Periodic Table of the Elements

Key to categories: Alkali metals · Alkaline earth metals · Transition metals · Lanthanides · Actinides · Other Metals · Metalloids · Nonmetals · Halogens · Noble gases

Key:
11 — Atomic number
Na — Element symbol
Sodium — Element name
22.99 — Standard atomic weight in amu (no data given for elements without stable nuclides)

The element symbol indicates the state at 0 °C and 1 atm:
Solid
Liquid
Gas
Unknown

PERIOD	(1) IA	(2) IIA	(3) IIIB	(4) IVB	(5) VB	(6) VIB	(7) VIIB	(8) VIIIB	(9) VIIIB	(10) VIIIB	(11) IB	(12) IIB	(13) IIIA	(14) IVA	(15) VA	(16) VIA	(17) VIIA	(18) VIIIA
1	1 H Hydrogen 1.008																	2 He Helium 4.00
2	3 **Li** Lithium 6.94	4 **Be** Beryllium 9.01											5 **B** Boron 10.81	6 **C** Carbon 12.01	7 N Nitrogen 14.01	8 O Oxygen 16.00	9 F Fluorine 19.00	10 Ne Neon 20.18
3	11 **Na** Sodium 22.99	12 **Mg** Magnesium 24.31											13 **Al** Aluminum 26.98	14 **Si** Silicon 28.09	15 **P** Phosphorus 30.97	16 **S** Sulfur 32.06	17 Cl Chlorine 35.45	18 Ar Argon 39.95
4	19 **K** Potassium 39.10	20 **Ca** Calcium 40.08	21 **Sc** Scandium 44.96	22 **Ti** Titanium 47.87	23 **V** Vanadium 50.94	24 **Cr** Chromium 52.00	25 **Mn** Manganese 54.94	26 **Fe** Iron 55.85	27 **Co** Cobalt 58.93	28 **Ni** Nickel 58.69	29 **Cu** Copper 63.55	30 **Zn** Zinc 65.38	31 **Ga** Gallium 69.72	32 **Ge** Germanium 72.63	33 **As** Arsenic 74.92	34 **Se** Selenium 78.97	35 **Br** Bromine 79.90	36 Kr Krypton 83.80
5	37 **Rb** Rubidium 85.47	38 **Sr** Strontium 87.62	39 **Y** Yttrium 88.91	40 **Zr** Zirconium 91.22	41 **Nb** Niobium 92.91	42 **Mo** Molybdenum 95.95	43 **Tc** Technetium	44 **Ru** Ruthenium 101.07	45 **Rh** Rhodium 102.91	46 **Pd** Palladium 106.42	47 **Ag** Silver 107.87	48 **Cd** Cadmium 112.41	49 **In** Indium 114.82	50 **Sn** Tin 118.71	51 **Sb** Antimony 121.76	52 **Te** Tellurium 127.60	53 **I** Iodine 126.90	54 Xe Xenon 131.29
6	55 **Cs** Cesium 132.91	56 **Ba** Barium 137.33	57-71 Lanthanides	72 **Hf** Hafnium 178.49	73 **Ta** Tantalum 180.95	74 **W** Tungsten 183.84	75 **Re** Rhenium 186.21	76 **Os** Osmium 190.23	77 **Ir** Iridium 192.22	78 **Pt** Platinum 195.08	79 **Au** Gold 196.97	80 **Hg** Mercury 200.59	81 **Tl** Thallium 204.38	82 **Pb** Lead 207.2	83 **Bi** Bismuth 208.98	84 **Po** Polonium	85 **At** Astatine	86 Rn Radon
7	87 **Fr** Francium	88 **Ra** Radium	89-103 Actinides	104 *Rf* Rutherfordium	105 *Db* Dubnium	106 *Sg* Seaborgium	107 *Bh* Bohrium	108 *Hs* Hassium	109 *Mt* Meitnerium	110 *Ds* Darmstadtium	111 *Rg* Roentgenium	112 *Cn* Copernicium	113 *Nh* Nihonium	114 *Fl* Flerovium	115 *Mc* Moscovium	116 *Lv* Livermorium	117 *Ts* Tennessine	118 *Og* Oganesson

57 **La** Lanthanum 138.91	58 **Ce** Cerium 140.12	59 **Pr** Praseodymium 140.91	60 **Nd** Neodymium 144.24	61 **Pm** Promethium	62 **Sm** Samarium 150.36	63 **Eu** Europium 151.96	64 **Gd** Gadolinium 157.25	65 **Tb** Terbium 158.93	66 **Dy** Dysprosium 162.50	67 **Ho** Holmium 164.93	68 **Er** Erbium 167.26	69 **Tm** Thulium 168.93	70 **Yb** Ytterbium 173.05	71 **Lu** Lutetium 174.97
89 **Ac** Actinium	90 **Th** Thorium 232.04	91 **Pa** Protactinium 231.04	92 **U** Uranium 238.03	93 **Np** Neptunium	94 **Pu** Plutonium	95 **Am** Americium	96 **Cm** Curium	97 **Bk** Berkelium	98 **Cf** Californium	99 **Es** Einsteinium	100 **Fm** Fermium	101 **Md** Mendelevium	102 **No** Nobelium	103 *Lr* Lawrencium

FIRST EDITION

Fundamentals of Chemistry for Today

General, Organic, and Biochemistry

Spencer L. Seager
Weber State University

Tiffiny D. Rye-McCurdy
The Ohio State University at Marion

Ryan J. Yoder
The Ohio State University at Marion

Australia • Brazil • Canada • Mexico • Singapore • United Kingdom • United States

Fundamentals of Chemistry for Today: General, Organic, and Biochemistry,
First Edition
Spencer L. Seager, Tiffiny D. Rye-McCurdy, and Ryan J. Yoder

SVP, Product: Cheryl Costantini

VP, Product: Thais Alencar

Portfolio Product Director: Maureen McLaughlin

Portfolio Product Manager: Roxanne Wang

Product Assistant: Ellie Purgavie

Learning Designer: Susan Pashos

Content Manager: Meaghan Ford

Subject Matter Expert: Dakin Sharum

Developmental Editor: John Murdzek

Digital Project Manager: Nikkita Kendrick

Technical Content Program Manager: Ivan Corriher

VP, Product Marketing: Jason Sakos

Director, Product Marketing: Danae April

Product Marketing Manager: Andrew Stock

Product Development Researcher: Nicole Hurt

Content Acquisition Analyst: Nichole Nalenz

Production Service: Lumina Datamatics, Inc.

Designer: Chris Doughman

Cover Image Source: MR.Cole_Photographer /Moment/Getty Images oxygen/Moment/Getty Images Westend61/Getty Images PM Images /DigitalVision/Getty Images

Library of Congress Control Number: 2023917528

ISBN: 978-0-357-45342-1

Cengage
5191 Natorp Boulevard
Mason, OH 45040
USA

Cengage is a leading provider of customized learning solutions. Our employees reside in nearly 40 different countries and serve digital learners in 165 countries around the world. Find your local representative at **www.cengage.com**.

To learn more about Cengage platforms and services, register or access your online learning solution, or purchase materials for your course, visit **www.cengage.com**.

Printed in the United States of America
Print Number: 03 Print Year: 2024

To Rem McCurdy and Steffi Yoder:
Without your love and encouragement, we might never have been able to start or finish this journey as authors.

To Maxine and Riggs, and Hannah and Noah:
Though you may never read the contents of this book, we hope you will develop a lifelong love and passion for learning and education.

About the Authors

Spencer L. Seager

Spencer L. Seager retired from Weber State University in 2013 after serving for 52 years as a faculty member of the chemistry department. He served as department chairman from 1969 until 1993 and taught general and physical chemistry at the university. Dr. Seager was also active in projects to help improve chemistry and other science education in local elementary schools. He received his B.S. in chemistry and Ph.D. in physical chemistry from the University of Utah.

Tiffiny D. Rye-McCurdy

Tiffiny D. Rye-McCurdy is the Administrative Manager of the Academic Success Center and a lecturer in chemistry at The Ohio State University at Marion, where she assists students in learning the concepts of chemistry and biology both within and outside the classroom. Dr. Rye-McCurdy currently teaches GOB chemistry, general chemistry, organic chemistry, and biochemistry. Prior to this position she taught introductory biology, and physiology courses. She received her B.A. in ACS certified biochemistry from Ohio Wesleyan University and received her Ph.D. from The Ohio State University Biochemistry program under the mentorship of Dr. Karin Musier-Forsyth, where she studied the mechanisms of retrovirology (predominately Rous sarcoma virus and HIV). Dr. Rye-McCurdy is involved in community outreach as the co-coordinator of Ohio State Marion's science and engineering camps for high school and middle school students.

Ryan J. Yoder

Ryan J. Yoder is an associate professor at The Ohio State University, serving the regional campus in Marion, OH. Dr. Yoder previously taught GOB chemistry at Marion before joining the full-time faculty in 2013. He currently teaches organic chemistry lecture and laboratory courses in addition to serving the campus and university community. He received his B.A. in chemistry from Ohio Wesleyan University and received his Ph.D. from The Ohio State University under the mentorship of Dr. Christopher Hadad, where he studied computational modeling and organic chemistry. Dr. Yoder mentors undergraduate research students at Marion and Columbus, examining protein-ligand interactions towards therapeutics against threats from chemical weapons and cancer. As a passionate teacher, Dr. Yoder additionally pursues interests in chemical education research. He lives in central Ohio with his wife and two children where he enjoys family time, travel, cooking, golfing, and following sports from around the world.

Brief Contents

Contents

Preface

Approach

One-semester courses in general, organic, and biological chemistry (GOB) are growing in frequency and importance compared to two-semester courses. For a course to cover such a vast array of topics successfully in a single semester, students must find the textbook associated with the course to be approachable and relevant to their educational journey. Students must also learn the material in sufficient depth to be beneficial in their future studies. The tension between breadth and depth is a challenge in currently available one-semester GOB textbooks. In *Fundamentals of Chemistry for Today: General, Organic, and Biochemistry*, we seek to thread this needle by integrating examples relevant to future health professionals at an appropriate level while gradually increasing broader chemical knowledge.

The goal of any GOB course is to give students a foundation in the principles of general chemistry that connect directly to the organic and biochemistry content encountered in health-related fields. Faculty must prepare students with quantitative reasoning skills as well as the ability to think critically and solve problems, as is necessary for the career paths these students are pursuing. By carefully covering the core principles of chemistry at the appropriate depth, our text provides instruction in the most essential analytical calculations where they matter most.

Student Audience

Most students taking a GOB course are preparing for careers in the health professions. We have assumed no previous college-level science or mathematics prerequisite. High school chemistry and algebra would be helpful but are not required since all concepts and skills are fully introduced. Ample problem-solving opportunities are found throughout the text in examples, Learning Checks, and exercises, giving students the ability to hone their critical thinking skills. Conscious of the one-semester time frame, we have presented the examples efficiently so that students are not bogged down with excessive repetition.

Each chapter begins with a Career Focus that exposes students to different potential professional pathways. There are also various boxed essays throughout the text that make crucial connections to health and the environment. An expanded collection of photos and art relevant to careers in health professions add context and color to the core narrative.

Organization and Integration of Topics

Since the average semester is 15 weeks long, this text contains 15 chapters and is intentionally structured for coverage of one chapter per week. Faculty may certainly adapt the Table of Contents to suit their needs, but these 15 chapters provide a basic framework suitable for one semester of instruction. There is sufficient depth that instructors who wish to delve more intensely into selected topics will have the resources to do so, while also covering all the topics essential to the GOB curriculum.

That said, many instructors want to "get to the biochemistry faster." With this factor in mind, we have sought to integrate concepts that showcase the intersection of fundamental themes in chemistry and relevant topics in health fields. Instructors who have the needs of future medical technologists, nurses, and other health professionals in mind may want to note the following items as examples (not all-inclusive) of the integration of career-relevant topics:

- Medical dosage calculations in Chapter 1 (Matter, Measurements, and Calculations)
- Health effects and medical uses of radioisotopes in Chapter 2 (Atomic Structure and the Periodic Table)

- Examples and figures throughout Chapter 3 (Chemical Bonds: Molecule Formation) that add context to important ionic and covalent compounds in medicine.
- Osmosis and dialysis in Chapter 6 (Gases and Solutions)
- Biological buffers in Chapter 7 (Chemical Equilibrium: Acids, Bases, and Buffers)
- Respiratory and urinary control of pH in Chapter 7 (Chemical Equilibrium: Acids, Bases, and Buffers)
- Amines as neurotransmitters in Chapter 9 (Alcohols, Ethers, and Amines)
- Amphetamines and alkaloids in Chapter 9 (Alcohols, Ethers, and Amines)
- Protein functions, mechanism of action of ampicillin, and enzymes in medical applications and disease in Chapter 12 (Amino Acids, Proteins, and Enzymes)
- Virology, vaccines, and gene therapy in Chapter 13 (Nucleic Acids and Protein Synthesis)
- Drug delivery in biological systems in Chapter 14 (Lipids)
- Biological functions of macronutrients, vitamins, and minerals as well as the highlights of the most important parts of metabolism in Ch. 15 (Nutrition and Metabolism)

Features of This Text

We have preserved the best features of an earlier two-semester text by Dr. Seager while rewriting much of the narrative and pedagogy to make the text more efficient for students and faculty to use in a one-semester format.

Career Focus/Career Description. This feature stimulates student interest and inquiry by introducing the diverse opportunities across a variety of health-related fields at the beginning of each chapter. The Career Focus is a vignette written in the first person describing the appeal of each highlighted career and some of the ways the skills and content from the chapter are relevant to that field. The Career Description at the close of each chapter describes job requirements and qualifications.

Health Career Focus

Psychosocial Oncologist

I remember back to my early nursing education when I discovered I could contribute to the mental well-being of my patients, and also help them through whatever struggles they may be experiencing with their physical well-being. So many of us have personally been affected by cancer or have loved ones or colleagues who have faced the mountain that is cancer. It's so important for anyone who is battling serious illness to have a support system.

As a professional in psychosocial oncology, I am able to use my training in a multitude of areas to help cancer patients deal with the mental aspects of fighting cancer via counseling and support. Although there are many causes and many types of cancer, one

Learning Objectives. At the beginning of each chapter, a list of learning objectives provides students with a convenient overview of what they should gain by studying the chapter. The relevant objectives are repeated at the beginning of each section to remind students of the main takeaways from that part of the chapter. Each chapter also begins with a brief overview, which provides a big picture view of the concepts discussed throughout that part of the text.

Learning Objectives

When you have completed your study of Chapter 2, you should be able to:

1. Locate elements in the periodic table on the basis of their group and period designations. **(Section 2.1)**
2. Describe the charge, relative mass, and location of the three subatomic particles. **(Section 2.2)**
3. Write the atomic symbol ($^A_Z X$) for a given set of subatomic particles. **(Section 2.3)**
4. Define the terms *isotope* and *atomic weight*. **(Section 2.4)**
5. Write balanced equations for radioactive decay and other nuclear processes. **(Section 2.5)**
6. Solve problems using the half-life concept. **(Section 2.5)**
7. Describe the applications of radiation in health and medicine. **(Section 2.5)**
8. Write correct electron configurations for each element through atomic number 56. **(Section 2.6)**
9. Describe periodic trends in atomic radius, first ionization energy, and electronegativity. **(Section 2.7)**

Key Terms. Identified within the text by the use of bold type, key terms are defined in the margin near the place where they are introduced. Students reviewing a chapter can quickly identify the important concepts on each page with this marginal glossary. A full glossary of key terms and concepts appears at the end of the text.

Examples. To reinforce students in their problem-solving skill development, complete step-by-step solutions for numerous examples are included throughout each chapter.

Learning Checks. Immediately following most examples is a Learning Check, which allows students to apply their knowledge and problem-solving skills from the worked-out examples. A complete set of solutions is included in Appendix C in order for students to measure their understanding and progress.

Health Connections. These boxed essays describe current chemistry-related health issues in order for students to further appreciate the application of the chemistry they are learning to general health and wellness.

Health Connections 14.1

Consider the Mediterranean Diet

If you are looking for a heart-healthy eating plan, the Mediterranean diet might be right for you. The Mediterranean diet incorporates the basics of healthy eating—plus a splash of flavorful olive oil and perhaps even an occasional glass of red wine. These components characterize the traditional cooking style of countries that border the Mediterranean Sea.

Most healthful diets include fruits, vegetables, fish, and whole grains, and they limit unhealthy fats. While these parts of a healthful diet remain tried and true, subtle variations or differences in the proportions of certain foods might make a difference in your risk for heart disease.

Research has shown that the traditional Mediterranean diet reduces the risk of heart disease. In fact, an analysis of more than 1.5 million healthy adults demonstrated that following a Mediterranean diet was associated with a reduced risk of death from heart disease and cancer, as well as a reduced incidence of Parkinson's and Alzheimer's diseases.

The Mediterranean diet emphasizes the following dietary practices:

1. Eating primarily plant-based foods, such as fruits, vegetables, whole grains, legumes, and nuts
2. Replacing butter with healthy fats such as olive oil
3. Using herbs and spices instead of salt to flavor foods
4. Limiting red meat to no more than a few times a month
5. Eating fish and poultry at least twice a week
6. Drinking red wine in moderation (optional)

The Mediterranean diet incorporates a variety of healthy foods.

Environmental Connections. These boxed essays contain spotlight areas where chemistry intersects with the natural world around us.

Environmental Connections 6.1

CO_2 Emissions: A Blanket around Earth

In the same way that a greenhouse heats up when radiant heat energy from the sun is trapped as sunlight passes through its glass or plastic windows, the surface of Earth warms when radiant solar energy passes through the atmosphere and is trapped. Scientists call this heating of Earth the greenhouse effect. Most climate scientists agree that one main cause of current observed climate change is *human activity* that enhances this "greenhouse effect." Since the Industrial Revolution in the nineteenth century, human activity has influenced the climate.

Certain gases, particularly carbon dioxide (CO_2), water vapor (H_2O), methane (CH_4), and nitrous oxide (N_2O), are called greenhouse gases because they act like the windows of a greenhouse and let visible light in but prevent heat from radiating back out. As a result of this effect and the presence of these gases in Earth's atmosphere, the average temperature at the surface of Earth is a life-supporting 59 °F or 15 °C. In naturally occurring amounts, these gases make life possible on Earth, but too much of a good thing can create problems.

The consequences of changing the natural atmospheric concentration of greenhouse gases are difficult to predict and controversial, but one thing is certain. During the past century, based on actual measurements, the average surface temperature of Earth has increased by up to 0.8 °C. Although that might not sound like much, a tiny increase in temperature can have enormous consequences. Glaciers, sea ice, and ice sheets around the world are melting. As the ice melts, sea levels rise. Also, as seawater warms, it expands and sea levels rise even more. Only humans can change the amount of greenhouse gases produced by human activity, and thereby minimize these negative environmental effects.

An increase in the rate of breakup of glaciers (calving) as they approach the oceans is one indication of global warming.

Study Tools. Most chapters contain a Study Tools feature that may prove useful to students. These features offer various strategies in order to aid in the development of students' critical thinking skills as they approach the material throughout the text.

Study Tools 1.1

Help with Calculations

Many students feel uneasy about working chemistry problems that involve the use of mathematics. One tip that will help you solve such problems in this textbook is to remember that almost all of these problems are one of two types: those for which a specific formula applies and those where dimensional analysis is used. When you encounter a math-type problem, your first task should be to decide which type of problem it is: one you solve with an equation or one that you can do using dimensional analysis.

In this chapter, the percentage calculations in Examples 1.15 and 1.16 are formula-based problems. The temperature conversions in Example 1.7 may seem formula-based, too, but they are a special case, because they use a formula to perform a unit conversion.

The dimensional analysis discussed in **Section 1.8** is used for many problems that require mathematical calculations. The beauty of this method is that it mimics your natural, everyday way of solving problems. This real-life method usually involves identifying where you are, where you want to go, and how to get there.

A key to working story problems is to see through all the words and find what is given (number and units). Then look for what is wanted by focusing on key words or phrases like "how much," "what is," and "calculate." Finally, use one of the two methods, formula or dimensional analysis, to solve the problem.

The steps used to solve both types of problems are summarized in the following flow charts.

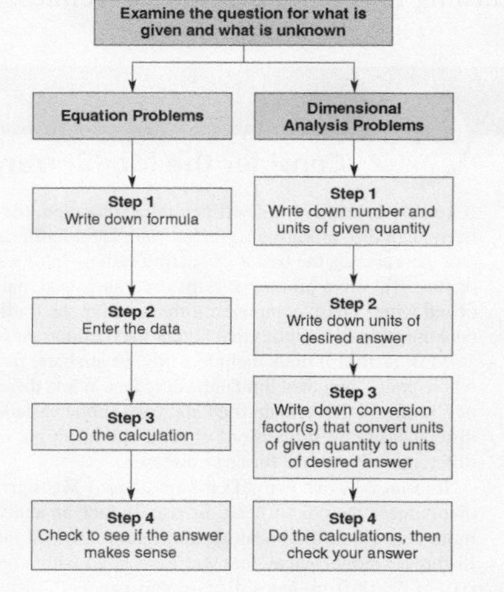

Concept Summary. Located at the end of each chapter, this feature provides a concise review of the concepts and challenges students to check their achievement of the learning objectives related to the concepts.

Key Terms and Concepts. The key terms throughout the text are listed in alphabetical order along with the section of the chapter in which the term can be found.

Key Equations. This feature provides a useful summary of general equations and reactions from the chapter, which is particularly useful for the organic chemistry chapters.

Exercises. At the end of each chapter, there are numerous exercises arranged by section. Within the exercises are a wide variety of question types to provide a well-rounded challenge for students, including a significant number of clinical and other applications. Solutions and answers to all exercises are provided in the Instructor's Manual. Brief answers to selected problems can be found in Appendix B.

Online Learning Platform: WebAssign

Built by educators, *WebAssign* provides flexible settings at every step to customize your course with online activities and secure testing to meet learners' unique needs. Students get everything in one place, including rich content and study resources designed to fuel deeper understanding, plus access to a dynamic, interactive eTextbook. Proven to help hone problem-solving skills, *WebAssign* helps you help learners in any course format. Learn more at https://www.cengage.com/webassign.

Instructor and Student Resources

Additional instructor and student resources for this title are available online. Instructor assets include an *Instructor's Manual with Solutions to Exercises*, Educator's Guide, PowerPoint® slides prepared by Drs. Rye-McCurdy and Yoder, and a Test Bank powered by Cognero®. Student assets include a *Student Study Guide and Solutions Manual*. Sign up or sign in at www.cengage.com to search for and access this title and its online resources.

The Educator's Guide explains key features of the WebAssign course for this title including types of homework questions, help tools available for students, course packs, and assets on the WebAssign Resources tab.

The *Student Study Guide and Solutions Manual* by Dr. Katherine Thomas contains the following content for each chapter: chapter outline, table of learning objectives correlated to examples and exercises, expanded solutions to even-numbered exercises, and a practice test.

Acknowledgments

We express our sincere appreciation to the following instructors who responded to questionnaires, participated in online discussion groups, or commented on early drafts of this text:

Corina E. Brown
University of Northern Colorado

Emma Chow
Palm Beach State College

Dr. Gerard G. Dumancas
The University of Scranton

Heather D. Hollandsworth
Harding University

Jason M. Hudzik
County College of Morris

Bruce Kowiatek
Blue Ridge Community and Technical College

Ganesh Kumar
Illinois Central College

Bethany Melroe Lehrman
Dakota Wesleyan University

William J. Magilton
Northampton Community College

Janet Maxwell, PhD
Angelo State University

Vinod Kumar Mishra, PhD
Snead State Community College

Dr. Carson Prevatte
Middle Tennessee State University

Osvaldo Rodriguez
El Paso Community College

Dr. Jessica L. Sardo
Mesa College

Amar S. Tung
Lincoln University

Dr. E. Viltchinskaia
NMMI

Linda Waldman
Cerritos College

Julie Wondergem
University of Wisconsin Green Bay

We also give special thanks to the staff of Cengage Learning, including Roxanne Wang, Portfolio Product Manager; Susan Pashos, Senior Learning Designer; Meaghan Ford, Senior Content Manager; Ellie Purgavie, Product Assistant; Andrew Stock, Product Marketing Manager; Dakin Sharum, Subject Matter Expert; Ivan Corriher, Technical Content Program Manager; and Nikkita Kendrick, Digital Project Manager.

We would like to especially thank our development editor, John Murdzek. His thoughtful comments and guidance have been indispensable in shaping this text, while his good humor and cheer buoyed the entire writing process. We would also like to thank Katherine Thomas, the author of our *Student Study Guide and Solutions Manual* and a valued friend and mentor, for her tireless efforts in helping craft end-of-chapter exercises. Redding Morse deserves our thanks as well, for her patience guiding us through the final production of this text. Lastly, we would like to thank our families for their love and sacrifice, our friends and colleagues for their support, and our students for their inspiration.

Tiffiny D. Rye-McCurdy
Ryan J. Yoder

1 Matter, Measurements, and Calculations

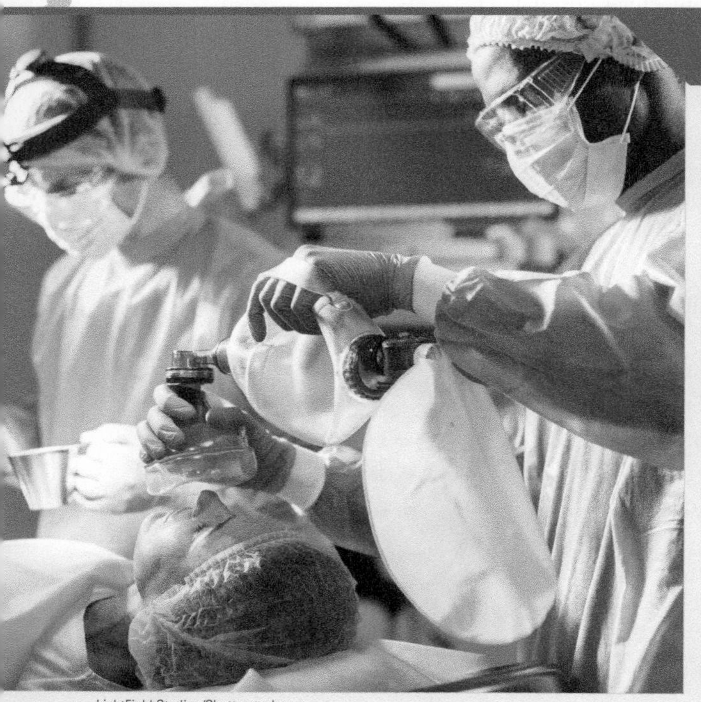

LightField Studios/Shutterstock.com

Health Career Focus

Anesthesiologist Assistant

Anesthesia, as we know it, has existed for less than 200 years! Can you imagine the world without it?"

By working as an anesthesiologist assistant, I feel that every day I give the greatest possible gift to my patients. It's awe inspiring to consider what the scientific application of a few gases and medications applied with precise skill and measurement makes possible—pain-free procedures and surgeries. What an amazing gift!

I considered premed in college, but I thought a doctor's schedule and responsibilities just weren't for me. I found that I could get the same sense of satisfaction working as an anesthesiologist assistant. I know I'm an important part of the surgical team. Eliminating pain—what could be better?

Follow-up to this Career Focus appears at the end of the chapter before the *Concept Summary*.

Learning Objectives

When you have completed your study of Chapter 1, you should be able to:

1 Explain what matter is. **(Section 1.1)**

2 Explain the difference between the terms *physical* and *chemical* as they apply to the properties of matter and changes in matter. **(Section 1.2)**

3 Classify matter as an element, compound, homogenous mixture, or heterogeneous mixture. **(Section 1.3)**

4 Describe four measurement units used in everyday activities. **(Section 1.4)**

5 Convert measurements within the metric system into related units. **(Section 1.5)**

6 Convert temperatures measured in Fahrenheit to Celsius and vice versa. **(Section 1.5)**

7 Express numbers using scientific notation. **(Section 1.6)**

8 Perform calculations with numbers expressed in scientific notation. **(Section 1.6)**

9 Express measurements and calculations using the correct number of significant figures. **(Section 1.7)**

10 Use dimensional analysis to solve numerical problems. **(Section 1.8)**

11 Perform calculations involving percentages. **(Section 1.9)**

12 Perform calculations involving densities. **(Section 1.10)**

CHEMISTRY is often described as the scientific study of matter. In a way, almost every study is a study of matter, because matter is the substance of everything. When chemists study matter, however, they attempt to understand it from nearly every possible point of view.

The chemical nature of all matter makes an understanding of chemistry useful and necessary for individuals who are studying in a wide variety of areas (**Figure 1.1**), including the health sciences, natural sciences, education, environmental science, and law enforcement.

iStock.com/SolStock

Figure 1.1 Chemistry is essential to research in multidisciplinary areas.

Matter comes in many shapes, sizes, and colors that are interesting to look at and describe. Early chemists did little more than describe what they observed, and their chemistry was severely limited in scope. It became a much more useful science when chemists began to make quantitative measurements, do calculations, and incorporate the results into their descriptions. Some fundamental ideas about matter are presented in this chapter, along with some ideas about quantitative measurement, the scientific measurement system, and calculations.

1.1 What Is Matter?

Learning Objective 1 Explain what matter is.

Definitions are useful in all areas of knowledge. You will be expected to learn a number of definitions as you study chemistry, and the first one is a definition of *matter*. Although matter is the substance of everything, that isn't a very scientific definition, even though we think we know what it means. If you stop reading for a moment and look around, you will see a number of objects that might include people, potted plants, walls, furniture, books, windows, and a television or computer. The objects you see have at least two things in common: Each one has mass, and each one occupies space. These two common characteristics provide the basis for the scientific definition of matter. **Matter** is anything that has mass and occupies space. You probably understand what is meant by an object occupying space, especially if you have tried to occupy the same space as some other object.

matter Anything that has mass and occupies space.

You might not understand the meaning of the term *mass* quite as well, but it can also be illustrated "painfully." Imagine walking into a very dimly lit room and being able to just barely see two large objects of equal size on the floor. You know that one is a bowling ball and the other is an inflated plastic ball, but you can't visually identify which is which. However, a hard kick delivered to either object allows you to distinguish between them. The bowling ball resists being moved much more strongly than does the inflated ball. Resistance to movement depends on the amount of matter in an object, and mass is an actual measurement of the amount of matter present.

The term *weight* is probably more familiar to you than *mass*, but the two are related. All objects are attracted to each other by gravity, and the greater their mass, the stronger the attraction between them. The **weight** of an object on Earth is a measurement of the gravitational force pulling the object toward Earth. An object with twice the mass of a second object is attracted with twice the force, and therefore has a weight twice the weight of the second object. The **mass** of an object is constant, no matter where it is located (even if it is in a "weightless" condition in outer space). However, the weight of an object depends on the strength of the gravitational attraction to which it is subjected. For example, a rock that weighs 16 pounds on Earth would weigh about 2.7 pounds on the moon (see **Figure 1.2**) because the gravitational attraction on the moon is only about

weight A measurement of the gravitational force acting on an object.

mass A measurement of the amount of matter in an object.

Moon/NASA

Figure 1.2 Astronauts weigh less on the moon than on Earth, making it relatively easy to jump several feet high on the moon!

one-sixth of that on Earth. However, the rock contains the same amount of matter and thus has the same mass whether it is located on Earth or on the moon.

Despite the difference in meaning between mass and weight, the determination of mass is commonly called "weighing." We will follow that practice in this book, but we use the correct term *mass* when referring to an amount of matter.

1.2 Physical and Chemical Properties and Changes

Learning Objective 2 Explain the difference between the terms *physical* and *chemical* as they apply to the properties of matter and changes in matter.

When you looked at your surroundings earlier, you probably didn't have much trouble identifying the various things you saw. For example, you could likely tell the difference between a TV and a potted plant by observing characteristics such as shape, color, and size. Our ability to identify objects or materials and discriminate between them depends on such characteristics. Scientists prefer to use the term *property* instead of *characteristic*, and they classify properties as either physical or chemical.

physical properties Properties of matter that can be observed or measured without trying to change the composition of the matter being studied.

Physical properties are those that can be observed or measured without changing or trying to change the composition of the matter—that is, no original substances are destroyed, and no new substances appear. For example, you can observe the color or measure the size of a sheet of paper without attempting to change the paper into anything else. Color and size are physical properties of the paper.

chemical properties Properties that matter demonstrates when attempts are made to change it into new substances.

Chemical properties are the properties that matter demonstrates when attempts are made to change it into other kinds of matter (see **Figure 1.3**). For example, burning a sheet of paper changes it into a new substance. On the other hand, attempts to burn a piece of glass under similar conditions fail. The ability of paper to burn is a chemical property, as is the inability of glass to burn.

physical changes Changes matter undergoes without changing composition.

You can change the size of a sheet of paper by cutting off a piece of it. The paper sheet is not converted into any new substance by this change, but it is simply made smaller. **Physical changes** can be carried out without changing the composition of a substance. However, there is no way you can burn a sheet of paper without changing it into new

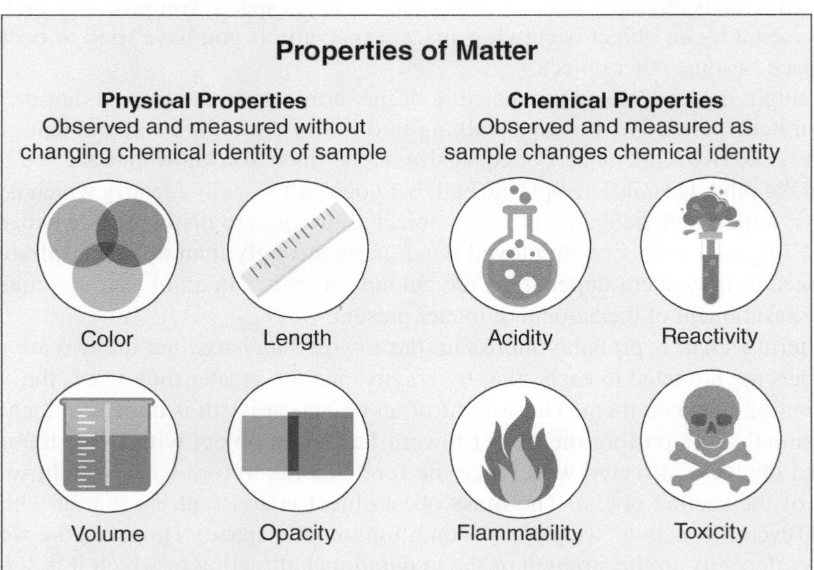

Figure 1.3 Examples of physical and chemical properties of matter.

Health Connections 1.1

A Central Science

Chemistry is often referred to as a "central science" because it serves as a necessary foundation for many other scientific disciplines. Regardless of which scientific field you are interested in, every single substance you will discuss or work with is made up of chemicals. Also, many processes important to those fields will be based on an understanding of chemistry.

Chemistry → Health sciences, Environmental sciences, Microbiology, Physiology, Nutrition

Chemistry is the foundation for many other scientific disciplines.

We also consider chemistry a central science because of its crucial role in responding to the needs of society. We use chemistry to discover new processes, develop new sources of energy, produce new products and materials, provide more food, and ensure better health.

As you read this text, you will encounter chapter-opening applications of chemistry in the health care professions. Within the chapters, Environmental Connections and Health Connections boxes focus on specific topics that play essential roles in meeting the needs of society.

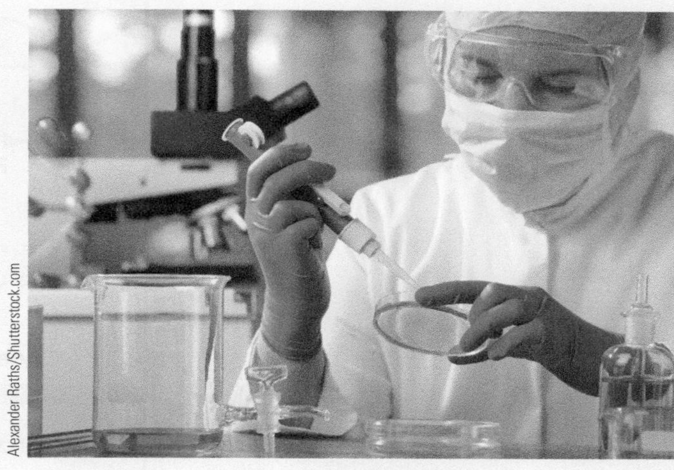

Alexander Raths/Shutterstock.com

Chemistry is essential to microbiology and the research performed by microbiologists help us understand viruses that may lead to global pandemics, such as the virus that causes COVID-19.

substances. Thus, the change that occurs when paper burns is called a **chemical change**. The burning of magnesium metal (**Figure 1.4**) is a chemical change, because magnesium is converted to magnesium oxide. The light magnesium produces when it is burned is so bright that it was used in the flash powder in early photography. Magnesium is still used in fireworks to produce a brilliant white light.

chemical changes Changes matter undergoes that involve changes in composition.

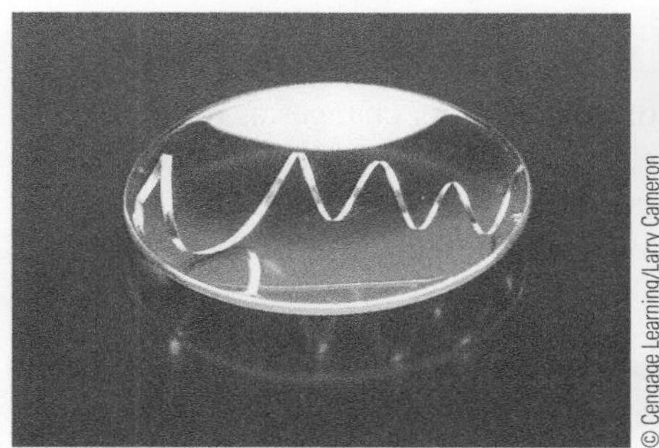

© Cengage Learning/Larry Cameron

1. A strip of magnesium metal.

© Cengage Learning/Larry Cameron

2. After being ignited with a flame, the magnesium burns with a blinding white light.

© Cengage Learning/Larry Cameron

3. The white ash of magnesium oxide from the burning of several magnesium strips.

Figure 1.4 A chemical change occurs when magnesium metal burns.

Among the most common physical changes are changes in state (see Chapter 5), such as the melting of solids to form liquids, the sublimation of solids to form gases, and the condensation of gases to form liquids. These changes take place when heat is added to or removed from matter (**Figure 1.5**).

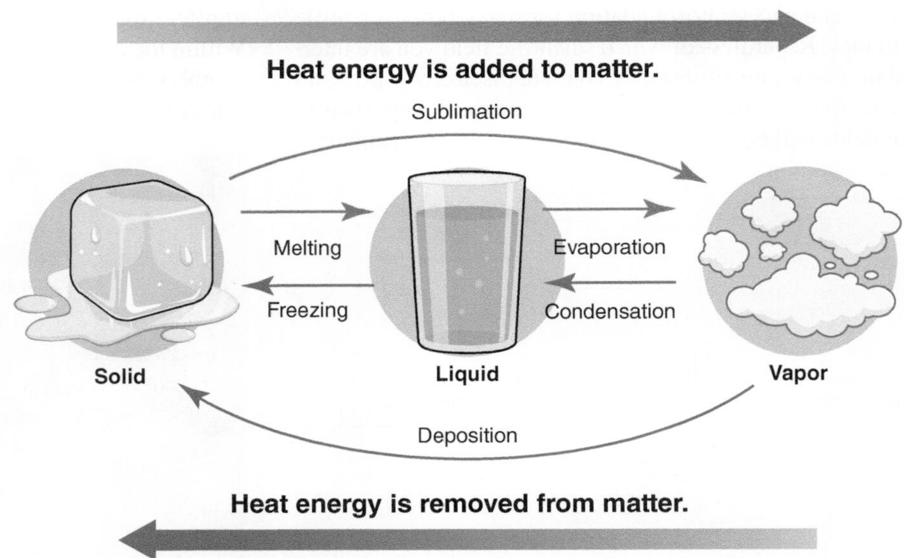

Figure 1.5 Description of changes of state.

Example 1.1 Classifying Changes as Physical or Chemical

Classify each of the following changes as physical or chemical: (a) a match is burned; (b) iron is melted; (c) limestone is crushed; (d) limestone is heated, producing lime and carbon dioxide; (e) an antacid seltzer tablet is dissolved in water; and (f) a rubber band is stretched.

Solution

Changes b, c, and f are physical changes because no composition changes occur and no new substance forms.

The others are chemical changes because new substances do form. When a match is burned, combustion gases are given off, and matchstick wood is converted to ashes. The lime and carbon dioxide produced when limestone is heated are new substances. The fizzing that results when a seltzer tablet is dissolved in water is evidence that at least one new material (a gas) is produced.

✔ **Learning Check 1.1** Classify each of the following changes as physical or chemical, and, in the cases of chemical change, describe one observation or test that indicates new substances have been formed: (a) milk sours, (b) a wet handkerchief dries, (c) fruit ripens, (d) a stick of dynamite explodes, (e) air is compressed into a steel container, and (f) water boils.

1.3 Classifying Matter

Learning Objective 3 Classify matter as an element, compound, homogenous mixture, or heterogeneous mixture.

In order to better understand matter, we must be able to classify it in a variety of ways. First, matter can broadly be classified as either a **pure substance** or a **mixture**. Things like water, sugar, and salt (see **Figure 1.6**) are each pure substances with a constant composition. Pure substances have a fixed set of physical and chemical properties.

pure substance Matter that has a constant composition and fixed properties.

mixture A blend of matter that can theoretically be physically separated into its separate pure components.

For example, water contains two hydrogen atoms for every one oxygen atom and freezes at a specific temperature. A mixture of sugar and water, on the other hand, varies in composition, and its properties will be different for each possible combination of the two pure substances. For example, a glass of water with only a few crystals of sugar will taste much less sweet than a glass of water with several spoonfuls of sugar. Other properties, such as the freezing point of these solutions, will change as the mixture changes. Because mixtures are combinations of matter, they can theoretically be separated through physical processes, such as heating a mixture of sugar water, until all the water evaporates, leaving behind the sugar. Some separations of mixtures are more difficult to achieve than others.

Figure 1.6 A pure substance such as salt has a constant composition.

Mixtures that have their own consistent properties and uniform appearance, like sugar water, are examples of **homogeneous matter**. All pure substances are also, by definition, homogeneous. Homogeneous mixtures like sugar water are called **solutions** (see **Figure 1.7**). Solutions of gases and solids do exist, such as the air in our atmosphere, which contains primarily oxygen and nitrogen. Metal alloys like steel (iron and carbon) and brass (copper and zinc) are solutions, too. Typically, the word *solution* is used to imply a homogeneous liquid mixture. We discuss solutions in more detail in Chapter 6.

homogeneous matter Matter that has the same properties throughout the sample.

solutions Homogeneous mixtures of two or more pure substances.

Mixtures in which the properties and appearance are not uniform throughout are examples of **heterogeneous matter**. An ice cream sundae, where all the components are layered separately, is a heterogeneous mixture. A milkshake, where all the various ingredients are blended together will ultimately appear as a homogeneous mixture. Most mixtures in nature are heterogeneous mixtures when classified on the basis of their appearance.

heterogeneous matter Matter with properties that are not the same throughout the sample.

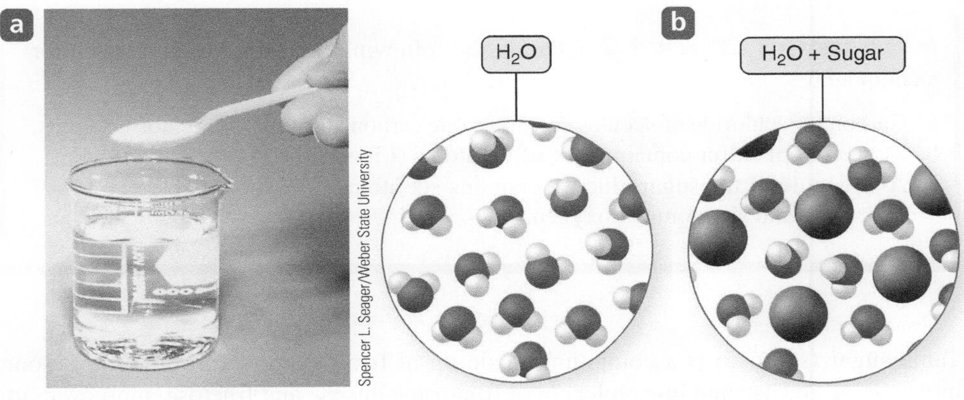

Figure 1.7 Sugar and water (a) form a solution when mixed (b).

Pure substances can be further classified as being made up of either elements or compounds. **Elements** cannot be separated chemically into simpler, smaller pure substances or components. Thus, elements (such as carbon, nitrogen, oxygen, and hydrogen) are often considered to be the "building blocks" of all matter. As we explain in Chapter 2, these elements are organized within the periodic table. If elements are the "building blocks" of matter, then **atoms** (from the Greek for "indivisible") are the building blocks of elements. Atoms are the smallest part of an element that retains all the characteristics of that element. When two or more elements combine through chemical bonding, the resulting pure substance with a definite composition is called a **compound**.

Molecules can be made up of atoms of the same element, such as oxygen gas, or they can be made up of atoms of two or more different elements, such as water. The oxygen gas in Earth's atmosphere is a **diatomic molecule**, because two individual oxygen atoms are bound through chemical bonds. Molecules like oxygen that contain atoms of only one kind of element are called **homoatomic molecules**. The greenhouse gas ozone (**Figure 1.8**), another homoatomic molecule, is a **triatomic molecule**, because it is composed of three oxygen atoms linked through chemical bonds. Molecules with more than three atoms are generally referred to as **polyatomic molecules**.

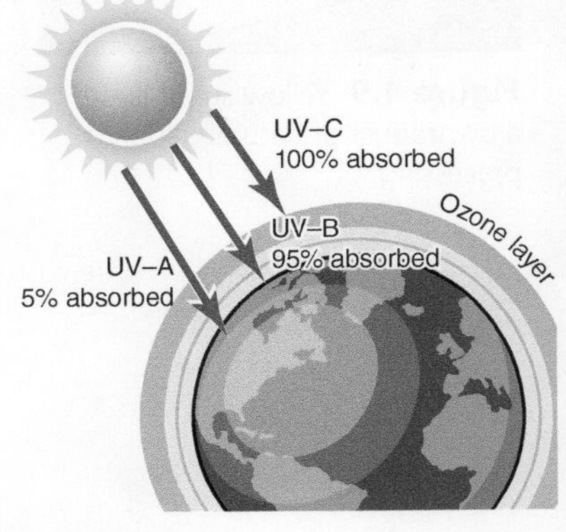

Figure 1.8 The ozone layer protects life on Earth from ultraviolet (UV) radiation emitted by the sun.

element A pure substance consisting of only one kind of atom in the form of homoatomic molecules or individual atoms.

atom The limit of chemical subdivision for matter.

compound A pure substance consisting of two or more elements, with a definite composition, that can be broken down into simpler substances only by chemical methods.

molecule The smallest particle of a pure substance that has the properties of that substance and is capable of a stable independent existence. Alternatively, a molecule is the limit of physical subdivision for a pure substance.

diatomic molecules Molecules that contain two atoms.

homoatomic molecules Molecules that contain only one kind of atom.

triatomic molecules Molecules that contain three atoms.

polyatomic molecules Molecules that contain more than three atoms.

heteroatomic molecules Molecules that contain two or more kinds of atoms.

Water, made up of two hydrogen atoms and one oxygen atom, is a **heteroatomic molecule** because it is made up of atoms of more than one element. It is also triatomic. Therefore, we can also define compounds as pure substances made up of heteroatomic molecules. Compounds can be separated through chemical processes into smaller components—either smaller molecules or atoms themselves.

Example 1.2 Classifying Molecules

Classify the following molecules as an element or a compound.

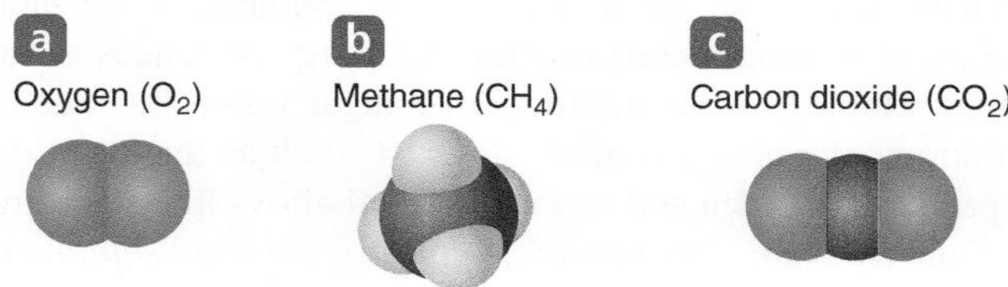

Solution

a. Oxygen gas (O_2) is a diatomic element, because it contains two identical atoms.
b. Methane (CH_4) consists of five total atoms, but more importantly it contains two types of atoms, carbon and hydrogen, so it is a compound.
c. Carbon dioxide (CO_2) consists of three total atoms and contains two types, oxygen and carbon, so it is a compound.

✔ Learning Check 1.2 Classify the following molecules as an element or a compound.

a. Carbon tetrachloride molecules consist of one carbon atom and four chlorine atoms.
b. Molecules of sulfur contain eight sulfur atoms (**Figure 1.9**).
c. A molecule of the sugar glucose contains six atoms of carbon, twelve atoms of hydrogen, and six atoms of oxygen.

Figure 1.9 Yellow sulfur is a by-product of natural gas processing.

Table sugar (sucrose) is a compound made up of two different carbohydrates—one molecule of glucose and one molecule of fructose. Glucose and fructose, moreover, are made up of carbon, oxygen, and hydrogen atoms. Table sugar, then, can be separated into individual glucose and fructose molecules through chemical means. And if broken down further, table sugar can be separated through chemical processes into the individual atoms or molecules of carbon, oxygen, and hydrogen that make up its chemical structure. **Figure 1.10** illustrates a classification scheme for matter based on the ideas we have discussed so far.

Example 1.3 Classifying Substances

Classify the substances italicized in the following scenarios as those of an element, compound, homogenous mixture, or heterogeneous mixture.

a. The IV bags used to treat a patient contained *9% saline in water*.
b. A patient with high blood pressure was asked to reduce their *salt (NaCl)* intake.

Solution

a. The *9% saline in water* is a solution that is uniform throughout, so it is a homogenous mixture (a solution).
b. *Salt (NaCl)* is a pure substance made of two different elements (sodium and chlorine) and is therefore a compound.

✔ **Learning Check 1.3** Classify the substances italicized in the following scenarios as those of an element, compound, homogenous mixture, or heterogeneous mixture.

a. Your dietician prescribed more *salads* in your diet.
b. An emergency medical technician encounters a patient exhibiting the symptoms of *carbon monoxide (CO)* poisoning (e.g., headache, dizziness, and confusion, amongst others).

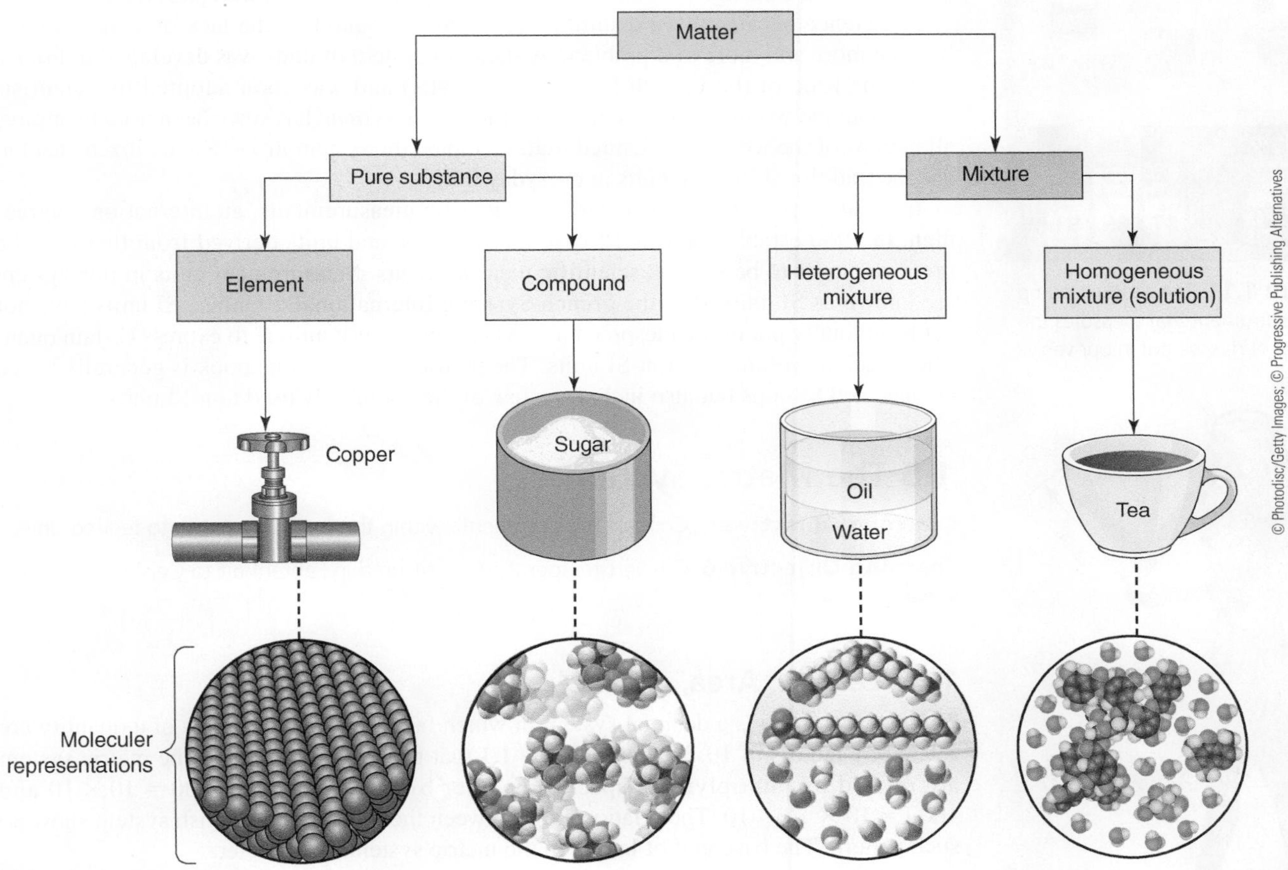

Figure 1.10 A classification scheme for matter.

1.4 Measurement Units

Learning Objective 4 Describe four measurement units used in everyday activities.

Matter can be classified, and some physical or chemical properties can be observed without making any measurement. However, the use of quantitative measurements and calculations greatly expands our ability to understand the chemical nature of the world around us. A measurement consists of two parts—a number and an identifying unit. A number expressed without a unit is generally useless, especially in scientific work. We constantly make and express measurements in our daily lives. We measure the gallons of gasoline put into our cars (**Figure 1.11**), the time it takes to drive a certain distance, and the temperature on a hot or cold day. In some of our daily measurements, the units might be implied or understood. For example, if someone said the temperature outside was 39, you might assume this was 39 degrees Fahrenheit if you lived in the United States, but in most other parts of the world, it would be 39 degrees Celsius. Because there is a large difference between the two temperatures, confusion is avoided by expressing both the number and the unit of a measurement.

Pexels

Figure 1.11 The gasoline pump is an instrument that measures the gallons of gas we put in our vehicles.

All measurements are based on units agreed on by those making and using the measurements. When a measurement is made in terms of an accepted unit, the result is expressed as some multiple of that unit. For example, when you purchase 10 pounds of potatoes, you are buying a quantity of potatoes equal to 10 times the standard quantity called 1 pound. Similarly, 3 feet of string is a length of string 3 times as long as the standard length that has been agreed on and called 1 foot.

The earliest units used for measurements were based on the dimensions of the human body. For example, the foot was the length of some important person's foot, and the biblical cubit (**Figure 1.12**) was the length along the forearm from the elbow to the tip of the middle finger. Unfortunately, the size of the units changed when the person on whom they were based changed because of death, change in political power, and so on.

As science became more quantitative, scientists found that the lack of standard units became more and more of a problem. A standard system of units was developed in France about the time of the French Revolution (1790s) and was soon adopted by scientists throughout the world. This system, called the *metric system*, has since been used by almost all nations of the world. The United States adopted the system in 1975, but citizens tend to use the traditional English units in everyday life.

In an attempt to further standardize scientific measurements, an international agreement in 1960 established certain base metric units, and units derived from them, as the preferred units to be used in scientific measurements. Measurement units in this system are known as SI units after the French Système International d'Unitès. SI units have not yet been totally put into widespread use. Many scientists continue to express certain quantities, such as volume, in non-SI units. The metric system in this book is generally based on accepted SI units but also includes a few of the commonly used non-SI units.

Figure 1.12 In ancient cultures, a cubit could vary from 18 to 21 inches, depending on the length of the arm.

1.5 The Metric System

Learning Objective 5 Convert measurements within the metric system into related units.

Learning Objective 6 Convert temperatures measured in Fahrenheit to Celsius and vice versa.

1.5A Length, Area, and Volume

The metric system is a decimal system in which larger and smaller units of a quantity are related by factors of 10. Notice in **Table 1.1** that the units of length in the metric system are related by multiplying a specific number by 10—remember, $100 = 10 \times 10$ and $1000 = 10 \times 10 \times 10$. The relationships between the units of the English system show no such pattern. The base unit of length in the metric system is the meter.

Table 1.1 Metric and English Units of Length

	Base Unit	Larger Unit	Smaller Unit
Metric	1 meter	1 kilometer = 1000 meters	10 decimeters = 1 meter 100 centimeters = 1 meter 1000 millimeters = 1 meter
English	1 yard	1 mile = 1760 yards	3 feet = 1 yard 36 inches = 1 yard

base unit of measurement A specific unit from which other units for the same quantity are obtained by multiplication or division.

The relationships between units of the metric system that are larger or smaller than a **base** (defined) **unit** are indicated by prefixes attached to the name of the base unit. Thus, 1 kilometer (km) (**Figure 1.13**) is a unit of length that is 1000 times longer than the base unit of 1 meter (m), and a millimeter (mm) is only $\frac{1}{1000}$ the length of 1 m. Some commonly used prefixes are listed in **Table 1.2**.

Table 1.2 Common Prefixes of the Metric System

Prefix[a]	Abbreviation	Relationship to Base Unit	Exponential Relationship to Base Unit[b]
mega-	M	1,000,000 × base unit	10^6 × base unit
kilo-	k	1000 × base unit	10^3 × base unit
deci-	d	1/10 × base unit	10^{-1} × base unit
centi-	c	1/100 × base unit	10^{-2} × base unit
milli-	m	1/1000 × base unit	10^{-3} × base unit
micro-	μ	1/1,000,000 × base unit	10^{-6} × base unit
nano-	n	1/1,000,000,000 × base unit	10^{-9} × base unit
pico-	p	1/1,000,000,000,000 × base unit	10^{-12} × base unit

[a] Prefixes in boldface (heavy) type are the most common ones.

[b] Use of exponents to express large and small numbers is discussed in **Section 1.6**.

Ian Melvin/Shutterstock.com

Figure 1.13 Olympic athletes in the triathlon compete over the Olympic standard distance of a 1.5 km swim, 40 km bike, and 10 km run!

Area and volume are examples of **derived units of measurement**. They are obtained or derived from the base unit of length as follows:

$$\text{area} = (\text{length})(\text{length}) = (\text{length})^2 \tag{1.1}$$

$$\text{volume} = (\text{length})(\text{length})(\text{length}) = (\text{length})^3 \tag{1.2}$$

The unit used to express an area or volume depends on the unit of length used.

derived unit of measurement A unit obtained by multiplication or division of one or more basic units.

Example 1.4 Calculating Areas

Calculate the area of a rectangle that has sides of 1.5 m and 2.0 m. Express the answer in units of square meters and square centimeters.

Solution

When working examples, especially exercises with calculations, it is important to use a stepwise approach. Learning the strategy here will help you in the future to deal with more challenging questions.

Step 1: Examine the question for what is given and what is unknown.

Given: sides are 1.5 m and 2.0 m Unknown: area = ?

Step 2: Identify an equation that links the given and unknown terms (Equation 1.1).

$$\text{area} = (\text{length})(\text{length})$$

Step 3: Enter the data into the formula.

$$\text{area} = (1.5 \text{ m})(2.0 \text{ m}) = 3.0 \text{ m}^2$$

According to **Table 1.1**, 1 m = 100 cm. In terms of centimeters, the lengths are 150 cm and 200 cm.

$$\text{area} = (150 \text{ cm})(200 \text{ cm}) = 30{,}000 \text{ cm}^2$$

✔ **Learning Check 1.4** The area of a circle is given by the formula $A = \pi r^2$, where r is the radius and $\pi = 3.14$. Calculate the area of a circle that has a radius of 3.5 cm.

Figure 1.14 A quart (left) is slightly smaller than a liter (right).

A volume could have units such as cubic meters (m^3), cubic decimeters (dm^3), or cubic centimeters (cm^3), depending on the unit of length used. The abbreviation cc is also used to represent cubic centimeters, especially in medical fields. The liter (L), a non-SI unit of volume, has been used by chemists for many years (see **Figure 1.14**). For all practical purposes, 1 L and 1 dm^3 are equal volumes. This also means that 1 milliliter (mL) is equal to 1 cm^3 or 1 cc in the medical field. Most laboratory glassware is calibrated in liters or milliliters.

Example 1.5 Expressing Volume in Metric Units

Laboratory tests for substances present in blood are often reported in mass per deciliter. According to **Table 1.2**, 1 deciliter (dL) = 0.1 liter (L). Express a volume of 1.36 L in terms of dL.

Solution

Because 1 dL = 0.1 L, 1.36 L can be converted to deciliters as

$$1.36\ \cancel{\text{L}} \times \frac{1\ \text{dL}}{0.1\ \cancel{\text{L}}} = 13.6\ \text{dL}$$

The units of the original quantity (1.36 L) were cancelled, and the desired units (dL) remain.

✔ **Learning Check 1.5** A clinical technician uses a micropipet to measure a 5.0 mL (5.0 milliliter) sample of blood serum for analysis. Express the sample volume in microliters (μL).

1.5B Mass

The base unit of mass in the metric system is the kilogram (kg), where 1 kg is equal to about 2.2 pounds in the English system. A kilogram is too large to be conveniently used in many laboratory and medical applications, so it is subdivided into smaller units. Two of these smaller units that are often used in chemistry are the gram (g) and milligram (mg) (see **Figure 1.15**). The prefixes *kilo-* (k) and *milli-* (m) indicate the following relationships between these units:

$$1\ \text{kg} = 1000\ \text{g}$$

$$1\ \text{g} = 1000\ \text{mg}$$

$$1\ \text{kg} = 1{,}000{,}000\ \text{mg}$$

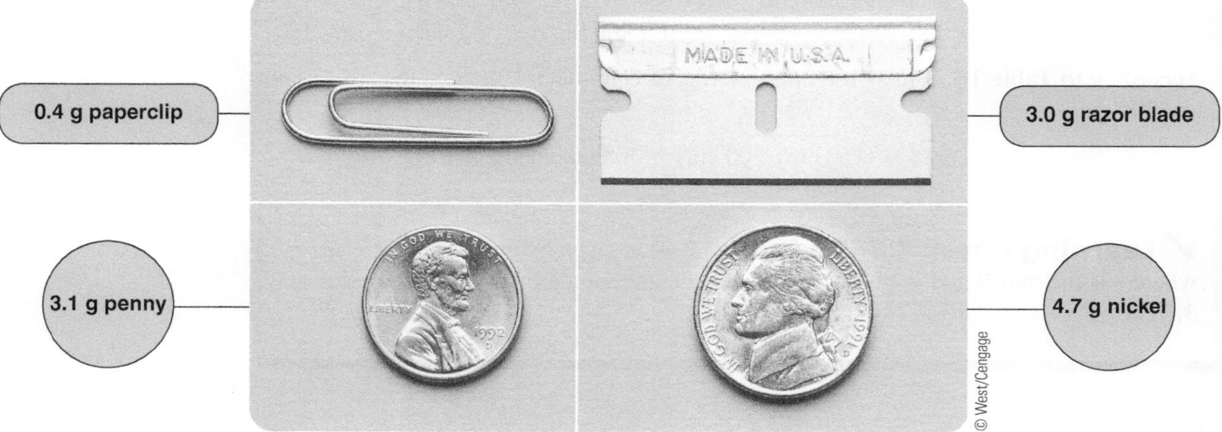

Figure 1.15 Metric masses of some common items.

Example 1.6 Expressing Measurements in Metric Units

Saline solution for intravenous infusion is used to treat dehydration and has a number of other important uses in medicine. A partial bag of saline solution has a mass of 600 grams. Express this mass in kilograms and milligrams.

Solution

Because 1 kg = 1000 g, 600 g can be converted to kilograms as follows:

$$600\ \cancel{g} \times \frac{1\ kg}{1000\ \cancel{g}} = 0.600\ kg$$

Also, because 1 g = 1000 mg,

$$600\ \cancel{g} \times \frac{1000\ mg}{1\ \cancel{g}} = 600{,}000\ mg$$

Once again, the units of the original quantity (600 g) cancel, and the desired units remain.

✔ **Learning Check 1.6** The javelin thrown by male competitors in track and field meets must have a minimum mass of 0.800 kg. A javelin is weighed and found to have a mass of 0.819 kg. Express the mass of the weighed javelin in grams.

1.5C Temperature

Temperature is difficult to define but easy for most of us to measure—we just read a thermometer (**Figure 1.16**). However, thermometers can have temperature scales that represent different units. For example, a temperature of 293 seems quite high, but it is just room temperature when measured using the Kelvin temperature scale. Temperatures on this scale are given in kelvins, K. (Note that the abbreviation is K, not °K.)

The Celsius scale (formerly known as the centigrade scale) is the temperature scale used in most scientific work. On this scale, water freezes at 0 °C and boils at 100 °C under normal atmospheric pressure. A Celsius degree (division) is the same size as a kelvin of the Kelvin scale, but the two scales have different zero points. **Figure 1.17** compares the two scientific temperature scales and the Fahrenheit scale, which is most commonly used in the United States and is not a metric unit. There are 100 Celsius degrees (divisions) between the freezing point (0 °C) and the boiling point (100 °C) of water. On the Fahrenheit scale, these same two temperatures are 180 degrees (divisions) apart (the freezing point is 32 °F and the boiling point is 212 °F). Readings on these two scales are related by Equations 1.3 and 1.4:

$$°C = \frac{5}{9}(°F - 32) \qquad (1.3)$$

$$°F = \frac{9}{5}(°C) + 32° \qquad (1.4)$$

Figure 1.16 Thermometer used to measure temperature on the Fahrenheit and Celsius scales.

As mentioned, the difference between the Kelvin and Celsius scales is simply the zero point. Consequently, readings on the two scales are related by Equations 1.5 and 1.6:

$$°C = K - 273 \qquad (1.5)$$

$$K = °C + 273 \qquad (1.6)$$

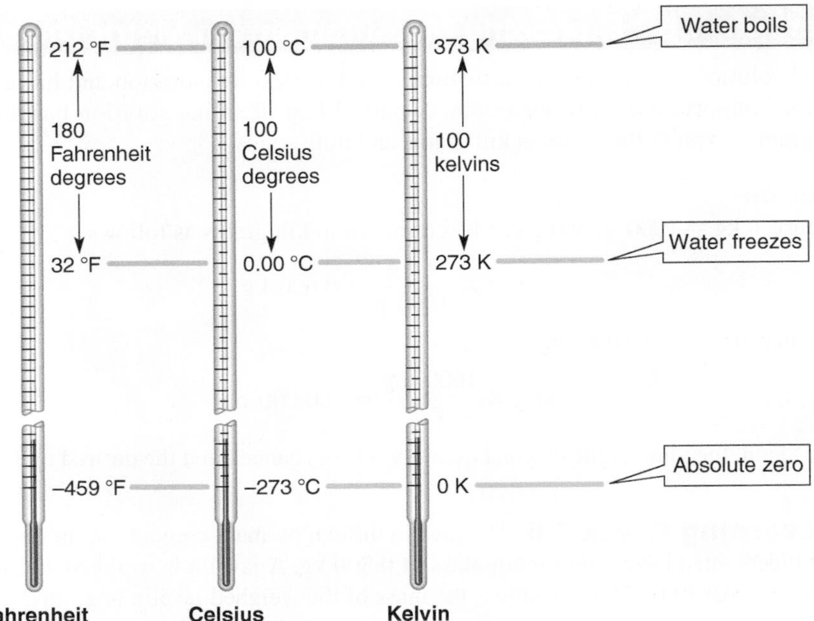

Figure 1.17 Fahrenheit, Celsius, and Kelvin temperature scales. The lowest temperature possible is absolute zero, 0 K.

Notice that Equation 1.4 can be obtained by solving Equation 1.3 for Fahrenheit degrees, and Equation 1.6 can be obtained by solving Equation 1.5 for kelvins. Thus, you need to remember only Equations 1.3 and 1.5, rather than all four.

Example 1.7 Converting Fahrenheit and Celsius Temperatures

A parent is worried because their child has a temperature of 100.8 °F. What is this reading in degrees Celsius?

Solution

Step 1: Examine the question for what is given and what is unknown.

Given: 100.8 °F Unknown: °C = ?

Step 2: Identify a formula that links the given and unknown terms.

$$\text{Equation 1.3: } ^\circ\text{C} = \frac{5}{9}(^\circ\text{F} - 32)$$

Step 3: Enter the data into the formula.

$$^\circ\text{C} = \frac{5}{9}(100.8^\circ - 32^\circ) = \frac{5}{9}(68.8^\circ) = 38.2\ ^\circ\text{C}$$

✔ **Learning Check 1.7** Body temperature is considered to be 37.0 °C when measured orally. What is this temperature in degrees Fahrenheit?

Health Connections 1.2

Effects of Temperature on Body Functions

The human body has the ability to remain at a relatively constant temperature, even when the surrounding temperature increases or decreases. Because of this characteristic, we humans are classified as warm-blooded. In reality, our body temperature varies over a significant range depending on the time of day and the temperature of our surroundings. "Normal" body temperature is considered to be 37.0 °C when measured orally. However, this "normal" value can fluctuate between a low of 36.1 °C for an individual just waking up in the morning to a high of 37.2 °C in the late evening.

In addition to this regular variation, our body temperature fluctuates in response to extremes in the temperature of our surroundings. In extremely hot environments, the capacity of our perspiration-based cooling mechanism can be overtaxed, and our body temperature will increase. Body temperatures more than 3.5 °C above normal begin to interfere with bodily functions. Body temperatures above 41.1 °C can cause convulsions and can result in permanent brain damage, especially in children.

Hypothermia occurs when the body's internal heat generation is insufficient to balance the heat lost to very cold surroundings. As a result, the body temperature decreases, and at 28.5 °C the afflicted person appears pale and might have an irregular heartbeat. Unconsciousness usually results if the body temperature drops below 26.7 °C. At these low temperatures, respiration also slows down and becomes shallow, so less oxygen is delivered to body tissues.

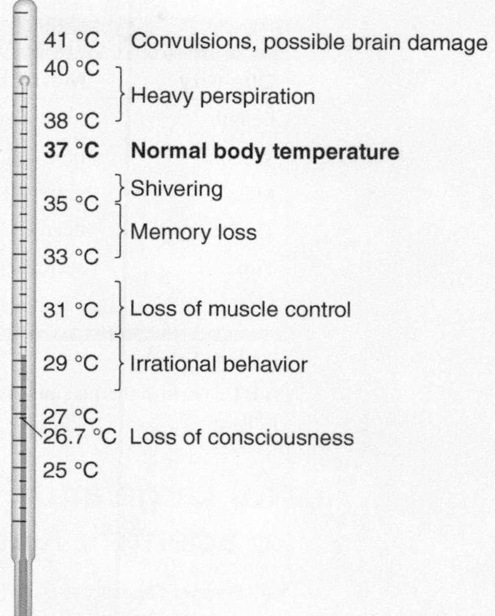

The effects of body temperature on body functions.

1.5D Energy

The last units discussed at this point are derived units of energy, a unit central to all life on Earth (**Figure 1.18**). Other units are introduced later in the book as they are needed. The metric system unit of energy is the joule (J), pronounced "jewl." A joule is quite small—a 50 watt light bulb uses 50 J of energy every second. A typical household in the United States uses several billion joules of electrical energy every month.

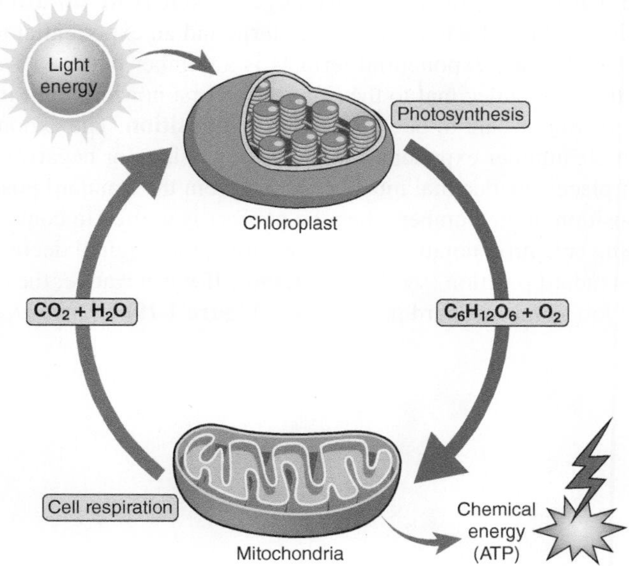

Figure 1.18 Energy cycle. Plants absorb energy from the sun and produce glucose ($C_6H_{12}O_6$) and oxygen (O_2) during photosynthesis. Animals consume plants and break down the glucose to produce chemical energy in the form of ATP during a process called cellular respiration. The by-products of cellular respiration are carbon dioxide (CO_2) and water (H_2O). These processes are discussed in greater detail in Chapter 15.

The calorie (cal), a slightly larger unit of energy, is sometimes used by chemists. One calorie is the amount of heat energy required to increase the temperature of 1 g of water by 1 °C. The calorie and joule are related as follows:

$$1 \text{ cal} = 4.184 \text{ J}$$

The nutritional calorie is actually 1000 scientific calories, or 1 kcal. It is represented by writing *calorie* with a capital C (Calorie, abbreviated Cal). **Table 1.3** lists the commonly used metric units, their relationship to basic units, and their relationship to English units.

Table 1.3 Commonly Used Units in Multiple Systems of Measurement

Quantity	Metric Unit	SI Unit	English Unit
Length	meter (m)	meter (m)	inches (in) or feet (ft)
Volume	liter (L)	cubic meter (m^3)	gallon (gal)
Mass	gram (g)	kilogram (kg)	pound (lb)
Temperature	degree Celsius (°C)	kelvin (K)	degree Fahrenheit (°F)
Time	second (s)	second (s)	minutes (min) or hours (h)
Energy	calories (cal) or joules (J)	joules (J)	British thermal unit (BTU)

[a]1 mL = 1 cm^3.

[b]A BTU (British thermal unit) is the amount of heat required to increase the temperature of 1 pound of water by 1 °F.

1.6 Large and Small Numbers: An Introduction to Scientific Notation

Learning Objective 7 Express numbers using scientific notation.

Learning Objective 8 Perform calculations with numbers expressed in scientific notation.

Numbers are used in all measurements and calculations. Many numbers are readily understood and represented because of common experience with them. A price of 10 dollars, a height of 7 feet, a weight of 165 pounds, and a time of 40 seconds are examples of such numbers. But how do we handle numbers like the diameter of a hydrogen atom (about one hundred-millionth of a centimeter) or the distance light travels in 1 year—a light-year (about 6 trillion miles)? These numbers are so small and large, respectively, that they defy understanding in terms of relationships to familiar distances. Even if we can't totally relate to them, it is important in scientific work to be able to conveniently represent and work with such numbers.

scientific notation A way of representing numbers consisting of a product between a nonexponential number and 10 raised to a whole-number exponent that may be positive or negative.

standard position for a decimal In scientific notation, the position to the right of the first nonzero digit in the nonexponential number.

Scientific notation provides a method for conveniently representing any number, including those that are very small or very large. In **scientific notation**, numbers are represented as the product of a nonexponential term and an exponential term in the general form $M \times 10^n$. The nonexponential term M is a number between 1 and 10 (but not equal to 10) written with a decimal to the right of the first nonzero digit in the number. A decimal in this position is said to be in the **standard position**. The exponential term is a 10 raised to a whole number exponent n that may be positive or negative. The value of n is the number of places the decimal must be moved from the standard position in M to be at the original position in the number when the number is written in conventional notation (i.e., without using scientific notation). If n is positive, the original decimal position is to the right of the standard position (see **Figure 1.19a**). If n is negative, the original decimal position is to the left of the standard position (see **Figure 1.19b**).

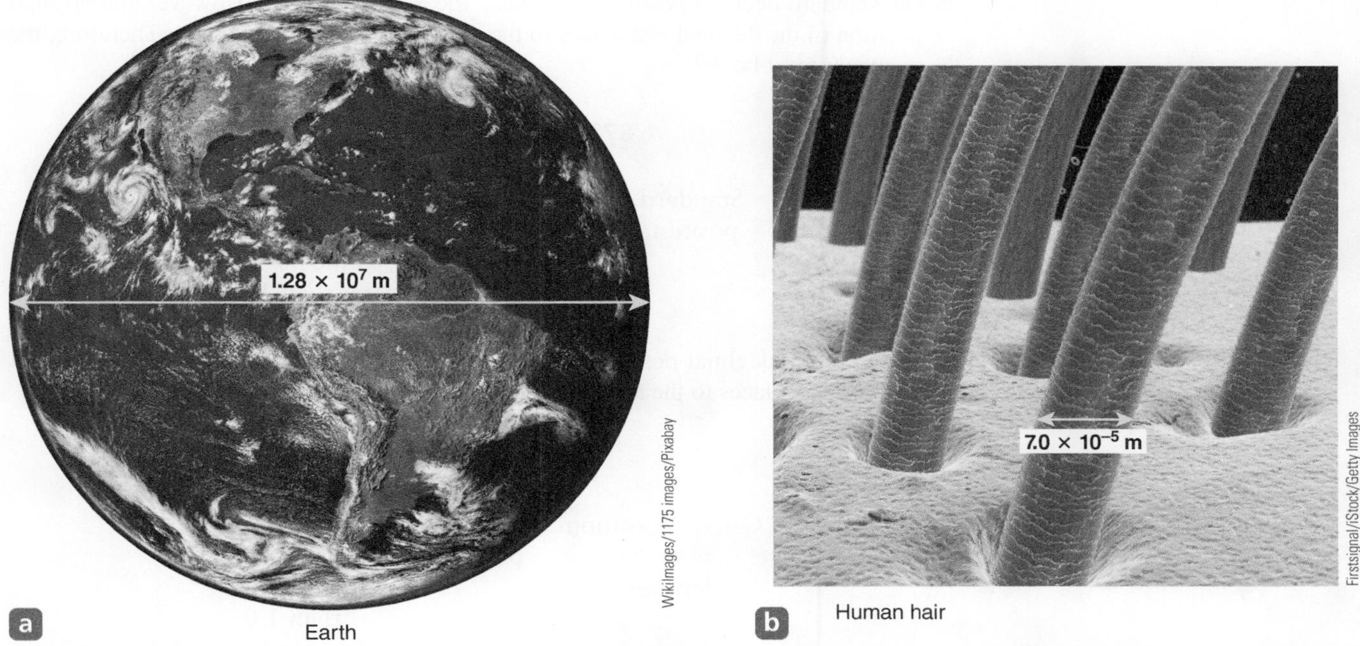

Figure 1.19 (a) The equatorial diameter of Earth is 1.28×10^7 meters. (b) The width of human hair varies depending on thickness. For average hair, the width is about 7.0×10^{-5} meters.

Example 1.8 Interconverting Scientific Notation to Nonscientific Notation

Very small numbers such as the diameter of a bacterium and very large numbers such as the number of viral particles in a sample are expressed using scientific notation. Interconvert the following numbers between scientific notation and conventional notation.

a. 3.72×10^5 b. 8.513×10^{-7} c. 8725.6 d. 0.000729

Solution

a. The exponent 5 indicates that the new position of the decimal should be located 5 places to the right of the standard position. Zeros are added to accommodate this change:

$$3.72 \times 10^5 = 372{,}000. = 372{,}000$$

Standard position (after the 3); New position (after 372,000)

b. The exponent −7 indicates that the new position of the decimal should be 7 places to the left of the standard position. Again, zeros are added as needed:

$$8.513 \times 10^{-7} = 0.0000008513$$

New position (after the 0); Standard position (after the 8)

c. The standard decimal position is between the 8 and 7: 8.7256. However, the original position of the decimal is 3 places to the right of the standard position. Therefore, the exponent must be +3:

$$8725.6 = 8.7256 \times 10^3$$

Standard position — Original position is to the right (+3) — Exponent is positive for numbers greater than 1.0

d. The standard decimal position is between the 7 and 2: 7.29. However, the original position is 4 places to the left of standard. Therefore, the exponent must be −4:

$$0.000729 = 7.29 \times 10^{-4}$$

Original position is to the left (−4) — Standard position — Exponent is negative for numbers less than 1.0

✔ **Learning Check 1.8** Some of the following numbers are written using scientific notation, and some are written using conventional notation. In each case, rewrite the numbers using the notation in which it is not written.

a. 5.88×10^2
b. 0.000439
c. 3.915×10^{-4}
d. 9870
e. 36.77
f. 0.102

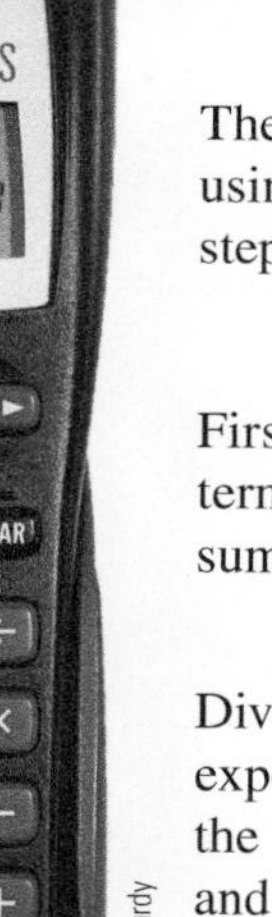

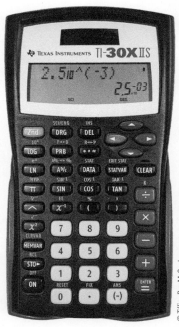

Figure 1.20 A typical scientific calculator used to express numbers in scientific notation.

The multiplication and division of numbers written in scientific notation can be done by using the characteristics of exponentials. The following multiplication can be done in two steps:

$$(a \times 10^y)(b \times 10^z)$$

First, the nonexponential terms a and b are multiplied in the usual way. The exponential terms 10^y and 10^z are multiplied by adding the exponents y and z and using the resulting sum as a new exponent of 10. Thus, we can write

$$(a \times 10^y)(b \times 10^z) = (a \times b)(10^{y+z})$$

Division is done similarly. The nonexponential terms are divided in the usual way, and the exponential terms are divided by subtracting the exponent of the bottom term from that of the top term. The final answer is then written as a product of the resulting nonexponential and exponential terms:

$$\frac{a \times 10^y}{b \times 10^z} = \left(\frac{a}{b}\right)(10^{y-z})$$

Multiplication and division calculations involving scientific notation can also be done using a calculator (**Figure 1.20**). **Table 1.4** lists the steps, the typical procedures (buttons to press), and the typical calculator readout or display for the division of 7.2×10^{-3} by 1.2×10^4.

Table 1.4 Using a Calculator for Scientific Notation Calculations

Step	Procedure		Calculator Display
1. Enter 7.2	Press buttons 7, ., 2	7 . 2	7.2
2. Enter 10^{-3}	Press button that activates exponential mode (EE, Exp, etc.)	EE or EXP	7.2^{00}
	Press 3	3	7.2^{03}
	Press change-sign button (±, etc.)	+/−	7.2^{-03}
3. Divide	Press divide button (÷)	÷	7.2^{-03}
4. Enter 1.2	Press buttons 1, ., 2	1 . 2	1.2
5. Enter 10^4	Press button that activates exponential mode (EE, Exp, etc.)	EE or EXP	1.2^{00}
	Press 4	4	1.2^{04}
6. Obtain answer	Press equals button (=)	=	$6.^{-07}$

Example 1.9 Calculations with Scientific Notation

Do the following operations.

a. $(3.5 \times 10^4)(2.0 \times 10^2)$ c. $(4.6 \times 10^{-7})(5.0 \times 10^3)$

b. $\dfrac{3.8 \times 10^5}{1.9 \times 10^2}$ d. $\dfrac{1.2 \times 10^3}{3.0 \times 10^{-2}}$

Solution

a. $(3.5 \times 10^4)(2.0 \times 10^2) = (3.5 \times 2.0)(10^4 \times 10^2) = (7.0)(10^{4+2}) = 7.0 \times 10^6$

b. $\dfrac{3.8 \times 10^5}{1.9 \times 10^2} = \dfrac{3.8}{1.9} \times \dfrac{10^5}{10^2} = (2.0)(10^{5-2}) = 2.0 \times 10^3$

c. $(4.6 \times 10^{-7})(5.0 \times 10^3) = (4.6 \times 5.0)(10^{-7} \times 10^3) = (23)(10^{-7+3}) = 23 \times 10^{-4}$

To get the decimal into the standard position, as required for correct scientific notation, move it one place to the left. This changes the exponent from −4 to −3, so the final result is 2.3×10^{-3}.

d. $\dfrac{1.2 \times 10^3}{3.0 \times 10^{-2}} = \dfrac{1.2}{3.0} \times \dfrac{10^3}{10^{-2}} = (0.40)(10^{3-(-2)}) = 0.40 \times 10^5$

Adjust the decimal to the standard position and get 4.0×10^4.

You may find it beneficial to repeat these four calculations using a calculator to make sure you get the same answers.

✔ Learning Check 1.9 Perform the following operations, and express the result in correct scientific notation.

a. $(2.4 \times 10^3)(1.5 \times 10^4)$ b. $\dfrac{6.3 \times 10^5}{2.1 \times 10^3}$ c. $\dfrac{(1.7 \times 10^4)(2.9 \times 10^3)}{3.4 \times 10^2}$

1.7 Significant Figures

Learning Objective 9 Express measurements and calculations using the correct number of significant figures.

Every measurement contains an uncertainty that is characteristic of the device used to make the measurement. These uncertainties are represented by the numbers used to record the measurement. Consider the square in **Figure 1.21**.

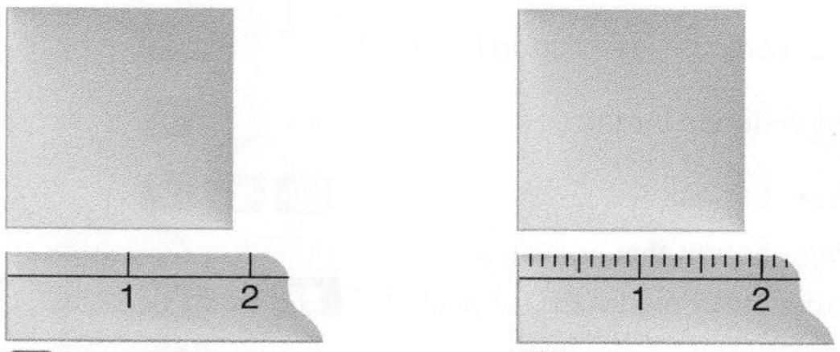

Figure 1.21 Measurements using two different rulers.

a The ruler is divided into centimeters.

b The ruler is divided into tenths of centimeters (mm).

In **Figure 1.21a**, the length of one side of the square is measured with a ruler divided into centimeters. Notice how the length is greater than 1 cm, but not quite 2 cm. The length is recorded by writing the number that is known with certainty to be correct (the 1) and writing an estimate for the uncertain number. The result is 1.9 cm, where the .9 is the estimate. In **Figure 1.21b**, the ruler is divided into tenths of centimeters. Notice that the length is at least 1.8 cm, but not quite 1.9 cm. Once again the certain numbers (1.8) are written, and an estimate is made for the uncertain part. The result is 1.86 cm.

When measurements are recorded this way, the numbers representing the certain measurement plus one number representing the estimate are called **significant figures**. Thus, the first measurement of 1.9 cm contains two significant figures, whereas the second measurement of 1.86 cm contains three significant figures.

significant figures The numbers in a measurement that represent the certainty of the measurement, plus one number representing an estimate.

The maximum number of significant figures possible in a measurement is determined by the design of the measuring device and cannot be changed by expressing the measurement in different units. Thus, the 1.9 cm length can also be represented in terms of meters and millimeters as follows:

$$1.9 \text{ cm} = 0.019 \text{ m} = 19 \text{ mm}$$

In this form, it may appear that the length expressed as 0.019 m contains four significant figures, but this is impossible; a measurement made with a device doesn't become more certain simply by changing the unit used to express the number. Thus, the zeros are not significant figures—their only function is to locate the correct position for the decimal. Zeros located to the left of nonzero numbers, such as the two zeros in 0.019 cm, are never considered to be significant. Thus, 12.5 mg, 0.0125 g, and 0.000125 kg all represent the same measured mass, and all contain three significant figures.

Significant figures

Leading zero — Buried zero — Nonzero digits — Trailing zero

0.0040710

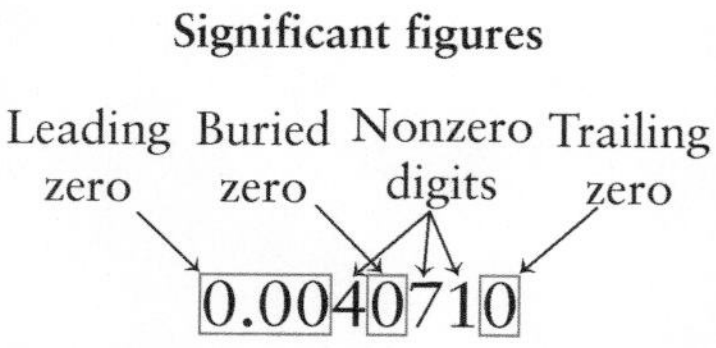

Figure 1.22 Five significant figures. Leading zeros are not significant, but buried zeros and trailing zeros to the right of a decimal point are.

Zeros located between nonzero numbers are considered significant. Thus, 2055 μL, 2.055 mL, and 0.002055 L all represent the same volume measurement, and all contain four significant figures. Trailing zeros at the end of a number are significant if there is a decimal point present in the number. Otherwise, trailing zeros are not significant. The rules for determining the significance of zeros can be summarized as follows (see **Figure 1.22**):

1. Zeros not preceded by nonzero numbers are not significant figures. These zeros are sometimes called *leading zeros*.
2. Zeros located between nonzero numbers are significant figures. These zeros are sometimes called *buried* or *confined zeros*.
3. Zeros located at the end of a number are significant figures if there is a decimal point present in the number. These zeros are sometimes called *trailing zeros*. If no decimal point is present, trailing zeros are not significant.

In this book, large numbers are always represented by scientific notation. In scientific notation, the correct number of significant figures is used in the nonexponential term, and the location of the decimal is determined by the exponent.

Example 1.10 Significant Figures and Scientific Notation

Determine the number of significant figures in each of the following measurements and use scientific notation to express each measurement using the correct number of significant figures.

a. 24.6 °C b. 0.036 g c. 15.0 mL d. 0.0020 m

Solution

a. All the numbers are significant: three significant figures, 2.46×10^1 °C.
b. The leading zeros are not significant: two significant figures, 3.6×10^{-2} g.
c. The trailing zero is significant because there is a decimal point: three significant figures, 1.50×10^1 mL.
d. The leading zeros are not significant, but the trailing zero is: two significant figures, 2.0×10^{-3} m.

✔ **Learning Check 1.10** Determine the number of significant figures in each of the following measurements and use scientific notation to express each of the following measurements using the correct number of significant figures.

a. 250 mg
b. 18.05 mL
c. 0.0108 kg
d. 37 °C
e. 0.001 mm
f. 101.0 K

Most measurements that are made do not stand as final answers. Instead, they are usually used to make calculations involving multiplication, division, addition, or subtraction. The answer obtained from such a calculation cannot have more certainty than the least certain measurement used in the calculation. It should be written to reflect an uncertainty equal to that of the most uncertain measurement. This is accomplished by using the following rules:

1. The answer obtained by multiplication or division must contain the same number of significant figures as the quantity with the fewest significant figures used in the calculation.

$$(14.00)(25.00)/(10.0) = 35.0$$

   The final answer has three significant figures because 10.0 has three, whereas 14.00 and 25.00 have four.

2. The answer obtained by addition or subtraction must contain the same number of places to the right of the decimal as the quantity in the calculation with the fewest number of places to the right of the decimal.

$$12.250 + 4.75 = 17.00$$

   The final answer has two zeros to the right of the decimal, not three, because 4.75 has two places to the right of the decimal, whereas 12.250 has three.

To follow these rules, it is often necessary to reduce the number of significant figures by rounding answers. The following are the rules for rounding:

1. If the first of the nonsignificant figures to be dropped from an answer is 5 or greater, all the nonsignificant figures are dropped, and the last significant figure is increased by 1. For example, rounding 37.38 to three significant figures gives 37.4.

Figure 1.23 Calculators usually display the sum of 4.362 and 2.638 as 7 (too few figures), and the product of 0.67 and 10.14 as 6.7938 (too many figures).

2. If the first of the nonsignificant figures to be dropped from an answer is less than 5, all nonsignificant figures are dropped, and the last significant figure is left unchanged. For example, rounding 38.383 to four significant figures gives 37.38.

When you use a calculator, it will often express answers with too few or too many significant figures (see **Figure 1.23**). It will be up to you to determine the proper number of significant figures to use and to round the calculator answer correctly.

Example 1.11 Significant Figures in Multiplication and Division

On a typical day, nurses make numerous measurements and calculations. Reporting data in the most meaningful way is critical. Do the following calculations, and round the answers to the correct number of significant figures.

a. $(4.95)(12.10)$ b. $\dfrac{3.0954}{0.0085}$ c. $\dfrac{(9.15)(0.9100)}{3.117}$

Solution

All calculations are done with a calculator, and the calculator answer is written first. Appropriate rounding is done to get the final answer.

a. Calculator answer: 59.895

The number 4.95 has three significant figures, whereas 12.10 has four. This is a multiplication, so the answer must have three significant figures:

59.895

Significant figures (59.8) Nonsignificant figures (95)

The first of the nonsignificant figures to be dropped is 9, so after both are dropped, the last significant figure is increased by 1. The final answer containing three significant figures is 59.9.

b. Calculator answer: 364.16471

The number 3.0954 has five significant figures, whereas 0.0085 has two. This is a division, so the answer must have only two significant figures:

364.16471

Significant figures (36) Nonsignificant figures (4.16471)

The first of the nonsignificant figures to be dropped is 4, so the last significant figure remains unchanged after the nonsignificant figures are dropped. The correct answer then is 360, which contains two significant figures. The answer also can be written with the proper number of significant figures by using scientific notation, 3.6×10^2.

c. Calculator answer: 2.6713186

The number 9.15 has three significant figures, 0.9100 has four, and 3.117 has four. Thus, the answer must have only three:

2.6713186

Significant figures (2.67) Nonsignificant figures (13186)

Appropriate rounding gives 2.67 as the correct answer.

✔ **Learning Check 1.11** Do the following calculations, and round the answers to the correct number of significant figures.

a. (0.0019)(21.39) b. $\frac{8.321}{4.1}$ c. $\frac{(0.0911)(3.22)}{1.379}$

Example 1.12 Significant Figures in Addition and Subtraction

Do the following additions and subtractions, then write the answers with the correct number of significant figures.

a. 1.9 + 18.65 b. 15.00 − 8.0 c. 1500 + 10.9 + 0.005

Solution

In each case, the numbers are arranged vertically with the decimal points aligned. The answer is then rounded so that it contains the same number of places to the right of the decimal as the quantity with the fewest.

a.
$$\begin{array}{r} 1.9 \\ 18.65 \\ \hline 20.55 \end{array}$$

The answer must be expressed with one place to the right of the decimal to match the one in 1.9.

Correct answer: 20.6 (Why was the final 5 increased to 6?)

b.
$$\begin{array}{r} 15.00 \\ -8.0 \\ \hline 7.0 \end{array}$$

The answer must be expressed using one place to the right of the decimal to match the 8.0. A typical calculator answer would probably be 7, requiring that a zero be added to the right of the decimal to provide the correct number of significant figures.

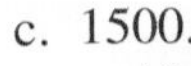

c.
$$\begin{array}{r} 1500. \\ 10.9 \\ 0.005 \\ \hline 1510.905 \end{array}$$

The answer must be expressed with no places to the right of the decimal to match the 1500.

Correct answer: 1511 (Why was the final 0 of the answer increased to 1? (See **Figure 1.24**.)

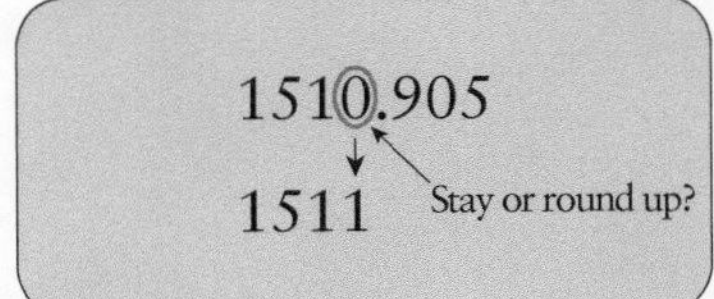

Figure 1.24 Should you round up or round down? The digit to the right of the zero that is rounded is a 5 or greater, so the zero must be rounded up to 1.

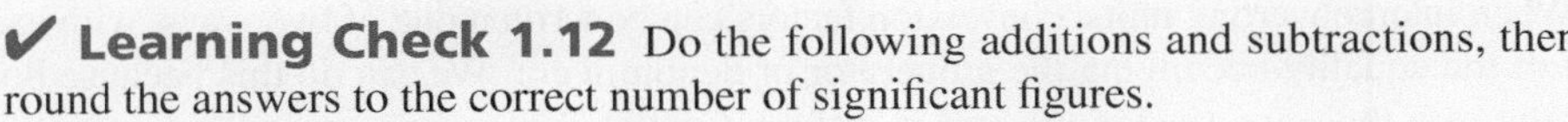

✔ **Learning Check 1.12** Do the following additions and subtractions, then round the answers to the correct number of significant figures.

a. 8.01 + 3.2
b. 3000 + 20.3 + 0.009
c. 4.33 − 3.12
d. 6.023 − 2.42

Some numbers used in calculations are exact. **Exact numbers** have no uncertainty associated with them and are considered to contain an unlimited number of significant trailing zeros. Such numbers are not used when the appropriate number of significant figures is determined for calculated answers. In other words, exact numbers do not limit the number of significant figures in calculated answers. One kind of exact number is a number used as part of a defined relationship between quantities. For example, 1 m contains exactly 100 cm:

$$1 \text{ m} = 100 \text{ cm}$$

Thus, the numbers 1 and 100 are exact. A second kind of exact number is a counting number obtained by counting individual objects. A dozen eggs contains exactly 12 eggs, not 11.8 and not 12.3. The 12 is an exact counting number. A third kind of exact number is one that is part of a simple fraction, such as $\frac{1}{4}$, $\frac{2}{3}$, or the $\frac{5}{9}$ used in Equation 1.3 to convert Fahrenheit temperatures into Celsius temperatures.

exact numbers Numbers that have no uncertainty; numbers from defined relationships, counting numbers, and numbers that are part of simple fractions.

1.8 Using Units in Calculations: An Introduction to Dimensional Analysis

Learning Objective 10 Use dimensional analysis to solve numerical problems.

A large number of different units are used to measure the same quantities, so we must be able to interconvert between these units. Sometimes it is necessary to convert between English units and SI or metric units (yd to m, kg to lb, etc.). Other times we must convert between different units within metric or English units (m to cm, ft to yd, etc.). **Table 1.5** lists some of the most common relationships between metric units and between metric and English units. Because these are standard values, we know the amount of one unit that is equal to another unit (e.g., 1 m = 1.094 yd and 1000 g = 1 kg). When these equalities are rewritten as ratios (fractions), they become **conversion factors**. We can use these conversion factors to convert one unit into another through **dimensional analysis**, a systematic, mathematical approach to this unit interconversion. In fact, you may not have realized it at the time, but we have already used dimensional analysis to solve Examples 1.5 and 1.6. Another practical application of the mathematical method of dimensional analysis can be seen when dosages must be altered based on body mass. An amount of a drug in mg may need to be altered based on the mass in kg of the patient in order for the medication to be effective (**Figure 1.25**).

conversion factors Fractions obtained from numerical relationships between quantities.

dimensional analysis The systematic, mathematical method of converting a quantity from one unit to another.

iStock.com/SDI Productions

Figure 1.25 It is vital that pharmacists correctly calculate the dosage of medicine based on a patient's body mass. An error in unit conversion could prove to be fatal.

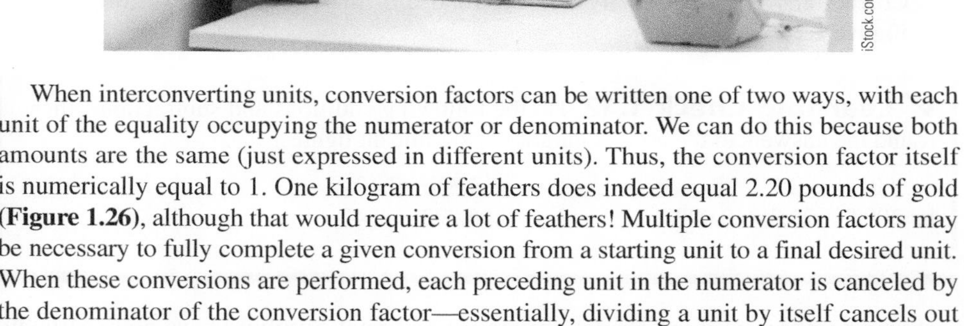

When interconverting units, conversion factors can be written one of two ways, with each unit of the equality occupying the numerator or denominator. We can do this because both amounts are the same (just expressed in different units). Thus, the conversion factor itself is numerically equal to 1. One kilogram of feathers does indeed equal 2.20 pounds of gold (**Figure 1.26**), although that would require a lot of feathers! Multiple conversion factors may be necessary to fully complete a given conversion from a starting unit to a final desired unit. When these conversions are performed, each preceding unit in the numerator is canceled by the denominator of the conversion factor—essentially, dividing a unit by itself cancels out that unit. The following is a stepwise approach to these dimensional analysis problems.

vetre/Adobe Stock Photos

Figure 1.26 A single kilogram of feathers is equal to 2.20 pounds of gold!

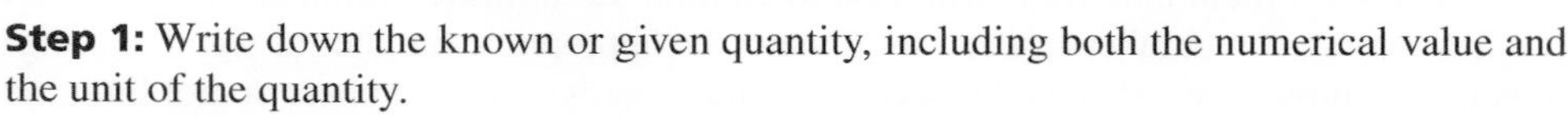

Step 1: Write down the known or given quantity, including both the numerical value and the unit of the quantity.

Step 2: Determine the unit required by the desired final product.

Step 3: Use as many conversion factors as necessary to turn the known quantity's units into the units of the desired final product. One or more conversion factors may be necessary to link the starting quantity's units to the units of the unknown final product, with units canceling each step along the way.

Step 4: Beginning with the starting quantity, multiply by each conversion factor so that you end with the units of the desired final product. Because the conversion factors are derived from equalities, it is as if we are multiplying the starting quantity each time by 1—that is, we aren't changing the *amount*, we are just changing the *unit*.

Table 1.5 Commonly Used Conversion Factors

Quantity	Metric	English	Metric-English
Length	100 cm = 1 m 1000 mm = 1 m 1 km = 1000 m	1 ft = 12 in 1 yd = 3 ft 1 mi = 5280 ft	1 m = 1.094 yd 1 cm = 0.394 in. 1 mm = 0.0394 in. 1 km = 0.621 mi
Volume	[a]1 cm^3 = 1 mL 1 L = 1 dm^3 1000 mL = 1 dm^3	1 qt = 4 cups 1 qt = 2 pints 1 gal = 4 qt	1 dm^3 = 1.057 qt 1 cm^3 = 0.0338 fl oz 1 L = 1.057 qt 1 mL = 0.0338 fl oz
Mass	1000 g = 1 kg 1,000,000 mg = 1 kg	1 lb = 16 oz	1 g = 0.035 oz 1 mg = 0.015 grain 1 kg = 2.20 lb
Energy	1 cal = 4.184 J 1 kcal = 4184 J	[b]1 ft lbf = 0.0013 BTU[c]	252 cal = 1 BTU 1055 J = 1 BTU
Time	1 h = 60 min 1 min = 60 s	1 h = 60 min 1 min = 60 s	Same unit used

[a] The cm^3 unit is commonly abbreviated "cc."

[b] A foot-pound is the energy it takes to push with one pound-force (lbf) for a distance of one foot.

[c] A BTU (British thermal unit) is the amount of heat required to increase the temperature of 1 pound of water by 1 °F.

Example 1.13 Dimensional Analysis Calculations

As part of an annual physical exam, Brent's weight was found to be 145 lb. What would Brent's weight be in kilograms? Use the dimensional analysis method and numerical relationships from **Table 1.5**.

Solution

The known quantity is 145 lb, and the unit of the unknown quantity is kilograms (kg).

Step 1: 145 lb

Step 2: 145 lb = ____ kg

Step 3: $145\ \cancel{\text{lb}} \times \dfrac{1\ \text{kg}}{2.20\ \cancel{\text{lb}}} = ?\ \text{kg}$

The factor $\dfrac{1\ \text{kg}}{2.20\ \text{lb}}$ came from the numerical relationship 1 kg = 2.20 lb found in **Table 1.5**.

Step 4: $\dfrac{(145)(1)\ \text{kg}}{2.20} = 65.9090\ \text{kg} = 65.9\ \text{kg}$

Step 5: Check to see whether the answer makes sense.

Yes. The answer should be rounded to 65.9 kg, an answer that contains three significant figures, just as 145 lb does. The 1 kg in the conversion factor is an exact number used as part of a defined relationship, so it doesn't influence the number of significant figures in the answer.

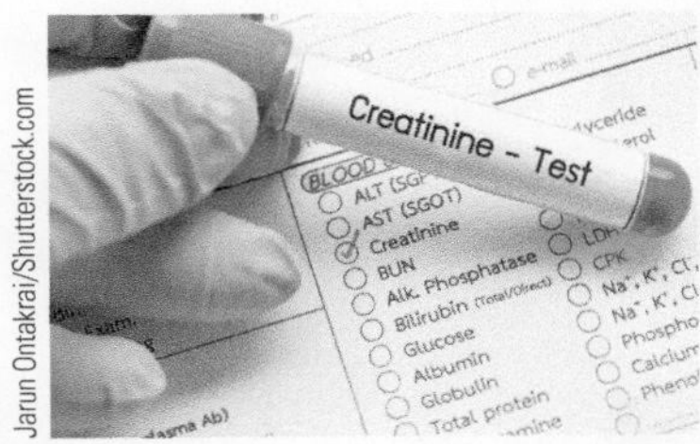

Jarun Ontakrai/Shutterstock.com

Figure 1.27 Blood tests can detect very low levels of creatinine, as well as a variety of other chemicals in the body.

✔ **Learning Check 1.13** Creatinine is a substance found in the blood (**Figure 1.27**). An analysis of a blood serum sample detected 1.1 mg of creatinine. Express this amount in grams by using dimensional analysis.

Example 1.14 Complex Dimensional Analysis Calculations

One of the fastest-moving nerve impulses in the body travels at a speed of 405 feet per second (ft/s). What is the speed in miles per hour?

Solution

The known quantity is 405 ft/s, and the unit of the unknown quantity is miles per hour (mi/h).

Step 1: $\dfrac{405 \text{ ft}}{\text{s}}$

Step 2: $\dfrac{405 \text{ ft}}{\text{s}} = ? \dfrac{\text{mi}}{\text{h}}$

Step 3: $\left(\dfrac{405 \cancel{\text{ft}}}{\cancel{\text{s}}}\right)\left(\dfrac{1 \text{ mi}}{5280 \cancel{\text{ft}}}\right)\left(\dfrac{60 \cancel{\text{s}}}{1 \cancel{\text{min}}}\right)\left(\dfrac{60 \cancel{\text{min}}}{1 \text{ h}}\right) = ? \dfrac{\text{mi}}{\text{h}}$

Figure 1.28 Jamaican sprinter Usain Bolt captured the Olympic gold medal in the 100 m race three times (2008, 2012, and 2016), making him the "fastest person on the planet." As of 2023, Bolt still held the world record of 9.58 s.

Everett Collection Inc/Alamy Stock Photo

The conversion factors came from the following numerical relationships: 1 mi = 5280 ft, 1 min = 60 s, and 1 h = 60 min. All numbers in these conversion factors are exact numbers based on definitions. The dimensional analysis leaves the desired units of miles in the numerator and hours in the denominator.

Step 4: $\dfrac{(405)(1)(60)(60) \text{ mi}}{(5280)(1)(1) \text{ h}} = 276.1 \dfrac{\text{mi}}{\text{h}}$

Rounding to three significant figures, the same as in 405 ft/s, gives 276 mi/h.

✔ **Learning Check 1.14** A world-class sprinter can run 100 m in 10.0 s (**Figure 1.28**). This corresponds to a speed of 10.0 m/s. Convert this speed to miles per hour. Use information from **Table 1.5**.

1.9 Calculating Percentages

Learning Objective 11 Perform calculations involving percentages.

The word *percent* literally means per one hundred. It is the number of specific items in a group of 100 such items. Because items are seldom found in groups of exactly 100, we usually have to calculate the number of specific items that would be in the group if it did contain exactly 100 items. This number is the percentage, and it is calculated as follows:

$$\text{Percent} = \frac{\text{number of specific items}}{\text{total items in the group}} \times 100\%$$
$$\% = \frac{\text{part}}{\text{total}} \times 100\% \qquad (1.7)$$

In Equation 1.7, the word *part* is used to represent the number of specific items included in the total.

Example 1.15 Percentage Calculations

On the advice of his doctor, Mark lost 18 pounds in body weight. If Mark's initial weight was 208 pounds, what percentage of his weight was lost?

Solution

Step 1: Examine the question for what is given and what is unknown.

Given: Beginning weight = 208 pounds

Unknown: percent = ?

Weight lost = 18 pounds

Step 2: Identify a formula that links the given and unknown terms.

$$\text{Equation 1.7: } \% = \frac{\text{part}}{\text{total}} \times 100\%$$

Step 3: Enter the data into the formula.

$$\% = \frac{18 \text{ pounds}}{208 \text{ pounds}} \times 100\% = 8.7\% (\text{rounded})$$

Step 4: Check to see whether the answer makes sense. Yes, Mark lost a little less than one-tenth of his weight.

Figure 1.29 Regular exercise is typically a part of any healthy weight-loss program.

> ✔ **Learning Check 1.15** In an effort to lose weight, Debra went from a beginning weight of 184 pounds to a new weight of 167 pounds (**Figure 1.29**). What percent of Debra's initial weight was lost?

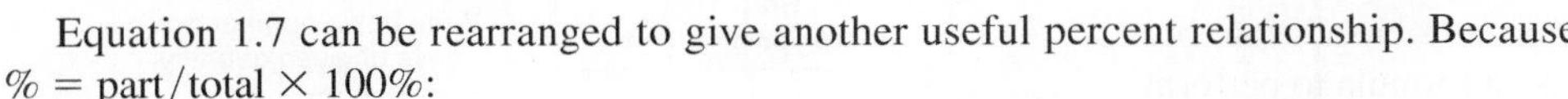

Equation 1.7 can be rearranged to give another useful percent relationship. Because $\% = \text{part}/\text{total} \times 100\%$:

$$\text{Part} = \frac{(\%)(\text{total})}{100\%} \tag{1.8}$$

According to Equation 1.8, the number of specific items corresponding to a percentage can be calculated by multiplying the percentage and total, then dividing their product by 100%.

Example 1.16 More Percentage Calculations

The human body is approximately 64% water by mass (**Figure 1.30**). What is the mass of water in a 170 pound (lb) person?

Solution

Step 1: Examine the question for what is given and what is unknown.

Given: 64% water

Unknown: mass of water in 170 lb person

Step 2: Identify a formula that links the given and unknown terms.

$$\text{Equation 1.8: Part} = \frac{(\%)(\text{total})}{100}$$

Step 3: Enter the data into the formula.

$$\text{Part} = \frac{(64\%)(170 \text{ 1b})}{100\%} = 108.8 \text{ lb}$$

We should round our answer to only two significant figures to match the two in the 64%. The answer is 110 lb, where the trailing zero is not significant.

Step 4: Check to see whether the answer makes sense.

Yes. If 64% of the body is water, you would expect the weight of water to be greater than half of the body weight.

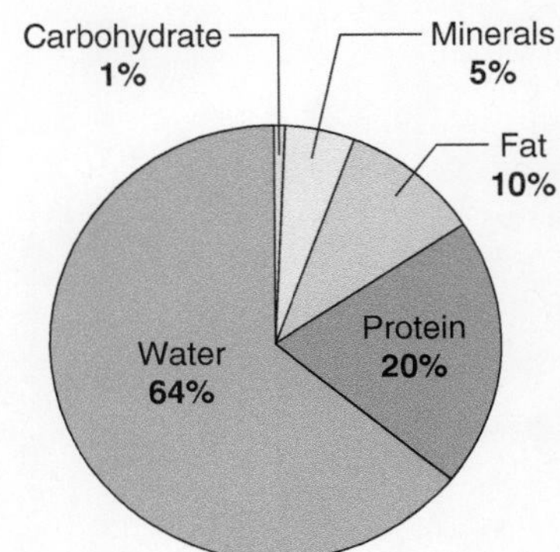

Figure 1.30 The human body, much like Earth itself, is composed primarily of water.

> ✔ **Learning Check 1.16**
>
> a. A student has saved \$988 toward a computer that will cost a total of \$1200. What percentage of the purchase price has been saved so far?
>
> b. In a chemistry class of 83 students, 90.4% voted to cancel the final exam. How many students wanted to have the exam?

Study Tools 1.1

Help with Calculations

Many students feel uneasy about working chemistry problems that involve the use of mathematics. One tip that will help you solve such problems in this textbook is to remember that almost all of these problems are one of two types: those for which a specific formula applies and those where dimensional analysis is used. When you encounter a math-type problem, your first task should be to decide which type of problem it is: one you solve with an equation or one that you can do using dimensional analysis.

In this chapter, the percentage calculations in Examples 1.15 and 1.16 are formula-based problems. The temperature conversions in Example 1.7 may seem formula-based, too, but they are a special case, because they use a formula to perform a unit conversion.

The dimensional analysis discussed in **Section 1.8** is used for many problems that require mathematical calculations. The beauty of this method is that it mimics your natural, everyday way of solving problems. This real-life method usually involves identifying where you are, where you want to go, and how to get there.

A key to working story problems is to see through all the words and find what is given (number and units). Then look for what is wanted by focusing on key words or phrases like "how much," "what is," and "calculate." Finally, use one of the two methods, formula or dimensional analysis, to solve the problem.

The steps used to solve both types of problems are summarized in the following flow charts.

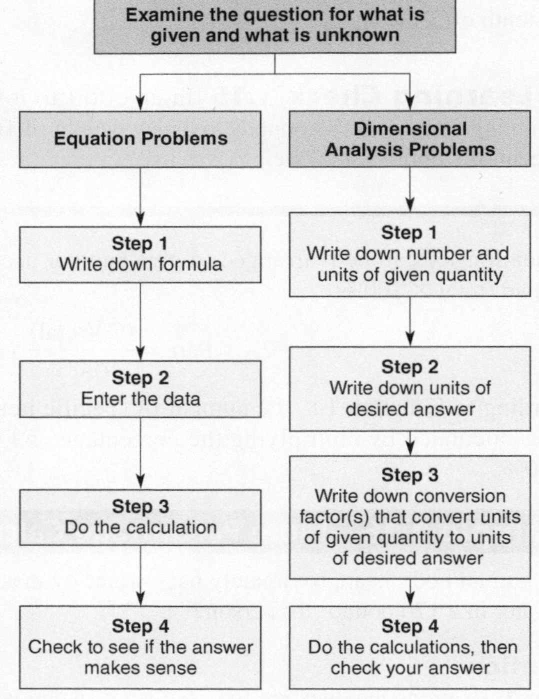

1.10 Density and Its Applications

Learning Objective 12 Perform calculations involving densities.

density The number obtained when the mass of a sample of a substance is divided by its volume.

Density is a physical property of matter, so it can be measured without changing the composition of the sample of matter under investigation. **Density** (*d*) is the number obtained by dividing the mass (*m*) of a sample of matter by the volume (*v*) of the same sample:

$$\text{Density} = \frac{\text{mass}}{\text{volume}}$$

$$d = \frac{m}{V} \qquad (1.9)$$

According to Equation 1.9, once we have obtained a numerical value for the density, we can use conversion factors that relate mass and volume, as shown in Example 1.17.

Example 1.17 Using Density in Calculations

a. Low-density lipoprotein (LDL) cholesterol is often referred to as the "bad" cholesterol (**Figure 1.31**). If a blood test reveals that a 0.365 g sample of LDL had a volume of 0.353 mL, what is the density of the LDL sample in g/mL?

b. The density of iron metal has been determined to be 7.20 g/cm^3. Use the density value to calculate the mass of an iron sample that has a volume of 35.0 cm^3.

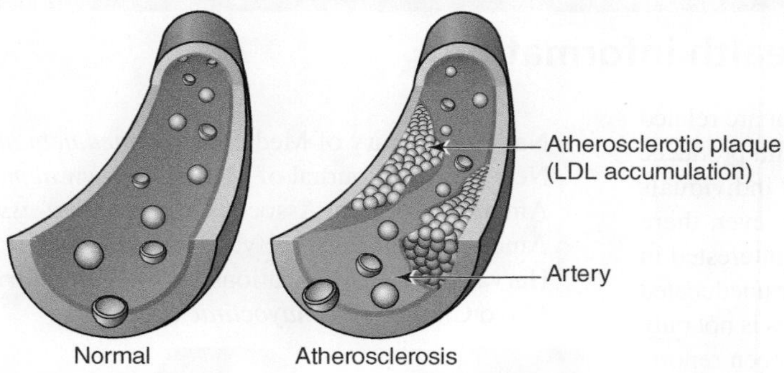

Figure 1.31 Low-density lipoprotein (LDL) cholesterol is referred to as the "bad" cholesterol because it can lead to atherosclerosis (hardening of the arteries).

Solution

a. **Step 1:** Examine the question for what is given and what is unknown.

Given: $m = 0.365$ g and $V = 0.353$ mL
Unknown: $d = ?$

Step 2: Identify a formula that links the given and unknown terms.

$$\text{Equation 1.9: } d = \frac{m}{V}$$

Step 3: Enter the data into the formula.

$$d = \frac{0.365 \text{ g}}{0.353 \text{ mL}} = 1.03 \text{ g/mL}$$

b. **Step 1:** Examine the question for what is given and what is unknown.

Given: $d = 7.20$ g/cm^3 and V = 35.0 cm^3
Unknown: $m = ?$

Step 2: Identify a formula that links the given and unknown terms.

$$\text{Equation 1.9: } d = \frac{m}{V}$$

Step 3: Enter the data into the formula.

$$7.20 \text{ g/cm}^3 = \frac{m}{35.0 \text{ cm}^3}$$

Multiply both sides of the equation by the denominator, 35.0 cm^3

$$(7.20 \text{ g/cm}^3)(35.0 \text{ cm}^3) = \left(\frac{m}{\cancel{35.0 \text{ cm}^3}}\right)(\cancel{35.0 \text{ cm}^3})$$

$$m = 252 \text{ g}$$

✔ Learning Check 1.17

a. A patient receives a transfusion of 1.20 L of type O blood. If the density of the blood is 1.04 g/mL, how many grams of blood did the patient receive?
b. Aluminum metal has a density of 2.7 g/cm^3. Calculate the volume of an aluminum sample that has a mass of 98.5 g.

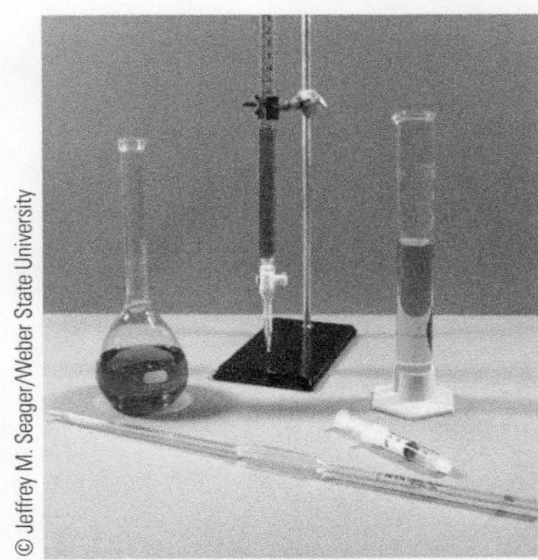
© Jeffrey M. Seager/Weber State University

Figure 1.32 Glassware for measuring volumes of liquids. Clockwise from top center: buret, graduated cylinder, syringe, pipet, and volumetric flask.

For some substances, density can be determined experimentally by direct measurement. The mass of a sample is obtained by weighing the sample. The sample volume can be calculated if the sample is a regular solid such as a cube. If the sample is a liquid or an irregular solid, the volume can be measured by using various volumetric apparatuses such as those shown in **Figure 1.32**. Densities of solids are often given in units of g/cm^3, and those of liquids in units of g/mL because of the different ways the volumes are determined. Recall from **Table 1.5**, however, that 1 cm^3 = 1 mL, so the numerical value is the same regardless of which of the two volume units is used.

Health Connections 1.3

Choose Wisely for Health information

The Internet contains more than 3 million sites that are related to health issues or health problems or that sell health products. Many of these sites are very good resources for individuals concerned about such topics. Unfortunately, however, there are also sites run by scam artists who are simply interested in trying to make money at the expense of gullible or uneducated web surfers. The information provided by these sites is not only useless but might also be dangerous. There have been reports of sites run by individuals claiming to have a "miracle cure" for serious diseases. The sites encourage individuals to stop taking their prescription medication and instead buy the new miracle product. Another characteristic of a site to be avoided is one that claims to have a physician who will diagnose or treat you without requiring you to have a proper examination and consultation. As a general rule, you should use common sense. Another good idea is to find websites that are already linked with organizations you are familiar with or recognize as being legitimate. The following sites are very helpful and contain accurate and complete information.

National Cancer Institute: *www.cancer.gov*
Centers for Disease Control and Prevention: *www.cdc.gov*
National Library of Medicine: *pubmed.ncbi.nlm.nih.gov*
New England Journal of Medicine: *www.nejm.org*
American Medical Association: *www.ama-assn.org/*
American Cancer Society: *www.cancer.org*
Harvard Health Publications: *www.health.harvard.edu*
Mayo Clinic: *www.mayoclinic.org*

iStockphoto.com/Daniel Laflor

An enormous amount of health information is available on the Internet. Telehealth appointments with legitimate health care providers are also becoming more common.

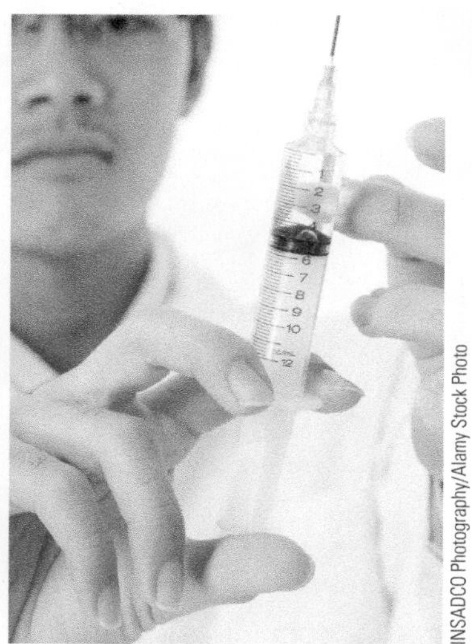

INSADCO Photography/Alamy Stock Photo

Figure 1.33 Hypodermic syringes are used to measure and administer medicines to patients.

Example 1.18 More Use of Density in Calculations

a. A hypodermic syringe (**Figure 1.33**) was used to deliver 5.0 cc (cm^3) of alcohol into an empty container that had a mass of 25.12 g when empty. The container with the alcohol sample weighed 29.08 g. Calculate the density of the alcohol.

b. A cube of copper metal measures 2.00 cm on each edge and weighs 71.36 g. What is the density of the copper sample?

Solution

a. **Step 1:** Examine the question for what is given and what is unknown.

Given: $V = 5.0$ cc, $m = 25.12$ g (empty), and $m = 29.08$ g (full)

Unknown: $d = ?$

Step 2: Identify a formula that links the given and unknown terms.

$$\text{Equation 1.9: } d = \frac{m}{V}$$

Step 3: Enter the data into Equation 1.9.

According to **Table 1.5**, 1 cc (or cm^3) = 1 mL, so the volume is 5.0 mL. The sample mass is equal to the difference between the mass of the container with the sample inside and the mass of the empty container.

$$m = 29.08 \text{ g} - 25.12 \text{ g} = 3.96 \text{ g}$$

$$d = \frac{m}{V} = \frac{3.96 \text{ g}}{5.0 \text{ mL}} = 0.79 \text{ g/mL (rounded to 2 sig figs)}$$

b. **Step 1:** Examine the question for what is given and what is unknown.

Given: A cube 2.00 cm each side and mass = 71.36 g

Unknown: $d = ?$

Step 2: Identify a formula that links the given and the unknown terms.

$$\text{Equation 1.9: } d = \frac{m}{V}$$

Step 3: Enter the data into the formula.

The volume of a cube is equal to the product of the three sides:

$$V = (2.00\text{ cm})(2.00\text{ cm})(2.00\text{ cm}) = 8.00\text{ cm}^3$$

$$d = \frac{m}{V} = \frac{71.36\text{ g}}{8.00\text{ cm}^3} = 8.92\text{ g/cm}^3 \text{(rounded to 3 sig figs)}$$

✔ Learning Check 1.18

a. A pipet was used to transfer 10.00 mL of a liquid into an empty container. The empty container weighed 51.22 g, and the container with the liquid sample weighed 64.93 g. Calculate the density of the liquid in g/mL.

b. A box of small irregular pieces of metal was found in a storage room. It was known that the metal was either nickel or chromium. A 35.66 g sample of the metal was weighed and put into a graduated cylinder that contained 21.2 mL of water (**Figure 1.34**, left). The water level after the metal was added was 25.2 mL (**Figure 1.34**, right). Was the metal nickel (density = 8.9 g/cm^3) or chromium (density = 7.2 g/cm^3)?

Figure 1.34 Measuring the volume of irregular metal pieces.

Health Career Description

Anesthesiologist Assistant

Anesthesiologist assistants work under the direct supervision of an anesthesiologist. They ensure continuity of care by introducing the patient to the anesthesia plan and prepping them for the procedure. Anesthesia assistants are trained in lifesaving skills such as cardiopulmonary resuscitation (CPR). They may work in hospitals or surgical centers. Options are available for typical work hours or on-call and weekend shifts. This career field promises to remain in high demand. Educational requirements include a bachelor's degree in biological or health sciences followed by a 2-year master's degree in anesthesia studies. Anesthesiologist assistants differ from nurse anesthetists in that they do not need to have a nursing license to practice. They take a premedical-type undergraduate program then move on to an anesthesiology master's degree program.

The website explorehealthcareers.org lists the duties of an anesthesiologist assistant as:

- Taking a complete health history of the patient and performing physical exams to identify any issues that may affect the anesthesia care plan
- Administering necessary diagnostic and laboratory tests (such as taking blood)
- Preparing the patient to be monitored, using noninvasive and invasive methods, as determined by the physician
- Assisting with preparatory procedures, such as pulmonary artery catheterization, electroencephalographic spectral analysis, and echocardiography
- Pretesting and calibrating anesthesia delivery systems and monitors
- Inducing, sustaining, and adjusting anesthesia levels
- Ensuring continuity of care through the postoperative recovery period
- Assisting with life support where required, including airway management
- Performing functions in the intensive care unit and pain clinic
- Performing administrative duties, research, and clinical instruction

This career, like so many in health sciences, requires the quantitative skills introduced here in Chapter 1.

Anesthesiologist assistants help provide pain-free procedures in hospitals and surgical centers, thereby providing essential patient care. Because they work with inhaled anesthetics such as nitrous oxide, halothane, isoflurane, desflurane, and sevoflurane, in conjunction with oxygen and intravenous anesthetics like midazolam and propofol, proficiency in chemistry and knowledge of human physiology are critical to success in this field.

Concept Summary

1.1 What Is Matter?

Learning Objective: **Explain what matter is.**

- Matter is the substance of everything.
- Matter is defined as anything that has mass and occupies space.
- Mass is a measurement of the amount of matter present in an object.
- Weight is a measure of the gravitational force pulling down on an object.

1.2 Physical and Chemical Properties and Changes

Learning Objective: **Explain the difference between the terms *physical* and *chemical* as they apply to the properties of matter and changes in matter.**

- Chemical properties are determined by attempts to change one kind of matter into another.
- Physical properties are determined without attempting such changes in composition.
- Any change in matter that is accompanied by a composition change is a chemical change.
- Physical changes take place without the occurrence of any composition changes.

1.3 Classifying Matter

Learning Objective: **Classify matter as an element, compound, homogenous mixture, or heterogeneous mixture.**

- Matter can be classified as either a pure substance or a mixture.
- Mixtures are physical blends of matter that can be physically separated.
- Mixtures are homogeneous if they have a uniform appearance and consistent properties. If not, they are heterogeneous. Homogeneous mixtures are often referred to as solutions.
- Pure substances are always homogeneous.
- Pure substances are made up of either elements or compounds.
- Elements cannot be separated into simpler pure substances. They are the building blocks of molecules. Atoms are the limit to which matter can be chemically subdivided and still serve as the building blocks of the elements.
- A compound is formed when two or more atoms combine through chemical bonds to form a pure substance with a definite composition.
- Diatomic molecules consist of two atoms. Triatomic molecules consist of three atoms. Polyatomic molecules consist of more than three atoms.
- Homoatomic molecules consist of only one type of element, while heteroatomic molecules consist two or more kinds of elements.

1.4 Measurement Units

Learning Objective: **Describe four measurement units used in everyday activities.**

- All measurements are based on standard units that have been agreed on and adopted.
- The earliest measurements were based on human body dimensions.
- The changeable nature of such basic units made the adoption of a worldwide standard necessary.
- A standard system of units called the metric system was developed.

1.5 The Metric System

Learning Objectives: **Convert measurements within the metric system into related units. Convert temperatures measured in Fahrenheit to Celsius and vice versa.**

- The metric system of measurement is used by most scientists worldwide and in all major nations except the United States.
- It is a decimal system in which larger and smaller units of a quantity are related by factors of 10.
- Prefixes are used to designate relationships between the base unit and larger or smaller units of a quantity.

1.6 Large and Small Numbers: An Introduction to Scientific Notation

Learning Objectives: **Express numbers using scientific notation. Perform calculations with numbers expressed in scientific notation.**

- A system of scientific notation has been devised because working with very large or very small numbers in calculations can be awkward.
- In scientific notation, numbers are represented as products of a nonexponential number and 10 raised to some power.
- The nonexponential number is always written with the decimal in the standard position (to the right of the first nonzero digit in the number).
- Numbers written in scientific notation can be manipulated in calculations by following a few rules.

1.7 Significant Figures

Learning Objective: **Express measurements and calculations using the correct number of significant figures.**

- Significant figures are the numbers representing the part of the measurement that is certain, plus one number representing an estimate.
- The maximum number of significant figures possible in a measurement is determined by the design of the measuring device.
- Results of calculations made using numbers from measurements can be expressed with the proper number of significant figures by following one set of rules for multiplication and division and another set for addition and subtraction.

1.8 Using Units in Calculations: An Introduction to Dimensional Analysis

Learning Objective: **Use dimensional analysis to solve numerical problems.**

- The dimensional analysis method for doing calculations involves the use of conversion factors that are obtained from fixed numerical relationships between quantities.

- The units of the conversion factor must always cancel the units of the known quantity and generate the units of the unknown or desired quantity.

1.9 Calculating Percentages

Learning Objective: Perform calculations involving percentages.

- The word *percent* means per one hundred.
- A percentage is literally the number of specific items contained in a group of 100 items.
- Percentage is calculated by dividing the number of specific items by the total number of items in a group, then multiplying by 100%.

1.10 Density and Its Applications

Learning Objective: Perform calculations involving densities.

- The density of a substance is the number obtained by dividing the mass of a sample by the volume of the same sample.
- Using the density equation, the mass of a substance can be calculated if the volume is known, or the volume can be calculated if the mass is known.

Key Terms and Concepts

Atom (1.3)
Base unit of measurement (1.5)
Chemical changes (1.2)
Chemical properties (1.2)
Compound (1.3)
Conversion factor (1.8)
Density (1.10)
Derived unit of measurement (1.5)
Diatomic molecules (1.3)
Dimensional analysis (1.8)
Element (1.3)
Exact numbers (1.7)
Heteroatomic molecules (1.3)
Heterogeneous matter (1.3)
Homoatomic molecules (1.3)
Homogeneous matter (1.3)
Mass (1.1)
Matter (1.1)
Mixture (1.3)
Molecule (1.3)
Physical changes (1.2)
Physical properties (1.2)
Polyatomic molecules (1.3)
Pure substance (1.3)
Scientific notation (1.6)
Significant figures (1.7)
Solutions (1.3)
Standard position for a decimal (1.6)
Triatomic molecules (1.3)
Weight (1.1)

Key Equations

1. Calculation of area (**Section 1.5**)	$\text{Area} = (\text{length})(\text{length}) = (\text{length})^2$	Equation 1.1
2. Calculation of volume (**Section 1.5**)	$\text{Volume} = (\text{length})(\text{length})(\text{length}) = (\text{length})^3$	Equation 1.2
3. Conversion of temperature readings from one scale to another (**Section 1.5**)	$°\text{C} = \frac{5}{9}(°\text{F} - 32)$	Equation 1.3
	$°\text{F} = \frac{9}{5}(°\text{C}) + 32°$	Equation 1.4
	$°\text{C} = \text{K} - 273$	Equation 1.5
	$\text{K} = °\text{C} + 273$	Equation 1.6
4. Calculation of percentage (**Section 1.9**)	$\text{Percent} = \frac{\text{number of specific items}}{\text{total items in the group}} \times 100\%$ $\% = \frac{\text{part}}{\text{total}} \times 100\%$	Equation 1.7
5. Calculation of number of items representing a specific percentage of a total (**Section 1.9**)	$\text{Part} = \frac{(\%)(\text{total})}{100\%}$	Equation 1.8
6. Calculation of density from mass and volume measurements (**Section 1.10**)	$d = \frac{m}{V}$	Equation 1.9

Exercises

Even-numbered exercises are answered in Appendix B.

Exercises with an asterisk (*) are more challenging.

What Is Matter? (Section 1.1)

1.1 A heavy steel ball is suspended by a thin wire. The ball is hit from the side with a hammer but hardly moves. Describe what you think would happen if this identical experiment were carried out on the moon.

1.2 Explain how matter, mass, and weight are related to each other.

1.3 Tell how you would try to prove to a doubter that air is matter.

1.4 Which of the following do you think is likely to change the most when done on Earth and then on the moon? Carefully explain your reasoning.

- **a.** The distance you can throw a bowling ball
- **b.** The distance you can roll a bowling ball on a flat, smooth surface

1.5 The attractive force of gravity for objects near Earth's surface increases as you move toward Earth's center. Suppose you are transported from a deep mine to the top of a tall mountain.

- **a.** How would your mass be changed by the move?
- **b.** How would your weight be changed by the move?

1.6 Earth's rotation causes it to bulge at the equator. How would the weights of people of equal mass differ when one was determined at the equator and one at the North Pole? (See Exercise 1.5.)

Physical and Chemical Properties and Changes (Section 1.2)

1.7 Classify each of the following as a physical or chemical change, and give at least one observation, fact, or reason to support your answer.

- **a.** A plum ripens.
- **b.** Water boils.
- **c.** A glass window breaks.
- **d.** Food is digested.

1.8 Classify each of the following as a physical or chemical change, and give at least one observation, fact, or reason to support your answer.

- **a.** A stick is broken into two pieces.
- **b.** A candle burns.
- **c.** Rock salt is crushed by a hammer.
- **d.** Tree leaves change color in autumn.

1.9 Classify each of the following properties as physical or chemical. Explain your reasoning in each case.

- **a.** Unsaturated fats are liquid at room temperature.
- **b.** Oxygen gas binds with hemoglobin.
- **c.** The metal used in artificial hip joint implants is not corroded by body fluids.
- **d.** An antacid tablet neutralizes stomach acid.
- **e.** Melanin pigment is brown in color.

1.10 Classify each of the following properties as physical or chemical. Explain your reasoning in each case.

- **a.** Mercury metal is a liquid at room temperature.
- **b.** Sodium metal reacts vigorously with water.
- **c.** Water freezes at 0 °C.
- **d.** Gold does not rust.
- **e.** Chlorophyll molecules are green in color.

1.11 Classify each of the following properties as physical or chemical. Explain your reasoning in each case.

- **a.** You dye your hair.
- **b.** Sugar is spun to make cotton candy.
- **c.** Bubbles form when hydrogen peroxide is used as a disinfectant on a wound.
- **d.** Magnesium chloride is used as a deicer for roads and airport runways.

Classifying Matter (Section 1.3)

1.12 Oxygen gas and solid carbon are both made up of homoatomic molecules. The two react to form a single substance, carbon dioxide. Use the term *homoatomic* or *heteroatomic* to classify molecules of carbon dioxide. Explain your reasoning.

1.13 Under appropriate conditions, hydrogen peroxide can be changed to water and oxygen gas. Use the term *homoatomic* or *heteroatomic* to classify molecules of hydrogen peroxide. Explain your reasoning.

1.14 Water can be decomposed to hydrogen gas and oxygen gas by passing electricity through it. Use the term *homoatomic* or *heteroatomic* to classify molecules of water. Explain your reasoning.

1.15 Glucose, commonly known as blood sugar, is chemically broken down in the process of cellular respiration, which requires oxygen. The only products of the process are water and carbon dioxide. Use the term *homoatomic* or *heteroatomic* to classify molecules of glucose. Explain your reasoning.

1.16 Classify each pure substance represented below by a capital letter as an element or a compound. Indicate when such a classification cannot be made and explain why.

- **a.** Substance A is composed of heteroatomic molecules.
- **b.** Substance D is composed of homoatomic molecules.
- **c.** Substance E is changed into substances G and J when it is heated.

1.17 Classify each pure substance represented below by a capital letter as an element or a compound. Indicate when such a classification cannot be made and explain why.

a. Two elements when mixed combine to form only substance L.

b. An element and a compound when mixed form substances M and Q.

c. Substance X is not changed by heating it.

1.18 Consider the following experiments, and answer the questions pertaining to classification.

a. A pure substance R is heated, cooled, put under pressure, and exposed to light but does not change into anything else. Is substance R an element or a compound? Explain your reasoning.

b. Upon heating, solid pure substance T gives off a gas and leaves another solid behind. Is substance T an element or a compound? Explain your reasoning.

c. Is the solid left in part b an element or a compound? Explain your reasoning.

1.19 Early scientists incorrectly classified calcium oxide (lime) as an element for a number of years. Discuss one or more reasons you think they might have done this.

1.20 Classify each of the following as homogeneous or heterogeneous.

a. Blood

b. Liquid eye drops

c. An aspirin tablet

d. A urine sample with kidney stones

e. Intravenous saline

f. An antibiotic ointment

g. Curdled milk

1.21 Classify each of the following as homogeneous or heterogeneous.

a. Muddy flood water

b. Gelatin dessert

c. Normal urine

d. Smog-filled air

e. An apple

f. Mouthwash

g. Petroleum jelly

1.22 Classify as a pure substance or solution each of the materials of Exercise 1.20 that were classified as homogeneous.

1.23 Classify as a pure substance or solution each of the materials of Exercise 1.21 that were classified as homogeneous.

Measurement Units (Section 1.4)

1.24 Briefly discuss why a system of measurement units is an important part of our modern society.

1.25 In the distant past, 1 in was defined as the length resulting from laying a specific number of grain kernels such as corn in a row. Discuss the disadvantages of such a system.

1.26 An old British unit used to express weight is a stone. It is equal to 14 lb. What sort of weighings might be expressed in stones? Suggest some standard that might have been used to establish the unit.

The Metric System (Section 1.5)

1.27 Which of the following quantities are expressed in metric units?

a. The amount of aspirin in a tablet: 5 grams

b. Approximate blood volume of an adult human: 5 L

c. Approximate length of a human small intestine: 20 ft

d. The time for a race: 4 min, 5.2 s

e. Human body surface area: 1.9 m^2

f. The temperature of someone with a fever: 103 °F

1.28 Which of the following quantities are expressed in metric units?

a. Normal body temperature: 37 °C

b. The amount of soft drink in a bottle: 2 L

c. The height of a ceiling in a room: 8.0 ft

d. The amount of aspirin in a tablet: 81 mg

e. The volume of a cooking pot: 4 qt

f. The time for a short race to be won: 10.2 s

1.29 Referring to **Table 1.5**, suggest an appropriate metric system unit for each nonmetric unit in Exercise 1.27.

1.30 Referring to **Table 1.5**, suggest an appropriate metric system unit for each nonmetric unit in Exercise 1.28.

1.31 Referring only to **Table 1.2**, answer the following questions.

a. A computer has 12 megabytes of memory storage. How many bytes of storage is this?

b. A 10 km race is 6.2 mi long. How many meters long is it?

c. A chemical balance can detect a mass as small as 0.1 mg. What is this detection limit in grams?

d. A micrometer is a device used to measure small lengths. If it lives up to its name, what is the smallest metric length that could be measured using a micrometer?

1.32 Referring only to **Table 1.2**, answer the following questions.

a. Devices are available that allow liquid volumes as small as one microliter (mL) to be measured. How many microliters would be contained in 1.00 liter?

b. Electrical power is often measured in kilowatts. How many watts would equal 75 kilowatts?

c. Ultrasound is sound at such high frequency that it cannot be heard. The frequency is measured in hertz (vibrations per second). How many hertz correspond to 15 megahertz?

d. A chlorine atom has a diameter of 200 picometers. How many meters is this diameter?

1.33 One inch is equal to 2.54 cm. Express this length in millimeters and meters.

1.34 In a cookbook that uses metric units, 1 cup equals 240 mL. Express 1 cup in terms of liters and cubic centimeters.

1.35 Two cities in Germany are located 25 km apart. What is the distance in miles?

1.36 The shotput used by female track and field athletes has a mass of 4.0 kg. What would be the weight of such a shotput in pounds?

1.37 Use **Table 1.5** to answer the following questions.

a. Which is longer, a centimeter or an inch of surgical tape?

b. How many milliliters are in a quart of blood?

c. How many grams are in an ounce of fat?

1.38 Use **Table 1.5** to answer the following questions.

a. Approximately how many inches longer is a meter stick than a yardstick?

b. A temperature increases by 65 °C. How many kelvins would this increase be?

c. You have a 5.00 lb bag of sugar. Approximately how many kilograms of sugar do you have?

1.39 Do the following, using appropriate values from **Table 1.5**.

a. Calculate the area in square meters of a circular skating rink that has a 12.5 m radius. For a circle, the area (A) is related to the radius (r) by $A = \pi r^2$, where $\pi = 3.14$.

b. Calculate the floor area and volume of a rectangular room that is 5.0 m long, 2.8 m wide, and 2.1 m high. Express your answers in square meters and cubic meters (meters cubed).

c. A model sailboat has a triangular sail that is 25 cm high (h) and has a base (b) of 15 cm. Calculate the area (A) of the sail in square centimeters. $A = \frac{(b)(h)}{2}$

1.40 Use appropriate values from **Table 1.5**, and answer the following questions.

a. One kilogram of water has a volume of 1.0 dm^3. What is the mass of 1.0 cm^3 of water?

b. One quart is 32 fl oz. How many fluid ounces are contained in 2.0 L of saline solution?

c. Approximately how many milligrams of aspirin are contained in a 5 grain tablet?

1.41 The weather report says the temperature is 23 °F. What is this temperature on the Celsius scale? On the Kelvin scale?

1.42 Recall from Health Connections 1.1 that a normal body temperature might be as low as 36.1 °C in the morning and as high as 37.2 °C at bedtime. What are these temperatures on the Fahrenheit scale?

1.43 One pound of body fat releases approximately 4500 kcal of energy when it is metabolized. How many joules of energy is this? How many BTUs?

Large and Small Numbers: An Introduction to Scientific Notation (Section 1.6)

1.44 Which of the following numbers are written using scientific notation correctly? For those that are not, explain what is wrong.

a. 02.7×10^{-3}

b. 4.1×10^2

c. 71.9×10^{-6}

d. 10^3

e. $.0405 \times 10^{-2}$

f. 0.119

1.45 Which of the following numbers are written using scientific notation correctly? For those that are not, explain what is wrong.

a. 4.2×10^3

b. 6.8^4

c. 202×10^{-3}

d. 0.026×10^{-2}

e. 10^{-2}

f. 74.5×10^5

1.46 Write each of the following numbers using scientific notation.

a. 14 thousand

b. 365

c. 0.00204

d. 461.8

e. 0.00100

f. 911

1.47 Write each of the following numbers using scientific notation.

a. Three hundred

b. 4003

c. 0.682

d. 91.86

e. Six thousand

f. 400

1.48 The speed of light is about 186 thousand mi/s, or 1100 million km/h. Write both numbers using scientific notation.

1.49 A human hair has a diameter of 0.0651 mm, or 0.00256 inches. Write both numbers using scientific notation.

1.50 A single copper atom has a mass of 1.05×10^{-22} g. Write this number in a decimal form without using scientific notation.

1.51 In 2.0 g of hydrogen gas, there are approximately 6.02×10^{23} hydrogen molecules. Write this number without using scientific notation.

1.52 Do the following multiplications, and express each answer using scientific notation.

a. $(8.2 \times 10^{-3})(1.1 \times 10^{-2})$

b. $(2.7 \times 10^2)(5.1 \times 10^4)$

c. $(3.3 \times 10^{-4})(2.3 \times 10^2)$

d. $(9.2 \times 10^{-4})(2.1 \times 10^4)$

e. $(4.3 \times 10^6)(6.1 \times 10^5)$

1.53 Do the following multiplications, and express each answer using scientific notation.

a. $(6.3 \times 10^5)(4.2 \times 10^{-8})$

b. $(2.8 \times 10^{-3})(1.4 \times 10^{-4})$

c. $(8.6 \times 10^2)(6.4 \times 10^{-3})$

d. $(9.1 \times 10^4)(1.4 \times 10^3)$

e. $(3.7 \times 10^5)(6.1 \times 10^{-3})$

1.54 Express each of the following numbers using scientific notation, then carry out the multiplication. Express each answer using scientific notation.

a. (144)(0.0876)

b. (751)(106)

c. (0.0422)(0.00119)

d. (128,000)(0.0000316)

1.55 Express each of the following numbers using scientific notation, then carry out the multiplication. Express each answer using scientific notation.

a. (538)(0.154)

b. (600)(524)

c. (22.8)(341)

d. (23.6)(0.047)

1.56 Do the following divisions, and express each answer using scientific notation.

a. $\frac{3.1 \times 10^{-3}}{1.2 \times 10^{2}}$

b. $\frac{7.9 \times 10^{4}}{3.6 \times 10^{2}}$

c. $\frac{4.7 \times 10^{-1}}{7.4 \times 10^{2}}$

d. $\frac{0.00229}{3.16}$

e. $\frac{119}{3.8 \times 10^{3}}$

1.57 Do the following divisions, and express each answer using scientific notation.

a. $\frac{223}{1.67}$

b. $\frac{6.7 \times 10^{3}}{4.2 \times 10^{4}}$

c. $\frac{8.7 \times 10^{-4}}{2.3 \times 10^{-2}}$

d. $\frac{6.8 \times 10^{3}}{2.7 \times 10^{-4}}$

e. $\frac{1.8 \times 10^{-2}}{6.5 \times 10^{4}}$

1.58 Do the following calculations, and express each answer using scientific notation.

a. $\frac{(5.3)(0.22)}{(6.1)(1.1)}$

b. $\frac{(3.8 \times 10^{-4})(1.7 \times 10^{-2})}{6.3 \times 10^{3}}$

c. $\frac{4.8 \times 10^{6}}{(7.4 \times 10^{3})(2.5 \times 10^{-4})}$

d. $\frac{5.6}{(0.022)(109)}$

e. $\frac{(4.6 \times 10^{-3})(2.3 \times 10^{2})}{(7.4 \times 10^{-4})(9.4 \times 10^{-5})}$

1.59 Do the following calculations, and express each answer using scientific notation.

a. $\frac{(7.4 \times 10^{-3})(1.3 \times 10^{4})}{5.5 \times 10^{-2}}$

b. $\frac{6.4 \times 10^{5}}{(8.8 \times 10^{3})(1.9 \times 10^{-4})}$

c. $\frac{(6.4 \times 10^{-2})(1.1 \times 10^{-8})}{(2.7 \times 10^{-4})(3.4 \times 10^{-4})}$

d. $\frac{(963)(1.03)}{(0.555)(412)}$

e. $\frac{1.15}{(0.12)(0.73)}$

Significant Figures (Section 1.7)

1.60 Indicate to what decimal position readings should be estimated and recorded (nearest 0.1, 0.01, etc.) for measurements made with the following devices.

a. A ruler with a smallest scale marking of 0.1 cm

b. A measuring telescope with a smallest scale marking of 0.1 mm

c. A protractor with a smallest scale marking of 1°

d. A tire pressure gauge with a smallest scale marking of 1 lb/in^2

1.61 Indicate to what decimal position readings should be estimated and recorded (nearest 0.1, 0.01, etc.) for measurements made with the following devices.

a. A buret with a smallest scale marking of 0.1 mL

b. A graduated cylinder with a smallest scale marking of 1 mL

c. A thermometer with a smallest scale marking of 0.1 °C

d. A barometer with a smallest scale marking of 1 torr

1.62 Write the following measured quantities as a health professional would record them, using the correct number of significant figures based on the device used to make the measurement.

a. Exactly 6 mL of water measured with a syringe whose smallest scale marking is 1 mL

b. A temperature that appears to be exactly 37 °C using a thermometer with a smallest scale marking of 0.1 °C

c. A time of exactly 9 s measured with a stopwatch whose smallest scale marking is 0.1 s

d. Fifteen and one-half degrees measured on an X-ray machine that has 1 degree scale markings

1.63 Write the following measured quantities using the correct number of significant figures based on the device used to make the measurements.

a. A length of two and one-half centimeters measured with a measuring telescope with a smallest scale marking of 0.1 mm

b. An initial reading of exactly 0 for a burette with a smallest scale marking of 0.1 mL

c. A length of four and one-half centimeters measured with a ruler that has a smallest scale marking of 0.1 cm

d. An atmospheric pressure of exactly 690 torr measured with a barometer that has a smallest scale marking of 1 torr

1.64 In each of the following, identify the measured numbers and the exact numbers. Do the indicated calculation, and write your answer using the correct number of significant figures.

a. A bag of potatoes is found to weigh 5.06 lb. The bag contains 16 potatoes. Calculate the weight of an average potato.

b. The foul-shooting percentages for the five starting players of a women's basketball team are 71.2%, 66.9%, 74.1%, 80.9%, and 63.6%. What is the average shooting percentage of the five players?

1.65 In each of the following, identify the measured numbers and the exact numbers. Do the indicated calculation, and write your answer using the correct number of significant figures.

a. An individual has a job of counting the number of people who enter a store between 1 p.m. and 2 p.m. each day for 5 days. The counts were 19, 24, 17, 31, and 40. What was the average number of people entering the store per day for the 5 day period?

b. The starting five members of a women's basketball team have the following heights: 6′9″, 5′8″, 5′6″, 5′1″, and 4′11″. What is the average height of the starting five?

1.66 Determine the number of significant figures in each of the following.

a. 0.0400

b. 309

c. 4.006

d. 4.4×10^{-3}

e. 1.002

f. 255.02

1.67 Determine the number of significant figures in each of the following.

a. 0.040

b. 11.91

c. 2.48×10^2

d. 149.1

e. 10.003

f. 148.67

1.68 Do the following calculations and use the correct number of significant figures in your answers. Assume all numbers are the results of measurements.

a. (3.71)(1.4)

b. (0.0851)(1.2262)

c. $\dfrac{(0.1432)(2.81)}{(0.7762)}$

d. $(3.3 \times 10^4)(3.09 \times 10^{-3})$

e. $\dfrac{(760)(2.00)}{6.02 \times 10^{20}}$

1.69 Do the following calculations and use the correct number of significant figures in your answers. Assume all numbers are the results of measurements.

a. (1.21)(3.2)

b. $(6.02 \times 10^{23})(0.220)$

c. $\dfrac{(0.023)(1.1 \times 10^{-3})}{100}$

d. $\dfrac{(365)(7.000)}{60}$

e. $\dfrac{(810)(3.1)}{8.632 \times 10^{-1}}$

1.70 Do the following calculations and use the correct number of significant figures in your answers. Assume all numbers are the results of measurements.

a. 0.208 + 4.9 + 1.11

b. 228 + 0.999 + 1.02

c. 8.543 − 7.954

d. $(3.2 \times 10^{-2}) + (5.5 \times 10^{-1})$
(**HINT:** Write in decimal form first, then add.)

e. 336.86 − 309.11

f. 21.66 − 0.02387

1.71 Do the following calculations and use the correct number of significant figures in your answers. Assume all numbers are the results of measurements.

a. 2.1 + 5.07 + 0.119

b. 0.051 + 8.11 + 0.02

c. 4.337 − 3.211

d. $(2.93 \times 10^{-1}) + (6.2 \times 10^{-2})$
(**HINT:** Write in decimal form first, then add.)

e. 471.19 − 365.09

f. 17.76 − 0.0479

1.72 Do the following calculations and use the correct number of significant figures in your answers. Assume all numbers are the results of measurements. In calculations involving both addition/subtraction and multiplication/division, it is usually better to do additions/subtractions first.

a. $\dfrac{(0.0267 + 0.0019)(4.626)}{28.7794}$

b. $\dfrac{212.6 - 21.88}{86.37}$

c. $\dfrac{27.99 - 18.07}{4.63 - 0.88}$

d. $\dfrac{18.87}{2.46} - \dfrac{18.07}{0.88}$
(**HINT:** Do divisions first, then subtract.)

e. $\dfrac{(8.46 - 2.09)(0.51 + 0.22)}{(3.74 + 0.07)(0.16 + 0.2)}$

f. $\dfrac{12.06 - 11.84}{0.271}$

1.73 Do the following calculations and use the correct number of significant figures in your answers. Assume all numbers are the results of measurements. In calculations involving both addition/subtraction and multiplication/division, it is usually better to do additions/subtractions first.

a. $\dfrac{132.15 - 32.16}{87.55}$

b. $\dfrac{(0.0844 + 0.1021)(7.174)}{19.1101}$

c. $\dfrac{(2.78 - 0.68)(0.42 + 0.4)}{(1.058 + 0.06)(0.22 + 0.2)}$

d. $\dfrac{27.65 - 21.71}{4.97 - 0.36}$

e. $\dfrac{12.47}{6.97} - \dfrac{203.4}{201.8}$

(**HINT:** Do divisions first, then subtract.)

f. $\dfrac{19.37 - 18.49}{0.822}$

1.74 The following measurements were obtained for the length and width of a series of rectangles. Each measurement was made using a ruler with a smallest scale marking of 0.1 cm.

Black rectangle: $l = 12.00$ cm, $w = 10.40$ cm
Red rectangle: $l = 20.20$ cm, $w = 2.42$ cm
Green rectangle: $l = 3.18$ cm, $w = 2.55$ cm
Orange rectangle: $l = 13.22$ cm, $w = 0.68$ cm

a. Calculate the area (length × width) and perimeter (sum of all four sides) for each rectangle and express your results in square centimeters and centimeters, respectively, and give the correct number of significant figures in the result.

b. Change all measured values to meters and then calculate the area and perimeter of each rectangle. Express your answers in square meters and meters, respectively, and give the correct number of significant figures.

c. Does changing the units used change the number of significant figures in the answers?

Using Units in Calculations: An Introduction to Dimensional Analysis (Section 1.8)

1.75 It is important that health care professionals and others be able to convert English units to metric units and vice versa. Determine a single equality from **Table 1.5** that could be used as a conversion factor to make each of the following conversions.

a. 3.4 lb to kilograms
b. 3.0 yd to meters
c. 1.5 oz to grams
d. 40. cm to inches

1.76 It is important that health care professionals and others be able to convert English units to metric units and vice versa. Determine a single equality from **Table 1.5** that could be used as a conversion factor to make each of the following conversions.

a. 20 mg to grains
b. 350 mL to fl oz
c. 4 qt to liters
d. 5 yd to meters

1.77 Obtain a factor from **Table 1.5** and calculate the number of liters in 1.00 gal (4 qt) by using dimensional analysis.

1.78 A patient receives 45 mL of a drug. Obtain a conversion factor from **Table 1.5** and use dimensional analysis to calculate how much the patient receives in fluid ounces (fl oz).

1.79 A state health department recommends that certain medications be stored between 35 °F and 40 °F. If the refrigerator used has a Celsius thermometer, what settings should you use?

1.80 A nurse's direction is to administer 55 mg of Garamycin (gentamicin) every 6 hours. If a vial is labeled 80 mg/2 mL, how many mL should be given?

1.81 A single tablet of baby aspirin contains 81 mg of aspirin. If someone takes 7 tablets in 1 day, how many grams of aspirin have been consumed?

1.82 If an intravenous solution is infused at the rate of 0.70 mL per minute, how many minutes will it take to administer 200.0 mL of solution?

1.83 A patient weighs 74.6 kg. How much is that weight in pounds?

1.84 During a glucose tolerance test, the serum glucose concentration of a patient was found to be 131 mg/dL. Convert the concentration to grams per liter.

Calculating Percentages (Section 1.9)

1.85 Retirement age is 65 years in many companies. What percentage of the way from birth to retirement is a 55-year-old person?

1.86 A salesperson made a sale of $467.80 and received a commission of $25.73. What percent commission was paid?

1.87 After drying, 140 lb of grapes yields 32 lb of raisins. What percentage of the grapes' mass was lost during the drying process?

1.88 The recommended daily intake of thiamin is 1.4 mg for a male adult. Suppose such a person takes in only 1.0 mg/day. What percentage of the recommended intake are they receiving?

1.89 The recommended daily caloric intake for a 20-year-old female is 2000. How many calories should her breakfast contain if she wants it to be 45% of her recommended daily total?

1.90 Immunoglobulin antibodies occur in five forms. A sample of serum is analyzed with the following results. Calculate the percentage of total immunoglobulin represented by each type.

Type:	IgG	IgA	IgM	IgD	IgE
Amount (mg):	987.1	213.3	99.7	14.4	0.1

Density and Its Applications (Section 1.10)

1.91 Calculate the density of the following materials for which the mass and volume of samples have been measured. Express the density of liquids in g/mL, the density of solids in g/cm^3, and the density of gases in g/L.

a. 250 mL of liquid mercury metal (Hg) has a mass of 3400 g.

b. 500 mL of concentrated liquid sulfuric acid (H_2SO_4) has a mass of 925 g.

c. 5.00 L of oxygen gas has a mass of 7.15 g.

d. A 200 cm^3 block of magnesium metal (Mg) has a mass of 350 g.

1.92 Calculate the density of the following materials for which the mass and volume of samples have been measured. Express the density of liquids in g/mL, the density of solids in g/cm^3, and the density of gases in g/L.

a. A 50.0 mL sample of liquid acetone has a mass of 39.6 g.

b. A 1.00 cup (236 mL) sample of homogenized milk has a mass of 243 g.

c. 20.0 L of dry carbon dioxide gas (CO_2) has a mass of 39.54 g.

d. A 25.0 cm^3 block of nickel metal (Ni) has a mass of 222.5 g.

1.93 Osteoporosis is a disease in which the density and quality of bone are reduced. If a bone sample has a volume of 1.35 cm^3 and a mass of 2.10 g, what is the density in g/cm^3?

1.94 Calculate the volume and density of a lead (Pb) cube with a mass of 718.3 g and side-length measuring 3.98 cm.

1.95* The volume of an irregularly shaped solid can be determined by immersing the solid in a liquid and measuring the volume of liquid displaced. Find the volume and density of the following.

a. An irregular piece of the mineral quartz weighs 12.4 g. It is then placed into a graduated cylinder that contains some water. The quartz does not float. The water in the cylinder was at a level of 25.2 mL before the quartz was added and at 29.9 mL afterward.

b. The volume of a sample of lead shot is determined using a graduated cylinder, as in part a. The cylinder readings are 16.3 mL before the shot is added and 21.7 mL after. The sample of shot weighs 61.0 g.

c. A sample of coarse rock salt has a mass of 11.7 g. The volume of the sample is determined by the graduated-cylinder method described in part a, but kerosene is substituted for water because the salt will not dissolve in kerosene. The cylinder readings are 20.7 mL before adding the salt and 26.1 mL after.

1.96 If the density of high-density lipoproteins (HDL) or "good" cholesterol is 1.069 g/mL, what is the volume in mL of 3.056 g of HDL?

1.97 A urine sample has a density of 1.107 g/mL. What is the mass in grams of 0.250 L of this urine?

Additional Exercises

1.98 Do the following metric system conversions by changing only the power of 10. For example, convert 2.5 L to mL: 2.5 L $= 2.5 \times 10^3$ mL.

a. Convert 4.5 km to mm.

b. Convert 6.0×10^6 mg to g.

c. Convert 9.86×10^{15} m to km.

d. Convert 1.91×10^{-4} kg to mg.

e. Convert 5.0 ng to mg.

1.99 A single water molecule has a mass of 2.99×10^{-23} g. Each molecule contains two hydrogen atoms that together make up 11.2% of the mass of the water molecule. What is the mass in grams of a single hydrogen atom?

1.100 It has been found that fat makes up 14% of a 170 lb person's body weight. One pound of body fat provides 4500 kcal of energy when it is metabolized. The person requires 2000 kcal of energy per day to survive. Assume all the energy needed for survival comes from the metabolism of body fat, and calculate the number of days the person could survive without eating before depleting the entire body fat reserve.

1.101 Cooking oil has a density of 0.812 g/mL. What is the mass in grams of 1.00 quart of cooking oil? Use **Table 1.5** for any necessary conversion factors.

1.102 A 175 lb patient is to undergo surgery and will be given an anesthetic intravenously. The safe dosage of anesthetic is 12 mg/kg of body weight. Determine the maximum dose of anesthetic in mg that should be used.

1.103 At 4.0 °C, pure water has a density of 1.00 g/mL. At 60.0 °C, the density is 0.98 g/mL. Calculate the volume in mL of 1.00 g of water at each temperature, and then calculate the percentage increase in volume that occurs as water is heated from 4.0 °C to 60.0 °C.

Chemistry for Thought

1.104 The following pairs of substances represent heterogeneous mixtures. For each pair, describe the steps you would follow to separate the components and collect them.

a. Wood sawdust and sand

b. Sugar and sand

c. Iron filings and sand

d. Sand soaked with oil

1.105 Explain why a bathroom mirror becomes foggy when someone takes a hot shower. Classify any changes that occur as physical or chemical.

1.106 A 20-year-old student was weighed and found to have a mass of 44.5 kg. In converting this weight to pounds, an answer of 20.2 lb was obtained. Describe the mistake made in doing the calculation.

1.107 Liquid mercury metal freezes to a solid at a temperature of −38.9 °C. Suppose you want to measure a temperature that is at least as low as −45 °C. Can you use a mercury thermometer? If not, propose a way to make the measurement.

1.108 Show how dimensional analysis can be used to prepare an oatmeal breakfast for 27 guests at a family reunion. The directions on the oatmeal box say that 1 cup of dry oatmeal makes 3 servings.

1.109 Explain what is meant by the following statement: All matter contains chemicals.

1.110 A small solid figurine is brought to a chemist. The owner wants to know if it is made of silver but doesn't want it damaged during the analysis. The chemist decides to determine the density, knowing that silver has a density of 10.5 g/mL. The figurine is put into a graduated cylinder that contains 32.6 mL of water. The reading while the figurine is in the water is 60.1 mL. The mass of the figurine is 240.8 g. Is the figurine made of silver? Explain your reasoning.

2 Atomic Structure and the Periodic Table

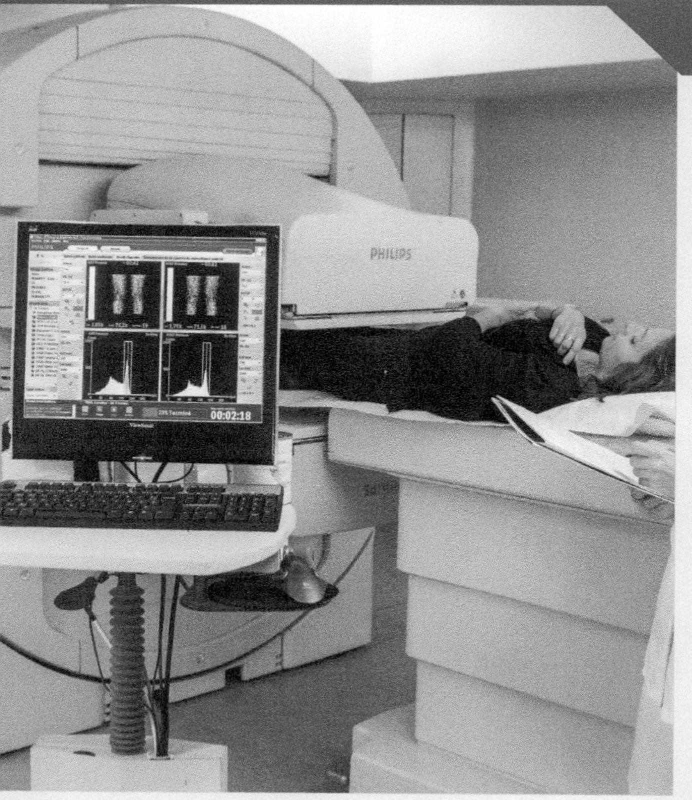

Phanie/Superstock.com

Health Career Focus

Nuclear Medicine Technologist

"It feels a little like a superpower," I explain to the classroom of radiology students. "I put on a protective suit to pick up the radioactive medication; then I connect it to the IV, or in some cases 'port,' and the medicine produces a 'contrast,' allowing the radiologists to see minute details. Without these medications, doctors couldn't make the accurate diagnoses needed for cancer treatment."

"Some of the medications I administer are for diagnosis, and some are for treatment, and most are radioactive. That's why I'm called a 'nuclear medicine' technologist."

Students pay keen attention when I show images of scans comparing normal and diseased patients. I feel proud when scans document improvement over time from treatments I've administered.

"I know my work gives many patients days, or even years, of life they might not otherwise enjoy—that really feels like a super power."

Follow-up to this Career Focus appears at the end of the chapter before the *Concept Summary*.

Learning Objectives

When you have completed your study of Chapter 2, you should be able to:

1 Locate elements in the periodic table on the basis of their group and period designations. **(Section 2.1)**

2 Describe the charge, relative mass, and location of the three subatomic particles. **(Section 2.2)**

3 Write the atomic symbol ($^{A}_{Z}X$) for a given set of subatomic particles. **(Section 2.3)**

4 Define the terms *isotope* and *atomic weight*. **(Section 2.4)**

5 Write balanced equations for radioactive decay and other nuclear processes. **(Section 2.5)**

6 Solve problems using the half-life concept. **(Section 2.5)**

7 Describe the applications of radiation in health and medicine. **(Section 2.5)**

8 Write correct electron configurations for each element through atomic number 56. **(Section 2.6)**

9 Describe periodic trends in atomic radius, first ionization energy, and electronegativity. **(Section 2.7)**

Chapter 1 introduced fundamental ideas about matter, atoms, molecules, measurements, and calculations. Here in Chapter 2, we further explore foundational principles regarding the nature of the atom itself, with a special focus on all the information contained within the periodic table. The chapter begins with an explanation of elemental symbols and the classification of elements into groups sharing common chemical characteristics. Next, we dive into subatomic structure and define the term *isotope*. This description makes it possible to explore the concepts of atomic weights, radioisotopes, and the applications of radiation in healthcare. We finish the chapter by examining the location of electrons within the atom and learning about periodic trends in three important properties of the elements.

World History Archive/Alamy Stock Photo

Figure 2.1 Dimitri Mendeleev (1834–1907) organized the elements into a classification scheme in 1869.

2.1 The Periodic Table

Learning Objective 1 Locate elements in the periodic table on the basis of their group and period designations.

By the early nineteenth century, detailed studies of the elements known at that time had produced an abundance of chemical information. Chemists like Antoine Lavoisier (1743–1794) had contributed many foundational ideas (e.g., the components of water and the importance of systematic measurement) that we take for granted today. As knowledge about chemical processes and the elements involved grew, scientists looked for order in these observations, with the hope of providing a systematic approach to the study of chemistry. In 1869, Dimitri Mendeleev (**Figure 2.1**), first produced a classification scheme that arranged elements with similar chemical properties at regular (periodic) intervals. And thus, the periodic table of the elements was born (see **Figure 2.2**). At the time of Mendeleev's breakthrough, only a little more than half of the 118 elements that currently make up the periodic table were known. As a result, Mendeleev left gaps in his table that correctly predicted elements that had not yet been discovered!

Each element has been assigned both a name and an abbreviation known as its **elemental symbol**. The symbol is based on the element's name and consists of a single capital letter or a capital letter followed by a lowercase letter. The organization of the elements within the periodic table places ones with similar chemical properties in the same vertical columns. These vertical columns of elements are referred to as **groups**.

In this book, the groups are designated in two ways. In the U.S. system, a Roman numeral and letter, such as IIA, is used. In the International Union of Pure and Applied Chemistry (IUPAC) system, the numbers 1–18 are used. Both designations appear on the periodic table in **Figure 2.2**, with the IUPAC number in parentheses, and both designations may be used in this text, either together or separately. The horizontal rows in the table are called **periods** and are numbered from top to bottom (1 to 7). Thus, each element belongs to both a period and a group of the periodic table.

More broadly, the periodic table also separates elements into one of three categories: **metals**, **nonmetals**, or **metalloids** (**Figure 2.2**). Notice that most elements are classified as metals. Metals (**Figure 2.3**) tend to have high thermal conductivity (transmit heat readily), high electrical conductivity (transmit electricity readily), and luster (a "metallic" appearance). Metals also tend to be malleable (can be hammered or pressed into thin sheets) and ductile (can be pulled into wires). Nonmetals, on the other hand, often appear as either brittle, powdery solids or as gases. Metalloids cut a diagonal pattern in the periodic table between the metals and the nonmetals. These seven elements—boron (B), silicon (Si), germanium (Ge), arsenic (As), antimony (As), tellurium (Te), and polonium (Po)—have properties somewhat between metals and nonmetals, exhibiting some characteristics of each. This is what makes them such good semiconductors.

elemental symbol A one- or two-letter designation assigned to an element based on the name of the element, consisting of one capital letter or a capital letter followed by a lowercase letter.

group of the periodic table A vertical column of elements that have similar chemical properties.

period of the periodic table A horizontal row of elements.

metals Elements found in the left two-thirds of the periodic table. Common properties of metals are: high thermal and electrical conductivities, high malleability and ductility, and a luster.

nonmetals Elements found in the right one-third of the periodic table. They often occur as brittle, powdery solids or gases and have properties generally opposite to those of metals.

metalloids Elements that form a narrow diagonal band in the periodic table between metals and nonmetals. They have properties somewhat between those of metals and nonmetals.

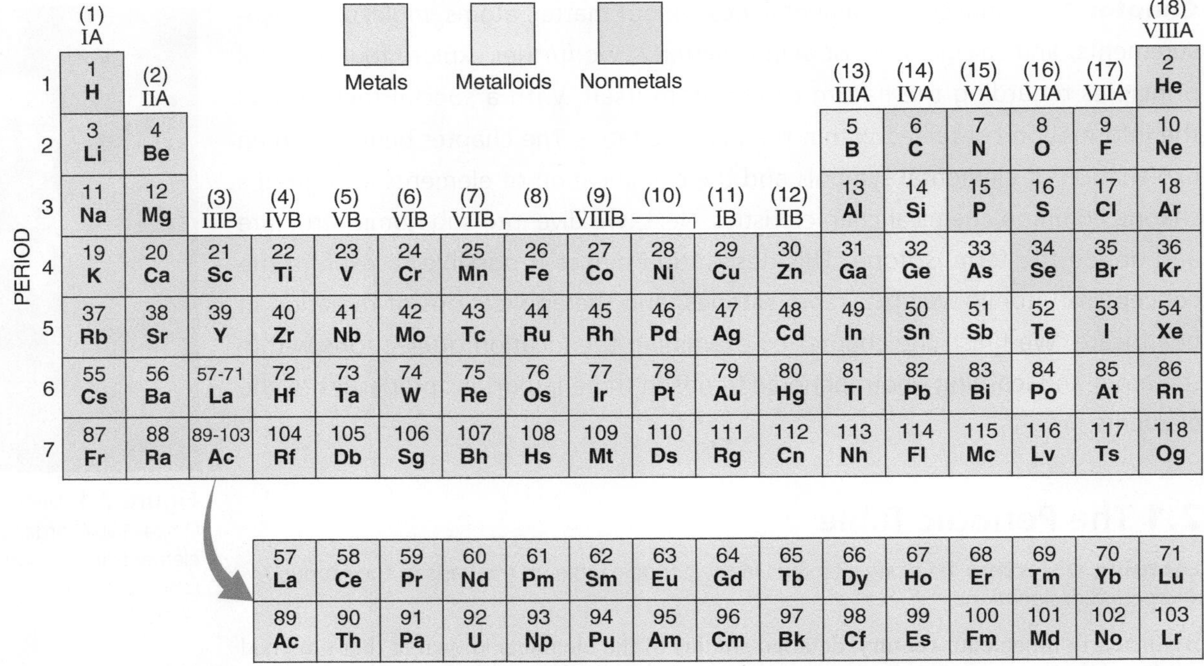

Figure 2.2 Locations of metals, nonmetals, and metalloids in the periodic table of the elements.

David Švihovec/Unsplash.com

Figure 2.3 Metals such as yttrium (Y) in the color screen and lanthanum (La) in the circuitry play important roles in your smartphone.

representative elements Elements found within Groups 1–2 and 13–18 in the periodic table.

transition metals All metallic elements in the central block (Groups 3–12).

inner-transition metals The lanthanides (elements 57–71) and actinides (elements 89–103).

The elements within Groups IA(1), IIA(2), and IIIA(13)–VIIIA(18), are referred to as the **representative elements**. **Figure 2.4** shows that certain individual groups within the periodic table are also known by unique names. For example, Group IA(1) is the *alkali metals*, Group IIA(2) is the *alkaline earth metals*, Group VIIA(17) is the *halogens*, and Group VIIIA(18) is the *noble gases*.

Elements in the middle of the periodic table that comprise the B groups (e.g., groups 3–12 in the IUPAC system) are known as **transition metals**. Many of the names of the transition metals may be familiar to you due to their numerous applications in everyday life (e.g., silver, gold, and iron). Elements 57–71 (the lanthanides) and elements 89–103 (the actinides) are placed beneath the rest of the periodic table to keep it from being too wide. The elements in these two rows are known as the **inner-transition metals** (see **Figure 2.4**).

Example 2.1 Distinguishing Element Classifications

Use the periodic table in **Figure 2.4** to classify the following elements into their specific families.

a. Neon (Ne) b. Cesium (Cs) c. Gold (Au) d. Germanium (Ge)

Solution

a. Ne: noble gas
b. Cs: alkali metal
c. Au: transition metal (**Figure 2.5**)
d. Ge: metalloid

✔ **Learning Check 2.1** Use the periodic table in **Figure 2.4** to classify the following elements into their specific families.

a. Xe b. As c. Hg d. Ba e. Br

Periodic Table of the Elements

Legend: Alkali metals · Alkaline earth metals · Transition metals · Lanthanides · Actinides · Other Metals · Metalloids · Nonmetals · Halogens · Noble gases

Key: 11 — Atomic number; Na — Element symbol; Sodium — Element name; 22.99 — Standard atomic weight in amu (no data given for elements without stable nuclides)

The element symbol indicates the state at 0 °C and 1 atm: **Solid** **Liquid** Gas *Unknown*

PERIOD	(1) IA	(2) IIA	(3) IIIB	(4) IVB	(5) VB	(6) VIB	(7) VIIB	(8) VIIIB	(9) VIIIB	(10) VIIIB	(11) IB	(12) IIB	(13) IIIA	(14) IVA	(15) VA	(16) VIA	(17) VIIA	(18) VIIIA
1	1 H Hydrogen 1.008																	2 He Helium 4.00
2	3 Li Lithium 6.94	4 Be Beryllium 9.01											5 B Boron 10.81	6 C Carbon 12.01	7 N Nitrogen 14.01	8 O Oxygen 16.00	9 F Fluorine 19.00	10 Ne Neon 20.18
3	11 Na Sodium 22.99	12 Mg Magnesium 24.31											13 Al Aluminum 26.98	14 Si Silicon 28.09	15 P Phosphorus 30.97	16 S Sulfur 32.06	17 Cl Chlorine 35.45	18 Ar Argon 39.95
4	19 K Potassium 39.10	20 Ca Calcium 40.08	21 Sc Scandium 44.96	22 Ti Titanium 47.87	23 V Vanadium 50.94	24 Cr Chromium 52.00	25 Mn Manganese 54.94	26 Fe Iron 55.85	27 Co Cobalt 58.93	28 Ni Nickel 58.69	29 Cu Copper 63.55	30 Zn Zinc 65.38	31 Ga Gallium 69.72	32 Ge Germanium 72.63	33 As Arsenic 74.92	34 Se Selenium 78.97	35 Br Bromine 79.90	36 Kr Krypton 83.80
5	37 Rb Rubidium 85.47	38 Sr Strontium 87.62	39 Y Yttrium 88.91	40 Zr Zirconium 91.22	41 Nb Niobium 92.91	42 Mo Molybdenum 95.95	43 Tc Technetium	44 Ru Ruthenium 101.07	45 Rh Rhodium 102.91	46 Pd Palladium 106.42	47 Ag Silver 107.87	48 Cd Cadmium 112.41	49 In Indium 114.82	50 Sn Tin 118.71	51 Sb Antimony 121.76	52 Te Tellurium 127.60	53 I Iodine 126.90	54 Xe Xenon 131.29
6	55 Cs Cesium 132.91	56 Ba Barium 137.33	57-71 Lanthanides	72 Hf Hafnium 178.49	73 Ta Tantalum 180.95	74 W Tungsten 183.84	75 Re Rhenium 186.21	76 Os Osmium 190.23	77 Ir Iridium 192.22	78 Pt Platinum 195.08	79 Au Gold 196.97	80 Hg Mercury 200.59	81 Tl Thallium 204.38	82 Pb Lead 207.2	83 Bi Bismuth 208.98	84 Po Polonium	85 At Astatine	86 Rn Radon
7	87 Fr Francium	88 Ra Radium	89-103 Actinides	104 *Rf* Rutherfordium	105 *Db* Dubnium	106 *Sg* Seaborgium	107 *Bh* Bohrium	108 *Hs* Hassium	109 *Mt* Meitnerium	110 *Ds* Darmstadtium	111 *Rg* Roentgenium	112 *Cn* Copernicium	113 *Nh* Nihonium	114 *Fl* Flerovium	115 *Mc* Moscovium	116 *Lv* Livermorium	117 *Ts* Tennessine	118 *Og* Oganesson

57 La Lanthanum 138.91	58 Ce Cerium 140.12	59 Pr Praseodymium 140.91	60 Nd Neodymium 144.24	61 Pm Promethium	62 Sm Samarium 150.36	63 Eu Europium 151.96	64 Gd Gadolinium 157.25	65 Tb Terbium 158.93	66 Dy Dysprosium 162.50	67 Ho Holmium 164.93	68 Er Erbium 167.26	69 Tm Thulium 168.93	70 Yb Ytterbium 173.05	71 Lu Lutetium 174.97
89 Ac Actinium	90 Th Thorium 232.04	91 Pa Protactinium 231.04	92 U Uranium 238.03	93 Np Neptunium	94 Pu Plutonium	95 Am Americium	96 Cm Curium	97 Bk Berkelium	98 Cf Californium	99 Es Einsteinium	100 Fm Fermium	101 Md Mendelevium	102 No Nobelium	103 *Lr* Lawrencium

Figure 2.4 Elements are classified into specific families with similar properties.

2.2 Subatomic Particles

Learning Objective 2 Describe the charge, relative mass, and location of the three subatomic particles.

In **Section 1.3**, we introduced the atom as the "building block" of the elements, because the atom is the smallest particle that still maintains the characteristics of that element. Several decades of research and experimentation have shown, however, that atoms are made up of *subatomic particles.* The study of the makeup of atoms is known today as *atomic theory.* Atomic theory is a true testament to the scientific method in action because many years of scientific observations have led to hypotheses which were tested through experimentation until a coherent theory could be established. The generally accepted atomic theory at the time then became consistently challenged as new experiments presented critical information that tested that current theory's validity and completeness. Even today,

Figure 2.5 Au represents the precious metal gold, which is categorized as a transition metal.

Proton A stable subatomic particle with an electric charge of +1 and a mass of 1.67×10^{-24} g that resides in the center of an atom.

Neutron A stable subatomic particle with an electric charge of 0 and a mass of 1.67×10^{-24} g that resides in the center of an atom.

Electron A stable subatomic particle with an electric charge of −1 and a mass of 9.07×10^{-28} g that orbits the center of an atom.

scientists still conduct experiments to understand the atom to the deepest level humanly possible in order to establish a universal atomic theory.

In this text, we focus on the three fundamental subatomic particles: the **proton**, **neutron**, and **electron** (see **Table 2.1**).

Table 2.1 Characteristics of the Fundamental Subatomic Particles

		Characteristics		
Particle	**Common Symbols**	**Charge (±)**	**Mass (g)**	**Location**
Electron	e^-	−1	9.07×10^{-28}	Outside nucleus
Proton	p, p^+, H^+	+1	1.67×10^{-24}	Inside nucleus
Neutron	n	0	1.67×10^{-24}	Inside nucleus

As mentioned in **Section 1.3**, the term *atom* comes from the Greek for "indivisible." Ancient Greek philosophers like Democritus viewed the existence of what we know today as the atom from more of a philosophical perspective than one based on experimentation and observation. Early chemists like Lavoisier began to revive the foundational principles of what would become atomic theory as they sought to explain the behavior of gases like oxygen and hydrogen. British scientist John Dalton (1766–1844) then furthered this knowledge by proposing a complete atomic theory in the early 1800s. Among Dalton's crucial tenets was the proposition that atoms were themselves tiny particles of matter that could not be created or destroyed in a reaction; that atoms of each element are similar to,

Health Connections 2.1

Take Care of Your Bones

Osteoporosis, the abnormal thinning of bones that often accompanies aging, may lead to bone fractures, disability, and even death. While women are most susceptible, this serious condition also affects men, but usually at a more advanced age than women. A number of significant risk factors have been identified, including the following:

1. A poor diet, especially one low in calcium
2. Advanced age
3. The onset of menopause (or having had ovaries removed)
4. A sedentary lifestyle
5. Smoking
6. A family history of osteoporosis or hip fracture
7. Heavy drinking
8. The long-term use of certain steroid drugs
9. Vitamin D deficiency

Ninety-nine percent of the calcium found in the body is located in the skeleton and teeth, so it is not surprising that the behavior of this metallic element in the body plays a central role in a number of these risk factors for osteoporosis.

The best insurance against developing osteoporosis in later life is to build and strengthen as much bone as possible during the first 25 to 35 years of life. Two essential components of this process are eating a healthful diet containing adequate amounts of calcium and vitamin D, and following a healthful lifestyle that includes regular weight-bearing exercise such as walking, jogging, weight lifting, stair climbing, and/or physical labor. Dietary calcium supplements provide an additional way to enhance calcium intake, especially for individuals who are at risk to develop osteoporosis.

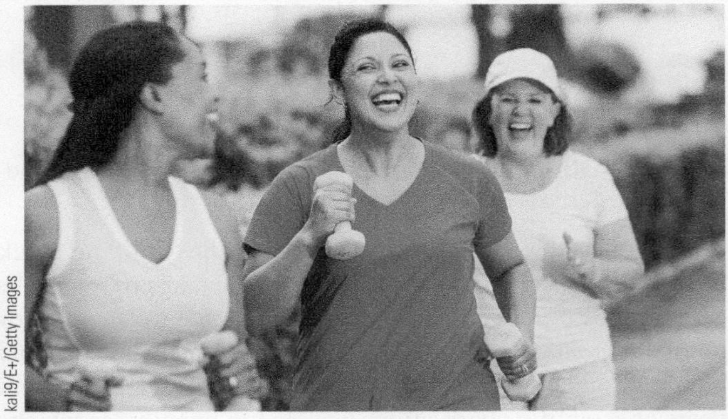
kali9/E+/Getty Images

An active lifestyle is an essential component of building and strengthening bones.

yet different from, atoms of other elements; and that molecules are formed when atoms of two or more elements combine. Dalton's theory was remarkably accurate, but it needed some tweaking when later experiments showed that there were charged subspecies within the atom.

It wasn't until the turn of the twentieth century that an advancement of Dalton's atomic theory was introduced to account for these charged subatomic particles. Another British scientist named J.J. Thomson (1856–1940) proposed the "plum pudding" model of the atom in which negatively charged particles (electrons—the "plums") were scattered about the atom (**Figure 2.6**). The "pudding" would have to consist of some positively charged matrix because the atom must be of neutral (zero) charge overall. Scientists understood at the time that these newly discovered electrons were remarkably small—namely, thousands of times smaller than the hydrogen atom.

A colleague of Thomson's, Ernest Rutherford (1871–1937), set out to test this new atomic model through his "gold foil experiment" (**Figure 2.7**). In that experiment, Rutherford shot positively charged particles, called alpha particles, at a thin piece of gold foil and monitored what happened. If Thomson's model were accurate, the positively charged particles would pass harmlessly through the atoms of gold to the other side. Most particles did pass through, but some of them were significantly deflected off course. In order to explain these unexpected results, Rutherford proposed in 1911 that there must exist a dense, positively charged focus of mass in the center of the atom, which he called the **nucleus** (**Figure 2.8**).

Soon, the presence of both positively charged protons and neutral neutrons within an atom's nucleus was established. The positively charged subatomic particle (the proton)

Figure 2.6 This plum pudding was used to describe the picture of the atom proposed by Thomson.

nucleus The central core of atoms that contains protons, neutrons, and most of the mass of atoms.

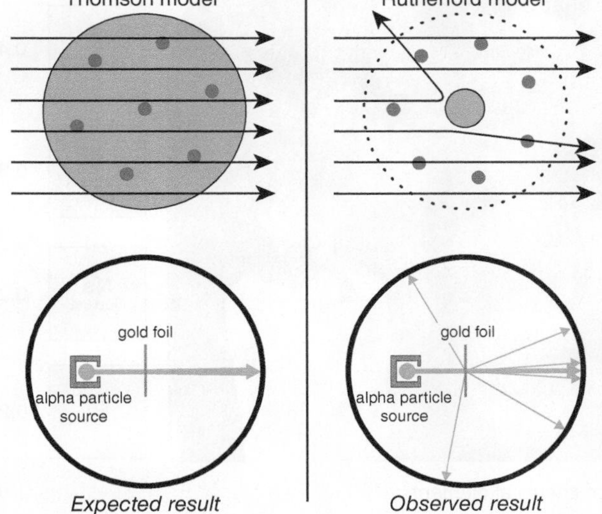

Figure 2.7 A schematic of the gold foil experiment showing what Thomson's plum pudding model would have produced (left) and the actual results Rutherford observed (right).

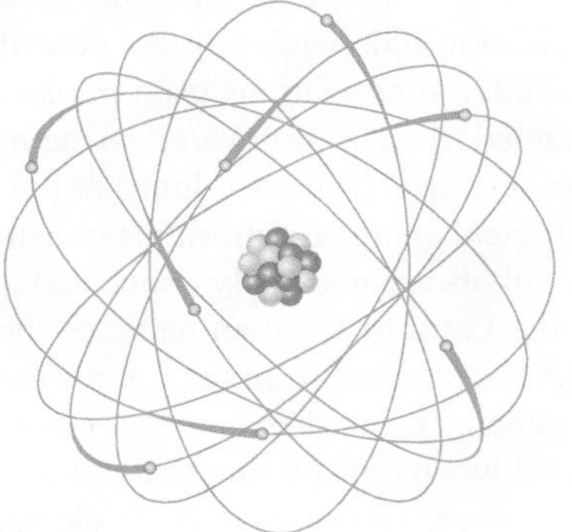

Figure 2.8 Rutherford's model of the atom consists of a small positively charged nucleus surrounded by negatively charged electrons. We now know that electrons do not follow orbits in precise paths as this diagram suggests. This figure is not drawn to scale. For a nucleus of the size shown, the closest electrons would be at least 80 m away.

was found to be equal in magnitude and opposite in sign relative to the charge of the electron. And relative to the electron, most of an atom's mass is contributed by the protons and neutrons that make up the nucleus. However, most of the space within an atom is taken up by the electrons. Thus, each of the foundational subatomic particles (electrons, protons, and neutrons) were now a part of atomic theory, though much more about the behavior and characteristics of these subatomic particles was yet to be learned. The unique attributes of each element, including its chemical properties, have their root in the number and type of subatomic particles present in atoms of that element. In the next section, we explain the notation that chemists have devised to represent atoms, with particular attention to their subatomic components.

2.3 Atomic Symbols

Learning Objective 3 Write the atomic symbol ($^{A}_{Z}X$) for a given set of subatomic particles.

The periodic table consists of 118 different elements, meaning that 118 different kinds of atoms are currently known to exist. Of these 118, there are 88 that occur naturally within Earth's crust, oceans, or atmosphere. The remaining 30 elements are synthetic, meaning they can only be produced in a laboratory setting. The elements most commonly found in the human body are shown in **Figure 2.9**.

Figure 2.9 The human body, like all matter, is made up of chemical elements.

Before we learn more about the individual elements themselves, we need to be able to name and identify each element independently. As described in **Section 2.1**, elemental symbols are one- or two-letter abbreviations used in place of an element's full name. Many elements have been named by their discoverer, so the names vary widely.

Some names are based on scientists (rutherfordium) or places (californium), while others are derived from mythological creatures (thorium) or astronomical bodies (uranium). However, some elemental symbols are not easily connected to the element they represent because they are based on the Latin or German name of the element. For instance, the elemental symbol for gold (Au) is derived from the Latin word *aurum*. Elemental symbols can represent the elements generally, but they can also represent a single atom of an element. This is especially useful when considering that compounds, as we defined them in Chapter 1, are pure substances made up of two or more elements.

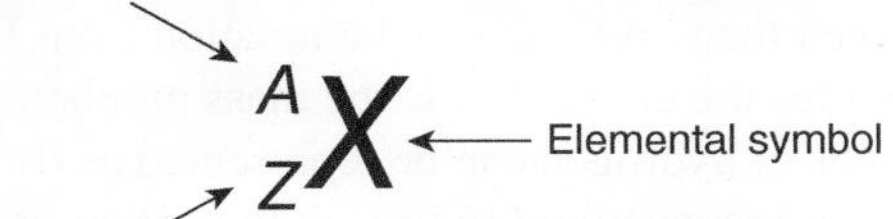

Figure 2.10 Atomic symbol of an atom, where X = elemental symbol, Z = the number of protons, and A = the number of protons plus the number of neutrons in an atom's nucleus.

Each element is more fully represented by its **atomic symbol** (see **Figure 2.10**), which provides not only the name and elemental symbol (sometimes represented generically by the letter X), but also the number of subatomic particles and the mass of each atom. The **atomic number** (Z) is a whole number that identifies the number of protons within each atom of an element. For instance, carbon has an atomic number of 6 because every carbon atom has 6 protons within its nucleus. Any atom with something other than 6 protons in its nucleus is not carbon.

atomic symbol The full representation of an element, which communicates both the name (abbreviation) of the element, and also information about the number of subatomic particles present.

atomic number The number of protons in the nucleus of an atom. Symbolically represented by Z.

Because each element is neutral in its naturally occurring state, the number of protons in an atom equals the number of electrons. That is, the positive charges of the protons cancel the negative charges of the electrons. Thus, a carbon atom contains not just 6 protons, but 6 electrons as well. Now that we can account for the number of protons and electrons in an atom, what about the number of neutrons? The **mass number** (A) is the sum total of the number of protons and neutrons in the nucleus of an atom. Thus, the number of neutrons = $A - Z$.

mass number The total number of protons and neutrons in a nucleus.

Example 2.2 Atomic Symbols

Use the periodic table to answer the following questions.

a. What are the mass number, atomic number, and atomic symbol ($^{A}_{Z}X$) for an atom that contains 7 protons and 8 neutrons?

b. Nickel is part of the stainless steel used in surgical instruments (**Figure 2.11**). How many neutrons are contained in an atom of nickel with a mass number of 60?

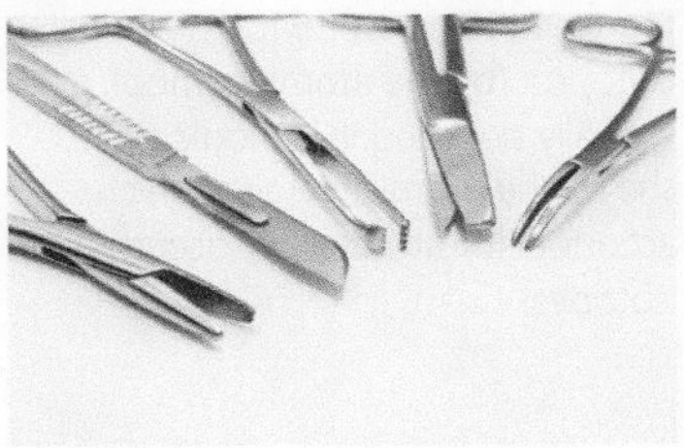

Solution

a. The mass number is the sum of the protons and neutrons in an atom, so $7 + 8 = 15$. The atomic number is the number of protons in an atom, and because there are 7 protons in this element, its atomic number is 7. The element corresponding to an atomic number of 7 is nitrogen ($^{15}_{7}N$).

b. According to the periodic table, nickel has the symbol Ni, and an atomic number, Z, of 28. The mass number, $A = 60$, is equal to the sum of the number of protons and the number of neutrons. The number of protons is equal to the atomic number, 28. Therefore, the number of neutrons is $A - Z = 60 - 28 = 32$. The atom contains 32 neutrons.

YJPTO/Shutterstock.com

Figure 2.11 Both regular stainless steel and surgical steel, a medical-grade version of stainless steel, contain up to 25% nickel.

✔ **Learning Check 2.2** Use the periodic table to answer the following question.
What are the atomic number, mass number, and atomic symbol for an atom that contains 4 protons and 5 neutrons?

2.4 Isotopes and Atomic Weights

Learning Objective 4 Define the terms *isotope* and *atomic weight*.

Although the number of protons and electrons is fixed for each element, atoms of the same element can have different numbers of neutrons! Atoms that have the same atomic number, but different mass numbers (i.e., a different number of neutrons in the nucleus) are called **isotopes**. Most elements consist of naturally occurring mixtures of two or more isotopes—even hydrogen, the smallest element of all, has three isotopes. Each hydrogen isotope has an atomic number of 1 (1 proton, 1 electron), but they differ in the number of neutrons present

isotopes Atoms that have the same atomic number but different mass numbers. That is, they are atoms of the same element that contain different numbers of neutrons in their nuclei.

Bokeh Art Photo/Shutterstock.com

Figure 2.12 Silicon has 23 known isotopes.

atomic weight The mass of the weighted average of the atoms (naturally occurring isotopes) of an element.

atomic mass unit (amu) A unit used to express the relative masses of atoms. One amu is equal to 1.661×10^{-24} g, which is 1/12 of the mass of a single carbon-12 atom.

in the nucleus. Their mass numbers are 1 (0 neutrons), 2 (1 neutron), and 3 (2 neutrons). In order to distinguish between them, we can use the notation from **Figure 2.10**: ($^{A}_{Z}X$), where X is the elemental symbol for the element, A is the mass number, and Z is the atomic number. Thus, the three isotopes of hydrogen can be represented as $^{1}_{1}H$, $^{2}_{1}H$, $^{3}_{1}H$.

When these symbols are inconvenient to use, as in written or spoken references to the isotopes, the elemental name followed by the mass number is used. Thus, the three hydrogen isotopes are hydrogen-1 (spoken as "hydrogen one"), hydrogen-2, and hydrogen-3. Unlike most other isotopes, these three isotopes have specific names: protium ($A = 1$), deuterium ($A = 2$), and tritium ($A = 3$), respectively. Silicon (see **Figure 2.12**), one of the major elements in Earth's crust, has 23 known isotopes!

If there are multiple possible isotopes for the majority of the elements in the periodic table, then how do we accurately, but concisely, communicate the mass of an atom? We can't use mass number because that limits us to looking at only a single isotope at a time. Furthermore, the mass of atoms is so small, with protons and neutrons each on the order of 10^{-24} g, it is difficult to comprehend the actual mass of atoms in any quantitative sense. Therefore, the values provided in the periodic table are a weighted average of all naturally occurring isotopes of an element (**Figure 2.13**). This value, weighted according to the abundance of each isotope, is referred to as an element's **atomic weight**. Atomic weight is reported in units of **atomic mass units**, represented by the symbol "amu" throughout the text. This particular unit is defined as 1/12th of the mass of a single carbon-12 atom, which equals 1.661×10^{-24} g.

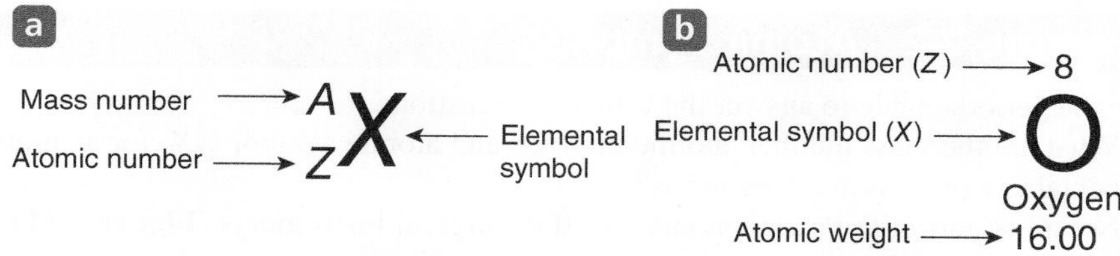

Figure 2.13 (a) The atomic symbol, as introduced in **Figure 2.10**, is useful for individual isotopes. (b) The atomic symbol, as typically depicted in periodic table, shows the atomic weight, which accounts for all naturally occurring isotopes.

Let us use chlorine to illustrate how atomic weights are calculated using the weighted averages of an element's isotopes. Naturally occurring chlorine consists of 75.53% chlorine-35 (atomic mass = 34.97 amu) and 24.47% chlorine-37 (atomic mass = 36.97 amu), so we can calculate the atomic weight as follows:

$$\begin{aligned}\text{Atomic weight} &= (\%\ \text{chlorine-35})(\text{mass chlorine-35}) + \\ &\quad (\%\ \text{chlorine-37})(\text{mass chlorine-37}) \\ &= (0.7553)(34.97\ \text{amu}) + (0.2447)(36.97\ \text{amu}) \\ &= 35.4594 = 35.46\ \text{amu}(\text{rounded answer})\end{aligned}$$

This result is slightly different from the periodic table atomic weight value of 35.45 because of slight errors introduced in rounding the isotope masses to four significant figures.

2.5 Radioactive Nuclei

Learning Objective 5 Write balanced equations for radioactive decay and other nuclear processes.

Learning Objective 6 Solve problems using the half-life concept.

Learning Objective 7 Describe the applications of radiation in health and medicine.

While our focus in subsequent chapters centers on the role electrons play in atoms and in chemical bonding, we now turn our attention to the nucleus of the atom. Specifically, we explore radioactivity, a topic of great importance to human health and modern medicine.

We've now seen that isotopes of the same element differ only in the number of neutrons present in their nuclei. However, while it has been found that some combinations of neutrons and protons in the nucleus are stable and do not change spontaneously, others are not so stable. In fact, these unstable nuclei emit radiation. Nuclei that emit radiation are said to be **radioactive nuclei**. Isotopes of elements that emit nuclear radiation are called

radioactive nuclei Nuclei that undergo spontaneous changes and emit energy in the form of radiation.

radioisotopes. In 1896, Henri Becquerel (1852–1908), a French physicist, discovered radioactivity by chance when observing that radiation is emitted by any compound of uranium. Later studies by other investigators showed that the radiation emitted by uranium, and by other radioactive elements discovered later, can be separated into three types by an electrical or a magnetic field (see **Figure 2.14**). The three types have different electrical charges: one is positive, one is negative, and one carries no charge. The types of radiation were given names that are still used today: alpha rays (positive), beta rays (negative), and gamma rays (uncharged).

radioisotope An isotope of an element that emits nuclear radiation.

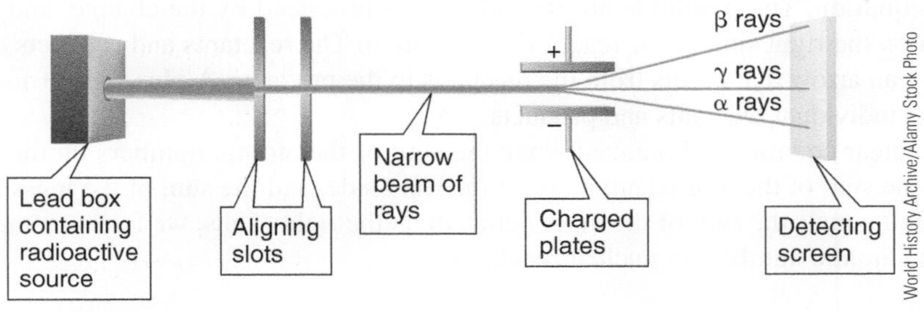

Figure 2.14 The three types of emitted radiation.

Today it is known that other types of particles, such as neutrons and positrons, are also emitted by radioactive nuclei. However, alpha, beta, and gamma rays are the most common. **Table 2.2** summarizes the characteristics of these forms. In this book, we usually use the symbol given first for each form; the symbols in parentheses are alternatives you might find elsewhere. Except for high-energy gamma radiation, the radiation emitted by radioactive nuclei consists of streams of particles. The emission of radiation by unstable nuclei is often called **radioactive decay**.

radioactive decay A process in which an unstable nucleus changes energy states and in the process emits radiation.

Table 2.2 Characteristics of Nuclear Radiation

Type of Radiation	Symbols	Mass Number	Charge	Composition
Alpha	${}^{4}_{2}\alpha$ (α, ${}^{4}_{2}He$, He^{2+})	4	+2	Helium nuclei, 2 protons + 2 neutrons
Beta	${}^{0}_{-1}\beta$ (β, β^-, ${}^{0}_{-1}e$, e^-)	0	−1	Electrons produced in nucleus and ejected
Gamma	γ (${}^{0}_{0}\gamma$)	0	0	Electromagnetic radiation
Neutron	${}^{1}_{0}n$ (n)	1	0	Neutrons
Positron	${}^{0}_{1}\beta$ (β^+, ${}^{0}_{1}e$, e^+)	0	+1	Positively charged electrons

As shown in **Table 2.2**, the particles that make up alpha rays are identical to the nuclei of helium atoms; that is, they are clusters containing two protons and two neutrons, and they have a charge of +2 from the two protons. **Alpha particles** are the most massive particles emitted by radioactive materials, and they have the highest charge. Because of their mass and charge, they collide often with the molecules of any matter through which they travel. As a result of these collisions, their energy is quickly dissipated, and they do not travel far. They cannot even penetrate through a few sheets of writing paper or the outer cells of the skin. However, exposure to intense alpha radiation can result in severe burns. Despite its side effects, radiation therapy is commonly used in treating various types of cancers (**Figure 2.15**).

alpha particle The particle that makes up alpha rays. It is identical to the helium nucleus and is composed of two protons and two neutrons.

Beta particles are actually electrons, but they are produced within the nucleus and then emitted. They are not part of the group of electrons moving around the nucleus that is responsible for chemical characteristics of the atoms. Beta particles have a charge of −1 and are 7000 times less massive than alpha particles. They undergo far fewer collisions with the molecules of matter through which they pass, so they are much more penetrating than alpha particles.

beta particle The particle that makes up beta rays. It is identical to an electron but is produced in the nucleus when a neutron is changed into a proton and an electron.

Gamma rays are not streams of particles but rays of electromagnetic radiation similar to X rays. Gamma rays, though, have higher energies than X rays. Like X rays, gamma

gamma ray A high-energy ray that is like an X ray, but with a higher energy.

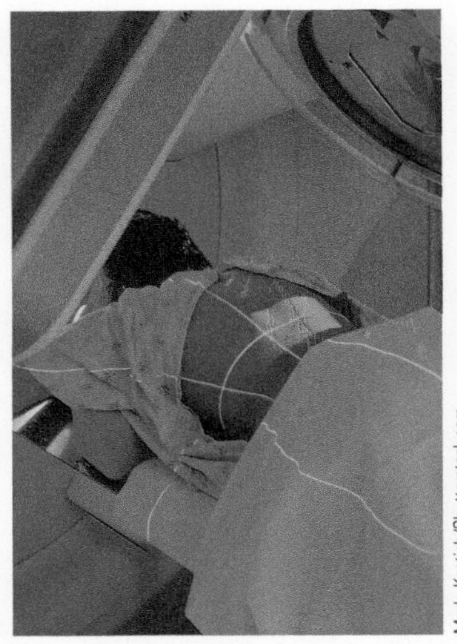

Mark_Kostich/Shutterstock.com

Figure 2.15 Alpha and beta particle emitters are frequently used in radiation therapy to treat cancer.

nuclear reaction A process where two atomic nuclei or an atomic nucleus and another subatomic particle collide, resulting in one or more new nuclei.

reactants of a reaction The substances that undergo chemical change during the reaction. They are written on the left side of the equation, representing the reaction.

products of a reaction The substances produced as a result of the reaction taking place. They are written on the right side of the equation, representing the reaction.

daughter nuclei The new nuclei produced when unstable nuclei undergo radioactive decay.

Source: Science Stock Photography

Figure 2.16 The radioisotope americium-241 is present in smoke detectors. Am-241 emits alpha particles, which conduct electricity. The presence of smoke reduces electrical conductivity, thus tripping the alarm.

radiation is very penetrating and dangerous to living organisms. Adequate shielding from gamma rays is provided only by thick layers of heavy materials like metallic lead or concrete.

2.5A Equations for Nuclear Reactions

The transformation of one or more atomic nuclei into one or more new nuclei is called a **nuclear reaction**. The reactants are the substances that undergo the change; by convention, the **reactants** are written on the left side of the reaction, which can also be referred to as the equation. The **products** are the substances produced by the change, and they are written on the right side of the reaction, or equation. The reactants and products are separated by an arrow that points from the reactants to the products. A plus sign (+) is used to separate individual reactants and products.

A nuclear equation is balanced when the sum of the atomic numbers on the left side equals the sum of the atomic numbers on the right side, and the sum of the mass numbers on the left equals the sum of the mass numbers on the right. Thus, we focus on mass numbers and atomic numbers in nuclear reactions.

Example 2.3 Writing Equations for Nuclear Reactions

Americium-241 nuclei undergo radioactive decay by emitting an alpha particle and changing into the nucleus of another element (**Figure 2.16**).

a. Write appropriate symbols for americium-241 using the ${}^{A}_{Z}X$ symbolism.
b. Write a balanced nuclear equation for the radioactive decay of americium-241 that produces one alpha particle.

Solution

a. **Find the elemental symbol:**
Americium-241 is an isotope of americium, so the symbol X is replaced by Am.

Find the mass number:
The mass number is 241 according to the rules learned earlier for designating isotopes, so $A = 241$.

Find the atomic number:
The atomic number of americium obtained from the periodic table is 95, so $Z = 95$.

Atomic symbol: ${}^{241}_{95}\text{Am}$

b. **Identify the radioactive particle emitted:**
Because americurium-241 is undergoing alpha particle decay, we know the following:

$$ {}^{241}_{95}\text{Am} \rightarrow {}^{4}_{2}\alpha + \text{daughter nucleus} $$

Find the daughter nucleus mass number:
The mass number of the new nucleus, called a **daughter nucleus**, is $241 - 4 = 237$ because an emitted alpha particle has a mass number of 4.

Find the daughter nucleus atomic number:
The atomic number of the new daughter nucleus is $95 - 2 = 93$ because the emitted alpha particle has an atomic number of 2.

Find the atomic symbol of the daughter nucleus:
The daughter nucleus has a mass number of 237 and an atomic number of 93, and based on this atomic number, the identity of the element produced is neptunium, ${}^{237}_{93}\text{Np}$.

Write the balanced nuclear equation:
Processes such as this where one alpha particle is ejected can be represented by the following general equation, where Y is the elemental symbol of the daughter nucleus:

$$ {}^{A}_{Z}X \rightarrow {}^{4}_{2}\alpha + {}^{A-4}_{Z-2}Y \qquad (2.1) $$

Taken together this information gives us the balanced equation:

$$^{241}_{95}\text{Am} \rightarrow {}^{4}_{2}\alpha + {}^{237}_{93}\text{Np}$$

> ✔ **Learning Check 2.3** Thorium ($Z = 90$) exists in a number of isotopic forms. One form, thorium-234, decays by emitting a beta particle.
>
> a. Write appropriate symbols for thorium-234 using the $^{A}_{Z}X$ symbolism.
> b. Write a balanced nuclear equation for the radioactive decay of thorium-234.
> c. Write a general equation for decays in which one beta particle is ejected.

Because gamma rays have no mass or atomic numbers, they do not enter into the balancing process of nuclear reactions. However, they do represent energy and should be included in balanced equations when they are known to be emitted.

Although beta particles are electrons ejected from the nucleus of a radioactive atom, we know from our discussions of atomic theory that atomic nuclei contain only protons and neutrons. Where, then, do the beta particles originate? According to a simplified theory of nuclear behavior, a beta particle (or electron) is produced in the nucleus when a neutron changes into a proton. The balanced equation for the nuclear reaction is

$$^{1}_{0}\text{n} \rightarrow {}^{1}_{1}\text{p} + {}^{0}_{-1}\beta \qquad (2.2)$$

Note that the criteria for balance in this equation are satisfied; that is, mass numbers balance: $1 = 1 + 0$, and atomic numbers balance $0 = 1 + (-1)$. Thus, a nuclear neutron is converted to a proton during beta emission, so the daughter nucleus has an atomic number *higher* by 1 than the decaying nucleus. Similar changes in nuclear particles occur when nuclei decay by positron emission. A **positron** is a positively charged electron, $^{0}_{1}\beta$ (see **Table 2.2**). When a nucleus emits a positron, a nuclear proton is changed to a neutron, so the daughter nucleus has an atomic number *lower* by 1 than the decaying nucleus. Positron-emitting isotopes are used in diagnostic positron emission tomography (PET) scans of the brain (see **Figure 2.17**).

positron A positively charged electron.

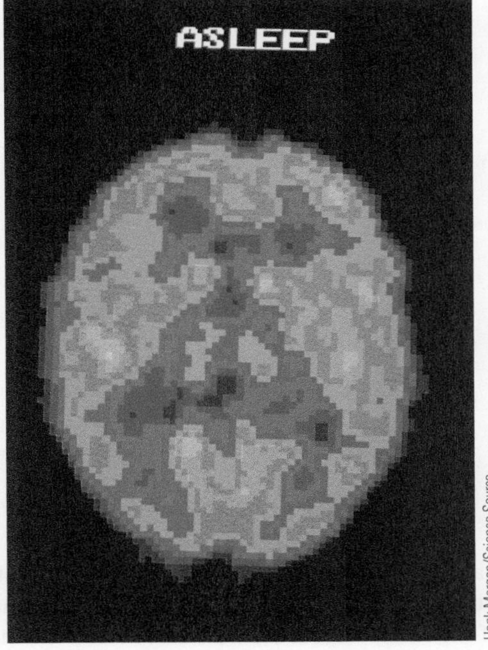

Figure 2.17 A positron emission tomography (PET) scan shows normal brain activity during sleep. A radioactive isotope that emits positrons is made into a chemical compound that is absorbed by active areas of the brain. The emitted positrons collide with nearby electrons and produce gamma rays that pass through the skull to detectors surrounding the patient's head. A computer uses the detector data to construct the image.

electron capture A mode of decay for some unstable nuclei in which an electron from outside the nucleus is drawn into the nucleus, where it combines with a proton to form a neutron.

Certain nuclear changes occur when an electron from outside the nucleus is drawn into the nucleus, where it converts a proton into a neutron. This process, called **electron capture**, is not as common as other decay processes. The daughter nucleus in electron capture has an atomic number 1 less than the decaying nucleus.

2.5B Isotope Half-Life and Units for Radiation

half-life The time required for one-half the unstable nuclei in a sample to undergo radioactive decay.

Some radioactive isotopes are more stable than others. The more stable isotopes undergo radioactive decay more slowly than the less stable isotopes. The **half-life** of an isotope is used to indicate stability, and it is equal to the time required for one-half (50%) of the atoms of a sample of the isotope to decay. **Table 2.3** lists isotopes with a wide range of half-lives.

Table 2.3 Half-Lives of Selected Radioactive Isotopes

Isotope	Half-Life	Source
$^{238}_{92}U$	4.5×10^9 years	Naturally occurring
$^{40}_{19}K$	1.3×10^9 years	Naturally occurring
$^{226}_{88}Ra$	1600 years	Naturally occurring
$^{14}_{6}C$	5600 years	Naturally occurring
$^{239}_{94}Pu$	24,000 years	Synthetically produced
$^{90}_{38}Sr$	28 years	Synthetically produced
$^{131}_{53}I$	8 days	Synthetically produced
$^{24}_{11}Na$	15 hours	Synthetically produced
$^{15}_{8}O$	2 minutes	Synthetically produced
$^{5}_{3}Li$	10^{-21} seconds	Synthetically produced

The fraction of original radioactive atoms remaining in a sample after a specific time has passed can be determined from the half-life. After one half-life has passed, $\frac{1}{2}$ of the original number have decayed, so $\frac{1}{2}$ remain. During the next half-life, $\frac{1}{2}$ of the remaining $\frac{1}{2}$ decay, so $\frac{1}{4}$ of the original atoms remain. After three half-lives, $\frac{1}{2} \times \frac{1}{2} \times \frac{1}{2} = \frac{1}{8}$ of the original atoms remain undecayed (see **Figure 2.18**).

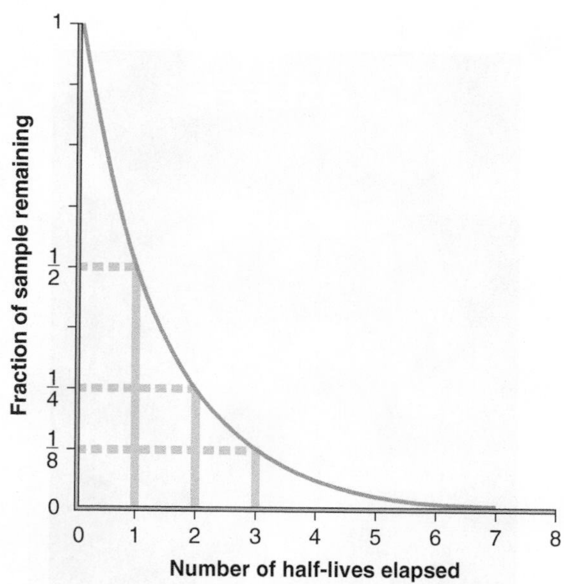

Figure 2.18 A radioactive decay curve (bars give values at 1, 2, and 3 half-lives). What fraction of the sample will be remaining after 4 half-lives?

The amount of radiation given off by a sample of radioactive material is proportional to the number of radioactive atoms present in the sample. The amount of radiation given off is called the *activity* of the sample and is measured by various instruments of detection.

Example 2.4 Calculating Half-Lives

The activity (amount of radiation emitted) of a radioactive substance is measured at 9 a.m. At 5 p.m. the activity is found to be only $\frac{1}{16}$ of the original (9 a.m.) value. What is the half-life of the radioactive material?

Solution

Because only $\frac{1}{16}$ of the original radioactive material is present, four half-lives have passed: $\frac{1}{2} \times \frac{1}{2} \times \frac{1}{2} \times \frac{1}{2} = \frac{1}{16}$. The time between measurements is 8 hours (9 a.m. to 5 p.m.). If 8 hours is four half-lives, then one half-life is 2 hours.

✔ **Learning Check 2.4** Iodine-123 is a radioisotope used to diagnose the function of the thyroid gland. It has a half-life of 13.3 hours. What fraction of a diagnostic dose of iodine-123 would be present in a patient 79.8 hours (a little over 3 days) after it was administered?

IanDagnall Computing/Alamy Stock Photo

Figure 2.19 Marie Curie (1867–1934) discovered the elements polonium and radium and is the only person to win Nobel Prizes in two separate sciences (physics and chemistry).

Radiation can be described in either physical or biological units (**Table 2.4**). **Physical units** indicate the activity of a source of radiation, while **biological units** are related to the damage caused by radiation. Physical units include the *becquerel*, named after Henri Becquerel, and the *curie,* named after the Polish-French chemist and physicist Marie Curie (see **Figure 2.19**). Biological units account for the fact that a given quantity of a type of radiation doesn't have the same effect on tissue as a given quantity of another type of radiation. *Rad* and *gray* describe the amount of energy transferred from the radiation to the tissue through which it passes. Lastly, the *sievert* and *rem* are biological units devised to account for differences that various types of radiation have on health. The advantage of the *rem* is that individuals exposed to low levels of various types of radiation have to keep track only of the rems absorbed from each type of radiation and add them up to get the total health effect.

physical unit of radiation A radiation measurement unit indicating the activity of the source of the radiation; for example, the number of nuclear decays per minute.

biological unit of radiation A radiation measurement unit indicating the damage caused by radiation in living tissue.

Geiger-Müller tube A radiation-detection device operating on the principle that ions form when radiation passes through a tube filled with low-pressure gas.

scintillation counter A radiation-detection device operating on the principle that certain substances give off light when struck by radiation.

Table 2.4 Units for Measuring Radiation

Measurement	Common Unit	SI Unit	Relationships between Units
Physical	curie (Ci) 1 Ci = 3.7×10^{10} disintegrations/s	becquerel (Bq) 1 Bq = 1 disintegrations/s	1 Ci = 3.7×10^{10} Bq
Biological	rad (D)	gray (Gy)	1 Gy = 100 D
Biological	rem	sievert (Sv)	1 Sv = 100 rem

The presence of radiation can be detected in several ways. A **Geiger-Müller tube** (a "Geiger counter") measures radiation as it passes between charged electrodes in the presence of a low-pressure gas. These instruments are very useful for detecting radiation in a variety of environments but cannot distinguish between types of radiation. They also have difficulty retaining accuracy at high levels of exposure. A **scintillation counter** measures flashes when radiation reacts with certain substances to emit visible light. Many workers in medical settings use simple film badges to track exposure to radiation over time (**Figure 2.20**).

Public Health England/Science Source

Figure 2.20 Film badges provide a convenient way to monitor the total amount of radiation received during a specific time period.

2.5C The Health Effects and Medical Uses of Radioisotopes

Radiation can be hazardous to living organisms, but do you know why? What does radiation do that is so dangerous? The greatest danger to living organisms results from the ability of high-energy or ionizing radiation to knock electrons out of compounds and generate

radical or **free radical** An electron-deficient particle that is very reactive.

acute radiation syndrome The condition that can result from short-term exposure to intense radiation.

particles called **radicals** or **free radicals** in the tissue through which the radiation passes. These electron-deficient particles are very reactive. They cause reactions to occur among more stable materials in the cells of living organisms. When these reactions involve genetic materials such as genes and chromosomes, the changes might lead to genetic mutations, cancer, or other serious consequences.

Long-term exposure to low-level radiation is more likely to cause these kinds of problems than are short exposures to intense radiation. Short-term intense radiation tends to destroy tissue rapidly in the area exposed and can cause symptoms of so-called **acute radiation syndrome**. Some of these symptoms are listed in **Table 2.5**. The rapid destruction of tissue makes relatively intense radiation a useful tool in the treatment of some cancers (**Figure 2.21**).

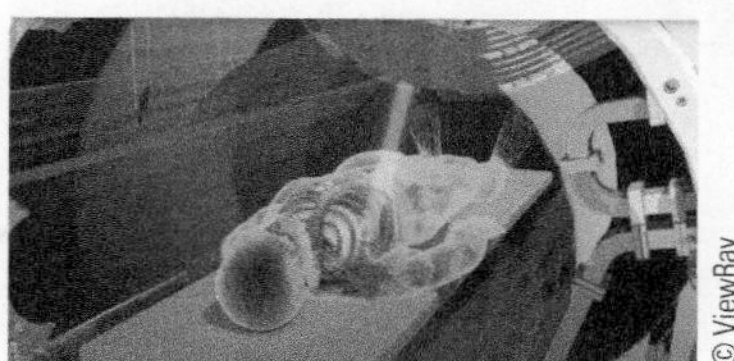

© ViewRay

Figure 2.21 Short-term intense radiation exposure is useful in fighting cancerous tissues. This image depicts Hershey Medical Center's MRI-magnetic resonance imaging guided radiation therapy.

Table 2.5 The Effects on Humans of Short-Term, Whole-Body Radiation Exposure

Dose (rems)[a]	Effects
0–25	No detectable clinical effects.
25–100	Slight short-term reduction in number of some blood cells; disabling sickness not common.
100–200	Nausea and fatigue, vomiting if dose is greater than 125 rems, longer-term reduction in number of some blood cells.
200–300	Nausea and vomiting first day of exposure, up to a 2-week latent period followed by appetite loss, general malaise, sore throat, pallor, diarrhea, and moderate emaciation. Recovery in about 3 months unless complicated by infection or injury.
300–600	Nausea, vomiting, and diarrhea in first few hours. Up to a 1-week latent period followed by loss of appetite, fever, and general malaise in the second week, followed by hemorrhage, inflammation of mouth and throat, diarrhea, and emaciation. Some deaths in 2–6 weeks. Eventual death for 50% if exposure is above 450 rems; others recover in about 6 months.
600 or more	Nausea, vomiting, and diarrhea in first few hours. Rapid emaciation and death as early as second week. Eventual death rate of nearly 100%.

[a] The rem, a biological unit of radiation, is defined in **Section 2.5B (see Table 2.4)**.

SOURCE: U.S. Atomic Energy Commission.

The health hazards presented by radiation make it imperative to minimize exposures. This is especially true for individuals with occupations that continually present opportunities for exposure. One approach is to absorb radiation by shielding. As shown in **Table 2.6**, alpha and beta rays are the easiest to stop. Gamma radiation and X rays require the use of very dense materials such as lead to provide protection. If the type of radiation is known, a careful choice of shielding materials can provide effective protection (see **Figure 2.22**).

Table 2.6 Penetrating Abilities of Alpha, Beta, and Gamma Rays

	Depth of Penetration into		
Type of Radiation	**Dry Air**	**Tissue**	**Lead**
Alpha	4 cm	0.05 mm	0
Beta[a]	6 to 300 cm	0.06 to 4 mm	0.005 to 0.3 mm
Gamma[b]	400 m	50 cm	300 mm

[a] Depth depends on energy of rays.

[b] Depth listed is that necessary to reduce initial intensity by 10%.

Another protection against radiation is distance. Because radiation spreads in all directions from a source, the amount falling on a given area decreases the farther that area is from the source. If you double the distance to the source, then the amount of radiation that reaches the area is only one-fourth of what it was before.

Radioisotopes undergo the same chemical reactions as nonradioactive isotopes of the same element. This characteristic, together with the fact that the location of radioisotopes

Health Connections 2.2

Diagnostic Radiation

We are continually exposed to radiation from naturally occurring radioisotopes, sunlight, and cosmic rays from outer space. The discovery of X rays in 1895 by Wilhelm Roentgen contributed to this exposure when X rays were introduced as a diagnostic tool for patients. Today, X rays are used in numerous medical and dental diagnostic tests and procedures.

In the early 1970s, X rays were integrated with computers to create a diagnostic tool called *computed tomography,* or CT scans. In this procedure, the intensity of X rays after they pass through the body is interpreted using a computer, and the computer results are used to create a cross-sectional image of the part of the body through which the X rays passed.

In 1946, the property called *nuclear magnetic resonance* was discovered. Today, this property is widely used medically in the diagnostic procedure called *magnetic resonance imaging* (MRI). During an MRI procedure, the patient is placed in a very strong magnetic field, and radiation in the form of powerful radio waves is passed through the patient's body.

The energy of the radio waves is absorbed by hydrogen atoms in the body and then re-emitted. The re-emitted radio waves are interpreted by a powerful computer program that then produces cross-sectional images of the body.

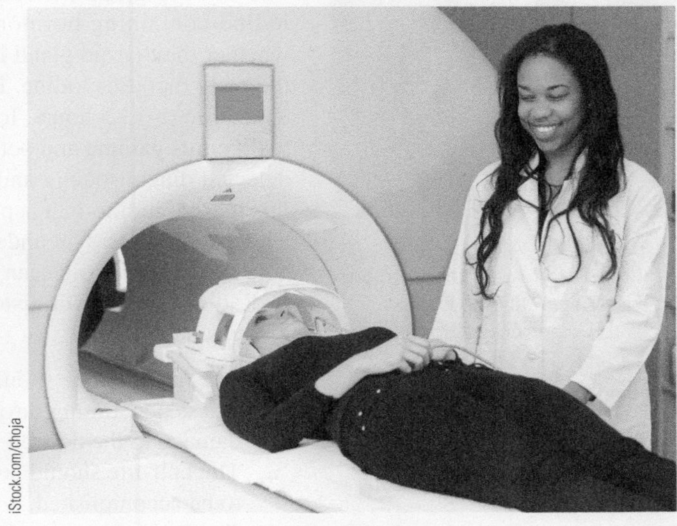

iStock.com/choja

Magnetic resonance imaging (MRI) scans enable medical personnel to view soft tissues in detail.

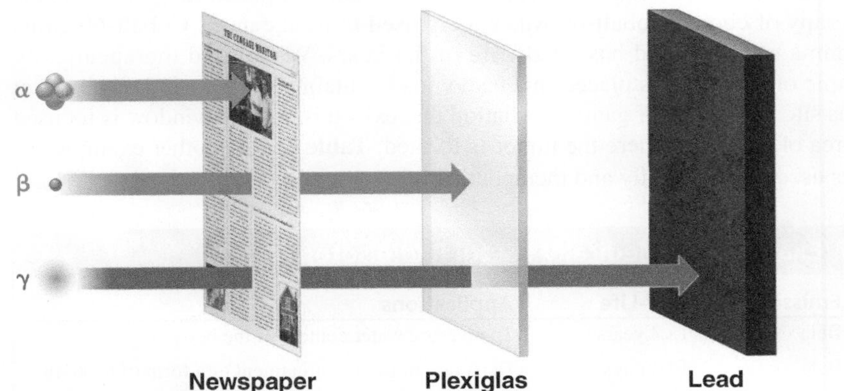

Figure 2.22 Penetration of materials by ionizing radiation.

in the body can be readily detected, makes them useful for diagnostic and therapeutic medical applications. In diagnostic applications, radioisotopes are used as **tracers**, whose progress through the body or localization in specific organs can be followed.

tracer A radioisotope used medically because its progress through the body or localization in specific organs can be followed.

To minimize the risks to patients from exposure to radiation, radioisotopes used as diagnostic tracers should have as many of the following characteristics as possible:

1. Tracers should have short half-lives so they will decay while the diagnosis is being done but will give off as little radiation as possible after the diagnostic procedure is completed.
2. The daughter nuclei produced by the decaying radioisotope should be nontoxic and give off little or no radiation of their own. Ideally, they will be stable.
3. The radioisotope should have a long enough half-life to allow it to be prepared and administered conveniently.

hot spot Tissue in which a radioactive tracer concentrates.

cold spot Tissue from which a radioactive tracer is excluded or rejected.

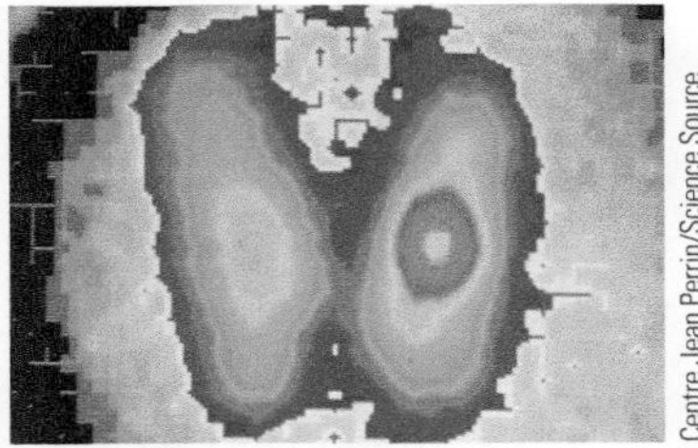

Centre Jean Perrin/Science Source

Figure 2.23 A thyroid scan produced after the administration of a radioactive iodine isotope. Does this scan show a hot spot or cold spot?

4. The radiation given off by the radioisotope should be penetrating gamma rays, if possible, to ensure that they can be detected readily by detectors located outside the body.
5. The radioisotope should have chemical properties that make it possible for the tissue being studied to either concentrate it in diseased areas and form a **hot spot** or essentially reject it from diseased areas to form a **cold spot**.

The thyroid gland concentrates iodine and uses it in the production of thyroxine, an iodine-containing hormone. Iodine is an excellent diagnostic tracer for determining whether the thyroid gland is functioning properly because the thyroid is the only tissue in the body that uses iodine. Iodine-123 is a good tracer because it is a gamma-emitter with a half-life of 13.3 hours. Iodine-131 is also radioactive and has a short half-life (8 days), but it emits gamma and beta radiation. The beta radiation cannot penetrate the tissue to be detected diagnostically and therefore is simply an added radiation risk to the patient. For this reason, iodine-123 is preferred. An overactive thyroid absorbs more iodine and forms a hot spot, whereas an underactive or nonactive gland absorbs less iodine and may show a cold spot on a thyroid scan (see **Figure 2.23**).

Radioisotopes administered internally for therapeutic use should ideally have the following characteristics:

1. The radioisotope should be an alpha or beta emitter because radiation in these forms is less penetrating through tissue than gamma emitters and thus restricts the extent of damage to the desired tissue.
2. The half-life should be long enough to allow sufficient time for the desired therapy to be accomplished.
3. The decay products should be nontoxic and give off little or no radiation.
4. The target tissue should concentrate the radioisotope so that the radiation damage is restricted to the target tissue.

Iodine-131, for example, is used therapeutically to treat thyroid cancer and hyperthyroidism. Its primary advantage is the ability of thyroid tissue to absorb it and localize its effects.

Until the mid-1950s, when improvements in the generation of penetrating X rays made them the therapy of choice, cobalt-60 was widely used to treat cancer. Cobalt-60 emits beta and gamma radiation and has a half-life of 5.3 years. When used therapeutically today, a sample of cobalt-60 is placed in a heavy lead container with a window aimed at the cancerous site. The beam of gamma radiation that exits through the window is focused on a small area of the body where the tumor is located. **Table 2.7** lists other examples of radioisotopes used diagnostically and therapeutically.

Table 2.7 Examples of Medically Useful Radioisotopes

Isotope	Emission	Half-Life	Applications
${}^{3}_{1}H$	Beta	12.3 years	To measure water content of the body
${}^{32}_{15}P$	Beta	14.3 days	Detection of tumors, treatment of a form of leukemia
${}^{51}_{24}Cr$	Gamma	27.8 days	Diagnosis: size and shape of spleen, gastrointestinal disorders
${}^{59}_{26}Fe$	Beta	45.1 days	Diagnosis: anemia, bone marrow function
${}^{60}_{27}Co$	Beta, gamma	5.3 years	Therapy: cancer treatment
${}^{67}_{31}Ga$	Gamma	78.1 hours	Diagnosis: various tumors
${}^{75}_{34}Se$	Beta	120.4 days	Diagnosis: pancreatic scan
${}^{81}_{36}Kr$	Gamma	2.1×10^5 years	Diagnosis: lung ventilation scan
${}^{85}_{38}Sr$	Gamma	64 days	Diagnosis: bone scan
${}^{99}_{43}Tc$	Gamma	6 hours	Diagnosis: brain, liver, kidney, bone, and heart muscle scans
${}^{123}_{53}I$	Gamma	13.3 hours	Diagnosis: thyroid cancer
${}^{131}_{53}I$	Beta, gamma	8.1 days	Diagnosis and treatment: thyroid cancer
${}^{197}_{80}Hg$	Gamma	65 hours	Diagnosis: kidney scan

2.6 Where Are the Electrons?

Learning Objective 8 Write correct electron configurations for each element through atomic number 56.

Now that we have discussed the nucleus and the radiation emitted during nuclear chemistry, let us now examine electrons in more detail. Electrons are largely responsible for the fundamental properties of the atoms and the interactions of atoms with one another. As we have seen, in the early twentieth century, scientists learned that electrons reside outside the nucleus. A further understanding of where electrons reside and how they behave was elucidated, in part, due to research on the energy that atoms absorb or emit. We have discussed various types of nuclear radiation, including high-energy rays of electromagnetic radiation like X rays (**Figure 2.24**) and gamma rays. However, this is only a part of the larger *electromagnetic spectrum* (**Figure 2.25**).

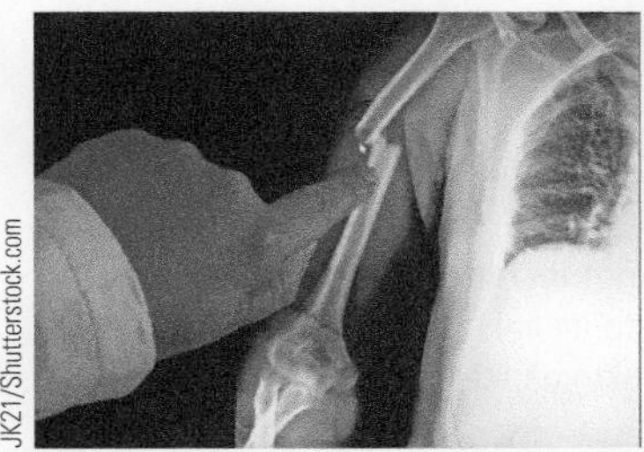

Figure 2.24 X rays are just one of the many ways energy from the electromagnetic spectrum is harnessed in medicine.

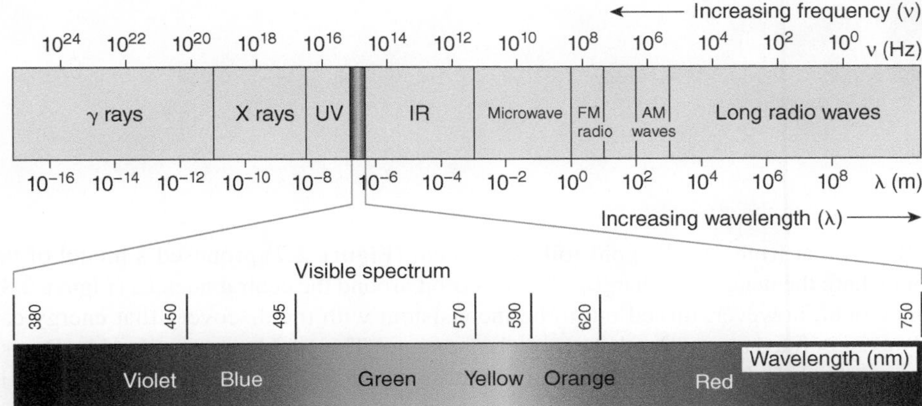

Figure 2.25 The full electromagnetic (EM) spectrum from low-energy radio waves (right) to high-energy gamma rays (left). The range of visible light is only a small part of the entire spectrum. It falls between the ultraviolet (UV) and infrared (IR) regions.

We interact over and over again with the different parts of the electromagnetic spectrum. Radio waves, for example, are used in magnetic resonance imaging (MRI), a powerful medical technology (**Figure 2.26**). Infrared waves are used for thermal imaging, and UV waves are the source of energy for things like tanning beds, although exposure must be monitored carefully (**Figure 2.27**). These different types of electromagnetic radiation all have different **wavelengths** (the distance from one wave peak to the next) and **frequencies** (the number of waves per unit of time). Their particular wavelengths and frequencies relate to their energies, where higher-frequency and shorter-wavelength waves are the most energetic. The visible light spectrum, the light and color we see with our naked eye, makes up the smallest part of the larger electromagnetic spectrum, from approximately 380 to 750 nm.

wavelength The distance between any given point and the same point in the next wave.

frequency The number of waves that pass a fixed point per unit of time.

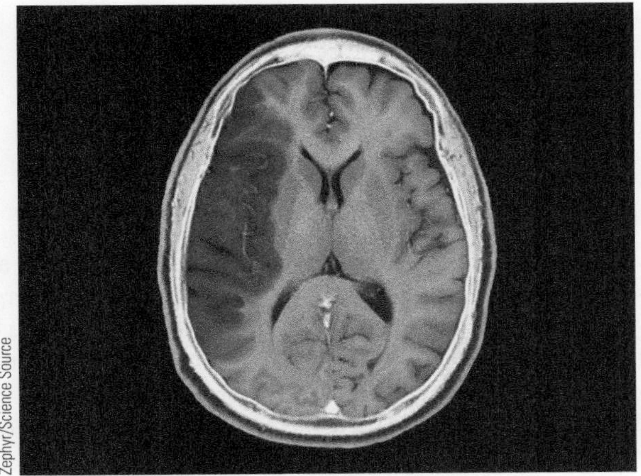

Figure 2.26 MRI is a diagnostic technique within medicine that uses, in part, high-powered radio waves. This particular image shows the use of a gadolinium (Gd) contrast to emphasize a defect in the blood–brain barrier after a stroke.

Viktor Cap/Alamy Stock Photo

Figure 2.27 UV radiation is the source of energy in most tanning beds. According to the U.S. Food and Drug Administration (FDA), the use of artificial tanning devices is not recommended for anyone due to its link to the development of skin cancer.

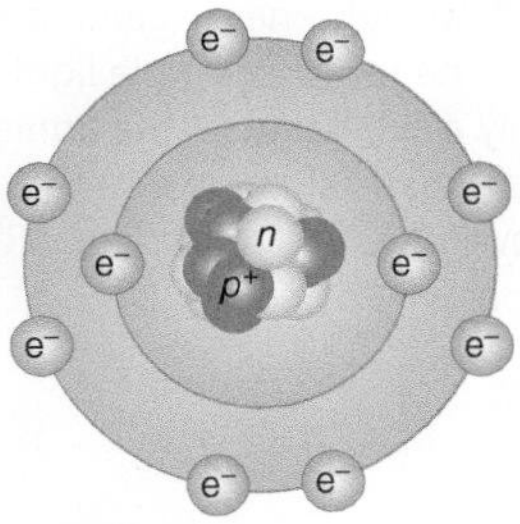

e^- e^- e^- e^- e^- n p^+ e^- e^- e^- e^- e^-

Neon, Ne
Atomic number: 10
Mass number: 20
(10 protons + 10 neutrons)
10 electrons

Figure 2.28 A Bohr model showing the subatomic particles of the nucleus and electron energy levels in neon.

Recall that Rutherford's gold foil experiment (**Figure 2.7**) proposed a model of the atom where the negatively charged electrons orbit around the central nucleus (**Figure 2.8**). This model, however, turned out to be inconsistent with the discovery that energy can only exist in fixed amounts—that energy is *quantized.* This problem was solved by Danish physicist Niels Bohr (1885–1962), who was awarded the Nobel Prize in Physics in 1922 in recognition of these efforts.

In the Bohr model of the atom, electrons are confined into clearly defined energy levels spaced at specific distances away from the nucleus (**Figure 2.28**). The electrons can climb up or down these energy levels, almost like the rungs of a ladder (**Figure 2.29**), but they can't move freely inward or outward. In order for electrons to transition between these energy levels, they must absorb or release a specific amount of energy that matches the energy gap between the different levels.

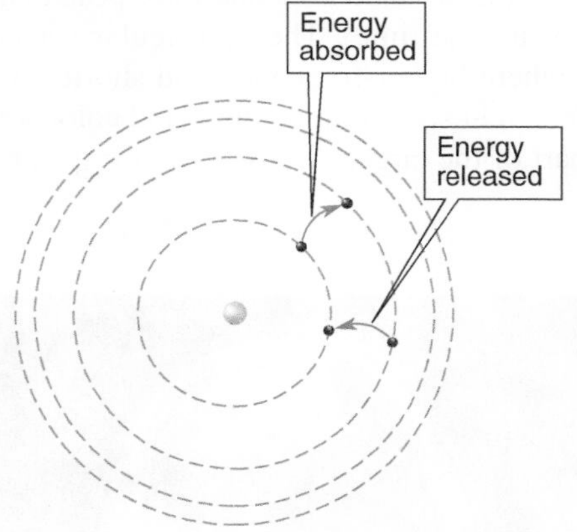

Figure 2.29 A diagram of the energy levels in an atom (not drawn to scale; the orbits are actually much larger than the nucleus). An electron is elevated to a higher-energy level when the correct amount of energy is absorbed, and it drops to a lower-energy level when the correct amount of energy is released.

shell The location and energy of electrons around a nucleus are designated by a value for n, where $n = 1, 2, 3$ ....

The energy levels described in the Bohr model that electrons occupy are often referred to as **shells** and assigned the letter n, where $n = 1, 2, 3, ...$ The $n = 1$ energy level, or shell, is lowest in energy, while energy levels farther away from the nucleus are higher in energy. This is consistent with the idea that opposite charges are attracted to one another. Negatively charged electrons in the $n = 1$ shell are closer to the positively charged nucleus, so they are more stable and lower in energy. Electrons in higher-numbered shells are less stable and higher in energy because they are farther from the nucleus, so their attraction to the nucleus is not as strong. As a result, the electrons in the $n = 4$ shell are less stable and higher in energy than the electrons in the $n = 1, 2,$ or 3 shells.

We can be even more specific about where electrons reside in an atom because each shell consists of one or more subshells, and each subshell consists of one or more atomic orbitals. Thus, we can precisely locate an electron by its shell, subshell, and orbital in the same way that addresses identify places by country or state, city, and street.

The **subshells** within a shell are designated by one of the letters *s, p,* or *d*. Because all subshells are designated by one of these letters regardless of the shell in which the subshell is found, a combination of both shell number and subshell letter is used to identify subshells clearly. Thus, a *p* subshell in shell number 2 is referred to as a 2*p* subshell. The number of subshells found in a shell is the same as the value of *n* for the shell. Thus, shell number 1 ($n = 1$) contains one subshell, the 1*s* subshell. Shell number 2 ($n = 2$) contains two subshells, the 2*s* and 2*p* subshells. Shell number 3 ($n = 3$) contains three subshells, the 3*s*, 3*p*, and 3*d* subshells. All electrons within a specific subshell have the same energy.

subshell A subdivision of a shell that is designated by *s, p,* or *d*.

The description of the location and energy of electrons moving around a nucleus is completed by specifying an orbital. Each subshell consists of one or more **atomic orbitals**, which are specific volumes of space around nuclei in which electrons may reside at any given time. Each orbital can hold two electrons. The shape of the orbital represents the probability of finding an electron, not a path the electron follows (**Figure 2.30**).

atomic orbital A volume of space around atomic nuclei in which electrons of the same energy may reside at any given time. Groups of orbitals with the same *n* value form subshells. Each orbital can hold two electrons.

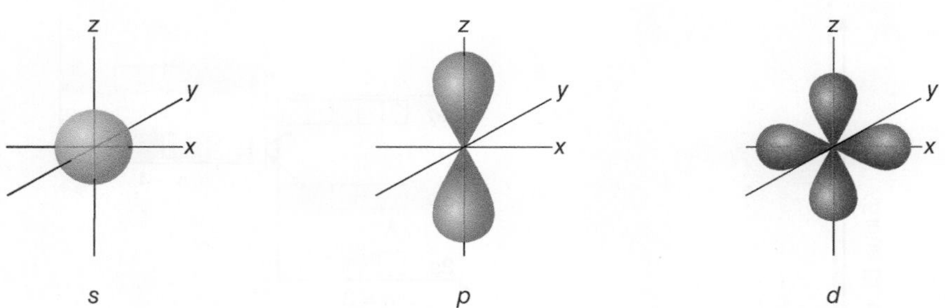

Figure 2.30 Shapes of typical *s, p,* and *d* orbitals.

All *s* subshells consist of a single orbital that is also designated by the letter *s* and further identified by the *n* value of the shell to which the subshell belongs. Thus, the 1*s* subshell consists of a single 1*s* orbital that can hold a maximum of two electrons. All *p* subshells consist of three *p* orbitals that also carry the *n* value of the shell. Thus, the 2*p* subshell consists of three individual 2*p* orbitals. Each 2*p* orbital can hold a maximum of two electrons, so the 2*p* subshell can hold a maximum total of six electrons. Because all the electrons in a subshell have the same energy, an electron in any one of the three 2*p* orbitals has the same energy, regardless of which orbital of the three it occupies. All *d* subshells contain five orbitals (which can hold a maximum of two electrons each, for a maximum total of 10).

The relationships between shells, subshells, orbitals, and electrons are summarized in **Table 2.8** for the first three shells (or energy levels) of an atom. These relationships exist beyond the $n = 3$ shell, but our discussion is usually limited to the first three energy levels.

Table 2.8 Relationships between Shells, Subshells, Orbitals, and Electrons

Shell Number (*n*)	Number of Subshells in Shell	Subshell Designation	Number of Orbitals in Subshell	Maximum Number of Electrons in Subshell	Maximum Number of Electrons in Shell
1	1	1*s*	1	2	2
2	2	2*s*	1	2	
		2*p*	3	6	8
3	3	3*s*	1	2	
		3*p*	3	6	
		3*d*	5	10	18

Understanding the arrangement of electrons in their shells, subshells, and orbitals is critical to recognizing where electrons reside in the atom. As we explain in Chapter 3, this then helps explain how atoms come together to form chemical bonds. Therefore, chemists have devised a notation that quickly communicates that exact information. These so-called **electron configurations** are a short-hand method for representing all of the electrons in a particular atom (**Figure 2.31**).

electron configurations The detailed arrangement of electrons indicated by a specific notation, such as $1s^22s^22p^4$.

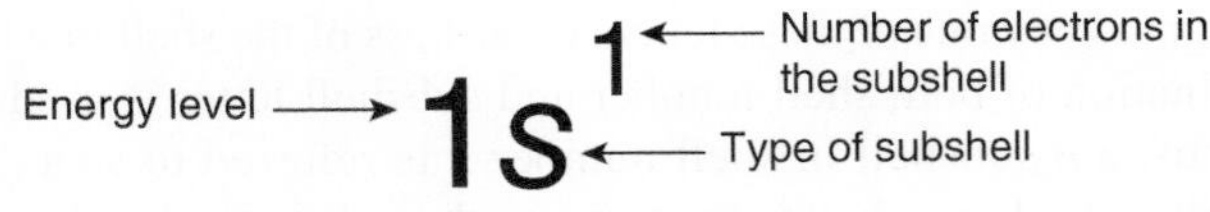

Heinz-Jürgen Göttert/picture alliance/Getty Images

Figure 2.31 An electron configuration, such as this one for hydrogen, consists of a coefficient number representing the shell (energy level), a letter representing the subshell, and a superscripted number representing the number of electrons in that subshell.

The electron configuration for a particular element is unique to that element and lists all of the electrons that occupy each shell, subshell, and orbital. For example, the proper electron configuration for carbon (atomic number = 6) is $1s^22s^22p^2$. Notice that the subshells are written in the correct filling order (i.e., by increasing energy; see **Figure 2.32**), that the number of electrons in each subshell is indicated by a superscript, and that the total number of electrons equals the atomic number.

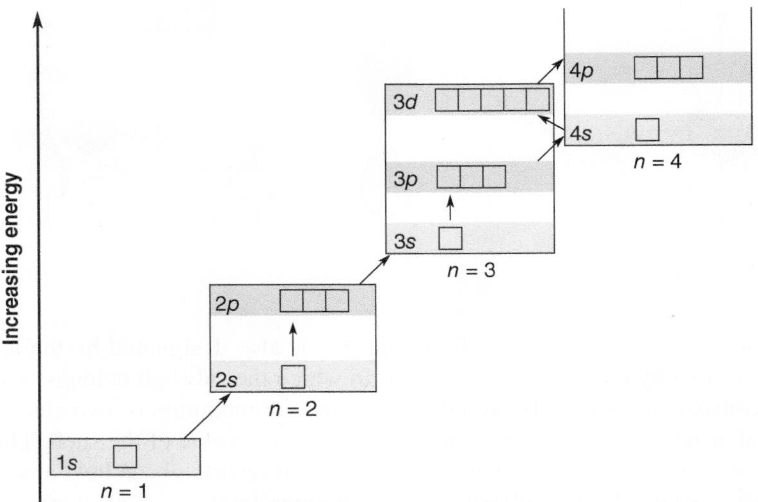

Figure 2.32 The relative energies and electron-filling order for select shells and subshells.

Thankfully, we can use the periodic table to quickly help us determine an element's electron configuration. As shown in **Figure 2.33**, the periodic table can be broken up into blocks on the basis of the type of subshell the electrons are placed into as we proceed through them in order of increasing energy. So, adding an electron to the correct subshell and orbital essentially is like moving a piece on a game board one square at a time! To go from carbon (atomic number = 6; electron configuration = $1s^22s^22p^2$) to nitrogen (atomic number = 7), we simply add a single electron to the $2p$ sublevel that nitrogen also occupies, thus making its electron configuration $1s^22s^22p^3$. Note that nitrogen is the third element inside the $2p$ area of the periodic table. Also note that the *s* area is 2 columns (elements) wide, the *p* area is 6 columns wide, and the *d* area is 10 columns wide—exactly the number of electrons required to fill the *s*, *p*, and *d* subshells, respectively. The *f* block of lanthanides and actinides is beyond the scope of this text.

What emerges from dividing the areas of the periodic table in this way is a notable pattern in the electron configurations of elements that occupy the same group (column). This pattern, which helps explain why elements in the same group display similar properties, can be seen in **Table 2.9**, which lists the electron configurations of the first four elements of Group IA(1)—namely, Li, Na, K, and Rb.

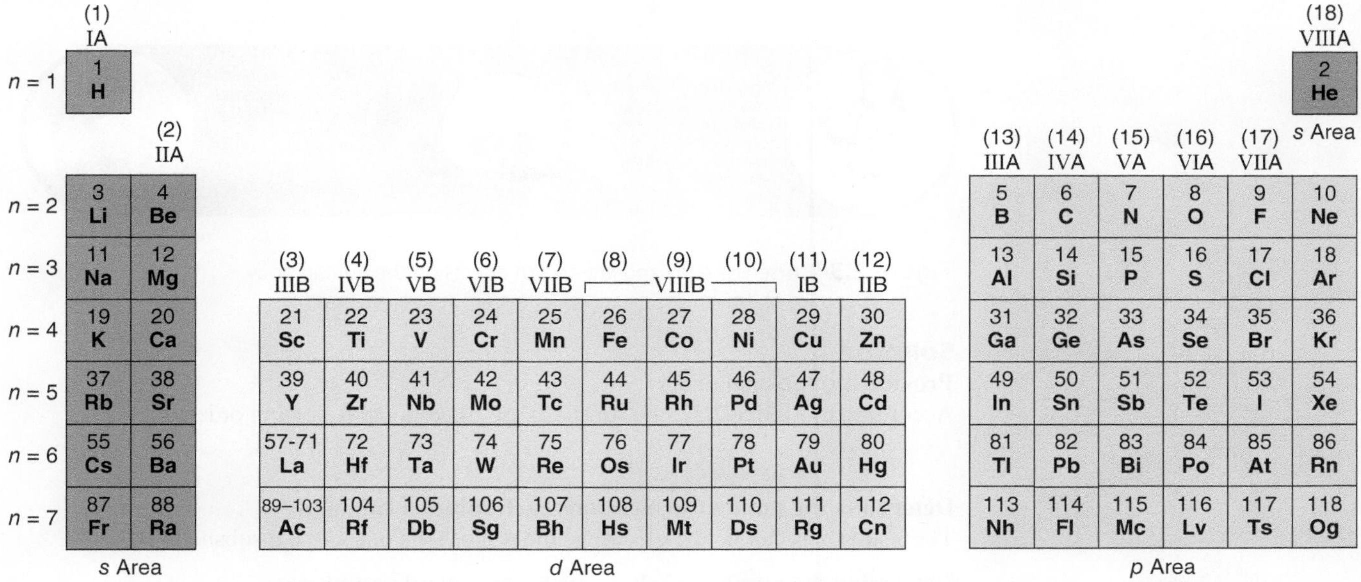

	(1) IA	(2) IIA	(3) IIIB	(4) IVB	(5) VB	(6) VIB	(7) VIIB	(8) VIIIB	(9) VIIIB	(10) VIIIB	(11) IB	(12) IIB	(13) IIIA	(14) IVA	(15) VA	(16) VIA	(17) VIIA	(18) VIIIA
n = 1	1 H																	2 He
n = 2	3 Li	4 Be											5 B	6 C	7 N	8 O	9 F	10 Ne
n = 3	11 Na	12 Mg											13 Al	14 Si	15 P	16 S	17 Cl	18 Ar
n = 4	19 K	20 Ca	21 Sc	22 Ti	23 V	24 Cr	25 Mn	26 Fe	27 Co	28 Ni	29 Cu	30 Zn	31 Ga	32 Ge	33 As	34 Se	35 Br	36 Kr
n = 5	37 Rb	38 Sr	39 Y	40 Zr	41 Nb	42 Mo	43 Tc	44 Ru	45 Rh	46 Pd	47 Ag	48 Cd	49 In	50 Sn	51 Sb	52 Te	53 I	54 Xe
n = 6	55 Cs	56 Ba	57-71 La	72 Hf	73 Ta	74 W	75 Re	76 Os	77 Ir	78 Pt	79 Au	80 Hg	81 Tl	82 Pb	83 Bi	84 Po	85 At	86 Rn
n = 7	87 Fr	88 Ra	89–103 Ac	104 Rf	105 Db	106 Sg	107 Bh	108 Hs	109 Mt	110 Ds	111 Rg	112 Cn	113 Nh	114 Fl	115 Mc	116 Lv	117 Ts	118 Og

s Area (Groups 1, 2 and He) — *d* Area (Groups 3–12) — *p* Area (Groups 13–18)

Figure 2.33 The periodic table divided by subshell blocks. The lanthanides and actinides form a separate block, the *f* area, which is not shown.

Table 2.9 Electron Configurations of Select Group IA(1) Elements

Element Symbol	Electron Configuration
Li	$1s^22s^1$
Na	$1s^22s^22p^63s^1$
K	$1s^22s^22p^63s^23p^64s^1$
Rb	$1s^22s^22p^63s^23p^64s^23d^{10}4p^65s^1$

Notice that the electron configuration for each of these elements ends with a single electron in the highest numbered shell. This shell is the outermost occupied shell, the one farthest from the nucleus, and the one highest in energy. Electrons in this outermost, highest-energy shell are referred to as **valence electrons**. Thus, each Group IA(1) element has one valence electron. For elements in the A groups of the periodic table—IA(l), IIA(2), and IIIA(13) through VIIIA(18)—the group number matches the number of valence electrons. Thus, the halogens in Group VIIA(17) each contain seven valence electrons. It is these electrons farthest from the nucleus that are responsible for chemistry—namely, the making of chemical bonds.

valence electrons The electrons found in the outermost (highest-energy) shell of an atom

We can further condense electron configurations to highlight the valence electrons. For example, the electron configuration of sodium (Na), $1s^22s^22p^63s^1$, can also be written as [Ne]$3s^1$. Here, [Ne] represents the electron configuration of the preceding noble gas, which for sodium happens to be neon (Ne). In this **noble gas configuration** or **abbreviated electron configuration**, only the valence electrons ($3s^1$ for Na) are listed explicitly.

noble gas configurations or **abbreviated electron configurations** The arrangement of electrons that starts with the noble gas prior to the valence shell and ends with specific notation representing the electrons in the valence shell, for example, [He]$2s^22p^4$ instead of $1s^22s^22p^4$.

Example 2.5 Electron Configurations for Atoms

Arsenic (As) is known for its toxicity and may play a role in the development of a number of serious conditions (**Figure 2.34**). For arsenic, provide the full electron configuration and the abbreviated electron configuration.

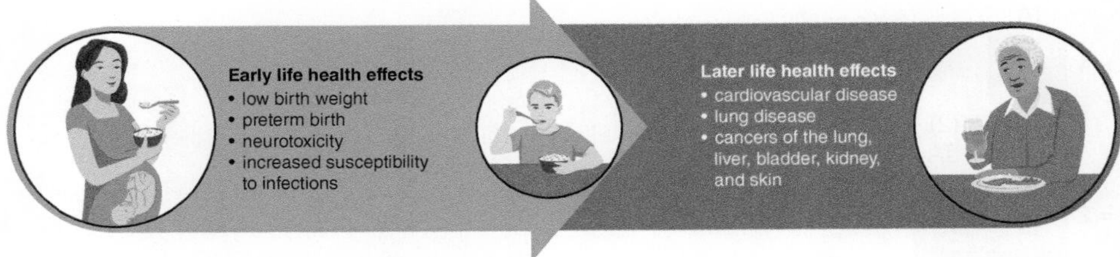

Early life health effects
• low birth weight
• preterm birth
• neurotoxicity
• increased susceptibility to infections
Later life health effects
• cardiovascular disease
• lung disease
• cancers of the lung, liver, bladder, kidney, and skin

Figure 2.34 Arsenic's early and late health effects on the human body.

Solution

Provide the subshell order:
According to **Figure 2.32**, the subshells are filled in the following order:

$$1s, 2s, 2p, 3s, 3p, 4s, 3d, 4p$$

Determine the number of electrons each sublevel can hold:
The *s* subshells can hold 2 electrons, the *p* subshells 6, and the *d* subshells 10.

Determine the number of electrons for the atomic number:
The atomic number for arsenic is 33, so only enough subshells to hold 33 electrons will be used.

Write the full electron configuration:
Therefore, the full configuration is $1s^22s^22p^63s^23p^64s^23d^{10}4p^3$.

Determine the noble gas preceding the element:
Argon (Ar) is the noble gas prior to arsenic (Ar).

Write the abbreviated electron configuration:
The electron configuration for Ar is $1s^22s^22p^63s^23p^6$, so the abbreviated electron configuration for As is $[Ar]4s^23d^{10}4p^3$.

✔ **Learning Check 2.5** Write the full electron configurations and the abbreviated electron configurations for the following.

a. Element number 9
b. Mg
c. The element found in Group VIA(16) and Period 3

2.7 Trends within the Periodic Table

Learning Objective 9 Describe periodic trends in atomic radius, first ionization energy, and electronegativity.

The periodic table is vitally important to chemistry because it illuminates so many trends in the properties of the elements, as we have just seen with valence electrons. In this section, we describe the periodic trends in atomic size, first ionization energy, and electronegativity.

2.7A Atomic Radii

atomic radius The distance extending from the center of the nucleus of the atom to the location of the valence shell electrons.

The size of an atom is determined by its **atomic radius**. It is defined as the radius of a sphere extending from the center of the nucleus of the atom to the location of the outermost electrons around the nucleus. As shown in **Figure 2.35**, atomic radii increase from the top to the bottom of each group, and increase from right to left across a period.

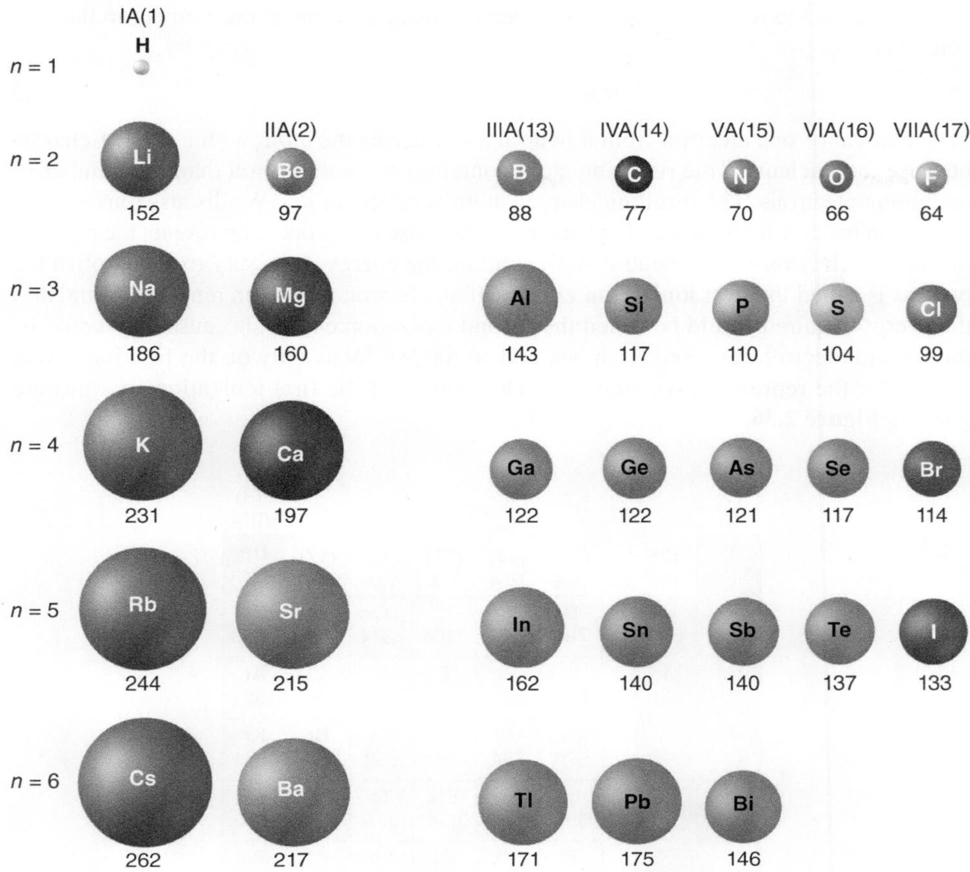

Figure 2.35 Scale drawings of the atoms of some representative elements in Groups IA(1), IIA(2), and IIIA(13)–VIIA(17). The numbers are the atomic radii in picometers (10^{-12} m), so these drawings are enlarged about 60 million times.

When we go from element to element down a group, a new electronic shell is being filled. In Group IA(1), for example, the second shell ($n = 2$) is being filled for lithium (Li), the third shell ($n = 3$) is being filled for sodium (Na), and so on for each element in the group until the sixth shell ($n = 6$) begins to fill for cesium (Cs). As n increases, the distance of the electrons from the nucleus in the shell designated by n increases, so the atomic radius also increases.

The decrease in atomic radius across a period can also be understood in terms of the outermost electrons around the nucleus. In Period 3, for example, the abbreviated electron configuration for sodium (Na) is [Ne]$3s^1$, so the outermost (valence) electron is in the third shell ($n = 3$). For magnesium (Mg), the electron structure is [Ne]$3s^2$, in which case its outermost electrons are still in the third shell, too. For aluminum (Al), the electron structure is [Ne]$3s^23p^1$, so once again, the outermost electrons lie in the third shell. Because all of these electrons are in the same third shell, they should all be the same distance from the nucleus, and the atoms should all be the same size. However, each time another electron is added to the third shell of these elements, another positively charged proton is added to the nucleus. The increasing nuclear charge attracts electrons in the same shell, thereby pulling all the electrons of the shell closer to the nucleus, which causes the atomic radii to decrease as the nuclear charge increases with increasing atomic number.

2.7B Ionization Energy

The chemical reactivity of the elements depends on the behavior of their electrons, especially their valence electrons. A property that is related to the behavior of the electrons of atoms is the ionization energy. The ionization energy of an element (e.g., sodium) is the energy required to remove an outermost electron from an atom of the element in the gaseous state.

$$Na(g) \rightarrow Na^+(g) + 1\ e^- \qquad 2.3$$

The removal of one electron from a neutral atom leaves the atom with a net +1 charge, because the nucleus of the resulting atom contains one more proton than the number of remaining electrons. The resulting charged atom is called an *ion.* We discuss ions and the ionization process in more detail in Chapter 3. Because this process represents the removal of the first electron from a neutral sodium atom, the energy necessary to accomplish the process is called the **first ionization energy**. If an electron were then removed from Na^+, the energy required would be called the second ionization energy, because that would be the second electron removed from Na, and so on. We focus only on the first ionization energy for the representative elements. The values for the first ionization energies are given in **Figure 2.36**.

first ionization energy The energy required to remove the first outermost electron from a neutral atom.

(1) IA	(2) IIA	(13) IIIA	(14) IVA	(15) VA	(16) VIA	(17) VIIA	(18) VIIIA
H 1311							He 2370
Li 521	Be 899	B 799	C 1087	N 1404	O 1314	F 1682	Ne 2080
Na 496	Mg 737	Al 576	Si 786	P 1052	S 1000	Cl 1245	Ar 1521
K 419	Ca 590	Ga 576	Ge 784	As 1013	Se 939	Br 1135	Kr 1351
Rb 402	Sr 549	In 559	Sn 704	Sb 834	Te 865	I 1007	Xe 1170
Cs 375	Ba 503	Tl 590	Pb 716	Bi 849	Po 791	At 926	Rn 1037

Figure 2.36 First ionization energies for the representative elements. The values are given in kJ/mol. (Source: Data from Bard, A. J.; Parsons, R.; Jordan, J. *Standard Potentials in Aqueous Solution.* New York: Dekker, 1985, pp. 24–27.)

According to **Figure 2.36**, the first ionization energies increase from the bottom to the top in a group, and increase from left to right across a period. The higher the value of the ionization energy, the more difficult it is to remove an electron from the atoms of an element. In general, therefore, the electrons of metals [i.e., Groups IA(1), IIA(2), and IIA(13)] are easier to remove than are the electrons of nonmetals [i.e., Groups IVA(14)–VIIIA(18)]. Also, the farther down a group a metal is located, the easier it is to remove one of its electrons.

The trends in ionization energy (i.e., the trends related to how easy it is to remove an electron) can be explained using arguments similar to those that explain the trends in atomic size. As we move down a group, the valence electrons are located farther and farther away from the nucleus because they are located in higher-energy shells. The farther the electrons are from the nucleus, the weaker is the attraction of the positively charged nucleus toward the negatively charged electrons, and the easier it is to pull the electrons away.

Similarly, as we proceed from left to right across a period, the valence electrons are placed into the same shell, in which case they are located at the same distance from the nucleus. But, as we saw with atomic size, the nuclear charge increases, as does the number of electrons. And as the nuclear charge increases, it gets harder and harder to remove a valence electron from an atom as we move farther to the right in a row.

The general trends in atomic size and first ionization energy are summarized in **Figure 2.37**. Note that they increase in opposite directions, both across a period and down a group.

2.7C Electronegativity

Electronegativity A measure of the tendency of an atom to attract electrons.

Electronegativity, another important periodic trend, refers to the preference of certain elements to hold on to their electrons or to be stable with extra electrons.

Electronegativity trends follow those observed for first ionization energies—that is, electronegativity increases left to right across a period and increases going up a group (**Figure 2.37**). As we explain in Chapter 3, electronegativity, along with trends in atomic size and ionization energy, are critical to understanding how electrons are distributed when atoms come together to form chemical bonds.

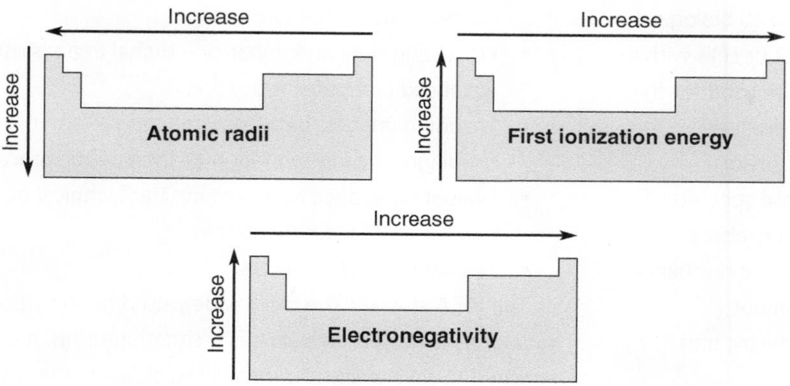

Figure 2.37 A summary of the periodic trends in atomic radii, first ionization energy, and electronegativity.

Example 2.6 Atomic Trends

Rank the elements Ba, Ne, Mo, and Br based on:

a. Increasing atomic radii
b. Increasing first ionization energy
c. Increasing electronegativity

Solution

As atomic radii increase (right to left across a period and top to bottom down a group), first ionization energy and electronegativity decrease.

a. Thus, to rank the elements by increasing atomic radii, we must start with the one in the list that has the smallest valence shell (i.e., the one nearest the top of a column) and that has the most protons (i.e., the one farthest to the right in a row). Of the elements provided, the smallest atom is Ne ($n = 2$ for this noble gas). The next largest atom is Br ($n = 4$ for this halogen), followed by Mo ($n = 5$ for this transition metal), and the largest atom is Ba ($n = 6$ for this alkaline earth metal).

$$\text{Ne} < \text{Br} < \text{Mo} < \text{Ba}$$

b. When ranking the elements by increasing first ionization energy, the trend is the exact opposite of the atomic radii:

$$\text{Ba} < \text{Mo} < \text{Br} < \text{Ne}$$

c. As with first ionization energy, electronegativity increases from left to right across a period and from bottom to top in a group:

$$\text{Ba} < \text{Mo} < \text{Br} < \text{Ne}$$

✔ Learning Check 2.6 Rank the elements Al, Fe, F, and Cs based on:

a. Decreasing atomic radii
b. Decreasing first ionization energy
c. Decreasing electronegativity

Health Career Description

Nuclear Medicine Technologist

According to the International Atomic Energy Agency (IAEA), "Radiopharmaceuticals are radioisotopes bound to biological molecules able to target specific organs, tissues or cells within the human body. These radioactive drugs can be used for the diagnosis and, increasingly, for the therapy of diseases."

The nuclear medicine technologist is a healthcare professional who prepares and administers these specialized medications. Nuclear medicine technologists may also operate computed tomography (CT) and magnetic resonance imaging (MRI) scanners. This field combines patient care, imaging, computer science, physics, chemistry, and mathematics.

The website explorehealthcareers.org lists the nuclear medicine technologist's responsibilities as:

- Putting the patient at ease, obtaining pertinent history, describing the procedure, and answering the patient's questions
- Administering radiopharmaceuticals and medications for patient imaging and therapeutic procedures
- Monitoring the patient's physical condition during the course of the procedure
- Processing data and enhancing digital images using advanced computer technology
- Providing images, data analysis, and patient information for diagnostic interpretation or therapeutic procedures
- Evaluating images to determine the technical quality and calibration of instrumentation
- Evaluating new protocols

The IAEA states, "The most widely used radioisotope in diagnostic nuclear medicine is technetium-99m. It can be attached to several specific molecules, allowing the diagnosis of many diseases, including certain types of cancers. For instance, technetium-99m-MDP (methylene diphosphonate) is widely used to detect bone metastasis associated with cancer." Nuclear medicine is a growing and important career field. Nuclear medicine technologists must have a specialized technical associate's degree or a bachelor's degree in radiology.

Sources: American Cancer Society. Retrieved April 12, 2020 from: https://www.cancer.org/treatment/treatments-and-side-effects/treatment-types/radiation/systemic-radiation-therapy.html; Diagnostic Radiopharmaceuticals. Retrieved April 12, 2020 from: https://www.iaea.org/topics/diagnostic-radiopharmaceuticals; Nuclear Medicine Technologist. Retrieved April 5, 2020 from: https://explorehealthcareers.org/career/allied-health-professions/nuclear-medicine-technologist/#academic

Concept Summary

2.1 The Periodic Table

Learning Objective: **Locate elements in the periodic table on the basis of their group and period designations.**

- The chemical properties of the elements tend to repeat in a regular (periodic) way when the elements are arranged in order of increasing atomic numbers.
- This periodic law is the basis for the arrangement of the elements called the periodic table.
- In the periodic table, each element belongs to a column called a group.
- Each element also belongs to a row called a period.
- All elements in a group have similar chemical properties.
- The elements can be classified on the basis of other properties as metals, nonmetals, and metalloids.
- Metals can be further classified into alkali metals, alkaline earth metals, transition metals, other metals, lanthanides, and actinides.
- Nonmetals can be further classified into metalloids, halogens, and noble gases.

2.2 Subatomic Particles

Learning Objective: **Describe the charge, relative mass, and location of the three subatomic particles.**

- Atoms are made up of smaller particles.
- The advancement of atomic theory led to the discovery of these subatomic particles.
- Rutherford's gold foil experiment, along with others, led to the current model of the atom, which has a dense center of mass called the nucleus, where the protons and neutrons reside, surrounded by a cloud of electrons.
- The nucleus constitutes most of the mass of an atom, whereas the electrons constitute most of the volume.

2.3 Atomic Symbols

Learning Objective: **Write the atomic symbol ($^{A}_{Z}X$) for a given set of subatomic particles.**

- Symbols based on names have been assigned to every element.
- Most elements consist of a single capital letter followed by a lowercase letter.

- A few elements consist of a single capital letter.
- An atomic symbol incorporating atomic number (Z), mass number (A), and elemental symbol is used to represent an element.

2.4 Isotopes and Atomic Weights

Learning Objective: **Define the terms *isotope* and *atomic weight*.**

- Most elements in their natural state are made up of more than one kind of atom.
- These different kinds of atoms of a specific element are called isotopes.
- They differ from one another only in the number of neutrons in their nuclei.
- Atomic weights have been assigned to each element and are tabulated in the periodic table.
- The units used are atomic mass units, abbreviated amu.
- The atomic weights measured for elements are average weights that depend on the percentages and masses of the isotopes in the naturally occurring element.
- If the isotope percent abundances and isotope masses are known for an element, its atomic weight can be calculated.

2.5 Radioactive Nuclei

Learning Objective: **Write balanced equations for radioactive decay and other nuclear processes. Solve problems using the half-life concept. Describe the applications of radiation in health and medicine.**

- Some nuclei are unstable and undergo radioactive decay.
- The common types of radiation emitted during decay processes are alpha, beta, and gamma.
- These types can be distinguished by their different masses and charges.
- Nuclear reactions can be represented by balanced equations in which the total mass numbers and atomic numbers balance on each side.
- In order to make it convenient to balance mass numbers and atomic numbers, atomic symbols are used in nuclear equations.
- Different radioisotopes generally decay at different rates, which are indicated by half-lives.
- One half-life is the time required for one-half of the unstable nuclei in a sample to undergo radioactive decay.
- Two systems, physical and biological, are used to describe quantities of radiation.
- Physical units indicate the number of nuclei of radioactive material that decay per unit of time.
- Biological units are related to the damage caused by radiation in living tissue.
- The common physical units are the curie, its fractions, and the becquerel.
- Biological units include the sievert, the rad, the gray, and the rem.
- Radiation generates free radicals in tissue as it passes through.
- Radiation is hazardous even at low intensity if there is long-term exposure.
- Radiation sickness is caused by short-term intense radiation.
- Those working around radioactive sources can minimize exposure by using shielding or distance as a protection.
- Radioisotopes behave chemically like nonradioisotopes of the same element and can be used diagnostically and therapeutically.
- Diagnostically, radioisotopes are used as tracers whose movement or location in the body can be followed.
- Therapeutic radioisotopes localize in diseased areas of the body, where their radiation can destroy diseased tissue.

2.6 Where Are the Electrons?

Learning Objective: **Write correct electron configurations for each element through atomic number 56.**

- The electromagnetic (EM) spectrum shows various types of emitted energy from low-energy (low frequency, long wavelength) radio waves to high-energy (high frequency, short wavelength) gamma rays.
- Visible light is only a small part of the entire electromagnetic spectrum.
- Niels Bohr proposed a revision to the Rutherford model of the atom, which placed electrons in quantized energy levels called shells.
- Electrons change shells only when they absorb or release a specific amount of energy.
- The energy and location of electrons can be further specified in terms of shells, subshells, and orbitals.
- The s sublevel contains one orbital, the p sublevel contains three orbitals, and the d sublevel contains five orbitals.
- Each orbital holds a maximum of two electrons, and they exist with opposite spins.
- The arrangements of electrons in orbitals, subshells, and shells are called electron configurations.
- The electron configurations of the elements correlate to their placement in the periodic table.
- Electron configurations reveal the number of valence electrons, the electrons that exist in the highest-energy shell.
- Electron configurations can be represented in an abbreviated form by using noble gas symbols to represent the inner electrons.
- Elements with similar chemical properties turn out to be those with identical numbers and types of electrons in their valence shells.

2.7 Trends within the Periodic Table

Learning Objective: **Describe periodic trends in atomic radius, first ionization energy, and electronegativity.**

- Chemical and physical properties of the elements follow trends within the periodic table.
- Atomic radii, the size of an atom, increase from right to left across a row and down a column of the periodic table.
- First ionization energy, the energy required to remove the outermost electron, increases from left to right across a row and up a column of the periodic table.
- Electronegativity, the preference of an atom to attract electrons, increases from left to right across a row and up a column of the periodic table.

Key Terms and Concepts

Abbreviated electron configurations (2.6)
Acute radiation syndrome (2.5)
Atomic radius (2.7)
Alpha particle (2.5)
Atomic mass unit (amu) (2.4)
Atomic number (2.3)
Atomic orbital (2.6)
Atomic symbol (2.3)
Atomic weight (2.4)
Biological unit of radiation (2.5)
Beta particle (2.5)
Cold spot (2.5)
Daughter nuclei (2.5)
Electron (2.2)
Electron capture (2.5)
Electron configurations (2.6)
Electronegativity (2.7)
Elemental symbol (2.1)
First ionization energy (2.7)
Frequency (2.6)
Gamma ray (2.5)
Geiger-Müller tube (2.5)
Group of the periodic table (2.1)
Half-life (2.5)
Hot spot (2.5)
Inner-transition metal (2.1)
Isotopes (2.4)
Mass number (2.3)
Metalloids (2.1)
Metals (2.1)
Neutron (2.2)
Noble gas configuration (2.6)
Nonmetals (2.1)
Nuclear reaction (2.5)
Nucleus (2.2)
Period of the periodic table (2.1)
Physical unit of radiation (2.5)
Positron (2.5)
Products of a reaction (2.5)
Proton (2.2)
Radical or free radical (2.5)
Radioactive dating (2.5)
Radioactive decay (2.5)
Radioactive nuclei (2.5)
Radioisotope (2.5)
Reactants of a reaction (2.5)
Representative element (2.1)
Scintillation counter (2.5)
Shell (2.6)
Subshell (2.6)
Tracer (2.5)
Transition metal (2.1)
Valence electrons (2.6)
Wavelength (2.6)

Key Equations

1. General equation for alpha particle decay of a radioactive nuclei (**Section 2.5**)	${}^{A}_{Z}X \rightarrow {}^{4}_{2}\alpha + {}^{A-4}_{Z-2}Y$	Equation 2.1
2. Conversion of a neutron to a proton and electron in the nucleus, leading to beta emission (**Section 2.5**)	${}^{1}_{0}\text{n} \rightarrow {}^{1}_{1}\text{p} + {}^{0}_{-1}\beta$	Equation 2.2
3. Equation associated with the first ionization energy of sodium (**Section 2.7**)	$\text{Na(g)} \rightarrow \text{Na}^{+}\text{(g)} + 1\ \text{e}^{-}$	Equation 2.3

Exercises

Even-numbered exercises are answered in Appendix B.

Exercises with an asterisk (*) are more challenging.

You will find it useful to refer to **Table 2.1** and the periodic table inside the front cover as you work through these exercises.

The Periodic Table (Section 2.1)

2.1 Identify the group and period to which each of the following elements belongs.

a. Si

b. element number 21

c. zinc

d. element number 35

2.2 Identify the group and period to which each of the following elements belongs:

a. element number 27

b. Pb

c. arsenic

d. Ba

2.3 **a.** How many elements are located in Group VIIB(7) of the periodic table?

b. How many elements are found in Period 5 of the periodic table?

c. How many total elements are in Groups IVA(14) and IVB(4) of the periodic table?

2.4 **a.** How many elements are in Group VIIB(7) of the periodic table?

b. How many total elements are found in Periods 1 and 2 of the periodic table?

c. How many elements are found in Period 5 of the periodic table?

2.5 The following statements either define or are closely related to the terms *period* and *group.* Match the terms to the appropriate statements.

a. This is a vertical arrangement of elements in the periodic table.

b. Elements 4 and 12 belong to this arrangement.

2.6 The following statements either define or are closely related to the terms *period* and *group.* Match the terms to the appropriate statements.

a. This is a horizontal arrangement of elements in the periodic table.

b. Element 11 begins this arrangement in the periodic table.

2.7 Classify the following elements as representative, transition metals, inner-transition metals, or noble gases.

a. Iron, found in hemoglobin

b. Element 15, found in bones, ATP, DNA, and many other biological materials

c. U, radioactive isotopes used for medical diagnosis and research

d. Bismuth, used in an antacid medication

2.8 Classify the following elements as representative, transition metals, inner-transition metals, or noble gases.

a. W, used in oncology treatment instruments

b. Cm, used in medical power sources

c. Element 10, used as a diagnostic tracer in lung diffusion tests

d. Helium, used in MRI cooling systems

e. Barium, sulfate used in digestive system CT scans

2.9 Classify the following as metals, nonmetals, or metalloids.

a. Argon, used in a cauterizing surgical tool

b. Element 3, used to help treat bipolar disorder

c. Ge, despite serious safety concerns, used by some people as a medicine

d. Boron, essential for the growth and maintenance of bone

2.10 Classify the following as metals, nonmetals, or metalloids.

a. Rubidium, used in positron emission tomographic (PET) scans to help diagnose heart disease

b. Arsenic, despite safety concerns, used in some diluted homeopathic remedies

c. Element 49, a radioactive isotope used in nuclear medicine tests

d. S, the third most abundant mineral in the human body

e. Br, ions used in sedatives, antiepileptics, and tranquilizers

Subatomic Particles (Section 2.2)

2.11 Determine the charge of nuclei made up of the following particles. Determine the atomic mass of the nuclei in amu.

a. 8 protons and 9 neutrons

b. 20 protons and 25 neutrons

c. 52 protons and 78 neutrons

2.12 Determine the charge of nuclei made up of the following particles. Determine the atomic mass of the nuclei in amu.

a. 9 protons and 10 neutrons

b. 20 protons and 23 neutrons

c. 47 protons and 60 neutrons

2.13 Determine the number of electrons that would have to be associated with each nucleus described in Exercise 2.11 to produce a neutral atom.

2.14 Determine the number of electrons that would have to be associated with each nucleus described in Exercise 2.12 to produce a neutral atom.

Atomic Symbols (Section 2.3)

2.15 Write the atomic symbol and name for the elements located in the periodic table as follows.

a. Belongs to Group VIA(16) and Period 3

b. The first element (reading down) in Group VIB(6)

c. The fourth element (reading left to right) in Period 3

d. Belongs to Group IB(11) and Period 5

2.16 Write the atomic symbol and name for the elements located in the periodic table as follows.

a. The noble gas belonging to Period 4

b. The fourth element (reading down) in Group IVA(14)

c. Belongs to Group VIB(6) and Period 5

d. The sixth element (reading left to right) in Period 6

2.17 Determine the number of electrons and protons contained in an atom of the following elements.

a. sulfur

b. As

c. element number 24

2.18 Determine the number of electrons and protons contained in an atom of the following elements.

a. silicon

b. Sn

c. element number 74

Isotopes and Atomic Weights (Section 2.4)

2.19 Determine the number of protons, neutrons, and electrons in atoms of the following isotopes.

a. ${}^{3}_{2}He$

b. ${}^{9}_{4}Be$

c. ${}^{235}_{92}U$

2.20 Determine the number of protons, neutrons, and electrons in atoms of the following isotopes.

a. ${}^{34}_{16}S$

b. ${}^{91}_{40}Zr$

c. ${}^{131}_{54}Xe$

2.21 Write the atomic symbols (${}^{A}_{Z}X$) showing elemental symbol (X), mass number (A), and atomic number (Z) for the following isotopes.

a. cadmium-110

b. cobalt-60

c. uranium-235

2.22 Write the atomic symbols (${}^{A}_{Z}X$) showing elemental symbol (X), mass number (A), and atomic number (Z) for the following isotopes.

a. silicon-28

b. argon-40

c. strontium-88

2.23 Write the atomic symbols (${}^{A}_{Z}X$) showing elemental symbol (X), mass number (A), and atomic number (Z) for the following isotopes.

a. 6 protons and 6 neutrons

b. 8 protons and 9 neutrons

c. 20 protons and 25 neutrons

2.24 Write the atomic symbols (^A_ZX) showing elemental symbol (X), mass number (A), and atomic number (Z) for the following isotopes.

a. 9 protons and 10 neutrons

b. 20 protons and 23 neutrons

c. 47 protons and 60 neutrons

2.25 Write the atomic symbols (^A_ZX) showing elemental symbol (X), mass number (A), and atomic number (Z) for the following neutral atoms.

a. Contains 20 electrons and 20 neutrons

b. Contains 1 electron and 2 neutrons

c. A magnesium atom that contains 14 neutrons

2.26 Write the atomic symbols (^A_ZX) showing elemental symbol (X), mass number (A), and atomic number (Z) for the following neutral atoms.

a. Contains 17 electrons and 20 neutrons

b. A copper atom with a mass number of 65

c. A zinc atom that contains 36 neutrons

2.27 Naturally occurring sodium has a single isotope. Determine the following for the naturally occurring atoms of sodium.

a. The number of neutrons in the nucleus

b. The mass (in amu) of the nucleus (to three significant figures)

2.28 Naturally occurring aluminum has a single isotope. Determine the following for the naturally occurring atoms of aluminum.

a. The number of neutrons in the nucleus

b. The mass (in amu) of the nucleus (to three significant figures)

2.29 Calculate the atomic weight of lithium on the basis of the following percent composition and atomic weights of the naturally occurring isotopes. Compare the calculated value with the atomic weight listed for lithium in the periodic table.

lithium-6 = 7.42% (6.0151 amu)

lithium-7 = 92.58% (7.0160 amu)

2.30 Calculate the atomic weight of boron on the basis of the following percent composition and atomic weights of the naturally occurring isotopes. Compare the calculated value with the atomic weight listed for boron in the periodic table.

boron-10 = 19.78% (10.0129 amu)

boron-11 = 80.22% (11.0093 amu)

2.31 Calculate the atomic weight of silicon on the basis of the following percent composition and atomic weights of the naturally occurring isotopes. Compare the calculated value with the atomic weight listed for silicon in the periodic table.

silicon-28 = 92.21% (27.9769 amu)

silicon-29 = 4.70% (28.9765 amu)

silicon-30 = 3.09% (29.9738 amu)

2.32 Calculate the atomic weight of copper on the basis of the following percent composition and atomic weights of the naturally occurring isotopes. Compare the calculated value with the atomic weight listed for copper in the periodic table.

copper-63 = 69.09% (62.9298 amu)

copper-65 = 30.91% (64.9278 amu)

Radioactive Nuclei (Section 2.5)

2.33 Define the term *radioactive*; then criticize the following statements.

a. Beta rays are radioactive.

b. Radon is a stable radioactive element.

2.34 Group the common nuclear radiations (**Table 2.3**) into the following categories.

a. Those with a mass number of 0

b. Those with a positive charge

c. Those with a charge of 0

2.35 Group the common nuclear radiations (**Table 2.3**) into the following categories.

a. Those with a negative charge

b. Those with a mass number greater than 0

c. Those that consist of particles

2.36 Characterize the following nuclear particles in terms of the fundamental particles—protons, neutrons, and electrons.

a. A beta particle

b. An alpha particle

c. A positron

2.37 Discuss how the charge and mass of particles that comprise radiation influence the range or ability of the radiation to penetrate matter.

2.38 Summarize how the atomic number and mass number of daughter nuclei compare with the original nuclei after:

a. An alpha particle is emitted

b. A beta particle is emitted

c. An electron is captured

d. A gamma ray is emitted

e. A positron is emitted

2.39 Write appropriate symbols for the following particles using the ^A_ZX symbolism.

a. A tin-117 nucleus

b. A nucleus of the chromium (Cr) isotope containing 26 neutrons

c. A nucleus of element number 20 that contains 24 neutrons

2.40 Write appropriate symbols for the following particles using the ^A_ZX symbolism.

a. A nucleus of the element in Period 5 and Group VB(5) with a mass number of 96

b. A nucleus of element number 37 with a mass number of 80

c. A nucleus of the calcium (Ca) isotope that contains 18 neutrons

2.41 Complete the following equations using appropriate notations and formulas.

a. $^{10}_{4}Be \rightarrow ? + ^{10}_{5}B$

b. $^{210}_{83}Bi \rightarrow \ + ^{4}_{2}\alpha + ?$

c. $^{15}_{8}O \rightarrow ? + ^{15}_{7}N$

d. $^{44}_{22}Te + ^{0}_{-1}e \rightarrow ?$

e. ${}^{8}_{4}Be \rightarrow ? + {}^{4}_{2}He$

f. ${}^{46}_{23}V \rightarrow ? + {}^{46}_{22}Ti$

2.42 Complete the following equations using appropriate notations and formulas.

a. ${}^{204}_{82}Pb \rightarrow ? + {}^{4}_{2}\alpha$

b. ${}^{84}_{35}Br \rightarrow ? + {}^{0}_{-1}\beta$

c. $? + {}^{0}_{-1}e \rightarrow {}^{41}_{19}K$

d. ${}^{149}_{62}Sm \rightarrow {}^{145}_{60}Nd + ?$

e. $? \rightarrow {}^{34}_{15}P + {}^{0}_{-1}\beta$

f. ${}^{15}_{8}O \rightarrow {}^{0}_{1}\beta + ?$

2.43 Write balanced equations to represent decay reactions of the following isotopes. The decay process or daughter isotope is given in parentheses.

a. ${}^{121}_{50}Sn$ (beta emission)

b. ${}^{55}_{26}Fe$ (electron capture)

c. ${}^{22}_{11}Na$ (daughter = neon-22)

d. ${}^{190}_{78}Pt$ (alpha emission)

e. ${}^{67}_{28}Ni$ (beta emission)

f. ${}^{67}_{31}Ga$ (daughter = Zinc-67)

2.44 Write balanced equations to represent decay reactions of the following isotopes. The decay process or daughter isotope is given in parentheses.

a. ${}^{157}_{63}Eu$ (beta emission)

b. ${}^{190}_{78}Pt$ (daughter = osmium-186)

c. ${}^{138}_{62}Sm$ (electron capture)

d. ${}^{188}_{80}Hg$ (daughter = Au-188)

e. ${}^{234}_{90}Th$ (beta emission)

f. ${}^{218}_{85}At$ (alpha emission)

2.45 What is meant by a half-life?

2.46 Describe half-life in terms of something familiar, such as a cake or pizza.

2.47 The radioisotope rubidium-84 has been used to measure cardiac output. If Rb-84 has a half-life of 33 days, how many milligrams of a 12.0 mg sample remains after 99 days?

2.48 Technetium-99 has a half-life of 6 hours. This isotope is used diagnostically to perform brain scans. A patient is given a 9.0 ng dose. How many nanograms will be present in the patient 24 hours later?

2.49 An archaeologist sometime in the future analyzes the iron used in an old building. The iron contains tiny amounts of nickel-63, with a half-life of 92 years. On the basis of the amount of nickel-63 and its decay products found, it is estimated that about 0.78% (1/128) of the original nickel-63 remains. If the building was constructed in 1980, in what year did the archaeologist make the discovery?

2.50 An archaeologist unearths the remains of a wooden box, analyzes for the carbon-14 content, and finds that about 93.75% of the carbon-14 initially present has decayed. Estimate the age of the box. The half-life of carbon-14 is 5600 years.

2.51 Germanium-66 decays by positron emission, with a half-life of 2.5 hours. What mass of germanium-66 remains in a sample after 10.0 hours if the sample originally weighed 50.0 mg?

2.52 A grain sample was found in a cave. The ratio of ${}^{14}_{6}C/{}^{12}_{6}C$ was 1/8 the value in a fresh grain sample. How old was the grain in the cave? The half-life of C-14 is 5600 years.

2.53 The element strontium is located just below calcium in the periodic table. The toxicity of radioactive strontium is due to its ability to chemically replace calcium in the human body. Write the nuclear equation for the beta emission of Sr-90.

2.54 Compare and contrast the general health effects of long-term exposure to low-level radiation and short-term exposure to intense radiation.

2.55 Explain why the rem is the best unit to use when evaluating the radiation received by a person working in an area where exposure to several types of radiation is possible.

2.56 Explain the difference between physical and biological units of radiation and give an example of each type.

2.57 Explain what a diagnostic tracer is and list the ideal characteristics one should have.

2.58 An individual receives a short-term whole-body dose of 3.1 rads of beta radiation. How many grays of X rays would represent the same health hazard?

2.59 List the ideal characteristics of a radioisotope that is to be administered internally for therapeutic use.

2.60 Describe the importance of hot and cold spots in diagnostic work using tracers.

2.61 Gold-198 is a β emitter used to treat leukemia. It has a half-life of 2.7 days. The dosage is 1.0 mCi/kg body weight. How long would it take for a 70.0 mCi dose to decay to an activity of 2.2 mCi?

2.62 Chromium-51 is used medically to monitor kidney activity. Chromium-51 decays by electron capture. Write a balanced equation for the decay process and identify the daughter that is produced.

2.63 Explain why carbon-14, with a half-life of about 5600 years, is not a good radioisotope to use if you want to determine the age of a coal bed thought to be several million years old.

2.64 A mixture of water (H_2O) and hydrogen peroxide (H_2O_2) will give off oxygen gas when solid manganese dioxide is added as a catalyst. Describe how you could use a tracer to determine if the oxygen comes from the water or the peroxide.

2.65 Radioactive isotopes of strontium were produced by the explosion of nuclear weapons. They were considered serious health hazardous because they were incorporated into the bones of animals that ingested them. Explain why strontium would be likely to be deposited in bones.

Where Are the Electrons? (Section 2.6)

2.66 What particles in the nucleus cause the nucleus to have a positive charge?

2.67 According to the Bohr theory, which of the following would have the higher energy?

a. An electron in an orbit close to the nucleus

b. An electron in an orbit located farther from the nucleus

2.68 Which form of electromagnetic radiation has a greater amount of energy?

a. X rays from the dentist's office or radiation emitted from your microwave

b. Red light from a stoplight or radio waves from a television broadcast station

2.69 Which form of electromagnetic radiation has a greater amount of energy?

a. A magnetic resonance image (MRI) of your knee or UV radiation from the sun

b. An abdominal ultrasound with a frequency of 10^7 Hz or gamma rays from a supernova explosion

2.70 What is the maximum number of electrons that can be contained in each of the following?

a. A $2p$ orbital

b. A $2p$ subshell

c. The second shell

2.71 What is the maximum number of electrons that can be contained in each of the following?

a. A $3d$ orbital

b. A $3d$ subshell

c. The third shell

2.72 How many orbitals are found in the second shell? Write designations for the orbitals.

2.73 How many orbitals are found in the fourth shell? Write designations for the orbitals.

2.74 How many orbitals are found in a $4p$ subshell? What is the maximum number of electrons that can be accommodated in this subshell?

2.75 How many orbitals are found in a $3d$ subshell? What is the maximum number of electrons that can be located in this subshell?

2.76 Write an electron configuration for each of the following elements, using the form $1s^22s^22p^6$, and so on. Indicate how many electrons are unpaired in each case.

a. element number 37

b. Si

c. Ti

d. Ar

2.77 Write an electron configuration for each of the following elements, using the form $1s^22s^22p^6$, and so on. Indicate how many of the electrons are unpaired in each case.

a. K

b. manganese

c. element number 33

d. Ti

2.78 Write electron configurations and answer the following.

a. How many total s electrons are found in magnesium?

b. How many unpaired electrons are in nitrogen?

c. How many subshells are completely filled in Al?

2.79 Write electron configurations and answer the following.

a. How many total electrons in Ge have a number designation (before the letters) of 4?

b. How many unpaired p electrons are found in sulfur? What is the number designation of these unpaired electrons?

c. How many $3d$ electrons are found in tin?

2.80 Write the symbol and name for each of the elements described. More than one element will fit some descriptions.

a. Contains only two $2p$ electrons

b. Contains an unpaired $3s$ electron

2.81 Write the symbol and name for each of the elements described. More than one element will fit some descriptions.

a. Contains one unpaired $5p$ electron

b. Contains a half-filled $5s$ subshell

2.82 Write abbreviated electron configurations for the following.

a. arsenic

b. An element that contains 25 electrons

c. silicon

d. element number 53

2.83 Write abbreviated electron configurations for the following.

a. An element that contains 24 electrons

b. element number 21

c. iodine

d. copper

2.84 Refer to the periodic table and write abbreviated electron configurations for all elements in which the noble gas symbol used will be [Ne].

2.85 Refer to the periodic table and determine how many elements have the symbol [Kr] in their abbreviated electron configurations.

2.86 Which Period 6 element has chemical properties most like those of sodium? How many valence-shell electrons does this element have? How many valence-shell electrons does sodium have?

2.87 Which Period 5 element has chemical properties most like those of silicon? How many valence shell electrons does this element have? How many valence shell electrons does silicon have?

2.88 Look at the periodic table and tell how many electrons are in the valence shell of the following elements.

a. element number 54

b. The first element (reading down) in Group VA(15)

c. Sn

d. The fourth element (reading left to right) in Period 3

2.89 Look at the periodic table and tell how many electrons are in the valence shell of the following elements.

a. element number 35

b. Zn

c. strontium

d. The second element of Group VA(15)

2.90 Classify each of the following elements into the s, p, or d area of the periodic table on the basis of the distinguishing electron.

a. Lead, poison to many organisms

b. Element 27, used in detection of tumors and metastases

c. Rb, used in PET scans to help diagnose heart disease

2.91 Classify each of the following elements into the s, p, or d area of the periodic table on the basis of the distinguishing electron.

a. Kr, radioactive isotope used in heart shunt studies

b. Tin, although not well researched, taken by some people by mouth for cancer

c. Element 40, used in the manufacture of prosthetics

Trends within the Periodic Table (Section 2.7)

2.92 If you discover an ore deposit containing copper, which other two elements might you also expect to find in the ore? Explain your reasoning completely.

2.93 Use trends within the periodic table and indicate which member of each of the following pairs has the larger atomic radius.

a. Mg or Sr

b. Rb or Ca

c. S or Te

d. I or Sn

2.94 Use trends within the periodic table and indicate which member of each of the following pairs has the larger atomic radius.

a. Ga or Se

b. N or Sb

c. O or C

d. Te or S

2.95 Use trends within the periodic table and indicate which member of each of the following pairs gives up one electron more easily.

a. Mg or Al

b. Ca or Be

c. S or Al

d. Te or O

2.96 Use trends within the periodic table and indicate which member of each of the following pairs gives up one electron more easily.

a. Li or K

b. C or Sn

c. Mg or S

d. Li or N

2.97 Use trends within the periodic table and indicate which member of each of the following pairs gives has a greater electronegativity.

a. Li or Br

b. F or Cl

c. Sn or C

d. Cl or Al

2.98 Use trends within the periodic table and indicate which member of each of the following pairs gives has a greater electronegativity.

a. K or Cl

b. Cl or I

c. N or P

d. Be or B

Additional Exercises

2.99 Refer to **Figure 2.33** and predict what would happen to the density of the metallic elements (purple color) as you go from left to right across a period of the periodic table. Explain your reasoning.

2.100 The mass of a single carbon-12 atom is 1.99×10^{-23} g. What is the mass in grams of a single carbon-14 atom?

2.101 Bromine (Br) and mercury (Hg) are the only elements that are liquids at room temperature. All other elements in the periodic group that contains mercury are solids. Explain why mercury and bromine are not in the same group.

2.102 How would you expect the chemical properties of isotopes of the same element to compare to each other? Explain your answer.

2.103 a. Explain how atoms of different elements differ from one another.

b. Explain how atoms of different isotopes of the same element differ from one another.

Chemistry for Thought

2.104 The atomic weight of aluminum is 26.98 amu, and the atomic weight of nickel is 58.69 amu. All aluminum atoms have a mass of 26.98 amu, but not a single atom of nickel has a mass of 58.69 amu. Explain.

2.105 Nuclear accidents, such as the one at Chernobyl, can release radioactive iodine and other radioisotopes into the atmosphere. Why might the distribution of potassium iodide (nonradioactive) tablets to the population be effective in countering radiation sickness?

2.106 One (unrealized) goal of ancient alchemists was to change one element into another (such as lead to gold). Do such changes occur naturally? Explain your reasoning.

2.107 Diagnostic procedures rely on radioisotopes that are gamma, beta, or positron emitters. Why are alpha emitters not useful for diagnostic purposes?

2.108 Consider the concept of half-life and decide if, in principle, a radioactive isotope ever completely disappears by radioactive decay. Explain your reasoning.

2.109 Nuclear wastes typically have to be stored for at least 20 half-lives before they are considered safe. This can be a time of hundreds or thousands of years for some isotopes. With that in mind, would you consider sending such wastes into outer space a responsible solution to the nuclear waste disposal problem? Explain your answer.

2.110 Plutonium-239 decays by alpha emission, and iodine-131 decays by beta emission and emits a gamma ray.

a. Write the balanced equations for radioactive decay for both reactions.

b. Iodine-131 can be hazardous if ingested. Where do the radioisotopes accumulate in the body?

2.111 Uranium-238 is the most abundant naturally occurring isotope of uranium. It undergoes radioactive decay to form other isotopes. In the first three steps of this decay, uranium-238 is converted to thorium-234, which is converted to protactinium-234, which is then converted to uranium-234. Assume that only one particle in addition to the daughter is produced in each step, and use equations for the decay processes to determine the type of radiation emitted during each step.

3 Chemical Bonds: Molecule Formation

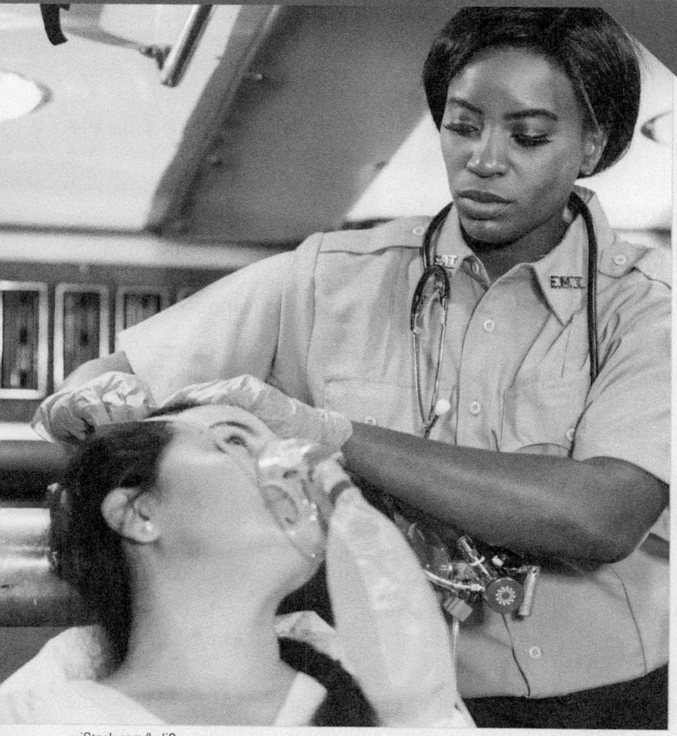

iStock.com/kali9

Health Career Focus

Paramedic

I took a "strengths test" to help me determine careers that fit my personality and interests. When the results suggested "paramedic," I wondered why I hadn't thought of that—it fit me perfectly! I love helping people and solving problems as well as working under pressure.

I scheduled a tour of the paramedic program at my local community college. Their "lab" was an actual ambulance! I liked the idea that I could use my physical strength in emergency situations. My interest in physics, chemistry, and anatomy mesh well with this career. From saline solutions in IVs to oxygen, many of the substances I administer to patients in need are either the covalent or the ionic compounds I first learned about in my chemistry course. I thrive in the high-stress environment. We respond to accidents, overdoses, and cardiac events; we make split-second decisions. Working as a paramedic is never boring!

Follow-up to this Career Focus appears at the end of the chapter before the *Concept Summary*.

Learning Objectives

When you have completed your study of Chapter 3, you should be able to:

1. Draw correct Lewis dot structures for atoms of the representative elements. **(Section 3.1)**
2. Use electron configurations to determine the number of electrons gained or lost by atoms as they achieve noble gas electron configurations. **(Section 3.2)**
3. Use electron configurations to determine the series of ions that are isoelectronic with the noble gases. **(Section 3.2)**
4. Use the octet rule to correctly predict the ions formed during the formation of ionic compounds. **(Section 3.3)**
5. Write correct formula units for ionic compounds containing a representative metal and a representative nonmetal. **(Section 3.3)**
6. Determine formula weights for ionic compounds in atomic mass units. **(Section 3.3)**
7. Correctly name binary ionic compounds. **(Section 3.4)**
8. Draw correct Lewis structures for covalent molecules. **(Section 3.5)**
9. Determine molecular weights for covalent compounds in atomic mass units. **(Section 3.5)**
10. Correctly name binary covalent compounds. **(Section 3.6)**
11. Draw correct Lewis structures for polyatomic ions. **(Section 3.7)**
12. Write correct formulas for ionic compounds containing representative metals and polyatomic ions. **(Section 3.8)**
13. Correctly name binary ionic compounds containing polyatomic ions. **(Section 3.8)**

In the discussion up to this point, we have emphasized that matter is composed of tiny particles: protons, neutrons, and electrons. Chapter 2 helped us understand where the electrons reside in an element, but now it is time to examine how the valence electrons of atoms interact to form chemical bonds. We begin Chapter 3 by explaining how to depict valence electrons using Lewis structures. Lewis dot structures make it possible to understand ions and how ionic bonds form between a metal and nonmetal. Once we are able to predict a compound formed between ions, we can properly name it and determine its formula weight. Next, we examine how the valence electrons of two nonmetals are shared in a covalent bond and how full Lewis structures can depict the discrete molecules of covalent compounds. We then learn how to calculate molecular weights and determine the names of covalent compounds. Chapter 3 concludes with a discussion of polyatomic ions.

3.1 An Introduction to Lewis Structures

Learning Objective 1 Draw correct Lewis dot structures for atoms of the representative elements.

Lewis dot structure A representation of an atom or ion in which the elemental symbol represents the atomic nucleus and all but the valence-shell electrons. The valence-shell electrons are represented by dots arranged around the elemental symbol.

In Chapter 2, we described the location of electrons within an atom, using electron configurations to list electrons in order of energy shells, sublevels, and orbitals. Abbreviated electron configurations, which list the immediately preceding noble gas and the remaining electrons in the outermost energy level, were introduced to emphasize the number of valence electrons. Another way to represent the valence electrons of atoms was invented by the American chemist G. N. Lewis (**Figure 3.1**). In Lewis's representations, called electron-dot structures or **Lewis dot structures**, the symbol for an element represents the nucleus and all electrons around the nucleus except those in the valence shell. Valence-shell electrons are shown as dots around the symbol.

For example, the electron configuration of sodium is $1s^22s^22p^63s^1$, and the abbreviated electron configuration is $[Ne]3s^1$, which indicates that sodium has one valence electron. Thus, the Lewis dot structure of sodium is represented as Na·. As mentioned in **Section 2.6**, the number of valence electrons in the representative elements—that is, Groups IA(1), IIA(2), and IIIA(13) through VIIIA(18)—is the same as the number of their group in the periodic table. Thus, the elements of Group IA(1) all have one valence electron and would have a similar Lewis dot structure to sodium. Throughout Chapter 3, we show how Lewis dot structures lead to full Lewis structures, which can be used to represent covalent bonds in molecules. Before that, it is important to clearly understand the relationship between Lewis dot structures and the electron configurations for atoms. Example 3.1 and Learning Check 3.1 will help you review this relationship.

Bettmann/Getty Images

Figure 3.1 G. N. Lewis (1875–1946) was nominated for the Nobel Prize in Chemistry 35 times, but he never won.

Example 3.1 Electron Configurations and Lewis Dot Structures of Atoms

Compounds of fluorine (F) are used in toothpaste and in drinking water to prevent dental cavities (**Figure 3.2**).

Represent the following using abbreviated electronic configurations and Lewis dot structures.

a. F b. K c. Mg d. Si

Solution

a. Fluorine (F), which resides in Group VIIA(17) of the periodic table, contains nine electrons, seven of which are classified as valence electrons (those in the 2*s* and 2*p* subshells). The abbreviated electron configuration and Lewis dot structure are:

$$[He]2s^22p^5 \qquad \text{and} \qquad :\ddot{\underset{\cdot\cdot}{F}}\cdot$$

b. Potassium (K), of Group IA(1), contains 19 electrons, one of which is classified as a valence electron (the one in the 4*s* subshell). The abbreviated electron configuration and Lewis dot structure are:

$$[Ar]4s^1 \qquad \text{and} \qquad K\cdot$$

c. Magnesium (Mg), of Group IIA(2), contains 12 electrons, two of which are classified as valence electrons (the two in the 3*s* subshell). The abbreviated electron configuration and Lewis dot structure are:

$$[Ne]3s^2 \qquad \text{and} \qquad \dot{M}g\cdot$$

d. Silicon (Si), of Group IVA(14), contains 14 electrons, four of which are classified as valence electrons (those in the 3*s* and 3*p* subshells). The abbreviated electron configuration and Lewis dot structure are:

$$[Ne]3s^23p^2 \qquad \text{and} \qquad \cdot\dot{\underset{\cdot}{Si}}\cdot$$

✔ **Learning Check 3.1** Write abbreviated electron configurations and Lewis dot structures for the following atoms.

a. Li b. Br c. Sr d. S

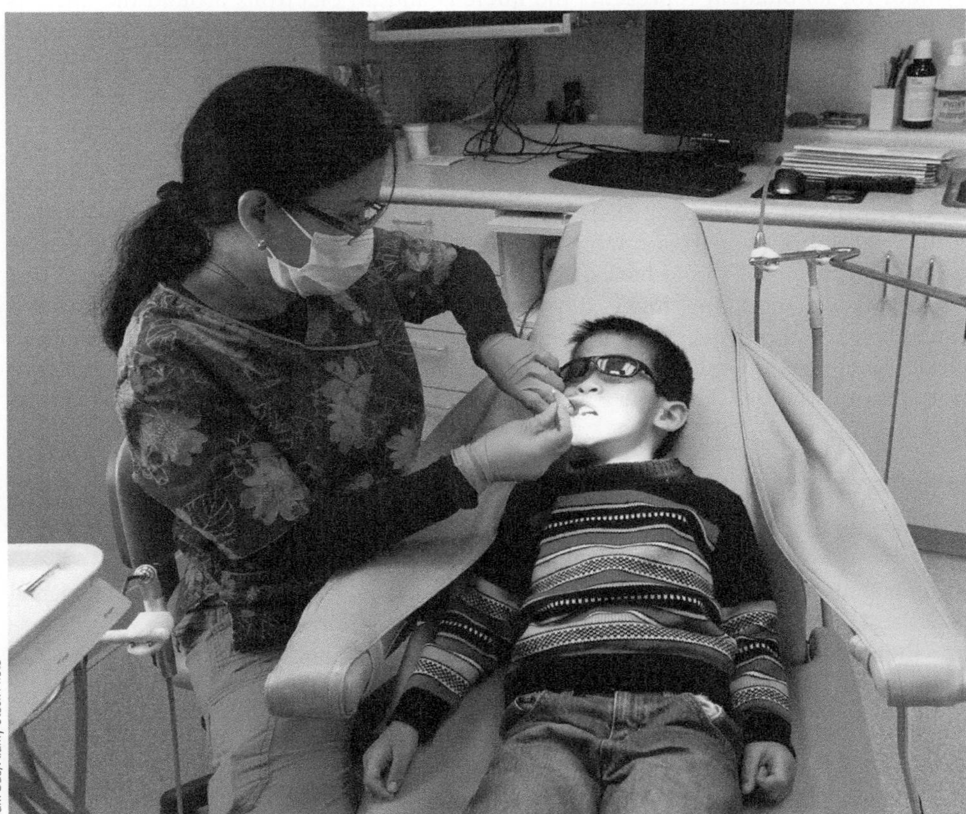

Figure 3.2 Fluoride treatments at the dentist's office are common in order to fight tooth decay. In Section 3.2, we explain that fluoride is the ionized version of the neutral element fluorine.

Sam Dao/Alamy Stock Photo

3.2 The Formation of Ions

Learning Objective 2 Use electron configurations to determine the number of electrons gained or lost by atoms as they achieve noble gas electron configurations.

Learning Objective 3 Use electron configurations to determine the series of ions that are isoelectronic with the noble gases.

The elements in Group VII(18)—the noble gases—only rarely react with other chemical species. They are so unreactive because their outermost energy shell is completely full of electrons, which just so happens to give these elements an inherent state of stability. This state is so energetically favorable that the atoms of other elements actually lose, gain, or share electrons with other atoms to reach the nearest noble gas electron configuration themselves. Chemists refer to this behavior as the **octet rule**. There are limited exceptions to the octet rule, but it sufficiently describes how most of the representative elements from the periodic table behave.

When atoms lose or gain electrons to satisfy the octet rule and reach a noble gas configuration, they become charged species because the number of protons and electrons is no longer balanced. These charged species are called **simple ions**. When atoms lose electrons, they become positively charged ions, which are called **cations**. When atoms gain electrons, they become negatively charged ions, which are called **anions**. For instance, oxygen, with its six valence electrons, gains two electrons to satisfy the octet rule, which gives the oxygen anion a −2 charge. Likewise, the calcium ion, with its two valence electrons, loses two electrons to satisfy the octet rule and achieve a noble gas configuration, which gives the calcium cation a +2 charge. As a general rule, metals lose electrons, whereas nonmetals gain electrons, and the loss or gain is usually no more than three electrons (**Figure 3.3**).

octet rule A rule for predicting electron behavior in reacting atoms—namely, atoms will lose, gain, or share sufficient electrons to achieve an outer electron arrangement identical to that of a noble gas. This arrangement usually consists of eight electrons in the valence shell.

simple ion An atom that has acquired a net positive or negative charge by losing or gaining electrons, respectively.

cation An atom that has acquired a net positive charge by losing one or more electrons.

anion An atom that has acquired a net negative charge by gaining one or more electrons.

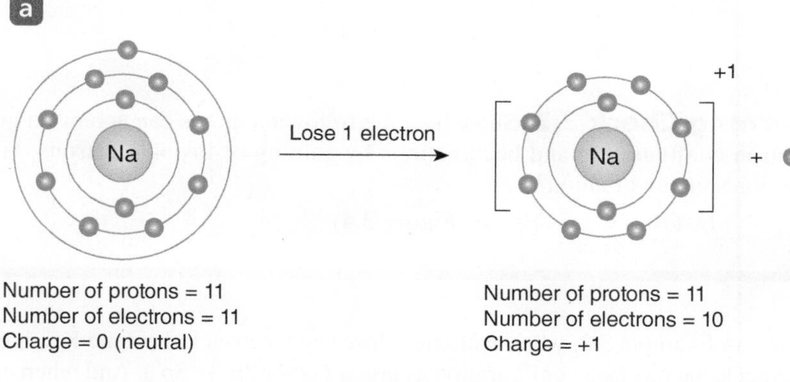

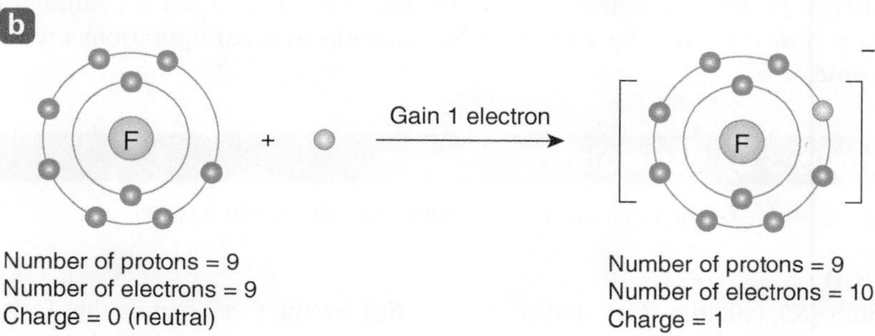

Figure 3.3 (a) Formation of the Na^+ cation from the neutral Na atom and (b) formation of the F^- anion from the neutral F atom.

Example 3.2 Ions and Noble Gas Configurations

Show how the following atoms can achieve a noble gas configuration and become ions by gaining or losing electrons.

a. K   b. Cl

Solution

a. The electronic structure of potassium (K) can be represented using the following abbreviated electron configuration and Lewis structure:

$$[Ar]4s^1 \quad \text{and} \quad K\cdot$$

The abbreviated electron configuration shows that the Ar electron configuration with eight electrons in the valence shell would result if the K atom loses the single electron located in the 4*s* subshell. The loss is represented by the following equation:

$$K \rightarrow K^+ + e^-$$

The removal of a single electron from a neutral K atom leaves the atom with 19 positive protons in the nucleus and 18 negative electrons. This gives the atom a net positive charge. The atom has become a cation with a +1 charge.

b. The electron structure for chlorine (Cl) can be represented using the following abbreviated electron configuration and Lewis structure:

$$[Ne]3s^2 3p^5 \quad \text{and} \quad :\ddot{\underset{\cdot\cdot}{Cl}}\cdot$$

The Cl atom can achieve the electron configuration of Ne by losing the seven valence electrons. However, it is energetically much more favorable to achieve the configuration of argon (Ar) by adding one electron to the valence shell. This electron would complete an octet and change a Cl atom (17 protons, 17 electrons) into a chloride anion (17 protons, 18 electrons) with a −1 charge. The gain can be represented by the following equation:

$$Cl + e^- \rightarrow Cl^-$$

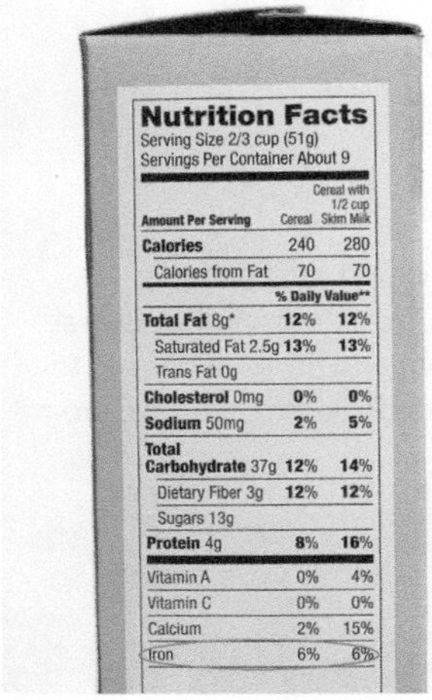

Chris Sadowski/E+/Getty Images

Figure 3.4 Acid in the stomach converts metallic iron (Fe), present in fortified cereals, into Fe^{2+} ions.

✔ **Learning Check 3.2** Show how the following atoms can achieve a noble gas electron configuration and become ions by gaining or losing electrons. Is the resulting ion a cation or anion?

a. Li   b. F   c. Fe (see **Figure 3.4**)

As shown in Example 3.2, when potassium loses an electron to become K^+, it ends up with the exact same electron configuration as argon ($1s^2 2s^2 2p^6 3s^2 3p^6$). And when chloride gains an electron to become Cl^-, it ends up with the exact same electron configuration as argon ($1s^2 2s^2 2p^6 3s^2 3p^6$). K^+ and Cl^- are not Ar, however, because they still differ in the number of protons in their respective nuclei. Instead, it is just the number of electrons that is now the same. Species with the same electron configurations are said to be **isoelectronic**.

isoelectronic Literally "same electronic." The term is used to describe atoms or ions that have identical electron configurations.

Example 3.3 Isoelectronic Series

Provide the series of ions that are isoelectronic with the noble gas Ar.

Solution

Potassium (K), calcium (Ca), and scandium (Sc) are the three metals that follow Ar in the periodic table. If these three elements lose one, two, and three electrons, respectively, then they will achieve electron configurations that are isoelectronic with Ar. This can be demonstrated explicitly with the formation of calcium ion:

Abbreviated electron configuration of calcium (Ca):

$$[Ar]4s^2$$

Equation showing the formation of the ion:

$$Ca \rightarrow Ca^{2+} + 2e^-$$

Abbreviated electron configuration of calcium ion (Ca^{2+}):

$$[Ar]$$

Applying the same logic to K and Sc shows that K^+ and Sc^{3+} are also isoelectronic with Ar. We encourage you to do this on your own.

Phosphorus (P), sulfur (S), and chlorine (Cl) are the three nonmetals that precede Ar in the periodic table. If these three elements gain three, two, and one electron, respectively, then they will achieve electron configurations that are isoelectronic with Ar. A closer look at phosphorus shows this is true:

Abbreviated electron configuration of phosphorus (P):

$$[Ne]3s^2 3p^3$$

Equation showing the formation of the ion:

$$P + 3e^- \rightarrow P^{3-}$$

Abbreviated electron configuration of phosphide ion (P^{3-}):

$$[Ne]3s^2 3p^6 = [Ar]$$

Applying this same logic to Cl and S shows that Cl^- and S^{2-} are also isoelectronic with Ar. Once again, we encourage you to do this on your own.

The complete series of ions isoelectronic with Ar is as follows:

$$P^{3-},\ S^{2-},\ Cl^-,\ K^+,\ Ca^{2+},\ Sc^{3+}$$

✔ **Learning Check 3.3** Different colors are produced during electric discharge in each noble gas (**Figure 3.5**). Krypton is one of the rarest gases on Earth! Provide the series of ions isoelectronic with Kr.

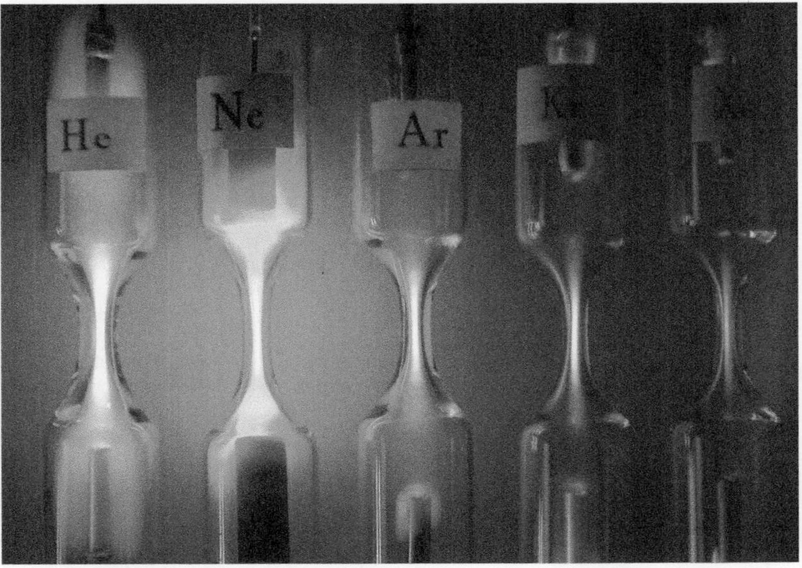

Figure 3.5 Shown here are the colors and spectra produced from electric discharge in flasks filled with noble gases.

3.3 Ionic Compounds

Learning Objective 4 Use the octet rule to correctly predict the ions formed during the formation of ionic compounds.

Learning Objective 5 Write correct formula units for ionic compounds containing a representative metal and a representative nonmetal.

Learning Objective 6 Determine formula weights for ionic compounds in atomic mass units.

So far the discussion has focused on the electron-transfer process as it occurs for isolated atoms. In reality, no atom can lose electrons unless another atom is available to accept them (see **Figure 3.6**). And, as mentioned in **Section 3.2**, metals tend to lose electrons, whereas nonmetals tend to gain them, so when a metal reacts with a nonmetal, the electrons lost by the metal are the same ones gained by the nonmetal. Substances formed from such reactions are called **ionic compounds**, and the attractive forces between the oppositely charged ions are called **ionic bonds**.

ionic compound A substance consisting of ions of opposite charge held together by an electrostatic attraction.

ionic bond The attractive force that holds together ions of opposite charge.

formula unit The simplest whole number ratio of ions that make up an ionic compound.

The formulas of ionic compounds must be neutral overall, so the total positive charge of the cation must balance the total negative charge of the anion. As a result, the ratio of cations to anions in an ionic compound is determined by the charges on the ions, which are determined, in turn, by the number of electrons transferred. This ratio that governs the number of atoms in an ionic compound is called a **formula unit**. When more than one atom of an element is present in an ionic compound, a subscript is used to indicate the number. For instance, the formula unit $CaCl_2$ indicates broadly that for every one calcium atom ("Ca") present in the ionic compound, there are two chlorine atoms ("Cl"). The subscript "1" is never written in these formulas; instead, it is always assumed.

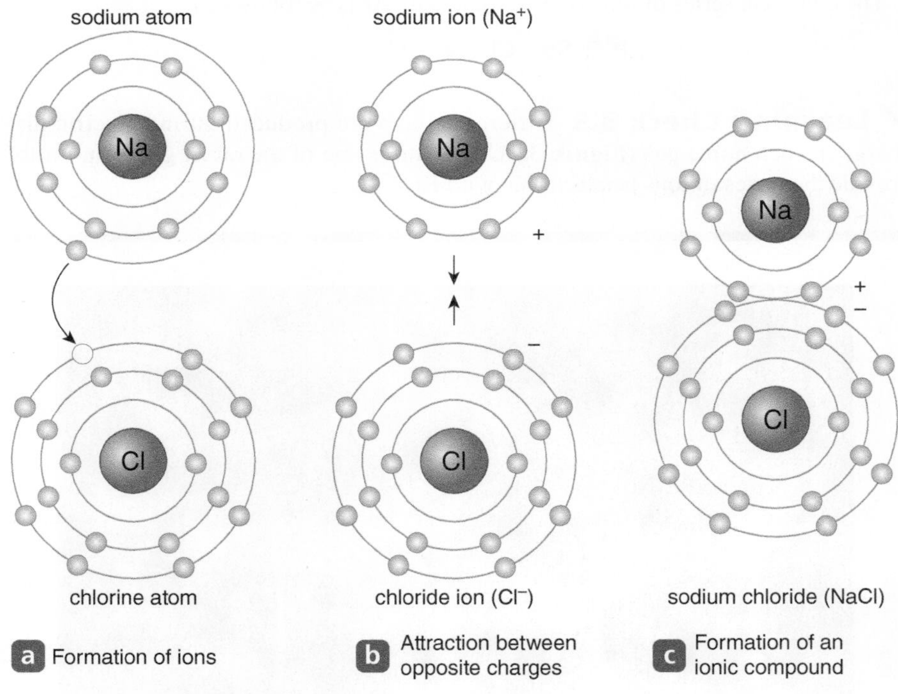

Figure 3.6 Sodium loses an electron that chloride gains in order to form the ionic compound NaCl. Both atoms now have a full octet in their valence shell.

Example 3.4 Electron-Transfer Processes and Formula Units

Represent the electron-transfer process that takes place when the ions of the following elements interact. Determine the formula unit for each resulting ionic compound.

a. Li and Br   b. Mg and F

Solution

a. Lithium (Li) is in Group IA(1), so it tends to lose one electron to form Li^+:

$$Li \rightarrow Li^+ + e^-$$

Bromine (Br), in Group VIIA(17), tends to gain one electron to form Br^-:

$$Br + e^- \rightarrow Br^-$$

The formula unit for ionic compounds must be neutral overall, so if the resulting ions, Li^+ and Br^-, combine in a 1:1 ratio, then the total positive and total negative charges in the final formula unit, LiBr, add up to zero. Note that the metal (Li) is written first in the formula unit and that the formula unit is written using the simplest whole number ratio (i.e., LiBr, not Li_2Br_2 or Li_3Br_3, etc.).

The following representation emphasizes the electron-transfer process and how the ions achieve their octets:

$$\begin{array}{lcll} \text{Li} & + & \text{Br} & \rightarrow \text{Li}^+ \quad \text{Br}^- \\ [\text{He}]2s^1 & + & [\text{Ar}]4s^2 3d^{10} 4p^5 & \rightarrow [\text{He}] + [\text{Kr}] \end{array}$$

b. Magnesium (Mg) of Group IIA(2) tends to lose two electrons per atom, whereas fluorine (F) of Group VIIA(17) tends to gain one electron per atom:

$$Mg \rightarrow Mg^{2+} + 2e^-$$

$$F + e^- \rightarrow F^-$$

$$\begin{array}{lcll} \text{Mg} & + & \text{F} & \rightarrow \text{Mg}^{2+} + \text{F}^- \\ [\text{Ne}]3s^2 & + & [\text{He}]2s^2 2p^5 & \rightarrow [\text{Ne}] + [\text{Ne}] \end{array}$$

The formula unit for ionic compounds must be neutral overall, so two fluorine atoms will be required to accept the electrons from one magnesium atom. That is, two F^- ions will be needed to balance the charge of a single Mg^{2+} ion. The formula unit for the resulting ionic compound, then, will be MgF_2. Notice how subscripts are used to indicate the number of ions involved in the formula. The subscript 1, on Mg, is understood and never written.

✔ Learning Check 3.4 Write equations to represent the formation of ions for each of the following pairs of elements. Write the formula unit for the ionic compound that would form in each case.

a. Mg and O   b. K and S   c. Ca and Br

The stable form of an ionic compound is not an individual molecule, but rather a crystal in which many ions of opposite charge occupy **lattice sites** in a rigid three-dimensional arrangement called a **crystal lattice**. For example, the lattice of NaCl (ordinary table salt) depicted in **Figure 3.7** is the stable form of the pure ionic compound.

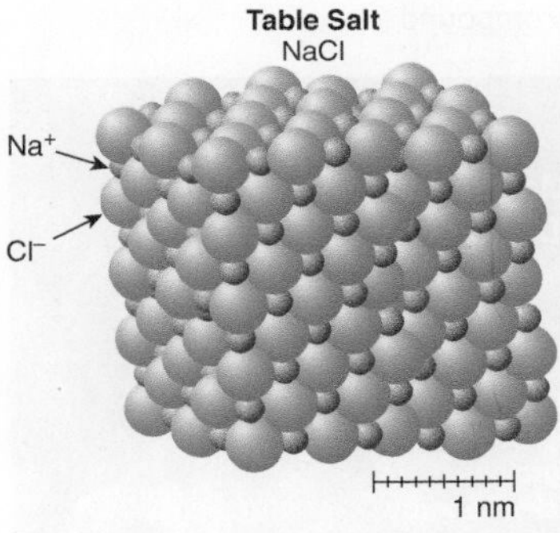

Figure 3.7 The crystal lattice for NaCl (table salt), an ionic compound. Notice how each Na^+ is surrounded by Cl^- ions and how each Cl^- is surrounded by Na^+ ions.

lattice site The individual location occupied by a particle in a crystal lattice.

crystal lattice A rigid three-dimensional arrangement of particles.

Health Connections 3.1

Watch the Salt

An average adult in the United States consumes about 3400 milligrams of sodium each day in the form of table salt. Only about one in 10 Americans does not exceed the recommended daily limit set by the American Heart Association of 2300 milligrams of sodium. The overconsumption of sodium increases the risk of developing high blood pressure, a leading cause of cardiovascular disease that accounts, in turn, for two-thirds of all strokes and half of the heart disease in the United States.

Salt is often present in processed and prepared foods such as pizza, bread, and cold cuts. A fast-food breakfast sandwich can contain as much as 50% of the recommended daily limit of salt in one food item alone. Make it a practice to read labels of items you buy at the grocery store. When you eat at home, be aware of the amount of salt you use.

Increase the amount of potassium in your diet: Bananas, yogurt, baked potatoes, salmon, and avocado are among the tasty foods that are rich in potassium. Dietary potassium helps relax blood vessels, decreases blood pressure, and helps eliminate excess sodium from your body.

Salt is often present in processed foods such as pizza, bread, and cold cuts.

formula weight The sum of the atomic weights of the ions present in the formula unit of an ionic compound.

Even though the formula units of ionic compounds do not represent the entire crystal lattice, they are the simplest, most useful way to identify what ions are present in the lattice. The sum total of the atomic weights of the ions in the formula unit of an ionic compound is called its **formula weight**.

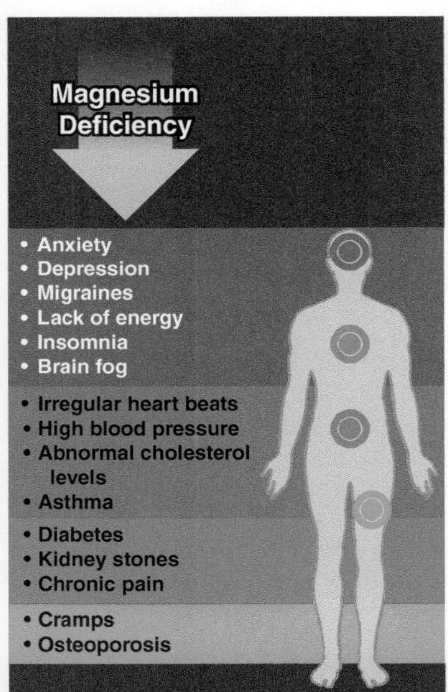

Magnesium Deficiency
• Anxiety
• Depression
• Migraines
• Lack of energy
• Insomnia
• Brain fog
• Irregular heart beats
• High blood pressure
• Abnormal cholesterol levels
• Asthma
• Diabetes
• Kidney stones
• Chronic pain
• Cramps
• Osteoporosis

Figure 3.8 Symptoms of magnesium deficiency in patients.

Example 3.5 Calculating Formula Weights

$MgCl_2$, an ionic compound, is a mineral supplement used to prevent and treat low amounts of magnesium in the blood. The main symptoms of a magnesium deficiency are listed in **Figure 3.8**. Determine the formula weight for $MgCl_2$ in atomic mass units.

For $MgCl_2$, the formula weight (FW) is equal to the sum of the atomic weights of the atoms in the formula unit.

$$\begin{aligned} FW &= (1)(\text{at. wt. Mg}) + (2)(\text{at. wt. Cl}) = (1)(24.3 \text{ amu}) + (2)(35.5 \text{ amu}) \\ &= 95.3 \text{ amu} \end{aligned}$$

✔ **Learning Check 3.5** Determine the formula weight for the ionic compound CaO.

3.4 Naming Binary Ionic Compounds

Learning Objective 7 Correctly name binary ionic compounds.

binary compound A compound made up of two different elements.

Ionic compounds of the types used in Example 3.4 are called **binary compounds** because each contains only two different kinds of atoms—namely, Li and Br in one case and Mg and F in the other. The reaction of sodium metal with chlorine gas to form the binary ionic compound NaCl is shown in **Figure 3.9**.

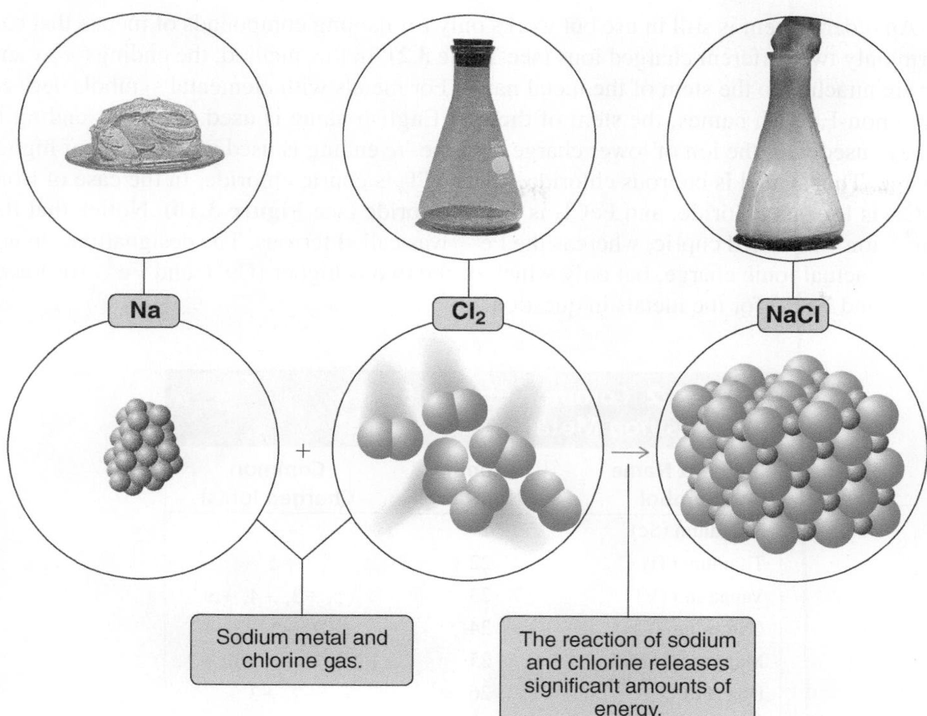
Na
Cl_2
NaCl
+
→
Sodium metal and chlorine gas.
The reaction of sodium and chlorine releases significant amounts of energy.

Figure 3.9 The reaction of sodium metal and chlorine gas produces the solid binary ionic compound sodium chloride (NaCl).

Names for binary ionic compounds are based on the names of the two elements involved. The name of the metallic element is given first, followed by the stem of the nonmetallic elemental name to which the suffix *-ide* has been added.

Name = metal + nonmetal stem + *-ide*

The stem of the name of a nonmetal is the name of the nonmetal with the ending dropped. **Table 3.1** lists the stem of the names of some common nonmetallic elements. For example, NaCl is sodium chloride.

Table 3.1 Stem Names and Ion Formulas of Common Nonmetallic Elements

Element	Stem	Formula of Ion
Bromine	*brom-*	Br^-
Chlorine	*chlor-*	Cl^-
Fluorine	*fluor-*	F^-
Iodine	*iod-*	I^-
Nitrogen	*nitr-*	N^{3-}
Oxygen	*ox-*	O^{2-}
Phosphorus	*phosph-*	P^{3-}
Sulfur	*sulf-*	S^{2-}

Names for the constituent ions of a binary compound are obtained in the same way as the compound name. Thus, Na^+ is a sodium ion, whereas Cl^- is a chloride ion.

Some metal atoms, especially those of transition and inner-transition elements, form more than one type of charged ion. Copper, for example, forms both Cu^+ and Cu^{2+}, and iron forms Fe^{2+} and Fe^{3+}. The names of ionic compounds containing such elements must indicate which ion is present in the compound. A nomenclature system that does this well indicates the ionic charge of the metal ion by a Roman numeral in parentheses following the name of the metal. Thus, CuCl is copper(I) chloride, and $CuCl_2$ is copper(II) chloride. These names are expressed verbally as “copper one chloride” and “copper two chloride,” respectively.

An older system is still in use but works only for naming compounds of metals that can form only two different charged ions (see **Table 3.2**). In this method, the endings *-ous* and *-ic* are attached to the stem of the metal name. For metals with elemental symbols derived from non-English names, the stem of the non-English name is used. The *-ous* ending is always used with the ion of lower charge, and the *-ic* ending is used with the ion of higher charge. Thus, CuCl is cuprous chloride, and $CuCl_2$ is cupric chloride. In the case of iron, $FeCl_2$ is ferrous chloride, and $FeCl_3$ is ferric chloride (see **Figure 3.10**). Notice that the Cu^{2+} ion was called cupric, whereas the Fe^{2+} was called ferrous. The designations do not tell the actual ionic charge, but only which of the two is higher (Cu^{2+} and Fe^{3+}) or lower (Cu^{+} and Fe^{2+}) for the metals in question.

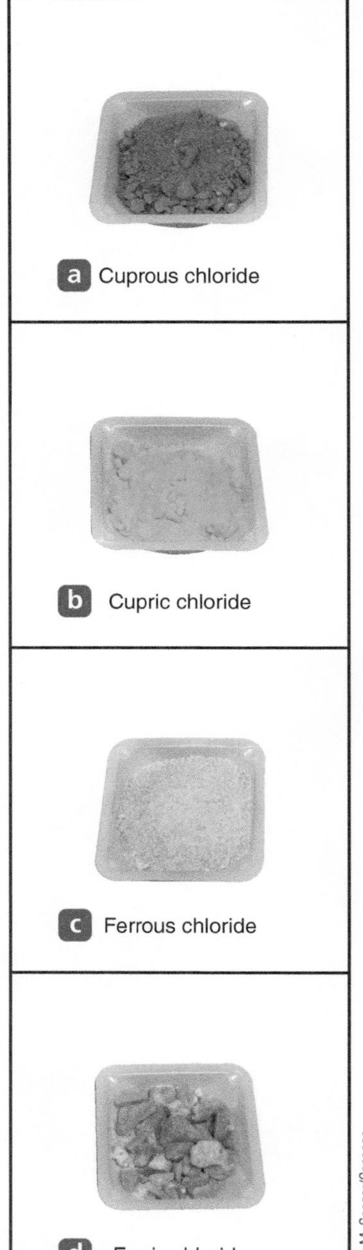

Figure 3.10 Chloride compounds of copper and iron:
(a) Cuprous chloride, CuCl
(b) Cupric chloride, $CuCl_2$
(c) Ferrous chloride, $FeCl_2$
(d) Ferric chloride, $FeCl_3$

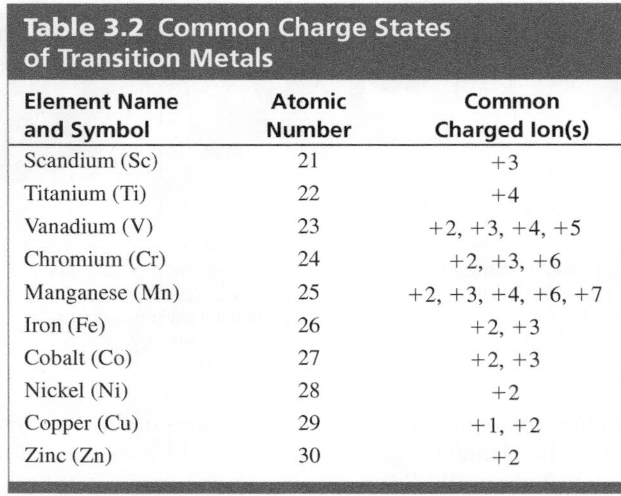

Table 3.2 Common Charge States of Transition Metals

Element Name and Symbol	Atomic Number	Common Charged Ion(s)
Scandium (Sc)	21	+3
Titanium (Ti)	22	+4
Vanadium (V)	23	+2, +3, +4, +5
Chromium (Cr)	24	+2, +3, +6
Manganese (Mn)	25	+2, +3, +4, +6, +7
Iron (Fe)	26	+2, +3
Cobalt (Co)	27	+2, +3
Nickel (Ni)	28	+2
Copper (Cu)	29	+1, +2
Zinc (Zn)	30	+2

Example 3.6 Naming Binary Ionic Compounds

The following examples include ionic compounds and ions that can combine to form binary ionic compounds. In each case, name either the compound provided or the compound formed from the electrostatic attraction of the ions.

a. KCl b. SrO c. Cr^{2+} and S^{2-} d. Cr^{3+} and S^{2-}

Solution

a. The metal is potassium (K), and the nonmetal is chlorine (Cl). Thus, the compound name is potassium chloride (the stem of the nonmetallic name is underlined). Potassium in ionic form is one of the most important minerals in the body, helping to regulate fluid balance, muscle contractions, and nerve signals.
b. Similarly, strontium (Sr) and oxygen (O) give the compound named strontium oxide.
c. The metal ion is Cr^{2+} and the nonmetal is S^{2-}. Because the ionic charges are equal in magnitude but opposite in sign, the ions will combine in a 1:1 ratio. The formula is CrS. Chromium forms simple ions with 2+ and 3+ charges (**Table 3.2**). This one is 2+, the lower of the two; thus, the names are chromium(II) sulfide and chromous sulfide.
d. The charges on the combining ions are 3+ and 2−. The smallest combining ratio that balances the charges is two Cr^{3+} (a total of 6+ charge) and three S^{2-} (a total of 6− charge). The formula is Cr_2S_3. The name is chromium(III) sulfide.

✔ **Learning Check 3.6** The following examples include ionic compounds and ions that can combine to form binary ionic compounds. In each case, name either the compound provided or the compound formed from the electrostatic attraction of the ions.

a. NaI b. MgF_2 c. Co^{2+} and Br^{-} d. Ca^{2+} and S^{2-}

3.5 Covalent Bonding

Learning Objective 8 Draw correct Lewis structures for covalent molecules.

Learning Objective 9 Determine molecular weights for covalent compounds in atomic mass units.

Based on what we've learned thus far, ionic compounds form when a metal loses electrons, a nonmetal gains those electrons, and the resulting ions form ionic bonds to each other. How, then, do nonmetals that cannot easily gain or lose electrons form chemical bonds? Carbon, a nonmetal from Group IVA(14), would need to gain or lose four electrons in order to form ionic bonds, which generally does not occur. Nevertheless, carbon is known to form many compounds with other nonmetals like hydrogen, nitrogen, and oxygen. These nonmetals can also bond with one another, or to themselves (see **Figure 3.11**). In fact, molecules of elements containing two atoms, such as chlorine (Cl_2), hydrogen (H_2), oxygen (O_2), and nitrogen (N_2), are known to represent the stable form in which these elements occur in nature.

So, what is the bonding of these compounds like? Instead of losing or gaining electrons, these nonmetals *share* electrons with one another to form what are called **covalent compounds**. Many of the compounds most commonly known to us, like water, ammonia, and methane, are examples of covalent compounds. When more than one atom of an element is present in a covalently bonded molecule, a subscript is used to indicate the number, just as in the formula units of ionic compounds. However, instead of formula units, the representation of covalently bond molecules is called a **molecular formula**. For instance, the molecular formula for water is H_2O, which indicates that every water molecule consists of one oxygen atom (O) and two hydrogen atoms (H). Likewise, a fluorine molecule, F_2, is made up of two fluorine atoms (F).

covalent compound A compound formed when two nonmetals share electrons to form a chemical bond.

molecular formula A representation of the molecule of a covalent compound, consisting of the symbols of the elements found in the molecule. Elements present in numbers greater than one have the number indicated by a subscript that follows the elemental symbol.

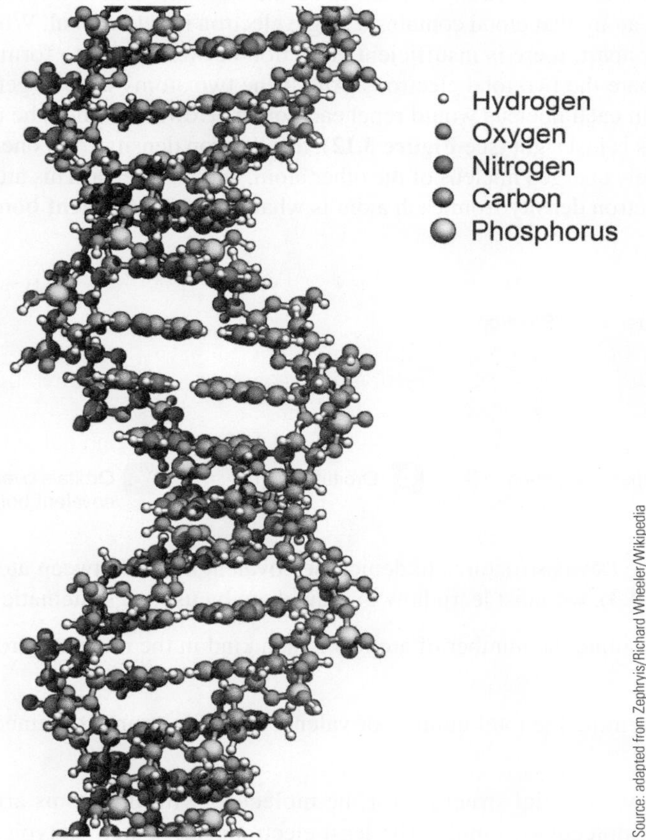

Figure 3.11 Carbon atoms, along with hydrogen, oxygen, nitrogen, and phosphorus atoms, are bonded in the structures that make up the DNA double helix.

Lewis structure A diagram that shows the bonding between atoms of a molecule, as well as the nonbonding pairs of electrons that may exist in the molecule.

G. N. Lewis suggested that the valence-shell electrons of the atoms in such molecules are shared in a way that satisfies the octet rule for each of the atoms. This process is symbolized as follows for fluorine, F_2, using full **Lewis structures**.

:F̤̈· + ·F̤̈: ⟶ :F̤̈:F̤̈:

Shared pair (Counted in the octet of each atom)

Covalent bonds are generally represented by straight lines, with each line (i.e., each covalent bond) being made up of two shared electrons. Thus, the full Lewis structure of F_2 could be represented as follows:

:F̤̈–F̤̈: ⟵ Lone pair

Shared pair

lone pairs Nonbonding pairs of electrons that serve to depict the completion of an atom's octet.

The nonbonding pairs of electrons (i.e., those represented in the Lewis structure of F_2 as pairs of electron dots instead of straight lines) are called **lone pairs**. Each F atom in F_2 has three lone pairs of electrons. In some of the Lewis structures that follow, we explicitly draw each individual electron, but the vast majority of covalent bonds are represented by straight lines.

Let's use the H_2 molecule and the concept of atomic orbital overlap to understand the origin of the attractive force between atoms that results from electron sharing. Recall that an atomic orbital (**Figure 2.30**) represents the space where an electron can be found at any given point in time. Chemists also generally may refer to orbitals as electron density (not to be confused with the density we have previously defined as the mass of a substance divided by the volume it occupies), because orbitals are where an atom's electrons are concentrated. While sometimes it is helpful to focus on electrons existing in defined shells, sublevels, and orbitals, as we have done previously, it can be helpful for understanding covalent bonding to imagine the valence electrons that are to be shared as existing in a cloud of electrons. In the hydrogen atom, that cloud contains a single electron in a 1*s* orbital. When two hydrogen atoms are far apart, there is insufficient attraction between them to form a covalent bond that would share the two total electrons. And if the two atoms were to get really close, the like charges in each nucleus would repel each other! However, when the distance between the two atoms is just right (see **Figure 3.12**), the electron density from one atom is attracted to the positively charged nucleus of the other atom, and vice versa. This attraction and overlap of the electron density from each atom is what forms the **covalent bond**.

covalent bond The attractive force that results between two atoms that are both attracted by a shared pair of electrons.

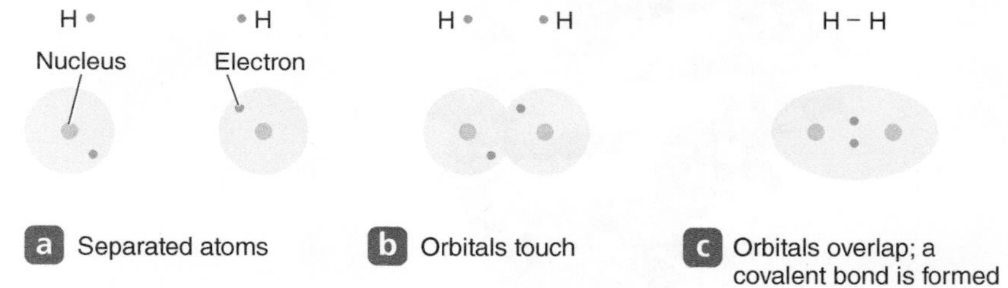

Figure 3.12 Orbital overlap during covalent bond formation.

In order to use Lewis structures to depict the covalent bonds between atoms within molecules (**Table 3.3**), we must learn how to draw them by using a systematic approach:

Step 1: Determine the number of atoms of each kind in the molecule from the molecular formula.

Step 2: Determine the total number of valence-shell electrons contained in the atoms of the molecule.

Step 3: Draw an initial structure for the molecule, with the atoms arranged properly, based on the molecular formula. The least electronegative atom is typically found in the middle, and more electronegative atoms (as well as hydrogen) are found on the outside. Electronegativity values of the representative elements are shown in **Figure 3.13**.

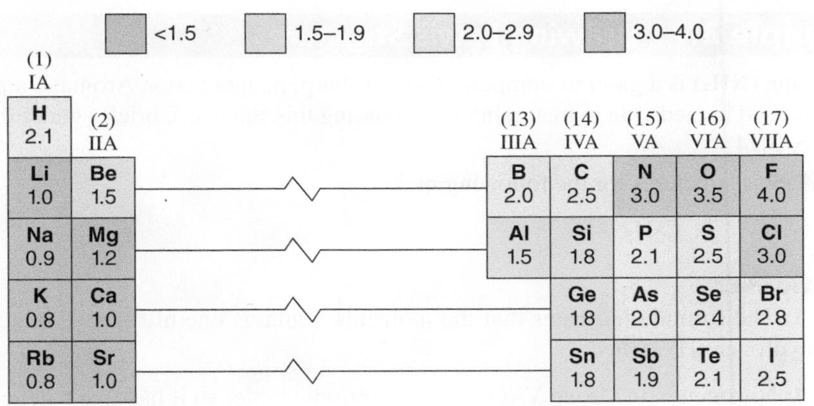

Figure 3.13 Pauling electronegativity values for the common representative elements. The values increase from left to right across a period and from bottom to top in a group. As a result, fluorine [at the top of Group VIIIA(18)] has the highest electronegativity value, and rubidium [at the bottom of Group IA(1)] has the lowest.

Step 4: Connect all the atoms in the initial structure drawn in Step 3 by a single covalent bond using a pair of electrons. Subtract the number of electrons used in this step from the total number determined in Step 2. Use the remaining electrons to complete the octets of all atoms in the structure (done typically by adding pairs of electrons), beginning with the atoms that are present in greatest number in the molecule. Remember, hydrogen atoms require only one pair (a duet) to achieve the electron configuration of helium.

Step 5: If all octets cannot be satisfied with the available electrons, move nonbonding pairs (i.e., lone pairs—those that are not between bonded atoms) to positions between bonded atoms to complete octets. This will create **double bonds** or **triple bonds** between some atoms.

double bonds The covalent bonds resulting from the sharing of two pairs of electrons.

triple bonds The covalent bonds resulting from the sharing of three pairs of electrons.

Table 3.3 Examples of Covalent Bonding

Molecule	Atomic Lewis Structure	Electronic Configuration	Lewis Structure with All Electrons Shown	Lewis Structure with Covalent Bonds and Lone Pairs Shown
Nitrogen gas (N_2)	·N̈·	($1s^22s^22p^3$)	:N⋮⋮N:	:N≡N:
Carbon dioxide (CO_2) (each O is bonded to the C)	·Ċ· :Ö·	($1s^22s^22p^2$) ($1s^22s^22p^4$)	:O::C::O:	:O=C=O:
Formaldehyde (H_2CO) (each H and the O are bonded to the C)	·Ċ· :Ö· H·	($1s^22s^22p^2$) ($1s^22s^22p^4$) ($1s^1$)	H:C::O: with H	H–C=Ö with C–H
Methane (CH_4) (each H is bonded to the C)	·Ċ· H·	($1s^22s^22p^2$) ($1s^1$)	H:C:H with H above and below	H–C–H with H above and below
Ammonia (NH_3) (each H is bonded to the N)	·N̈· H·	($1s^22s^22p^3$) ($1s^1$)	H:N̈:H with H below	H–N̈–H with H below
Water (H_2O) (each H is bonded to the O)	:Ö· H·	($1s^22s^22p^4$) ($1s^1$)	:Ö:H with H below	H–Ö–H
Oxygen difluoride (OF_2) (each F is bonded to the O)	:Ö· :F̤:	($1s^22s^22p^4$) ($1s^22s^22p^5$)	:Ö:F: with :F: below	:F–Ö–F:

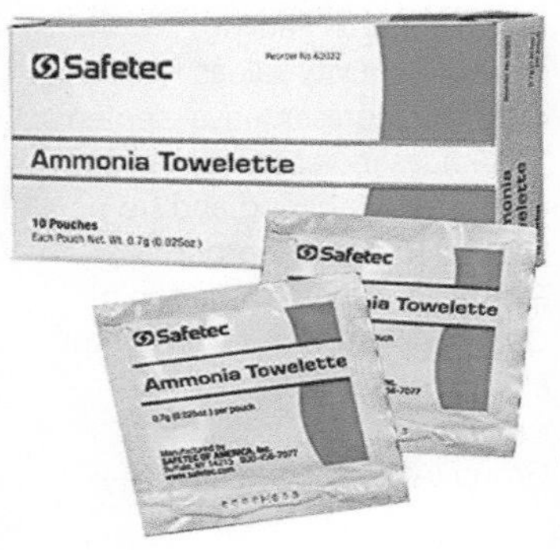

Source: Safetec

Figure 3.14 Ammonia towelette to treat fainting spells.

Example 3.7 Drawing Lewis Structures

Ammonia (NH_3) is a gaseous compound with a sharp, pungent odor. Aromatic ammonia spirit is used in medicine to treat fainting by passing this substance briefly under the nose (**Figure 3.14**).

Draw Lewis structures for the following molecules.

a. NH_3 b. C_2H_2

Solution a.

Step 1: The formula indicates that the molecule contains one nitrogen (N) atom and three hydrogen (H) atoms.

Step 2: Nitrogen is in Group VA(15) of the periodic table, so it has five valence electrons; hydrogen is in Group IA(1), so each one contributes one valence electron. The total, then, is eight (five from one N atom and three from the three H atoms).

Step 3: Nitrogen must be the central atom because hydrogen can form only one bond, so it is always on the outside of Lewis structures. This is confirmed in **Table 3.3**, which shows that each H atom is connected to the N atom; thus, we draw

$$\begin{matrix}\mathrm{H} & \mathrm{N} & \mathrm{H}\\ & \mathrm{H} & \end{matrix}$$

Step 4: We put one pair of electrons between each H atom and the N atom in the initial structure drawn in Step 2:

$$\begin{matrix}\mathrm{H:} & \mathrm{N} & \mathrm{:H}\\ & \ddot{\mathrm{H}} & \end{matrix}$$

This required six of the eight available electrons. The remaining pair is used to complete the octet of nitrogen. Remember, hydrogen achieves the noble gas configuration of helium with only two electrons:

$$\begin{matrix}\mathrm{H:} & \ddot{\mathrm{N}} & \mathrm{:H}\\ & \ddot{\mathrm{H}} & \end{matrix} \quad \text{or} \quad \begin{matrix}\mathrm{H}- & \ddot{\mathrm{N}} & -\mathrm{H}\\ & | & \\ & \mathrm{H} & \end{matrix}$$

Step 5: All octets are satisfied by Step 4, so nothing more needs to be done.

Solution b.

Step 1: The formula indicates two carbon (C) atoms and two hydrogen (H) atoms per C_2H_2 molecule.

Step 2: Carbon is in Group IVA(14), so each one contributes four valence electrons; hydrogen is in Group IA(1), so each one contributes one valence electron. Thus, the total number of valence electrons available is 10 (eight from the two C atoms and two from the two H atoms).

Step 3: The C atoms are bonded to each other, and one H atom is bonded to each C atom.

$$\mathrm{H\ C\ C\ H}$$

Step 4: We put one pair of electrons between the two C atoms and between each C atom and H atom.

$$\mathrm{H:C:C:H}$$

This required six of the available 10 electrons. The remaining four electrons are used to complete the C octets (remember, hydrogen needs only two electrons):

$$\mathrm{H:C:}\underset{\cdot\cdot}{\ddot{\mathrm{C}}}\mathrm{:H}$$

Step 5: All electrons are used in Step 4, but the octet of one C atom is still incomplete. It needs two more pairs. If both nonbonding pairs on the second C atom were shared with the first C atom, both carbons would have complete octets. The result is:

H:C:::C:H or H—C≡C—H

This molecule is called acetylene, and it contains a triple bond (three shared pairs) between the C atoms.

> ✔ **Learning Check 3.7** Draw Lewis structures for the following molecules.
>
> a. HNO_3 (each O atom is bonded to the N atom and one O atom is also bonded to the H atom)
>
> b. Formaldehyde, H_2CO (the O atom and two H atoms are each bonded to the C atom), which is used to preserve organs for medical studies (**Figure 3.15**)

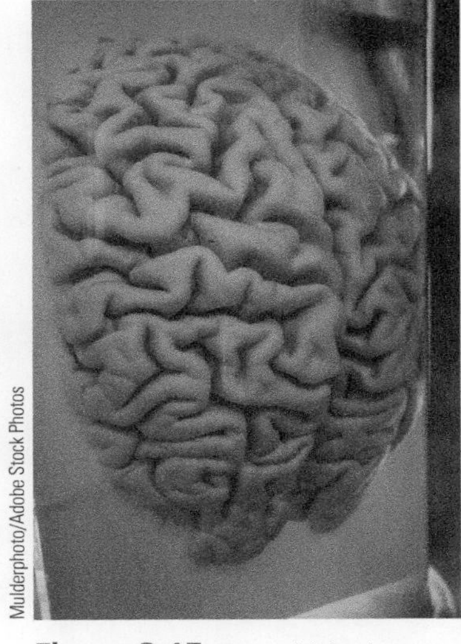

Mulderphoto/Adobe Stock Photos

Figure 3.15 Formaldehyde is used in laboratories and at universities to preserve tissues and organs for study purposes.

When drawing Lewis structures, how can we be certain we are placing electrons on the right atoms? *Formal charges* are a way of keeping track of the electrons on each atom in a Lewis structure. Formal charges for any atom can be calculated by subtracting the number of covalent bonds and the total number of lone pair electrons from its number of valence electrons.

$$\text{Formal charge (FC)} = \begin{matrix}\text{number of}\\ \text{valence } e^-\end{matrix} - \begin{matrix}\text{number of}\\ \text{bonds}\end{matrix} - \begin{matrix}\text{number of lone-pair}\\ \text{electrons}\end{matrix} \quad (3.1)$$

The sum of all the formal charges on atoms in a Lewis structure should equal the overall charge of the compound. Thus, for neutral compounds, the sum of the formal charges should be zero. Generally speaking, compounds are most stable when the number of atoms with formal charges is minimized (**Figure 3.16**). Some Lewis structures do not adhere to the octet rule, and formal charges help explain why these exceptions are allowed. There are some atoms (e.g., boron) that are stable with *fewer* electrons than a full octet. There are other atoms (e.g., sulfur and phosphorus) that are stable with *more* electrons than a full octet. Elements that expand their octet in this way are generally found in the third row of the periodic table or lower (i.e., $n \geq 3$ shell). Example 3.8 shows that these atoms, while not satisfying the octet rule, still carry a neutral formal charge!

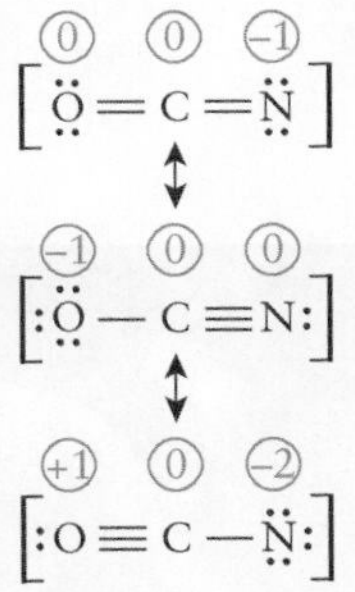

(0) (0) (−1)
[Ö=C=N̈]
↕
(−1) (0) (0)
[:Ö—C≡N:]
↕
(+1) (0) (−2)
[:O≡C—N̈:]

Figure 3.16 Lewis structure representations of the cyanate ion (CNO^-), with formal charges indicated in red circles. We can use formal charges to determine the most stable and therefore the most important Lewis structure. The bottom structure is not favored because its charge is not minimized. The middle structure is more stable than the top structure because the −1 formal charge is on the more electronegative oxygen atom!

Example 3.8 Exceptions to the Octet Rule: Lewis Structures

Draw Lewis structures for the following molecules.

a. BH_3 b. SO_3

Solution a.

Step 1: The formula indicates that the molecule contains one boron (B) atom and three hydrogen (H) atoms.

Step 2: Boron is in Group IIIA(13) of the periodic table, so it contributes three valence electrons; hydrogen is in Group IA(1), so each one contributes one valence electron. The total is six (three from the one B atom and three from the three H atoms)

Step 3: The connecting pattern is similar to ammonia (see **Table 3.3**) because hydrogens can only appear on the outside of Lewis structures; thus, we draw

H B H
H

Step 4: We put one pair of electrons between each H atom and the B atom in the initial structure drawn in Step 3. At this point, all available electrons are used up, even though boron has not achieved an octet.

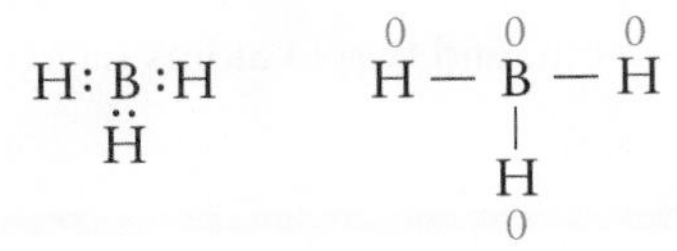

H:B:H (with H below B) H—B—H (with H below B; formal charges 0 on each atom)

$FC_B = 3 - 3 - 0 = 0$
$FC_H = 1 - 1 - 0 = 0$

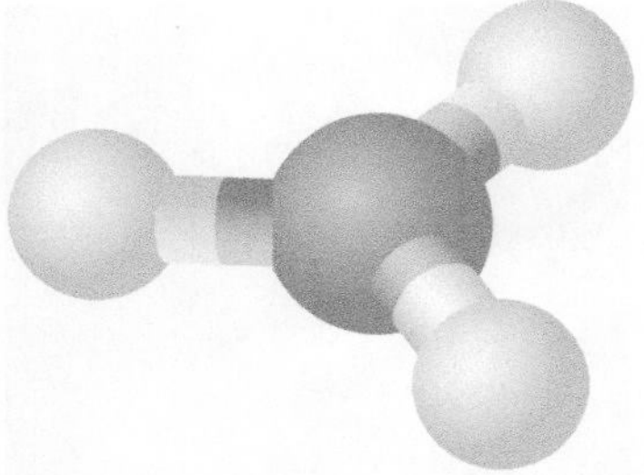

Figure 3.17 Ball-and-stick model of BH_3.

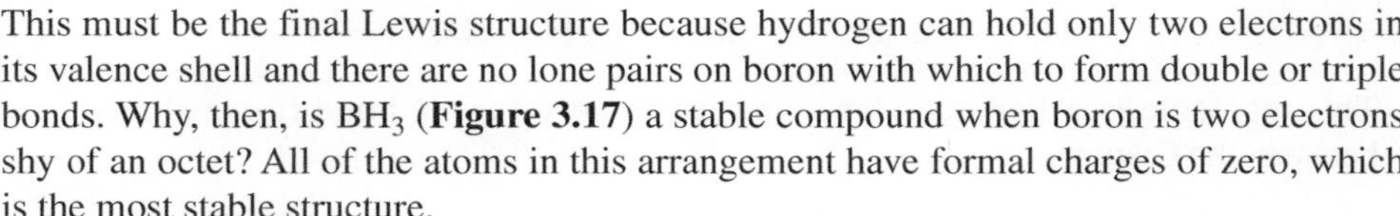

This must be the final Lewis structure because hydrogen can hold only two electrons in its valence shell and there are no lone pairs on boron with which to form double or triple bonds. Why, then, is BH_3 (**Figure 3.17**) a stable compound when boron is two electrons shy of an octet? All of the atoms in this arrangement have formal charges of zero, which is the most stable structure.

Solution b.

Step 1: The formula indicates that one sulfur (S) and three oxygen (O) atoms are present in the molecule.

Step 2: Sulfur and oxygen are both in Group VIA(16), so each of them contributes 6 valence electrons. The total is 24 (6 from the one S atom and 18 from the three O atoms).

Step 3: Sulfur is directly below oxygen in the periodic table, so S is the less electronegative atom. We therefore arrange the atoms as follows:

O S O
  O

Step 4: We begin by putting one pair of electrons between each O atom and the S atom:

O : S : O
  Ö

This uses 6 of the 24 available electrons. The remaining 18 are used to complete the octets, beginning with the O atoms:

:Ö: S :Ö:
  :Ö:

Step 5a: At this point, sulfur has only six electrons, even though all available electrons have been used. If one nonbonding pair of any one of the three O atoms is moved to a bonding position between the oxygen and sulfur, it will help satisfy the octet of both atoms:

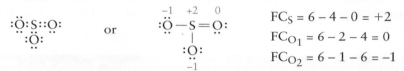

:Ö:S::O:
  :Ö:

or

−1 +2 0
:Ö—S=O:
  |
  :O:
  −1

$FC_S = 6 - 4 - 0 = +2$
$FC_{O_1} = 6 - 2 - 4 = 0$
$FC_{O_2} = 6 - 1 - 6 = -1$

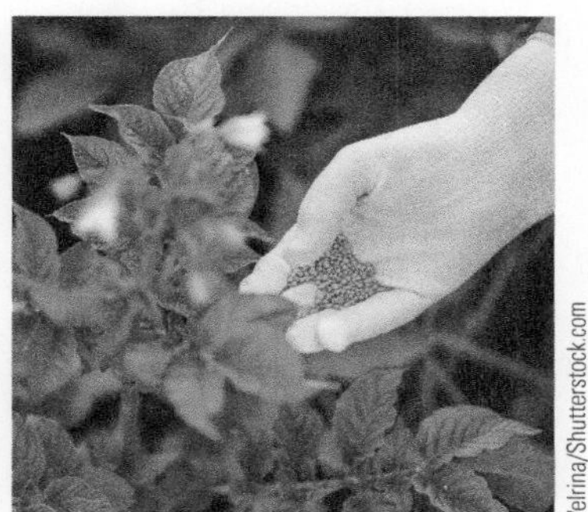

Figure 3.18 The compound SO_3 is used in fertilizer for potato plants.

Sulfur now has a full octet, but the structure is unfinished. A sulfur bonded to three atoms by two single bonds and one double bond results in a formal charge of +2, which is highly unfavorable for an electronegative atom.

Step 5b: To rid sulfur of its unfavorable formal charge, it can expand its octet and share 4 electrons with each oxygen, resulting in 12 shared electrons in a total of three double bonds. By expanding sulfur's octet, it now has a neutral formal charge. All three oxygen atoms have neutral formal charges, too, and the octet rule is satisfied for each of them. Note that the sum of the formal charges in both steps 5a and 5b match the overall charge of zero on SO_3 (**Figure 3.18**).

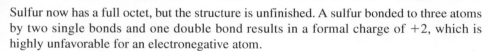

Ö::S::Ö
  :Ö:

or

  0
  :O:
  ‖
  S 0
 // \\
:O: :O:
0   0

$FC_S = 6 - 6 - 0 = 0$
$FC_O = 6 - 2 - 4 = 0$

✔ **Learning Check 3.8** Draw Lewis structures for the following molecules.

a. PF_5 (Each F atom is bonded to the P atom.)
b. H_2SO_4 (Each O atom is bonded to the S atom, and two O atoms have one H atom bonded on the exterior.)

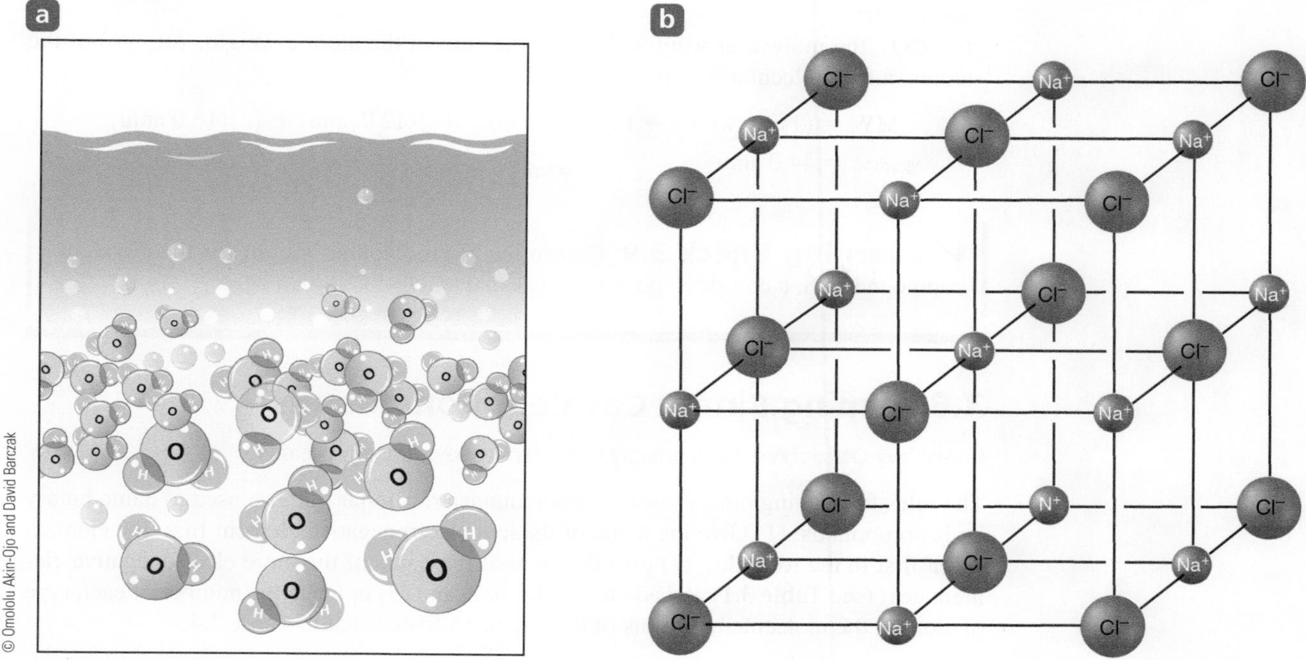

Figure 3.19 (a) Covalently bonded water molecules associate with one another in space in a less organized way than (b) the crystal lattice of ionic compounds like NaCl. Nevertheless, relatively weak attractive forces exist between the water molecules (see Chapter 5).

Covalent bonds give rise to discrete molecules that associate with each other in whatever phase they exist (**Figure 3.19a**). They are not arranged in an organized crystal lattice like ionic compounds (**Figure 3.19b**). We discuss the forces that attract covalent molecules to one another in Chapter 5.

Much as we saw with the formula weight of ionic compounds (**Section 3.3**), we can quantify the mass of covalently bonded molecules. When the atomic weights of the elements shown in the molecular formula of a covalent compound are added, the total is called the **molecular weight**.

molecular weight The sum of the atomic weights of the elements in the molecular formula of a covalent compound.

Example 3.9 Calculating Molecular Weights

CO_2 is a covalent compound that is produced and exhaled as a part of human respiration (**Figure 3.20**). Determine the molecular weight for CO_2.

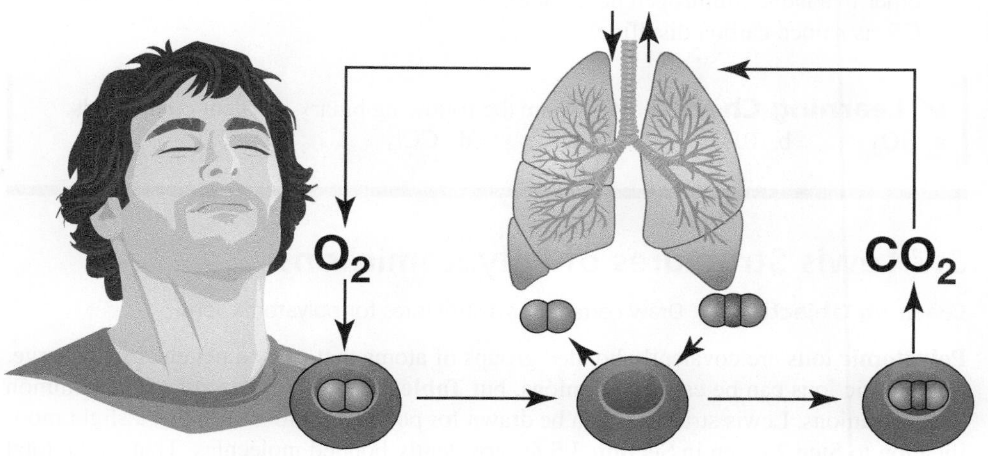

Figure 3.20 Humans breathe in oxygen (O_2), which is used up in cellular respiration. One of the by-products is carbon dioxide (CO_2) (Chapter 15).

Solution

For CO_2, the molecular weight (MW) is the sum of the atomic weights (at. wt.) of the atoms in the molecular formula:

$$MW = (1)(\text{at. wt. C}) + (2)(\text{at. wt. O}) = (1)(12.0\text{ amu}) + (2)(16.0\text{ amu})$$
$$= 44.0\text{ amu}$$

✔ **Learning Check 3.9** Determine the molecular weight for the covalent compound H_2S, a colorless, poisonous gas that smells like rotten eggs.

3.6 Naming Binary Covalent Compounds

Learning Objective 10 Correctly name binary covalent compounds.

The rules for naming binary covalent compounds are similar to those used to name binary ionic compounds: (1) Give the name of the less electronegative element first (the element given first in the formula), (2) give the stem of the name of the more electronegative element next (see **Table 3.1**) and add the suffix *-ide,* and (3) indicate the number of each type of atom in the molecule by means of the Greek prefixes listed in **Table 3.4**.

Table 3.4 Greek Numerical Prefixes

Number	Prefix
1	*mono-*
2	*di-*
3	*tri-*
4	*tetra-*
5	*penta-*
6	*hexa-*
7	*hepta-*
8	*octa-*
9	*nona-*
10	*deca-*

Example 3.10 Naming Binary Covalent Compounds

Name the following binary covalent compounds.

a. CO_2 b. CO c. NO_2 d. N_2O_5 e. CS_2

Solution

a. The elements are carbon (C) and oxygen (O), and carbon is less electronegative than oxygen (see **Figure 3.13**). Because the stem of oxygen is *ox-,* the two portions of the name will be carbon and oxide. Only one C atom is found in the molecule, but the prefix *mono-* is not used when it appears at the beginning of a name. The two O atoms in the molecule are indicated by the prefix *di-*. The correct name, therefore, is carbon dioxide.
b. Applying similar logic to CO, the correct name is carbon monoxide. Patients with carbon monoxide poisoning are often treated with hyperbaric oxygen therapy (**Figure 3.21**).
c. Nitrogen is less electronegative than oxygen, and there are two oxygen atoms, so the correct name for NO_2 is nitrogen dioxide.
d. N_2O_5 is the first example where the less electronegative element (N) needs a Greek prefix (*di-*, in this case, because there are two nitrogens). The correct name, then, is dinitrogen pentoxide. *Note:* The *a* is dropped from the *penta-* prefix for oxygen; this is done to avoid the pronunciation problem created by having two vowels next to each other in a name (dinitrogen pentoxide).
e. CS_2 is named carbon disulfide.

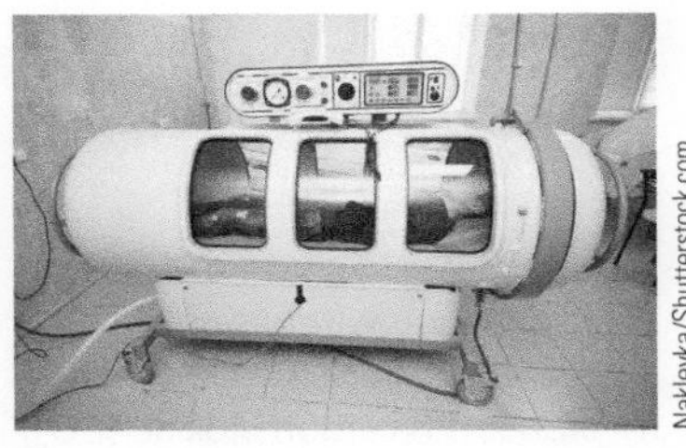

Nakleyka/Shutterstock.com

Figure 3.21 Hyperbaric oxygen therapy is often used to treat patients with carbon monoxide poisoning.

✔ **Learning Check 3.10** Name the following binary covalent compounds.

a. SO_3 b. BF_3 c. S_2O_7 d. CCl_4

3.7 Lewis Structures of Polyatomic Ions

Learning Objective 11 Draw correct Lewis structures for polyatomic ions.

polyatomic ions Covalently bonded groups of atoms that carry a net electrical charge.

Polyatomic ions are covalently bonded groups of atoms that carry a net electrical charge. Polyatomic ions can be cations or anions, but **Table 3.5** shows that most of the common ones are anions. Lewis structures can be drawn for polyatomic ions with just a slight modification to Step 2 given in **Section 3.5** for covalently bonded molecules. That is, the total number of electrons available, which is obtained by summing the number of valence-shell electrons contained in the atoms of the ion, does not account for the charge on the ion. For anions, we must *add* one electron to this number for each *negative* charge. For cations, we must *subtract* one electron for each *positive* charge.

For the SO_4^{2-} ion, for example, the total number of valence-shell electrons is 30, with six coming from the sulfur atom and six from each of the four oxygen atoms. To this number we must add two, because two additional electrons are required to give the group of neutral atoms the charge of 2− found on the ion. This gives 32 total electrons that must be accounted for in its Lewis structure.

Table 3.5 Some Common Polyatomic Ions

Very Common		Common	
NH_4^+	ammonium	CrO_4^{2-}	chromate
$C_2H_3O_2^-$	acetate	$Cr_2O_7^{2-}$	dichromate
CO_3^{2-}	carbonate	NO_2^-	nitrite
ClO_3^-	chlorate	MnO_4^-	permanganate
CN^-	cyanide	SO_3^{2-}	sulfite
HCO_3^-	hydrogen carbonate (bicarbonate)	ClO^-	hypochlorite
OH^-	hydroxide	HPO_4^{2-}	hydrogen phosphate
NO_3^-	nitrate	$H_2PO_4^-$	dihydrogen phosphate
PO_4^{3-}	phosphate	HSO_4^-	hydrogen sulfate (bisulfate)
SO_4^{2-}	sulfate	HSO_3^-	hydrogen sulfite (bisulfite)

For the positively charged ammonium ion (NH_4^+), on the other hand, the number of electrons available is the total number of valence electrons $[5 + 4(1) = 9]$ minus one (for a total of 8), because one electron has to be removed from the neutral group of atoms to produce the 1+ charge on the ion.

Example 3.11 Lewis Structures for Polyatomic Ions

Sulfate ions (SO_4^{2-}) in combination with zinc (zinc sulfate) are used medically as a dietary supplement for treating zinc deficiency (**Figure 3.22**).

Draw Lewis structures for the following polyatomic ions. The connecting patterns of the atoms to each other are indicated.

a. SO_4^{2-} (Each O atom is bonded to the S atom.)
b. NO_3^- (Each O atom is bonded to the N atom.)

Solution a.

The only difference between drawing Lewis structures for molecules and polyatomic ions comes in Step 2, where the total number of valence-shell electrons is determined.

Step 1: The formula indicates that the ion contains one sulfur (S) atom and four oxygen (O) atoms.

Step 2: Both the S and O atoms are in Group VIA(16), so each atom contributes six valence electrons. The total number of valence-shell electrons from the atoms is therefore $5 \times 6 = 30$. However, the ion has a 2− charge, so two additional electrons must be included in the total valence-shell electrons. The total number of electrons, then, is $30 + 2 = 32$.

Step 3: The given bonding relationships lead to the following initial structure:

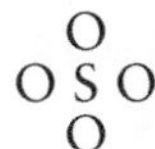

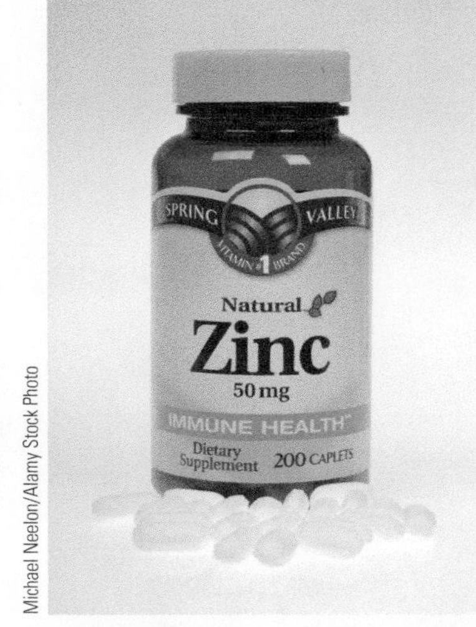

Michael Neelon/Alamy Stock Photo

Figure 3.22 Zinc sulfate is used as a dietary supplement to support immune system function.

Step 4: The following is obtained when the 32 electrons are distributed among the atoms of the initial structure of Step 2:

$$:\ddot{O}:\ddot{S}:\ddot{O}: \text{ (with } :\ddot{O}: \text{ above and below S)} \quad \text{or} \quad {}^{-1}:\ddot{O}-\overset{+2}{S}-\ddot{O}:^{-1} \text{ (with } {}^{-1}:\ddot{O}: \text{ single-bonded above and below S)}$$

$$FC_S = 6 - 4 - 0 = +2$$
$$FC_O = 6 - 1 - 6 = -1$$

Step 5: All octets are satisfied in the Lewis structure shown in Step 4, but the formal charges suggest it can be improved. A sulfur bonded to four atoms via single bonds has a formal charge of +2. Each attached oxygens has a formal charges of −1, giving the entire polyatomic ion a correct net charge of −2. The +2 charge on sulfur is highly unfavorable for an electronegative atom, however, so sulfur will expand its octet and share four electrons with two oxygens and two electrons with two oxygens, resulting in 12 shared electrons in total between sulfur and the four oxygen atoms:

$$\left[:\ddot{O}:\ddot{S}:\ddot{O}: \text{ (with } :O: \text{ above and below S)}\right]^{2-} \quad \text{or} \quad \left[{}^{-1}:\ddot{O}-S-\ddot{O}:^{-1} \text{ (with } \overset{0}{:\ddot{O}:} \text{ double-bonded above and below S, S}^{0}\text{)}\right]^{2-}$$

$$FC_S = 6 - 6 - 0 = 0$$
$$FC_{O_1} = 6 - 2 - 4 = 0$$
$$FC_{O_2} = 6 - 1 - 6 = -1$$

This leaves two sulfur–oxygen single bonds and two sulfur–oxygen double bonds. It does not matter which oxygens have the single bonds and which have the double bonds. This final Lewis structure also gives the minimum number of charged atoms and places negative charges on the most electronegative atoms (i.e., the oxygen atoms). It is conventional to enclose Lewis structures of polyatomic ions in brackets and indicate the net charge on the ion.

Solution b.

Step 1: Three oxygen (O) atoms and one nitrogen (N) atom are found in the nitrate ion (**Figure 3.23**).

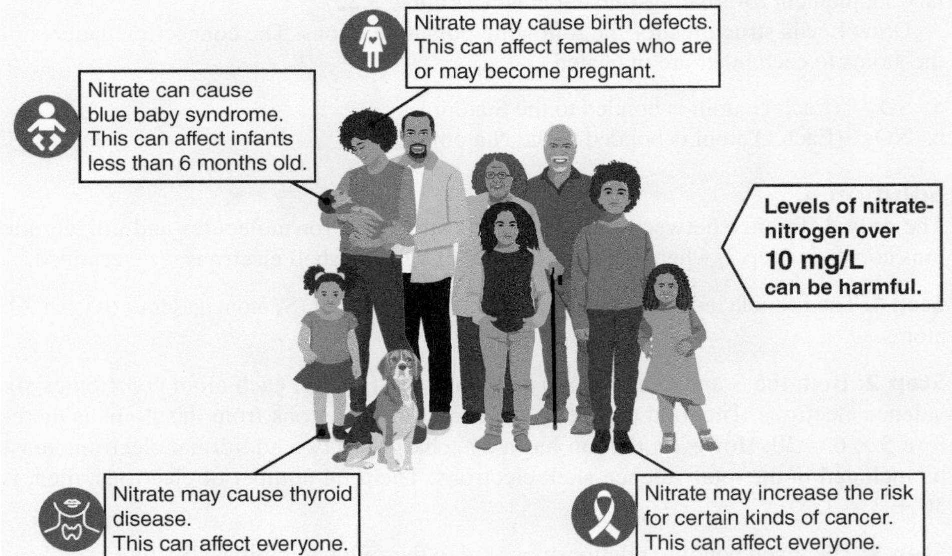

Figure 3.23 High levels of nitrate in drinking water can lead to a variety of harmful health conditions.

Step 2: Electrons from oxygen = 3 × 6 = 18.

Electrons from nitrogen = 1 × 5 = 5.

One electron comes from the ionic charge = 1.

Total electrons = 18 + 5 + 1 = 24.

Step 3:

```
  O
O N O
```

Step 4:

```
   ··
  :O:
 ·· ·· ··
:O:N:O:
 ··    ··
```

Step 5: The octet of nitrogen is not satisfied by Step 4, so an unshared pair of electrons on an O atom must be shared with the N atom. This will form a double bond in the ion:

```
[   ··   ]⁻           [       -1      ]⁻
[  :O:   ]            [       ··      ]
[ ·· ··  ]     or     [      :O:      ]
[:O:N::O ]            [  -1   |    0  ]
[ ··  ·· ]            [ :O — N = O:   ]
                      [  ··  +1   ··  ]
```

In this Lewis structure, the N atom has a +1 formal charge, but it cannot be reduced by expanding the octet because nitrogen is in Period 2 of the periodic table, and Period 2 elements can accommodate no more than 8 electrons in their valence shells.

> ✔ **Learning Check 3.11** Draw Lewis structures for the following polyatomic ions.
>
> a. PO_4^{3-} (Each O atom is bonded to the P atom.)
>
> b. NH_4^{+} (Each H atom is bonded to the N atom.)

3.8 Compounds Containing Polyatomic Ions

Learning Objective 12 Write correct formulas for ionic compounds containing representative metals and polyatomic ions.

Learning Objective 13 Correctly name binary ionic compounds containing polyatomic ions.

Using **Table 3.5**, the formulas and names for compounds containing polyatomic ions can be written. The rules are essentially the same as those used previously for binary ionic compounds (**Section 3.4**). In the formulas, the metal (or ammonium ion) is written first, followed by the polyatomic anion. The positive and negative charges must add up to zero, and parentheses are used around the polyatomic ions if more than one of a certain kind is present. For example, barium hydroxide is $Ba(OH)_2$. In the names, the positive metal (or ammonium) ion is given first, followed by the name of the negative polyatomic ion (see **Figure 3.24**). No numerical prefixes are used except where they are a part of the polyatomic ion name. Names and formulas of acids (compounds in which hydrogen is bound to simple or polyatomic ions) are discussed in Chapter 7.

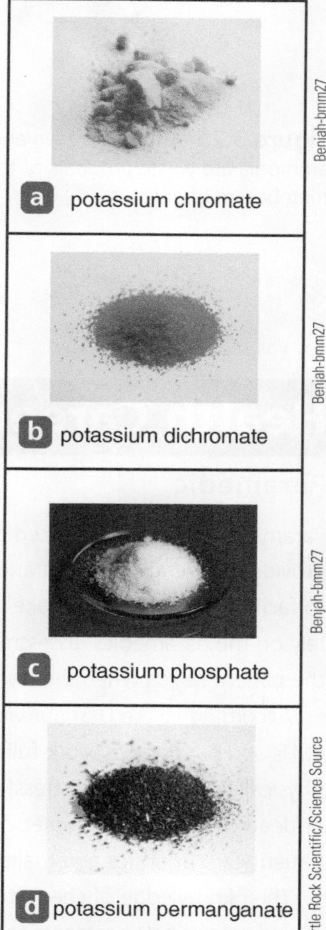

Figure 3.24 Examples of potassium compounds that contain polyatomic ions, resulting in strikingly different colors. The samples are shown as solids. Referring to Table 3.5, write formulas for: (a) potassium chromate, (b) potassium dichromate, (c) potassium phosphate, and (d) potassium permanganate.

Example 3.12 Writing Compound Formulas and Names

Ammonium ions (NH_4^+) (**Figure 3.25**) are a waste product of the metabolism of animals. Fish excrete it directly into water, while in mammals it is converted into less toxic urea.

Write formulas and names for compounds composed of ions of the following metals and the polyatomic ions indicated.

a. Na and NO_3^-	c. K and HPO_4^{2-}
b. Ca and ClO_3^-	d. NH_4^+ and NO_3^-

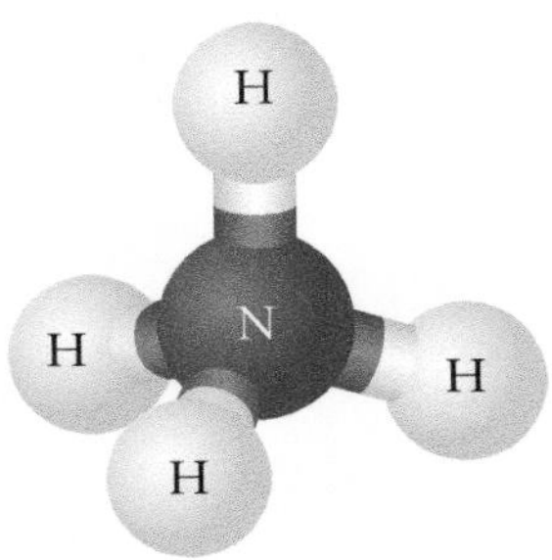

ammonium ion

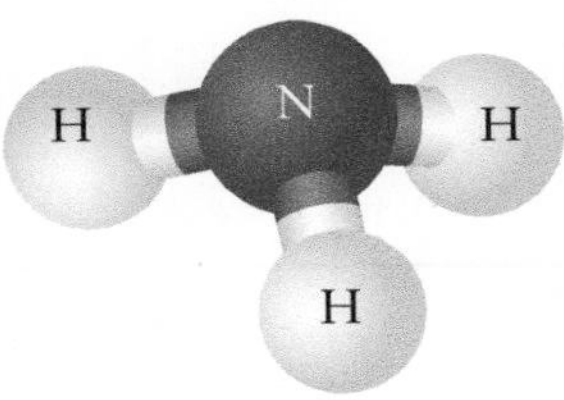

ammonia

Figure 3.25 Ammonium ion and ammonia are waste products of the metabolism of animals.

Solution

a. Sodium (Na) is a Group IA(1) metal and forms Na^+ ions. Electrical neutrality requires a combining ratio of one Na^+ for one NO_3^-. The formula is $NaNO_3$. The name is given by the metal name plus the polyatomic ion name: sodium nitrate.

b. Calcium (Ca), a Group IIA(2) metal, forms Ca^{2+} ions. Electrical neutrality therefore requires that one Ca^{2+} ion combines with two ClO_3^- ions. The formula is $Ca(ClO_3)_2$. (*Note:* The use of parentheses around the polyatomic ion prevents the confusion resulting from writing $CaClO_{32}$, which implies that there are 32 oxygen atoms in the formula. Parentheses are always used when multiples of a specific polyatomic ion are present in a formula and are indicated by a subscript.) The name is calcium chlorate.

c. Potassium (K), a Group IA(1) metal, forms K^+ ions, whereas hydrogen phosphate is a polyatomic ion with a -2 charge. Therefore, K^+ and HPO_4^{2-} combine in a 2:1 ratio, which gives a formula of K_2HPO_4. The name is potassium hydrogen phosphate.

d. The ammonium ion, NH_4^+, is not a metallic ion, but it behaves like one in numerous compounds. It combines in a 1:1 ratio with nitrate ion, NO_3^-, to give a formula of NH_4NO_3. (*Note:* The polyatomic ions are written separately, and the nitrogen atoms are not grouped to give a formula such as $N_2H_4O_3$.) The name is ammonium nitrate.

✔ **Learning Check 3.12** Write formulas and names for compounds containing ions of the following metals and the polyatomic ions indicated.

a. Ca and HPO_4^{2-}
b. Mg and PO_4^{3-}
c. K and MnO_4^-
d. NH_4^+ and $Cr_2O_7^{2-}$

Health Career Description

Paramedic

Paramedics work in fast-paced critical-care situations. They provide lifesaving treatment and stabilize patients for transport to facilities where doctors take over the patients' care. Doctors rely on the paramedics' assessment to quickly understand what the patient needs when the transport is complete.

According to the U.S. Bureau of Labor Statistics, "Most EMTs and paramedics work full time. Their work can be physically strenuous and stressful, sometimes involving life-or-death situations." Paramedics must be able to establish patient-care priorities and make quick decisions.

Their knowledge of chemistry and biochemistry is essential in understanding treatments for stabilizing patients and which medications are appropriate for various situations. A paramedic's proficiency in chemistry serves to ensure proper patient care, guard against drug interactions, and recognize potentially dangerous situations such as ketoacidosis or respiratory alkalosis.

Physical fitness is a job requirement. Paramedics are often expected to work on physical fitness at a gym during work hours. Career preparation varies from technical-college certificate programs to university associate degree programs. After completing the education for paramedic training, students will take a state licensure exam. Paramedics require state licenses in order to work—requirements vary by state. Employment opportunities for paramedics are growing at a rate of 7% per year.

Good paramedics should have excellent overall communication skills and knowledge of psychology and special populations (e.g., elderly, children, and disabled), and they should be trained in substance abuse issues. Paramedics are primary communicators with patients and their families. They're often called on to diffuse emotional situations and bring a sense of calm and confidence when a scene is fraught with fear and anxiety. Communities rely on paramedics as frontline responders to keep their people safe and secure.

Sources: AAS in Paramedic Studies, GradMaps, Weber State University. Retrieved from: https://apps.weber.edu/gradmaps/getAttachment.aspx7MAP=2703&Att=703. March 13, 2020. Bureau of Labor Statistics, U.S. Department of Labor, Occupational Outlook Handbook, EMTs and Paramedics. Retrieved from: https://www.bls.gov/ooh/healthcare/emts-and-paramedics.htm March 13, 2020.

Concept Summary

3.1 An Introduction to Lewis Structures

Learning Objective: **Draw correct Lewis dot structures for atoms of the representative elements.**

- Noble gases are stable and unreactive.
- Noble gas electron configurations usually consist of eight electrons in the valence shell.

3.2 The Formation of Ions

Learning Objectives: **Use electron configurations to determine the number of electrons gained or lost by atoms as they achieve noble gas electron configurations. Use electron configurations to determine the series of ions that are isoelectronic with the noble gases.**

- Ionic compounds are formed when reacting atoms gain or lose electrons to achieve a noble gas configuration of eight electrons in the valence shell. This is the "octet rule."
- The octet rule predicts that atoms will be changed into charged particles called simple ions in order to obtain a completely filled valence shell.
- Ions and elements with the same number of electrons are said to be isoelectronic.

3.3 Ionic Compounds

Learning Objectives: **Use the octet rule to correctly predict the ions formed during the formation of ionic compounds. Write correct formula units for ionic compounds containing a representative metal and a representative nonmetal. Determine formula weights for ionic compounds in atomic mass units.**

- Ions of opposite charge are attracted to each other.
- This attractive force is called an ionic bond.
- Oppositely charged ions group together to form compounds in ratios determined by the positive and negative charges of the ions.
- The formula units representing these ratios contain the lowest number of each ion possible in a proportion such that the total positive charges and total negative charges used are equal.
- Ionic compounds do not exist in the form of discrete molecules; instead, they form three-dimensional arrangements of oppositely charged ions called lattices. Their formula units represent the simplest whole-number ratio of ions that repeats throughout the crystal lattice.
- The sum of the atomic weights of the elements in the formula of ionic compounds is called the formula weight.

3.4 Naming Binary Ionic Compounds

Learning Objective: **Correctly name binary ionic compounds.**

- Binary ionic compounds contain a metal and a nonmetal.
- They are named by naming the metal, then adding the suffix *-ide* to the stem of the nonmetal.

3.5 Covalent Bonding

Learning Objectives: **Draw correct Lewis structures for covalent molecules. Determine molecular weights for covalent compounds in atomic mass units.**

- Elements with little or no tendency to gain or lose electrons often react and achieve noble gas electronic configurations by sharing electrons.
- Lewis structures can be used to represent electron sharing.
- Shared pairs of electrons exert an attractive force on both of the atoms that share them.
- The attractive force that holds these atoms together is called a covalent bond.
- Covalent compounds exist as discrete molecules. Their molecular formulas represent the precise number of atoms of each element present in each covalently bonded molecule.
- The sum of the atomic weights of the elements in the formula of covalent compounds is called the molecular weight.

3.6 Naming Binary Covalent Compounds

Learning Objective: **Correctly name binary covalent compounds.**

- Binary covalent compounds contain two nonmetals.
- Binary covalent compounds are named using the name of the less electronegative element first, followed by the stem of the more electronegative element plus the suffix *-ide.*
- Greek prefixes (see **Table 3.4**) are used to represent the number of each type of atom in molecules of the compounds.

3.7 Lewis Structures of Polyatomic Ions

Learning Objective: **Draw correct Lewis structures for polyatomic ions.**

- Polyatomic ions (see **Table 3.5**) are groups of two or more covalently bonded atoms that carry a net electrical charge.
- They are conveniently represented using Lewis structures.

3.8 Compounds Containing Polyatomic Ions

Learning Objectives: **Write correct formulas for ionic compounds containing representative metals and polyatomic ions. Correctly name binary ionic compounds containing polyatomic ions.**

- As with simple binary ionic compounds, compounds containing representative metal ions and polyatomic ions form in ratios determined by the positive and negative charges of the ions.
- The formulas representing these ratios contain the lowest number of each ion necessary for the total positive charges and total negative charges used to be equal.
- Ionic compounds that contain a metal ion (or ammonium ion) plus a polyatomic ion are named by first naming the metal (or ammonium ion) followed by the name of the polyatomic ion.

Key Terms and Concepts

Anion (3.2)
Binary compound (3.4)
Cation (3.2)
Covalent bond (3.5)
Covalent compound (3.5)
Crystal lattice (3.3)
Double bond (3.5)
Formula unit (3.3)
Formula weight (3.3)
Ionic bond (3.3)
Ionic compound (3.3)
Isoelectronic (3.2)
Lattice site (3.3)
Lewis dot structure (3.1)
Lewis structure (3.5)
Lone pairs (3.5)
Molecular formula (3.5)
Molecular weight (3.5)
Octet rule (3.2)
Polyatomic ion (3.7)
Simple ion (3.2)
Triple bond (3.5)

Key Equation

1. Calculation of the formal charge of an atom (**Section 3.5**)	$\text{Formal charge (FC)} = \text{number of valence } e^- - \text{number of bonds} - \text{number of lone-pair electrons}$	Equation 3.1

Exercises

Even-numbered exercises are answered in Appendix B.

Exercises with an asterisk (*) are more challenging.

An Introduction to Lewis Structures (Section 3.1)

3.1 Refer to the group numbers of the periodic table and draw Lewis dot structures for atoms of the following.

a. potassium

b. barium

c. aluminum

d. bromine

3.2 Refer to the group numbers of the periodic table and draw Lewis dot structures for atoms of the following.

a. iodine

b. strontium

c. tin

d. sulfur

3.3 For each element, write the abbreviated electron configuration and Lewis dot structure.

a. fluorine

b. element 37

c. selenium

d. silicon

3.4 For each element, write the abbreviated electron configuration and Lewis dot structure.

a. germanium

b. Cs

c. element number 49

d. calcium

3.5 Draw Lewis dot structures of the elements represented by the following condensed electron configurations.

a. $[He]2s^2 2p^2$

b. $[Ne]3s^2 3p^3$

c. $[Ne]3s^1$

d. $[Ar]5s^1$

3.6 Draw Lewis dot structures of the elements represented by the following condensed electron configurations.

a. $[He]2s^1$

b. $[Ne]3s^2 3p^5$

c. $[Ar]4s^2 3d^2$

d. $[Ar]4s^2 3d^{10} 4p^3$

3.7 Use the symbol *E* to represent an element in a general way and draw Lewis dot structures for atoms of the following.

a. Any Group IIIA(13) element

b. Any Group VIA(16) element

3.8 Use the symbol *E* to represent an element in a general way and draw Lewis dot structures for atoms of the following.

a. Any Group IIA(2) element

b. Any Group VA(15) element

The Formation of Ions (Section 3.2)

3.9 Indicate the minimum number of electrons that would have to be added and the minimum number that would have to be removed to change the electron configuration of each element listed in **Exercise 3.3** to attain a noble gas configuration.

3.10 Indicate the minimum number of electrons that would have to be added and the minimum number that would have to be removed to change the electron configuration of each element listed in **Exercise 3.4** to attain a noble gas configuration.

3.11 Using the periodic table, predict the number of electrons lost or gained by the following elements as they change into simple ions. Next, write an equation using elemental symbols, ionic symbols, and electrons to represent each change.

a. Mg

b. silicon

c. element 53

d. sulfur

3.12 Using the periodic table, predict the number of electrons lost or gained by the following elements as they change into simple ions. Next, write an equation using elemental symbols, ionic symbols, and electrons to represent each change.

a. Cs

b. oxygen

c. element number 7

d. iodine

3.13 Write the charged elemental symbol for each of the following ions.

a. A bromine atom that has gained one electron

b. A sodium atom that has lost one electron

c. A sulfur atom that has gained two electrons

3.14 Write the charged elemental symbol for each of the following ions.

a. A selenium atom that has gained two electrons

b. A rubidium atom that has lost one electron

c. An aluminum atom that has lost three electrons

3.15 Identify the element in Period 2 that would form each of the following ions. *E* is used as a general symbol for an element.

a. E^-

b. E^{2+}

c. E^{3-}

d. E^+

3.16 Identify the element in Period 3 that would form each of the following ions. *E* is used as a general symbol for an element.

a. E^{2-}

b. E^{3+}

c. E^+

d. E^-

3.17 Identify the noble gas that is isoelectronic with each of the following ions.

a. Mg^{2+}

b. Te^{2-}

c. N^{3-}

d. Be^{2+}

3.18 Identify the noble gas that is isoelectronic with each of the following ions.

a. Li^+

b. I^-

c. S^{2-}

d. Sr^{2+}

3.19 Provide the series of ions that are isoelectronic for the following noble gases.

a. Kr

b. Xe

c. Rn

3.20 How many protons and electrons are present in each ion?

a. Cu^+

b. S^{2-}

c. Sc^{3+}

Ionic Compounds (Section 3.3)

3.21 Represent the electron-transfer process when the ions of the following elements interact. Then determine the formula for the resulting ionic compound.

a. Ca and S

b. Mg and N

c. element numbers 19 and 17

3.22 Represent the electron-transfer process when the ions of the following elements interact. Then determine the formula for the resulting ionic compound.

a. Ca and Cl

b. lithium and bromine

c. element numbers 12 and 16

3.23 Write the formulas for the ionic compound formed from Ca^{2+} and each of the following ions.

a. Br^-

b. O^{2-}

c. S^{2-}

d. N^{3-}

3.24 Write the formulas for the ionic compound formed from Ba^{2+} and each of the following ions.

a. Se^{2-}

b. P^{3-}

c. I^-

d. As^{3-}

3.25 What must be the charge on the cation X in each of the following ionic compounds in order for the compound to be neutral?

a. XO

b. XBr_2

c. X_3P

d. XI

3.26 What must be the charge on the anion X in each of the following ionic compounds in order for the compound to be neutral?

a. BaX

b. MgX_2

c. Al_2X_3

d. Na_2X

3.27 Determine the formula weight in atomic mass units for each of the following ionic compounds.

a. Na_2O

b. FeO

c. PbS_2

d. $AlCl_3$

3.28 Determine the formula weight in atomic mass units for each of the following ionic compounds.

a. NaBr

b. CaF_2

c. Cu_2S

d. Li_3N

Naming Binary Ionic Compounds (Section 3.4)

3.29 Classify each of the following as a binary compound or not a binary compound.

a. HF

b. OF_2

c. H_2SO_4

d. H_2S

e. $MgBr_2$

3.30 Classify each of the following as a binary compound or not a binary compound.

a. PbO_2

b. $CuCl_2$

c. KNO_3

d. Be_3N_2

e. $CaCO_3$

3.31 Name the following metal ions.

a. Ca^{2+}

b. K^+

c. Al^{3+}

d. Rb^+

3.32 Name the following metal ions.

a. Li^+

b. Mg^{2+}

c. Ba^{2+}

d. Cs^+

3.33 Name the following nonmetal ions.

a. Cl^-

b. N^{3-}

c. S^{2-}

d. Se^{2-}

3.34 Name the following nonmetal ions.

a. Br^-

b. O^{2-}

c. P^{3-}

d. F^-

3.35 Name the following binary ionic compounds.

a. Na_2O

b. $CaCl_2$

c. Al_2S_3

d. LiF

e. Ba_3N_2

3.36 Name the following binary ionic compounds.

a. SrS

b. CaF_2

c. $BaCl_2$

d. Li_2O

e. MgO

3.37 Write formulas for the following binary ionic compounds.

a. magnesium chloride

b. lithium oxide

c. potassium fluoride

d. sodium nitride

3.38 Write formulas for the following binary ionic compounds.

a. cesium iodide

b. calcium sulfide

c. aluminum chloride

d. beryllium phosphide

3.39 Name the following binary ionic compounds, using a Roman numeral to indicate the charge on the metal ion.

a. $CrCl_2$ and $CrCl_3$

b. CoS and Co_2S_3

c. FeO and Fe_2O_3

d. $PbCl_2$ and $PbCl_4$

3.40 Name the following binary ionic compounds, using a Roman numeral to indicate the charge on the metal ion.

a. PbO and PbO_2

b. CuCl and $CuCl_2$

c. Au_2S and Au_2S_3

d. CoO and Co_2O_3

3.41 Name the binary compounds in **Exercise 3.39** by adding the endings *-ous* and *-ic* to indicate the lower and higher ionic charges of the metal ion in each pair of compounds. The non-English root for lead (Pb) is *plumb-*.

3.42 Name the binary compounds in **Exercise 3.40** by adding the endings *-ous* and *-ic* to indicate the lower and higher ionic charges of the metal ion in each pair of compounds. The non-English root for gold (Au) is *aur-*, and that of lead (Pb) is *plumb-*.

3.43 Write formulas for the following binary ionic compounds.

a. manganese(II) chloride

b. iron(III) sulfide

c. chromium(II) oxide

d. iron(II) bromide

e. tin(II) chloride

3.44 Write formulas for the following binary ionic compounds.

a. mercury(I) oxide

b. lead(II) oxide

c. platinum(IV) iodine

d. copper(I) nitride

e. cobalt(II) sulfide

Covalent Bonding (Section 3.5)

3.45 Represent the following reaction using Lewis structures.

$$I + I \rightarrow I_2$$

3.46 Represent the following reaction using Lewis structures.

$$8S \rightarrow S_8 \qquad \text{(The atoms form a ring.)}$$

3.47 Represent the following molecules by Lewis structures.

a. HF

b. IBr

c. PH_3 (Each H atom is bonded to the P atom.)

d. $HClO_2$ (The O atoms are each bonded to the Cl, and the H is bonded to one of the O atoms.)

3.48 Represent the following molecules by Lewis structures.

a. CH_4 (Each H atom is bonded to the C atom.)

b. CO_2 (Each O atom is bonded to the C atom.)

c. H_2Se (Each H atom is bonded to the Se atom.)

d. NH_3 (Each H atom is bonded to the N atom.)

3.49 Represent the following molecules by Lewis structures.

a. CCl_2O (Each Cl and O atom is bonded to the C atom.)

b. SiF_4 (Each F atom is bonded to the Si atom.)

c. PF_3 (Each F atom is bonded to the P atom.)

d. C_2H_2 (Each H atom is bonded to one C atom.)

3.50 Draw correct Lewis structures for the following compounds and use formal charges to explain why they violate the octet rule.

a. BCl_3 (Each Cl atom is bonded to the B atom.)

b. SF_6 (Each F atom is bonded to the S atom.)

c. SO_2 (Each O atom is bonded to the S atom.)

d. SeO_3 (Each Se atom is bonded to the Se atom.)

3.51 Draw Lewis structures for the following molecules.

a. O_3 (The O atoms are bonded together, like beads on a string.)

b. CS_2 (Each S atom is bonded to the C atom.)

c. SeO_2 (Each O atom is bonded to the Se atom.)

3.52 Determine the molecular weight in atomic mass units for each of the following binary compounds.

a. Chlorine dioxide, used to disinfect water.

b. Dinitrogen monoxide, "laughing gas" anesthetic.

c. Sulfur dioxide, used to preserve dried foods.

d. Carbon tetrachloride, once a popular solvent, now phased out because of health concerns.

3.53 Determine the molecular weight in atomic mass units for each of the following binary compounds.

a. Sulfur hexafluoride, an ultrasound contrast agent

b. Silicon tetrachloride, an intermediate in the production of solar cells

c. Dinitrogen trioxide, a toxic gas

d. Sulfur trioxide, used in making sulfuric acid and explosives

3.54 Determine the molecular weight in atomic mass units for each of the following covalent compounds.

a. propane gas (C_3H_8)

b. carbon monoxide (CO)

3.55 Determine the molecular weight in atomic mass units for each of the following covalent compounds.

a. glucose ($C_6H_{12}O_6$)

b. the anesthetic lidocaine ($C_{14}H_{22}N_2O$)

c. the amino acid methionine ($C_5H_{11}NO_2S$)

Naming Binary Covalent Compounds (Section 3.6)

3.56 Name the following binary covalent compounds.

a. PCl_3, used in making plasticizers and pesticides

b. N_2O, "laughing gas" anesthetic

c. CCl_4, once a popular solvent, now phased out because of health concerns

d. BF_3, a polymerization reaction catalyst

e. CS_2, used as an insecticide fumigant for grains and cereal mills

3.57 Name the following binary covalent compounds.

a. SiO_2, has many uses in tablet making in the pharmaceutical industry

b. SiF_4, found in volcanic plumes

c. B_2O_3, used in making borosilicate glass

d. NO, a vasodilator

e. CBr_4, can damage liver and kidneys

3.58 For each of the following compounds, indicate whether it is covalent or ionic and give the correct name.

a. SO_2

b. $MgBr_2$

c. BaS

d. N_2O_4

3.59 For each of the following compounds, indicate whether it is covalent or ionic and give the correct name.

a. CS_2

b. SiF_4

c. Cu_2O

d. NaF

Lewis Structures of Polyatomic Ions (Section 3.7)

3.60 Draw Lewis structures and provide names for the following polyatomic ions.

a. SO_3^{2-} (Each O atom is bonded to the S atom.)

b. SO_4^{2-} (Each O atom is bonded to the S atom.)

3.61 Draw Lewis structures and provide names for the following polyatomic ions.

a. NO_3^- (Each O atom is bonded to the N atom.)

b. NO_2^- (Each O atom is bonded to the N atom.)

3.62 Draw Lewis structures for the following polyatomic ions.

a. NH_4^+ (Each H atom is bonded to the N atom.)

b. PO_4^{3-} (Each O atom is bonded to the P atom.)

c. SO_3^{2-} (Each O atom is bonded to the S atom.)

3.63 Draw Lewis structures for the following polyatomic ions.

a. Hypochlorite (ClO^-) is the anion in bleach (NaClO), a common disinfectant. In ClO^-, each O atom is bonded to the Cl atom.

b. The cyanide ion (CN^-) is the molecule responsible for cyanide poisoning in human and animal patients.

c. Bicarbonate (HCO_3^-) plays a key role in the blood buffering system. In HCO_3^-, each O atom is bonded to the C atom, and one oxygen is also bonded to an H.

Compounds Containing Polyatomic Ions (Section 3.8)

3.64 Write a formula for the following compounds, using *M* with appropriate charges to represent the metal ion.

a. Any Group IA(1) element and SO_3^{2-}

b. Any Group IA(1) element and $C_2H_3O_2^-$

c. Any Group IIA(2) element and $Cr_2O_7^{2-}$

d. Any Group IIIA(3) element and PO_4^{3-}

e. Any Group IIIA(3) element and NO_3^-

3.65 Write a formula for the following compounds, using any appropriate metal with the following polyatomic ions specified.

a. Any metal that forms M^+ ions and SO_4^{2-}

b. Any metal that forms M^{3+} ions and OH^-

c. Any metal that forms M^{2+} ions and HPO_4^{2-}

3.66 Write the formulas and names for compounds composed of ions of the following metals and the indicated polyatomic ions.

a. calcium and hypochlorite ion

b. cesium and nitrite ion

c. Mg and SO_3^{2-}

d. K and $Cr_2O_7^{2-}$

3.67 Write the formulas and names for compounds composed of ions of the following metals and the indicated polyatomic ions.

a. calcium and carbonate ion

b. sodium and sulfate ion

c. K^+ and PO_4^{3-}

d. Mg^{2+} and NO_3^-

3.68 Write formulas for the following compounds.

a. Magnesium sulfite, used to remove SO_2 gas from industrial flue gases

b. Barium hydroxide, used to titrate weak acids

c. Calcium carbonate, used as a calcium dietary supplement

d. Ammonium sulfate, used in the purification of water and vaccine preparations

e. Lithium hydrogen carbonate, used to treat bipolar disorders

3.69 Write formulas for the following compounds.

a. Potassium hydroxide, used to dissolve warts and cuticles

b. Sodium carbonate, found in some toothpastes as an abrasive

c. Ammonium chloride, used as an expectorant in cough medicine

d. Sodium phosphate, commonly used to treat constipation

e. Calcium nitrate, used in wastewater treatments

3.70 Correct the incorrect formulas for the following ionic compounds.

a. Na_2NO_3

b. CaOH

c. MgBr

d. KCO_3

3.71 Correct the incorrect formulas for the following ionic compounds.

a. $CsSO_3$

b. $LiPO_4$

c. $K(MnO_4)_2$

d. $AlClO_3$

3.72 Determine whether each of the following is a covalent or ionic compound and give the correct name.

a. $FeSO_4$, a common iron supplement for anemia

b. NO, a potent gas used to dilate blood vessels

c. CF_4, was a useful refrigerant but now is classified as a greenhouse gas

3.73 Determine whether each of the following is a covalent or ionic compound and give the correct name.

a. $NaHCO_3$, the active ingredient in baking soda

b. $CaC_2H_3O_2$, a compound used to treat patients with too much phosphate in the blood

c. KI, a compound taken to protect a patient's thyroid from radioactive materials

Chemistry for Thought

3.74 Three atoms with an electronic configuration of $1s^1$ are covalently bonded to one atom with an electronic configuration of $1s^2 2s^2 2p^3$ to form a molecule. What is the formula of this molecule?

3.75 Suppose an element from Group IIA(2) and Period 3 of the periodic table form an ionic compound with the element that has an electronic configuration of $1s^2 2s^2 2p^5$. What would be the formula and name of the compound?

3.76 Three compounds in **Figure 3.24** are highly colored (other than white). All the compounds consist of potassium and a polyatomic ion. If you have not yet done so, write formulas for the compounds and see if you can find a characteristic of the polyatomic ions of the colored compounds that are not found in the white compounds.

3.77 Refer to **Figure 3.9**. What other metals and nonmetals would you predict might react similarly?

3.78 Recall how a nonmetal atom such as chlorine changes to form a negatively charged ion. How do you think the size of a nonmetal atom and a negatively charged ion of the same nonmetal will compare?

3.79 Recall how a metal atom changes to form a positively charged metal ion. How do you think the sizes of a metal atom and a positive ion of the same metal will compare?

3.80 Below is the stimulant methamphetamine. Include all nonbonding electron pairs on atoms that contain them.

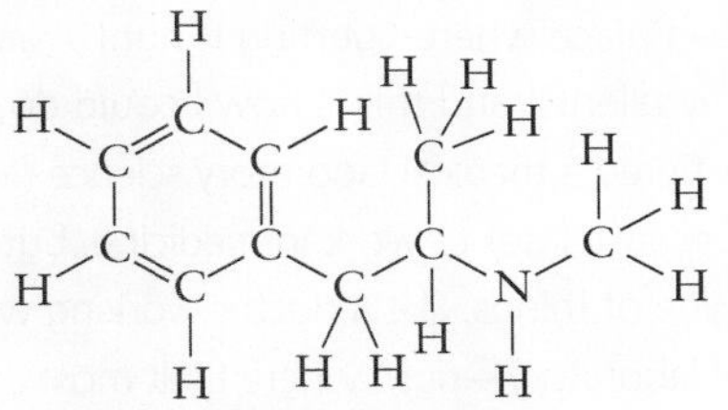

methamphetamine

3.81 Neon atoms do *not* combine to form Ne_2 molecules. Explain.

3.82 Below is the amino acid glycine. Include all nonbonding electron pairs on atoms that contain them.

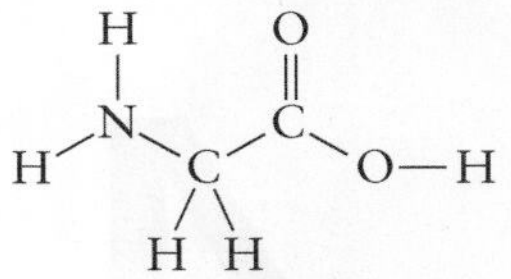

glycine

3.83 Hydrogen cyanide is a colorless gas with a bitter, almond-like odor and is extremely poisonous. Draw in all nonbonding electron pairs and calculate formal charges for each atom.

$$H-C\equiv N$$

hydrogen cyanide

3.84 Below is the compound propanoic acid, which is responsible for foot odor that smells like vinegar. Include all nonbonding electron pairs on atoms that contain them.

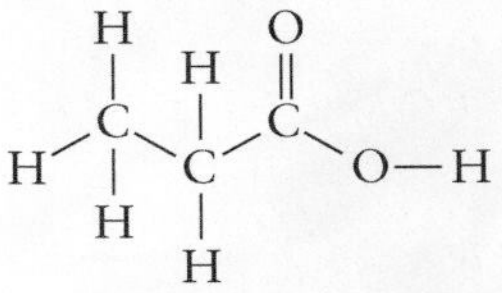

propanoic acid

4 The Mole and Chemical Reactions

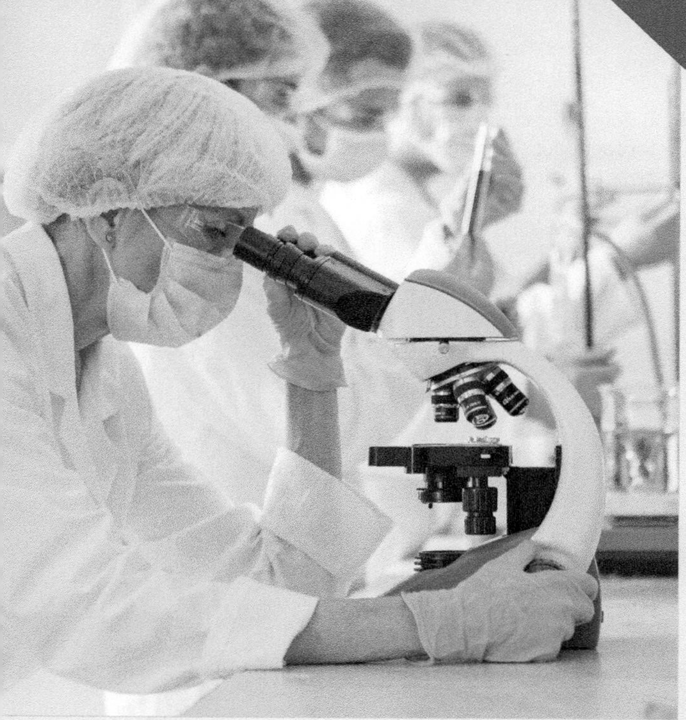

Nastasic/E+/Getty Images

Health Career Focus

Clinical Laboratory Technician

I'm naturally meticulous. Some might call me a perfectionist. I like working on puzzles and getting the right answer. I like math—the world of numbers, measurement, and predictability. I have this knack for precision. It really comes in handy at work.

I've always liked watching science and true crime shows that showed people working in laboratories wearing their protective gear, working with racks of tiny tubes and fancy droppers, and looking at slides under microscopes. Somehow, I knew a lab was the place for me—a place where "getting it right" really mattered. I wanted to use my talents, and this is how I could do it.

My university offered a medical laboratory science (MLS) major. It was a perfect fit because I got to work in medicine, but not so much on the "people" side of things, like a doctor working with patients, but in the medical laboratory—right where I felt most comfortable.

Follow-up to this Career Focus appears at the end of the chapter before the *Concept Summary*.

Learning Objectives

When you have completed your study of Chapter 4, you should be able to:

1. Convert between the number of moles, number of grams, and number of atoms of elements. **(Section 4.1)**
2. Convert between the number of moles, number of grams, number of compounds, and number of atoms in compounds. **(Section 4.2)**
3. Calculate the percent by mass of atoms in compounds. **(Section 4.2)**
4. Write a balanced chemical equation when provided with a chemical reaction. **(Section 4.3)**
5. Use the mole concept to perform calculations based on the stoichiometry of balanced chemical equations. **(Section 4.4)**
6. Use the mole concept to calculate percent yields. **(Section 4.5)**
7. Identify the oxidizing and reducing agents in oxidation–reduction (redox) reactions when given the oxidation states of each element in the reactants and products. **(Section 4.6)**
8. Classify reactions first as redox or nonredox, then as decomposition, combination, single replacement, double replacement, or combustion. **(Section 4.6)**

In Chapters 1–3, we have learned about elements, ionic compounds, covalent molecules, and chemical change. Here in Chapter 4, we explain how to better quantify elements and molecules through the use of the mole, the foundational unit of chemistry. Next, we express chemical reactions using chemical equations and apply the mole concept to balance them. Then we tie these concepts together as we use balanced chemical equations to calculate reaction yields. We conclude the chapter by discussing several of the most prominent classes of reactions that you may encounter.

4.1 Avogadro's Number: The Mole

Learning Objective 1 Convert between the number of moles, number of grams, and number of atoms of elements.

You probably know that there are 12 donuts in a *dozen*, but did you know that the "four score and seven years" in Abraham Lincoln's Gettysburg Address amounted to 87 years, because a *score* is 20? Terms like *dozen* and *score*, and *gross* (144) and *grand* (1000) are used to identify particular amounts of items. Given how small the atoms are in elements and molecules, similar terminology is needed to relate the millions and millions and millions of atoms on the micro level to quantities we can easily understand, measure, and describe on the macro level. Without it, we would be unable to quantify elements, molecules, and chemical reactions without constantly relying on numbers that are incredibly large or infinitesimally small.

The critical unit we use for this purpose in chemistry is the **mole** (**Figure 4.1**). A mole (mol) was originally defined as the same number of particles as there are atoms in 12.01 g of carbon. Since 2019, it has been defined precisely as $6.02214076 \times 10^{23}$ entities, whether those are atoms, molecules, ions, or something else. In any case, this quantity is equal to **Avogadro's number**, which is named in honor of Italian scientist Amedeo Avogadro (1776–1856), who made important contributions to the concept of atomic weights.

mole The number of particles (atoms or molecules) contained in a sample of an element or a compound with a mass in grams equal to the atomic or molecular weight, respectively. Numerically, 1 mol is equal to 6.022×10^{23} particles.

Avogadro's number The number of atoms or molecules in one mole of a substance, equal to 6.022×10^{23}.

Zephyr/Science Source

Figure 4.1 A collection of beakers, each containing one mole of a substance. One mole of any sample contains the same number of molecules or atoms, but the weight and volume are different. From left to right, the substances are sucrose, nickel(II) chloride, copper(II) sulfate, potassium permanganate, copper turnings, and iron filings.

One mole is a specific number of particles, but in chemistry it is customary to also let 1 mol stand for the mass of a sample of element or compound that contains an Avogadro's number of particles. This relationship is very helpful when we quantify elements and compounds using either their atomic weight or their molecular weight. Recall from **Section 2.4** that atomic weight is expressed in atomic mass units (amu), where 1 amu is equal to the mass of one atom of carbon-12. That probably doesn't sound very useful, but relationships introduced by Avogadro's number and the mole mean that 1 amu is actually equal to the amount (in grams) of 1 mole of an atom. For carbon, we can demonstrate this conversion as follows:

$$\frac{12.01\ \cancel{\text{amu}}}{1\ \cancel{\text{atom}}} \times \frac{1.6605 \times 10^{-24}\ \text{g}}{1\ \cancel{\text{amu}}} \times \frac{6.022 \times 10^{23}\ \cancel{\text{atoms}}}{1\ \text{mole}}$$

$$= 12.01\ \text{g/mole} = \text{molar mass}$$

molar mass The mass of 1 mol of a substance in grams.

Thus, the atomic mass of a substance (in amu) equals the mass of 1 mol of the substance (in grams), where the mass of 1 mol of a substance (in grams) is called its **molar mass**. In the development of these ideas to this point, we have used four significant figures for atomic weights, molecular weights, and Avogadro's number to minimize the introduction of rounding errors. However, in calculations throughout the remainder of the book, two decimal places are generally used because they are usually sufficient.

Example 4.1 Calculations Relating Grams, Moles, and Atoms for Sulfur

Sulfur is a vital element for all forms of life because covalent bonds between sulfur atoms help maintain the overall structure of proteins in all organisms (**Figure 4.2**). Calculate the following values for sulfur (S).

a. The mass in grams of 1.35 mol of S
b. The number of S atoms in 98.6 g of S

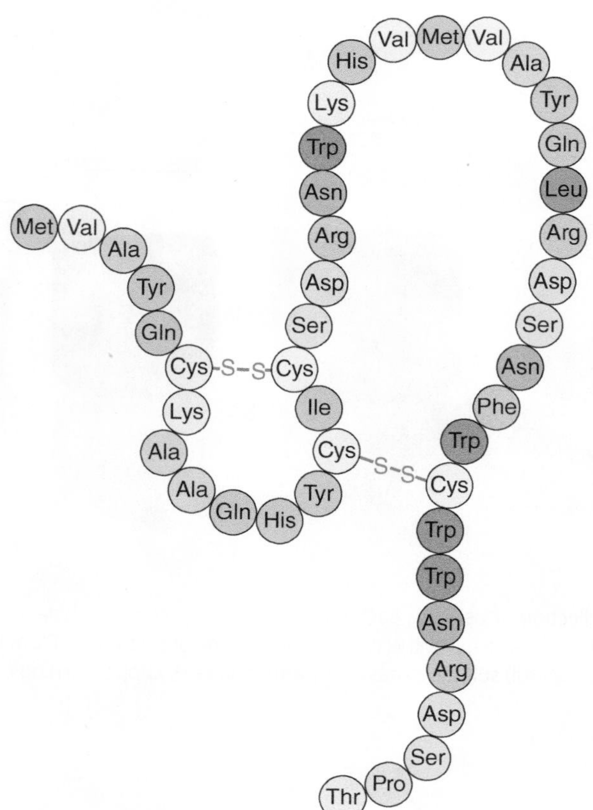

Figure 4.2 Proteins consist of amino acid subunits (circles shown in the depiction to the right). The overall structure of a protein is stabilized by covalent bonds between sulfur atoms (shown in red) protruding from amino acids called cysteine. We discuss amino acids and proteins in much greater detail in Chapter 12.

Solution

a. Use the four steps of the dimensional analysis method (presented in Study Tools 1.1). The known quantity is 1.35 mol of S, and the unit of the unknown quantity is grams of S.

Step 1: 1.35 mole S atoms

Step 2: 1.35 mole S atoms = ? g S

Step 3: $(1.35 \text{ \sout{mol S atoms}})\left(\frac{32.1 \text{ g S}}{1 \text{ \sout{mol S atoms}}}\right) = 43.3 \text{ g S}$

The conversion factor comes from the relationship 1 mol S atoms = 32.1 g S, the molar mass of sulfur.

Step 4: Check to see whether the answer makes sense.
Yes. The answer should be greater than 32.1 g S (the mass of 1 mol of S) because we began with greater than 1 mol of S.

b. The known quantity is 98.6 g S, and the unit of the unknown quantity is atoms of S.

Step 1: 98.6 g S

Step 2: 98.6 g S = ? S atoms

Step 3: $(98.6 \text{ \sout{g S}})\left(\frac{1 \text{ \sout{mol S atoms}}}{32.1 \text{ \sout{g S}}}\right)\left(\frac{6.02 \times 10^{23} \text{ S atoms}}{1 \text{ \sout{mol S atoms}}}\right) = 1.85 \times 10^{24} \text{ S atom}$

The factor comes from the relationships 32.1 g S = 1 mol S atoms, the molar mass of sulfur, and 6.02×10^{23} S atoms = 1 mol S atoms, Avogadro's number.

Step 4: Check to see whether the answer makes sense.
Yes. The answer seems reasonable because the value should be larger than Avogadro's number.

✔ **Learning Check 4.1** The element oxygen (O) is present in everything from simple compounds like water, to more complex molecules like glucose, to biopolymers such as DNA.

a. Determine the number of moles of O atoms in 98.6 g of O.
b. Determine the mass in grams of one atom of O.

4.2 The Mole and Chemical Formulas

Learning Objective 2 Convert between the number of moles, number of grams, number of compounds, and number of atoms in compounds.

Learning Objective 3 Calculate the percent by mass of atoms in compounds.

The mole concept can also be applied when the particles are molecules instead of atoms. The compound carbon dioxide (**Figure 4.3**) consists of molecules that contain one carbon atom (C) and two oxygen atoms (O). The formula for the molecule is CO_2. The molecular weight of the molecule is calculated by adding the atomic weight of one carbon atom and the atomic weight of two oxygen atoms:

$$\text{MW} = 1(\text{at. wt. C}) + 2(\text{at. wt. O}) = 1(12.0 \text{ amu}) + 2(16.0 \text{ amu}) = 44.0 \text{ amu}$$

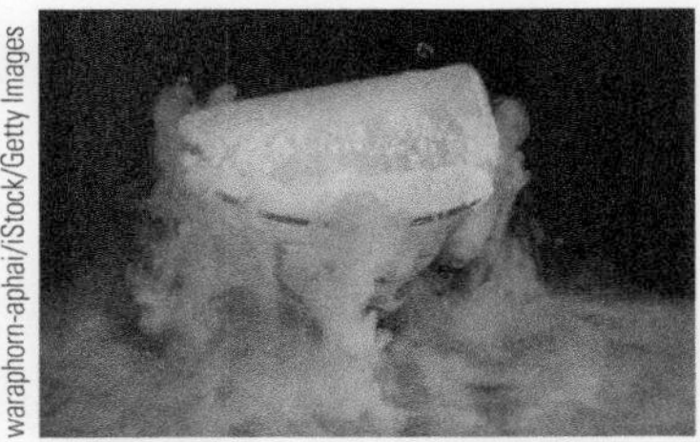

waraphorn-aphai/iStock/Getty Images

Figure 4.3 Dry ice is a solid (and very cold) form of carbon dioxide.

Applying the mole concept to CO_2 molecules gives the following relationships:

$$1 \text{ mol } CO_2 \text{ molecules} = 6.02 \times 10^{23} \text{ } CO_2 \text{ molecules} = 44.0 \text{ g } CO_2$$

These equalities lead to two conversion factors that can be used in dimensional analysis (unit conversion) calculations involving CO_2:

$$1 \text{ mol } CO_2 \text{ molecules} = 6.02 \times 10^{23} \text{ } CO_2 \text{ molecules}$$

$$1 \text{ mol } CO_2 \text{ molecules} = 44.0 \text{ g } CO_2$$

Paul Burns/DigitalVision/Getty Images

Figure 4.4 Along with water vapor, we exhale carbon dioxide.

Example 4.2 Calculations Relating Grams, Moles, and Molecules of a Compound

Atmospheric carbon dioxide (see **Figure 4.4**) is the primary carbon source for life on Earth. Use dimensional analysis and conversion factors to determine the following.

a. The mass in grams of 1.62 mol of CO_2
b. The number of CO_2 molecules in 63.9 g of CO_2

Solution

a. Use the four steps of dimensional analysis shown in Study Tools 1.1.
The known quantity is 1.62 mol of CO_2, and the unit of the unknown quantity is g of CO_2.

Step 1: 1.62 mol CO_2 molecules

Step 2: 1.62 mol CO_2 molecules = ? g CO_2

Step 3: $1.62\ \cancel{\text{mol } CO_2 \text{ molecules}} \left(\dfrac{44.0 \text{ g } CO_2}{1\ \cancel{\text{mol CO molecules}}}\right) = 71.3 \text{ g } CO_2$

The conversion factor comes from the relationship 1 mol CO_2 molecules = 44.0 g CO_2.

Step 4: Check to see whether the answer makes sense.
Yes. The answer seems reasonable because if 1 mol is 44.0 g, then 1.62 mol should be more than that.

b. The known quantity is 63.9 g of CO_2, and the unit of the unknown is molecules of CO_2.

Step 1: 63.9 g CO_2

Step 2: 63.9 g CO_2 = ? molecules CO_2

Step 3: $63.9\ \cancel{\text{g } CO_2} \left(\dfrac{1\ \cancel{\text{mol } CO_2 \text{ molecules}}}{44.0\ \cancel{\text{g } CO_2}}\right)\left(\dfrac{6.02 \times 10^{23}\ CO_2 \text{ molecules}}{1\ \cancel{\text{mol } CO_2 \text{ molecules}}}\right)$

$= 8.74 \times 10^{23}\ CO_2$ molecules

The conversion factors come from the relationships 1 mol CO_2 molecules = 44.0 g CO_2 and 6.02×10^{23} CO_2 molecules = 1 mole CO_2 molecules.

Step 4: Check to see if the answer makes sense.
Yes. The answer seems reasonable because 63.6 g is greater than 1 mol of CO_2, so the number of molecules must be greater than 6.02×10^{23} (the number of molecules in 1 mol).

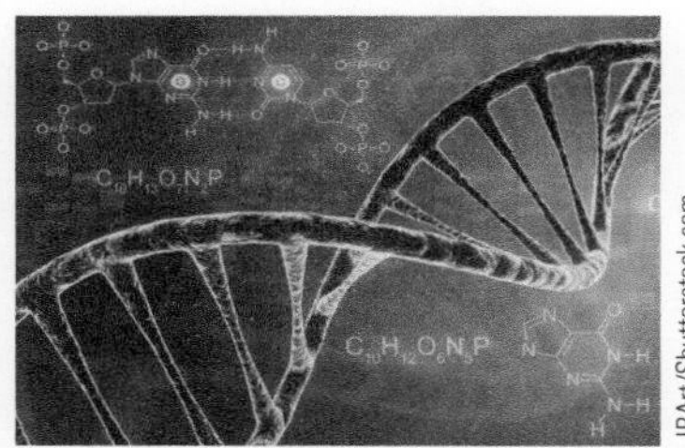
JBArt/Shutterstock.com

Figure 4.5 Oxygen is one of the essential elements in the structure of our DNA. We discuss DNA in much greater detail in Chapter 13.

✔ Learning Check 4.2 Deoxyribose is one of the components that make up DNA, the biopolymer responsible for carrying the genetic code (**Figure 4.5**).

a. Determine the number of moles of deoxyribose ($C_5H_{10}O_4$) molecules in 63.9 g of $C_5H_{10}O_4$.
b. Determine the mass in grams of one molecule of deoxyribose ($C_5H_{10}O_4$).

Recall from Chapter 3 that the formula for a compound is made up of the symbols for each element present and that subscripts following the elemental symbols indicate the number of each type of atom represented in the molecule. Thus, chemical formulas reflect the numerical (i.e., molar) relationships that exist among the atoms in a compound.

The formula for water (H_2O), for example, indicates there is a 2:1 ratio of hydrogen atoms to oxygen atoms. Because this ratio is fixed, the following statements are true:

1. 2 H_2O molecules contain 4 H atoms and 2 O atoms.
2. 10 H_2O molecules contain 20 H atoms and 10 O atoms.
3. 100 H_2O molecules contain 200 H atoms and 100 O atoms.
4. 6.02×10^{23} H_2O molecules contain 1.204×10^{24} H atoms and 6.02×10^{23} O atoms.

Statement 4 is significant because 6.02×10^{23} particles is 1 mol. Thus, Statement 4 can be restated as Statement 5:

5. 1 mol of H_2O molecules contains 2 mol of H atoms and 1 mol of O atoms.

Figure 4.6 shows another example of this concept.

Figure 4.6 Liquid carbon disulfide (CS_2) is composed of the solid elements carbon (left) and sulfur (right). How many moles of sulfur atoms would be contained in 1.5 mol of CS_2 molecules?

Example 4.3 Calculating Moles of Atoms from Moles of Compound

How many moles of each type of atom are contained in 1 mol of vitamin C ($C_6H_8O_6$)?

Solution

Each vitamin C molecule contains six C atoms, eight H atoms, and six O atoms. Therefore, 1 mol of $C_6H_8O_6$ contains 6 mol of C atoms, 8 mol of H atoms, and 6 mol of O atoms.

> ✔ **Learning Check 4.3** How many moles of each type of atom would be contained in 0.50 mol of glucose ($C_6H_{12}O_6$)?

This approach becomes even more useful when we use the mass relationships of the mole concept.

$$1\ \cancel{\text{mol H}_2\text{O}} \times \frac{1\ \cancel{\text{mol O}}}{1\ \cancel{\text{mol H}_2\text{O}}} \times \frac{16.00\text{ g O}}{1\ \cancel{\text{mol O}}} = 16.00\text{ g O}$$

and

$$1\ \cancel{\text{mol H}_2\text{O}} \times \frac{2\ \cancel{\text{mol H}}}{1\ \cancel{\text{mol H}_2\text{O}}} \times \frac{1.01\text{ g}}{1\ \cancel{\text{mol H}}} = 2.02\text{ g H}$$

Thus, Statement 5, for water, yields Statement 6:

6. 1 mol H_2O = 18.02 g H_2O.

Mass relationships such as this make it possible to calculate percent compositions. This process is shown in Example 4.4.

Example 4.4 Mass Percentage Calculations

The human body makes ammonia (NH_3) when proteins are broken down, then converts that ammonia into urea. Calculate the mass percentage of nitrogen (N) in ammonia.

Solution

Step 1: Examine the question for what is given and what is unknown.

Given: ammonia is NH_3
Unknown: mass % of N = ?

Step 2: Identify a formula that links the given and unknown terms.

$$\text{Equation 1.7: } \% = \frac{\text{part}}{\text{total}} \times 100\%$$

Step 3: Enter the data into the formula.

Use 1 mol each of NH_3 and N because the mass in grams of 1 mol of compound and the mass in grams of N in the 1 mol of compound are readily determined. One mol of NH_3 weighs 17.0 g and contains 1 mol of N atoms, which weighs 14.0 g.

$$\%\text{ N} = \frac{14.0\text{ g}}{17.0\text{ g}} \times 100\% = 82.4\%$$

Step 4: Check to see whether the answer makes sense.

Yes. Nitrogen should contribute the majority of the mass in NH_3.

> ✔ **Learning Check 4.4** Determine the mass percentage of carbon in carbon dioxide (CO_2) and carbon monoxide (CO) (**Figure 4.7**).

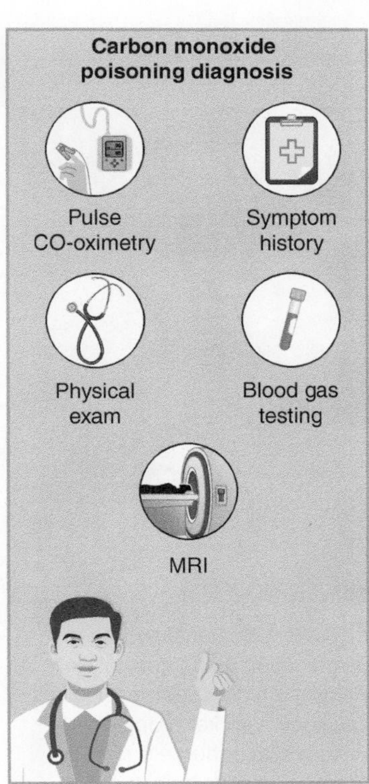

Figure 4.7 Physicians must use a combination of analysis techniques to diagnose carbon monoxide poisoning in patients.

Study Tools 4.1

Help with Mole Calculations

Problems involving the use of the mole may seem challenging at first. The good news is that problems involving the use of moles, atoms, molecules, and grams are made easier by using the dimensional analysis method. This method focuses your attention on the goal of eliminating the unit of the known, or given, quantity in order to generate the unit of the answer, or the unknown quantity. Remember, Step 1 is to write down the number and unit of the given quantity. In Step 2, write down the unit of the answer. In Step 3, multiply the known quantity by a factor whose units will cancel out the units of the known quantity and will generate the unit of the answer. In Step 4, obtain the answer by doing the required arithmetic using the numbers that were introduced in Steps 1–3. Remember, anytime you see the word *moles*, think of the method of dimensional analysis.

These steps are summarized in the flow chart to the right. Referring to it often will help you solve these kinds of problems.

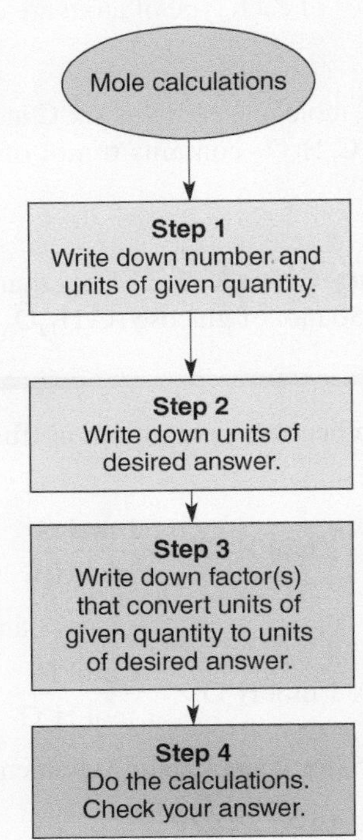
Mole calculations
Step 1
Write down number and units of given quantity.
Step 2
Write down units of desired answer.
Step 3
Write down factor(s) that convert units of given quantity to units of desired answer.
Step 4
Do the calculations. Check your answer.

NASA/Science Source

Figure 4.8 Liquid hydrogen in combination with liquid oxygen yields the highest efficiency of any known rocket propellant. Here, a space shuttle is shown about 5 seconds before launch, after the hydrogen-burning main engines have ignited but before the solid rocket boosters begin firing.

4.3 Chemical Equations

Learning Objective 4 Write a balanced chemical equation when provided with a chemical reaction.

In **Section 2.5**, we showed how to represent nuclear reactions with balanced nuclear equations. The same concepts apply to conventional chemical equations.

That is, the reactants are the substances that undergo the chemical change, and the products are the substances produced by the chemical change. By convention, the reactants are written on the *left* side of the equation, whereas the products are written on the *right* side of the equation. The reactants and products are separated by an arrow that points from the reactants to the products. A plus sign (+) is used to separate individual reactants and products.

A simple chemical reaction between elemental hydrogen and oxygen has been used to power the engines of a number of spacecraft, including the space shuttle, which was operated by the United States from 1981 to 2011 (see **Figure 4.8**). The products are water and a lot of heat. For now, let's focus only on the substances involved; we discuss the heat in Chapter 5. The reaction can be represented by the following word equation:

$$\text{hydrogen} + \text{oxygen} \rightarrow \text{water}$$

The corresponding *chemical* equation is:

$$2H_2(g) + O_2(g) \rightarrow 2H_2O(\ell)$$

In this chemical equation, the reactants and products are represented by molecular formulas that reveal much more than the names used in the word equation. Note, for example, that both hydrogen and oxygen molecules are diatomic. The equation is also consistent with a fundamental law of nature called the **law of conservation of matter**. According to this law, atoms are neither created nor destroyed in chemical reactions, but are rearranged to form new substances. Thus, atoms are conserved in chemical reactions, but molecules are not. The numbers written as coefficients to the left of the molecular formulas make the equation consistent with this law by making the total number of each kind of atom equal in the reactants and products. Note that coefficients of 1 are never written but are understood. Equations written this way are said to be **balanced equations**. In addition, the symbol in parentheses to the right of each formula indicates the state or form in which the substance exists. Thus, (g) indicates the reactants hydrogen and oxygen are both in the form of gases, and (ℓ) indicates that the product water is in the form of a liquid. Other common symbols are (s) to designate a solid and (aq) to designate a substance dissolved in water. The symbol (aq) comes from the first two letters of *aqua*, the Latin word for water.

law of conservation of matter Atoms are neither created nor destroyed in chemical reactions.

balanced equation An equation in which the number of atoms of each element in the reactants is the same as the number of atoms of each element in the products.

The coefficient 2 written to the left of H_2 means that two hydrogen molecules with the formula H_2 are reacted. Because each molecule contains two hydrogen atoms, a total of four hydrogen atoms is represented. The coefficient to the left of O_2 is 1, even though it is not written in the equation. The oxygen molecule contains two atoms, so a total of two oxygen atoms is represented. The coefficient 2 written to the left of H_2O means that two molecules of H_2O are produced. Each molecule contains two hydrogen atoms and one oxygen atom. Therefore, the total number of hydrogen atoms represented is four, and the total number of oxygen atoms is two. These are the same as the number of hydrogen and oxygen atoms represented in the reactants, so this chemical equation is balanced, and it satisfies the law of conservation of matter. When the identity and formulas of the reactants and products of a reaction are known, the reaction can be balanced by applying the law of conservation of matter.

Example 4.5 Writing Balanced Equations

Nitrogen dioxide (NO_2) is an air pollutant (**Figure 4.9**) that is produced in part when nitric oxide (NO) reacts with oxygen gas (O_2). Write a balanced equation for the production of NO_2 by this reaction.

darksoul72/Shutterstock.com

Figure 4.9 Nitrogen dioxide is produced through a variety of human activities and, per molecule, has a much more pronounced effect on warming the atmosphere than carbon dioxide does.

Solution

The reactants written to the left of the arrow are NO and O_2. The product, written to the right of the arrow, is NO_2:

$$NO(g) + O_2(g) \rightarrow NO_2(g)$$

A quick inspection shows that the reactants contain three oxygen atoms, whereas the product has just two. The number of nitrogen atoms is the same in both the reactants and the products. An incorrect approach to balancing the equation would be to change the formula of oxygen gas to O by changing the subscript:

$$NO(g) + O(g) \rightarrow NO_2(g)$$

This is *not* allowed. The natural molecular formulas of any compounds or elements *cannot* be adjusted; they are fixed by the principles of chemical bond formation described in Chapter 3. All that can be done to balance a chemical equation is to change the coefficients of the reactants and products. In this case, inspection reveals that the following is a balanced form of the equation:

$$2NO(g) + O_2(g) \rightarrow 2NO_2(g)$$

The following form of the equation is also balanced:

$$4NO(g) + 2O_2(g) \rightarrow 4NO_2(g)$$

However, only the lowest possible whole-number coefficients are used in balanced equations. Thus, $2NO(g) + O_2(g) \rightarrow 2NO_2(g)$ is the correct form.

✔ **Learning Check 4.5** Write and balance an equation that represents the reaction of (a) sulfur dioxide (SO_2) with oxygen gas (O_2) to give sulfur trioxide (SO_3) and (b) magnesium (Mg) with nitrogen gas (N_2) to form magnesium nitride (Mg_3N_2).

4.4 The Mole and Chemical Equations

Learning Objective 5 Use the mole concept to perform calculations based on the stoichiometry of balanced chemical equations.

stoichiometry The study of mass relationships in chemical reactions.

The mole concept introduced in **Section 4.1** and applied to chemical formulas in **Section 4.2** can now be used to calculate mass relationships in chemical reactions. The study of these mass relationships is called **stoichiometry**, a word derived from the Greek *stoicheion* (element) and *metron* (measure).

The chemical equations used in stoichiometry calculations must be balanced. Consider, for example, the following equation for the combustion of methane (CH_4), a type of reaction we discuss in greater detail in **Section 4.6** (**Figure 4.10**):

$$CH_4(g) + 2O_2(g) \rightarrow CO_2(g) + 2H_2O(\ell)$$

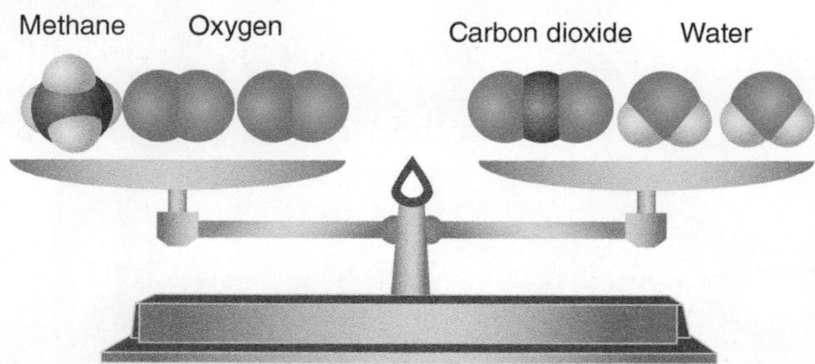

Figure 4.10 The balanced equation for the combustion of methane gas satisfies the law of conservation of mass.

Remember, coefficients in balanced equations refer to the formula that follows the coefficient. With this in mind, note that the following statement is consistent with this balanced equation:

1. 1 CH_4 molecule + 2 O_2 molecules → 1 CO_2 molecule + 2 H_2O molecules

According to this statement, one molecule of CH_4 gas reacts with two molecules of O_2 gas and produces one molecule of CO_2 gas plus two molecules of liquid H_2O. Put another way, the number of O_2 molecules that react with CH_4 will be equal to twice the number of

CH_4 molecules that react. As a result, the coefficients in balanced chemical equations can be interpreted as the number of moles of reactants and products involved in the reaction, which leads to the following statement for the reaction of CH_4 with O_2:

2. $1 \text{ mol } CH_4(g) + 2 \text{ mol } O_2(g) \rightarrow 1 \text{ mol } CO_2(g) + 2 \text{ mol } H_2O(\ell)$

In general, any balanced chemical equation can be interpreted this way, where the coefficients of the equation become the number of moles of reactants and products involved in the reaction.

Because 1 mol of any substance is equal to 6.02×10^{23} particles of the substance, the following statement is also true for the reaction of CH_4 with O_2:

3. $6.02 \times 10^{23} \text{ } CH_4 \text{ molecules} + 2(6.02 \times 10^{23}) \text{ } O_2 \text{ molecules} \rightarrow$
$6.02 \times 10^{23} \text{ } CO_2 \text{ molecules} + 2(6.02 \times 10^{23}) \text{ } H_2O \text{ molecules}$

Finally, because 1 mol of a compound has a mass in grams equal to the molecular weight of the compound, the following statement also applies to the reaction:

4. $16.0 \text{ g } CH_4 + 2(32.0) \text{ g } O_2 \rightarrow 44.0 \text{ g } CO_2 + 2(18.0) \text{ g } H_2O$

The relationships expressed in Statements 1–4 are all based on the definition of the mole. As a result, they are very useful in solving numerical problems involving balanced reaction equations and unit conversions utilizing the method of dimensional analysis described in **Section 1.8**. Any two quantities from Statements 2, 3, and 4 can be used to create conversion factors that can be used to solve problems. For example, the following factors are just four of the many that can be obtained from Statements 2, 3, and 4 by combining various quantities from the statements:

$$\frac{2 \text{ mol } O_2}{1 \text{ mol } CH_4} \qquad \frac{16.0 \text{ g } CH_4}{44.0 \text{ g } CO_2} \qquad \frac{2(18.0) \text{ g } H_2O}{6.02 \times 10^{23} \text{ } CH_4 \text{ molecules}} \qquad \frac{1 \text{ mol } CH_4}{44.0 \text{ g } CO_2}$$

Example 4.6 shows how to use these kinds of conversion factors to determine the amounts (grams or moles) of reactants required or products obtained in chemical reactions.

Example 4.6 Stoichiometric Relationships

The combustion of methane gas (**Figure 4.11**) can be represented by following balanced chemical equation:

$$CH_4(g) + 2O_2(g) \rightarrow CO_2(g) + 2H_2O(\ell)$$

a. How many moles of oxygen gas (O_2) will be required to react with 1.5 mol of methane (CH_4)?
b. How many grams of H_2O will be produced when 2.0 mol of CH_4 reacts with an excess of O_2?
c. How many grams of O_2 must react with excess CH_4 to produce 8.0 g of carbon dioxide (CO_2)?

Spencer Platt/Getty Images News/Getty Images

Figure 4.11 The combustion of methane gas, a commonly used fuel that is a major component of natural gas. Like NO_2 (**Figure 4.9**), it has a more pronounced effect on warming the atmosphere than CO_2 does.

Solution a.

Use the unit conversion method of dimensional analysis from **Section 1.8**. The steps are as follows:

1. Write down the known or given quantity. It will include both a number and a unit.
2. Leave working space, and set the known quantity equal to the unit of the unknown quantity.
3. Multiply the known quantity by a factor such that the unit of the known quantity is canceled and the unit of the unknown quantity is generated.
4. After the units on each side of the equation match, do the necessary arithmetic to get the final answer.

In this problem, the known quantity is 1.5 mol of CH_4, and the unknown quantity has the unit of mol O_2.

Step 1: 1.5 mol CH_4

Step 2: 1.5 mol CH_4 = ? mol O_2

Step 3: $1.5\ \cancel{\text{mol CH}_4} \times \dfrac{2\ \text{mol O}_2}{1\ \cancel{\text{mol CH}_4}} = ?\ \text{mol O}_2$

The factor $\dfrac{2\ \text{mol O}_2}{1\ \cancel{\text{mol CH}_4}}$ was used because it properly canceled the unit of the known and generated the unit of the unknown.

Step 4: $(1.5\ \cancel{\text{mol CH}_4})\left(\dfrac{2\ \text{mol O}_2}{1\ \cancel{\text{mol CH}_4}}\right) = 3.0\ \text{mol O}_2$

The answer contains the proper number of significant figures (two) because the 1 and 2 in the conversion factor are exact counting numbers and therefore do not affect the number of significant figures. (To review significant figures, see **Section 1.7**.)

Solution b.

Step 1: 2.0 mol CH_4

Step 2: 2.0 mol CH_4 = ? g H_2O

Step 3: $2.0\ \cancel{\text{mol CH}_4} \times \dfrac{2\ \cancel{\text{mol H}_2\text{O}}}{1\ \cancel{\text{mol CH}_4}} \times \dfrac{18.0\ \text{g H}_2\text{O}}{1\ \cancel{\text{mol H}_2\text{O}}} = ?\ \text{g H}_2\text{O}$

Step 4: $2.0\ \cancel{\text{mol CH}_4} \times \dfrac{2\ \cancel{\text{mol H}_2\text{O}}}{1\ \cancel{\text{mol CH}_4}} \times \dfrac{18.0\ \text{g H}_2\text{O}}{1\ \cancel{\text{mol H}_2\text{O}}} = 72\ \text{g H}_2\text{O}$

The answer was rounded to two significant figures to match the two significant figures in 2.0 mol CH_4. The 1 and 2 in the conversion factor are exact counting numbers and therefore do not affect the number of significant figures.

Solution c.

Step 1: 8.0 g CO_2

Step 2: 8.0 g CO_2 = ? g O_2

Step 3: $8.0\ \cancel{\text{g CO}_2} \times \dfrac{1\ \cancel{\text{mol CO}_2}}{44.0\ \cancel{\text{g CO}_2}} \times \dfrac{2\ \cancel{\text{mol O}_2}}{1\ \cancel{\text{mol CO}_2}} \times \dfrac{32.0\ \text{g O}_2}{1\ \cancel{\text{mol O}_2}} = ?\ \text{g}$

Step 4: $8.0\ \cancel{\text{g CO}_2} \times \dfrac{1\ \cancel{\text{mol CO}_2}}{44.0\ \cancel{\text{g CO}_2}} \times \dfrac{2\ \cancel{\text{mol O}_2}}{1\ \cancel{\text{mol CO}_2}} \times \dfrac{32.0\ \text{g O}_2}{1\ \cancel{\text{mol O}_2}} = 12\ \text{g O}_2$

The answer was rounded to two significant figures to match the two significant figures in 8.0. The 1 and 2 in the mole:mole conversion factor are exact counting numbers and therefore do not affect the number of significant figures in the final answer.

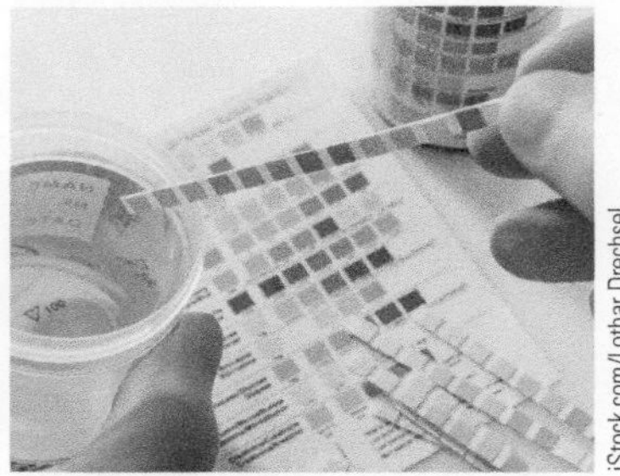

Figure 4.12 A clinical laboratory technician prepares to test the acidity of a patient's urine.

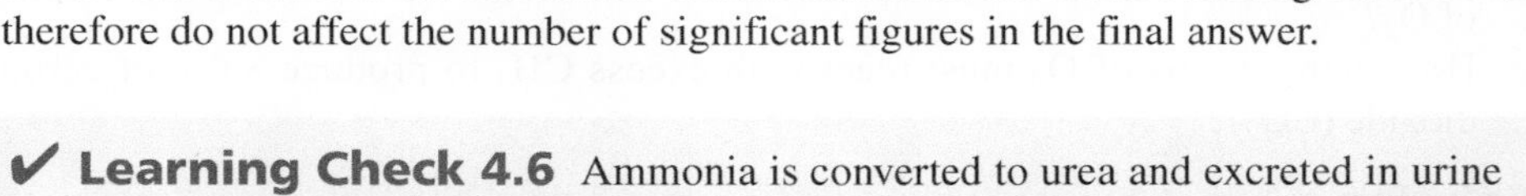

✔ Learning Check 4.6 Ammonia is converted to urea and excreted in urine in mammals (**Figure 4.12**). The balanced chemical equation for the formation of ammonia is:

$$N_2(g) + 3H_2(g) \rightarrow 2NH_3(g)$$

a. Calculate the mass of NH_3 (in g) that would be produced if 2.0 mol of H_2 were reacted with excess N_2.

b. Calculate the number of grams of N_2 required to react with 10.0 g H_2.

4.5 Reaction Yields

Learning Objective 6 Use the mole concept to calculate percent yields.

When reactions are carried out in the laboratory, less product is always obtained than the amount predicted from stoichiometric relationships like the ones calculated in Example 4.6. Does this mean that matter has been destroyed when reactions are actually done in the laboratory? It can't be because that would violate the law of conservation of matter, a fundamental law of nature. Sometimes less product is obtained because some of the reactants form compounds other than the desired product. Such reactions are called **side reactions**. Other times product is lost in the process of isolating it from the reaction mixture.

side reactions Reactions that do not give the desired product of a reaction.

The mass of product obtained in an experiment is called the *actual yield*, and the mass calculated according to the methods of **Section 4.4** is called the *theoretical yield*. The **percent yield** is the actual yield divided by the theoretical yield multiplied by 100% (see **Section 1.9**):

percent yield The percentage of the theoretical amount of a product that is actually produced by a reaction.

$$\% \text{ yield} = \frac{\text{actual yield}}{\text{theoretical yield}} \times 100\% \qquad (4.1)$$

Example 4.7 Calculating Percent Yields

As we've discovered, urea (**Figure 4.13**) is a waste product of many living organisms and is the major carbon compound in human urine. A chemist wants to produce urea (CH_4N_2O) by reacting ammonia (NH_3) and carbon dioxide (CO_2). The balanced equation for the reaction is

$$2NH_3(g) + CO_2(g) \rightarrow CH_4N_2O(s) + H_2O(\ell)$$

The chemist reacts 5.0 g of NH_3 with excess CO_2 and isolates 5.0 g of solid CH_4N_2O. Calculate the percent yield of the experiment.

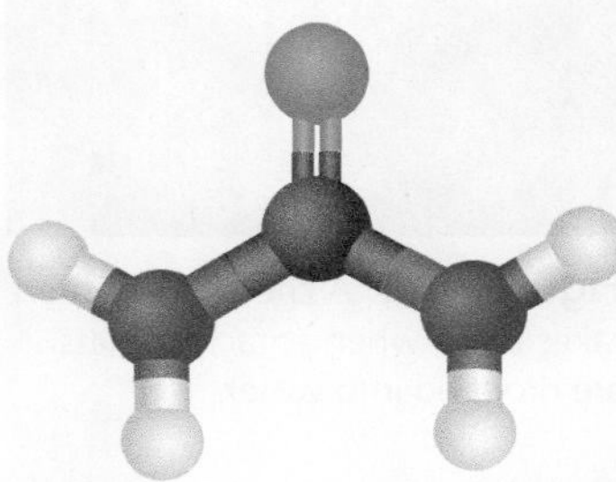

Figure 4.13 A ball-and-stick model of urea (CH_4N_2O). Notice the double bond between carbon and oxygen.

Solution

First, the theoretical yield must be calculated. Because this involves the stoichiometric ratio of reactants and products, the balanced equation will be used:

$$2NH_3(g) + CO_2(g) \rightarrow CH_4N_2O(s) + H_2O(\ell)$$

The known quantity is 5.0 g of NH_3, and the unit of the unknown quantity of CH_4N_2O is g.

Step 1: 5.0 g NH_3

Step 2: 5.0 g NH_3 = ? g CH_4N_2O

Step 3: $5.0 \text{ g } \cancel{NH_3} \times \frac{1 \text{ mol } \cancel{NH_3}}{17.0 \text{ g } \cancel{NH_3}} \times \frac{1 \text{ mol } \cancel{CH_4N_2O}}{2 \text{ mol } \cancel{NH_3}} \times \frac{60.0 \text{ g } CH_4N_2O}{1 \text{ mol } \cancel{CH_4N_2O}}$

$= 8.8 \text{ g } CH_4N_2O$

The actual yield was 5.0 g, so the percent yield is

$$\% \text{ yield} = \frac{\text{actual yield}}{\text{theoretical yield}} \times 100\% = \frac{5.0 \text{ g}}{8.8 \text{ g}} \times 100\% = 56.8\% = 57\%$$

✔ Learning Check 4.7

a. A chemist isolates 17.5 g of product from a reaction that has a calculated theoretical yield of 21.0 g. What is the percent yield?

b. Lime (CaO) is produced by heating calcium carbonate. The balanced equation for the reaction is $CaCO_3(s) \rightarrow CaO(s) + CO_2(g)$. Suppose 510.0 g of $CaCO_3$ is heated, and 235 g of CaO is isolated after the reaction mixture cools. What is the percent yield for the reaction?

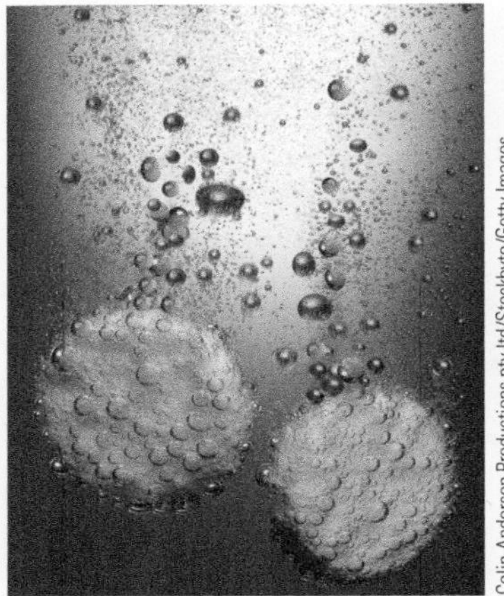

Colin Anderson Productions pty ltd/Stockbyte/Getty Images

Figure 4.14 A chemical reaction takes place when antacid tablets are dropped into water.

4.6 Types of Reactions

Learning Objective 7 Identify the oxidizing and reducing agents in oxidation–reduction (redox) reactions when given the oxidation states of each element in the reactants and products.

Learning Objective 8 Classify reactions first as redox or nonredox, then as decomposition, combination, single replacement, double replacement, or combustion.

Now that we have been able to quantify chemical reactions, let's discuss some of the many classes of chemical reactions that are most prominent, a few of which we have already seen in this chapter. A large number of chemical reactions are known to occur (**Figure 4.14**), but only a relatively small number of them are discussed in this book. Studying reactions can be made easier by classifying them based on their characteristics. We first classify reactions as being either oxidation–reduction (redox) reactions or nonredox reactions. Redox reactions are very important in numerous areas of study, including metabolism. Once reactions are classified as redox or nonredox, many can be further classified into one of several other categories, as shown in **Figure 4.15**. Notice in **Figure 4.15** that single-replacement (substitution) and combustion reactions are redox reactions, whereas double-replacement (metathesis) reactions are nonredox. Combination and decomposition reactions can be either redox or nonredox.

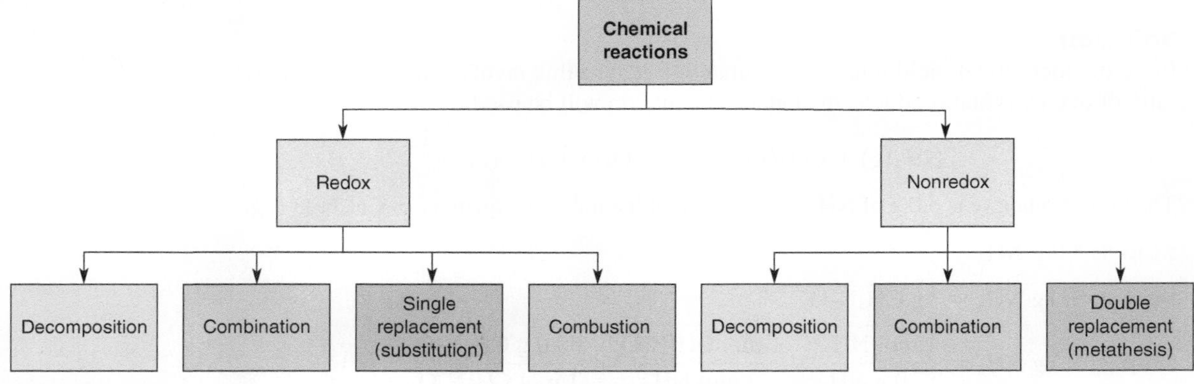

Figure 4.15 The scheme used in this text for classifying chemical reactions.

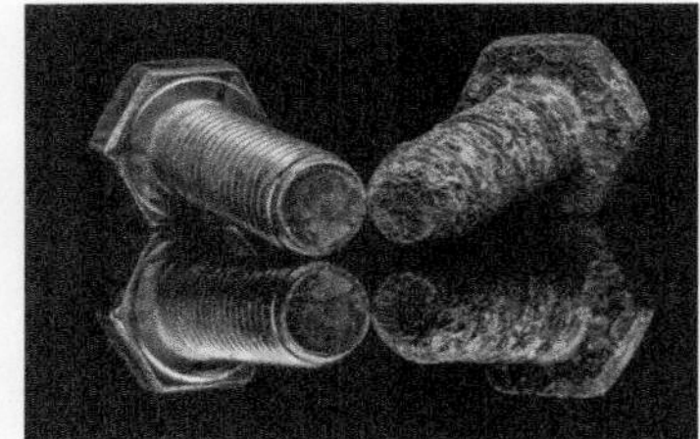

Gza Blint Ujvrosi/EyeEm/Getty Images

Figure 4.16 Corrosion (which includes the rusting of iron) costs the U.S. economy about $276 billion a year.

4.6A Redox Reactions

Almost all elements react with oxygen to form oxides. The process is so common that the word *oxidation* was coined to describe it. One example is the rusting of iron (**Figure 4.16**),

$$4Fe(s) + 3O_2(g) \rightarrow 2Fe_2O_3(s)$$

The reverse process, *reduction*, originally referred to the technique of removing oxygen from metal oxide ores to produce the free metal. An example is

$$CuO(s) + H_2(g) \rightarrow Cu(s) + H_2O(\ell)$$

Today, the words **oxidation** and **reduction** are used in a broader sense. **Table 4.1** lists most of the common meanings. To understand oxidation and reduction in terms of electron transfer or a change in oxidation number (O.N.), you must become familiar with the concept of oxidation numbers. **Oxidation numbers**, sometimes called **oxidation states**, are numbers assigned to an element in a chemical combination that represents the number of electrons lost or gained by an atom of that element in the compound. Positive oxidation numbers result when electrons are lost, and negative numbers result when electrons are gained. Let us examine the reaction of sulfur burning in oxygen represented by the following balanced chemical equation, where the oxidation numbers are indicated beneath the elements of each compound:

$$\underset{0}{S(s)} + \underset{0}{O_2(g)} \rightarrow \underset{+4\ -2}{SO_2(g)}$$

There are three indications that oxidation has taken place (see **Table 4.1**):

1. Sulfur has combined with oxygen.
2. The oxidation number of sulfur increases from 0 in S(s) to +4 in the SO_2 product.
3. This change in charge results when the sulfur atom releases four electrons that have a −1 charge each (i.e., electrons are lost).

There are two indications that reduction has taken place:

1. The oxidation number of oxygen decreases from 0 in O_2 to −2 in SO_2.
2. This change occurs when each uncharged oxygen atom in the O_2 molecule accepts two electrons (i.e., electrons are gained).

It is no coincidence that the number of electrons lost by sulfur as it was oxidized (4) is the same as the number accepted by the oxygen as it was reduced. An oxidation process is always accompanied by a reduction process, and all the electrons released during oxidation are accepted during reduction—hence the term *redox*. In redox reactions, the substance that is oxidized (that releases electrons) is called the **reducing agent** because it is responsible for reducing another substance. Similarly, the substance that is reduced (that accepts electrons) is called the **oxidizing agent** because it is responsible for oxidizing another material (see **Figure 4.17**). These characteristics are summarized in **Table 4.2**. By convention, the "agents" are always the entire reactant containing the element that is oxidized or reduced.

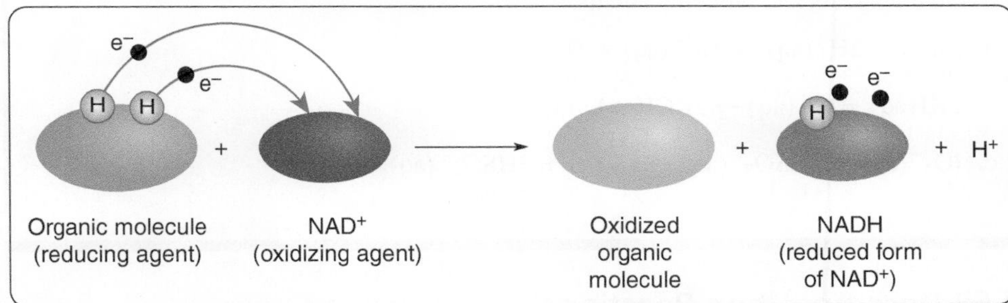

Figure 4.17 An essential step in energy production inside your body is the electron transport chain. It relies on an oxidizing agent called NAD^+ that gets reduced to a species called NADH. An organic molecule serves as the reducing agent. It gives up electrons to NAD^+ and becomes oxidized in the process.

Table 4.2 Properties of Oxidizing and Reducing Agents

Oxidizing Agent	Reducing Agent
Gains electrons	Loses electrons
Oxidation number decreases	Oxidation number increases
Becomes reduced	Becomes oxidized

In covalent substances like SO_2, the electrons are actually shared (see **Section 3.5**). The oxidation numbers of S and O are obtained by arbitrarily assigning the shared electrons of the covalent compound to the more electronegative element sharing them (i.e., O of SO_2).

oxidation Originally, the term referred to a process involving a reaction with oxygen. Today, it means a number of things, including a process in which electrons are given up, hydrogen is lost, or an oxidation number increases.

reduction Originally, the term referred to a process in which oxygen was lost. Today, it means a number of things, including a process in which electrons are gained, hydrogen is accepted, or an oxidation number decreases.

oxidation numbers or **oxidation states** The number assigned to an element within a compound after its atoms have lost or gained electrons.

Table 4.1 Common Meanings of the Terms *Oxidation* and *Reduction*

Term	Meaning
Oxidation	To combine with oxygen
	To lose hydrogen
	To lose electrons
	To increase in oxidation number
Reduction	To lose oxygen
	To combine with hydrogen
	To gain electrons
	To decrease in oxidation number

reducing agent The substance in a redox reaction that reduces another substance. The reducing agent itself is oxidized.

oxidizing agent The substance in a redox reaction that oxidizes another substance. The oxidizing agent itself is reduced.

Keep in mind, however, that none of the atoms in covalent molecules actually acquires a net charge. That is, SO_2 does not consist of sulfur and oxygen ions.

Example 4.8 Identifying Oxidizing and Reducing Agents

Use the oxidation numbers provided to identify the oxidizing agents and reducing agents.

a. $4Al(s) + 3O_2(g) \rightarrow 2Al_2O_3(s)$
   0   0   +3 −2

b. $CO(g) + 3H_2(g) \rightarrow H_2O(g) + CH_4(g)$
   +2 −2   0   +1 −2   −4 +1

c. $S_2O_8^{2-}(aq) + 2I^-(aq) \rightarrow I_2(aq) + 2SO_4^{2-}(aq)$
   +7 −2   −1   0   +6 −2

Solution

a. $4Al + 3O_2 \rightarrow 2Al_2O_3$
   0   0   +3 −2

   The O.N. of Al has changed from 0 to +3. Therefore, Al has been oxidized and is the reducing agent. The O.N. of O has decreased from 0 to −2, so oxygen has been reduced and O_2 is the oxidizing agent.

b. $CO + 3H_2 \rightarrow H_2O + CH_4$
   +2−2   0   +1−2   −4+1

   The O.N. of H has increased from 0 to +1. Hydrogen has been oxidized, so H_2 is the reducing agent. The O.N. of C has decreased from +2 to −4. Carbon has been reduced and could be called the oxidizing agent. However, when one element in a molecule or ion is the oxidizing or reducing agent, the convention is to refer to the entire molecule or ion by the appropriate term. Thus, carbon monoxide (CO) is the oxidizing agent.

c. $S_2O_8^{2-} + 2I^- \rightarrow I_2 + 2SO_4^{2-}$
   +7 −2   −1   0   +6 −2

   The O.N. of I has increased from −1 to 0, so the I^- ion is the reducing agent. The O.N. of S in $S_2O_8^{2-}$ has decreased from +7 to +6. Thus, sulfur has been reduced, and $S_2O_8^{2-}$ is the oxidizing agent.

✔ Learning Check 4.8 Use the oxidation numbers provided to identify the oxidizing agents and reducing agents.

a. $Zn(s) + 2H^+(aq) \rightarrow Zn^{2+}(aq) + H_2(g)$
   0   +1   +2   0

b. $2KI(aq) + Cl_2(aq) \rightarrow 2KCl(aq) + I_2(aq)$
   +1 −1   0   +1 −1   0

c. $IO_3^-(aq) + 3HSO_3^-(aq) \rightarrow I^-(aq) + 3HSO_4^-(aq)$
   +5 −2   +1 +4 −2   −1   +1 +6 −2

decomposition reaction
A chemical reaction in which a single substance reacts to form two or more simpler substances.

4.6B Decomposition Reactions

In **decomposition reactions**, a single substance is broken down to form two or more simpler substances. In **Figure 4.18** and other "box" representations of reactions throughout the chapter, the number of molecules in the boxes will not match the coefficients of the reaction equation. However, they will be in the correct proportions. In **Figure 4.18**, for example, the box on the left contains the same number of H_2O_2 molecules as the number of H_2O molecules on the right, as required by the balanced chemical equation beneath the space-filling models of the respective molecules. Also, the number of H_2O molecules on the right is twice the number of O_2 molecules. The general form of the equation for a decomposition reaction is

$$A \rightarrow B + C \tag{4.2}$$

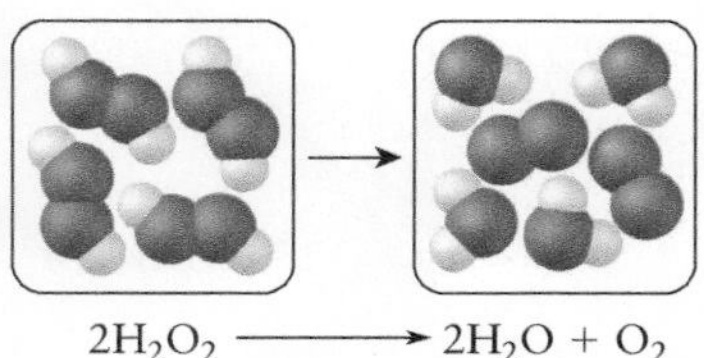

$2H_2O_2 \longrightarrow 2H_2O + O_2$

Figure 4.18 The decomposition reaction of hydrogen peroxide to form water and oxygen.

Some decomposition reactions are also redox reactions, whereas others are not. The following are examples of decomposition reactions:

$$2HgO(s) \xrightarrow{\text{heat}} 2Hg(\ell) + O_2(g)$$

$$CaCO_3(s) \xrightarrow{\text{heat}} CaO(s) + CO_2(g)$$

Health Connections 4.1

Antioxidants in Diet

Scientific evidence accumulated during the 1990s suggested that diets rich in fruits and vegetables have a protective effect against a number of different types of cancer. Studies showed that simply increasing the levels of vitamins and minerals in the diet did not provide the increased protection. This led to research into the nature of other substances found in fruits and vegetables that are important for good health. As a result of this research, a number of chemical compounds found in plants called *phytonutrients* have been shown to be involved in the maintenance of healthy tissues and organs. The mechanism for their beneficial action in the body is not understood for all phytonutrients, but a significant number are known to work as antioxidants that stop harmful oxidation reactions from occurring.

Fruits and vegetables are often rich in beneficial compounds called phytonutrients.

The first equation represents the redox reaction that takes place when mercury(II) oxide (HgO) is heated. Mercury metal (Hg) and oxygen gas (O_2) are the products. This reaction was used by Joseph Priestley (1733–1804) in 1774 when he discovered oxygen. The second equation represents a nonredox reaction used commercially to produce lime (CaO) by heating limestone ($CaCO_3$) to a high temperature. The decomposition of H_2O_2 represented in **Figure 4.18** is shown in **Figure 4.19**.

Figure 4.19 The decomposition of hydrogen peroxide. A solution of hydrogen peroxide (H_2O_2) in water (left) does not decompose rapidly until an enzyme catalyst from a piece of freshly cut potato is added (right). How can you tell that one of the products of the reaction is a gas?

4.6C Combination Reactions

combination reaction A chemical reaction in which two or more substances react to form a single substance.

Combination reactions are sometimes called *addition* or *synthesis reactions* because two or more substances react to form a single substance (see **Figure 4.20**). The reactants can be any combination of elements or compounds, but the product is always a compound. The general form of the equation is

$$A + B \rightarrow C \qquad (4.3)$$

$$H_2 + F_2 \longrightarrow 2HF$$

Figure 4.20 The combination reaction of hydrogen with fluorine to produce HF.

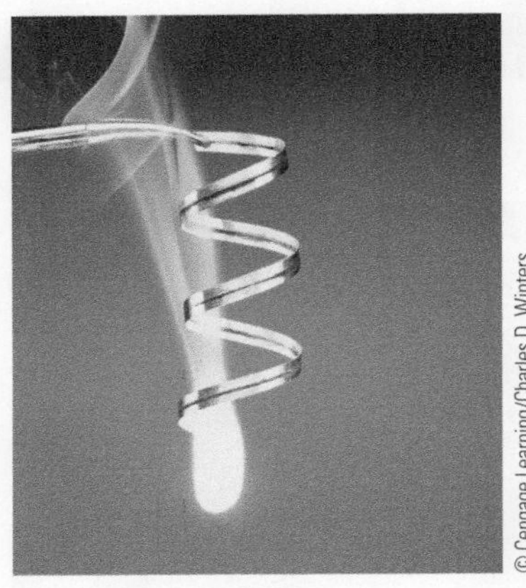

© Cengage Learning/Charles D. Winters

Figure 4.21 Magnesium metal burns in air to form magnesium oxide in a combination reaction that is also a redox reaction.

At high temperatures, a number of metals will burn and give off very bright light (**Figure 4.21**). This burning is a redox combination reaction, represented for magnesium metal by the following equation:

$$2Mg(s) + O_2(g) \rightarrow 2MgO(s)$$

A nonredox combination reaction that takes place in the atmosphere contributes to the acid rain problem. An air pollutant called sulfur trioxide (SO_3) reacts with water vapor and forms sulfuric acid. The reaction is represented by the following balanced chemical equation:

$$SO_3(g) + H_2O(\ell) \rightarrow H_2SO_4(aq)$$

4.6D Replacement Reactions

single-replacement reaction A redox chemical reaction in which an element reacts with a compound and displaces another element from the compound.

Single-replacement reactions, also called *substitution reactions*, are always redox reactions and take place when one element reacts with a compound and displaces another element from the compound. **Figure 4.22** shows the single-replacement reaction that occurs when a copper wire is immersed in an aqueous (water) solution of silver nitrate.

$$2AgNO_3(aq) + Cu(s) \rightarrow Cu(NO_3)_2(aq) + 2Ag(s)$$

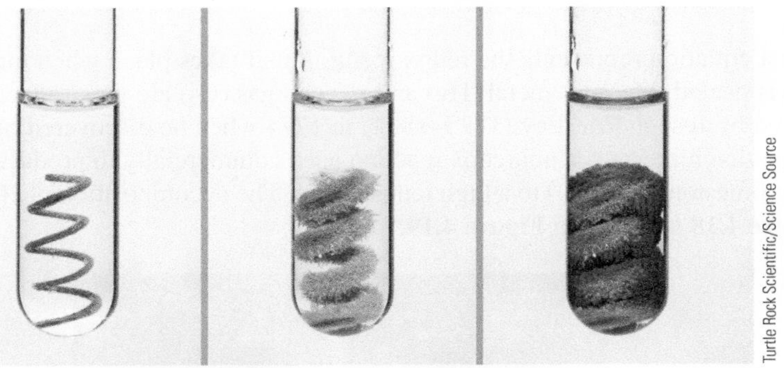

Turtle Rock Scientific/Science Source

Figure 4.22 When a piece of copper wire (Cu) is placed in a solution of silver nitrate ($AgNO_3$) in water, crystals of silver metal (Ag) form on the wire, and the liquid solution that was originally colorless turns blue as copper nitrate [$Cu(NO_3)_2$] forms in solution.

The general equation for a single-replacement reaction is shown in Equation 4.4.

$$A + BX \rightarrow B + AX \quad (4.4)$$

This type of reaction is used in a number of processes to obtain metals from their oxide ores. Iron, for example, can be obtained by reacting iron(III) oxide ore (Fe_2O_3) with carbon. The carbon displaces the iron from the oxide, and carbon dioxide is formed. The balanced chemical equation for the reaction is

$$3C(s) + 2Fe_2O_3(s) \rightarrow 4Fe(s) + 3CO_2(g)$$

double-replacement reaction A chemical reaction in which two compounds react and exchange partners to form two new compounds.

Double-replacement reactions, also called *metathesis reactions*, are never redox reactions. They take place when two compounds exchange partners to form two new compounds. These reactions, depicted in **Figure 4.23** for the formation of NaF and water from NaOH and HF, often take place between substances dissolved in water. The general form of a double-replacement reaction is shown in Equation 4.5.

$$AX + BY \rightarrow BX + AY \quad (4.5)$$

The reaction between NaOH and HCl to form NaCl and water is another example of a double-replacement reaction:

$$HCl(aq) + NaOH(aq) \rightarrow NaCl(aq) + H_2O(\ell)$$

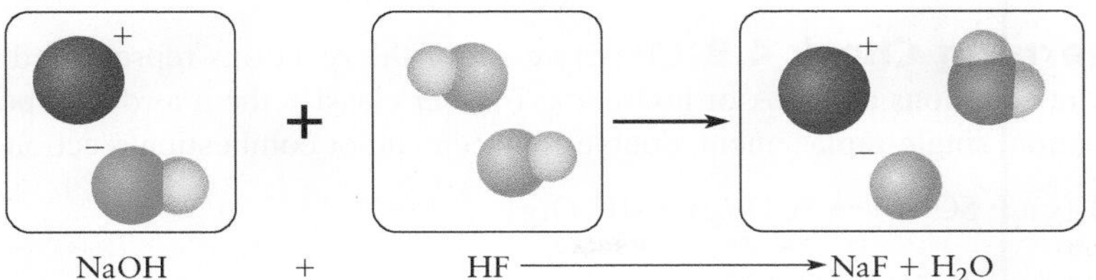

Figure 4.23 A double-replacement, or metathesis, reaction.

4.6E Combustion Reactions

combustion A chemical reaction in which hydrocarbons are oxidized in the presence of oxygen.

hydrocarbon An organic compound that contains only carbon and hydrogen.

Combustion is the rapid oxidation of **hydrocarbons**, molecules made of exclusively carbon and hydrogen that we further explore in Chapter 8. In the presence of ample oxygen, hydrocarbons burn to form carbon dioxide and water, liberating large quantities of heat. The general form of the equation representing the combustion of the simplest hydrocarbons, called alkanes (see Chapter 8), is shown as follows:

$$C_xH_y + zO_2 \rightarrow xCO_2 + (y/2)H_2O + \text{energy} \qquad (4.6)$$

It is combustion that accounts for the wide use of hydrocarbons as fuels. Natural gas contains 80% to 95% methane (CH_4), some ethane (C_2H_6), and small amounts of other hydrocarbons. The balanced chemical equation representing the combustion of methane produces 890.4 kJ/mol.

$$CH_4(g) + 2O_2(g) \rightarrow CO_2(g) + 2H_2O(\ell) + 890.4\ \text{kJ/mol}$$

Propane (C_3H_8) and butane (C_4H_{10}) are extracted from natural gas and sold in pressurized metal containers (**Figure 4.24**). In this form, they are used for heating and cooking in campers, trailers, boats, and rural homes. Gasoline is a mixture of hydrocarbons that contain 5 to 12 carbon atoms per molecule (**Figure 4.25**). Diesel fuel is a similar mixture, except the molecules contain 12 to 16 carbon atoms. The hot CO_2 and water vapor generated during combustion in an internal combustion engine have a much greater volume than the air and fuel mixture. It is this sudden increase in gaseous volume and pressure that pushes the pistons and delivers power to the crankshaft in some motors.

David/Adobe Stock Photos

Figure 4.24 Propane is used to heat campers and homes.

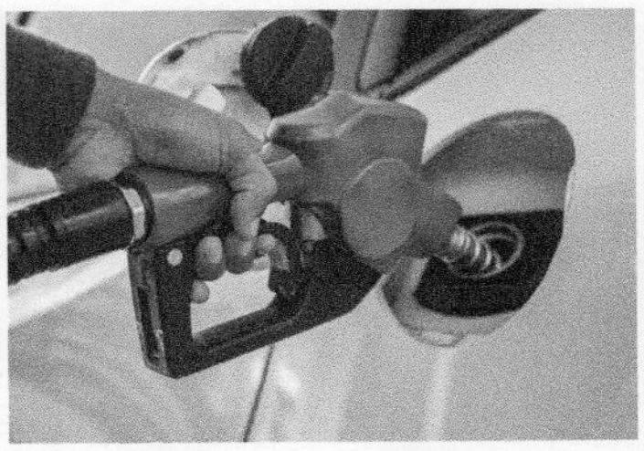

15Studio/Shutterstock.com

Figure 4.25 Gasoline consists of hydrocarbons that provide energy through combustion reactions.

Example 4.9 Classifying Reactions

Use the oxidation numbers provided to classify each of the reactions represented by the following equations as redox or nonredox. Further classify them as decomposition, combination, single-replacement, double-replacement, or combustion reactions.

a. $\underset{+4\,-2}{SO_2} + \underset{+1\,-2}{H_2O} \rightarrow \underset{+1\,+4\,-2}{H_2SO_3}$

b. $\underset{0}{2K} + \underset{+1\,-2}{2H_2O} \rightarrow \underset{+1\,-2\,+1}{2KOH} + \underset{0}{H_2}$

c. $\underset{-2.5\,+1}{2C_4H_{10}} + \underset{0}{13O_2} \rightarrow \underset{+4\,-2}{8CO_2} + \underset{+1\,-2}{10H_2O}$

d. $\underset{+2\,-1}{BaCl_2} + \underset{+1\,+4\,-2}{Na_2CO_3} \rightarrow \underset{+2\,+4\,-2}{BaCO_3} + \underset{+1\,-1}{2NaCl}$

Solution

a. No O.N. changes take place, so the reaction is nonredox. Because two substances combine to form a third, this is a combination reaction.

b. The O.N. of K increases from 0 to +1 and that of H decreases from +1 to 0, so this is a redox reaction. Because K displaces H, it is a single-replacement reaction.

c. The O.N.s of both C and O change, so this is a redox reaction. A hydrocarbon reacts with oxygen to produce carbon dioxide and water, so it is a combustion reaction.

d. No changes in O.N. occur. The reaction is nonredox and is an example of a double-replacement, or metathesis, reaction.

✔ **Learning Check 4.9** Classify each of the reactions represented by the following equations as redox or nonredox. Further classify them as decomposition, combination, single-replacement, double-replacement, or combustion reactions.

a. $\underset{-2.7\,+1}{C_3H_8(g)} + \underset{0}{5O_2(g)} \rightarrow \underset{+4\,-2}{3CO_2(g)} + \underset{+1\,-2}{4H_2O(g)}$

b. $\underset{+1\,-1}{2H_2O_2(aq)} \rightarrow \underset{+1\,-2}{2H_2O(\ell)} + \underset{0}{O_2(g)}$

c. $\underset{+1\,-1}{NaCl(aq)} + \underset{+1\,+5\,-2}{AgNO_3(aq)} \rightarrow \underset{+1\,-1}{AgCl(s)} + \underset{+1\,+5\,-2}{NaNO_3(aq)}$

d. $\underset{0}{4P(s)} + \underset{0}{5O_2(g)} \rightarrow \underset{+5\,-2}{2P_2O_5(s)}$

e. $\underset{+1\,-1}{2NaI(aq)} + \underset{0}{Cl_2(aq)} \rightarrow \underset{+1\,-1}{2NaCl(aq)} + \underset{0}{I_2(aq)}$

Health Career Description

Clinical Laboratory Technician

Clinical laboratory technicians typically work under clinical technologists, who may work, in turn, under a medical laboratory scientist. Clinical laboratory technician is a career that may be entered with an associate of science degree, whereas the laboratory technologist and medical laboratory scientist careers require a bachelor's degree.

According to the Bureau of Labor Statistics, clinical laboratory technologists and technicians typically do the following:

- Analyze body fluids, such as blood, urine, and tissue samples, and record normal or abnormal findings
- Study blood samples for use in transfusions by identifying the number of cells, the cell morphology or the blood group, blood type, and compatibility with other blood types
- Operate sophisticated laboratory equipment, such as microscopes and cell counters
- Use automated equipment and computerized instruments capable of performing a number of tests at the same time
- Log data from medical tests and enter results into a patient's medical record
- Discuss results and findings of laboratory tests and procedures with physicians

Both technicians and technologists perform tests and procedures that physicians and surgeons, or other healthcare personnel, prescribe. However, technologists perform more complex tests and laboratory procedures than technicians do. For example, technologists may prepare specimens and perform detailed manual tests, whereas technicians perform routine tests that may be more automated. Clinical laboratory technicians usually work under the general supervision of clinical laboratory technologists or laboratory managers.

Clinical laboratory technicians and technologists are close partners with physicians in the diagnosis of disease. They provide essential feedback about patient status through laboratory testing. Their precise work is crucial to the proper diagnosis and treatment of patients.

Concept Summary

4.1 Avogadro's Number: The Mole

Learning Objective: **Convert between the number of moles, number of grams, and number of atoms of elements.**

- An Avogadro's number of the atoms of an element is called a mole of atoms of that element.
- The atomic mass of an element in atomic mass unit (amu) is equal to the amount in grams of 1 mole of that element.

4.2 The Mole and Chemical Formulas

Learning Objectives: **Convert between the number of moles, number of grams, number of compounds, and number of atoms in compounds. Calculate the percent by mass of atoms in compounds.**

- An Avogadro's number of compounds or molecules is called a mole of those compounds or molecules.
- When applied to molecular formulas, the mole concept gives numerous relationships that yield useful conversion factors for dimensional analysis calculations.

4.3 Chemical Equations

Learning Objective: **Write a balanced chemical equation when provided with a chemical reaction.**

- Chemical reactions can be represented by equations.
- Reacting substances are called reactants, and they are written on the left side of the chemical equation.
- The substances produced are called products, and they are written on the right side of the chemical equation.
- The reactants and products are written as chemical formulas.
- Coefficients are placed before reactant and product formulas to balance the equation.
- A balanced equation satisfies the law of conservation of matter.

4.4 The Mole and Chemical Equations

Learning Objective: **Use the mole concept to perform calculations based on the stoichiometry of balanced chemical equations.**

- The mole concept, when applied to chemical equations, yields relationships that can be used to obtain factors for converting units via dimensional analysis.

4.5 Reaction Yields

Learning Objective: **Use the mole concept to calculate percent yields.**

- The mass of product isolated after a reaction is always less than the mass that is theoretically possible.
- The percent yield of a reaction is the ratio of the actual isolated mass (actual yield) to the calculated theoretical yield multiplied by 100%.

4.6 Types of Reactions

Learning Objectives: **Identify the oxidizing and reducing agents in oxidation–reduction (redox) reactions when given the oxidation states of each element in the reactants and products. Classify reactions first as redox or nonredox, then as decomposition, combination, single replacement, double replacement, or combustion.**

- To facilitate their study, reactions are classified first as redox or nonredox.
- They are then further classified as decomposition, combination, single-replacement, double-replacement, or combustion reactions.
- Reactions in which reactants undergo oxidation or reduction are called redox reactions.
- Oxidation and reduction can be identified by the changes in oxidation number that occur as reactants are converted to products.
- A substance is oxidized when the oxidation number of a constituent element increases.
- A substance is reduced when the oxidation number of a constituent element decreases.
- Decomposition reactions are characterized by one substance reacting to give two or more products.
- Decomposition reactions can be redox or nonredox.
- Combination reactions are characterized by two or more reactants that form a single compound as a product.
- Combination reactions can be redox or nonredox.
- Combination reactions can also be called addition or synthesis reactions.
- Single-replacement reactions, sometimes called substitution reactions, are always redox reactions.
- In a single-replacement reaction, one element reacts with a compound and displaces another element from the compound.
- Double-replacement reactions, also called metathesis reactions, are always nonredox reactions.
- Double-replacement reactions can be recognized by their partner-swapping characteristics.
- Combustion reactions occur when a hydrocarbon reacts with oxygen to produce carbon dioxide and water vapor.
- Combustion reactions release large amounts of heat.

Key Terms and Concepts

Avogadro's number (4.1)
Balanced equation (4.3)
Combination reaction (4.6)
Combustion (4.6)
Decomposition reaction (4.6)
Double-replacement reaction (4.6)
Hydrocarbon (4.6)
Law of conservation of matter (4.3)
Molar mass (4.1)
Mole (4.1)
Oxidation (4.6)
Oxidation number (oxidation state) (4.6)
Oxidizing agent (4.6)
Percent yield (4.5)
Reducing agent (4.6)
Reduction (4.6)
Side reactions (4.5)
Single-replacement reaction (4.6)
Stoichiometry (4.4)

Key Equations

1. Percent yield **(Section 4.5)**	$\% \text{ yield} = \frac{\text{actual yield}}{\text{theoretical yield}} \times 100\%$	Equation 4.1
2. Decomposition reaction—one substance changes into two or more new substances **(Section 4.6)**	$A \rightarrow B + C$	Equation 4.2
3. Combination reaction—two or more substances react to produce one new substance **(Section 4.6)**	$A + B \rightarrow C$	Equation 4.3
4. Single-replacement reaction—one element reacts with a compound to produce a new compound and new element **(Section 4.6)**	$A + BX \rightarrow B + AX$	Equation 4.4
5. Double-replacement reaction (metathesis reaction)—two compounds exchange partners to form two new compounds **(Section 4.6)**	$AX + BY \rightarrow BX + AY$	Equation 4.5
6. Combustion reaction—hydrocarbons react with ample O_2 to form CO_2, H_2O, and heat **(Section 4.6)**	$C_xH_y + zO_2 \rightarrow xCO_2 + (y/2)H_2O + \text{energy}$	Equation 4.6

Exercises

Even-numbered exercises are answered in Appendix B.

Exercises with an asterisk (*) are more challenging.

Avogadro's Number: The Mole (Section 4.1)

4.1 Using the periodic table and Avogadro's number:

a. How many atoms are in 3.10 g of phosphorus?

b. How many grams of sulfur would contain that same number of atoms of phosphorus?

4.2 Using the periodic table and Avogadro's number:

a. How many atoms are in 1.60 g of oxygen?

b. How many grams of fluorine would contain that same number of atoms of oxygen?

4.3 Using the periodic table and the dimensional analysis method, determine the following.

a. The number of moles of beryllium atoms in a 10.0 g sample of beryllium

b. The number of lead atoms in a 2.00 mol sample of lead

c. The number of sodium atoms in a 50.0 g sample of sodium

4.4 Using the periodic table and the dimensional analysis method, determine the following.

a. The mass in grams of one phosphorus atom

b. The number of grams of aluminum in 1.65 mol of aluminum

c. The total mass in grams of one-fourth Avogadro's number of krypton atoms

The Mole and Chemical Formulas (Section 4.2)

4.5 Fructose ($C_6H_{12}O_6$), a fruit sugar, is approximately 1.2 to 1.8 times sweeter than table sugar. What is the molar mass of fructose?

4.6 Creatine ($C_4H_9N_3O_2$) is a natural energy source for muscle contractions. What is the molar mass of creatine?

4.7 The opioid Olinvyk ($C_{22}H_{30}N_2O_2S$) is used as an intravenous medication to treat moderate to severe acute pain. What is the molar mass of Olinvyk?

4.8 The drug Camzyos ($C_{15}H_{19}N_3O_2$) is used to treat adults with hypertrophic cardiomyopathy. What is the molar mass of Camzyos?

4.9 Use the periodic table to calculate the molecular weights for PH_3 and SO_2. How many grams of PH_3 contain the same number of molecules as 6.41 g of SO_2?

4.10 Use the periodic table to calculate the molecular weights for BF_3 and H_2S. How many grams of BF_3 contain the same number of molecules as 0.34 g of H_2S?

4.11 Answer the following questions by using the mole and mole relationships.

a. How many moles of oxygen atoms are contained in 1 mol of CO_2 molecules?

b. How many grams of carbon are contained in 1.00 mol of $C_3H_6O_3$?

c. What is the mass percentage of oxygen in $C_6H_{12}O_6$?

4.12 Answer the following questions by using the mole and mole relationships.

a. How many carbon atoms are contained in 0.25 mol of C_2H_6O?

b. How many moles of hydrogen atoms are contained in 0.50 mol of diethyl ether ($C_4H_{10}O$)?

c. How many grams of hydrogen are contained in 2.00 mol of halothane ($C_2HBrClF_3$)?

4.13 An intravenous solution of normal saline may contain 1.70 mol of sodium chloride (NaCl). How many sodium atoms are present in the solution?

4.14 If an adult female has a mass of 60.0 kg and contains 0.00172% iron, how many moles of iron are in her body?

4.15 Determine the mass percentage of carbon in glucose ($C_6H_{12}O_6$) and ethanol (C_2H_6O).

4.16 Determine the mass percentage of hydrogen in CH_4 and C_2H_6.

4.17* Nitrophenol ($C_6H_5NO_3$) is used to make drugs, insecticides, and dyes.

a. How many grams of nitrogen are contained in 70.0 g of $C_6H_5NO_3$?

b. How many moles of oxygen atoms are contained in 1.50 mol of $C_6H_5NO_3$?

c. How many carbon atoms are contained in 9.00×10^{22} molecules of $C_6H_5NO_3$?

4.18 Fructose ($C_6H_{12}O_6$) is a common sweetener in various foods.

a. How many grams of oxygen are contained in 43.5 g of $C_6H_{12}O_6$?

b. How many moles of hydrogen atoms are contained in 1.50 mol of $C_6H_{12}O_6$?

c. How many carbon atoms are contained in 7.50×10^{22} molecules of $C_6H_{12}O_6$?

4.19 Which compound contains the higher mass percentage of nitrogen—ammonium sulfate [$(NH_4)_2SO_4$], which is used to isolate proteins, or urea (CH_4N_2O), which is a product of protein metabolism?

4.20 Epinephrine ($C_9H_{13}NO_3$) is a hormone released in response to stress. What is the mass in grams of 6.40×10^{-3} mol of epinephrine?

4.21 Cortisone ($C_{21}H_{28}O_5$) is used as an anti-inflammatory drug. How many moles of cortisone are present in one 10.0 mg tablet?

4.22 Chloroquine ($C_{18}H_{26}ClN_3$) is used to treat malaria and rheumatoid arthritis. How many grams are in 5.5 moles?

4.23 Lysine is an essential amino acid that must be obtained from a proper diet and has been shown to inhibit the replication of herpes viruses.

$$\overset{+}{H_3N}-CH-COO^-$$
$$|$$
$$CH_2$$
$$|$$
$$CH_2$$
$$|$$
$$CH_2$$
$$|$$
$$CH_2$$
$$|$$
$$NH_3^+$$

lysine

a. Calculate the molecular formula of lysine.

b. Determine the molar mass of lysine.

c. Calculate the percentage of nitrogen by mass in lysine.

d. How many moles are in the recommended daily amount of 2.5 grams of lysine?

4.24 The most common neurotransmitter in the brain is glutamate. It regulates learning and memory but can be toxic in large quantities.

$$\overset{+}{H_3N}-CH-COO^-$$
$$|$$
$$CH_2$$
$$|$$
$$CH_2$$
$$|$$
$$COO^-$$

glutamate

a. Calculate the molecular formula of glutamate.

b. Determine the molar mass of glutamate.

c. Calculate the percentage of oxygen by mass in glutamate.

d. The high-risk level of consuming glutamates is around 80 mg/kg daily for toddlers. Should you contact the pediatrician if a toddler ingested 5.4 g of glutamate?

Chemical Equations (Section 4.3)

4.25 Identify the reactants and products in each of the following reaction equations.

a. $CH_4(g) + 2O_2(g) \rightarrow CO_2(g) + 2H_2O(g)$

b. $2Al(s) + 3Cl_2(g) \rightarrow 2AlCl_3(s)$

c. methane + water → carbon monoxide + hydrogen

d. copper(II) oxide + hydrogen → copper + water

4.26 Identify the reactants and products in each of the following reaction equations.

a. $H_2(g) + Cl_2(g) \rightarrow 2HCl(g)$

b. $2KClO_3(s) \rightarrow 2KCl(s) + 3O_2(g)$

c. magnesium oxide + carbon → magnesium + carbon monoxide

d. ethane + oxygen → carbon dioxide + water

4.27 Identify which of the following are consistent with the law of conservation of matter. For those that are not, explain why they are not.

a. $Fe(s) + O_2(g) \rightarrow Fe_2O_3(s)$

b. $2Na_3PO_4(aq) + 3MgCl_2(aq) \rightarrow Mg_3(PO_4)_2(s) + 6NaCl(aq)$

c. 3.20 g oxygen + 3.21 g sulfur → 6.41 g sulfur dioxide

d. $CH_4(g) + 2O_2(g) \rightarrow CO_2(g) + 2H_2O(g)$

4.28 Identify which of the following are consistent with the law of conservation of matter. For those that are not, explain why they are not.

a. $ZnS(s) + O_2(g) \rightarrow ZnO(s) + SO_2(g)$

b. $Cl_2(aq) + 2I^-(aq) \rightarrow I_2(aq) + 2Cl^-(aq)$

c. 1.50 g oxygen + 1.50 g carbon → 2.80 g carbon monoxide

d. $2C_2H_6(g) + 7O_2(g) \rightarrow 4CO_2(g) + 6H_2O(g)$

4.29 Determine the number of atoms of each element on each side of the following equations and decide which equations are balanced.

a. $3Pb(NO_3)_2(aq) + 2AlBr_3(aq) \rightarrow 3PbBr_2(s) + 2Al(NO_3)_3(aq)$

b. $K(s) + H_2O(\ell) \rightarrow KOH(aq) + H_2(g)$

c. $CaCO_3(s) \rightarrow CaO(s) + CO_2(g)$

d. $Ba(ClO_3)_2(aq) + H_2SO_4(aq) \rightarrow 2HClO_3(aq) + BaSO_4(s)$

4.30 Determine the number of atoms of each element on each side of the following equations and decide which equations are balanced.

a. $Ag(s) + Cu(NO_3)_2(aq) \rightarrow Cu(s) + AgNO_3(aq)$

b. $2N_2O(g) + 3O_2(g) \rightarrow 4NO_2(g)$

c. $Mg(s) + O_2(g) \rightarrow 2MgO(s)$

d. $H_2SO_4(aq) + Ca(OH)_2(aq) \rightarrow CaSO_4(s) + 2H_2O(\ell)$

4.31 Balance the following equations.

a. $Li_3N(s) \rightarrow Li(s) + N_2(g)$

b. $Cu_2O(s) + O_2(g) \rightarrow CuO(s)$

c. $Zn(s) + HNO_3(aq) \rightarrow Zn(NO_3)_2(aq) + H_2(g)$

d. $K_2SO_4(aq) + BaCl_2(aq) \rightarrow BaSO_4(s) + KCl(aq)$

e. dinitrogen monoxide → nitrogen + oxygen

f. dinitrogen pentoxide → nitrogen dioxide + oxygen

g. $P_4O_{10}(s) + H_2O(\ell) \rightarrow H_4P_2O_7(aq)$

h. $CaCO_3(s) + HCl(aq) \rightarrow CaCl_2(aq) + H_2O(\ell) + CO_2(g)$

4.32 Balance the following equations.

a. $KClO_3(s) \rightarrow KCl(s) + O_2(g)$

b. $C_2H_6(g) + O_2(g) \rightarrow CO_2(g) + H_2O(\ell)$

c. nitrogen + oxygen → dinitrogen pentoxide

d. $MgCl_2(s) + H_2O(g) \rightarrow MgO(s) + HCl(g)$

e. $CaH_2(s) + H_2O(\ell) \rightarrow Ca(OH)_2(s) + H_2(g)$

f. $Al(s) + Fe_2O_3(s) \rightarrow Al_2O_3(s) + Fe(s)$

g. aluminum + bromine → aluminum bromide

h. $Hg_2(NO_3)_2(aq) + NaCl(aq) \rightarrow Hg_2Cl_2(s) + NaNO_3(aq)$

The Mole and Chemical Equations (Section 4.4)

4.33 For the reactions represented by the following equations, write statements equivalent to Statements 2, 3, and 4 given in **Section 4.4**.

a. $2SO_2(g) + O_2(g) \rightarrow 2SO_3(g)$

b. $4HCl(g) + O_2(g) \rightarrow 2Cl_2(g) + 2H_2O(g)$

c. $Fe_2O_3(s) + 3C(s) \rightarrow 2Fe(s) + 3CO(g)$

d. $2H_2O_2(aq) \rightarrow 2H_2O(\ell) + O_2(g)$

e. $2C_3H_6(g) + 9O_2(g) \rightarrow 6CO_2(g) + 6H_2O(g)$

4.34 For the reactions represented by the following equations, write statements equivalent to Statements 2, 3, and 4 given in **Section 4.4**.

a. $S(s) + O_2(g) \rightarrow SO_2(g)$

b. $Sr(s) + 2H_2O(\ell) \rightarrow Sr(OH)_2(s) + H_2(g)$

c. $2H_2S(g) + 3O_2(g) \rightarrow 2H_2O(g) + 2SO_2(g)$

d. $4NH_3(g) + 5O_2(g) \rightarrow 4NO(g) + 6H_2O(g)$

e. $CaO(s) + 3C(s) \rightarrow CaC_2(s) + CO(g)$

4.35 For the following equation:

$$H_2O(\ell) + SnCl_4(s) \rightarrow Sn(OH)(s) + HCl(aq)$$

a. How many moles of $H_2O(\ell)$ will react with 1.5 moles of $SnCl_4(s)$?

b. How many grams of $H_2O(\ell)$ will be produced when 2.0 moles of $SnCl_4(s)$ react with excess $H_2O(\ell)$?

4.36 For the following equation, write statements equivalent to Statements 2, 3, and 4 given in **Section 4.4**. Then write at least six factors (including numbers and units) that could be used to solve problems by the dimensional analysis method.

$$2SO_2(g) + O_2(g) \rightarrow 2SO_3(g)$$

4.37 Calculate the number of grams of SO_2 that must react to produce 350.0 g of SO_3. Use the statements mentioned in **Exercise 4.36** and express your answer using the correct number of significant figures.

4.38 Calculate the mass of limestone ($CaCO_3$) that must be decomposed to produce 500.0 g of lime (CaO). The equation for the reaction is

$$CaCO_3(s) \rightarrow CaO(s) + CO_2(g)$$

4.39 Calculate the number of moles of CO_2 generated by the reaction of **Exercise 4.38** when 500.0 g of CaO is produced.

4.40 Calculate the number of grams of bromine (Br_2) needed to react exactly with 50.1 g of aluminum (Al). The equation for the reaction is

$$2Al(s) + 3Br_2(\ell) \rightarrow 2AlBr_3(s)$$

4.41 Given the following equation for the reaction used to clean tarnish from silver:

$$3Ag_2S(s) + 2Al(s) \rightarrow 6Ag(s) + Al_2S_3(s)$$

a. How many grams of aluminum would need to react to remove 0.250 g of Ag_2S tarnish?

b. How many moles of Al_2S_3 would be produced by the reaction described in part a?

4.42 Pure titanium metal is produced by reacting titanium(IV) chloride with magnesium metal. The equation for the reaction is

$$TiCl_4(s) + 2Mg(s) \rightarrow Ti(s) + 2MgCl_2(s)$$

How many grams of Mg would be needed to produce 1.00 kg of pure titanium?

4.43 An important metabolic process of the body is the oxidation of glucose to water and carbon dioxide. The equation for the reaction is

$$C_6H_{12}O_6(aq) + 6O_2(aq) \rightarrow 6CO_2(aq) + 6H_2O(\ell)$$

a. What mass of water in grams is produced when the body oxidizes 1.00 mol of glucose?

b. How many grams of oxygen are needed to oxidize 1.00 mol of glucose?

4.44 Caproic acid ($C_6H_{12}O_2$) is oxidized in the body as follows:

$$C_6H_{12}O_2(aq) + 8O_2(aq) \rightarrow 6CO_2(aq) + 6H_2O(\ell)$$

How many grams of oxygen are needed to oxidize 1.00 mol of caproic acid?

Reaction Yields (Section 4.5)

4.45 A product weighing 14.37 g was isolated from a reaction. The amount of product possible according to a calculation was 17.55 g. What was the percent yield?

4.46 Upon heating, mercury(II) oxide undergoes a decomposition reaction:

$$2HgO(s) \rightarrow 2Hg(\ell) + O_2(g)$$

A sample of HgO weighing 7.22 g was heated. The collected mercury weighed 5.95 g. What was the percent yield of the reaction?

4.47 A sample of calcium metal with a mass of 2.00 g was reacted with excess oxygen. The following equation represents the reaction that took place:

$$2Ca(s) + O_2(g) \rightarrow 2CaO(s)$$

The isolated product (CaO) weighed 2.26 g. What was the percent yield of the reaction?

Types of Reactions (Section 4.6)

4.48 For each of the following equations, indicate whether the blue element has been oxidized, reduced, or neither oxidized nor reduced.

a. $4Al(s) + 3O_2(g) \rightarrow 2Al_2O_3(s)$

0 0 +3 −2

b. $SO_2(g) + H_2O(\ell) \rightarrow H_2SO_3(aq)$

+4 −2 +1 −2 +1 +4 −2

c. $2KClO_3(s) \rightarrow 2KCl(s) + 3O_2(g)$

+1 +5 −2 +1 −1 0

d. $2CO(g) + O_2(g) \rightarrow 2CO_2(g)$

+2 −2 0 +4 −2

e. $2Na(s) + 2H_2O(\ell) \rightarrow 2NaOH(aq) + H_2(g)$

0 +1 −2 +1 −2 +1 0

4.49 For each of the following equations, indicate whether the blue element has been oxidized, reduced, or neither oxidized nor reduced.

a. $2Mg(s) + O_2(g) \rightarrow 2MgO(s)$

0 0 +1 −2

b. $CuO(s) + H_2(g) \rightarrow Cu(s) + H_2O(g)$

+2 −2 0 0 +1 −2

c. $Ag^+(aq) + Cl^-(aq) \rightarrow AgCl(s)$

+1 −1 +1 −1

d. $BaCl_2(aq) + H_2SO_4(aq) \rightarrow BaSO_4(s) + 2HCl(aq)$

+2 −1 +1 +6 −2 +2 +6 −2 +1 −1

e. $Zn(s) + 2H^+(aq) \rightarrow Zn^{2+}(aq) + H_2(g)$

0 +1 +2 0

4.50 Identify the oxidizing and reducing agents, given the oxidation states provided.

a. $2Cu(s) + O_2(g) \rightarrow 2CuO(s)$

0 0 +2 −2

b. $Cl_2(aq) + 2KI(aq) \rightarrow 2KCl(aq) + I_2(aq)$

0 +1 −1 +1 −1 0

c. $3MnO_2(s) + 4Al(s) \rightarrow 2Al_2O_3(s) + 3Mn(s)$

+4 −2 0 +3 −2 0

d. $2H^+(aq) + 3SO_3{}^{2-}(aq) + 2NO_3{}^-(aq) \rightarrow$

+1 +4 −2 +5 −2

$2NO(g) + H_2O(\ell) + 3SO_4{}^{2-}(aq)$

+2 −2 +1 −2 +6 −2

e. $Mg(s) + 2HCl(aq) \rightarrow MgCl_2(aq) + H_2(g)$

0 +1 −1 +2 −1 0

f. $4NO_2(g) + O_2(g) \rightarrow 2N_2O_5(g)$

+4 −2 0 +5 −2

4.51 Identify the oxidizing and reducing agents, given the oxidation states provided:

a. $H_2(g) + Cl_2(g) \rightarrow 2HCl(g)$

0 0 +1 −1

b. $H_2O(g) + CH_4(g) \rightarrow CO(g) + 3H_2(g)$

+1 −2 −4 +1 +2 −2 0

c. $CuO(s) + H_2(g) \rightarrow Cu(s) + H_2O(g)$

+2 −2 0 0 +1 −2

d. $B_2O_3(s) + 3Mg(s) \rightarrow 2B(s) + 3MgO(s)$

+3 −2 0 0 +2 −2

e. $Fe_2O_3(s) + CO(g) \rightarrow 2FeO(s) + CO_2(g)$

+3 −2 +2 −2 +2 −2 +4 −2

f. $Cr_2O_7{}^{2-}(aq) + 2H^+(aq) + 3Mn^{2+}(aq) \rightarrow$

+6 −2 0 +2

$2Cr^{3+}(aq) + 3MnO_2(s) + H_2O(\ell)$

+3 +4 −2 +1 −2

4.52 Aluminum metal reacts rapidly with fundamental solutions to liberate hydrogen gas and a large amount of heat. This reaction is utilized in a popular solid drain cleaner composed primarily of lye (sodium hydroxide) and aluminum granules. When wet, the mixture reacts as follows:

$$6NaOH(aq) + 2Al(s) \rightarrow 3H_2(g) + 2Na_3AlO_3(aq) + \text{heat}$$

+1 −2 +1 0 0 +1 +3 −2

The liberated H_2 provides agitation that, together with the heat, breaks the drain stoppage loose. What are the oxidizing and reducing agents in the reaction?

4.53 The overall reaction for photosynthesis is

$$6CO_2 + 6H_2O \rightarrow C_6H_{12}O_6 + 6O_2$$

+4 −2 +1 −2 0 +1 −2 0

Is this a redox reaction? If so, identify the reducing and oxidizing agents.

4.54 Classify each of the reactions represented by the following chemical equations as a decomposition, single-replacement, combination, double-replacement, or combustion reaction.

a. $K_2CO_3(s) \rightarrow K_2O(s) + CO_2(g)$

b. $Ca(s) + 2H_2O(\ell) \rightarrow Ca(OH)_2(s) + H_2(g)$

c. $BaCl_2(aq) + H_2SO_4(aq) \rightarrow BaSO_4(s) + 2HCl(aq)$

d. $SO_2(g) + H_2O(\ell) \rightarrow H_2SO_3(aq)$

e. $2NO(g) + O_2(g) \rightarrow 2NO_2(g)$

f. $2Zn(s) + O_2(g) \rightarrow 2ZnO(s)$

g. $C_3H_8(g) + 5O_2(g) \rightarrow 3CO_2(g) + 4H_2O(g)$

4.55 Classify each of the reactions represented by the following chemical equations as a decomposition, single-replacement, combination, double-replacement, or combustion reaction.

a. $N_2O_5(g) + H_2O(\ell) \rightarrow 2HNO_3(aq)$

b. $Cr_2O_3(s) + 2Al(s) \rightarrow 2Cr(s) + Al_2O_3(s)$

c. $CaO(s) + SiO_2(s) \rightarrow CaSiO_3(s)$

d. $H_2CO_3(aq) \rightarrow CO_2(g) + H_2O(\ell)$

e. $PbCO_3(s) \rightarrow PbO(s) + CO_2(g)$

f. $Zn(s) + Cl_2(g) \rightarrow ZnCl_2(s)$

g. $2C_2H_2(g) + 5O_2(g) \rightarrow 4CO_2(g) + 2H_2O(g)$

4.56 Cooking fires are the number one cause of home fires and home fire injuries. Baking soda ($NaHCO_3$) can serve as an emergency fire extinguisher for grease fires in the kitchen. When heated, it liberates CO_2, which smothers the fire. The equation for the reaction is

$$2NaHCO_3(s) \xrightarrow{\text{Heat}} Na_2CO_3(s) + H_2O(g) + CO_2(g)$$

Classify the reaction as a decomposition, single-replacement, combination, double-replacement, or combustion reaction.

4.57 Baking soda may serve as a source of CO_2 in bread dough. It causes the dough to rise. The CO_2 is released when $NaHCO_3$ reacts with an acidic substance:

$$\underset{+1\,+1\,+4\,-2}{NaHCO_3(aq)} + \underset{+1}{H^+(aq)} \rightarrow \underset{+1}{Na^+(aq)} + \underset{+1\,-2}{H_2O(\ell)} + \underset{+4\,-2}{CO_2(g)}$$

Classify the reaction as redox or nonredox.

4.58 Since the 1800s, natural gas has replaced coal as a heating and electrical-power-generating source. This change has reduced air pollution emissions linked to lung diseases. Today, many homes are heated by the energy released when natural gas (represented by CH_4) reacts with oxygen. The equation for the reaction is

$$\underset{-4\,+1}{CH_4(g)} + \underset{0}{2O_2(g)} \rightarrow \underset{+4\,-2}{CO_2(g)} + \underset{+1\,-2}{2H_2O(g)}$$

Classify the reaction as redox or nonredox.

4.59 Hydrogen peroxide solutions are used as an antiseptic to treat wounds. Hydrogen peroxide will react with and liberate oxygen gas. In commercial solutions, the addition of an inhibitor prevents the reaction to a larger degree. The equation for the oxygen-liberating reaction is

$$\underset{+1\,-1}{2H_2O_2(aq)} \rightarrow \underset{+1\,-2}{2H_2O(\ell)} + \underset{0}{O_2(g)}$$

Classify the reaction as redox or nonredox.

4.60 Chlorine, used to treat drinking water, undergoes the reaction in water represented by the following equation:

$$\underset{0}{Cl_2(aq)} + \underset{+1\,-2}{H_2O(\ell)} \rightarrow \underset{+1\,-2\,+1}{HOCl(aq)} + \underset{+1\,-1}{HCl(aq)}$$

Classify the reaction as redox or nonredox.

4.61 Triple superphosphate, an ingredient of some fertilizers, is prepared by reacting rock phosphate (calcium phosphate) and phosphoric acid. The equation for the reaction is

$$Ca_3(PO_4)_2(s) + 4H_3PO_4(aq) \rightarrow 3Ca(H_2PO_4)_2(s)$$

Classify the reaction as a decomposition, single-replacement, combination, double-replacement, or combustion reaction.

4.62 Farmers rely on nitrogen fertilizers produced through the Haber process to feed the billions of people on this planet. This process produces ammonia from elemental nitrogen and hydrogen. The ammonia is then further processed. The equation for the reaction is

$$\underset{0}{N_2(g)} + \underset{0}{3H_2(g)} \rightarrow \underset{-3\,+1}{2NH_3(g)}$$

Identify the oxidizing and reducing agents in the Haber process.

Additional Exercises

4.63 About one billion (1.0×10^9) peas can fit into a railroad car. How many moles of peas is this?

4.64 The mass of a single carbon-12 atom is 1.99×10^{-23} g. What is the mass in grams of a single carbon-14 atom?

4.65 One mole of water molecules, H_2O, has a mass of 18.0 g. What would be the mass in grams of one mole of heavy water molecules, D_2O, where D represents the ^2H isotope?

4.66 Would you expect lithium, Li, which is a trace element in biological systems, to be used in a redox reaction? Explain your reasoning.

4.67 Would you expect argon, Ar, to be involved in a redox reaction? Explain your reasoning.

4.68 Assuming a 100% reaction yield, it was calculated that 6.983 g of naturally occurring elemental iron would be needed to react with another element to form 18.149 g of product. If iron composed only of the iron-60 isotope was used to form the product, how many grams of iron-60 would be required?

4.69 The element with an electron configuration of $1s^22s^22p^63s^1$ undergoes a combination reaction with the element that has 15 protons in its nucleus. Assume the product of the reaction consists of simple ions and write a balanced chemical equation for the reaction.

4.70 The puffer fish has a potent neurotoxin, tetrodotoxin ($C_{11}H_{17}N_3O_8$), in the liver that blocks sodium channels and causes respiratory paralysis. It is estimated that the lethal dose in an adult human is 2.4 mg.

a. Calculate the number of moles of tetrodotoxin in 2.4 mg of $C_{11}H_{17}N_3O_8$.

b. Calculate the number of molecules of tetrodotoxin in 2.4 mg of $C_{11}H_{17}N_3O_8$.

Chemistry for Thought

4.71* Many vitamin supplements are candy-like in appearance, which can lead to iron poisoning in small children. Symptoms will appear in a 3-year-old at doses greater than 10 mg/kg.

a. Should you contact the pediatrician if a 15 kg pediatric patient ingested five multivitamin gummies that contained 12 mg of iron per gummy?

b. The treatment for iron poisoning is to administer 15 mg/kg/hr of deferoxamine for up to 24 hours. How many grams of deferoxamine would the 15 kg pediatric patient receive in 24 hours?

4.72 In experiments where students prepare compounds by precipitation from water solutions, they often report yields of dry products greater than 100%. Propose an explanation for this.

4.73 The decomposition of a sample of a compound produced 1.20×10^{24} atoms of nitrogen and 80.0 grams of oxygen atoms. What was the formula of the sample that was decomposed? What is the correct name of the decomposed sample?

4.74 In an ordinary flashlight battery, an oxidation reaction and a reduction reaction occur at different locations to produce an electrical current consisting of electrons. In one of the reactions, the zinc content of the battery slowly dissolves as it is converted into zinc ions. Is this the oxidation or the reduction reaction? Is this reaction the source of electrons, or are electrons used to carry out the reaction? Explain your answers with a reaction equation.

5 Heating and Changes of State: A Story of Polarity and Intermolecular Forces

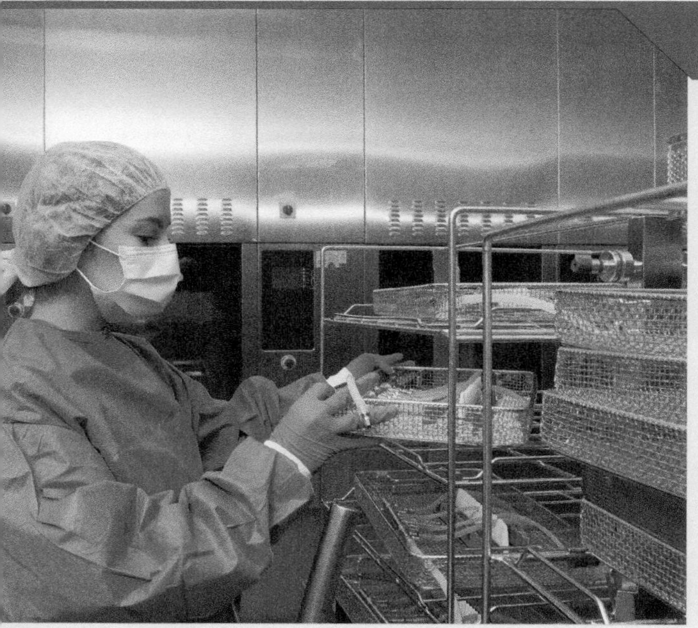
lortie/iStock/Getty Images

Health Career Focus

Sterile Processing Technician

"Without us, nothing happens in this hospital. We're essential to every treatment, procedure, and surgery that occurs in this facility."

"There, behind that window, we receive trays of used surgical equipment. We meticulously wash the blood, bone, and tissue off and get them ready for steam-pressure sterilization. This job isn't for the squeamish."

"One of the most important parts of my job is actually counting the instruments, making sure each item returns from surgery and is recorded carefully in the database. I also monitor equipment for proper temperatures and settings to ensure we meet protocols."

Follow-up to this Career Focus appears at the end of the chapter before the *Concept Summary*.

Learning Objectives

When you have completed your study of Chapter 5, you should be able to:

1. Use VSEPR theory to predict the geometries of molecules and polyatomic ions. **(Section 5.1)**
2. Use electronegativities to classify the covalent bonds in molecules. **(Section 5.2)**
3. Determine whether covalent molecules are polar or nonpolar. **(Section 5.2)**
4. Identify the major attractive force between molecules in a given substance. **(Section 5.3)**
5. Distinguish phenomena that demonstrate kinetic energy from those that demonstrate potential energy. **(Section 5.4)**
6. Identify the states of matter using five properties: density, shape, compressibility, particle interaction, and molecular movement. **(Section 5.4)**
7. Classify changes of state as exothermic or endothermic. **(Section 5.5)**
8. Use the factors that affect evaporation and condensation to rank substances in order of increasing vapor pressure. **(Section 5.5)**
9. Use the factors that affect boiling and melting to rank substances in order of increasing boiling point and melting point. **(Section 5.5)**
10. Calculate energy changes that accompany heating, cooling, or changing the state of a substance. **(Section 5.6)**

In Chapter 2, we ended with a discussion of periodic trends, including electronegativity. Here in Chapter 5, we examine how electronegativity and the three-dimensional shape of compounds combine to determine polarity, and in turn how polarity determines the strength of attractive forces between molecules. Next we will apply our knowledge of attractive and disruptive forces to understand states of matter; the energy required to transition between those states of matter; and lastly, the energy involved in heating or cooling a substance.

narikan/Shutterstock.com

Figure 5.1 The amount of water in the human body ranges from 70% in infants to 55% to 60% in adults. The bent (rather than linear) shape of water molecules gives rise to properties that support life.

5.1 Geometries of Molecules and Polyatomic Ions

Learning Objective 1 Use VSEPR theory to predict the geometries of molecules and polyatomic ions.

Most molecules and polyatomic ions do not have the flat, two-dimensional shapes implied by their molecular Lewis structures (**Section 3.5**). Instead, they have distinct three-dimensional geometries. Being able to predict those geometries is important because the shapes of molecules and ions contribute to their properties (see **Figure 5.1**).

To predict geometries, first draw Lewis structures for the molecules or ions using the methods discussed in **Sections 3.5** and **3.7**. Next, apply *valence-shell electron-pair repulsion theory* (**VSEPR theory**—sometimes pronounced "vesper" theory) to the Lewis structure. According to VSEPR theory, electron pairs in the valence shell of an atom are repelled by other electron pairs and arrange themselves as far away from one another as possible. Any atom in a molecule or ion that is bonded to two or more other atoms is called a *central atom*. When VSEPR theory is applied to the valence-shell electrons of central atoms, the shape of the molecule or ion containing the atoms can be predicted. Two rules are followed.

VSEPR theory A theory based on the mutual repulsion of electron pairs. It is used to predict molecular shapes.

1. All valence-shell electron pairs around the central atom are counted equally, regardless of whether they are bonding or nonbonding pairs.
2. Double or triple bonds between atoms are treated like a single pair of electrons when predicting shape. As a result, single, double, and triple bonds are all considered to be a single **electron domain**. Electron domains that are due to bonds are called *bonding domains*, whereas lone pairs of electrons are called *nonbonding domains*.

electron domain Any lone pair of electrons or shared electrons resulting in a single, double, or triple bond.

Two electron domains around a central atom will arrange themselves in a linear fashion (**Figure 5.2a**)—that is, the two electron domains will be on opposite sides of the central atom. Three electron domains will form a triangle around the central atom (**Figure 5.2b**), and four electron domains will be located at the corners of a regular tetrahedron with the central atom in the center (**Figure 5.2c**). The shapes shown in **Figure 5.2**, which result from the repulsion of electron domains (bonding *and* nonbonding), are called

© Jeffrey M. Seager

a Two balloons; linear　b Three balloons; trigonal planar　c Four balloons; tetrahedral

Figure 5.2 When balloons of the same size and shape are tied together, they will assume positions like those taken by electron domains around a central atom. Where would the central atom be located in these balloon models?

electron-domain geometry The general shape that a molecule assumes when repulsions are minimized between electron domains.

molecular geometry The three-dimensional shape of bonded atoms in a molecule.

electron-domain geometries. VSEPR theory can be used to predict the shapes of molecules and ions with five and six electron domains around the central atom, but we limit the discussion to four electron domains in this book.

The three balloon models in **Figure 5.2** represent the three electron-domain geometries that we discuss—namely, linear, trigonal planar, and tetrahedral. Notice in **Table 5.1** that these three electron-domain geometries lead to six different molecular geometries. The **molecular geometry** reflects the shape exhibited by only the bonding domains. The molecular geometry is the same as the electron-domain geometry when all the electron domains are bonding domains, but it differs when one or more of the electron domains are nonbonding domains (i.e., lone pairs of electrons). **Table 5.1** provides examples of the molecular geometries possible within the linear, trigonal planar, and tetrahedral electron domain geometries.

The only molecular geometry possible for the linear electron-domain geometry is also linear, with two bonding domains and zero lone pairs of electrons. The bond angle

Table 5.1 Electron-Domain and Molecular Geometries for Two, Three, and Four Electron Domains around a Central Atom

Number of Electron Domains	Electron-Domain Geometry	Bonding Domains	Nonbonding Domains	Molecular Geometry	Bond Angle	Example
2	Linear	2	0	Linear	180°	H—C≡N:
3	Trigonal planar	3	0	Trigonal planar	120°	BH_3 (H–B–H, three H)
		2	1	Bent	<120°	NO_2^- (:Ö–N̈=Ö:)
4	Tetrahedral	4	0	Tetrahedral	109.5°	CF_4
		3	1	Trigonal pyramidal	<109.5°	PBr_3
		2	2	Bent	<109.5°	H_2S (H–S̈–H)

around the central atom here is 180°. For the trigonal planar electron-domain geometry, there are two different molecular geometries possible: trigonal planar (3 bonding domains, 0 lone pairs of electrons) and bent (2 bonding domains, 1 lone pair of electrons). The bond angles around the central atom are 120° for the trigonal planar molecular geometry, but they are less than 120° for the bent molecular geometry because the lone pair of electrons occupies more space than a covalent bond. As a result, the lone pair of electrons tends to compress the bond angle below its idealized value.

Lastly, three molecular geometries are possible within the tetrahedral electron-domain geometry: tetrahedral (4 bonding domains, 0 lone pairs of electrons), trigonal pyramidal (3 bonding domains, 1 lone pair of electrons), and bent (2 bonding domains, 2 lone pairs of electrons). The bond angle is 109.5° for the tetrahedral molecular geometry, but it once again decreases as covalent bonds are replaced with lone pairs of electrons.

Example 5.1 Applications of VSEPR Theory

Carbon dioxide (CO_2) is commonly used as an insufflation gas for minimally invasive surgery (such as laparoscopy) to enlarge body cavities to make it easier to see the surgical area (**Figure 5.3**).

Use VSEPR theory to predict the electron-domain geometry and molecular geometry of each of the following molecules.

a. CO_2 (each O atom is bonded to the central C atom via a double bond)
b. Formaldehyde, CH_2O (the central C atom is bonded to two H atoms via single bonds and to the oxygen via a double bond)
c. Ammonia (**Figure 5.4**), NH_3 (each H atom is bonded to the central N atom via a single bond)
d. Water (**Figure 5.5**), H_2O (each H atom is bonded to the central O atom via a single bond)

Solution

a. **Draw the Lewis structure:**

Before we can apply VSEPR theory, we must draw the Lewis structure. For CO_2, the Lewis structure is

$$:\ddot{O}=C=\ddot{O}:$$

Count the electron domains:

The central carbon atom has two electron domains—namely, the two double bonds to oxygen. Remember, a double bond or triple bond is counted as a single electron domain (a bonding domain) in VSEPR theory.

Determine the electron-domain geometry:

When there are just two electron domains around the central atom, the electron-domain geometry is linear.

Determine the molecular geometry:

Because both electron domains are bonding domains, the bond angle is 180°, and the molecular geometry is linear.

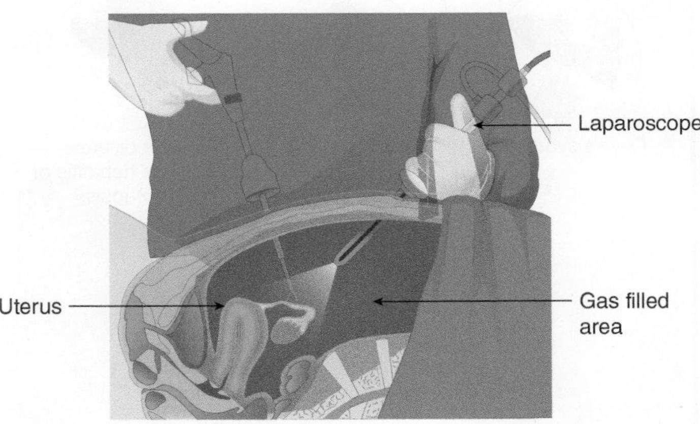

Figure 5.3 Gases such as carbon dioxide are used to inflate the body cavity during laparoscopic surgery.

b. **Draw the Lewis structure:**
The Lewis structure of formaldehyde is

$$\begin{array}{ccc} & :\ddot{\mathrm{O}}: & \\ & \| & \\ & \mathrm{C} & \\ \diagup & & \diagdown \\ \mathrm{H} & & \mathrm{H} \end{array}$$

Count the electron domains:
The Lewis structure of formaldehyde shows that the central C atom has three electron domains—namely, two single bonds to H and one double bond to O.

Determine the electron-domain geometry:
When there are three electron domains around the central atom, the electron-domain geometry is trigonal planar.

Determine the molecular geometry: Because all three electron domains are bonding domains, bond angles are 120°, and the molecular geometry is trigonal planar.

c. High levels of ammonia in an individual may indicate a variety of diseases (**Figure 5.4**).

Draw the Lewis structure:
The Lewis structure of ammonia is

$$\begin{array}{c} \mathrm{H}-\ddot{\mathrm{N}}-\mathrm{H} \\ | \\ \mathrm{H} \end{array}$$

Count the electron domains:
The Lewis structure of ammonia shows that the central N atom has four electron domains—namely, a lone pair of electrons and single bonds to three H atoms.

Determine the electron-domain geometry:
When the central atom is surrounded by four electron domains, the electron-domain geometry is tetrahedral.

Determine the molecular geometry:
When the tetrahedral electron-domain geometry results from three bonding domains and a single lone pair of electrons, the molecular geometry is trigonal pyramidal, with bond angles that are less than 109.5°. In fact, the actual bond angles are 107°.

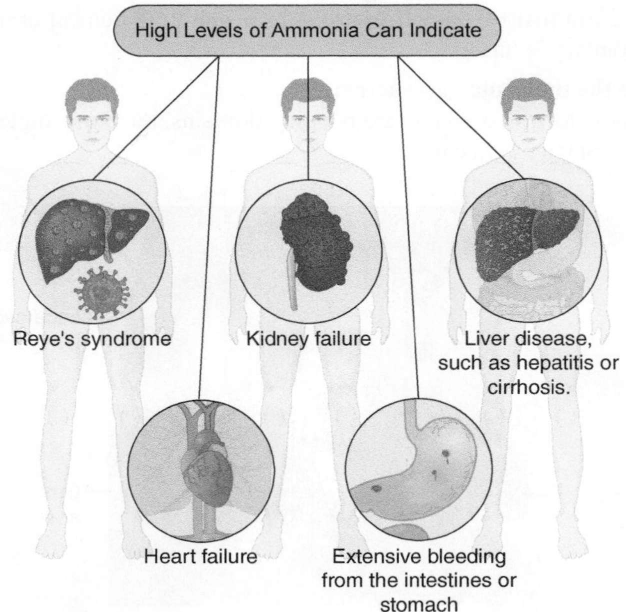

Figure 5.4 High levels of ammonia can indicate diseases such as Reye's syndrome, cirrhosis of the liver, or even heart failure.

d. Water is used to dissolve medicines so they can be administered to patients via intravenous therapy (**Figure 5.5**).

Draw the Lewis structure:

The Lewis structure of water is

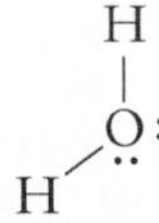

Count the electron domains:

The Lewis structure of water shows that the central oxygen atom has four electron domains—namely, two lone pairs of electrons and single bonds to two H atoms.

Determine the electron-domain geometry:

When the central atom is surrounded by four electron domains, the electron-domain geometry is tetrahedral.

Determine the molecular geometry:

When the tetrahedral electron-domain geometry results from two bonding domains and two lone pairs of electrons, the molecular geometry is bent, with bond angles that are less than 109.5°. The actual bond angle is 104.5°.

Jose Luis Pelaez/Media Bakery

Figure 5.5 The medicines delivered by intravenous therapy are often dissolved in water.

✔ **Learning Check 5.1** Use VSEPR theory to predict the electron-domain geometry and the molecular geometry of the following molecules.

a. BF_3 (each F atom is bonded to fluorine via a single bond)
b. SO_4^{2-} (two O atoms are bonded to the central S atom via single bonds, and two O atoms are bonded to the central S atom via single bonds)
c. CS_2 (each S atom is bonded to the central C atom via a double bond)
d. NO_3^- (one O atom is bonded to the central N atom via a double bond, and two O atoms are bonded to the central N atom via a single bond)

5.2 The Polarity of Covalent Molecules

Learning Objective 2 Use electronegativities to classify the covalent bonds in molecules.

Learning Objective 3 Determine whether covalent molecules are polar or nonpolar.

Chlorine (Cl—Cl) and hydrogen chloride (H—Cl) consist of two atoms each, making both of them diatomic molecules. Both are held together by a single covalent bond. The chlorine molecule consists of two atoms of the same element, whereas the hydrogen chloride molecule consists of two atoms of different elements. This difference influences the distribution of the shared electrons in the two molecules. The distribution of electrical charge over the atoms joined by a chemical bond is called **polarity**.

An electron pair shared by two atoms of the same element is attracted equally to each of the atoms, so the electrons are distributed equally between the two atoms. Thus, on average, the electrons spend exactly the same amount of time associated with each atom. Covalent bonds of this type are called **nonpolar covalent bonds**.

Different atoms generally have different tendencies to attract the shared electrons of a covalent bond. A measurement of this tendency is called electronegativity (**Section 2.7**). The electronegativity of the elements increases from left to right across a period of the periodic table and increases up a group. These trends are shown in **Figure 5.6** for Groups IA–VIIA. As a result of electronegativity differences, atoms of two different elements are shared unequally. That is, the electrons spend more of their time near the atom with the higher electronegativity. The resulting shift in average location of the bonding electrons is called **bond polarization**, and the resulting covalent bond is called a **polar covalent bond**.

As a result of the unequal electron sharing, the more electronegative atom acquires a partial negative charge (δ^-), while the less electronegative atom acquires a partial

polarity The distribution of electrical charge between atoms bonded together.

nonpolar covalent bond A covalent bond in which the bonding pair of electrons is shared equally by the bonded atoms.

bond polarization A result of shared electrons being attracted to the more electronegative atom of a bonded pair of atoms.

polar covalent bond A covalent bond that shows bond polarization; that is, the bonding electrons are shared unequally.

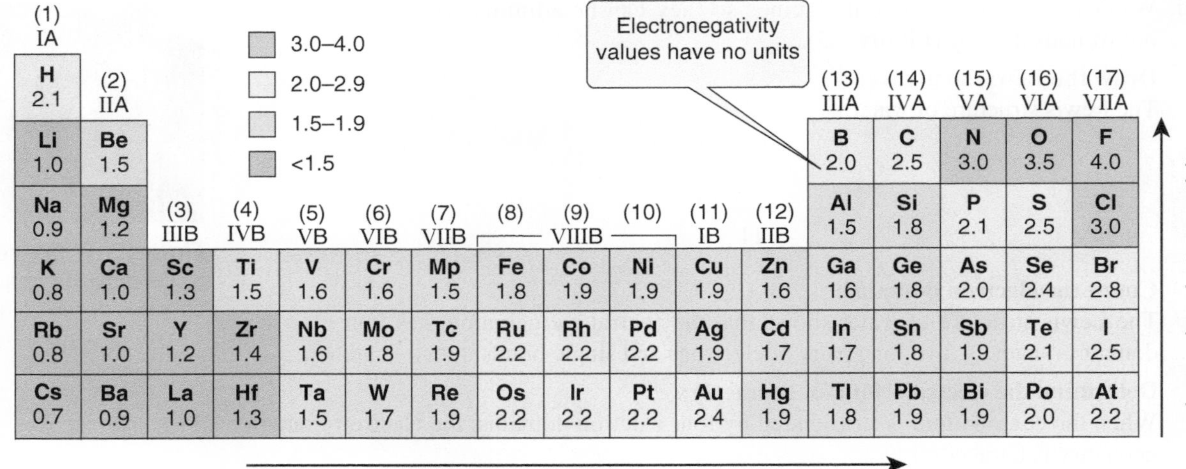

(1) IA	(2) IIA	(3) IIIB	(4) IVB	(5) VB	(6) VIB	(7) VIIB	(8)	(9) VIIIB	(10)	(11) IB	(12) IIB	(13) IIIA	(14) IVA	(15) VA	(16) VIA	(17) VIIA
H 2.1																
Li 1.0	Be 1.5											B 2.0	C 2.5	N 3.0	O 3.5	F 4.0
Na 0.9	Mg 1.2											Al 1.5	Si 1.8	P 2.1	S 2.5	Cl 3.0
K 0.8	Ca 1.0	Sc 1.3	Ti 1.5	V 1.6	Cr 1.6	Mp 1.5	Fe 1.8	Co 1.9	Ni 1.9	Cu 1.9	Zn 1.6	Ga 1.6	Ge 1.8	As 2.0	Se 2.4	Br 2.8
Rb 0.8	Sr 1.0	Y 1.2	Zr 1.4	Nb 1.6	Mo 1.8	Tc 1.9	Ru 2.2	Rh 2.2	Pd 2.2	Ag 1.9	Cd 1.7	In 1.7	Sn 1.8	Sb 1.9	Te 2.1	I 2.5
Cs 0.7	Ba 0.9	La 1.0	Hf 1.3	Ta 1.5	W 1.7	Re 1.9	Os 2.2	Ir 2.2	Pt 2.2	Au 2.4	Hg 1.9	Tl 1.8	Pb 1.9	Bi 1.9	Po 2.0	At 2.2

Figure 5.6 Pauling electronegativity values for the first six periods of elements. The values increase from left to right across a period and from bottom to top in a group. As a result, fluorine [at the top of Group VIIA(17)] has the highest electronegativity value, and cesium [at the bottom of Group IA(1)] has the lowest.

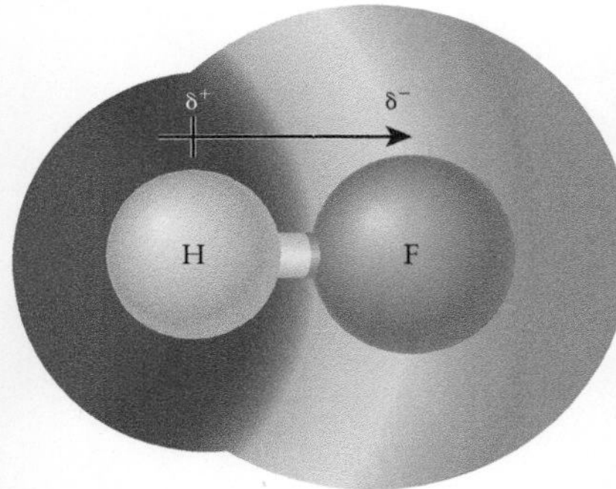

Figure 5.7 The polar covalent nature of HF. The color red indicates greater electron density, and the color blue indicates lesser electron density.

positive charge (δ^+) (**Figure 5.7**). These partial charges should make intuitive sense. The more electronegative atom becomes partially negative because it is surrounded by a greater amount of electron density, while the less electronegative atom becomes partially positive because it has lost some of its electron density. The molecule as a whole still has a net charge of zero, so it remains neutral overall—it just has an uneven charge distribution.

Example 5.2 Electronegativity and Bond Polarization

Using only the periodic table, determine (1) the more electronegative element and (2) the direction of bond polarization resulting from the polarization of the following diatomic covalent molecules.

a. I—Cl b. Br—Br c. C≡O

Solution

a. **Determine the more electronegative element:**
Both chlorine (Cl) and iodine (I) belong to Group VIIA(17). Chlorine is higher in the group, so Cl is more electronegative than I.

Determine the direction of bond polarization:
The bond polarization (the shift in average bonding-electron location) will be toward chlorine, as indicated by the arrow shown below the bond. The result is a partial negative charge on Cl and a partial positive charge on I:

$$\overset{\delta^+}{\mathrm{I}}-\overset{\delta^-}{\mathrm{Cl}}$$
$$\longmapsto$$

b. **Determine the more electronegative element:**
Both bromine (Br) atoms have the same electronegativity.

Determine the direction of bond polarization:
Because there is no difference in electronegativity between the two atoms, there is no bond polarization. Instead, the charge distribution between the two atoms is equal.

The molecule is nonpolar covalent.

c. Carbon monoxide (CO) is harmful when breathed in because it displaces oxygen in the blood, depriving the heart, brain, and other vital organs of oxygen (**Figure 5.8**).

Determine the more electronegative element:
Oxygen (O) and carbon (C) both belong to the second period. Oxygen, however, is located farther to the right in the period, so it is more electronegative than carbon.

Determine the direction of bond polarization:
The bond polarization will be toward oxygen, as indicated by the arrow shown below the bond. The result is a partial negative charge on O and a partial positive charge on C:

$$\overset{\delta^+}{C}\equiv\overset{\delta^-}{O}$$
$$\longmapsto$$

What Is Carbon Monoxide Poisoning?

It's caused by excessive exposure to carbon monoxide, which builds up in the bloodstream

Symptoms include nausea, chest pain, and irregular heartbeats, but unconsciousness and death can occur

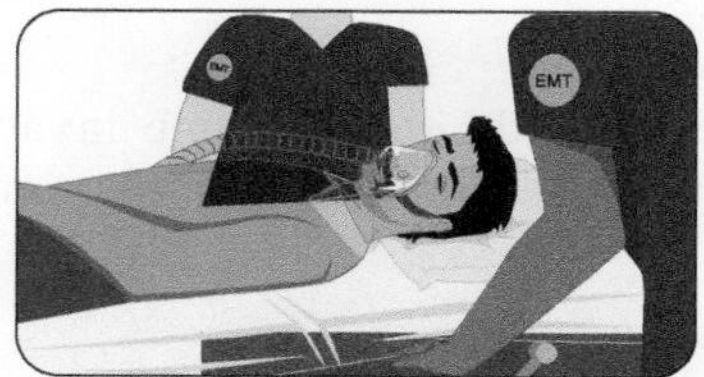

It's commonly treated with pressurized oxygen to clear the blood

It's most easily prevented with a carbon monoxide alarm

Figure 5.8 Carbon monoxide poisoning is often fatal. CO is odorless and colorless, so it may render a person unconscious before the danger is realized. It can be prevented, however, with a functioning carbon monoxide detector.

> ✔ **Learning Check 5.2** Using only the periodic table, show the direction of any bond polarization and the resulting charge distribution in the following molecules.
>
> a. N≡N b. I—Br c. H—Br

The extent of bond polarization depends on the electronegativity difference (ΔEN) between the bonded atoms and forms one basis for the classification of bonds (**Figure 5.9**).

- $\Delta EN = 0$: The electrons are shared in a perfectly even manner; this is a *pure covalent bond*.
- $\Delta EN < 0.4$: The electrons are shared evenly enough that the bond is called a *nonpolar covalent bond*.
- $\Delta EN = 0.4$ to 1.8: The electrons are shared unevenly enough that one atom is partially positive and the other is partially negative; this is a *polar covalent bond*.
- $\Delta EN > 1.8$: The electrons are fully lost or gained instead of shared; this is an *ionic bond*.

Generally, bonds between a metal atom and nonmetal atom are ionic, while those between two nonmetal atoms are some type of covalent bond.

Figure 5.9 The electronegativity difference (ΔEN) increases as the sharing of the bonding electrons becomes more and more unequal.

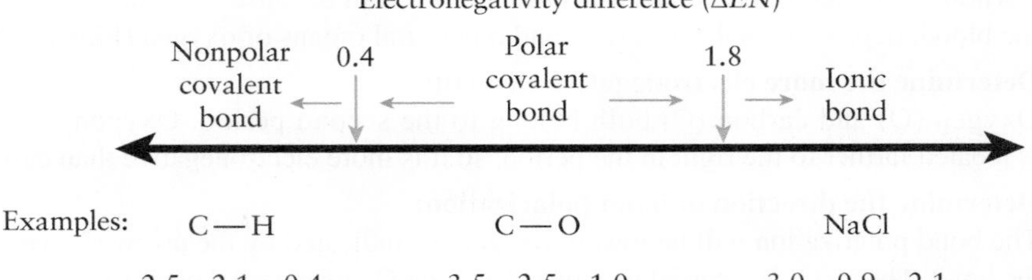

Example 5.3 Classifying Covalent Bonds

Use **Figures 5.6** and **5.9** to classify the bonds in the following compounds as nonpolar covalent, polar covalent, or ionic.

a. ClF b. MgO c. PI_3

Solution

a. The electronegativities of chlorine (Cl) and fluorine (F) are 3.0 and 4.0, respectively. Obtain ΔEN by subtracting the smaller electronegativity value from the larger, regardless of the order of the elements in the formula. Thus,

$$\Delta EN = 4.0 - 3.0 = 1.0$$

According to **Figure 5.9**, this bond is polar covalent.

b. Magnesium oxide (MgO) is used in some antacids and in some laxatives (**Figure 5.10**). The electronegativity values of magnesium (Mg) and oxygen (O) are 1.2 and 3.5, respectively. Thus,

$$\Delta EN = 3.5 - 1.2 = 2.3$$

So, the bond is ionic.

c. The electronegativity values of phosphorus (P) and iodine (I) are 2.1 and 2.5, respectively. For each phosphorus–iodine bond, then,

$$\Delta EN = 2.5 - 2.1 = 0.4$$

Thus, this bond is classified as nonpolar covalent.

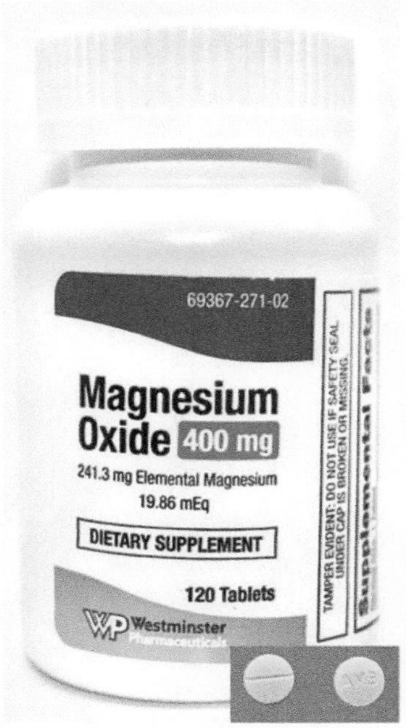

Source: Westminster Research

Figure 5.10 Magnesium oxide can be used as a laxative to empty the bowels prior to surgery.

✔ **Learning Check 5.3** Use **Figures 5.6** and **5.9** to classify the bonds in the following compounds as nonpolar covalent, polar covalent, or ionic.

a. KF b. NO c. AlN d. K_2O

polar molecule A molecule that contains polarized bonds and in which the resulting charges are distributed nonsymmetrically throughout the molecule.

nonpolar molecule A molecule that contains no polarized bonds, or a molecule containing polarized bonds in which the resulting charges are distributed symmetrically throughout the molecule.

Being able to predict the polarity of bonds within molecules is the first step toward being able to predict the polar nature of the molecules themselves. The terms *polar* and *nonpolar*, when used to describe molecules, indicate their overall electron charge distribution. In a **polar molecule**, one or more of the bonds are polar, and the charge distribution resulting from these polar bonds is nonsymmetric, as indicated by the color gradient for CH_3F in **Figure 5.11a**. In a **nonpolar molecule**, either all of the bonds are nonpolar, as in H_2 (**Figure 5.11b**), or any charges resulting from individual polar bonds are symmetrically distributed throughout the molecule, in which case they cancel each other, as in CO_2 (**Figure 5.11c**).

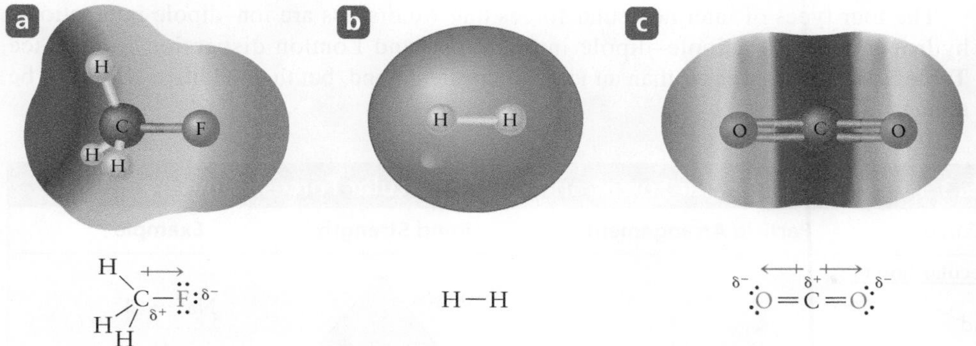

Figure 5.11 These models of electron density show the equal and unequal sharing of electrons, from blue (partial positive) to green (neutral) to red (partial negative). (a) CH_3F is a polar molecule due to the polar C—F bond. (b) H_2 is nonpolar because the H—H bond is nonpolar. (c) Both C=O bonds in CO_2 are polar, but they point in opposite directions, so they cancel each other, making CO_2 a nonpolar molecule.

Table 5.2 lists several polar and nonpolar molecules. Notice in particular that all of the bonds in CO_2, BF_3, and CCl_4 are polar, but all three molecules are nonpolar. This occurs because their molecular geometries, as determined using VSEPR theory, are so symmetrical that their individual polar covalent bonds cancel each other.

Table 5.2 Molecular Geometry and the Polarity of Molecules

Molecular Formula	Lewis Structure	Electron-Domain Geometry	Molecular Geometry	Bond Polarization and Molecular Charge Distribution	Is the Molecule Polar or Nonpolar?
H_2	H—H	Linear	Linear	H—H	Nonpolar
HCl	H—Cl:	Linear	Linear	$\delta+$ H ⇸ Cl $\delta-$	Polar
CO_2	:O=C=O:	Linear	Linear	$\delta-$ O ⇸ $\delta+$ C ⇸ O $\delta-$	Nonpolar
BF_3	:F: — B — :F: :F:	Trigonal planar	Trigonal planar	$\delta-$ F, $\delta+$ B, $\delta-$ F, $\delta-$ F	Nonpolar
N_2O	:N≡N—O:	Linear	Linear	N≡$\overset{+}{N}$ ⇸ O^-	Polar
H_2O	H—O:—H	Tetrahedral	Bent	$\delta+$ H ⇸ O $\delta-$ ⇷ H $\delta+$	Polar
CCl_4	:Cl: — C — :Cl: :Cl: :Cl:	Tetrahedral	Tetrahedral	$\delta-$ Cl, $\delta+$ C, $\delta-$ Cl, $\delta-$ Cl, Cl	Nonpolar

5.3 Intermolecular Forces

Learning Objective 4 Identify the major attractive force between molecules in a given substance.

The covalent bonds in covalent molecules and the ionic bonds in ionic compounds are classified as **intramolecular forces** because they occur *within* the molecules and compounds. There are also relatively weak forces *between* individual covalent molecules that are classified as **intermolecular forces**.

intramolecular forces The attractive forces that exist *within* molecules.

intermolecular forces The attractive forces that exist *between* molecules.

The four types of intermolecular forces that we discuss are ion–dipole interactions, hydrogen bonding, dipole–dipole interactions, and London dispersion forces (see **Table 5.3**). Each is weaker than an ionic or covalent bond, but their additive effect can be quite strong.

Table 5.3 How Intramolecular Forces and Intermolecular Forces Compare

Type of Force	Particle Arrangement	Bond Strength	Example
Intramolecular Forces			
Ionic bond	+ − + / − + −	Strongest	$Na^+ \cdots Cl^-$
Covalent bond (X = nonmetal)	X:X		Cl — Cl
Intermolecular Forces			
Ion–dipole interaction	$\overset{\delta+}{Y}\overset{\delta-}{X} \cdots +$		$\overset{\delta+}{H}—\overset{\delta-}{F}\cdots Na^+$
Hydrogen bonds (X = F, O, or N)	$\overset{\delta+}{H}\overset{\delta-}{X} \cdots \overset{\delta+}{H}\overset{\delta-}{X}$		$\overset{\delta+}{H}—\overset{\delta-}{F}\cdots\overset{\delta+}{H}—\overset{\delta-}{F}$
Dipole–dipole attractions (X and Y = different nonmetals)	$\overset{\delta+}{Y}\overset{\delta-}{X} \cdots \overset{\delta+}{Y}\overset{\delta-}{X}$		$\overset{\delta+}{Br}—\overset{\delta-}{Cl}\cdots\overset{\delta+}{Br}—\overset{\delta-}{Cl}$
London dispersion forces	$\overset{\delta+}{X}:\overset{\delta-}{X} \cdots \overset{\delta+}{X}:\overset{\delta-}{X}$	Weakest	$\overset{\delta+}{F}—\overset{\delta-}{F}\cdots\overset{\delta+}{F}—\overset{\delta-}{F}$

ion–dipole interaction The attraction between a charged ion (cation or anion) and a polar molecule.

The strongest of the four intermolecular forces is the **ion–dipole interaction**. This interaction occurs when a cation or anion is attracted to the partial negative or partial positive end of a polar molecule, respectively. **Figure 5.12** shows the two ion–dipole interactions that occur in saline solution, which is NaCl dissolved in water.

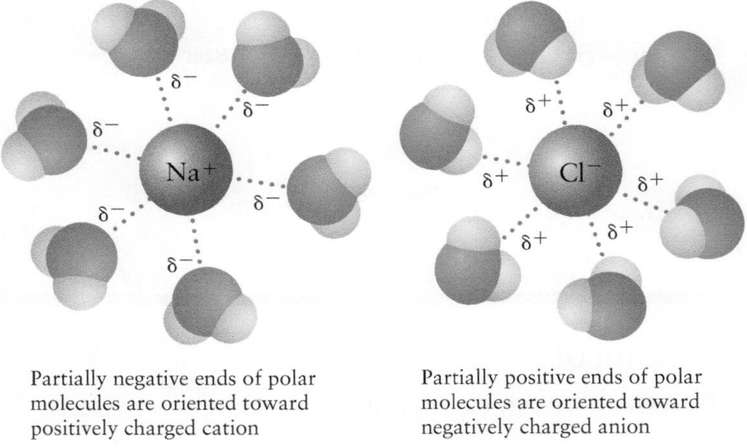

δ−
δ−
δ−
Na+
δ−
δ−
δ−
δ+
δ+
δ+
Cl−
δ+
δ+
δ+
Partially negative ends of polar molecules are oriented toward positively charged cation
Partially positive ends of polar molecules are oriented toward negatively charged anion

Figure 5.12 NaCl dissolves in water because both Na^+ and Cl^- ions form ion–dipole interactions with polar water molecules. Na^+ interacts with the negative end (the oxygen end) of the water molecules, whereas Cl^- interacts with the positive end (the hydrogen end) of them.

hydrogen bonding The attraction between a hydrogen atom that is bonded to a F, N, or O atom and any F, N, or O atom on another molecule or a distant region of the same molecule.

The second strongest intermolecular force is **hydrogen bonding**. Despite the name, a hydrogen bond is an attractive force between molecules, not a covalent bond. A hydrogen bond is the relatively strong attraction between a hydrogen atom that is bonded to a F, N, or O and a pair of electrons on a F, N, or O atom on another molecule. For example, hydrogen bonds readily form between individual water molecules (**Figure 5.13**). The partially positive hydrogen of one water molecule becomes attracted to the oxygen lone

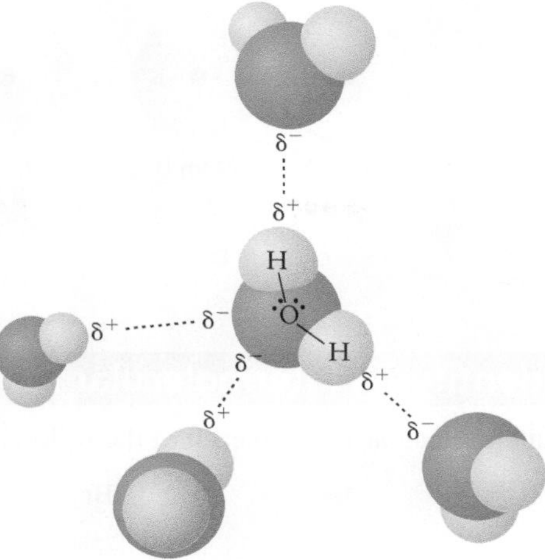

Figure 5.13 Hydrogen bonding in water. The dotted lines represent hydrogen bonds, whereas the solid lines represent covalent bonds.

pairs of a second water molecule, and so on until an entire network of hydrogen bonds is created. This network is responsible for the surface tension of water (**Figure 5.14**) and the unusually high melting and boiling points of water (**Section 5.5**). Hydrogen bonding is also responsible for the unique double-helical structure of DNA (see **Section 13.3**). Hydrogen bonds can also form within distant reaches of the same molecule.

Figure 5.14 The surface tension of water, created by hydrogen bonding, is so strong that some animals can run along its surface without sinking.

The next strongest intermolecular force is **dipole–dipole interactions**. These interactions occur between two or more molecules that have permanent dipoles themselves. Thus, the partially positive end of one molecule is attracted to the partially negative end of another molecule. This can be seen in **Figure 5.15**, which depicts the interaction between two molecules of H—F.

dipole–dipole interactions The attractive force that exists between the positive end of one polar molecule and the negative end of another.

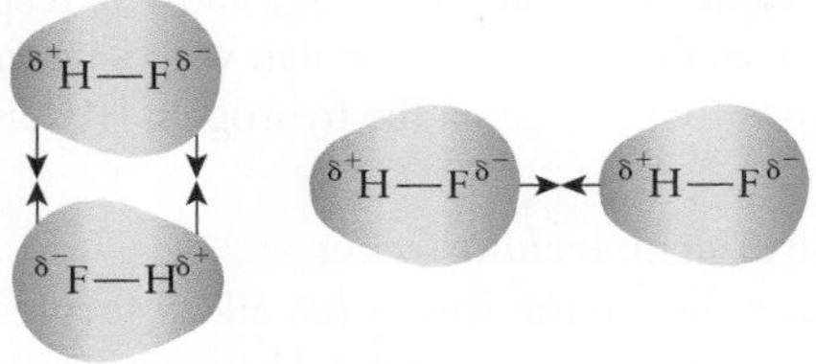

Figure 5.15 Dipole–dipole interactions between polar compounds.

London dispersion forces are the weakest intermolecular force. They occur between all molecules because they result from the temporary shift of the electron clouds in nonpolar bonds when they come within close proximity of one another (**Figure 5.16**). This type of intermolecular force often dominates in the long chains of hydrocarbons (molecules made up exclusively of bonds between carbon and hydrogen atoms; see **Section 8.3**). This is in part because the clouds of electron density of large hydrocarbons are easily distorted—that is, they are *polarizable*. Dispersion forces tend to increase with increasing molecule size for this reason. In addition to size, the shape of a molecule affects the strength of the forces. Long, narrow molecules tend to have stronger forces than short, globular molecules (**Figure 5.17**).

London dispersion forces Very weak attractive forces acting between the particles of all matter. They result from momentary nonsymmetric electron distributions in molecules or atoms.

Figure 5.16 Formation of an induced dipole between two helium atoms.

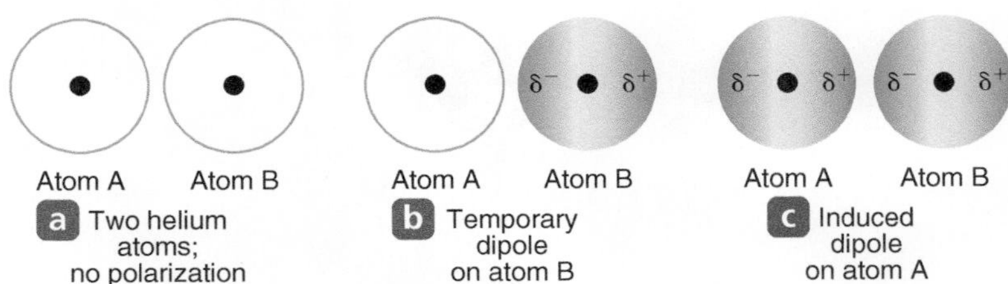

Large contact area; strong attraction

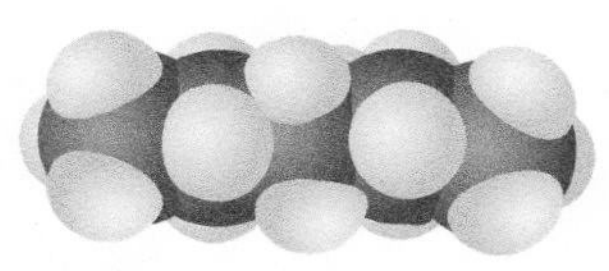

a *n*-pentane

Less surface area; less attraction

b isopentane

Small contact area; weakest attraction

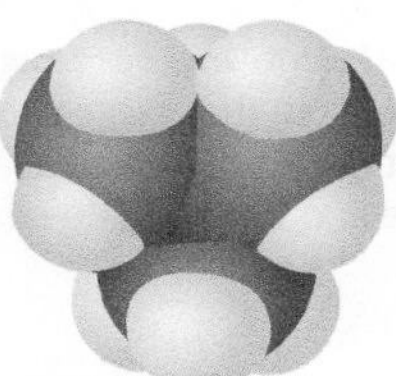

c neopentane

Figure 5.17 The strength of London dispersion forces depends on the shape of the molecule. (a) They tend to be strongest for long, linear molecules because they have large surface areas. (b) and (c) Because the surface area decreases, London dispersion forces decrease in strength as the molecule becomes more and more compact.

Example 5.4 Identifying Intermolecular Forces

Identify the strongest intermolecular force in each of the following substances.

a. NI_3 b. HF c. Br_2

Solution

a. **Draw the Lewis structure:**

The Lewis structure of NI_3 consists of a central nitrogen with one lone pair and three covalently bonded iodine atoms.

$$\overset{\delta^-}{\ddot{\mathrm{N}}}(\overset{}{\mathrm{I}}^{\delta^+})_3$$

Determine the polarity:

The electron-domain geometry is tetrahedral, but the molecular geometry of NI_3 is trigonal pyramidal. The nitrogen has an electronegativity of 3.0, and each iodine has an electronegativity of 2.5. Each bond, then, is polarized toward nitrogen, resulting in a partial negative charge on the nitrogen and partial positive charges on the iodine atoms. NI_3 has a permanent dipole, and is therefore polar.

Determine the major intermolecular force:

All molecules exhibit London dispersion forces, but a polar molecule like NI_3 also exhibits dipole–dipole interactions, which are stronger, making them the major force.

b. **Draw the Lewis structure:**

The Lewis structure of HF consists of a hydrogen covalently bonded to fluorine with three lone pairs.

$$\overset{\delta^+}{\mathrm{H}}-\overset{\delta^-}{:\!\ddot{\mathrm{F}}\!:}$$

Determine the polarity:

The electronegativity values for H and F are 2.1 and 4.0, respectively. This difference in electronegativity makes the H—F bond polar, with a partial negative charge on the fluorine and a partial positive charge on the hydrogen. HF has a permanent dipole and is polar.

Determine the major intermolecular force:

HF will exhibit London dispersion forces (as all molecules do), and it will exhibit dipole–dipole interactions because it is polar. However, neither of these are the strongest intermolecular force present. Instead, HF molecules can form hydrogen bonds when the hydrogen of one molecule interacts with the lone pair on the fluorine of another molecule. As a result, the strongest intermolecular force is hydrogen bonding.

c. **Draw the Lewis structure:**

The Lewis structure of Br_2 consists of two bromines, each with three lone pairs, connected via a single covalent bond.

$$:\!\ddot{\underset{..}{\mathrm{Br}}}-\ddot{\underset{..}{\mathrm{Br}}}\!:$$

Determine the polarity:

The electronegativity of each bromine atom is 2.8, so the Br—Br bond is nonpolar.

Determine the major intermolecular force:
Because Br_2 is nonpolar, the only intermolecular forces exhibited by the molecule are London dispersion forces.

✔ **Learning Check 5.4** Identify the strongest intermolecular force in each of the following substances.

a. NH_3 b. O_2 c. H_2S d. KCl in water

5.4 Energy and Properties of Matter

Learning Objective 5 Distinguish phenomena that demonstrate kinetic energy from those that demonstrate potential energy.

Learning Objective 6 Identify the states of matter using five properties: density, shape, compressibility, particle interaction, and molecular movement.

In order to understand how these forces shape matter, we must discuss in more detail the relationship between matter and energy. While we have already defined matter as anything that occupies space and has mass (see **Section 1.1**), energy is the ability to transfer heat or do work. That is, energy is something that must be expended to perform some type of task. It takes energy to get out of bed in the morning, just as it takes energy to run a marathon. It even takes energy to read a textbook!

The energy of matter in motion is called **kinetic energy**. Kinetic energy describes the motion of particles on the molecular level as well as objects on the macro level. Kinetic energy is related to **disruptive forces**, which tend to scatter particles and make them independent of one another. **Cohesive forces**, on the other hand, such as intermolecular forces, attract particles. Cohesive forces can best be related to **potential energy**, which is energy that is stored (**Figure 5.18**). On the macro level, potential energy describes a boulder sitting at the top of a hill, whereas on the micro level, it describes the energy stored in chemical bonds.

kinetic energy The energy a particle has as a result of its motion.

disruptive forces The forces resulting from particle motion; they are associated with kinetic energy.

cohesive forces The attractive forces between particles; they are associated with potential energy.

potential energy The energy that is stored in particles due to their position, composition, or arrangement.

Figure 5.18 When a person skates up and down a ramp, they are using kinetic energy. When a person is perched at the top of the ramp preparing to skate down it, they are exhibiting potential energy.

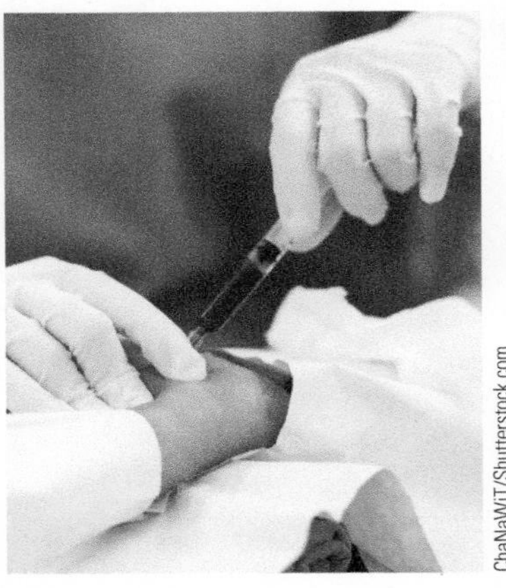
ChaNaWiT/Shutterstock.com

Figure 5.19 Arterial blood gas analysis is a method used to determine the acidity of a patient's blood.

Example 5.5 Determining the Type of Energy

The acidity of our blood is maintained within a very narrow range by the bicarbonate buffer system (**Figure 5.19** and **Section 7.7**). Is the energy stored in the ionic bond between the sodium cation and the bicarbonate anion in $NaHCO_3$ kinetic or potential?

Solution

While Na^+ is bonded to HCO_3^-, the energy is stored and no work is being done, so the ionic bond represents potential energy.

✔ Learning Check 5.5 Do the following examples represent kinetic energy or potential energy?

a. The reaction of ATP to ADP to perform cellular work
b. Energy stored in the covalent bond between carbon and chlorine in carbon tetrachloride (CCl_4)
c. The chemical digestion that takes place in the stomach

On the molecular level, the state of matter of a substance depends on the relative strengths of the intermolecular (cohesive) forces that hold the particles together and the kinetic (disruptive) forces that tend to separate them. The three states of matter—namely, solids, liquids, and gases—are shown for water in **Figure 5.20**. They can be identified and distinguished by the differences in their physical properties. The five properties that we use for this purpose are density, shape, compressibility, particle interaction, and molecule movement.

Ralph Lee Hopkins/National Geographic Image Collection/Getty Images

Figure 5.20 This winter scene at Yellowstone National Park shows water in all three forms (solid, liquid, and gaseous vapor) simultaneously.

Density (**Section 1.10**) is the mass of a sample of matter divided by its volume. As shown in **Table 5.4**, the density of solids and liquids tends to be high, whereas the density of gases tends to be low. **Figure 5.21** shows how the *shape* of matter depends on its physical state. Solids do not conform to the shape of their container, whereas liquids and gases do. In the case of liquids, the shape depends on the extent to which the container is filled. Gases always fill the container completely.

© Spencer L. Seager

a Solids have a shape and volume that is independent of the container.

© Spencer L. Seager

b Liquids take the shape of the part of the container that they fill. Each sample shown here has the same volume.

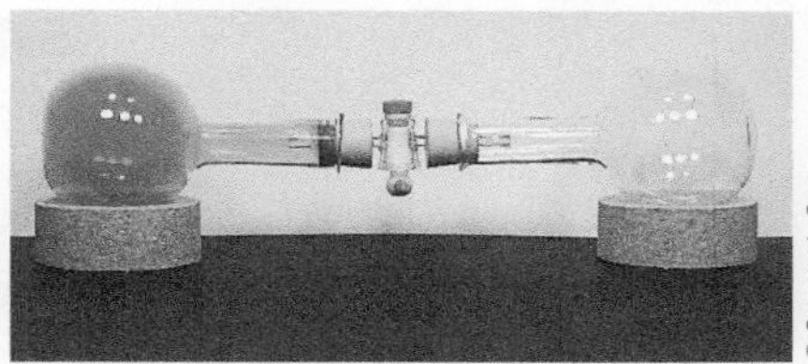

© Spencer L. Seager

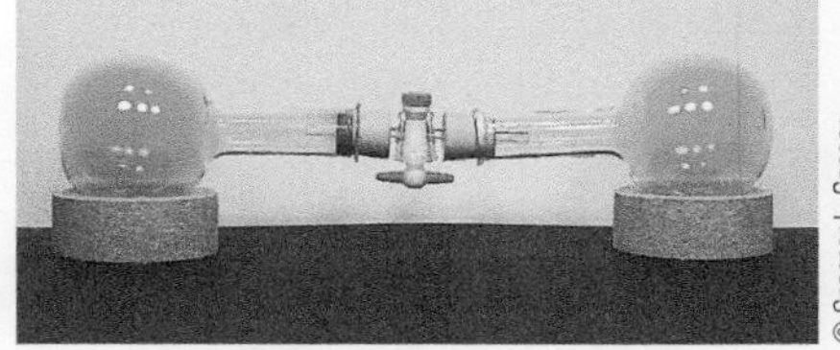

© Spencer L. Seager

c Gases completely fill and take the shape of their container. When the valve separating the two parts of the container is opened, the gas fills the entire container volume (bottom photo).

Figure 5.21 Shape characteristics of the (a) solid, (b) liquid, and (c) gaseous states of matter.

compressibility The change in volume of a sample resulting from a pressure change acting on the sample.

Compressibility, the change in volume resulting from a pressure change, is quite high for gases. Compressibility allows a lot of gas to be squeezed into a small volume if the gas is put under sufficient pressure—think of automobile tires, or a tank of compressed oxygen used to treat a patient with respiratory issues. Compressibility for liquids is quite low because liquid particles are already quite close together, and it is even lower for solids.

Particle interaction consists of the attractive forces between molecules, mainly intermolecular forces. The strength of these intermolecular forces directly affects the *movement of molecules* in each state. Because gases have little to no intermolecular forces (the molecules are too far apart), their particles can move freely past one another. Gas particles, then, can diffuse and flow easily in space. Attractive forces increase in strength from gases to liquids to solids because the particles get closer together. Therefore, particle interaction is closely linked to the fifth principle, *molecule movement*. The movement of liquid molecules is more restricted than that of gases, and the diffusion or flow of solid molecules is extremely limited.

These five properties are compared for the three states of matter in **Table 5.4**. **Sections 5.4A** to **5.4C** provide a detailed explanation of how these properties shape the states of matter, then **Section 5.5** focuses on the effects of energy on changes of state.

Table 5.4 Physical Properties of Solids, Liquids, and Gases

	State		
Property	**Solid**	**Liquid**	**Gas**
Density	High	High—usually lower than that of corresponding solid	Low
Shape	Definite	Indefinite—takes shape of container to the extent it is filled	Indefinite—takes shape of container it fills
Compressibility	Low	Low—usually greater than that of corresponding solid	High
Particle interaction	Strong attractive forces	Intermediate attractive forces	Weak attractive forces
Molecule movement	Does not flow	Intermediate flow	Flows readily

5.4A Solids

In solids, the cohesive forces are stronger than the disruptive forces (see **Figure 5.22a**). Disruptive kinetic energy causes the particles to vibrate about their fixed positions in the crystal lattice, but strong cohesive forces (i.e., intermolecular forces) prevent the solid from breaking down. The properties of solids in **Table 5.4** are explained as follows:

- *High density*. The particles of solids are located as closely together as possible. Therefore, large numbers of particles are contained in a small volume, resulting in a high density.
- *Definite shape*. The strong cohesive forces hold the particles of solids in essentially fixed positions, resulting in a definite shape.
- *Low compressibility*. Because there is already very little space between the particles of solids, increased pressure cannot push them any closer together, so it has little effect on the volume of the solid.
- *Strong particle interactions*. The intermolecular forces in solids are very strong, so the particles "stick together."
- *Very little molecule movement*. The strong attractive forces in solids restrict the movement of molecules, thereby preventing the diffusion and flow of particles.

Figure 5.22 A particle-based view of solids, liquids, and gases.

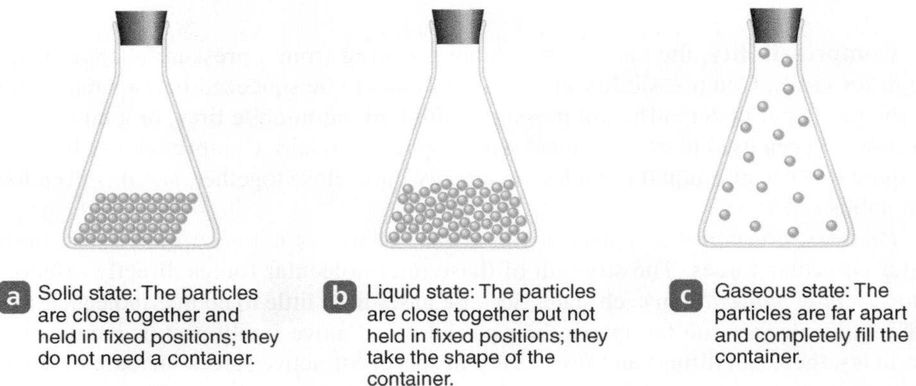

a Solid state: The particles are close together and held in fixed positions; they do not need a container.

b Liquid state: The particles are close together but not held in fixed positions; they take the shape of the container.

c Gaseous state: The particles are far apart and completely fill the container.

5.4B Liquids

Particles in the liquid state are randomly packed and relatively close to each other (see **Figure 5.22b**). They are in constant, random motion, sliding freely over one another, but they lack sufficient kinetic energy to separate completely from each other. In liquids, cohesive forces are slightly favored over disruptive forces. The characteristic properties of liquids can be explained as follows:

- *High density*. The particles of liquids are not widely separated; they essentially touch each other. Therefore, there will be a large number of particles per unit volume, which results in a high density.
- *Indefinite shape*. Although not completely independent of each other, the particles in a liquid are free to move over and around each other in a random manner, so they are limited only by the container walls and the extent to which the container is filled.
- *Low compressibility*. Because the particles in a liquid essentially touch each other, there is very little space between them. Therefore, increased pressure cannot squeeze the particles much more closely together. Nevertheless, a substance in the liquid state is more compressible than when it is in the solid state.
- *Intermediate particle interactions*. The intermolecular forces in liquids are intermediate in strength. The attractive forces compete with increasing disruptive forces compared to solids.
- *Substantial molecule movement*. The intermediate intermolecular forces in liquids allow molecules to move, which increases flow and causes particles to diffuse freely.

5.4C Gases

Disruptive forces completely overcome cohesive forces between particles in the gaseous (vapor) state. As a result, the particles of a gas move essentially independently of one another in a totally random way (see **Figure 5.22c**). Under ordinary pressure, the particles are relatively far apart except when they collide with each other. Between collisions with each other or with the container walls, gas particles travel in straight lines. The particle velocities and resultant collision frequencies are quite high for gases. The characteristic properties of gases can be explained as follows:

- *Low density*. The particles of a gas are widely separated. There are relatively few of them in a given volume, which means there is little mass per unit volume.
- *Indefinite shape*. The forces of attraction between particles have been overcome by kinetic energy, so the particles are free to travel in all directions. The particles, therefore, completely fill the container and assume its inner shape.
- *High compressibility*. The gas particles are widely separated, so a gas sample is mostly empty space. When pressure is applied, the particles are easily pushed closer together, decreasing the amount of empty space and the gas volume, as shown in **Figure 5.23**.
- *Weak particle interactions*. The intermolecular forces in gases are extremely weak, and disruptive forces dominate over attractive forces.
- *Extreme molecule movement*. There are very few attractive forces in gases, so the molecules move quickly past one another, which causes the particles to diffuse and flow freely.

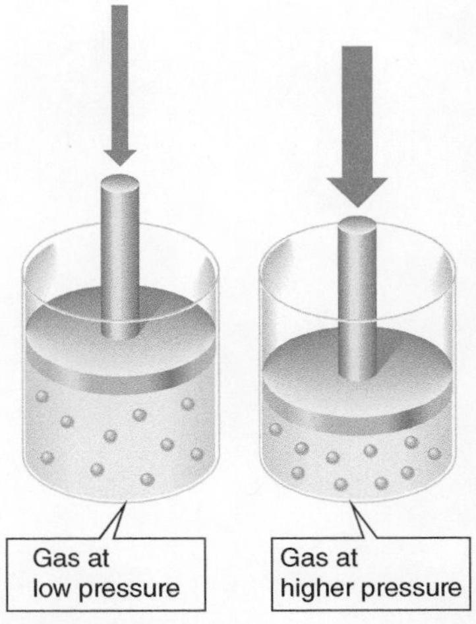

Figure 5.23 The compression of a gas.

Example 5.6 Determining the State of Matter

Identify the state of matter of the following substances printed in italics.

a. The *nitrous oxide* administered to sedate a patients during dental procedures, has a low density and diffuses freely into the nasal hood (**Figure 5.24**).
b. *Ice chips*, which have definite shape and low compressibility, are provided to individuals who are in the process of giving birth.

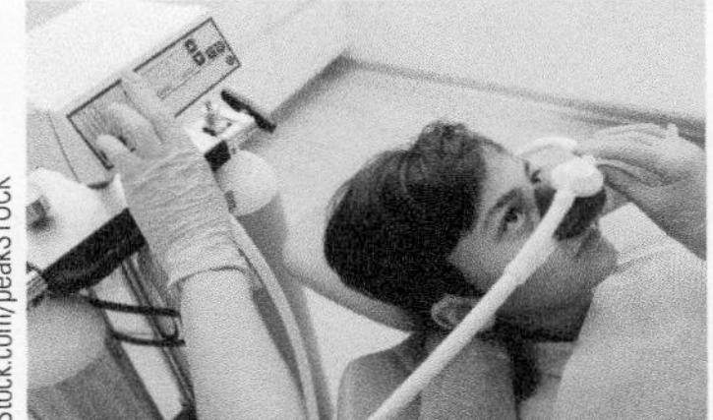

Figure 5.24 Child receiving nitrous oxide sedation during a visit to the dentist.

Solution

a. Nitrous oxide has low density and flows freely, so it must be a gas.
b. Ice is a solid because it has a definite shape and low compressibility.

✔ Learning Check 5.6 Identify the state of matter of the following substances printed in italics.

a. *Dimethylsulfoxide (DMSO)*, used to treat painful bladder syndrome, has an indefinite shape and intermediate attractive forces.
b. A patient experiencing *carbon monoxide* poisoning is brought to the emergency department for treatment. Carbon monoxide has a very low density, no definite shape, and weak attractive forces.

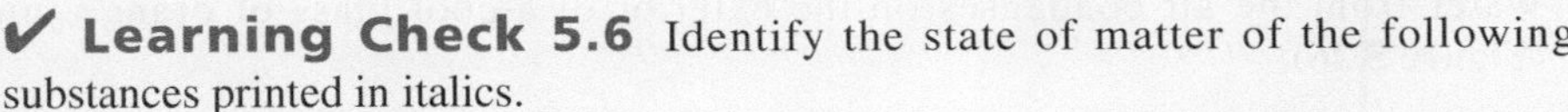

5.5 Changes of State

Learning Objective 7 Classify changes of state as exothermic or endothermic.

Matter can be changed from one state into another by processes such as heating, cooling, or changing the pressure. Heating and cooling are the processes most often used. A change in state that requires an input of heat is said to be **endothermic**, whereas one in which heat is given up (or removed) is said to be **exothermic**. Endothermic changes are those in which particles are moved farther apart as disruptive forces overcome cohesive intermolecular

endothermic A process that absorbs heat.

exothermic A process that liberates heat.

forces, such as the change of a solid to a liquid or a liquid to a gas (**Figure 5.25**). Changes in the opposite direction—from a gas to a liquid to a solid—are exothermic. In exothermic processes, particles are moved closer together as cohesive forces overcome disruptive forces.

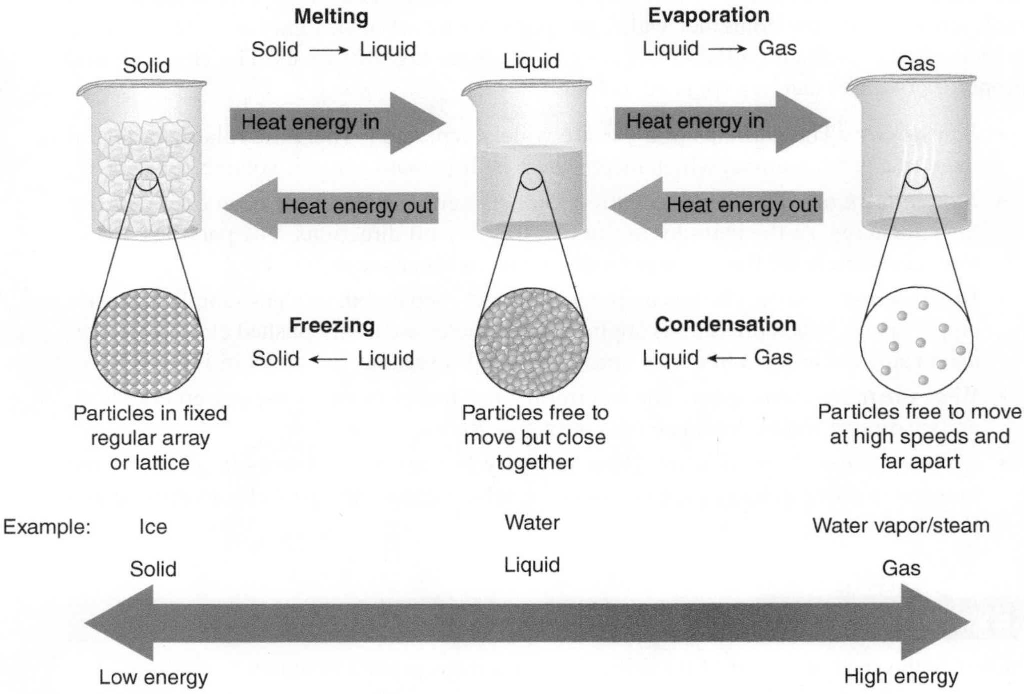

Figure 5.25 Endothermic and exothermic changes of state.

Example 5.7 Classifying Processes as Endothermic or Exothermic

Classify the following processes as endothermic or exothermic.

a. Ice chips melt at room temperature.
b. Water from the air condenses on the exterior of a cool glass of orange juice (**Figure 5.26**).

Solution

a. When solid ice melts to liquid water, heat energy must be absorbed to overcome the hydrogen bonding (i.e., the cohesive forces) that hold water molecules together. Melting is an *endothermic* process.
b. When gaseous water condenses to liquid water, heat energy is released to reduce disruptive forces (i.e., the kinetic energy of the molecules). Condensation is an *exothermic* process.

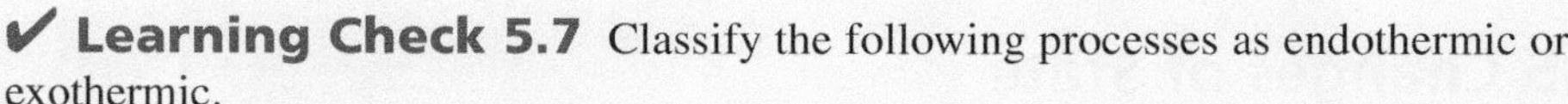

✔ **Learning Check 5.7** Classify the following processes as endothermic or exothermic.

a. Isopropyl alcohol evaporates from a sanitizing wipe.
b. Water freezes on a sidewalk in winter.
c. Dry ice is used to keep blood donations and organs for transplant cool during transport. At room temperature, dry ice changes directly from a solid to a gas without first becoming a liquid.

pixelman/Shutterstock.com

Figure 5.26 Condensation forming on a cool glass of orange juice is an exothermic process.

5.5A Vapor Pressure, Evaporation, and Condensation

Learning Objective 8 Use the factors that affect evaporation and condensation to rank substances in order of increasing vapor pressure.

Water in a container will soon disappear if the container is left uncovered. **Evaporation**, or **vaporization**, is an endothermic process that takes place when molecules leave the surface of a liquid. The rate of evaporation depends on the temperature of the liquid and the surface area from which the molecules can escape. Temperature is an important factor because it is directly related to the speed and kinetic energy of the molecules and thus their ability to break away from the attractive forces present at the liquid's surface. No covalent bonds are broken in evaporation—only the cohesive forces between molecules.

evaporation or **vaporization** An endothermic process in which a liquid is changed to a gas.

Evaporating molecules carry significant amounts of kinetic energy away from the liquid, and as a result, the temperature of the remaining liquid will drop unless heat flows in from the surroundings. This principle is the basis for all evaporative cooling processes, including evaporative coolers for homes, the cooling of the human body by perspiration, and the cooling of a panting dog by the evaporation of saliva from the mucous membranes of its mouth.

Evaporation in an open container occurs until the container is dry. Evaporation occurs in a closed container (see **Figure 5.27**), too, but the liquid level eventually stops decreasing and becomes constant.

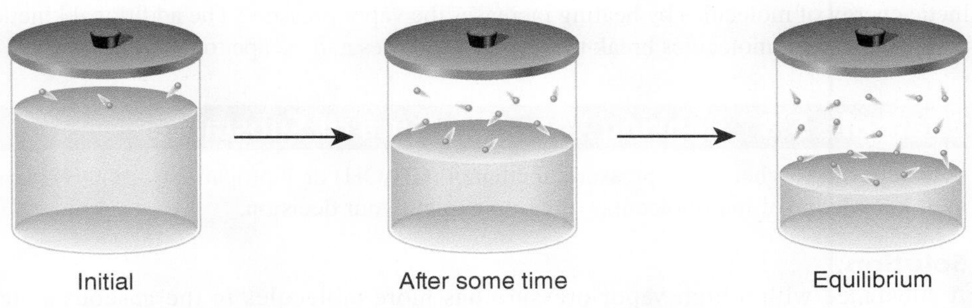

Figure 5.27 Liquid evaporation in a closed container. The drop in liquid level is greatly exaggerated for emphasis.

What explains this behavior? In a closed container, the molecules of liquid that go into the vapor (gaseous) state are unable to move completely away from the liquid surface as they do in an open container. Instead, the vapor molecules are confined to a space immediately above the liquid, where they have many random collisions with the container walls, other vapor molecules, and the liquid surface. Occasionally, their collisions with the liquid result in condensation, and they are recaptured by the liquid. **Condensation** is an exothermic process in which a gas (a vapor) is converted to a liquid (see **Figure 5.28**). Thus, two processes—evaporation (escape) and condensation (recapture)—actually take place in the closed container. Initially, the rate of evaporation exceeds that of condensation, so the liquid level decreases. The rates of the two processes eventually become equal, however, so the liquid level stops decreasing. At this point, the number of molecules that escape in a given time is the same as the number recaptured.

sethink/ 10 images/Pixabay

Figure 5.28 The formation of dew on a spider web is a common example of condensation.

A system in which two opposite processes take place at equal rates is said to be in equilibrium (**Section 7.2**). Under the equilibrium conditions just described, the number of molecules in the vapor state remains constant. This constant number of molecules will exert a constant pressure on the liquid surface and the container walls. This pressure exerted by a vapor in equilibrium with a liquid is called the **vapor pressure** of the liquid. Vapor pressure is often expressed in units of **standard atmosphere** (atm) or **torr**. The magnitude of a vapor pressure depends on the nature of the liquid (e.g., its molecular polarity and mass) and the temperature of the liquid—see **Tables 5.5** and **5.6**.

condensation An exothermic process in which a gas or vapor is changed to a liquid.

vapor pressure The pressure exerted by vapor that is in equilibrium with its liquid.

standard atmosphere The pressure needed to support a 760-mm column of mercury in a barometer tube.

torr The pressure needed to support a 1-mm column of mercury in a barometer tube.

Table 5.5 The Vapor Pressure of Various Liquids at 20 °C

Liquid	Molecular Weight (amu)	Polarity	Vapor Pressure (torr)
Pentane (C_5H_{12})	72	Nonpolar	414.5
Hexane (C_6H_{14})	86	Nonpolar	113.9
Heptane (C_7H_{16})	100	Nonpolar	37.2
Ethanol (C_2H_5—OH)	46	Polar (hydrogen bonds)	43.9
1-Propanol (C_3H_7—OH)	60	Polar (hydrogen bonds)	17.3
1-Butanol (C_4H_9—OH)	74	Polar (hydrogen bonds)	7.1

Table 5.6 Vapor Pressure of Water at Various Temperatures

Temperature (°C)	Vapor Pressure (torr)
0	4.6
20	17.5
40	55.3
60	149.2
80	355.5
100	760.0

The effect of molecular mass on vapor pressure is seen in **Table 5.5** for both nonpolar compounds (pentane, hexane, and heptane) and polar compounds (ethanol, 1-propanol, and 1-butanol). Within each series of compounds, the polarities are the same, but the vapor pressure *decreases* as the molecular weight *increases*. Vapor pressure decreases because London dispersion forces increase as the molecular weight increases.

Pentane (nonpolar) and 1-butanol (polar) have similar molecular weights (72 g/mol vs. 74 g/mol, respectively), but 1-butanol has a much lower vapor pressure because it can form hydrogen bonds. Hydrogen bonds require much more energy to break than the London dispersion forces in pentane. The data for water in **Table 5.6** show that increasing the kinetic energy of molecules by heating increases the vapor pressure. The additional kinetic energy helps more molecules break their cohesive forces and escape to the gaseous state.

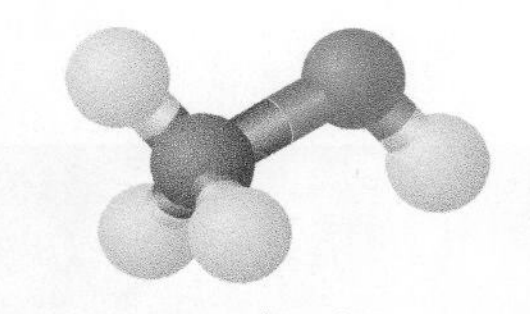

methanol

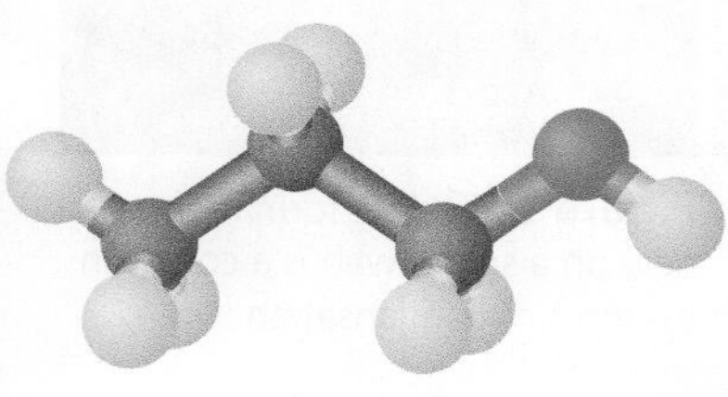

1-propanol

Figure 5.29 Ball-and-stick models of methanol and 1-propanol.

Example 5.8 Ranking Vapor Pressure of Substances

Which has the higher vapor pressure, methanol (CH_3OH) or 1-propanol (C_3H_7OH)? Use your knowledge of intermolecular forces to explain your decision.

Solution

A substance with a high vapor pressure has more molecules in the gaseous state. Increased numbers of molecules in the gaseous state is a result of lower intermolecular forces, because the molecules in the liquid state are able to move apart and escape into the gaseous state. While both methanol and 1-propanol are able to hydrogen bond through their –OH groups, it is the remainder of the molecules that will determine the strength of attractions between molecules. 1-Propanol has stronger London dispersion forces because it is larger than methanol (**Figure 5.29**). As a result, methanol has a higher vapor pressure.

✔ Learning Check 5.8 Which member of each of the following pairs of compounds has the higher vapor pressure? Use your knowledge of intermolecular forces to explain your decision.

a. Liquid helium (He) and liquid nitrogen (N_2)
b. Liquid hydrogen fluoride (HF) and liquid neon (Ne)

5.5B Boiling, Melting, and Sublimation

Learning Objective 9 Use the factors that affect boiling and melting to rank substances in order of increasing boiling point and melting point.

As a liquid is heated, its vapor pressure increases, as shown for water in **Table 5.6**. If the liquid's temperature is increased enough, its vapor pressure will reach a value equal to that of the prevailing atmospheric pressure. When bubbles of vapor form and rise rapidly to the surface, where the vapor escapes, the liquid is boiling (see **Figure 5.30**). The **boiling point** of a liquid is the temperature at which the vapor pressure of the liquid is equal to the atmospheric pressure above the liquid. Thus, the boiling point of water decreases as the atmospheric pressure decreases (**Figure 5.31**).

The increase in boiling point caused by an increase in pressure is the principle used to cook food faster in a pressure cooker. Increasing the pressure inside the cooker causes the boiling point of the water in it to rise, as shown in **Table 5.7**. It then becomes possible to increase the temperature of the food plus water in the cooker above 100 °C. An increase of just 10 °C makes the food cook approximately twice as fast. This same concept is used to increase the boiling point of water in an autoclave, a piece of equipment that sterilizes medical instruments for surgery (see **Figure 5.32**).

rommex/iStock/Getty Images

Figure 5.30 As the temperature of a liquid reaches the boiling point, bubbles of vapor form within the liquid and rise to the surface.

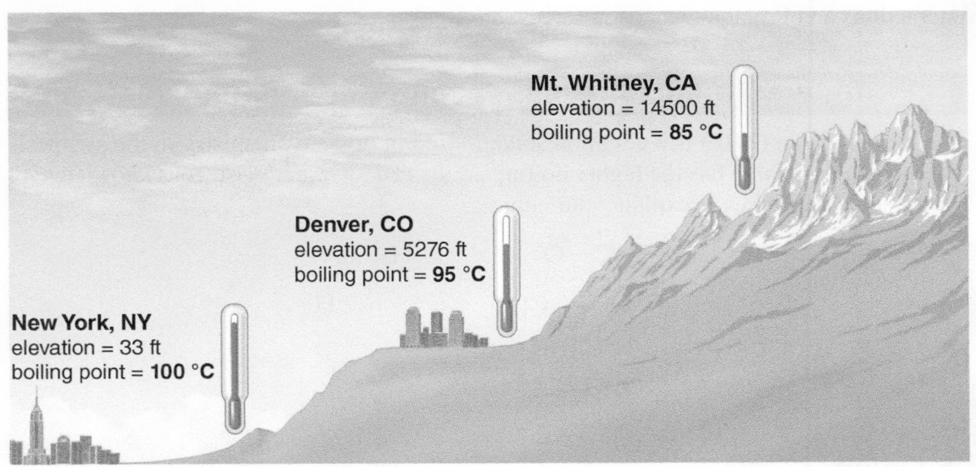

Figure 5.31 The boiling point of water decreases as the elevation increases.

boiling point The temperature at which the vapor pressure of a liquid is equal to the prevailing atmospheric pressure.

Table 5.7 The Boiling Point of Water in a Pressure Cooker

Pressure above Atmospheric (torr)	Boiling Point of Water (°C)
259	108
517	116
776	121

Solids, like liquids, have vapor pressures. Although the motion of particles is much more restricted in solids, particles at the surface can escape into the vapor state if they acquire sufficient energy. However, the strong cohesive forces characteristic of the solid state usually cause the vapor pressures of solids to be quite low.

Similar to the vapor pressures of liquids, the vapor pressures of solids increase with temperature as well. When the vapor pressure of a solid is high enough to allow the escaping molecules to go directly into the vapor state without passing through the liquid state, the process is called **sublimation** (see **Figure 5.33**). Sublimation is characteristic of materials such as solid carbon dioxide (dry ice) and naphthalene (moth crystals). Frozen water also sublimes under appropriate conditions. Freeze drying, a technique based on this process, is used to remove water from materials that would be damaged by heating (e.g., freeze-dried foods). The reverse process, when molecules go directly into the solid state from the gaseous state, is referred to as **deposition**. The most common example of deposition is frost. Water in the gaseous state deposits onto a cool surface, thus skipping the liquid state.

sublimation The endothermic process in which a solid is changed directly to a gas without first becoming a liquid.

deposition The exothermic process in which a gas is changed directly to a solid without first becoming a liquid.

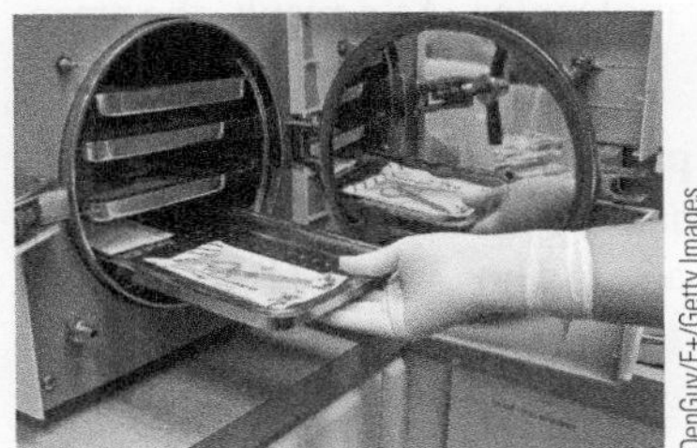
DenGuy/E+/Getty Images

Figure 5.32 An autoclave, used to sterilize medical instruments, functions like a pressure cooker, with higher temperatures and shorter times required to accomplish its task.

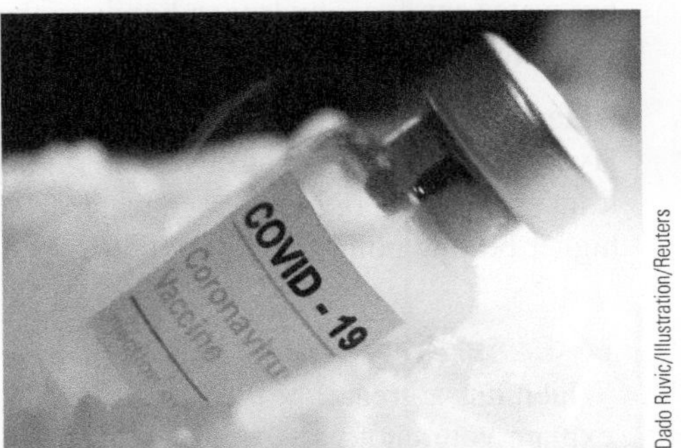

Dado Ruvic/Illustration/Reuters

Figure 5.33 Solid CO_2 (dry ice) is used to keep COVID-19 vaccines cold during transport.

melting point The temperature at which a solid changes to a liquid; the solid and liquid have the same vapor pressure.

Even though solids have vapor pressures, most pure substances in the solid state melt before appreciable sublimation takes place. Melting involves the breakdown of a rigid, orderly solid structure into a mobile, disorderly liquid state. This collapse of the solid structure occurs at a characteristic temperature called the **melting point**. At the melting point, the kinetic energy of solid particles is large enough to partially overcome the strong cohesive forces holding the particles together, and the solid and liquid states have the same vapor pressure.

No chemical bonds are broken when a substance undergoes a change of state. Only the intermolecular forces that attract molecules are broken. It takes energy to break these intermolecular forces, which is reflected in the melting points and boiling points. That is, a higher melting point or boiling point means that more energy is required because the intermolecular forces (like hydrogen bonding) are stronger. A lower melting point or boiling point means less energy is required because the intermolecular forces (like dipole–dipole interactions and London dispersion forces) are weaker.

Example 5.9 Ranking Boiling Points and Melting Points

Tetrahydrofuran (THF) is a common solvent used in organic chemistry in the synthesis of molecules. Which has the higher boiling point, THF or water? Use your knowledge of intermolecular forces to explain your choice.

THF

water

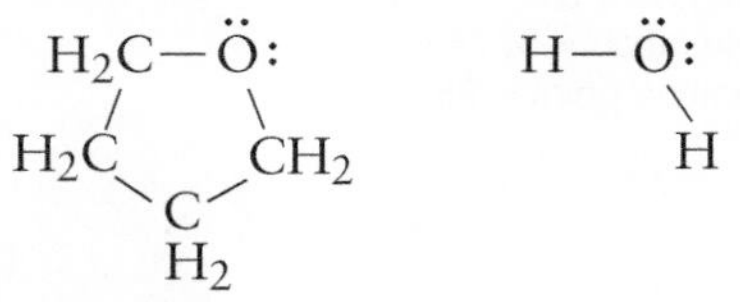

Solution

Determine the attractive forces in THF:

THF is polar and has an overall dipole moment that points in the direction of oxygen. It is unable to form hydrogen bonds to other THF molecules, so its strongest intermolecular forces are dipole–dipole interactions.

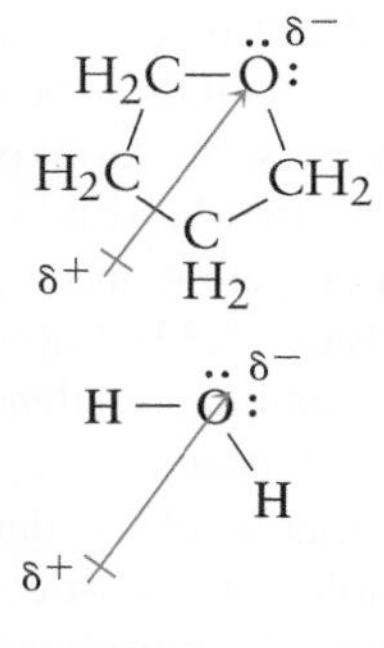

Determine the attractive forces in water:

Water is also polar with an overall dipole moment that points in the direction of oxygen, but water can form hydrogen bonds to the other water molecules. As a result, the strongest intermolecular force is hydrogen bonding.

Determine which compound has the higher boiling point:

Hydrogen bonding is a stronger intermolecular force than dipole–dipole interactions, so it takes more energy to change water from a liquid to a gas. Thus, water (100 °C) has a higher boiling point than THF (66 °C).

✔ Learning Check 5.9 Which has a higher melting point, F_2 or diethyl ether? Which has a greater boiling point? Use your knowledge of intermolecular forces to explain your choices.

5.6 Energetics: Changes of State vs. Specific Heat

Learning Objective 10 Calculate energy changes that accompany heating, cooling, or changing the state of a substance.

Kinetic energy, the energy of particle motion, is related to heat. In fact, temperature is a measurement of the average kinetic energy of the particles in a system. Potential energy, in contrast, is related to the extent to which particles are separated, rather than to their motion. Thus, an increase in temperature on adding heat corresponds to an increase in kinetic energy of the particles, whereas their potential energy remains unchanged.

Imagine a system composed of 1 g of ice at an initial temperature of −20 °C. Heat is added at a constant rate until the ice is converted into 1 g of steam at 120 °C. The atmospheric pressure is assumed to be 760 torr throughout the experiment. The changes in the system take place in several steps, as shown in the heating curve in **Figure 5.34**.

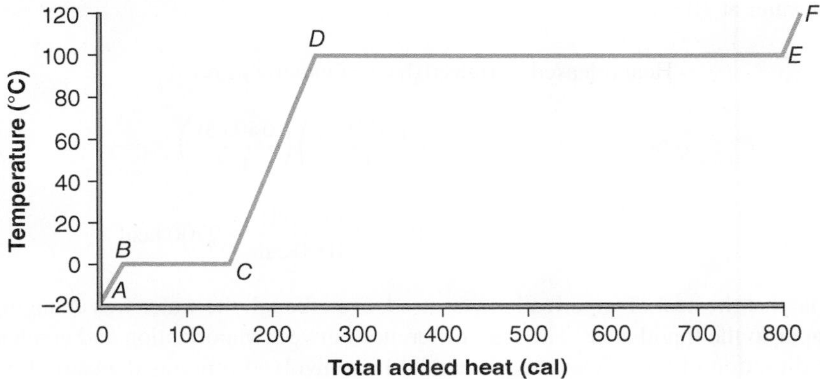

Figure 5.34 The temperature behavior of a system composed of water as it is heated from −20 °C to 120 °C. The horizontal segments (i.e., *B* to *C* and *D* to *E*) correspond to melting (0 °C) and boiling (100 °C), respectively.

The solid is first heated from −20 °C to the melting point at 0 °C (line *AB*). The temperature increase indicates that most of the added heat causes an increase in the kinetic energy of the molecules. Along line *BC*, the temperature remains constant at 0 °C until all of the solid melts. The temperature remains constant during melting because the added heat goes toward breaking attractive forces, thus increasing the potential energy of the molecules. The added heat does not increase their kinetic energy; the molecules are moved farther apart, but their motion is not increased. Once all of the solid has melted, the addition of more heat warms the liquid water from 0 °C to 100 °C along line *CD*. At 100 °C, the normal boiling point (**Figure 5.35**), another change of state occurs as heat is added—namely, the liquid is converted into vapor (steam; line *DE*). The constant temperature during boiling indicates once again that the added heat goes to an increase in potential energy. Once all of the liquid has been converted into vapor, line *EF* represents heating the steam from 100 °C to 120 °C by increasing the kinetic energy of the molecules.

Figure 5.35 Boiling water kills or inactivates bacteria and other pathogens. Boil water notices are used by health agencies in response to biological contamination of drinking water.

The amount of heat required to change 1 g of a substance at its melting point from solid to liquid is called the **heat of fusion**, whereas the amount of heat required to change 1 g of a substance at its boiling point from liquid to gas is the **heat of vaporization** (as shown in Equations 5.1 and 5.2). The units used for these quantities of heat are calories (or joules) per gram, where 1 calorie = 4.184 joules.

heat of fusion The amount of heat energy required to melt exactly 1 g of a solid substance at constant temperature.

heat of vaporization The amount of heat energy required to vaporize exactly 1 g of a liquid substance at constant temperature.

Heat = (mass)(heat of fusion)

$$q = m \cdot \Delta H_{\text{fus}} \quad (5.1)$$

Heat = (mass)(heat of vaporization)

$$q = m \cdot \Delta H_{\text{vap}} \quad (5.2)$$

Image Source/Photodisc/Getty Images

Figure 5.36 Whether you are working with steam pipes, or in the kitchen, steam can cause severe burns.

The heats of fusion and vaporization for water are 80 and 540 cal/g, respectively. This explains why a burn caused by steam at 100 °C is more severe than one caused by water at 100 °C. Vaporization has added 540 cal to each gram of steam, and each gram will release these 540 cal when it condenses on the skin. Liquid water at 100 °C would not have this extra heat with which to burn the skin (see **Figure 5.36**).

Example 5.10 Heat of Vaporization and Fusion Calculations

Steam rooms can be used to break up the congestion in the sinuses and lungs, helping to treat colds and aid breathing. Calculate the heat released when 5.0 kg of steam condenses to water at 100 °C in a system.

Solution

Use Equation 5.2 to calculate the amount of heat released when the steam condenses to liquid water at 100 °C:

$$\text{Heat released} = (\text{mass})(\text{heat of vaporization})$$

$$q = m \cdot \Delta H_{\text{vap}} = (5.0\ \cancel{\text{kg}})\left(\frac{1000\ \cancel{\text{g}}}{1\ \cancel{\text{kg}}}\right)\left(\frac{540\ \text{cal}}{1\ \cancel{\text{g}}}\right)$$

$$= 2.7 \times 10^6\ \cancel{\text{cal}}\frac{(1\ \text{kcal})}{1000\ \cancel{\text{cal}}} = 2700\ \text{kcal}$$

Note that the heat of vaporization was used, even though the water was changing from the vapor to the liquid state. The only difference between vaporization and condensation is the direction of heat flow; the amount of heat involved remains the same for a specific quantity of material. (Accordingly, 1 g of water vapor will release 540 cal when it condenses.)

✔ **Learning Check 5.10** How much heat in joules is needed to melt 25.0 g of ice (H_2O) at 0 °C at constant pressure?

Table 5.8 Specific Heats for Selected Substances

Substance and State	Specific Heat J/g·°C	cal/g·°C
Aluminum (solid)	1.0	0.24
Copper (solid)	0.39	0.093
Ethylene glycol (liquid)	2.43	0.58
Helium (gas)	5.23	1.25
Hydrogen (gas)	14.2	3.39
Lead (solid)	0.13	0.031
Mercury (liquid)	0.14	0.033
Nitrogen (gas)	1.1	0.25
Oxygen (gas)	0.92	0.22
Sodium (solid)	1.2	0.29
Sodium (liquid)	1.3	0.32
Water (solid)	2.1	0.51
Water (liquid)	4.18	1.00
Water (gas)	2.0	0.48

specific heat The amount of heat energy required to raise the temperature of exactly 1 g of a substance by exactly 1°C.

Along lines *AB*, *CD*, and *EF* of **Figure 5.34**, the kinetic energy added to the system increased the temperature. The amount of heat energy required to raise the temperature of 1 g of the substance by 1 °C is called the **specific heat** of the substance. In scientific work, this is often given in units of calories per gram degree Celsius (cal/g·°C) or joules per gram degree Celsius (J/g·°C). Equation 5.3 represents the relationship between specific heat and the amount of heat required to increase the temperature of a sample of substance.

$$\text{Heat} = (\text{sample mass})(\text{specific heat})(\text{temperature change})$$

$$q = m \cdot c \cdot \Delta T \tag{5.3}$$

Specific heats for a number of substances in various states are listed in **Table 5.8**. A substance with a high specific heat (e.g., liquid water) is capable of absorbing more heat with a small temperature change than are substances with lower specific heats. The high specific heat of water is crucial in helping our bodies maintain physiological temperature (around 98.6 °F). If this weren't the case, our body temperature would fluctuate drastically at the smallest change in ambient temperature!

Health Connections 5.1

Therapeutic Uses of Oxygen Gas

A steady supply of oxygen is essential for the human body to function properly. The most common source of this gas is inhaled air, which consists of about 21% oxygen gas. In a healthy individual, this amount of oxygen is sufficient to be transported into the blood and distributed throughout the body. Patients suffering from a lung disease, such as pneumonia or chronic obstructive pulmonary disease (COPD), often cannot transport sufficient oxygen to the blood from the air they breathe unless the amount of oxygen in the air is increased. This is done by mixing an appropriate amount of oxygen with the air by using an oxygen mask or nasal cannula.

The effective concentration of oxygen present in inhaled air can be increased for other clinical applications based on a technique called *hyperbaric oxygenation*. In one application, patients infected by anaerobic bacteria, such as those that cause tetanus and gangrene, are placed in a hyperbaric chamber in which the relative amount of oxygen is substantially increased, which causes body tissues to pick up large amounts of oxygen and kill the bacteria.

Hyperbaric oxygenation may also be used to treat other abnormal conditions or injuries, such as certain heart disorders, carbon monoxide poisoning, crush injuries, certain hard-to-treat bone infections, smoke inhalation, near-drowning, asphyxia, and burns.

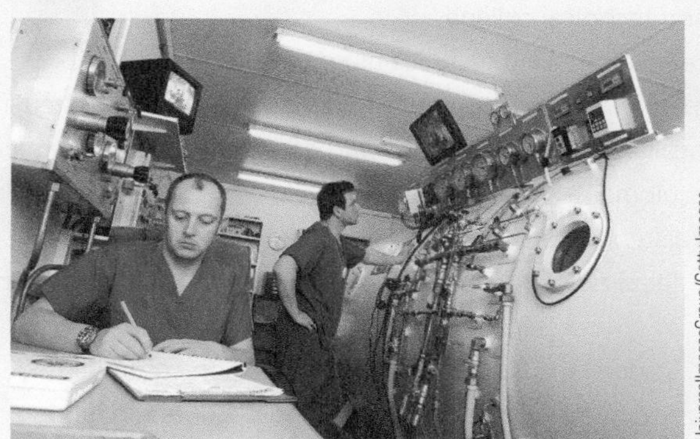

UniversalImagesGroup/Getty Images

Hyperbaric chambers must be strongly built to resist significant internal gas pressure.

Example 5.11 Specific Heat Calculations

Therapeutic hypothermia is a method in which medical professionals use ice or ice packs to lower a patient's body temperature after cardiac arrest in order to reduce damage to the brain (**Figure 5.37**). How much heat has been removed from a patient's body when a mass of 6.0 kg of ice applied to the patient was warmed from an initial temperature of −18 °C to a final temperature of −12 °C?

Solution

The patient's temperature will drop as heat is transferred from the patient to the ice in contact with the body. The specific heat of water (solid) is 0.51 cal/g·°C.

Heat absorbed by the patient from the ice = (mass)(specific heat)(temp. change)

$$\text{Heat absorbed} = (6.0\ \cancel{\text{kg}})\frac{(1000\ \cancel{\text{g}})}{1\ \cancel{\text{kg}}}\left(\frac{0.51\ \text{cal}}{\cancel{\text{g}}\cdot\cancel{°\text{C}}}\right)(6\ \cancel{°\text{C}})$$

$$= 18{,}360\ \cancel{\text{cal}}\ \frac{(1\ \text{kcal})}{1000\ \cancel{\text{cal}}} = 18\ \text{kcal}$$

✔ Learning Check 5.11 Some nuclear reactors are cooled by gases. Calculate the number of calories that 1.00 kg of helium gas will absorb when it is heated from 25 °C to 700 °C. See **Table 5.8** for the specific heat of He.

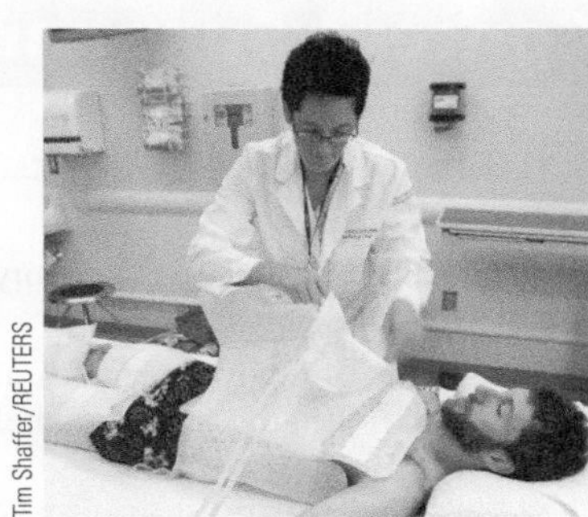

Tim Shaffer/REUTERS

Figure 5.37 Nurse administering the therapeutic hypothermia technique as they apply cooling pads to a patient after cardiac arrest.

Health Career Description

Sterile Processing Technician

Sterile processing technicians maintain the cleanliness and functionality of medical and surgical equipment. They may work in hospitals, surgery centers, cosmetic surgery clinics, or laboratories. Training programs are available through universities and technical colleges.

Sterile processing technicians certify that equipment destined for use in procedures and surgeries is reliably sterile and available when physicians and other professionals require the equipment.

In a sterile processing technician program, students learn to:

- Sterilize instrumentation and equipment through a series of critical steps
- Understand microbiology and infection control as it pertains to sterile processing and decontamination procedures
- Implement infection control practices to ensure that patients avoid infections
- Provide instrumentation and equipment access to doctors, nurses, and allied health professionals as required

In order to practice as a sterile process technician, you must pass a Certified Registered Central Service Technician exam (CRCST) and complete 400 hours of hands-on experience. The CRCST exam tests applicants in the following areas:

- Cleaning, decontamination, and disinfection
- Preparation and packaging
- Sterilization process
- Patient care equipment
- Sterile storage and inventory management
- Documentation and record maintenance
- Customer relations

Sterile processing technicians work behind the scenes, but their work is indispensable in maintaining the reliability and safety of health care delivery. Understanding the actions of chemical disinfectants, the proper handling of chemicals, and laboratory safety protocols is important in this career. The work of these technicians stops the spread of bacterial and viral infection and ensures the safety of patients, medical personnel, and the public.

Sources: American Institute of Medical Sciences and Education (AIMS). How to become a Sterile Processing Technician. Retrieved May 2, 2020 from: https://www.aimseducation.edu/blog/how-do-i-become-a-sterile-processing-technician/; International Association of Healthcare Central Service Material Management (IAHCSMM). Certified Registered Central Service Technician (CRCST). Retrieved May 2, 2020 from: https://www.iahcsmm.org/certification-menu/crcst-certification.html; Portland Community Hospital. Sterile Processing Technician. Retrieved May 2, 2020 from: https://www.youtube.com/watch?v=GAOCDMbDvRQ

Concept Summary

5.1 Geometries of Molecules and Polyatomic Ions

Learning Objective: Use VSEPR theory to predict the geometries of molecules and polyatomic ions.

- The shapes of many molecules and polyatomic ions can be predicted by using valence-shell electron-pair repulsion (VSEPR) theory.
- According to VSEPR theory, electron domains in the valence shell of the central atom of a molecule or ion repel one another and arrange themselves so as to maximize their separation distances.
- The resulting arrangement determines the geometry when one or all of the electron pairs involved form bonds between the central atom and other atoms.

5.2 The Polarity of Covalent Molecules

Learning Objectives: Use electronegativities to classify the covalent bonds in molecules. Determine whether covalent molecules are polar or nonpolar.

- Electrons in covalent bonds may be shared equally, or they may be attracted more strongly to one of the atoms forming the bond.
- The tendency of a covalently bonded atom to attract shared electrons is called the electronegativity of the atom.
- Unequally shared bonding-electron pairs form polar covalent bonds.
- The extent of bond polarization can be estimated from the electronegativity differences between the bonded atoms.
- The higher the electronegativity difference, the more polar (or ionic) the bond is.

- Polar covalent bonds cause partial positive and partial negative charges to form within molecules.
- When these partial charges are symmetrically distributed in the molecule, it is said to be nonpolar.
- An unsymmetric distribution gives rise to a polar molecule.

5.3 Intermolecular Forces

Learning Objective: **Identify the major attractive force between molecules in a given substance.**

- Forces other than ionic and covalent bonds are also known to hold the particles of some pure substances together in the solid and liquid states.
- These intermolecular forces include ion–dipole interactions, hydrogen bonds, dipole–dipole interactions, and London dispersion forces.

5.4 Energy and Properties of Matter

Learning Objectives: **Distinguish phenomena that demonstrate kinetic energy from those that demonstrate potential energy. Identify the states of matter using five properties: density, shape, compressibility, particle interaction, and molecular movement.**

- Kinetic energy is the energy of motion and thus is a disruptive force.
- Potential energy is energy that is stored (i.e., in chemical bonds) and thus is a cohesive force.
- In the solid state, cohesive forces between particles of matter are stronger than disruptive forces.
- As a result, the particles of solids are held in rigid three-dimensional shapes in which the particles' kinetic energy only takes the form of vibrations of the atoms.
- In the liquid state, cohesive forces between particles slightly exceed disruptive forces.
- As a result, particles of liquids are randomly arranged but relatively close to each other and are in constant random motion, sliding freely over each other but without enough kinetic energy to become separated.
- In the gaseous state, disruptive forces dominate and particles move randomly, essentially independent of each other.
- Under ordinary pressure, the particles of gases are separated from each other by relatively large distances except when they collide.

5.5 Changes of State

Learning Objectives: **Classify changes of state as exothermic or endothermic. Use the factors that affect evaporation and condensation to rank substances in order of increasing vapor pressure. Use the factors that affect boiling and melting to rank substances in order of increasing boiling point and melting point.**

- Most matter can be changed from one state to another by heating, cooling, or a change in pressure.
- State changes that give up heat are said to be exothermic.
- Those that absorb heat are said to be endothermic.
- The evaporation of a liquid is an endothermic process. As a result, it is involved in many cooling processes.
- In a closed container, evaporation takes place only until the rate of escape of molecules from the liquid is equal to the rate at which they return to the liquid.
- The pressure exerted by the vapor, which is in equilibrium with the liquid, is called the vapor pressure of the liquid.
- Liquid vapor pressures increase as the liquid temperature increases.
- At the boiling point of a liquid, its vapor pressure equals the prevailing atmospheric pressure.
- Bubbles of vapor form within the liquid and rise to the surface as the liquid boils.
- The boiling point of a liquid decreases as the prevailing atmospheric pressure decreases.
- Solids, like liquids, have vapor pressures that increase with temperature.
- Some solids have vapor pressures high enough to allow them to change to vapor without first becoming a liquid, a process called sublimation.
- The reverse process, when gases become solids without first becoming liquids, is called deposition.
- Most solids change to the liquid state before they change to the vapor state.
- The temperature at which solids change to liquids is called the melting point.
- The strength of the predominant intermolecular force acting in a substance is indicated by the melting points and boiling points of the substance. High melting points and boiling points indicate strong intermolecular forces (e.g., hydrogen bonds), whereas low ones indicate weak intermolecular forces (e.g., London dispersion forces).

5.6 Energetics: Changes of State vs. Specific Heat

Learning Objective: **Calculate energy changes that accompany heating, cooling, or changing the state of a substance.**

- Energy is absorbed or released when the temperature of matter is changed or when matter changes from one state to another.
- For melting or freezing changes, the amount of heat required or released is called the heat of fusion; for boiling or condensing, the amount of heat required or released is called the heat of vaporization.
- The amount of heat energy required to raise the temperature of 1 g of a substance by 1 °C is called the specific heat of the substance.

Key Terms and Concepts

Boiling point (5.5)
Bond polarization (5.2)
Cohesive forces (5.4)
Compressibility (5.4)
Condensation (5.5)
Decomposition (5.5)
Deposition (5.5)
Dipole–dipole interactions (5.3)
Disruptive forces (5.4)
Endothermic (5.5)
Electron domain (5.1)
Electron-domain geometry (5.1)
Evaporation or vaporization (5.5)
Exothermic (5.5)
Heat of fusion (5.6)
Heat of vaporization (5.6)
Hydrogen bonding (5.3)
Intermolecular forces (5.3)
Intramolecular forces (5.3)
Ion–dipole interactions (5.3)
Kinetic energy (5.4)
London dispersion forces (5.3)
Melting point (5.5)
Molecular geometry (5.1)
Nonpolar covalent bond (5.2)
Nonpolar molecule (5.2)
Polar covalent bond (5.2)
Polar molecule (5.2)
Polarity (5.2)
Potential energy (5.4)
Specific heat (5.6)
Standard atmosphere (5.5)
Sublimation (5.5)
Torr (5.5)
Vapor pressure (5.5)
VSEPR theory (5.1)

Key Equations

1. Heat of fusion (**Section 5.6**)	Heat = (sample mass)(heat of fusion) $q = (m)(\Delta H_{fus})$	Equation 5.1
2. Heat of vaporization (**Section 5.6**)	Heat = (sample mass)(heat of vaporization) $q = (m)(\Delta H_{vap})$	Equation 5.2
3. Specific heat calculation (**Section 5.6**)	Heat = (sample mass)(specific heat)(temperature change) $q = (m)(c)(\Delta T)$	Equation 5.3

Exercises

Even-numbered exercises are answered in Appendix B.

Exercises with an asterisk (*) are more challenging.

Geometries of Molecules and Polyatomic Ions (Section 5.1)

5.1 For the following molecules, draw a proper Lewis structure and use **Table 5.1** to identify the number of bonding and nonbonding domains.

a. CH_4 (each H atom is bonded to the C atom)

b. SO_2 (each O atom is bonded to the S atom)

c. $AlCl_3$ (each Cl atom is bonded to the Al atom)

5.2 For the following molecules, draw a proper Lewis structure and use **Table 5.1** to identify the number of bonding and nonbonding domains.

a. NH_3 (each H atom is bonded to the N atom)

b. $BeCl_2$ (each Cl atom is bonded to the Be atom)

c. ClCN (the Cl and N atoms are bonded to the C atom)

5.3 Predict the molecular geometry of each of the following molecules by first drawing a Lewis structure and then applying VSEPR theory.

a. BF_3 (each F atom is around the B atom)

b. PH_3 (each H atom is around the P atom)

c. SCl_2 (each Cl atom is bonded to the S atom)

5.4 Predict the molecular geometry of each of the following molecules by first drawing a Lewis structure, then applying VSEPR theory.

a. H_2S (each H atom is bonded to the S atom)

b. PCl_3 (each Cl atom is bonded to the P atom)

c. OF_2 (each F atom is bonded to the O atom)

5.5 Predict the molecular geometry of each of the following molecules by first drawing a Lewis structure, then applying VSEPR theory.

a. O_3

b. SeO_2

c. PH_3

d. SO_3 (each O atom is bonded to the S atom)

5.6 Predict the molecular geometry of each of the following polyatomic ions by first drawing a Lewis structure, then applying VSEPR theory.

a. NO_2^-

b. ClO_3^- (each O is bonded to Cl)

c. CO_3^{2-}

d. H_3O^+ (note the positive charge and compare with NH_4^+)

5.7 Predict the molecular geometry of each of the following polyatomic ions by first drawing a Lewis structure, then applying VSEPR theory.

a. NH_2^- (each H is bonded to N)

b. PO_3^{3-} (each O is bonded to P)

c. $BeCl_4^{2-}$ (each Cl is bonded to Be)

d. ClO_4^- (each O is bonded to Cl)

5.8 What is the molecular geometry around the atom in red?

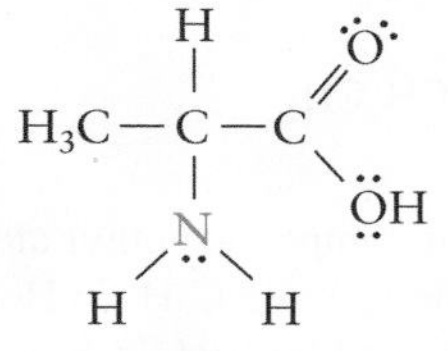

alanine

5.9 What is the molecular geometry around the atom in red?

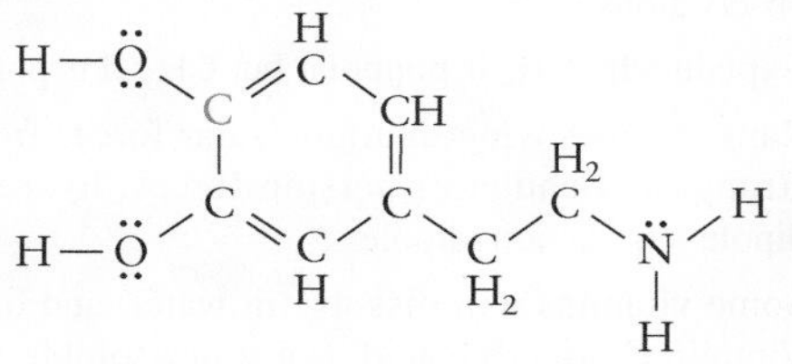

dopamine

5.10 What is the molecular geometry around the atom in red?

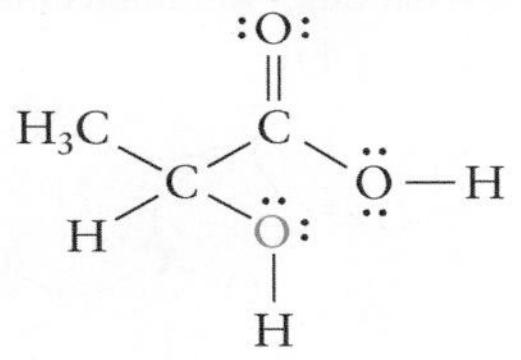

lactic acid

5.11 What is the molecular geometry around the atom in red?

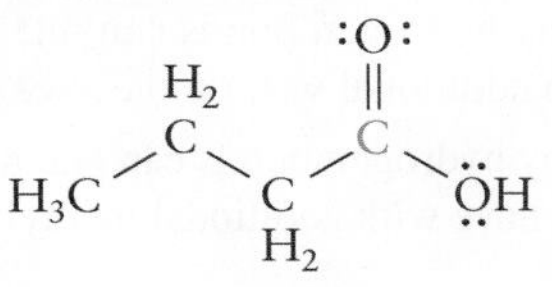

butyric acid

The Polarity of Covalent Molecules (Section 5.2)

5.12 Identify the more electronegative atom in each of the following pairs.

a. O and S

b. Si and F

5.13 Identify the more electronegative atom in each of the following pairs.

a. F and Cl

b. C and N

5.14 Identify the more polar bond in each of the following pairs by using the electronegativity values given in **Figure 5.6**.

a. N—C or N—F

b. Si—S or Si—Cl

5.15 Identify the more polar bond in each of the following pairs by using the electronegativity values given in **Figure 5.6**.

a. C—O or C—S

b. Br—Cl or Br—F

5.16 Determine the direction of the bond polarization for the following bonds.

a. H—N

b. F—Cl

c. C—F

5.17 Determine the direction of the bond polarization for the following bonds.

a. H—O

b. Br—I

c. P—Cl

5.18 Use the periodic table and **Figure 5.9** to determine which bonds will be polarized. Show the resulting charge distribution in those molecules that contain polarized bonds.

a. H—I

b. 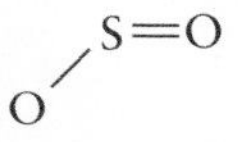 O—S=O

c. 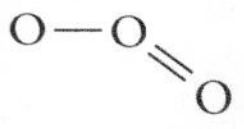 O—O=O

5.19 Use the periodic table and **Figure 5.9** to determine which bonds will be polarized. Show the resulting charge distribution in those molecules that contain polarized bonds.

a. Cl—F

b. 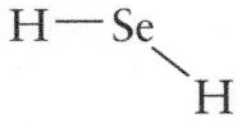 H—Se—H

c. 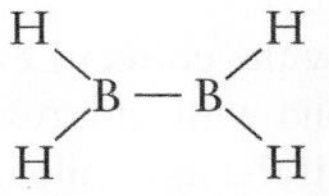 $H_2B—BH_2$

5.20 Use **Figure 5.9** to classify the bonds in the following compounds as nonpolar covalent, polar covalent, or ionic.

a. LiBr

b. HCl

c. PH_3 (each H is bonded to P)

d. SO_2 (each O is bonded to S)

e. CsF

5.21 Use **Figure 5.9** to classify the bonds in the following compounds as nonpolar covalent, polar covalent, or ionic.

a. MgI_2 (each I is bonded to Mg)

b. NCl_3 (each Cl is bonded to N)

c. H_2S (each H is bonded to S)

d. RbF

e. SrO

5.22 Based on the charge distributions you drew for the molecules in Exercise 5.18, classify each of the molecules as polar or nonpolar.

5.23 Based on the charge distributions you drew for the molecules in Exercise 5.19, classify each of the molecules as polar or nonpolar.

5.24 Use **Figures 5.6 and 5.9** to predict the type of bond you would expect to find in compounds formed from the following elements.

a. Magnesium and chlorine

b. Carbon and hydrogen

c. Phosphorus and hydrogen

5.25 Use **Figures 5.6 and 5.9** to predict the type of bond you would expect to find in compounds formed from the following elements.

a. C and Br

b. Aluminum and chlorine

c. Nitrogen and oxygen

5.26 Show the charge distribution in the following molecules, and predict which are polar molecules.

a. C≡O

b. H—Se—H (Se bonded to two H atoms)

c.

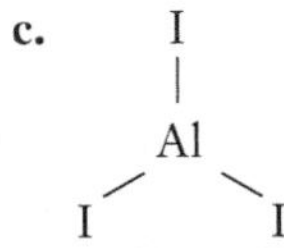

5.27 Show the charge distribution in the following molecules, and predict which are polar molecules.

a. S=C=S

b. H—C≡N

c. F—O—F

5.28 For the following molecules, draw the correct Lewis structure, label each polar covalent bond with an arrow to show bond polarization, and determine if the molecule is polar or nonpolar.

a. $SOCl_2$ (S atom in the middle)

b. NF_3 (N atom in the middle)

c. CH_4 (C atom in the middle)

5.29 For the following molecules, draw the correct Lewis structure, label each polar covalent bond with an arrow to show bond polarization, and determine if the molecule is polar or nonpolar.

a. $AlCl_3$ (Al atom in the middle)

b. OF_2 (O atom in the middle)

c. H_2S (S atom in the middle)

Intermolecular Forces (Section 5.3)

5.30 Identify the strongest intermolecular force in each of the following substances.

a. CF_4

b. CS_2

c. HCl

5.31 Identify the strongest intermolecular force in each of the following substances.

a. CH_3OH

b. $CH_3CH_2CH_2CH_3$

c. CH_2Cl_2

5.32 The covalent compounds ethyl alcohol and dimethyl ether both have the formula C_2H_6O. However, the alcohol melts at −117.3 °C and boils at 78.5 °C, whereas the ether melts at −138.5 °C and boils at −23.7 °C. How could differences in forces between molecules be used to explain these observations?

5.33 Explain why CH_4 is nonpolar but CH_3Cl is polar.

5.34 Rank the following intermolecular forces from weakest to strongest: London dispersion forces, hydrogen bonding, dipole–dipole, ion–dipole.

5.35 Some vitamins will dissolve in water, and others will not. Vitamin C, ascorbic acid, is a water-soluble vitamin that is important for maintaining good health. What is the most important type of intermolecular force vitamin C experiences with water? What other intermolecular forces are present?

ascorbic acid

5.36 How many hydrogen bonds can one water molecule (H_2O) have with additional water molecules?

5.37 How many hydrogen bonds can one methylamine (CH_3NH_2) molecule have with additional methylamine molecules?

Energy and Properties of Matter (Section 5.4)

5.38 Explain each of the following observations.

a. A liquid takes the shape, but not necessarily the volume, of its container.

b. Solids and liquids are practically incompressible.

c. A gas always exerts uniform pressure on all walls of its container.

5.39 Explain each of the following observations.

a. Gases have low densities.

b. The densities of a substance in the solid and liquid states are nearly identical.

c. Solids, liquids, and gases all expand when heated.

5.40 Explain how liquids are similar to gases. Explain how liquids are different from gases.

5.41 Identify the states of matter of the following substances in italics below.

a. *Phosgene* is a toxic nerve agent that smells like fresh-cut grass and causes irritation to the eyes; dry, burning throat; and difficulty breathing. It is shipped in a compressed steel cylinder.

b. *Benzocaine* is a topical anesthetic used to relieve pain and itching caused by insect bites and other minor burns. It has a high boiling point, a low compressibility, and a density that is heavier than water.

5.42 Discuss differences in kinetic and potential energy of the constituent particles for a substance in the solid, liquid, and gaseous states.

Changes of State (Section 5.5)

5.43 The following statements are best associated with the solid, liquid, or gaseous states of matter. Match each statement to the appropriate state of matter.

a. This state is characterized by the lowest density of the three.

b. This state is characterized by an indefinite shape and a high density.

c. In this state, disruptive forces prevail over cohesive forces.

d. In this state, cohesive forces are most dominant.

5.44 The following statements are best associated with the solid, liquid, or gaseous state of matter. Match each statement to the appropriate state of matter.

a. Temperature changes influence the volume of this state substantially.

b. In this state, constituent particles are less free to move about than in other states.

c. Pressure changes influence the volume of this state more than that of the other two states.

d. This state is characterized by an indefinite shape and a low density.

5.45 Classify each of the following processes as endothermic or exothermic.

a. Freezing

b. Sublimation

c. Vaporization

5.46 Classify each of the following processes as endothermic or exothermic.

a. Condensation

b. Liquefaction

c. Boiling

5.47 Classify each of the following processes as endothermic or exothermic.

a. Frost appears on a car window after a cold evening.

b. Coffee is freeze-dried before shipping.

5.48 Classify each of the following processes as endothermic or exothermic.

a. Rubbing alcohol disappears after being wiped on the skin before an injection.

b. Frost forms on food in the freezer.

5.49 Classify each of the following processes as endothermic or exothermic.

a. A cold pack is used to reduce muscle swelling due to an injury.

b. A heat pack is used to relax a muscle cramp.

5.50 Classify each of the following processes as endothermic or exothermic.

a. An ice cube melts after being left out on the table.

b. Cellular metabolism breaks chemical bonds and generates heat.

5.51 The following are all nonpolar liquid hydrocarbon compounds derived from petroleum: butane (C_4H_{10}), pentane (C_5H_{12}), hexane (C_6H_{14}), and heptane (C_7H_{16}). Arrange these compounds in order of increasing vapor pressure (lowest first, highest last) and explain how you arrived at your answer.

5.52 An autoclave is used to sterilize surgical equipment. In it, water is heated to become high-pressure steam. High temperature and high-pressure steam effectively inactivate bacteria and viruses. At the end of the sterilization cycle, the steam is cooled back to liquid water and ambient pressure before the autoclave is opened. Name the changes of state that occurred to the water and identify them as endothermic or exothermic.

5.53 Arrange the following substances in order of increasing vapor pressure (lowest first, highest last) and explain how you arrived at your answer: CH_4, CCl_4, CH_2Cl_2, CH_3Cl.

5.54 In the past, dentists used methylene chloride (CH_2Cl_2) as a local anesthetic. It was sprayed onto the area to be anesthetized. Propose an explanation for how it worked.

5.55 Determine which substance will have the higher vapor pressure and explain how you arrived at your answer: water (H_2O) or hydrogen sulfide (H_2S).

5.56 Each of two glass containers contains a clear, colorless, odorless liquid that has been heated until it is boiling. One liquid is water (H_2O) and the other is ethylene glycol ($HOCH_2CH_2OH$). Explain how you could make one measurement of each boiling liquid, using the same device, and tell which liquid was which.

5.57 Autoclaves are used to sterilize equipment in hospitals. They typically operate at a pressure of 776 torr above ambient pressure. Autoclaves also require water in the form of steam to operate. Explain what happens to the boiling point of water under 776 torr pressure. The normal boiling point of water is 100 °C. What would the temperature inside an autoclave be?

5.58 Identify which pure substance will have a lower vapor pressure, water (H_2O) or 1-propanol ($CH_3CHOHCH_3$), and explain your reasoning.

5.59 Identify which pure substance will have the higher boiling point, pentane ($CH_3CH_2CH_2CH_2CH_3$) or octane ($CH_3CH_2CH_2CH_2CH_2CH_2CH_2CH_3$), and explain your reasoning.

5.60 Arrange the following substances in order of increasing boiling point (lowest to highest) and explain your answer: C_6H_{14}, $C_{10}H_{22}$, C_5H_{12}, C_8H_{18}.

5.61 Arrange the following substances in order of increasing boiling point (lowest to highest) and explain your answer: CF_4, CCl_4, CBr_4, CI_4.

5.62 Predict which substance will have the higher melting point, the antacid $Mg(OH)_2$ or palmitic acid $CH_3(CH_2)_{14}COOH$ (the primary fatty acid used in processed foods).

5.63 Predict which substance will have the higher melting point, the primary component of eggshells $CaCO_3$ or cetyl palmitate $C_{32}H_{64}O_2$ (the primary wax found in sperm whales).

5.64 When moderately heated, solid iodine readily sublimes without melting. The hot vapor will condense back to the solid state when it cools. Describe a method that could be used to obtain pure solid iodine from a mixture of solid iodine and sand. Explain your reasoning.

5.65 In a clinical procedure, a pathologist may choose to put tissue samples that need to remain frozen and dry in a dry ice (solid CO_2) cooler rather than a regular water ice cooler. Why?

5.66 A mixture was made of pure water and ice. The mixture was allowed to come to a constant temperature of 0.0 °C, the melting point of solid water. The vapor pressure of the water was measured and found to be 4.58 torr. What is the vapor pressure of the ice in torr? Explain your answer.

Energetics: Changes of State vs. Specific Heat (Section 5.6)

5.67 Calculate the kJ for the following.

a. A protein bar produces 280 kilocalories when burned.

b. Two apples produce 144 kilocalories when burned.

5.68 Calculate the kJ for the following.

a. An IV of dextrose produces 170 kilocalories.

b. One packet of electrolyte powder produces 25 kilocalories.

5.69 How much heat is released when 2.5 g of $H_2O(s)$ is melted to $H_2O(\ell)$ at constant pressure? The ΔH_{fus} of water is 80 cal/g.

5.70 How much heat is needed when 66 g of $H_2O(\ell)$ is vaporized to $H_2O(g)$ at constant pressure? The ΔH_{vap} of water is 540 cal/g.

5.71 Using the specific heat data from **Table 5.8**, calculate the amount of heat (in calories) needed to increase the temperature of the following.

a. 210 g of copper from 40.°C to 95.°C

b. 150 g of mercury from 120.°C to 300.°C

c. 2.50×10^3 g of helium gas from 250.°C to 900.°C

5.72 Using the specific heat data from **Table 5.8**, calculate the amount of heat (in calories) needed to increase the temperature of the following.

a. 50. g of aluminum from 25 °C to 55 °C

b. 2.50×10^3 g of ethylene glycol from 80.°C to 85 °C

c. 500. g of steam from 110.°C to 120.°C

5.73 For solar energy to be effective, collected heat must be stored for use during periods of decreased sunshine. One proposal suggests that heat can be stored by melting solids that, upon solidification, would release the heat. Calculate the heat that could be stored by melting 1000. kg of each of the following solids. (*Note:* The water in each formula is included in the molecular weight.)

a. calcium chloride ($CaCl_2 \cdot 6H_2O$): melting point = 30.2 °C, heat of fusion = 40.7 cal/g

b. lithium nitrate ($LiNO_3 \cdot 3H_2O$): melting point = 29.9 °C, heat of fusion = 70.7 cal/g

c. sodium sulfate ($Na_2SO_4 \cdot 10H_2O$): melting point = 32.4 °C, heat of fusion = 57.1 cal/g

5.74* Why wouldn't a solid such as K_2SO_4 (melting point = 1069 °C, heat of fusion = 50.3 cal/g) be suitable for use in a solar heat storage system? (See Exercise 5.73.)

5.75 Calculate the total amount of heat needed to change 0.500 kg of ice at −10.°C into 0.500 kg of steam at 120.°C. Do this by calculating the heat required for each of the following steps and adding to get the total.

Step 1. Ice (−10.°C) → ice (0.°C)

Step 2. Ice (0.°C) → water (0.°C)

Step 3. Water (0.°C) → water (100.°C)

Step 4. Water (100.°C) → steam (100.°C)

Step 5. Steam (100.°C) → steam (120.°C)

5.76 Refrigerators and freezers are an integral part of maintaining human health both at home and in hospitals. Liquid Freon (CCl_2F_2) was once used as a refrigerant. It was circulated inside the cooling coils of older refrigerators and freezers. As it vaporized, it absorbed heat. How much heat could be removed by 2.00 kg of Freon as it vaporized inside the coils of a refrigerator? The heat of vaporization of Freon is 38.6 cal/g.

Additional Exercises

5.77 Sevoflurane, an inhalation anesthesia, has a boiling point of 58.5 °C. As the liquid is heated past the boiling point, it changes to gas. Explain this in terms of the effect of temperature on cohesive and disruptive forces.

5.78 If one atom of oxygen reacted with two atoms of nitrogen to form a molecule, what would be the formula of the molecule? Use **Figures 5.6 and 5.9** to determine if the bond between the atoms is ionic or covalent. Also, name the compound that is formed.

5.79 Water is sometimes made safe to drink by boiling. Explain why this might not work if you attempted to do it in an open pan on the summit of Mount Everest.

5.80 Suppose a sample of water was heated to the boiling point in a glass beaker, using a single burner. What would happen to the temperature of the boiling water if a second burner was added to help with the heating? Explain.

5.81 Suppose you put four 250 mL bottles of water into an ice chest filled with crushed ice. If the bottles were initially at a temperature of 23 °C, calculate the number of grams of ice that would have to melt in order to cool the water in the bottles to 5 °C. Assume the density of water is 1.00 g/mL.

5.82 Using **Table 5.8**, determine which metal—aluminum, copper, or lead—would need the least amount of heat added to it to reach 50.0 °C if you began with 5.0 g of each at 10.0 °C?

5.83 An ice bag treats muscle injuries and releases 43.4 kJ of heat when all ice melts at 0 °C. How many grams of ice were placed in the bag? The ΔH_{fus} of water is 334 J/g.

5.84 When an egg is hard-boiled, the main protein solidifies at around 180 °C. If a 68.8 g egg is dropped into a pot of boiling water at 175 °C and removed when the water reaches 183 °C, how much heat is required for the transformation? (*Note:* The specific heat of an egg is 3.32 kJ/kg °C.)

5.85 The specific heat of a protein is 1.48 J/g·K, and it denatures over a temperature range of 5 K. How much heat is released if 1 mg of protein is denatured?

6 Gases and Solutions

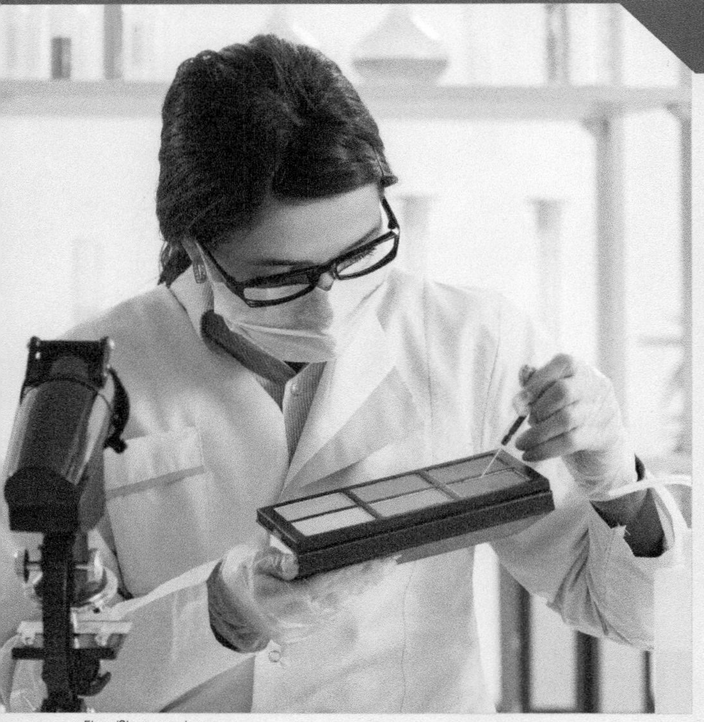
Elnur/Shutterstock.com

Health Career Focus

Personal Care Product Developer

I've loved cosmetics—shampoos, makeup, and lotions—for as long as I can remember. In reading product catalogs, I discovered the fascinating career of "product formulation." Eventually, I landed a job as an assistant developer at a nutritional supplement laboratory. Not only must I know chemistry to understand active ingredients, but I must also understand the product bases. Understanding solutions in order to control the smoothness, fluffiness, thickness, or silkiness of a product is critical.

To meet my approval, a product must smell amazing, contain natural botanicals and no animal products, should be environmentally conscientious, and promote overall health. Every product I develop solves a complex puzzle addressing multiple questions. I love when all the parts come together—when a client opens a bottle, smells the product, and can't wait to go home and try it.

Follow-up to this Career Focus appears at the end of the chapter before the *Concept Summary*.

Learning Objectives

When you have completed your study of this Chapter 6, you should be able to:

1 Calculate the kinetic energy of moving particles. **(Section 6.1)**

2 Convert pressure and temperature values into various units. **(Section 6.1)**

3 Do calculations based on Boyle's law, Charles's law, and the combined gas law. **(Section 6.2)**

4 Do calculations based on the ideal gas law. **(Section 6.3)**

5 Do calculations based on Dalton's law. **(Section 6.4)**

6 Apply your knowledge of intermolecular forces to determine the solubility of substances in a given solvent. **(Section 6.5)**

7 Write molecular equations in total ionic and net ionic forms. **(Section 6.6)**

8 Calculate solution concentrations in units of molarity, weight/weight percent, weight/volume percent, and volume/volume percent. **(Section 6.7)**

9 Describe how to prepare solutions of specific concentration by diluting concentrated solutions. **(Section 6.8)**

10 Describe the process of osmosis. **(Section 6.9)**

11 Describe the role of osmosis and diffusion in dialysis. **(Section 6.9)**

In Chapter 5, we learned about attractive and repulsive forces between molecules and how they affect the energetics of changes of state. Here in Chapter 6, we expand on that discussion as we explore how intermolecular forces affect the properties of gases and solutions. The first half of the chapter focuses on the relationships between pressure, temperature, volume, and moles of a gas. Then we transition to solubility, with a focus on the dissolution of ionic compounds. Next, we quantify solutions as we learn about molarity, dilution preparation, % (w/v), and % (v/v). We end the chapter by applying these concepts of solution chemistry to osmosis and dialysis.

6.1 Properties of Gases

Learning Objective 1 Calculate the kinetic energy of moving particles.

Learning Objective 2 Convert pressure and temperature values into various units.

The kinetic molecular theory of gases is the name scientists have given to a model, or theory, used to explain the behavior of matter in the gaseous state. Some theories, including this one, are based on a group of generalizations or postulates. This useful practice makes it possible to study and understand each postulate individually, instead of the entire theory.

The postulates of the kinetic molecular theory of gases are as follows:

1. Matter is composed of tiny particles called molecules that move in constant straight-line motion and therefore possess kinetic energy.
2. Gas particles are far apart, so there are little to no attractive or repulsive forces between them.
3. Gas particles transfer energy from one to another during collisions in which no net energy is lost from the system.
4. The combined volume of the gas particles is very small compared to the total volume of the container they are in.
5. The average gas particle speed increases as the temperature increases.

Volker Meyer/Pexels

Figure 6.1 Kinetic energy is associated with mass and speed. A skier in motion possesses kinetic energy.

As mentioned in Chapter 5, kinetic energy is the energy a particle has as a result of its motion (**Figure 6.1**). Mathematically, kinetic energy is calculated as

$$\text{K.E.} = \frac{1}{2}mv^2 \tag{6.1}$$

where m is the particle mass and v is its velocity. Thus, if two particles of different mass are moving at the same velocity, the heavier particle will possess more kinetic energy than the other particle. Similarly, the faster moving of two particles with equal masses will have more kinetic energy.

Example 6.1 Calculating Kinetic Energies

Calculate the kinetic energy of two particles with masses of 2.00 g and 3.00 g if they are both moving at a velocity of 15.0 cm/s.

Solution

The kinetic energy of the 2.00 g particle is

$$\text{K.E.} = \frac{1}{2}(2.00\text{ g})\left(15.0\frac{\text{cm}}{\text{s}}\right)^2 = 15\frac{\text{g}\cdot\text{cm}^2}{\text{s}^2}$$

The kinetic energy of the 3.00 g particle is

$$\text{K.E.} = \frac{1}{2}(3.00\text{ g})\left(15.0\frac{\text{cm}}{\text{s}}\right)^2 = 22.5\frac{\text{g}\cdot\text{cm}^2}{\text{s}^2}$$

Rounding gives 23 $\text{g}\cdot\text{cm}^2/\text{s}^2$; thus, the more massive particle has more kinetic energy.

✔ **Learning Check 6.1** Calculate the kinetic energy of two 3.00 g particles if one has a velocity of 10.0 cm/s and the other has a velocity of 20.0 cm/s.

So far, the states of matter have been discussed in qualitative terms, not quantitative ones. We have pointed out, for example, that solids, liquids, and gases expand when heated, but the amounts by which they expand have not been calculated. Such calculations for liquids and solids are beyond the intended scope of this text, but gases obey relatively simple quantitative relationships. These relationships, called **gas laws**, describe in mathematical terms the behavior of gases as they are mixed, subjected to pressure or temperature changes, or allowed to diffuse. As a result, we describe gases in terms of four properties: pressure, volume, amount, and temperature (**Table 6.1**).

gas laws Mathematical relationships that describe the behavior of gases as they are mixed, subjected to pressure or temperature changes, or allowed to diffuse.

pressure A force per unit area of surface on which the force acts. In measurements and calculations involving gases, it is often expressed in units related to measurements of atmospheric pressure.

Table 6.1 Properties of Gases

Property	Definition in the Context of Gases	Common Units of Measurement
Pressure (*P*)	The force a gas exerts on the walls of its container	atmospheres (atm), torr (torr), millimeters of mercury (mm Hg), pounds per square inch (psi), bar (bar), pascal (Pa)
Volume (*V*)	The space a gas occupies	milliliter (mL), liter (L), cubic centimeter (cm^3 or cc)
Amount (*n*)	The amount of gas particles measured in moles	grams (g), moles (n)
Temperature (*T*)	A measure of the kinetic energy of the gas particles	degrees Celsius (°C) or kelvin (K)

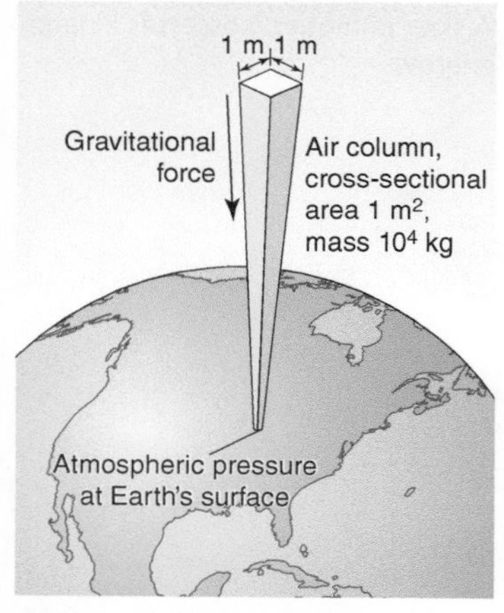

Figure 6.2 Atmospheric pressure is the force per unit area exerted by the weight of the atmosphere.

To use the gas laws, you must clearly understand the units used to express pressure. **Pressure** is defined as force per unit area. However, most of the units commonly used to express pressure reflect a relationship to barometric measurements of atmospheric pressure (**Figure 6.2**). Despite attempts to standardize measurement units, you will probably encounter a number of different pressure units in your future studies and employment. Some of the more common are standard atmosphere (atm), torr, millimeters of mercury (mm Hg), inches of mercury (in. Hg), pounds per square inch (psi), bar, and kilopascals (kPa). Recall from **Section 5.5A** that one standard atmosphere (atm) is the pressure needed to support a 760 mm column of mercury in a barometer tube, and 1 torr is the pressure needed to support a 1 mm column of mercury in a barometer tube. The relationships of the various units to the standard atmosphere are listed in **Table 6.2**. Note that the values of 1 atm and 760 torr (or 760 mm Hg) are exact numbers based on definitions, so they do not limit the number of significant figures in calculated numbers.

Table 6.2 Units of Pressure

Unit	Relationship to One Standard Atmosphere	Typical Application
Atmosphere	—	Gas laws
Torr	760 torr = 1 atm	Gas laws
Millimeters of mercury	760 mm Hg = 1 atm	Gas laws
Pounds per square inch	14.7 psi = 1 atm	Compressed gases
Bar	1.01 bar = 29.9 in. Hg = 1 atm	Meteorology
Kilopascal	101 kPa = 1 atm	Gas laws

Example 6.2 Units of Gas Pressure

Hyperbaric therapy uses oxygen gas to help heal wounds (**Figure 6.3**). The gauge on a cylinder of compressed oxygen gas reads 1500. psi. Recall from **Section 1.7** that the decimal point present at the end of the number makes the trailing zeros significant. Express this pressure in (a) atm, (b) torr, and (c) mm Hg.

Solution

Use the dimensional analysis method of calculation from **Section 1.8**, and the necessary factors from **Table 6.2**.

a. According to **Table 6.2**, 14.7 psi = 1 atm. Therefore, the known quantity, 1500. psi, is multiplied by the factor atm/psi in order to generate units of atm. The result is

$$1500.\ \cancel{\text{psi}}\left(\frac{1\ \text{atm}}{14.7\ \cancel{\text{psi}}}\right) = 1.02 \times 10^2\ \text{atm}$$

b. **Table 6.2** does not provide a direct relationship between psi and torr. However, both psi and torr are related to atm. Therefore, the atm is used somewhat like a bridge between psi and torr:

$$(1500.\ \cancel{\text{psi}})\left(\frac{1\ \cancel{\text{atm}}}{14.7\ \cancel{\text{psi}}}\right)\left(\frac{760\ \text{torr}}{1\ \cancel{\text{atm}}}\right) = 77{,}551\ \text{torr} = 7.76 \times 10^4\ \text{torr}$$

Note how the bridge term canceled out.

c. Because the torr and mm Hg are identical, the problem is worked as it was in part b:

$$(1500.\ \cancel{\text{psi}})\left(\frac{1\ \cancel{\text{atm}}}{14.7\ \cancel{\text{psi}}}\right)\left(\frac{760\ \text{mm Hg}}{1\ \cancel{\text{atm}}}\right) = 7.76 \times 10^4\ \text{mm Hg}$$

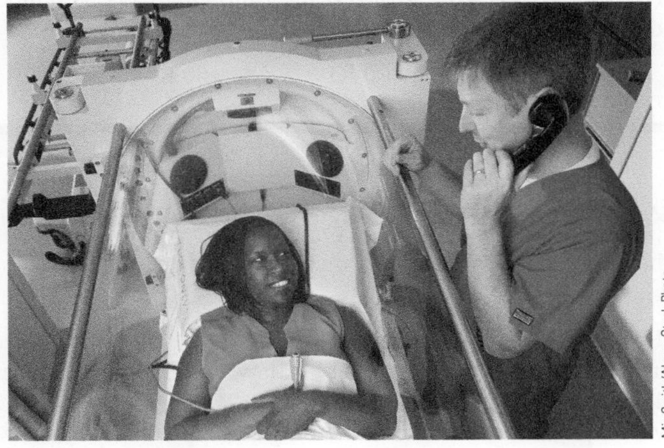

Mic Smith/Alamy Stock Photo

Figure 6.3 A patient about to undergo hyperbaric oxygen therapy.

✔ **Learning Check 6.2** A barometer has a pressure reading of 670. torr. Convert this reading into (a) atm, (b) psi, and (c) mm Hg.

volume The amount of three-dimensional space an object or a substance occupies.

Volume, one of the four properties of gases we intend to discuss, is the amount of three-dimensional space a gas occupies. A gas will diffuse to fill the entire space available in its container. The volume of gases is most commonly measured in liters (L) and milliliters (mL). According to the kinetic molecular theory, a gas expands when it is heated at constant pressure, because the gaseous particles move faster at the higher temperature. It makes no difference to the particles what temperature scale is used to describe the heating. However, gas law calculations are based on the Kelvin temperature scale (**Section 1.5**) rather than the Celsius or Fahrenheit scales. The only apparent difference between the Kelvin and Celsius scales is the location of the zero reading (see **Figure 1.17**). The 0 reading on the Kelvin scale has a great deal of significance. It is the lowest possible temperature and is called **absolute zero**. It represents the temperature at which particles have no kinetic energy because all motion stops. The Kelvin and Celsius temperature scales have the same size degree, but the 0 of the Kelvin scale is −273 °C. Thus, a Celsius reading can be converted to a Kelvin reading by adding 273 (see Equation 1.6).

absolute zero The temperature at which all motion stops; a value of 0 on the Kelvin scale.

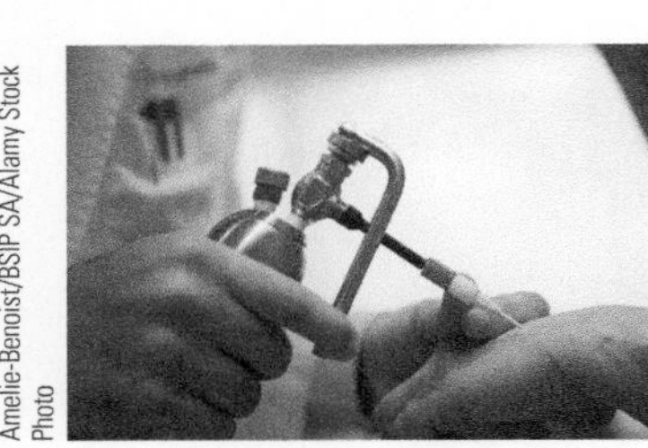

Amelie-Benoist/BSIP SA/Alamy Stock Photo

Figure 6.4 The temperature of liquid nitrogen used in cryogenic surgery is between 63 K and 77 K (i.e., between −210 °C and −196 °C).

Example 6.3 Converting Temperature Values

The Kelvin temperature scale is convenient for recording the very low temperatures of liquid helium [used to cool magnets for magnetic resonance imaging (MRI)] and liquid nitrogen (used in cryosurgery; see **Figure 6.4**). Refer to **Section 1.5C** for the necessary conversion factors, and make the following temperature conversions (remember, Kelvin temperatures are expressed in kelvins, K).

a. 37 °C (body temperature) to K b. −50 °C to K c. 400 K to °C

Solution

a. Celsius is converted to Kelvin by adding 273: K = °C + 273. Therefore,

$$\text{K} = 37\ °\text{C} + 273 = 310\ \text{K}\ \ \text{(body temperature)}$$

b. Again, 273 is added to the Celsius reading. However, in this case, the Celsius reading is negative, so the addition must be done with the signs in mind. Therefore,

$$\text{K} = -50\ °\text{C} + 273 = 223\ \text{K}$$

c. Because K = °C + 273, an algebraic rearrangement shows that °C = K − 273. Therefore,

$$°\text{C} = 400\ \text{K} - 273 = 127\ °\text{C}$$

✔ **Learning Check 6.3** Convert the following Celsius temperatures into kelvins, and the Kelvin (K) temperatures into degrees Celsius:

a. 27 °C b. 0 °C c. 0.00 K d. 100 K

6.2 Pressure, Temperature, and Volume Relationships

Learning Objective 3 Do calculations based on Boyle's law, Charles's law, and the combined gas law.

Experimental investigations into the behavior of gases as they were subjected to changes in temperature and pressure led to several gas laws that could be expressed by simple mathematical equations. In 1662, Robert Boyle (1627–1691) discovered an inverse relationship between the pressure and volume of a gas sample kept at constant temperature. The change in one is in a direction opposite (inverse) to the change in the other. Thus, if the pressure on a gas sample increases, the volume of the gas sample decreases. Similarly, if the pressure decreases, then the volume increases.

If the temperature and amount of gas are held constant, the product of the initial pressure (P_i) and initial volume (V_i) will be equal to the product of the final pressure (P_f) and final volume (V_f). This relationship, known as **Boyle's law**, is shown in Equation 6.2:

Boyle's law A gas law describing the inverse relationship between the pressure and volume of a gas at constant temperature.

$$P_i V_i = P_f V_f \tag{6.2}$$

Breathing is a real-world application of Boyle's law. Each time you breathe, your lungs expand, increasing their volume and reducing the pressure of the gas inside your lungs. The atmospheric pressure is higher than the pressure of your expanded lungs, causing air to flood your empty lungs. Inhaling air has changed not only the pressure and volume of the gas in your lungs, but also the total amount of gaseous particles contained in them. For instance, when inhaling, the concentration of oxygen in your lungs increases (**Figure 6.5**).

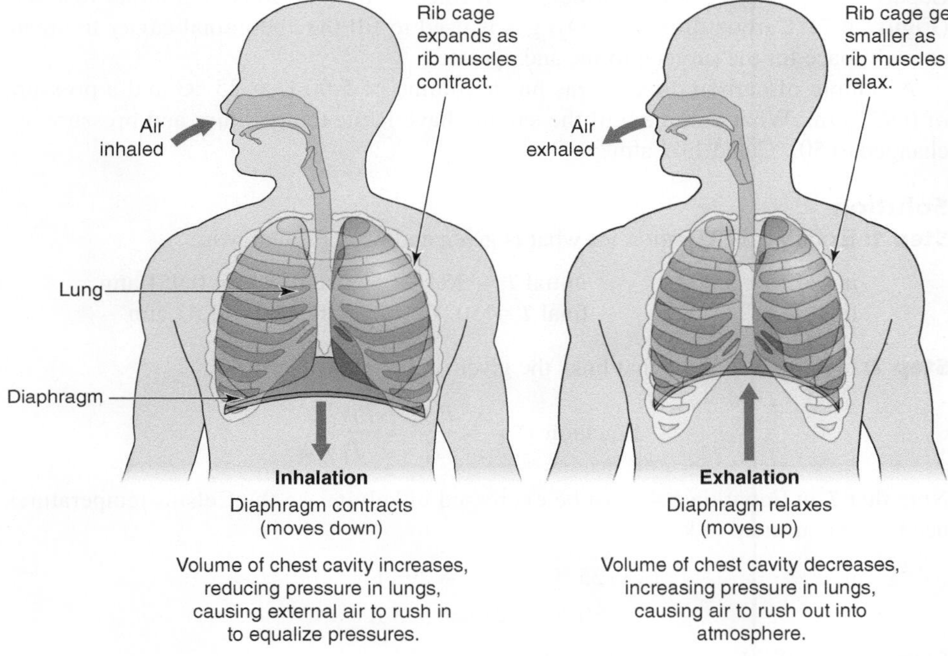
Rib cage expands as rib muscles contract.
Rib cage gets smaller as rib muscles relax.
Air inhaled
Air exhaled
Lung
Diaphragm
Inhalation
Diaphragm contracts (moves down)
Volume of chest cavity increases, reducing pressure in lungs, causing external air to rush in to equalize pressures.
Exhalation
Diaphragm relaxes (moves up)
Volume of chest cavity decreases, increasing pressure in lungs, causing air to rush out into atmosphere.

Figure 6.5 The diaphragm controls the volume of the chest cavity, which regulates the pressure in the lungs and thus controls breathing.

In 1787, Jacques Charles (1746–1823) found that when pressure remains constant, the volume of a gas sample is directly proportional to its temperature expressed in kelvins. In other words, if the temperature is doubled, the sample volume doubles as long as the pressure is kept constant. Similarly, if the temperature is halved, then the volume is halved as well (see **Figure 6.6**).

If the pressure and amount of gas are held constant, the quotient of initial volume (V_i) and initial temperature (T_i) will be equal to the quotient of final volume (V_f) and final temperature (T_f). This behavior, known as **Charles's law**, is represented mathematically in Equation 6.3:

Charles's law A gas law describing the direct relationship between the volume and temperature of a gas at constant pressure.

$$\frac{V_i}{T_i} = \frac{V_f}{T_f} \tag{6.3}$$

Boyle's law and Charles's law can be combined to give a single law called the combined gas law, which provides a relationship between the pressure, volume, and temperature of gases. The **combined gas law** is shown in Equation 6.4:

combined gas law A gas law that describes the pressure, volume, and temperature behavior of a gas sample when the amount of the gas is held constant.

$$\frac{P_i V_i}{T_i} = \frac{P_f V_f}{T_f} \tag{6.4}$$

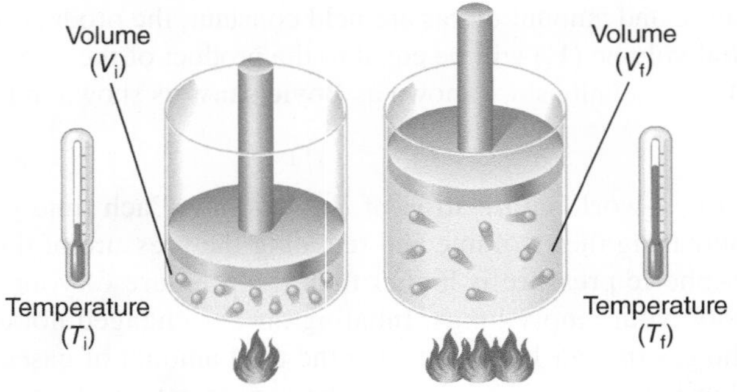

Figure 6.6 A sealed container fitted with a movable piston demonstrates that the volume of a gas is directly proportional to the temperature (at constant pressure).

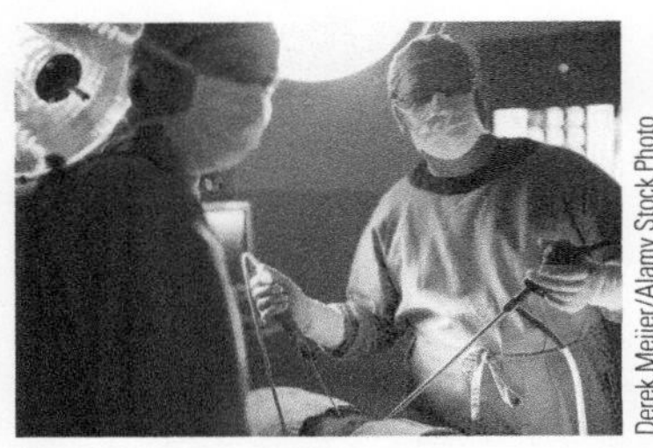

Figure 6.7 Laparoscopic surgery is a minimally invasive technique that allows surgeons to work through smaller incisions and reduce the recovery times for patients.

Example 6.4 Gas Law Calculations

Laparoscopic surgery is now widely performed to treat various abdominal diseases (**Figure 6.7**). Carbon dioxide (CO_2) gas is used to fill the abdominal cavity to create enough space for the surgeon to see and operate.

A sample of carbon dioxide gas has a volume of 5.00 L at 25 °C and a pressure of 0.951 atm. What volume will the sample have if the temperature and pressure are changed to 50. °C and 1.41 atm?

Solution

Step 1: Examine the question for what is given and what is unknown.

initial V = 5.00 L	initial T = 25 °C	initial P = 0.951 atm
final V = ?	final T = 50.°C	final P = 1.41 atm

Step 2: Identify a formula that links the given and unknown terms.

$$\text{Equation 6.4:} = \frac{P_i V_i}{T_i} = \frac{P_f V_f}{T_f}$$

Note that T in Equation 6.4 must be expressed in kelvins, so the Celsius temperatures need to be converted to kelvins.

$$25\ °\text{C} + 273 = 298\ \text{K}$$
$$50.°\text{C} + 273 = 323\ \text{K}$$

Step 3: Enter the data into the formula.

$$\frac{(0.951\ \text{atm})(5.00\ \text{L})}{298\ \text{K}} = \frac{(1.41\ \text{atm})V_f}{323\ \text{K}}$$

$$V_f = \frac{(323\ \cancel{\text{K}})(0.951\ \cancel{\text{atm}})(5.00\ \text{L})}{(298\ \cancel{\text{K}})(1.41\ \cancel{\text{atm}})} = 3.66\ \text{L}$$

✔ Learning Check 6.4

a. A patient is receiving hyperbaric therapy using oxygen gas to heal an infection. The hyperbaric chamber is filled with 650 L of O_2. If the pressure inside the chamber is 3.00 atm at 28 °C, what is the pressure at 20 °C?

b. A sample of argon gas is confined in a 10.0 L container at a pressure of 1.90 atm and a temperature of 30. °C. What volume would the sample have at 1.00 atm and 210.2 °C?

Environmental Connections 6.1

CO_2 Emissions: A Blanket around Earth

In the same way that a greenhouse heats up when radiant heat energy from the sun is trapped as sunlight passes through its glass or plastic windows, the surface of Earth warms when radiant solar energy passes through the atmosphere and is trapped. Scientists call this heating of Earth the greenhouse effect. Most climate scientists agree that one main cause of current observed climate change is *human activity* that enhances this "greenhouse effect." Since the Industrial Revolution in the nineteenth century, human activity has influenced the climate.

Certain gases, particularly carbon dioxide (CO_2), water vapor (H_2O), methane (CH_4), and nitrous oxide (N_2O), are called greenhouse gases because they act like the windows of a greenhouse and let visible light in but prevent heat from radiating back out. As a result of this effect and the presence of these gases in Earth's atmosphere, the average temperature at the surface of Earth is a life-supporting 59 °F or 15 °C. In naturally occurring amounts, these gases make life possible on Earth, but too much of a good thing can create problems.

The consequences of changing the natural atmospheric concentration of greenhouse gases are difficult to predict and controversial, but one thing is certain. During the past century, based on actual measurements, the average surface temperature of Earth has increased by up to 0.8 °C. Although that might not sound like much, a tiny increase in temperature can have enormous consequences. Glaciers, sea ice, and ice sheets around the world are melting. As the ice melts, sea levels rise. Also, as seawater warms, it expands and sea levels rise even more. Only humans can change the amount of greenhouse gases produced by human activity, and thereby minimize these negative environmental effects.

iStock.com/Nancy Nehring

An increase in the rate of breakup of glaciers (calving) as they approach the oceans is one indication of global warming.

6.3 The Ideal Gas Law

Learning Objective 4 Do calculations based on the ideal gas law.

The combined gas law (Equation 6.4) works only for samples in which the mass of gas remains constant during changes in temperature, pressure, and volume. However, it is often useful to work with situations in which the amount of gas varies. The foundation for this kind of work was proposed in 1811 by Amadeo Avogadro. According to his proposal, which is now known as **Avogadro's law** (Equation 6.5), equal volumes of different gases (V) measured at the same temperature and pressure contain equal numbers of gas molecules (n).

Avogadro's law Equal volumes of gases measured at the same temperature and pressure contain equal numbers of molecules.

$$\frac{V_i}{n_i} = \frac{V_f}{n_f} \tag{6.5}$$

According to this idea, two identical compressed gas cylinders of helium and oxygen at the same pressure and temperature would contain identical numbers of molecules of the respective gases. However, the mass of gas in the cylinders would *not* be the same because the molecules of the two gases have different molecular weights.

The actual temperature and pressure used do not influence the validity of Avogadro's law, but it is convenient to specify a standard set of values for use when tying all the properties of gases together. Chemists have chosen 0 °C (273 K) and 1.00 atm to represent what are called **standard conditions** for gas measurements. These conditions are often abbreviated as STP (standard temperature and pressure). Experiments show that 1.00 mol of any gas molecules has a volume of 22.4 L at STP (see **Figure 6.8**).

standard conditions (STP) A set of specific temperature and pressure values used for gas measurements.

© Spencer L. Seager

Figure 6.8 The ball has a volume of 22.4 L, the volume of 1 mol of any gas at STP. The bottle has a volume of 1.00 quart (0.946 L). How many moles of gas would the bottle contain at STP?

ideal gas law A gas law that relates the pressure, volume, temperature, and number of moles in a gas sample. Mathematically, it is *PV = nRT*.

universal gas constant The constant that relates pressure, volume, temperature, and number of moles of gas in the ideal gas law.

A combination of Boyle's, Charles's, and Avogadro's laws leads to the **ideal gas law**, which includes the quantity of gas in a sample as well as the temperature, pressure, and volume of the sample:

$$PV = nRT \qquad (6.6)$$

In this equation, P, V, and T are pressure, volume, and temperature, respectively, just as they were in the gas laws given in Equations 6.2 to 6.5. The symbol n stands for the number of moles of gas in the sample being used, and R is a constant known as the **universal gas constant**. The measured value for the volume of 1.00 mol of gas at STP allows R to be evaluated by substituting the values into Equation 6.6. After rearrangement to isolate R:

$$R = \frac{PV}{nT} = \frac{(1.00 \text{ atm})(22.4 \text{ L})}{(1.00 \text{ mol})(273 \text{ K})} = 0.0821 \frac{\text{L} \cdot \text{atm}}{\text{mol} \cdot \text{K}}$$

This value of R is the same for all gases under any conditions of temperature, pressure, and volume.

Example 6.5 Ideal Gas Law Calculations

Nitrous oxide (N_2O) is an inhaled gas that is used as a pain medication during childbirth (**Figure 6.9**), following trauma, and as a part of end-of-life care.

Use the ideal gas law to calculate the volume of 0.41 mol of nitrous oxide gas at a temperature of 20. °C and a pressure of 1.20×10^3 torr.

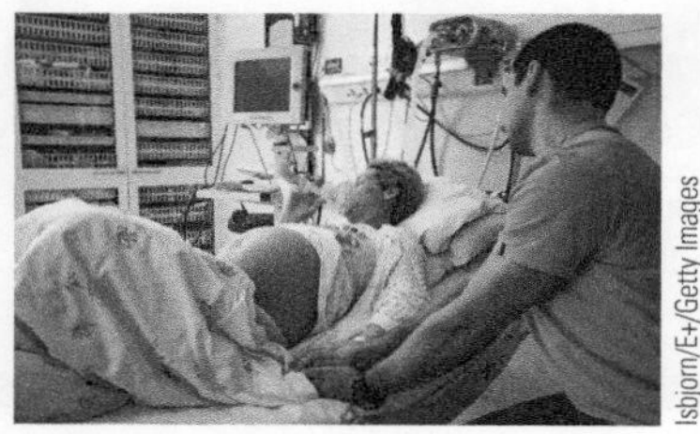

Isbjorn/E+/Getty Images

Figure 6.9 A patient using nitrous oxide to manage pain during labor.

Solution

Step 1: Examine the question for what is given and what is unknown.

$$\text{Given: } n = 0.41 \text{ mol}, T = 20.°\text{C}, P = 1.20 \times 10^3 \text{ torr}$$

$$\text{Unknown: } V = ?$$

Step 2: Identify a formula that links the given and unknown terms.

$$\text{Equation 6.6: } PV = nRT \text{ (the ideal gas law)}$$

Because volume is the desired quantity, we isolate it by dividing both sides of the ideal gas law by P:

$$V = \frac{nRT}{P}$$

To use the ideal gas law (Equation 6.6), it is necessary that all units match those of R. Thus, the temperature will have to be expressed in kelvins and the pressure in atmospheres:

$$\text{K} = °\text{C} + 273 = 20.°\text{C} + 273 = 293 \text{ K}$$

$$P = (1.20 \times 10^3 \cancel{\text{torr}})\left(\frac{1 \text{ atm}}{760 \cancel{\text{torr}}}\right) = 1.58 \text{ atm}$$

Step 3: Enter the data into the formula:

$$V = \frac{(0.41 \cancel{\text{mol}})\left(0.0821 \frac{\text{L} \cdot \cancel{\text{atm}}}{\cancel{\text{mol}} \cdot \cancel{\text{K}}}\right)(293 \cancel{\text{K}})}{(1.58 \cancel{\text{atm}})} = 6.2 \text{ L}$$

✔ **Learning Check 6.5** A 2.15 mol sample of sulfur dioxide gas (SO_2) occupies a volume of 12.6 L at 30. °C. What is the pressure of the gas in atm?

The gas laws presented in Equations 6.2 to 6.6 apply only to gases that are ideal, but no ideal gases actually exist. If they did exist, ideal gases would behave exactly as predicted by the gas laws at all temperatures and pressures. Instead, real gases deviate from the behavior predicted by the gas laws, but under normally encountered temperatures and pressures, the deviations are small for many real gases. This fact allows the gas laws to be used to predict the behavior of real gases. Intermolecular forces tend to make gases behave less ideally. Thus, the gas laws work best for gases in which such forces are weak—that is, those made up of single atoms [the noble gases or nonpolar molecules (O_2, N_2, etc.)]. Highly polar molecules such as water vapor, hydrogen chloride, and ammonia deviate significantly from ideal behavior.

6.4 Dalton's Law

Learning Objective 5 Do calculations based on Dalton's law.

John Dalton made a number of important contributions to chemistry. Some of his experiments led to the law of partial pressures, also called **Dalton's law**. According to Dalton's law, the total pressure exerted by a mixture of different gases kept at a constant volume and temperature is equal to the sum of the partial pressures of the gases in the mixture. The **partial pressure** of each gas in such mixtures is the pressure each gas would exert if it were confined alone under the same temperature and volume conditions as the mixture.

Dalton's law of partial pressures The total pressure exerted by a mixture of gases is equal to the sum of the partial pressures of the gases in the mixture.

partial pressure The pressure an individual gas of a mixture would exert if it were in the container alone at the same temperature as the mixture.

Imagine you have the four identical gas containers shown in **Figure 6.10**. Samples of three different gases (represented by △, ○, and □) are placed into three of the containers, one to a container, and the pressure exerted by each sample is measured. Then all three samples are placed into the fourth container and the total pressure (P_t) exerted is measured. The result is a statement of Dalton's law:

$$P_t = P_\triangle + P_\bigcirc + P_\square \qquad (6.7)$$

where $P_\triangle$, $P_\bigcirc$, and $P_\square$ are the partial pressures of gases △, ○, and □, respectively.

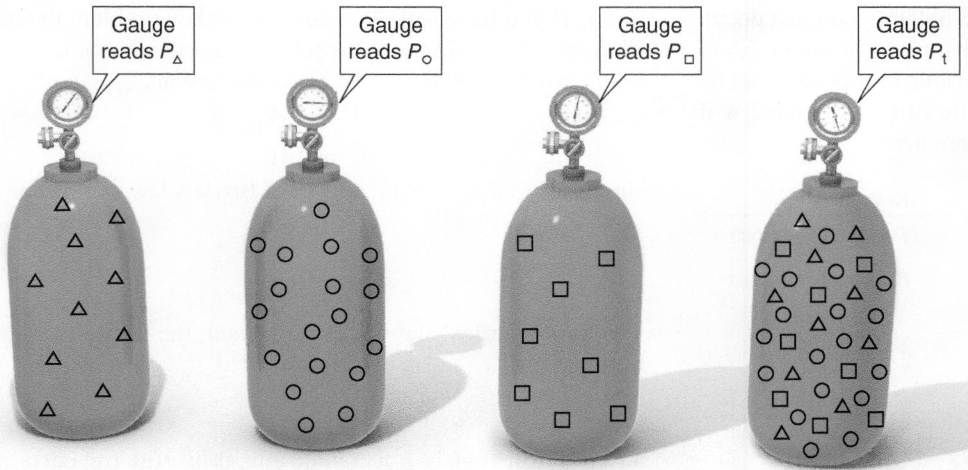

Figure 6.10 Dalton's law of partial pressures.

Example 6.6 Partial Pressure Calculations

Nitrox is a mixture made of nitrogen (N_2) and oxygen (O_2) that is used as breathing gas in pressurized tanks for scuba divers (**Figure 6.11**). Hyperoxia, caused by an excess of oxygen in tissues and organs, occurs when divers breathe in oxygen at partial pressures greater than 1.6 atm. At a recreational dive depth of 99 feet below sea level, the partial pressure of nitrogen in a common nitrox mix is 2.72 atm, and the total pressure exerted on a diver by the combined atmosphere and water is 4.00 atm. Given this information, solve for the partial pressure of oxygen at this depth. Has the diver exceeded the oxygen partial pressure limit that can lead to hyperoxia?

Carlos Villoch - MagicSea.com/Alamy Stock Photo

Figure 6.11 Scuba divers use pressured gas stored in tanks to breathe underwater.

Solution

The total pressure of the compressed nitrox is 4 atm and according to Dalton's law, $P_t = P_{N_2} + P_{O_2}$. Therefore,

$$4.00 \text{ atm} = 2.72 \text{ atm} + P_{O_2}$$

and

$$P_{O_2} = 4.00 - 2.72 = 1.28 \text{ atm}$$

The partial pressure of oxygen at this depth does *not* exceed the safety limit of 1.6 atm. However, the diver should monitor their depth because they will exceed the limit at a depth of 132 feet below sea level.

✔ **Learning Check 6.6** A sample of medical grade compressed air is collected when the atmospheric pressure is 742 torr. The partial pressures of nitrogen and oxygen in the sample are found to be 581 torr and 141 torr, respectively. If water vapor is the only other gas present in the air sample, what is its partial pressure?

Study Tools 6.1

Which Gas Law to Use

For many students, the biggest challenge in this chapter is solving problems using the gas laws. Once you recognize that you are faced with a gas law problem, you must decide which of the six gas laws will work to solve the problem. One aid to selecting the appropriate law is to look for keywords, phrases, or ideas that are often associated with specific laws. Some of these are given here:

Gas Law	Equation	Key
Boyle's law	$P_iV_i = P_fV_f$	T and n are constant
Charles's law	$\frac{V_i}{T_i} = \frac{V_f}{T_f}$	P and n are constant
Combined gas law	$\frac{P_iV_i}{T_i} = \frac{P_fV_f}{T_f}$	n is constant
Avogadro's law	$\frac{V_i}{n_i} = \frac{V_f}{n_f}$	T and n are constant
Dalton's law	$P_t = P_\triangle + P_\bigcirc + P_\square$	Two or more different gases
Ideal gas law	$PV = nRT$	Interrelates all variables

The first three gas laws (Boyle's, Charles's, and the combined) are very similar in that they all use the symbols P, V, and T. If you have narrowed down a gas law problem to one of these three, a simple approach is to use the combined gas law. It works in all three cases. If T is constant,

$$\frac{P_iV_i}{T_i} = \frac{P_fV_f}{T_f}$$

simplifies to $P_iV_i = P_fV_f$, a form of Boyle's law. If P is constant, it simplifies to

$$\frac{V_i}{T_i} = \frac{V_f}{T_f}$$

a form of Charles's law. If V is constant, the combined law becomes

$$\frac{P_i}{T_i} = \frac{P_f}{T_f}$$

an equation that is not named but is useful for certain problems.

Remember too, that temperatures must be expressed in kelvins to get correct answers using any of the gas laws that involve temperature.

6.5 Solutions and Solubility

Learning Objective 6 Apply your knowledge of intermolecular forces to determine the solubility of substances in a given solvent.

Solutions, first mentioned in **Section 1.3**, are homogeneous mixtures of two or more substances in which the components are present as atoms, molecules, or ions. In solutions, one substance is usually present in a larger amount compared to the other components. This most abundant substance in a solution is called the **solvent**, whereas the other components are called **solutes**. Most people think of solutions as liquids, but solutions in solid (see **Figure 6.12**) and gaseous forms (e.g., nitrox, the mixture of O_2 and N_2 in Example 6.6) are known as well. The state of a solution is often the same as the state of the solvent. This is illustrated in **Table 6.3**, which lists examples of solutions in various states. The original states of the solvents and solutes are given in parentheses. Solutions made with water as the solvent are called **aqueous solutions** and are denoted (aq) in chemical equations.

Table 6.3 Solutions in Various States

Solution	Solution State	Solvent	Solute
Saltwater	Liquid	Water (liquid)	Sodium chloride (solid)
Alcoholic beverage	Liquid	Water (liquid)	Alcohol (liquid)
Carbonated water	Liquid	Water (liquid)	Carbon dioxide (gas)
Gold alloy (jewelry)	Solid	Gold (solid)	Copper (solid)
Hydrogen in palladium	Solid	Palladium (solid)	Hydrogen (gas)
Air	Gaseous	Nitrogen (gas)	Oxygen (gas)
Humid oxygen	Gaseous	Oxygen (gas)	Water (liquid)

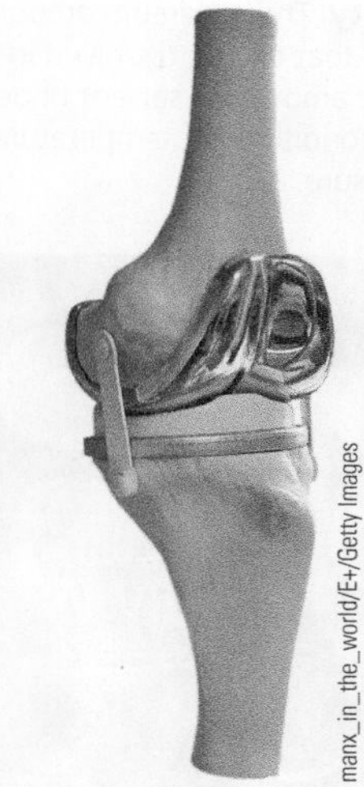
manx_in_the_world/E+/Getty Images

Figure 6.12 Stainless steel is a solid solution of several metals. It is used in the construction of surgical implants and instruments where corrosion resistance, reliability, and hygiene are particularly important.

Solution formation takes place when one or more solutes **dissolve** in a solvent. To better understand this process, **Figure 6.13** shows a solid colored drink mix and oil being poured into two separate glasses of water and photographed over time. In the final photo of **Figure 6.13a**, the solid colored drink and water appear as a single phase, because they have formed a homogeneous mixture (a solution). Substances such as these that completely dissolve in a solvent are defined as **soluble substances**.

In the final photograph of **Figure 6.13b**, the oil and water form a two-layer (heterogeneous) mixture. Substances that do not dissolve in a solvent are defined as **insoluble substances**. The term **immiscible** is used to describe a liquid solute (e.g., the cooking oil) that does not dissolve in a liquid solvent (e.g., the water).

solvent The substance present in a solution in the largest amount.

solute One or more substances present in a solution in amounts less than that of the solvent.

aqueous solution A solution in which the solvent is water. The state of an aqueous solution is indicated by (aq) in chemical reactions.

dissolving The process of solution formation when one or more solutes are dispersed in a solvent to form a homogeneous mixture.

soluble substance A substance that dissolves to a significant extent in a solvent.

insoluble substance A substance that does not dissolve to a significant extent in a solvent.

immiscible A term used to describe liquids that are insoluble in each other.

Turtle Rock Scientific/Science Source

a Drink mix dissolves in water to form a homogenous mixture (a solution).

Giphotostock/Science Source

b Oil and water form a heterogeneous mixture, as evidenced by their separation into distinct layers.

Figure 6.13 Comparison of homogeneous and heterogeneous mixtures. (a) The dissolution of purple drink mix in water. (b) A mixture of oil and water, which together are immiscible.

solubility The maximum amount of solute that can be dissolved in a specific amount of solvent under specific conditions of temperature and pressure.

Substances that are soluble at low concentrations can become insoluble at high concentrations. The term **solubility** refers to the maximum amount of solute that can be dissolved in a specific amount of solvent at a specific temperature. Assume, for example, that 200 g of drink mix can be dissolved in the 100 mL of water at 20 °C (this is denoted as 200 g/100 g H_2O because 100 mL of water = 100 g water). This means that any drink mix in excess of 200 g will sink to the bottom of the glass of 100 mL of water at 20 °C and remain undissolved.

The solubilities of a number of solutes in water are listed in **Table 6.4**. The use of specific units, such as grams of solute per 100 g of water, makes it possible to compare solubilities precisely.

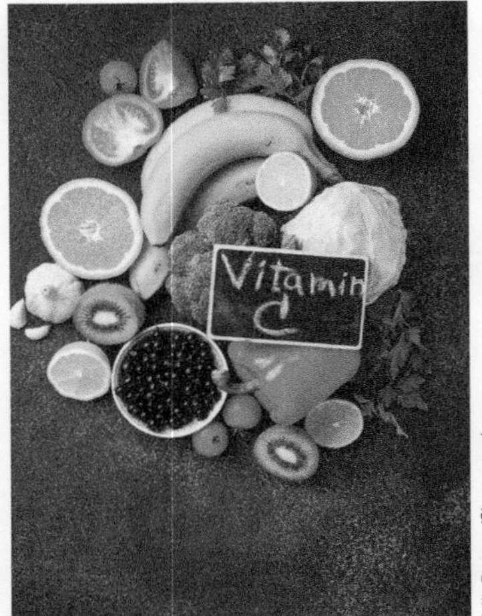

Yulia Furman/Shutterstock.com

Figure 6.14 Ascorbic acid (vitamin C) is found in the foods shown here. If you consume more vitamin C than your body can absorb at one time, the excess will be excreted in your urine, because vitamin C is soluble in the water of your urine.

Table 6.4 Examples of Solute Solubilities in Water (0 °C)

Solute		Solubility
Name	**Formula**	**(g solute/100 g H_2O)**
Ammonium chloride	NH_4Cl	29.7
Ammonium nitrate	NH_4NO_3	118.3
Calcium carbonate	$CaCO_3$	0.0012
Calcium chloride	$CaCl_2$	53.3
Potassium carbonate	K_2CO_3	101
Potassium chloride	KCl	29.2
Sodium bicarbonate	$NaHCO_3$	6.9
Sodium bromide	NaBr	111
Sodium carbonate	Na_2CO_3	7.1
Sodium chloride	NaCl	35.7
Ascorbic acid (vitamin C) (**Figure 6.14**)	$C_6H_8O_6$	33
Ethyl alcohol	C_2H_5OH	∞[a]
Ethylene glycol	$C_2H_4(OH)_2$	∞[a]
Sucrose (table sugar)	$C_{12}H_{22}O_{11}$	179.2

[a]Soluble in all proportions.

saturated solution A solution that contains the maximum amount possible of dissolved solute in a stable situation under the prevailing conditions of temperature and pressure.

supersaturated solution An unstable solution that contains an amount of solute greater than the solute solubility under the prevailing conditions of temperature and pressure.

A solution in which the maximum amount of solute has been dissolved in a quantity of solvent is called a **saturated solution**. Solutions in which the amount of solute dissolved is greater than the solute solubility are called **supersaturated solutions**. Supersaturated solutions are usually prepared by forming a nearly saturated solution at a high temperature and then cooling the solution to a lower temperature at which the solubility is lower. Supersaturated solutions are unstable. The addition of a small amount of solid solute (or even a dust particle) will usually cause the excess solute to crystallize out of the solution until the solution becomes saturated (see **Figure 6.15**). The temperature dependence of solute solubility is illustrated in **Figure 6.16**.

Turtle Rock Scientific/Science Source

1 A supersaturated solution just before adding a small seed crystal.

Turtle Rock Scientific/Science Source

2 Rapid crystallization begins on the seed crystal.

Turtle Rock Scientific/Science Source

3 The excess solute has all crystallized, leaving behind a saturated solution.

Figure 6.15 Crystallization converts a supersaturated solution to a saturated solution.

Whereas the solubility of most liquids and solids in water increases with temperature, the solubility of most gases in water decreases as the temperature increases (see the SO_2 curve in **Figure 6.16**). For instance, as a cold carbonated beverage warms up, more and more of the CO_2 gas escapes. The solubility of gaseous solutes is also influenced significantly by pressure, whereas the effect of pressure on liquid or solid solutes is minimal. The solubility of many gases is directly proportional to the pressure of the gas above the solution at a constant temperature. Thus, if the gas pressure is doubled, the solubility doubles.

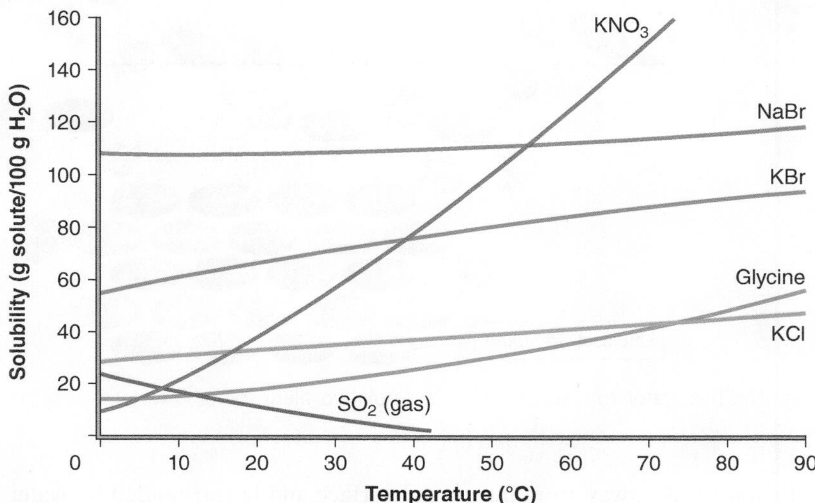

Figure 6.16 The effect of temperature on solute solubility.

The pressure dependence of gas solubility provides the "sparkle" for carbonated beverages. The cold beverage is saturated with CO_2 and capped under pressure. When the bottle is opened, the pressure is relieved, and the gas, now less soluble, comes out of the solution as fine bubbles (see **Figure 6.17**). A similar effect sometimes takes place in the bloodstream of deep-sea divers. While submerged, they inhale air under pressure that causes nitrogen to be more soluble in the blood than it is under normal atmospheric pressure. If the diver is brought to the lower pressure on the surface too quickly, the excess dissolved nitrogen comes out of the solution and forms bubbles in the blood and joints. The result, called decompression sickness or the bends, is painful and dangerous. The chances of getting the bends are decreased by breathing a mixture of oxygen and helium rather than air (oxygen and nitrogen), because helium is less soluble in the body fluids than nitrogen.

Image Source/Getty Images

Figure 6.17 As the pressure is released in a carbonated beverage, the solubility of CO_2 decreases, allowing gaseous bubbles of CO_2 to rise to the surface.

During solution formation, how are solute particles removed from the bulk solute and uniformly distributed throughout the solvent? Why are some solutes very soluble, whereas others are not? Consider the formation of a saltwater solution. Recall from **Section 3.3** that solid ionic compounds are collections of ions held together by attractions between the cations and anions. When an ionic compound dissolves, the orderly ionic arrangement is disrupted as the interionic attractions are overcome. This occurs because the attractive forces between the water molecules and the individual ions are stronger than the attractions of the ions within the crystal lattice.

The solution-forming process for an ionic solute such as NaCl is depicted in **Figure 6.18a**. When the solid NaCl is placed in water, the polar water molecules become oriented so that the partial negative charge of the oxygen points toward the positive sodium ions, and the partial positive charges of the hydrogens point toward the negative chloride ions. The attraction between a polar molecule and a charged ion described here is called an *ion–dipole interaction*. As the polar water molecules begin to surround ions on the crystal surface, they tend to create a shielding effect that reduces the attraction between the ion and the remainder of the crystal.

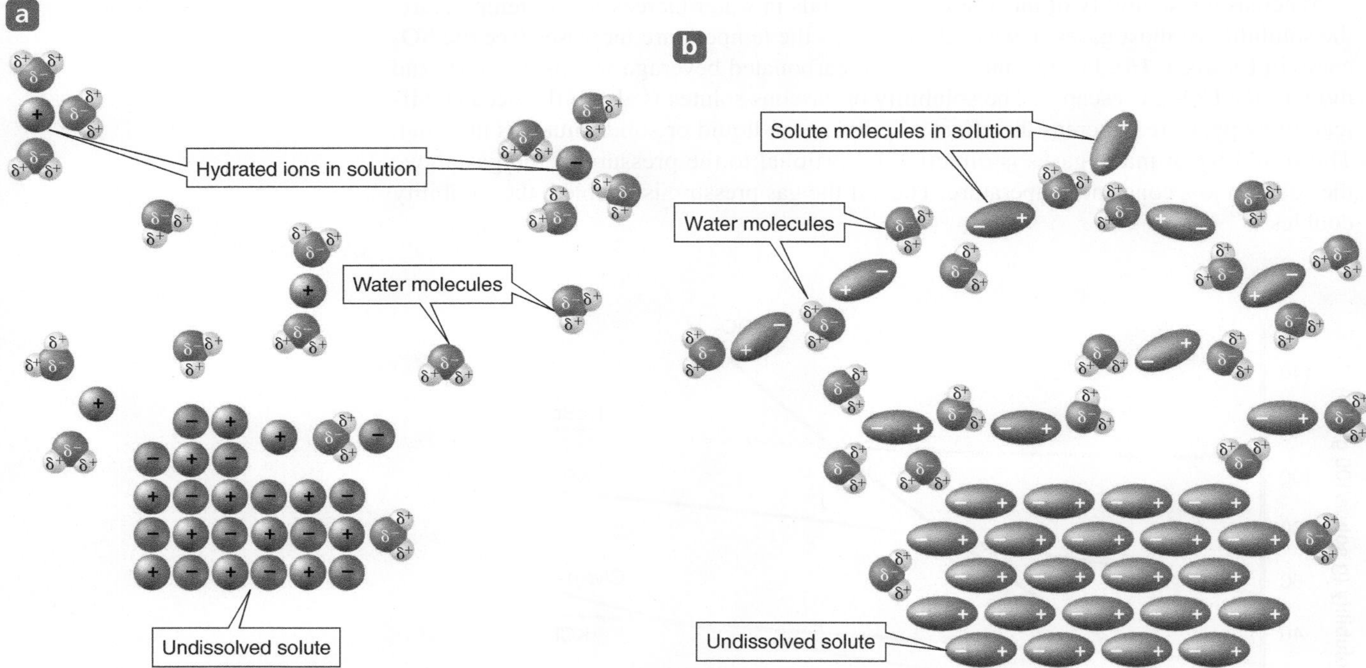

Figure 6.18 (a) The dissolving of an ionic substance (e.g., NaCl) in water. (b) The dissolving of a polar covalent solute in water.

hydrated ions Ions in solution that are surrounded by water molecules.

electrolyte A solute that when dissolved in water forms a solution that conducts electricity.

nonelectrolyte A solute that when dissolved in water forms a solution that does not conduct electricity.

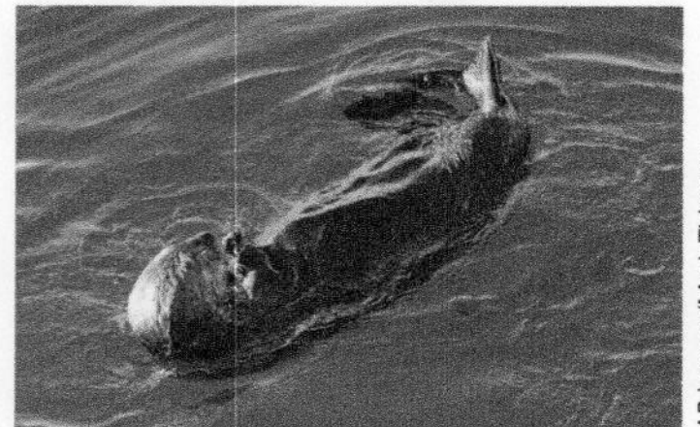

Figure 6.19 Wildlife can be severely impacted by oil spills because oil is insoluble in water. Most oils float, so creatures such as sea otters and seabirds are particularly affected.

As a result, the ion breaks away from the crystal surface and is surrounded by water molecules. Ions surrounded by water molecules in solution are called **hydrated ions**. As each ion leaves the surface, others are exposed to the water, and the crystal is picked apart ion by ion. Once in solution, the hydrated ions are uniformly distributed by stirring or by random collisions with other molecules or ions. Polar covalent solutes, such as sugar or methanol (CH_3OH), dissolve in water in much the same way as ionic solutes do through *dipole–dipole interactions*. The difference is the attraction of polar water molecules to both poles of the solute molecules. The process is represented in **Figure 6.18b**. Solutes that dissociate into hydrated ions in solution, as shown in **Figure 6.18a**, are called **electrolytes**, whereas solutes that do not form ions, as shown in **Figure 6.18b**, are called **nonelectrolytes**. Electrolytes form solutions that conduct electricity, while nonelectrolytes form solutions that do not.

A solute will not dissolve in a solvent if (1) the forces between solute particles are too strong to be overcome by interactions with solvent particles or (2) the solvent particles are more strongly attracted to each other than to solute particles. This phenomenon is why oil spills are so problematic to marine wildlife (**Figure 6.19**). If we think back to the oil in **Figure 6.13b**, it did not dissolve in water because the polar water molecules were attracted to each other more strongly than they were to the nonpolar oil molecules. Cooking oil will dissolve, however, in a nonpolar solvent such as gasoline or carbon tetrachloride (CCl_4). In these solvents, the weak London dispersion forces between oil molecules and solvent molecules are no stronger than the weak London dispersion forces between nonpolar solvent molecules. A good rule of thumb is that "like dissolves like." Thus, polar solvents will dissolve polar or ionic solutes, whereas nonpolar solvents will dissolve nonpolar solutes.

Example 6.7 Predicting Solute Solubilities

To be effective, drugs must be soluble in the aqueous pathways of the body (**Figure 6.20**). Predict whether the following solutes are soluble or insoluble in the solvent indicated.

a. Ammonia gas (NH_3) in water
b. Oxygen gas (O_2) in water
c. $Ca(NO_3)_2$ in water

Solution

a. Soluble: Both NH_3 and H_2O are polar—like dissolves like. The actual solubility at 20 °C is 51.8 g/100 g H_2O.
b. Insoluble: O_2 is nonpolar, whereas water is polar. The actual solubility at 20 °C is 4.3×10^{-3} g/100 g H_2O.
c. Soluble: Calcium nitrate will dissociate into calcium cations and nitrate anions, which will interact with water through ion–dipole interactions to form hydrated ions. The actual solubility at 20 °C is 121.2 g/100 mL.

Grace Cary/Moment/Getty Images

Figure 6.20 The solubility of a drug is crucial in achieving the desired concentration of medicine in the body.

✔ **Learning Check 6.7** Sea birds end up soaked with crude oil (**Figure 6.21**) after disasters such as British Petroleum's Deepwater Horizon oil spill in the Gulf of Mexico in 2010. Why is dish soap incredibly effective at removing crude oil from contaminated feathers? Explain your answer.

Justin Sullivan/Getty Images News/Getty Images

Figure 6.21 Animal rescuers used dish soap to remove crude oil from sea birds after major oil spills.

6.6 Electrolytes and Net Ionic Equations

Learning Objective 7 Write molecular equations in total ionic and net ionic forms.

There are no simple rules for quantitatively predicting the water solubility of ionic compounds. That is, we can't easily predict how many grams of a compound will dissolve in a specific quantity of water. However, ionic compounds are generally insoluble in nonpolar solvents. The solubility guidelines for ionic compounds when water is the solvent are given in **Table 6.5**.

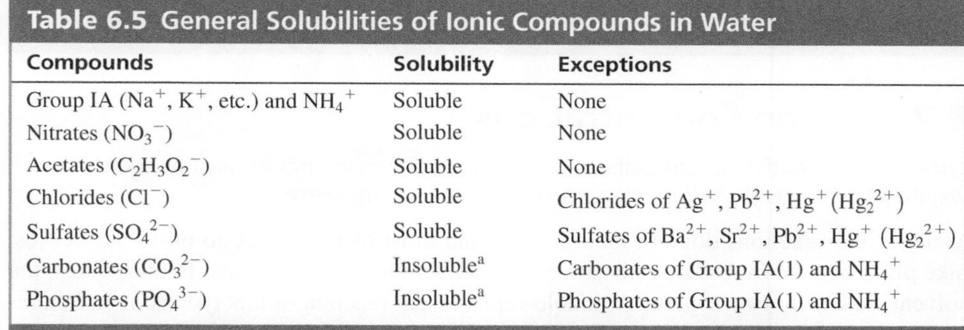

Table 6.5 General Solubilities of Ionic Compounds in Water

Compounds	Solubility	Exceptions
Group IA (Na^+, K^+, etc.) and NH_4^+	Soluble	None
Nitrates (NO_3^-)	Soluble	None
Acetates ($C_2H_3O_2^-$)	Soluble	None
Chlorides (Cl^-)	Soluble	Chlorides of Ag^+, Pb^{2+}, Hg^+ (Hg_2^{2+})
Sulfates (SO_4^{2-})	Soluble	Sulfates of Ba^{2+}, Sr^{2+}, Pb^{2+}, Hg^+ (Hg_2^{2+})
Carbonates (CO_3^{2-})	Insoluble[a]	Carbonates of Group IA(1) and NH_4^+
Phosphates (PO_4^{3-})	Insoluble[a]	Phosphates of Group IA(1) and NH_4^+

[a]Many hydrogen carbonates (HCO_3^-) and phosphates (HPO_4^{2-} and $H_2PO_4^-$) are soluble.

Chemical equations for reactions between substances that form ions in solution can be written in several ways. For example, the following equation is written in the form of a **molecular equation** because each compound is represented by its formula.

$$HCl(aq) + NaOH(aq) \rightarrow NaCl(aq) + H_2O(\ell)$$

This same reaction can also be represented by a **total ionic equation**:

$$H^+(aq) + Cl^-(aq) + Na^+(aq) + OH^-(aq) \rightarrow Na^+(aq) + Cl^-(aq) + H_2O(\ell)$$

In a total ionic equation, each ionic compound is shown dissociated into its ions, the form it takes when it is dissolved in water. Ions that appear as both reactants and products (Cl^- and Na^+ in our example) do not undergo any changes in the reaction, so they are called **spectator ions**. When spectator ions are eliminated from the equation, what's left is called a **net ionic equation**:

$$H^+(aq) + OH^-(aq) \rightarrow H_2O(\ell)$$

This particular net ionic equation obscures the double-replacement nature of the molecular equation, but it does emphasize the actual chemical changes that take place. **Figure 6.22** shows an experiment in which a double-replacement reaction occurs.

molecular equation An equation written with each compound represented by its formula.

total ionic equation An equation written with all soluble ionic substances represented by the ions they form in solution.

spectator ions The ions in a total ionic reaction that are not changed as the reaction proceeds. They appear in identical forms on the left and right sides of the equation.

net ionic equation An equation that contains only unionized or insoluble materials and ions that undergo changes as the reaction proceeds. All spectator ions are eliminated.

© Mike Slabaugh

Figure 6.22 The liquid in the large container is a solution of solid sodium chloride (NaCl) dissolved in water. The liquid being added is a solution of solid silver nitrate ($AgNO_3$) dissolved in water. When the two aqueous solutions are mixed, an insoluble white solid forms. The solid is silver chloride (AgCl). Write the molecular, total ionic, and net ionic equations for the reaction. Identify the spectator ions.

Example 6.8 Writing Reaction Equations

Sodium sulfate (Na_2SO_4) in combination with other salts is used to cleanse the colon (bowel) before a colonoscopy. Write total ionic and net ionic equations for the following double-replacement reaction, and identify any spectator ions. Note that barium sulfate ($BaSO_4$) does not dissolve in water and should not be written in dissociated form. All other compounds are ionic and soluble in water.

$$Na_2SO_4(aq) + BaCl_2(aq) \rightarrow BaSO_4(s) + 2NaCl(aq)$$

Solution

Total ionic: All compounds except $BaSO_4$ dissociate into ions.

$$2Na^+(aq) + SO_4^{2-}(aq) + Ba^{2+}(aq) + 2Cl^-(aq) \rightarrow BaSO_4(s) + 2Na^+(aq) + 2Cl^-(aq)$$

Net ionic: The Na and Cl ions appear in equal numbers as reactants and products. Thus, they are spectator ions and are not shown in the net ionic equation.

$$SO_4^{2-}(aq) + Ba^{2+}(aq) \rightarrow BaSO_4(s)$$

✔ **Learning Check 6.8** Write total and net ionic equations for the following reactions, and identify any spectator ions. Assume all ionic compounds are soluble except $CaCO_3$ and $BaSO_4$, and remember that covalent molecules do not form ions when they dissolve.

a. $2NaI(aq) + Cl_2(aq) \rightarrow 2NaCl(aq) + I_2(aq)$
b. $CaCl_2(aq) + Na_2CO_3(aq) \rightarrow 2NaCl(aq) + CaCO_3(s)$
c. $Ba(OH)_2(aq) + H_2SO_4(aq) \rightarrow 2H_2O(\ell) + BaSO_4(s)$

6.7 Solution Concentrations

Learning Objective 8 Calculate solution concentrations in units of molarity, weight/weight percent, weight/volume percent, and volume/volume percent.

Many of the reactions done in laboratories, and most of those that go on in our bodies, take place between substances dissolved in a solvent to form solutions. In our bodies, the solvent is almost always water. A double-replacement reaction of this type done in laboratories is represented by the following equation:

$$2AgNO_3(aq) + Na_2CO_3(aq) \rightarrow Ag_2CO_3(s) + 2NaNO_3(aq)$$

Hemera Technologies/PhotoObjects.net/ Getty Images

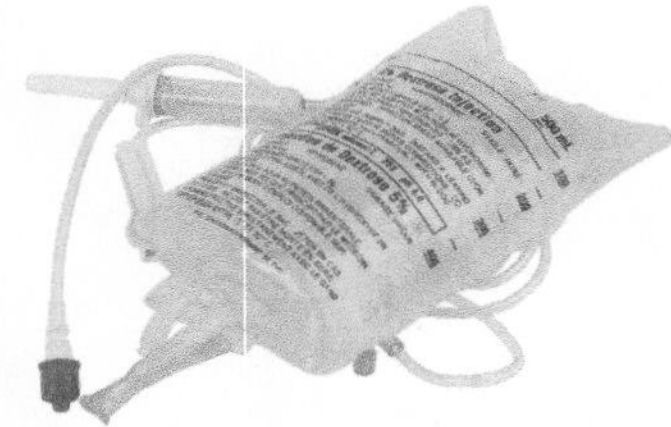

Figure 6.23 The concentration of every component in an intravenous solution is shown on the label.

concentration The relationship between the amount of solute and the specific amount of solution in which it is contained.

molarity (M) A solution concentration expressed as the number of moles of solute contained in one liter of solution.

Recall from Chapter 4 that the coefficients in equations allow the relative number of moles of pure reactants and products involved in the reaction to be determined. These relationships, coupled with the mole definition in terms of masses, then yield factors that can be used to solve stoichiometric problems involving the reactants and products. Similar calculations can be done for reactions that take place between the solutes of solutions if the amount of solute contained in a specific quantity of the reacting solutions is known. Such relationships are known as solution **concentrations** (see **Figure 6.23**). Solution concentrations may be expressed in a variety of units, but only two, molarity and percentage, are discussed at this time.

The **molarity (M)** of a solution expresses the number of moles of solute contained in exactly 1 L of the solution:

$$M = \frac{\text{moles of solute}}{\text{liters of solution}} \quad (6.8)$$

Even though a concentration in molarity expresses the number of moles contained in 1 L of solution, the molarity of solutions that have total volumes different from 1 L can also be calculated. We simply determine the number of moles of solute contained in a specified volume of solution, then express that volume in liters before substituting the values into Equation 6.8.

Example 6.9 Calculating Molarity of Solutions

Express the concentration of each of the following solutions in terms of molarity.

a. 2.00 L of solution contains 1.50 mol of solute.
b. 315 mL of solution contains 10.3 g of isopropyl alcohol, C_3H_7OH (**Figure 6.24**).

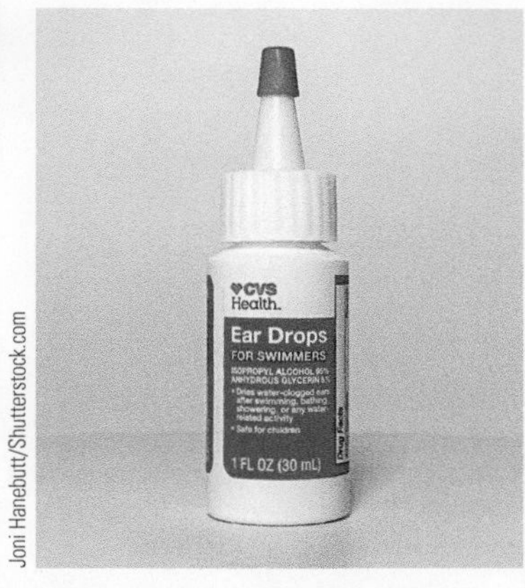

Figure 6.24 Isopropyl alcohol is the active ingredient in medicines used to dry excess water in swimmers' ears.

Solution a.

Step 1: Examine the question for what is given and what is unknown.

Given: $V = 2.00$ L and moles = 1.50 mol
Unknown: M (molarity) = ?

Step 2: Identify a formula that links the given and unknown terms.

$$\text{Equation 6.8: M (molarity)} = \frac{\text{moles of solute}}{\text{liters of solution}}$$

Step 3: Enter the data into the formula.

$$\text{M} = \frac{1.50 \text{ mol solute}}{2.00 \text{ L solution}} = 0.750 \frac{\text{mol solute}}{\text{L solution}}$$

The solution is 0.750 molar, or 0.750 M.

Solution b.

Step 1: Examine the question for what is given and what is unknown.

Given: $V = 315$ mL and mass = 10.3 g
Unknown: M (molarity) = ?

Step 2: Identify a formula that links the given and unknown terms.

$$\text{Equation 6.8: M (molarity)} = \frac{\text{moles of solute}}{\text{liters of solution}}$$

Before Equation 6.8 can be used, the number of moles of solute (isopropyl alcohol) must be determined. This is done as follows, where the factor

$$\frac{1 \text{ mol alcohol}}{60.1 \text{ g alcohol}}$$

comes from the calculated molecular weight of 60.1 g/mol for isopropyl alcohol:

$$(10.3 \text{ g \cancel{alcohol}}) \frac{1 \text{ mol alcohol}}{60.1 \text{ g \cancel{alcohol}}} = 0.171 \text{ mol alcohol}$$

The solution volume expressed in liters is

$$315 \cancel{\text{mL}} \times \frac{1 \text{ L}}{1000 \cancel{\text{mL}}} = 0.315 \text{ L}$$

Step 3: Enter the data into the formula.

$$\text{M} = \frac{0.171 \text{ mol alcohol}}{0.315 \text{ L solution}} = 0.543 \frac{\text{mol alcohol}}{\text{L solution}}$$

The solution is 0.543 molar, or 0.543 M.

✔ Learning Check 6.9 Express the concentrations of each of the following solutions in terms of molarity.

a. 2.50 L of solution contains 1.25 mol of solute.
b. 225 mL of solution contains 0.486 mol of solute.
c. 100 mL of solution contains 2.60 g of NaCl solute.

Sometimes a detailed knowledge of the actual stoichiometry of a process involving solutions is not needed, but some information about the solution concentrations would be useful. When this is true, solution concentrations are often expressed as percentages. In general, a concentration in percent gives the amount of solute contained in 100 parts of solution. It is convenient to use a formula for percentage calculations because we seldom work with exactly 100 units of anything. The general formula used is:

$$\% = \frac{\text{part}}{\text{total}} \times 100\%$$

weight/weight percent A concentration that expresses the mass of solute contained in 100 mass units of solution. The mass units of the solute and solution must be identical.

Three different percent concentrations for solutions are used. A **weight/weight percent**, abbreviated % (w/w), is the mass of solute contained in 100 mass units of solution. Thus, a 12.0% (w/w) sugar solution contains 12.0 grams of sugar in each 100 g of solution. The general formula for calculating this kind of percent concentration is

$$\% \text{ (w/w)} = \frac{\text{solute mass}}{\text{solution mass}} \times 100\% \tag{6.9}$$

Any mass units may be used, but the mass of solute and solution must be expressed in the same units.

weight/volume percent A concentration that expresses the grams of solute contained in 100 mL of solution.

A more commonly used percent concentration is **weight/volume percent**, abbreviated % (w/v), which is the grams of solute contained in 100 mL of solution (see **Figure 6.25**). In these units, a 12.0% (w/v) sugar solution would contain 12.0 g of sugar in each 100 mL of solution. This percent concentration is normally used when the solute is a solid, the solvent is a liquid, and resulting solutions are liquids. The general formula for calculating percent concentrations in these units is

$$\% \text{ (w/v)} = \frac{\text{solute mass in grams}}{\text{solution volume in mL}} \times 100\% \tag{6.10}$$

In weight/volume percent calculations, the solute amount is always given in grams, and the solution volume is always given in milliliters.

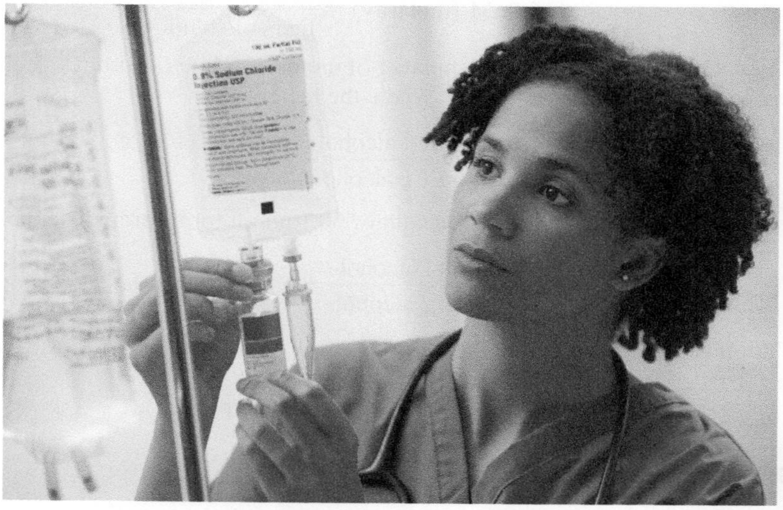

Figure 6.25 A 0.9% (w/v) sodium chloride solution in water is given intravenously to replace lost fluids and to provide carbohydrates to the body.

Jose Luis Pelaez Inc/Getty Images

volume/volume percent A concentration that expresses the volume of solute contained in 100 volume units of solution.

A percent concentration that is useful when the solute and solvent are either both liquids or both gases is **volume/volume percent**, abbreviated % (v/v). Concentrations given in these units express the number of volumes of solute found in 100 volume units of solution (see **Figure 6.26**). For these units, the general percentage equation becomes

$$\% \text{ (v/v)} = \frac{\text{solute volume}}{\text{solution volume}} \times 100\% \tag{6.11}$$

Any volume units may be used, but they must be the same for both the solute and the solution.

Example 6.10 Calculating % (w/v) and % (v/v) of Solutions

The majority of drug concentrations are expressed as a unit of mass per volume or % (w/v).

a. A 500. mL IV bag of 0.90% saline solution (NaCl in water) is hung for a medical patient. Calculate the mass of NaCl (in grams) required to produce the solution.
b. A solution is made by mixing 90.0 mL of alcohol with enough water to give 250. mL of solution. What is the % (v/v) concentration of alcohol in the solution?

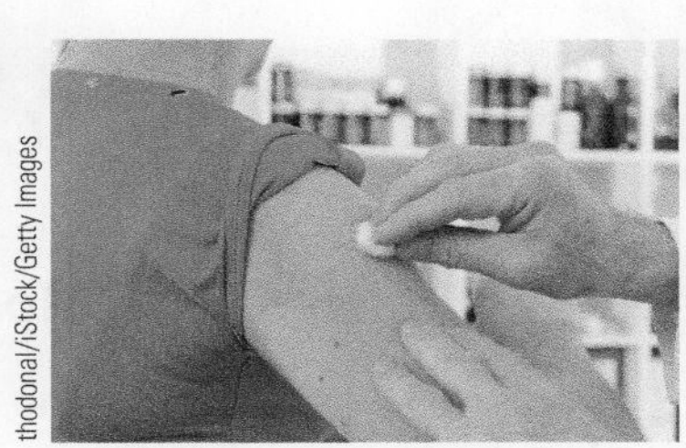

thodonal/iStock/Getty Images

Figure 6.26 Isopropyl rubbing alcohol is generally marketed as 70% alcohol (v/v), which is the concentration that works best as a topical antimicrobial against bacteria, fungi, and viruses. This concentration means there are 70 mL of solute (isopropyl alcohol) for every 100 mL of solvent (water).

Solution a.

Step 1: Examine the question for what is given and what is unknown.

Given: % (w/v) = 0.90% and volume = 500. mL
Unknown: mass in grams = ?

Step 2: Identify a formula that links the given and unknown terms.

$$\text{Equation 6.10: } \% \text{ (w/v)} = \frac{\text{solute mass in grams}}{\text{solution volume in mL}} \times 100\%$$

or

$$\text{Solute mass in grams} = \frac{\% \text{ (w/v)} \times (\text{solution volume in mL})}{100\%}$$

Step 3: Enter the data into the formula.

$$\frac{0.90\% \times 500. \text{ mL}}{100\%} = 4.5 \text{ g NaCl}$$

Although not shown, note that mL do cancel out of the numerator, because % (w/v) is in units of g/mL.

Solution b.

Step 1: Examine the question for what is given and what is unknown.

Given: volume of solute = 90.0 mL and volume of solution = 250. mL
Unknown: % (v/v) = ?

Step 2: Identify a formula that links the given and unknown terms.

$$\text{Equation 6.11: (v/v)} = \frac{\text{solute volume}}{\text{solution volume}} \times 100\%$$

Step 3: Enter the data into the formula.

$$\% \text{ (v/v)} = \frac{90.0 \text{ mL}}{250. \text{ mL}} \times 100\% = 36.0\% \text{ (v/v)}$$

✔ Learning Check 6.10

a. A solution is made by dissolving 0.900 g of salt in 100.0 mL of water. Assume that each milliliter of water weighs 1.00 g and that the final solution volume is 100.0 mL. Calculate the % (w/w) and % (w/v) for the solution, using the assumptions as necessary.
b. An alcoholic beverage is labeled 90 proof, which means the alcohol concentration is 45% (v/v). How many milliliters of pure alcohol would be present in 1 oz (30.0 mL) of the beverage?

Study Tools 6.2

Getting Started with Molarity Calculations

Knowing where to start or what to do first is a critical part of working with any math problem. First, you must identify a problem as a molarity problem. This can be done by looking for a keyword or phrase (e.g., *molar*, *molarity*, or *moles/liter*) or the abbreviation M. Second, remember that you have a formula for molarity:

$$M = \frac{\text{moles of solute}}{\text{liters of solution}}$$

Look for the given numbers and their units, and see if they match the units of the formula. If they do, put them into the formula and do the calculations. For example, if you are asked to calculate the molarity of a solution that contains 0.0856 mol NaCl dissolved in enough water to give 0.100 L of solution, you could put the numbers and their units directly into the formula:

$$M = \frac{0.0856\ \text{mol NaCl}}{0.100\ \text{L solution}} = 0.856\ \text{mol NaCl/L solution}$$

However, if you must calculate the molarity of a solution that contains 5.00 g of NaCl in enough water to give 100 mL of solution, the units of the numbers do not match those of the formula. Dimensional analysis can be used to convert each quantity into the units needed by the formula:

$$5.00\ \cancel{g\ \text{NaCl}} \times \frac{1\ \text{mol NaCl}}{58.4\ \cancel{g\ \text{NaCl}}} = 0.0856\ \text{mol NaCl}$$

$$100\ \cancel{\text{mL solution}} \times \frac{1\ \text{L solution}}{1000\ \cancel{\text{mL solution}}} = 0.100\ \text{L solution}$$

These numbers and units can then be put into the formula:

$$M = \frac{0.0856\ \text{mol NaCl}}{0.100\ \text{L solution}} = 0.856\ \text{mol NaCl/L solution}$$

6.8 Solution Preparation

Learning Objective 9 Describe how to prepare solutions of specific concentration by diluting concentrated solutions.

Solutions are usually prepared by mixing together proper amounts of solute and solvent or by diluting a concentrated solution with solvent (usually water) to produce a solution of lower concentration. A pharmacist, for example, might dilute a concentrated solution of amoxicillin to the proper concentration for treating an infant (see **Figure 6.27**).

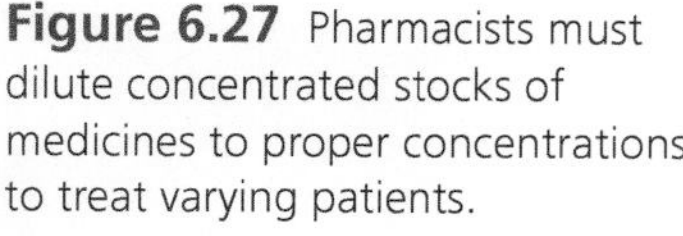

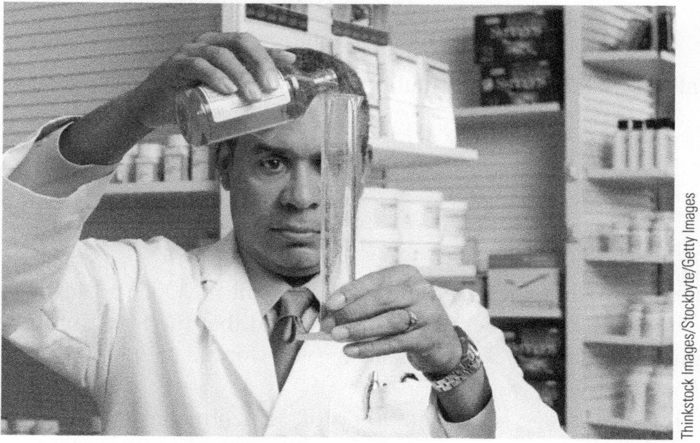

Thinkstock Images/Stockbyte/Getty Images

Figure 6.27 Pharmacists must dilute concentrated stocks of medicines to proper concentrations to treat varying patients.

Suppose that you want to prepare 250 mL of 0.100 M NaCl solution using a 2.00 M NaCl solution as the source of NaCl. First, use Equation 6.8 to calculate the number of moles of NaCl that would be contained in 250 mL of 0.100 M solution.

$$M = \frac{\text{moles of solute}}{\text{liters of solution}}$$

This equation can be rearranged to

$$(\text{M})(\text{liters of solution}) = \text{moles of solute}$$

So, the number of moles of NaCl contained in the desired solution is

$$\left(0.100\ \frac{\text{mol}}{\cancel{\text{L}}}\right)(0.250\ \cancel{\text{L}}) = 0.0250\ \text{mol}$$

This is the number of moles of NaCl that must be obtained from the 2.00 M NaCl solution. The volume of this solution that contains the desired number of moles can be obtained by rearranging Equation 6.8 in a different way:

$$\text{Liters of solution} = \frac{\text{moles of solute}}{\text{M}}$$

or

$$\text{Liters of solution} = \frac{0.0250\ \cancel{\text{mol}}}{2.00\ \cancel{\text{mol}}/\text{L}} = 0.0125\ \text{L} = 12.5\ \text{mL}$$

Thus, the solution is prepared by putting 12.5 mL of 2.00 M NaCl solution into a 250 mL volumetric flask and adding water up to the mark. **Figure 6.28** shows the steps in this process using NaCl solution dyed orange to make it easier to see.

Richard Megna/Fundamental Photographs

1 12.5 mL of 2.00 M NaCl solution is withdrawn from a beaker using a pipette.

Richard Megna/Fundamental Photographs

2 The 12.5 mL of 2.00 M NaCl solution is put into a 250 mL volumetric flask.

Richard Megna/Fundamental Photographs

3 Water is added (while the solution is swirled) to fill the flask to the mark.

Figure 6.28 Preparation of a 0.100 M NaCl solution by dilution of a 2.00 M NaCl solution.

The same result can be obtained by using a simplified calculation. Note that the product of concentration in molarity (M) and solution volume in liters will give the number of moles of solute in a sample of solution. In a dilution such as the one just described for the NaCl solution, the number of moles of solute taken from the concentrated solution and diluted with water is the same as the number of moles of solute in the resulting more dilute solution (see **Figure 6.29**). Thus, the following equation can be written:

$$\underset{\substack{\text{solute moles in}\\ \text{concentrated solution}}}{M_1V_1} = \underset{\substack{\text{solute moles in}\\ \text{dilute solution}}}{M_2V_2} \qquad (6.12)$$

Science History Images/Alamy Stock Photo

Figure 6.29 The concentrated solution (on the left) contains the same amount of solute as the dilute solution (on the right).

where the subscripts 1 and 2 refer to the more concentrated and dilute solutions, respectively, and M is used to represent any appropriate concentration. An advantage of using this equation is that any volume units may be used, as long as the same one is used for both V_1 and V_2. Also, Equation 6.12 is true for any solution concentration based on volume. This means that M, % (v/v), or % (w/v) concentrations can be used as long as M_1 and M_2 are expressed in the same units.

Example 6.11 Preparing Solutions by Dilution

To save preparation time and reduce storage space, some medical solutions are purchased in concentrated form that will be diluted to some lower concentration for actual use.

Use Equation 6.12 and describe how to prepare 250. mL of 5.0% (w/v) dextrose solution using a 25% (w/v) dextrose solution as the source of dextrose.

Solution

Step 1: Examine the question for what is given and what is unknown.

Given: $M_2 = 5.0\%$ (w/v), $V_2 = 250.$ mL, and $M_1 = 25\%$ (w/v)
Unknown: $V_1 = ?$

Step 2: Identify a formula that links the given and unknown terms.

Equation 6.12: $(M_1)(V_1) = (M_2)(V_2)$

Step 3: Enter the data into the formula.

$$[25\% \text{ (w/v)}](V_1) = [5.0\% \text{ (w/v)}](250. \text{ mL})$$

or

$$V_1 = \frac{[5.0\% \text{ (w/v)}](250. \text{ mL})}{[25\% \text{ (w/v)}]} = 50. \text{ mL}$$

Thus, we quickly find a volume of 50 mL of 25% (w/v) dextrose needs to be put into a 250 mL volumetric flask, and then water must be added to the mark as before.

✔ Learning Check 6.11 Using Equation 6.12, show how to prepare 500 mL of a 0.250 M NaOH solution from a 6.00 M NaOH solution.

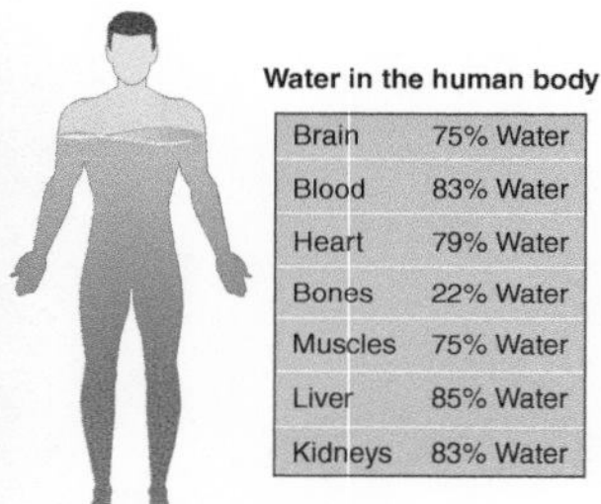
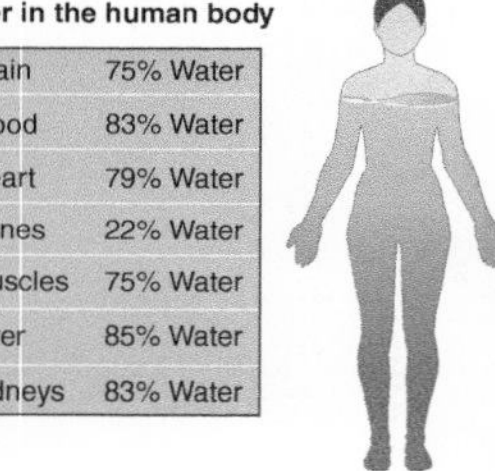

Water in the human body

Brain	75% Water
Blood	83% Water
Heart	79% Water
Bones	22% Water
Muscles	75% Water
Liver	85% Water
Kidneys	83% Water

Figure 6.30 Percent of water in various organs of the human body.

6.9 Osmosis and Dialysis

Learning Objective 10 Describe the process of osmosis.

Learning Objective 11 Describe the role of osmosis and diffusion in dialysis.

The human body is roughly 60% water (**Figure 6.30**), and the blood in our veins and the cytoplasm in our cells are aqueous solutions! So how are the physiological concentrations of these solutions maintained? The answer is quite complex, but osmosis plays a critical role in this regulation. Osmosis is the means by which your cells take in water and some essential nutrients, as well as the method by which some wastes are removed on the cellular level. To help better understand this concept, consider the experiment shown in **Figure 6.31**, in which pure water is separated from a sugar solution by a **semipermeable membrane**. The membrane has pores large enough to allow small molecules (such as water) to pass through, but small enough to prevent the passage of larger molecules (such as sugar) or hydrated ions.

semipermeable membrane A biological or synthetic membrane that will allow only certain molecules to pass through via diffusion in the case of solutes and osmosis in the case of solvent particles.

Figure 6.31 Osmosis. A semipermeable membrane separates a sugar solution from pure water. Although water can pass through the membrane in both directions, there is a net flow of water from the side with the least concentrated solution (pure water) to the side with the most concentrated solution (sugar solution).

In this experiment, a semipermeable membrane separates a sugar solution from pure water, which creates a concentration difference between the liquids. Water molecules move through the membrane in both directions, but the membrane prevents the **diffusion** of sugar molecules into the pure water to equalize the concentration. As a result, the net flow of water through the membrane is into the sugar solution. This increases the volume of the sugar solution, which decreases its concentration to a value closer to that of the pure water. The movement of water creates a difference in liquid levels (h), which causes hydrostatic pressure against the membrane. This pressure increases until it becomes high enough to balance the tendency for net water to flow through the membrane into the sugar solution. From that time on, the flow of water in both directions through the membrane is equal, and the volume of liquid on each side of the membrane no longer changes.

The pressure exerted by water that is required to prevent the net flow of water through a semipermeable membrane into a solution is called the **osmotic pressure** of the solution. The process in which solvent molecules move through semipermeable membranes is called **osmosis**. The same is true for solvent molecules other than water, meaning solvent molecules will also flow osmotically through semipermeable membranes separating solutions of different concentrations. The net flow of solvent is always from the more dilute solution into the more concentrated solution.

A real-life example of osmosis affects the shape of red blood cells (**Figure 6.32**).

- A **hypertonic** blood plasma solution has a higher concentration of solutes than the interior of the blood cell. In a hypertonic solution, the osmotic pressure on the inside of the cell is greater than on the outside of the cell, resulting in an overall net flow of water out of the cell. When this happens, the blood cell shrinks and **crenation** (plasmolysis) is observed.
- An **isotonic** blood plasma solution has the same concentration of solutes as the interior of the blood cell. In this case, the osmotic pressure on either side of the cell membrane is equal. As a result, water molecules diffuse into and out of the blood cell at the same rate. There is no net flow of water in either direction.
- A **hypotonic** plasma solution has a lower solute concentration than the inside of the blood cell. In this case, the osmotic pressure is greater outside the cell, causing an overall net flow of water into the cell. As a result, the blood cell swells or in extreme cases undergoes **hemolysis** (bursts).

diffusion The movement of individual molecules of a substance from an area of higher concentration to an area of lower concentration.

osmotic pressure The hydrostatic pressure required to prevent the net flow of solvent through a semipermeable membrane into a solution.

osmosis The process in which solvent molecules flow through a semipermeable membrane into a solution.

hypertonic solution A solution that contains a higher concentration of solutes compared to another solution.

crenation (plasmolysis) The shrinking of a cell after exposure to a hypertonic solution due to a net movement of water out of the cell via osmosis.

isotonic solution A solution that contains the same concentration of solutes as another solution.

hypotonic solution A solution that contains a lower concentration of solutes compared to another solution.

hemolysis The swelling of a cell after exposure to a hypotonic solution due to a net movement of water into the cell via osmosis.

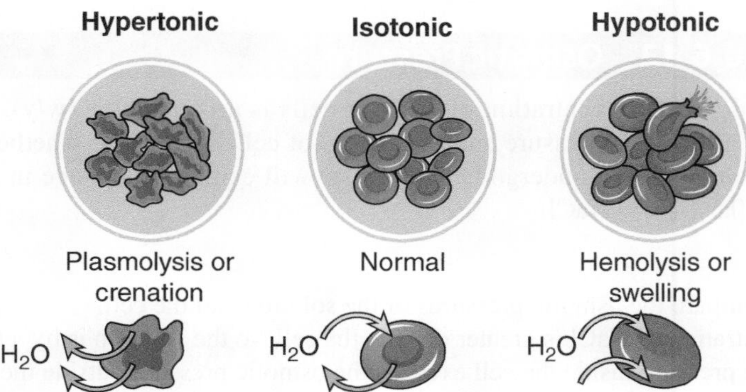

Figure 6.32 Red blood cells in solution. A hypertonic plasma solution has a higher concentration of solutes than the blood cells. An isotonic plasma solution has an equal concentration of solutes as the blood cells. A hypotonic plasma solution has a lower concentration of solutes than the blood cells.

Health Connections 6.1

Juices and Sports Drinks

Many people believe that fruit juice and sparkling juice are a healthy alternative to soda. This is a mistake, because juices have as many calories and as much sugar as non-diet soda, and they are just as detrimental to your dental health and overall body weight. Unlike soda, juice does contribute some vitamins and other nutrients, especially vitamin C. Therefore, adding just a splash of juice to plain water or sparkling water can make the water more palatable, which may encourage you to consume more water.

The consumption of high-sugar drinks can cause health problems such as weight gain, as well as increased risk of type 2 diabetes, cardiovascular disease, and gout. The average can of fruit punch or soda contains about 150 calories. Consuming one can of soda per day over the course of a year contributes to a weight gain of 5 pounds for the average person. Conversely, eliminating 150 calories per day (or one can of soda) could result in a 5 pound weight reduction (not to mention reduced in your chances of developing diabetes).

Sports drinks are closely related to soda. They contain a similar number of calories and are designed to deliver carbohydrates and electrolytes to athletes during high-intensity physical workouts. If you are not an athlete burning calories and losing fluid rapidly, sports drinks may not be for you.

Another popular choice is energy drinks. These drinks contain large amounts of sugar and enough caffeine to influence your blood pressure. They also contain additives that produce unknown health effects. If you need a pick-me-up, plain coffee or tea is a better choice. Diet drinks, fruit juice, and milk should also be limited or consumed in moderation. Alcoholic drinks should be limited to no more than one drink per day for females and no more than two drinks per day for males ("one drink" equals 5 oz of wine, 12 oz of beer, or 1 oz of hard liquor). Clean, refreshing, plain water is always the healthiest choice.

mimagephotography/Shutterstock.com

Plain water is always a healthy choice.

Example 6.12 Osmotic Pressure

The physiological concentration of NaCl in cells is around 0.1% (w/v). Using your knowledge of osmotic pressure and the shapes of cells, determine whether a cell will undergo crenation, will undergo hemolysis, or will exhibit no change in volume in a solution of 0.5% (w/v) NaCl.

Solution

Step 1: Compare the osmotic pressures of the solution and the cell.
The concentration of NaCl is greater outside the cell, so the solution is hypertonic. Thus, the osmotic pressure inside the cell exceeds the osmotic pressure outside the cell.

Step 2: Determine the direction of net flow.
The osmotic pressure inside the cell is greater, so there is a net flow of water via osmosis out of the cell.

Step 3: Describe the change in cell volume.
The net flow of water out of the cell will cause the cell to shrink. In extreme cases, it will undergo crenation.

✔ Learning Check 6.12 The physiological concentration of glucose in cells is around 7 mM. Using your knowledge of osmotic pressure and the shape of cells, determine if a cell will undergo crenation, will undergo hemolysis, or will exhibit no change in volume in a solution of 1 mM glucose.

Dialyzing membranes are semipermeable membranes with larger pores than the osmotic membranes discussed so far. They hold back large molecules but allow solvent, hydrated ions, and small molecules to pass through. Recall that the net flow of particles down their concentration gradient (i.e., from high solute concentration to low solute concentration) is called diffusion. The diffusion of these ions and small molecules through such membranes is called **dialysis**. An application of dialysis called hemodialysis (**Figure 6.33**) is used to remove waste particles from the blood of people with failing kidneys. Hemodialysis takes advantage of the fact that dialysis can be used to separate particles from large molecules.

dialyzing membrane A semipermeable membrane with pores large enough to allow solvent molecules, other small molecules, and hydrated ions to pass through.

dialysis A process in which solvent molecules, other small molecules, and hydrated ions pass from a solution through a membrane.

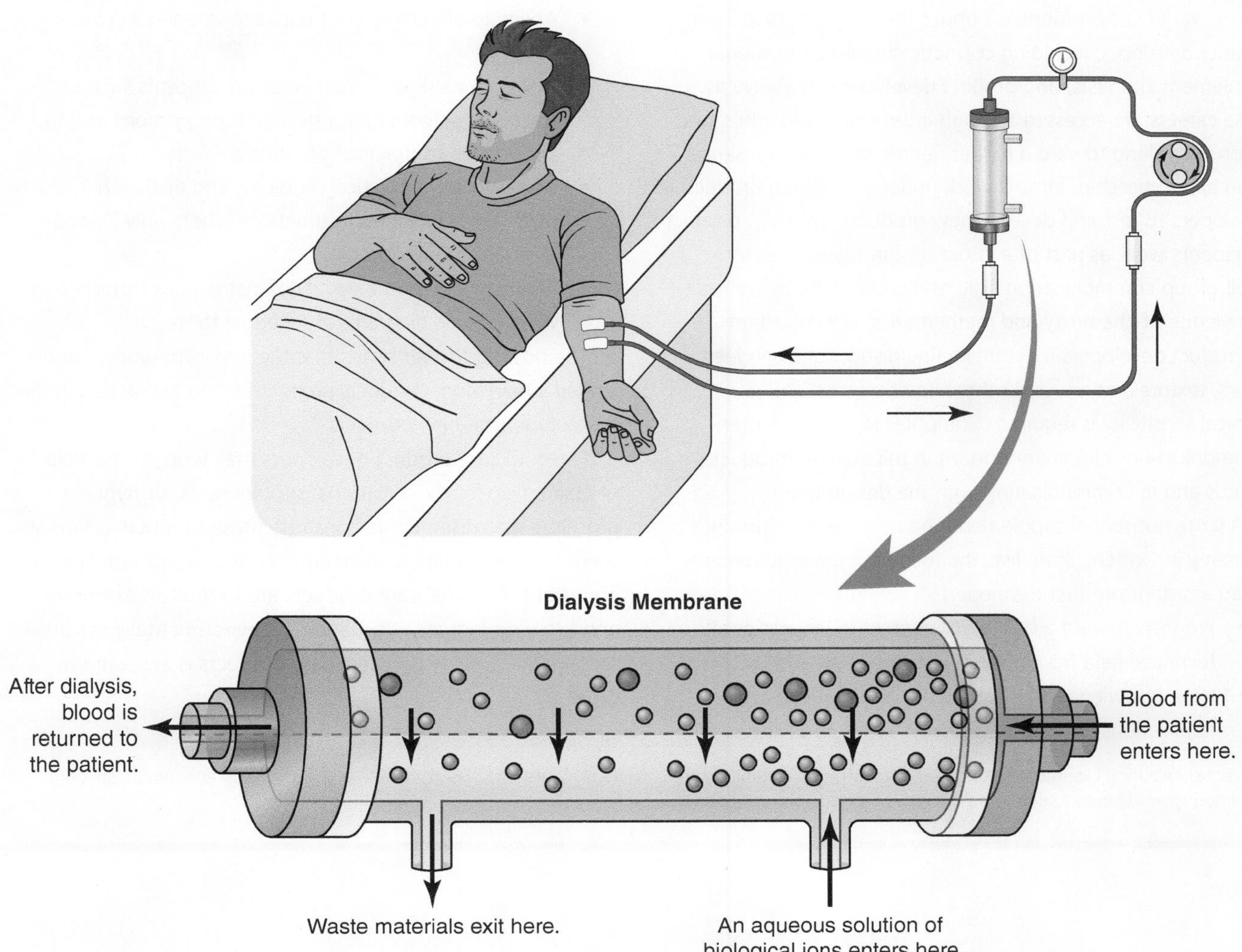
Dialysis Membrane
After dialysis, blood is returned to the patient.
Blood from the patient enters here.
Waste materials exit here.
An aqueous solution of biological ions enters here.

Figure 6.33 Hemodialysis, the dialysis method used to treat acute kidney injury.

The blood from a patient undergoing hemodialysis contains medium-molecular-weight particles, such as proteins, that must be retained and low-molecular-weight impurities, such as urea, that must be removed. As the patient's blood enters the dialysis chamber, an aqueous solution with essential biological ions enters it, too. The purpose of the aqueous solution is to maintain the biological ion concentration while also carrying away the low-molecular-weight impurities. The impurities (shown in orange in **Figure 6.33**) pass through the membrane down their concentration gradient and out of the chamber. Water molecules move through the membrane in both directions, but the medium-molecular-weight particles (i.e., blood proteins) remain inside the chamber and are returned to the patient's bloodstream.

Health Career Description

Personal Care Product Developer

A wide variety of positions fall under the title of personal care product developer, including cosmetic chemists, nutritional supplement chemists, and product developers. In many cases, these careers are accessed through internships held while the intern is working toward a higher degree in chemistry. Similar to an apprenticeship, interns work under established product developers to test and develop new products. Typically, product developers work as part of a larger design team; therefore, small-group communication skills and social skills, as well as knowledge of chemistry and mathematics, are important.

Product developers must make fine distinctions between colors, textures, odors, and other characteristics, so physical sensitivity is required. Computer skills and written communication skills are important in maintaining product records and in communicating with the design team.

A large nutritional supplement and personal care product company in Ogden, Utah, lists the following job requirements for an assistant product developer:

- Progress toward a bachelor's degree (BS) in a scientific or technical field from a four-year college or university
- Have knowledge of a variety of computer software applications, including Microsoft Word and Excel
- Ability to effectively communicate verbally and in writing
- Ability to work with mathematical concepts such as fractions, percentages, ratios, and proportions and to apply them to practical situations
- Ability to solve practical problems and deal with a variety of concrete variables in situations where only limited standardization exists
- Ability to interpret a variety of instructions furnished in written, oral, diagram, or schedule form

The posting further states that the applicant works "under limited supervision, conducts product testing and assists in the development of new products."

Personal care product developers may work in the field of cosmetics, foods, nutritional supplements, or hygiene products. Because people consume these products or apply them to their bodies, knowledge of toxicity and nutrition is important. Personal care products are part of an extensive and growing industry. Demand for environmentally sensitive, sustainable, quality personal care products is expected to grow.

Source: Nutraceutical Laboratories. Retrieved from: https://recruiting2.ultipro.com/NUT1002NUTRC/JobBoard/e686a42b-cd18-4509-8047-8b5a75be06fe/OpportunityDetail?opportunityId=11a1715f-47f2-43fd-ad30-b275659d46af.

Concept Summary

6.1 Properties of Gases

Learning Objectives: **Calculate the kinetic energy of moving particles. Convert pressure and temperature values into various units.**

- Gases are defined in terms of their pressure, volume, amount, and temperature.
- Matter possesses kinetic energy because it is composed of tiny particles called molecules that move in constant straight-line motion.
- Gas particles are far apart, so there are little to no attractive or repulsive forces between them.
- Gas particles transfer energy from one to another during collisions in which no net energy is lost from the system.
- Gas particles have very small volumes compared to the containers they occupy.
- The average particle speed increases as the temperature increases.
- Mathematical relationships called gas laws describe the observed behavior of gases when they are mixed or subjected to changes in pressure, volume, or temperature.
- When the gas laws are used, it is necessary to express the volume and pressure in consistent units and the temperature in kelvins.

6.2 Pressure, Temperature, and Volume Relationships

Learning Objective: **Do calculations based on Boyle's law, Charles's law, and the combined gas law.**

- Boyle's law relates changes in pressure and volume, whereas Charles's law relates changes in volume and temperature.
- The combined gas law relates the temperature, pressure, and volume changes for gases.

6.3 The Ideal Gas Law

Learning Objective: **Do calculations based on the ideal gas law.**

- The ideal gas law relates the pressure and volume of a gas sample to its temperature and amount (number of moles).

6.4 Dalton's law

Learning Objective: **Do calculations based on Dalton's law.**

- Dalton's law states that the total pressure exerted by a mixture of gases is equal to the sum of the partial pressure of the individual gases in the mixture.

6.5 Solutions and Solubility

Learning Objective: **Apply your knowledge of intermolecular forces to rank the solubility of substances in a given solvent.**

- The amount of solute that will dissolve in a quantity of solvent to form a saturated solution is the solubility of the solute.
- Solubility depends on the similarities in polarities of the solvent and solute—namely, "like dissolves like."
- Solubility generally increases with temperature for solid and liquid solutes, but decreases with temperature for gaseous solutes.
- The solution process of a solid solute in a liquid solvent can be thought of in terms of the solvent molecules attracting the solute particles away from the solute crystal lattice.
- The solute is picked apart and solute particles are hydrated as a result of attractions between water molecules and solute particles.

6.6 Electrolytes and Net Ionic Equations

Learning Objective: **Write molecular equations in total ionic and net ionic forms.**

- Many water-soluble compounds separate into ions when dissolved in water.
- Ionic or highly polar solutes that dissociate in solution are called electrolytes. They result in solutions that conduct electricity.
- Solutes that do *not* dissociate in solution are called nonelectrolytes. Their solutions do *not* conduct electricity.
- Molecular equations depict each compound with its molecular or ionic formula.
- Total ionic equations depict all soluble ionic substances as the ions they form in solution.
- Net ionic equations depict only the ions and insoluble compounds that undergo changes in the reaction. All spectator ions have been eliminated.

6.7 Solution Concentrations

Learning Objective: **Calculate solution concentrations in units of molarity, weight/weight percent, weight/volume percent, and volume/volume percent.**

- Relationships between the amount of solute and the amount of solution containing the solute are called concentrations.
- They may be expressed as molarity, weight/weight percent, weight/volume percent, or volume/volume percent.

6.8 Solution Preparation

Learning Objective: **Describe how to prepare solutions of specific concentration by diluting concentrated solutions.**

- Solutions of specific concentration can be prepared by diluting a more concentrated solution with solvent.

6.9 Osmosis and Dialysis

Learning Objectives: **Describe the process of osmosis. Describe the role of osmosis and diffusion in dialysis.**

- Diffusion is the movement of individual molecules of a substance from an area of higher concentration to an area of lower concentration.
- Osmosis is the process in which solvent molecules flow through a semipermeable membrane into a solution.
- Osmotic pressure is the hydrostatic pressure required to prevent the net flow of solvent through a semipermeable membrane into a solution.
- Dialyzing membranes are semipermeable but with pores large enough to allow solvent molecules, hydrated ions, and small molecules to pass through in a process called dialysis.
- Dialysis is used to remove impurities from the blood of people suffering from kidney malfunction.

Key Terms and Concepts

Absolute zero (6.1)
Aqueous solution (6.5)
Avogadro's law (6.3)
Boyle's law (6.2)
Charles's law (6.2)
Combined gas law (6.2)
Concentration (6.7)
Crenation (6.9)
Dalton's law of partial pressures (6.4)
Dialysis (6.9)
Dialyzing membrane (6.9)
Diffusion (6.9)
Dissolving (6.5)
Electrolyte (6.5)
Gas laws (6.1)
Hemolysis (6.9)
Hydrated ions (6.5)
Hypertonic solution (6.9)
Hypotonic solution (6.9)
Ideal gas law (6.3)
Immiscible (6.5)
Insoluble substance (6.5)
Isotonic solution (6.9)
Molarity (M) (6.7)
Molecular equation (6.6)
Net ionic equation (6.6)
Nonelectrolyte (6.5)
Osmosis (6.9)
Osmotic pressure (6.9)
Partial pressure (6.4)
Pressure (6.1)
Saturated solution (6.5)
Semipermeable membrane (6.9)
Solubility (6.5)
Soluble substance (6.5)
Solute (6.5)
Solvent (6.5)
Spectator ions (6.6)
Standard conditions (STP) (6.3)
Supersaturated solution (6.5)
Total ionic equation (6.6)
Universal gas constant (6.3)
Volume (6.1)
Volume/volume percent (6.7)
Weight/volume percent (6.7)
Weight/weight percent (6.7)

Key Equations

1. Calculation of kinetic energy of particles in motion (**Section 6.1**)	$\text{K.E.} = \frac{1}{2}mv^2$	Equation 6.1
2. Boyle's law (**Section 6.2**)	$P_iV_i = P_fV_f$	Equation 6.2
3. Charles's law (**Section 6.2**)	$\frac{V_i}{T_i} = \frac{V_f}{T_f}$	Equation 6.3
4. Combined gas law (**Section 6.2**)	$\frac{P_iV_i}{T_i} = \frac{P_fV_f}{T_f}$	Equation 6.4
5. Avogadro's law (**Section 6.2**)	$\frac{V_i}{n_i} = \frac{V_f}{n_f}$	Equation 6.5
6. Ideal gas law (**Section 6.3**)	$PV = nRT$	Equation 6.6
7. Dalton's law (**Section 6.4**)	$P_t = P_{\triangle} + P_{\bigcirc} + P_{\square}$	Equation 6.7
8. Solution concentration molarity (**Section 6.7**)	$\text{M} = \frac{\text{moles of solute}}{\text{liters of solution}}$	Equation 6.8
9. Solution concentration weight/weight % (**Section 6.7**)	$\%\ (\text{w/w}) = \frac{\text{solute mass}}{\text{solution mass}} \times 100\%$	Equation 6.9
10. Solution concentration weight/volume % (**Section 6.7**)	$\%\ (\text{w/v}) = \frac{\text{solute mass in grams}}{\text{solution volume in mL}} \times 100\%$	Equation 6.10
11. Solution concentration volume/volume % (**Section 6.7**)	$\%\ (\text{v/v}) = \frac{\text{solute volume}}{\text{solution volume}} \times 100\%$	Equation 6.11
12. Dilution of a concentrated solution to make a less concentrated solution (**Section 6.8**)	$M_1V_1 = M_2V_2$	Equation 6.12

Exercises

Even-numbered exercises are answered in Appendix B.

Exercises with an asterisk (*) are more challenging.

Properties of Gases (Section 6.1)

6.1 Suppose a 180 lb (81.8 kg) halfback running at a speed of 8.0 m/s collides head-on with a 260 lb (118.2 kg) linebacker running at 3.0 m/s. Which one will be pushed back? That is, which one has less kinetic energy?

6.2 Nitrous oxide (N_2O, M.W. = 44.0 amu) and sevoflurane ($C_4H_3F_7O$, M.W. = 200.1 amu) are commonly used in outpatient inhalation anesthesia. At 25.0 °C, nitrous oxide molecules have an average velocity of 3.78×10^4 cm/s, and sevoflurane molecules have an average velocity of 1.77×10^4 cm/s. Calculate the kinetic energy of each type of molecule at 25.0 °C and determine whether they are significantly different. Express molecular masses in amu for this calculation.

6.3 What is a gas law?

6.4 A weather reporter on TV reports the barometer pressure as 28.6 inches of mercury. Calculate this pressure in the following units.

a. atm

b. torr

c. psi

d. bars

6.5 An anesthesiologist reads the pressure of an anesthesia administered during an operation as 16.8 cm Hg. Calculate this pressure in the following units.

a. atm

b. mm Hg

c. torr

d. psi

6.6 An engineer reads the pressure gauge of a boiler as 210 psi. Calculate this pressure in the following units.

a. atm

b. bars

c. mm Hg

d. in. Hg

6.7 Which have the greater kinetic energy, hydrogen molecules traveling with a velocity of $2v$, or helium molecules traveling with a velocity of v? Express molecular masses in amu.

6.8 Convert each of the following temperatures from the unit given to the unit indicated.

a. Liquid helium used to cool MRI magnets has a boiling point of −268.9 °C. Convert this to kelvins.

b. Convert the freezing point of liquid hydrogen, 14.1 K, to degrees Celsius.

c. An inhaled anesthesia boils at 63.7 °C, at which point it can be inhaled. Convert this to kelvins.

6.9 Convert each of the following temperatures from the unit given to the unit indicated.

a. The melting point of gold, 1337.4 K, to degrees Celsius

b. The melting point of tungsten, 3410 °C, to kelvins

c. The melting point of tin, 505 K, to degrees Celsius

Pressure, Temperature, and Volume Relationships (Section 6.2)

6.10 Use the combined gas law (Equation 6.4) to calculate the unknown quantity for each gas sample described in the following table.

	Sample		
	A	B	C
P_i	1.50 atm	2.35 atm	9.86 atm
V_i	2.00 L	1.97 L	11.7 L
T_i	300 K	293 K	500. K
P_f	?	1.09 atm	5.14 atm
V_f	3.00 L	?	9.90 L
T_f	450. K	310 K	?

6.11 A 200. mL sample of oxygen gas is collected at 26.0 °C and a pressure of 690. torr. What volume will the gas occupy at STP (0 °C and 760 torr)?

6.12 A 200. mL sample of nitrogen gas is collected at 45.0 °C and a pressure of 610. torr. What volume will the gas occupy at STP (0 °C and 760 torr)?

6.13 A 3.00 L sample of helium at 0.00 °C and 1.00 atm is compressed into a 0.50 L cylinder. What pressure will the gas exert in the cylinder at 50.0 °C?

6.14 A 2.50 L sample of neon at 0.00 °C and 1.00 atm is compressed into a 0.75 L cylinder. What pressure will the gas exert in the cylinder at 30.0 °C?

6.15 What volume (in liters) of air measured at 1.00 atm would have to be put into a bicycle tire with a 1.00 L volume if the pressure in the bike tire is to be 65.0 psi? Assume the temperature of the gas remains constant.

6.16 What volume (in liters) of air measured at 1.00 atm would have to be put into a car tire with a volume of 14.5 L if the pressure in the car tire is to be 32.0 psi? Assume the temperature of the gas remains constant.

6.17 A patient is diagnosed with a 1000 mL pneumothorax (air trapped between the lung and chest wall, causing the lung to collapse) at sea level (air pressure, 760 torr). If the patient is to be transported by helicopter at an altitude of 5000 ft (air pressure, 620 torr), what will the size of the pneumothorax be during the transport?

6.18 A 5.0 L portable oxygen tank used by people with respiratory ailments is pressurized to 2000 psi with oxygen. Assuming no temperature change, how many liters of oxygen are available to the patient at 14.0 psi?

6.19 A scuba diving regulator is designed to reduce the pressure of the air in a scuba tank to the pressure that the diver is experiencing, thus allowing the diver to breathe. If the diver breathes in 300 mL of air from the tank through the regulator, which is at 4 °C, what is the volume of the air in her lungs, which are at 37 °C? Assume constant pressure.

6.20 A 3.8 L sample of gas at 1.0 atm and 20. °C is heated to 85 °C. Calculate the gas volume in liters at the higher temperature if the pressure remains constant at 1.0 atm.

6.21 If an individual draws a breath of 1.00 L at 16 °C, what is the new volume of air as it warms to 37 °C? Assume the pressure and amount of gas remain constant.

6.22 What volume of gas in liters at 120. °C must be cooled to 40. °C if the gas volume at constant pressure and 40. °C is to be 2.0 L?

6.23 A person undergoing a laparoscopic abdominal surgery procedure will have their abdominal cavity inflated with carbon dioxide for the procedure. If 180 mL of CO_2 at 2500 torr and 20. °C is put into the abdominal cavity of the patient, which exerts 940 torr of pressure on the gas at a temperature of 37 °C, what is the new volume of the insufflated carbon dioxide?

6.24 When stressed, people sometimes swallow air, which causes them to burp. A researcher is studying the relationship between volume of burps and stress. Your diaphragm detects pressure as discomfort at 210.0 kPa (body temperature is 37 °C), leading to a burp. If you feel discomfort and burp into the researcher's balloon, the balloon slightly inflates to 7.64 mL at 94.6 kPa and 22 °C. What volume did the burp occupy in your stomach?

6.25 A 6.25 L tank contains the inhalation anesthesia sevoflurane at 3.25 atm and 10.0 °C. It is to be given to a patient at 0.250 atm of pressure and 37 °C. What volume of sevoflurane (in liters) will be available to the patient?

6.26 A helium balloon was partially filled with 8.00×10^3 ft^3 of helium when the atmospheric pressure was 0.98 atm and the temperature was 23 °C. The balloon rose to an altitude where the atmospheric pressure was 400 torr and the temperature was 5.3 °C. What volume did the helium occupy at this altitude?

6.27 You have a 1.50 L balloon full of air at 30. °C. To what Celsius temperature would you have to heat the balloon to double its volume if the pressure remained unchanged?

6.28 An oyster diver completely fills her lungs with 4.00 L of air at the ocean's surface (1.00 atm pressure). She descends to a depth of 20 m, where the water exerts 3.00 atm of pressure on her. What is the volume of air in her lungs now, assuming no temperature change?

6.29 What minimum pressure would a 250 mL aerosol can have to withstand if it were to contain 2.00 L of gas measured at 700 torr? Assume constant temperature.

6.30 A 2.00 L sample of nitrogen gas at 760 torr and 0.0 °C weighs 2.50 g. The pressure on the gas is increased to 4.00 atm at 0.0 °C. Calculate the gas density at the new pressure in g/L.

The Ideal Gas Law (Section 6.3)

6.31 Use the ideal gas law and calculate the following.

a. The pressure exerted by 2.00 mol of oxygen confined to a volume of 400 mL at 20.0 °C

b. The volume of hydrogen gas in a steel cylinder if 0.525 mol of the gas exerts a pressure of 3.00 atm at a temperature of 15.0 °C

c. The temperature (in degrees Celsius) of a nitrogen gas sample that has a volume of 2.25 L and a pressure of 300 torr and contains 0.100 mol

6.32 Use the ideal gas law and calculate the following.

a. The number of moles of argon in a gas sample that occupies a volume of 400 mL at a temperature of 90.0 °C and has a pressure of 735 torr

b. The pressure in atm exerted by 0.750 mol of hydrogen gas confined to a volume of 2.60 L at 50.0 °C

c. The volume in liters of a tank of nitrogen if 1.50 mol of the gas exerts a pressure of 4.32 atm at 20.0 °C

6.33 Suppose 0.156 mol of SO_2 gas is compressed into a 0.750 L steel cylinder at a temperature of 27 °C. What pressure in atmospheres is exerted by the gas?

6.34 Suppose 10.0 g of SO_2 gas is compressed into a 1.25 L steel cylinder at a temperature of 38.0 °C. What pressure in atm is exerted by the gas?

6.35 Calculate the volume occupied by 8.75 g of oxygen gas (O_2) at a pressure of 0.890 atm and a temperature of 35.0 °C.

6.36 The pressure gauge of a steel cylinder of methane gas (CH_4) reads 400. psi. The cylinder has a volume of 1.50 L and is at a temperature of 30.0 °C. How many grams of methane does the cylinder contain?

6.37 Suppose 12.0 g of dry ice (solid CO_2) was placed in an empty 400. mL steel cylinder. What pressure would develop if all the solid sublimed at a temperature of 35.0 °C?

6.38 An experimental chamber has a volume of 60 L. How many moles of oxygen gas will be required to fill the chamber at STP?

6.39 How many molecules of nitrogen gas (N_2) are present in a sample that fills a 10.0 L tank at STP?

6.40 A sample of gaseous inhalation anesthesia desflurane has a mass of 29.7 g and occupies a volume of 3.96 L at STP. What is the molecular weight of desflurane?

6.41 If an individual draws a breath of 1.00 L at 1.00 atm and 35 °C, what mass of air is contained in the breath? Assume that air has an average molar mass of 28.6 g/mol.

6.42 A sample of gas weighs 0.176 g and has a volume of 114.0 mL at a pressure and temperature of 640 torr and 20. °C. Determine the molecular weight of the gas, and identify it as CO, CO_2, or O_2.

6.43 A 2.00 g sample of gas has a volume of 1.12 L at STP. Calculate its molecular weight and identify it as He, Ne, or Ar.

Dalton's Law (Section 6.4)

6.44 Heliox is a supplemental breathing gas mixture of helium and oxygen that is used to alleviate symptoms of severe asthma, emphysema, and COPD (chronic obstructive pulmonary disease). The total pressure in the tank is 2100 torr. The pressure exerted by the helium is 1680 torr. What is the partial pressure in torr of the oxygen in the mixture? What is the percent composition of this mixture (percent helium/percent oxygen)?

6.45 A 250 mL sample of oxygen gas is collected by water displacement. As a result, the oxygen is saturated with water vapor. The partial pressure of water vapor at the prevailing temperature is 22 torr. Calculate the partial pressure of the oxygen if the total pressure of the sample is 720 torr.

Solutions and Solubility (Section 6.5)

6.46 Many solutions are found in the home. Some are listed below, with the composition as printed on the label. When no percentage is indicated, components are usually given in order of decreasing amount. When water is present, it is often not mentioned on the label or it is included in the inert ingredients. Identify the solvent and solutes of the following solutions.

a. Liquid laundry bleach: sodium hypochlorite 5.25%, inert ingredients 94.75%

b. Rubbing alcohol: isopropyl alcohol 70%

c. Hydrogen peroxide: 3% hydrogen peroxide

d. Aftershave: SD alcohol, water, glycerin, fragrance, menthol, benzophenone-1, coloring

6.47 Many solutions are found in the home. Some are listed below, with the composition as printed on the label. When no percentage is indicated, components are usually given in order of decreasing amount. When water is present, it is often not mentioned on the label or it is included in the inert ingredients. Identify the solvent and solutes of the following solutions.

a. Antiseptic mouthwash: alcohol 25%, thymol, eucalyptol, methyl salicylate, menthol, benzoic acid, boric acid

b. Paregoric: alcohol 45%, opium 0.4%

c. Baby oil: mineral oil, lanolin (there happens to be no water in this solution—why?)

d. Distilled vinegar: acetic acid 5%

6.48 Use the term *soluble*, *insoluble*, or *immiscible* to describe the behavior of the following pairs of substances when they are shaken together.

a. 25 mL of cooking oil and 25 mL of vinegar—the resulting mixture is cloudy and gradually separates into two layers.

b. 25 mL of water and 10 mL of rubbing alcohol—the resulting mixture is clear and colorless.

c. 25 mL of chloroform and 1 g of roofing tar—the resulting mixture is clear but dark brown in color.

6.49 Use the term *soluble*, *insoluble*, or *immiscible* to describe the behavior of the following pairs of substances when they are shaken together.

a. 25 mL of water and 1 g of salt—the resulting mixture is clear and colorless.

b. 25 mL of water and 1 g of solid silver chloride—the resulting mixture is cloudy and solid settles out.

c. 25 mL of water and 5 mL of mineral oil—the resulting mixture is cloudy and gradually separates into two phases.

6.50 Classify the following solutions as unsaturated, saturated, or supersaturated.

a. A solution to which a small piece of solute is added, and it dissolves

b. A solution to which a small piece of solute is added, and much more solute comes out of solution

c. The final solution resulting from the process in part b

6.51 Ammonium sulfate is used in pharmacologic preparations to "salt out" unwanted proteins and contaminants. Suppose a technician puts 35.8 g of ammonium sulfate into a flask and adds 100 g of water at 0 °C. After stirring to dissolve as much solute as possible, will the technician have a saturated or an unsaturated solution? Explain your answer. Refer to **Table 6.4**.

6.52 Rubbing alcohol is used as a topical antiseptic. It is typically a mixture of isopropyl alcohol and water. Because isopropyl alcohol is miscible in water, it can be purchased in a variety of concentrations. Define the term *miscible*. It is *not* defined in the text.

6.53 What is the difference between a nonhydrated ion and a hydrated ion? Draw a sketch using the Cl^- ion to help illustrate your answer.

6.54 Suppose you have a saturated solution that is at room temperature. Discuss how it could be changed into a supersaturated solution without using any additional solute.

6.55 Indicate which of the following substances (with geometries as given) would be soluble in water (a polar solvent) and in benzene (a nonpolar solvent).

a. H—S—H (bent)

b. HCl

c. H—O—O—H

d. N≡N

6.56 Indicate which of the following substances (with geometries as given) would be soluble in water (a polar solvent) and in benzene (a nonpolar solvent).

a. 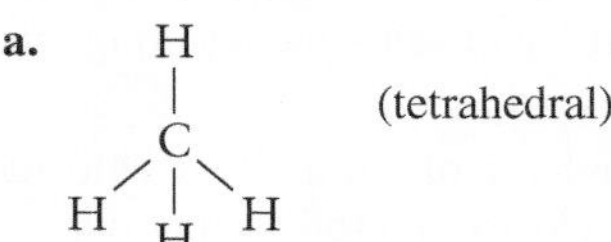

(tetrahedral)

b. Ne

c. N, H H H (trigonal pyramidal)

d. F, B, F F (trigonal planar)

6.57 Suppose you put a piece of a solid into a beaker that contains water and stir the mixture briefly. You find that the solid does not immediately dissolve completely. Describe three things you might do to try to get the solid to dissolve.

6.58 Halothane is a general-use inhalation anesthetic. The molecular structure is:

F F
F—C—C—H
F Br

Explain why halothane would or would not be soluble in water.

Electrolytes and Net Ionic Equations (Section 6.6)

6.59 Consider all of the following ionic compounds to be water soluble, and write the formulas of the ions that would be formed if the compounds were dissolved in water. **Table 6.5** will be helpful.

a. Na_2SO_4

b. $CaCl_2$

c. $(NH_4)_3PO_4$

d. NaOH

e. KNO_3

f. $NaMnO_4$

6.60 Consider all of the following ionic compounds to be water soluble, and write the formulas of the ions that would be formed if the compounds were dissolved in water. **Table 6.5** will be helpful.

a. $LiNO_3$

b. Na_2HPO_4

c. $Ca(ClO_3)_2$

d. KOH

e. $MgBr_2$

f. $(NH_4)_2SO_4$

6.61 Reactions represented by the following equations take place in aqueous solutions. Write each molecular equation in total ionic form, then identify spectator ions and write the equations in net ionic form. Solids that do not dissolve are designated by (s), gases that do not dissolve are designated by (g), and substances that dissolve but do not dissociate appear in blue.

a. $H_2O(\ell) + Na_2SO_3(aq) + SO_2(aq) \rightarrow 2NaHSO_3(aq)$

b. $3Cu(s) + 8HNO_3(aq) \rightarrow 3Cu(NO_3)_2(aq) + 2NO(g) + 4H_2O(\ell)$

c. $2HCl(aq) + CaO(s) \rightarrow CaCl_2(aq) + H_2O(\ell)$

d. $CaCO_3(s) + 2HCl(aq) \rightarrow CaCl_2(aq) + CO_2(aq) + H_2O(\ell)$

e. $MNO_2(s) + 4HCl(aq) \rightarrow MnCl_2(aq) + Cl_2(aq) + 2H_2O(\ell)$

f. $2AgNO_3(aq) + Cu(s) \rightarrow Cu(NO_3)_2(aq) + 2Ag(s)$

6.62 Reactions represented by the following equations take place in aqueous solutions. Write each molecular equation in total ionic form, then identify spectator ions and write the equations in net ionic form. Solids that do not dissolve are designated by (s), gases that do not dissolve are designated by (g), and substances that dissolve but do not dissociate appear in blue.

a. $SO_2(aq) + H_2O(\ell) \rightarrow H_2SO_3(aq)$

b. $CuSO_4(aq) + Zn(s) \rightarrow Cu(s) + ZnSO_4(aq)$

c. $2KBr(aq) + 2H_2SO_4(aq) \rightarrow Br_2(aq) + SO_2(aq) + K_2SO_4(aq) + 2H_2O(\ell)$

d. $AgNO_3(aq) + NaOH(aq) \rightarrow AgOH(s) + NaNO_3(aq)$

e. $BaCO_3(s) + 2HNO_3(aq) \rightarrow Ba(NO_3)_2(aq) + CO_2(g) + H_2O(\ell)$

f. $N_2O_5(aq) + H_2O(\ell) \rightarrow 2HNO_3(aq)$

Solution Concentrations (Section 6.7)

6.63 Calculate the molarity of the following solutions.

a. 2.50 L of solution contains 0.860 mol of solute.

b. 400 mL of solution contains 0.304 mol of solute.

c. 0.120 mol of solute is put into a container, and enough distilled water is added to give 250 mL of solution.

6.64 Calculate the molarity of the following solutions.

a. 1.25 L of solution contains 0.455 mol of solute.

b. 250 mL of solution contains 0.215 mol of solute.

c. 0.175 mol of solute is put into a container, and enough distilled water is added to give 100. mL of solution.

6.65 Calculate the molarity of the following solutions.

a. A sample of solid Na_2SO_4 weighing 0.140 g is dissolved in enough water to make 10.0 mL of solution.

b. A 4.50 g sample of glucose ($C_6H_{12}O_6$) is dissolved in enough water to give 150. mL of solution.

c. A 43.5 g sample of K_2SO_4 is dissolved in a quantity of water, and the solution is stirred well. A 250.0 mL sample of the resulting solution is evaporated to dryness and leaves behind 2.18 g of solid K_2SO_4.

6.66 Calculate the molarity of the following solutions.

a. A sample of solid KBr weighing 11.9 g is dissolved in enough distilled water to give 200.0 mL of solution.

b. A 14.2 g sample of solid Na_2SO_4 is dissolved in enough water to give 500.0 mL of solution.

c. A 10.0 mL sample of solution is evaporated to dryness and leaves 0.29 g of solid residue that is identified as Li_2SO_4.

6.67 Because concentrations are relatively dilute, intravenous solutions given in hospitals have concentrations listed as mmol/L (millimoles per liter; a millimole is 0.001 of a mole), rather than M (moles per liter). So a 250. mmol/L concentration is the same as 0.250 M. Calculate the following.

a. How many moles of dextrose (blood sugar) are contained in 1.25 L of a 278 mmol/L dextrose solution?

b. How many moles of sodium chloride are contained in 200. mL of a 154 mmol/L saline solution?

c. What volume of a 154 mmol/L saline solution contains 0.500 mol of sodium chloride?

6.68 Calculate the following.

a. How many moles of solute is contained in 1.75 L of 0.215 M solution?

b. How many moles of solute is contained in 250. mL of 0.300 M solution?

c. How many milliliters of 0.350 M solution contains 0.200 mol of solute?

6.69 Calculate the following.

a. How many moles of solute is contained in 2.50 L of 0.125 M solution?

b. How many moles of solute is contained in 150. mL of 0.265 M solution?

c. How many milliliters of 0.450 M solution contains 0.100 mol of solute?

6.70 Calculate the following.

a. How many grams of solid would be left behind if 20.0 mL of a 0.550 M KCl solution was evaporated to dryness?

b. What volume of a 0.315 M HNO_3 solution is needed to provide 0.0410 mol HNO_3?

c. What volume of 1.21 M NH_4NO_3 contains 50.0 g of solute?

6.71 Calculate the following.

a. How many grams of solid $AgNO_3$ will be needed to prepare 200. mL of a 0.200 M solution?

b. How many grams of vitamin C ($C_6H_8O_6$) would be contained in 25.0 mL of a 1.00 M solution?

c. How many moles of HCl are contained in 250. mL of a 6.0 M solution?

6.72 Calculate the concentration in % (w/w) of the following solutions. Assume water has a density of 1.00 g/mL.

a. 5.3 g of sugar and 100. mL of water

b. 5.3 g of any solute and 100. mL of water

c. 5.3 g of any solute and 100. g of any solvent

6.73 Calculate the concentration in % (w/w) of the following solutions. Assume water has a density of 1.00 g/mL.

a. 6.5 g of table salt and 100 mL of water

b. 6.5 g of any solute and 100 mL of water

c. 6.5 g of any solute and 100 g of any solvent

6.74 Calculate the concentration in % (w/w) of the following solutions. Assume water has a density of 1.00 g/mL.

a. 20.0 g of salt is dissolved in 250. mL of water.

b. 0.100 mol of solid glucose ($C_6H_{12}O_6$) is dissolved in 100. mL of water.

c. 120. g of solid is dissolved in 100 mL of water.

d. 10.0 mL of ethyl alcohol (density = 0.789 g/mL) is mixed with 10.0 mL of water.

6.75 Calculate the concentration in % (w/w) of the following solutions. Assume water has a density of 1.00 g/mL.

a. 5.20 g of $CaCl_2$ is dissolved in 125 mL of water.

b. 0.200 mol of solid KBr is dissolved in 200. mL of water.

c. 50.0 g of solid is dissolved in 250. mL of water.

d. 10.0 mL of ethyl alcohol (density = 0.789 g/mL) is mixed with 10.0 mL of ethylene glycol (density = 1.11 g/mL).

6.76 Calculate the concentration in % (w/w) of the following solutions.

a. 20.0 g of solute is dissolved in enough water to give 150. mL of solution. The density of the resulting solution is 1.20 g/mL.

b. A 10.0 mL sample with a density of 1.10 g/mL leaves 1.18 g of solid residue when evaporated.

c. A 25.0 g sample of solution on evaporation leaves a 1.87 g residue of $MgCl_2$.

6.77* Calculate the concentration in % (w/w) of the following solutions.

a. 424 g of solute is dissolved in enough water to give 1.00 L of solution. The density of the resulting solution is 1.18 g/mL.

b. A 50.0 mL solution sample with a density of 0.898 g/mL leaves 12.6 g of solid residue when evaporated.

c. A 25.0 g sample of solution on evaporation leaves a 2.32 g residue of NH_4Cl.

6.78 Calculate the concentration in % (v/v) of the following solutions.

a. 200. mL of solution contains 15 mL of alcohol.

b. 200. mL of solution contains 15 mL of any soluble liquid solute.

c. 8.0 fluid ounces of oil is added to 2.0 gallons (256 fluid ounces) of gasoline.

d. A solution of alcohol and water is separated by distillation. A 200. mL sample gives 85.9 mL of alcohol.

6.79 Calculate the concentration in % (v/v) of the following solutions.

a. 250. mL of solution contains 15.0 mL of acetone.

b. 250. mL of solution contains 15.0 mL of any soluble liquid solute.

c. 1.0 quart of acetic acid is put into a 5 gallon container, and enough water is added to fill the container.

d. A solution of acetone and water is separated by distillation. A 300. mL sample gives 109 mL of acetone.

6.80 Consider the blood volume of an adult to be 5.0 L. A blood alcohol level of 0.50% (v/v) can cause a coma. What volume of pure ethyl alcohol, if consumed in one long drink and assumed to be absorbed completely into the blood, would result in this critical blood alcohol level?

6.81* The blood serum acetone level for a person is determined to be 1.8 mg of acetone per 100. mL of serum. Express this concentration as % (v/v) if liquid acetone has a density of 0.79 g/mL.

6.82 Calculate the concentration in % (w/v) of the following intravenous solutions used in hospitals.

a. 150. mL of a dextrose solution contains 7.50 g of dissolved solid dextrose (glucose).

b. 150. mL of solution contains 7.50 g of any dissolved solid solute.

c. 350. mL of normal saline solution contains 3.15 g of dissolved solid sodium chloride.

6.83 Calculate the concentration in % (w/v) of the following solutions.

a. 26.5 g of solute is dissolved in 200. mL of water to give a solution with a density of 1.10 g/mL.

b. A 30.0 mL solution sample on evaporation leaves a solid residue of 0.38 g.

c. On analysis for total protein, a blood serum sample of 15.0 mL is found to contain 1.15 g of total protein.

6.84 A saturated solution of KBr in water is formed at 20.0 °C. Consult **Figure 6.16** and calculate the concentration of the solution in % (w/w).

6.85 Phenol (C_6H_5OH) is an antiseptic used in some mouthwashes, throat sprays, and lozenges. What is the molarity of phenol in a mouthwash that contains 1.4 g of phenol per 100. mL of solution?

Solution Preparation (Section 6.8)

6.86 Explain how you would prepare the following solutions using pure solute and water. Assume water has a density of 1.00 g/mL.

a. 200. mL of 0.150 M Na_2SO_4 solution

b. 250.0 mL of 0.250 M $Zn(NO_3)_2$ solution

c. 150.0 g of 2.25% (w/w) NaCl solution

d. 125.0 mL of 0.75% (w/v) KCl solution

6.87 Explain how you would prepare the following solutions using pure solute and water. Assume water has a density of 1.00 g/mL.

a. 250. mL of a 2.00 M NaOH solution

b. 500. mL of a 40.0% (v/v) alcohol solution (C_2H_5OH)

c. 100. mL of a 15.0% (w/v) glycerol solution. Glycerol is a liquid with a density of 1.26 g/mL. Describe two ways to measure the amount of glycerol needed.

d. Approximately 50. mL of a normal saline solution, 0.89% (w/w) NaCl

6.88* A saline solution used for intravenous injections is a 0.16 M solution of sodium chloride. Calculate the mass in grams of sodium chloride needed to prepare 250. mL of solution.

6.89 Calculate the following for these common solutions found in clinics and hospitals.

a. The number of moles of dextrose (glucose) in 250. mL of a 0.278 M dextrose solution

b. The number of grams of NaCl in 120. mL of a 0.154 M normal saline solution

c. The number of grams of iodine in 20.0 mL of a 3.00% (w/v) tincture of iodine solution

d. The number of milliliters of alcohol in 250. mL of a 90.0% (v/v) disinfectant solution

6.90 Calculate the following.

a. The number of grams of Li_2CO_3 in 250. mL of 1.75 M Li_2CO_3 solution

b. The number of moles of NH_3 in 200. mL of 3.50 M NH_3 solution

c. The number of mL of alcohol in 250. mL of 12.5% (v/v) solution

d. The number of grams of $CaCl_2$ in 50.0 mL of 4.20% (w/v) $CaCl_2$ solution

6.91 Explain how you would prepare the following dilute solutions from the more concentrated ones.

a. 200. mL of 0.500 M HCl from a 6.00 M HCl solution

b. 50.0 mL of 2.00 M H_2SO_4 from a 6.00 M H_2SO_4 solution

c. 100. mL of normal saline solution, 0.89% (w/v) NaCl, from a 5.0% (w/v) NaCl solution

d. 250. mL of 5.00% (v/v) acetone from 20.5% (v/v) acetone

6.92 Explain how you would prepare the following dilute solutions from the more concentrated ones.

a. 5.00 L of 6.00 M H_2SO_4 from a 18.0 M H_2SO_4 solution

b. 250. mL of 0.500 M $CaCl_2$ from a 3.00 M $CaCl_2$ solution

c. 200. mL of 1.50% (w/v) KBr from a 10.0% (w/v) KBr solution

d. 500. mL of 10.0% (v/v) alcohol from 50.0% (v/v) alcohol

6.93 What is the molarity of the solution prepared by diluting 25.0 mL of 0.412 M $Mg(NO_3)_2$ solution to each of the following final volumes?

a. 40.0 mL

b. 0.100 L

c. 1.10 L

d. 350.0 mL

6.94 What is the molarity of the solution prepared by diluting 50.0 mL of 0.195 M KBr to each of the following volumes?

a. 1.50 L

b. 200 mL

c. 500 mL

d. 700 mL

Osmosis and Dialysis (Section 6.9)

Note: **In Exercises 6.95 to 6.99, assume the temperature is 25 °C, and express your answer in torr, mm Hg, and atm.**

6.95 Suppose you have a bag made of a semipermeable membrane. Inside the bag is a solution containing water and dissolved small molecules. Describe the behavior of the system when the bag functions as an osmotic membrane and when it functions as a dialyzing membrane.

6.96 Suppose an osmotic membrane separates a 5.00% sugar solution from a 10.0% sugar solution. In which direction will water flow? Which solution will become diluted as osmosis takes place?

6.97 Refer to **Figure 6.8** and answer the question. Would a quart bottle full of gas actually contain the number of moles you calculated? Explain.

6.98 Each of the following mixtures was placed in a dialysis bag similar to the membrane in **Figure 6.33**. The bag was immersed in pure water. Which substances will pass through the bag into the surrounding water?

a. NaCl solution and starch (colloid)

b. Urea (small organic molecule) and starch (colloid)

c. Albumin (colloid), KCl solution, and glucose solution (small organic molecule)

Additional Exercises

6.99 Explosives react very rapidly and produce large quantities of heat and gaseous products. When nitroglycerine explodes, several gases are produced:

$$4C_3H_5O_9N_3(\ell) \rightarrow 12CO_2(g) + O_2(g) + 6N_2(g) + 10H_2O(g)$$

Suppose 11 g of nitroglycerine was sealed inside a 1.0 L soda bottle and detonated. Assume the bottle would not break, and the temperature immediately after detonation was 750 K. Calculate the pressure of the gases inside the bottle in atmospheres.

6.100 How many liters of oxygen gas, O_2, will it take to completely react with 2.31 L of hydrogen gas, H_2, to produce water? Assume both gases are at the same pressure and temperature. What type of reaction is this?

6.101 Which of the following gases would you expect to behave most ideally: He, Ar, or HCl? Explain.

7 Chemical Equilibrium: Acids, Bases, and Buffers

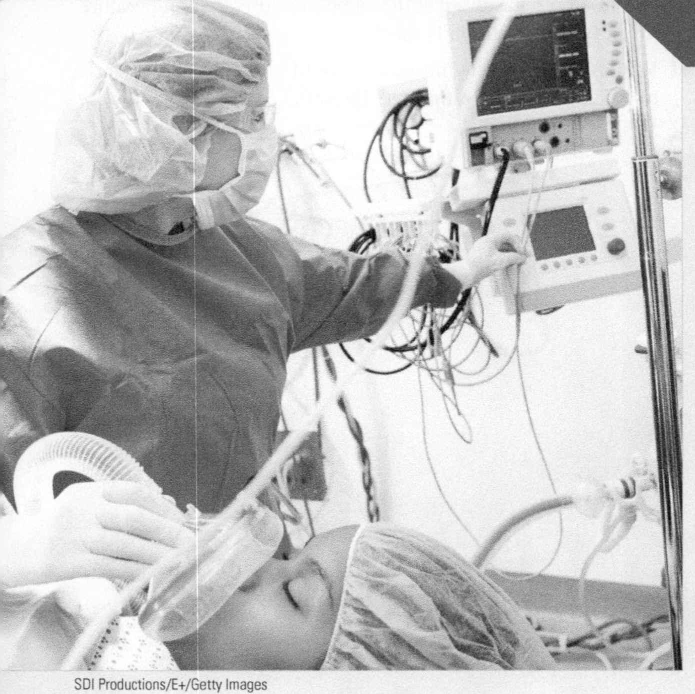
SDI Productions/E+/Getty Images

Health Career Focus

Respiratory Therapist

Respiratory therapy is ranked among the most competitive programs at the university. At first I was wait-listed, but eventually I was accepted to the program and now, I work at a large community hospital.

The other night, while checking a patient on a ventilator, I noticed his hands and feet spasming. I recognized this sign of respiratory alkalosis and, based on my observation, the doctor ordered an arterial blood gas measurement. I recalled my nervousness in performing that test for the first time; now, I proudly conduct it with confidence. The test results showed a blood pH imbalance, with a pH greater than 7.5. I increased the breaths per minute setting on the ventilator to correct it.

How satisfying to know that my observations help to keep the patients safe!

Follow-up to this Career Focus appears at the end of the chapter before the *Concept Summary*.

Learning Objectives

When you have completed your study of Chapter 7, you should be able to:

1. Use energy diagrams to represent and interpret the energy relationships for reactions. **(Section 7.1)**
2. Explain how factors such as reactant concentrations, temperature, and catalysts influence reaction rates. **(Section 7.1)**
3. Write equilibrium expressions based on reaction equations. **(Section 7.2)**
4. Use Le Châtelier's principle to predict the influence of changes in concentration and reaction temperature on the position of equilibrium for reactions. **(Section 7.2)**
5. Write reaction equations that illustrate Brønsted-Lowry acid–base behavior. **(Section 7.3)**
6. Name acids. **(Section 7.3)**
7. Classify solutions as acidic, basic, or neutral based on hydroxide ion or hydronium ion concentrations. **(Section 7.4)**
8. Calculate the pH of solutions given hydroxide ion or hydronium ion concentrations. **(Section 7.4)**
9. Use equilibrium constants to rank the strengths of acids and bases. **(Section 7.5)**
10. Write reaction expressions for the neutralization reactions between acids and bases. **(Section 7.6)**
11. Perform titration calculations using balanced acid–base neutralization expressions. **(Section 7.6)**
12. Write reaction expressions to illustrate how buffers work. **(Section 7.7)**
13. Determine the pH of buffers. **(Section 7.7)**
14. Describe respiratory and urinary control of blood pH. **(Section 7.7)**

In Chapter 4, we discussed types of chemical reactions in detail, but here in Chapter 7, we explore how chemical reactions behave. This includes how energetically favorable reactions are, a concept known as *thermodynamics*, and how fast reactions take place, a concept known as *kinetics*. From there, we examine reactions that take place in both forward and reverse directions, a concept known as *chemical equilibrium*. Le Châtelier's principle then provides a way to understand how some reaction variables can affect equilibrium. Next, we explore acid–base chemistry, the most prevalent class of equilibrium reactions. We learn to name acids, express acidity in terms of pH, perform pH calculations, and explore the strength of acids and bases. Lastly, we discuss pH in the context of buffers using blood pH as a real-life example.

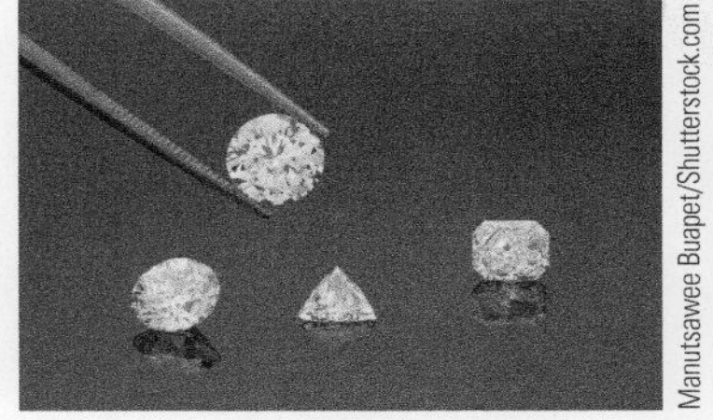
Manutsawee Buapet/Shutterstock.com

Miriam Doerr Martin Frommherz/Shutterstock.com

Figure 7.1 Though diamond and graphite have vastly different appearances, both are elemental forms of carbon atoms.

7.1 Principles of Chemical Reactions

Learning Objective 1 Use energy diagrams to represent and interpret the energy relationships for reactions.

Learning Objective 2 Explain how factors such as reactant concentrations, temperature, and catalysts influence reaction rates.

Why do some reactions happen immediately, whereas others take place over thousands of years? Before we can answer that question, we must understand the difference between kinetics and thermodynamics.

7.1A Kinetics and Thermodynamics

According to calculations, carbon in the form of a diamond will spontaneously change into graphite, if given enough time (**Figure 7.1**). This is because graphite is a more energetically stable form of carbon than diamond. The study of these kinds of changes in energy that occur during the course of a chemical reaction is called **chemical thermodynamics**. So, should you be concerned if you own a diamond? Diamond does change to graphite, but it happens so slowly that it is undetectable over many human lifetimes. The study of the *speed* at which chemical reactions go to completion is called **chemical kinetics**. All chemical reactions are governed by the relationship between thermodynamics (the energy of reaction) and kinetics (the rate of reaction). The chemical process of diamond turning into graphite is controlled, or dominated, by kinetics. But there are many other reactions where thermodynamics takes precedent.

chemical thermodynamics The study of the changes in energy that take place when processes such as chemical reactions occur.

chemical kinetics The study of the rates of chemical reactions.

The thermodynamic property of molecules and reactions that is used most often in chemistry is **enthalpy**, which is defined as the sum of internal energies plus the product of pressure and volume. Rather than focusing on individual enthalpies of chemical reactants and products, most of the time chemists are interested in how the thermodynamics change during the course of a reaction, which is defined as the **reaction enthalpy**. Energy taken in or given off by a chemical reaction can be closely related to the heat accepted or released during a chemical reaction. Chemical reactions that give off heat as they proceed, where the reactants are at a higher enthalpy than the products, are referred to as being *exothermic*. Conversely, chemical reactions that require heat being added, where the products have a higher enthalpy than the reactants, are called *endothermic*. We first introduced these terms in **Section 5.5** when discussing changes of state, but they actually apply to all chemical reactions (**Figure 7.2**).

enthalpy The sum of internal energies plus the product of pressure and volume.

reaction enthalpy The difference in enthalpy between the products and the reactants in a chemical reaction.

Figure 7.2 Melting ice to produce liquid water requires the input of heat and is thus endothermic. The combustion of wood gives off heat and is therefore exothermic.

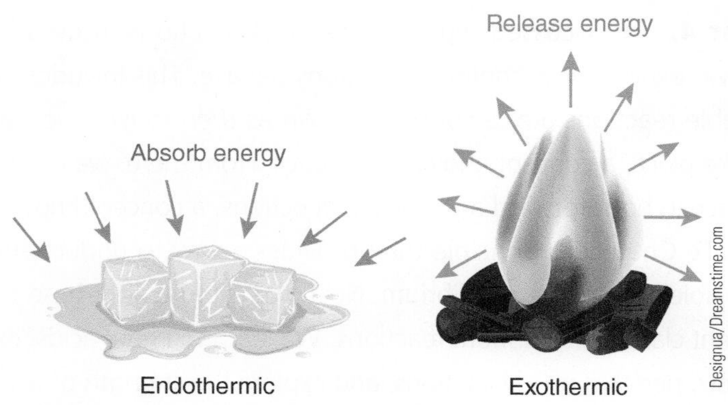

activation energy Energy needed to start some spontaneous processes. Once started, the processes continue without further stimulus or energy from an outside source.

The reaction enthalpy can often be shown in a graph that depicts the change in energy of a reaction over the duration of the reaction. This type of graph, shown in **Figure 7.3a** for a typical exothermic reaction and in **Figure 7.3b** for a typical endothermic reaction, is known as a *reaction energy diagram*. In both cases, there is a steep rise in energy between the reactants on the left and the products on the right. This energy barrier, which must be overcome in the transition from reactants to products in a chemical reaction, is known as the **activation energy**.

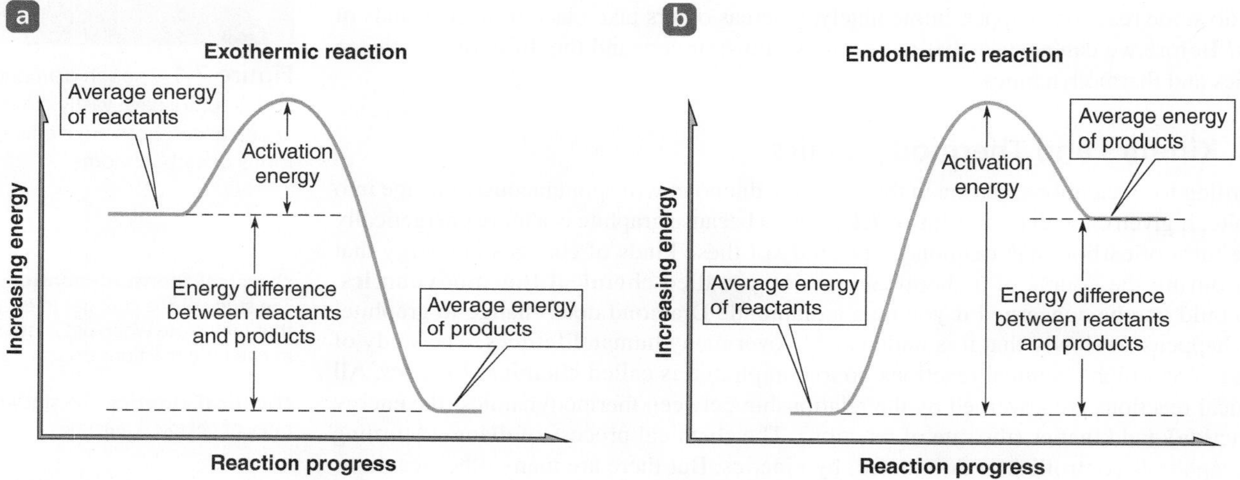

Figure 7.3 (a) A typical energy diagram for an exothermic chemical reaction. The average energy of the products is *lower* than the average energy of the reactants. (b) A typical energy diagram for an endothermic chemical reaction. The average energy of the products is *higher* than the average energy of the reactants.

collision model of chemical kinetics The idea that a chemical reaction can occur only when the reactant molecules, atoms, or ions collide in the proper orientation and exceed the activation energy.

The magnitude of this activation energy, which varies from reaction to reaction, is often the most prominent factor in governing the kinetics of a chemical reaction. In many cases, the average energy of the molecules is too low at the prevailing temperature for a reaction to proceed, so an increase in energy is needed to produce the molecular collisions necessary for a chemical reaction to occur. Before we discuss in earnest the speed of chemical reactions, let us take a moment to better understand the **collision model of chemical kinetics**, which outlines the conditions necessary on a molecular level for reactions to occur. This model is based on the following three assumptions:

1. Reactant particles must collide with one another in order for a reaction to occur.
2. Particles must collide with at least a certain minimum total amount of energy if the collision is to result in a reaction.
3. In some cases, colliding reactants must be oriented in a specific way if a reaction is to occur.

effective collision A collision that causes a reaction to occur between the colliding molecules.

If all three of these assumptions are met, then the collision is defined as an **effective collision**, and a reaction occurs.

Assumption 1 is necessary because two molecules cannot react with each other if they do not collide at some point. That is, in order to break bonds, exchange atoms, and form new bonds, they must come in contact. If collisions were the only factor, and if every collision resulted in a reaction, then most gaseous and liquid state reactions would take place almost instantaneously. Such high reaction rates are not observed, however, which brings us to assumptions 2 and 3.

Orientation effects are related to which side or end of a particle actually hits another particle during a collision. While a specific orientation is not essential to all reactions, it plays a key role in many bond formations. For instance, **Figure 7.4** shows that nitrogen dioxide and oxygen gas form from the collision of nitrogen monoxide and ozone only when the nitrogen atom collides with a terminal oxygen atom in ozone. Collision with the central oxygen atom is ineffective.

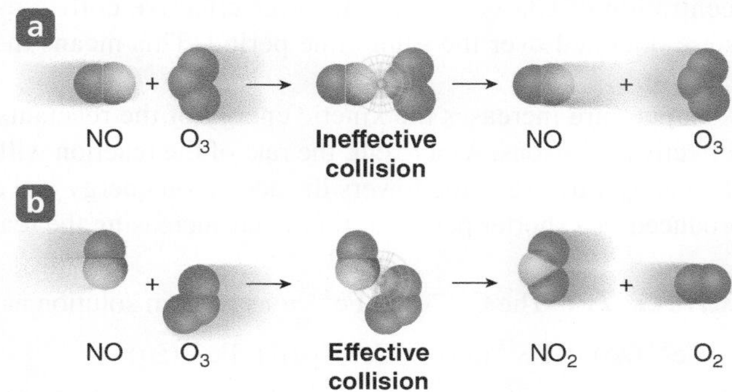

Figure 7.4 (a) An ineffective collision of nitrogen monoxide and ozone leads to a failed reaction. (b) An effective collision of nitrogen monoxide and ozone leads to the production of nitrogen dioxide and oxygen.

reaction rate The speed of a reaction.

catalyst A substance that changes (usually increases) reaction rates without being used up in the reaction.

inhibitor A substance that decreases reaction rates.

7.1B Reaction Rates

The speed of a reaction is called the **reaction rate**. It is determined experimentally as the change in concentration of a reactant or product divided by the time required for the change to occur. Three factors that affect the rates of all reactions are:

1. The concentration of the reactants.
2. The temperature of the reactants.
3. The presence of catalysts.

How do reactant concentration and temperature influence reaction rates? Essentially, any increase in reactant concentration results in more effective collisions in a given period of time and thus speeds up the rate of reaction. An increase in the temperature of a system increases the energy of the reacting molecules. The increased molecular speed (kinetic energy) causes more collisions to take place in a given time. Also, because the kinetic and internal energies of the colliding molecules are greater, a larger fraction of the collisions will exceed the activation energy and result in effective collisions, thus increasing the rate of reaction.

A **catalyst** is a substance that changes reaction rates without being used up in the reactions. Usually the term *catalyst* is used to describe substances that speed up reactions. A substance that slows reactions is known as an **inhibitor**. Catalysts enhance a reaction rate by providing an alternate reaction pathway that requires less activation energy than the normal pathway. This effect is depicted in the reaction energy diagram in **Figure 7.5**.

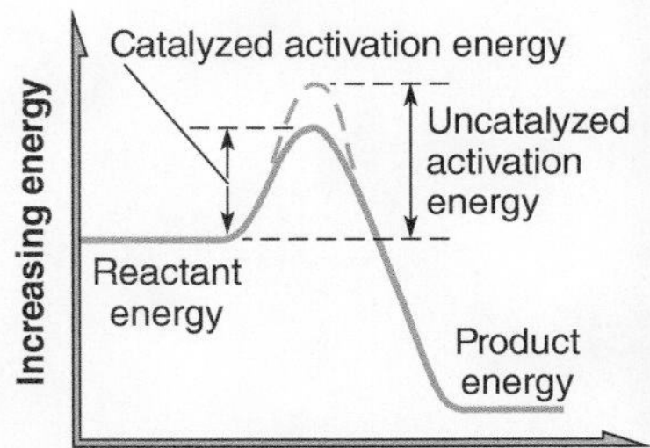

Figure 7.5 A catalyst increases the rate of the reaction by lowering the activation energy. The catalyst has no effect on the energies of the reactants or products.

Enzymes are proteins (large biomolecules we study in Chapter 12) that catalyze the chemical reactions that take place in your body. They do this by placing molecules in the exact orientation for them to collide in the quickest, most favorable way possible. The enzyme acetylcholinesterase (AChE), for example, regulates the levels of acetylcholine in the body. Acetylcholine is a molecule involved in signaling muscles to contract (**Figure 7.6**). The enzyme is so efficient that it can break down 14,000 molecules of acetylcholine every second! This reaction can be slowed, though, by exposure to specific inhibitors, either for good (as in the case of Alzheimer's treatments), or bad (as in the case of toxic nerve agents like sarin).

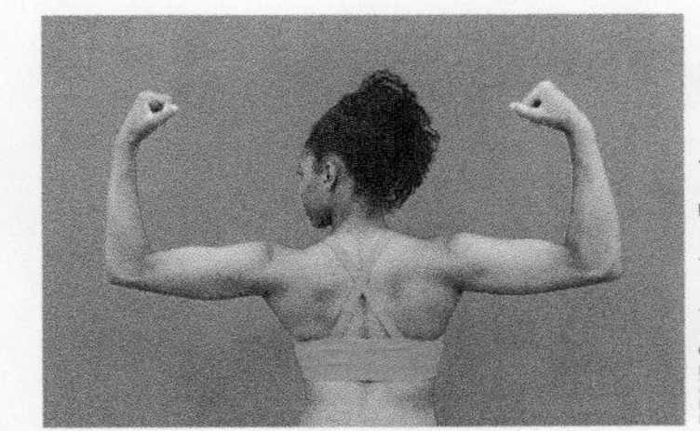

Figure 7.6 Acetylcholine is a signaling molecule that triggers muscles to contract.

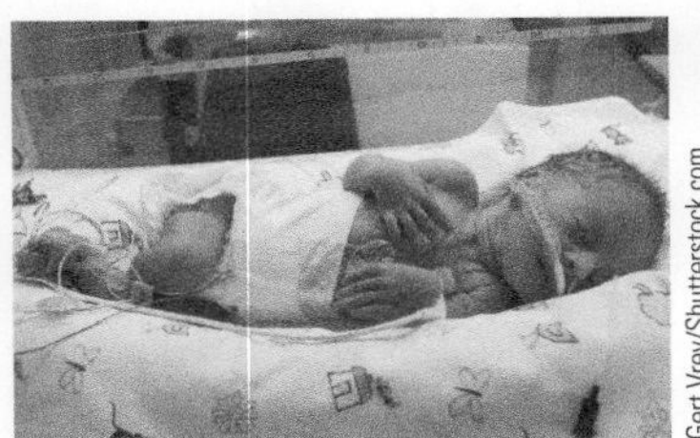
Gert Vrey/Shutterstock.com

Figure 7.7 Nitric oxide (NO) is used to dilate blood vessels in the lungs of premature babies to help them breathe.

Example 7.1 How Factors Affect Reaction Rates

Nitric oxide (**Figure 7.7**) is used in hospitals for lung vasodilation of preterm newborns and patients with pulmonary distress. The gases NO_2 and CO react to form NO as follows:

$$NO_2(g) + CO(g) \rightarrow NO(g) + CO_2(g)$$

Will the following changes in conditions increase or decrease the reaction rate of production of NO(g) and $CO_2(g)$ from $NO_2(g)$ and CO(g)?

a. Decreasing the concentration of CO(g) from 0.40 M to 0.20 M
b. Increasing the temperature from 25 °C to 37 °C
c. Adding a catalyst to the reaction

Solution

a. When the concentration of CO is decreased, fewer effective collisions occur, and fewer products are obtained over the same time period. This means the rate of the reaction will decrease.
b. An increase in temperature increases the kinetic energy of the reactants and causes more frequent effective collisions. As a result, the rate of the reaction will increase.
c. The addition of a catalyst to a reaction lowers the activation energy and allows more product to be produced in a shorter period of time, thus increasing the reaction rate.

✔ Learning Check 7.1 The Ce^{4+} and Fe^{2+} ions react in solution as follows:

$$Ce^{4+}(aq) + Fe^{2+}(aq) \rightarrow Ce^{3+}(aq) + Fe^{3+}(aq)$$

Will the following changes in conditions increase or decrease the reaction rate of the production of Ce^{3+} and Fe^{3+} from Ce^{4+} and Fe^{2+}?

a. Increasing the concentration of Fe^{2+} from 3.0 M to 5.0 M
b. Decreasing the temperature from 50 °C to 37 °C
c. Adding an inhibitor to the reaction

7.2 Equilibrium

Learning Objective 3 Write equilibrium expressions based on reaction equations.

Learning Objective 4 Use Le Châtelier's principle to predict the influence of changes in concentration and reaction temperature on the position of equilibrium for reactions.

So far, we have focused only on the reactants of a reaction. However, all reactions can (in principle) go in both directions, so the products can be viewed as reactants, too. Suppose equal amounts of gaseous H_2 and I_2 are placed in a closed container and allowed to react to form HI. Initially, no HI is present, so the only possible reaction is

$$H_2(g) + I_2(g) \rightarrow 2HI(g)$$

After a short time, however, some HI molecules are produced. Those HI molecules can collide with one another in a way that causes the reverse reaction to occur:

$$2HI(g) \rightarrow H_2(g) + I_2(g)$$

The low concentration of HI makes this reverse reaction slow at first, but as the concentration increases, so does the reaction rate. Meanwhile, the rate of the forward reaction decreases as the concentrations of H_2 and I_2 decrease. Eventually, the concentrations of H_2, I_2, and HI in the reaction mixture reach levels at which the rates of the forward and reverse reactions are equal. From that time on, the concentrations of H_2, I_2, and HI in the mixture remain *constant*, because both forward and reverse reactions take place at the same rate and because each substance is being produced as fast as it is used up. When the forward and reverse reaction rates are equal, the reaction is said to be at **equilibrium**, and the concentrations of the reactants and products are called **equilibrium concentrations**. The behavior of the reaction $2HI(g) \rightleftarrows H_2(g) + I_2(g)$ as it reaches equilibrium is depicted graphically in **Figure 7.8**. The majority of chemical reactions that govern life on Earth are at or near equilibrium!

state of equilibrium A condition in a reaction system when the rates of the forward and reverse reactions are equal.

equilibrium concentrations The unchanging concentrations of reactants and products in a reaction system that is in a state of equilibrium.

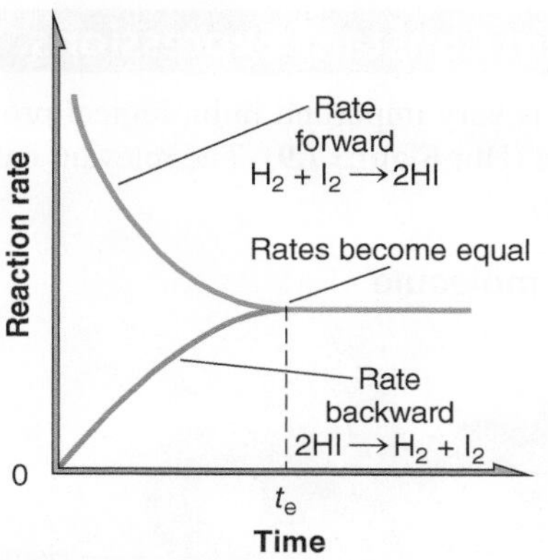

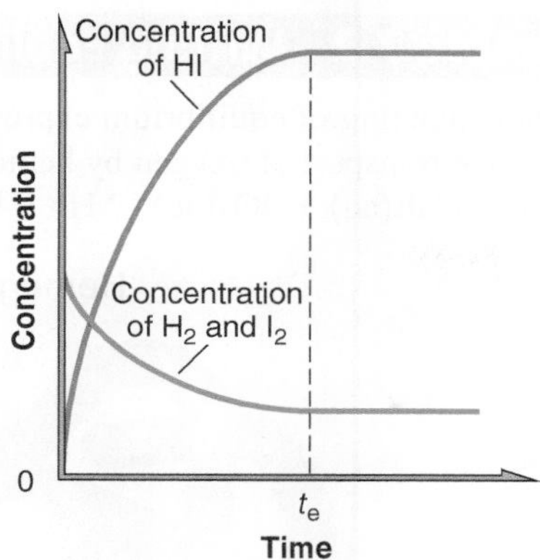

Figure 7.8 Variation of reaction rates and reactant concentrations as equilibrium is established (t_e is the time needed to reach equilibrium).

Instead of writing separate equations for both the forward and reverse reactions, the usual practice is to represent reversibility by double arrows:

$$2HI(g) \rightleftarrows H_2(g) + I_2(g)$$

The **position of equilibrium** for a reaction indicates the relative amounts of reactants and products present at equilibrium. When the position is described as being far to the right, it means that at equilibrium, the concentration of products is much higher than the concentration of reactants (the products predominate). A position far to the left means the concentration of reactants is much higher than that of products (the reactants predominate).

position of equilibrium An indication of the relative amounts of reactants and products present at equilibrium.

The position of equilibrium can be represented numerically with an **equilibrium constant** (K_{eq}). Any reaction that establishes an equilibrium can be represented by the following general reaction expression:

equilibrium constant A numerical relationship between reactant and product concentrations in a reaction at equilibrium.

$$aA + bB \rightleftarrows cD + dD$$

In this expression, the capital letters stand for reactants and products, whereas the lowercase letters are the coefficients in the balanced chemical equation. For a reaction at equilibrium, the following **equilibrium expression** is valid:

equilibrium expression An equation relating the equilibrium constant and reactant and product equilibrium concentrations.

$$K_{eq} = \frac{[C]^c[D]^d}{[A]^a[B]^b} \tag{7.1}$$

In Equation 7.1, the brackets [ ] represent molar concentrations of the reactants (A and B) and the products (C and D). K_{eq} is the equilibrium constant, and the exponents on each bracket are the coefficients from the balanced chemical equation for the reaction.

According to Equation 7.1, the product of equilibrium concentrations of products (raised to appropriate powers) divided by the product of the equilibrium concentration of reactants (also raised to appropriate powers) gives a number that does not change with time (the equilibrium constant). It does not change because the concentrations of reactants and products remain constant at equilibrium. K_{eq} values are constant at any given temperature, but they will change as temperature changes. Because they only involve reactants and products that can change concentration during a chemical reaction, the concentrations of pure liquids and solids (e.g., water) are omitted from the K_{eq} expression.

For a reaction with an equilibrium position far to the right, the product concentrations $[C]$ and $[D]$ will be much higher than the reactant concentrations $[A]$ and $[B]$ and should result in a large K_{eq} value (greater than 1). For reactions with equilibrium positions far to the left, the K_{eq} values should be small (less than 1). Thus, the value of the equilibrium constant indicates the position of equilibrium for a reaction. In fact, some K_{eq} values are so small (e.g., 1.1×10^{-36}) that for all practical purposes, no products are present at equilibrium. Others are so large (e.g., 1.2×10^{40}) that the reaction can be considered to go completely to products. For reactions with K_{eq} values between 10^{-3} and 10^3, the position of equilibrium is not extremely favorable to either side, and significant concentrations of both reactants and products can be detected in the equilibrium mixtures.

Example 7.2 Writing Equilibrium Constant Expressions

An understanding of equilibrium expressions is very important in biological processes such as the transport of oxygen by hemoglobin (Hb; **Figure 7.9**). The relevant balanced equation is $Hb(aq) + 4O_2(aq) \rightleftarrows Hb \cdot 4O_2(aq)$

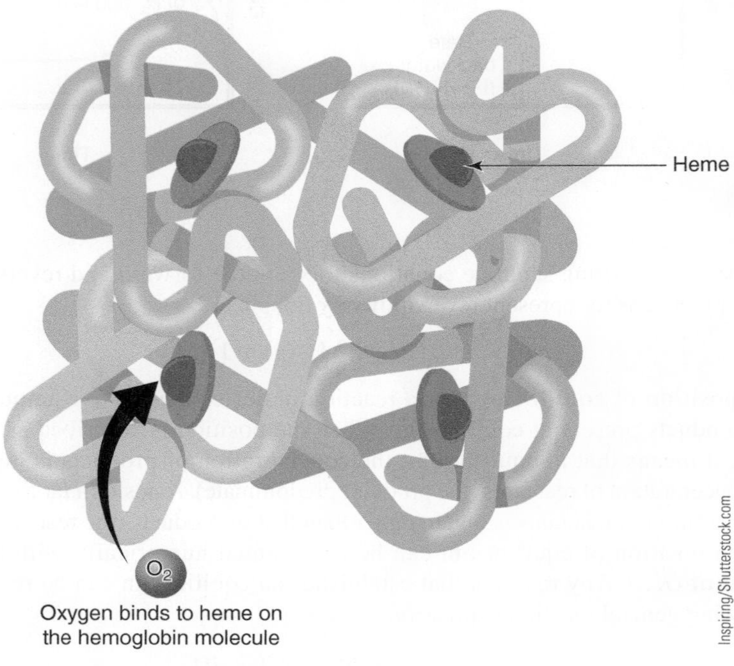

Figure 7.9 Oxygen is transported by our blood cells using a protein called hemoglobin.

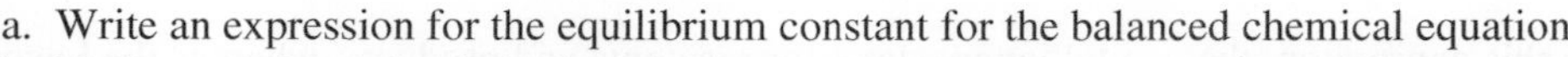

a. Write an expression for the equilibrium constant for the balanced chemical equation.
b. If $K_{eq} = 3.3$ at physiological body temperature, does oxygenated hemoglobin ($Hb \cdot 4O_2$) or deoxygenated hemoglobin (Hb) predominate at equilibrium?

Solution

a. The concentration of the product $Hb \cdot 4O_2$ goes on top of the K_{eq} expression and is raised to the power 1 (which is not written but assumed by convention). The concentrations of the reactants Hb and O_2 go on the bottom, with Hb raised to the power 1, and O_2 raised to the power 4:

$$K_{eq} = \frac{[Hb \cdot 4O_2]}{[Hb][O_2]^4}$$

b. $K_{eq} = 3.3$, so you can think of the equilibrium constant as the ratio of 3.3/1. This means that there are 3.3 oxygenated hemoglobins ($Hb \cdot 4O_2$) for every 1 set of deoxygenated hemoglobin and oxygen (Hb and O_2). As a result, oxygenated hemoglobin predominates at equilibrium.

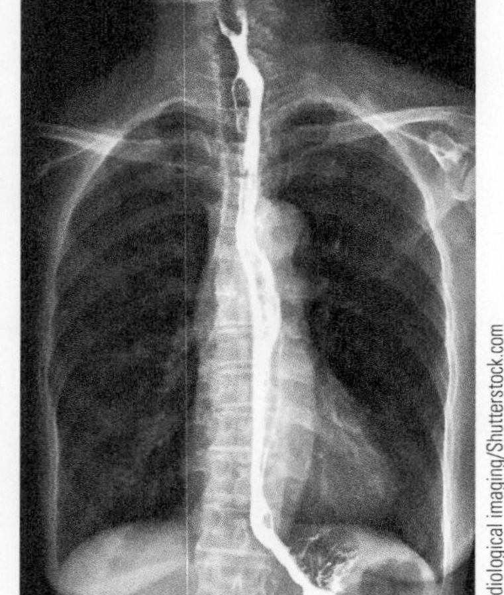

Figure 7.10 X-ray of a patient who has ingested barium sulfate for a diagnostic test of the stomach and intestinal tract.

✔ **Learning Check 7.2** Suspensions of solid barium sulfate ($BaSO_4$) are routinely swallowed by patients undergoing diagnostic X-ray photography of the stomach and intestinal tract (**Figure 7.10**). When $BaSO_4$ dissolves, Ba^{2+} and SO_4^{2-} ions form via the following reaction:

$$BaSO_4(s) + H_2O(\ell) \rightleftarrows Ba^{2+}(aq) + SO_4^{2-}(aq)$$

a. Write an expression for the equilibrium constant for the balanced chemical equation.
b. If $K_{eq} = 1.5 \times 10^{-9}$ for this reaction at physiological body temperature, do the ions (Ba^{2+} and SO_4^{2-}) or the solid ($BaSO_4$) predominate at equilibrium?

A number of factors can change the position of an established equilibrium. The influence of such factors can be predicted by using a concept known as Le Châtelier's principle. According to **Le Châtelier's principle**, when a change is made to an established equilibrium, the position of equilibrium will shift in a direction that will minimize or oppose the change. The factors we are most concerned with are concentrations of reactants and products, and reaction temperature.

Le Châtelier's principle The position of an equilibrium shifts in response to changes made in factors of the equilibrium.

The effect of concentration changes can be illustrated by once again using the reaction of Hb with O_2:

$$Hb(aq) + 4O_2(aq) \rightleftarrows Hb \cdot 4O_2(aq)$$

Suppose an equilibrium mixture of Hb, O_2, and $Hb \cdot 4O_2$ is formed. What happens if some additional O_2 is added to the equilibrium mixture? The chances for favorable collisions between Hb and O_2 molecules are increased, so the rate of formation of $Hb \cdot 4O_2$ is increased, and more $Hb \cdot 4O_2$ is formed than disappears. The new equilibrium mixture will contain more $Hb \cdot 4O_2$ than the original mixture, but the amounts of the reactants will also be different such that the equilibrium constant remains unchanged. The original equilibrium of the reaction has been shifted toward the right.

Le Châtelier's principle can also be used to predict the influence of temperature on an equilibrium by treating heat as a product or reactant. Consider the following reaction expression for a hypothetical exothermic (i.e., heat out) reaction:

$$A + B \rightleftarrows \text{products} + \text{heat}$$

If the temperature is increased (by adding more heat, which appears on the right side of the equation), the equilibrium shifts to the left in an attempt to use up the added heat. In the new equilibrium position, the concentrations of *A* and *B* will be higher, whereas the chemical product concentrations will be lower than those in the original equilibrium. In this case, the value of the equilibrium constant is changed by the change in temperature (see **Figure 7.11**).

Michael C. Slabaugh

Figure 7.11 The effect of temperature on the position of equilibrium for the reaction

$$N_2O_4 \rightleftarrows 2NO_2$$

Left: A sealed tube containing an equilibrium mixture of red–brown NO_2 and colorless N_2O_4 is cooled in an ice bath. *Right:* The same sealed tube is heated in a hot water bath. On which side of the equation will heat appear? Is the reaction exothermic or endothermic?

Example 7.3 Using Le Châtelier's Principle

The large-scale production of ammonia is important in meeting the agricultural demand for nitrogen-based fertilizers. Ammonia is made from hydrogen and atmospheric nitrogen:

$$N_2(g) + 3H_2(g) \rightleftarrows 2NH_3(g) + \text{heat}$$

Use Le Châtelier's principle to determine if the following change in conditions will shift the equilibrium right (toward products) or left (toward reactants).

a. The reaction is cooled.
b. H_2 is removed from the equilibrium mixture

Solution

a. Heat is removed by cooling. Heat is replenished when the equilibrium shifts to the right. Thus, more NH_3 will be present at equilibrium in the cooler mixture.
b. The equilibrium will shift to the left in an attempt to replenish the H_2. At the new equilibrium position, less NH_3 will be present.

✔ **Learning Check 7.3** A saturated solution of ammonium nitrate is formed as follows. The solution process is endothermic.

$$\text{Heat} + NH_4NO_3(s) \rightleftarrows NH_4^+(aq) + NO_3^-(aq)$$

Use Le Châtelier's principle to determine if the following changes in conditions will shift the equilibrium right (toward products) or left (toward reactants).

a. Heat is added to the equilibrium solution.
b. $NH_4^+(aq)$ is added to the equilibrium solution.

Table 7.1 summarizes the effects of concentration and temperature changes on the position of chemical equilibria.

Table 7.1 Effects of Concentration and Temperature Changes on the Position of Chemical Equilibria

Factor	Stress (Change)	Reaction Shifts in the Direction of
Concentration	Remove reactant	Reactants
	Add reactant	Products
	Remove product	Products
	Add product	Reactants
Temperature	Lower temperature of exothermic reaction	Products
	Raise temperature of exothermic reaction	Reactants
	Lower temperature of endothermic reaction	Reactants
	Raise temperature of endothermic reaction	Products

Svante Arrhenius

Figure 7.12 Svante Arrhenius (1859–1927) was a pioneer in physical chemistry and was awarded the Nobel Prize in Chemistry in 1903.

7.3 Acid–Base Equilibria

Learning Objective 5 Write reaction equations that illustrate Brønsted-Lowry acid–base behavior.

Learning Objective 6 Name acids.

Now that we have examined the key principles of thermodynamics, kinetics, and equilibrium, it's time to apply what we've learned to acid–base chemistry. Acid–base chemistry is fundamental to many processes in organic chemistry and biochemistry, helping to maintain equilibrium in a number of biological processes.

7.3A Defining Acids and Bases

Arrhenius acid Any substance that provides H^+ ions when dissolved in water.

Arrhenius base Any substance that provides OH^- ions when dissolved in water.

In 1887, Svante Arrhenius (**Figure 7.12**) defined an **acid** as a substance that dissociates when dissolved in water and produces hydrogen ions (H^+). Similarly, a **base** is a substance that dissociates and releases hydroxide ions (OH^-). Hydrogen chloride (HCl) and sodium hydroxide (NaOH) are examples of an Arrhenius acid and Arrhenius base, respectively. They dissociate in water as follows:

$$HCl(aq) \rightarrow H^+(aq) + Cl^-(aq)$$
$$NaOH(aq) \rightarrow Na^+(aq) + OH^-(aq)$$

hydronium ion The ion resulting from the addition of a proton to water.

Note that the hydrogen ion is a bare proton, the nucleus of a hydrogen atom. Arrhenius did not know that free hydrogen ions cannot exist in water. Instead, they covalently bond with water molecules to form **hydronium ions** (H_3O^+). The water molecules provide both electrons used to form the covalent bond:

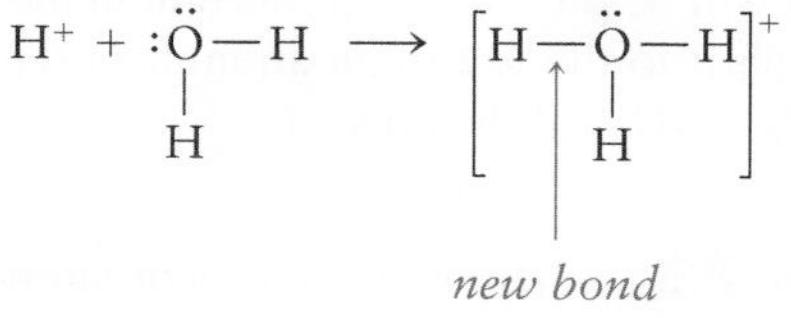

Or, more simply,

$$H^+(aq) + H_2O(\ell) \rightarrow H_3O^+(aq)$$

Brønsted-Lowry acid Any hydrogen-containing substance that is capable of donating a proton (H^+) to another substance.

Brønsted-Lowry base Any substance that is capable of accepting a proton (H^+) from another substance.

In 1923, Johannes Brønsted (1879–1947) and Thomas Lowry (1874–1936) proposed a different acid–base theory that took into account this behavior of hydrogen ions. They defined an **acid** as any hydrogen-containing substance that donates a proton (a hydrogen ion) to another substance and a **base** as any substance that accepts a proton.

According to this theory, the acidic behavior of covalently bonded HCl molecules in water is written as follows:

$$HCl(aq) + H_2O(\ell) \rightleftharpoons H_3O^+(aq) + Cl^-(aq)$$

The HCl behaves as a Brønsted-Lowry acid by donating a proton to a water molecule. The water molecule, by accepting the proton, behaves as a Brønsted-Lowry base.

The double arrows of unequal length in the reaction equation indicate that the reaction is reversible, with the equilibrium lying far to the right. In actual aqueous solutions, essentially 100% of the dissolved HCl is in the ionic form at equilibrium. Remember, both the forward and the reverse reactions are taking place at equilibrium (see **Section 7.2**). When the reverse reaction occurs, hydronium ions donate protons to chloride ions to form HCl and H_2O molecules. Thus, H_3O^+ behaves as a Brønsted-Lowry acid, and Cl^- behaves as a Brønsted-Lowry base.

When a substance like HCl behaves as a Brønsted-Lowry acid by donating a proton, the species that remains (Cl^-) is a Brønsted-Lowry base (**Figure 7.13**). The Cl^- is called the **conjugate base** of HCl. Every Brønsted-Lowry acid and the base formed when it donates a proton is called a **conjugate acid–base pair**. Thus, in the HCl reaction equation, HCl and Cl^- form a conjugate acid–base pair, as do H_3O^+ and H_2O for the reverse reaction. Notice that the acid and base in a conjugate acid–base pair differ only by a proton, H^+.

conjugate base The species remaining when a Brønsted-Lowry acid donates a proton.

conjugate acid–base pair A Brønsted-Lowry acid and its conjugate base.

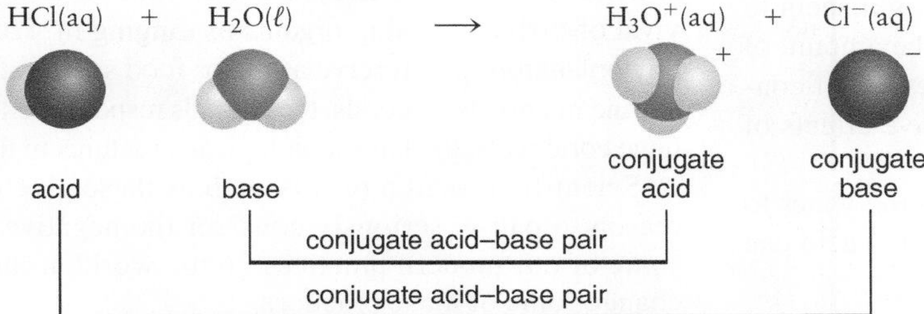

Figure 7.13 In a Brønsted-Lowry acid–base reaction, the acid donates a proton, the base accepts a proton, and the conjugate acid–base pairs differ by a single proton.

Example 7.4 Identifying Brønsted-Lowry Acids and Bases

Identify all Brønsted-Lowry acids, bases, and conjugate acid–base pairs in the reactions represented by the following chemical equations.

a. $HNO_3(aq) + H_2O(\ell) \rightleftharpoons H_3O^+(aq) + NO_3^-(aq)$ (**Figure 7.14**)
b. $NH_3(aq) + H_2O(\ell) \rightleftharpoons OH^-(aq) + NH_4^+(aq)$

Solution

a. HNO_3 is a Brønsted-Lowry acid because it donates a proton to H_2O. The conjugate base of HNO_3 is the result, NO_3^-.

H_2O is a Brønsted-Lowry base because it accepts a proton from HNO_3. The conjugate acid of H_2O is the result, H_3O^+.

$$HNO_3(aq) + H_2O(\ell) \rightleftharpoons H_3O^+(aq) + NO_3^-(aq)$$

acid base acid base
conjugate pair
conjugate pair

b. H_2O is a Brønsted-Lowry acid because it donates a proton to NH_3. The conjugate base of H_2O is the result, OH^-.

NH_3 is a Brønsted-Lowry base because it accepts a proton from H_2O. The conjugate acid of NH_3 is the result, NH_4^+:

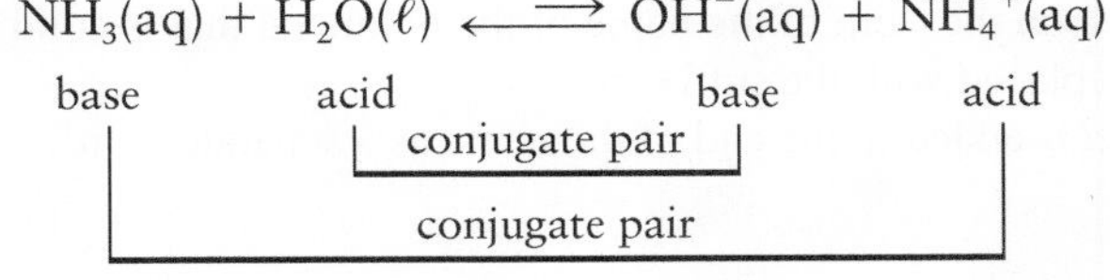

Andia/Alamy Stock Photo

Figure 7.14 Nitric acid is sometimes used by law enforcement agents in the field to detect illegal drugs.

✔ **Learning Check 7.4** Identify all Brønsted-Lowry acids, bases, and acid–base conjugate pairs in the reactions represented by the following chemical equations.

a. $HC_2H_3O_2(aq) + H_2O(\ell) \rightleftarrows H_3O^+(aq) + C_2H_3O_2^-(aq)$
b. $NO_2^-(aq) + H_2O(\ell) \rightleftarrows HNO_2(aq) + OH^-(aq)$

Environmental Connections 7.1

Carbon Dioxide in the Oceans

Carbon dioxide (CO_2) is an atmospheric greenhouse gas that contributes to global warming. Concerns are now being raised about the potential adverse impact of dissolved CO_2 on the oceans of the world.

It is estimated that about one-third of the CO_2 produced by burning fossil fuels in our highly industrialized, modern world ends up dissolved in our oceans. This might seem to be a positive development, because it reduces the amount of CO_2 contributing to global warming, but the results of continuing studies raise questions about the negative effects of this dissolved CO_2 on ocean ecosystems.

When CO_2 dissolves in water, it reacts with the water to form carbonic acid, H_2CO_3 (Reaction 1). Carbonic acid can dissociate to form H^+ and HCO_3^- (Reaction 2).

1. $CO_2(g) + H_2O(\ell) \rightleftarrows H_2CO_3(aq)$
2. $H_2CO_3(aq) \rightleftarrows H^+(aq) + HCO_3^-(aq)$

The H^+ ions increase the acidity of the water. This has the side effect of reducing the concentration of carbonate ions (CO_3^{2-}), which are the building blocks of shells (**Figure 7.15**) and skeletons. This, in turn, interferes with the growth and survival of carbonate-using organisms ranging in size from tiny phytoplankton, which serve as major food sources for fish and marine mammals, to corals, the animals responsible for building huge coral reefs, the largest biological structures in the world.

Scientific research results such as these give us another reason to pause, seriously consider the negative impact of some of our modern practices on the world around us, and change some of those practices.

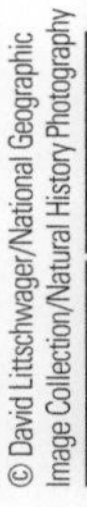

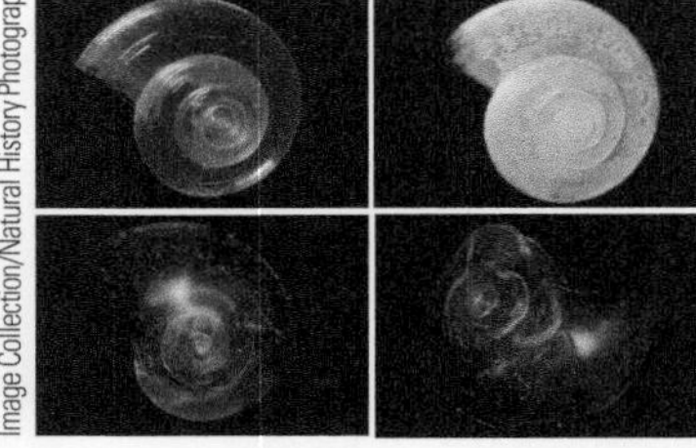

Figure 7.15 Ocean acidification caused by burning fossil fuels can damage the shells of sea animals. These images show the damage to shells after 45 days in the acidic sea water projected for the year 2100.

7.3B Naming Acids

The rules for naming binary ionic compounds were given in **Section 3.4**. Moreover, the rules for naming binary covalent compounds and ionic compounds that contain polyatomic ions were discussed in **Sections 3.6** and **3.8**, respectively. We now conclude our discussion of inorganic nomenclature by presenting the rules used to name hydrogen-containing compounds that behave as acids. Acids can be grouped into two different types.

Acids of the first type, represented by HCl, are compounds in which hydrogen is covalently bonded to a nonmetal. In acids of the second type, represented by HNO_3, hydrogen is covalently bonded to a polyatomic ion.

HCl is a binary covalent compound that is named hydrogen chloride by the rules given in **Section 3.6**. In fact, that is the correct name for HCl when it has not been dissolved in water, and it is represented in chemical equations by HCl(g). Compounds that have not been dissolved in water are said to be anhydrous (without water). However, when the gas is dissolved in water and represented in chemical equations by HCl(aq), it behaves as an acid and is given another name. The following rules are used to name acidic, aqueous solutions of such compounds:

1. The word *hydrogen* in the anhydrous compound name is dropped.
2. The prefix *hydro-* is attached to the stem of the name of the nonmetal that is combined with hydrogen.
3. The suffix *-ide* on the stem of the name of the nonmetal that is combined with hydrogen is replaced with the suffix *-ic*.
4. The word *acid* is added to the end of the name as a separate word.

Example 7.5 Naming Acids

Determine the name that would be given to each of the following binary covalent compounds in the anhydrous form and in the aqueous solution (hydrous) form.

a. HCl (stomach acid)
b. H_2S (a gas produced naturally in hot springs; see **Figure 7.16**)

Yongyut Kumsri/Shutterstock.com

Figure 7.16 H_2S is found naturally in hot springs, crude petroleum, and natural gas.

Solution

a. The name of the anhydrous compound was just given as hydrogen chloride. The name of the aqueous solution is obtained by dropping *hydrogen* from the anhydrous compound name and adding the prefix *hydro-* to the stem *chlor*. The suffix *-ide* on the stem *chlor* is replaced by the suffix *-ic* to give the name *hydrochloric*. The word *acid* is added, giving the final name *hydrochloric acid* for the aqueous solution.
b. According to the rules of **Section 3.6**, the anhydrous compound name is hydrogen sulfide. The first two steps in obtaining the name of the aqueous solution are to drop *hydrogen* and to add the prefix *hydro-* to the stem *sulf*. However, in acids involving sulfur as the nonmetal combined with hydrogen, the stem *sulf* is replaced by the entire name *sulfur* for pronunciation reasons. The next steps involve dropping the suffix *-ide*, adding the suffix *-ic*, and adding the word *acid*. The resulting name of the aqueous solution is *hydrosulfuric acid*.

✔ **Learning Check 7.5** Determine the name that would be given to each of the following binary covalent compounds in the hydrous and anhydrous forms.

a. HI b. HBr c. HF

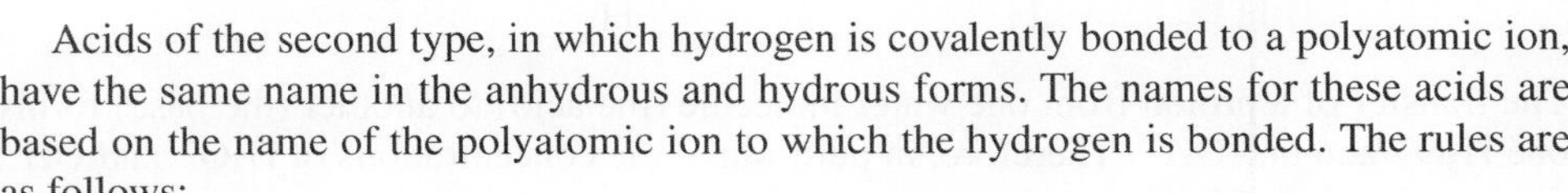

Acids of the second type, in which hydrogen is covalently bonded to a polyatomic ion, have the same name in the anhydrous and hydrous forms. The names for these acids are based on the name of the polyatomic ion to which the hydrogen is bonded. The rules are as follows:

1. All hydrogens that are written as the first part of the formula of the acid are removed. The hydrogens are removed in the form of H^+ ions.
2. The polyatomic ion that remains after the H^+ ions are removed is named by referring to sources such as **Table 3.5**.
3. When the remaining polyatomic ion has a name ending in the suffix *-ate*, the suffix is replaced by the suffix *-ic*, and the word *acid* is added.
4. When the remaining polyatomic ion has a name ending in the suffix *-ite*, the suffix is replaced by the suffix *-ous*, and the word *acid* is added.
5. If the polyatomic ion contains sulfur or phosphorus, the stems *-sulf* or *-phosph* that remain in Steps 3 or 4, when the suffixes *-ate* or *-ite* are replaced, are expanded for pronunciation reasons to *-sulfur* and *-phosphor* before the suffixes *-ic* or *-ous* are added.

Example 7.6 Naming Acids Containing Polyatomic Ions

Compounds derived from the acid H_3PO_4 serve numerous important functions in the body, including the control of the acidity of urine (**Figure 7.17**) and body cells and the storage of energy in the form of ATP. Name this important acid.

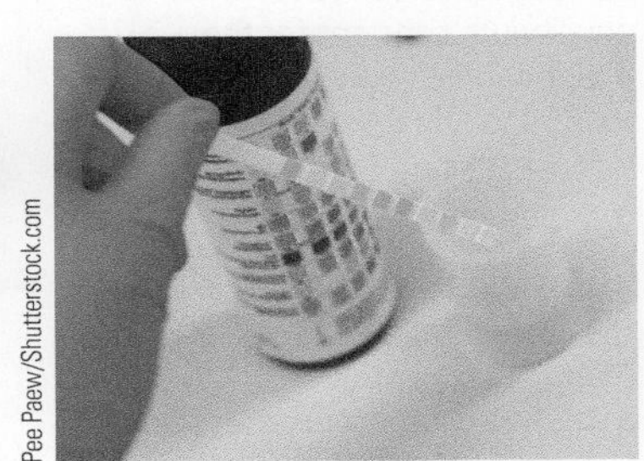
Pee Paew/Shutterstock.com

Figure 7.17 Compounds derived from H_3PO_4 help control the acidity of urine.

Solution

The removal of the three H^+ ions leaves behind the PO_4^{3-} polyatomic ion. This ion is named the phosphate ion in **Table 3.5**. According to Rules 1–4, the suffix *-ate* is replaced by the suffix *-ic* to give the name *phosphic acid*. However, Rule 5 must be used for this phosphorus-containing acid. The stem is expanded to *phosphor* to give the final name, *phosphoric acid*.

✔ **Learning Check 7.6** The acid H_2CO_3 is involved in many processes in the body, including removal of CO_2 gas produced by cellular metabolism and control of the acidity of various body fluids. Name this important acid.

7.4 The pH Concept

Learning Objective 7 Classify solutions as acidic, basic, or neutral based on hydroxide ion or hydronium ion concentrations.

Learning Objective 8 Calculate the pH of solutions given hydroxide ion or hydronium ion concentrations.

In Example 7.4a, water behaved as a Brønsted-Lowry base. In Example 7.4b, it was a Brønsted-Lowry acid. But what happens when only pure water is present? It turns out that water behaves as both an acid and a base and undergoes a self- or autoionization. The reaction expression representing this self-ionization is:

$$H_2O(\ell) + H_2O(\ell) \rightleftharpoons H_3O^+(aq) + OH^-(aq)$$

or

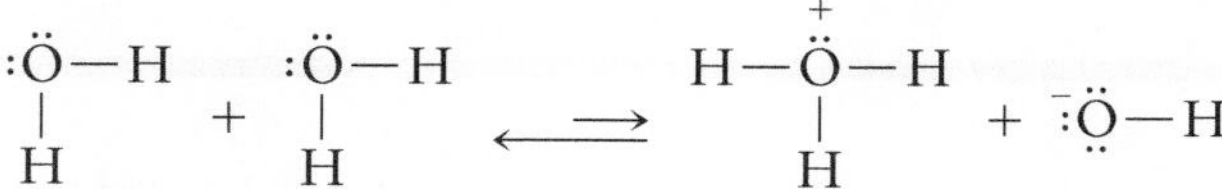

Figure 7.18 Pure water is considered to be neutral, neither an acid nor a base.

The transfer of a proton from one water molecule (the acid) to another (the base) forms one H_3O^+ and one OH^-. Therefore, in pure water, the concentrations of H_3O^+ and OH^- must be equal. At 25 °C, these concentrations are 10^{-7} mol/L (M). Thus, the equilibrium position is far to the left, as indicated by the unequal arrows in the reaction expressions. Unless noted otherwise, all concentrations and related terms in this chapter are given at 25 °C.

The term **neutral** is used to describe an aqueous solution (see **Figure 7.18**) in which the concentrations of H_3O^+ and OH^- are equal. Thus, pure water is neutral because each liter of pure water contains 10^{-7} mol H_3O^+ and 10^{-7} mol OH^- at equilibrium. Although all aqueous solutions are not necessarily neutral, it is true that in any solution that contains water, the product of the molar concentrations of H_3O^+ and OH^- is a constant. This becomes apparent by writing the equilibrium expression for the autoionization of water:

neutral A term used to describe any water solution in which the concentrations of H_3O^+ and OH^- are equal. Also, an aqueous solution with pH = 7.

$$K_w = [H_3O^+][OH^-]$$

where K_w is a new equilibrium constant called the **ion product of water**. As a pure liquid, water is kept out of the expression. At 25 °C, K_w can be evaluated from the measured values of $[H_3O^+]$ and $[OH^-]$ in pure water.

ion product of water The equilibrium constant for the dissociation of pure water into H_3O^+ and OH^-.

$$K_w = [H_3O^+][OH^-] = (1.0 \times 10^{-7}\,M)(1.0 \times 10^{-7}\,M) = 1.0 \times 10^{-14}\,(M)^2 \quad (7.2)$$

Equation 7.2 is valid not only for pure water but for any solution in which water is the solvent. Note that we are including units for equilibrium constants in this discussion. This makes the calculation of concentrations easier to follow and understand.

A solution is classified as **acidic** when the concentration of H_3O^+ is greater than the concentration of OH^-. In a **basic** or **alkaline solution**, the concentration of OH^- is greater than that of H_3O^+. However, the product of the molar concentrations of H_3O^+ and OH^- will be 1.0×10^{-14} $(M)^2$ in either case. Many acidic and basic materials are found in the home (see **Figure 7.19**).

acidic solution A solution in which the concentration of H_3O^+ is greater than the concentration of OH^-. Also, a solution in which pH is less than 7.

basic or alkaline solution A solution in which the concentration of OH^- is greater than the concentration of H_3O^+. Also, a solution in which pH is greater than 7.

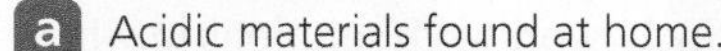
a Acidic materials found at home.

b Basic or alkaline materials found at home.

Figure 7.19 Acidic and basic materials are common in the home. Based on the photos, there is a common category to which most of the acidic materials belong, and one to which most of the basic materials belong. What are these two categories?

Example 7.7 Solution Classifications and Calculations

Calculate the molar concentration of the hydroxide ion when given the hydronium ion concentration and calculate the hydronium ion concentration when the hydroxide ion concentration is given. Classify each solution as acidic, basic, or neutral.

a. $[H_3O^+] = 1.0 \times 10^{-5}\,M$
b. $[OH^-] = 1.0 \times 10^{-6}\,M$

Solution a.

Step 1: Examine the question for what is given and what is unknown.

$$\text{Given: } [H_3O^+] = 1.0 \times 10^{-5}\,M \qquad \text{Unknown: } [OH^-] = ?$$

Step 2: Identify a formula that links the given and unknown terms.

$$\text{Equation 7.2: } [H_3O^+][OH^-] = 1.0 \times 10^{-14}\,(M)^2$$

Step 3: Enter the data into the formula.

$$(1.0 \times 10^{-5}\,M)[OH^-] = 1.0 \times 10^{-14}\,(M)^2$$

Rearranging gives

$$[OH^-] = \frac{1.0 \times 10^{-14}\,(M)^2}{1.0 \times 10^{-5}\,M} = 1 \times 10^{-9}\,M$$

The solution is acidic because $[H_3O^+] = 1.0 \times 10^{-5}$ M, which is greater than $[OH^-] = 1.0 \times 10^{-9}$.

Solution b.

Step 1: Examine the question for what is given and what is unknown.

$$\text{Given: } [OH^-] = 1 \times 10^{-6}\,M \qquad \text{Unknown: } [H_3O^+] = ?$$

Step 2: Identify a formula that links the given and the unknown terms.

$$\text{Equation 7.2: } [H_3O^+][OH^-] = 1.0 \times 10^{-14}\,(M)^2$$

Step 3: Enter the data into the formula.

$$[H_3O^+](1.0 \times 10^{-6}\,M) = 1.0 \times 10^{-14}\,(M)^2$$

Rearranging gives

$$[H_3O^+] = \frac{1.0 \times 10^{-14}\,(M)^2}{1.0 \times 10^{-6}\,M} = 1.0 \times 10^{-8}\,M$$

The solution is basic because $[OH^-] = 1.0 \times 10^{-6}$ M is greater than $[H_3O^+] = 1.0 \times 10^{-8}$ M.

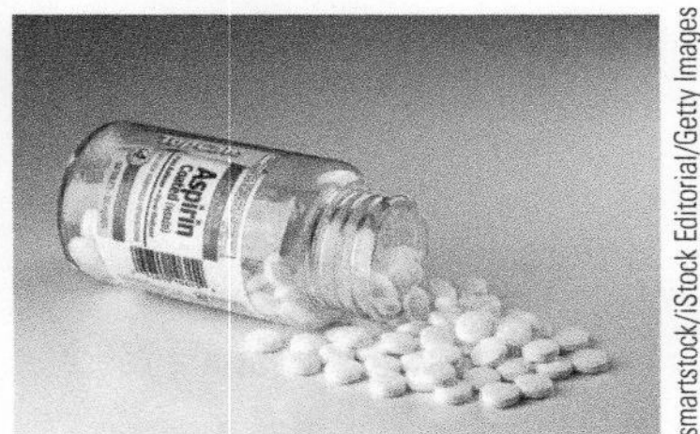

smartstock/iStock Editorial/Getty Images

Figure 7.20 Aspirin is a nonsteroidal anti-inflammatory drug (NSAID).

✔ Learning Check 7.7 Calculate the molar concentration of the ion whose concentration is not given and then classify each of the following solutions as acidic, basic, or neutral.

a. Milk of magnesia, used to relieve indigestion and heartburn, has $[OH^-] = 1 \times 10^{-5}$ M.
b. The nonsteroidal anti-inflammatory drug (NSAID) aspirin (**Figure 7.20**) in water gives $[H_3O^+] = 1 \times 10^{-5}$ M.

pH The negative logarithm of the molar concentration of H_3O^+ in a solution.

Chemists, technologists, and other laboratory personnel routinely work with solutions in which the H_3O^+ concentration may be anywhere from 10 to 10^{-14} M. Because of the inconvenience of working with numbers that extend over such a wide range, chemists long ago adopted a notation known as the pH. Mathematically, the **pH** is the negative logarithm (log) of $[H_3O^+]$:

$$pH = -\log[H_3O^+] \quad (7.3)$$

$$[H_3O^+] = 10^{-pH} \quad (7.4)$$

Thus, pH is simply the negative of the exponent used to express the hydronium ion concentration in moles per liter. If we know the pH but want to know $[H_3O^+]$, we can use Equation 7.4. **Figure 7.21** summarizes the mathematical relationships between pH, $[H_3O^+]$, and $[OH^-]$.

$$pH \underset{pH = -\log[H_3O^+]}{\overset{[H_3O^+] = 10^{-pH}}{\rightleftarrows}} [H_3O^+] \underset{[H_3O^+] = \frac{1 \times 10^{-14}}{[OH^-]}}{\overset{[OH^-] = \frac{1 \times 10^{-14}}{[H_3O^+]}}{\rightleftarrows}} [OH^-]$$

Figure 7.21 Diagram providing the mathematical relationships between pH, $[H_3O^+]$, and $[OH^-]$.

JGI/Jamie Grill/Getty Images

Figure 7.22 Milk has a pH of around 6.5 to 6.7, which makes it slightly acidic.

Example 7.8 Calculating pH Values

Cow's milk is slightly acidic (pH 6.5 to 6.7; see **Figure 7.22**). Calculate the pH of the following common beverages.

a. Nondairy milks like rice, soy, and almond have $[H_3O^+] = 1 \times 10^{-6}$ M.
b. Orange juice has $[OH^-] = 1 \times 10^{-10}$ M.

Solution a.

Step 1: Examine the question for what is given and what is unknown.

Given: $[H_3O^+] = 1 \times 10^{-6}$ M   Unknown: pH = ?

Step 2: Identify a formula that links the given and unknown terms.

Equation 7.3: $pH = -\log[H_3O^+]$

Step 3: Enter the data into the formula.

$pH = -\log(1 \times 10^{-6}\ M)$; pH = 6

Solution b.

Step 1: Examine the question for what is given and what is unknown.

Given: $[OH^-] = 1 \times 10^{-10}$ M   Unknown: pH = ?

To find pH, $[H_3O^+]$ must first be calculated.

Step 2: Identify formulas that link the given and unknown terms.

Equation 7.2: $[H_3O^+][OH^-] = 1.0 \times 10^{-14} (M)^2$

Equation 7.3: $pH = -\log[H_3O^+]$

Step 3: Enter the data into the formulas.

$$[H_3O^+](1 \times 10^{-10}\,M) = 1.0 \times 10^{-14}\,(M)^2$$

Rearranging gives

$$[H_3O^+] = \frac{1.0 \times 10^{-14}\,(M)^2}{1 \times 10^{-10}\,M} = 1 \times 10^{-4}\,M$$

Now enter that value for $[H_3O^+]$ into Equation 7.3.

$$pH = -\log(1 \times 10^{-4}) = 4$$

✔ Learning Check 7.8 Calculate the pH of the following solutions.

a. A urine sample has $[H_3O^+] = 1 \times 10^{-6}\,M$.
b. A household ammonia cleaner has $[OH^-] = 1 \times 10^{-3}\,M$.

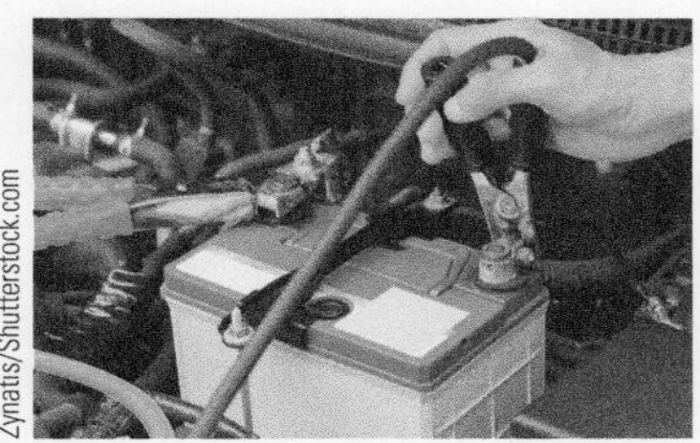

Zynatis/Shutterstock.com

Figure 7.23 The acid in automobile lead-acid batteries is sulfuric acid, H_2SO_4, and is highly corrosive.

The pH value of 6 obtained in Example 7.8a for nondairy milk corresponds to a solution in which the H_3O^+ concentration is greater than the OH^- concentration. Thus, the solution is acidic. Any solution with a pH less than 7 is classified as acidic. Any solution with a pH greater than 7 is classified as basic or alkaline. The pH values of some familiar solutions are listed in **Table 7.2**. Among them, the most acidic is battery acid (see **Figure 7.23**).

Table 7.2 Relationship between $[H_3O^+]$, $[OH^-]$, and pH

$[H_3O^+]$	$[OH^-]$	pH	Examples (Solids Are Dissolved in Water)
10^0	10^{-14}	0	Battery acid
10^{-1}	10^{-13}	1	Gastric juice
10^{-2}	10^{-12}	2	Lemon juice
10^{-3}	10^{-11}	3	Vinegar, carbonated drinks, aspirin
10^{-4}	10^{-10}	4	Orange juice, apple juice
10^{-5}	10^{-9}	5	Black coffee
10^{-6}	10^{-8}	6	Normal urine, milk, liquid detergent
10^{-7}	10^{-7}	7	Saliva, pure water, blood
10^{-8}	10^{-6}	8	Soap, baking soda
10^{-9}	10^{-5}	9	Milk of magnesia
10^{-10}	10^{-4}	10	Phosphate-free detergent
10^{-11}	10^{-3}	11	Household ammonia
10^{-12}	10^{-2}	12	Liquid household cleaner
10^{-13}	10^{-1}	13	Oven cleaner
10^{-14}	10^0	14	Liquid drain cleaner

Example 7.9 Calculating $[H_3O^+]$ and $[OH^-]$ from pH

Perspiration (**Figure 7.24**) is found at moderately acidic to neutral pH levels, typically between 4.5 and 7.0. Determine the H_3O^+ and OH^- molar concentrations that correspond to a perspiration pH of 5.0.

iStock.com/wundervisuals

Figure 7.24 Body sweat, or perspiration, is often slightly acidic, having a pH in the range of 4.5 to 7.0.

Solution

Step 1: Examine the question for what is given and what is unknown.

Given: pH = 5.0 Unknown: $[H_3O^+] = ?$ and $[OH^-] = ?$

Step 2: Identify a formula that links the given and unknown terms.
Calculate $[H_3O^+]$ first, using Equation 7.4: $[H_3O^+] = 10^{-pH}$
To calculate $[OH^-]$, use Equation 7.2: $[H_3O^+][OH^-] = 1.0 \times 10^{-14}\ (M)^2$

Step 3: Enter the data into the formulas.

$$[H_3O^+] = 1.0 \times 10^{-pH} = 1.0 \times 10^{-5}\ M$$

Enter the value of $[H_3O^+]$ into Equation 7.2.

$$(1.0 \times 10^{-5}\ M)\ [OH^-] = 1.0 \times 10^{-14}\ (M)^2$$

Rearranging gives

$$[OH^-] = \frac{1.0 \times 10^{-14}\ (M)^2}{1.0 \times 10^{-5}\ M} = 1.0 \times 10^{-9}\ M$$

✔ **Learning Check 7.9** Determine the values of $[H_3O^+]$ and $[OH^-]$ that correspond to the following pH values.

a. Gastric juice has pH = 1.0
b. A laundry detergent has pH = 10.0

7.5 The Strengths of Acids and Bases

Learning Objective 9 Use equilibrium constants to rank the strengths of acids and bases.

strong acids and **strong bases** Acids and bases that dissociate (ionize) completely when dissolved to form a solution.

weak (or **moderately weak**) **acids** and **bases** Acids and bases that dissociate (ionize) less than completely when dissolved to form a solution.

When salts dissolve in water, they generally dissociate completely, but this is not true for all acids and bases. The acids and bases that do dissociate almost completely are classified as **strong acids** and **strong bases** (they are also strong electrolytes). Those that dissociate to a much smaller extent are called **weak** or **moderately weak**, depending on the degree of dissociation (they are also weak or moderately weak electrolytes). Examples of strong and weak acids are listed in **Table 7.3**.

Table 7.3 Some Common Strong and Weak Acids

Name	Formula	% Dissociation[a]	K_a	Classification
Hydrochloric acid	HCl	100	Very large	Strong
Hydrobromic acid	HBr	100	Very large	Strong
Nitric acid	HNO_3	100	Very large	Strong
Sulfuric acid	H_2SO_4	100	Very large	Strong
Phosphoric acid	H_3PO_4	28	7.5×10^{-3}	Moderately weak
Hydrofluoric acid	HF	8.1	6.6×10^{-4}	Moderately weak
Sulfurous acid[b]	H_2SO_3	34	1.5×10^{-2}	Moderately weak
Acetic acid	$HC_2H_3O_2$	1.3	1.8×10^{-5}	Weak
Boric acid	H_3BO_3	0.01	7.3×10^{-10}	Weak
Carbonic acid[b]	H_2CO_3	0.2	4.3×10^{-7}	Weak
Nitrous acid[b]	HNO_2	6.7	4.6×10^{-4}	Weak

[a] Based on dissociation of one proton in 0.1 M solution at 25 °C.

[b] Unstable acid.

According to **Table 7.3**, 100% of the HCl dissociates into H_3O^+ and Cl^-. Thus, the concentration of H_3O^+ in a 0.10 M HCl solution is 0.10 M, and the pH is 1.00. The strength of acids and bases is shown quantitatively by the value of the equilibrium constant for the dissociation reaction in aqueous solutions. The reaction expression for the dissociation of an acid (HA) in water is as follows:

$$HA(aq) + H_2O(\ell) \rightleftarrows H_3O^+(aq) + A^-(aq)$$

where A^- represents the conjugate base of the acid. The equilibrium constant for this reaction is called the **acid dissociation constant (K_a)**:

acid dissociation constant (K_a) The equilibrium constant for the dissociation of an acid.

$$K_a = \frac{[H_3O^+][A^-]}{[HA]} \tag{7.5}$$

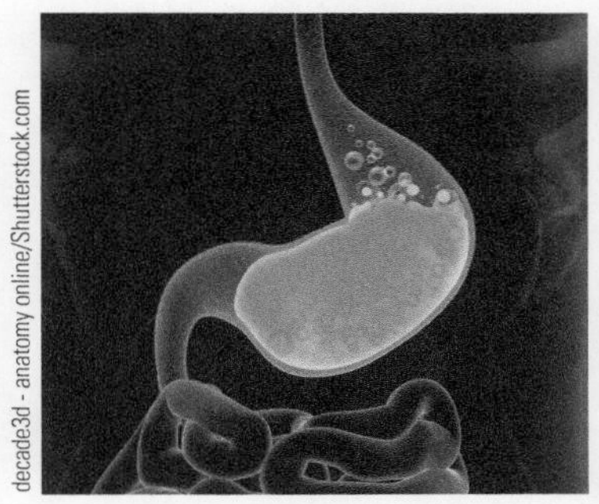

Figure 7.25 Gastric juice, or stomach acid, is approximately 0.05% HCl.

In Equation 7.5, the brackets, again, represent molar concentrations of the species in the solution, and $[H_3O^+]$ and $[A^-]$ are, respectively, the equilibrium concentrations of the hydronium ion and the conjugate base of the acid. [HA] represents the concentration of that part of the dissolved acid that remains undissociated in the equilibrium mixture. Recall that water is a pure liquid, so it is not included in the equilibrium expression.

K_a values for strong acids are quite large, while K_a values for weak acids are quite small. In solutions of strong acids, $[H_3O^+]$ and $[A^-]$ values are quite large, while [HA] has a value near 0, so K_a is relatively bigger. In weak acids, [HA] has larger values, while $[H_3O^+]$ and $[A^-]$ are smaller, so K_a is relatively smaller. Thus, the larger a K_a value, the stronger the acid it represents. Remember that the terms *weak* and *strong* apply to the extent of dissociation, not to the concentration of an acid or base. For example, gastric juice (0.05% HCl) is a dilute (not weak) solution of a strong acid (**Figure 7.25**).

Acid behavior is linked to the loss of protons. Thus, acids must contain hydrogen atoms that can be removed to form H^+. **Monoprotic acids** can lose only one proton per molecule, whereas **diprotic** and **triprotic** acids can lose two and three, respectively. For example, HCl is monoprotic, H_2SO_4 is diprotic, and H_3PO_4 is triprotic (see **Figure 7.26**). Diprotic and triprotic acids dissociate in steps, as shown for H_2SO_4 in the following reaction expressions:

monoprotic acid An acid that gives up only one proton (H^+) per molecule when dissolved.

diprotic acid An acid that gives up two protons (H^+) per molecule when dissolved.

triprotic acid An acid that gives up three protons (H^+) per molecule when dissolved.

$$H_2SO_4(aq) + H_2O(\ell) \rightleftharpoons H_3O^+(aq) + HSO_4^-(aq)$$

$$HSO_4^-(aq) + H_2O(\ell) \rightleftharpoons H_3O^+(aq) + SO_4^{2-}(aq)$$

The second proton is not as easily removed as the first because it must be pulled away from a negatively charged species, HSO_4^-. Accordingly, HSO_4^- is a weaker acid than H_2SO_4.

The number of ionizable hydrogens cannot always be determined from the molecular formula for an acid. For example, acetic acid (CH_3COOH) is monoprotic even though the molecule contains four hydrogen atoms. Only the hydrogen bonded to the oxygen is ionizable. Because the O—H bond is polar (**Section 5.2**), the less electronegative hydrogen is partially positive, making it easier to be removed by the base. **Table 7.4** provides other examples, with the ionizable hydrogens shown in color.

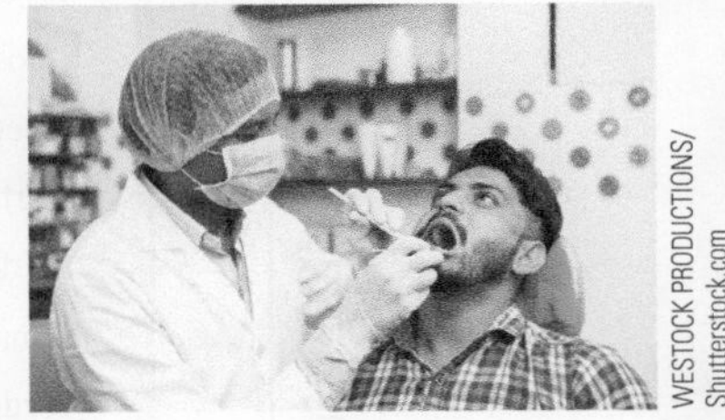

Figure 7.26 Phosphoric acid, H_3PO_4, is used in dental cements.

We have focused our attention on acids, using the extent of dissociation as a way to determine whether they are relatively strong or weak (**Table 7.5**). However, most acid dissociations are reversible to some degree. Thus, the anions produced by the forward reaction (the conjugate base) can themselves behave as Brønsted-Lowry bases in the reverse reaction. So, it would be helpful to also understand the equilibrium between acids and bases from the perspective of the base, even just qualitatively. For example, a strong acid completely dissociates to its conjugate base. But if this process is in equilibrium, then the reverse reaction must not occur to any measurable extent. This implies, then, that the conjugate base of a strong acid is weak (i.e., unreactive); it is stable as a conjugate base and does not use a pair of electrons to add a proton.

Table 7.4 Examples of Monoprotic, Diprotic, and Triprotic Acids

Name	Formula	Structural Formula	Classification
Butyric acid	$HC_4H_7O_2$	H H H O |		‖ H—C—C—C—C—O—H |	| H H H	Monoprotic
Nitric acid	HNO_3	O ‖ $^+$N—O—H | O_-	Monoprotic
Carbonic acid	H_2CO_3	O ‖ H—O—C—O—H	Diprotic
Phosphoric acid	H_3PO_4	O ‖ H—O—P—O—H | O | H	Triprotic

On the other hand, a weak acid dissociates incompletely to its conjugate base. In this equilibrium, the reverse reaction occurs to a greater extent than the forward reactions. As a result, the conjugate base of a weak acid is strong and reactive. It readily uses its electrons to add a proton, thereby acting as a Brønsted-Lowry base.

Table 7.5 Relative Strengths of Conjugate Acid–Base Pairs

(Increasing acid strength ↑ — Increasing base strength ↓)

Acid		Conjugate Base	
Name	**Formula**	**Formula**	**Name**
Hydrochloric acid	HCl	Cl^-	chloride ion
Sulfuric acid	H_2SO_4	HSO_4^-	hydrogen sulfate ion
Nitric acid	HNO_3	NO_3^-	nitrate ion
Hydronium ion	H_3O^+	H_2O	water
Phosphoric acid	H_3PO_4	$H_2PO_4^-$	dihydrogen phosphate
Acetic acid	CH_3COOH	CH_3COO^-	acetate ion
Carbonic arid	H_2CO_3	HCO_3^-	hydrogen carbonate
Hydrogen sulfide	H_2S	HS^-	hydrogen sulfide ion
Ammonium ion	NH_4^+	NH_3	ammonia
Hydrogen cyanide	HCN	CN^-	cyanide ion
Water	H_2O	OH^-	hydroxide ion
Ammonia	NH_3	NH_2^-	amide ion
Hydrogen	H_2	H^-	hydride ion
Methane	CH_4	CH_3^-	methide ion

Ammonia (NH_3) and the anions of moderately strong acids are all weak bases. Ammonia in particular acts as a Brønsted-Lowry base in an aqueous environment (**Figure 7.27**) by accepting a proton from water:

$$NH_3(aq) + H_2O(\ell) \rightleftarrows NH_4^+(aq) + OH^-(aq)$$

While the reaction of ammonia with water produces a measurable amount of ammonium (NH_4^+) and hydroxide (OH^-) ions, the same is not true for the anions of the strongest acids. For example, chloride (Cl^-) reacts with water to produce hydrochloric acid (HCl) and hydroxide (OH^-). HCl dissociates 100% of the time, however, so it is inappropriate to write it as a product. By convention, then, we write:

$$Cl^-(aq) + H_2O \rightarrow \text{no reaction}$$

In contrast, strong bases produce large amounts of hydroxide (OH^-). The most common strong bases are the hydroxides of the Group IA(1) metals (NaOH, KOH, etc.), the hydroxides of the Group IIA(2) metals [$Mg(OH)_2$, $Ca(OH)_2$, etc.], hydride (H^-), and many compounds containing anions of carbon. The following dissociation of NaOH (an Arrhenius base) in water is representative of how hydroxides of the Groups IA(1) and IIA(2) metals react with water to produce hydroxide (OH^-):

$$NaOH(s) + H_2O(\ell) \rightleftarrows Na^+(aq) + OH^-(aq)$$

When hydride or compounds with negatively charged nitrogen or carbon atoms (e.g., amide or methide) react with water, they also act as Brønsted-Lowry bases, as is shown here for the reaction of hydride (H^-) in water:

$$H^-(aq) + H_2O(\ell) \rightleftarrows H_2(g) + OH^-(aq)$$

skynesher/iStock/Getty Images

Figure 7.27 Ammonia (NH_3) solutions are good household cleaning agents. Because ammonia evaporates quickly, it is often used in glass-cleaning products to help avoid streaking.

Example 7.10 Strengths of Acids and Bases

Which acid or base in each of the following pairs is stronger?

a. H_3PO_4 and $H_2PO_4^-$ (acid)
b. CH_3NH_2 and $Sr(OH)_2$ (base)
c. Ascorbic acid, C_6H_6O ($K_a = 7.9 \times 10^{-5}$) (**Figure 7.28**), and HNO_2 ($K_a = 4.0 \times 10^{-4}$) (acid)

Solution

a. H_3PO_4 is the stronger acid. In diprotic and triprotic acids, the second and third protons are sequentially harder to remove, so the corresponding acid is weaker.
b. The stronger bases are Group IA(1) and Group IIA(2) metal hydroxides, such as $Sr(OH)_2$, whereas the most common weak bases are derivatives of NH_3.
c. HNO_2 is the stronger acid because its K_a is larger than that of ascorbic acid, indicating HNO_2 dissociates more readily into H_3O^+ and the conjugate base.

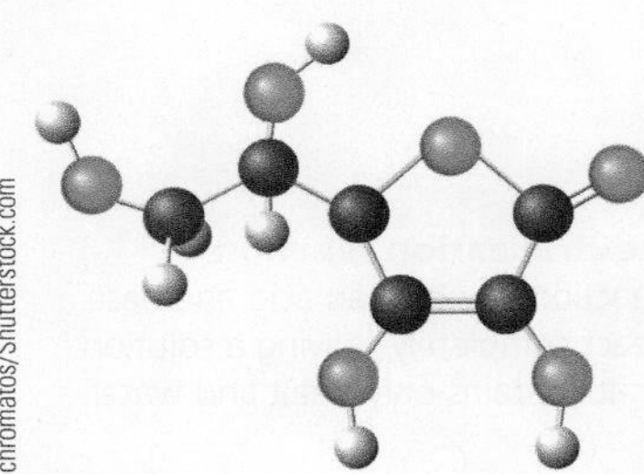

chromatos/Shutterstock.com

Figure 7.28 Ascorbic acid (vitamin C) is vital for blood vessel, cartilage, muscle, and collagen formation in the body. Red atoms represent oxygen, white atoms represent hydrogen, and grey atoms represent carbon in this ball-and-stick model of ascorbic acid.

✔ Learning Check 7.10 Which acid or base in each of the following pairs is stronger? Use **Table 7.5** as needed.

a. H_2SO_4 and HSO_4^- (acid)
b. NaOH and H^- (base)
c. CH_3COO^- and NO_3^- (base)

Health Connections 7.1

Heartburn Alert

The medical name for acid reflux disease is *gastroesophageal reflux disease* (GERD). The disease is often mistaken for occasional heartburn and treated with over-the-counter remedies. GERD is the result of a malfunctioning muscle [the lower esophageal sphincter (LES)] located at the bottom of the esophagus, just above the stomach. When operating normally, this muscle relaxes and opens to allow food to pass from the esophagus down into the stomach, then contracts to close the opening and prevent the acidic contents of the stomach from backing up into the esophagus.

Doctors often recommend lifestyle and dietary changes for most GERD patients, including the avoidance of foods and beverages that weaken the LES. These foods include chocolate, peppermint, fatty foods, coffee, and alcoholic beverages. The use of foods and beverages that can irritate a damaged esophageal lining, such as acidic fruits and juices, pepper, and tomato products, is also discouraged. GERD symptoms in overweight individuals often diminish when some weight is lost. Smokers who quit also generally gain some relief.

Two types of prescription medications that reduce the amount of acid in the stomach are also available. Both types, called H2 blockers and proton (acid) pump inhibitors, decrease the amount of acid secreted into the stomach, but they do so by different mechanisms.

GERD has many degrees of severity.

7.6 Acid–Base Titrations

Learning Objective 10 Write reaction expressions for the neutralization reactions between acids and bases.

Learning Objective 11 Perform titration calculations using balanced acid–base neutralization expressions.

neutralization reaction A reaction in which an acid and base react completely, leaving a solution that contains only a salt and water.

In most of the earliest acid–base reactions studied, the complete reaction of an acid with a base produced a neutral solution. For this reason, these reactions were often called **neutralization reactions**. Although it is now known that many "neutralization" reactions do not produce neutral solutions, the name is still used. The most common example of a neutralization reaction is between sodium hydroxide (NaOH) and hydrochloric acid (HCl).

Molecular equation:

$$HCl(aq) + NaOH(aq) \rightarrow NaCl(aq) + H_2O(\ell)$$

Total ionic equation:

$$H^+(aq) + Cl^-(aq) + Na^+(aq) + OH^-(aq) \rightarrow Na^+(aq) + Cl^-(aq) + H_2O(\ell)$$

Net ionic equation:

$$H^+(aq) + OH^-(aq) \rightarrow H_2O(\ell)$$

Thus, when an acid and a base are combined, a salt and water are always produced because the H^+ ions (from the acid) and the OH^- ions (from the base) combine to form water. The remaining (spectator) ions combine to form the salt.

The analysis of solutions for the total amount of acid or base they contain is a regular activity in many laboratories. The total amount of acid in a solution is indicated by its capacity to neutralize a base, and vice versa. We can take advantage of neutralization reactions to analyze how acidic or basic solutions are through a common laboratory procedure called **titration** (**Figure 7.29**).

Suppose the acidity of an unknown acid solution needs to be determined. To do this, a basic solution of known concentration (a so-called standard solution) is added via a buret to a known volume of the acid solution until the **equivalence point of the titration** is reached. This is the point in the neutralization reaction where all of the acid has been completely reacted with base. The volume of base required is determined from the difference between the initial and final readings of the buret.

To successfully complete a titration, the equivalence point must somehow be detected. One way to do this is to add an indicator (**Figure 7.30**) to the solution being titrated (i.e., the acid in our example). An indicator is a chemical that changes color depending on the pH of its surroundings. When the indicator changes color during the titration, the end of the titration has been reached and is referred to as the **end point of the titration**. Different indicators change colors at different pH values, which is convenient because not all neutralization reactions produce neutral solutions (pH = 7). The goal is to choose an indicator that changes color at the pH corresponding to the equivalence point of your particular acid–base titration. If indicators are not practical for a titration, then pH meters, which contain electrodes that can detect the pH of an unknown solution directly, can be used.

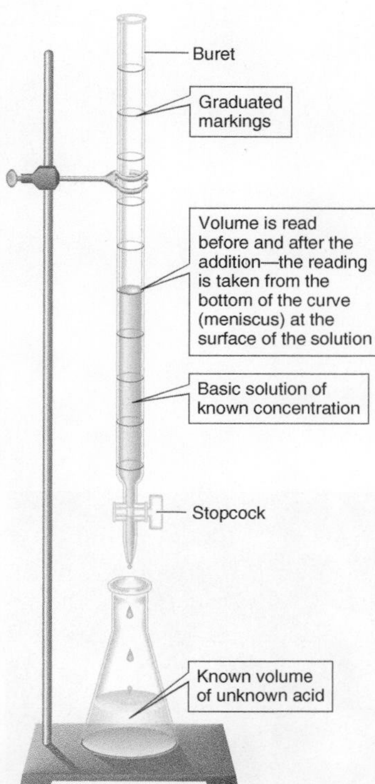

Figure 7.29 Representation of an acid–base titration apparatus.

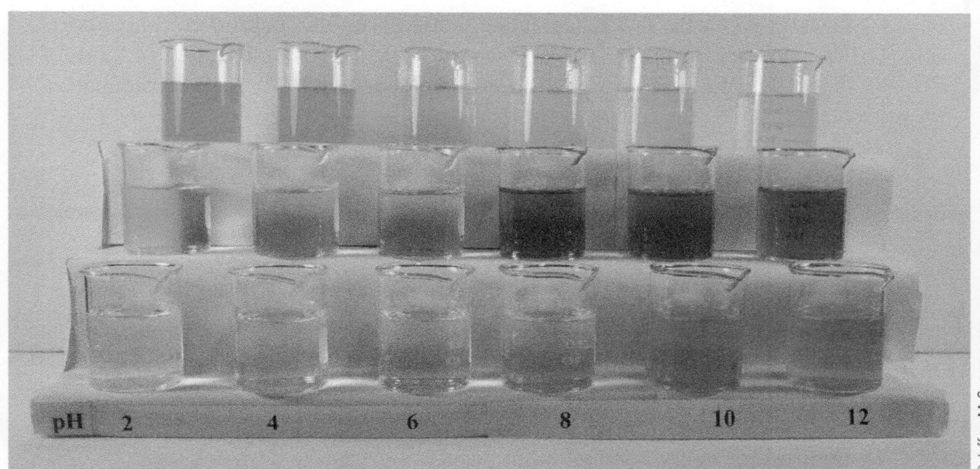

Figure 7.30 Indicators change color with changes in pH (the numbers across the bottom). Methyl red (top row) goes from red at low pH to yellow at higher pH. Bromothymol blue (middle row) goes from yellow to blue, and phenolphthalein (bottom row) goes from colorless to red. Which of the three indicators would be the best to use to differentiate between the solutions with pH values of 4 and 6?

titration An analytical procedure in which one solution (often a base) of known concentration is slowly added to a measured volume of an unknown solution (often an acid). The volume of the added solution is measured with a buret.

equivalence point of a titration The point at which the unknown solution has exactly reacted with the known solution. Neither is in excess.

end point of a titration The point at which the titration is stopped on the basis of an indicator color change or pH meter reading.

Example 7.11 Titration Calculations

A 25.0 mL sample of an HNO_3 solution requires the addition of 16.3 mL of 0.200 M NaOH to reach the equivalence point. Calculate the molarity of the HNO_3 solution. The chemical equation for the reaction is

$$HNO_3(aq) + NaOH(aq) \rightarrow NaNO_3(aq) + H_2O(\ell)$$

Solution

Step 1: Examine the equation for what is given and what is unknown.

Given: 25.0 mL of HNO_3 reacts with 16.3 mL of 0.200 M NaOH.

Unknown: M of HNO_3 = ?

Step 2: Identify a formula that links the given and unknown terms. We know the volume of acid solution reacted. If we also knew the number of moles of acid in the volume of reacted solution, we could calculate the solution molarity by using Equation 6.8:

$$M = \frac{\text{moles of solute}}{\text{liters of solution}}$$

Therefore, our task will be to calculate the number of moles of acid reacted. The balanced equation for the reaction shows that 1 mol of HNO_3 requires 1 mol of NaOH. Thus, by calculating the moles of NaOH, we know the moles of HNO_3 reacted.

Step 3: Calculate the moles of NaOH using the method of dimensional analysis.

$$16.3\ \cancel{\text{mL NaOH}}\ \text{solution} \times \frac{1\ \cancel{\text{L}}}{1000\ \cancel{\text{mL}}} \times \frac{0.200\ \cancel{\text{mol NaOH}}}{1\ \cancel{\text{L NaOH}}} \times \frac{1\ \text{mol HNO}_3}{1\ \cancel{\text{mol NaOH}}}$$

$$= 0.00326\ \text{mol HNO}_3$$

Singkham/Shutterstock.com

Figure 7.31 Phosphoric acid is reacted to form the phosphate salts used in fertilizers.

Step 4: Calculate the molarity of HNO_3. Remember to convert the HNO_3 volume to liters in your calculation.

$$\frac{0.00326\ \text{mol HNO}_3}{25.0\ \cancel{\text{mL HNO}_3}} \times \frac{1000\ \cancel{\text{mL HNO}_3}}{1\ \text{L HNO}_3} = 0.130\ \text{M HNO}_3$$

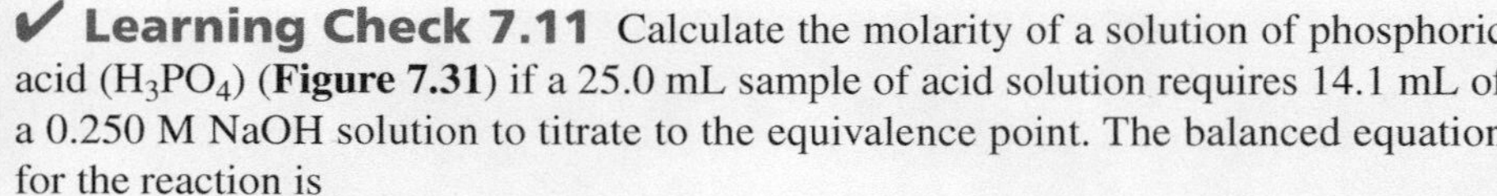

✔ **Learning Check 7.11** Calculate the molarity of a solution of phosphoric acid (H_3PO_4) (**Figure 7.31**) if a 25.0 mL sample of acid solution requires 14.1 mL of a 0.250 M NaOH solution to titrate to the equivalence point. The balanced equation for the reaction is

$$H_3PO_4(aq) + 3NaOH(aq) \rightarrow Na_3PO_4(aq) + 3H_2O(aq)$$

7.7 Maintaining pH: The Balancing Act

Learning Objective 12 Write reaction expressions to illustrate how buffers work.

Learning Objective 13 Determine the pH of buffers.

Learning Objective 14 Describe respiratory and urinary control of blood pH.

buffer A solution with the ability to resist changing pH when acids (H^+) or bases (OH^-) are added.

acidosis An abnormally low blood pH.

alkalosis An abnormally high blood pH.

7.7A Biological Buffers

A **buffer** is a solution that has the ability to resist changes in pH when acids or bases are added. This is particularly important in biological systems like you! For example, the human circulatory and respiratory processes operate only within a very narrow range of blood pH—namely, 7.35 to 7.45 (**Figure 7.32**). A decrease in pH, called **acidosis**, or an increase in pH, called **alkalosis**, is serious and requires prompt attention. Death can result when the pH value falls below 6.8 or rises above 7.8.

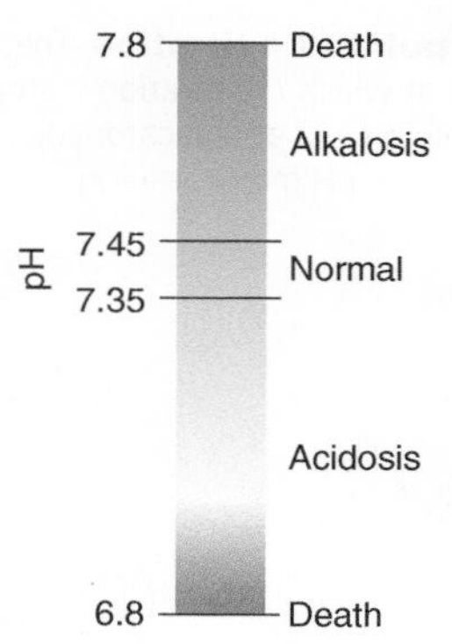

Figure 7.32 Human circulatory and respiratory processes operate only within a very narrow range of blood pH (i.e., 7.35 to 7.45).

Large amounts of acids and smaller amounts of bases normally enter the blood. Some mechanisms must neutralize or eliminate these substances if the blood pH is to remain constant. In practice, a constant pH is maintained by the interactive operation of buffers, the respiratory system, and the urinary system. Only when all three parts of this complex mechanism function in tandem can the acid–base balance be maintained.

Most buffers consist of a pair of compounds, one with the ability to react with and neutralize H_3O^+ and the other with the ability to react with and neutralize OH^-. In this way, the buffer solution resists changes in pH in either direction. The amount of

H_3O^+ or OH^- that a buffer system can absorb without allowing significant pH changes to occur is called its **buffer capacity**. Three major buffer systems prevent the buffer capacity from being exceeded within blood: the bicarbonate buffer, the phosphate buffer, and the plasma proteins (**Figure 7.33**). The most important of these is the bicarbonate buffer system, consisting of a mixture of bicarbonate ions (HCO_3^-) and carbonic acid (H_2CO_3).

buffer capacity The amount of acid (H^+) or base (OH^-) that can be absorbed by a buffer without causing a significant change in pH.

FatCamera/E+/Getty Images

Figure 7.33 Blood buffers enable the body to maintain normal blood pH levels under a range of human activities.

As an example of buffers in action, consider the fate of lactic acid (HLac) entering the bloodstream. When lactic acid dissolves in the blood, it dissociates to produce H_3O^+ and lactate ions (Lac^-). The liberated H_3O^+ ions drive down the pH of the blood:

$$\text{HLac(aq)} + \text{H}_2\text{O}(\ell) \rightleftharpoons \text{H}_3\text{O}^+\text{(aq)} + \text{Lac}^-\text{(aq)}$$

The bicarbonate ions (HCO_3^-) of the bicarbonate buffer system react with the excess H_3O^+ and return the pH to the normal range:

$$\text{HCO}_3^-\text{(aq)} + \text{H}_3\text{O}^+\text{(aq)} \rightleftharpoons \text{H}_2\text{CO}_3\text{(aq)}$$

In cases where the scenario is reversed and base is introduced to the bloodstream, the carbonic acid protects against added OH^-:

$$\text{H}_2\text{CO}_3\text{(aq)} + \text{OH}^-\text{(aq)} \rightleftharpoons \text{HCO}_3^-\text{(aq)} + \text{H}_2\text{O}(\ell)$$

If large amounts of H_3O^+ or OH^- are added to a buffer, the buffer capacity can be exceeded, in which case the buffer system is overwhelmed, and the pH changes. For example, if large amounts of H_3O^+ were added to the bicarbonate ion–carbonic acid buffer, bicarbonate (HCO_3^-) would react with hydronium (H_3O^+) until it was depleted. The pH would then drop as additional H_3O^{++} ions were added. In blood, the concentration of HCO_3^- is 10 times the concentration of H_2CO_3. Thus, this buffer has a greater capacity against added acid than against bases. This is consistent with the normal functions of the body that allow larger amounts of acidic substances rather than basic substances to enter the blood.

To calculate the pH of buffers, we begin with Equation 7.5:

$$K_a = \frac{[\text{H}_3\text{O}^+][\text{A}^-]}{[\text{HA}]}$$

This equation can be rearranged to give:

$$[\text{H}_3\text{O}^+] = K_a \frac{[\text{HA}]}{[\text{A}^-]}$$

Application of the logarithm concept gives:

$$\text{pH} = \text{p}K_a + \log \frac{[\text{A}^-]}{[\text{HA}]} \qquad (7.6)$$

In this equation, $\text{pH} = -\log[\text{H}_3\text{O}^{++}]$ (Equation 7.4), and $\mathbf{p}\boldsymbol{K}_\mathbf{a}$ $= -\log K_a$. Equation 7.6 is known as the **Henderson-Hasselbalch equation**. It is often used by biologists, biochemists, and others who frequently work with buffers. We see from Equation 7.6 that when the concentrations of a weak acid [HA] and its conjugate base $[A^-]$ are equal in a solution, the pH of the solution is equal to pK_a. When it is desired to produce a buffer with a pH different from the exact pK_a, Equation 7.6 indicates that it can be done. An acid with a pK_a near the desired pH is selected, and the ratio of the concentrations of conjugate base (anion of the acid) and acid is adjusted to give the desired pH. **Table 7.6** lists some weak acids, together with values of K_a and pK_a as a reference guide.

pK_a The negative logarithm of K_a.

Henderson-Hasselbalch equation A relationship between the pH of a buffer, pK_a, and the concentrations of acid and salt in the buffer.

Table 7.6 K_a and pK_a Values for Selected Weak Acids

Name	Formula	K_a	pK_a
Acetic acid[a]	CH_3COOH	1.8×10^{-5}	4.74
Ammonium ion	NH_4^+	5.6×10^{-10}	9.25
Boric acid	H_3BO_3	7.3×10^{-10}	9.14
Carbonic acid	H_2CO_3	4.3×10^{-7}	6.37
Bicarbonate ion	HCO_3^-	5.6×10^{-11}	10.25
Citric acid[a]	$C_3H_4(OH)(COOH)_3$	8.4×10^{-4}	3.08
Dihydrogen citrate ion[a]	$C_3H_4(OH)(COOH)_2COO^-$	1.8×10^{-5}	4.74
Hydrogen citrate ion[a]	$C_3H_4(OH)(COOH)(COO)_2{}^{2-}$	4.0×10^{-6}	5.40
Formic acid[a]	$HCOOH$	1.8×10^{-4}	3.74
Lactic acid[a]	$C_2H_4(OH)COOH$	1.4×10^{-4}	3.85
Nitrous acid	HNO_2	4.6×10^{-4}	3.33
Phosphoric acid	H_3PO_4	7.5×10^{-3}	2.12
Dihydrogen phosphate ion	$H_2PO_4^-$	6.2×10^{-8}	7.21
Hydrogen phosphate ion	HPO_4^{2-}	2.2×10^{-13}	12.66
Sulfurous acid	H_2SO_3	1.5×10^{-2}	1.82
Bisulfite ion	HSO_3^-	1.0×10^{-7}	7.00

[a]The hydrogen that ionizes in organic acids and ions is a part of a carboxylic acid group, represented by COOH (Chapter 10).

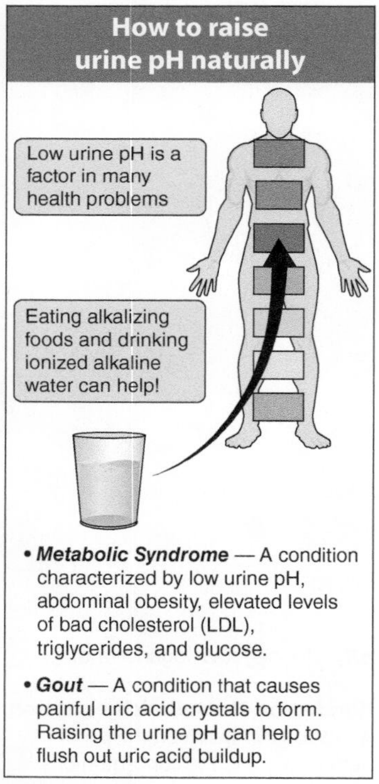

How to raise urine pH naturally

Low urine pH is a factor in many health problems

Eating alkalizing foods and drinking ionized alkaline water can help!

- ***Metabolic Syndrome*** — A condition characterized by low urine pH, abdominal obesity, elevated levels of bad cholesterol (LDL), triglycerides, and glucose.
- ***Gout*** — A condition that causes painful uric acid crystals to form. Raising the urine pH can help to flush out uric acid buildup.

Figure 7.34 Low urine pH can contribute to health problems such as metabolic syndrome and gout.

Example 7.12 Buffer Solution Calculations

A buffer system consisting of the ions $H_2PO_4^-$ and HPO_4^{2-} helps to control the pH of urine (**Figure 7.34**). What is the pH of a buffer in which the concentration of NaH_2PO_4 is 0.10 M and that of Na_2HPO_4 is 0.50 M? The pK_a for $H_2PO_4^-$ is 7.21.

Solution

Step 1: Examine the question for what is given and what is unknown.

Given: $[Na_2H_2PO_4] = 0.10$ M and $[Na_2HPO_4] = 0.05$M and
pK_a of $H_2PO_4^- = 7.21$ Unknown: pH = ?

Step 2: Identify a formula that links the given and unknown terms.
Both compounds are salts that produce the $H_2PO_4^-$ and HPO_4^{2-} ions. In such situations, the ion with more hydrogens will be the acid, and the one with fewer will be the anion or conjugate base. Therefore, the acid is $H_2PO_4^-$, and the conjugate base or anion is HPO_4^{2-}. Equation 7.6 links the given and unknown terms.

$$\mathrm{pH} = \mathrm{p}K_a + \log\frac{[A^-]}{[HA]} \text{ or } \mathrm{pH} = \mathrm{p}K_a + \log\frac{[HPO_4^{2-}]}{[H_2PO_4^-]}$$

Step 3: Enter the data into the formula.

$$\mathrm{pH} = 7.21 + \log\left[\frac{0.50\,\cancel{\mathrm{mol/L}}}{0.10\,\cancel{\mathrm{mol/L}}}\right]$$

$$\mathrm{pH} = 7.91$$

✔ Learning Check 7.12

a. What is the pH of a buffer solution in which formic acid (HCOOH) and sodium formate (HCOONa) are both at a concentration of 0.22 M?
b. What is the pH of a buffer solution that is 0.25 M in sulfurous acid (H_2SO_3) and 0.10 M in sodium bisulfite ($NaHSO_3$)?
c. A buffer system in the blood consists of the bicarbonate ion (HCO_3^-) and carbonic acid (H_2CO_3). What value of the ratio $[HCO_3^-]/[H_2CO_3]$ would be required to maintain blood at the average normal pH of 7.40?

7.7B Respiratory and Urinary Control of Blood pH

The respiratory system helps control the acidity of blood by regulating the elimination of carbon dioxide and water molecules. These molecules, exhaled in every breath, come from carbonic acid as follows:

$$\underset{\text{carbonic acid}}{H_2CO_3} \rightleftarrows H_2O + CO_2$$

As more CO_2 and H_2O are exhaled, more carbonic acid is removed from the blood, thus elevating the blood pH to a more alkaline level.

The respiratory mechanism for controlling blood pH begins in the brain with respiratory center neurons that are sensitive to blood CO_2 levels and pH. If the concentration of CO_2 in arterial blood increases, or if the pH of arterial blood decreases below 7.38, then breathing increases both in rate and depth, resulting in **hyperventilation**. This increased ventilation eliminates more carbon dioxide, reduces carbonic acid and hydronium ion concentrations, and increases the blood pH back toward the normal level (see **Figure 7.35**).

hyperventilation Rapid, deep breathing.

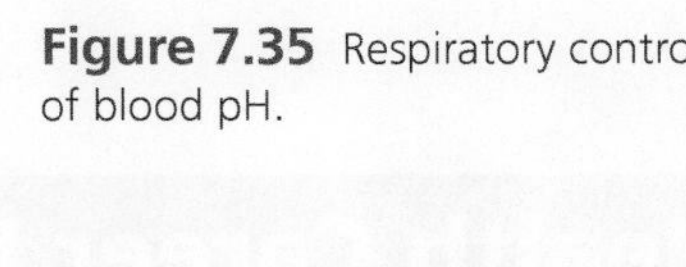

Figure 7.35 Respiratory control of blood pH.

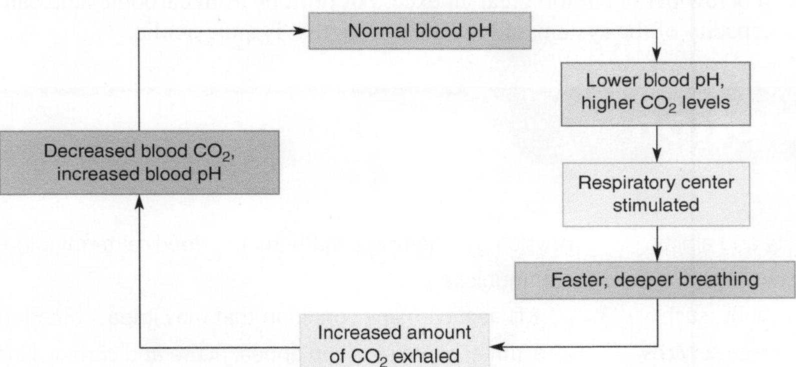

In the opposite situation, an increase in blood pH above normal causes **hypoventilation**, a reduced rate of respiration. Less CO_2 is exhaled, and the higher concentration of carbonic acid remaining in the blood lowers the pH back to normal.

hypoventilation Slow, shallow breathing.

In addition to respiratory control, the kidneys play a vital role in pH control because they can excrete varying amounts of acid and base. For example, with acidic blood (acidosis), the excretion of hydronium ions by the kidneys decreases the urine pH and simultaneously increases the blood pH back toward normal.

Figure 7.36 traces the reactions involved in the excretion of hydronium ions by the kidneys:

1. Carbon dioxide diffuses from the blood capillaries into the kidney distal tubule cells.
2. Catalyzed by the enzyme carbonic anhydrase, water and carbon dioxide react to give carbonic acid.
3. The carbonic acid ionizes to give hydronium ions and bicarbonate ions. The hydronium ions diffuse into the developing urine.
4. For every hydronium ion entering the urine, a sodium ion passes into the tubule cells to balance charge.
5. The sodium ions and bicarbonate ions enter the bloodstream capillaries.

Figure 7.36 Urinary control of blood pH in the kidney.

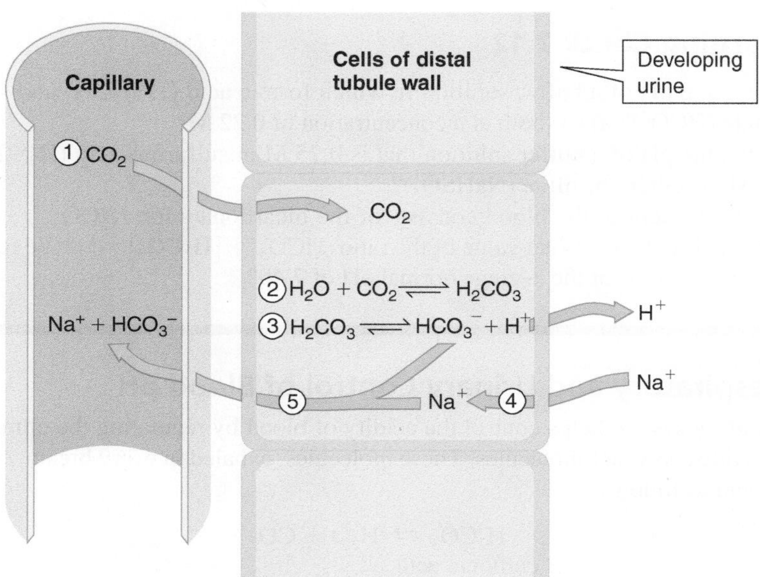

The net result of these reactions is the conversion of CO_2 to HCO_3^- within the blood. Both the decrease in CO_2 and the increase in HCO_3^- tend to increase blood pH levels back to normal. The developing urine has picked up the hydrogen ions, which now react with buffering ions such as hydrogen phosphate that are present in urine:

$$H^+ + HPO_4^{2-} \rightarrow H_2PO_4^-$$

The presence of the phosphate buffer system $H_2PO_4^-/HPO_4^{2-}$ usually keeps urine from going much below pH 6, but too great an excess of protons from carbonic acid can exceed the buffer capacity of the system and result in urine that is quite acidic.

Health Career Description

Respiratory Therapist

Respiratory therapists (RTs) have the responsibility and ability to keep patients breathing freely. RTs ensure patient health with an emphasis on breathing in a variety of settings, such as hospitals, care facilities, sleep centers, and hospice services. They diagnose disease; interview patients and conduct physical chest exams; consult with physicians and other health care personnel; and administer medications that enhance breathing such as albuterol or steroids, often referred to as "breathing treatments," via nebulizers and other medication routes.

RTs are responsible for managing the airway of patients who cannot breathe on their own. They provide airways in a variety of ways: RTs might "bag a patient," squeezing an apparatus that forces air into the lungs, or intubate or use manual maneuvers to maintain a patent airway. They analyze breath, blood, and tissue samples to diagnose, manage, and treat disease. They are expected to respond to emergency and urgent care calls as well as high-risk births. RTs have knowledge in a wide range of subjects, including anatomy, physiology, chemistry, mathematics, medical terminology, and communication skills.

RTs deal with any condition that may impact breathing, such as asthma, allergies, sleep apnea, acute and chronic illness, and trauma. They are often involved in providing community education on subjects such as smoking or air-quality issues.

In order to become a respiratory therapist, students must be accepted into an accredited respiratory therapy program. An associate's degree in health professions or in respiratory therapy with a GPA above 3.5 is helpful in gaining acceptance to a program. RT programs are demanding. One such program at Weber State University in Ogden, Utah, requires students to take 20 credit hours per semester, and working outside of the program is discouraged. After completion of the program, students will hold a bachelor's degree but must also pass a state licensing exam in order to practice. Master's degrees in respiratory therapy are also available and equip students to deal with special populations such as neonatal, pediatric, or intensive care.

Concept Summary

7.1 Principles of Chemical Reactions

Learning Objectives: **Use energy diagrams to represent and interpret the energy relationships for reactions. Explain how factors such as reactant concentrations, temperature, and catalysts influence reaction rates.**

- Explanations of how reactions take place are called reaction mechanisms.
- Most mechanisms are based on three assumptions:
  - Molecules must collide with one another.
  - The collision must involve a certain minimum of energy.
  - Some colliding molecules must be oriented in a specific way during collision in order to react.
- Energy relationships for reactions can be represented by energy diagrams, in which energy is plotted versus the reaction progress.
- Energy diagrams represent the concepts of exothermic and endothermic reactions and activation energy.
- Spontaneous processes take place naturally, with no apparent cause or stimulus.
- The speed of a reaction is called a reaction rate.
- Reaction rate can be determined by measuring how fast reactants are consumed or products are formed.
- Three factors affect the rates of all reactions:
  - Reactant concentrations
  - Reactant temperature
  - The presence of catalysts

7.2 Equilibrium

Learning Objectives: **Write equilibrium expressions based on reaction equations. Use Le Châtelier's principle to predict the influence of changes in concentration and reaction temperature on the position of equilibrium for reactions.**

- Reactions are in equilibrium when the rate of the forward reaction is equal to the rate of the reverse reaction.
- Equilibrium is emphasized in equations for reactions by writing double arrows pointing in both directions between reactants and products.
- The relative amounts of reactants and products present in a system at equilibrium define the position of equilibrium.
- The equilibrium position is toward the right when a large amount of product is present and toward the left when a large amount of reactant is present.
- The equilibrium position is indicated by the value of the equilibrium constant.
- Factors known to influence the position of equilibrium include changes in the amount of reactants and/or products and changes in temperature.
- The influence of such factors can be predicted by using Le Châtelier's principle.
- Catalysts cannot change the position of an equilibrium.

7.3 Acid–Base Equilibria

Learning Objectives: **Write reaction equations that illustrate Brønsted-Lowry acid–base behavior. Name acids.**

- Arrhenius acids dissociate in water to provide hydrogen ions (H^+).
- Arrhenius bases dissociate in water to provide hydroxide ions (OH^-).
- Johannes Brønsted and Thomas Lowry proposed that acids are hydrogen-containing substances capable of donating protons to other substances, whereas bases are substances that accept protons and form covalent bonds with them.
- When a substance behaves as a Brønsted-Lowry acid by donating a proton, it becomes a conjugate base.
- When a substance behaves as a Brønsted-Lowry base by accepting a proton, it becomes a conjugate acid.
- There are two ways to name acids.
  - Water solutions of binary covalent compounds containing hydrogen and a nonmetal are named following the pattern of hydro(stem)ic, where (stem) is the name of the nonmetal bonded to hydrogen.
  - Acids in which hydrogen is bonded to polyatomic ions have names based on the name of the polyatomic ion to which the hydrogen is attached.

7.4 The pH Concept

Learning Objectives: **Classify solutions as acidic, basic, or neutral based on hydroxide ion or hydronium ion concentrations. Calculate the pH of solutions given hydroxide ion or hydronium ion concentrations.**

- Water can behave both as a Brønsted-Lowry acid and a Brønsted-Lowry base.
- In pure water, water molecules will transfer protons from one water molecule (the acid) to another (the base). This process is called the autoionization of water.
- The pH is the negative logarithm of the molar H_3O^+ concentration of a solution.
- Solutions with pH values lower than 7 are acidic.
- Solutions with pH values higher than 7 are basic or alkaline.
- Solutions with a pH value of 7 are neutral.

7.5 The Strengths of Acids and Bases

Learning Objective: **Use equilibrium constants to rank the strengths of acids and bases.**

- Acids and bases that dissociate completely when dissolved in solution are called strong.
- Acids and bases that do *not* dissociate completely are called weak or moderately weak, depending on the degree of dissociation they undergo.
- Acid strength is indicated by the value of the equilibrium constant K_a.
- In general, polyprotic acids become weaker during each successive dissociation reaction.

- The Brønsted-Lowry base produced by the dissociation of an acid has a strength opposite that of the acid. That is, strong bases are produced by weak acids, and weak bases are produced by strong acids.
- The acid strengths of cations produced by the dissociation of bases follow a similar pattern.

7.6 Acid–Base Titrations

Learning Objectives: **Write reaction expressions for the neutralization reactions between acids and bases. Perform titration calculations using balanced acid–base neutralization expressions.**

- The neutralization reaction of acids and bases is used in titration, an experimental technique used to analyze acids and bases.
- During a typical titration, a base solution of known concentration is added slowly to a known volume of an acid solution of unknown concentration.
- The titration is stopped at the end point when a color change occurs in an indicator.
- The volumes of acid and base required are used to calculate the acid concentration.
- Titrations are used to determine the total amount of acid or base in solutions.

7.7 Maintaining pH: The Balancing Act

Learning Objectives: **Write reaction expressions to illustrate how buffers work. Determine the pH of buffers. Describe respiratory and urinary control of blood pH.**

- Solutions that maintain constant pH values when acid (H_3O^+) or base (OH^-) are added are called buffers.
- All buffers have a limit to the amount of acid or base they can absorb without changing pH. This limit is called buffer capacity.
- Blood pH is normally in the range of 7.35 to 7.45.
- Acidosis and alkalosis occur when the blood pH becomes lower or higher, respectively, than the normal values.
- The blood contains three major buffer systems to maintain normal pH levels: the bicarbonate system, the phosphate system, and the plasma protein system.
- The most important is the bicarbonate system, which consists of a mixture of bicarbonate ions and carbonic acid.
- The respiratory system influences blood pH by regulating the elimination of carbon dioxide and water.
- Because carbon dioxide and water are produced during the decomposition of carbonic acid, the greater the elimination of carbon dioxide and water from the body, the greater is the elimination of carbonic acid from the blood.
- The kidneys influence blood pH by varying the amount of hydrogen ions excreted in the urine.
- When the blood is too acidic, carbonic acid is converted to bicarbonate ions and hydrogen ions. When the hydrogen ions are excreted, and the bicarbonate ions enter the blood.

Key Terms and Concepts

Acid dissociation constant (K_a) (7.5)
Acidic solution (7.4)
Acidosis (7.7)
Activation energy (7.1)
Alkalosis (7.7)
Arrhenius acid (7.3)
Arrhenius base (7.3)
Basic or alkaline solution (7.4)
Brønsted-Lowry acid (7.3)
Brønsted-Lowry base (7.3)
Buffer (7.7)
Buffer capacity (7.7)
Catalyst (7.1)
Chemical equilibrium (7.2)
Chemical kinetics (7.1)
Chemical thermodynamics (7.1)
Collision model of chemical kinetics (7.1)
Conjugate acid–base pair (7.3)
Conjugate base (7.3)
Diprotic acid (7.5)
Effective collision (7.1)
End point of a titration (7.6)
Enthalpy (7.1)
Equilibrium concentrations (7.2)
Equilibrium constant (7.2)
Equilibrium expression (7.2)
Equivalence point of a titration (7.6)
Henderson-Hasselbalch equation (7.7)
Hydronium ion (7.3)
Hyperventilation (7.7)
Hypoventilation (7.7)
Inhibitor (7.1)
Ion product of water (7.4)
Le Châtelier's principle (7.2)
Monoprotic acid (7.5)
Neutral (7.4)
Neutralization reaction (7.6)
pH (7.4)
pK_a (7.7)
Position of equilibrium (7.2)
Reaction enthalpy (7.1)
Reaction rate (7.1)
State of equilibrium (7.2)
Strong acids and strong bases (7.5)
Titration (7.6)
Triprotic acid (7.5)
Weak (or moderately weak) acids and bases (7.5)

Key Equations

1. Equilibrium expression for general reaction (**Section 7.2**)	$aA + bB \rightleftarrows cD + dD$ $K_{eq} = \frac{[C]^c[D]^d}{[A]^a[B]^b}$	Generic Reaction Expression Equation 7.1
2. Relationship between $[H_3O^+]$ and $[OH^-]$ in water solutions (**Section 7.4**)	$K_w = [H_3O^+][OH^-] = (1.0 \times 10^{-7}\text{ M})(1.0 \times 10^{-7}\text{ M}) = 1.0 \times 10^{-14}(\text{M})^2$	Equation 7.2
3. Relationships between pH and $[H_3O^+]$ (**Section 7.4**)	$\text{pH} = -\log[H_3O^+]$ $[H_3O^+] = 1 \times 10^{-\text{pH}}$	Equation 7.3 Equation 7.4
4. Dissociation constant for acids (**Section 7.5**)	$HA(aq) + H_2O(\ell) \rightleftarrows H_3O^+(aq) + A^-(aq)$ $K_a = \frac{[H_3O^+][A^-]}{[HA]}$	Weak Acid Reaction Expression Equation 7.5
5. Calculation of buffer pH, Henderson-Hasselbalch equation (**Section 7.7**)	$\text{pH} = \text{p}K_a + \log\frac{[A^-]}{[HA]}$	Equation 7.6

Exercises

Even-numbered exercises are answered in Appendix B.

Exercises with an asterisk (*) are more challenging.

Principles of Chemical Reactions (Section 7.1)

7.1 Sketch energy diagrams to represent each of the following. Label the diagrams completely and tell how they are similar to each other and how they are different.

a. Exothermic reaction with activation energy

b. Exothermic reaction without activation energy

7.2 Sketch energy diagrams to represent each of the following. Label the diagrams completely and tell how they are similar to each other and how they are different.

a. Endothermic reaction with activation energy

b. Endothermic reaction without activation energy

7.3 Use energy diagrams to compare catalyzed and uncatalyzed reactions.

7.4 Describe two ways by which an increase in temperature increases a reaction rate.

7.5 Classify the following processes as spontaneous or nonspontaneous. Explain your answers in terms of whether energy must be continually supplied to keep the process going.

a. Water is decomposed into hydrogen and oxygen gas by passing electricity through the liquid.

b. An explosive detonates after being struck by a falling rock.

c. A coating of magnesium oxide forms on a clean piece of magnesium exposed to air.

d. A lightbulb emits light when an electric current is passed through it.

e. A cube of sugar dissolves in a cup of hot coffee.

7.6 Classify the following processes as spontaneous or nonspontaneous. Explain your answers in terms of whether energy must be continually supplied to keep the process going.

a. The space shuttle Orion leaves its pad and goes into orbit.

b. The fuel in a booster rocket of the space shuttle burns.

c. Water boils at 100 °C and 1 atm pressure.

d. Water temperature increases to 100 °C at 1 atm pressure.

e. Your bedroom becomes orderly.

7.7 Classify the following processes as exothermic or endothermic. Explain your answers.

a. Any combustion process

b. Perspiration evaporating from the skin

c. Melted lead solidifying

d. An explosive detonating

e. An automobile being pushed up a slight hill (from point of view of the automobile)

7.8 Classify the following processes as exothermic or endothermic. Explain your answers.

a. An automobile being pushed up a slight hill (from point of view of the one pushing)

b. Ice melting (from point of view of the ice)

c. Ice melting (from point of view of surroundings of the ice)

d. Steam condensing to liquid water (from point of view of the steam)

e. Steam condensing to liquid water (from point of view of surroundings of the steam)

7.9 Describe the observations or measurements that could be made to allow you to follow the rate of the following processes.

a. The diffusion of ink from a drop placed in a pan of quiet, undisturbed water

b. The loss of water from a pan of boiling water

c. The growth of a corn plant

7.10 Describe the observations or measurements that could be made to allow you to follow the rate of the following processes.

a. The melting of a block of ice

b. The setting (hardening) of concrete

c. The burning of a candle

7.11 A reaction is started by mixing reactants. As time passes, the rate decreases. Explain this behavior that is characteristic of most reactions.

7.12 Suppose you are running a reaction and want to speed it up. Describe three things you might try to do this.

Equilibrium (Section 7.2)

7.13 Describe the observation or measurement result that would indicate when each of the following had reached equilibrium.

a. $2CO + O_2 \rightleftarrows 2CO_2$
(2CO: colorless gas; O_2: colorless gas; $2CO_2$: colorless gas (apply Dalton's law))

b. $LiOH + CO_2 \rightleftarrows LiHCO_3$
(LiOH: colorless solid; CO_2: colorless gas; $LiHCO_3$: colorless solid)

c. paycheck → checking account → checks to pay bills

7.14 Describe the observation or measurement result that would indicate when each of the following had reached equilibrium.

a. $H_2 + I_2 \rightleftarrows 2HI$
(H_2: colorless gas; I_2: violet gas; 2HI: colorless gas)

b. solid sugar + water ⇄ sugar solution

c. $N_2 + 2O_2 \rightleftarrows 2NO_2$
(N_2: colorless gas; $2O_2$: colorless gas; $2NO_2$: red-brown gas)

7.15 Write an equilibrium expression for each of the following gaseous reactions.

a. $3O_2 \rightleftarrows 2O_3$

b. $COCl_2 \rightleftarrows CO + Cl_2$

c. $CS_2 + 4H_2 \rightleftarrows CH_4 + 2H_2S$

d. $2SO_2 + O_2 \rightleftarrows 2SO_3$

e. $CO + H_2O \rightleftarrows CO_2 + H_2$

7.16 Write an equilibrium expression for each of the following gaseous reactions.

a. $2CO + O_2 \rightleftarrows 2CO_2$

b. $N_2O_4 \rightleftarrows 2NO_2$

c. $2C_2H_6 + 7O_2 \rightleftarrows 4CO_2 + 6H_2O$

d. $2NOCl \rightleftarrows 2NO + Cl_2$

e. $2Cl_2O_5 \rightleftarrows O_2 + 4ClO_2$

7.17 The following equilibria are established in aqueous solutions. Write an equilibrium expression for each reaction.

a. $Ni^{2+} + 6NH_3 \rightleftarrows Ni(NH_3)_6{}^{2+}$

b. $Sn^{2+} + 2Fe^{3+} \rightleftarrows Sn^{4+} + 2Fe^{2+}$

c. $F_2 + 2Cl^- \rightleftarrows 2F^- + Cl_2$

7.18 The following equilibria are established in aqueous solutions. Write an equilibrium expression for each reaction.

a. $Fe^{3+} + 6CN^- \rightleftarrows Fe(CN)_6{}^{3-}$

b. $Ag^+ + 2NH_3 \rightleftarrows Ag(NH_3)_2{}^+$

c. $Au^{3+} + 4Cl^- \rightleftarrows AuCl_4{}^-$

7.19 Write a reaction equation that corresponds to each of the following equilibrium expressions.

a. $K = \dfrac{[PH_3][F_2]^3}{[HF]^3[PF_3]}$

b. $K = \dfrac{[O_2]^7[NH_3]^4}{[NO_2]^4[H_2O]^6}$

c. $K = \dfrac{[O_2][ClO_2]^4}{[Cl_2O_5]^2}$

d. $K = \dfrac{[N_2][H_2O]^2}{[NO]^2[H_2]^2}$

7.20 Write a reaction equation that corresponds to each of the following equilibrium expressions.

a. $K = \dfrac{[CO_2][H_2O]^2}{[CH_4][O_2]^2}$

b. $K = \dfrac{[CH_4][H_2O]}{[H_2]^3[CO]}$

c. $K = \dfrac{[O_2]^3}{[O_3]^2}$

d. $K = \dfrac{[NH_3]^4[O_2]^7}{[NO_2]^4[H_2O]^6}$

7.21 Consider the following equilibrium constants. Describe how you would expect the equilibrium concentrations of reactants and products to compare with each other (larger than, smaller than, etc.) for each case.

a. $K = 2.1 \times 10^{-6}$

b. $K = 0.15$

c. $K = 1.2 \times 10^8$

d. $K = 0.00036$

7.22 Consider the following equilibrium constants. Describe how you would expect the equilibrium concentrations of reactants and products to compare with each other (larger than, smaller than, etc.) for each case.

a. $K = 5.9$

b. $K = 3.3 \times 10^6$

c. $K = 2.7 \times 10^{-4}$

d. $K = 0.0000558$

7.23 Use Le Châtelier's principle to predict the direction of equilibrium shift and the changes that will be observed (color, amount of precipitate, etc.) in the following equilibria when the indicated stress is applied:

a. $heat + Co^{2+}(aq) + 4Cl^-(aq) \rightleftarrows CoCl_4{}^{2-}(aq)$;
(Co^{2+}: pink; Cl^-: colorless; $CoCl_4{}^{2-}$: blue)
The equilibrium mixture is heated.

b. $heat + Co^{2+}(aq) + 4Cl^-(aq) \rightleftarrows CoCl_4{}^{2-}(aq)$;
(Co^{2+}: pink; Cl^-: colorless; $CoCl_4{}^{2-}$: blue)
Co^{2+} is added to the equilibrium mixture.

c. $Fe^{3+}(aq) + 6SCN^-(aq) \rightleftarrows Fe(SCN)_6{}^{3-}(aq)$;
(Fe^{3+}: brown; SCN^-: colorless; $Fe(SCN)_6{}^{3-}$: red)
SCN^- is added to the equilibrium mixture.

d. $Pb^{2+}(aq) + 2Cl^-(aq) \rightleftarrows PbCl_2(s) + heat$;
(Pb^{2+}: colorless; Cl^-: colorless; $PbCl_2$: white solid)
Pb^{2+} is added to the equilibrium mixture.

e. $C_2H_4 + I_2 \rightleftarrows C_2H_4I_2 + heat$;
(C_2H_4: colorless gas; I_2: violet gas; $C_2H_4I_2$: colorless gas)
A catalyst is added to the equilibrium mixture.

7.24 Use Le Châtelier's principle to predict the direction of equilibrium shift and the changes that will be observed (color, amount of precipitate, etc.) in the following equilibria when the indicated stress is applied

a. $Cu^{2+}(aq) + 4NH_3(aq) \rightleftarrows Cu(NH_3)_4^{2+}(aq)$;
blue; colorless; dark purple
Some NH_3 is added to the equilibrium mixture.

b. $Pb^{2+}(aq) + 2Cl^-(aq) \rightleftarrows PbCl_2(s) + \text{heat}$;
colorless; colorless; white solid
The equilibrium mixture is cooled.

c. $C_2H_4 + I_2 \rightleftarrows C_2H_4I_2 + \text{heat}$;
colorless gas; violet gas; colorless gas
Some $C_2H_4I_2$ is removed from the equilibrium mixture.

d. $C_2H_4 + I_2 \rightleftarrows C_2H_4I_2 + \text{heat}$;
colorless gas; violet gas; colorless gas
The equilibrium mixture is cooled.

e. $\text{heat} + 4NO_2 + 6H_2O \rightleftarrows 7O_2 + 4NH_3$;
brown gas; colorless gas; colorless gas; colorless gas
A catalyst is added, and NH_3 is added to the equilibrium mixture.

7.25* Tell what will happen to each equilibrium concentration in the following when the indicated stress is applied and a new equilibrium position is established.

a. $H^+(aq) + HCO_3^-(aq) \rightleftarrows H_2O(\ell) + CO_2(g)$;
HCO_3^- is added.

b. $CO_2(g) + H_2O(\ell) \rightleftarrows H_2CO_3(aq) + \text{heat}$;
CO_2 is removed.

c. $CO_2(g) + H_2O(\ell) \rightleftarrows H_2CO_3(aq) + \text{heat}$;
The system is cooled.

7.26* Tell what will happen to each equilibrium concentration in the following when the indicated stress is applied and a new equilibrium position is established.

a. $LiOH(s) + CO_2(g) \rightleftarrows LiHCO_3(s) + \text{heat}$;
CO_2 is removed.

b. $2NaHCO_3(s) + \text{heat} \rightleftarrows Na_2O(s) + 2CO_2(g) + H_2O(g)$;
The system is cooled.

c. $CaCO_3(s) + \text{heat} \rightleftarrows CaO(s) + CO_2(g)$;
The system is cooled.

7.27 The gaseous reaction $2HBr(g) \rightleftarrows H_2(g) + Br_2(g)$ is endothermic. Tell which direction the equilibrium will shift for each of the following.

a. Some H_2 is added.

b. The temperature is increased.

c. Some Br_2 is removed.

d. A catalyst is added.

e. Some HBr is removed.

f. The temperature is decreased, and some HBr is removed.

7.28 The gaseous reaction $N_2(g) + 3H_2(g) \rightleftarrows 2NH_3(g)$ is exothermic. Tell which direction the equilibrium will shift for each of the following.

a. Some N_2 is added.

b. The temperature is increased.

c. Some NH_3 is removed.

d. Some H_2 is removed.

e. A catalyst is added.

f. The temperature is increased, and some H_2 is removed.

7.29 The equilibrium constant for the reaction $PCl_5 \rightleftarrows PCl_3 + Cl_2$ is 0.0245 at 250 °C. What molar concentration of PCl_5 would be present at equilibrium if the concentrations of PCl_3 and Cl_2 were both 0.250 M?

7.30 At 448 °C, the equilibrium constant for the reaction $H_2 + I_2 \rightleftarrows 2HI$ is 50.5. What concentration of I_2 would be found in an equilibrium mixture in which the concentrations of HI and H_2 were 0.500 M and 0.050 M, respectively?

7.31 A sample of gaseous BrCl is allowed to decompose in a closed container at 25 °C.

$$2BrCl(g) \rightleftarrows Br_2(g) + Cl_2(g)$$

When the reaction reaches equilibrium, the following concentrations are measured: $[BrCl] = 0.38$ M, $[Cl_2] = 0.26$ M, $[Br_2] = 0.26$ M. Evaluate the equilibrium constant for the reaction at 25 °C.

7.32 A mixture of the gases NOCl, Cl_2, and NO is allowed to reach equilibrium at 25 °C. The measured equilibrium concentrations are $[NO] = 0.92$ mol/L, $[NOCl] = 1.31$ mol/L, and $[Cl_2] = 0.20$ mol/L. What is the value of the equilibrium constant at 25 °C for the following reaction?

$$2NOCl(g) \rightleftarrows 2NO(g) + Cl_2(g)$$

Acid–Base Equilibria (Section 7.3)

7.33 Write the dissociation equations for the following that emphasize their behavior as Arrhenius acids.

a. HI

b. HBrO

c. HCN

d. $HClO_2$

7.34 Write the dissociation equations for the following that emphasize their behavior as Arrhenius acids.

a. $HBrO_2$

b. HS^-

c. HBr

d. $HC_2H_3O_2$ (only the first listed H dissociates)

7.35 Identify each Brønsted-Lowry acid and base, and conjugate acid–base pair, in the following equations. Note that the reactions are assumed to be reversible.

a. $HBr(aq) + H_2O(\ell) \rightleftarrows H_3O^+(aq) + Br^-(aq)$

b. $H_2O(\ell) + N_3^-(aq) \rightleftarrows HN_3(aq) + OH^-(aq)$

c. $H_2S(aq) + H_2O(\ell) \leftrightarrows H_3O^+(aq) + HS^-(aq)$

d. $SO_3^{2-}(aq) + H_2O(\ell) \rightleftarrows HSO_3^-(aq) + OH^-(aq)$

e. $HCN(aq) + H_2O(\ell) \leftrightarrows H_3O^+(aq) + CN^-(aq)$

7.36 Identify each Brønsted-Lowry acid and base, and conjugate acid–base pair. Note that the reactions are assumed to be reversible.

a. $HC_2O_4^-(aq) + H_2O(\ell) \rightleftharpoons H_3O^+(aq) + C_2O_4^{2-}(aq)$

b. $HNO_2(aq) + H_2O(\ell) \rightleftharpoons H_3O^+(aq) + NO_2^-(aq)$

c. $PO_4^{3-}(aq) + H_2O(\ell) \rightleftharpoons HPO_4^{2-}(aq) + OH^-(aq)$

d. $H_2SO_3(aq) + H_2O(\ell) \rightleftharpoons HSO_3^-(aq) + H_3O^+(\ell)$

e. $F^-(aq) + H_2O(\ell) \rightleftharpoons HF(aq) + OH^-(aq)$

7.37 Write equations to represent the Brønsted-Lowry acid behavior for each of the following acids in water solution. Remember to represent the reactions as being reversible.

a. HI **b.** HBrO

c. HCN **d.** HSe^-

7.38 Write equations to represent the Brønsted-Lowry acid behavior for each of the following acids in water solution. Remember to represent the reactions as being reversible.

a. HF **b.** $HClO_3$

c. HClO **d.** HS^-

7.39 Write a formula for the conjugate base formed when each of the following behaves as a Brønsted-Lowry acid.

a. HSO_3^- **b.** HPO_4^{2-}

c. $HClO_3$ **d.** $CH_3NH_3^+$

e. $H_2C_2O_4$

7.40 Write a formula for the conjugate base formed when each of the following behaves as a Brønsted-Lowry acid.

a. HSO_4^- **b.** $CH_3NH_3^+$

c. $HClO_4$ **d.** NH_4^+

e. HCl

7.41 Write a formula for the conjugate acid formed when each of the following behaves as a Brønsted-Lowry base.

a. NH_2^- **b.** CO_3^{2-}

c. OH^- **d.** $(CH_3)_2NH$

e. NO_2^-

7.42 Write a formula for the conjugate acid formed when each of the following behaves as a Brønsted-Lowry base.

a. HCO_3^- **b.** S^{2-}

c. HS^- **d.** $HC_2O_4^-$

e. $HN_2O_2^-$

7.43 The following reactions illustrate Brønsted-Lowry acid–base behavior. Complete each equation.

a. $HI(aq) + ? \rightarrow H_3O^+(aq) + I^-(aq)$

b. $NH_3(\ell) + ? \rightarrow NH_4^+ + NH_2^-$

c. $H_2C_2O_4(aq) + H_2O(\ell) \rightarrow ? + HC_2O_4^-(aq)$

d. $H_2N_2O_2(aq) + H_2O(\ell) \rightarrow H_3O^+(aq) + ?$

e. $? + H_2O(\ell) \rightarrow H_3O^+(aq) + CO_3^{2-}(aq)$

7.44 The following reactions illustrate Brønsted-Lowry acid–base behavior. Complete each equation.

a. $H_2AsO_4^-(aq) + NH_3(aq) \rightarrow NH_4^+(aq) + ?$

b. $? + H_2O(\ell) \rightarrow C_6H_5NH_3^+(aq) + OH^-(aq)$

c. $S^{2-}(aq) + ? \rightarrow HS^-(aq) + OH^-(aq)$

d. $? + HBr(aq) \rightarrow (CH_3)_2NH_2^+(aq) + Br^-(aq)$

e. $CH_3NH_2(aq) + HCl(aq) \rightarrow ? + Cl^-(aq)$

7.45* Write equations to illustrate the acid–base reaction when each of the following pairs of Brønsted-Lowry acids and bases are combined.

	Acid	Base
a.	HOCl	H_2O
b.	$HClO_4^-$	NH_3
c.	H_2O	NH_2^-
d.	H_2O	OCl^-
e.	HC_2O_4	H_2O

7.46* Write equations to illustrate the acid–base reaction when each of the following pairs of Brønsted-Lowry acids and bases are combined.

	Acid	Base
a.	H_3O^+	NH_2^-
b.	$H_2PO_4^-$	NH_3
c.	$HS_2O_3^-$	ClO^-
d.	H_2O	ClO_4^-
e.	H_2O	NH_3

7.47 Name the following acids. Refer to **Table 3.5** as needed.

a. $H_2Se(aq)$ **b.** $HClO_3$

c. H_2SO_4 **d.** HNO_3

7.48 Name the following acids. Refer to **Table 3.5** as needed.

a. $H_2Te(aq)$ **b.** HClO

c. H_2SO_3 **d.** HNO_2

7.49 The acid $H_3C_6H_5O_7$ forms the citrate ion, $C_6H_5O_7^{3-}$, when all three hydrogens are removed. Citrate salts are a class of drugs known as urinary alkalinizers, which are used to make urine less acidic. Name the acid.

7.50 The acid $H_2C_4H_4O_4$ forms the succinate ion, $C_4H_4O_4^{2-}$, when both hydrogens are removed. Succinate is used in treating symptoms of menopause and as a topical aid for joint pain. Name $H_2C_4H_4O_4$ as an acid.

The pH Concept (Section 7.4)

7.51 Classify solutions with the following characteristics as acidic, basic, or neutral.

a. Phosphate-free detergent, pH = 10.0

b. Tomato juice, pH = 4.0

c. Saliva, pH = 6.8

d. Household ammonia, pH = 11.6

7.52 Classify solutions with the following characteristics as acidic, basic, or neutral.

a. Vinegar, pH = 2.8

b. Bile, pH = 8

c. Black coffee, pH = 5

d. Household bleach, pH = 11

7.53 Below are the H_3O^+ concentrations of common foods. Calculate the OH^- concentrations of these items.

a. Artichokes, 1.0×10^{-6}

b. Jonathan apples, 4.8×10^{-4}

c. Frozen strawberries, 0.0051

d. Fresh beef, 1.0×10^{-7}

e. Crackers, 6.3×10^{-9}

7.54 Below are the H_3O^+ concentrations of common foods. Calculate the OH^- concentrations of these items.

a. Crab, 1.0×10^{-7}

b. Fruit cocktail, 1.3×10^{-4}

c. Fresh lemon juice, 0.0063

d. Cottage cheese, 1.0×10^{-5}

e. Egg whites, 5.0×10^{-9}

7.55 Below are the OH^- concentrations of common foods. Calculate the H_3O^+ concentrations of these items.

a. Frozen whole eggs, 1.0×10^{-6}

b. Devil's food cake, 5.0×10^{-7}

c. Angel's food cake, 2.6×10^{-9}

d. Fresh grapefruit, 1.3×10^{-11}

e. Egg white solids, 1.0×10^{-7}

7.56 Below are the OH^- concentrations of common foods. Calculate the H_3O^+ concentrations of these items.

a. Pomegranates, 1×10^{-11}

b. Mangos, 1.3×10^{-10}

c. Most freshwater fish, 1.0×10^{-7}

d. Sponge cake, 3.2×10^{-7}

e. Canned tomatoes, 5.0×10^{-11}

7.57 The pH values listed in **Table 7.2** are generally the average values for the listed materials. Most natural materials, such as body fluids and fruit juices, have pH values that cover a range for different samples. Some measured pH values for specific body fluid samples are given below. Convert each one to $[H_3O^+]$, and classify the fluid as acidic, basic, or neutral.

a. Blood, pH = 7.41

b. Gastric juice, pH = 1.60

c. Urine, pH = 5.93

d. Saliva, pH = 6.85

e. Pancreatic juice, pH = 7.85

7.58 The pH values listed in **Table 7.2** are generally the average values for the listed materials. Most natural materials, such as body fluids and fruit juices, have pH values that cover a range for different samples. Some measured pH values for specific body fluid samples are given below. Convert each one to $[H_3O^+]$, and classify the fluid as acidic, basic, or neutral.

a. Bile, pH = 8.05

b. Vaginal fluid, pH = 3.93

c. Semen, pH = 7.38

d. Cerebrospinal fluid, pH = 7.40

e. Perspiration, pH = 6.23

7.59 The ideal pH for a public swimming pool is considered by many to be pH 7.5. Outside the range of pH 7.2 to 7.8, exposed eyes will become bloodshot, and other adverse effects may be evident. Calculate the following.

a. The $[H_3O^+]$ value of pH 7.5

b. The $[H_3O^+]$ value of pH 8.5

c. The multiple differences there are in $[H_3O^+]$ between pH 7.5 and pH 8.5

d. $\frac{[H_3O^+]\text{ pH } 7.5}{[H_3O^+]\text{ pH } 8.5}$

7.60 The pH for a public swimming pool is usually pH 7.2 to 7.8. Outside that range, exposed eyes will become bloodshot and other adverse effects may be evident. Calculate the following.

a. The $[H_3O^+]$ value of pH 7.2

b. The $[H_3O^+]$ value of pH 7.8

c. $\frac{[H_3O^+]\text{ pH } 7.2}{[H_3O^+]\text{ pH } 7.8}$

7.61 Convert the following pH values into both $[H_3O^+]$ and $[OH^-]$ values.

a. pH = 9.00

b. pH = 6.27

c. pH = 3.10

7.62 Convert the following pH values into both $[H_3O^+]$ and $[OH^-]$ values.

a. pH = 3.95

b. pH = 4.00

c. pH = 11.86

The Strengths of Acids and Bases (Section 7.5)

7.63 The K_a values for four weak acids are as follows:

$$\text{Acid } A\text{: } (K_a = 5.6 \times 10^{-5})$$
$$\text{Acid } B\text{: } (K_a = 1.8 \times 10^{-5})$$
$$\text{Acid } C\text{: } (K_a = 1.3 \times 10^{-4})$$
$$\text{Acid } D\text{: } (K_a = 1.1 \times 10^{-3})$$

a. Arrange the four acids in order of increasing acid strength (weakest first, strongest last).

b. Arrange the conjugate bases of the acids (identify as base *A*, etc.) in order of increasing base strength (weakest base first, strongest last).

7.64 K_a values for four weak acids are given below:

$$\text{Acid } A\text{: } (K_a = 2.6 \times 10^{-4})$$
$$\text{Acid } B\text{: } (K_a = 3.7 \times 10^{-5})$$
$$\text{Acid } C\text{: } (K_a = 5.8 \times 10^{-4})$$
$$\text{Acid } D\text{: } (K_a = 1.5 \times 10^{-3})$$

a. Arrange the four acids in order of increasing acid strength (weakest first, strongest last).

b. Arrange the conjugate bases of the acids (identify as base *A*, etc.) in order of increasing base strength (weakest base first, strongest last).

7.65 Write dissociation reactions and K_a expressions for the following weak acids.

a. hypobromous acid, $HBrO$

b. sulfurous acid, H_2SO_3 (first H only)

c. hydrogen sulfite ion, HSO_3^-

d. hydroselenic acid, H_2Se (first H only)

e. arsenic acid, H_3AsO_4 (first H only)

7.66 Write dissociation reactions and K_a expressions for the following weak acids.

a. Hydrogen selenide ion, HSe^-

b. Dihydrogen borate ion, $H_2BO_3^-$ (first H only)

c. Hydrogen borate ion, HBO_3^{2-}

d. Hydrogen arsenate ion, $HAsO_4^{2-}$

e. Hypochlorous acid, $HClO$

7.67 Equal molar solutions are made of three monoprotic acids, HA, HB, and HC. The pH values of the solutions are, respectively, 4.82, 3.16, and 5.47. Rank the acids in order of increasing acid strength and explain your reasoning.

7.68 If someone asked you for a weak acid solution, which of the following would you provide, according to definitions in this chapter?

a. 0.05 M HCl

b. 20% acetic acid

If the individual really wanted the other solution, what term should have been used instead of *weak*?

7.69 Arsenic acid (H_3AsO_4) is a moderately weak triprotic acid. Write equations showing its stepwise dissociation. Which of the three anions formed in these reactions will be the strongest Brønsted-Lowry base? Which will be the weakest Brønsted-Lowry base? Explain your answers.

Acid–Base Titrations (Section 7.6)

7.70* Write balanced molecular, total ionic, and net ionic equations to represent neutralization reactions between RbOH and the following acids. Use all H's possible for each acid.

a. H_3PO_4

b. $H_2C_2O_4$ (oxalic acid)

c. $HC_2H_3O_2$

7.71* Write balanced molecular, total ionic, and net ionic equations to represent neutralization reactions between RbOH and the following acids. Use all H's possible for each acid.

a. HCl

b. HNO_3

c. H_2SO_4

7.72 Aluminum hydroxide, $Al(OH)_3$, is an ingredient in some antacids. Write a reaction for the neutralization of stomach acid, HCl, with $Al(OH)_3$.

7.73 In one of the mechanisms for protection of the stomach, the cells lining the stomach release HCO_3^-, bicarbonate ions, to react with stomach acid, HCl. Write a reaction for $NaHCO_3$ and HCl.

7.74 Oxalic acid ($H_2C_2O_4$) and phosphoric acid (H_3PO_4) are polyprotic acids that can form more than one salt, depending on the number of H's that react with base. Oxalic acid is found in plants such as spinach, cabbage, broccoli, and brussel sprouts. Phosphoric acid is added to foods and some cola beverages to provide a tangy, sour taste. Write balanced molecular, total ionic, and net ionic equations to represent the following neutralization reactions between KOH and

a. H_3PO_4 (react two H's)

b. H_3PO_4 (react three H's)

c. $H_2C_2O_4$ (react one H)

7.75 Carbonic acid (H_2CO_3) and phosphoric acid (H_3PO_4) are polyprotic acids that can form more than one salt, depending on the number of H's that react with base. In the body, carbonic acid plays an important role in maintaining acid–base homeostasis via the bicarbonate-buffer system. Phosphoric acid is added to foods and some cola beverages to provide a tangy, sour taste. Write balanced molecular, total ionic, and net ionic equations to represent the following neutralization reactions between KOH and

a. H_2CO_3 (react only one H)

b. H_2CO_3 (react both H's)

c. H_3PO_4 (react only one H)

7.76 Determine the number of moles of NaOH that could be neutralized by each of the following.

a. 250.0 mL of 0.400 M HBr

b. 750.0 mL of 0.300 M $HClO_4$

7.77 Determine the number of moles of NaOH that could be neutralized by each of the following.

a. 1.00 L of 0.250 M HCl

b. 500. mL of 0.300 M HNO_3

7.78 Oxalic acid, $H_2C_2O_4$, is present in a number of foodstuffs but can be toxic if ingested in larger amounts. For this reason, rhubarb leaves are not consumed. A 25.00 mL sample of $H_2C_2O_4$ solution required 43.88 mL of 0.1891 M NaOH solution to titrate it to the equivalence point. Calculate the molarity of the $H_2C_2O_4$ solution.

7.79* A 25.00 mL sample of gastric juice is titrated with a 0.0210 M NaOH solution. The titration to the equivalence point requires 26.4 mL of NaOH solution. If the equation for the reaction is

$$HCl(aq) + NaOH(aq) \rightarrow NaCl(aq) + H_2O(\ell)$$

what is the molarity of HCl in the gastric juice?

7.80 The following acid solutions were titrated to the equivalence point with the base listed. Use the titration data to calculate the molarity of each acid solution.

a. 5.00 mL of dilute H_2SO_4 required 29.88 mL of a 1.17 M NaOH solution.

b. 10.00 mL of vinegar (acetic acid) required 35.62 mL of a 0.250 M KOH solution.

c. 10.00 mL of muriatic acid (HCl) used to clean brick and cement required 20.63 mL of a 6.00 M NaOH solution.

7.81* The following acid solutions were titrated to the equivalence point with the base listed. Use the titration data to calculate the molarity of each acid solution.

a. 25.00 mL of HI solution required 27.15 mL of a 0.250 M NaOH solution.

b. 20.00 mL of H_2SO_4 solution required 11.12 mL of a 0.109 M KOH solution.

c. 25.00 mL of gastric juice (HCl) required 18.40 mL of a 0.0250 M NaOH solution.

Maintaining pH: The Balancing Act (Section 7.7)

7.82 Why are two components needed in buffering systems?

7.83 Could a mixture of ammonia (NH_3), a weak base, and ammonium chloride (NH_4Cl) behave as a buffer when dissolved in water? Use reaction equations to justify your answer.

7.84 Write a reaction expression to illustrate how a mixture of sodium hydrogen phosphate (Na_2HPO_4) and sodium dihydrogen phosphate (NaH_2PO_4) could function as a buffer when dissolved in water. Remember that phosphoric acid (H_3PO_4) ionizes in three steps.

7.85 Calculate the pH of a buffer made by dissolving 1 mol formic acid (HCOOH) and 1 mol sodium formate (HCOONa) in 1 L of solution (see **Table 7.6**).

7.86 **a.** Calculate the pH of a buffer that is 0.1 M in lactic acid, $C_2H_4(OH)COOH$, and 0.1 M in sodium lactate, $C_2H_4(OH)COONa$.

b. What is the pH of a buffer that is 1 M in lactic acid and 1M in sodium lactate?

c. What is the difference between the buffers described in parts a and b?

7.87 Which of the following acids and its conjugate base would you use to make a buffer with a pH of 3.00? Explain your reasons. Formic acid, lactic acid, nitrous acid.

7.88 Calculate the pH of buffers that contain the acid and conjugate base concentrations listed below.

a. $[CH_3COOH] = 0.40$ M, $[CH_3COO^-] = 0.25$ M

b. $[H_2PO_4^-] = 0.10$ M, $[HPO_4^{2-}] = 0.40$ M

c. $[HSO_3^-] = 1.50$ M, $[SO_3^{2-}] = 0.20$ M

7.89 Calculate the pH of buffers that contain the acid and conjugate base concentrations listed below.

a. $[HPO_4^{2-}] = 0.33$ M, $[PO_4^{3-}] = 0.52$ M

b. $[HNO_2] = 0.029$ M, $[NO_2^-] = 0.065$ M

c. $[HCO_3^-] = 0.50$ M, $[CO_3^{2-}] = 0.15$ M

7.90 List three systems that cooperate to maintain blood pH in an appropriate narrow range. Which one of these provides the most rapid response to changes in blood pH?

7.91 A citric acid–citrate buffer has a pH of 3.20. You want to increase the pH to a value of 3.35. Would you add citric acid or sodium citrate to the solution? Explain.

7.92 Can the capacity of the blood buffering systems be exceeded? Explain your answer.

7.93 What is meant by the terms *acidosis* and *alkalosis*?

7.94* Hemoglobin (Hb) is a weak acid that forms the Hb^- anion. Write equations to show how the Hb/Hb^- buffer would react to added H^+ and added OH^-.

7.95 Explain how the reaction of a strong acid with bicarbonate serves to buffer the blood.

7.96 Hyperventilation is a symptom of what type of acid–base imbalance in the body?

7.97 Write a short summary description of the role of the kidneys and urine in the control of blood pH.

7.98 What happens to CO_2 in the blood when the kidneys function to control blood pH?

7.99 What happens to blood pH as HCO_3^- increases?

7.100 What buffer system in the urine usually keeps the urine pH from dropping lower than 6?

7.101 What ionic shift maintains electron charge balance when the kidneys function to increase the pH of blood?

Additional Exercise

7.102 Consider the following dissociation reaction of a weak acid, HA:

$$HA(aq) + H_2O(\ell) \rightleftarrows H_3O^+(aq) + A^-(aq)$$

It was determined that 2.63% of the acid in a 0.150 M solution of HA in water was dissociated. Calculate the pH of the 0.150 M solution.

7.103 What happens if the amount of H^+ ions in the urine exceeds urine's buffering capacity?

7.104 Ordinary water may be the most effective "sports drink." Explain why this might be true.

7.105 Refer to **Figure 7.24**. Which of the three indicators would be the best to use to differentiate between the solutions with pH values of 4 and 6?

7.106 Use the concept of reaction rates to explain why no smoking is allowed in hospital areas where patients are being administered oxygen gas.

7.107 Suppose you have two identical unopened bottles of carbonated beverage. The contents of both bottles appear to be perfectly clear. You loosen the cap of one of the bottles and hear a hiss as gas escapes, and at the same time gas bubbles appear in the liquid. The liquid in the unopened bottle still appears to be perfectly clear. Explain these observations using the concept of equilibrium and Le Châtelier's principle. Remember, a carbonated beverage contains carbon dioxide gas dissolved in a liquid under pressure. What happens to the pH of a carbonated beverage as CO_2 escapes?

7.108 Consider the following reaction equation:

$$Cl^-(aq) + H_3O^+(aq) \rightarrow HCl(aq) + H_2O(\ell)$$

Draw Lewis structures (electron-dot formulas) for all the substances in the reaction equation, identify each reactant as a Brønsted-Lowry acid or base, and then propose a definition for an acid and a base based on the ability of the substance to accept or donate a pair of electrons to form a covalent bond.

7.109 In the blood, both oxygen (O_2) and poisonous carbon monoxide (CO) can react with hemoglobin (Hb). The respective compounds are HbO_2 and HbCO. When both gases are present in the blood, the following equilibrium is established:

$$HbCO + O_2 \rightleftarrows HbO_2 + CO$$

Use Le Châtelier's principle to explain why pure oxygen is often administered to victims of CO poisoning.

8 Introduction to Organic Chemistry: Hydrocarbons

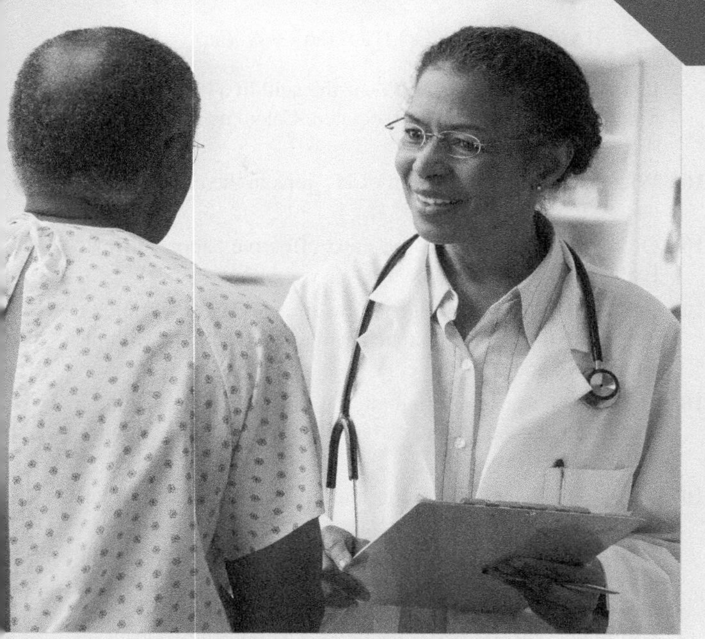

Jose Luis Pelaez/The Image Bank/Getty Images

Health Career Focus

Psychosocial Oncologist

I remember back to my early nursing education when I discovered I could contribute to the mental well-being of my patients, and also help them through whatever struggles they may be experiencing with their physical well-being. So many of us have personally been affected by cancer or have loved ones or colleagues who have faced the mountain that is cancer. It's so important for anyone who is battling serious illness to have a support system.

As a professional in psychosocial oncology, I am able to use my training in a multitude of areas to help cancer patients deal with the mental aspects of fighting cancer via counseling and support. Although there are many causes and many types of cancer, one thing is for certain: Nobody should have to endure it alone.

Follow-up to this Career Focus appears at the end of the chapter before the *Concept Summary*.

Learning Objectives

When you have completed your study of Chapter 8, you should be able to:

1. Distinguish between organic and inorganic chemicals. **(Section 8.1)**
2. Write condensed, expanded, and skeletal structural representations for compounds. **(Section 8.2)**
3. Identify functional groups within larger molecules. **(Section 8.3)**
4. Use structural formulas to identify compounds that are isomers of each other. **(Section 8.4)**
5. Assign IUPAC names to alkanes. **(Section 8.5)**
6. Draw structural formulas for alkanes when given the IUPAC name. **(Section 8.5)**
7. Assign IUPAC names to cycloalkanes. **(Section 8.6)**
8. Write the IUPAC names of alkenes and alkynes from their molecular structures. **(Section 8.7)**
9. Write chemical equations for addition reactions of alkenes. **(Section 8.8)**
10. Use Markovnikov's rule to predict the major products of certain alkene addition reactions. **(Section 8.8)**
11. Write chemical equations for addition polymerization. **(Section 8.8)**
12. Describe the key physical properties of alkanes, alkenes, and alkynes. **(Section 8.9)**
13. Name aromatic compounds. **(Section 8.10)**
14. Draw structural formulas for aromatic compounds. **(Section 8.10)**

In previous chapters, we described matter and how to quantify chemical bonding, intermolecular forces between molecules, thermodynamics and kinetics of chemical reactions, and chemical equilibrium. With this solid foundation in place, we can now apply these concepts to the compounds responsible for living matter—that is, organic compounds. We begin our discussion of organic chemistry by explaining how to represent carbon-containing compounds in condensed, expanded, and skeletal formulas. This then leads to the introduction of organic functional groups, a discussion of isomers, and a tutorial in naming organic compounds. Then we examine the major reactions of hydrocarbons and their unique properties before wrapping up the chapter with a discussion of aromatic compounds.

8.1 Organic Chemistry: The Story of Carbon

Learning Objective 1 Distinguish between organic and inorganic chemicals.

Figure 8.1 Grocery stores usually have a separate produce area for fruits and vegetables labeled organic. "Organic" is a designation used by the U.S. Department of Agriculture to certify food that is produced without the use of antibiotics, hormones, synthetic chemicals or fertilizers, most pesticides, bioengineering, ionizing radiation, or sewage sludge.

Early scientists struggled to determine what was responsible for the difference between living and non-living organisms. Originally, chemists described this mysterious entity that separated living organisms, or organic matter, from nonliving organisms as the "vital force," something that could not be synthesized. This idea was central to the study of organic chemistry, a field originally envisioned to classify these compounds that are responsible for living matter. Through the early 1800s, scientists had been unable to synthesize an organic compound from its elements or from naturally occurring minerals. In 1828, however, Friedrich Wöhler (1800–1882) heated an inorganic salt called ammonium cyanate and produced urea. This compound, normally found in the blood and urine of living organisms, was unquestionably organic, yet it had come from an inorganic source. The reaction is:

$$\underset{\text{ammonium cyanate}}{NH_4NCO} \xrightarrow{\text{Heat}} \underset{\text{urea}}{H-\overset{\displaystyle H}{\overset{|}{N}}-\overset{\displaystyle O}{\overset{\|}{C}}-\overset{\displaystyle H}{\overset{|}{N}}-H}$$

organic compound A compound that contains the element carbon.

organic chemistry The study of carbon-containing compounds.

Wöhler's urea synthesis was one of the first experiments to discredit the "vital force" theory, and his success prompted other chemists to attempt to synthesize organic compounds. Organic chemistry has rapidly evolved in the time since as chemists have harnessed their growing knowledge to synthesize everything from polymers to perfumes to pesticides to life-saving therapeutics. Organic compounds are now synthesized in thousands of laboratories across the world, from small academic labs to large industrial settings.

Today, **organic compounds** share one unique feature: They all contain atoms of carbon. Therefore, **organic chemistry** is presently defined as the study of carbon-containing compounds. The use of "organic" as a way of describing produce and other household goods is an entirely different definition (**Figure 8.1**).

The importance of carbon compounds to life on Earth cannot be overemphasized. If all carbon compounds were removed from Earth, it would be like the barren surface of the moon (see **Figure 8.2**). There would be no animals, plants, or any other form of life. If carbon-containing compounds were removed from the human body, all that would remain would be water, a very brittle skeleton, and a small residue of minerals. Many of the essential constituents of living matter—such as carbohydrates, fats, proteins, nucleic acids, enzymes, and hormones—are organic chemicals.

Figure 8.2 Organic chemistry makes a tremendous difference when comparing the physical makeup of Earth and the moon.

Organic compounds can be found around us throughout everyday life. The principal components of food (with the exception of water) are organic. This includes both the

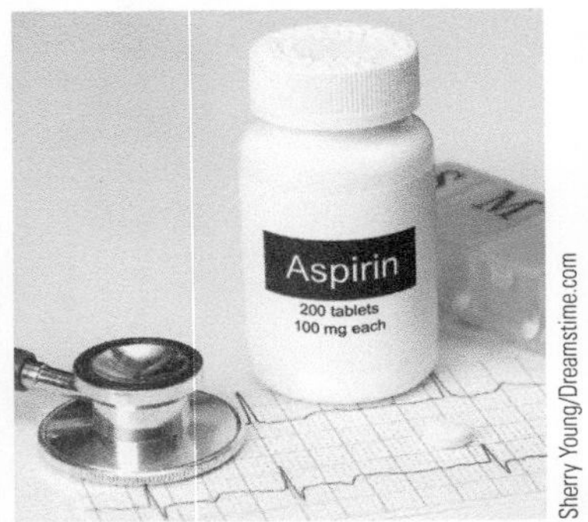

Sherry Young/Dreamstime.com

Figure 8.3 Many organic compounds, such as drugs like aspirin, have relatively low melting points compared to inorganic molecules. What does this observation reveal about the forces between organic molecules?

naturally occurring components of food and additives like artificial colorings, flavorings, and preservatives. The fuels we use (e.g., wood, coal, petroleum, and natural gas) are mixtures of organic compounds.

An estimated 500,000 inorganic compounds have been identified, but more than 9 million organic compounds are known, and thousands of new ones are synthesized or isolated each year. One of the reasons for the large number of organic compounds is the unique ability of carbon atoms to form stable covalent bonds with other carbon atoms and with atoms of other elements. Generally, organic and inorganic compounds differ physically (see **Figure 8.3**) and chemically, as shown in **Table 8.1**.

Table 8.1 Properties of Typical Organic and Inorganic Compounds

Property	Organic Compounds	Inorganic Compounds
Bonding within molecules	Covalent	Often ionic
Forces between molecules	Generally weak	Quite strong
Normal physical state	Gases, liquids, or low-melting-point solids	Usually high-melting-point solids
Flammability	Often flammable	Usually nonflammable
Solubility in water	Often low	Often high
Conductivity of water solutions	Nonconductor	Conductor
Rate of chemical reactions	Usually slow	Usually fast

Example 8.1 Classify Compounds as Organic or Inorganic

Classify each of the following compounds as organic or inorganic.

a. NaCl b. C_6H_6

Solution

a. NaCl consists of ionic bonds between a metal (Na) and a nonmetal (Cl). The nonmetal is not carbon, so the compound is inorganic.
b. C_6H_6 is a hydrocarbon that consists of carbon covalently bonded to other carbon and hydrogen atoms, so the molecule is organic.

✔ **Learning Check 8.1** Classify each of the following compounds as organic or inorganic.

a. CH_3OH b. $Mg(NO_3)_2$

8.2 Representations of Organic Molecules

Learning Objective 2 Write condensed, expanded, and skeletal structural representations for compounds.

Organic chemists have several ways they can represent organic molecules, ranging from ones that are very explicit, to ones that have several assumptions. You must be comfortable with these different structural representations because they are used interchangeably throughout various sources, including this text, and because you may encounter them in your career. For large organic molecules such as carbohydrates and amino acids, drawing structures in a very explicit way becomes cumbersome. However, as we first learn about organic structures, these detailed representations can be very helpful.

expanded structural representation A structural molecular formula showing all the covalent bonds.

condensed structural representation A structural molecular formula showing the general arrangement of atoms but without all the covalent bonds.

The most detailed representation is the **expanded structural representation**. In this form, all covalent bonds are shown explicitly (**Figure 8.4a**). Other structural representations get less and less specific. For instance, the **condensed structural representations** (**Figure 8.4b**) omit all the covalent bonds and simply present the structural formula in a way that reflects the bonding to each carbon atom in sequential order. In the condensed representation, it is also possible to omit just the covalent bonds between carbon and hydrogen, explicitly showing only the covalent bonds between carbon atoms and other carbons or other nonhydrogen atoms.

In the **skeletal structural representation** (**Figure 8.4c**), the elemental symbol C is not used for any carbon atoms. Instead, carbon atoms are depicted as points. Chains of carbon are drawn in an extended zig-zag in order to more accurately represent the geometry at each carbon. The hydrogens bonded to carbon are not even shown! Instead, they are simply *assumed* to be present in the proper number to satisfy carbon's octet.

skeletal structural representation A structural molecular formula showing carbon atoms as points, with hydrogens bonded to carbon assumed. Other atoms (e.g., nitrogen, oxygen, and chlorine) are shown explicitly, as well as hydrogens bonded to these noncarbon atoms.

```
    H   H   H
    |   |   |
H — C — C — C — H        CH3CH2CH3          /\
    |   |   |
    H   H   H
```

a Expanded **b** Condensed ($CH_3CH_2CH_3$) **c** Skeletal

Figure 8.4 Three different structural representations of propane (C_3H_8): (a) expanded, (b) condensed, and (c) skeletal.

How are other, noncarbon, atoms depicted in skeletal structures? These atoms, such as nitrogen, oxygen, and chlorine, are shown explicitly using their elemental symbol. Hydrogens bonded to these atoms are also shown explicitly in order to distinguish them from hydrogens bonded to carbon, as shown in ethylamine (**Figures 8.5** and **8.6**). If these atoms have lone pairs of electrons, they can be shown explicitly or assumed to be present to complete the octet of the noncarbon atom. The nitrogen atom in the structure of ethylamine does indeed have a full octet because it is bonded to 3 atoms (1 carbon and 2 hydrogens), and the presence of one lone pair of electrons is *assumed*.

ADAM GAULT/Science Source

Figure 8.5 Ethylamine is a stabilizer in latex rubber, which is used in sterile gloves.

```
    H   H
    |   |
H — C — C — N — H        CH3CH2NH2          /\NH2
    |   |   |
    H   H   H
```

a Expanded **b** Condensed ($CH_3CH_2NH_2$) **c** Skeletal

Figure 8.6 (a) Expanded, (b) condensed, and (c) skeletal structures of ethylamine (C_2H_7N).

These three structural representations are used interchangeably throughout the remainder of the text, although we most commonly use condensed structures in Chapters 8–10, with more of an emphasis on skeletal structures as we move into cyclic compounds and biochemical structures, which are often larger and more complex.

Example 8.2 Writing Structural Representations

Recognizing and writing structural formulas is an important skill in biological areas of study because structure determines chemical and physical properties.

Write a condensed and skeletal structural formula for the following compound, which is called 2-methylpentane (**Figure 8.7**):

```
               H
               |
    H      H — C — H   H   H   H
    |          |6      |   |   |
H — C ———————— C ————— C — C — C — H
    |1         |2      |3  |4  |5
    H          H       H   H   H
```

iStock.com/Koonsiri Boonnak

Figure 8.7 The compound 2-methylpentane is one of many hydrocarbons in gasoline.

Solution

In a condensed structural formula, the hydrogens belonging to a carbon are grouped to the right. Thus, the groups

$$\begin{array}{c} \text{H} \\ | \\ \text{H}-\text{C}- \\ | \\ \text{H} \end{array} \text{ and } \begin{array}{c} \text{H} \\ | \\ -\text{C}- \\ | \\ \text{H} \end{array} \text{ and } \begin{array}{c} | \\ -\text{C}- \\ | \\ \text{H} \end{array}$$

condense to CH_3–, –CH_2–, and $\overset{|}{-CH-}$, respectively. The formula then condenses to $(CH_3)_2CH(CH_2)_2CH_3$. Other acceptable condensed structures are

$$\begin{array}{c} CH_3 \\ | \\ CH_3-CH-CH_2-CH_2-CH_3 \end{array} \quad \text{or} \quad \begin{array}{c} CH_3 \\ | \\ CH_3CHCH_2CH_2CH_3 \end{array}$$

In the first condensed structure, where none of the covalent bonds are shown, the first set of parentheses indicates two identical CH_3 groups attached to the same carbon atom (carbon 2 in the original structure), and the second set of parentheses denotes two CH_2 groups bonded in sequence.

In a skeletal structure, each point is a carbon, and only carbon–carbon bonds are shown (the carbon–hydrogen bonds are not shown but are assumed). This particular compound contains five carbons bonded together in a straight chain and a sixth carbon branching off that linear chain at the second carbon in the chain. Thus, the compound is represented with a five-point zig-zag line with a single line branching off at the second carbon. The correct number of hydrogens are assumed to be at each carbon in order to fulfill each carbon's octet.

6
2 4
1 3 5

✔ **Learning Check 8.2** Write a condensed and skeletal structural formula for each of the following compounds.

a. $\begin{array}{ccccc} & \text{H} & \text{OH} & \text{H} & \text{H} & \text{H} \\ & | & | & | & | & | \\ \text{H}- & \text{C}- & \text{C}- & \text{C}- & \text{C}- & \text{C}-\text{H} \\ & | & | & | & | & | \\ & \text{H} & \text{H} & \text{H} & \text{H} & \text{H} \end{array}$

b. $\begin{array}{ccccc} & \text{H} & \text{H} & \text{H} & \text{O} \\ & | & | & | & \| \\ \text{H}- & \text{C}- & \text{C}- & \text{C}- & \text{C}-\text{OH} \\ & | & | & | & \\ & \text{H} & \text{H}-\text{C}-\text{H} & \text{H} & \\ & & | & & \\ & & \text{H} & & \end{array}$

8.3 Functional Groups

Learning Objective 3 Identify functional groups within larger molecules.

functional group A unique reactive combination of atoms that differentiates molecules of organic compounds of one class from those of another.

Because of the enormous number of possible organic compounds, the study of organic chemistry might appear to be unbelievably disorganized. However, the arrangement of organic compounds into a relatively small number of classes can simplify the study a great deal. This organization is done on the basis of characteristic structural features called **functional groups**. For example, compounds with a carbon–carbon double bond

$$\begin{array}{c} \diagdown \quad \diagup \\ C=C \\ \diagup \quad \diagdown \end{array}$$

are classified as alkenes. The major classes and functional groups are listed in **Table 8.2**. We can reliably group these patterns of bonding together because they provide distinct patterns of reactivity. For instance, small alcohols like methanol (CH_3OH) behave in many of the same ways as much larger alcohols. Notice that each functional group in **Table 8.2** contains a multiple bond or at least one oxygen or nitrogen atom. These functional groups are the focus of our study in Chapters 8, 9, and 10. In fact, many organic compounds contain multiple functional groups within their structure, as shown in **Example 8.3**.

Hydrocarbons, the simplest of all organic compounds, contain only carbon and hydrogen. **Saturated hydrocarbons**, also called **alkanes**, are organic compounds in which carbon is bonded to four other atoms by single bonds; there are no double or triple bonds in the molecule. Alkanes are generally not considered to be an actual functional group because the presence of C–H single bonds is so ubiquitous throughout organic molecules.

saturated hydrocarbon Another name for an alkane.

alkane A hydrocarbon that contains only carbon–hydrogen bonds and carbon–carbon single bonds.

Unsaturated hydrocarbons contain either carbon–carbon double bonds or carbon–carbon triple bonds (**Figure 8.7**). Aromatics, which consist of a ring of six carbons that alternates carbon–carbon single and double bonds, are unsaturated hydrocarbons, too (**Section 8.10**).

unsaturated hydrocarbon A hydrocarbon that contains one or more carbon–carbon multiple bonds.

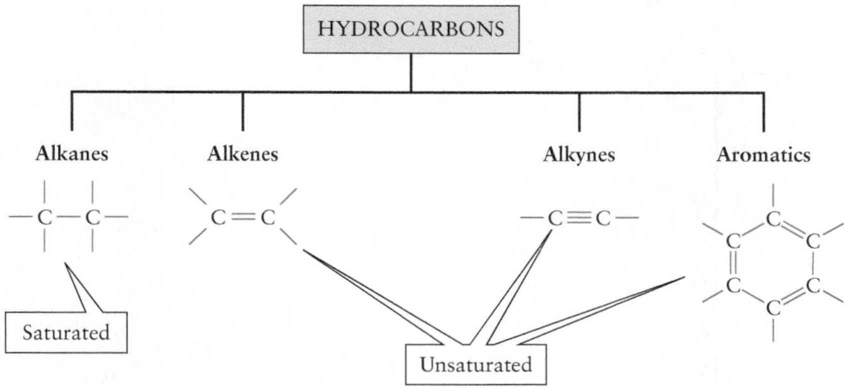

HYDROCARBONS

Alkanes | Alkenes | Alkynes | Aromatics

Saturated

Unsaturated

Figure 8.7 Classification of hydrocarbons.

Although alkanes do not play a direct role in biochemistry, some biomolecules, such as the long-chain fatty acids that make up the cell membrane, do contain a series of saturated or unsaturated hydrocarbons (**Figure 8.8**). So, some understanding of the structure, physical properties, and chemical behavior of hydrocarbons is beneficial.

Hydrocarbons play a crucial role in modern industrial society. We use naturally occurring hydrocarbons as important sources of raw materials for the manufacture of plastics, synthetic fibers, drugs, and hundreds of other compounds that are used daily. We also use hydrocarbons as primary sources of energy. In fact, the burning of hydrocarbons for energy releases the huge amount of CO_2 that is a major cause of global climate change.

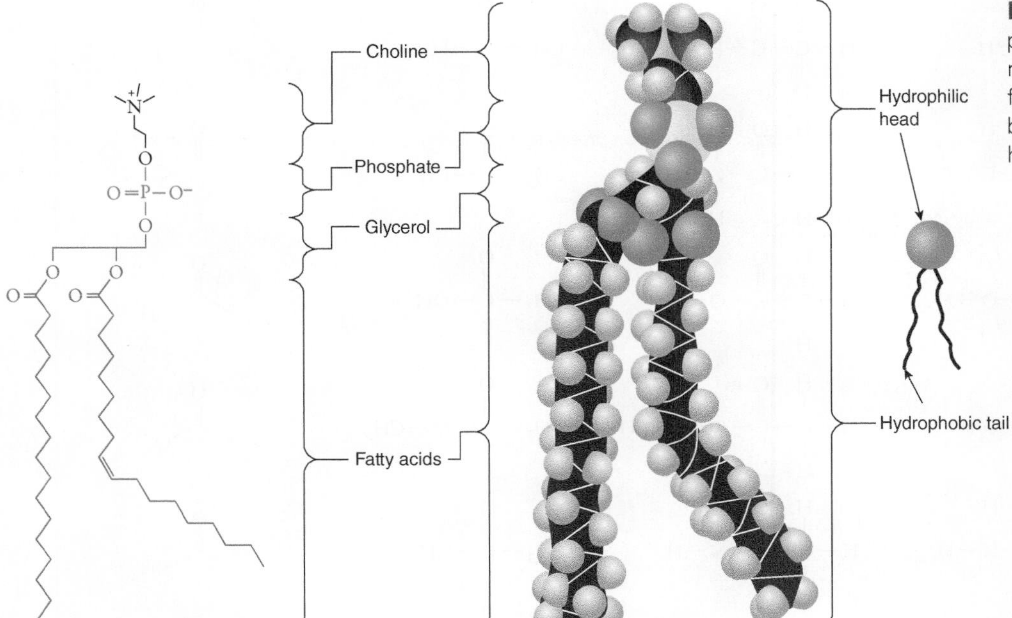

Figure 8.8 Depiction of a phospholipid found in the cell membrane lipid bilayer. The long fatty chains in this structure may be saturated or unsaturated hydrocarbons.

Table 8.2 Classes and Functional Groups of Organic Compounds

Class	Functional Group	Example of Expanded Structural Formula	Example of Condensed Structural Formula	Example of Skeletal Structural Formula
Alkane	None	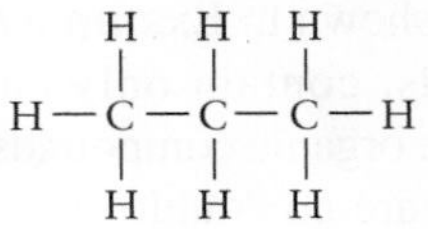 H—C—C—C—H	$CH_3CH_2CH_3$	
Alkene	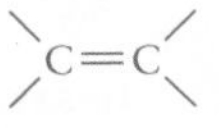 C=C	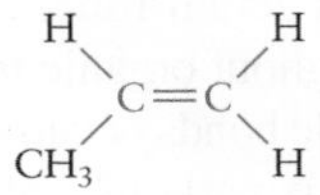 C=C (H, H, CH_3, H)	$CH_3CH{=}CH_2$	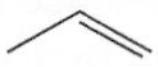
Alkyne	—C≡C—	CH_3—C≡C—H	$CH_3C{\equiv}CH$	
Aromatic	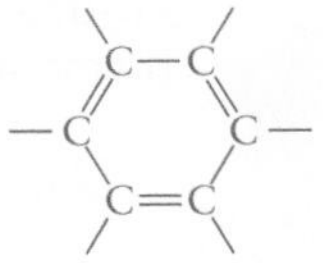 —C C— (ring: C—C, C=C)	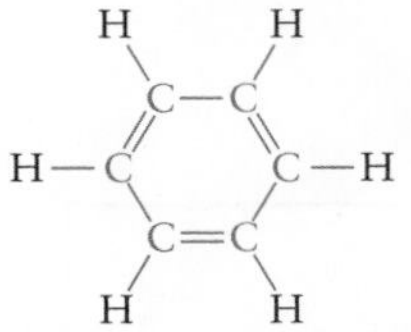 H—C C—H (ring with H on each C)	C_6H_6	
Alcohol	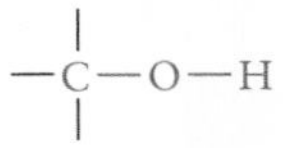 —C—O—H	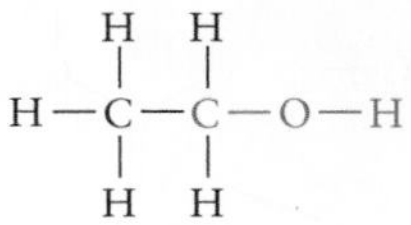 H—C—C—O—H	CH_3CH_2—OH	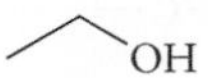 OH
Thiol	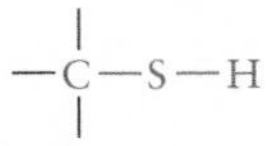 —C—S—H	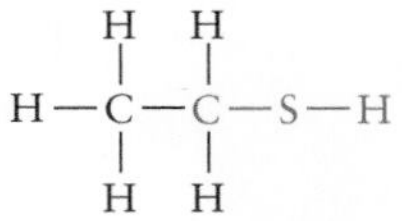 H—C—C—S—H	CH_3CH_2—SH	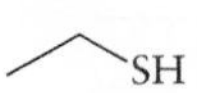 SH
Ether	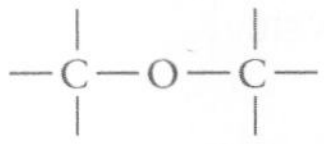 —C—O—C—	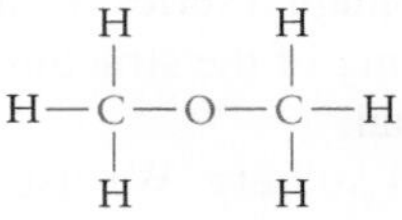 H—C—O—C—H	CH_3—O—CH_3	 O
Amine	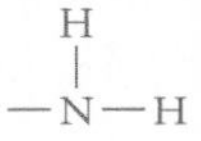 —N—H (H)	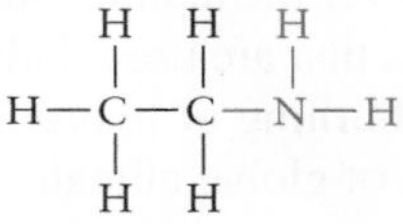 H—C—C—N—H	CH_3CH_2—NH_2	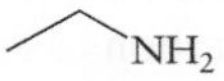 NH_2
Aldehyde	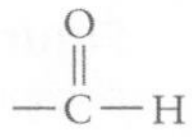 —C—H (O)	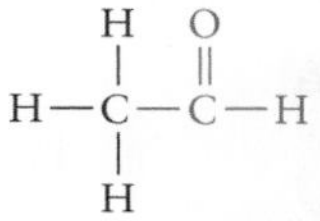 H—C—C—H (O)	CH_3—C—H (O)	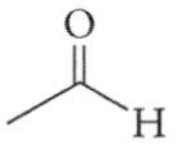 O, H
Ketone	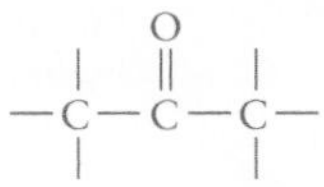 —C—C—C— (O)	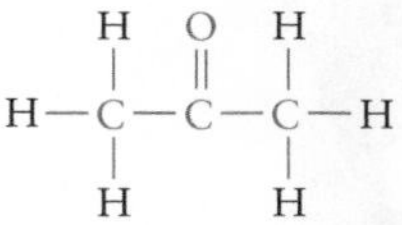 H—C—C—C—H (O)	CH_3—C—CH_3 (O)	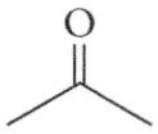 O
Carboxylic acid	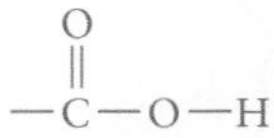 —C—O—H (O)	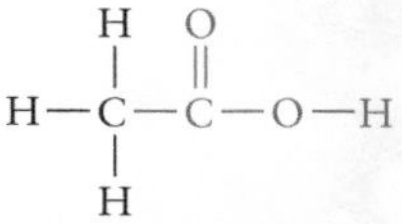 H—C—C—O—H (O)	CH_3—C—OH (O)	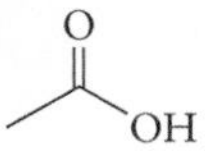 O, OH
Ester	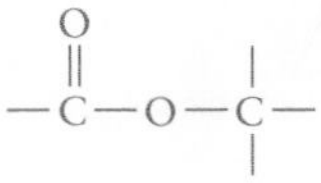 —C—O—C— (O)	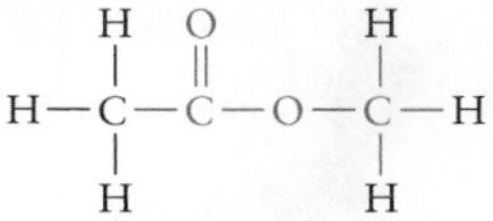 H—C—C—O—C—H (O)	CH_3—C—O—CH_3 (O)	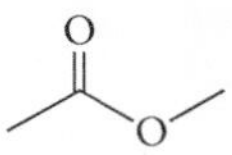 O, O
Amide	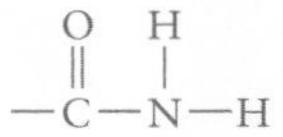 —C—N—H (O, H)	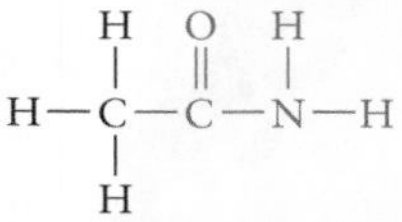 H—C—C—N—H (O, H)	CH_3—C—NH_2 (O)	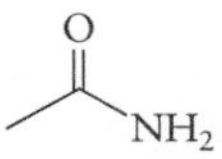 O, NH_2

Example 8.3 Identifying Functional Groups

Identify by name the functional groups in aspartame (**Figure 8.9**), an artificial sweetener in foods and beverages. Please do *not* consider alkane a functional group for this exercise.

aspartame

Solution

Ester, Amine, Amide, Carboxylic acid, Aromatic

Figure 8.9 The artificial sweetener aspartame is about 200 times sweeter than sucrose (table sugar).

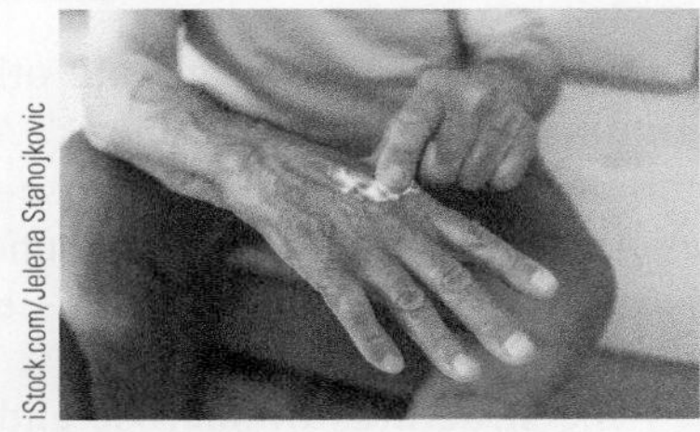

Figure 8.10 Capsaicin is a key ingredient in creams used to treat arthritis pain. It is also added to food products to make them taste more spicy.

✔ **Learning Check 8.3** Identify by name the functional groups in capsaicin (**Figure 8.10**), a compound found in creams and patches used to treat pain.

capsaicin

Table 8.3 Molecular Formulas and Possible Structural Isomers of Alkanes

Molecular Formula	Number of Possible Structural Isomers
C_4H_{10}	2
C_5H_{12}	3
C_6H_{14}	5
$C_{10}H_{22}$	75
$C_{20}H_{42}$	366,319
$C_{30}H_{62}$	4,111,846,763

8.4 Isomers and Conformations

Learning Objective 4 Use structural formulas to identify compounds that are isomers of each other.

In principle, there is no limit to the number of carbon atoms that can bond covalently. Thus, organic molecules range from simple molecules such as methane (CH_4) to very complicated molecules containing over a million carbon atoms such as DNA. The variety of possible carbon atom arrangements is even more important than the size range of the resulting molecules. The number of possible structural isomers increases dramatically with the number of carbon atoms in an alkane (**Table 8.3**). The carbon atoms in all but the very simplest organic molecules can bond in more than one arrangement, giving rise to different compounds with different structures and properties. This property, called **isomerism**, is characterized by compounds that have identical molecular formulas but different arrangements of atoms. **Structural isomers** have the same molecular formula, but their atoms are connected differently.

isomerism A property in which two or more compounds have the same molecular formula but different arrangements of atoms.

structural isomers Compounds that have the same molecular formula but in which the atoms bond in different patterns.

Health Connections 8.1

How Sweet Is Too Sweet?

Aspartame is one of the most commonly used artificial sweeteners in a host of food and beverages. You will see its presence on the nutrition label of many foods labeled "diet" or "sugar free." It may be more familiar to you under names such as Equal or NutraSweet. The component that aspartame is replacing in food and beverages, the "sugar" in this instance, is sucrose. Sucrose, a carbohydrate (Chapter 11), has a caloric value when it is ingested into the body. Artificial sweeteners, like aspartame, however, have little to no caloric value when they are broken down in the body.

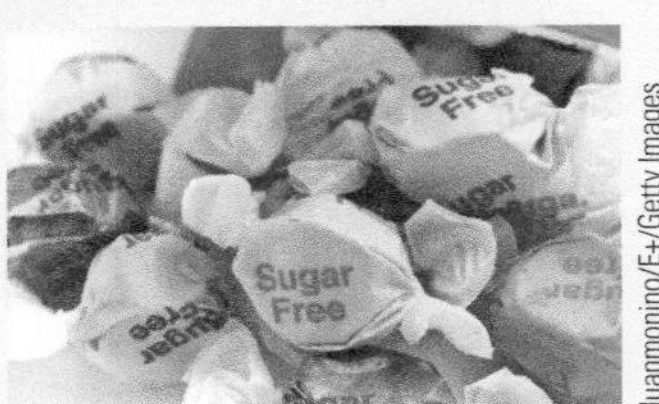

Juanmonino/E+/Getty Images

So, how do aspartame and other artificial sweeteners mimic the sweetness of sugars? Actually, many don't just mimic the sweetness of sucrose—they greatly exceed it. Aspartame, for example, is around 200 times sweeter than sucrose, so very little of it is needed to make food and beverages taste sweet.

The structures of sucrose and aspartame appear very different. Nevertheless, the way aspartame interacts with our taste receptors evokes the "sweet" response the same way sucrose does. Aspartame is derived from two amino acids (aspartic acid and phenylalanine), which we discuss in more detail in Chapter 12.

Artificial sweeteners have come under great scrutiny for their purported safety in foods. Attempts have been made to link aspartame specifically to such maladies as cancer, Alzheimer's, Parkinson's, and other afflictions. In fact, aspartame was labeled by the International Agency for Research on Cancer (IARC), a part of the World Health Organization, as "possibly carcinogenic to humans" in July 2023. However, no links have been conclusively proven, so aspartame is approved for use in foods and beverages by numerous government agencies around the world, including the U.S. Food & Drug Administration (FDA), to be consumed at so-called normal levels.

sucrose | aspartame

Skeletal structures of sucrose and aspartame.

8.4A Isomers

straight-chain alkane Any alkane in which all the carbon atoms are aligned in a continuous chain.

branched alkane An alkane in which at least one carbon atom is not part of a continuous chain.

Butane is a saturated hydrocarbon with the molecular formula C_4H_{10}. It can form two structural isomers (**Figure 8.11**). *n*-Butane is a **straight-chain alkane** because the four carbon atoms are bonded sequentially in a "straight line," with the appropriate number of hydrogens bonded to each carbon. Thus, there are three hydrogens on each end (*terminal*) carbon and two hydrogens on each middle (*internal*) carbon. However, there are other combinations of bonding between carbon atoms that have the same chemical formula. Isobutane is a **branched alkane** because one of *n*-butane's terminal carbons and its three hydrogens (known as a *methyl* group, CH_3–) is removed from the straight chain and attached to a different internal carbon. Thus, isobutane has three methyl groups all bonded to the same carbon, which also has a single hydrogen attached.

$CH_3—CH_2—CH_2—CH_3$ | $CH_3—CH(CH_3)—CH_3$

n-butane | isobutane

Figure 8.11 Condensed and skeletal structural representations of the two isomeric butanes.

Example 8.4 Drawing Structural Isomers

Draw two structural isomers with the molecular formula C_2H_6O, showing all covalent bonds. Identify the functional groups within each isomer.

Solution

Carbon forms four covalent bonds by sharing its four valence-shell electrons. Similarly, oxygen should form two covalent bonds, and hydrogen a single bond. On the basis of these bonding relationships, two structural isomers are possible:

```
    H  H                     H       H
    |  |                     |       |
 H—C—C—O—H              H—C—O—C—H
    |  |                     |       |
    H  H                     H       H
```

ethyl alcohol (an alcohol) — dimethyl ether (an ether)

✔ Learning Check 8.4 Identify the functional groups in compounds a–d (some structures may contain more than one). Identify which of structures a–d represents a structural isomer of the following compound.

```
    H  O  H
    |  ‖  |
 H—C—C—C—H
    |     |
    H     H
```

```
      H  H  O—H
      |  |  |
a. H—C=C—C—H
            |
            H

      H  H  O
      |  |  ‖
b. H—C=C—C—H
```

c. (skeletal structure) O, OH

d. (skeletal structure) O

Even though two molecules may have the same molecular formula, their properties may be quite different if they have different functional groups. The two isomers in Example 8.4 are quite different. Ethyl alcohol (grain alcohol; see **Figure 8.12**) is a liquid at room temperature, whereas dimethyl ether is a gas. Thus, molecular formulas such as C_2H_6O provide much less information about a compound than do structural formulas. **Figure 8.13** shows ball-and-stick models of these two molecules.

smereka/Shutterstock.com

Figure 8.12 A number of grains can be fermented and distilled to produce alcohol, including wheat, corn, rice, and rye. Grain alcohol is considered a "neutral spirit," meaning it has no added flavor.

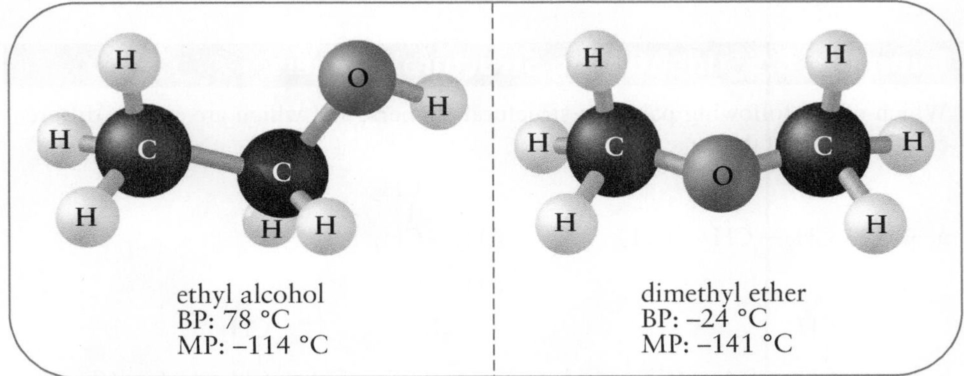

Figure 8.13 Ball-and-stick models, boiling points (BP) and melting points (MP) of the structural isomers of C_2H_6O. Ethyl alcohol is a liquid at room temperature and completely soluble in water, whereas dimethyl ether is a gas at room temperature and only partially soluble in water. What does this suggest about the kind of intermolecular forces each compound exhibits?

8.4B Conformations of Alkanes

Planar representations of organic molecules such as $CH_3—CH_2—CH_3$ or

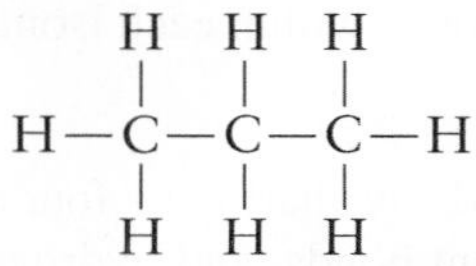

$$\begin{array}{ccccccc} & & H & & H & & H & \\ & & | & & | & & | & \\ H & — & C & — & C & — & C & — & H \\ & & | & & | & & | & \\ & & H & & H & & H & \end{array}$$

make no attempt to accurately portray correct bond angles or molecular geometries. Recall from **Section 5.1** that a carbon with four bonding groups has a tetrahedral geometry with 109.5° bond angles. Condensed and expanded structural formulas are usually written horizontally simply because it is convenient to do so. Actual organic molecules are in constant motion—twisting, turning, vibrating, and bending. Groups connected by a single bond are capable of rotating about that bond much like a wheel rotates around an axle (see **Figure 8.14**). As a result of this rotation about single bonds, a molecule can exist in many different orientations in three-dimensional space, called **conformations**.

conformations The different arrangements of atoms in three-dimensional space achieved by rotation about single bonds.

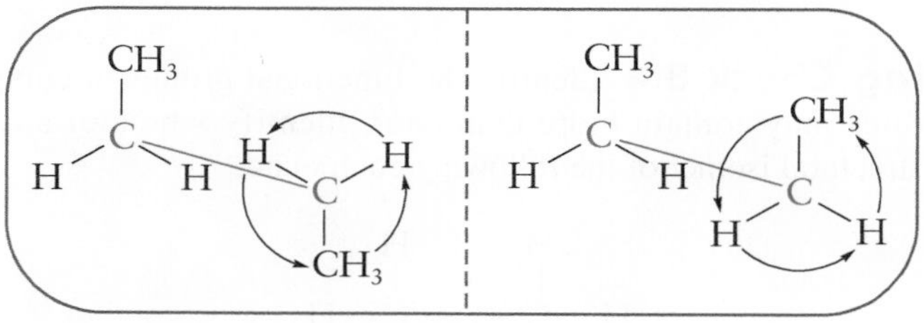

Figure 8.14 Rotation occurs readily about the carbon–carbon single bonds, as shown here for butane.

In a sample of butane containing billions of identical molecules, there are countless conformations present at any instant, and each conformation is rapidly changing into another. Two of the possible conformations of butane are shown in **Figure 8.15**. *These different conformations do not represent different structural isomers.* In each case, the four carbon atoms are bonded in a continuous (unbranched) chain. Because the order of bonding is identical, the conformations correspond to the same molecule. Two structures would be structural isomers only if bonds had to be broken and remade to convert one into the other.

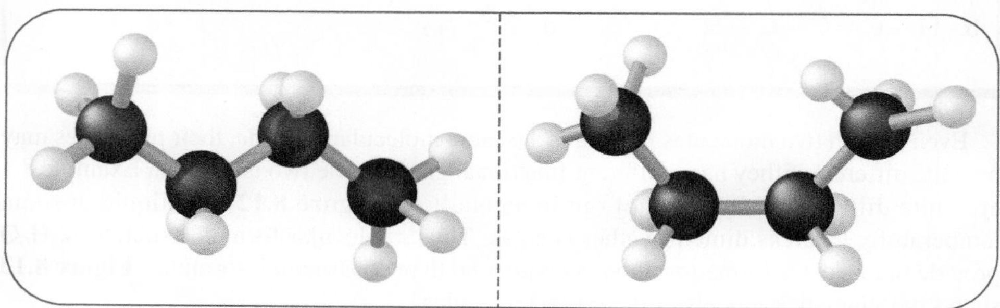

Figure 8.15 Ball-and-stick models of two conformations of *n*-butane.

Example 8.5 Identifying Structural Isomers

Which of the following pairs are structural isomers, and which are simply different conformations of the same molecule?

a. $CH_3—CH_2—CH_2—CH_3$

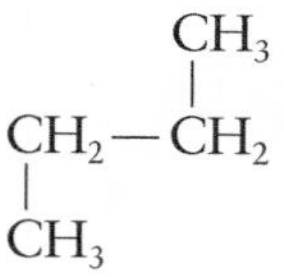

$$\begin{array}{ccc} & & CH_3 \\ & & | \\ CH_2 & — & CH_2 \\ | & & \\ CH_3 & & \end{array}$$

b. $\begin{array}{l} \quad\;\; CH_3 \\ \quad\;\;\; | \\ CH_3—CH—CH_2—CH_2—CH_3 \end{array}$ $\qquad$ $\begin{array}{l} \qquad\qquad\quad\; CH_3 \\ \qquad\qquad\quad\;\; | \\ CH_3—CH_2—CH—CH_2—CH_3 \end{array}$

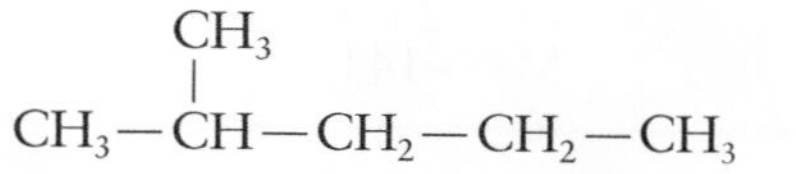

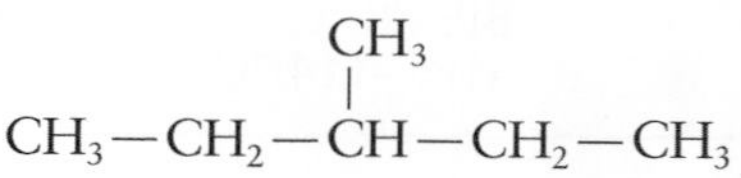

Solution

a. Different conformations of the same molecule: In both molecules, the four carbons are bonded in a continuous chain.

$$\overset{1}{CH_3}-\overset{2}{CH_2}-\overset{3}{CH_2}-\overset{4}{CH_3} \qquad \begin{array}{l} \quad\ \ {}^{4}CH_3 \\ \quad\quad\ | \\ {}^{2}CH_2-\underset{3}{CH_2} \\ \ \ | \\ {}^{1}CH_3 \end{array}$$

b. Structural isomers: Both molecules have a continuous chain of five carbons, but the branch is located at different positions.

$$\begin{array}{c} \quad CH_3 \\ \quad | \\ \underset{1}{CH_3}-\underset{2}{CH}-\underset{3}{CH_2}-\underset{4}{CH_2}-\underset{5}{CH_3} \end{array} \qquad \begin{array}{c} CH_3 \\ | \\ \underset{1}{CH_3}-\underset{2}{CH_2}-\underset{3}{CH}-\underset{4}{CH_2}-\underset{5}{CH_3} \end{array}$$

✔ **Learning Check 8.5** Which of the following pairs represent structural isomers, and which are simply different conformations of the same compound?

a.
$$\begin{array}{c} CH_3 \\ | \\ CH_3-CH-CH_2-CH_2-CH_3 \end{array} \qquad \begin{array}{c} CH_3 \\ | \\ CH_3-CH_2-CH_2-CH-CH_3 \end{array}$$

b.
$$\begin{array}{c} CH_3 \\ | \\ CH_3-CH-CH_2-CH_3 \end{array} \qquad \begin{array}{c} CH_2-CH_3 \\ | \\ CH_3-CH-CH_3 \end{array}$$

8.5 Classification and Naming of Alkanes

Learning Objective 5 Assign IUPAC names to alkanes.

Learning Objective 6 Draw structural formulas for alkanes when given the IUPAC name.

When only a relatively few organic compounds were known, chemists gave them what are today called common names, such as methane (see **Figure 8.16**), ethane, propane (see **Figure 8.17**), and butane. The names for the larger alkanes were derived from the Greek prefixes that indicate the number of carbon atoms in the molecule. Thus, *pent*ane contains five carbons, *hex*ane has six, *hept*ane has seven, and so forth, as shown in **Table 8.4**. The general molecular formula for each alkane is C_nH_{2n+2}, where n is the number of carbon atoms. This formula applies to straight-chain and branched alkanes.

Antranias/4323 images/Pixabay

Figure 8.16 Microorganisms in the guts of cows produce some of the methane in Earth's atmosphere.

Table 8.4 Names of Alkanes

Number of Carbon Atoms	Name	Molecular Formula	Structure of Straight-Chain Isomer
1	Methane	CH_4	CH_4
2	Ethane	C_2H_6	CH_3CH_3
3	Propane	C_3H_8	$CH_3CH_2CH_3$
4	Butane	C_4H_{10}	$CH_3CH_2CH_2CH_3$
5	Pentane	C_5H_{12}	$CH_3CH_2CH_2CH_2CH_3$
6	Hexane	C_6H_{14}	$CH_3CH_2CH_2CH_2CH_2CH_3$
7	Heptane	C_7H_{16}	$CH_3CH_2CH_2CH_2CH_2CH_2CH_3$
8	Octane	C_8H_{18}	$CH_3CH_2CH_2CH_2CH_2CH_2CH_2CH_3$
9	Nonane	C_9H_{20}	$CH_3CH_2CH_2CH_2CH_2CH_2CH_2CH_2CH_3$
10	Decane	$C_{10}H_{22}$	$CH_3CH_2CH_2CH_2CH_2CH_2CH_2CH_2CH_2CH_3$

Karol Franks/Moment/Getty Images

Figure 8.17 Hot air balloons use liquid propane to fuel their flight, keeping the balloon filled with air near 120 °C (250 °F).

As more compounds and isomers were discovered, however, it became increasingly difficult to devise unique names and much more difficult to commit them to memory; so, a systematic naming method was needed. Although a systematic method is now in use for larger, more complex molecules, in this text (and in many outside sources), the well-established common names are predominately used for smaller organic compounds.

The names for the two isomeric butanes (*n*-butane and isobutane) illustrate the important features of the common nomenclature system used for alkanes. The stem *but-* indicates that four carbons are present in the molecule. The *-ane* ending signifies the alkane family. The prefix *n*-indicates that all carbons form an unbranched chain. The prefix *iso-* refers to compounds in which all carbons except one are in a continuous chain and in which that one carbon is branched from a next-to-the-end carbon:

$CH_3CH_2CH_2CH_3$ *n*-butane	CH_3 on C2 of CH_3CHCH_3 isobutane
$CH_3CH_2CH_2CH_2CH_3$ *n*-pentane	CH_3 on C2 of $CH_3CHCH_2CH_3$ isopentane
$CH_3CH_2CH_2CH_2CH_2CH_3$ *n*-hexane	CH_3 on C2 of $CH_3CHCH_2CH_2CH_3$ isohexane

This common naming system has limitations. Pentane has three isomers, and hexane has five. The more complicated the compound, the greater the number of isomers, and the greater the number of special prefixes needed to name all the isomers. It would be extremely difficult and time-consuming to try to use a unique prefix or name to identify each of the 75 isomeric alkanes containing 10 carbon atoms!

To devise a system of nomenclature that could be used for even the most complicated compounds, committees of chemists have developed the IUPAC (International Union of Pure and Applied Chemistry) system. This system is much the same for all classes of organic compounds. The IUPAC name for an organic compound consists of three component parts:

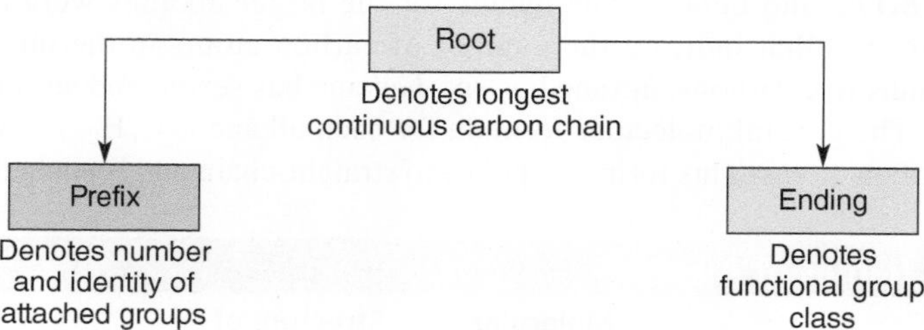

The *root* of the IUPAC name specifies the longest continuous chain of carbon atoms in the compound. The roots for the first 10 normal hydrocarbons are based on the names given in **Table 8.4**: C_1 *meth-*, C_2 *eth-*, C_3 *prop-*, C_4 *but-*, C_5 *pent-*, C_6 *hex-*, C_7 *hept-*, C_8 *oct-*, C_9 *non-*, and C_{10} *dec-*.

The *ending* of an IUPAC name specifies the major functional group of the compound. The ending *-ane* specifies an alkane. Each of the other functional groups has a characteristic ending; for example, *-ene* is the ending for alkenes, and the *-ol* ending designates alcohols.

alkyl group A group differing by one hydrogen from an alkane.

Prefixes are used to specify the identity, number, and location of atoms or groups of atoms that are attached to the longest carbon chain. **Table 8.5** lists several common carbon-containing groups referred to as **alkyl groups**. Each alkyl group is a collection of atoms that can be thought of as an alkane where one hydrogen has been replaced by branching off the root chain. Alkyl groups are named simply by dropping *-ane* from the name of the corresponding alkane and replacing it with *-yl*. For example, CH_3— is called a methyl group, and CH_3—CH_2— is an ethyl group:

CH_4	CH_3-	CH_3-CH_3	CH_3-CH_2-
methane	methyl group	ethane	ethyl group

Two different alkyl groups can be derived from propane, depending on which hydrogen is replaced by substituting to the root carbon chain. Replacing a hydrogen from an end carbon results in a propyl group:

$CH_3-CH_2-CH_3$	$CH_3-CH_2-CH_2-$
propane	propyl group

Replacing a hydrogen from the center carbon results in an isopropyl group:

$CH_3-CH_2-CH_3$	| $CH_3-CH-CH_3$
propane	isopropyl group

Wavebreakmedia/iStock/Getty Images

Figure 8.18 Camping wouldn't be nearly as fun without canister stoves, which typically use an isobutane–propane mixture as fuel.

An isopropyl group also can be represented by $(CH_3)_2CH$—. As shown in **Table 8.5**, four butyl groups can be derived from butane; two from the straight-chain, or normal, butane; and two from the branched-chain isobutane (see **Figure 8.18**). A number of nonalkyl groups are also commonly used in naming organic compounds (see **Table 8.6**).

Table 8.5 Common Alkyl Groups

Parent Alkane	Structure of Parent Alkane	Structure of Alkyl Group	Name of Alkyl Group
Methane	CH_4	CH_3-	Methyl
Ethane	CH_3CH_3	CH_3CH_2-	Ethyl
Propane	$CH_3CH_2CH_3$	$CH_3CH_2CH_2-$	Propyl
		| CH_3CHCH_3	Isopropyl
n-Butane	$CH_3CH_2CH_2CH_3$	$CH_3CH_2CH_2CH_2-$	Butyl
		| $CH_3CH_2CHCH_3$	*sec*-Butyl
Isobutane	CH_3 | CH_3CHCH_3	CH_3 | CH_3CHCH_2-	Isobutyl
		CH_3 | CH_3CCH_3 |	*tert*-Butyl

Table 8.6 Common Nonalkyl Groups

Group	Name
—F	Fluoro
—Cl	Chloro
—Br	Bromo
—I	Iodo
$—NO_2$	Nitro
$—NH_2$	Amino

The following steps are useful when the IUPAC name of an alkane is written on the basis of its structural formula:

Step 1: *Number the longest chain.* The carbon atoms in the longest chain are numbered consecutively from the end that will give the lowest possible number to any carbon to which a group is attached.

If two or more alkyl groups are attached to the longest chain and more than one numbering sequence is possible, the chain is numbered to get the lowest series of numbers. An easy way to follow this rule is to number from the end of the chain nearest a branch:

Step 2: *Name the longest chain.* The longest continuous carbon-atom chain is chosen as the basis for the name of the root chain. The names are those given in **Table 8.4**.

$$\begin{array}{llllll} ^{1} & ^{2} & ^{3} & ^{4} & ^{5} & ^{6} \\ CH_3- & CH- & CH_2- & CH- & CH_2- & CH_3 \\ & | & & | & & \\ & CH_3 & & CH_2-CH_3 & & \end{array}$$

Groups are located at positions 2 and 4.

$$\begin{array}{llllll} ^{6} & ^{5} & ^{4} & ^{3} & ^{2} & ^{1} \\ CH_3- & CH- & CH_2- & CH- & CH_2- & CH_3 \\ & | & & | & & \\ & CH_3 & & CH_2-CH_3 & & \end{array}$$

Groups are *not* located at positions 3 and 5.

Step 3: *Locate and name the attached alkyl groups.* Each group is located by the number of the carbon atom to which it is attached on the chain.

$$\begin{array}{llllll} ^{1} & ^{2} & ^{3} & ^{4} & ^{5} & ^{6} \\ CH_3- & CH- & CH_2- & CH- & CH_2- & CH_3 \\ & | & & | & & \\ & CH_3 & & CH_2-CH_3 & & \end{array}$$

Groups are located at positions 2 and 4.

The one-carbon group at position 2 is a methyl group. The two-carbon group at position 4 is an ethyl group.

Step 4: *Combine the longest chain and indicate the number and position of attached alkyl groups to give the full IUPAC name.*

$$\begin{array}{llllll} ^{1} & ^{2} & ^{3} & ^{4} & ^{5} & ^{6} \\ CH_3- & CH- & CH_2- & CH- & CH_2- & CH_3 \\ & | & & | & & \\ & CH_3 & & CH_2-CH_3 & & \end{array}$$

4-ethyl-2-methylhexane

If two or more of the same alkyl group occur as branches, the number of them is indicated by the prefixes *di-*, *tri-*, *tetra-*, *penta-*, and so on, and the location of each is again indicated by a number (see Example 8.6). These position numbers, separated by commas, are put just before the name of the group, with hyphens before and after the numbers when necessary.

If two or more different alkyl groups are present, their names are alphabetized and added to the name of the basic alkane, again as one word. Thus, the compound we just named is 4-ethyl-2-methylhexane, not 2-methyl-4-ethylhexane. For purposes of alphabetizing, the prefixes *di-*, *tri-*, and so on are ignored, as are the italicized prefixes *sec* and *tert*. The prefix *iso-* is an exception and is used for alphabetizing.

Example 8.6 Naming Alkanes Given the Structure

Give the correct IUPAC name for the following alkane.

$$\begin{array}{llllllll} & & & CH_2-CH_3 & & & & \\ & & & | & & & & \\ CH_3- & CH_2- & CH_2- & CH- & CH_2- & CH- & CH_2- & CH_3 \\ & & & & & | & & \\ & & & & & CH_2-CH_3 & & \end{array}$$

Solution

Step 1: *Number the longest chain.* In this case, the longest continuous chain contains 8 carbons. Counting right to left gives the branching point nearest to an end of the chain the lowest number possible.

$$\begin{array}{llllllll} & & & CH_2-CH_3 & & & & \\ ^{8} & ^{7} & ^{6} & ^{5}| & ^{4} & ^{3} & ^{2} & ^{1} \\ CH_3- & CH_2- & CH_2- & CH- & CH_2- & CH- & CH_2- & CH_3 \\ & & & & & | & & \\ & & & & & CH_2-CH_3 & & \end{array}$$

Step 2: *Name the longest chain.* In this case, the longest continuous chain contains 8 carbons, so the root name is octane.

Step 3: *Locate and name the attached alkyl groups.* Numbering the chain as shown here places an ethyl substituent on carbons 3 and 5.

$$\overset{8}{CH_3}-\overset{7}{CH_2}-\overset{6}{CH_2}-\overset{5}{\underset{}{CH}}(\text{Ethyl: } CH_2-CH_3)-\overset{4}{CH_2}-\overset{3}{CH}(\text{Ethyl: } CH_2-CH_3)-\overset{2}{CH_2}-\overset{1}{CH_3}$$

kali9/Getty Images

Figure 8.19 Many chemical engineers study branched hydrocarbons because they are crucial to the fuel industry.

Step 4: *Combine the longest chain and indicate the number and position of attached alkyl groups to give the full IUPAC name.* Putting it all together with the prefix *di-* because there are two ethyl groups, the correct IUPAC name is 3,5-diethyloctane (**Figure 8.19**).

✔ Learning Check 8.6 Give the correct IUPAC name for each of the following alkanes.

a. $CH_3-CH_2-CH_2-CH(CH_3)-CH_3$

b. $CH_3-CH_2-CH_2-CH(CH(CH_3)-CH_3)-CH_2-CH(CH_3)-CH_3$

Naming compounds is a very important skill, as is the reverse process of using IUPAC nomenclature to specify a chemical structure. The two processes are very similar. To obtain a chemical structure from a name, first determine the longest chain and draw it out. Then number the chain and add any attached groups specified in the IUPAC name to the appropriate carbons of the longest chain. Lastly, fill in the appropriate number of hydrogens for each carbon.

Example 8.7 Drawing Condensed Structural Formulas, Given the Name

There are thousands of biological and medicinal compounds, so being able to draw and name structures is an important skill, leaving no ambiguity concerning which chemical compound is being used.

Draw a condensed structural formula for 3-ethyl-2-methylhexane.

Solution

Step 1: *Use the root of the name to draw the longest chain.* Use the last part of the name (i.e., *hexane*) to determine the longest chain. Draw a chain of six carbons. Then number the carbon atoms.

$$\underset{1}{C}-\underset{2}{C}-\underset{3}{C}-\underset{4}{C}-\underset{5}{C}-\underset{6}{C}$$

Step 2: *Use the alkyl group positions in the name* to attach a methyl group at position 2 and an ethyl group at position 3.

$$\underset{1}{C}-\underset{2}{C}(CH_3)-\underset{3}{C}(CH_2CH_3)-\underset{4}{C}-\underset{5}{C}-\underset{6}{C}$$

Step 3: *Fulfill the octet.* Complete the structure by adding enough hydrogen atoms so that each carbon has four bonds.

$$\underset{1}{CH_3}-\underset{2}{CH}(CH_3)-\underset{3}{CH}(CH_2CH_3)-\underset{4}{CH_2}-\underset{5}{CH_2}-\underset{6}{CH_3}$$

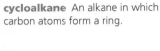

✔ **Learning Check 8.7** Draw a condensed structural formula for each of the following compounds.

a. 2,2,4-trimethylpentane
b. 3-ethyl-2,4-dimethylheptane

8.6 Cycloalkanes

Learning Objective 7 Assign IUPAC names to cycloalkanes.

cycloalkane An alkane in which carbon atoms form a ring.

So far, we have studied straight-chain alkanes and branched alkanes. However, there is another class of alkanes, called **cycloalkanes**, where the two carbon atoms at the ends of the alkane circle back and bond to each other to form a ring, or cyclic structure. As a result, the two terminal carbons must have one less hydrogen in order for each carbon to have 4 bonds. Thus, the general molecular formulas of cycloalkanes is C_nH_{2n}.

Like the other alkanes, cycloalkanes are not found in human cells in isolation. However, several important molecules in human cells do contain rings of five or six atoms (**Figure 8.20**), so the study of cycloalkanes will help you better understand the chemical behavior of these complex molecules.

According to IUPAC rules, cycloalkanes are named by placing the prefix *cyclo-* before the name of the corresponding alkane with the same number of carbon atoms (**Table 8.7**). When substituted cycloalkanes (those with attached groups) are named, the position of a single attached group does not need to be specified in the name because all positions in the ring are equivalent. However, when two or more groups are attached, their positions of attachment are indicated by numbers, just as they were for branched alkanes. The ring numbering begins with the carbon attached to the first group alphabetically and proceeds around the ring in the direction that will give the lowest numbers for the locations of the other attached groups.

a

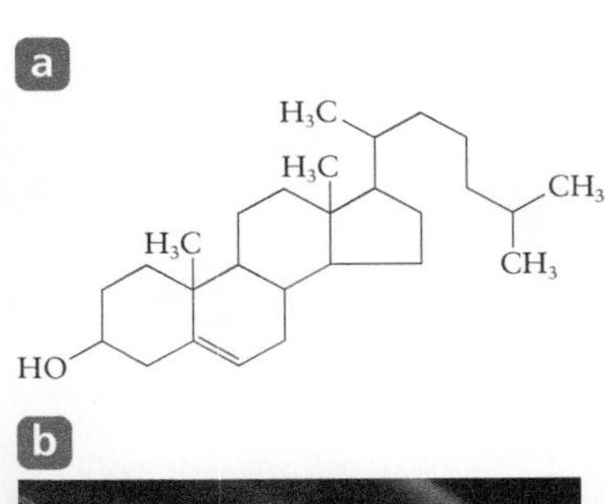

b

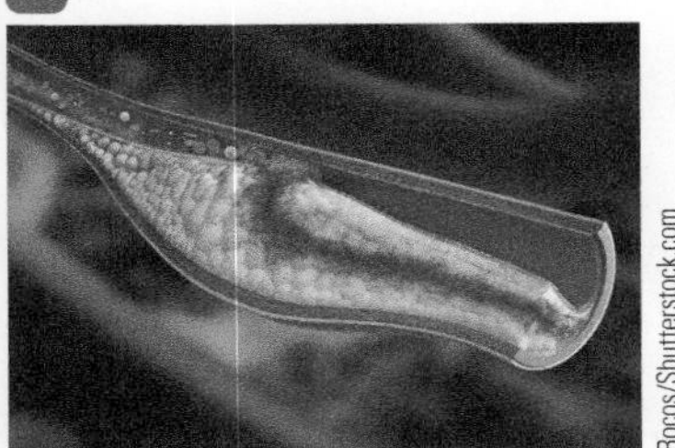

Rocos/Shutterstock.com

Figure 8.20 (a) Cholesterol, an essential biomolecule that is important for hormone synthesis and cell membranes, consists of several cycloalkanes as part of its structure. (b) Even though it is essential to the body, the build-up of excess cholesterol in the bloodstream can cause arteries to narrow and become blocked.

Table 8.7 Structural Formulas and Symbols for Common Cycloalkanes

Name	Condensed Structure	Skeletal Representation	Ball-and-Stick Model
Cyclopropane	CH_2 $H_2C—CH_2$		
Cyclobutane	$H_2C—CH_2$ $H_2C—CH_2$		
Cyclopentane	H_2 C H_2C CH_2 $H_2C—CH_2$		
Cyclohexane	H_2 C H_2C CH_2 H_2C CH_2 C H_2		

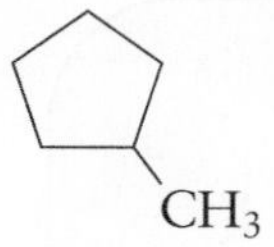

methylcyclopentane

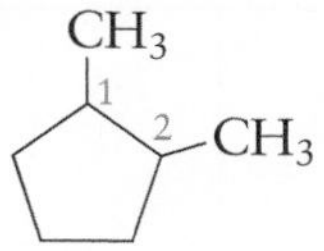

1,2-dimethylcyclopentane
not 1,5-dimethylcyclopentane

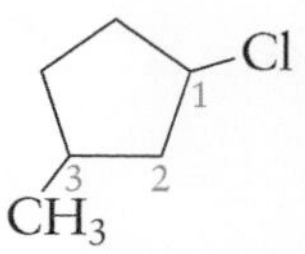

1-chloro-3-methylcyclopentane
not 1-chloro-4-methylcyclopentane
not 3-chloro-1-methylcyclopentane

Example 8.8 Naming Cycloalkanes

Provide the correct IUPAC name for the following cycloalkane.

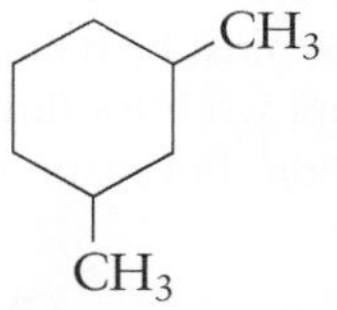

Solution

Step 1: *Number the longest chain.* In this case, two methyl groups are attached to a six-carbon ring. To properly indicate the positions of the two alkyl groups, the ring is numbered beginning with a carbon to which a methyl group is attached, counting in the direction that gives the lowest numbers.

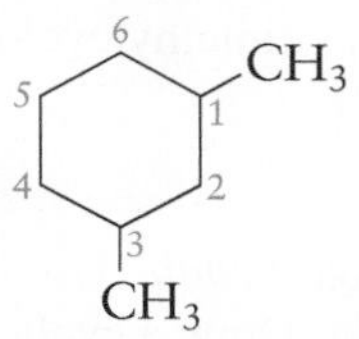

not

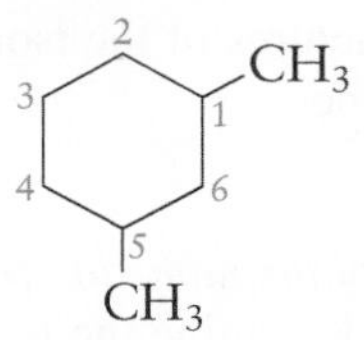

Step 2: *Name the longest chain.* In this case, a hexagon represents a six-carbon ring, which is called cyclohexane.

Step 3: *Locate and name the attached alkyl groups.* Because there are two methyl groups, one each at positions 1 and 3 in the cyclic ring, the correct name is 1,3-dimethylcyclohexane.

✔ Learning Check 8.8 Provide the correct IUPAC name for the following cycloalkane.

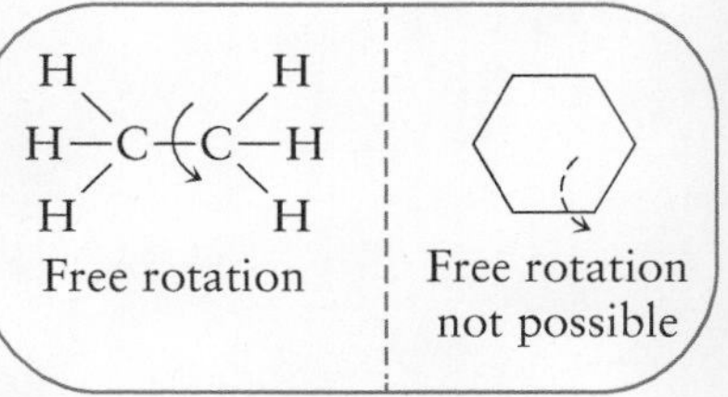

Figure 8.21 Rotation about C–C single bonds occurs in open-chain compounds but not within rings.

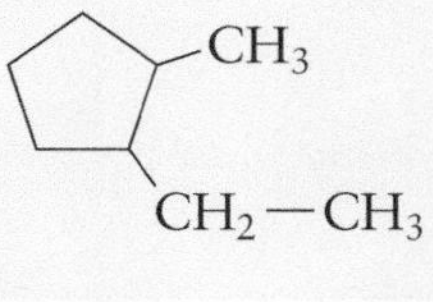

The lack of free rotation around C–C bonds in cycloalkanes (**Figure 8.21**) leads to an extremely important kind of isomerism called *stereoisomerism*. Two different compounds that have the same molecular formula and the same structural formula but different spatial arrangements of atoms are called **stereoisomers**. In 1,2-dimethylcyclopentane (**Figure 8.22**), for example, the methyl groups attached to the ring project above or below the plane of the ring.

stereoisomers Compounds with the same structural formula but different spatial arrangements of atoms.

Two stereoisomers are possible: Either both groups may project in the same direction from the plane, or they may project in opposite directions from the plane of the ring. Because the methyl groups cannot rotate from one side of the ring to the other due to the inherent restriction of cycloalkanes, the two compounds represented in **Figure 8.22** are distinct. These two compounds have physical and chemical properties that are quite different, so they can be separated from each other.

Figure 8.22 The cis–trans isomers of 1,2-dimethylcyclopentane. Note that the other possible cis representation (both methyl groups up) is identical to the one shown if the molecule is simply flipped like a pancake. The same is true for the trans stereoisomer. Also note that rings with leading edges that are bolded like these are not meant to communicate that these cycloalkanes are flat. Instead, they are meant to emphasize the location of subsituents above or below the plane of the ring.

cis- Two groups attached to the *same* side of an organic molecule whose rotation is restricted (as applied to cis–trans isomers).

trans- Two groups attached to the *opposite* side of an organic molecule whose rotation is restricted (as applied to cis–trans isomers).

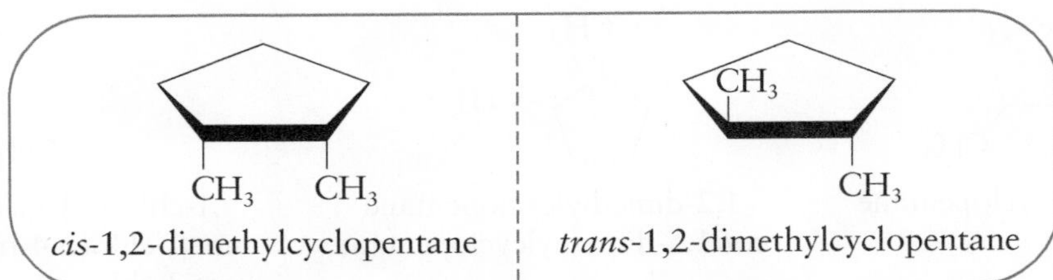

Stereoisomers of this type, in which the spatial arrangement or geometry of their groups is maintained by rings, are called **cis–trans isomers**. The prefix *cis-* denotes the isomer in which both groups are on the same side of the ring, whereas the prefix *trans-* denotes the isomer in which they are on opposite sides. To exist as cis–trans isomers, a disubstituted (and only a disubstituted) cycloalkane must be bonded to groups at two different carbons of the ring. This nomenclature does not work for three or more attached groups or for two groups bonded to the same ring carbon. For example, there are no cis–trans isomers of 1,1-dimethylcyclohexane:

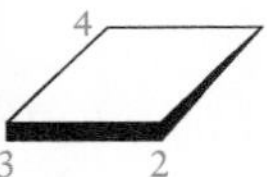

Example 8.9 Identifying Cis–Trans Isomers

Draw the skeletal structures of the isomers of 1,3 dimethylcyclobutane. Indicate which ones are cis–trans isomers.

Solution

Step 1: *Use the root of the name to draw the longest chain.* Use the last part of the name (i.e., cyclobutane) to determine the longest chain. Draw a cyclic chain of four carbons. Then number the carbon atoms.

Step 2: *Use the alkyl group positions in the name* to attach a methyl group at position 1 and 3. Notice you can add both methyl groups on the same side (cis) or opposite sides (trans) of the ring.

cis-1,3-dimethylcyclobutane *trans*-1,3-dimethylcyclobutane

✔ Learning Check 8.9

a. Identify each of the following cycloalkanes as a cis or trans compound.

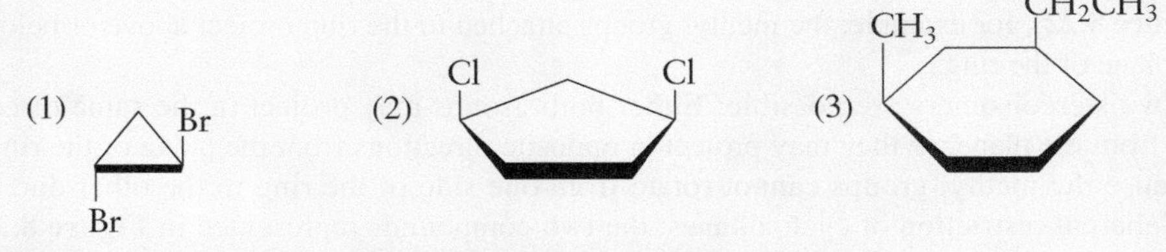

b. Draw the structural formula for *cis*-1,2-dichlorocyclobutane.

Now that you have a good grasp of stereoisomers, you might be wondering, "Why should I care?" The answer is that stereoisomerism is extremely important in drug design. For example, thalidomide, a drug administered to pregnant females to ease morning sickness in the late 1950s and early 1960s, led to a reported 5000–7000 babies born with disabilities, with a 60% fatality rate (**Figure 8.23**). Unfortunately, what wasn't known at the time the drug was first administered was that only one of the stereoisomers of thalidomide aided symptoms of morning sickness. The other stereoisomer led to negative effects on fetal development. Later research found that even if the desired stereoisomer is isolated and administered as a drug, the body is capable of interconverting the desired isomer to the one that causes fetal side effects!

Figure 8.23 Thalidomide was used as a drug to treat females with morning sickness. (a) Children affected by thalidomide. (b) The skeletal structure of thalidomide.

8.7 Alkenes and Alkynes

Learning Objective 8 Write the IUPAC names of alkenes and alkynes from their molecular structures.

alkene A hydrocarbon containing one or more carbon–carbon double bonds.

alkyne A hydrocarbon containing one or more carbon–carbon triple bonds.

aromatic hydrocarbon Any organic compound that contains a benzene ring.

As discussed in **Section 8.3**, unsaturated hydrocarbons contain one or more double or triple bonds between carbon atoms and belongs to one of three classes: alkenes, alkynes, or aromatic hydrocarbons. An **alkene** contains one or more carbon–carbon double bonds; an **alkyne** contains one or more carbon–carbon triple bonds; and an **aromatic hydrocarbon** contains a benzene ring, with its three carbon–carbon double bonds alternating with three carbon–carbon single bonds in a six-carbon ring, and its derivatives. Ethylene is the simplest alkene, acetylene is the simplest alkyne, and benzene is the simplest aromatic:

$H_2C{=}CH_2$
ethylene
(an alkene)

$H{-}C{\equiv}C{-}H$
acetylene
(an alkyne)

C_6H_6
benzene
(an aromatic)

8.7A Alkenes

The general formula for alkenes is C_nH_{2n} (the same as that for cycloalkanes). Instead of forming a ring, though, two hydrogens from the generic alkane formula are replaced by

Robert Fried/Alamy Stock Photo

Figure 8.24 Ethylene (ethene), a plant hormone that controls growth and fruit development, is used commercially to artificially ripen fruits.

a double bond between two carbon atoms. The simplest alkenes are well known by their common names, ethylene (see **Figure 8.24**) and propylene:

$$\underset{\text{ethylene, } C_2H_4}{CH_2{=}CH_2} \qquad \underset{\text{propylene, } C_3H_6}{CH_3{-}CH{=}CH_2}$$

Three alkene structural isomers have the formula C_4H_8:

$$CH_3{-}CH_2{-}CH{=}CH_2 \qquad CH_3{-}CH{=}CH{-}CH_3 \qquad CH_3{-}\underset{\underset{CH_3}{|}}{C}{=}CH_2$$

As discussed in **Section 8.4A**, the number of structural isomers increases rapidly as the number of carbons increases because of variations in chain length and branching and, for alkenes, because variations occur in the position of the double bond. IUPAC nomenclature is extremely useful in differentiating among these many alkene compounds.

The IUPAC rules for naming alkenes are similar to those used for the alkanes, with a few additions to indicate the presence and location of the double bonds.

Step 1: *Number the longest chain* of carbon atoms so that the carbon atoms joined by the double bond have the lowest numbers possible.

Step 2: *Name the longest chain* that contains the double bond. The suffix name of the compound is *-ene*.

Step 3: *Locate the double bond* by the lower-numbered carbon atom in the double bond.

Step 4: *Locate and name any attached groups.*

Step 5: *Combine the names for the attached groups and the longest chain into the name.*

Example 8.10 Naming Alkenes

Give the IUPAC name for the following alkenes.

a. $CH_3{-}CH{=}CH{-}CH_3$ b. $CH_3{-}\underset{\underset{CH_3}{|}}{CH}{-}CH{=}CH_2$

Solution

a. **Step 1:** *Number the longest chain of carbon atoms.* The longest chain containing both carbons of the double bond has four carbon atoms. The chain can be numbered from either end because the double bond will be between carbons 2 and 3 either way.

$$\overset{1}{C}H_3{-}\overset{2}{C}H{=}\overset{3}{C}H{-}\overset{4}{C}H_3$$

Step 2: *Name the longest chain.* The four-carbon alkane is butane. Thus, the compound is a butene.

Step 3: *Locate the double bond.* The position of the double bond is indicated by the lower-numbered carbon atom that is part of the double bond (carbon 2 in this case).

Step 4: *Locate and name attached groups.* No groups are attached to the main chain in this case.

Step 5: *Combine the names for the attached groups and the longest chain into the name.* In this case, the name is 2-butene.

b. **Step 1:** *Number the longest chain of carbon atoms.* The longest chain containing both carbons of the double bond has four carbon atoms. To give the lowest numbers possible to the carbons of the double bond, the chain is numbered from right to left.

$$\overset{4}{C}H_3{-}\underset{\underset{CH_3}{|}}{\overset{3}{C}H}{-}\overset{2}{C}H{=}\overset{1}{C}H_2 \qquad \textit{not} \qquad \overset{1}{C}H_3{-}\underset{\underset{CH_3}{|}}{\overset{2}{C}H}{-}\overset{3}{C}H{=}\overset{4}{C}H_2$$

Step 2: *Name the longest chain.* The four-carbon alkane is butane. Thus, the compound is a butene.

Step 3: *Locate the double bond.* The position of the double bond is indicated by the lower-numbered carbon atom that is part of the double bond (carbon 1 in this case).

Step 4: *Locate and name attached groups.* There is an attached methyl group at position 3 of the main carbon chain.

Step 5: *Combine the names for the attached groups and the longest chain into the name.* In this case, the name is 3-methyl-1-butene.

✔ Learning Check 8.10 Give the IUPAC name for each of the following.

a. $Br-CH_2-CH=CH_2$ (Br attached to the CH_2)

b. $CH_2=C(CH_2CH_3)-CH_2-CH_2-CH_3$ (with CH_2-CH_3 attached to C-2)

c. H_3C CH_3 (cyclohexene ring bearing two adjacent methyl groups)

Some compounds contain more than one double bond per molecule (see **Figure 8.25**). Molecules of this type are important components of natural and synthetic rubber and other useful materials. The nomenclature of these compounds is the same as for the alkenes with one double bond, except that the endings *-diene*, *-triene*, and the like are used to denote the number of double bonds. Also, the locations of all the multiple bonds must be indicated in all molecules, including those within rings:

valentinrussanov/iStock/Getty Images

Figure 8.25 β-Carotene, the compound responsible for the orange color of carrots and yams, contains 11 double bonds.

$$\overset{1}{C}H_2=\overset{2}{C}H-\overset{3}{C}H=\overset{4}{C}H_2$$

1,3-butadiene

1,3-cyclohexadiene (ring carbons numbered 1, 2, 3, 4)

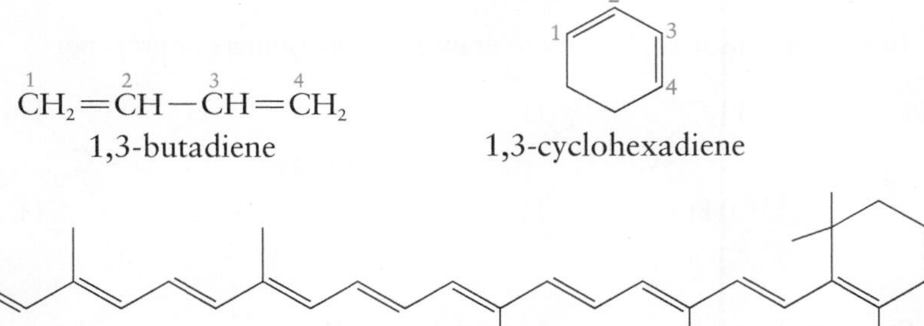

β-carotene

Because of the double bond, alkenes have a different geometry than alkanes. With only three bonded groups—namely, one double bond and two other groups (H or C)—the carbon atoms of the double bond have a trigonal planar geometry with a 120° bond angle (see **Table 5.1**). In addition to geometry, alkenes also differ from open-chain alkanes in that the double bonds prevent the relatively free rotation that is characteristic of carbon atoms bonded by single bonds. As a result, alkenes can exhibit cis–trans isomerism, the same type of stereoisomerism seen in **Section 8.6** for cycloalkanes. For instance, there are two cis–trans isomers of 2-butene:

H H C=C H_3C CH_3	H_3C H C=C H CH_3
cis-2-butene	*trans*-2-butene

coffeekai/iStock/Getty Images

Figure 8.26 The unsaturated components in vegetable oils contain cis double bonds, which contribute to making the material a liquid at room temperature.

Once again, the cis isomer has the two similar or identical groups on the same side of the double bond, whereas the trans isomer has them on opposite sides. The two isomers, *cis*- and *trans*-2-butene, represent distinctly different compounds with different physical properties (see **Table 8.8** and **Figure 8.26**).

Table 8.8 Physical Properties of a Pair of Cis–Trans Isomers

Isomer	Melting Point (°C)	Boiling Point (°C)	Density (g/mL)
cis-2-Butene	−139.9	3.7	0.62
trans-2-Butene	−105.6	0.9	0.60

Not all double-bonded compounds show cis–trans stereoisomerism. Cis–trans stereoisomerism is found only in alkenes that have two different groups attached to each double-bonded carbon atom. In 2-butene, the two different groups are a methyl and a hydrogen for each double-bonded carbon:

Two groups are different. (H and H_3C) — $C{=}C$ — (H and CH_3) *Two groups are different.*

cis-2-butene

Two groups are different. (H and H_3C) — $C{=}C$ — (CH_3 and H) *Two groups are different.*

trans-2-butene

If either double-bonded carbon is attached to identical groups, no cis–trans isomers are possible. Thus, there are no cis–trans isomers of ethene or propene:

(H, H) $C{=}C$ (H, H) — *Two groups are the same.*

ethene

(H_3C, H) $C{=}C$ (H, H) — *Two groups are the same.*

propene

To see why this is so, let's try to draw cis–trans isomers of propene:

(H_3C, H) $C{=}C$ (H, H) (A) is the same as (H, H_3C) $C{=}C$ (H, H) (B)

Structure B can be converted into A by just flipping it over, so they are identical, not isomers.

Example 8.11 Drawing Cis–Trans Isomers

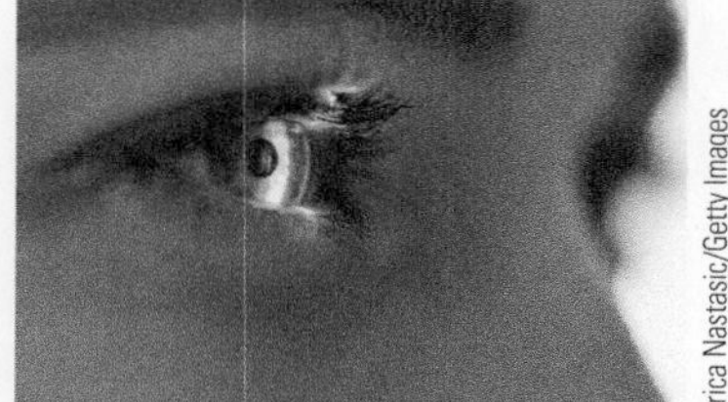

Zorica Nastasic/Getty Images

Figure 8.27 The isomerization of *cis*-retinal to *trans*-retinal is a vital process involved in vision.

Cis–trans isomerism plays an essential role in many biological processes, including the response to light by the retina of the eye (**Figure 8.27**). Determine which of the following molecules can exhibit cis–trans isomerism, and draw structural formulas to illustrate your conclusions.

a. $Cl{-}CH{=}CH{-}Cl$
b. $CH_2{=}CH{-}Cl$

Solution

a. **Step 1:** Begin by drawing the carbon–carbon double bond with two additional bonds about each carbon atom:

$$\rangle C{=}C \langle$$

Step 2: Complete the structure and analyze each carbon of the double bond to see if it is attached to two different groups:

```
                          H         H
 H and Cl are      ↗   \       /   ↖     H and Cl are
 different groups.      C=C           different groups.
                   ↘   /       \   ↙
                         Cl        Cl
```

Step 3: In this case, each carbon is attached to two different groups, and cis–trans isomers are possible:

```
  H          H            Cl          H
    \       /                \       /
      C=C                      C=C
    /       \                /       \
  Cl         Cl            H          Cl
```

cis-1,2-dichloroethene *trans*-1,2-dichloroethene

b. **Step 1:** Draw the carbon–carbon bond as in Step 1 of solution a.

Step 2: Complete the structure and analyze each carbon of the double bond to see if it is attached to two different groups. No cis–trans isomers are possible because one carbon contains two identical groups:

```
                     H          H
 Two identical  ↗     \       /
   groups               C=C
                ↘     /       \
                     H          Cl
```

Figure 8.28 Alkynes have been isolated from several plant species, marine sponges, and corals.

ifish/E+/Getty Images

> ✔ **Learning Check 8.11** Determine which of the following can exhibit cis–trans isomerism, and draw structural formulas for the cis and trans isomers of those that can.
>
> a. $CH_3-CH-CH=CH-CH_3$ (with CH_3 on the second carbon)
>
> b. $CH_3-C=CH-CH_3$ (with CH_3 on the second carbon)

8.7B Alkynes

The characteristic feature of alkynes is the presence of a triple bond between carbon atoms. Thus, alkynes are also unsaturated hydrocarbons. Although relatively rare, compounds containing the carbon–carbon triple bond are found in nature (see **Figure 8.28**). The simplest and most important compound of this series is ethyne, more commonly called acetylene (C_2H_2):

$$H-C\equiv C-H$$

acetylene

Acetylene is used in torches for welding steel and in making plastics and synthetic fibers. The systematic IUPAC naming of alkynes is identical to the naming of alkenes, except the ending suffix is *-yne* instead of *-ene*.

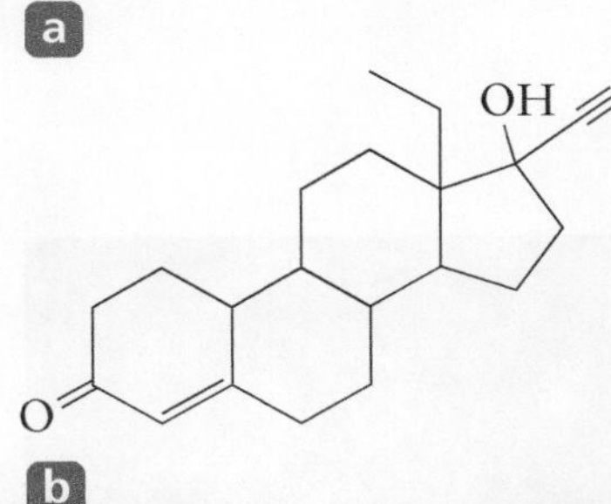

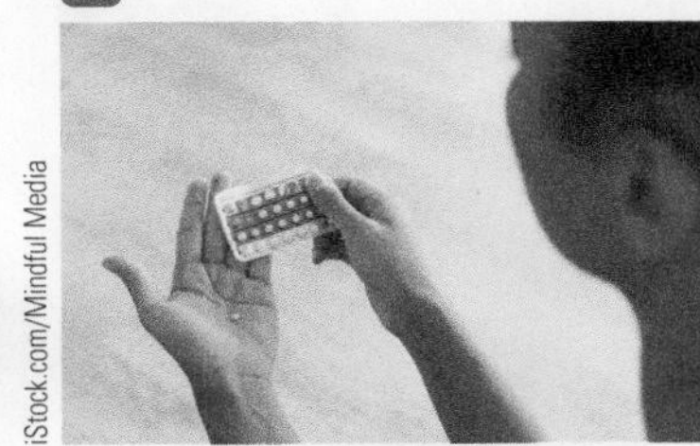

iStock.com/Mindful Media

Figure 8.29 Norgestrel is an oral contraceptive (birth control pill) that contains a carbon–carbon triple bond (an alkyne). (a) Chemical structure of Norgestrel. (b) Woman taking one of her monthly birth control pills.

Example 8.12 Naming Alkynes

Norgestrel, a compound once found in some birth control pills, has a carbon–carbon triple bond (**Figure 8.29**). There are over a thousand biomolecules containing carbon–carbon triple bonds, so understanding how to name them is important. Give the IUPAC name for the following alkynes.

a. $CH_3CH_2CH_2-C\equiv C-H$

b. $H-C\equiv C-CH-CH_3$ (with CH_3 on the third carbon)

Solution

a. Alkynes are named in exactly the same way as alkenes, except the ending *-yne* is used.

Step 1: *Number the longest chain of carbon atoms.* The longest chain containing both carbons of the triple bond has five carbon atoms. To give the lowest possible numbers to the carbons in the triple bond, the chain is numbered from the right.

$$\overset{5}{C}H_3\overset{4}{C}H_2\overset{3}{C}H_2-\overset{2}{C}\equiv\overset{1}{C}-H$$

Step 2: *Name the longest chain.* The five-carbon alkane is pentane. Thus, the alkyne is a pentyne.

Step 3: *Locate the triple bond.* The position of the triple bond is indicated by the lower-numbered carbon atom in the triple bond—carbon 1 in this case.

Step 4: *Locate and name attached groups.* There are no groups attached to the main carbon chain.

Step 5: *Combine the names for the attached groups and the longest chain into the name.* In this case the name is 1-pentyne.

b. **Step 1:** *Number the longest chain of carbon atoms.* The longest chain containing both carbons of the triple bond has four carbon atoms. To give the lowest numbers to the carbons in the triple bond, the chain is numbered from left to right:

$$H-\overset{1}{C}\equiv\overset{2}{C}-\underset{\displaystyle CH_3}{\overset{3}{C}H}-\overset{4}{C}H_3 \quad \text{not} \quad H-\overset{4}{C}\equiv\overset{3}{C}-\underset{\displaystyle CH_3}{\overset{2}{C}H}-\overset{1}{C}H_3$$

Step 2: *Name the longest chain.* The four-carbon alkane is butane. Thus, the alkyne is a butyne.

Step 3: *Locate the triple bond.* The position of the triple bond is indicated by the lower-numbered carbon atom in the triple bond—carbon 1 in this case.

Step 4: *Locate and name attached groups.* There is a methyl group attached at position 3 of the main carbon chain.

Step 5: *Combine the names for the attached groups and the longest chain into the name.* In this case, the name is 3-methyl-1-butyne.

✔ **Learning Check 8.12** Give the IUPAC name for each of the following.

a. $CH_3-CH_2-C\equiv C-CH_3$

b. $CH_3-\overset{\displaystyle CH_3}{\overset{|}{C}H}-CH_2-C\equiv C-CH_3$

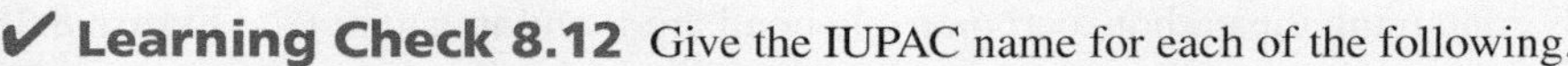

8.8 Reactions of Hydrocarbons

Learning Objective 9 Write chemical equations for addition reactions of alkenes.

Learning Objective 10 Use Markovnikov's rule to predict the major products of certain alkene addition reactions.

Learning Objective 11 Write chemical equations for addition polymerization.

Danita Delimont/Shutterstock.com

Figure 8.30 A mixture of solid alkanes makes up paraffin wax, a substance used to seal homemade jam jars.

Alkanes are the least reactive of all organic compounds. In general, they do not react with strong acids (e.g., sulfuric acid), strong bases (e.g., sodium hydroxide), most oxidizing agents (e.g., potassium dichromate), or most reducing agents (e.g., sodium metal).

This lack of reactivity is reflected in the name *paraffins* (from Latin words that mean "little affinity"). Paraffin wax, sometimes used to seal jars of homemade preserves, is a mixture of solid alkanes (**Figure 8.30**). Paraffin wax is also used in the preparation of milk cartons and wax paper. The inertness of the compounds in the wax makes it ideal for these uses. Alkanes do undergo reactions with halogens such as chlorine and bromine, but these are not important for our purposes. The combustion of alkanes was previously discussed in **Section 4.6E**.

The reactions of alkenes, and to a lesser extent alkynes, are much more varied because of the presence of the carbon–carbon multiple bond. Because both alkenes and alkynes rely on the reactivity of a multiple bond, their reactions are quite similar. In this section, we focus on alkenes, which are typically easier to control and more useful because they contain only the double bond. The reactions presented in this section are just a very limited selection of the ones alkenes and alkynes are known to undergo.

8.8A Alkene Addition

The first reaction we consider is the addition of two components across the carbon–carbon double bond of an alkene:

$$\mathrm{>C{=}C<} \; + \; A{-}B \longrightarrow \; -\underset{|}{\overset{\overset{A}{|}}{\mathrm{C}}}-\underset{|}{\overset{\overset{B}{|}}{\mathrm{C}}}- \qquad (8.1)$$

addition reaction A reaction in which a compound adds to a multiple bond.

These are called **addition reactions**, because a substance is added across the double bond. Addition reactions are characterized by two reactants (one of which is the alkene) that combine to form one product. In **hydrogenation**, two hydrogen atoms are added across the double bond of the alkene. A transition metal such as platinum (Pt) is needed to catalyze the reaction. One of the bonds in the double bond is replaced by two new C–H bonds. Thus, this reaction converts alkenes into alkanes.

hydrogenation A reaction in which the addition of hydrogen takes place. A precious metal catalyst (e.g., Pt, Ni, or Pd) is typically needed to make the reaction work.

polyunsaturated A term usually applied to molecules with several double bonds.

$$\mathrm{>C{=}C<} \; + \; H_2 \xrightarrow{\text{Pt}} \; -\underset{\underset{H}{|}}{\overset{|}{\mathrm{C}}}-\underset{\underset{H}{|}}{\overset{|}{\mathrm{C}}}- \qquad (8.2)$$

an alkane

Specific example: $CH_3CH{=}CHCH_3 + H_2 \xrightarrow{\text{Pt}} CH_3\underset{\underset{H}{|}}{CH}-\underset{\underset{H}{|}}{CH}CH_3$

2-butene → butane

Vegetable oils, such as soybean and cottonseed oil, are composed of long-chain organic molecules that contain many alkene bonds. The high degree of unsaturation characteristic of these oils is why they are said to be **polyunsaturated**. Upon hydrogenation, the melting point of the oils is raised, and the oils become low-melting-point solids. These products are used in the form of margarine (see **Figure 8.31**) and shortening.

George Doyle/Stockbyte/Getty Images

Figure 8.31 Margarines are prepared by the hydrogenation of vegetable oils.

Example 8.13 Writing Hydrogenation Reactions

Hydrogenation is a key step in the industrial synthesis of L-DOPA, a precursor to various neurotransmitters and a key treatment for Parkinson's disease (**Figure 8.32**).

[Structure: HO, HO on benzene ring; side chain CH₂–CH(NH₂)–C(=O)OH]

l-3,4-dihydroxyphenylalanine
(L-DOPA)

FG Trade/Getty Images

Figure 8.32 Parkinson's causes nerve cell death, which results in symptoms that include tremors and muscle stiffness.

Write the structural formula for the product of each of the following hydrogenation reactions.

a. (cyclohexene) $+ H_2 \xrightarrow{Pt} ?$ b. $CH_3-\underset{\displaystyle CH_3}{\underset{|}{C}}=CH_2 + H_2 \xrightarrow{Pt} ?$

Solution

a. In an addition reaction, the carbon–carbon double bond is converted to a single bond. The two hydrogen atoms attach to the carbon atoms that were part of the double bond. In the skeletal structure of the product, none of the hydrogen atoms attached to ring carbon atoms are shown.

(cyclohexene) $+ H_2 \xrightarrow{Pt}$ (cyclohexane with H, H adding) $\longrightarrow$ (cyclohexane with H, H) written as (cyclohexane)

product

b. The carbon–carbon double bond becomes a carbon–carbon single bond with two hydrogen atoms attached.

$$CH_3-\underset{\displaystyle CH_3}{\underset{|}{C}}=CH_2 + H_2 \xrightarrow{Pt} CH_3-\underset{\displaystyle CH_3}{\underset{|}{\overset{H}{C}}}-\overset{H}{CH_2} \longrightarrow CH_3-\underset{\displaystyle CH_3}{\underset{|}{\overset{\displaystyle H}{\overset{|}{C}}}}-CH_3$$

✔ **Learning Check 8.13** Write the structural formula for the product of each of the following hydrogenation reactions.

a. $CH_3-\underset{\displaystyle CH_3}{\underset{|}{C}}=CH-CH_3+H_2 \xrightarrow{Pt} ?$ b. (1-methylcyclopentene, CH_3) $+ H_2 \xrightarrow{Pt} ?$

hydration The addition of water to a multiple bond.

In the absence of a catalyst, water does not react with alkenes. But if a catalyst such as sulfuric acid is added, then water adds to carbon–carbon double bonds to give alcohols. In this reaction, which is called **hydration**, a water molecule is effectively split in such a way that –H attaches to one carbon of the double bond, and –OH attaches to the other carbon. In Equation 8.3, H_2O is written H–OH to emphasize the portions that add to the double bond.

$$\gt C=C \lt \; + \; H-OH \xrightarrow{H_2SO_4} -\underset{\displaystyle H}{\underset{|}{\overset{|}{C}}}-\underset{\displaystyle OH}{\underset{|}{\overset{|}{C}}}- \quad (8.3)$$

an alcohol

$$\underset{\text{2-butene}}{CH_3CH=CHCH_3} + H-OH \xrightarrow{H_2SO_4} \underset{\text{2-butanol}}{CH_3\underset{\displaystyle H}{\underset{|}{CH}}-\underset{\displaystyle OH}{\underset{|}{CH}}CH_3}$$

The addition reaction of H_2 yielded only one product because the same group (H and H) adds to each double-bonded carbon. With H–OH, however, a different group can add to each carbon, and for certain alkenes, there are two possible products. For example, in the reaction of H_2O with propene, both 2-propanol and 1-proponal might be formed. However, we only observe the formation of 2-propanol as a product.

$$CH_3CH{=}CH_2 + H{-}OH \xrightarrow{H_2SO_4} \underset{\text{2-propanol}}{CH_3\underset{OH}{CH}{-}\underset{H}{CH_2}} \quad \textit{not} \quad \underset{\text{1-propanol}}{CH_3\underset{H}{CH}{-}\underset{OH}{CH_2}}$$

propene

Markovnikov's rule In the addition of H—X to an alkene, the hydrogen becomes attached to the carbon atom that is already bonded to more hydrogens.

This fact, first reported in 1869 by chemist Vladimir Markovnikov (1837–1904), gave rise to a rule for predicting which product will be exclusively or predominantly formed. According to **Markovnikov's rule**, when a molecule of water adds to an alkene, the hydrogen becomes attached to a double-bonded carbon atom that is already bonded to more hydrogens. In other words, "H's go where H's are." Applying this rule to propene, we find:

$$CH_3 - CH{=}CH_2$$

One hydrogen attached (arrow to CH)

Three hydrogens, but they are not attached to the double-bonded carbons (arrow to CH_3)

Two hydrogens attached (arrow to CH_2)

Figure 8.33 Hydration of alkenes is crucial to carbohydrate metabolism.

Therefore, H attaches to the end carbon of the double bond, OH attaches to the second carbon, and 2-propanol is the observed product.

The hydration of alkenes provides a convenient method for preparing alcohols on a large scale. The reaction is also important in living organisms, but the catalyst is an enzyme rather than sulfuric acid. For example, one of the steps in the body's utilization of carbohydrates for energy (**Figure 8.33**) involves the hydration of fumaric acid, which is catalyzed by the enzyme fumarase:

$$\underset{\text{fumaric acid}}{HO{-}\overset{O}{\overset{\|}{C}}{-}CH{=}CH{-}\overset{O}{\overset{\|}{C}}{-}OH} + H_2O \xrightarrow{\text{Fumarase}} \underset{\text{malic acid}}{HO{-}\overset{O}{\overset{\|}{C}}{-}\overset{H}{\overset{|}{CH}}{-}\overset{OH}{\overset{|}{CH}}{-}\overset{O}{\overset{\|}{C}}{-}OH}$$

Example 8.14 Writing Hydration Reactions

Hydration is one of the steps in the citric acid cycle (Chapter 15). Draw structural formulas for the major organic product of each of the following reactions.

a. (cyclohexene) $+ H_2O \xrightarrow{H_2SO_4}$

b. $CH_3{-}CH{=}\overset{CH_3}{\overset{|}{C}}{-}CH_3 + H_2O \xrightarrow{H_2SO_4}$

Solution

a. In addition reactions, the carbon–carbon double bond is converted to a single bond. The –H and –OH of water attach to the carbon atoms that were part of the double bond. In the skeletal structure of the product, the additional hydrogen atom is not shown; nor are the other hydrogen atoms attached to the ring carbon atoms. It isn't necessary to apply Markovnikov's rule because both ends of the alkene are the same.

(cyclohexene) $+ H_2O \xrightarrow{H_2SO_4}$ (ring with H and OH adding) $\longrightarrow$ (cyclohexane with H and OH) written as (cyclohexanol, OH) product

b. With a different number of hydrogen atoms attached to the carbon atoms connected by a double bond, Markovnikov's rule must be followed—namely, the hydrogen adds to the carbon that is already bonded to more hydrogens.

$$CH_3-\underset{\substack{\uparrow \\ \textit{One hydrogen attached}}}{CH}=\underset{\substack{\uparrow \\ \textit{No hydrogen attached}}}{\overset{\overset{\displaystyle CH_3}{|}}{C}}-CH_3$$

The carbon–carbon double bond is converted to a single bond.

$$CH_3-CH=\overset{\overset{\displaystyle CH_3}{|}}{C}-CH_3+H_2O \xrightarrow{H_2SO_4} CH_3\underset{\substack{\uparrow \\ \textit{H attaches here.}}}{CH}-\underset{\substack{\uparrow \\ \textit{OH attaches here.}}}{\overset{\overset{\displaystyle CH_3}{|}}{C}}-CH_3 \longrightarrow CH_3CH_2-\underset{\underset{\displaystyle OH}{|}}{\overset{\overset{\displaystyle CH_3}{|}}{C}}-CH_3$$

iStock.com/SDI Productions

Figure 8.34 Keeping a physical or digital reaction card file is a great tactic for organizing and learning organic chemistry reactions.

> ✔ **Learning Check 8.14** Draw structural formulas for the major organic product of each of the following reactions.
>
> a. $CH_3CH_2CH_2CH=CH_2+H_2O \xrightarrow{H_2SO_4}$
>
> b.  $+ H_2O \xrightarrow{H_2SO_4}$

Study Tools 8.1

Keeping a Reaction Card File

Remembering organic reactions for exams is challenging for most students. Because the number of reactions being studied increases rapidly over the semester, it is a good idea to develop a systematic way to organize them for easy and effective review.

One way to do this is to focus on the functional group concept. When an exam question asks you to complete a reaction by identifying the product, your first step should be to identify the functional group of the reactant. Usually, only the functional group portion of a molecule undergoes reaction. In addition, a particular functional group usually undergoes the same characteristic reactions regardless of the other features of the organic molecule to which it is bonded. Thus, by remembering the behavior of a functional group under specific conditions, you can predict the reactions of many compounds, no matter how complex the structures look, as long as they contain the same functional group. For example, any structure containing a C=C will undergo reactions typical of alkenes. Other functional group reactions are introduced in later chapters.

Keeping a reaction card file (**Figure 8.34**) based on the functional group concept is a good way to organize reactions for review. Write the structures and names of the reactants on one side of an index card with an arrow showing any catalyst or special conditions. Write the product structure and name on the back of the card. We recommend that you do this for the general reaction (like those in the Key Reactions section at the end of most chapters) and for a specific example. Review your cards every day (this can even be done while waiting for a bus, etc.), and add to them as new reactions are studied. As an exam approaches, put aside the reactions you know well, and concentrate on the others in what should be a dwindling deck. This is an effective way to focus on learning what you don't know.

8.8B Polymer Formation

Certain alkenes undergo an addition reaction with one another in the presence of specific catalysts. In this reaction, the double bonds of the reacting alkenes are converted to single bonds as hundreds or thousands of molecules bond and form long chains. For example, several ethylene molecules react as follows:

$$\underset{\text{ethylene molecules}}{CH_2{=}CH_2 + CH_2{=}CH_2 + CH_2{=}CH_2 + CH_2{=}CH_2} \xrightarrow[\text{catalysts}]{\text{Heat, pressure}}$$

$$\underset{\text{polyethylene}}{-CH_2-CH_2-CH_2-CH_2-CH_2-CH_2-CH_2-CH_2-}$$

Figure 8.35 Sneeze guards, often used during the COVID-19 pandemic, are frequently made of plexiglass, a polymer.

The product is commonly called polyethylene, even though there are no longer any double bonds present. The newly formed bonds in this long chain are shown in color. This type of reaction is called a **polymerization**, and the long-chain product made up of repeating units is a **polymer** (*poly* = many, *mer* = parts). The trade names of common polymers include Orlon, Plexiglas, Lucite, and Teflon (see **Figure 8.35**). These products are referred to as **addition polymers** because the addition reaction between double-bonded compounds is used to produce them. The starting materials that make up the repeating units of polymers are called **monomers** (*mono* = one, *mer* = part). Polymers are used extensively in medicine as suture materials (**Figure 8.36**), tissue adhesives, contact lenses, and vascular grafts.

polymerization A reaction that produces a polymer.

polymer A very large molecule made up of repeating units.

addition polymer A polymer formed by the linking together of many alkene molecules through addition reactions.

monomer The starting material that becomes the repeating units of polymers.

It is not possible to give an exact formula for a polymer produced by a polymerization reaction because the individual polymer molecules vary in size. We could represent polymerization reactions as we just did for the formation of polyethylene from ethylene. This type of reaction is inconvenient, however, so instead we represent the polymer by a simple repeating unit based on the monomer. For polyethylene, the unit is $\text{+}CH_2{-}CH_2\text{+}$. The large number of units making up the polymer is denoted by *n*, a whole number that varies from several hundred to several thousand. The polymerization reaction of ethylene is then written as follows:

$$\underset{\text{ethylene}}{nCH_2{=}CH_2} \xrightarrow[\text{catalysts}]{\text{Heat, pressure}} \underset{\text{polyethylene}}{\text{+}CH_2-CH_2\text{+}_n} \qquad (8.4)$$

The lowercase *n* in Equation 8.4 represents a large, unspecified number. Based on this reaction, polyethylene is essentially a very long chain alkane. As a result, it has the chemical inertness of alkanes, a characteristic that makes polyethylene suitable for food storage containers, garbage bags, eating utensils, laboratory apparatus, and hospital equipment (see **Figure 8.37**). Polymer characteristics can be modified by using alkenes with different groups attached to either or both of the double-bonded carbons. For example, the polymerization of vinyl chloride gives the polymer polyvinyl chloride (PVC):

$$\underset{\text{vinyl chloride}}{nCH_2{=}\overset{\overset{\large Cl}{|}}{C}H} \xrightarrow[\text{catalysts}]{\text{Heat}} \underset{\text{polyvinyl chloride}}{\text{+}CH_2-\overset{\overset{\large Cl}{|}}{C}H\text{+}_n}$$

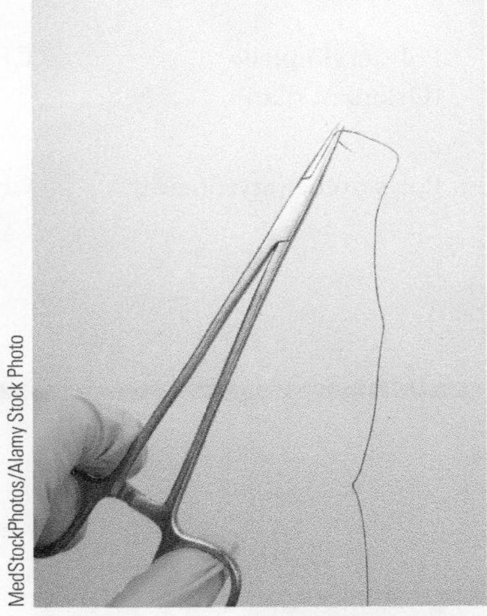
Figure 8.36 Sutures composed of synthetic polymers are ideal for soft tissue.

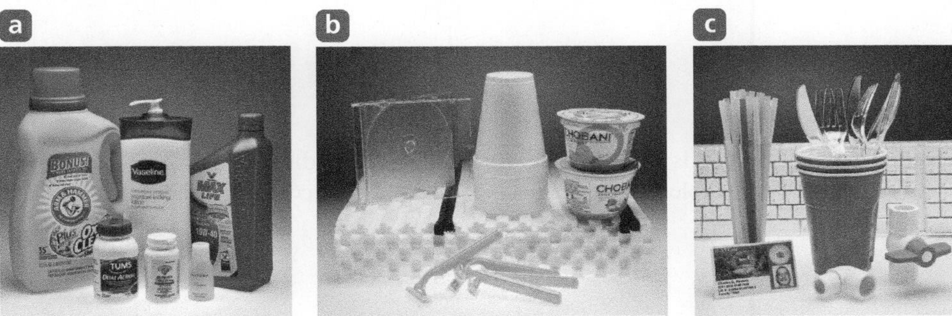

Figure 8.37 Common polymer-based consumer products include (a) high-density polyethylene, (b) polystyrene, and (c) polyvinyl chloride.

A number of the more important addition polymers are shown in **Table 8.9**. Based on the typical uses, addition polymers have become nearly indispensable in modern life.

Table 8.9 Common Addition Polymers

Chemical Name and Trade Name(s)	Monomer	Polymer	Typical Uses
Polyethylene	$CH_2{=}CH_2$	$\text{+}CH_2-CH_2\text{+}_n$	Bottles, plastic bags, film
Polypropylene	$CH_2{=}CH(CH_3)$	$\text{+}CH_2-CH(CH_3)\text{+}_n$	Carpet fiber, pipes, bottles, artificial turf
Polyvinyl chloride (PVC)	$CH_2{=}CH(Cl)$	$\text{+}CH_2-CH(Cl)\text{+}_n$	Synthetic leather, floor tiles, garden hoses, water pipe
Polytetrafluoroethylene (Teflon)	$CF_2{=}CF_2$	$\text{+}CF_2-CF_2\text{+}_n$	Pan coatings, plumbers' tape, heart valves, fabrics
Poly(methyl methacrylate) (Lucite, Plexiglas)	$CH_2{=}C(CH_3)(C({=}O)-O-CH_3)$	$\text{+}CH_2-C(CH_3)(C({=}O)-O-CH_3)\text{+}_n$	Airplane windows, paint, contact lenses, fiber optics
Polyvinyl acetate	$CH_2{=}CH(O-C({=}O)-CH_3)$	$\text{+}CH_2-CH(O-C({=}O)-CH_3)\text{+}_n$	Adhesives, latex paint, chewing gum
Polyacrylonitrile (Orlon, Acrilan)	$CH_2{=}CH(CN)$	$\text{+}CH_2-CH(CN)\text{+}_n$	Carpets, fabrics
Polystyrene (Styrofoam)	$CH_2{=}CH(C_6H_5)$	$\text{+}CH_2-CH(C_6H_5)\text{+}_n$	Food coolers, drinking cups, insulation

Example 8.15 Drawing Structural Formulas for Addition Polymers

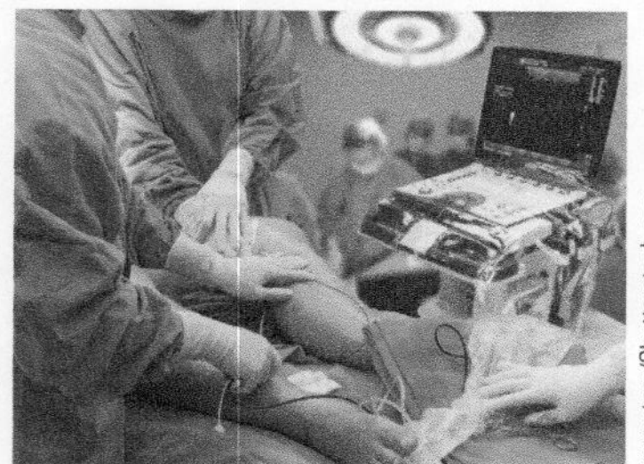

kung_tom/Shutterstock.com

Figure 8.38 Vascular grafts used during surgery are made from polymers.

The synthesis of specialty polymers for medical devices, such as vascular grafts (**Figure 8.38**), is a very active area of science. Draw the structural formula of a portion of polypropylene containing four repeating units of the monomer propylene.

$$CH_2{=}\underset{}{\overset{\displaystyle CH_3}{\overset{|}{CH}}}$$

Solution

Remove one of the bonds in the carbon–carbon double bond. Place new bonds on the left and right side of the molecule so that each carbon has four covalent bonds attached.

$$-CH_2-\overset{\displaystyle CH_3}{\overset{|}{CH}}-$$

This is the repeating unit in polypropylene. Now, attach three more units to the structure.

$$
\begin{array}{cccc}
\mathrm{CH_3} & \mathrm{CH_3} & \mathrm{CH_3} & \mathrm{CH_3} \\
| & | & | & | \\
\underbrace{-\mathrm{CH_2}-\mathrm{CH}}_{\text{Unit 1}} & \underbrace{-\mathrm{CH_2}-\mathrm{CH}}_{\text{Unit 2}} & \underbrace{-\mathrm{CH_2}-\mathrm{CH}}_{\text{Unit 3}} & \underbrace{-\mathrm{CH_2}-\mathrm{CH}-}_{\text{Unit 4}}
\end{array}
$$

✔ **Learning Check 8.15** Draw the structural formula of a portion of polyacrylonitrile containing four repeating units of C_3H_3N, given as follows.

$$
\begin{array}{l}
\mathrm{CH_2}{=}\mathrm{CH} \\
\quad\quad\ \ | \\
\quad\quad\ \mathrm{CN}
\end{array}
$$

8.9 Physical and Chemical Properties of Hydrocarbons

Learning Objective 12 Describe the key physical properties of alkanes, alkenes, and alkynes.

Because alkanes are composed of nonpolar carbon–carbon and carbon–hydrogen bonds, they are nonpolar molecules. Alkanes have lower melting points and boiling points than other organic compounds of comparable molecular weight (see **Table 8.10**) because they exert only very weak attractions (i.e., London dispersion forces) for each other. Alkanes are colorless, odorless compounds.

Table 8.10 Physical Properties of Some Normal Alkanes

Number of Carbon Atoms	IUPAC Name	Condensed Structural Formula	Melting Point (°C)	Boiling Point (°C)	Density (g/mL)
1	Methane	CH_4	−182.5	−164.0	0.55
2	Ethane	CH_3CH_3	−183.2	−88.6	0.57
3	Propane	$CH_3CH_2CH_3$	−189.7	−42.1	0.58
4	Butane	$CH_3CH_2CH_2CH_3$	−133.4	−0.5	0.60
5	Pentane	$CH_3CH_2CH_2CH_2CH_3$	−129.7	36.1	0.63
6	Hexane	$CH_3CH_2CH_2CH_2CH_2CH_3$	−95.3	68.9	0.66
7	Heptane	$CH_3CH_2CH_2CH_2CH_2CH_2CH_3$	−90.6	98.4	0.68
8	Octane	$CH_3CH_2CH_2CH_2CH_2CH_2CH_2CH_3$	−56.8	125.7	0.70
9	Nonane	$CH_3CH_2CH_2CH_2CH_2CH_2CH_2CH_2CH_3$	−53.5	150.8	0.72
10	Decane	$CH_3CH_2CH_2CH_2CH_2CH_2CH_2CH_2CH_2CH_3$	−29.7	174.1	0.73

The straight-chain alkanes make up what is called a **homologous series**. This term describes any series of compounds in which each member differs from a previous member only by having an additional $-CH_2-$ unit. The physical and chemical properties of compounds making up a homologous series are usually closely related and vary in a systematic and predictable way.

homologous series Compounds of the same functional class that differ by a $-CH_2-$ group.

For example, as the number of carbons in an alkane increase, the melting points and boiling points increase. This pattern results from increasing London dispersion forces as molecular weight increases. At ordinary temperatures and pressures, straight-chain alkanes with 1 to 4 carbon atoms are gases, those with 5 to 20 carbon atoms are liquids, and those with more than 20 carbon atoms are waxy solids. Because they are nonpolar, alkanes and other hydrocarbons are insoluble in water, which is a highly polar solvent. They are also less dense than water and thus float on it. These two properties of hydrocarbons are partly responsible for the adverse effects of oil spills from ships and drilling rigs.

The word **hydrophobic** (literally, "water fearing") is often used to refer to molecules or parts of molecules that are insoluble in water. Many biomolecules, the large organic molecules associated with living organisms, contain nonpolar (hydrophobic) parts. Thus, such molecules are insoluble in water. One such example can be found in the long hydrocarbon

hydrophobic Molecules or parts of molecules that repel (are insoluble in) water.

tails that make up the interior of cellular membranes (see Chapter 14). A common hydrocarbon present in our bodies is palmitic acid. The following condensed structure shows that palmitic acid contains a large nonpolar hydrophobic portion, so it is insoluble in water.

$$\overbrace{CH_3-CH_2-CH_2-CH_2-CH_2-CH_2-CH_2-CH_2-CH_2-CH_2-CH_2-CH_2-CH_2-CH_2-CH_2}^{\textit{Nonpolar portion}}-\overset{\overset{\displaystyle O}{\|}}{C}-OH$$

palmitic acid

8.10 Aromatic Compounds and Benzene

Learning Objective 13 Name aromatic compounds.

Learning Objective 14 Draw structural formulas for aromatic compounds.

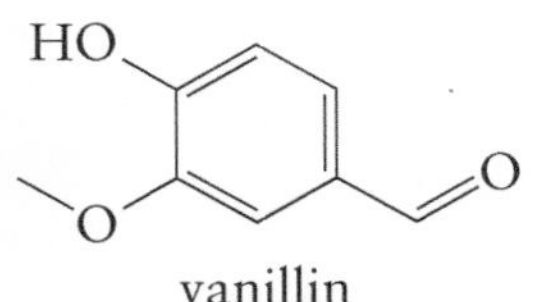

vanillin

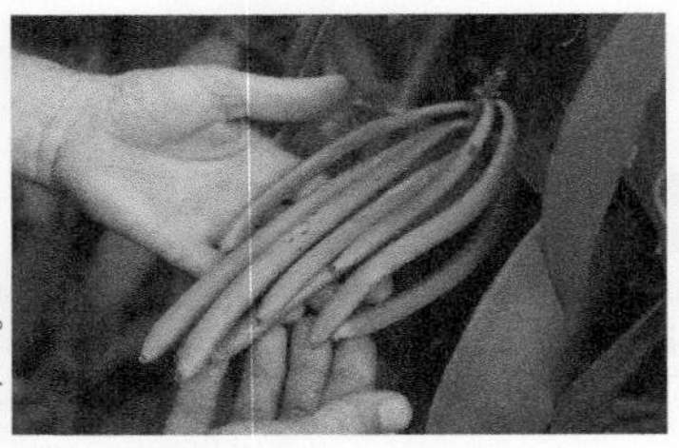

Owen Franken/Corbis Documentary/Getty Images

Figure 8.39 The coveted culinary spice vanilla comes from the bean of a vining orchid native to Mexico, *Vanilla planifolia*.

Aromatic compounds originally derived their name from the fragrant oils of plants, such as oil of wintergreen and vanillin (**Figure 8.39**), that seemed to have similar chemical properties. However, as chemists better understood the structure of benzene, the central ring of many aromatic compounds, our understanding of both benzene and aromatic compounds became more informed. Chemists learned that many compounds belonging to this class, based on their chemical properties and structures, were not at all fragrant. Conversely, there are many fragrant compounds that do not have the properties or structures of aromatic compounds. Today, the old class name is still used to group together benzene and its derivatives due to their inherent and similar chemical properties and reactivity.

8.10A The Structure of Benzene

The molecular structure of benzene presented chemists with an intriguing puzzle. The molecular formula C_6H_6 indicated that the molecule was highly unsaturated. However, the compound did not show the typical reactivity of unsaturated hydrocarbons. Benzene underwent relatively few reactions, and these proceeded slowly and often required heat and catalysts. This was in marked contrast to alkenes, which reacted rapidly with many reagents, in some cases almost instantaneously. This apparent discrepancy between structure and reactivity plagued chemists until 1865, when chemist Friedrich August Kekulé (1829–1896) suggested that the benzene molecule might be represented by a ring arrangement of carbon atoms with alternating single and double bonds between the carbon atoms:

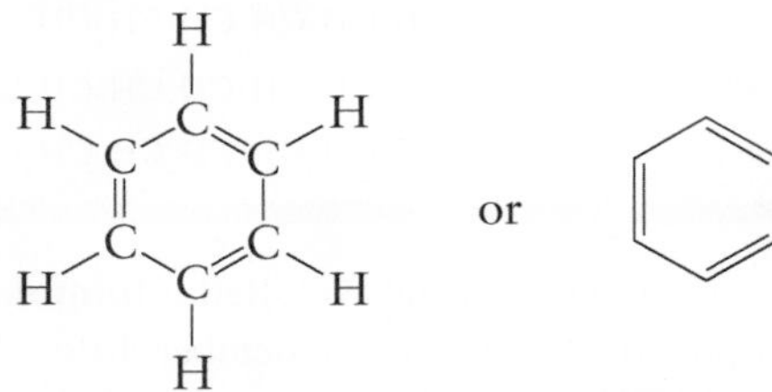

He later suggested that the double bonds alternate in their positions between carbon atoms to give two equivalent structures:

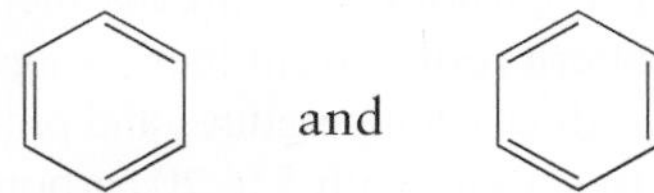

Kekulé structures

Because the two structures are equivalent, the benzene structure is often represented by the following symbol, which is an average of the two:

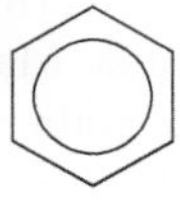

All six carbon and six hydrogen atoms in benzene molecules lie in the same plane. As a result, substituted aromatic compounds do not exhibit cis–trans isomerism.

As you draw the structure of aromatic compounds, remember that only one hydrogen atom or group can be attached to a particular position on the benzene ring. For example, compounds A and B exist, but C does not. The Kekulé structure of compound C shows that the carbon has five bonds attached to it in violation of the octet rule:

Cl | Cl, Cl | Cl Cl | Cl Cl — *Five bonds attached (not possible)*

(A) Exists   (B) Exists   (C) Does not exist   Kekulé structure

8.10B Naming Aromatic Compounds

The naming of aromatic compounds developed somewhat haphazardly for many years before systematic nomenclature schemes were developed. Some of the common names used previously are still used today—some have even been incorporated into the IUPAC nomenclature system (**Figure 8.40**).

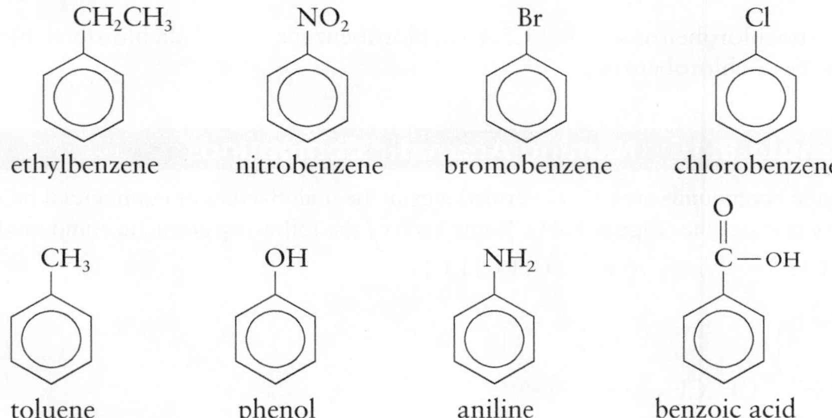

Figure 8.40 Some common aromatic compounds and their names.

Compounds formed by replacing a hydrogen of benzene with a more complex hydrocarbon group can be named by designating the benzene ring as a substituent. This is shown as the benzene ring C_6H_5–:

which is called a **phenyl group**:

phenyl group A benzene ring with one hydrogen removed, C_6H_5–.

$$CH_3-CH_2-CH_2-CH-CH_2-CH_2-CH_3$$

4-phenylheptane

C_6H_5 C_6H_5

1,1-diphenylcyclobutane

It's easy to confuse the words *phenyl* and *phenol*. The key to keeping them straight is the ending: *-ol* means an alcohol (C_6H_5—OH), whereas *-yl* means a *phenyl* group (C_6H_5–).

When two groups are attached to a benzene ring, three isomeric structures are possible. They can be designated by the prefixes *ortho* (*o*), *meta* (*m*), and *para* (*p*):

o-dibromobenzene | *m*-dibromobenzene | *p*-dibromobenzene

Their positions can be indicated by numbering the carbon atoms of the ring so as to obtain the lowest possible numbers for the attachment positions. Groups are arranged in alphabetical order. If there is a choice of identical sets of numbers, the group that comes first in alphabetical order is given the lower number. IUPAC-acceptable common names may be used:

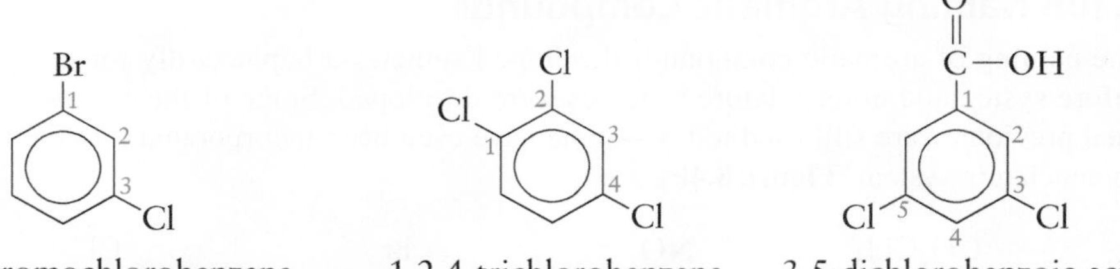

m-bromochlorobenzene or 1-bromo-3-chlorobenzene | 1,2,4-trichlorobenzene | 3,5-dichlorobenzoic acid

Example 8.16 Naming Aromatic Compounds

iStock.com/ThamKC

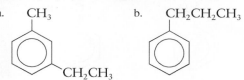

Figure 8.41 Polystyrene, otherwise known as Styrofoam, is manufactured from aromatic derivatives.

Aromatic compounds are often intermediates in the manufacture of commercial products such as polystyrene (**Figure 8.41**). Name each of the following aromatic compounds.

a. CH_3 / CH_2CH_3 b. $CH_2CH_2CH_3$

Solution

a. **Step 1:** *Identify the common aromatic compound* (use **Figure 8.40**). This compound is a substituted toluene, because it contains a benzene ring attached to a methyl group (toluene).

Step 2: *Number the benzene ring*. Position 1 is assigned to the methyl group that makes this compound toluene. In this case, we count clockwise around the ring because this gives the second attached group a lower number, 3, than if we counted counterclockwise.

Correct | Incorrect

Step 3: *Locate and identify attached groups*. An ethyl group is attached at position 3 of the benzene ring. This group is positioned *meta* to the methyl group.

Step 4: *Combine the names for the attached groups and the common aromatic compound into the name.* Both of the following are correct:

3-ethyltoluene and *m*-ethyltoluene

b. **Step 1:** *Identify the common aromatic compound* (use **Figure 8.40**). This compound is named as a substituted benzene, because it contains a benzene ring attached to a propyl group.

Step 2: *Number the benzene ring.* Position 1 is assigned to the only group attached to the benzene ring.

Step 3: *Locate and identify attached groups.* A propyl group is attached at position 1 of the benzene ring.

Step 4: *Combine the names for the attached groups and the common aromatic compound into the name.* The name of the compound is propylbenzene (the 1 is not stated but assumed because there are no other groups attached to the ring).

✔ Learning Check 8.16 Name the following aromatic compounds.

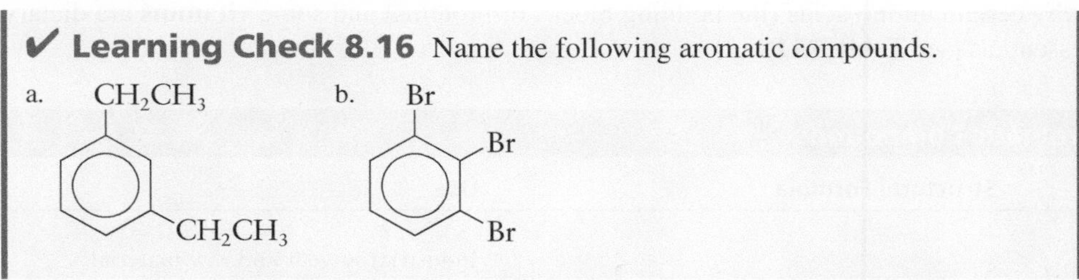

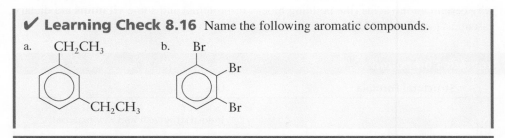

8.10C Properties and Applications of Aromatic Compounds

The physical properties of benzene and other aromatic hydrocarbons are similar to those of alkanes and alkenes. They are nonpolar and thus insoluble in water. Aromatic rings are relatively stable chemically. Because of this, benzene often reacts in such a way that the aromatic ring remains intact. Thus, benzene does not undergo the addition reactions that are characteristic of alkenes and alkynes. The predominant type of reaction of aromatic molecules is substitution, in which one of the ring hydrogens is replaced by some other group. Such aromatic reactions are of lesser importance for our purposes and are not shown here.

All aromatic compounds discussed to this point contain a single benzene ring. There are also substances called **polycyclic aromatic compounds**, which contain two or more benzene rings that share a common side. The rings are said to be "fused" together. The simplest of these compounds, naphthalene (**Figure 8.42**), is the original active ingredient in mothballs:

polycyclic aromatic compound A derivative of benzene in which carbon atoms are shared between two or more benzene rings.

Stephen Barnes/Alamy Stock Photo

Figure 8.42 Naphthalene is the original active ingredient used in mothballs to deter bugs and mice.

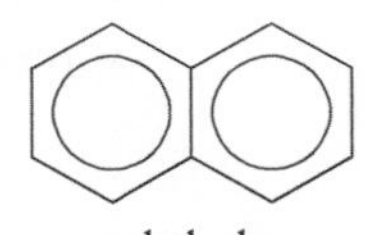

naphthalene

A number of more complex polycyclic aromatic compounds are known to be carcinogens—that is, they are chemicals that cause cancer. Two of these compounds are:

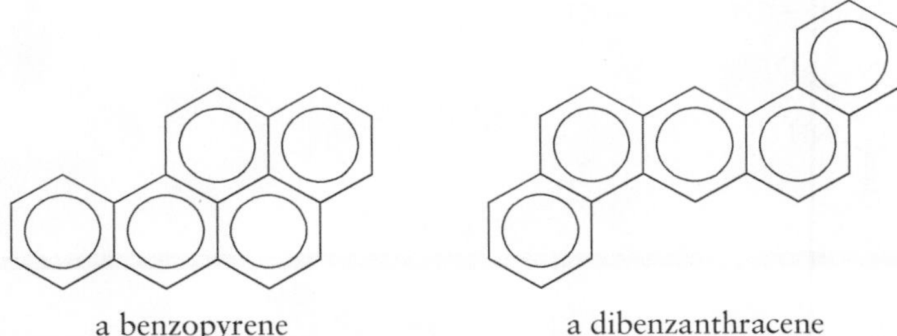

a benzopyrene a dibenzanthracene

These cancer-producing compounds are often formed as a result of heating organic materials to high temperatures. They are present in tobacco smoke (see **Figure 8.43**),

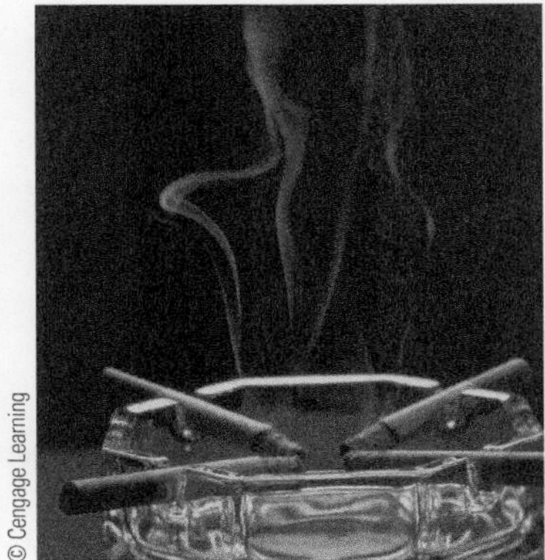

© Cengage Learning

Figure 8.43 Cigarette smoke contains carcinogenic polycyclic aromatic compounds, which is why "secondhand smoke" (i.e., smoke from another person's cigarette) can adversely affect your health, even if you don't smoke.

automobile exhaust, and sometimes in burned or heavily browned food. Such compounds are believed to be at least partially responsible for the high incidence of lung and lip cancer among cigarette smokers. Those who smoke heavily face an increased risk of getting cancer. Chemists have identified more than 4000 compounds in cigarette smoke, including 43 known carcinogens. The Environmental Protection Agency (EPA) considers tobacco smoke a Class A carcinogen.

The major sources of aromatic compounds are petroleum and coal tar, a sticky, dark-colored material derived from coal. As with many classes of organic compounds, the simplest structures are the most important commercial materials. Benzene and toluene are excellent laboratory and industrial solvents. In addition, they are the starting materials for the synthesis of hundreds of other valuable aromatic compounds that are intermediates in the manufacture of a wide variety of commercial products, including polystyrene (see **Table 8.11**).

A number of aromatic compounds must be present in our diet for proper nutrition. Unlike plants, which have the ability to synthesize the benzene ring from simpler materials, humans must obtain any necessary aromatic rings from their diet. This helps explain why certain amino acids (the building blocks of proteins) and some **vitamins** are dietary essentials (see **Table 8.11**).

vitamin An organic nutrient that the body cannot produce in the small amounts needed for good health.

Table 8.11 Some Important Aromatic Compounds

Name	Structural Formula	Use
Benzene	C_6H_6 (benzene ring)	Industrial solvent and raw material
Toluene	$C_6H_5-CH_3$	Industrial solvent and raw material
Phenol	C_6H_5-OH	Manufacture of Formica
Aniline	$C_6H_5-NH_2$	Manufacture of drugs and dyes
Styrene	$C_6H_5-CH{=}CH_2$	Preparation of polystyrene products
Phenylalanine	$C_6H_5-CH_2CH(NH_2)-C(=O)-OH$	An essential amino acid
Riboflavin	$CH_2-CH(OH)-CH(OH)-CH(OH)-CH_2OH$ attached at N of a ring system bearing two H_3C groups, N, N, NH and two C=O groups	Vitamin B_2

Health Career Description

Psychosocial Oncologist

According to the website for the American Psychosocial Oncology Society (APOS), professionals in this field come from many backgrounds. Those with degrees or training as social workers, psychologists and psychiatrists, nurses, and counselors can all eventually become a psychological oncologist. Psychological oncologists work with patients and their primary oncologist to develop a holistic treatment plan that targets not only the cancer in the body but also the strain on mental health that facing cancer often causes.

While fighting the physical attack of cancer on the body, cancer patients may deal with a range of afflictions, including stress, anxiety, and depression. There's also the upheaval that cancer brings to the patient's personal and professional life, from the workplace to relationships with friends and family.

At the James Cancer Hospital and Solove Research Institute at The Ohio State University in Columbus, Ohio, psychosocial oncologists use a variety of techniques to help patients cope with the stress of cancer on their mental health. Beyond counseling sessions, therapy based around art and music is also available. These nontraditional forms of therapy provide opportunities for patients to have a safe space to explore creativity and self-expression on their cancer journey.

The James Cancer Hospital and Solove Research Institute.

Concept Summary

8.1 Organic Chemistry: The Story of Carbon

Learning Objective: **Distinguish between organic and inorganic chemicals.**

- Organic compounds contain carbon.
- Organic chemistry is the study of those compounds.
- Inorganic chemistry is the study of the elements and all noncarbon compounds.
- Organic compounds are of tremendous everyday importance to life on Earth.
- Organic compounds are the basis of all life processes.
- The properties of organic and inorganic compounds often differ, largely as a result of bonding differences.
- Organic compounds contain primarily covalent bonds, whereas ionic bonding is more prevalent in inorganic compounds.

8.2 Representations of Organic Molecules

Learning Objective: **Write condensed, expanded, and skeletal structural representations for compounds.**

- Large numbers of organic compounds are possible because carbon atoms link to form chains and networks.
- Organic compounds are represented by three types of structural formulas: expanded structural formulas, condensed structural formulas, and skeletal structures.
- Expanded structural formulas show all covalent bonds.
- Condensed structural formulas show no covalent bonds or only selected bonds.
- Skeletal structures show carbon–carbon covalent bonds and covalent bonds between carbon and other elements, but carbon–hydrogen bonds are assumed.

8.3 Functional Groups

Learning Objective: **Identify functional groups within larger molecules.**

- All organic compounds are grouped into classes based on characteristic features called functional groups.
- Compounds containing double or triple bonds between carbon atoms are said to be unsaturated.

8.4 Isomers and Conformations

Learning Objective: **Use structural formulas to identify compounds that are isomers of each other.**

- Alkanes possess a three-dimensional geometry in which each carbon is surrounded by four bonds directed to the corners of a tetrahedron.
- Methane, the simplest alkane, is an important fuel (natural gas) and a chemical feedstock for the preparation of other organic compounds.

- Isomerism is one reason so many organic compounds exist.
- Isomers are compounds that have the same molecular formula but different arrangements of atoms.
- The number of structural isomers possible for an alkane increases dramatically with the number of carbon atoms present in the molecule.
- The straight-chain isomer is called a normal alkane.
- Others are called branched isomers.
- Rotation about the single bonds between carbon atoms allows alkanes to exist in many different conformations.
- When an alkane is drawn using only two dimensions, the structure can be represented in a variety of ways, as long as the order of bonding is not changed.

8.5 Classification and Naming of Alkanes

Learning Objectives: **Assign IUPAC names to alkanes. Draw structural formulas for alkanes when given the IUPAC name.**

- Some simple alkanes are known by common names.
- More complex compounds are usually named using the IUPAC system.
- The characteristic IUPAC ending for alkanes is *-ane*.

8.6 Cycloalkanes

Learning Objective: **Assign IUPAC names to cycloalkanes.**

- Cycloalkanes are alkanes in which the carbon atoms form a ring.
- The prefix *cyclo-* is used in the names of these compounds to indicate their cyclic nature.
- Because rotation about the single bonds in the ring is restricted, certain disubstituted cycloalkanes can exist as cis–trans isomers.

8.7 Alkenes and Alkynes

Learning Objective: **Write the IUPAC names of alkenes and alkynes from their molecular structures.**

- Alkenes contain carbon–carbon double bonds.
- Alkynes contain carbon–carbon triple bonds.
- In the IUPAC nomenclature system, alkene names end in *-ene*, whereas alkyne names end in *-yne*.
- In alkenes, the double-bonded carbons and the four groups attached to these carbons lie in the same plane.
- Because rotation about the double bond is restricted, alkenes may exist as cis–trans isomers.
- This type of stereoisomerism is possible when each double-bonded carbon is attached to two different groups.
- IUPAC names of stereoisomers contain the prefix *cis-* or *trans-*.
- In alkynes, the triple-bonded carbons and the two carbons attached to them possess a linear geometry.
- The physical and chemical properties of alkynes are very similar to those of alkenes.

8.8 Reactions of Hydrocarbons

Learning Objectives: **Write chemical equations for addition reactions of alkenes. Use Markovnikov's rule to predict the major products of certain alkene addition reactions. Write chemical equations for addition polymerization.**

- Alkanes are relatively unreactive and remain unchanged by most reagents.
- The physical properties of alkenes are very similar to those of alkanes.
- Alkanes and alkenes are nonpolar, insoluble in water, less dense than water, and soluble in nonpolar solvents.
- Alkenes are quite reactive, and their characteristic reaction is addition to the double bond.
- Two important addition reactions are hydrogenation (i.e., addition of H_2 to give an alkane) and hydration (i.e., addition of H_2O to produce an alcohol).
- Hydration is governed by Markovnikov's rule—namely, the hydrogen becomes attached to the carbon atom of the double bond that is already bonded to more hydrogens. The OH bonds to the other carbon of the double bond.
- Addition polymers are formed from alkene monomers that undergo repeated addition reactions with each other.
- Many familiar and widely used materials, such as fibers and plastics, are addition polymers.

8.9 Physical and Chemical Properties of Hydrocarbons

Learning Objective: **Describe the key physical properties of alkanes, alkenes, and alkynes.**

- The physical properties of alkanes are typical of all hydrocarbons:
  - Nonpolar
  - Insoluble in water
  - Less dense than water
  - Melting points and boiling points that increase with molecular weight

8.10 Aromatic Compounds and Benzene

Learning Objectives: **Name aromatic compounds. Draw structural formulas for aromatic compounds.**

- Benzene (C_6H_6) is the simplest aromatic compound. It consists of a six-membered ring with three double bonds.
- The other members of the aromatic class contain at least one of these six-membered ring with three double bonds.
- Several acceptable IUPAC names are possible for many benzene compounds.
- Some IUPAC names are based on widely used common names such as toluene and aniline.
- Other compounds are named as derivatives of benzene or by designating the benzene ring as a phenyl group.

Key Terms and Concepts

Addition polymer (8.8)
Addition reaction (8.8)
Alkane (8.3)
Alkene (8.7)
Alkyl group (8.5)
Alkyne (8.7)
Aromatic hydrocarbon (8.7)
Branched alkane (8.4)
cis- (8.6)
Condensed structural representation (8.2)
Conformations (8.4)
Cycloalkane (8.6)
Expanded structural representation (8.2)
Functional group (8.3)
Homologous series (8.9)
Hydration (8.8)
Hydrogenation (8.8)
Hydrophobic (8.9)
Isomerism (8.4)
Markovnikov's rule (8.8)
Monomer (8.8)
Organic chemistry (8.1)
Organic compound (8.1)
Phenyl group (8.10)
Polycyclic aromatic compound (8.10)
Polymer (8.8)
Polymerization (8.8)
Polyunsaturated (8.8)
Saturated hydrocarbon (8.3)
Skeletal structural representation (8.2)
Stereoisomers (8.6)
Straight-chain alkane (8.4)
Structural isomers (8.4)
trans- (8.6)
Unsaturated hydrocarbon (8.3)
Vitamin (8.10)

Key Equations

1. Generic addition reaction of an alkene (**Section 8.8**)	$>C=C< \; + \; A-B \longrightarrow -\overset{A}{\underset{|}{\overset{|}{C}}}-\overset{B}{\underset{|}{\overset{|}{C}}}-$	Equation 8.1
2. Hydrogenation of an alkene (**Section 8.8**)	$>C=C< \; + \; H-H \xrightarrow{Pt} -\underset{H}{\overset{|}{C}}-\underset{H}{\overset{|}{C}}-$	Equation 8.2
3. Hydration of an alkene (**Section 8.8**)	$>C=C< \; + \; H-OH \xrightarrow{H_2SO_4} -\underset{H}{\overset{|}{C}}-\underset{OH}{\overset{|}{C}}-$	Equation 8.3
4. Addition polymerization of an alkene (**Section 8.8**)	$\underset{\text{Ethylene}}{nCH_2=CH_2} \xrightarrow[\text{catalysts}]{\text{Heat, pressure,}} \underset{\text{Polyethylene}}{+CH_2-CH_2+_n}$	Equation 8.4

Exercises

Even-numbered exercises are answered in Appendix B.

Exercises with an asterisk (*) are more challenging.

Organic Chemistry: The Story of Carbon (Section 8.1)

8.1 Why were the compounds of carbon originally called organic compounds?

8.2 Name at least six items you recognize to be composed of organic compounds.

8.3 Describe what Wöhler did that made the vital force theory highly questionable.

8.4 What is the unique structural feature shared by all organic compounds?

8.5 Classify each of the following compounds as organic or inorganic.

a. KBr

b. C_2H_2

c. H_2O

d. LiOH

8.6 Classify each of the following compounds as organic or inorganic.

a. NH_3

b. CH_3NH_2

c. NH_4NCO

d. $AlPO_4$

8.7 What kind of bond between atoms is most prevalent among organic compounds?

8.8 Are the majority of all compounds that are insoluble in water organic or inorganic? Why?

8.9 Indicate for each of the following characteristics whether it more likely describes an inorganic or an organic compound. Give one reason for your answer.

a. This compound is a liquid that readily burns.

b. A white solid upon heating is found to melt at 735 °C.

c. A liquid added to water floats on the surface and does not dissolve.

d. This compound exists as a gas at room temperature and ignites easily.

8.10 Indicate for each of the following characteristics whether it more likely describes an inorganic or organic compound. Give one reason for each of your answers.

a. A solid substance melts at 65 °C.

b. This compound is soluble in water and does not burn.

c. A water solution of this substance conducts electricity.

d. A green solid resists melting when heated to 600 °C.

Representations of Organic Molecules (Section 8.2)

8.11 Complete the following structures by adding hydrogen atoms where needed.

a. 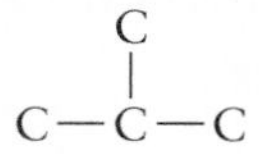
```
    C
    |
 C—C—C
```
b. 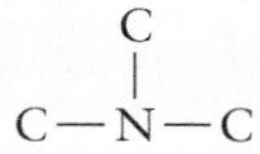
```
    C
    |
 C—N—C
```
c. C—C≡C—C

d. C—O—C

8.12 Complete the following structures by adding hydrogen atoms where needed.

a. C—C—C

b. C—C=C

c.
```
    O
    ‖
 C—C
```
d. C—C—N

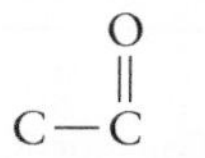

8.13 On the basis of the number of covalent bonds possible for each atom, determine which of the following structural formulas are correct. Explain what is wrong with the incorrect structures.

a. 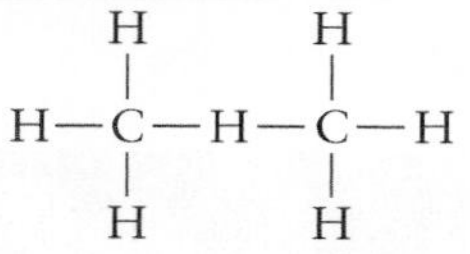
```
    H       H
    |       |
 H—C—H—C—H
    |       |
    H       H
```
b.
```
       H  H
       |  |
 H—N—C—C—H
    |  |  |
    H  H  H
```
c.
```
   H      O       H
   |      ‖       |
 H—C——————C———————C—H
   |      |       |
   H   H—C—H      H
          |
          H
```
d. $CH_3—C=C—CH_3$

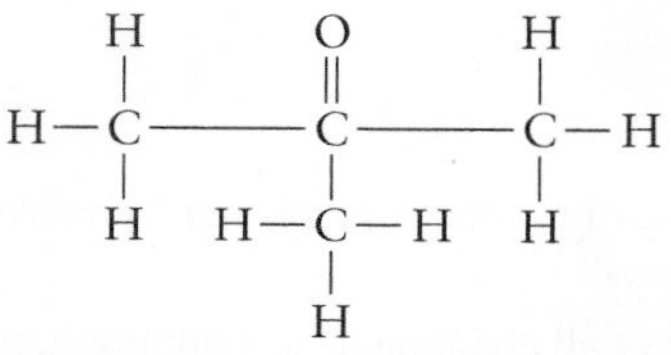

8.14 On the basis of the number of covalent bonds possible for each atom, determine which of the following structural formulas are correct. Explain what is wrong with the incorrect structures.

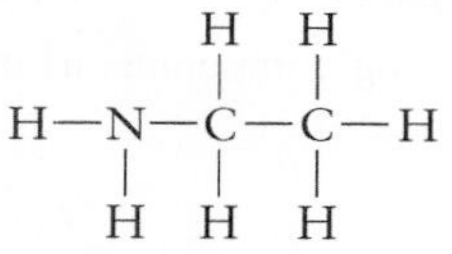

a. 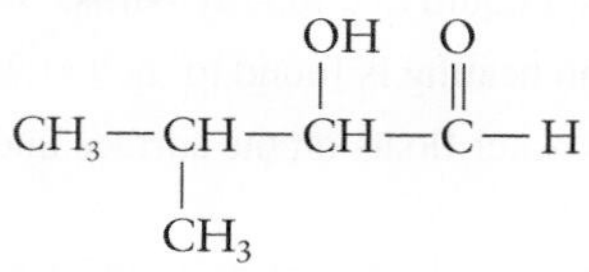
```
           OH   O
           |    ‖
 CH3—CH—CH—C—H
       |
       CH3
```
b. 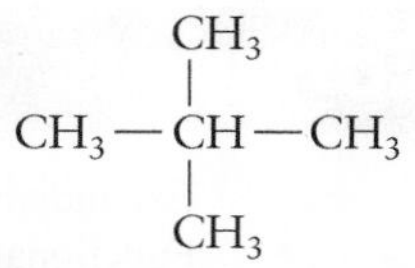
```
       CH3
       |
 CH3—CH—CH3
       |
       CH3
```
c. $CH_3—O—CH_2—CH_4$

d. 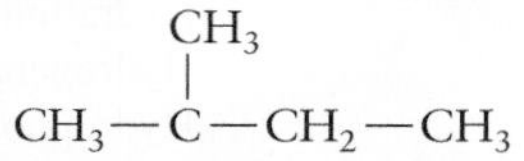
```
       CH3
       |
 CH3—C—CH2—CH3
```
8.15 Aspartic acid is an essential component of proteins. It has the basic structure shown here. Complete the structure by adding hydrogen atoms on the basis of the number of covalent bonds possible for each atom.

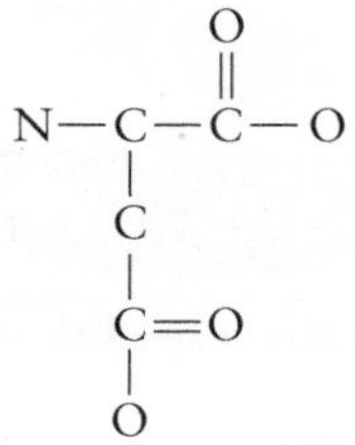

```
        O
        ‖
 N—C—C—O
   |
   C
   |
   C=O
   |
   O
```
8.16 Asparagine is an essential component of proteins. It has the basic structure shown here. Complete the structure by adding hydrogen atoms on the basis of the number of covalent bonds possible for each atom.

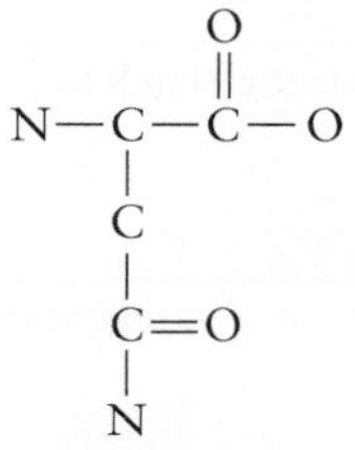

```
        O
        ‖
 N—C—C—O
   |
   C
   |
   C=O
   |
   N
```
8.17 Identify each of the following as a condensed structural formula, expanded structural formula, or molecular formula.

a. $CH_3CH_2CH_2—NH_2$

b. 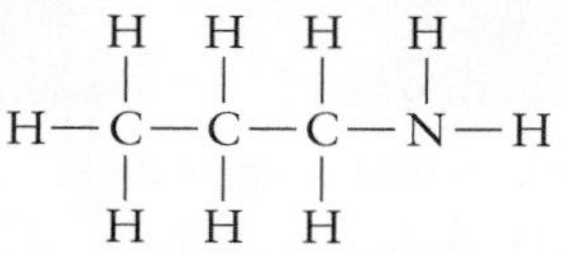
```
    H  H  H  H
    |  |  |  |
 H—C—C—C—N—H
    |  |  |
    H  H  H
```
c. $CH_3—CH_2—CH_2—NH_2$

d. C_3C_9N

8.18 Identify each of the following as a condensed structural formula, expanded structural formula, or molecular formula.

a. 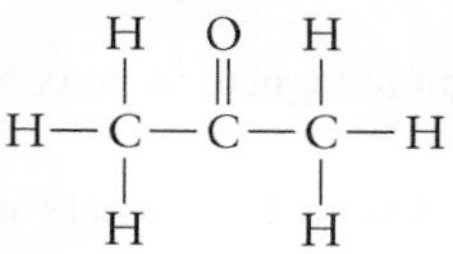
```
    H  O  H
    |  ‖  |
 H—C—C—C—H
    |     |
    H     H
```
b. 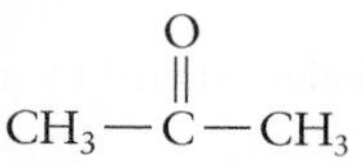
```
       O
       ‖
 CH3—C—CH3
```
c. C_3H_6O

d. 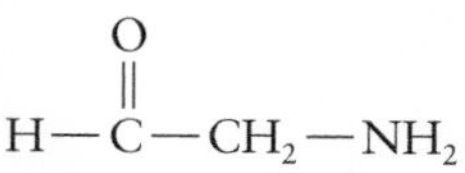
```
    O
    ‖
 H—C—CH2—NH2
```

8.19 Write a condensed structural formula for the following compounds.

a. 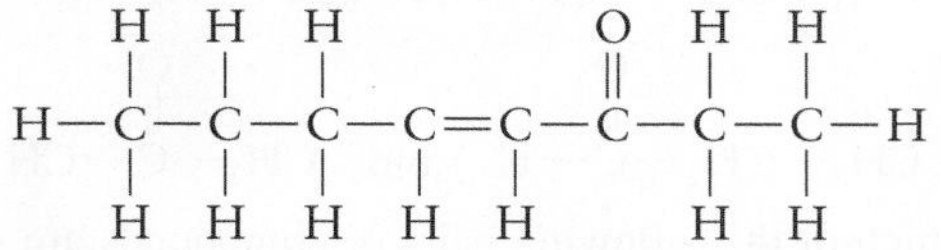
```
  H  H  H        O  H  H
  |  |  |        ‖  |  |
H—C—C—C—C=C—C—C—C—H
  |  |  |  |  |     |  |
  H  H  H  H  H     H  H
```

b. 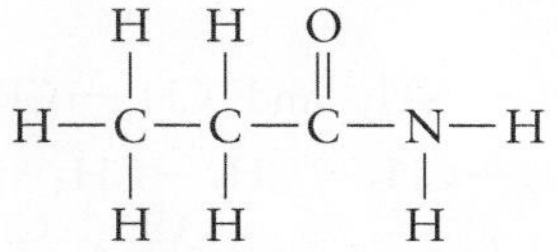
```
  H  H  O
  |  |  ‖
H—C—C—C—N—H
  |  |     |
  H  H     H
```

8.20 Write a condensed structural formula for the following compounds.

a. 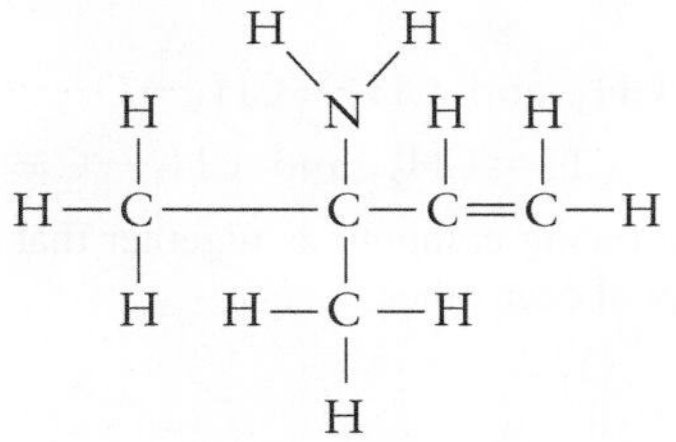
```
       H     H
        \   /
  H       N    H  H
  |       |    |  |
H—C———————C—C=C—H
  |       |
  H     H—C—H
          |
          H
```

b. 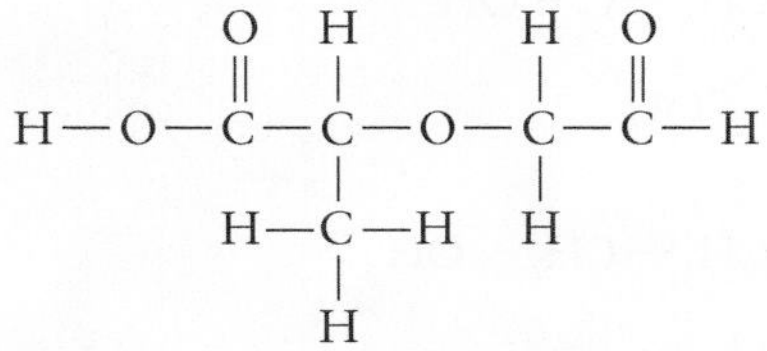
```
      O  H      H  O
      ‖  |      |  ‖
H—O—C—C—O—C—C—H
         |      |
       H—C—H    H
         |
         H
```

8.21* Write an expanded structural formula for the following.

a. 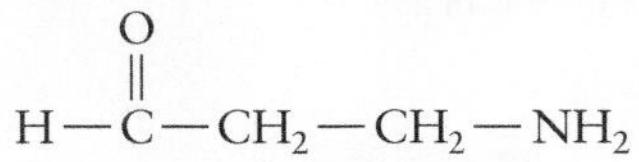 $$\mathrm{H-\overset{\overset{\displaystyle O}{\|}}{C}-CH_2-CH_2-NH_2}$$

b. 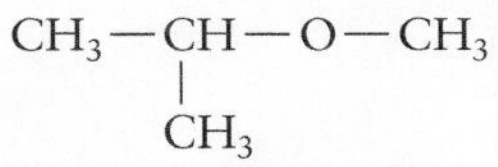 $$\mathrm{CH_3-\underset{\underset{\displaystyle CH_3}{|}}{CH}-O-CH_3}$$

8.22 Write an expanded structural formula for the following.

a. $CH_3-CH_2-O-CH_2-O-CH_3$

b. 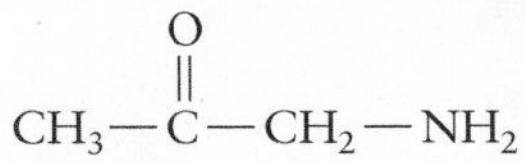 $$\mathrm{CH_3-\overset{\overset{\displaystyle O}{\|}}{C}-CH_2-NH_2}$$

8.23 Write skeletal structural formulas from the following expanded structural formulas.

a. 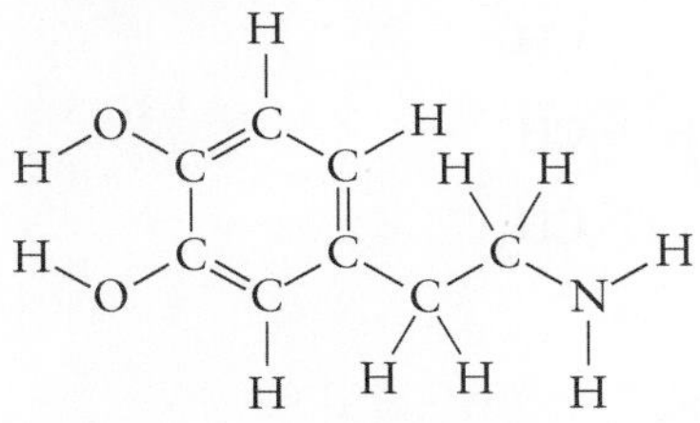
```
             H
             |
    H   O    C      H
     \ / \ // \    /
         C     C        H  H
         |     ‖         \ /
    H    C     C          C        H
     \  / \\  / \        / \      /
      O     C     C          N
            |    / \         |
            H   H   H        H
```
dopamine

b. 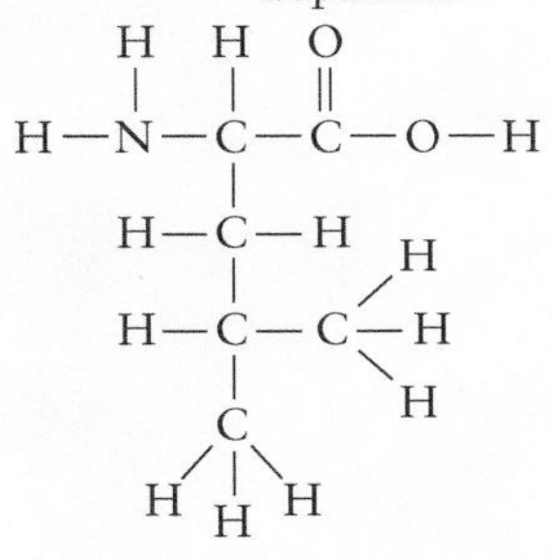
```
   H  H  O
   |  |  ‖
H—N—C—C—O—H
      |
    H—C—H    H
      |     /
    H—C—C—H
      |    \
      C     H
    / | \
   H  H  H
```
leucine

8.24 Write skeletal structural formulas from the following expanded structural formulas.

a. 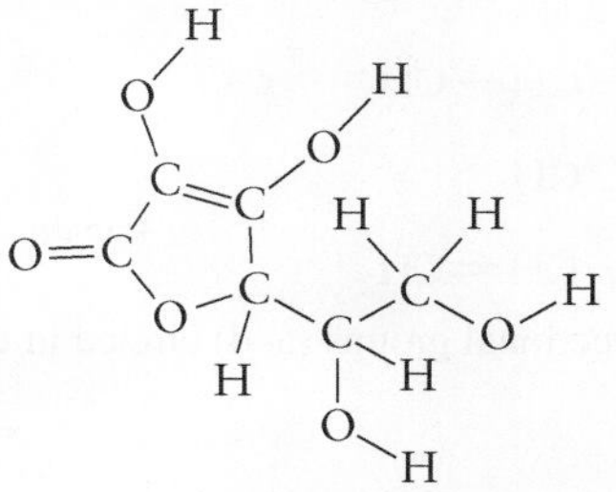
```
           H
          /
         O         H
          \       /
           C==C—O
          /    |      H    H
     O=C       |       \  /
         \     C        C       H
          O — C  \     /  \   /
              |    C        O
              H    |  \
                   O    H
                    \
                     H
```
ascorbic acid

b. 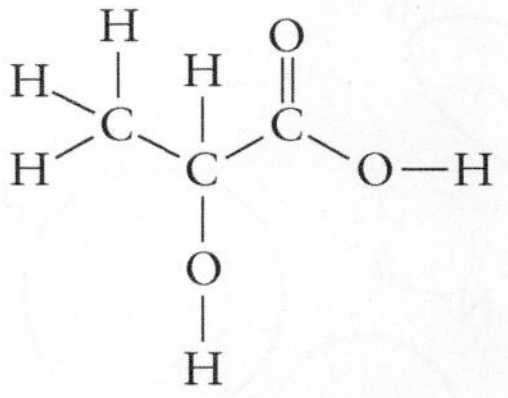
```
       H
   H   |   H   O
    \  |   |   ‖
      C —  C — C — O—H
    /      |
   H       O
           |
           H
```
lactic acid

8.25 Write a skeletal structural formula from the following.

a. CH_3CHNH_2COOH (alanine, an amino acid)

b. $CH_2CH_2CHCHCH_2CHO$ (3-hexenal, fresh-cut grass smell)

8.26 Write a skeletal structural formula from the following.

a. $(CH_3)_3COCH_3$ (MTBE, a gasoline additive)

b. $HSCH_2CHNH_2COOH$ (cysteine, an amino acid)

Functional Groups (Section 8.3)

8.27 What is the difference between an unsaturated hydrocarbon and a saturated hydrocarbon?

8.28 Classify each of the following hydrocarbons as saturated or unsaturated.

a. 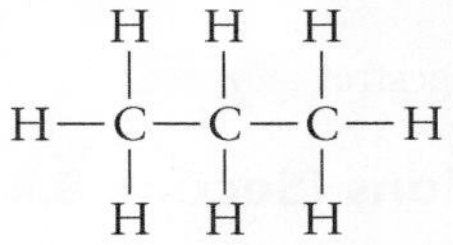
```
  H  H  H
  |  |  |
H—C—C—C—H
  |  |  |
  H  H  H
```

b. 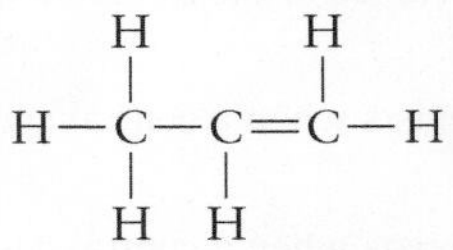
```
  H     H
  |     |
H—C—C=C—H
  |  |
  H  H
```

c. $CH_3-C\equiv C-CH_3$

d. $CH_3-CH_2-CH=CH_2$

8.29 Classify each of the following hydrocarbons as saturated or unsaturated.

a. 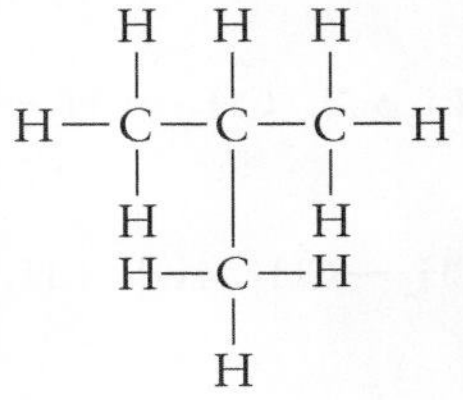
```
  H  H  H
  |  |  |
H—C—C—C—H
  |  |  |
  H  |  H
   H—C—H
     |
     H
```

b. 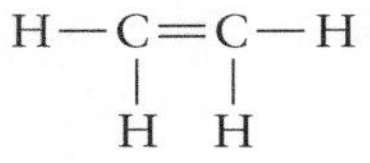 $H-C=C-H$ (each C also bonded to H: $\overset{}{\underset{H}{|}}$ $\overset{}{\underset{H}{|}}$)

c. $CH_3-CH_2-CH-CH_3$
$\qquad\qquad\qquad\;|$
$\qquad\qquad\quad\; CH_3$

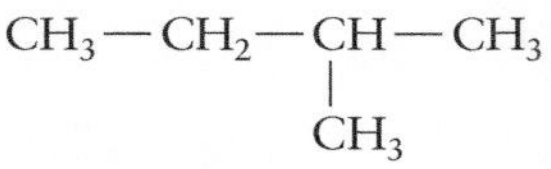

d. $CH_3-CH_2-CH=CH_2$

8.30 Name the four functional groups (a–d) circled in the antibiotic cephalexin.

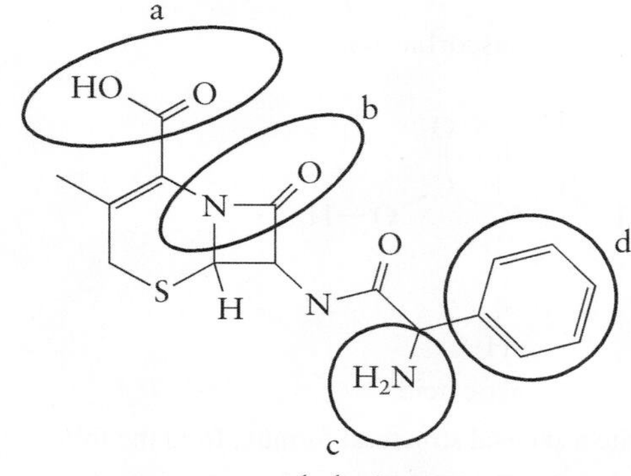

cephalexin

8.31 Name the four functional groups (a–d) circled in the steroid levonorgestrel.

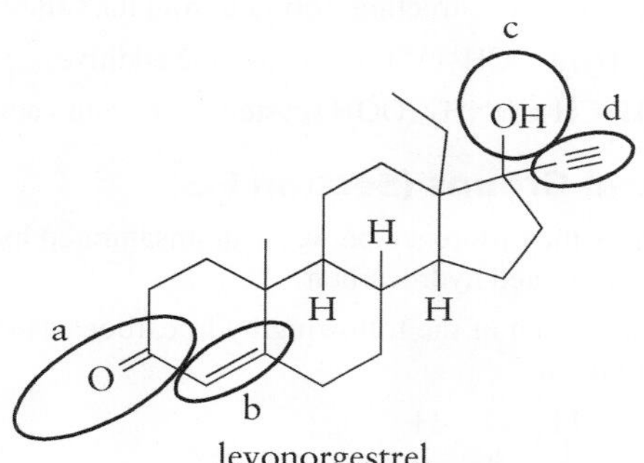

levonorgestrel

Isomers and Conformations (Section 8.4)

8.32 Give two reasons for the existence of the tremendous number of organic compounds.

8.33 What molecular geometry exists when a central carbon atom bonds to four other atoms?

8.34 What molecular geometry exists when a central carbon atom bonds to three other atoms?

8.35 Determine the number of covalent bonds typically formed by atoms of the following elements: carbon, hydrogen, oxygen, nitrogen, and bromine.

8.36 Which of the following pairs of compounds are structural isomers?

a. $CH_3-CH=CH-CH_3$ and $CH_3-CH_2-CH_2-CH_3$

b. 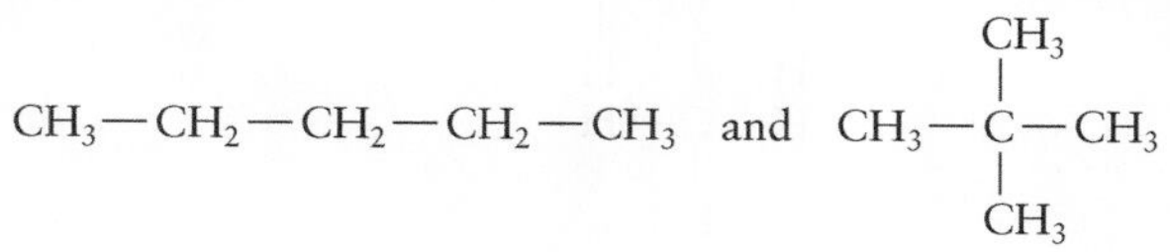 $CH_3-CH_2-CH_2-CH_2-CH_3$ and $CH_3-\underset{CH_3}{\overset{CH_3}{\overset{|}{\underset{|}{C}}}}-CH_3$

c. $CH_3-CH_2-\overset{CH_3}{\overset{|}{CH}}-OH$ and $CH_3-\overset{O}{\overset{\|}{C}}-CH_2-CH_3$

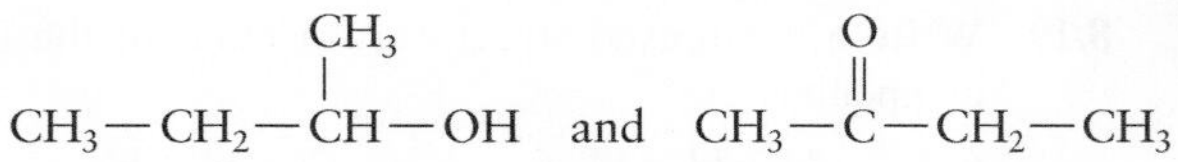

d. 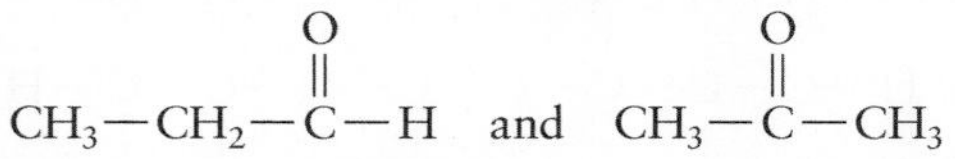 $CH_3-CH_2-\overset{O}{\overset{\|}{C}}-H$ and $CH_3-\overset{O}{\overset{\|}{C}}-CH_3$

8.37 Which of the following pairs of compounds are structural isomers?

a. $CH_3-CH_2-CH_2-NH_2$ and $CH_3-CH_2-NH-CH_3$

b. $CH_3-CH_2-CH_2-CH_2-CH_2-CH_3$ and $CH_3-\overset{CH_3}{\overset{|}{CH}}-\overset{CH_3}{\overset{|}{CH}}-CH_3$

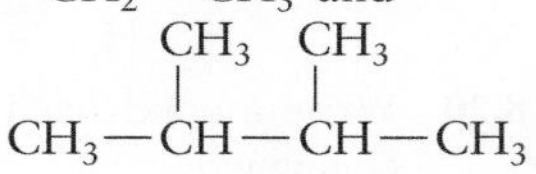

c. 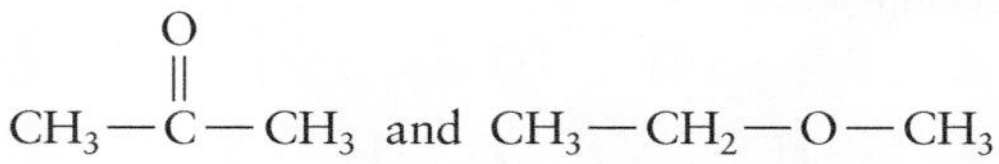 $CH_3-\overset{O}{\overset{\|}{C}}-CH_3$ and $CH_3-CH_2-O-CH_3$

d. $CH_3-CH_2-CH=CH_2$ and $CH_3-C\equiv C-CH_3$

8.38 Group all the following compounds together that represent structural isomers of each other.

a. 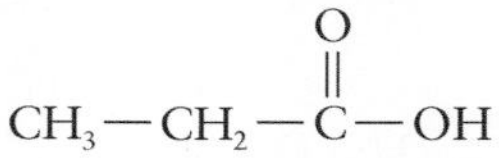 $CH_3-CH_2-\overset{O}{\overset{\|}{C}}-OH$

b. 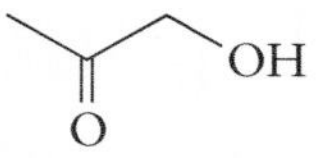 (skeletal structure) OH, O

c. $CH_3-CH_2-CH_2-OH$

d. 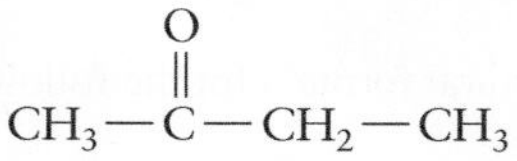 $CH_3-\overset{O}{\overset{\|}{C}}-CH_2-CH_3$

e. 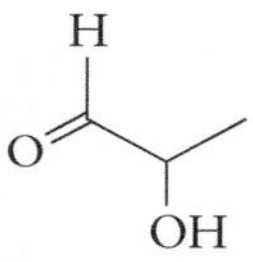 (skeletal structure) H, O, OH

f. 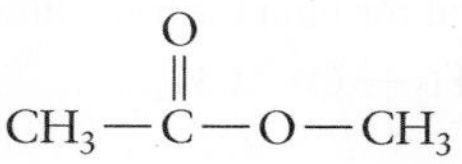 $CH_3-\overset{O}{\overset{\|}{C}}-O-CH_3$

g.  HO (skeletal structure) OH

8.39 Classify each of the following compounds as a normal alkane or a branched alkane.

a. 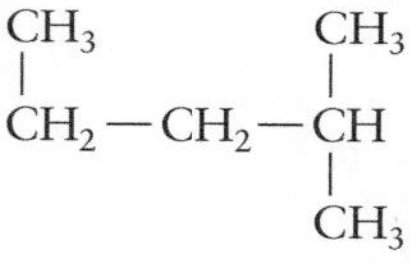 $\overset{CH_3}{\overset{|}{CH_2}}-CH_2-\underset{CH_3}{\underset{|}{\overset{CH_3}{\overset{|}{CH}}}}$

b. 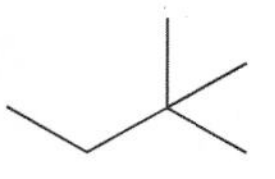 (skeletal structure)

c. 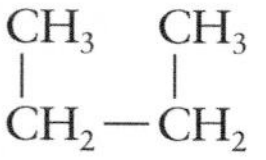 $\overset{CH_3}{\overset{|}{CH_2}}-\overset{CH_3}{\overset{|}{CH_2}}$

d. 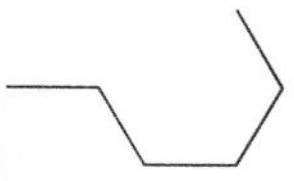 (skeletal structure)

8.40 Classify each of the following compounds as a normal alkane or a branched alkane.

a. 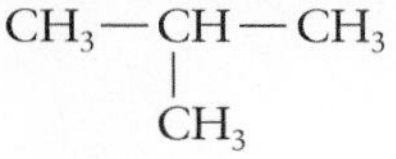 $CH_3-CH-CH_3$
$\quad\quad\quad | $
$\quad\quad CH_3$

b. 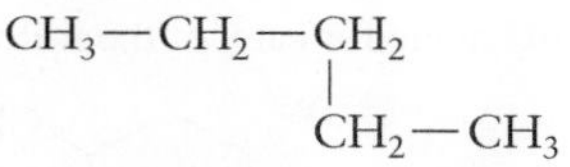 $CH_3-CH_2-CH_2$
$\quad\quad\quad\quad\quad\quad | $
$\quad\quad\quad\quad\quad CH_2-CH_3$

c.

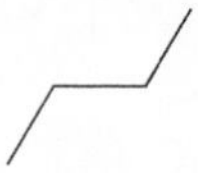

d.

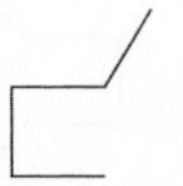

8.41* Which of the following pairs represent structural isomers, and which are simply the same compound?

a.  and $CH_3CH_2CH_2CH_3$

b. 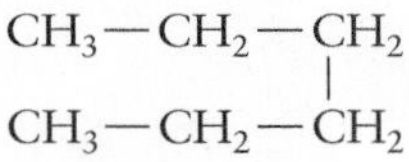 $CH_3-CH_2-CH_2$ and $CH_3CH_2CH_2CH_2CH_2CH_3$
$\quad\quad\quad\quad\quad\quad | $
$CH_3-CH_2-CH_2$

c. 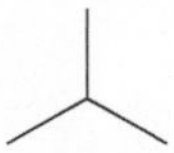 and $CH_3CH_2CH_2CH_3$

d. 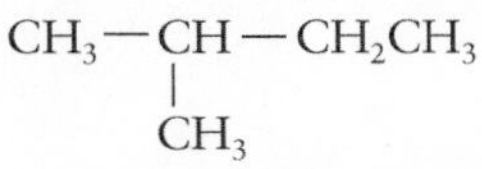 $CH_3-CH-CH_2CH_3$ and $CH_3-CH_2-CH-CH_3$
$\quad\quad | \quad\quad\quad\quad\quad\quad\quad\quad\quad\quad\quad\quad | $
$\quad\quad CH_3 \quad\quad\quad\quad\quad\quad\quad\quad\quad\quad CH_3$

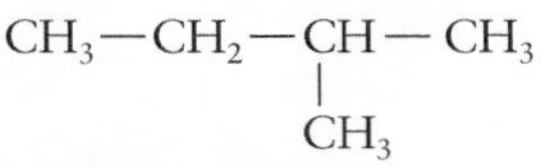

8.42 Which of the following pairs represents structural isomers, and which are simply the same compound?

a. 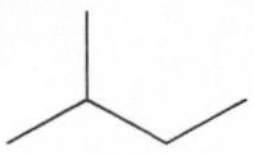 and

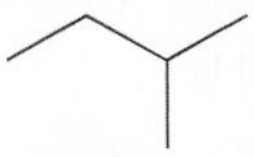

b. 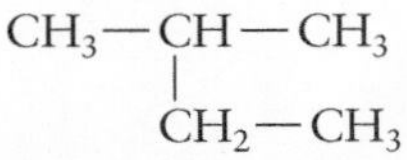 $CH_3-CH-CH_3$ and $CH_3-CH-CH_2-CH_3$
$\quad\quad | \quad\quad\quad\quad\quad\quad\quad\quad\quad\quad\quad | $
$\quad\quad CH_2-CH_3 \quad\quad\quad\quad\quad\quad\quad CH_3$

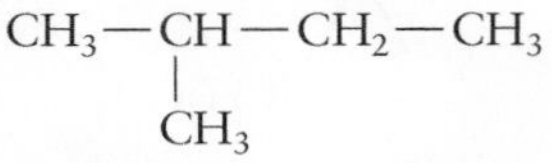

c. 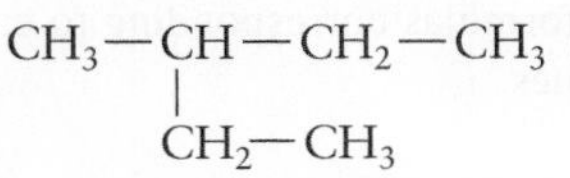 $CH_3-CH-CH_2-CH_3$ and $CH_3-CH-CH-CH_3$
$\quad\quad | \quad\quad\quad\quad\quad\quad\quad\quad\quad\quad\quad\quad | \quad\quad | $
$\quad\quad CH_2-CH_3 \quad\quad\quad\quad\quad\quad\quad CH_3 \quad CH_3$

d. and CH_3
$\quad\quad\quad\quad | $
$CH_3CH_2CH_2$

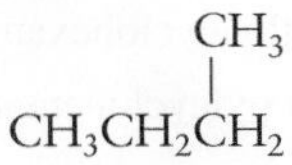

Classification and Naming of Alkanes (Section 8.5)

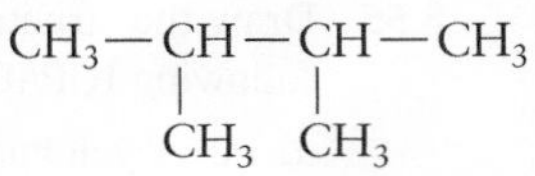

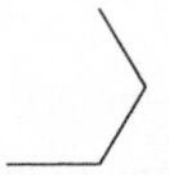

8.43 Identify the following alkyl groups.

a. 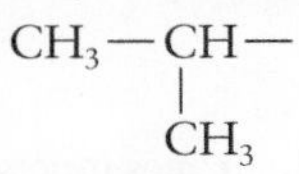 CH_3-CH-
$\quad\quad | $
$\quad\quad CH_3$

b. $CH_3-CH_2-CH_2-CH_2-$

c. 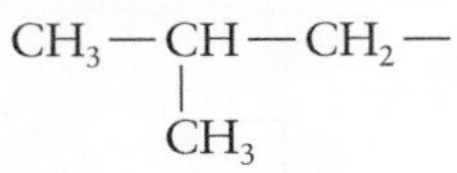 $CH_3-CH-CH_2-$
$\quad\quad | $
$\quad\quad CH_3$

d. CH_3-

8.44 Identify the following alkyl groups.

a. $CH_3-CH_2-CH_2-$

b. CH_3-CH_2-

c. 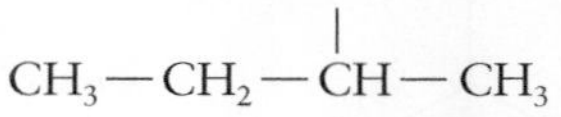 $\quad\quad\quad\quad\quad\quad | $
$CH_3-CH_2-CH-CH_3$

d. 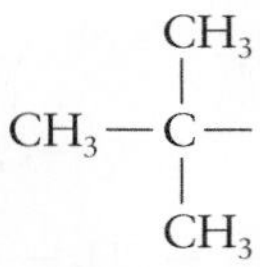 $\quad\quad CH_3$
$\quad\quad\quad | $
CH_3-C-
$\quad\quad\quad | $
$\quad\quad CH_3$

8.45 Give the correct IUPAC name for each of the following alkanes.

a. 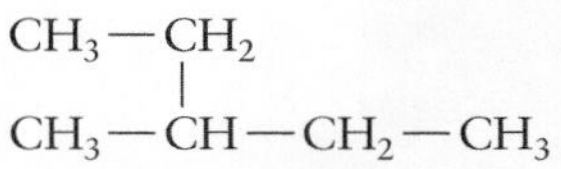 CH_3-CH_2
$\quad\quad\quad\quad | $
$CH_3-CH-CH_2-CH_3$

b. 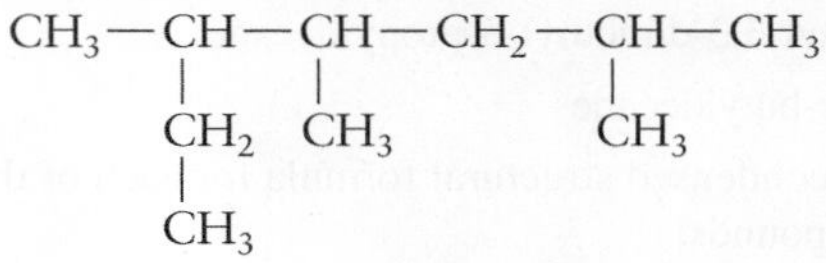 $CH_3-CH-CH-CH_2-CH-CH_3$
$\quad\quad | \quad\quad | \quad\quad\quad\quad\quad | $
$\quad\quad CH_2 \quad CH_3 \quad\quad\quad CH_3$
$\quad\quad | $
$\quad\quad CH_3$

c.

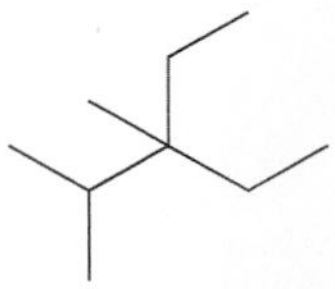

d.

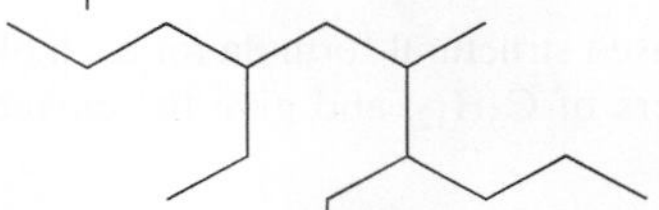

e. 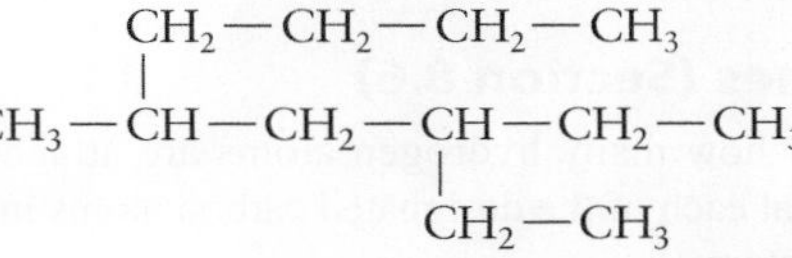 $\quad\quad CH_2-CH_2-CH_2-CH_3$
$\quad\quad\quad | $
$CH_3-CH-CH_2-CH-CH_2-CH_3$
$\quad\quad\quad\quad\quad\quad\quad\quad\quad | $
$\quad\quad\quad\quad\quad\quad\quad\quad CH_2-CH_3$

8.46 Give the correct IUPAC name for each of the following alkanes.

a.

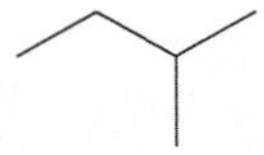

b. 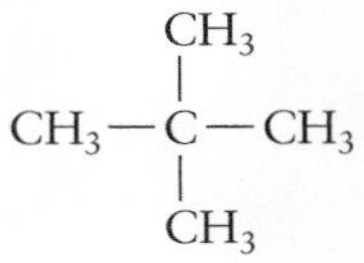 $\quad\quad CH_3$
$\quad\quad\quad | $
CH_3-C-CH_3
$\quad\quad\quad | $
$\quad\quad CH_3$

c. $\quad\quad CH_2-CH_3$
$\quad\quad\quad | $
$CH_3-CH-CH-CH_3$
$\quad\quad\quad\quad\quad\quad | $
$\quad\quad\quad\quad\quad CH_3$

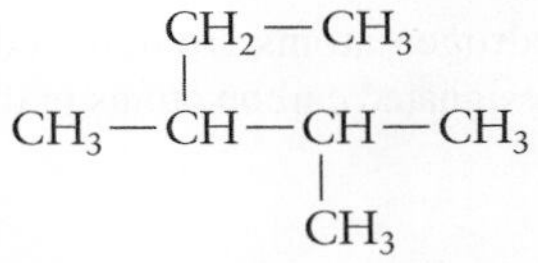

d.

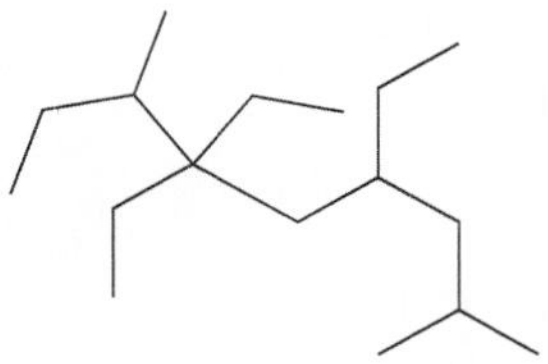

e.

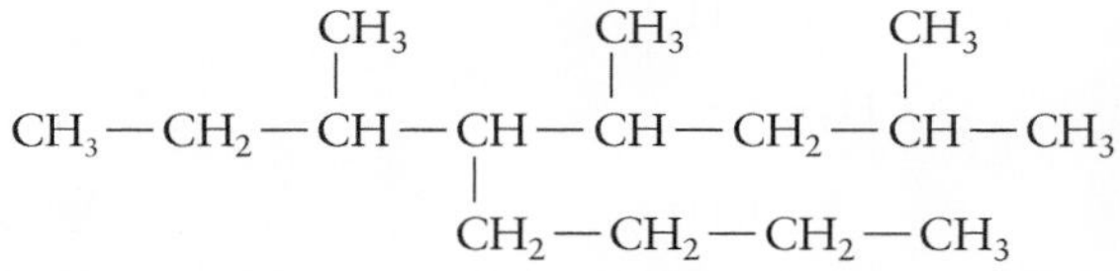

$$\begin{array}{c} \quad\quad CH_3 \quad\quad\quad CH_3 \quad\quad\quad\quad CH_3 \\ CH_3-CH_2-CH-CH-CH-CH_2-CH-CH_3 \\ \quad\quad\quad\quad | \\ \quad\quad\quad CH_2-CH_2-CH_2-CH_3 \end{array}$$

8.47 Draw a condensed structural formula for each of the following compounds.

a. 3-ethylpentane

b. 2,2-dimethylbutane

c. 4-ethyl-3,3-dimethyl-5-propyldecane

d. 5-*sec*-butyldecane

8.48 Draw a condensed structural formula for each of the following compounds.

a. 2,2,4-trimethylpentane

b. 4-isopropyloctane

c. 3,3-diethylhexane

d. 5-*tert*-butyl-2-methylnonane

8.49 Draw the condensed structural formula for each of the three structural isomers of C_5H_{12}, and give the correct IUPAC names.

8.50 Why are different conformations of an alkane not considered structural isomers?

Cycloalkanes (Section 8.6)

8.51 Identify how many hydrogen atoms are attached (but not shown) at each of the designated carbon atoms in the following structures.

a.

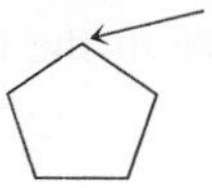

b. CH_3

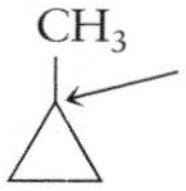

c. CH_3 CH_2CH_3

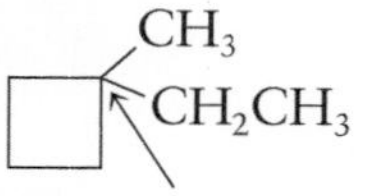

8.52 Identify how many hydrogen atoms are attached (but not shown) at each of the designated carbon atoms in the following structures.

a. CH_3

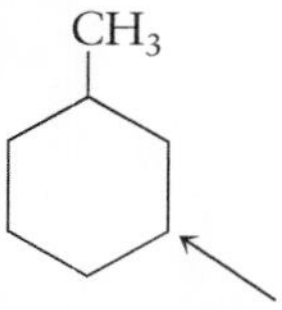

b. CH_3

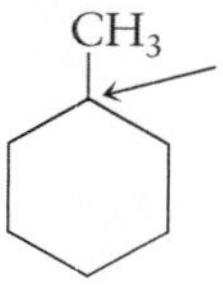

c. CH_3 CH_3 CH_3

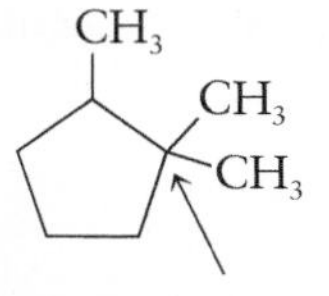

8.53 Write the correct IUPAC name for each of the following.

a.

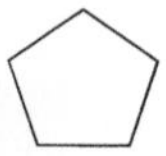

b. CH_3 CH_3

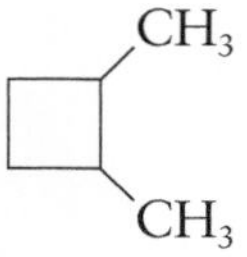

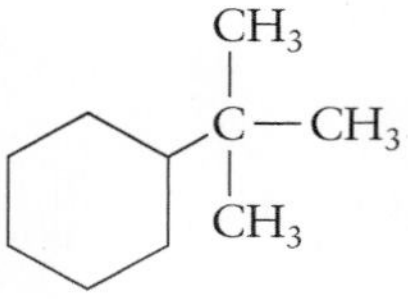

c. CH_3 CH_3

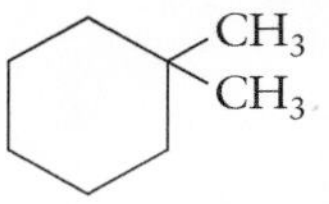

d. H_3C CH_3 CH_3

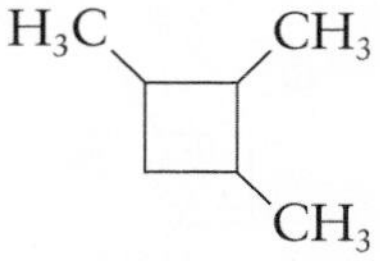

8.54 Write the correct IUPAC name for each of the following.

a. CH_3 CH_2CH_3

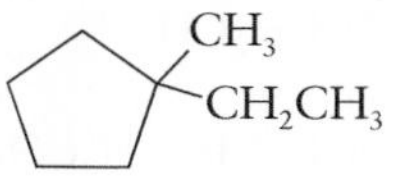

b. CH_3 $C-CH_3$ CH_3

c. CH_3 Cl CH_3

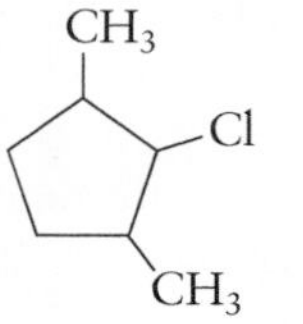

d. $CH_2CH_2CH_3$ CH_2CH_3

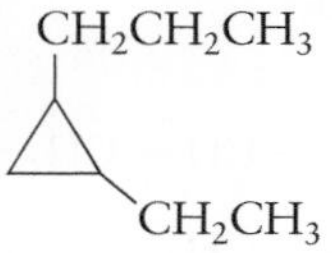

8.55 Draw the structural formulas corresponding to each of the following IUPAC names.

a. ethylcyclobutane

b. 1,1,2,5-tetramethylcyclohexane

c. 1-butyl-3-isopropylcyclopentane

8.56 Draw the structural formulas corresponding to each of the following IUPAC names.

a. 1-ethyl-1-propylcyclopentane

b. isopropylcyclobutane

c. 1,2-dimethylcyclopropane

8.57 Which of the following pairs of cycloalkanes represent structural isomers?

a. CH_3 CH_3 H_3C CH_3

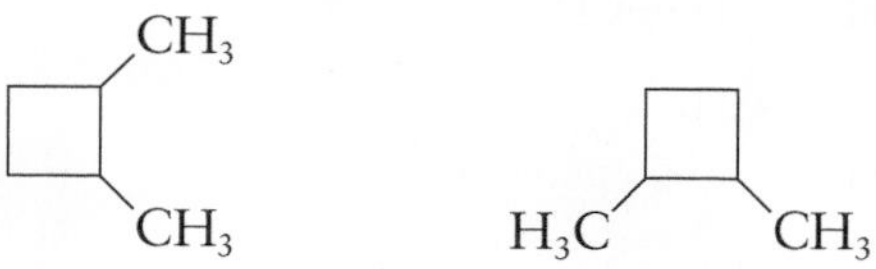

b. CH_2CH_3 CH_3 CH_2CH_3 H_3C

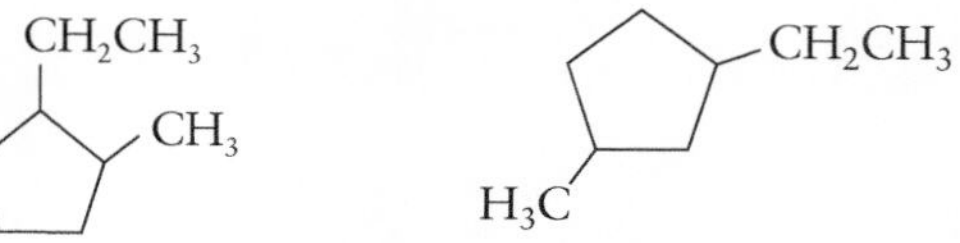

c. CH_2CH_3

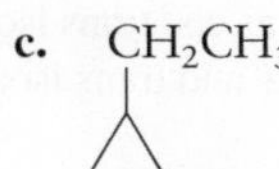

CH_3 CH_3

d. 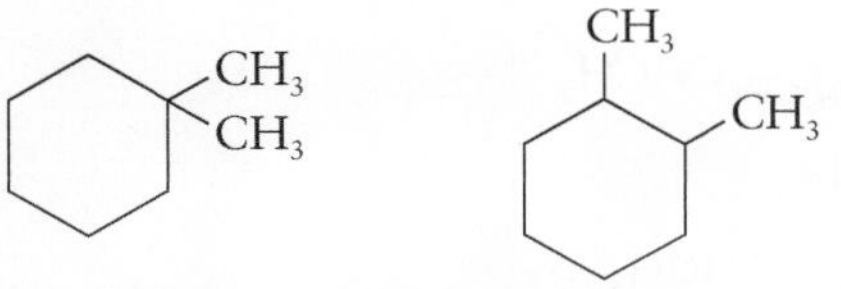 CH_3 CH_3 CH_3 CH_3

8.58* Draw structural formulas for the five structural isomers of C_5H_{10} that are cycloalkanes.

8.59 Explain the difference between cis–trans and structural isomers.

8.60 Using the prefix *cis-* or *trans-*, name each of the following.

a.

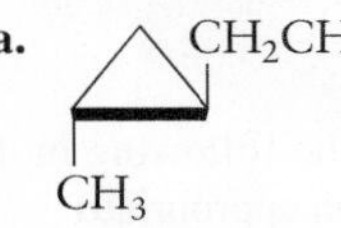

c.

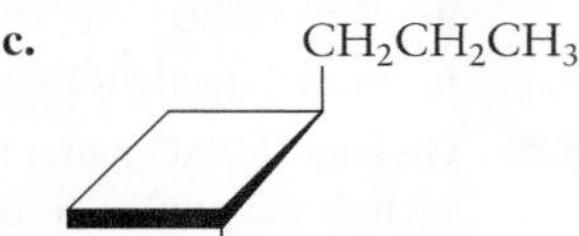

b.

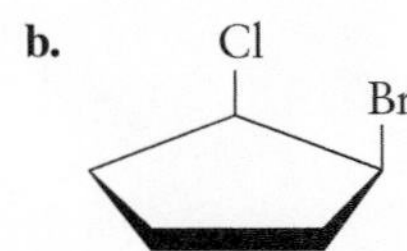

d.

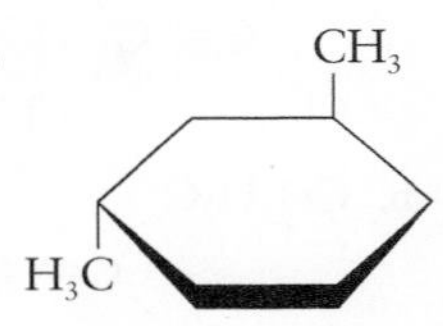

Alkenes and Alkynes (Section 8.7)

8.61 What is the definition of an unsaturated hydrocarbon?

8.62 Define the terms *alkene*, *alkyne*, and *aromatic hydrocarbon*.

8.63 Select those compounds that can be correctly called unsaturated and classify each one as an alkene or an alkyne.

a. $CH_3—CH_2—CH_3$

b. $CH_3CH{=}CHCH_3$

c. $H—C{\equiv}C—CH—CH_3$ (with CH_3 on the CH)

d. 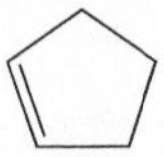 (cyclopentene)

e. 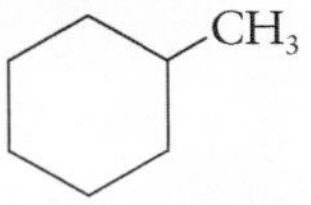 (cyclohexane with CH_3)

8.64 Select the compounds that can be correctly called unsaturated and classify each as an alkene or alkyne.

a. 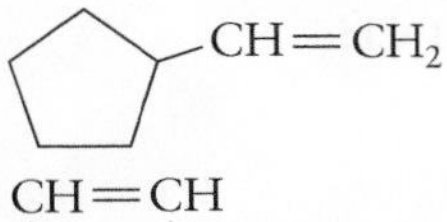 (cyclopentane with $CH{=}CH_2$)

b. $CH{=}CH$ / $CH_2—CH_2$ (ring)

c. $CH_2{=}CHCH_2CH_3$

d. CH_3CHCH_3 (with CH_3 on the CH)

8.65 Which of the following molecular structures cannot exist in nature? Explain.

a. $H_2C{=}C{=}CH_2$

b. $H_3C—CH_2{=}CH_3$

c. $H_3C—C{=}C—CH_3$ (each C bearing H)

d. $CH_2{=}CH—CH_2—CH_3$ (with CH_3 on the CH)

8.66 The IUPAC name of the compound below is 1,1-dibromo-4-methyl-2-pentene. Identify what substituents would be attached at points A, B, C, and D.

Br—C—C=C—C—(D)
(A) H (B) (C)

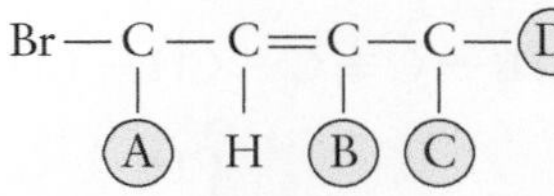

8.67 The IUPAC name of the compound below is 3,3-dichloro-2,3-difluoro-1-propene. Identify what substituents would be attached at points A, B, C, and D.

(C)
H—C=C—C—Cl
(A) (B) (D)

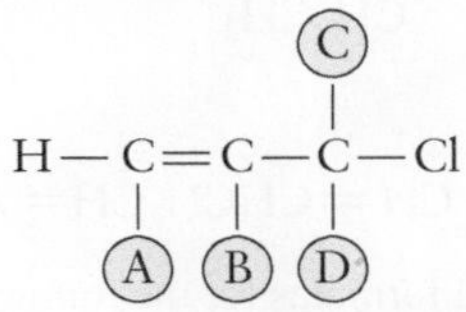

8.68 Give the IUPAC name for the following compounds.

a. $CH_3CHCH{=}CHCH_2CH_3$ (with CH_3 on the second C)

b. $CH_3CH{=}CHCH{=}CHCHCH_3$ (with CH_3 on the sixth C)

c. 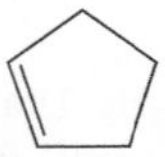 (cyclopentene)

d. $H_3C—C{\equiv}C—CH_2CH_3$

e. $CH_3C—CH_2CH_2—CCH_3$ (each with $=CH_2$)

f. 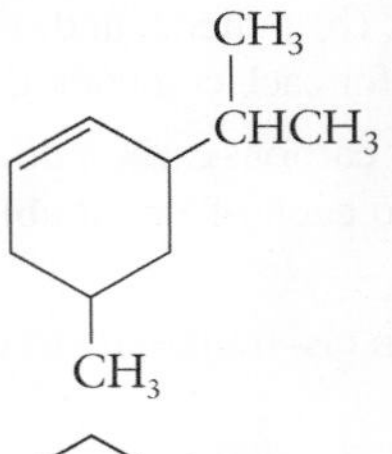 CH_3 / $CHCH_3$ (cyclohexene with $CH(CH_3)_2$ and CH_3)

g. 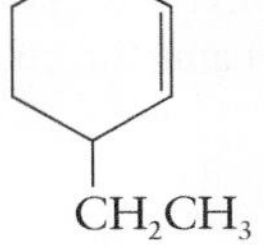 (cyclohexene with CH_2CH_3)

8.69 Give the IUPAC name for the following compounds.

a. $CH_3CH{=}CHCH_3$

b. 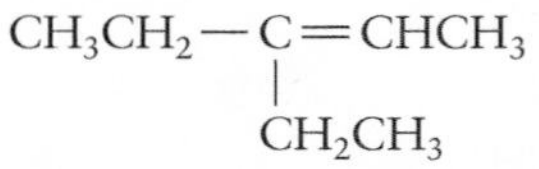 $CH_3CH_2{-}C{=}CHCH_3$
   (with CH_2CH_3 on the C)

$$
\begin{array}{l}
\,\mid \\
CH_3CH_2-C{=}CHCH_3 \\
\,\mid \\
CH_2CH_3
\end{array}
$$

c.

$$
\begin{array}{c}
CH_3 \\
\mid \\
CH_3-C{\equiv}C-C-CH_2CH_3 \\
\mid \\
CH_3
\end{array}
$$

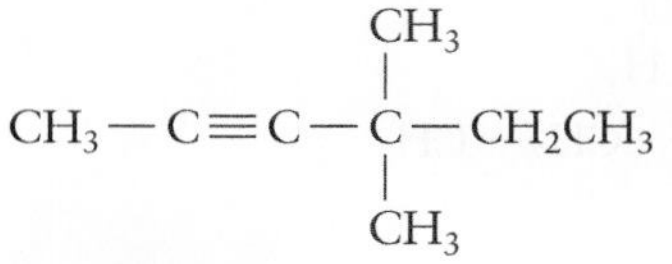

d. 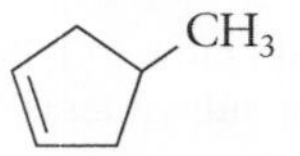 (cyclopentene ring) CH_3

e.

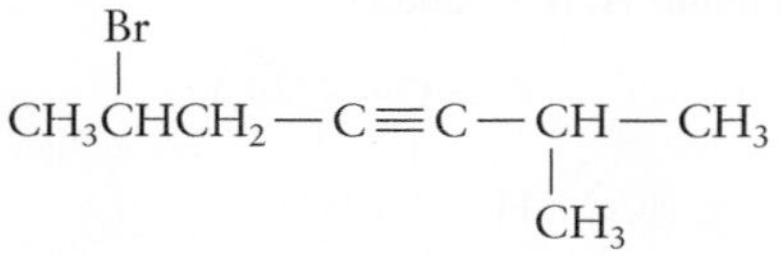

$$
\begin{array}{l}
Br \\
\mid \\
CH_3CHCH_2-C{\equiv}C-CH-CH_3 \\
\phantom{CH_3CHCH_2-C{\equiv}C-C}\mid \\
\phantom{CH_3CHCH_2-C{\equiv}C-}CH_3
\end{array}
$$

f. 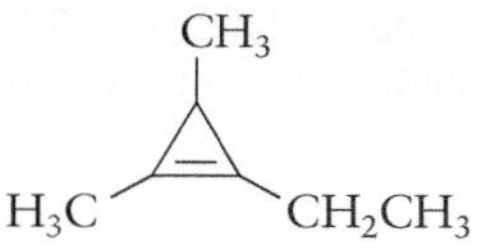 (cyclopropene ring with substituents) CH_3, H_3C, CH_2CH_3

g.

$$
\begin{array}{l}
CH_3 \\
\mid \\
CH_3CH-CH{=}CHCH_2CH{=}CH_2
\end{array}
$$

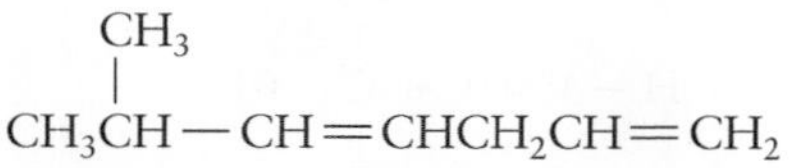

8.70 Draw structural formulas for the following compounds.

a. 4-methylcyclohexene

b. 2-ethyl-1,4-pentadiene

c. 4-isopropyl-3,3-dimethyl-1,5-octadiene

8.71 Draw structural formulas for the following compounds.

a. 3-ethyl-2-hexene

b. 3,4-dimethyl-1-pentene

c. 3-methyl-1,3-pentadiene

8.72 A compound has the molecular formula C_5H_8. Draw a structural formula for a compound with this formula that would be classified as (a) an alkyne, (b) a diene, and (c) a cyclic alkene. Give the IUPAC name for each compound.

8.73* A compound has the molecular formula C_4H_6. Draw a structural formula for a compound with this formula that would be classified as (a) an alkyne, (b) a diene, and (c) a cyclic alkene. Give the IUPAC name for each compound.

8.74 Describe the geometry of the carbon–carbon double bond and the two atoms attached to each of the double-bonded carbon atoms.

8.75* Explain the difference between cis–trans and structural isomers of alkenes.

8.76 Which of the following alkenes can exist as cis–trans isomers? Draw structural formulas and name the cis and trans isomers.

a. $CH_3CH_2CH_2CH_2CH{=}CH_2$

b. $CH_3CH_2CH{=}CHCH_2CH_3$

c.

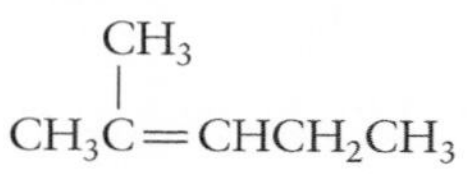

$$
\begin{array}{l}
CH_3 \\
\mid \\
CH_3C{=}CHCH_2CH_3
\end{array}
$$

8.77 Which of the following alkenes can exist as cis–trans isomers? Draw structural formulas and name the cis and trans isomers.

a. $H_2C{=}CH{-}CH_3$

b.

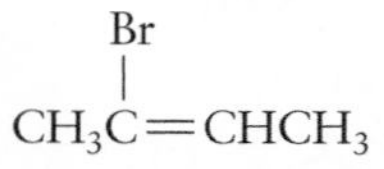

$$
\begin{array}{l}
Br \\
\mid \\
CH_3C{=}CHCH_3
\end{array}
$$

c.

$$
\begin{array}{l}
Cl \\
\mid \\
HC{=}CHCH_3
\end{array}
$$

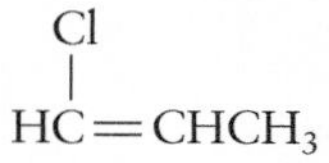

8.78 Draw structural formulas for the following.

a. *cis*-3-hexene

b. *trans*-3-heptene

8.79 Draw structural formulas for the following.

a. *cis*-4-methyl-2-pentene

b. *trans*-5-methyl-2-heptene

8.80 Give an IUPAC name for each of the following molecules. Include the prefix *cis*- or *trans*- when appropriate.

a. 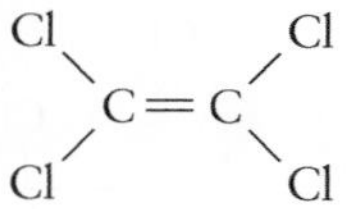 $Cl_2C{=}CCl_2$ (Cl, Cl on top; Cl, Cl on bottom)

b. 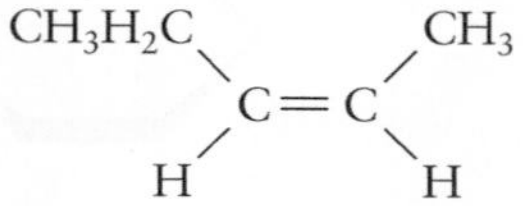 CH_3H_2C and H on left carbon; CH_3 (top) and H (bottom) on right carbon: $C{=}C$

c. H_3C (top) and H (bottom) on left carbon; H (top) and Cl (bottom) on right carbon: $C{=}C$

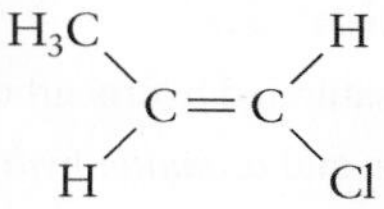

d. 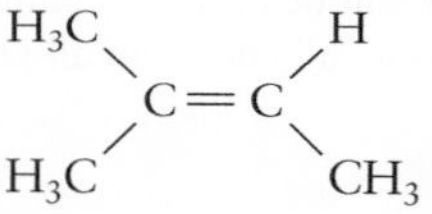 H_3C (top) and H_3C (bottom) on left carbon; H (top) and CH_3 (bottom) on right carbon: $C{=}C$

8.81 Give an IUPAC name for each of the following molecules. Include the prefix *cis*- or *trans*- when appropriate.

a. 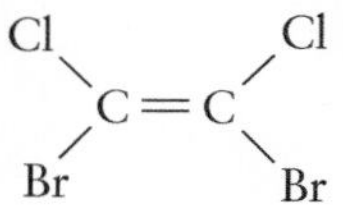 Cl (top) and Br (bottom) on left carbon; Cl (top) and Br (bottom) on right carbon: $C{=}C$

b. 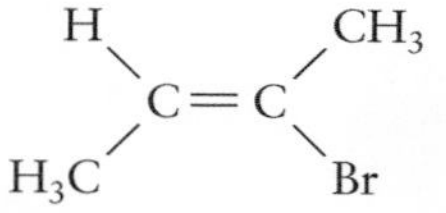 H (top) and H_3C (bottom) on left carbon; CH_3 (top) and Br (bottom) on right carbon: $C{=}C$

c. CH_3CH_2 (top) and H (bottom) on left carbon; H (top) and CH_2CH_3 (bottom) on right carbon: $C{=}C$

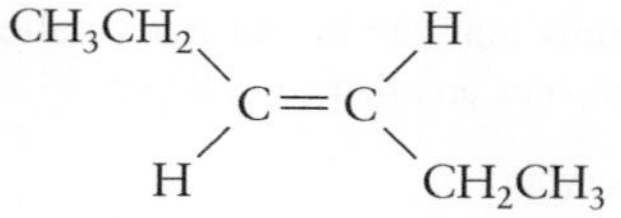

d. 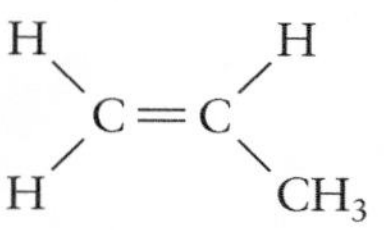 H (top) and H (bottom) on left carbon; H (top) and CH_3 (bottom) on right carbon: $C{=}C$

8.82 Explain why cis–trans isomerism is not possible in alkynes.

8.83 Give the common name and major uses of the simplest alkyne.

8.84 Describe the geometry in an alkyne of the carbon–carbon triple bond and the two attached atoms.

Reactions of Hydrocarbons (Section 8.8)

8.85 State Markovnikov's rule and write a reaction that illustrates its application.

8.86 Write a chemical equation showing reactants, products, and catalysts (if needed) for the reaction of 1-butene ($CH_2{=}CHCH_2CH_3$) with H_2. Use Markovnikov's rule as needed.

8.87 Write a chemical equation showing reactants, products, and catalysts (if needed) for the reaction of 1-butene ($CH_2{=}CHCH_2CH_3$) with H_2O. Use Markovnikov's rule as needed.

8.88 Complete the following reactions. Where more than one product is possible, show only the one expected according to Markovnikov's rule.

a. 1-methylcyclopentene (ring with CH_3) $+ H_2 \xrightarrow{Pt}$

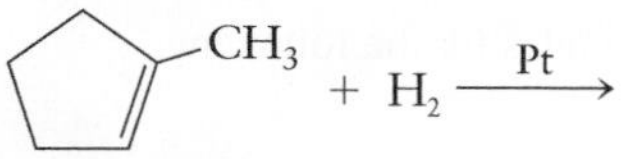

b.

$$CH_3CH_2\overset{\overset{CH_3}{|}}{C}{=}CH_2 + H_2O \xrightarrow{H_2SO_4}$$

8.89 Complete the following reactions. Where more than one product is possible, show only the one expected according to Markovnikov's rule.

a. cyclohexene $+ H_2O \xrightarrow{H_2SO_4}$

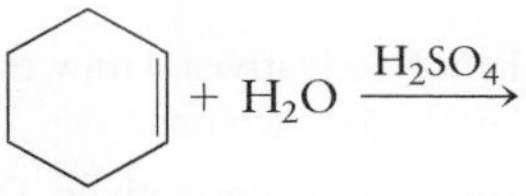

b.

$$CH_2{=}\overset{\overset{CH_3}{|}}{C}{-}CH{=}CH_2 + 2H_2 \xrightarrow{Pt}$$

8.90* What reagents would you use to prepare each of the following from 2-pentene?

a. $CH_3CH_2CH_2CH_2CH_3$

b.

$$CH_3\overset{\overset{OH}{|}}{C}HCH_2CH_2CH_3$$

8.91 Draw the "start" (the first three repeating units) of the structural formula of the addition polymers formed from the following monomers.

a. propene

b. 2-methyl-1-propene

c. 1,1,2,2-tetrafluoroethene

d. 1-butene

8.92 Draw the "start" (the first three repeating units) of the structural formula of the addition polymers formed from the following monomers.

a. ethene

b. chloroethene

c. 1,2-dichloroethene

d. 1,1,2-trichloroethene

8.93 Explain what is meant by each of the following terms: *monomer*, *polymer*, and *addition polymer*.

Physical and Chemical Properties of Hydrocarbons (Section 8.9)

8.94 The compound decane is a straight-chain alkane. Predict the following.

a. Is decane a solid, liquid, or gas at room temperature?

b. Is it soluble in water?

c. Is it soluble in hexane?

d. Is it more or less dense than water?

8.95 Which alkane in each of the following pairs of compounds has the higher boiling point?

a. ethane or propane

b. hexane or pentane

8.96 Which alkane in each of the following pairs of compounds has the higher boiling point?

a. butane or hexane

b. decane or octane

8.97 Identify the physical state (solid, liquid, or gas) at room temperature and pressure for each of the following alkanes.

a. hexane

b. butane

c. $C_{11}H_{24}$

8.98 Identify the physical state (solid, liquid, or gas) at room temperature and pressure for each of the following alkanes.

a. propane

b. $C_{24}H_{50}$

c. $C_{18}H_{30}$

8.99 Explain why alkanes of low molecular weight have lower melting and boiling points than water.

8.100 Terpin hydrate is used medicinally as an expectorant for coughs. It is prepared by the following addition reaction. What is the structure of terpin hydrate?

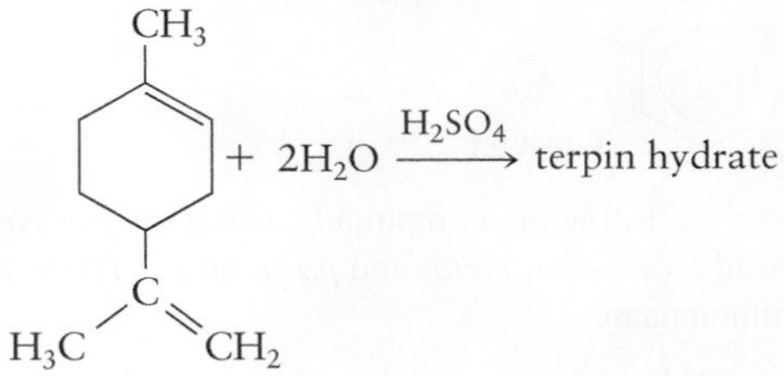

(cyclohexene ring bearing CH_3 on the double-bond carbon and a $C(=CH_2)$–H_3C group at position 4) $+ 2H_2O \xrightarrow{H_2SO_4}$ terpin hydrate

Aromatic Compounds and Benzene (Section 8.10)

8.101 What does the circle within the hexagon represent in the structural formula for benzene?

8.102 Define the term *aromatic*.

8.103 Limonene, which is present in citrus peelings, has a very pleasant lemon-like fragrance. However, it is not classified as an aromatic compound. Explain.

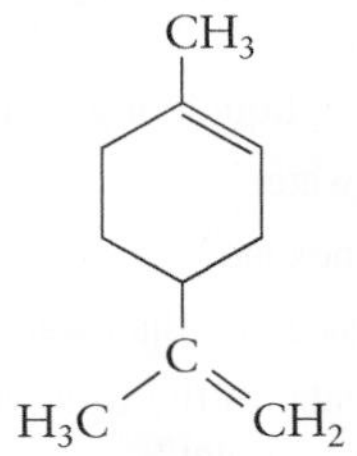

CH_3

C

H_3C CH_2

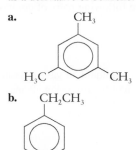

limonene

8.104 Give an IUPAC name for each of the following hydrocarbons as a derivative of benzene.

a.

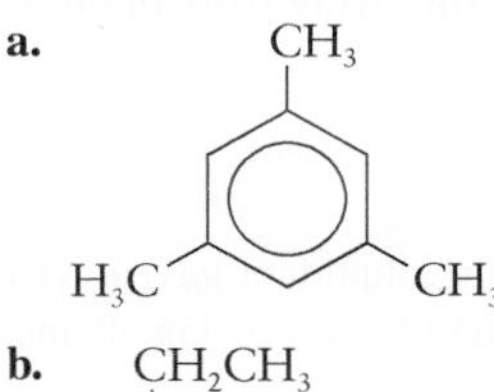

b. CH_2CH_3

CH_2CH_3

8.105 Give an IUPAC name for each of the following hydrocarbons as a derivative of benzene.

a. CH_2CH_3

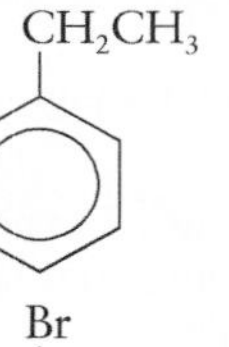

b. Br

CH_2CH_3

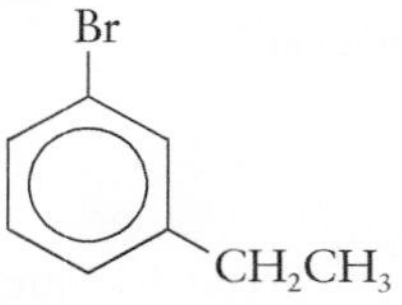

8.106 Name the following compounds, using the prefixed abbreviations for *ortho*, *meta*, and *para*, and IUPAC-acceptable common names.

a. OH

Br

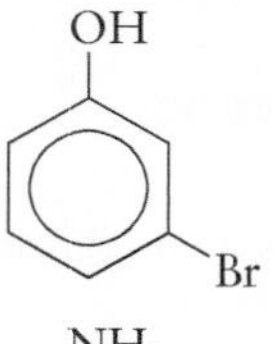

b. NH_2

CH_2CH_3

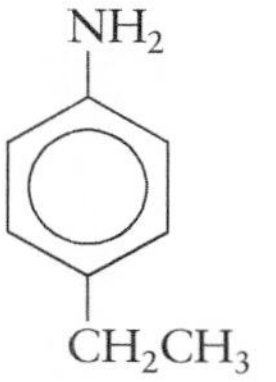

8.107 Name the following compounds, using the prefixed abbreviations for *ortho*, *meta*, and *para*, and IUPAC-acceptable common names.

a. CH_3

Br

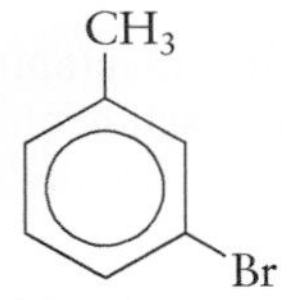

b. NH_2

CH_2CH_3

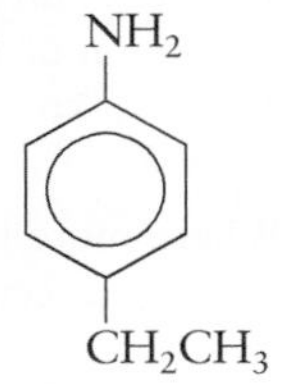

8.108 Write structural formulas for the following.

a. *o*-ethylphenol

b. *m*-chlorobenzoic acid

c. 3-methyl-3-phenylpentane

8.109 Draw structural formulas for the following.

a. *p*-chloroaniline

b. *m*-ethylphenol

c. 2-phenyl-1-pentene

8.110 There are three chlorophenol derivatives. Draw their structures and give an IUPAC name for each one.

8.111 There are three bromonitrobenzene derivatives. Draw their structures and give an IUPAC name for each.

8.112 Why does benzene not readily undergo addition reactions characteristic of other unsaturated compounds?

8.113 Describe the chief physical properties of aromatic hydrocarbons.

8.114 For each of the following uses, list an appropriate aromatic compound.

a. A solvent

b. A vitamin

c. An essential amino acid

d. Starting material for dyes

Chemistry for Thought

8.115 Compare the chemical behavior of benzene and cyclohexene.

8.116 Why might the study of organic compounds be important to someone interested in the health or life sciences?

8.117 Why do very few aqueous solutions of organic compounds conduct electricity?

8.118 The aspirin in **Figure 8.3** has a relatively low melting point when compared to inorganic compounds. What does that fact reveal about the forces between molecules of aspirin?

8.119 A semitruck loaded with cyclohexane overturns during a rainstorm, spilling its contents over the road embankment. If the rain continues, what will be the fate of the cyclohexane?

8.120 Oil spills along coastal shores can be disastrous to the environment. What physical and chemical properties of alkanes contribute to the consequences of an oil spill?

8.121 One of the fragrant components in mint plants is menthene, a compound whose IUPAC name is 1-isopropyl-4-methylcyclohexene. Draw a structural formula for menthene.

8.122 Why does propene not exhibit cis–trans isomerism?

8.123 In general, alkynes have slightly higher boiling points and densities than structurally equivalent alkanes. What intermolecular force would this be attributable to?

9 Alcohols, Ethers, and Amines

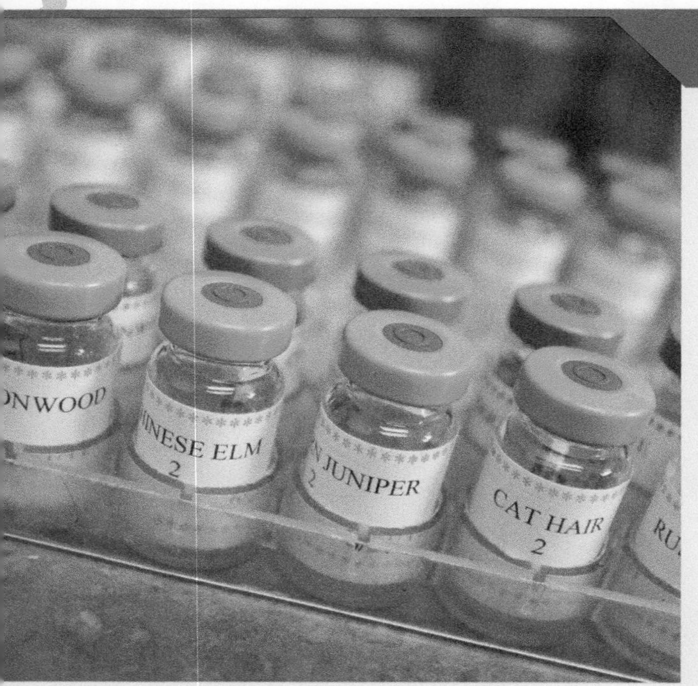

Julien McRoberts/Adobe Stock Photos

Health Career Focus

Allergy/Immunology Nurse

My son suffered from severe allergies. After his allergy test, the doctor said he reacted to every grass, tree, and animal on the skin test. On a scale of 1 to 8, most of his reactions approached 8. Since antihistamines, drugs that often contain amines, offered little relief, biweekly allergy shots were recommended.

Allergy shots must be taken on a regularly scheduled basis so that the dosage can be increased and resistance to the allergens can be developed.

An allergy nurse inspects the injection site and records the result. If the red spot from the shot stays within limits, they can increase the dosage next time. If the reaction is too large, they back off on the strength, or tell you to wait another week.

Follow-up to this Career Focus appears at the end of the chapter before the *Concept Summary*.

Learning Objectives

When you have completed your study of Chapter 9, you should be able to:

1. Name alcohols using common names and the IUPAC naming system. **(Section 9.1)**
2. Classify alcohols as primary, secondary, or tertiary on the basis of their structural formulas. **(Section 9.1)**
3. Explain how hydrogen bonding influences the physical properties of alcohols. **(Section 9.2)**
4. Predict the products when alcohols react with different acids and bases. **(Section 9.3)**
5. Write chemical equations for alcohol dehydration and oxidation reactions. **(Section 9.3)**
6. List uses for three specific alcohols. **(Section 9.4)**
7. Name ethers by using common names. **(Section 9.5)**
8. Describe the key physical and chemical properties of ethers. **(Section 9.5)**
9. Assign common names to simple amines. **(Section 9.6)**
10. Classify amines as primary, secondary, or tertiary based on their structural formulas. **(Section 9.6)**
11. Predict the products when amines react with different acids and bases. **(Section 9.7)**
12. Explain how hydrogen bonding influences the physical properties of amines. **(Section 9.7)**
13. Describe the uses of three different crucial biological amines. **(Section 9.8)**

Alcohols, ethers, and amines, the focus of this chapter, are important organic compounds that occur naturally and are produced synthetically in significant amounts. They are used in numerous industrial and pharmaceutical applications (see **Figure 9.1**). We begin by exploring the nomenclature, properties, reactions, and applications of alcohols. Then, after a brief discussion of ethers, we learn to name amines and study their properties. We conclude the chapter with an extensive look at biologically relevant amines.

inga spence/Alamy Stock Photo

Figure 9.1 Taxol, the most commonly used and effective chemotherapy drug for breast cancer, was originally isolated from the bark of the Pacific yew tree. It is now made via cell culture. Though taxol is a complex compound, what does its name suggest about its structure?

9.1 Nomenclature and Classification of Alcohols

Learning Objective 1 Name alcohols using common names and the IUPAC naming system.

Learning Objective 2 Classify alcohols as primary, secondary, or tertiary on the basis of their structural formulas.

Structurally, an **alcohol** is obtained from a hydrocarbon by replacing a hydrogen atom with a **hydroxy group** (–OH). The formula for an alcohol can be generalized as R—OH, where R– represents CH_3–, CH_3CH_2–, or any other alkyl group. Thus, R—OH can stand for CH_3—OH, CH_3CH_2—OH, and so on. If the replaced hydrogen was attached to an aromatic ring, the resulting compound is known as a **phenol**:

R—OH (alcohol) C_6H_5—OH (phenol)

alcohol A compound in which an –OH group is connected to a hydrocarbon.

hydroxy group The –OH functional group.

phenol A compound in which an –OH group is connected to a benzene ring. The parent compound is also called phenol.

Alcohols and phenols may also be considered to be derived from water by the replacement of one of its hydrogen atoms with an alkyl group or an aromatic ring:

$$\underset{\text{water}}{\text{H—OH}} \xrightarrow{\text{Replace H with R}} \underset{\text{alcohol}}{\text{R—OH}}$$

$$\underset{\text{water}}{\text{H—OH}} \xrightarrow{\text{Replace H with } C_6H_5} \underset{\text{phenol}}{C_6H_5\text{—OH}}$$

bogdanhoda/Shutterstock.com

Figure 9.2 The word *alcohol* is commonly used to refer to alcoholic drinks, which contain ethyl alcohol. Different types of beer have different concentrations of alcohol and the amount of alcohol a beer may contain varies by state.

The simpler alcohols are often known by common names, where the alkyl group name is followed by the word *alcohol,* or the suffix *-ol* is used with the root alkyl group (see **Figure 9.2**):

CH_3—OH	CH_3CH_2—OH	$CH_3CH_2CH_2$—OH	$CH_3CH(OH)CH_3$
methyl alcohol (methanol)	ethyl alcohol (ethanol)	propyl alcohol (propanol)	isopropyl alcohol (isopropanol)

As discussed in Chapter 8, this common naming convention quickly becomes unreasonable for the number of alcohols that are possible. Thus, the more systematic IUPAC naming procedure established for alkanes can also be used for alcohols. The IUPAC rules for naming alcohols that contain a single hydroxy group are as follows:

Sebastian Willnow/DDP/Getty Images

Figure 9.3 Ethylene glycol ($HOCH_2CH_2OH$), a diol, is used to de-ice aircraft.

Step 1: *Number the longest chain* to give the lower number to the carbon with the attached hydroxy group.

Step 2: *Name the longest chain* to which the hydroxy group is attached. The chain name is obtained by dropping the final *-e* from the name of the hydrocarbon that contains the same number of carbon atoms and adding the ending *-ol.*

Step 3: *Locate the position of the hydroxy group* by the number of the carbon atom to which it is attached.

Step 4: *Locate and name any groups attached to the chain.*

Step 5: *Combine the name and location for other groups, the hydroxy group location, and the longest chain into the final name.*

Alcohols containing two hydroxy groups are called *diols* (see **Figure 9.3**). Those containing three –OH groups are called *triols.* The IUPAC nomenclature rules for these compounds are essentially the same as those for the single hydroxy alcohols, except that the ending *-diol* or *-triol* is attached to the name of the parent hydrocarbon.

Substituted phenols are usually named as derivatives of the parent compound phenol:

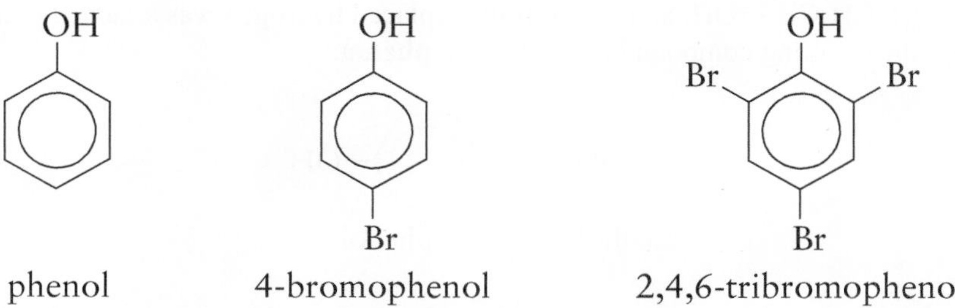

Example 9.1 Naming Alcohols

Name the following alcohols according to the IUPAC system.

a. $CH_3CH_2CH_2CHCH_2CH_3$ with CH_2-OH attached to the fourth carbon

b. A benzene ring bearing CH_2CH_3 and, on the adjacent carbon, OH

c. $CH_2-CH-CH_2$ with OH on each carbon

iStock.com/BrianAJackson

Figure 9.4 Glycerol is a component of syrups used to treat a patient's cough.

[The compound in part c is commonly known as glycerol. It is a component in most cough syrups (**Figure 9.4**) as a thickening agent, lubricant, and moistening agent.]

Solution

a. **Step 1:** *Number the longest chain* to give the lower number to the carbon with the attached hydroxy group. The longest chain containing the –OH group is numbered as follows:

$$\overset{5}{C}H_3\overset{4}{C}H_2\overset{3}{C}H_2\overset{2}{C}HCH_2CH_3 \quad (\text{C-2 bears } \overset{1}{C}H_2-OH)$$

There is a longer chain (six carbon atoms), but the –OH group is not directly attached to it.

Step 2: *Name the longest chain to which the hydroxy group is attached.* The five-carbon chain makes the alcohol a pentanol, with the ending *-ol* signifying the presence of the alcohol.

Step 3: *Locate the position of the hydroxy group.* The –OH group is given preference, so the carbon to which it is attached is carbon number 1 of the carbon chain.

Step 4: *Locate and name any groups attached to the chain.* There is an ethyl group attached at position 2 of the carbon chain.

Step 5: *Combine the name and location for other groups, the hydroxy group location, and the longest chain into the final name.* Taken together the final name is 2-ethyl-1-pentanol.

b. **Step 1:** *Number the longest chain* to give the lower number to the carbon with the attached hydroxy group.

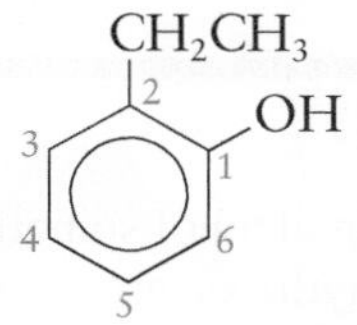

CH_2CH_3 (position 2), OH (position 1); benzene ring numbered 1–6

Step 2: *Name the longest chain to which the hydroxy group is attached.* The six-carbon aromatic ring (benzene) is attached to the –OH group, so phenol is the parent compound.

Step 3: *Locate the position of the hydroxy group.* The –OH group is at position number 1 on the benzene ring.

Step 4: *Locate and name any groups attached to the chain.* There is an ethyl group attached to the adjacent carbon, which is numbered 2 in order to provide the lowest numbering possible to the ethyl substituent.

Step 5: *Combine the name and location for other groups, the hydroxy group location, and the longest chain into the final name.* The complete name of the compound is 2-ethylphenol. An alternate name is *ortho*-ethylphenol because the benzene ring is disubstituted in a 1, 2 pattern (**Section 8.10**).

c. **Step 1:** *Number the longest chain* to give the lower number to the carbon with the attached hydroxy group.

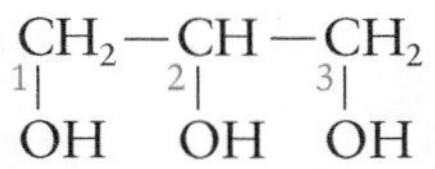

$$\underset{\text{OH}}{\overset{1}{CH_2}}-\underset{\text{OH}}{\overset{2}{CH}}-\underset{\text{OH}}{\overset{3}{CH_2}}$$

Step 2: *Name the longest chain to which the hydroxy group is attached.* The longest chain contains three carbons and three hydroxy groups, so the compound is a propanetriol.

Step 3: *Locate the position of the hydroxy group.* The –OH groups are attached to carbons 1, 2, and 3 of the carbon chain.

Step 4: *Locate and name any groups attached to the chain.* There are no additional groups attached to the carbon chain.

Step 5: *Combine the name and location for other groups, the hydroxy group location, and the longest chain into the final name.* The complete name of the compound is 1,2,3-propanetriol. Common names for this substance are glycerin and glycerol. Glycerol forms part of the structure of body fats and oils (see **Chapter 14**).

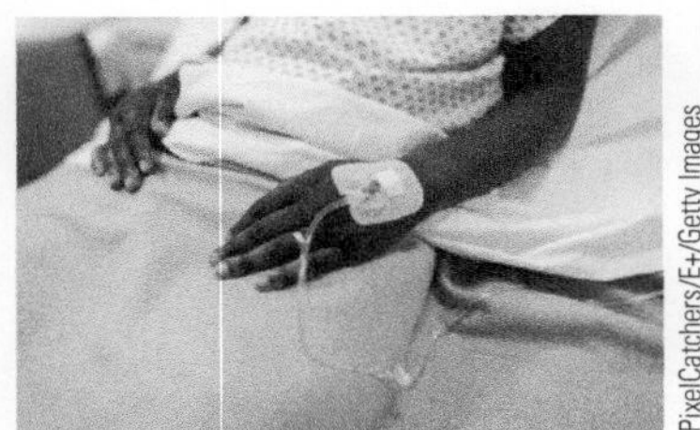

PixelCatchers/E+/Getty Images

Figure 9.5 Propofol is being delivered intravenously to a patient before surgery.

✔ **Learning Check 9.1** Propofol (part a and **Figure 9.5**) is an intravenous sedative that can be used for the initiation and maintenance of anesthesia. Provide IUPAC names for the following alcohols.

a.
```
 CH3     OH   CH3
 |       |     |
CH3CH ⟨benzene ring⟩ CHCH3
```

c.
```
   CH3
    |
⟨cyclopentane ring⟩—OH
```

b.
```
          OH
          |
CH3CHCHCH2CHCH3
     |        |
   CH2CH3   CH3
```

d.
```
      CH3
       |
CH2CHCH2CH2
|          |
OH         OH
```

primary alcohol An alcohol in which the –OH group is attached to CH_3 or to a carbon attached to one other carbon atom.

secondary alcohol An alcohol in which the carbon bearing the –OH group is attached to two other carbon atoms.

tertiary alcohol An alcohol in which the carbon bearing the –OH group is attached to three other carbon atoms.

The characteristic chemistry of an alcohol sometimes depends on the groups bonded to the carbon atom that bears the hydroxy group. Alcohols are classified as primary, secondary, or tertiary on the basis of these attached groups (see **Table 9.1**). In a **primary alcohol**, the hydroxy-bearing carbon atom is attached to one other carbon atom and two hydrogen atoms. Methanol (CH_3—OH), is also a primary alcohol. In a **secondary alcohol**, the hydroxy-bearing carbon atom is attached to two other carbon atoms and one hydrogen atom. In a **tertiary alcohol**, the hydroxy-bearing carbon atom is attached to three other carbon atoms. The use of R, R′, and R″ in **Table 9.1** signifies that each of those alkyl groups may (or may not) be different.

Table 9.1 Primary, Secondary, and Tertiary Alcohols

	Primary (1°)	Secondary (2°)	Tertiary (3°)
General formula	$R-CH_2-OH$	$R-CH(R')-OH$	$R'-C(R)(R'')-OH$
Specific example	$CH_3CH_2CH_2-OH$ 1-propanol (propyl alcohol)	$CH_3CH(CH_3)-OH$ 2-propanol (isopropyl alcohol)	$CH_3-C(CH_3)_2-OH$ 2-methyl-2-propanol (*t*-butyl alcohol)

Example 9.2 Classification of Alcohols

Propylene glycol is used as a solvent in many pharmaceuticals for drugs that are insoluble in water (**Figure 9.6**). Classify each of the two hydroxyl groups in propylene glycol as primary, secondary, or tertiary.

```
    OH
    |
CH3CHCH2—OH
```

propylene glycol

Solution

Classification depends on the number of carbon atoms bonded to the carbon atom that bears the hydroxyl group, so look carefully at the carbon skeleton.

```
     OH
     |
C—C—C—OH
    2    1
```

Carbon number 1 is bonded to one other carbon atom, which makes the group at position 1 a primary alcohol. The carbon atom at position 2 is bonded to two carbon atoms, making the group at position 2 a secondary alcohol.

Figure 9.6 Propylene glycol is the solvent for many drug compounds that aren't water soluble.

✔ Learning Check 9.2 Classify the following alcohols as primary, secondary, or tertiary.

a. $CH_3CH_2CH_2CH_2—OH$

b. (cyclohexane ring)—OH

c.
```
   OH
   |
CH3CCH3
   |
  CH2CH3
```

9.2 Physical Properties of Alcohols

Learning Objective 3 Explain how hydrogen bonding influences the physical properties of alcohols.

The replacement of one hydrogen of water with an organic group does not cause all the water-like properties to disappear. In fact, water could be considered the simplest alcohol of all, with two hydrogens bonded to oxygen instead of any alkyl groups. Thus, the low-molecular-weight alcohols—namely, methyl, ethyl, propyl, and isopropyl alcohol (**Figure 9.7**)—are soluble in water in all proportions. As the size of the alkyl group in an alcohol increases, the physical properties become less water-like and more alkane-like. Long-chain alcohols are less soluble in water (see **Figure 9.8**) and more soluble in nonpolar solvents such as benzene and carbon tetrachloride.

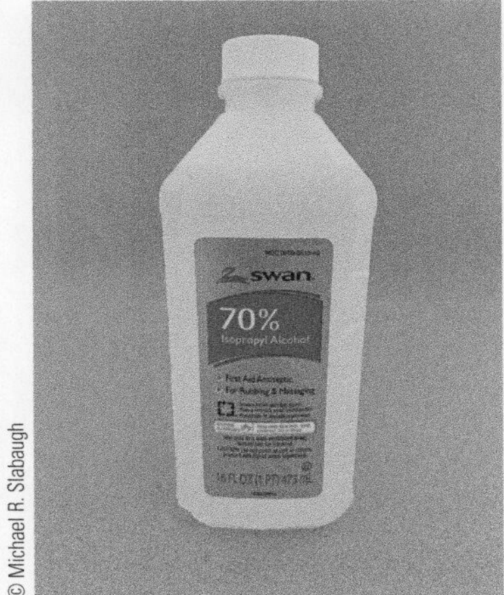

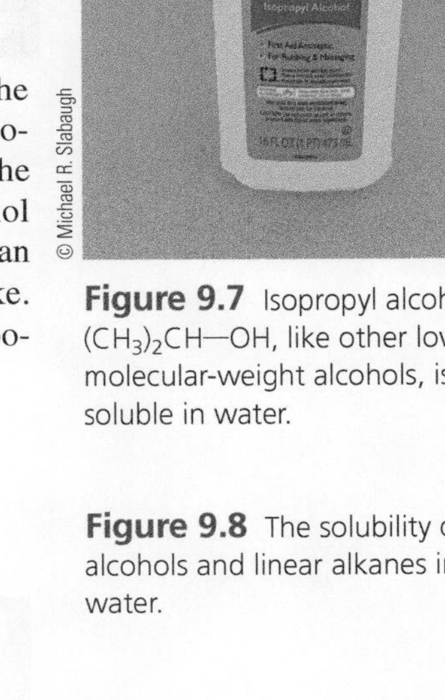

Figure 9.7 Isopropyl alcohol, $(CH_3)_2CH—OH$, like other low-molecular-weight alcohols, is highly soluble in water.

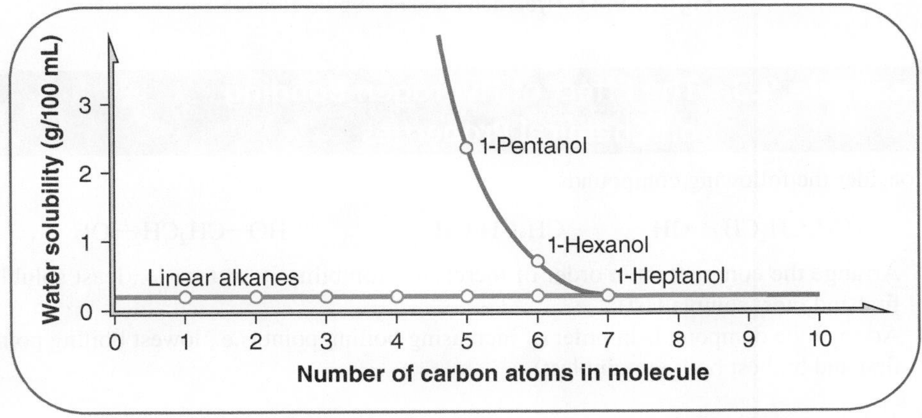

Figure 9.8 The solubility of alcohols and linear alkanes in water.

The solubility of alcohols in water depends on the number of carbon atoms per hydroxy group in the molecule. In general, one hydroxy group can solubilize three to four carbon atoms. The low-molecular-weight alcohols are highly soluble in water because hydrogen bonding between the alcohol and water molecules (see **Figure 9.9**) is more meaningful than the London dispersion forces present in the limited amount of the hydrocarbon component.

In long-chain alcohols, the hydrophobic alkyl group does not form hydrogen bonds with water. Instead, its predominant intermolecular interactions are the London dispersion forces between alkyl groups. Thus, in 1-heptanol, water molecules surround and form bonds with the –OH end but do not associate with the alkyl portion (see **Figure 9.10**). Because the alkyl portion is unsolvated by water molecules, 1-heptanol is insoluble in water. Its solubility in water is comparable to that of heptane (seven carbon atoms), as shown in **Figure 9.8**.

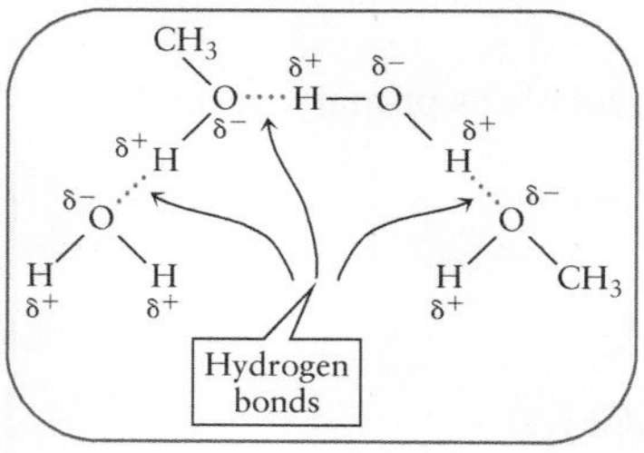

Figure 9.9 Hydrogen bonding in a water–methanol solution. A three-dimensional network of molecules and hydrogen bonds is formed.

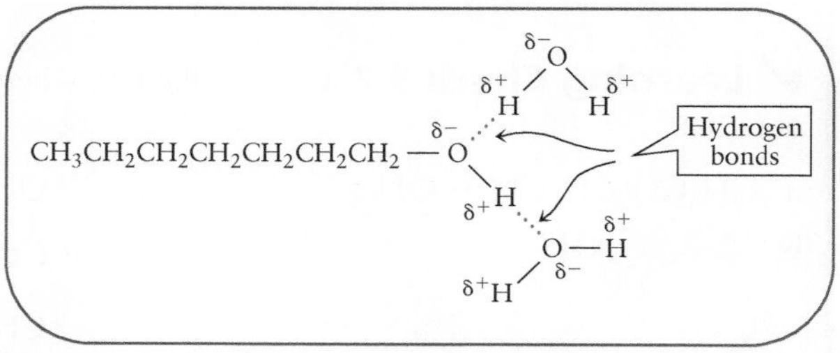

Figure 9.10 Water interacts only with the -OH group of 1-heptanol, which is insufficient to make it soluble in water, given its long hydrocarbon chain.

Hydrogen bonding also causes alcohols to have much higher boiling points than most other compounds of similar molecular weight. In this case, the hydrogen bonding is between alcohol molecules. Ethanol (CH_3CH_2OH) boils at 78 °C (**Figure 9.11**), whereas dimethyl ether ($CH_3—O—CH_3$), which has the same molecular weight, boils at −24 °C. Propane ($CH_3CH_2CH_3$) has nearly the same molecular weight but boils at −42 °C.

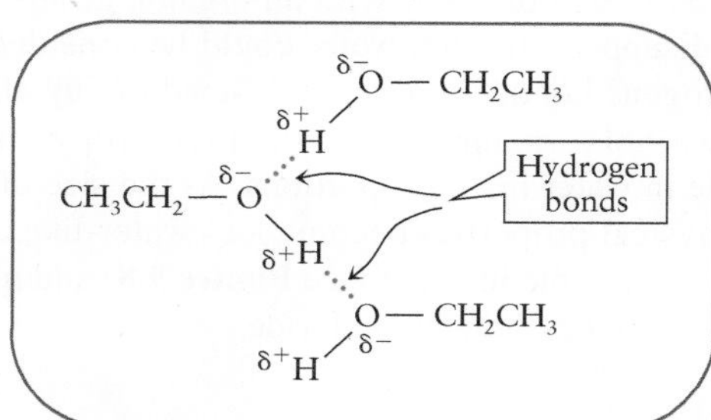

Figure 9.11 Hydrogen bonding in pure ethanol.

Example 9.3 The Effect of Hydrogen Bonding on Chemical Properties

Consider the following compounds.

$CH_3CH_2CH_2—OH$ $\quad$ $CH_3CH_2CH_3$ $\quad$ $HO—CH_2CH_2—OH$

a. Arrange the compounds in order of increasing solubility in water (i.e., least soluble first and most soluble last).

b. Arrange the compounds in order of increasing boiling point (i.e., lowest boiling point first and highest boiling point last).

Solution

a. The solubility of a compound in water largely depends on its ability to hydrogen bond, so the alkane ($CH_3CH_2CH_3$), which is unable to hydrogen bond due to the lack of –OH groups, is the least soluble in water. The next most soluble compound is $CH_3CH_2CH_2$—OH because it has a single –OH group. The most soluble in water is the diol (HO—CH_2CH_2—OH) because it has the most hydrogen bonding groups.

b. The boiling point of a pure compound is based on the intermolecular forces between molecules. The stronger the forces, the higher the boiling point. The dominant forces between alkanes are London dispersion forces, the weakest of the intermolecular forces. So, $CH_3CH_2CH_3$ has the lowest boiling point. The alcohols are both dominated by hydrogen bonding (the strongest intermolecular force), but the diol (HO—CH_2CH_2—OH) has two groups that can participate in hydrogen bonding, and $CH_3CH_2CH_2$—OH has only one –OH group. Thus, the diol has the highest boiling point.

Lolostock/Shutterstock.com

Figure 9.12 Butane gas is a common fuel source for portable stoves used during camping.

✔ **Learning Check 9.3** Explain why methanol has a higher boiling point (65 °C) than butane (−1 °C) (**Figure 9.12**), even though butane has a higher molecular weight.

9.3 Reactions of Alcohols

Learning Objective 4 Predict the products of alcohols with different acids and bases.

Learning Objective 5 Write chemical equations for alcohol dehydration and oxidation reactions.

9.3A Acid–Base Chemistry of Alcohols

The acid–base nature of alcohols is mixed because they are neither strongly acidic nor strongly basic. Their behavior in acid–base reactions greatly depends upon what other species are in solution. They are best described as being *amphoteric*, meaning they can be either acidic or basic. When reacting with a strong acid, the oxygen in an alcohol becomes protonated (Equation 9.1).

$$R-OH + H-X \rightleftharpoons R-{}^{+}OH_2 + X^- \qquad (9.1)$$

However, for an alcohol to be deprotonated (Equation 9.2), the base must be stronger than hydroxide, because hydroxide would be roughly equal in base strength to $R-O^-$, the conjugate base of the alcohol generated.

$$R-OH + {}^{-}Base \rightleftharpoons R-O^- + H\text{-}Base \qquad (9.2)$$

Recall from Chapter 7 that these acid–base reactions are at equilibrium, so if the conjugate acid–base pairs are of roughly equal strength, the equilibrium will not be strongly favored in one direction or another.

Unlike alcohols, phenols act as weak acids:

$$C_6H_5OH + H_2O \rightleftharpoons C_6H_5O^- + H_3O^+$$

phenol
(a weak acid)

$$C_6H_5OH + KOH \rightleftharpoons C_6H_5O^-K^+ + H_2O$$

Figure 9.13 Rashes associated with poison ivy are caused by phenolic compounds present in the leaves.

These weak acids can denature the proteins in the skin (see **Figure 9.13**), so they must be handled with care.

Example 9.4 Acid–Base Chemistry of Alcohols

Ethanol is often used as the solvent in medicines and has antimicrobial properties due to its abilities to denature proteins in organisms (**Figure 9.14**). Predict the major product of the following reactions of ethanol.

a. 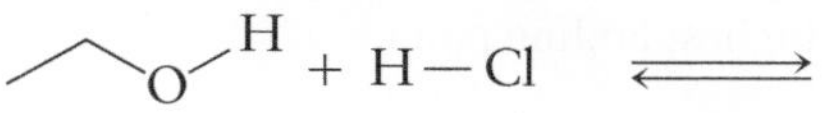 O–H + H—Cl $\rightleftharpoons$

b. 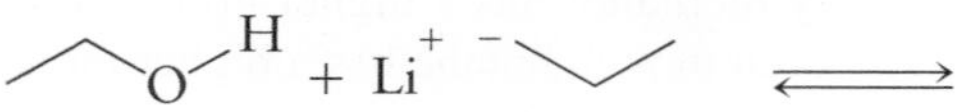 O–H + Li^{+} $^{-}$ $\rightleftharpoons$

Solution

a. Because hydrochloric acid is a strong acid, it will protonate the oxygen of ethanol to create two charged conjugate species, $CH_3CH_2OH_2^+$ and Cl^-. In this case, which is like Equation 9.1, the alcohol is acting as a base.

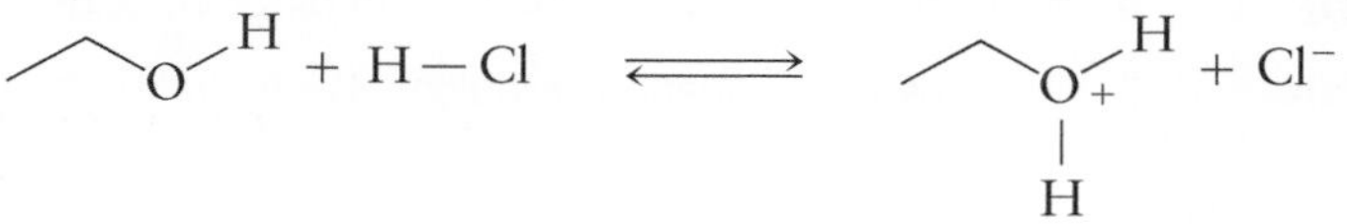

O–H + H—Cl $\rightleftharpoons$ O^+(H)–H + Cl^-

b. *n*-Propyl lithium is one of the strongest bases in existence—it's even stronger than sodium hydroxide. The ionic bond between Li and C places a very unstable, highly basic negative charge on the carbon atom. Thus, like in Equation 9.2, the alcohol is acting as an acid and is deprotonated.

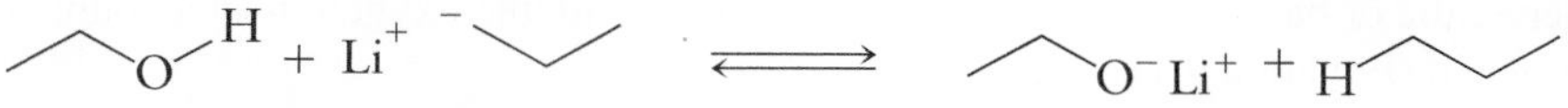

O–H + Li^{+} $^{-}$ $\rightleftharpoons$ O^- Li^+ + H

Figure 9.14 Many hand sanitizers contain >70% ethanol to increase the antimicrobial properties of the solution.

✔ Learning Check 9.4 Predict the major products of the following reactions.

a. 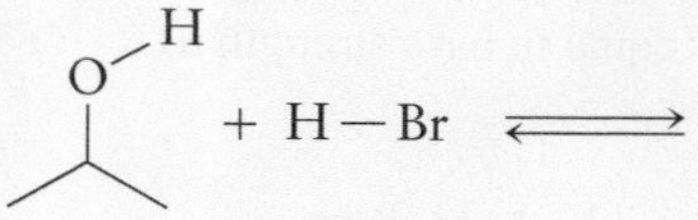 O–H + H—Br $\rightleftharpoons$

b. 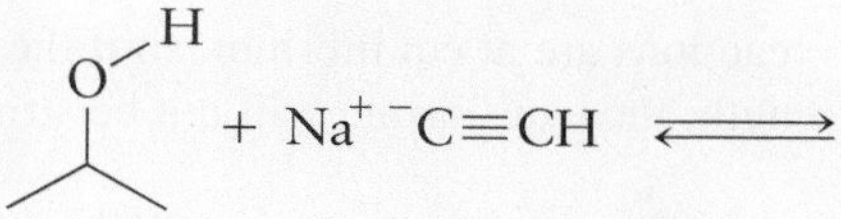 O–H + Na^{+} $^{-}C{\equiv}CH$ $\rightleftharpoons$

dehydration reaction A reaction in which water is chemically removed from a compound.

Alcohols undergo many reactions, but we consider only two important ones at this time—dehydration and oxidation. In an alcohol **dehydration reaction**, water is chemically removed from an alcohol. This reaction can occur in two different ways, depending on the reaction temperature. At 140 °C, the main product is an ether, whereas at 180 °C, alkenes are the predominant products.

9.3B Dehydration of Alcohols

elimination reaction A reaction in which two or more covalent bonds are broken and a new multiple bond is formed.

When the elements of a water molecule are removed at a higher temperature from a single alcohol molecule, an alkene is produced (see Equation 9.3). A reaction of this type, in which two or more covalent bonds are broken and a new multiple bond is formed, is called an **elimination reaction**. In this case, a water molecule is eliminated from an alcohol.

General reaction:

$$-\underset{\displaystyle H}{\overset{|}{\underset{|}{C}}}-\underset{\displaystyle OH}{\overset{|}{\underset{|}{C}}}- \xrightarrow[180\ °C]{H_2SO_4} \;\; >C{=}C< \; + H_2O \qquad (9.3)$$

alcohol → alkene

Specific example:

$$CH_3-\underset{\displaystyle H}{\underset{|}{CH}}-\underset{\displaystyle OH}{\underset{|}{CH}}-\underset{\displaystyle H}{\underset{|}{CH_2}} \xrightarrow[180\ °C]{H_2SO_4} \underset{\text{2-butene}}{(H)(H_3C)C{=}C(CH_3)(H)} + \underset{\text{1-butene}}{(H)(H_3CH_2C)C{=}C(H)(H)} + H_2O$$

2-butanol

2-butene: major product (90%) (two carbon groups on double bond)

1-butene: minor product (10%) (one carbon group on double bond)

In the specific example of 2-butanol, two alkenes can be produced, depending on which carbon next to the hydroxy-bearing carbon loses the hydrogen atom. The major product produced in such cases will be the one in which the higher number of carbon groups is bonded to the double-bonded carbon atoms. Alcohol dehydration is an important reaction in the human body, where enzymes (rather than sulfuric acid) function as catalysts. For example, the dehydration of citrate is part of the citric acid cycle, which plays a crucial role in the production of cellular energy discussed in Chapter 15.

$$^{-}OOC-\underset{\displaystyle H}{\overset{\displaystyle H}{\overset{|}{\underset{|}{C}}}}-\underset{\displaystyle COO^-}{\overset{\displaystyle OH}{\overset{|}{\underset{|}{C}}}}-CH_2-COO^- \xrightarrow{\text{Enzyme}} (H)(^{-}OOC)C{=}C(CH_2-COO^-)(COO^-) + H_2O$$

citrate → *cis*-aconitate

Note that the dehydration of citrate and 2-butanol both involve exactly the same changes in bonding. In each reaction, H and –OH are removed from adjacent carbons, leaving a C=C double bond. The other molecular bonds in all three of those compounds do not change.

Example 9.5 Dehydration of an Alcohol

Menthol is a component of ointments, like Vicks VapoRub© (**Figure 9.15**), which are used to ease minor muscle aches. Predict the major product of the following reaction of menthol.

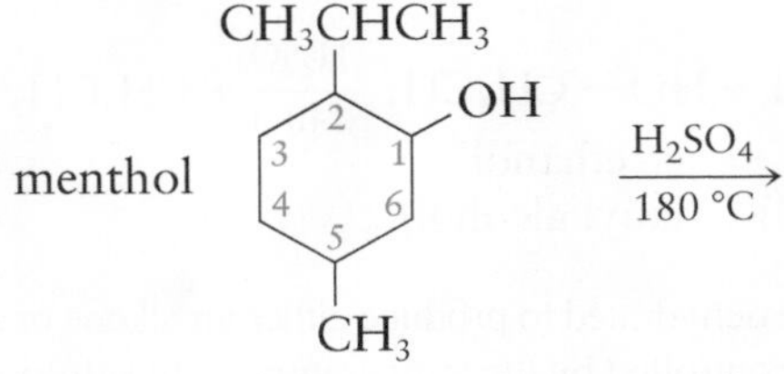

Enriscapes/Shutterstock.com

Figure 9.15 The alcohol menthol is present in many ointments used to relieve congestions and minor muscle aches.

Solution

This reaction is similar to Equation 9.3, where a molecule of water is removed and an alkene forms. The –OH group is removed along with a –H atom from an adjacent carbon atom (not the carbon to which the –OH is attached). There are –H atoms on both adjacent carbons (2 and 6).

CH_3CHCH_3 (at position 2 of a cyclohexane ring with –OH at position 1, –H at positions 2 and 6, and CH_3 at position 5) $\xrightarrow[180\ °C]{H_2SO_4}$ product A and product B $+ H_2O$

If the water molecule removed involves the –H at position 6, then product A is formed. If the formation of water involves the –H at position 2, then product B is formed.

The major product is the one that has the higher number of carbon atoms bonded to the carbons of the double bond. Each carbon bonded to C=C is screened in the following structures:

product A and product B

Product B has more alkyl substituents, so it is the major product.

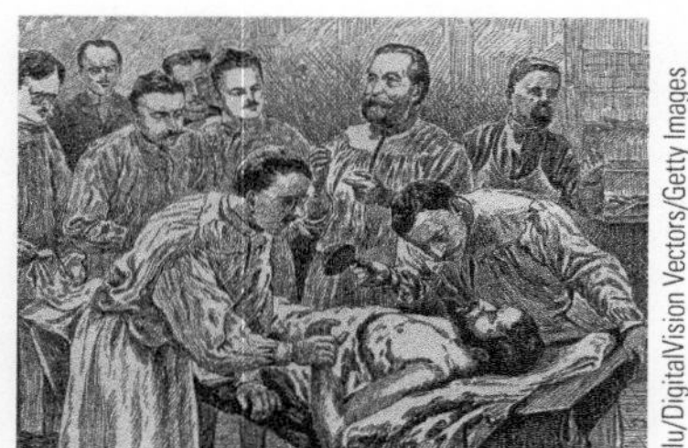

Figure 9.16 Historically, diethyl ether was used as an anesthetic during surgery. Diethyl ether is no longer used as an inhalation anesthetic because of its flammability.

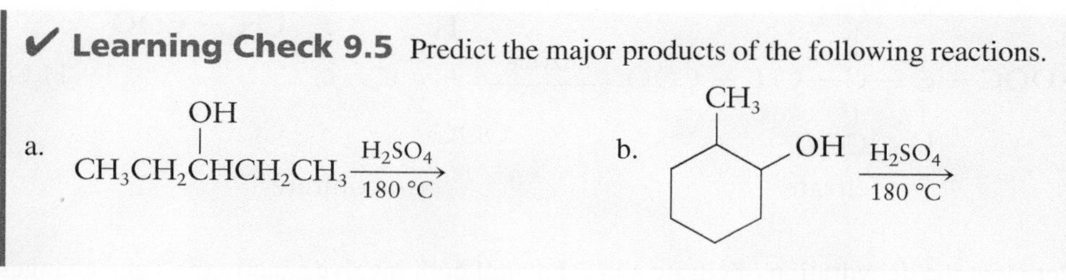

Alcohol dehydration can also occur when the elements of water are removed from two alcohol molecules. The residual molecular fragments of the alcohols bond and form an ether. This ether-forming reaction is useful mainly with primary alcohols. Equation 9.4 gives the general reaction, and the formation of diethyl ether (**Figure 9.16**) is the specific example.

General reaction:

$$\underset{\text{alcohol}}{R-O-H} + \underset{\text{alcohol}}{H-O-R} \xrightarrow[140\ °C]{H_2SO_4} \underset{\text{ether}}{R-O-R} + H_2O \qquad (9.4)$$

Specific example:

$$\underset{\substack{\text{ethanol}\\\text{(ethyl alcohol)}}}{CH_3CH_2-OH} + \underset{\substack{\text{ethanol}\\\text{(ethyl alcohol)}}}{HO-CH_2CH_3} \xrightarrow[140\ °C]{H_2SO_4} \underset{\text{diethyl ether}}{CH_3CH_2-O-CH_2CH_3} + H_2O$$

Amino acid (1) Amino acid (2)

N-terminus C-terminus

Protein Water

Figure 9.17 In the body, dehydration reactions are responsible for connecting amino acid monomers together to form proteins.

Thus, an alcohol can be dehydrated to produce either an alkene or an ether. The type of dehydration that takes place is controlled by the temperature, with sulfuric acid serving as a catalyst.

Although a mixture of alkene and ether products often results, we use 180 °C to indicate that the major dehydration product is an alkene and 140 °C to indicate when it is an ether. The reaction that produces ethers is an example of a dehydration synthesis in which two molecules are joined and a new functional group is generated by the removal of a water molecule. Dehydration synthesis is important in living organisms because it is part of the formation of carbohydrates, fats, proteins (**Figure 9.17**), and many other essential substances.

Example 9.6 Dehydration to Produce an Ether

Predict the major products of the following reaction.

$$2CH_3\underset{\displaystyle |}{\overset{\displaystyle OH}{C}}HCH_3 \xrightarrow[140\ °C]{H_2SO_4}$$

Solution

This reaction is similar to Equation 9.4, where two molecules of alcohol come together to form an ether.

$$CH_3CHCH_3-O-H \;+\; H-O-CH_3CHCH_3 \longrightarrow CH_3CHCH_3-O-CH_3CHCH_3 + H_2O$$

✔ Learning Check 9.6 Predict the major products of the following reaction.

$$2\ \text{(cyclopentyl)}-OH \xrightarrow[140\ °C]{H_2SO_4}$$

9.3C Oxidation of Alcohols

An oxidation reaction occurs when a molecule gains oxygen atoms or loses hydrogen atoms. Under appropriate conditions, alcohols can be oxidized (see **Figure 9.18**) in a controlled way (not combusted) by removing two hydrogen atoms per molecule and adding a second bond between carbon and oxygen to form a double bond. A number of oxidizing agents can be used, including potassium dichromate ($K_2Cr_2O_7$) and potassium permanganate ($KMnO_4$). The symbol (O) is used to represent an oxidizing agent in the reactions that follow. The three classes of alcohols (i.e., primary, secondary, and tertiary) behave differently toward oxidizing agents.

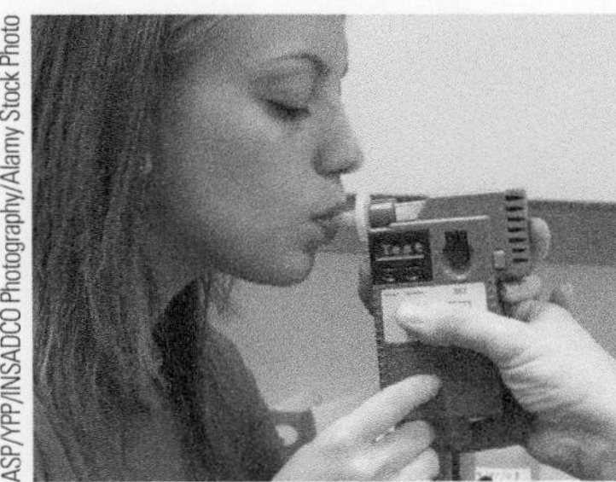

ASP/YPP/INSADCO Photography/Alamy Stock Photo

Figure 9.18 The breathalyzer test is based upon the oxidation of ethanol in a person's breath. Blood alcohol level is related to the concentration of alcohol in the breath.

Primary Alcohols

Equation 9.5 gives the general reaction, and the formation of acetic acid is a specific example (**Figure 9.19**):

General reaction:

$$\underset{\text{primary alcohol}}{R-\overset{\displaystyle OH}{\underset{\displaystyle H}{C}}-H} + (O) \longrightarrow \underset{\text{aldehyde}}{R-\overset{\displaystyle O}{\overset{\|}{C}}-H} + H_2O \qquad (9.5)$$

$$\text{aldehyde} \xrightarrow[(O)]{\text{Further oxidation}} \underset{\text{carboxylic acid}}{R-\overset{\displaystyle O}{\overset{\|}{C}}-OH}$$

Specific example:

$$\underset{\text{ethanol}}{CH_3CH_2-OH} + (O) \longrightarrow \underset{\text{acetaldehyde}}{CH_3-\overset{\displaystyle O}{\overset{\|}{C}}-H} + H_2O$$

$$\text{acetaldehyde} \xrightarrow[(O)]{\text{Further oxidation}} \underset{\text{acetic acid}}{CH_3-\overset{\displaystyle O}{\overset{\|}{C}}-OH}$$

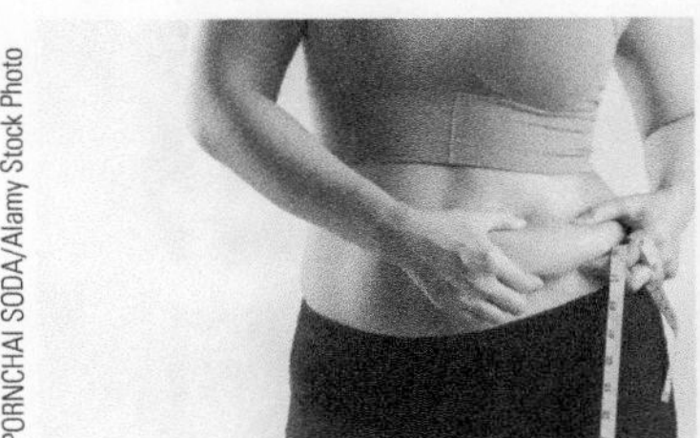

PORNCHAI SODA/Alamy Stock Photo

Figure 9.19 Apple cider vinegar is a solution of dilute acetic acid. Studies have demonstrated that acetic acid helps us regulate fat metabolism and storage.

The intermediate of the oxidation of a primary alcohol is an aldehyde. The aldehyde intermediate, however, is readily oxidized to give a carboxylic acid, a process known as overoxidation. Therefore, the oxidation of a primary alcohol normally results in the corresponding carboxylic acid as the product.

Secondary Alcohols

Equation 9.6 gives the general reaction, and the oxidation of 2-propanol is a specific example.

General reaction:

$$\underset{\text{secondary alcohol}}{R-\overset{\overset{\large OH}{|}}{\underset{\underset{\large H}{|}}{C}}-R'} + (O) \longrightarrow \underset{\text{ketone}}{R-\overset{\overset{\large O}{\|}}{C}-R'} + H_2O \qquad (9.6)$$

Specific example:

$$\underset{\text{2-propanol}}{CH_3-\overset{\overset{\large OH}{|}}{\underset{\underset{\large H}{|}}{C}}-CH_3} + (O) \longrightarrow \underset{\text{acetone}}{CH_3-\overset{\overset{\large O}{\|}}{C}-CH_3} + H_2O$$

Secondary alcohols are oxidized to ketones in exactly the same way primary alcohols are oxidized to aldehydes and by the same oxidizing agents. Unlike aldehydes, however, ketones resist further oxidation, so this reaction provides an excellent way to prepare ketones.

Tertiary Alcohols

$$R-\overset{\overset{\large OH}{|}}{\underset{\underset{\large R''}{|}}{C}}-R' + (O) \longrightarrow \text{No reaction} \qquad (9.7)$$

Tertiary alcohols do not have any hydrogen on the –OH-bearing carbon, so they do not react with oxidizing agents.

Example 9.7 Oxidation Reactions of Alcohols

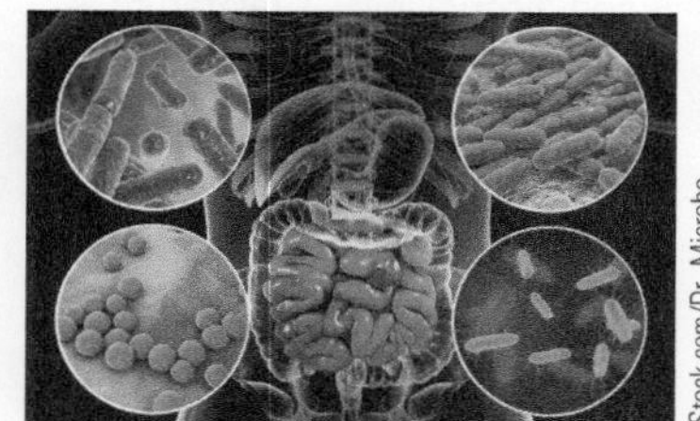

Figure 9.20 Isobutyric acid is a minor by-product of bacteria that live in the digestive tract in humans.

Draw the structural formulas for the final product of the following reactions.

a. $\underset{\underset{\large CH_3}{|}}{CH_3CH}CH_2-OH + (O) \longrightarrow$

b. 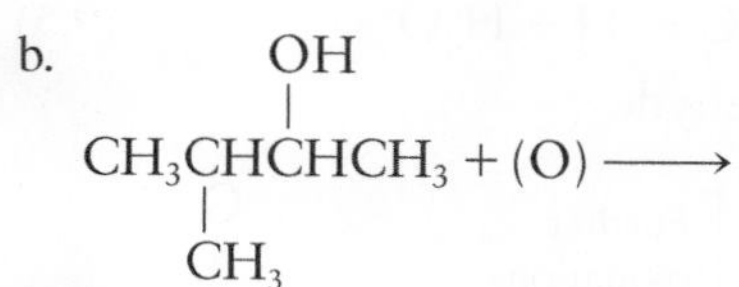

$CH_3\underset{\underset{\large CH_3}{|}}{CH}\overset{\overset{\large OH}{|}}{CH}CH_3 + (O) \longrightarrow$

Solution

a. This reaction is similar to Equation 9.5, where a primary alcohol is oxidized first to an aldehyde and then to a carboxylic acid. The carboxylic acid is isobutyric acid (**Figure 9.20**) in this case.

$$CH_3\underset{\underset{\large CH_3}{|}}{CH}-\overset{\overset{\large H}{|}}{\underset{\underset{\large H}{|}}{C}}-OH + (O) \longrightarrow CH_3\underset{\underset{\large CH_3}{|}}{CH}-\overset{\overset{\large H}{|}}{C}=O + H_2O \xrightarrow{\text{Further oxidation}} \underset{\text{isobutyric acid}}{CH_3\underset{\underset{\large CH_3}{|}}{CH}-\overset{\overset{\large O}{\|}}{C}-OH}$$

b. This reaction is similar to Equation 9.6, where a secondary alcohol is oxidized to a ketone.

$$\mathrm{CH_3CH(CH_3)-C(OH)(H)-CH_3 + (O) \longrightarrow CH_3CH(CH_3)-C(=O)-CH_3 + H_2O}$$

✔ **Learning Check 9.7** Draw the structural formulas of the final product of the following reactions.

a. (benzene ring)$-\mathrm{CH_2-OH}$ + (O) ⟶

b. (cyclohexane ring)$-\mathrm{OH}$ + (O) ⟶

c. (cyclopentane ring bearing $\mathrm{H_3C}$ and OH on the same carbon) + (O) ⟶

Study Tools 9.1

A Reaction Map for Alcohols

To solve a challenging question, using a stepwise approach is often helpful. First, decide what type of question is being asked. Is it a nomenclature, typical uses, physical properties, or reaction question? Second, if you recognize it as a reaction question, look for the functional group. Third, if there is an alcohol group in the starting material, identify the reagent and reaction conditions and use the following diagram to predict the right products.

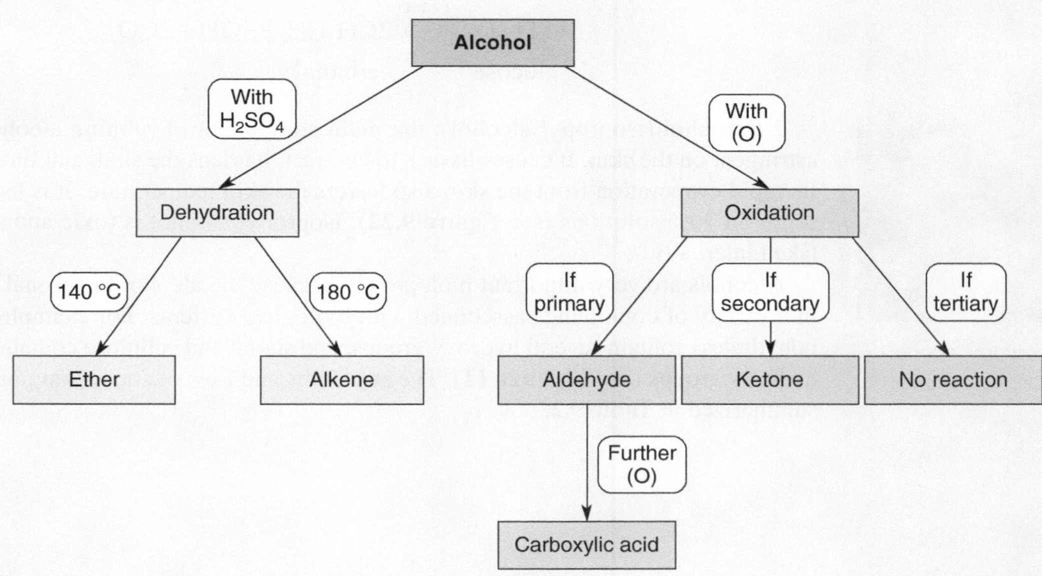

Figure 9.21 Some racing cars have used alcohols like methanol, ethanol, or mixtures of the two as fuel.

9.4 Applications of Alcohols

Learning Objective 6 List uses for three specific alcohols.

Methanol is a very important industrial chemical; more than 1 billion gallons are produced and used annually. It is sometimes known as wood alcohol because the principal source for many years was the distillation of wood. Today, it is synthetically produced in large quantities by reacting hydrogen gas with carbon monoxide (Equation 9.8):

$$CO + 2H_2 \xrightarrow[\text{Heat, pressure}]{\text{Catalysts}} CH_3{-}OH \qquad (9.8)$$

The industrial importance of methanol is due to its oxidation product, formaldehyde, which is a major starting material for the production of plastics. Methanol has been used as a fuel in Indy-style racing cars (see **Figure 9.21**) and is still used today in other forms of racing.

Methanol is highly toxic and causes permanent blindness if taken internally. Deaths and injuries have resulted as a consequence of mistakenly substituting methanol for ethanol in beverages, such as in homemade illicit moonshine. Ethanol, or ethyl alcohol, $CH_3CH_2{-}OH$, is probably the most familiar alcohol because it is the "active" ingredient in alcoholic beverages. It is also used in pharmaceuticals as a solvent (tinctures are ethanol solutions) and in aftershave lotions as an antiseptic and skin softener. It is an important industrial solvent as well. Ethanol has also become an emerging additive, and even alternative, to gasoline as a fuel.

Most ethanol used industrially is produced on a large scale by the direct hydration of ethylene obtained from petroleum:

$$\underset{\text{ethylene}}{H_2C{=}CH_2 + H{-}OH} \xrightarrow[300\ °C]{70\ atm} \underset{\text{ethanol}}{\begin{matrix} H_2C - CH_2 \\ | \quad\;\; | \\ H \quad OH \end{matrix}}$$

fermentation A reaction of sugars, starch, or cellulose to produce ethanol and carbon dioxide.

Ethanol for beverages, and increasingly for use in fuel, is produced by the yeast **fermentation** of carbohydrates such as sugars, starch, and cellulose. For example, a common method is the fermentation of glucose:

$$\underset{\text{glucose}}{C_6H_{12}O_6} \xrightarrow{\text{Yeast}} \underset{\text{ethanol}}{2CH_3CH_2{-}OH} + 2CO_2 \qquad (9.9)$$

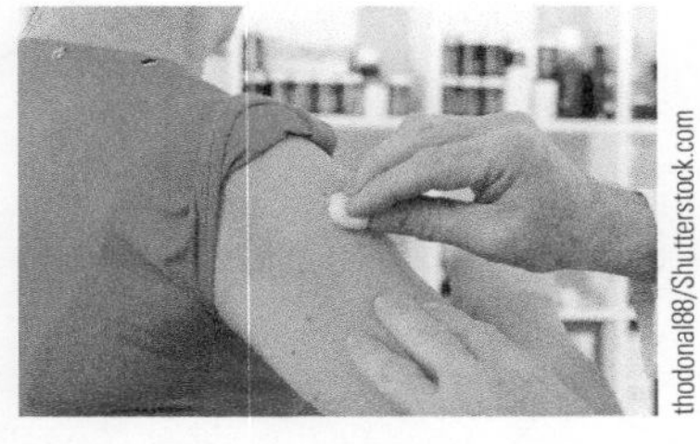

Figure 9.22 Isopropyl alcohol is used as an antiseptic before an injection.

2-Propanol (isopropyl alcohol), the main component of rubbing alcohol, acts as an astringent on the skin. It causes tissues to contract, hardens the skin, and limits secretions. Its rapid evaporation from the skin also lowers the skin temperature. It is used as an antiseptic in 70% solutions (see **Figure 9.22**). Isopropyl alcohol is toxic and should not be taken internally.

Alcohols are very important biologically because the alcohol functional group occurs in a variety of compounds associated with biological systems. For example, sugars (carbohydrates) contain several hydroxy groups, and starch and cellulose contain thousands of hydroxy groups (see **Chapter 11**). The structures and uses of some common alcohols are summarized in **Table 9.2**.

Health Connections 9.1

Alcohol and Antidepressants Don't Mix

Individuals experiencing depression often demonstrate one or more of the following symptoms: extreme feelings of sadness or melancholy, feelings of inadequacy in facing life's challenges, inability to make decisions, physical inactivity, lack of appetite, inability to sleep, and frequent crying.

Antidepressant medications are designed to counteract these symptoms. However, when alcohol, itself a depressant, is combined with antidepressant medications, serious consequences can result. For example, some antidepressants cause sedation and drowsiness. Alcohol has similar effects, and the combination of the two could easily cause a person to fall asleep while driving. Initially, alcohol might seem to improve the mood of a depressed person, but its overall effect is often to increase the symptoms of depression.

The combination of some types of antidepressants and alcohol influences coordination, judgment, reaction time, and other motor skills more than alcohol alone. When combined with certain types of alcoholic beverages and foods, antidepressants known as monoamine oxidase inhibitors (MAOIs) cause bodily reactions such as a dangerous increase in blood pressure.

It has been found that people experiencing depression have an increased risk of becoming addicted to alcohol or other addictive substances. When these factors are all taken into consideration, it is not worth the risks for people being treated for depression to mix their antidepressant medications with alcohol.

When alcohol is combined with antidepressants, serious consequences can result.

Table 9.2 Examples of Alcohols

Name	Structural Formula	Typical Uses
Methanol (methyl alcohol)	$CH_3{-}OH$	Solvent, making formaldehyde
Ethanol (ethyl alcohol)	$CH_3CH_2{-}OH$	Solvent, alcoholic beverages
2-Propanol (isopropyl alcohol)	CH_3CHCH_3 OH (on central C)	Rubbing alcohol, solvent
1-Butanol (butanol)	$CH_3CH_2CH_2CH_2{-}OH$	Solvent, hydraulic fluid
1,2-Ethanediol (ethylene glycol)	$HO{-}CH_2CH_2{-}OH$	Automobile antifreeze, polyester fibers
1,2-Propanediol (propylene glycol)	$CH_3CH{-}CH_2$ OH OH	Moisturizer in lotions and foods, automobile antifreeze
1,2,3-Propanetriol (glycerin, glycerol)	$CH_2{-}CH{-}CH_2$ OH OH OH	Moisturizer in foods, tobacco, and cosmetics
Menthol	CH_3CHCH_3 (cyclohexane ring with OH) CH_3	Cough drops, shaving lotion, mentholated tobacco

In dilute solutions, phenol is used as an antiseptic and disinfectant. Joseph Lister, a surgeon, introduced the use of phenol as a hospital antiseptic in the late 1800s. Before that time, antiseptics had not been used, and very few patients survived even minor surgery because of postoperative infections (**Figure 9.23**).

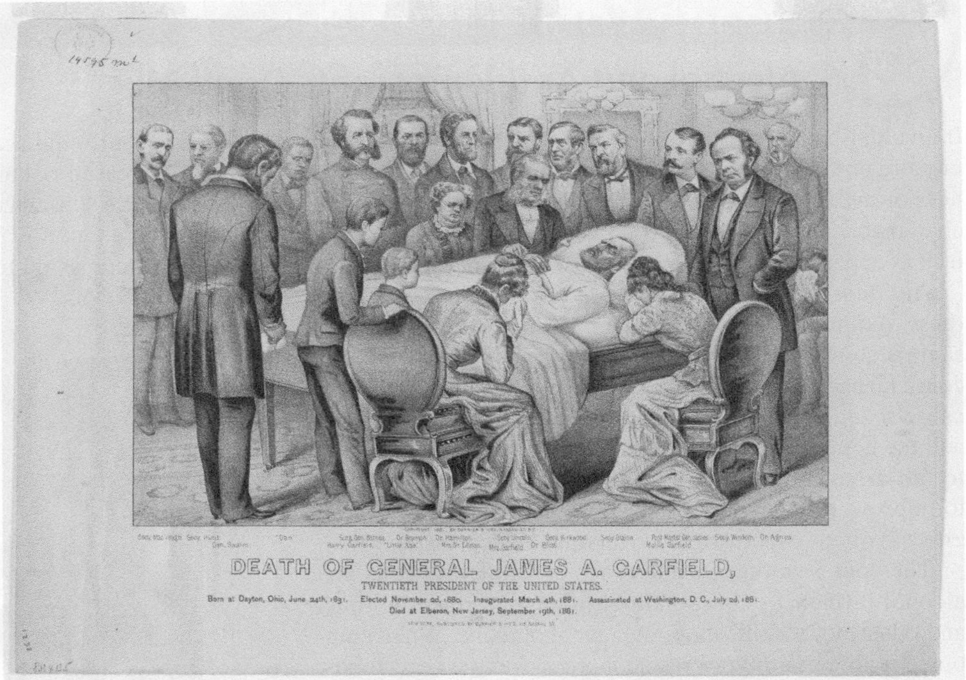

Figure 9.23 James A. Garfield (1831–1888), 20th president of the United States, died from a postoperative infection two months after being shot.

Figure 9.24 The phenol derivative 4-hexylresorcinol is used as soothing agents in mouthwashes and throat lozenges.

Today, there are effective derivatives used as antiseptics instead of phenol that are less irritating to the skin. Two such compounds, 4-chloro-3,5-dimethylphenol and 4-hexylresorcinol (**Figure 9.24**), are shown here.

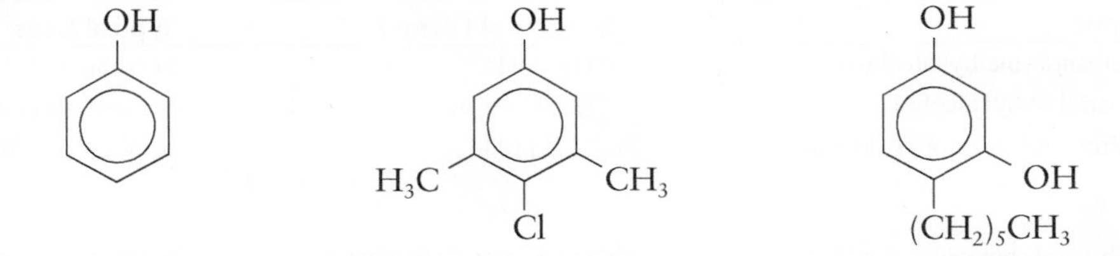

phenol (lister's original disinfectant)	4-chloro-3,5-dimethylphenol (nonirritating topical antiseptic)	4-hexylresorcinol (used in mouthwashes and throat lozenges)

OH; H_3C, CH_3, Cl; OH, $(CH_2)_5CH_3$

Two other derivatives of phenol, *o*-phenylphenol and 2-benzyl-4-chlorophenol, are ingredients in Lysol, a disinfectant for walls and furniture in hospitals and homes:

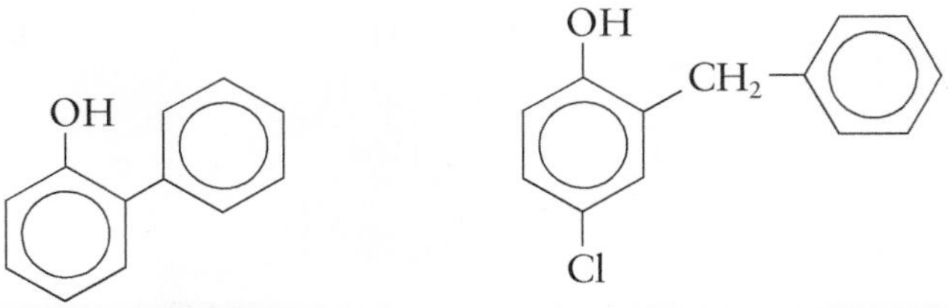

OH; OH, CH_2, Cl

o-phenylphenol   2-benzyl-4-chlorophenol

antioxidant A substance that prevents another substance from being oxidized.

Other important phenols act as **antioxidants** by interfering with oxidizing reactions. These phenols are useful in protecting foods from spoilage and a variety of other materials from unwanted reactions. Butylated hydroxy anisole (BHA) and butylated hydroxy toluene (BHT) are widely used as antioxidants in gasoline, lubricating oils, rubber, certain foods, and packaging materials for foods that might turn rancid (see **Figure 9.25**). In food or containers used to package food, the amount of these antioxidants is limited to 200 ppm (parts per million) or 0.02%, based on the fat or oil content of the food. Notice that industrial nomenclature for these compounds does not follow IUPAC rules.

(45mcg folic acid)

Vitamin B_{12}	20%	20
Phosphorus	6%	15
Magnesium	6%	8
Zinc	20%	20

* Amount in cereal. A serving of cereal plus skim milk provides 4.5g Total Fat (0.5g Saturated Fat), 5mg Cholesterol, 280mg Sodium, 39g Total Carbohydrate (19g Total Sugars), 6g Protein, 280mg Calcium, 270mg Potassium.

** The % Daily Value (DV) tells you how much a nutrient in a serving of food contributes to a daily diet. 2,000 calories a day is used for general nutrition advice.

Ingredients: Whole Grain Wheat, Sugar, Rice Flour, Canola Oil, Fructose, Maltodextrin, Dextrose, Salt, Cinnamon, Trisodium Phosphate, Soy Lecithin, Caramel Color. BHT Added to Preserve Freshness.

Vitamins and Minerals: Calcium

Figure 9.25 BHT is listed as a food or packaging ingredient that will "preserve freshness."

BHA (butylated hydroxy anisole)
2-*t*-butyl-4-methoxyphenol
(IUPAC name)

BHT (butylated hydroxy toluene)
2,6-di-*t*-butyl-4-methylphenol
(IUPAC name)

9.5 Ethers

Learning Objective 7 Name ethers by using common names.

Learning Objective 8 Describe the key physical and chemical properties of ethers.

An **ether** is an organic compound in which two alkyl groups are bonded to an oxygen atom, as if both hydrogen atoms of water were replaced by alkyl groups.

ether A compound that contains a $-\overset{|}{\underset{|}{C}}-O-\overset{|}{\underset{|}{C}}-$ functional group.

$$\underset{\text{water}}{H-O-H} \xrightarrow[\text{with R}]{\text{Replace both H's}} \underset{\text{ether}}{R-O-R'}$$

The R and R′ groups can be the same or different.

Common names for ethers are obtained by first naming the two groups attached to the oxygen and then adding the word *ether.* If the groups are different, they are listed alphabetically (e.g., ethyl before methyl). If the groups are the same, it is appropriate to use the prefix *di-*, as in diethyl ether, but sometimes the *di-* is omitted. As we have mentioned throughout the chapter, ethers have found various applications in the medical field in the past but are not generally a favored source of treatment today (**Figure 9.26**).

Brooks/Brown/Science Source

Figure 9.26 Hoffmann's Drops were a mixture of alcohol and ether used over a century ago to allegedly cure a variety of maladies.

$CH_3-O-CH_2CH_3$
ethyl methyl ether

$C_6H_5-O-CH(CH_3)-CH_3$
isopropyl phenyl ether

CH_3-O-CH_3
dimethyl ether
(methyl ether)

$CH_3CH_2-O-CH_2CH_3$
diethyl ether
(ethyl ether)

Example 9.8 Naming Ethers

Assign a common name to the following ethers.

a. CH_3—O—(benzene ring) b. $CH_3CH_2CH_2$—O—$CH_2CH_2CH_3$

Solution

a. Common names for ethers are formed by naming the two groups attached to the oxygen and adding the word *ether*. The two groups are different, so they must be listed alphabetically in the name. The common name is methyl phenyl ether.

b. The two groups attached to the oxygen are the same, so the prefix *di-* is used. The common name is dipropyl ether.

✔ Learning Check 9.8 Assign a common name to the following ethers.

a. CH_3CH_2—O—(benzene ring)

b. CH_3CH(—CH_3)—O—CH(—CH_3)CH_3

---

heterocyclic ring A ring containing at least one atom of an element other than carbon.

The ether linkage is also found in cyclic structures. A ring that contains elements in addition to carbon is called a **heterocyclic ring**. Two common structures with oxygen-containing heterocyclic rings are furan and pyran:

furan pyran

Many carbohydrates contain fundamental ring systems that are similar to furan and pyran (see **Chapter 11**).

The oxygen atom of ethers can form hydrogen bonds with water (see **Figure 9.26a**), so ethers are slightly more soluble in water than hydrocarbons but still *less* soluble than alcohols of comparable molecular weight. Ethers, however, cannot form hydrogen bonds with other ether molecules in the pure state (see **Figure 9.26b**), which results in low boiling points that are close to those of hydrocarbons with similar molecular weights.

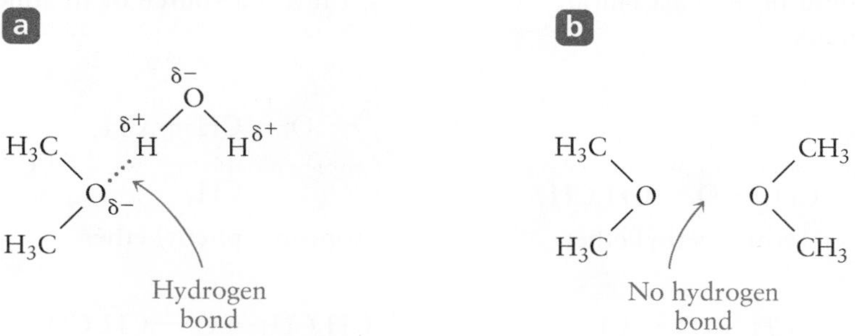

Figure 9.26 Ethers such as dimethyl ether can form hydrogen bonds with (a) water, but not with (b) other ether molecules.

Like alkanes, the chief chemical property of ethers is that they are inert and do not react with most reagents. It is this property that makes diethyl ether such a useful solvent. Like hydrocarbons, ethers are flammable, and diethyl ether is especially flammable because it is so highly volatile.

Health Connections 9.2

General Anesthetics

Anesthetics are compounds that induce a loss of sensation in a specific part (local anesthetic) or all (general anesthetic) of the body. A general anesthetic acts on the brain to produce unconsciousness as well as insensitivity to pain.

The word *ether* is often associated with general anesthetics because diethyl ether was once commonly used for this purpose. Diethyl ether became generally used as an anesthetic for surgical operations in about 1850. Before that time, surgery was an agonizing procedure. The patient was strapped to a table and (if lucky) soon fainted from the pain. Diethyl ether was the first general anesthetic used in surgery. For many years, it was the most important compound used as an anesthetic, but it has now been largely replaced by other compounds. Divinyl ether (Vinethene) is an anesthetic that acts more rapidly and is less nauseating than diethyl ether:

$$CH_2{=}CH-O-CH{=}CH_2$$

divinyl ether

Another common ether anesthetic is enflurane:

```
    F         F   F
    |         |   |
H—C—O—C—C—H      enflurane
    |         |   |
    H         H   Cl
```

Numerous inhalation anesthetics are not ethers at all. Nitrous oxide, for example, is a simple inorganic compound with the formula N_2O. Also known as "laughing gas," some dentists use it as a general anesthetic because its effects wear off quickly.

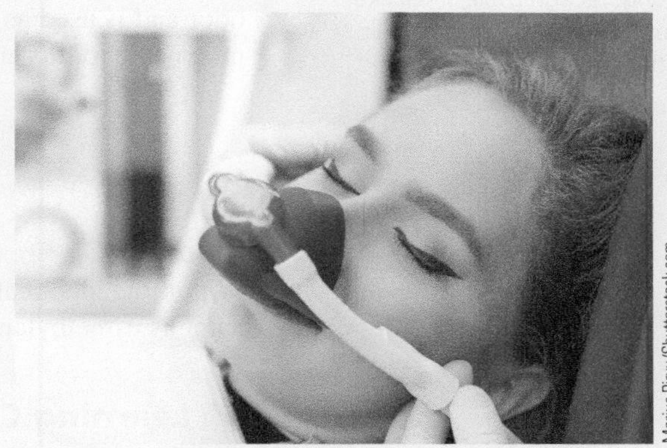

Marius Pirvu/Shutterstock.com

A patient receives an anesthetic prior to dental surgery.

9.6 The Nomenclature and Classification of Amines

Learning Objective 9 Assign common names to simple amines.

Learning Objective 10 Classify amines as primary, secondary, or tertiary based on the basis of their structural formulas.

The effectiveness of a wide variety of important medicines depends either partly or entirely on the presence of a nitrogen-containing group in their molecules. The simplest of the nitrogen-containing functional groups are the amines. **Amines** are organic derivatives of ammonia (NH_3) in which one or more of the hydrogens are replaced by an aromatic or alkyl group (R). Amines, then, are the nitrogen analogue of alcohols and ethers. Because of the unique bonding nature of nitrogen (preferring three covalent bonds and one lone pair of electrons), amines can have different combinations of hydrogen atoms and alkyl groups. The R′–, R″–, and R‴– in the following structures indicate that each of these three positions can be some type of alkyl group (or hydrogen, if indicated). The alkyl groups can be the same or different.

amine An organic compound derived by replacing one or more of the hydrogen atoms of ammonia with alkyl or aromatic groups, as in RNH_2, R_2NH, and R_3N.

```
   N          N          N          N
 / | \      / | \      / | \      / | \
H  H  H    R  H  H    R  R'  H   R  R'  R"
```

Several approaches are used in naming amines. Common names are used extensively for those with low molecular weights. These are named by alphabetically listing the names of the alkyl or aromatic groups bonded to the nitrogen and attaching the suffix *-amine.*

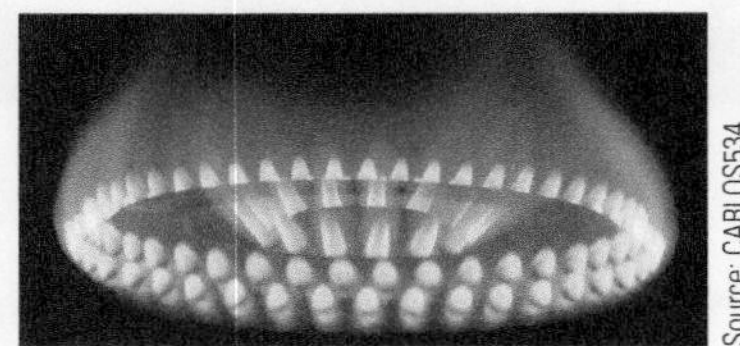

Figure 9.27 The natural gas that you might use on your stovetop, or to heat your home, is odorless; therefore, a chemical with a strong, recognizable smell like trimethylamine is added to help identify gas leaks.

The name is written as one word, and the prefixes *di-* and *tri-* are used when identical alkyl groups are present. Some examples are as follows:

CH_3-NH_2	$CH_3-NH-CH_3$	$CH_3CH_2-NH-CH_3$
methylamine	dimethylamine	ethylmethylamine

Example 9.9 Naming Amines

The nitrogen compound with three attached methyl groups is a gas with a strong fishy odor and is used as a warning (smell) agent in natural gas (**Figure 9.27**). Assign a common name to this amine.

$$\begin{array}{c} CH_3-N-CH_3 \\ | \\ CH_3 \end{array}$$

Solution

With three identical methyl groups attached to nitrogen, the prefix *tri-* can be used. The name is trimethylamine.

✔ Learning Check 9.9 Assign a common name to the following amines.

a. $CH_3CH_2CH_2-NH_2$ b. $CH_3CH_2CH_2-NH-CH_2CH_3$

primary amine An amine having one alkyl or aromatic group bonded to nitrogen, as in $R-NH_2$.

secondary amine An amine having two alkyl or aromatic groups bonded to nitrogen, as in R_2NH.

tertiary amine An amine having three alkyl or aromatic groups bonded to nitrogen, as in R_3N.

Like alcohols (**Section 9.1**), amines are classified as primary, secondary, or tertiary on the basis of their molecular structure. However, the classification is done differently. For amines, the number of groups (R) that have replaced hydrogens in the NH_3 is counted. The nitrogen atom in a **primary amine** is bonded to one R group. The nitrogen in **secondary amines** is bonded to two R groups, and that of **tertiary amines** is bonded to three R groups. These amine subclasses are summarized in **Table 9.3**.

Table 9.3 Subclasses of Amines

Subclass	General Formula	Example
Primary (1°)	$\begin{array}{c} R-N-H \\ \vert \\ H \end{array}$	$\begin{array}{c} CH_3-N-H \\ \vert \\ H \end{array}$
Secondary (2°)	$\begin{array}{c} R-N-R' \\ \vert \\ H \end{array}$	$\begin{array}{c} CH_3-N-CH_3 \\ \vert \\ H \end{array}$
Tertiary (3°)	$\begin{array}{c} R-N-R' \\ \vert \\ R'' \end{array}$	$\begin{array}{c} CH_3-N-CH_3 \\ \vert \\ CH_3 \end{array}$

Example 9.10 Classifying Amines

Carvedilol (**Figure 9.28**), an amine used to treat high blood pressure and heart failure, contains two amines. Classify each of the nitrogens in carvedilol as primary, secondary, or tertiary.

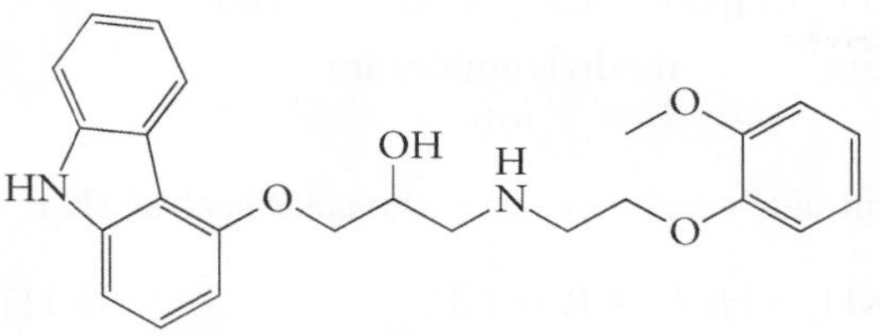

carvedilol

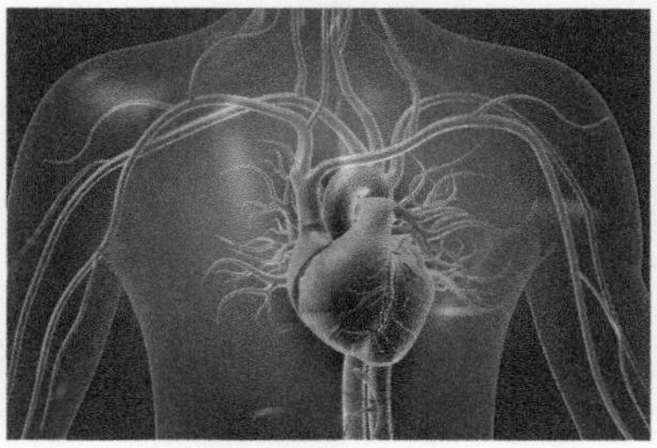

Explode/Shutterstock.com

Figure 9.28 Carvedilol is a drug containing amines used to treat high blood pressure and congestive heart failure.

Solution

The ring nitrogen is bonded to two carbon atoms, making it secondary. The R_2NH nitrogen atom is bonded to two carbon atoms, so it is secondary, too.

✔ **Learning Check 9.10** Classify each of the following amines as primary, secondary, or tertiary.

a. 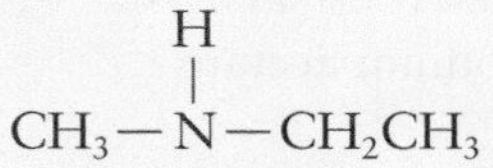 $CH_3-\underset{}{N}(H)-CH_2CH_3$

b. 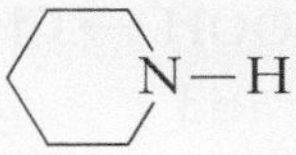 (piperidine ring) N—H

c. 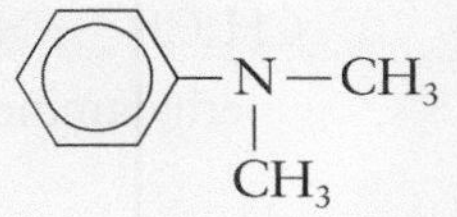 (benzene ring)—$N(CH_3)-CH_3$

9.7 Properties of Amines

Learning Objective 11 Predict the products of amines with different acids and bases.

Learning Objective 12 Explain how hydrogen bonding influences the physical properties of amines.

9.7A The Acid–Base Chemistry of Amines

Although amines undergo many reactions, their single most distinguishing feature is their behavior as weak bases. They are the most common organic bases. We know from inorganic chemistry that ammonia (**Figure 9.29**) behaves as a Brønsted base by accepting protons, and in the process becomes a conjugate acid. The following reaction of NH_3 with HCl gas illustrates this point:

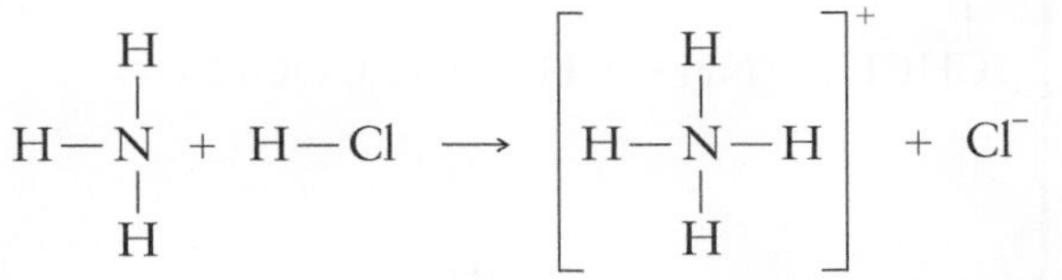

$$\underset{\text{ammonia}}{H-NH_2} + H-Cl \longrightarrow \underset{\text{ammonium ion}}{\left[H-NH_2-H\right]^+} + \underset{\text{chloride ion}}{Cl^-}$$

Recall from **Section 7.3** that Brønsted bases such as ammonia react with water to liberate OH^- ions:

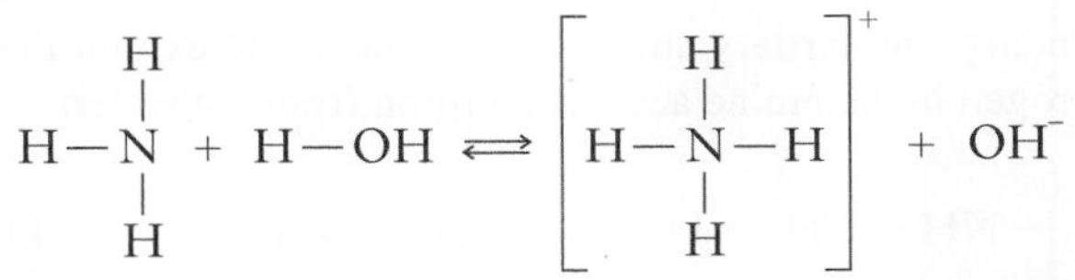

$$\underset{\text{ammonia}}{H-NH_2} + \underset{\text{water}}{H-OH} \rightleftarrows \underset{\text{ammonium ion}}{\left[H-NH_2-H\right]^+} + \underset{\text{hydroxide ion}}{OH^-}$$

REUTERS/Alamy Stock Photo

Figure 9.29 Approximately 14 million metric tons of ammonia were produced in the United States in 2021. Most of it was used in fertilizers.

Because amines are derivatives of ammonia, they react in similar ways. In water, they produce OH^- ions:

$$\textbf{General reaction:}\quad R-NH_2 + H_2O \rightleftharpoons R-NH_3^+ + OH^- \qquad (9.10)$$

$$\textbf{Specific example:}\quad \underset{\text{methylamine}}{CH_3-NH_2} + H_2O \rightleftharpoons \underset{\text{methylammonium ion}}{CH_3-NH_3^+} + OH^-$$

All amines behave as weak bases and form salts when they react with acids such as HCl:

$$\textbf{General reaction:}\quad \underset{\text{amine}}{R-NH_2} + \underset{\text{acid}}{HCl} \longrightarrow \underset{\text{amine salt}}{R-NH_3^+Cl^-} \qquad (9.11)$$

$$\textbf{Specific example:}\quad \underset{\text{methylamine}}{CH_3-NH_2} + \underset{\text{acid}}{HCl} \longrightarrow \underset{\text{methylammonium chloride}}{CH_3-NH_3^+Cl^-}$$

Other acids, such as sulfuric, nitric, phosphoric, and carboxylic acids (acetic acid is shown here), also react with amines to form salts:

$$\underset{\text{ethylamine}}{CH_3CH_2-NH_2} + \underset{\text{acetic acid}}{CH_3COOH} \longrightarrow \underset{\text{ethylammonium acetate}}{CH_3CH_2-NH_3^+CH_3COO^-}$$

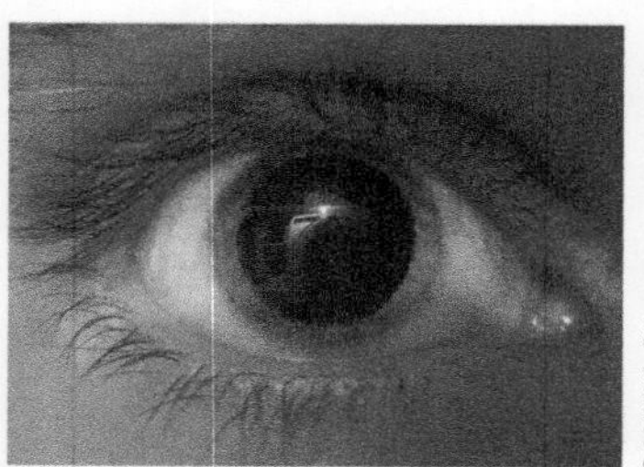

© MJH Life Sciences

Figure 9.30 Optometrists use phenylephrine on their patients to dilate pupils during eye exams.

Example 9.11 Acid–Base Chemistry of Amines

Phenylephrine (**Figure 9.30**) is a medication used as a decongestant, to dilate the pupil of the eye, and to increase blood pressure. Complete the following reactions.

a. $HO-C_6H_4-CH(OH)CH_2-NH-CH_3 + CH_3COOH \longrightarrow$

phenylephrine

b. $CH_3-NH-CH_3 + H_2O \longrightarrow$

Solution

a. In this reaction, an amine reacts with a carboxylic acid to produce a salt. The nitrogen in the amine accepts a proton from the carboxylic acid.

$$HO-C_6H_4-CH(OH)CH_2-NH-CH_3 + CH_3COOH \rightleftharpoons$$

$$HO-C_6H_4-CH(OH)CH_2-\overset{+}{N}H_2-CH_3 + CH_3COO^-$$

b. Primary, secondary, and tertiary amines are all basic and exhibit the same reaction in water. The nitrogen in the amine accepts a proton from the water.

$$CH_3-NH-CH_3 + H_2O \rightleftharpoons CH_3-\overset{+}{N}H_2-CH_3 + OH^-$$

✔ Learning Check 9.11 Complete the following reactions.

a. $CH_3-N(CH_3)-CH_2CH_3 + H_2O \rightleftarrows$

b. $C_6H_5-NH_2 + H_2O \rightleftarrows$

c. cyclopentyl$-NH_2 + CH_3CH_2COOH \rightleftarrows$

d. $C_6H_5-NH-CH_3 + HCl \rightleftarrows$

As we have seen, amine salts may be named by changing *amine* to *ammonium* and adding the name of the negative ion derived from the acid. The following are additional examples:

$CH_3-N(CH_3)(CH_2CH_3)-H^+Br^-$

ethyldimethylammonium bromide

$(CH_3CH_2)_3NH^+CH_3COO^-$

triethylammonium acetate

Amine salts have physical properties characteristic of other ionic compounds. They are white crystalline solids with high melting points. Because they are ionic, amine salts are more soluble in water than the parent amines. For this reason, amine drugs are often given in the form of salts so that they will dissolve in body fluids (**Figure 9.31**).

morphine (water insoluble) $+ H_2SO_4 \rightarrow$ morphine sulfate (water soluble)

Amine salts are easily converted back to amines by adding a strong base:

$$R{-}NH_3^+Cl^- + NaOH \longrightarrow R{-}NH_2 + H_2O + NaCl$$

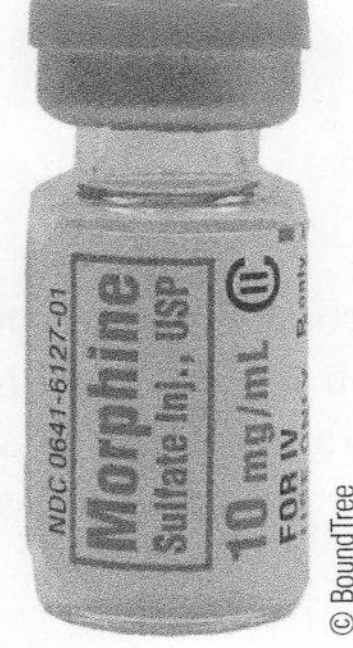

© BoundTree

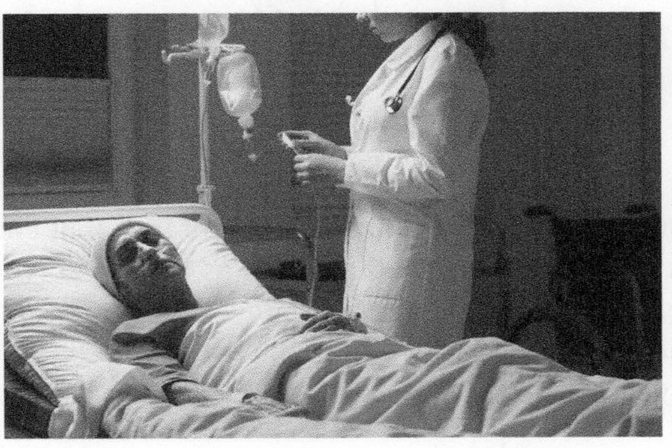

istock.com/KatarzynaBialasiewicz

Figure 9.31 Morphine sulfate, a drug containing an amine salt, is used to treat patients experiencing severe pain.

The amine cations in the salts we have discussed to this point have one, two, or three alkyl groups attached to the nitrogen:

$$R-NH_3^+ \qquad R-\overset{+}{N}H_2-R' \qquad R-\overset{+}{N}H(-R'')-R'$$

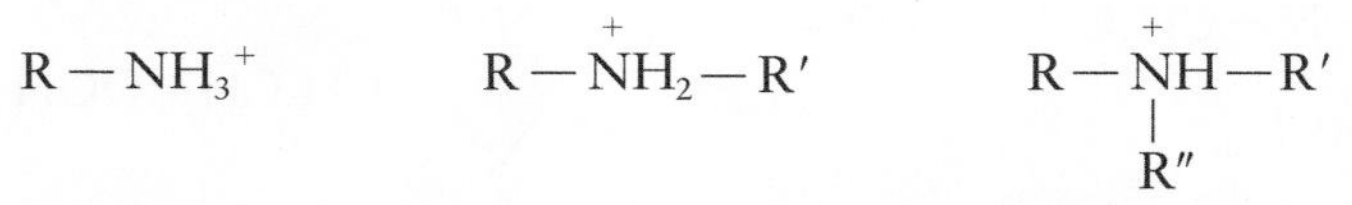

primary amine cation — secondary amine cation — tertiary amine cation

It is also possible to have amine cations in which four alkyl groups or benzene rings are attached to the nitrogen, as in triethylmethylammonium chloride:

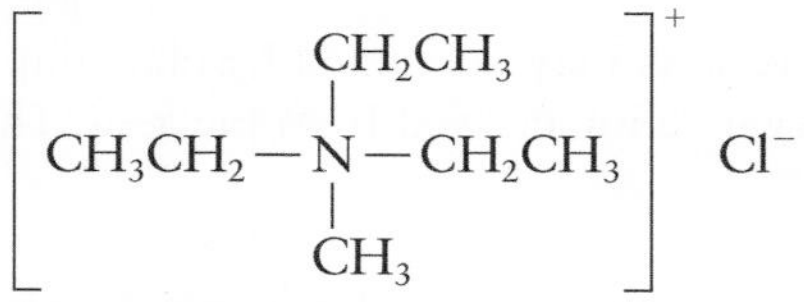

$$\left[CH_3CH_2-N(CH_2CH_3)(CH_3)-CH_2CH_3\right]^+ Cl^-$$

triethylmethylammonium chloride

quaternary ammonium salt An ionic compound containing a positively charged ion in which four alkyl or aromatic groups are bonded to nitrogen, as in R_4N^+.

These compounds are called **quaternary ammonium salts** and are named the same way as other amine salts. Unlike other amine salts, quaternary ammonium salts contain no hydrogen attached to the nitrogen that can be removed by adding a base. Thus, quaternary ammonium salts are present in solution in only one form, which is independent of the pH of the solution.

Choline is an important quaternary ammonium ion that is a component of certain lipids (see **Chapter 14**). Acetylcholine is involved in the transmission of nerve impulses from one cell to another (see **Section 9.8**):

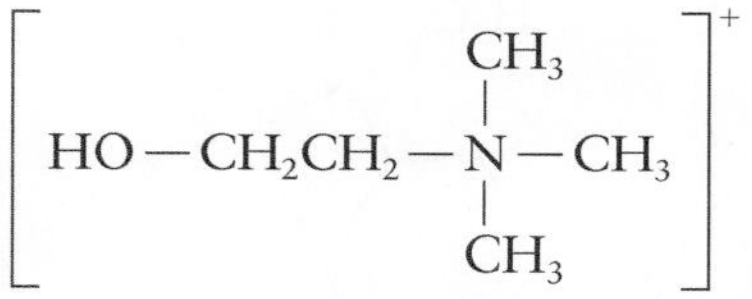

$$\left[HO-CH_2CH_2-N(CH_3)_2-CH_3\right]^+ \qquad \left[CH_3-\overset{O}{\overset{\|}{C}}-O-CH_2CH_2-N(CH_3)_2-CH_3\right]^+$$

choline cation 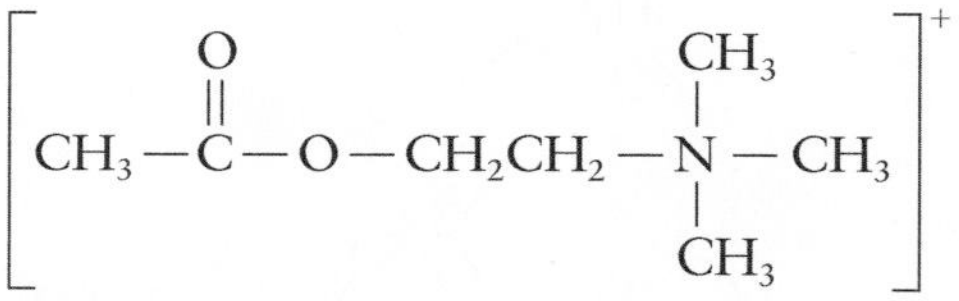 — acetylcholine cation

Figure 9.32 Quaternary ammonium compounds are used broadly in routine cleaning. In hospitals, quaternary ammonium disinfectants are used on noncritical surfaces such as floors, bed rails, and tray tables.

Cristian Gutu/Shutterstock.com

Some quaternary ammonium salts are important because they have disinfectant properties (see **Figure 9.32**). For example, benzalkonium chloride (Zephiran) is a well-known antiseptic compound that kills many pathogenic (disease-causing) bacteria and fungi on contact:

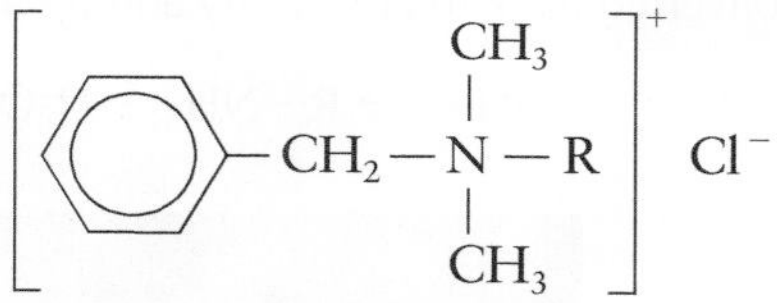

$$\left[C_6H_5-CH_2-N(CH_3)_2-R\right]^+ Cl^-$$

zephiran chloride
(R represents a long alkyl chain)

Its detergent action destroys the membranes that coat and protect the microorganisms. Zephiran chloride is recommended as a disinfectant solution for skin and hands prior to surgery and for the sterile storage of instruments. The trade names of some other anti-infectives that contain quaternary ammonium salts are Phemerol, Bactine, and Ceepryn.

9.7B Physical Properties of Amines

The simple, low-molecular-weight amines are gases at room temperature (see **Table 9.4**). Heavier, more complex compounds are liquids or solids.

Table 9.4 Properties of Some Amines

Name	Structure	Melting Point (°C)	Boiling Point (°C)
Methylamine	CH_3-NH_2	−94	−6
Ethylamine	$CH_3CH_2-NH_2$	−81	17
Dimethylamine	$CH_3-NH-CH_3$	−93	7
Diethylamine	$CH_3CH_2-NH-CH_2CH_3$	−48	56
Trimethylamine	CH_3-N-CH_3 $\vert$ CH_3	−117	3
Triethylamine	$CH_3CH_2-N-CH_2CH_3$ $\vert$ CH_2CH_3	−114	89

Like alcohols, primary and secondary amines form hydrogen bonds among themselves:

$$CH_3-\overset{\delta^-}{N}(H_{\delta^+})-\overset{\delta^+}{H}\cdots\overset{\delta^-}{N}(H_{\delta^+})(H_{\delta^+})-CH_3$$

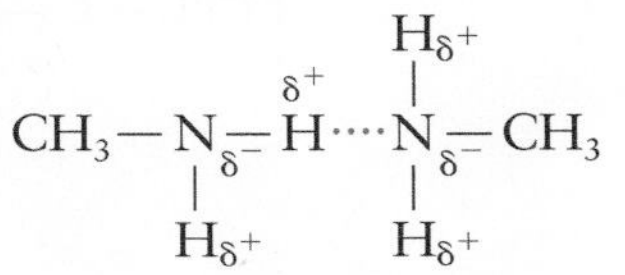

Because nitrogen is less electronegative than oxygen, the hydrogen bonds formed by amines are weaker than those formed by alcohols. For this reason, the boiling points of primary and secondary amines are usually somewhat lower than the boiling points of alcohols with similar molecular weights. Tertiary amines cannot form hydrogen bonds among themselves (because the nitrogen has no attached hydrogen atom), and their boiling points are similar to those of alkanes that have about the same molecular weights.

Amines with fewer than six carbon atoms are generally soluble in water as a result of hydrogen bond formation between amine functional groups and water molecules. This is true for tertiary as well as primary and secondary amines (see **Figure 9.33**). As we saw with alcohols (see **Section 9.2**), amines with longer carbon chains are insoluble in water because London dispersion forces between the alkyl groups dominate.

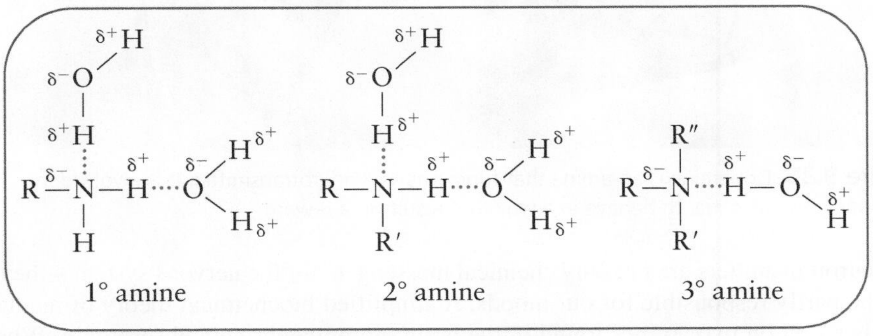

$$\overset{\delta^+}{H}-\overset{\delta^-}{O}-\overset{\delta^+}{H}\cdots R-\overset{\delta^-}{N}(H)-\overset{\delta^+}{H}\cdots\overset{\delta^-}{O}(H^{\delta^+})(H_{\delta^+})$$

1° amine

$$\overset{\delta^+}{H}-\overset{\delta^-}{O}-H^{\delta^+}\cdots R-N(R')-\overset{\delta^+}{H}\cdots\overset{\delta^-}{O}(H^{\delta^+})(H_{\delta^+})$$

2° amine

$$R-\overset{\delta^-}{N}(R'')(R')\cdots\overset{\delta^+}{H}-\overset{\delta^-}{O}-H^{\delta^+}$$

3° amine

Figure 9.33 Hydrogen bonding between amines in water.

9.8 Biologically Significant Amines

Learning Objective 13 Describe the uses of three different crucial biological amines.

9.8A Amines as Neurotransmitters

neurotransmitter A substance that acts as a chemical bridge in nerve impulse transmission between nerve cells.

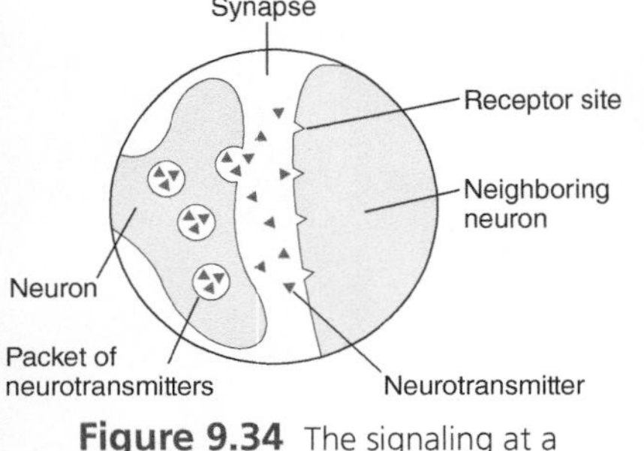

Figure 9.34 The signaling at a nerve synapse by neurotransmitters.

Neurotransmitters, the chemical messengers of the nervous system, carry nerve impulses from one nerve cell (neuron) to another (**Figure 9.34**). A neuron consists of many different parts, but the critical junction is at the synapse, the gap between the ends of two neurons. The released neurotransmitter molecules diffuse across the synapse and bind to receptors of the next neuron. Once the neurotransmitter binds to the receptor, the chemical message has been delivered. In the central nervous system—the brain and spinal cord—the most important neurotransmitters are acetylcholine, norepinephrine, dopamine (**Figure 9.35**), and serotonin:

HO, HO (ring) — $CH(OH)CH_2-NH_2$

norepinephrine

HO, HO (ring) — $CH_2CH_2-NH_2$

dopamine

HO (indole ring, N–H) — $CH_2CH_2-NH_2$

serotonin

$$\left[CH_3-\overset{\overset{O}{\|}}{C}-O-CH_2CH_2-\underset{\underset{CH_3}{|}}{\overset{\overset{CH_3}{|}}{N}}-CH_3\right]^+$$

acetylcholine

SDI Productions/Getty Images

Figure 9.35 Dopamine is an amine that functions as a neurotransmitter. It is involved in motivating an individual to behave in a way that results in a reward.

Neurotransmitters are not only chemical messengers for the nervous system—they may also be partly responsible for our moods. A simplified biochemical theory of mental illness is based on two amines found in the brain: norepinephrine and serotonin. When an excess of norepinephrine is formed in the brain, the result is a feeling of elation. Extreme excesses of norepinephrine can even induce a manic state, while low norepinephrine levels may be a cause of depression.

Several different receptors in the body can be activated by norepinephrine and related compounds (**Figure 9.36**). The major role of norepinephrine as a neurotransmitter is to work in tandem with the hormone epinephrine (**Section 9.8B**) to prepare the brain and body for action in high-stress scenarios. A class of drugs called beta blockers reduces the stimulant action of epinephrine and norepinephrine on various kinds of cells. Some beta blockers are used to treat cardiac arrhythmias, angina, and hypertension (high blood pressure). They function by slightly decreasing the force of each heartbeat. Depression is sometimes a side effect of using such drugs.

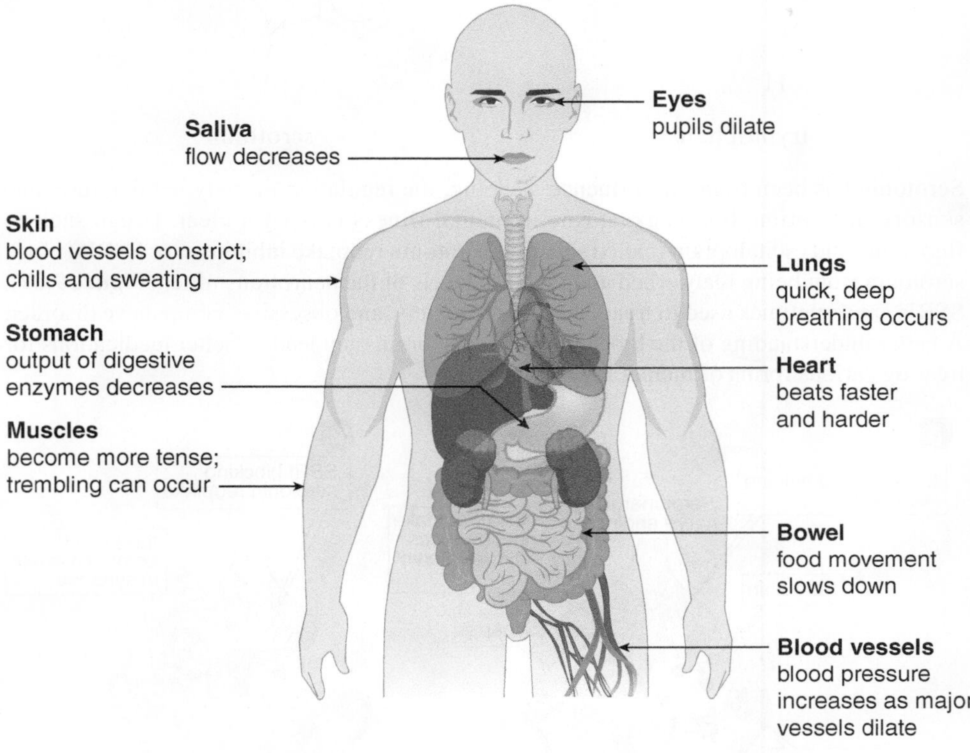

Figure 9.36 Norepinephrine is released in response to high-stress situations.

In a complex, multistep process, the amino acid tyrosine is used in the body to synthesize norepinephrine.

HO– C_6H_4 –CH_2–CH(COOH)–NH_2 ⟶ (HO)$_2$ C_6H_3 –CH_2–CH(COOH)–NH_2

tyrosine ⟶ dopa

⟶ (HO)$_2$ C_6H_3 –CH_2–CH_2–NH_2

dopamine

⟶ (HO)$_2$ C_6H_3 –CH(OH)–CH_2–NH_2

norepinephrine

Like norepinephrine, serotonin also functions as a neurotransmitter. It is produced in the body from the amino acid tryptophan:

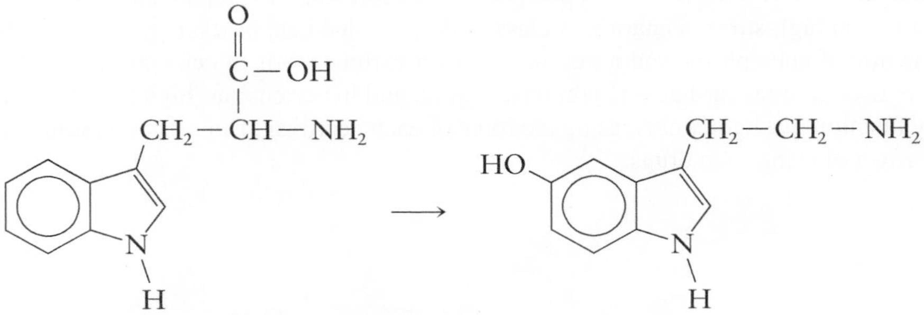

tryptophan serotonin

Serotonin has been found to influence sleeping, the regulation of body temperature, and sensory perception, but its exact role in mental illness is not yet clear. Drugs such as fluoxetine and escitalopram (called selective serotonin reuptake inhibitors, or SSRIs) block serotonin from being reabsorbed and increase levels of the neurotransmitter (**Figure 9.37**). SSRIs are sometimes used to treat depression, anxiety, and obsessive–compulsive disorder. A better understanding of the biochemistry of the brain may lead to better medications for treating various forms of mental illness.

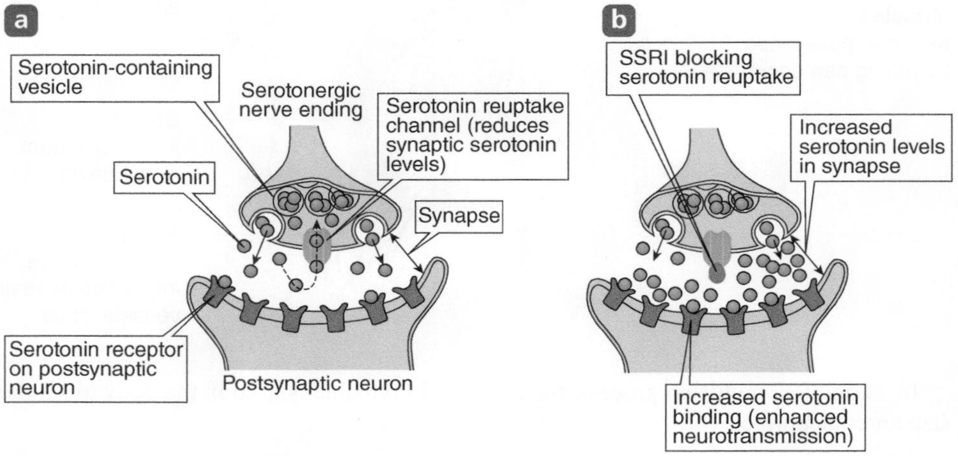

Figure 9.37 (a) The behavior of normal serotonin reuptake channels, which regulate the amount of serotonin in the synapse. (b) The use of selective serotonin reuptake inhibitors (SSRIs) block the reuptake of serotonin, thus increasing its concentration in the synapse, leading to enhanced binding.

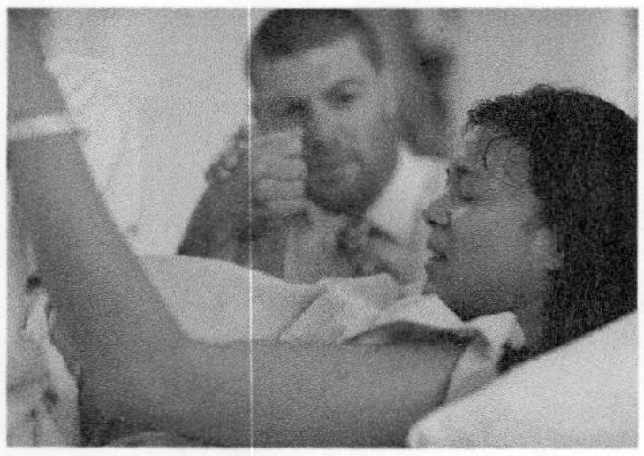

PeopleImages.com - Yuri A/Shutterstock.com

Figure 9.38 Epinephrine, or adrenaline as it is commonly known, is released during high-stress situations, such as when a woman is experiencing severe pain during labor. High adrenaline production can actually slow the process of labor.

9.8B Amphetamines

Epinephrine (adrenaline; **Figure 9.38**) has a molecular structure similar to that of norepinephrine. The difference is that epinephrine contains an *N*-methyl group:

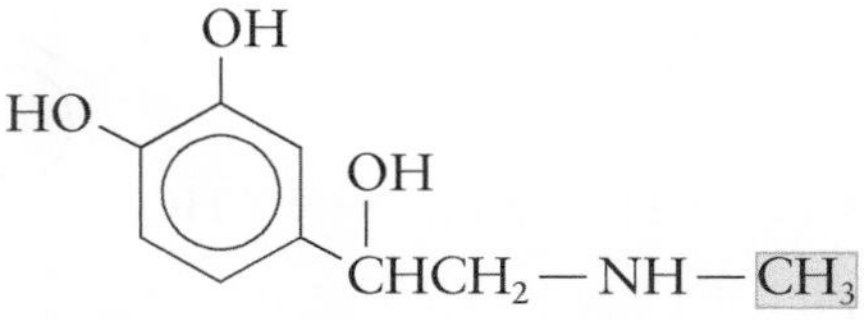

epinephrine
(adrenaline)

Epinephrine is more important as a hormone than as a neurotransmitter. It is synthesized in the adrenal gland and acts to increase the blood level of glucose, a key source of energy for body processes. The release of epinephrine—usually in response to pain, anger, or fear—provides glucose for a sudden burst of energy. For this reason, epinephrine has been called the fight-or-flight hormone.

Epinephrine also raises blood pressure. It does this by increasing the rate and force of heart contractions and by constricting the peripheral blood vessels. Injectable local anesthetics usually contain epinephrine because it constricts blood vessels in the vicinity of the injection. This prevents the blood from rapidly "washing" the anesthetic away and prolongs the anesthetic effect in the target tissue. Epinephrine is also used to reduce hemorrhage, treat asthma attacks, and combat anaphylactic shock (collapse of the circulatory system). Amphetamine (also known as Benzedrine) is a powerful nervous system stimulant with an amine structure similar to that of epinephrine:

$C_6H_5-CH_2CH(CH_3)-NH_2$
amphetamine
(benzedrine)

$(HO)_2C_6H_3-CH(OH)CH_2-NH-CH_3$
epinephrine
(adrenaline)

Both compounds contain an aromatic ring attached to a substituted ethylamine. That structural arrangement, sometimes referred to as a phenethylamine, is present in several physiologically active compounds. These structural similarities lead to similar behaviors in the body: Both compounds raise the glucose level in the blood and increase pulse rate and blood pressure. Other closely related phenethylamine compounds, such as Methedrine, also act as powerful nervous system stimulants. Compounds of this type can be thought of as being derived from amphetamine and as a result are sometimes collectively referred to as **amphetamines**:

amphetamines A class of drugs structurally similar to epinephrine, used to stimulate the central nervous system.

$C_6H_5-CH_2CH(CH_3)-NH-CH_3$
N-methylamphetamine
(methedrine, or "meth")

Amphetamines are used both legally (as prescribed by a physician) and illegally to elevate mood or reduce fatigue. The abuse of such drugs, like methamphetamine, results in severe detrimental effects on both the body and the mind. They are addictive and concentrate in the brain and nervous system. Abusers often experience hallucinations, long periods of sleeplessness, weight loss, and paranoia. Abusers of amphetamines often compound their problems by taking other drugs in order to prevent the "crash," or deep depression, brought on by discontinuation of the drug.

9.8C Alkaloids

Plants are the sources of some of the most powerful drugs known (see **Figure 9.39**). Numerous indigenous people in the world possess knowledge about the physiological effects that result from eating or chewing the leaves, roots, flowers, or bark of certain plants. The effects vary from plant species to plant species. Some cure diseases; others, such as opium, are addictive drugs. Still others are deadly poisons, such as the leaves of the belladonna plant and the hemlock herb. The substances responsible for these marked

Figure 9.39 Common daffodil bulbs are poisonous if eaten because of the presence of powerful alkaloids.

alkaloids A class of nitrogen-containing organic compounds obtained from plants.

physiological effects are often nitrogen-containing compounds called **alkaloids**. The name, which means alkali-like, reflects the fact that these amine compounds are weakly basic.

The molecular structures of alkaloids vary from simple to complex. Two common and relatively simple alkaloids are nicotine, which is found in tobacco, and caffeine, which is present in coffee and cola drinks:

N, N, CH_3

nicotine

O, CH_3, H_3C, N, N, N, O, N, CH_3

caffeine

Many people absorb nicotine into their bodies as a result of smoking or chewing tobacco. In small doses, nicotine behaves like a stimulant and is not especially harmful. It is habit-forming, however, and habitual tobacco users are exposed to other harmful substances such as tars, carbon monoxide, and polycyclic carcinogens.

Figure 9.40 Tea is a common source of caffeine.

Besides coffee and cola drinks, other sources of caffeine are tea (see **Figure 9.40**), chocolate, and cocoa. Caffeine is a mild stimulant of the respiratory and central nervous systems, the reason for its well-known side effects of nervousness and insomnia. These characteristics, together with its behavior as a mild diuretic, account for the use of caffeine in a wide variety of products, including pain relievers, cold remedies, diet pills, and energy drinks. Because caffeine is considered to be a drug, women who are pregnant should be prudent about how much caffeine they consume. Like most other drugs, caffeine enters the bloodstream, crosses the placental barrier, and reaches the fetus.

A number of alkaloids are used medicinally. Quinine, for example, is used to treat malaria. Atropine is used as a preoperative drug to relax muscles and reduce the secretion of saliva in surgical patients, and to dilate the pupil of the eye in patients undergoing eye examinations.

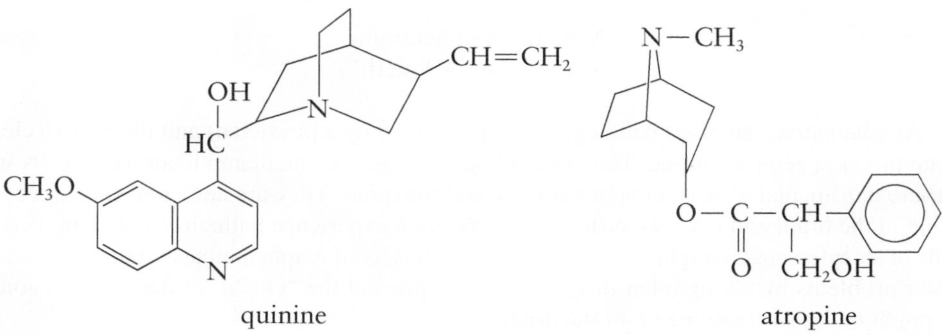

Figure 9.41 Since ancient times, the milky fluid that seeps from cuts on the unripe poppy seed pod has been air-dried to produce opium.

Opium, the dried juice of the poppy plant (see **Figure 9.41**), has been used for centuries as a painkilling drug. It contains numerous alkaloids, including morphine and its methyl ether, codeine. Both are central nervous system depressants, with codeine being much weaker. They exert a soothing effect on the body when administered in sufficient amounts. These properties make them useful as painkillers. Morphine is especially useful in this regard, being one of the most effective painkillers known. Codeine is used in some cough syrups to depress the action of the cough center in the brain. The major drawback to using these drugs is that they are all addictive.

Heroin, a derivative of morphine, is one of the more destructive illegally used drugs. Although intended at one time as a less-addictive substitute for morphine (**Figure 9.42**), heroin is actually much more addictive than morphine and has no accepted medical use in the United States.

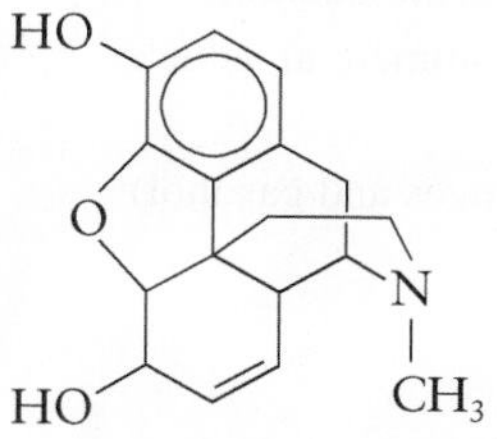

morphine 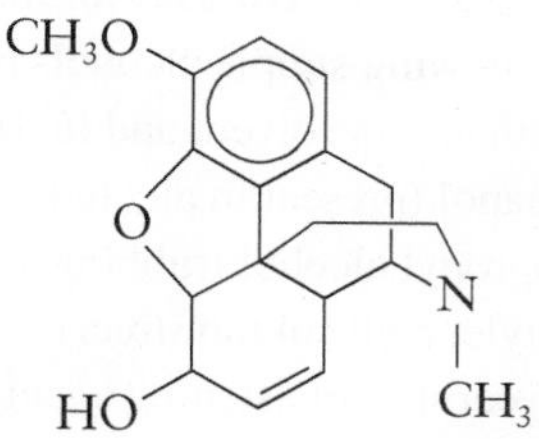 — codeine 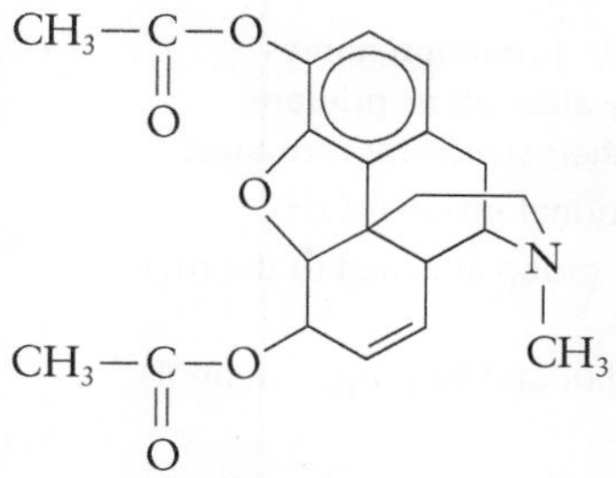 — heroin

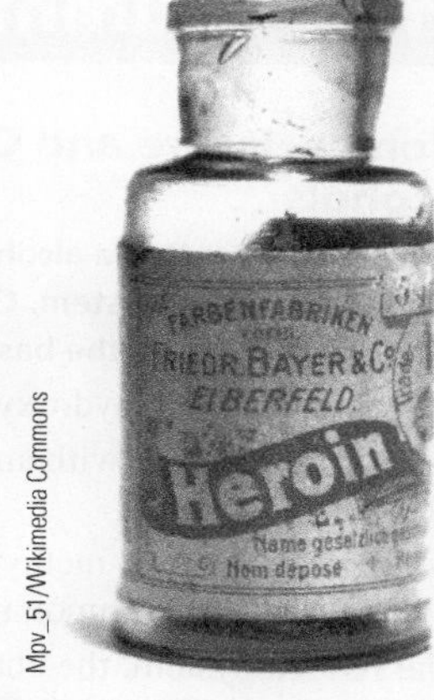

Mpv_51/Wikimedia Commons

Figure 9.42 Around the turn of the twentieth century, heroin was originally marketed as a "non-addictive cough suppressant."

Health Career Description

Allergy/Immunology Nurse

According to DailyNurse.org, "An allergy/immunology nurse focuses on the care of patients with chronic allergic conditions. These conditions include asthma, allergic rhinitis, urticaria, and atopic dermatitis. Duties include providing direct patient care and health education and, in most cases, administrative responsibilities such as those of an Allergy Office Manager."

Because allergies, in rare instances, may be life-threatening, allergy nurses must be certified in basic life support, in addition to holding R.N. certification. Many allergy nurses also hold certification in asthma education from the National Asthma Educator Certification Board.

Most allergy nurses work in clinics that conduct desensitization therapy (allergy shots) and allergen skin testing. Allergy nurses must possess acute observation skills, including the ability to discern slight differences in coloration and to discern slight changes in behavior or physical appearance. They must also be able to differentiate between normal and abnormal breathing sounds and respiration rates.

Allergy nurses must be able to work with patients of all ages and demographics. Communication skills are extremely important. Allergy nurses must observe strict attention to detail and maintain meticulous records.

Knowledge of chemistry and biochemistry is important for the allergy nurse in order to understand the reactions of the human body to proteins it perceives as threatening (as an allergen). Understanding the role of histamine in the body and the way antihistamines work is fundamental to working in the field of immunology.

Allergy nurses must have the ability to remain calm and act quickly in emergency situations and be able to impart calm to stressed patients or family members. Empathy and compassion along with precise technical skills are necessary for the immunological nurse.

Allergies run the gamut from somewhat irritating to highly dangerous. Immunological nurses help patients find their way to a healthier relationship with their environments.

Concept Summary

9.1 Nomenclature and Classification of Alcohols

Learning Objectives: **Name alcohols using common names and the IUPAC naming system. Classify alcohols as primary, secondary, or tertiary on the basis of their structural formulas.**

- Alcohols contain the hydroxy functional group (–OH).
- Aromatic compounds with an –OH group attached to the ring are called phenols.
- Several alcohols (e.g., methyl alcohol and isopropyl alcohol) are well known by common names.
- In the IUPAC system, the characteristic ending *-ol* is used to designate alcohols.
- Phenols are named as derivatives of the parent compound phenol.
- Alcohols are classified on the basis of the number of carbon atoms bonded to the carbon that is attached to the –OH group.
- In primary alcohols, the OH-bonded carbon is bonded to one other carbon.
- In secondary alcohols, the OH-bonded carbon is bonded to two other carbons.
- In tertiary alcohols, the OH-bonded carbon is bonded to three other carbons.

9.2 Physical Properties of Alcohols

Learning Objective: **Explain how hydrogen bonding influences the physical properties of alcohols.**

- Hydrogen bonding can occur between alcohol molecules and between alcohol molecules and water.
- As a result, alcohols have higher boiling points than hydrocarbons of similar molecular weight, and low-molecular-weight alcohols are soluble in water.

9.3 Reactions of Alcohols

Learning Objectives: **Predict the products when alcohols react with different acids and bases. Write chemical equations for alcohol dehydration and oxidation reactions.**

- Alcohols are amphoteric, meaning they can react as a base in the presence of an acid, or as an acid in the presence of a very strong base.
- Phenols are weak acids that are widely used for their antiseptic and disinfectant properties.
- Two important reactions of alcohols are dehydration and oxidation.
- Alcohols may be dehydrated in two different ways, depending on the reaction temperature.
- With H_2SO_4 at 180 °C, an alkene is produced, whereas at 140 °C, the product is an ether.
- The oxidation products obtained from the alcohols depend on the class of alcohol being oxidized.
- A primary alcohol produces an aldehyde that is further oxidized to a carboxylic acid.
- A secondary alcohol produces a ketone.
- A tertiary alcohol does not react with oxidizing agents.

9.4 Applications of Alcohols

Learning Objective: **List uses for three specific alcohols.**

- The following simple alcohols have commercial value:
  - Methanol (a solvent and fuel)
  - Ethanol (present in alcoholic beverages and gasohol)
  - Isopropyl alcohol (rubbing alcohol)
  - Ethylene glycol (antifreeze)
  - Glycerol (a moistening agent)
- Some phenols are used as antioxidants in foods and a variety of other materials.

9.5 Ethers

Learning Objectives: **Name ethers by using common names. Describe the key physical and chemical properties of ethers.**

- Ethers contain an oxygen attached to two carbons as the characteristic functional group.
- Ethers are commonly named by listing the alkyl groups attached to the oxygen in alphabetical order followed by the word *ether.*
- Like the alkanes, ethers are very unreactive.
- Pure ethers cannot form hydrogen bonds with other ether molecules.
- As a result, their boiling points are comparable to those of hydrocarbons of similar molecular weight and lower than those of alcohols of similar molecular weight.
- Ethers are much less polar than alcohols but are still slightly soluble in water because of their limited ability to form hydrogen bonds.

9.6 The Nomenclature and Classification of Amines

Learning Objectives: **Assign common names to simple amines. Classify amines as primary, secondary, or tertiary based on their structural formulas.**

- Amines are organic derivatives of ammonia in which one or more of the ammonia hydrogens are replaced by alkyl or aromatic groups.
- Common names are given to simple amines by adding the ending *-amine* to the names of the alkyl groups attached to the nitrogen.
- Amines are classified as primary, secondary, or tertiary, depending on the number of groups (one, two, or three) attached to the nitrogen.

9.7 Properties of Amines

Learning Objectives: **Predict the products when amines react with different acids and bases. Explain how hydrogen bonding influences the physical properties of amines.**

- Amines are weak bases.
- They react with water to liberate hydroxide ions.
- They react with acids to form salts.
- Primary and secondary amines have boiling points slightly lower than those of the corresponding alcohols.

- Tertiary amines have boiling points similar to those of alkanes.
- Low-molecular-weight amines are water soluble.

9.8 Biologically Significant Amines

Learning Objective: **Describe the uses of three different crucial biological amines.**

- Neurotransmitters are the chemical messengers of the nervous system.
- They carry nerve impulses from one nerve cell (neuron) to another.
- The most important neurotransmitters are acetylcholine and three other amines: norepinephrine, dopamine, and serotonin.
- Epinephrine is also known as the fight-or-flight hormone.
- The amphetamines have structures similar to that of epinephrine.
- Alkaloids are nitrogen-containing compounds isolated from plants.
- Alkaloids exhibit a variety of physiological effects on the body.
- Examples of alkaloids include nicotine, caffeine, quinine, atropine, morphine, and codeine.

Key Terms and Concepts

Alcohol (9.1)
Alkaloids (9.8)
Amine (9.6)
Amphetamines (9.8)
Antioxidant (9.4)
Dehydration reaction (9.3)
Elimination reaction (9.3)
Ether (9.5)
Fermentation (9.4)
Heterocyclic ring (9.5)
Hydroxy group (9.1)
Neurotransmitter (9.8)
Phenol (9.1)
Primary alcohol (9.1)
Primary amine (9.6)
Quaternary ammonium salt (9.7)
Secondary alcohol (9.1)
Secondary amine (9.6)
Tertiary alcohol (9.1)
Tertiary amine (9.6)

Key Equations

1. Reaction of an alcohol with a strong acid **(Section 9.3)**	$R{-}OH + H{-}X \rightleftharpoons R{-}{}^{+}OH_2 + X^-$	Equation 9.1
2. Reaction of an alcohol with a strong base **(Section 9.3)**	$R{-}OH + {}^{-}Base \rightleftharpoons R{-}O^- + H{-}Base$	Equation 9.2
3. Dehydration of alcohols to give alkenes **(Section 9.3)**	$-\overset{\vert}{\underset{H}{\underset{\vert}{C}}}-\overset{\vert}{\underset{OH}{\underset{\vert}{C}}}- \xrightarrow[180\ °C]{H_2SO_4} \;>C{=}C< + H_2O$ alcohol → alkene	Equation 9.3
4. Dehydration of alcohols to give ethers **(Section 9.3)**	$R{-}O{-}H + H{-}O{-}R \xrightarrow[140\ °C]{H_2SO_4} R{-}O{-}R + H_2O$	Equation 9.4
5. Oxidation of a primary alcohol to give an aldehyde and then a carboxylic acid **(Section 9.3)**	$R{-}\overset{OH}{\overset{\vert}{\underset{H}{\underset{\vert}{C}}}}{-}H + (O) \longrightarrow R{-}\overset{O}{\overset{\Vert}{C}}{-}H + H_2O$ primary alcohol → aldehyde Further oxidation (O) → $R{-}\overset{O}{\overset{\Vert}{C}}{-}OH$ carboxylic acid	Equation 9.5
6. Oxidation of a secondary alcohol to give a ketone **(Section 9.3)**	$R{-}\overset{OH}{\overset{\vert}{\underset{H}{\underset{\vert}{C}}}}{-}R' + (O) \longrightarrow R{-}\overset{O}{\overset{\Vert}{C}}{-}R' + H_2O$ secondary alcohol → ketone	Equation 9.6

7. Attempted oxidation of a tertiary alcohol gives no reaction **(Section 9.3)**	$R-\overset{\displaystyle OH}{\underset{\displaystyle R''}{C}}-R' + (O) \longrightarrow$ No reaction	Equation 9.7
8. Reaction of hydrogen gas with carbon monoxide gas to produce methanol **(Section 9.4)**	$CO + 2H_2 \xrightarrow[\text{Heat, pressure}]{\text{Catalysts}} CH_3-OH$	Equation 9.8
9. Production of alcohol from yeast fermentation of glucose **(Section 9.4)**	$\underset{\text{glucose}}{C_6H_{12}O_6} \xrightarrow{\text{Yeast}} \underset{\text{ethanol}}{2CH_3CH_2-OH} + 2CO_2$	Equation 9.9
10. Reaction of amines with water **(Section 9.7)**	$R-NH_2 + H_2O \rightleftharpoons R-NH_3^+ + OH^-$	Equation 9.10
11. Reaction of amines with acids **(Section 9.7)**	$R-NH_2 + HCl \longrightarrow R-NH_3^+ + Cl^-$	Equation 9.11

Exercises

Even-numbered exercises are answered in Appendix B.

Exercises with an asterisk (*) are more challenging.

Nomenclature and Classification of Alcohols (Section 9.1)

9.1 Draw general formulas for an alcohol and phenol, showing the functional group.

9.2 Assign IUPAC names to the following alcohols.

a. $CH_3\underset{}{\overset{\displaystyle OH}{C}}HCH_3$

b. $CH_3CH_2\overset{\displaystyle CH_2CH_3}{\underset{\displaystyle OH}{C}}CH_3$

c. (cyclopentane ring bearing OH on one carbon and two CH_3 groups on the adjacent carbon)

d. $CH_3CH_2\overset{\displaystyle OH}{C}HCH_2\overset{\displaystyle OH}{C}HCH_2-OH$

9.3 Assign IUPAC names to the following alcohols.

a. $CH_3CH_2CH_2-OH$

b. (benzene ring)$-CH_2CH_2CH_2-OH$

c. $CH_3\overset{\displaystyle CH_3}{C}H\underset{\displaystyle Cl}{C}HCH_2-OH$

d. (cyclopentane ring bearing Cl, OH, and CH_2CH_3 on adjacent carbons)

9.4 Several important alcohols are well known by common names. Give a common name for each of the following.

a. CH_3-OH

b. $CH_3\overset{\displaystyle OH}{C}HCH_3$

c. CH_3CH_2-OH

d. $\overset{\displaystyle OH}{CH_2}-\overset{\displaystyle OH}{CH_2}$

9.5 Draw structural formulas for each of the following.

a. 2-methyl-1-butanol

b. 2-bromo-3-methyl-3-pentanol

c. 1-methylcyclopentanol

9.6 Draw structural formulas for each of the following.

a. 2-methyl-2-pentanol

b. 1,3-butanediol

c. 1-ethylcyclopentanol

9.7 Name each of the following as a derivative of phenol.

a. (structure: benzene ring with $CHCH_3$ bearing CH_3, and OH ortho)

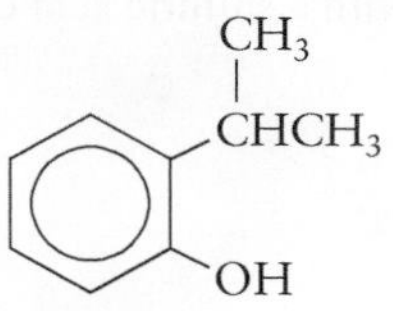

b. (structure: benzene ring with OH, CH_3, CH_3)

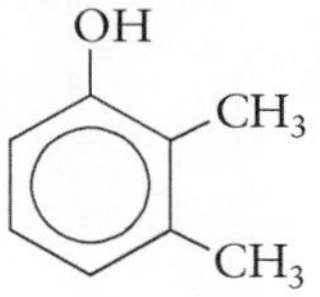

9.8 Name each of the following as a derivative of phenol.

a. (structure: benzene ring with CH_2CH_3 and OH ortho)

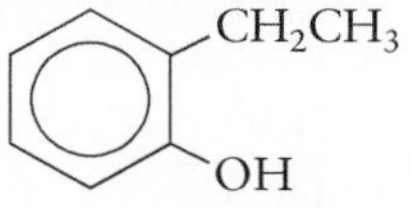

b. (structure: benzene ring with OH, H_3C, CH_2CH_3, CH_3)

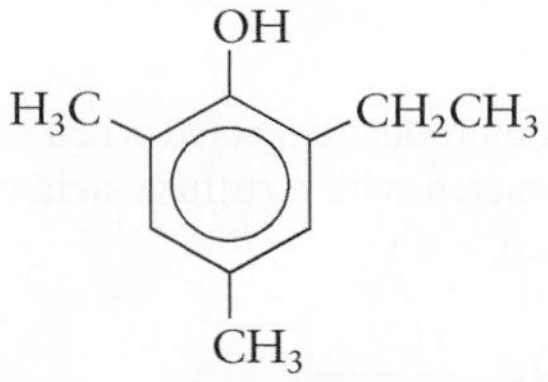

9.9 Draw structural formulas for each of the following.

a. *m*-methylphenol

b. 2,3-dichlorophenol

9.10 Draw structural formulas for each of the following.

a. *p*-chlorophenol

b. 2,5-diisopropylphenol

9.11 What is the difference between a primary, secondary, and tertiary alcohol?

9.12 Classify the following alcohols as primary, secondary, or tertiary.

a. (structure: cyclopentane with OH and CH_3 on same carbon)

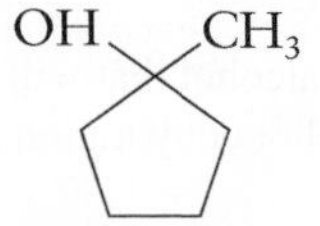

b. (structure: cyclopentane with OH)

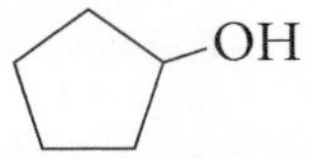

c. $CH_3CH_2CH{-}OH$ with CH_3 on the CH

9.13 Classify the following alcohols as primary, secondary, or tertiary.

a. $CH_3CH_2C{-}OH$ with two CH_3 groups on the C

b. $CH_3CH_2CH_2CH_2{-}OH$

c. (structure: cyclohexane with OH)

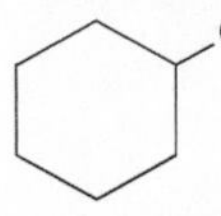

9.14 Draw structural formulas for the eight aliphatic alcohols with the molecular formula $C_5H_{12}O$. Name each compound using the IUPAC system and classify it as a primary, secondary, or tertiary alcohol.

9.15 Draw structural formulas for the four aliphatic alcohols with the molecular formula $C_4H_{10}O$. Name each compound using the IUPAC system and classify it as a primary, secondary, or tertiary alcohol.

9.16 Explain why the first of the following two compounds is a phenol, and the second is not.

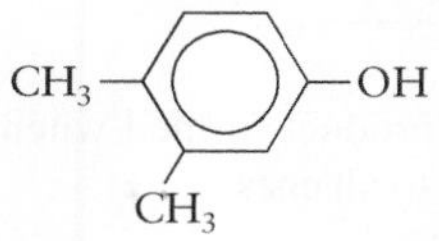

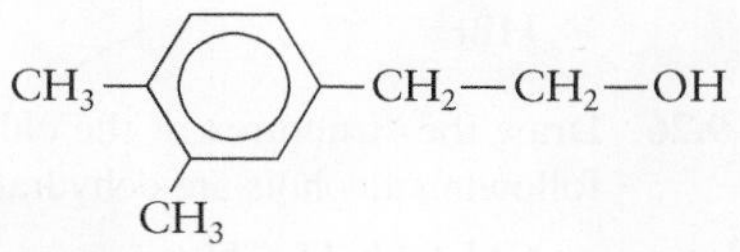

(structures: CH_3–benzene ring–OH with CH_3; CH_3–benzene ring–$CH_2{-}CH_2{-}OH$ with CH_3)

Physical Properties of Alcohols (Section 9.2)

9.17 Arrange the compounds of each group in order of increasing boiling point.

a. 1-butanol, 1-pentanol, 1-propanol

b. 1,2-butanediol, 2-butanol, pentane

9.18 Arrange the compounds of each group in order of increasing boiling point.

a. ethanol, 1-propanol, methanol

b. butane, ethylene glycol, 1-propanol

9.19 Which member of each of the following pairs would you expect to be more soluble in water? Briefly explain your choices.

a. 2,3-pentanediol or 2,3,4-pentanetriol

b. hexane or 3-hexanol

c. 1-butanol or 2-octanol

9.20 Which member of each of the following pairs would you expect to be more soluble in water? Briefly explain your choices.

a. butane or 2-butanol

b. 2-propanol or 2-pentanol

c. 2-butanol or 2,3-butanediol

9.21 Draw structural formulas for the following molecules and use a dotted line to show the formation of hydrogen bonds.

a. One molecule of 1-butanol and one molecule of ethanol

b. cyclohexanol and water

9.22 Draw structural formulas for the following molecules and use a dotted line to show the formation of hydrogen bonds.

a. One molecule of 3-pentanol and one molecule of 1-propanol

b. 2-propanol and water

9.23 Explain why the use of glycerol (1,2,3-propanetriol) in lotions helps retain water and keep the skin moist.

Reactions of Alcohols (Section 9.3)

9.24 Draw the structures of the chief product formed when the following alcohols are treated with a strong acid.

a. H_2SO_4 + (skeletal structure with methyl branch and OH) $\rightleftharpoons$

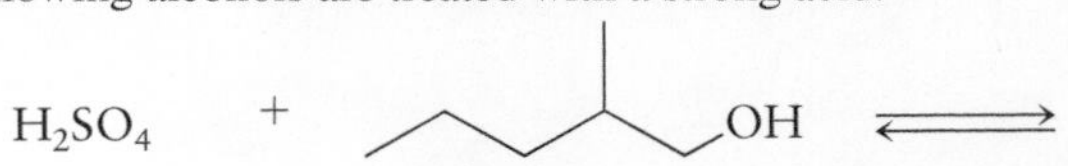

b. HNO_3 + (cyclohexanol, OH) $\rightleftharpoons$

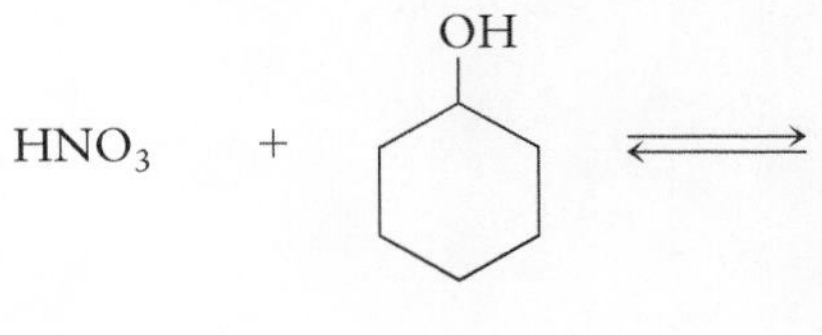

9.25 Draw the structures of the chief product formed when the following alcohols are treated with a strong acid.

a. HCl + (skeletal structure: 2-propanol, OH) $\rightleftharpoons$

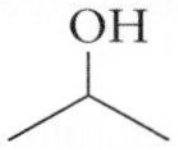

b. HBr + (skeletal structure: ethanol, OH) $\rightleftharpoons$

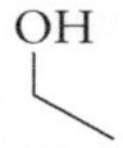

9.26 Draw the structures of the chief product formed when the following alcohols are dehydrated to alkenes.

a. $CH_3CH(OH)CH_2CH_3$

b. $CH_3CH(OH)CH(CH_3)CH_3$

9.27 Draw the structures of the chief product formed when the following alcohols are dehydrated to alkenes.

a. 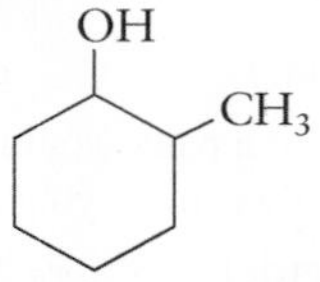 (cyclohexane ring bearing OH and an adjacent CH_3)

b. $CH_3C(CH_3)_2CH_2CH(OH)CH_3$

9.28 Draw the structures of the ethers that can be produced from the following alcohols.

a. $CH_3—OH$

b. $CH_3CH_2CH_2CH_2—OH$

c. 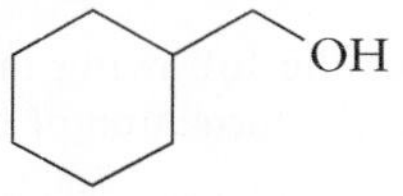 (cyclohexane ring–CH_2–OH)

9.29 Draw the structures of the ethers that can be produced from the following alcohols.

a. $CH_3CH_2CH_2—OH$

b. 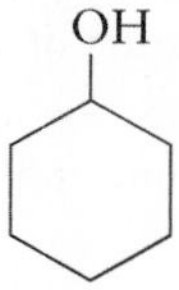 (cyclohexanol: cyclohexane ring with OH)

c. 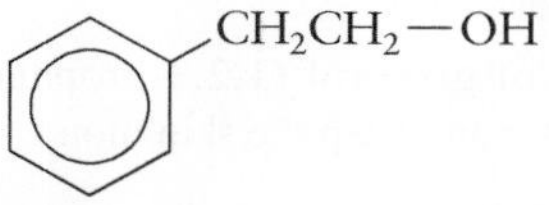 (benzene ring)–$CH_2CH_2—OH$

9.30 Draw the structures of the two organic compounds that can be obtained by oxidizing 1-propanol ($CH_3CH_2CH_2—OH$).

9.31 Draw the structure of the product expected when each of the following alcohols is reacted with a sulfuric acid catalyst at the temperature indicated.

a. $CH_3—CH(OH)—CH_3 \xrightarrow[180\ °C]{H_2SO_4}$

b. $CH_3—CH(CH_3)—CH_2—OH \xrightarrow[180\ °C]{H_2SO_4}$

c. $CH_3—CH(CH_3)—OH \xrightarrow[140\ °C]{H_2SO_4}$

d. $CH_3—CH(OH)—CH_2—CH_3 \xrightarrow[140\ °C]{H_2SO_4}$

9.32 Draw the structure of the product expected when each of the following alcohols is reacted with a sulfuric acid catalyst at the temperature indicated.

a. $CH_3—CH(OH)CH_2CH_3 \xrightarrow[180\ °C]{H_2SO_4}$

b. $CH_3—CH_2—CH(CH_3)—CH_2OH \xrightarrow[180\ °C]{H_2SO_4}$

c. $CH_3—CH_2—CH_2—OH \xrightarrow[140\ °C]{H_2SO_4}$

d. 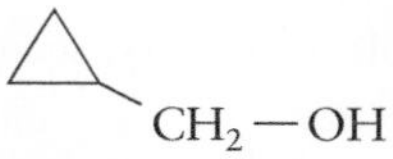 (cyclopropane ring)–$CH_2—OH \xrightarrow[140\ °C]{H_2SO_4}$

9.33 Draw the structural formula of the alcohol that will make the following products through an alcohol dehydration reaction.

a. alcohol $\xrightarrow[180\ °C]{H_2SO_4} CH_3—CH_2—CH{=}C(CH_3)—CH_3$

b. alcohol $\xrightarrow[180\ °C]{H_2SO_4} CH_3—CH_2—CH{=}CH_2$

c. alcohol $\xrightarrow[140\ °C]{H_2SO_4} CH_3—CH_2—O—CH_2—CH_3$

d. alcohol $\xrightarrow[140\ °C]{H_2SO_4} CH_3—CH(CH_3)—O—CH(CH_3)—CH_3$

9.34 Draw the structural formula of the alcohol that will make the following products through an alcohol dehydration reaction.

a. alcohol $\xrightarrow[180\ ^\circ C]{H_2SO_4}$ $CH_3-CH_2-\underset{\displaystyle CH_3}{\underset{|}{C}}=CH_2$

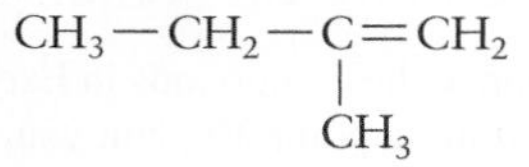

b. alcohol $\xrightarrow[180\ ^\circ C]{H_2SO_4}$ $CH_3-CH=CH_2$

c. alcohol $\xrightarrow[140\ ^\circ C]{H_2SO_4}$ CH_3-O-CH_3

d. alcohol $\xrightarrow[140\ ^\circ C]{H_2SO_4}$ $CH_3-\underset{\displaystyle CH_3}{\underset{|}{CH}}-CH_2-O-CH_2-\underset{\displaystyle CH_3}{\underset{|}{CH}}-CH_3$

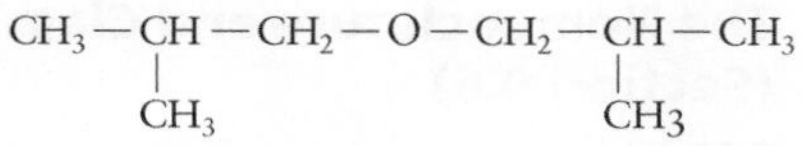

9.35 Identify whether a primary or secondary alcohol was oxidized to make each of the following compounds.

a. 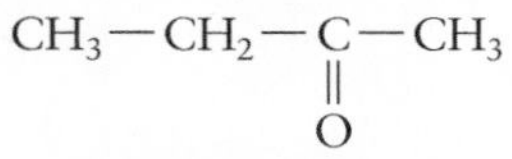 $CH_3-CH_2-\underset{\displaystyle O}{\underset{||}{C}}-CH_3$

b. $CH_3-CH_2-\underset{\displaystyle O}{\underset{||}{C}}-OH$

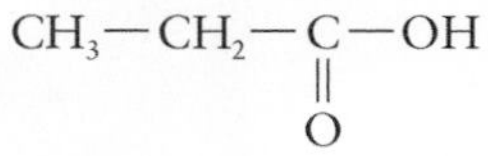

c. $CH_3-CH_2-\underset{\displaystyle O}{\underset{||}{C}}-H$

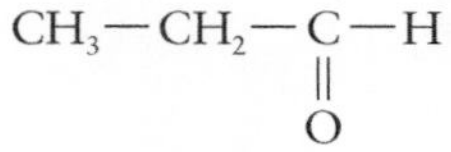

d. 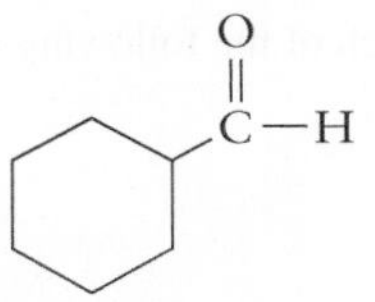 cyclohexane ring attached to $\overset{\displaystyle O}{\overset{||}{C}}-H$

9.36 Identify whether a primary or secondary alcohol was oxidized to make each of the following compounds.

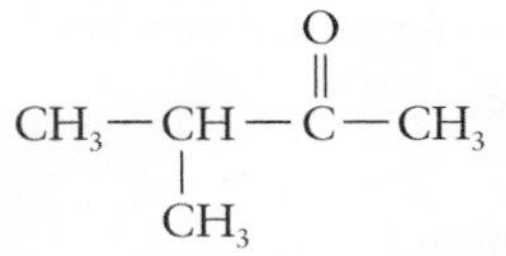

a. 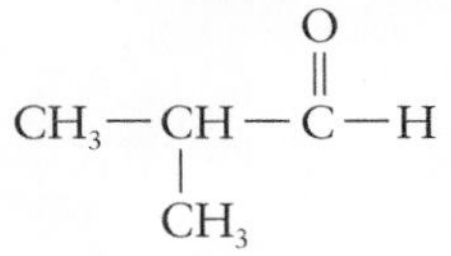 $CH_3-\underset{\displaystyle CH_3}{\underset{|}{CH}}-\overset{\displaystyle O}{\overset{||}{C}}-H$

b. $CH_3-\underset{\displaystyle CH_3}{\underset{|}{CH}}-\overset{\displaystyle O}{\overset{||}{C}}-CH_3$

c. $CH_3-\underset{\displaystyle CH_3}{\underset{|}{CH}}-CH_2-\underset{\displaystyle O}{\underset{||}{C}}-OH$

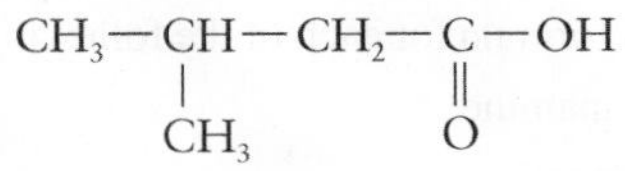

d. 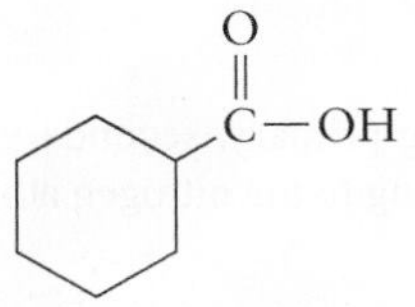 cyclohexane ring attached to $\overset{\displaystyle O}{\overset{||}{C}}-OH$

9.37 Draw the structures of the two organic compounds that can be obtained by oxidizing 2-methyl-1-propanol.

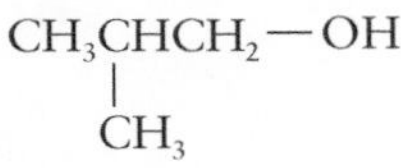

$CH_3\underset{\displaystyle CH_3}{\underset{|}{C}}HCH_2-OH$

9.38 Give the structure of an alcohol that could be used to prepare each of the following compounds.

a. 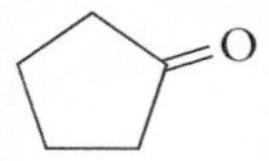 cyclopentane ring with =O (cyclopentanone)

b. 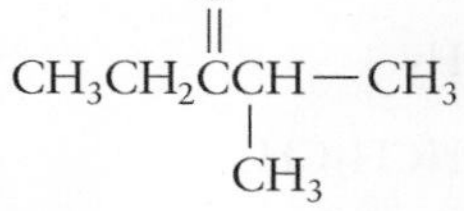 $CH_3CH_2\overset{\displaystyle O}{\overset{||}{C}}\underset{\displaystyle CH_3}{\underset{|}{C}}H-CH_3$

c. $CH_3CH_2CH_2-\overset{\displaystyle O}{\overset{||}{C}}-OH$

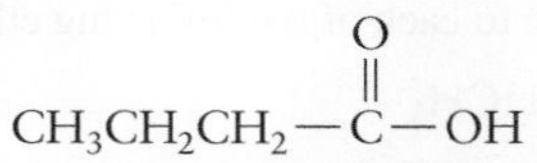

9.39 Give the structure of an alcohol that could be used to prepare each of the following compounds.

a. 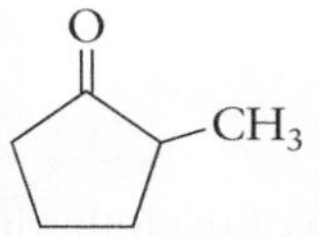 cyclopentane ring with =O and adjacent CH_3

b. 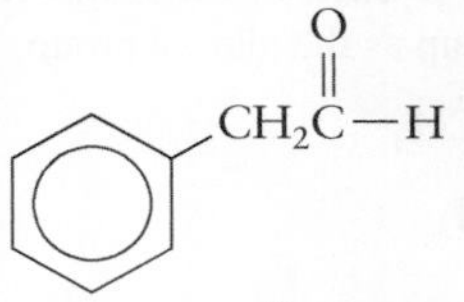 benzene ring attached to $CH_2\overset{\displaystyle O}{\overset{||}{C}}-H$

c. $CH_3CH_2-\overset{\displaystyle O}{\overset{||}{C}}-OH$

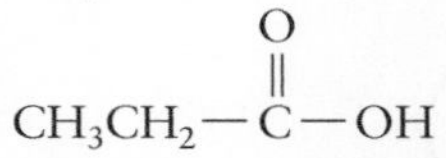

Applications of Alcohols (Section 9.4)

9.40 Methanol is fairly volatile and evaporates quickly if spilled. Methanol is also absorbed quite readily through the skin. If, in the laboratory, methanol accidentally spilled on your clothing, why would it be a serious mistake to just let it evaporate?

9.41 Suppose you are making some chocolate cordials (chocolate-coated candies with soft fruit filling). Why might you want to add a little glycerol to the filling?

9.42 Name an alcohol used in each of the following ways.

a. A moistening agent in many cosmetics

b. A solvent in solutions called tinctures

c. Automobile antifreeze

d. Rubbing alcohol

9.43 Name an alcohol used in each of the following ways.

a. A flavoring in cough drops

b. Present in gasohol

c. Active ingredient in alcoholic beverages

d. Industrially produced from CO and H_2

Ethers (Section 9.5)

9.44 Draw a general formula for an ether, emphasizing the functional group.

9.45 Name a phenol used in each of the following ways.

a. A disinfectant used for cleaning walls

b. An antiseptic found in some mouthwashes

c. An antioxidant used to prevent rancidity in foods

9.46 Assign a common name to each of the following ethers.

a. $CH_3-O-CH_2CH_3$

b. 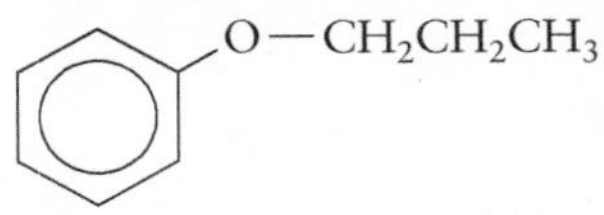 (benzene ring)$-O-CH_2CH_2CH_3$

c.

$$\begin{array}{c} CH_3 \qquad\qquad CH_3 \\ | \qquad\qquad\quad | \\ CH_3CH_2CH-O-CHCH_2CH_3 \end{array}$$

9.47 Assign a common name to each of the following ethers.

a.

$$\begin{array}{l} CH_3CH_2-O-CHCH_3 \\ \qquad\qquad\qquad | \\ \qquad\qquad\quad CH_3 \end{array}$$

b. $CH_3-O-CH_2CH_2CH_2CH_3$

c. 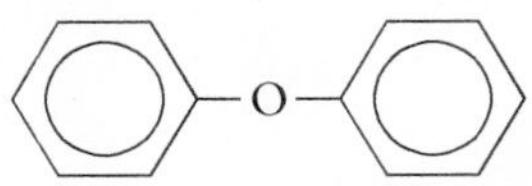 (benzene ring)$-O-$(benzene ring)

9.48 Assign the common name to each of the following ethers. Name the smaller alkyl group as the alkoxy group.

a. $CH_3-O-CH_2CH_2CH_3$

b. 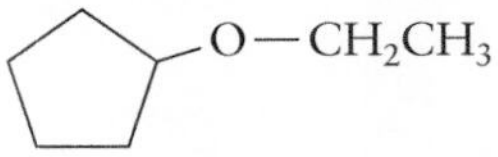 (cyclopentane ring)$-O-CH_2CH_3$

c.

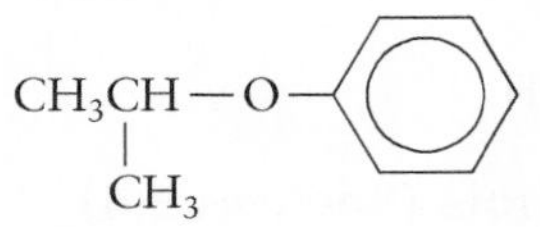

$$\begin{array}{l} CH_3CH-O-\text{(benzene ring)} \\ \quad\;\; | \\ \quad CH_3 \end{array}$$

9.49 Assign the common name to each of the following ethers. Name the smaller alkyl group as the alkoxy group.

a. $CH_3CH_2-O-CH_2CH_2CH_3$

b.

$$\begin{array}{l} CH_3CH_2-O-CHCH_3 \\ \qquad\qquad\qquad | \\ \qquad\qquad\quad CH_3 \end{array}$$

c. 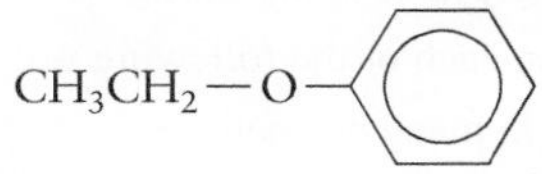 CH_3CH_2-O-(benzene ring)

9.50 Draw structural formulas for the following.

a. ethyl methyl ether

b. butyl phenyl ether

9.51 Draw structural formulas for the following.

a. methyl isopropyl ether

b. phenyl propyl ether

9.52 Arrange the following compounds in order of decreasing solubility in water. Explain the basis for your decision.

$CH_3CH_2-O-CH_2CH_3$

$$\begin{array}{c} OH \\ | \\ CH_3CHCH_2CH_3 \end{array}$$

$$\begin{array}{c} CH_3 \\ | \\ CH_3CHCH_2CH_3 \end{array}$$

9.53 Arrange the following compounds in order of decreasing solubility in water. Explain the basis for your decisions.

$CH_3CH_2CH_2-OH \quad CH_3CH_2CH_2CH_3 \quad CH_3CH_2-O-CH_3$

9.54 Arrange the compounds in Exercise 9.52 in order of decreasing boiling point. Explain your answer.

9.55 Arrange the compounds in Exercise 9.53 in order of decreasing boiling point. Explain your answer.

9.56 Why is diethyl ether hazardous to use as an anesthetic or as a solvent in the laboratory?

The Nomenclature and Classification of Amines (Section 9.6)

9.57 Assign the common name to each of the following amines.

a.

$$\begin{array}{c} NH_2 \\ | \\ CH_3CHCH_3 \end{array}$$

b.

$$\begin{array}{c} NH_2 \\ | \\ CH_3CCH_3 \\ | \\ CH_3 \end{array}$$

c.

$$\begin{array}{l} \qquad\qquad NH-CH_2CH_3 \\ \qquad\qquad | \\ CH_3CH_2CH_2 \end{array}$$

9.58 Assign the common name to each of the following amines.

a. 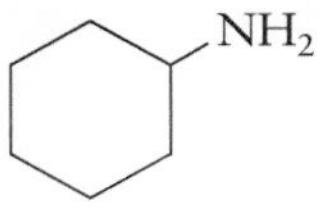 (cyclohexane ring)$-NH_2$

b. 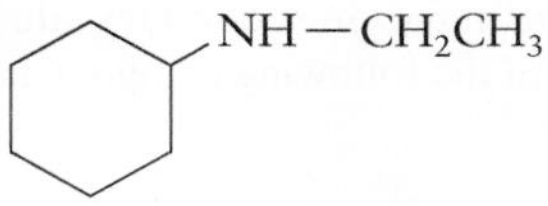 (cyclohexane ring)$-NH-CH_2CH_3$

c.

$$\begin{array}{l} CH_3CH_2CH-NH_2 \\ \qquad\quad\;\; | \\ \qquad\quad CH_3 \end{array}$$

9.59 Draw the structural formula for each of the following amines.

a. 3-ethyl-2-pentanamine

b. diethylpropylamine

c. 3-methylcyclohexanamine

9.60 Draw the structural formula for each of the following amines.

a. 3-chloro-2-pentanamine

b. 3-methyl-2-butanamine

c. ethylamine

9.61 What is the difference among primary, secondary, and tertiary amines in terms of bonding to the nitrogen atom?

9.62 Give a general formula for a primary amine, a secondary amine, and a tertiary amine.

9.63 Classify each of the following as a primary, secondary, or tertiary amine.

a. 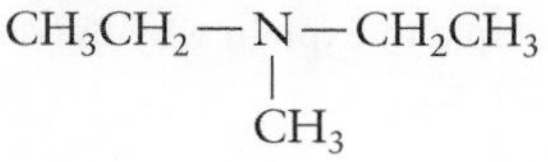 $CH_3CH_2-N(CH_3)-CH_2CH_3$

b. 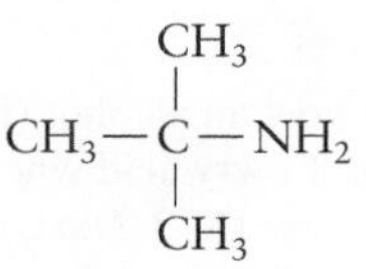 $CH_3-C(CH_3)_2-NH_2$ (a central C bonded to three CH_3 groups and NH_2)

c. $C_6H_5-NH-CH_3$ (benzene ring bonded to $NH-CH_3$)

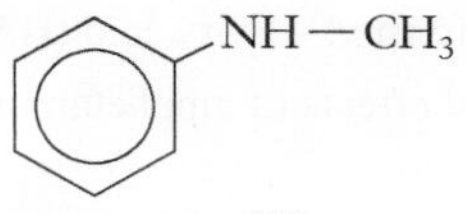

d. 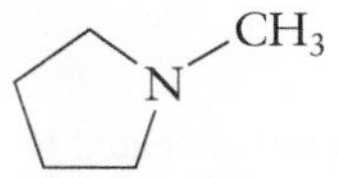 Five-membered ring containing N, with CH_3 on the N

9.64 Classify each of the following as a primary, secondary, or tertiary amine.

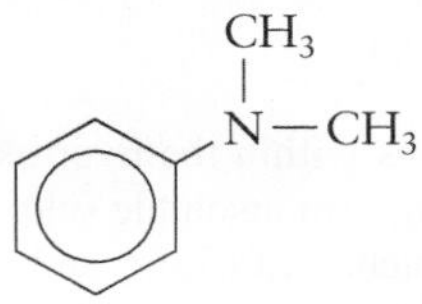

a. 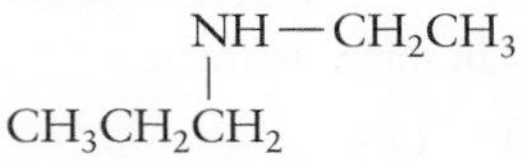 $CH_3CH_2CH_2-NH-CH_2CH_3$

b. $C_6H_5-N(CH_3)-CH_3$ (benzene ring bonded to N, which carries two CH_3 groups)

c. Cyclobutane ring bonded to NH_2

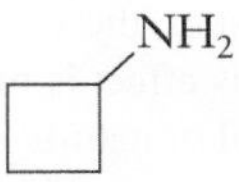

d. 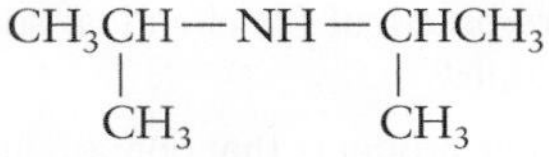 $CH_3CH(CH_3)-NH-CH(CH_3)CH_3$

Properties of Amines (Section 9.7)

9.65 Explain why CH_3-NH_2 is a Brønsted–Lowry base.

9.66 When diethylamine is dissolved in water, the solution becomes basic. Write an equation to account for this observation.

9.67 Draw the structural formula of the missing substance in each of the following reactions.

a. 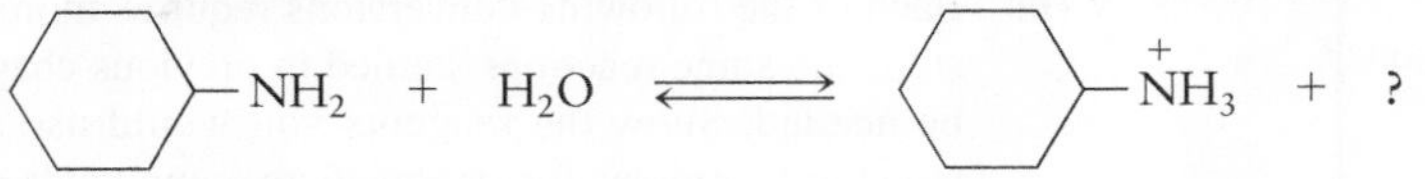 Cyclohexyl$-NH_2$ + H_2O $\rightleftharpoons$ Cyclohexyl$-\overset{+}{N}H_3$ + ?

b. 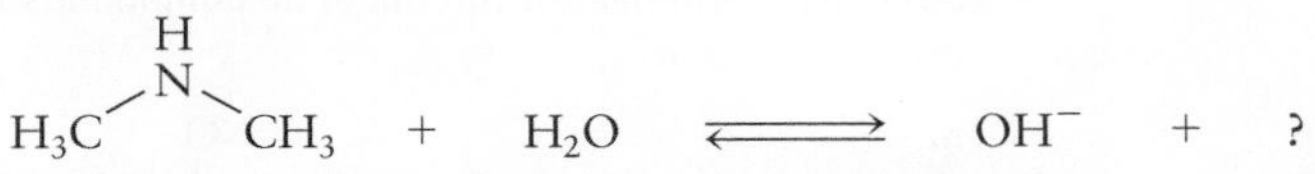 $H_3C-NH-CH_3$ + H_2O $\rightleftharpoons$ OH^- + ?

9.68 Draw the structural formula of the missing substance in each of the following reactions.

a. 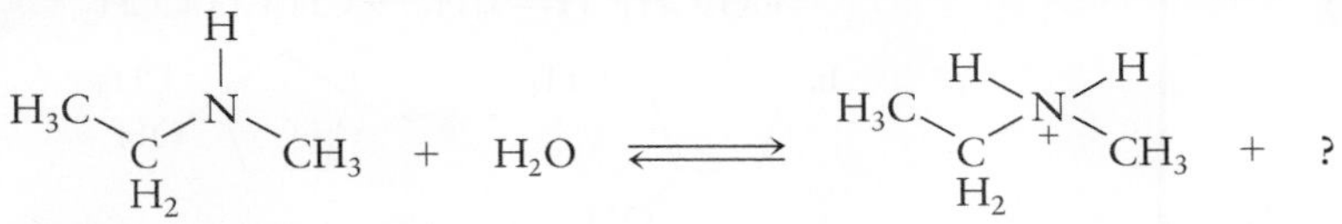 $H_3C-CH_2-NH-CH_3$ + H_2O $\rightleftharpoons$ $H_3C-CH_2-\overset{+}{N}H_2-CH_3$ + ?

b. 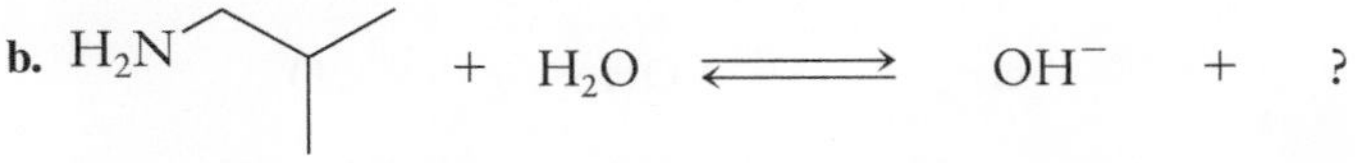 $H_2N-CH_2-CH(CH_3)_2$ (skeletal) + H_2O $\rightleftharpoons$ OH^- + ?

9.69 Why are amine drugs commonly administered in the form of their salts?

9.70 How does the structure of a quaternary ammonium salt differ from the structure of a salt of a tertiary amine?

9.71 Explain why all classes of low-molecular-weight amines are water soluble.

9.72* Why are the boiling points of amines lower than those of corresponding alcohols?

9.73* Why are the boiling points of tertiary amines lower than those of corresponding primary and secondary amines?

9.74 Write a chemical equation to show how the drug dextromethorphan, a cough suppressant, could be made more water soluble.

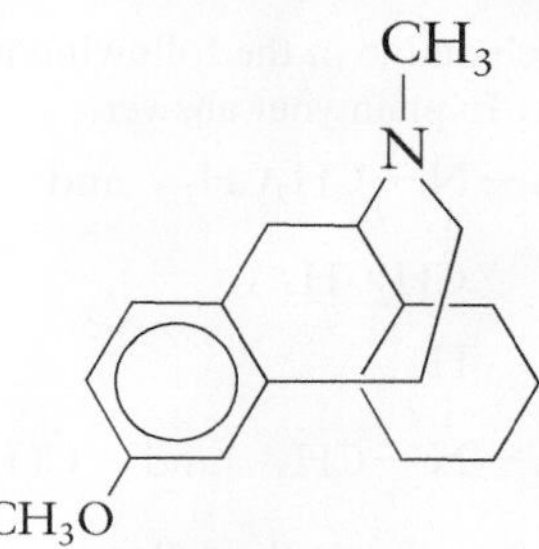

dextromethorphan

9.75 Draw diagrams similar to **Figure 9.33** to illustrate hydrogen bonding between the following compounds.

a. $CH_3CH_2-NH-CH_3$ and H_2O

b. 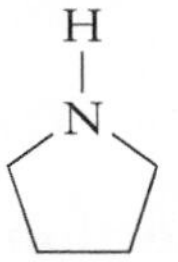 Pyrrolidine (five-membered ring with N–H) and pyrrolidine (five-membered ring with N–H)

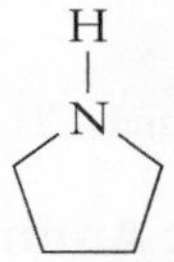

9.76 Draw diagrams similar to **Figure 9.33** to illustrate hydrogen bonding between the following compounds.

a. 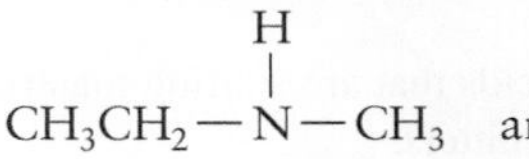 $CH_3CH_2-NH-CH_3$ and $CH_3CH_2-NH-CH_3$

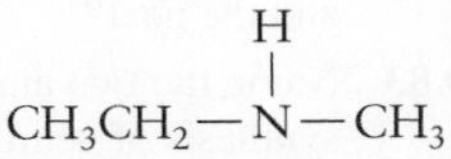

b. 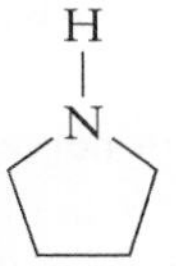 Pyrrolidine (five-membered ring with N–H) and H_2O

9.77 Arrange the following compounds by functional group in order of increasing boiling point.

$CH_3CH_2CH_2{-}NH_2$

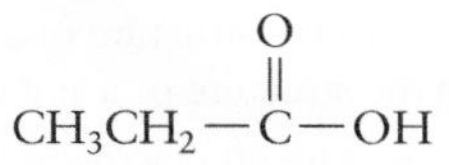

$CH_3CH_2{-}\overset{\overset{\large O}{||}}{C}{-}OH$

$CH_3CH{=}CHCH_3$

9.78 Arrange the following compounds by functional group in order of increasing boiling point.

$CH_3CH_2CH_2CH_3$

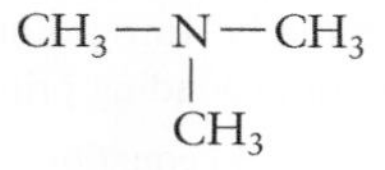

$CH_3{-}\underset{\underset{\large CH_3}{|}}{N}{-}CH_3$

$CH_3CH_2CH_2{-}OH$

9.79 Identify which amine in the following pairs would be more water soluble. Explain your answer.

a. 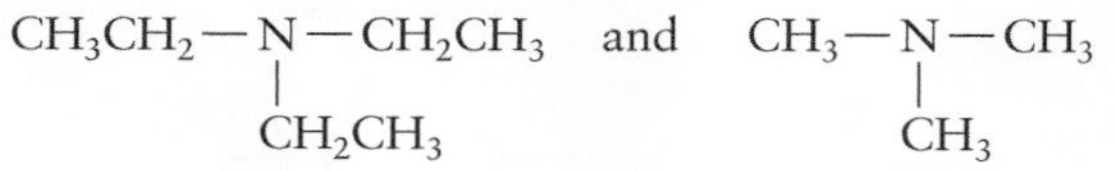 $CH_3CH_2{-}\underset{\underset{\large CH_2CH_3}{|}}{N}{-}CH_2CH_3$ and $CH_3{-}\underset{\underset{\large CH_3}{|}}{N}{-}CH_3$

b. 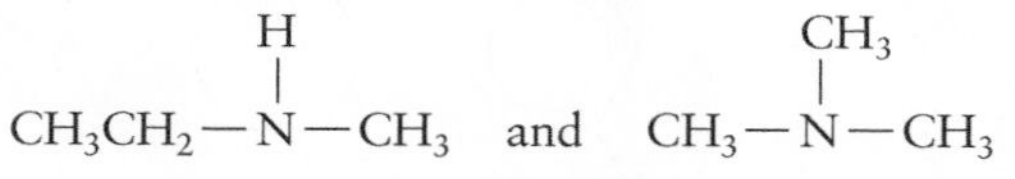 $CH_3CH_2{-}\overset{\overset{\large H}{|}}{N}{-}CH_3$ and $CH_3{-}\overset{\overset{\large CH_3}{|}}{N}{-}CH_3$

9.80 Identify which amine in the following pairs would be more water soluble. Explain your answer.

a. 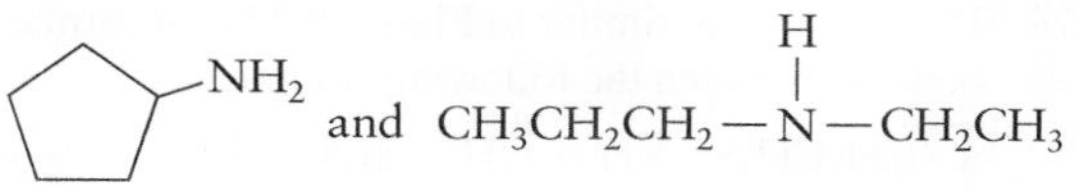 (cyclopentyl)$-NH_2$ and $CH_3CH_2CH_2{-}\overset{\overset{\large H}{|}}{N}{-}CH_2CH_3$

b. 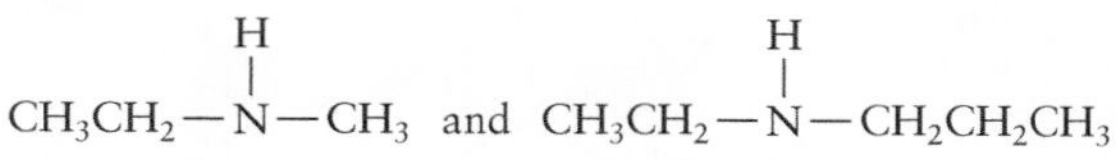 $CH_3CH_2{-}\overset{\overset{\large H}{|}}{N}{-}CH_3$ and $CH_3CH_2{-}\overset{\overset{\large H}{|}}{N}{-}CH_2CH_2CH_3$

Biologically Significant Amines (Section 9.8)

9.81 Describe the general structure of a neuron.

9.82 What term is used to describe the gap between one neuron and the next?

9.83 Name the two amino acids that are starting materials for the synthesis of neurotransmitters.

9.84 Name the two amines often associated with the biochemical theory of mental illness.

9.85 What role do neurotransmitters play in nerve impulse transmission?

9.86 List four neurotransmitters that are important in the central nervous system.

9.87 Why is epinephrine called the fight-or-flight hormone?

9.88 What is the source of alkaloids?

9.89 Why are alkaloids weakly basic?

9.90 Give the name of an alkaloid for the following.

a. Found in cola drinks

b. Used to reduce saliva flow during surgery

c. Present in tobacco

d. A cough suppressant

e. Used to treat malaria

f. An effective painkiller

9.91 In simplistic terms, it can be thought that an equilibrium exists in hydration–dehydration reactions:

$$\text{alkene} + \text{water} \underset{}{\overset{H_2SO_4}{\rightleftharpoons}} \text{alcohol} + \text{heat}$$

Use Le Châtelier's principle to explain why the dehydration reaction is favored at 180 °C.

Additional Problems

9.92 Use general reactions to show how an alcohol (R—O—H) can behave as a weak Brønsted–Lowry acid when it reacts with a strong Brønsted–Lowry base (X^-). Next, show how an alcohol can behave as a weak Brønsted–Lowry base in the presence of a strong Brønsted–Lowry acid (HA).

9.93 What are the physiological effects of amphetamines on the body?

Chemistry for Thought

9.94 A mixture of ethanol and 1-propanol is heated to 140 °C in the presence of sulfuric acid. Careful analysis reveals that three ethers are formed. Write formulas for these three products; name each; and explain why three form, rather than a single product.

9.95 The two-carbon alcohol in some antifreeze

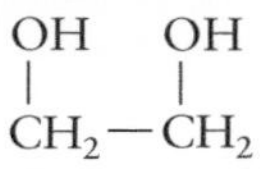

$$\overset{\overset{\large OH}{|}}{CH_2}{-}\overset{\overset{\large OH}{|}}{CH_2}$$

is toxic if ingested. Reactions within the liver oxidize the diol to oxalic acid, which can form insoluble salts and damage the kidneys. Give the structure of oxalic acid.

9.96 When diethyl ether is spilled on the skin, the skin takes on a dry appearance. Explain whether this effect is more likely due to a removal of water or a removal of natural skin oils.

9.97 Why might it be more practical for chemical suppliers to ship amines as amine salts?

9.98 Several over-the-counter products that contain amine salts are available. Sometimes the active ingredients are given common names ending in *hydrochloride*. What acid do you think is used to prepare an amine hydrochloride salt?

9.99 In **Section 13.2**, several components of nucleic acids, DNA and RNA, are identified as "bases." What molecular feature of these substances might explain why they are referred to as bases?

9.100 Each of the following conversions requires more than one step, and some reactions studied in previous chapters may be needed. Show the reagents you would use and draw structural formulas for intermediate compounds formed in each conversion.

a. 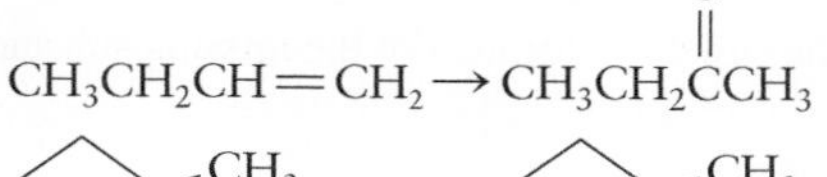 $CH_3CH_2CH{=}CH_2 \rightarrow CH_3CH_2\overset{\overset{\large O}{||}}{C}CH_3$

b. (2-methylcyclopentanol, ring with CH_3 and OH on adjacent carbons) $\longrightarrow$ (1-methylcyclopentanol, ring with CH_3 and OH on the same carbon)

c. $CH_3CH_2CH_2\overset{\overset{\large OH}{|}}{CH_2} \rightarrow CH_3CH_2\overset{\overset{\large O}{||}}{C}CH_3$

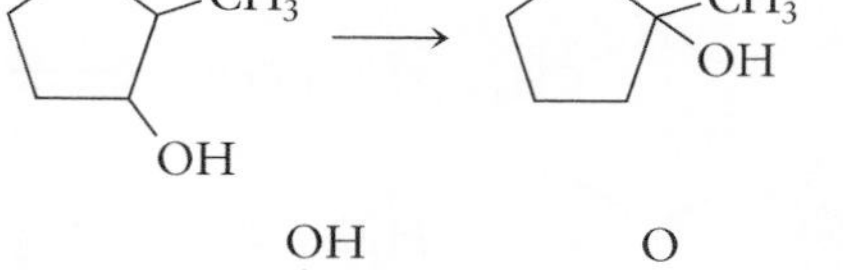

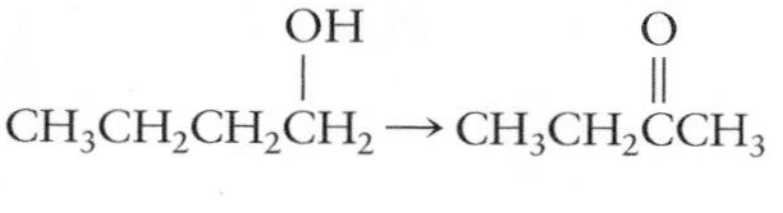

9.101* Each of the following conversions requires more than one step, and some reactions studied in previous chapters may be needed. Show the reagents you would use and draw structural formulas for intermediate compounds formed in each conversion.

a. $CH_2{=}CH_2 \rightarrow CH_3CH_2{-}O{-}CH_2CH_3$

b. cyclopentene $\longrightarrow$ cyclopentanone

c. $CH_3CH_2CH_2OH \rightarrow CH_3CH(OH)CH_3$

9.102 Alkaloids can be extracted from plant tissues using dilute hydrochloric acid. Why does the use of acid enhance their water solubility?

10 Carbonyl and Carboxyl Compounds

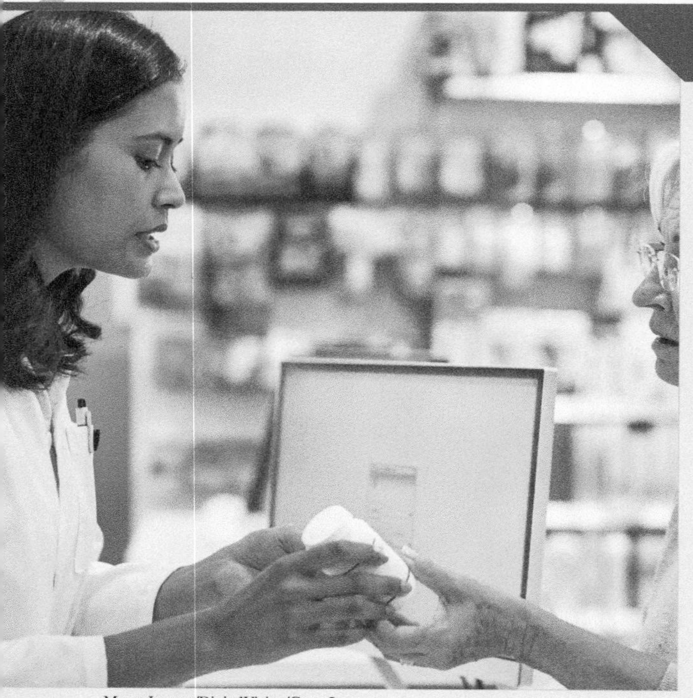
Morsa Images/DigitalVision/Getty Images

Health Career Focus

Pharmacy Technician

My junior year in high school I took Chem 1110 and discovered that I love chemistry! Something "just clicked" with the way I visualize chemical reactions—I just "got it." My high school counselor suggested I try the pharmacy program at the technical college.

I was really interested in chemical flavorings (esters) and how to compound medications to make them more palatable and effective for certain individuals. I liked learning how to improve the flavor of children's medicine so that children would comply more easily with treatment.

By the end of my senior year, I'd been certified as a pharmacy technician. After graduation, I stepped into a good job and started college without having to borrow much money. Now, I might go on to become a pharmacist!

Follow-up to this Career Focus appears at the end of the chapter before the *Concept Summary*.

Learning Objectives

When you have completed your study of Chapter 10, you should be able to:

1. Classify carbonyl compounds as aldehydes or ketones. **(Section 10.1)**
2. Identify specific uses for aldehydes and ketones. **(Section 10.1)**
3. Assign IUPAC names to aldehydes and ketones. **(Section 10.1)**
4. Explain how intermolecular forces influence the physical properties of aldehydes and ketones. **(Section 10.2)**
5. Predict the products of the hydrogenation, oxidation, reduction, and hydration reactions of aldehydes and ketones. **(Section 10.3)**
6. Assign IUPAC names to carboxylic acids. **(Section 10.4)**
7. Assign common and IUPAC names to carboxylate salts. **(Section 10.4)**
8. Describe uses for carboxylate salts. **(Section 10.4)**
9. Explain how hydrogen bonding affects the physical properties of carboxylic acids. **(Section 10.5)**
10. Write key acid–base reactions of carboxylic acids. **(Section 10.5)**
11. Assign common and IUPAC names to esters and amides. **(Section 10.6)**
12. Explain how hydrogen bonding affects the physical properties of esters and amides. **(Section 10.7)**
13. Write reactions for the formation and hydrolysis of esters and amides. **(Section 10.8)**
14. Write key reactions for saponification. **(Section 10.8)**

The carbonyl group is one of the most influential bonds in all of organic chemistry and biochemistry. A carbonyl group is characterized by a carbon double-bonded to an oxygen and single-bonded to two other atoms:

$$\begin{array}{c} \mathrm{O} \\ \| \\ -\mathrm{C}- \end{array}$$

carbonyl group

The carbonyl group is a feature of many functional groups that are the focus of the first part of this chapter. Namely, we examine the nomenclature, properties, and reactions of aldehydes and ketones. The nature of the atoms bonded to the carbonyl carbon (the carbon double-bonded to oxygen) dictates the identity of the specific functional group. In fact, another subset of functional groups containing the structurally similar **carboxyl group** is the focus of the second half of the chapter as we study the naming, properties, and reactions of functional groups like carboxylic acids and two of their derivatives: esters and amides. Many of the functional groups at the center of focus in this chapter play important roles in the biochemistry discussed in Chapters 11–15.

carbonyl group The $\overset{\mathrm{O}}{\overset{\|}{-\mathrm{C}-}}$ group.

carboxyl group The $\overset{\mathrm{O}}{\overset{\|}{-\mathrm{C}-\mathrm{OH}}}$ group.

aldehyde A compound that contains the $\overset{\mathrm{O}}{\overset{\|}{-\mathrm{C}-\mathrm{H}}}$ group; the general formula is $\mathrm{R}-\overset{\mathrm{O}}{\overset{\|}{\mathrm{C}}}-\mathrm{H}$, which can be written more concisely as RCHO.

ketone A compound that contains the $\begin{array}{c} \quad \mathrm{O} \\ | \quad \| \quad | \\ -\mathrm{C}-\mathrm{C}-\mathrm{C}- \\ | \qquad\quad | \end{array}$ group; the general formula is $\mathrm{R}-\overset{\mathrm{O}}{\overset{\|}{\mathrm{C}}}-\mathrm{R}'$, or RC(O)R′.

10.1 The Nomenclature of Aldehydes and Ketones

Learning Objective 1 Classify carbonyl compounds as aldehydes or ketones.

Learning Objective 2 Identify specific uses for aldehydes and ketones.

Learning Objective 3 Assign IUPAC names to aldehydes and ketones.

When a carbonyl group is directly bonded to at least one hydrogen atom, the compound is an **aldehyde**. In **ketones**, two carbon atoms are directly bonded to the carbonyl group. Structural models of the simplest aldehyde (formaldehyde) and the simplest ketone (acetone) are shown in **Figure 10.1**.

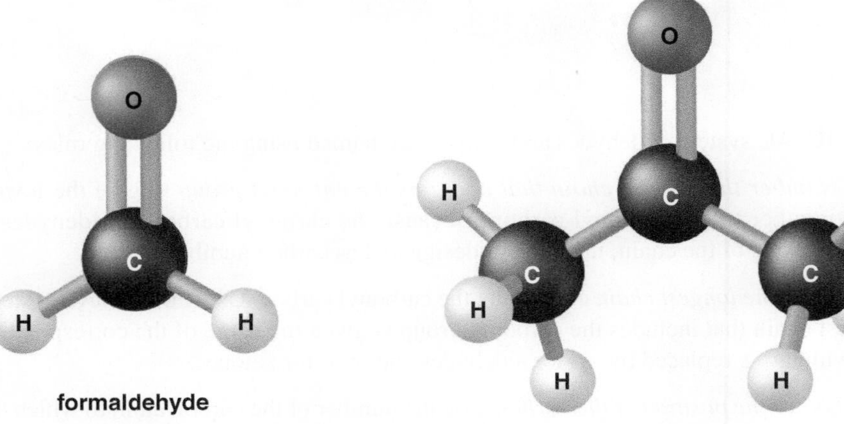

Figure 10.1 Ball-and-stick models of formaldehyde (an aldehyde) and acetone (a ketone).

$$\text{General formula:}\quad \underset{\text{aldehyde}}{-\overset{\overset{\displaystyle O}{\|}}{C}-H} \qquad \underset{\text{ketone}}{-\underset{|}{\overset{|}{C}}-\overset{\overset{\displaystyle O}{\|}}{C}-\underset{|}{\overset{|}{C}}-}$$

$$\text{Specific example:}\quad \underset{\text{formaldehyde}}{H-\overset{\overset{\displaystyle O}{\|}}{C}-H} \qquad \underset{\text{acetone}}{CH_3-\overset{\overset{\displaystyle O}{\|}}{C}-CH_3}$$

Several aldehydes and ketones, in addition to formaldehyde (see **Figure 10.2**) and acetone, are well known by common names (see **Table 10.1**). Notice that aldehyde and ketone common names have distinctive *-aldehyde* and *-one* endings, respectively.

As seen with other functional groups, the most important compound of the class is usually the simplest member. Formaldehyde, the simplest aldehyde, is a gas at room temperature, but it is often supplied and used in the form of a 37% aqueous solution called formalin. This solution kills microorganisms and effectively sterilizes surgical instruments. It is also used to embalm cadavers. Formaldehyde is a key industrial chemical in the production of plastics such as Bakelite, the first commercial polymer (invented in 1901), and Formica.

Judging from the quantity used (more than 1 billion pounds annually in the United States), acetone is by far the most important of the ketones. It is particularly useful as a solvent because it dissolves most organic compounds and yet it dissolves in water in all proportions (i.e., acetone and water are *miscible*).

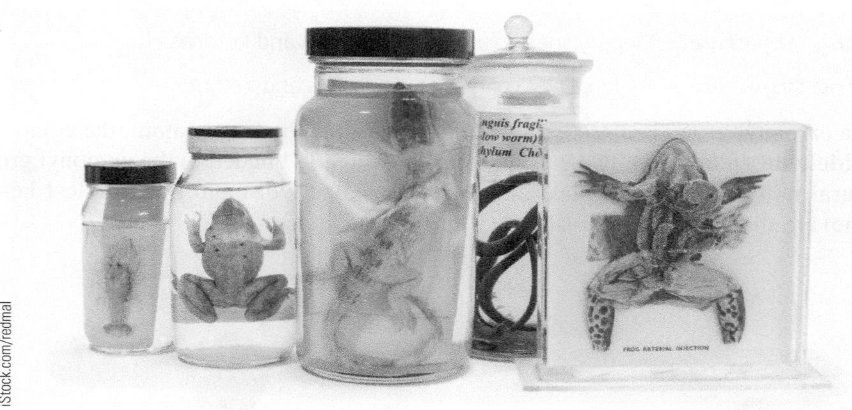

Figure 10.2 Formaldehyde is used to preserve biological specimens.

iStock.com/redmal

In the IUPAC system, aldehydes and ketones are named using the following rules:

Step 1: *Number the longest chain that contains the carbonyl group* to give the lowest possible number to the carbonyl carbon. Because the carbonyl carbon in aldehydes is always at the end of the chain, it is always designated as carbon number 1.

Step 2: *Name the longest chain* containing the carbonyl carbon. Using the IUPAC system, the longest chain that includes the carbonyl group is given the name of the corresponding alkane, with the *-e* replaced by *-al* for aldehydes and *-one* for ketones.

Step 3: *Locate the position of the carbonyl* by the number of the carbon atom to which the double-bonded oxygen is attached.

Step 4: *Locate and name any groups attached to the chain.*

Step 5: *Combine the name and location for other groups, the carbonyl group location, and the longest chain into the final name.*

Table 10.1 Some Common Aldehydes and Ketones

Structural Formula	IUPAC Name	Common Name	Boiling Point (°C)
Aldehydes			
$H-\overset{\overset{\Large O}{\|}}{C}-H$	methanal	formaldehyde	−21
$CH_3CH_2CH_2-\overset{\overset{\Large O}{\|}}{C}-H$	butanal	butyraldehyde	76
$C_6H_5-\overset{\overset{\Large O}{\|}}{C}-H$	benzaldehyde	benzaldehyde	178
Ketones			
$CH_3-\overset{\overset{\Large O}{\|}}{C}-CH_3$	propanone	acetone	56
(cyclohexane ring)$=O$	cyclohexanone	cyclohexanone	156

Example 10.1 Naming Aldehydes and Ketones

The common name for the two-carbon aldehyde, shown in part a, is acetaldehyde (**Figure 10.3**). It is a toxic substance produced when alcohol is broken down in the liver. Name the following aldehyde and ketones according to the IUPAC system.

a. $CH_3-\overset{\overset{\Large O}{\|}}{C}-H$

b. (skeletal structure: four-carbon chain with C=O on carbon 2)

c. (cyclohexane ring with $=O$ and adjacent CH_3)

iStock.com/Space_Cat

Figure 10.3 The fatigue, weakness, and headaches associated with excessive alcohol consumption are due, in part, to the acetaldehyde by-product that builds up when ethanol is metabolized.

Solution

a. **Step 1:** *Number the longest chain that contains the carbonyl group* to give the lowest possible number to the carbonyl carbon. The longest chain containing the carbonyl group is numbered as follows:

$$\underset{2}{CH_3}-\underset{1}{\overset{\overset{\Large O}{\|}}{C}}-H$$

Step 2: *Name the longest chain* containing the carbonyl carbon. This aldehyde contains two carbon atoms, so the name is ethanal (**Figure 10.4**).

Step 3: *Locate the position of the carbonyl* by the number of the carbon atom to which double-bonded oxygen is attached. The aldehyde group is always at position number 1, so a number designating the aldehyde position is unnecessary and is never included in the name.

Dids/Pexels

Figure 10.4 Ethanal, commonly known as acetaldehyde, is used to produce perfumes.

Step 4: *Locate and name any groups attached to the chain.* In this case, no other groups are attached.

Step 5: *Combine the name and location for other groups, the carbonyl group location, and the longest chain into the final name.* The final name is ethanal.

b. **Step 1:** *Number the longest chain that contains the carbonyl group* to give the lowest possible number to the carbonyl carbon. The longest chain containing the carbonyl group is numbered as follows:

O

1 2 3 4 5

Step 2: *Name the longest chain* containing the carbonyl carbon. There are five carbons in the longest chain containing the carbonyl group of the ketone. So the correct name is pentanone.

Step 3: *Locate the position of the carbonyl* by the number of the carbon atom to which the double-bonded oxygen is attached. The number 3 is used to identify the position of the carbonyl group on the longest carbon chain.

Step 4: *Locate and name any groups attached to the chain.* In this case, no other groups are attached.

Step 5: *Combine the name and location for other groups, the carbonyl group location, and the longest chain into the final name.* The final name is 3-pentanone.

c. **Step 1:** *Number the longest chain that contains the carbonyl group* to give the lowest possible number to the carbonyl carbon. The longest chain containing the carbonyl group is numbered as follows:

O

6 5 1 4 2 3

CH_3

Step 2: *Name the longest chain* containing the carbonyl carbon. This ketone contains six carbon atoms in a ring, so the name is cyclohexanone.

Step 3: *Locate the position of the carbonyl* by the number of the carbon atom to which the double-bonded oxygen is attached. In cyclic ketones, the carbonyl carbon will always be at position 1, so the number is omitted from the name.

Step 4: *Locate and name any groups attached to the chain.* In this case, there is a methyl group attached to the ring at carbon 2.

Step 5: *Combine the name and location for other groups, the carbonyl group location, and the longest chain into the final name.* The final name is 2-methylcyclohexanone (**Figure 10.5**).

Fotokostic/Depositphotos

Figure 10.5 2-Methylcyclohexanone, a ketone, is used as a solvent in varnish.

✔ Learning Check 10.1 Give IUPAC names for the following aldehyde and ketones.

a. $CH_3CH_2CH_2CH_2-\overset{\overset{\large O}{\|}}{C}-H$

b. CH_3 / O / CH_3

c. O

Several naturally occurring aldehydes and ketones have very fragrant odors and are used in flavorings and perfumes. A number of naturally occurring aldehydes and ketones play important roles within living systems, too. For example, progesterone and testosterone, female and male sex hormones, respectively, both contain ketones (**Figure 10.6**).

progesterone

testosterone

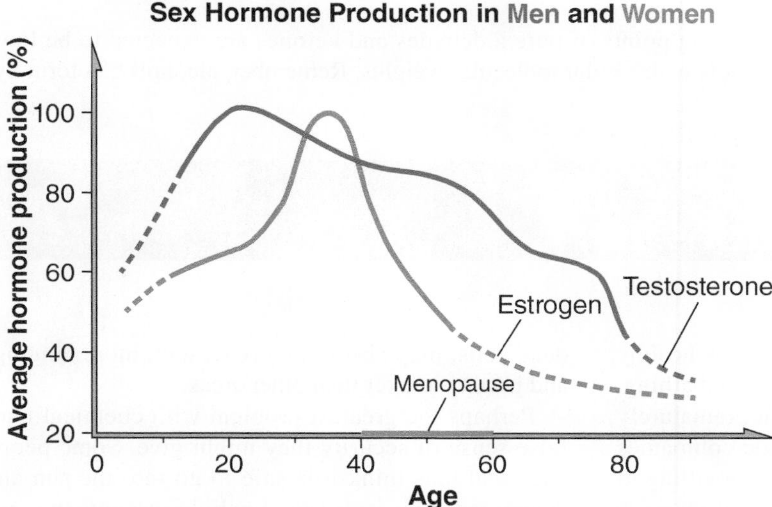

Figure 10.6 Estrogen and testosterone production in males and females as a function of age.

Moreover, these two carbonyl functional groups are found in the open-chain configuration of numerous carbohydrates, including glucose (an aldehyde) and fructose (a ketone). Glucose, a major source of energy in living systems, is found combined with fructose in cane sugar (**Figure 10.7**):

MK photograp55/Shutterstock.com

Figure 10.7 Cane sugar is made exclusively from sugarcane, a tall grass native to the tropics.

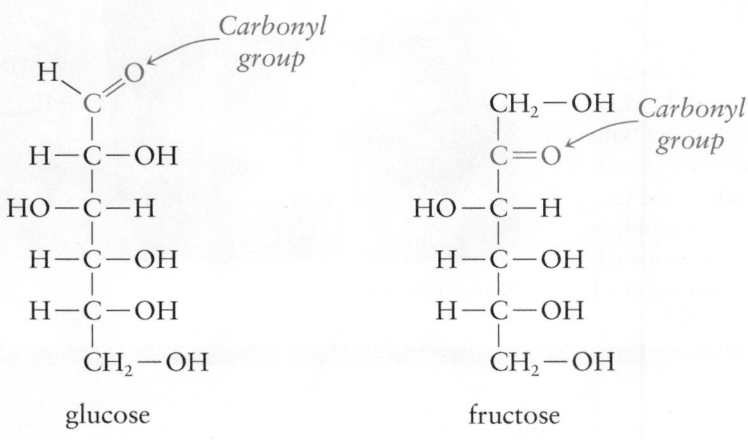

Thus, the naming of aldehydes and ketones, as well as some of their important reactions (see **Section 10.3**), figure prominently when we study the chemistry of carbohydrates in Chapter 11.

10.2 Properties of Aldehydes and Ketones

Learning Objective 4 Explain how intermolecular forces influence the physical properties of aldehydes and ketones.

The physical properties of aldehydes and ketones can be explained by examining their structures. First, the lack of a hydrogen on the oxygen prevents the formation of hydrogen bonds between molecules:

No H attached to the oxygen

$$\begin{array}{c} O \\ \| \\ R-C-H \end{array} \qquad \begin{array}{c} O \\ \| \\ R-C-R' \end{array}$$

Therefore, boiling points of pure aldehydes and ketones are expected to be lower than those of alcohols with similar molecular weights. Remember, alcohols can form hydrogen

Health Connections 10.1

Faking a Tan

Many people believe that a suntan makes them look healthy and attractive. Studies indicate, however, that sunbathing, especially when sunburn results, ages the skin prematurely and increases the risk of skin cancer. Cosmetic companies have developed a tanning alternative for those unwilling to risk using the sun but who want to be "fashionably" tan.

Tanning lotions and creams that chemically darken the skin are now available. The active ingredient in these "bronzers" is dihydroxyacetone (DHA), a colorless compound classified by the Food and Drug Administration as a safe skin dye.

$$\begin{array}{c} O \\ \| \\ HO-CH_2-C-CH_2-OH \end{array}$$

dihydroxyacetone (DHA)

Within several hours after application, DHA produces a brown skin color by reacting with the outer layer of the skin, which consists of dead cells. Only the dead cells react with DHA, so the color gradually fades as the dead cells slough off and are replaced. This process generally leads to the fading of chemical tans within a few weeks. Another problem with chemical tans is uneven skin color. Areas of skin such as the elbows and knees, which contain a thicker layer of dead cells, may absorb and react with more tanning lotion and become darker than other areas.

Perhaps the greatest problem with chemical tans is the false sense of security they might give. Some people with chemical tans think it is safe to go into the sun and get a deeper tan. This isn't true. Sunlight presents the same hazards to chemically tanned skin that it does to untanned skin.

© Maren Slabaugh

Some DHA-containing products.

bonds with one another. **Table 10.2** shows that the boiling points of propanal and acetone are 49 °C and 56 °C, respectively, whereas the alcohol of comparable molecular weight, 1-propanol, has a boiling point of 97 °C.

Table 10.2 A Comparison of Physical Properties

Class	Example	Formula	Molecular Weight (amu)	Boiling Point (°C)	Water Solubility
Alkane	butane	$CH_3CH_2CH_2CH_3$	58	0	Insoluble
Aldehyde	propanal	$CH_3CH_2-C(=O)-H$	58	49	Soluble
Ketone	propanone (acetone)	$CH_3-C(=O)-CH_3$	58	56	Soluble
Alcohol	1-propanol	$CH_3CH_2CH_2-OH$	60	97	Soluble

Also notice in **Table 10.2** that the boiling points of the aldehyde and ketone are higher than that of the comparable alkane. This can be explained by differences in polarity. Whereas alkanes are nonpolar, the carbonyl group of aldehydes and ketones is polar because the oxygen atom is more electronegative than the carbon atom:

$$\overset{\delta^-}{O}=\underset{\delta^+}{C}\!<$$

Water is a polar solvent and would be expected to be effective in dissolving compounds containing polar carbonyl groups. Besides exhibiting dipole–dipole attractions, the carbonyl oxygen also forms hydrogen bonds with water:

$$R-\underset{\delta^+}{C}(=\overset{\delta^-}{O}\cdots{}^{\delta^+}H-O-H)-H \qquad R-\underset{\delta^+}{C}(=\overset{\delta^-}{O}\cdots{}^{\delta^+}H-O-H)-R'$$

Hydrogen bond *Hydrogen bond*

As a result, low-molecular-weight aldehydes and ketones are water soluble (**Figure 10.8a**). As the length of the hydrocarbon chain (R and/or R′) increases, however, the water solubility of these aldehydes and ketones decreases because their hydrogen-bonding interactions with water become less and less important compared to the large amount of London dispersion forces between the alkane portions of the molecules (**Figure 10.8b**). And while the attractive forces between molecules possessing such dipoles are not as strong as hydrogen bonds, they do cause aldehydes and ketones to boil at higher temperatures than the nonpolar alkanes.

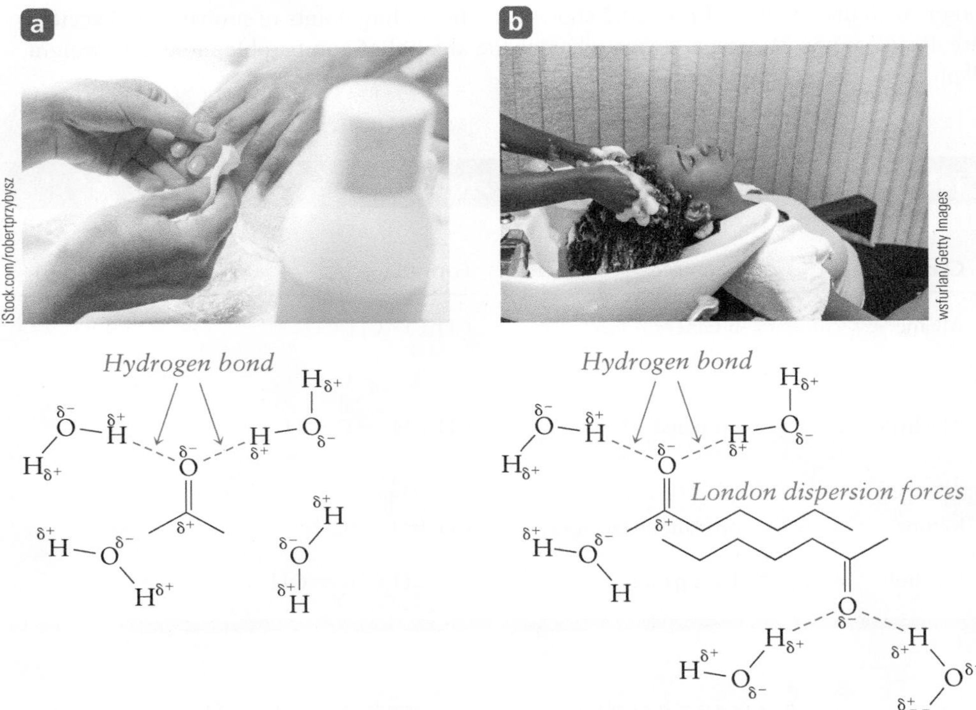

Figure 10.8 (a) Smaller aldehydes and ketones like acetaldehyde and acetone (used as nail polish remover) are water soluble. (b) Large aldehydes and ketones like 2-heptanone (which provides a fruity fragrance in shampoo) are insoluble in water.

10.3 Reactions of Aldehydes and Ketones

Learning Objective 5 Predict the products of the hydrogenation, oxidation, reduction, and hydration reactions of aldehydes and ketones.

10.3A Hydrogenation

In **Section 8.8**, we showed that addition reactions are common for compounds containing carbon–carbon double bonds. The same is true for the carbon–oxygen double bonds of carbonyl groups. An important example is the addition of H_2, known as hydrogenation, in the presence of catalysts such as nickel or platinum:

General reaction:

$$\underset{\text{aldehyde}}{R-\overset{\overset{\displaystyle O}{\|}}{C}-H} + H_2 \xrightarrow{Pt} \underset{\text{primary alcohol}}{R-\overset{\overset{\displaystyle OH}{|}}{\underset{\underset{\displaystyle H}{|}}{C}}-H} \qquad (10.1)$$

Specific example:

$$\underset{\text{propanal}}{CH_3CH_2-\overset{\overset{\displaystyle O}{\|}}{C}-H} + H_2 \xrightarrow{Pt} \underset{\text{1-propanol}}{CH_3CH_2\overset{\overset{\displaystyle OH}{|}}{C}H_2}$$

General reaction:

$$\underset{\text{ketone}}{R-\overset{\overset{\displaystyle O}{\|}}{C}-R'} + H_2 \xrightarrow{Pt} \underset{\text{secondary alcohol}}{R-\overset{\overset{\displaystyle OH}{|}}{\underset{\underset{\displaystyle H}{|}}{C}}-R'} \qquad (10.2)$$

Specific example:

$$\underset{\text{acetone}}{CH_3-\overset{\overset{\displaystyle O}{\|}}{C}-CH_3} + H_2 \xrightarrow{Pt} \underset{\text{2-propanol}}{CH_3-\overset{\overset{\displaystyle OH}{|}}{C}H-CH_3}$$

Notice that the addition of hydrogen to an aldehyde produces a primary alcohol (**Figure 10.9**), whereas the addition of hydrogen to a ketone gives a secondary alcohol.

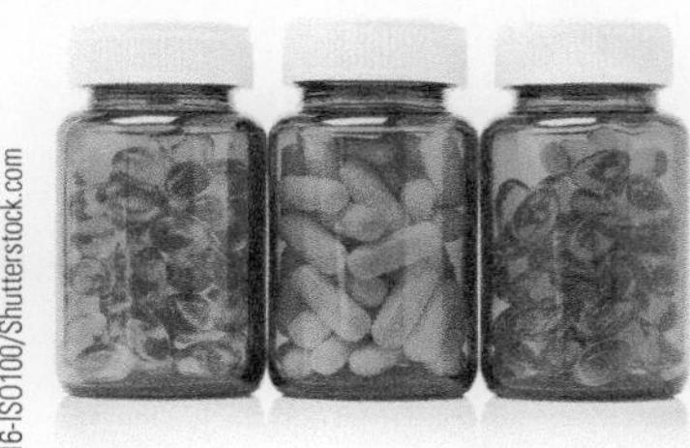
F16-ISO100/Shutterstock.com

Figure 10.9 1-Propanol, a primary alcohol, is used as an organic solvent in pharmaceutical laboratories during the synthesis of medicines.

Example 10.2 Writing Hydrogenation Reactions

Benzaldehyde, the main flavor component in almonds (**Figure 10.10**), can react with hydrogen. Show the formula for the product of the following reaction.

$$C_6H_5-\overset{O}{\overset{\|}{C}}-H + H_2 \xrightarrow{Pt}$$

Solution

Hydrogen adds across the carbon–oxygen double bond of an aldehyde to give a primary alcohol (see Equation 10.1).

$$C_6H_5-\overset{O}{\overset{\|}{C}}-H + H_2 \xrightarrow{Pt} C_6H_5-\overset{OH}{\overset{|}{CH_2}}$$

LookerStudio/Shutterstock.com

Figure 10.10 Benzaldehyde is used to provide artificial almond flavoring in pastries.

✔ Learning Check 10.2 Draw the structural formula for each product of the following reactions.

a. $CH_3CH_2-\overset{O}{\overset{\|}{C}}-CH_2CH_3 + H_2 \xrightarrow{Pt}$

b. $C_6H_5-CH_2-\overset{O}{\overset{\|}{C}}-H + H_2 \xrightarrow{Pt}$

10.3B Oxidation

Recall from **Section 9.3** that when aldehydes are prepared by oxidizing primary alcohols with $KMnO_4$ or $K_2Cr_2O_7$, the reaction may continue and produce carboxylic acids (Equation 10.3). Ketones are prepared similarly from secondary alcohols (Equation 10.4), but they are much less susceptible to further oxidation, especially by mild oxidizing agents:

General reaction:

$$\underset{\text{primary alcohol}}{R-CH_2-OH} \xrightarrow{(O)} \underset{\text{aldehyde}}{R-\overset{O}{\overset{\|}{C}}-H} \xrightarrow{(O)} \underset{\text{carboxylic acid}}{R-\overset{O}{\overset{\|}{C}}-OH} \qquad (10.3)$$

Specific example:

$$\underset{\text{benzaldehyde}}{C_6H_5-\overset{O}{\overset{\|}{C}}-H} + (O) \longrightarrow \underset{\text{benzoic acid}}{C_6H_5-\overset{O}{\overset{\|}{C}}-OH}$$

General reaction:

$$\mathrm{R{-}\underset{|}{\overset{OH}{CH}}{-}R' \xrightarrow{(O)} R{-}\overset{\overset{O}{\|}}{C}{-}R' \xrightarrow{(O)} \text{no reaction}} \qquad (10.4)$$

secondary alcohol — ketone

Specific example:

$$\mathrm{CH_3{-}\overset{\overset{O}{\|}}{C}{-}CH_3 + (O) \longrightarrow \text{no reaction}}$$

acetone

The difference in reactivity toward oxidation is the chief reason aldehydes and ketones are classified as separate functional groups.

The oxidation of aldehydes occurs so readily that even atmospheric oxygen causes the reaction. A few drops of liquid benzaldehyde placed on a watch glass will begin to oxidize to solid benzoic acid within an hour. In the laboratory, it is not unusual to find that bottles of aldehydes that have been opened for any length of time contain considerable amounts of carboxylic acids.

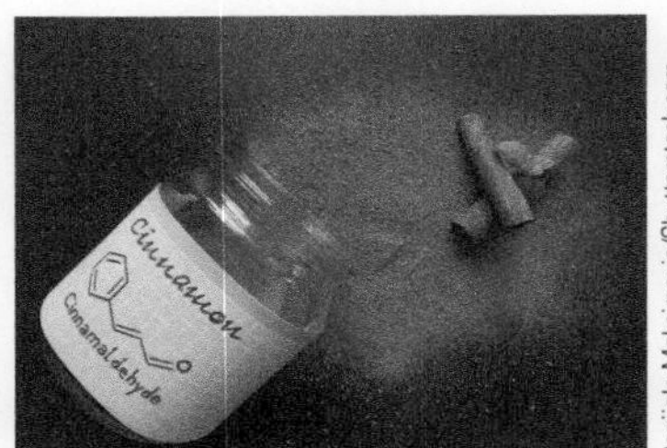

Danijela Maksimovic/Shutterstock.com

Figure 10.11 The taste and smell of cinnamon is due to the compound cinnamaldehyde, which contains three functional groups—namely an aldehyde, an alkene, and an aromatic.

Example 10.3 Writing Oxidation Reactions

Cinnamaldehyde (**Figure 10.11**) is the compound that gives cinnamon its characteristic aroma and taste. Draw the structural formula for the oxidation product of cinnamaldehyde.

$$\mathrm{C_6H_5{-}CH{=}CH{-}\overset{\overset{O}{\|}}{C}{-}H + (O) \longrightarrow}$$

Solution

This reaction is similar to Equation 10.3, in which an aldehyde is oxidized to a carboxylic acid.

$$\mathrm{C_6H_5{-}CH{=}CH{-}\overset{\overset{O}{\|}}{C}{-}H + (O) \longrightarrow C_6H_5{-}CH{=}CH{-}\overset{\overset{O}{\|}}{C}{-}OH}$$

✔ **Learning Check 10.3** Draw the structural formula for each product. Write "no reaction" if none occurs.

a. $\mathrm{CH_3\underset{}{\overset{\overset{CH_3}{|}}{C}H}CH_2{-}\overset{\overset{O}{\|}}{C}{-}H + (O) \longrightarrow}$

b. cyclopentanone $+ \mathrm{(O)} \longrightarrow$

10.3C Reduction

Reduction is the opposite of oxidation, meaning carbonyl compounds like aldehydes and ketones are converted back to their corresponding alcohol. While oxidation increases the number of bonds between carbon and oxygen, notice in Equations 10.5 and 10.6 that reduction increases the number of bonds between carbon and hydrogen. We use (R) to represent the reducing agent in these chemical equations.

General reaction: $R-\overset{\displaystyle O}{\overset{\|}{C}}-H \xrightarrow{(R)} R-\underset{\displaystyle H}{\overset{\displaystyle H}{C}}-OH$ (10.5)

Specific example: benzaldehyde $\xrightarrow{(R)}$ benzyl alcohol (**Figure 10.12**)

benzaldehyde

benzyl alcohol
(**Figure 10.12**)

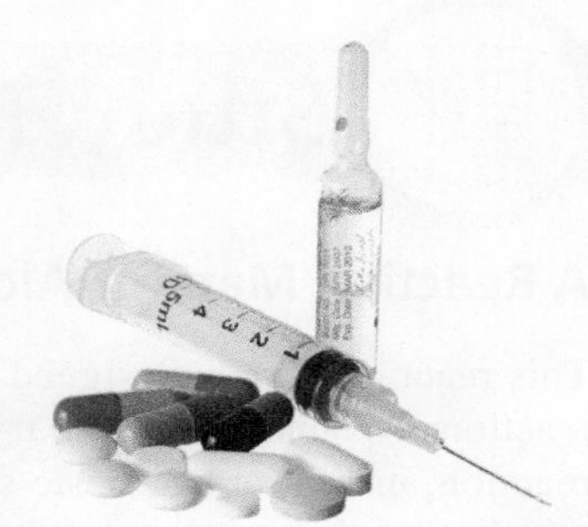
Jiad Durrani/Dreamstime

Figure 10.12 Benzyl alcohol has been used as a local anesthetic.

General reaction: $R-\overset{\displaystyle O}{\overset{\|}{C}}-R' \xrightarrow{(R)} R-\underset{\displaystyle H}{\overset{\displaystyle OH}{C}}-R'$ (10.6)

Specific example: $H_3C-\overset{\displaystyle O}{\overset{\|}{C}}-CH_3 \xrightarrow{(R)} H_3C-\underset{\displaystyle H}{\overset{\displaystyle OH}{C}}-CH_3$

acetone

2-propanol

The most effective reducing agents are $NaBH_4$ (sodium borohydride) and $LiAlH_4$ (lithium aluminum hydride). Less complex hydrogen-containing compounds, such as sodium hydride (NaH), are ineffective at carrying out these reductions. Tertiary alcohols ($R_3C–OH$) are impossible to synthesize by this reduction method because the carbonyl carbon of the ketone would have to have three alkyl groups bonded to it, which is impossible.

Example 10.4 Writing Reduction Reactions

Draw the structural formula for the reduction product of cyclohexanone (**Figure 10.13**).

cyclohexanone $\xrightarrow{(R)}$

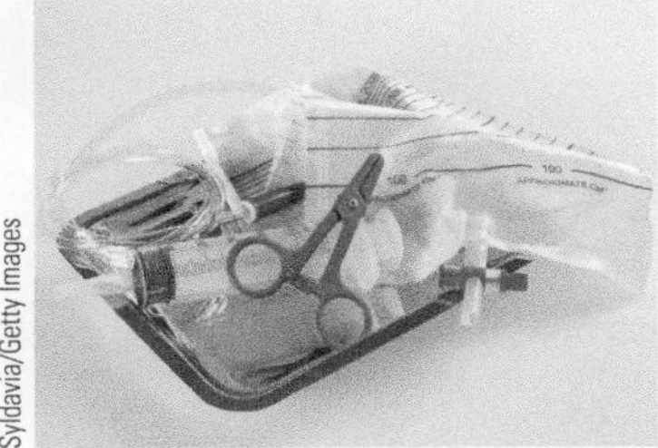
Syldavia/Getty Images

Figure 10.13 Cyclohexanone is often used as an adhesive in PVC (polyvinyl chloride) medical devices.

Solution

This reaction is a specific example of Equation 10.6, in which a ketone is reduced to a secondary alcohol.

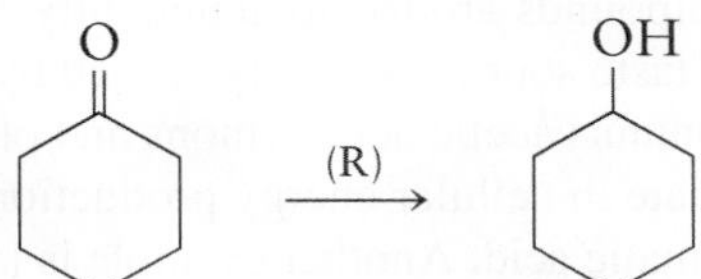

✔ Learning Check 10.4 Draw the structural formula for each product. Write "no reaction" if none occurs.

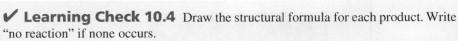

a. $H_3CH_2C-\overset{\displaystyle O}{\overset{\|}{C}}-CH_3 \xrightarrow{(R)}$

b. cyclopentyl$-\overset{\displaystyle O}{\overset{\|}{C}}-H \xrightarrow{(R)}$

Study Tools 10.1

A Reaction Map for Aldehydes and Ketones

This reaction map is designed to help you master organic reactions. Whenever you are trying to complete an organic reaction, use these two basic steps: (1) Identify the functional group that is to react and (2) identify the reagent that is to react with the functional group. If the reacting functional group is an aldehyde or a ketone, find the reagent in the summary diagram, and use the diagram to predict the correct products.

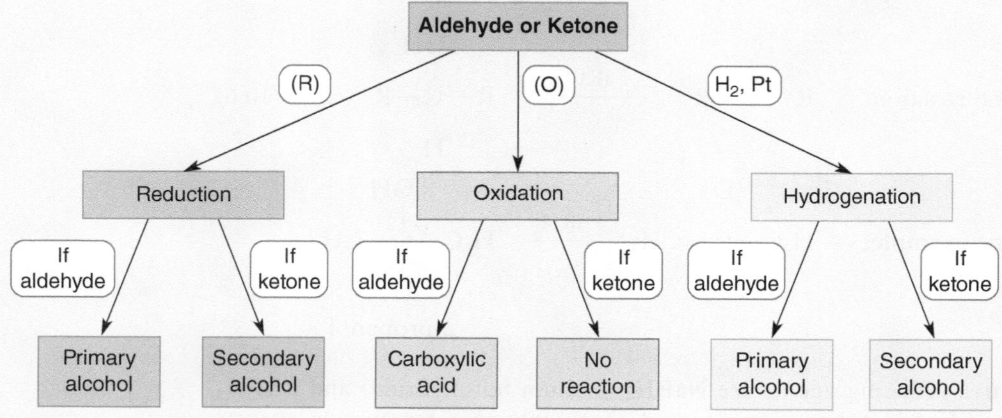

10.4 The Nomenclature of Carboxylic Acids and Carboxylate Salts

Learning Objective 6 Assign IUPAC names to carboxylic acids.

Learning Objective 7 Assign common and IUPAC names to carboxylate salts.

Learning Objective 8 Describe uses for carboxylate salts.

carboxylic acid An organic compound that contains the

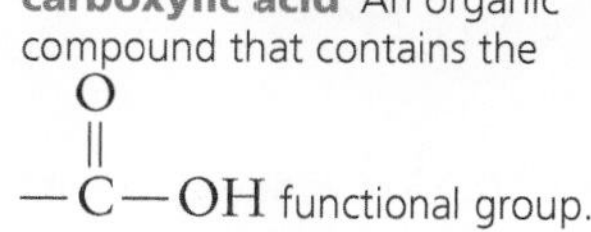

functional group.

The characteristic functional group of **carboxylic acids** is the carboxyl group:

$$\underset{\text{carboxyl group}}{-\overset{\overset{\displaystyle O}{\|}}{C}-OH} \qquad \underbrace{-COOH \quad \text{or} \quad -CO_2H}_{\text{abbreviated forms}}$$

Carboxyl-containing compounds are found abundantly in nature (see **Figure 10.14**). The tart flavor of foods that taste sour is generally caused by the presence of one or more carboxylic acids. Vinegar contains acetic acid; lemons and other citrus fruits contain citric acid, an important intermediate in cellular energy production (see **Figure 10.15**); and the tart taste of apples is due to malic acid**.** Another example is lactic acid, which is formed in muscle cells during vigorous exercise (see **Table 10.3**).

Mark Slabaugh

Figure 10.14 Rhubarb and spinach contain oxalic acid, a compound with two carboxyl groups. The nomenclature of dicarboxylic acids is not discussed here. What would you propose as the IUPAC ending for molecules with two carboxyl groups?

Mark Slabaugh

Figure 10.15 Citrus fruits and juices are rich in citric acid. What flavor sensation is characteristic of acids?

Table 10.3 Examples of Carboxylic Acids

Common Name	IUPAC Name	Structural Formula	Characteristics and Primary Uses
Formic acid	methanoic acid	$H-\overset{\overset{O}{\|\|}}{C}-OH$	Stinging agents of certain ants and nettles; used in food preservation
Acetic acid	ethanoic acid	$CH_3-\overset{\overset{O}{\|\|}}{C}-OH$	Active ingredient in vinegar; used in food preservation
Propionic acid	propanoic acid	$CH_3CH_2-\overset{\overset{O}{\|\|}}{C}-OH$	Used as mold inhibitor
Oxalic acid	ethanedioic acid	$HO-\overset{\overset{O}{\|\|}}{C}-\overset{\overset{O}{\|\|}}{C}-OH$	Present in leaves of some plants such as rhubarb and spinach; used as a cleaning agent for rust stains on fabric and porcelain
Citric acid	2-hydroxy-1,2,3-propane-tricarboxylic acid	$HO-\overset{\overset{O}{\|\|}}{C}-CH_2-\overset{\overset{OH}{\|}}{\underset{\underset{\underset{O}{\|\|}}{C-OH}}{\|}}{C}-CH_2-\overset{\overset{O}{\|\|}}{C}-OH$	Present in citrus fruits; used as a flavoring agent in foods; present in cells
Lactic acid	2-hydroxypropanoic acid	$CH_3\underset{\underset{OH}{\|}}{CH}-\overset{\overset{O}{\|\|}}{C}-OH$	Found in sour milk and sauerkraut; formed in muscles during exercise

Because of their abundance in nature, carboxylic acids were among the first organic compounds studied in detail. No systematic nomenclature system was then available, so the acids were usually named after some familiar source. Acetic acid,

$$CH_3-\overset{\overset{O}{\|\|}}{C}-OH$$

acetic acid

the sour constituent in vinegar (see **Figure 10.16**), is named after *acetum*, the Latin word for vinegar. Many of these common names are still widely used today (see **Table 10.3**). Carboxylic acids with long hydrocarbon chains, generally 12 to 20 carbon atoms, are called **fatty acids** (see **Figure 10.17** and Chapter 14) because they were first isolated from natural fats. These acids, such as stearic acid with 18 carbons, are known almost exclusively by their common names.

fatty acid A long-chain carboxylic acid found in fats.

In the IUPAC system, carboxylic acids are named using the following rules:

Step 1: *Number the longest chain containing the functional group* to give the lowest possible number to the carboxyl group carbon. Thus, in monocarboxylic acids, the carbonyl carbon is always at the end of the chain, and it is always designated as carbon number 1.

Step 2: *Name the longest chain* containing the carboxyl group. The final *-e* of the parent hydrocarbon chain is dropped, and the ending *-oic* is added, followed by the word *acid.*

Step 3: *Locate the position of the carboxyl group* by the number of the carbon atom to which the double-bonded oxygen is attached.

Step 4: *Locate and name any groups attached to the chain.*

Step 5: *Combine the name and location for other groups, the carboxyl group location, and the longest chain into the final name.*

Michael Neelon(misc)/Alamy stockphoto

Figure 10.16 Acetic acid is the active component in vinegar and a familiar laboratory weak acid.

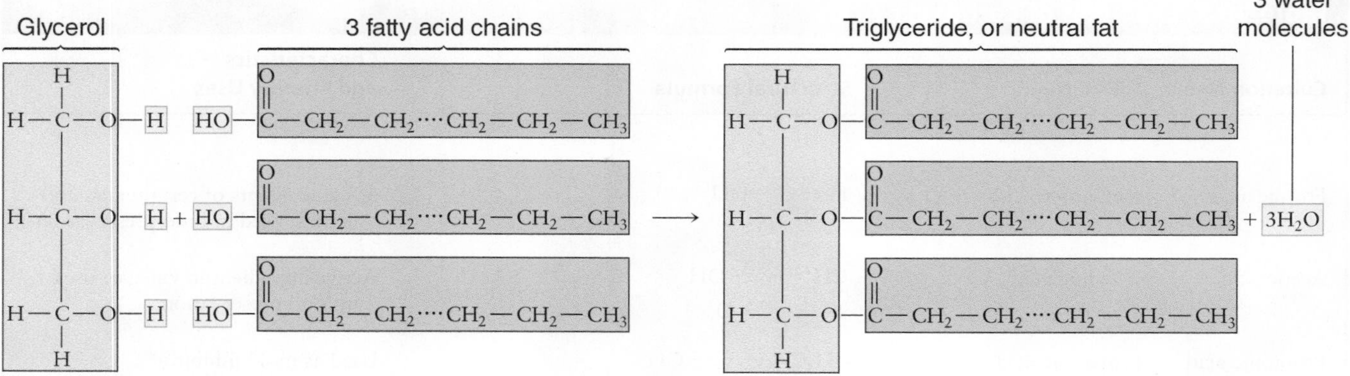

Figure 10.17 Our bodies store energy in several forms, one of which is triglycerides (fat). Triglycerides form when three fatty acid chains attach to a glycerol head group via a dehydration reaction (i.e., a reaction that releases water as a by-product).

Additionally aromatic acids are given names derived from the parent compound, benzoic acid (see **Figure 10.18**). Carboxylate salts are formed when the hydrogens of the –COOH group are removed and replaced by a metal cation (such as Na^+ or K^+). Both common and IUPAC names are assigned to carboxylate salts by naming the metal first and changing the *-ic* ending of the acid name to *-ate*.

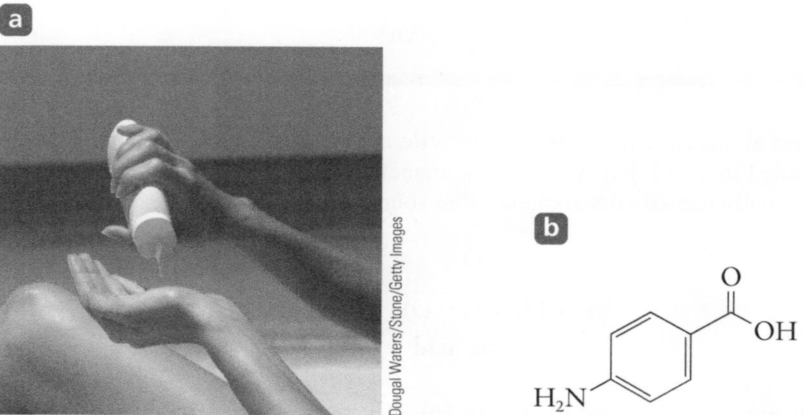

Figure 10.18 PABA (*p*-aminobenzoic acid) is a common ingredient in sunscreens. It absorbs ultraviolet (UV) radiation, protecting the skin from damage. (a) A person applying sunscreen. (b) Skeletal structure of PABA.

Example 10.5 Naming Carboxylic Acids and Carboxylate Salts

Use IUPAC rules to name the following carboxylic acids.

a. $CH_3CH_2-\overset{\overset{\large O}{\|}}{C}-OH$

b. $\underset{\underset{\large Br}{|}}{CH_3CH}CH_2-\overset{\overset{\large O}{\|}}{C}-OH$

c. $CH_3-\overset{\overset{\large O}{\|}}{C}-O^-Na^+$

Solution

a. **Step 1:** *Number the longest chain containing the functional group* to give the lowest possible number to the carboxyl group carbon.

$$\underset{3}{CH_3}\underset{2}{CH_2}-\underset{1}{\overset{\overset{\displaystyle O}{\|}}{C}}-OH$$

Step 2: *Name the longest chain* containing the carboxyl group. The carbon chain contains three carbons, so the parent hydrocarbon name is propane. Following the IUPAC rules, we must drop the *-e* and add *-oic acid*, giving the name propanoic acid.

Step 3: *Locate the position of the carboxyl group* by the number of the carbon atom to which the double-bonded oxygen is attached. The carboxyl group is at position 1.

Step 4: *Locate and name any groups attached to the chain.* There are no other groups attached the hydrocarbon chain.

Step 5: *Combine the name and location for other groups, the carboxyl group location, and the longest chain into the final name.* Because there are no additional groups, the final name is propanoic acid (**Figure 10.19**).

Aleksandar Malivuk/Shutterstock.com

Figure 10.19 Propanoic acid is used as a mold inhibitor in livestock feed.

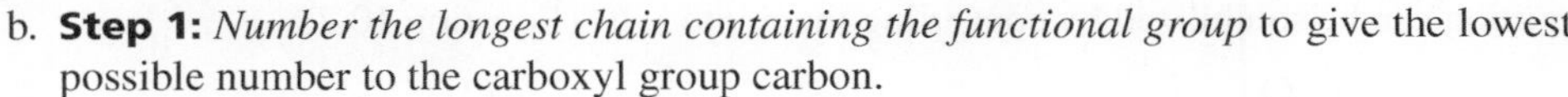

b. **Step 1:** *Number the longest chain containing the functional group* to give the lowest possible number to the carboxyl group carbon.

$$\underset{4}{CH_3}\underset{\underset{\displaystyle Br}{|3}}{CH}\underset{2}{CH_2}-\underset{1}{\overset{\overset{\displaystyle O}{\|}}{C}}-OH$$

Step 2: *Name the longest chain* containing the carboxyl group. The carbon chain contains four carbons, so the parent hydrocarbon name is butane. Following the IUPAC rules, we must drop the *-e* and add *-oic acid*, giving the name butanoic acid.

Step 3: *Locate the position of the carboxyl group* by the number of the carbon atom to which double-bonded oxygen is attached. The carboxyl group is at position 1.

Step 4: *Locate and name any groups attached to the chain.* There is a bromine attached to carbon number 3.

Step 5: *Combine the name and location for other groups, the carboxyl group location, and the longest chain into the final name.* The name is 3-bromobutanoic acid.

c. **Step 1:** *Number the longest chain containing the functional group* to give the lowest possible number to the carboxyl group carbon.

$$\underset{2}{CH_3}-\underset{1}{\overset{\overset{\displaystyle O}{\|}}{C}}-O^{-}Na^{+}$$

Step 2: *Name the longest chain* containing the carboxyl group. The carbon chain contains two carbons and following the IUPAC rules, we must drop the *-e* and add *-oic acid*, giving the name ethanoic acid. Alternatively, the common name for a carboxylic acid containing two carbons is acetic acid.

Step 3: *Locate the position of the carboxyl group* by the number of the carbon atom to which the double-bonded oxygen is attached. The carboxyl group is at position 1.

Step 4: *Locate and name any groups attached to the chain.* There are no groups attached to the chain.

Step 5: *Combine the name and location for other groups, the carboxyl group location, and the longest chain into the final name.* Because this compound is a carboxylate salt formed from ethanoic acid (or acetic acid), we must name the metal first and change the *-ic* ending of the acid name to *-ate*. Thus, the salt is sodium acetate (common name) or sodium ethanoate (IUPAC name).

✔ Learning Check 10.5 Give the IUPAC name of the following.

a. $CH_3CH_2CHCH_3$ with $-COOH$ on C2 (structure: $\overset{O}{\overset{\|}{C}}-OH$ attached to CH)

b. Benzene ring bearing $\overset{O}{\overset{\|}{C}}-OH$, with Br substituents at positions 3 and 4

c. Benzene ring bearing $\overset{O}{\overset{\|}{C}}-O^-K^+$

Carboxylic acid salts are solids at room temperature and are usually soluble in water because they are ionic. Even long-chain acids with an extensive nonpolar hydrocarbon portion can be solubilized by converting them into salts (Equation 10.7):

$$\underset{\substack{\text{stearic acid}\\ \text{(insoluble)}}}{CH_3(CH_2)_{16}-\overset{O}{\overset{\|}{C}}-O-H} + NaOH \longrightarrow \underset{\substack{\text{sodium stearate}\\ \text{(soluble)}}}{CH_3(CH_2)_{16}-\overset{O}{\overset{\|}{C}}-O^-Na^+} + H_2O \quad (10.7)$$

A number of acid salts are important around the home. Sodium stearate $[CH_3(CH_2)_{16}COO^-Na^+]$ and other sodium and potassium salts of long-chain carboxylic acids are used as soaps (see **Figure 10.20**). Calcium and sodium propanoate are used commercially as preservatives in bread, cakes, and cheese to prevent the growth of bacteria and molds (see **Figure 10.21**). The parent acid, CH_3CH_2COOH, occurs naturally in Swiss cheese. The labels for these bakery products often contain the common name propionate, rather than the IUPAC-acceptable name propanoate.

Cengage Learning/Charles D. Winters

Figure 10.20 Soap products enable us to keep ourselves and our homes clean.

Michael C. Slabaugh

Figure 10.21 Propanoates extend the shelf life of bread, preventing the formation of mold.

Sodium benzoate, another common food preservative, occurs naturally in many foods, especially cranberries and prunes. It is used in bakery products, ketchup, carbonated beverages, and a host of other foods:

$$\underset{\text{sodium benzoate}}{C_6H_5-\overset{O}{\overset{\|}{C}}-O^-Na^+}$$

Zinc 10-undecylenate, the zinc salt of $CH_2{=}CH(CH_2)_8COOH$, is commonly used to treat athlete's foot (**Figure 10.22**).

A mixture of sodium citrate and citric acid

$$HO-\overset{\overset{\displaystyle O}{\|}}{C}-CH_2-\overset{\overset{\displaystyle OH}{|}}{\underset{\underset{\underset{\displaystyle O}{\|}}{C-OH}}{C}}-CH_2-\overset{\overset{\displaystyle O}{\|}}{C}-OH$$

citric acid

$$Na^{+}\,{}^{-}O-\overset{\overset{\displaystyle O}{\|}}{C}-CH_2-\overset{\overset{\displaystyle OH}{|}}{\underset{\underset{\underset{\displaystyle O}{\|}}{C-O^{-}Na^{+}}}{C}}-CH_2-\overset{\overset{\displaystyle O}{\|}}{C}-O^{-}Na^{+}$$

sodium citrate

is widely used as a buffer to control pH. Products sold as foams or gels such as jelly, ice cream, candy, and whipped cream maintain their desirable characteristics only at certain pH values, which can be controlled by the citrate/citric acid buffer. This same buffer is used in medicines and in human blood that is used for transfusions. In blood, it also functions as an anticoagulant (**Figure 10.23**).

bymuratdeniz/Getty Images

Figure 10.22 Zinc 10-undecylenate is used as a treatment for athlete's foot.

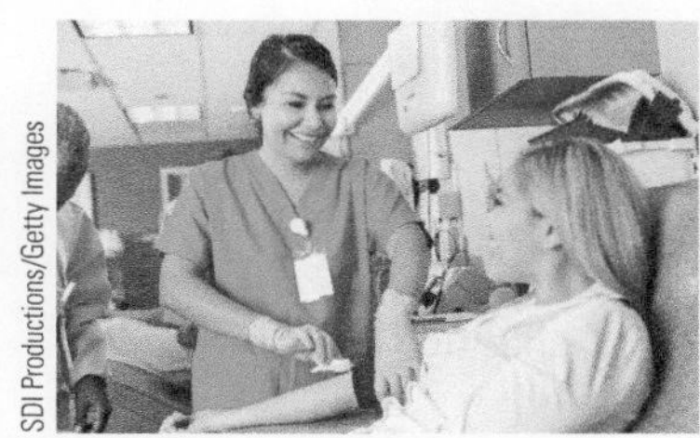

SDI Productions/Getty Images

Figure 10.23 Citrate/citric acid buffers are used as an anticoagulant and also to regulate the pH in blood transfusions.

10.5 Properties of Carboxylic Acids

Learning Objective 9 Explain how hydrogen bonding affects the physical properties of carboxylic acids.

Learning Objective 10 Write key acid–base reactions of carboxylic acids.

10.5A Physical Properties of Carboxylic Acids

Carboxylic acids with low molecular weights are liquids at room temperature and have characteristically sharp or unpleasant odors (see **Table 10.4**). Butyric acid, for example, is a component of perspiration and is partially responsible for the odor of locker rooms and unwashed socks. As the molecular weight of carboxylic acids increases, so does the boiling point. The heavier acids (containing more than 10 carbons) are waxlike solids. Stearic acid ($C_{18}H_{36}O_2$), for example, is mixed with paraffin to produce a wax used to make candles.

Table 10.4 Physical Properties of Some Carboxylic Acids

Common Name	Structural Formula	Boiling Point (°C)	Melting Point (°C)	Solubility (g/100 mL H_2O)
Formic acid	$H{-}COOH$	101	8	Infinite
Acetic acid	$CH_3{-}COOH$	118	17	Infinite
Butyric acid	$CH_3{-}(CH_2)_2{-}COOH$	164	−5	Infinite
Valeric acid	$CH_3{-}(CH_2)_3{-}COOH$	186	−34	5
Caproic acid	$CH_3{-}(CH_2)_4{-}COOH$	205	−3	1
Lauric acid	$CH_3{-}(CH_2)_{10}{-}COOH$	299	44	Insoluble
Myristic acid	$CH_3{-}(CH_2)_{12}{-}COOH$	Decomposes	58	Insoluble
Palmitic acid	$CH_3{-}(CH_2)_{14}{-}COOH$	Decomposes	63	Insoluble
Stearic acid	$CH_3{-}(CH_2)_{16}{-}COOH$	Decomposes	71	Insoluble

When boiling points of compounds with similar molecular weights are compared, the carboxylic acids have the highest boiling points of any organic compounds we have studied so far. See **Table 10.5** and **Figure 10.24**.

Table 10.5 Boiling Points of Compounds with Similar Molecular Weight

Class	Compound	Molecular Weight (amu)	Boiling Point (°C)
Alkane	pentane	72	35
Ether	diethyl ether	74	35
Aldehyde	butanal	72	76
Alcohol	1-butanol	74	118
Carboxylic acid	propanoic acid	74	141

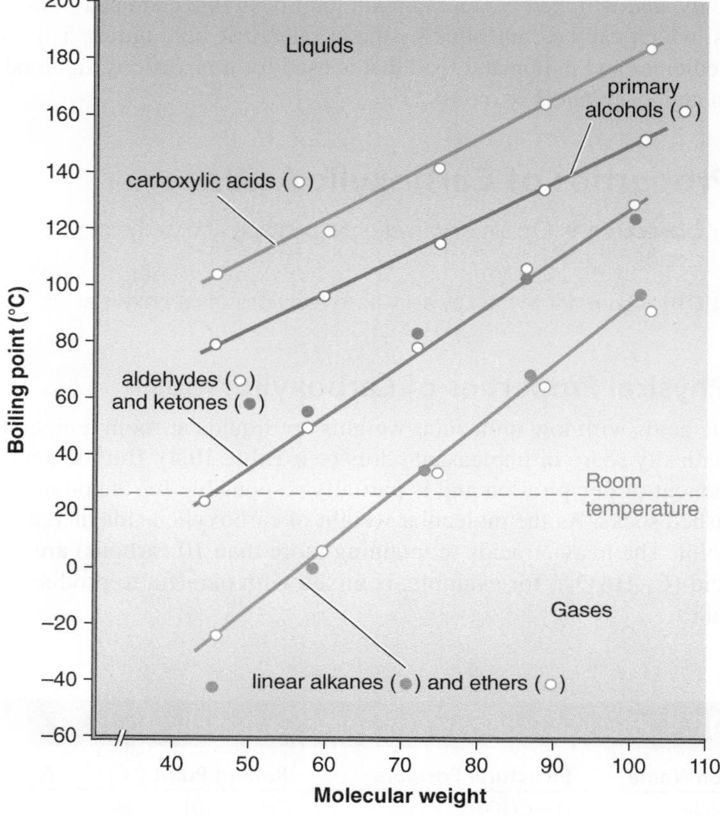

Figure 10.24 The boiling points of carboxylic acids are higher than those of all the other organic compounds studied thus far.

For simple nonaromatic compounds, the boiling points usually increase in the following order: hydrocarbons and ethers < aldehydes and ketones < alcohols < carboxylic acids. Carboxylic acids have such high boiling points because they, like alcohols, can form intermolecular hydrogen bonds. Acids have higher boiling points than alcohols with similar molecular weights because hydrogen-bonded acids are held together by two hydrogen bonds rather than the one hydrogen bond that is characteristic of alcohols. A structure where two identical molecules are joined together is called a **dimer**:

dimer Two identical molecules bonded together.

$$
\begin{array}{c}
\delta^- \quad \delta^+ \quad \delta^- \\
\mathrm{O}\cdots\mathrm{H}-\mathrm{O} \\
\mathrm{R}-\mathrm{C}^{\delta+} \qquad\qquad {}^{\delta+}\mathrm{C}-\mathrm{R} \\
\mathrm{O}-\mathrm{H}\cdots\mathrm{O} \\
\delta^- \quad \delta^+ \quad \delta^-
\end{array}
$$

Low-molecular-weight acids are very soluble in water (see **Table 10.4**) because of hydrogen bonds that form between the carboxyl group and water molecules (these are intermolecular hydrogen bonds; see margin to the right). As observed with aldehyde and ketones, the water solubility of carboxylic acids decreases as the length of the nonpolar hydrocarbon portion (the R group) of the carboxylic acid increases. Carboxylic acids containing eight or more carbon atoms are not appreciably soluble in water and are considered to be insoluble.

Organic compounds may generally be arranged according to increasing water solubility in the following order: hydrocarbons < ethers < aldehydes and ketones < alcohols < carboxylic acids. Because forces resulting from hydrogen bonding play important roles in determining both boiling points and water solubility, it should not be surprising that the same order is generally observed for these two properties.

$$\underset{\delta+}{\mathrm{R-C}}(\overset{\delta-}{=\mathrm{O}}\cdots\overset{\delta+}{\mathrm{H}}-\mathrm{O}^{\delta-}\mathrm{H}^{\delta+})-\overset{\delta-}{\mathrm{O}}-\overset{\delta+}{\mathrm{H}}\cdots\mathrm{O}^{\delta-}(\mathrm{H}^{\delta+})_2$$

carboxylate ion The $\mathrm{R-\overset{O}{\overset{\|}{C}}-O^-}$ ion that results from the dissociation of a carboxylic acid.

10.5B Chemical Properties of Carboxylic Acids

The most important chemical property of carboxylic acids is the acidic behavior implied by their name. The hydrogen attached to the oxygen of the carboxyl group gives carboxylic acids their acidic character (see **Figure 10.25**). In water, this proton may leave the acid, converting it into a **carboxylate ion**. Equation 10.8 gives the general reaction, and the reaction of acetic acid with water gives a specific example:

Clarence Holmes Wildlife/Alamy Stock Photo

Figure 10.25 As a defensive action, formicine ants can spray formic acid (methanoic acid, HCOOH), which causes skin irritation because of its acidic properties.

General reaction:

$$\underset{\text{carboxylic acid}}{\mathrm{R-\overset{O}{\overset{\|}{C}}-O-H}} + \mathrm{H_2O} \rightleftharpoons \underset{\substack{\text{carboxylate}\\ \text{ion}}}{\mathrm{R-\overset{O}{\overset{\|}{C}}-O^-}} + \underset{\substack{\text{hydronium}\\ \text{ion}}}{\mathrm{H_3O^+}} \qquad (10.8)$$

Specific example:

$$\underset{\text{acetic acid}}{\mathrm{CH_3-\overset{O}{\overset{\|}{C}}-O-H}} + \mathrm{H_2O} \rightleftharpoons \underset{\text{acetate ion}}{\mathrm{CH_3-\overset{O}{\overset{\|}{C}}-O^-}} + \mathrm{H_3O^+}$$

Generally, carboxylic acids behave as weak acids (see **Figure 10.26**). A 1 M solution of acetic acid, for example, is only about 0.5% dissociated. Moreover, the reaction of carboxylic acids with water (Equation 10.8) is reversible. According to Le Châtelier's principle, the addition of H_3O^+ (low pH) should favor formation of the carboxylic acid, whereas removal of H_3O^+ by adding base (high pH) should favor formation of the carboxylate ion:

$$\underset{\substack{\text{carboxylic}\\ \text{acid}}}{\mathrm{R-\overset{O}{\overset{\|}{C}}-O-H}} + \mathrm{H_2O} \underset{\substack{\text{Acidic conditions}\\ \text{(low pH)}}}{\overset{\substack{\text{Basic conditions}\\ \text{(high pH)}}}{\rightleftharpoons}} \underset{\substack{\text{carboxylate}\\ \text{ion}}}{\mathrm{R-\overset{O}{\overset{\|}{C}}-O^-}} + \mathrm{H_3O^+}$$

Thus, the pH of a solution determines the form in which a carboxylic acid exists in the solution. At pH 7.4, the normal pH of body fluids, the carboxylate form predominates. For example, citric acid in body fluids is often referred to as citrate. This is an important point for the upcoming chapters on biochemistry.

a

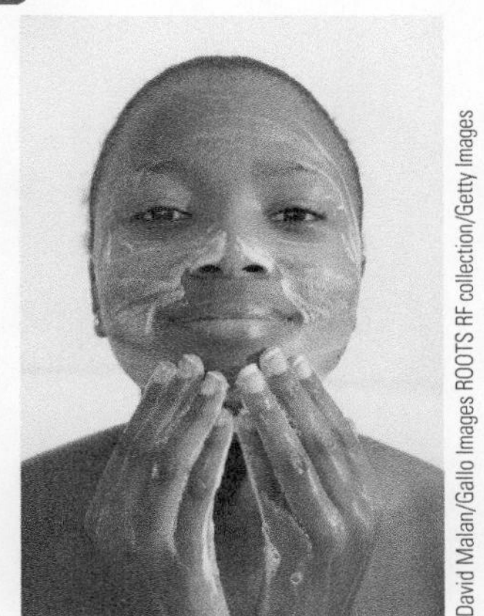
David Malan/Gallo Images ROOTS RF collection/Getty Images

b

HO–CH₂–C(=O)–OH

Figure 10.26 A common component of facial cleansers is glycolic acid, which helps shed dead skin cells, revealing the newer, brighter layers underneath. (a) Woman washing face with facial cleanser. (b) Skeletal structure of glycolic acid.

Although they are weak, carboxylic acids react readily with strong bases such as sodium hydroxide or potassium hydroxide to form salts. Equation 10.9 gives the general reaction, and the reaction of acetic acid with sodium hydroxide gives a specific example:

General reaction:

$$\underset{\text{carboxylic acid}}{R-\overset{\overset{\Large O}{\|}}{C}-O-H} + NaOH \rightleftarrows \underset{\text{sodium carboxylate}}{R-\overset{\overset{\Large O}{\|}}{C}-O^{-}Na^{+}} + H_2O \quad (10.9)$$

Specific examples:

$$\underset{\text{acetic acid}}{CH_3-\overset{\overset{\Large O}{\|}}{C}-O-H} + NaOH \rightleftarrows \underset{\text{sodium acetate}}{CH_3-\overset{\overset{\Large O}{\|}}{C}-O^{-}Na^{+}} + H_2O$$

Example 10.6 Writing Acid–Base Reactions of Carboxylic Acids

Cellular respiration is the process by which humans utilize oxygen to produce ATP (energy). A by-product of this process is carbonic acid (**Figure 10.27**), which reacts with water and in turn helps regulates oxygen levels in the blood.

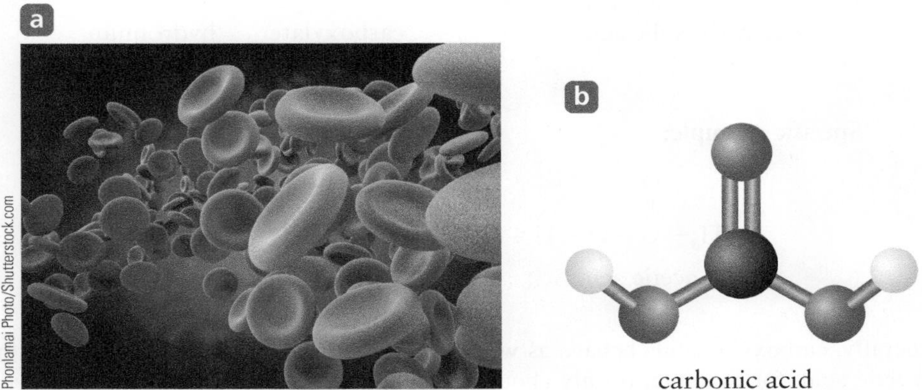

Figure 10.27 Carbonic acid is part of a buffer system that regulates the binding of oxygen to the hemoglobin protein in red blood cells. (a) Red blood cells, responsible for transporting oxygen throughout our bodies. (b) A ball-and-stick model of carbonic acid.

Predict the products of a reaction of carbonic acid and water.

$$\underset{\text{carbonic acid}}{HO-\overset{\overset{\Large O}{\|}}{C}-OH} + H_2O \rightleftarrows$$

Solution

Using Equation 10.8 as a guide:

$$HO-\overset{\overset{\Large O}{\|}}{C}-OH + H_2O \rightleftarrows \underset{\text{bicarbonate ion}}{HO-\overset{\overset{\Large O}{\|}}{C}-O^{-}} + H_3O^{+}$$

✔ Learning Check 10.6 Use Equation 10.9 to write the structural formulas for the products of the following reaction.

$$CH_3CH_2CH_2-\overset{\overset{\displaystyle O}{\|}}{C}-OH + NaOH \rightleftharpoons$$

10.6 The Nomenclature of Esters and Amides

Learning Objective 11 Assign common and IUPAC names to esters and amides.

Replacing the hydrogen atom of a carboxylic acid with another alkyl group forms a new carbonyl functional group called the **carboxylic esters**, or **esters** for short. Widely found in fruits and flowers, many esters are very fragrant and represent some of nature's most pleasant odors. Because of this characteristic, esters are commonly used as flavoring agents in foods and as scents in personal products (see **Figure 10.28**). The nitrogen analogs to esters are the **amides**. The characteristic functional group of amides is a carbonyl group attached to nitrogen. The single bond linking the carbonyl carbon and nitrogen atoms in the group is called an amide linkage:

$$-\overset{\overset{\displaystyle O}{\|}}{C}-\underset{|}{N}- \quad \leftarrow \textit{Amide linkage}$$

amide functional group

carboxylic ester or ester An organic compound having the functional group

$$-\overset{\overset{\displaystyle O}{\|}}{C}-OR.$$

amide An organic compound having the functional group

$$-\overset{\overset{\displaystyle O}{\|}}{C}-\underset{|}{N}-.$$

Esters are named by a common system as well as by IUPAC rules. The common names are used most often. It is helpful with both IUPAC and common names to think of the esters as being formed from a carboxylic acid and an alcohol:

$$R-\overset{\overset{\displaystyle O}{\|}}{C}-OH + H-OR' \rightleftharpoons \underbrace{R-\overset{\overset{\displaystyle O}{\|}}{C}}_{\text{Acid part}}-\overset{\text{Alcohol part}}{OR'} + H_2O$$

Step 1: *Name the alcohol group.* The first word of the name of an ester is the name of the alkyl or aromatic group (R′) contributed by the alcohol.

Step 2: *Name the carboxylate.* The second word is the carboxylic acid name, with the *-ic acid* ending changed to *-ate*. Thus, an ester of acetic acid becomes an acetate, one of butyric acid becomes a butyrate, one of lactic acid becomes a lactate, and so on. What, then, is the name of the following ester?

$$\underbrace{CH_3-\overset{\overset{\displaystyle O}{\|}}{C}}_{\substack{\textit{acetate (common name)} \\ \text{ethanoate (IUPAC name)}}}-O-\overset{\textit{Methyl}}{CH_3}$$

Step 3: *Combine the alcohol group and carboxylate into a final name.* The first word is methyl, which refers to the alcohol portion. The second word is derived from the name of the two-carbon carboxylic acid (acetic acid or ethanoic acid). Thus, we have methyl acetate (common name) and methyl ethanoate (IUPAC name).

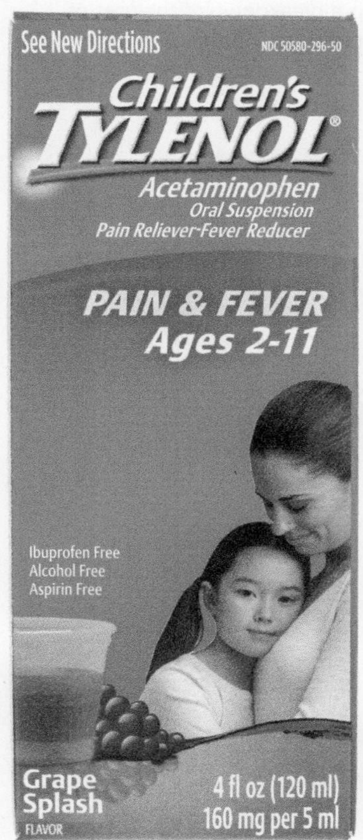

Keith Homan/Shutterstock.com

Figure 10.28 Esters are common in familiar flavorings. Grape flavoring, the ester methyl anthranilate, is added to children's Tylenol to make it tastier.

Similarly, simple amides are named after the corresponding carboxylic acids, using the following steps:

Step 1: *Name the carboxylic acid portion* then change the *-ic* ending (common names) or the *-oic* ending (IUPAC names) of the acid to *-amide*. Common names are used more often than IUPAC names. Some examples, such as acetamide (**Figure 10.29**), are as follows:

$$H-\overset{\overset{\displaystyle O}{\|}}{C}-OH \longrightarrow H-\overset{\overset{\displaystyle O}{\|}}{C}-NH_2$$

Common name	formic acid	formamide
IUPAC name	methanoic acid	methanamide

$$CH_3-\overset{\overset{\displaystyle O}{\|}}{C}-OH \longrightarrow CH_3-\overset{\overset{\displaystyle O}{\|}}{C}-NH_2$$

Common name	acetic acid	acetamide
IUPAC name	ethanoic acid	ethanamide

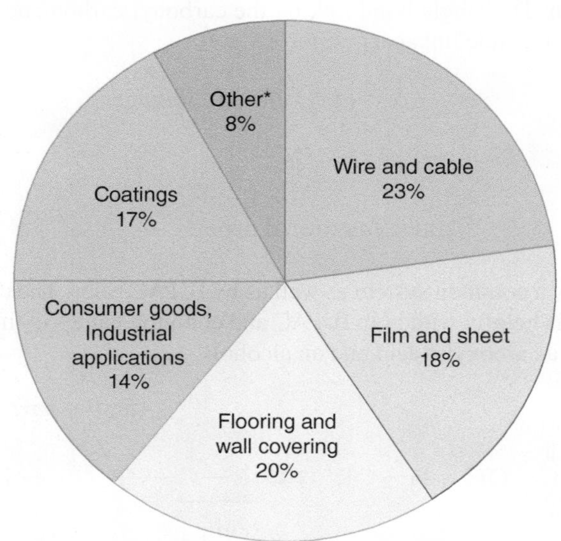

Figure 10.29 Acetamide can be added to materials to make them softer and more flexible. Substances such as these are known as plasticizers. Shown here is a breakdown of the uses of plasticizers in the United Kingdom.

Step 2: *Name alkyl groups attached to the nitrogen.* If alkyl groups are attached to the nitrogen atom of amides (substituted amides), the name of the alkyl group precedes the name of the amide, and an italic capital *N* is used to indicate that the alkyl group is bonded to the nitrogen. This style of naming is also used, sometimes, with amines.

Step 3: *Combine the carboxylic acid portion and any alkyl groups attached into a final name*, as shown in the following examples:

$$H-\overset{\overset{\displaystyle O}{\|}}{C}-NH-CH_3 \qquad CH_3-\overset{\overset{\displaystyle O}{\|}}{C}-\underset{\underset{\displaystyle CH_3}{|}}{N}-CH_3 \qquad C_6H_5-\overset{\overset{\displaystyle O}{\|}}{C}-\underset{\underset{\displaystyle CH_3}{|}}{N}-CH_2CH_3$$

Common name	*N*-methylformamide	*N*,*N*-dimethylacetamide	*N*-ethyl-*N*-methylbenzamide
IUPAC name	*N*-methylmethanamide	*N*,*N*-dimethylethanamide	*N*-ethyl-*N*-methylbenzamide

The most biologically relevant application of amides is arguably the **peptide bond** (which is actually an amide bond), a covalent bond that forms between amino acids, the building blocks of proteins (**Figure 10.30**). Proteins are one of the classes of biomolecules that we examine in great detail in Chapter 12.

peptide bond The amide linkage between amino acids that results when the amino group of one acid reacts with the carboxylate group of another.

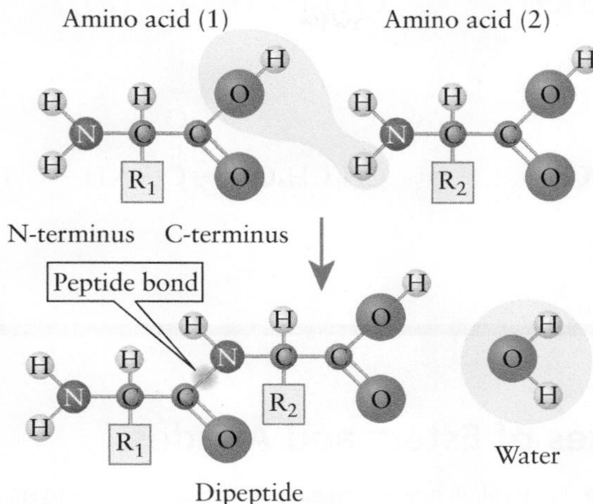

Figure 10.30 A peptide bond is formed via a dehydration reaction between two amino acids. The carboxylate group of one amino acid reacts with the amino group of another amino acid. The end result is long chains of amino acids that ultimately make up proteins.

Example 10.7 Naming Esters and Amides

Give common and IUPAC names for the following ester and amide.

a. $CH_3CH_2CH_2-\overset{\overset{\large O}{\|}}{C}-O-C_6H_5$ (phenyl ring)

b. $CH_3CH_2-\overset{\overset{\large O}{\|}}{C}-NH_2$

Solution

a. **Step 1:** *Name the alcohol group.* The alcohol portion is a phenyl group.

Step 2: *Name the carboxylate.* The carboxylate portion is four carbons long, so it is butanoate (or butyrate).

Step 3: *Combine the alcohol group and carboxylate into a final name.* The two names are phenyl butyrate (common name) and phenyl butanoate (IUPAC name).

$$\underbrace{CH_3CH_2CH_2}-\overset{\overset{\large O}{\|}}{C}-O-C_6H_5$$

Phenyl

phenyl butyrate (common name)
phenyl butanoate (IUPAC name)

b. **Step 1:** *Name the carboxylic acid portion* then change the *-ic* ending (common names) or the *-oic* ending (IUPAC names) of the acid to *-amide*. The main chain is three carbons long, so *propanoic acid* becomes *propanamide*.

$$\underset{3}{C}H_3\underset{2}{C}H_2-\underset{1}{\overset{\overset{\large O}{\|}}{C}}-NH_2$$

Step 2: *Name alkyl groups attached to the nitrogen.* There are no alkyl groups attached to the nitrogen.

Step 3: *Combine the carboxylic acid portion and any alkyl groups attached into a final name.* The final common and IUPAC names are the same: propanamide.

✔ Learning Check 10.7 Give both the common and the IUPAC names for each of the following esters and amides.

a. $H{-}\overset{\overset{\large O}{\|}}{C}{-}O{-}\underset{\underset{\large CH_3}{|}}{C}HCH_3$

b. $C_6H_5{-}\overset{\overset{\large O}{\|}}{C}{-}OCH_3$

c. $CH_3CH_2CH_2{-}\overset{\overset{\large O}{\|}}{C}{-}NH_2$

d. $CH_3CH_2CH_2{-}\overset{\overset{\large O}{\|}}{C}{-}NH{-}CH_3$

10.7 Properties of Esters and Amides

Learning Objective 12 Explain how hydrogen bonding affects the physical properties of esters and amides.

Because esters have replaced the hydrogen atom of carboxylic acids with a hydrocarbon chain, esters are incapable of intermolecular hydrogen bonding. As a result, esters are generally less water soluble and have a lower boiling point than carboxylic acids of similar molecular weight.

Amides, on the other hand, can have 0, 1, or 2 hydrogen atoms attached to the nitrogen. Unsubstituted amides (i.e., those with two hydrogens) can form a complex network of intermolecular hydrogen bonds similar to carboxylic acids (see **Figure 10.31a**). It is this characteristic that causes the melting points of these substances to be relatively high. For example, formamide is a liquid at room temperature, whereas all other unsubstituted amides are solids.

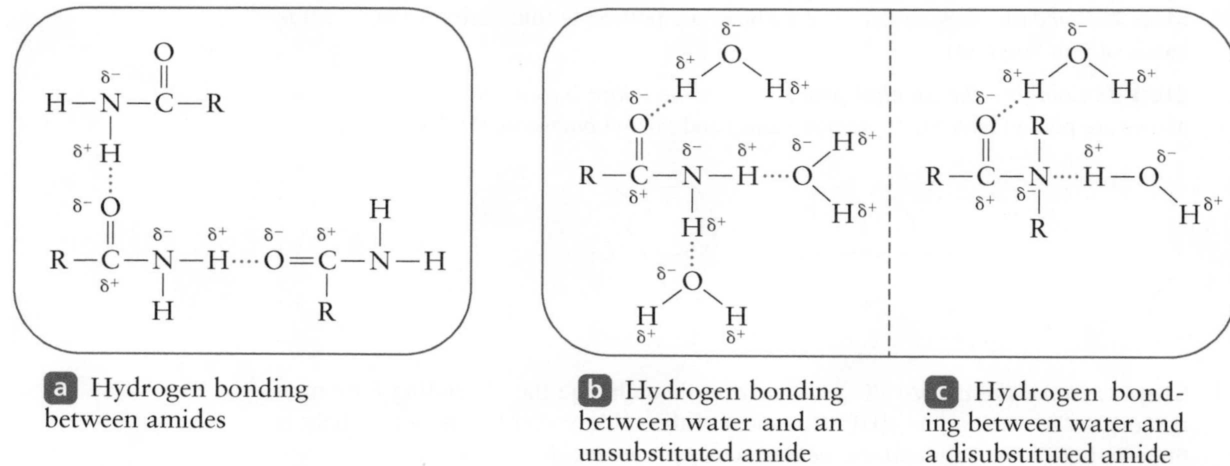

a Hydrogen bonding between amides

b Hydrogen bonding between water and an unsubstituted amide

c Hydrogen bonding between water and a disubstituted amide

Figure 10.31 Intermolecular hydrogen bonding between (a) amides, (b) water and an unsubstituted amide, and (c) water and a disubstituted amide. The nitrogen atom of the unsubstituted amide in part b can also form a hydrogen bond to a hydrogen atom of water (analogous to the nitrogen of the disubstituted amide in part c).

The substitution of alkyl or aromatic groups for hydrogen atoms on the nitrogen reduces the number of intermolecular hydrogen bonds that can form and causes the melting points to decrease. Thus, disubstituted amides often have lower melting points and boiling points than monosubstituted and unsubstituted amides. We show in Chapter 12 that hydrogen bonding between amide groups is important in maintaining the shape of protein molecules.

Amides are rather water soluble, especially those containing fewer than six carbon atoms. This solubility results from the ability of amides to form hydrogen bonds with water (**Figure 10.31b**). Even disubstituted amides can form hydrogen bonds with water (see **Figure 10.31c**).

Example 10.8 Showing Hydrogen Bonding

Phenacetin (**Figure 10.32**), a pain-relieving drug removed from medicinal use in the 1970s because it can cause severe kidney damage, has the following structure. Show how phenacetin forms hydrogen bonds with water molecules.

$$
\begin{array}{c}
\text{O} \\
\| \\
\text{NH}-\text{C}-\text{CH}_3 \\
| \\
\text{C}_6\text{H}_4 \\
| \\
\text{O}-\text{CH}_2\text{CH}_3
\end{array}
$$

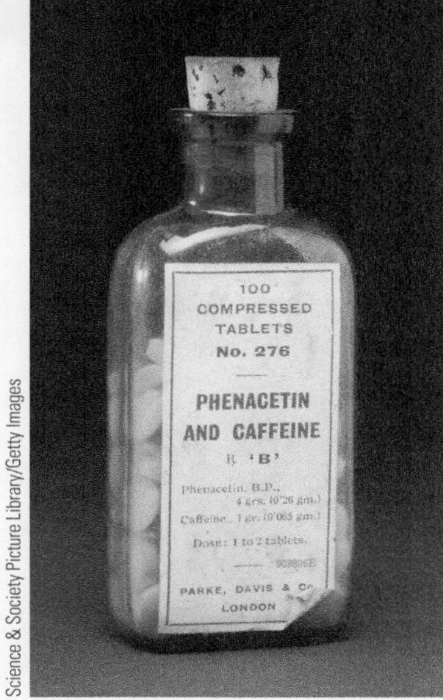

Figure 10.32 Phenacetin was an amide compound used as a pain killer prior to the 1970s.

Solution

$$
\begin{array}{c}
\overset{\delta+}{\text{H}}\ \ \overset{\delta-}{\text{O}}\ \ \overset{\delta+}{\text{H}} \\
\vdots \\
\overset{\delta-}{\text{O}}\diagup \text{H}_{\delta+}\ \cdots\ \overset{\delta+}{\text{H}}\ \ \overset{\delta-}{\text{O}}\cdots\overset{\delta+}{\text{H}}-\overset{\delta-}{\text{O}}-\text{H}^{\delta+} \\
\text{H}^{\delta+}\ \ \ \ \ \ | \ \ \ \ \ \ \| \\
\overset{\delta+}{\text{H}}\cdots\overset{\delta-}{\text{N}}-\overset{\delta+}{\text{C}}-\text{CH}_3 \\
| \\
\text{C}_6\text{H}_4 \\
| \\
\delta^-\,\text{O}-\text{CH}_2\text{CH}_3 \\
\vdots \\
\delta^+\,\text{H}-\text{O}\,\delta^- \\
| \\
\text{H}\,\delta^+
\end{array}
$$

✔ Learning Check 10.8 Show how *N*-methylacetamide can form the following.

a. Intermolecular hydrogen bonds with other amide molecules

b. Hydrogen bonds with water molecules

$$
\begin{array}{c}
\text{O} \\
\| \\
\text{CH}_3-\text{C}-\text{N}-\text{CH}_3 \\
| \\
\text{H}
\end{array}
$$

Basicity is the most important property of amines, so does that mean amides are also basic? The answer is no. Although they are formed from carboxylic acids and basic amines, amides are neither basic nor acidic; instead, they are neutral. The carbonyl group bonded to the nitrogen has destroyed the basicity of the original amine, and the nitrogen of the amine has replaced the acidic –OH of the carboxylic acid.

Figure 10.33 Esters are partially responsible for the fragrance of oranges, pears, bananas, pineapples, and strawberries.

10.8 Formation and Reactions of Esters and Amides

Learning Objective 13 Write reactions for the formation and hydrolysis of esters and amides.

Learning Objective 14 Write key reactions for saponification.

10.8A Formation of Esters and Amides

The most straightforward way to form an ester is to heat an alcohol (or phenol) in the presence of a carboxylic acid. At room temperature, the reaction will not go to completion, but in the presence of an acid catalyst and heat, an ester (**Figure 10.33**) will form.

General reaction:

$$\underset{\text{carboxylic acid}}{R-\overset{\overset{\displaystyle O}{\|}}{C}-OH} + \underset{\text{alcohol or phenol}}{H-O-R'} \xrightleftharpoons{H^+,\ \text{heat}} \underset{\text{carboxylic ester}}{R-\overset{\overset{\displaystyle O}{\|}}{C}-O-R'} + H_2O \qquad (10.10)$$

esterification The process of forming an ester.

As before, R′ indicates that the two variable alkyl groups can be the same or different. The process of ester formation is called **esterification**, and the carbonyl carbon–oxygen single bond of the ester group is called the *ester linkage*:

$$\underset{\text{ester group}}{-\overset{\overset{\displaystyle O}{\|}}{C}-O-\overset{|}{\underset{|}{C}}-}$$

Ester linkage (arrow pointing to the C–O bond)

The reaction of butanoic acid with ethyl alcohol illustrates the formation of a widely used ester.

Specific examples:

$$\underset{\text{butanoic acid}}{CH_3CH_2CH_2-\overset{\overset{\displaystyle O}{\|}}{C}-OH} + \underset{\text{ethyl alcohol}}{HO-CH_2CH_3} \xrightleftharpoons{H^+} \underset{\text{ethyl butanoate (strawberry flavoring)}}{CH_3CH_2CH_2-\overset{\overset{\displaystyle O}{\|}}{C}-O-CH_2CH_3} + H_2O$$

acid chloride An organic compound that contains the $-\overset{\overset{\displaystyle O}{\|}}{C}-Cl$ functional group.

The reaction of an amine with a carboxylic acid normally produces a salt, not an amide. For this reason, the preparation of amides requires a more reactive carbonyl functional group, such as an **acid chloride**. An acid chloride is like a carboxylic acid, except the OH bonded to the carbonyl carbon has been replaced with Cl.

General reaction:

$$\underset{\text{acid chloride}}{R-\overset{\overset{\displaystyle O}{\|}}{C}-Cl} + \underset{\text{amine}}{R'-\underset{\underset{\displaystyle H}{|}}{N}-H} \longrightarrow \underset{\text{amide}}{R-\overset{\overset{\displaystyle O}{\|}}{C}-\underset{\underset{\displaystyle H}{|}}{N}-R'} + HCl \qquad (10.11)$$

Specific examples:

$$\text{C}_6\text{H}_5-\overset{\overset{\displaystyle O}{\|}}{C}-Cl + H-\underset{}{\overset{\overset{\displaystyle H}{|}}{N}}-CH_3 \longrightarrow \text{C}_6\text{H}_5-\overset{\overset{\displaystyle O}{\|}}{C}-NH-CH_3 + HCl$$

benzoyl chloride — methylamine (a 1° amine) — *N*-methylbenzamide

Both primary and secondary amines can be used to prepare amides in the manner shown in Equation 10.11, a method similar to the formation of esters. Tertiary amines, however, do not form amides because they lack a hydrogen atom on the nitrogen.

Example 10.9 Synthesis of Esters and Amides

DEET (*N,N*-diethyl-meta-toluamide) is one of the few insecticides that is effective in warding off mosquitoes (**Figure 10.34**).

Figure 10.34 *N,N*-Diethyl-meta-toluamide (DEET) is used in bug sprays to prevent mosquito bites and the diseases they carry.

a Complete the following reaction to show the synthesis of DEET.

$$CH_3-C_6H_4-\overset{\overset{\displaystyle O}{\|}}{C}-Cl + CH_3CH_2-NH-CH_2CH_3 \longrightarrow$$

b. Give the structure of the ester or amide formed in the following reaction.

$$CH_3-\overset{\overset{\displaystyle O}{\|}}{C}-OH + CH_3-OH \xrightleftharpoons{H^+,\ heat}$$

Solution

a. This reaction is analogous to Equation 10.11, but with a secondary amine. The products are an amide and HCl.

$$CH_3-C_6H_4-\overset{\overset{\displaystyle O}{\|}}{C}-\underset{\underset{\displaystyle CH_2CH_3}{|}}{N}-CH_2CH_3 + HCl$$

b. This reaction is an example of Equation 10.10, so the products are an ester and water:

$$CH_3-\overset{\overset{\displaystyle O}{\|}}{C}-O-CH_3 + H_2O$$

✔ Learning Check 10.9 Give the structure of the products formed in the following reactions.

a.

$$CH_3-\overset{\overset{\displaystyle O}{\|}}{C}-OH + C_6H_5-OH \longrightarrow$$

b.

$$CH_3-\overset{\overset{\displaystyle O}{\|}}{C}-Cl + CH_3CH_2CH_2NH_2 \longrightarrow$$

Health Connections 10.2

Aspirin Substitutes

Even though aspirin is an excellent and useful drug, it occasionally produces adverse effects that some people cannot tolerate. These effects include gastrointestinal bleeding and allergic reactions such as skin rashes and asthmatic attacks. Children who are feverish with influenza or chickenpox should not be given aspirin because of a strong correlation with Reye's syndrome, a devastating illness characterized by liver and brain dysfunction.

The amide *acetaminophen*, marketed under trade names such as Tylenol, Excedrin Aspirin Free, Panadol, and Anacin-3, is an effective aspirin substitute available for use in such situations. Acetaminophen does not irritate the intestinal tract and yet has comparable analgesic (pain-relieving) and antipyretic (fever-reducing) effects. Acetaminophen is available in a stable liquid form that is suitable for administration to children and other patients who have difficulty taking tablets or capsules. Unlike aspirin, acetaminophen has virtually no anti-inflammatory effect and is therefore a poor substitute for aspirin in the treatment of inflammatory disorders such as rheumatoid arthritis. The overuse of acetaminophen has been linked to liver and kidney damage.

Ibuprofen, a carboxylic acid rather than an amide, is a drug with analgesic, antipyretic, and anti-inflammatory properties that was available for decades only as a prescription drug under the trade name Motrin. In 1984, it became available in an over-the-counter (OTC) form under trade names such as Advil, Ibuprin, Mediprin, and Motrin IB.

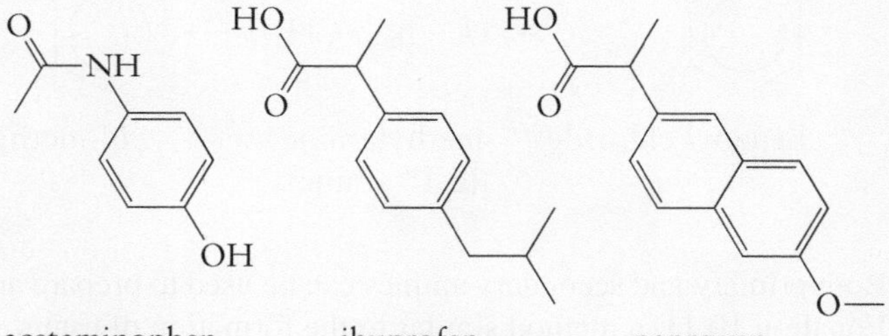

acetaminophen ibuprofen naproxen

Ibuprofen is thought to be somewhat superior to aspirin as an anti-inflammatory drug, but it has about the same effectiveness as aspirin in relieving mild pain and reducing fever.

In 1994, the FDA approved another compound for sale as an over-the-counter pain reliever. *Naproxen*, marketed under the trade names Aleve and Anaprox, exerts its effects for a longer time in the body (8 to 12 hours per dose) than ibuprofen (4 to 6 hours per dose) or aspirin and acetaminophen (4 hours per dose). However, the chances for causing slight intestinal bleeding and stomach upset are greater with naproxen than with ibuprofen, and naproxen is not recommended for use by children under age 12.

10.8B Ester and Amide Hydrolysis

The most important reaction of esters, both in commercial processes and in the body, involves breaking the ester linkage. The process is called either ester hydrolysis or saponification, depending on the reaction conditions. Ester hydrolysis is the reaction of an ester with water to break an ester linkage and produce an alcohol and a carboxylic acid. This process (Equation 10.12) is simply the reverse of the esterification reaction (Equation 10.10). Strong acids are frequently used as catalysts to hydrolyze esters.

General reaction:

$$\underset{\text{ester}}{R-\overset{\overset{\displaystyle O}{\|}}{C}-OR'} + H-OH \overset{H^+}{\rightleftarrows} \underset{\text{carboxylic acid}}{R-\overset{\overset{\displaystyle O}{\|}}{C}-OH} + \underset{\text{alcohol or phenol}}{R'-OH} \quad (10.12)$$

Specific example:

$$\underset{\text{ethyl acetate}}{CH_3-\overset{\overset{\displaystyle O}{\|}}{C}-O-CH_2CH_3} + H-OH \overset{H^+}{\rightleftarrows} \underset{\text{acetic acid}}{CH_3-\overset{\overset{\displaystyle O}{\|}}{C}-OH} + \underset{\text{ethyl alcohol}}{CH_3CH_2-OH}$$

When ester linkages break during hydrolysis, the components of water (–OH and –H) are attached to the ester fragments to form the carboxylic acid and alcohol. The breaking of a bond and the attachment of the components of water to the fragments are characteristic of all hydrolysis reactions.

The human body and other biological organisms are constantly forming and hydrolyzing a variety of esters (**Figure 10.35**). The catalysts used in these cellular processes are very efficient protein catalysts called enzymes (see **Section 12.5**). Animal fats and vegetable oils are esters. Their enzyme-catalyzed hydrolysis is an important process that takes place when they are digested.

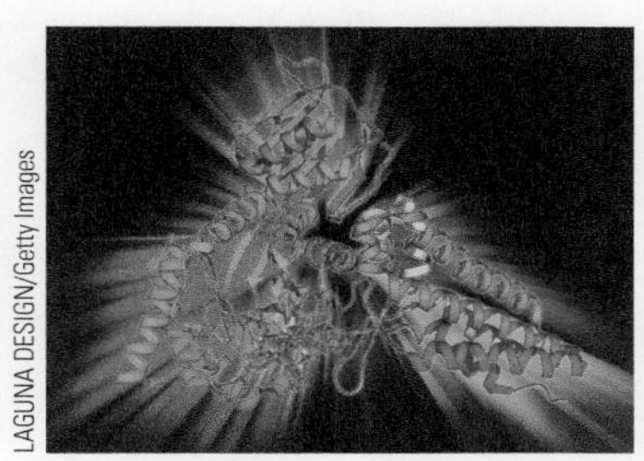

LAGUNA DESIGN/Getty Images

Figure 10.35 A molecular model of a glycoside hydrolase. Glycoside hydrolases are enzymes that break apart the bond between two or more carbohydrates.

The most important reaction of amides is also hydrolysis. This reaction corresponds to the reverse of amide formation; the amide is cleaved into two compounds, a carboxylic acid and an amine or ammonia:

Cleavage here (at the C–NH bond)

$$\underset{\text{amide}}{R-\overset{\overset{\displaystyle O}{\|}}{C}-NH-R'} + H_2O \xrightarrow[\text{Heat}]{\text{Acid or base}} \underset{\text{carboxylic acid}}{R-\overset{\overset{\displaystyle O}{\|}}{C}-OH} + \underset{\text{amine}}{R'-NH_2}$$

As with the hydrolysis of carboxylic esters, amide hydrolysis requires the presence of a strong acid or base, and the nature of the final products depends on which of these reactants is used. Under acidic conditions, the salt of the amine is produced along with the carboxylic acid (Equation 10.13), whereas under basic conditions, the salt of the carboxylic acid is formed along with the amine (Equation 10.14). Notice that heat is required for the hydrolysis of amides.

General reactions:

$$\underset{\text{amide}}{R-\overset{\overset{\displaystyle O}{\|}}{C}-NH-R'} + H_2O + HCl \xrightarrow{\text{Heat}} \underset{\text{carboxylic acid}}{R-\overset{\overset{\displaystyle O}{\|}}{C}-OH} + \underset{\text{amine salt}}{R'-NH_3^+Cl^-} \qquad (10.13)$$

Specific examples:

$$\underset{\text{acetamide}}{CH_3-\overset{\overset{\displaystyle O}{\|}}{C}-NH_2} + H_2O + HCl \xrightarrow{\text{Heat}} \underset{\text{acetic acid}}{CH_3-\overset{\overset{\displaystyle O}{\|}}{C}-OH} + \underset{\text{ammonium chloride}}{NH_4^+Cl^-}$$

General reactions:

$$\underset{\text{amide}}{R-\overset{\overset{\displaystyle O}{\|}}{C}-NH-R'} + NaOH \xrightarrow{\text{Heat}} \underset{\text{carboxylate salt}}{R-\overset{\overset{\displaystyle O}{\|}}{C}-O^-Na^+} + \underset{\text{amine}}{R'-NH_2} \qquad (10.14)$$

Specific examples:

$$\underset{N\text{-methylacetamide}}{CH_3-\overset{\overset{\displaystyle O}{\|}}{C}-NH-CH_3} + NaOH \xrightarrow{\text{Heat}} \underset{\text{sodium acetate}}{CH_3-\overset{\overset{\displaystyle O}{\|}}{C}-O^-Na^+} + \underset{\text{methylamine}}{CH_3-NH_2}$$

Because the form in which carboxylic acids and amines occur in solution depends on the solution pH, one of the hydrolysis products must always be in the form of a salt.

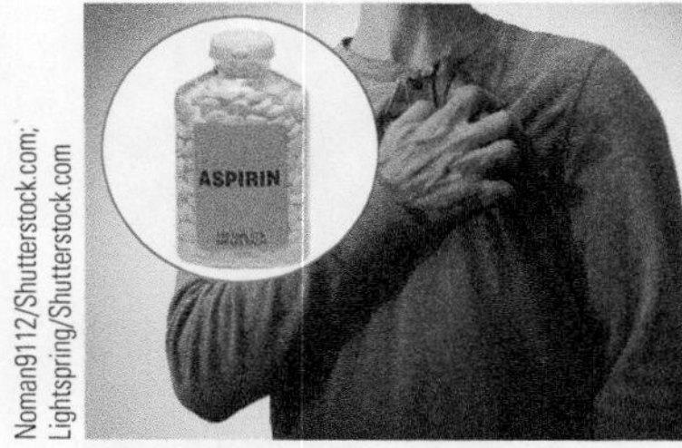

Noman9112/Shutterstock.com; Lightspring/Shutterstock.com

Figure 10.36 Low-dose aspirin has been prescribed by physicians to combat heart disease. Newer guidance has suggested this treatment is not without risks due to possible internal bleeding.

Example 10.10 Writing Hydrolysis Reactions

When aspirin (**Figure 10.36**) enters the stomach, a hydrolysis reaction takes place. Give the structure of the products formed in the following reactions.

a.

O
‖
C—OH
(benzene ring)
O—C—CH_3
‖
O

$+ H_2O \overset{H^+}{\rightleftharpoons}$

acetylsalicylic acid (aspirin)

b.

O
‖
NH—C—CH_3
(benzene ring)
OH

$+ NaOH \xrightarrow{\text{Heat}}$

Solution

a. Using Equation 10.12 as a guide, hydrolysis will produce an alcohol group (specifically a phenol) and a carboxylic acid.

O
‖
C—OH
(benzene ring)
O—C—CH_3
‖
O

$+ H_2O \overset{H^+}{\rightleftharpoons}$

O
‖
C—OH
(benzene ring)
OH ← alcohol

$+ CH_3—\overset{\overset{\displaystyle O}{\|}}{C}—OH$
carboxylic acid

b. Equation 10.14 is the model for this reaction, so the hydrolysis products are a carboxylate salt and an amine.

O
‖
NH—C—CH_3
(benzene ring)
OH

$+ NaOH \xrightarrow{\text{Heat}}$

NH_2
(benzene ring)
OH

$+ CH_3—\overset{\overset{\displaystyle O}{\|}}{C}—O^-Na^+$

✔ Learning Check 10.10 Give the structure of the products formed in the following hydrolysis reactions.

a.

O
‖
C—O—$CH_2CH_2CH_3$
(benzene ring)

$+ H_2O \overset{H^+}{\rightleftharpoons}$

b.

O
‖
C—NH—CH_2CH_3
(benzene ring)

$+ NaOH \overset{\text{Heat}}{\rightleftharpoons}$

Amide hydrolysis is a central reaction in the digestion of proteins and the breakdown of proteins within cells. However, most amide hydrolysis in the body is catalyzed by enzymes rather than by strong acids or bases. Because of their physiological properties, a number of amides play valuable roles in medicine. **Table 10.6** lists some of these important compounds (**Figures 10.37** and **10.38**).

katleho Seisa/E+/Getty Images

Figure 10.37 Your doctor may send a prescription to your pharmacist for antibiotics for a bacterial infection. Antibiotics like penicillin and its derivatives contain an amide within their structure.

Table 10.6 Some Important Amides in Medicine

Structure	Generic Name (Trade Name)	Use
$CH_3CH_2CH_2CH(CH_3)$— and CH_3CH_2— on a ring with N–H, C=O, C=S, N–H, C=O	thiopental (Pentothal)	Intravenous anesthesia
$CH_3CH(CH_3)CH_2CH_2$— and CH_3CH_2— on a ring with N–H, C=O, C=O, N–H, C=O	amobarbital (Amytal)	Treatment of insomnia
C_6H_5–$CH(NH_2)$–C(=O)–NH– on a ring system with N, C=O, S, CH_3, CH_3, COOH	ampicillin (Polycillin)	Antibiotic

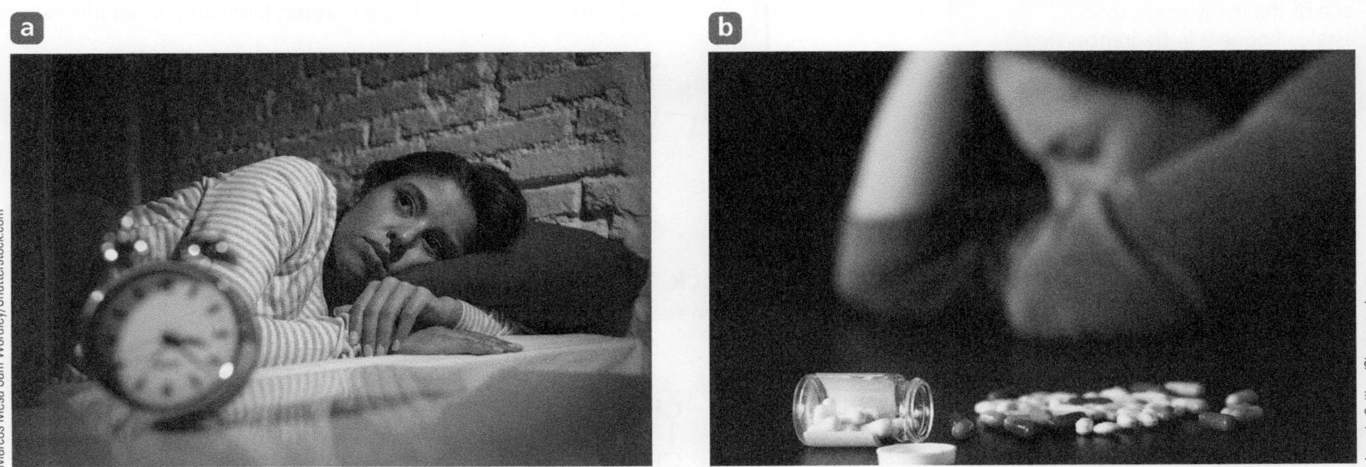

Marcos Mesa Sam Wordley/Shutterstock.com

Antonio Guillem/Shutterstock.com

Figure 10.38 Barbiturates and their derivatives are central nervous system (CNS) depressants. Their medical usefulness for (a) treating maladies like insomnia and anxiety has receded due to their propensity for (b) addiction and overdose.

10.8C Saponification

saponification The basic cleavage of an ester linkage.

Like ester hydrolysis, **saponification** results in a breaking of the ester linkage and produces an alcohol and a carboxylic acid. However, unlike ester hydrolysis, saponification is done in solutions containing strong bases such as potassium or sodium hydroxide. The carboxylic acid is converted to a salt under these basic conditions.

General reactions:

$$\underset{\text{ester}}{R-\overset{\overset{\displaystyle O}{\|}}{C}-O-R'} + NaOH \longrightarrow \underset{\text{carboxylic acid salt}}{R-\overset{\overset{\displaystyle O}{\|}}{C}-O^-Na^+} + \underset{\text{alcohol}}{R'-OH} \qquad (10.15)$$

Specific example:

$$\underset{\text{methyl benzoate}}{C_6H_5-\overset{\overset{\displaystyle O}{\|}}{C}-O-CH_3} + NaOH \longrightarrow \underset{\text{sodium benzoate}}{C_6H_5-\overset{\overset{\displaystyle O}{\|}}{C}-O^-Na^+} + \underset{\text{methyl alcohol}}{CH_3-OH}$$

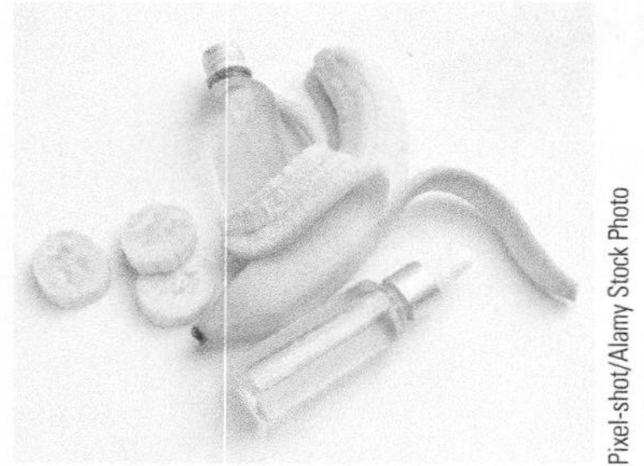

Pixel-shot/Alamy Stock Photo

Figure 10.39 Pentyl ethanoate, also known as amyl acetate, is one of the many esters used as a component in fragrances (especially those that smell like bananas).

Example 10.11 Writing Saponification Reactions

The ester pentyl ethanoate gives rise to the odor of bananas (**Figure 10.39**). Give structural formulas for the products of the following saponification reaction.

$$\underset{\text{pentyl ethanoate}}{CH_3-\overset{\overset{\displaystyle O}{\|}}{C}-O-CH_2(CH_2)_3CH_3} + NaOH \longrightarrow$$

Solution

Using Equation 10.15 as a guide, the products will be a carboxylate salt and an alcohol.

$$\underset{\text{salt}}{CH_3-\overset{\overset{\displaystyle O}{\|}}{C}-O^-Na^+} + \underset{\text{alcohol}}{HO-CH_2(CH_2)_3CH_3}$$

✔ **Learning Check 10.11** Give structural formulas for the products of the following saponification reactions.

a.

$$CH_3CH_2-\overset{\overset{\displaystyle O}{\|}}{C}-O-\underset{\underset{\displaystyle CH_3}{|}}{CH}CH_3 + NaOH \longrightarrow$$

b.

$$C_6H_5-\overset{\overset{\displaystyle O}{\|}}{C}-O-CH_2CH_3 + KOH \longrightarrow$$

Health Connections 10.3

Nitroglycerin in Dynamite and in Medicine

Nitroglycerin is a nitrate ester resulting from the reaction of nitric acid and glycerol:

$$
\begin{array}{l}
CH_2-OH \\
\;| \\
CH-OH \\
\;| \\
CH_2-OH \\
\text{glycerol}
\end{array}
\quad + \; \underset{\text{nitric acid}}{3HO-NO_2} \; \rightleftharpoons \;
\begin{array}{l}
CH_2-O-NO_2 \\
\;| \\
CH-O-NO_2 \\
\;| \\
CH_2-O-NO_2 \\
\text{nitroglycerin}
\end{array}
\; + 3H_2O
$$

First made in 1846 by chemist Ascanio Sobrero (1812–1888), nitroglycerin was discovered to be a powerful explosive with applications in both war and peace. However, as a shock-sensitive, unstable liquid, nitroglycerin proved to be extremely dangerous to manufacture and handle. The chemist Alfred Nobel (1833–1896) later perfected its synthesis and devised a safe method for handling it. Nobel mixed nitroglycerin with a clay-like absorbent material. The resulting solid, called dynamite, is an explosive that is much less sensitive to shock than liquid nitroglycerin. Dynamite is still one of the most important explosives used for mining, digging tunnels, and blasting hills for road building.

Surprising as it may seem, nitroglycerin is also an effective medicine. It is used to treat patients with angina pectoris—sharp chest pains caused by an insufficient supply of oxygen to the heart muscle. Angina pectoris is usually found in patients with coronary artery diseases, such as arteriosclerosis (hardening of the arteries). During overexertion or excitement, the partially clogged coronary arteries prevent the heart from getting an adequate supply of oxygenated blood, and pain results. Nitroglycerin relaxes the cardiac muscle and causes a dilation of the arteries, thus increasing blood flow to the heart and relieving the chest pains.

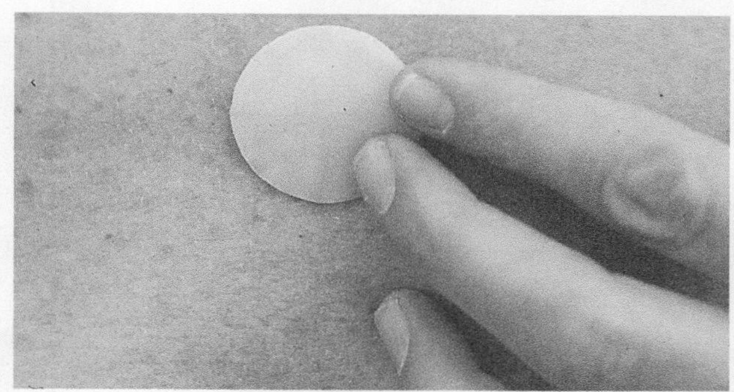

Van D. Bucher/Science Source

A transdermal patch allows nitroglycerin to be absorbed through the skin.

Health Career Description

Pharmacy Technician

Pharmacy technicians work under the supervision of pharmacists in a variety of settings. They may work in a traditional "drug-store" type pharmacy, a pharmacy within a "big box store," an online pharmacy, or a pharmacy within a hospital or clinic. Their time spent in customer interaction depends on the location of their work.

Pharm techs that work in stores or stand-alone pharmacies will serve as a liaison between the pharmacist and the public, answering questions about medications, store hours, insurance coverage, and so on. Those who work in a hospital setting will be more involved in compounding medications or making sure medications are correctly matched to patients.

Pharm techs are expected to perform office duties and must have excellent computer and organizational skills. They manage inventories, stock machines and organize orders, and collect payments. They have a high level of responsibility because there is no room for error in a pharmacy practice.

Pharmacists depend on pharmacy techs to keep their businesses running smoothly and to provide an extra level of safety by checking that orders are filled and delivered correctly. Pharmacy technicians are not expected to have the same education level or understanding of chemistry and medications that pharmacists have, but they are expected to have a working understanding of pharmaceuticals so as to operate safely within the pharmacy industry.

According to the Bureau of Labor Statistics, the median salary for a pharmacy technician with an associate degree is $36,740 per year, and the field is expected to grow at a rate of 7% annually. This is a career that is well suited to people with an interest in pharmaceuticals who enjoy interacting with people and want to help others.

Concept Summary

10.1 The Nomenclature of Aldehydes and Ketones

Learning Objectives: Classify carbonyl compounds as aldehydes or ketones. Identify specific uses for aldehydes and ketones. Assign IUPAC names to aldehydes and ketones.

- The functional groups characteristic of aldehydes and ketones are very similar because they both contain a carbonyl group.

$$\overset{\displaystyle O}{\overset{\|}{-C-}}$$

- Aldehydes have a hydrogen attached to the carbonyl carbon.

$$\overset{\displaystyle O}{\overset{\|}{-C}}-H$$

- Ketones have two carbons attached to the carbonyl carbon.

$$-\underset{|}{\overset{|}{C}}-\overset{\displaystyle O}{\overset{\|}{C}}-\underset{|}{\overset{|}{C}}-$$

- The IUPAC ending for aldehyde names is *-al*, whereas the IUPAC ending for ketones is *-one*.
- Some naturally occurring aldehydes and ketones are important in living systems.
- Some function as sex hormones; others are used as flavorings.

10.2 Properties of Aldehydes and Ketones

Learning Objective: Explain how intermolecular forces influence the physical properties of aldehydes and ketones.

- Molecules of aldehydes and ketones cannot form hydrogen bonds with each other.
- As a result, they have lower boiling points than alcohols of similar molecular weight.
- Aldehydes and ketones have higher boiling points than alkanes of similar molecular weight because of the presence of the polar carbonyl group.
- The polarity of the carbonyl group and the fact that aldehydes and ketones can form hydrogen bonds with water explain why the low-molecular-weight compounds of these organic classes are water soluble.

10.3 Reactions of Aldehydes and Ketones

Learning Objective: Predict the products of the hydrogenation, oxidation, reduction, and hydration reactions of aldehydes and ketones.

- A characteristic reaction of both aldehydes and ketones is the addition of hydrogen to the carbonyl double bond to form alcohols.
- Aldehydes are prepared by the oxidation of primary alcohols.
- Ketones are prepared by the oxidation of secondary alcohols.
- Aldehydes can be further oxidized to carboxylic acids, but ketones resist further oxidation.
- Reduction is the addition of a hydride to an aldehyde or ketone to make an alcohol.
- Aldehydes are reduced to primary alcohols, whereas ketones are reduced to secondary alcohols.

10.4 The Nomenclature of Carboxylic Acids and Carboxylate Salts

Learning Objectives: Assign IUPAC names to carboxylic acids. Assign common and IUPAC names to carboxylate salts. Describe uses for carboxylate salts.

- The characteristic functional group of carboxylic acids is the carboxyl group.

$$\overset{\displaystyle O}{\overset{\|}{-C}}-OH$$

- Many of the simpler carboxylic acids are well known by common names.
- In the IUPAC system, the *-oic acid* ending is used to name these compounds.
- Aromatic acids are named as derivatives of benzoic acid.
- The carboxylate salts are named by changing the *-ic* ending of the acid to *-ate*.
- The ionic nature of the salts makes them water soluble.
- A number of carboxylate salts are useful as food preservatives, soaps, and medicines.

10.5 Properties of Carboxylic Acids

Learning Objectives: Explain how hydrogen bonding affects the physical properties of carboxylic acids. Write key acid–base reactions of carboxylic acids.

- At room temperature, low-molecular-weight carboxylic acids are liquids with distinctively sharp or unpleasant odors. High-molecular-weight, long-chain acids are waxlike solids.
- Carboxylic acids are can form dimers: two molecules held together by hydrogen bonds.
- As a result, their boiling points are relatively high compared to alcohols, aldehydes, ketones, and alkanes of similar molecular weights. Carboxylic acids with low molecular weights are soluble in water.
- Soluble carboxylic acids behave as weak acids.
- They dissociate only slightly in water to form an equilibrium mixture with the carboxylate ion.
- The equilibrium concentrations of the carboxylic acid and the carboxylate ion depend on pH.
- At low pH, the acid form predominates, and at pH 7.4 (the pH of cellular fluids) and above, the carboxylate ion predominates.
- Carboxylic acids react with bases to produce carboxylate salts and water.

10.6 The Nomenclature of Esters and Amides

Learning Objective: Assign common and IUPAC names to esters and amides.

- Both common and IUPAC names for esters are formed by first naming the alkyl group of the alcohol portion followed by the name of the acid portion in which the *-ic acid* ending has been changed to *-ate*.

- Amides are named by changing the *-ic acid* or *-oic acid* ending of the carboxylic acid portion of the compound to *-amide*.
- Groups attached to the nitrogen of the amide are denoted by an italic capital *N* that precedes the name of the attached group.

10.7 Properties of Esters and Amides

Learning Objective: **Explain how hydrogen bonding affects the physical properties of esters and amides.**

- Low-molecular-weight amides are water soluble because they can form hydrogen bonds with water.
- The melting and boiling points of unsubstituted amides are higher than those of comparable substituted amides because of intermolecular hydrogen bonding.
- Many esters are very fragrant and represent some of nature's most pleasant odors.
- Because of this characteristic, esters are widely used as flavoring agents.
- Amides are neither basic nor acidic.

10.8 Formation and Reactions of Esters and Amides

Learning Objectives: **Write reactions for the formation and hydrolysis of esters and amides. Write key reactions for saponification.**

- Carboxylic acids react with alcohols in the presence of heat and an acid catalyst to produce esters.
- Amides are formed from the reaction of acid chlorides with amines.
- Esters can be converted back to carboxylic acids and alcohols under either acidic or basic conditions.
- Hydrolysis, the reaction of an ester with water in the presence of an acid, produces the carboxylic acid and alcohol.
- Saponification, the cleavage of an ester under basic conditions, produces the carboxylate salt and alcohol.
- Amides undergo hydrolysis in acidic conditions to yield a carboxylic acid and an amine salt. Amide hydrolysis requires heat.
- Hydrolysis of amides under basic conditions produces a carboxylate salt and an amine.

Key Terms and Concepts

Acid chloride (10.8)
Aldehyde (10.1)
Amide (10.6)
Carbonyl group (Introduction)
Carboxylate ion (10.5)
Carboxyl group (Introduction)
Carboxylic acid (10.4)
Carboxylic ester or ester (10.6)
Dimer (10.5)
Esterification (10.8)
Fatty acid (10.4)
Ketone (10.1)
Peptide bond (10.6)
Saponification (10.8)

Key Equations

1. Hydrogenation of an aldehyde to give a primary alcohol (**Section 10.3**)	$R-\overset{O}{\overset{\|\|}{C}}-H + H_2 \xrightarrow{Pt} R-\underset{H}{\underset{\|}{\overset{OH}{\overset{\|}{C}}}}-H$	Equation 10.1
2. Hydrogenation of a ketone to give a secondary alcohol (**Section 10.3**)	$R-\overset{O}{\overset{\|\|}{C}}-R' + H_2 \xrightarrow{Pt} R-\underset{H}{\underset{\|}{\overset{OH}{\overset{\|}{C}}}}-R'$	Equation 10.2
3. Overoxidation of a primary alcohol to give an aldehyde and then ultimately a carboxylic acid (**Section 10.3**)	$\underset{\text{primary alcohol}}{R-CH_2-OH} \xrightarrow{(O)} \underset{\text{aldehyde}}{R-\overset{O}{\overset{\|\|}{C}}-H} \xrightarrow{(O)} \underset{\text{carboxylic acid}}{R-\overset{O}{\overset{\|\|}{C}}-OH}$	Equation 10.3

4. Oxidation of a secondary alcohol to give a ketone (**Section 10.3**)	$\underset{\text{secondary alcohol}}{R-\overset{OH}{\overset{\vert}{C}H}-R'} \xrightarrow{(O)} \underset{\text{ketone}}{R-\overset{O}{\overset{\Vert}{C}}-R'} \xrightarrow{(O)}$ no reaction	Equation 10.4
5. Reduction of an aldehyde to give a primary alcohol (**Section 10.3**)	$R-\overset{O}{\overset{\Vert}{C}}-H \xrightarrow{(R)} R-\underset{H}{\underset{\vert}{\overset{H}{\overset{\vert}{C}}}}-OH$	Equation 10.5
6. Reduction of a ketone to give a secondary alcohol (**Section 10.3**)	$R-\overset{O}{\overset{\Vert}{C}}-R' \xrightarrow{(R)} R-\underset{H}{\underset{\vert}{\overset{OH}{\overset{\vert}{C}}}}-R'$	Equation 10.6
7. Reaction of a carboxylic acid and base to produce a carboxylate salt (**Section 10.4**)	$\underset{\text{stearic acid}}{CH_3(CH_2)_{16}-\overset{O}{\overset{\Vert}{C}}-O-H} + NaOH \longrightarrow CH_3(CH_2)_{16}-\overset{O}{\overset{\Vert}{C}}-O^-Na^+ + H_2O$	Equation 10.7
8. Reaction of a carboxylic acid to give a carboxylate ion (**Section 10.5**)	$R-\overset{O}{\overset{\Vert}{C}}-O-H + H_2O \rightleftharpoons R-\overset{O}{\overset{\Vert}{C}}-O^- + H_3O^+$	Equation 10.8
9. Reaction of a carboxylic acid with base to produce a carboxylate salt plus water (**Section 10.5**)	$R-\overset{O}{\overset{\Vert}{C}}-O-H + NaOH \rightleftharpoons R-\overset{O}{\overset{\Vert}{C}}-O^-Na^+ + H_2O$	Equation 10.9
10. Reaction of a carboxylic acid with an alcohol to produce an ester plus water (**Section 10.8**)	$\underset{\text{carboxylic acid}}{R-\overset{O}{\overset{\Vert}{C}}-OH} + \underset{\text{alcohol or phenol}}{H-O-R'} \xrightleftharpoons{H^+,\text{ heat}} \underset{\text{carboxylic ester}}{R-\overset{O}{\overset{\Vert}{C}}-O-R'} + H_2O$	Equation 10.10
11. Reaction of amines with acid chlorides to form amides (**Section 10.8**)	$R-\overset{O}{\overset{\Vert}{C}}-Cl + R'-\underset{H}{\underset{\vert}{N}}-H \longrightarrow R-\overset{O}{\overset{\Vert}{C}}-\underset{H}{\underset{\vert}{N}}-R' + HCl$	Equation 10.11
12. Ester hydrolysis to produce a carboxylic acid and an alcohol (**Section 10.8**)	$R-\overset{O}{\overset{\Vert}{C}}-OR' + H-OH \xrightleftharpoons{H^+} R-\overset{O}{\overset{\Vert}{C}}-OH + R'-OH$	Equation 10.12
13. Acid hydrolysis of amides to form carboxylic acids (**Section 10.8**)	$R-\overset{O}{\overset{\Vert}{C}}-NH-R' + H_2O + HCl \xrightarrow{\text{Heat}} R-\overset{O}{\overset{\Vert}{C}}-OH + R'-NH_3^+Cl^-$	Equation 10.13
14. Basic hydrolysis of amides to form carboxylate salts (**Section 10.8**)	$R-\overset{O}{\overset{\Vert}{C}}-NH-R' + NaOH \xrightarrow{\text{Heat}} R-\overset{O}{\overset{\Vert}{C}}-O^-\ Na^+ + R'-NH_2$	Equation 10.14
15. Ester saponification to give a carboxylate salt and alcohol (**Section 10.8**)	$R-\overset{O}{\overset{\Vert}{C}}-O-R' + NaOH \longrightarrow R-\overset{O}{\overset{\Vert}{C}}-O^-\ Na^+ + R'-OH$	Equation 10.15

Exercises

Even-numbered exercises are answered in Appendix B.

Exercises with an asterisk (*) are more challenging.

The Nomenclature of Aldehydes and Ketones (Section 10.1)

10.1 What is the structural difference between an aldehyde and a ketone?

10.2 Draw structural formulas for an aldehyde and a ketone that contain the fewest number of carbon atoms possible.

10.3 Oxybenzone is a useful component in sunscreens because of its UV light-absorbing property. What does the ending of the common name oxybenzone imply about its structure?

10.4 Indicate which of the following compounds contains a carbonyl group.

a. $CH_3-O-CH_2CH_3$

b. 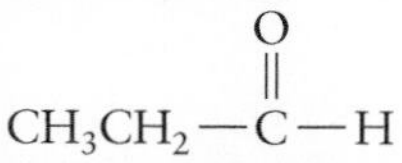 $CH_3CH_2-\overset{\displaystyle O}{\overset{\|}{C}}-H$

c. CH_3CH_2-OH

d. 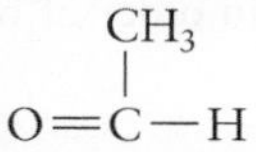 $O=\overset{\displaystyle CH_3}{\overset{|}{C}}-H$

10.5 Indicate which of the following compounds contains a carbonyl group.

a. 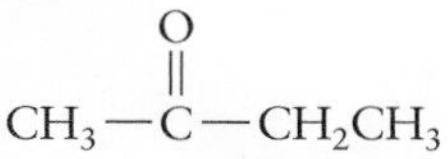 $CH_3-\overset{\displaystyle O}{\overset{\|}{C}}-CH_2CH_3$

b. $CH_3CH_2-\overset{\displaystyle OH}{\overset{|}{CH}}-CH_3$

c. $CH_3CH_2CH_2\overset{\displaystyle H}{\overset{|}{C}}=O$

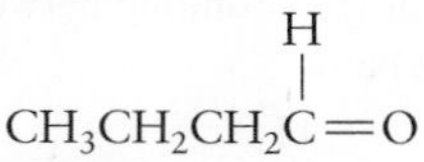

d. CH_3-O-CH_3

10.6 Identify each of the following compounds as an aldehyde, a ketone, or neither.

a. 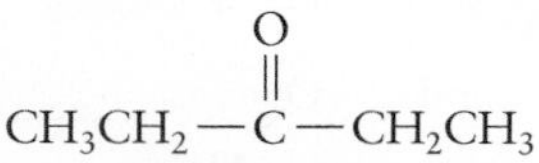 $CH_3CH_2-\overset{\displaystyle O}{\overset{\|}{C}}-CH_2CH_3$

b. (cyclohexane ring)$-\overset{\displaystyle O}{\overset{\|}{C}}-OH$

c. (six-membered ring containing O, with C=O adjacent to ring O)

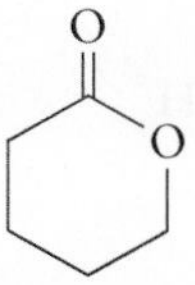

d. 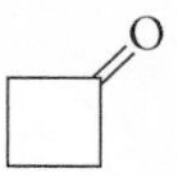 (cyclobutane ring)=O

e. 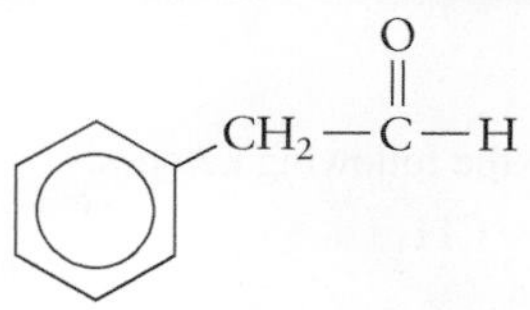 (benzene ring)$-CH_2-\overset{\displaystyle O}{\overset{\|}{C}}-H$

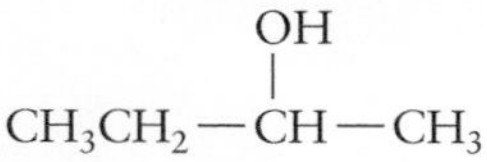

f.  $NH_2-\overset{\displaystyle O}{\overset{\|}{C}}-NH_2$

10.7 Identify each of the following compounds as an aldehyde, a ketone, or neither.

a. 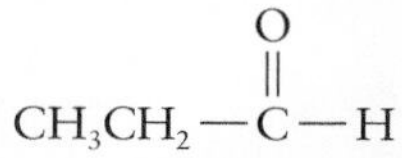 $CH_3CH_2-\overset{\displaystyle O}{\overset{\|}{C}}-H$

b. $CH_3-\overset{\displaystyle O}{\overset{\|}{C}}-O-CH_3$

c. (cyclopentane ring)=O

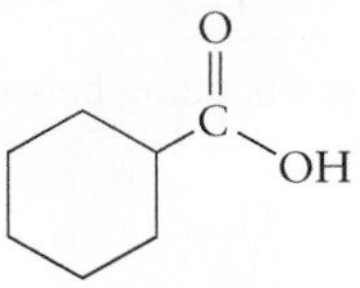

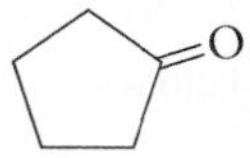

d. 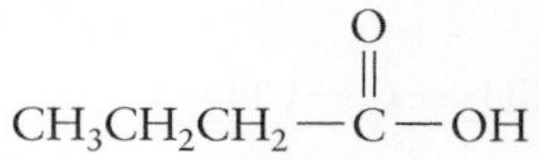 $CH_3CH_2CH_2-\overset{\displaystyle O}{\overset{\|}{C}}-OH$

e. 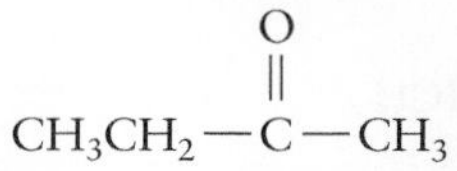 $CH_3CH_2-\overset{\displaystyle O}{\overset{\|}{C}}-CH_3$

f. 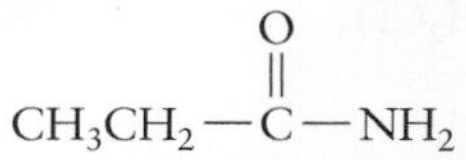 $CH_3CH_2-\overset{\displaystyle O}{\overset{\|}{C}}-NH_2$

10.8 Assign IUPAC names to the following aldehydes.

a. 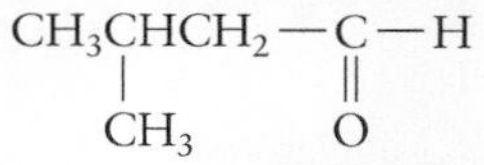 $CH_3\underset{\underset{\displaystyle CH_3}{|}}{CH}CH_2-\underset{\underset{\displaystyle O}{\|}}{C}-H$

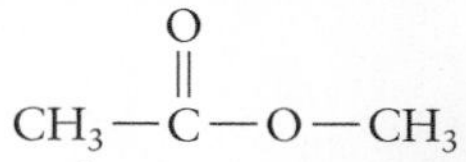

b. $CH_3-\underset{\underset{\displaystyle CH_2CH_2CH_3}{|}}{CH}-CH_2CH_2-\underset{\underset{\displaystyle O}{\|}}{C}-H$

c. (benzene ring)$-CH_2-\overset{\displaystyle O}{\overset{\|}{C}}-H$

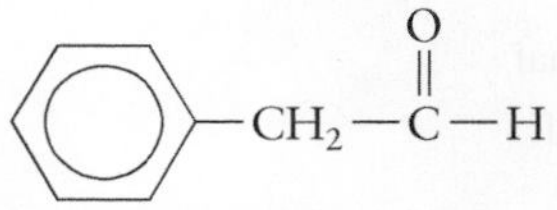

d. $CH_3CH_2CH_2CHO$

10.9 Assign IUPAC names to the following aldehydes.

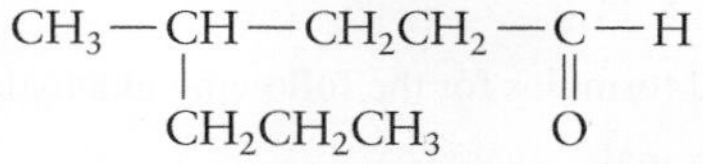

a. 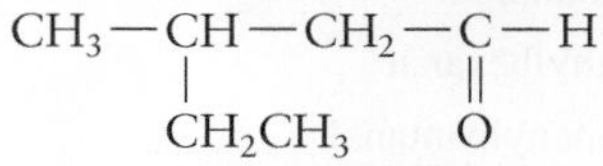 $CH_3-\underset{\underset{\displaystyle CH_2CH_3}{|}}{CH}-CH_2-\underset{\underset{\displaystyle O}{\|}}{C}-H$

b. $CH_3-\underset{\underset{\displaystyle CH_3}{|}}{CH}-CHO$

c. $CH_3\underset{\underset{\displaystyle \text{(benzene ring)}}{|}}{CH}CH_2-\overset{\displaystyle O}{\overset{\|}{C}}-H$

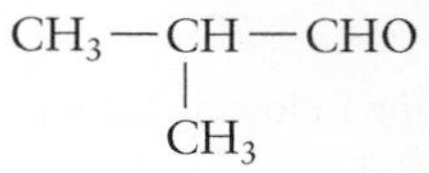

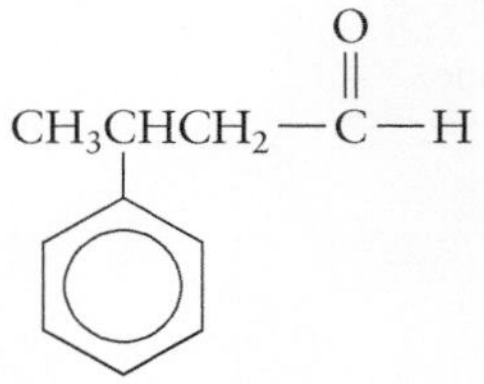

d. 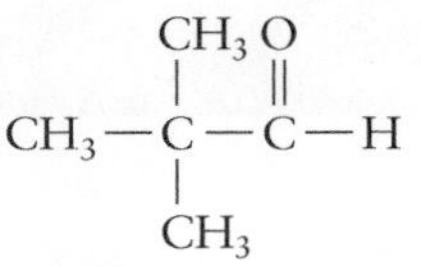 $CH_3-C(CH_3)_2-C(=O)-H$

$$\begin{array}{c} \quad CH_3\ O \\ \quad | \quad \| \\ CH_3-C-C-H \\ | \\ CH_3 \end{array}$$

10.10 Assign IUPAC names to the following ketones.

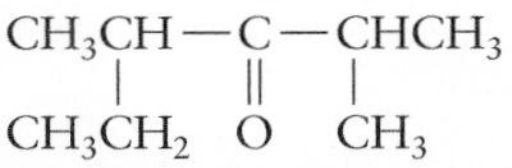

a. 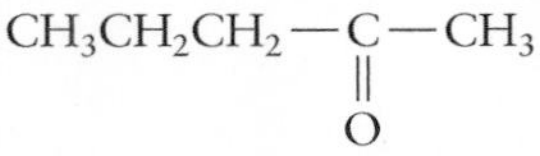 $CH_3CH_2CH_2-\underset{\underset{O}{\|}}{C}-CH_3$

b. $\underset{CH_3CH_2}{CH_3\underset{|}{C}H}-\underset{\underset{O}{\|}}{C}-\underset{\underset{CH_3}{|}}{C}HCH_3$

c. $CH_3CH_2-\underset{\underset{O}{\|}}{C}-CH_2CH_3$

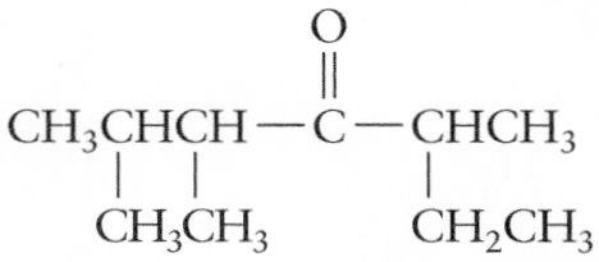

d. 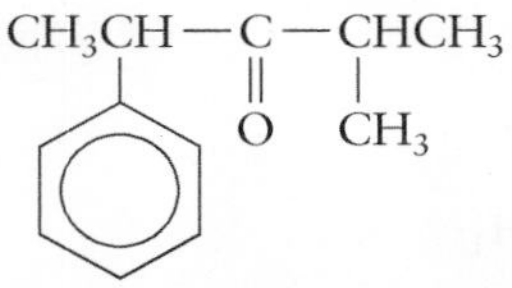 $CH_3\underset{|}{C}H-\underset{\underset{O}{\|}}{C}-\underset{\underset{CH_3}{|}}{C}HCH_3$ (with a phenyl ring on the first CH)

10.11 Assign IUPAC names to the following ketones.

a.  $CH_3CH_2CH_2CH_2CH_2-\overset{\overset{O}{\|}}{C}-CH_3$

b. $CH_3\underset{CH_3}{\underset{|}{C}H}\underset{CH_3}{\underset{|}{C}H}-\overset{\overset{O}{\|}}{C}-\underset{\underset{CH_2CH_3}{|}}{C}HCH_3$

c. (ring structure with =O, substituted with H_3C)

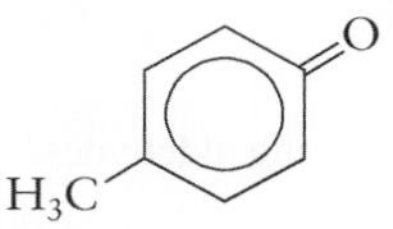

d. 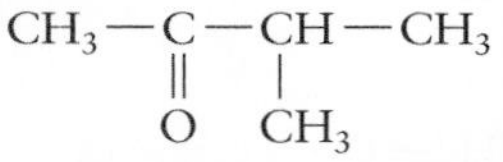 $CH_3-\underset{\underset{O}{\|}}{C}-\underset{\underset{CH_3}{|}}{C}H-CH_3$

10.12 Draw structural formulas for the following aldehydes.

a. 3-methylhexanal

b. 3,3-dimethylpentanal

c. 2,3-dibromobutanal

d. 2-phenylpropanal

10.13 Draw structural formulas for the following aldehydes.

a. 2,4,5-trimethyloctanal

b. 2-hydroxy-3-methylhexanal

c. 4,4-dimethyl-3-phenylpentanal

d. butanal

10.14 Draw structural formulas for the following ketones.

a. cyclopentanone

b. 1,2-dichloro-3-pentanone

c. 2-hexanone

d. 3-phenyl-2-pentanone

10.15 Draw structural formulas for the following ketones:

a. 1,5-dibromo-2-pentanone

b. 3-methylcyclohexanone

c. 1,3,4-trichloro-2-butanone

d. 4-hydroxy-5-phenyl-2-hetpanone

Properties of Aldehydes and Ketones (Section 10.2)

10.16 Why does hydrogen bonding *not* take place between molecules of aldehydes or ketones?

10.17 Most of the remaining water in washed laboratory glassware can be removed by rinsing the glassware with acetone (propanone). Explain how this process works (acetone is much more volatile than water).

10.18 Arrange the following compounds in order of increasing water solubility.

a. $CH_3CH_2CH_2-OH$

b. 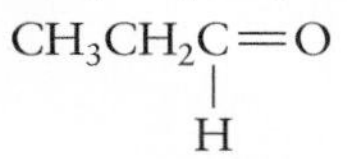 $CH_3CH_2\underset{\underset{H}{|}}{C}=O$

c. $CH_3CH_2CH_2CH_3$

10.19 Arrange the following compounds in order of increasing water solubility.

a. 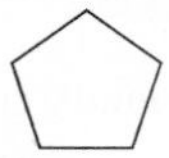 (cyclopentane)

b. 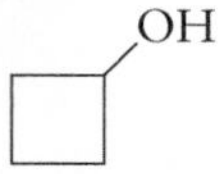 (cyclobutane with OH)

c. 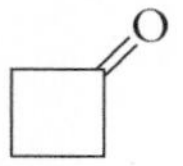 (cyclobutane with =O)

10.20 Use a dotted line to show hydrogen bonding between molecules in each of the following pairs.

a. 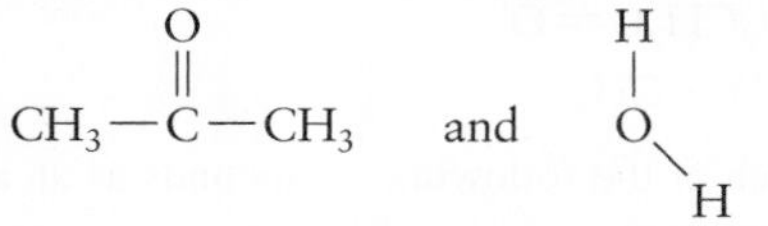 $CH_3-\overset{\overset{O}{\|}}{C}-CH_3$ and $H-O-H$

b. 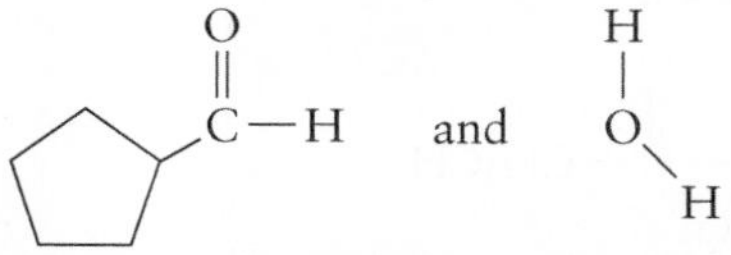 (cyclopentane)$-\overset{\overset{O}{\|}}{C}-H$ and $H-O-H$

10.21 Use a dotted line to show hydrogen bonding between molecules in each of the following pairs.

a. 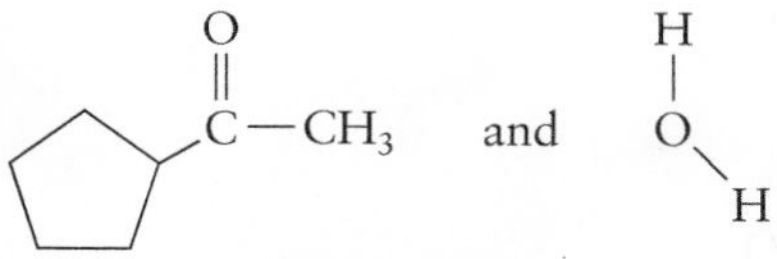 (cyclopentane)$-\overset{\overset{O}{\|}}{C}-CH_3$ and $H-O-H$

b. 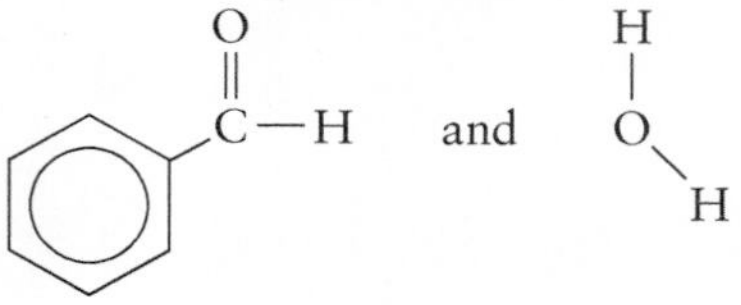 (benzene ring)$-\overset{\overset{O}{\|}}{C}-H$ and $H-O-H$

10.22 Arrange the following compounds in order of increasing boiling point.

a. $CH_3CH_2CH_2-OH$

b. $CH_3CH_2CH_2CH_3$

c. $CH_3CH_2-C(H)=O$

10.23 Arrange the following compounds in order of increasing boiling point.

a. 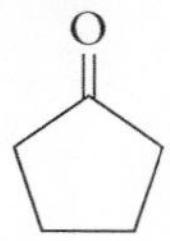 O (cyclopentanone)

b. 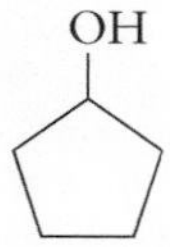 OH (cyclopentanol)

c. 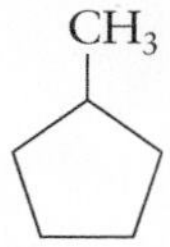 CH_3 (methylcyclopentane)

Reactions of Aldehydes and Ketones (Section 10.3)

10.24 Draw the structural formula of the aldehyde or ketone needed to make each of the following alcohols.

a. 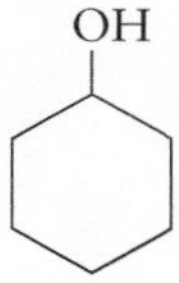 OH (cyclohexanol)

b. 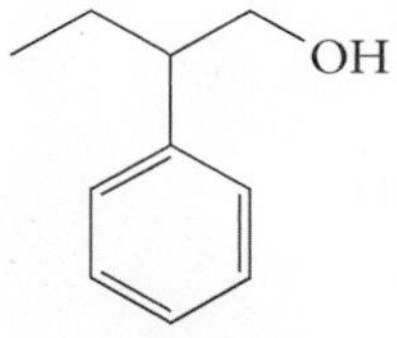 OH

c. 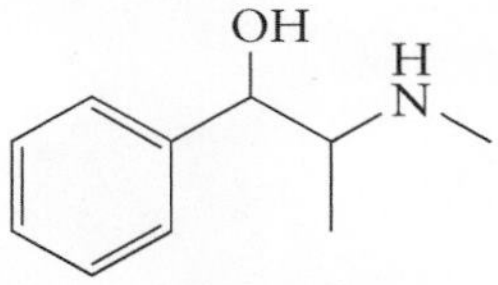 OH H N

pseudoephedrine

d.  OH

ethanol

10.25 Draw the structural formula of the aldehyde or ketone needed to make each of the following alcohols.

a. 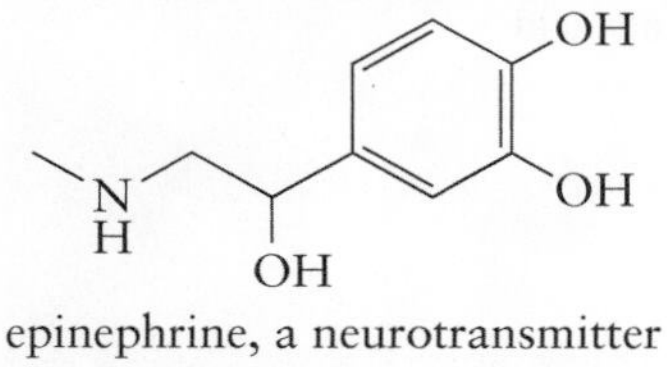 OH OH N H OH

epinephrine, a neurotransmitter

(Note: The phenol hydroxyl groups do not react)

b. 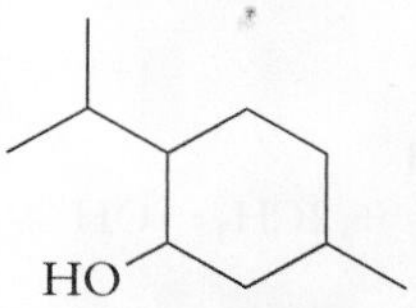 HO

menthol, a flavoring in cough drops

c. 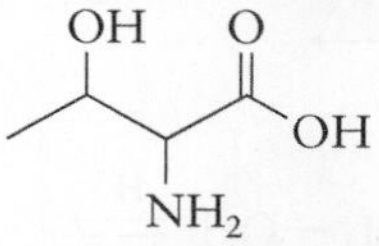 OH O OH NH_2

threonine, an amino acid

d. 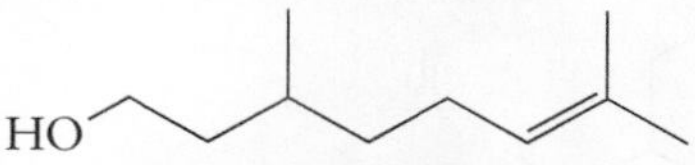 HO

citronellol, fragrance

10.26 Draw the structural formula of the aldehyde or ketone needed to make each of the following alcohols.

a. 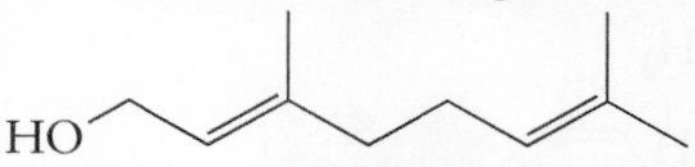 HO

geraniol, floral scent

b. 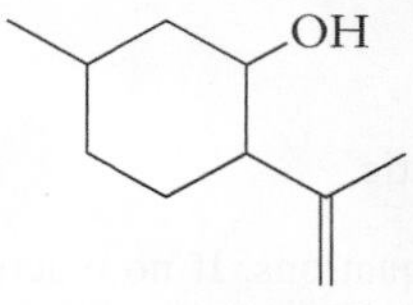 OH

isopulegol, minty aroma

c. 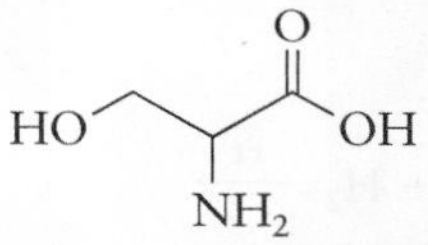 O HO OH NH_2

serine, an amino acid

d. 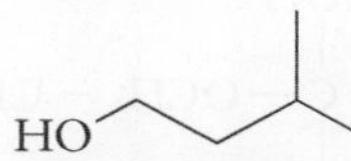 HO

isoamyl alcohol, a solvent used to make flavorings

10.27 Complete the following statements.

a. Oxidation of a secondary alcohol produces ___________.

b. Oxidation of a primary alcohol produces an aldehyde that can be further oxidized to a ___________.

c. Hydrogenation of a ketone produces ___________.

d. Hydrogenation of an aldehyde produces ___________.

e. Hydrolysis of an acetal produces ___________.

f. Reduction of a ketone produces ___________.

g. Reduction of an aldehyde produces ___________.

10.28 Complete the following equations. If no reaction occurs, write "no reaction."

a. $CH_3CH_2-\overset{O}{\overset{\|}{C}}-CH_3 + H_2 \xrightarrow{Pt}$

b. $CH_3\underset{CH_3}{\underset{|}{C}}H\overset{Br}{\overset{|}{C}}H-\overset{O}{\overset{\|}{C}}-H + (O) \longrightarrow$

c. [cyclohexanecarbaldehyde: cyclohexane ring with C(=O)–H] $+ 2CH_3-OH \xrightleftharpoons{H^+}$

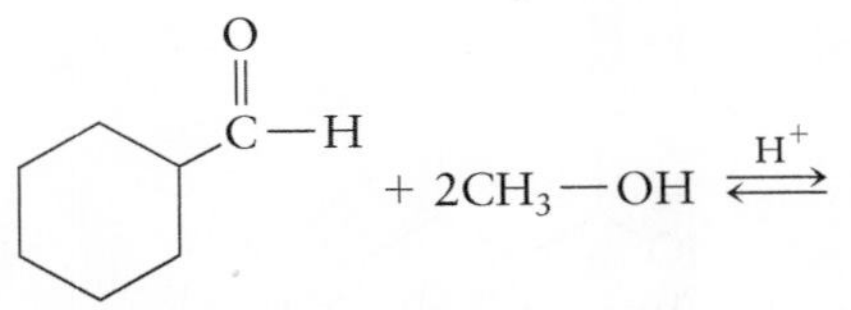

d. [cyclopentanone] $+ (O) \longrightarrow$

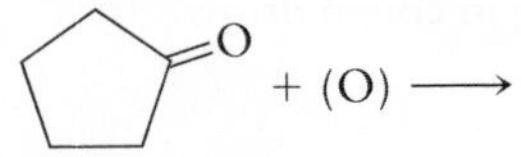

e. [four-membered ring containing O, with OH on adjacent carbon] $+ CH_3CH_2-OH \xrightleftharpoons{H^+}$

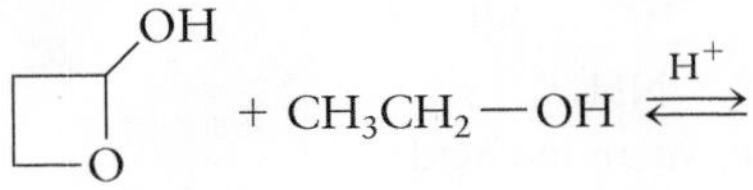

f. [carvone structure] $\xrightarrow{(R)}$

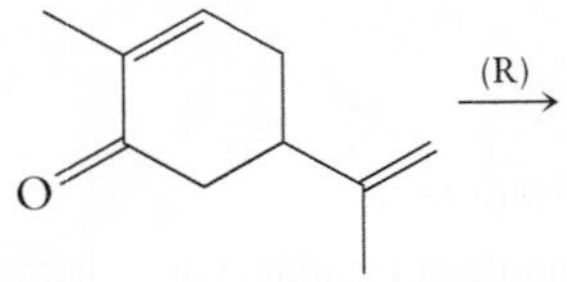

carvone, spearmint scent

g. [ketamine structure: cyclohexanone with NH–CH$_3$ and 2-chlorophenyl] $\xrightarrow{(R)}$

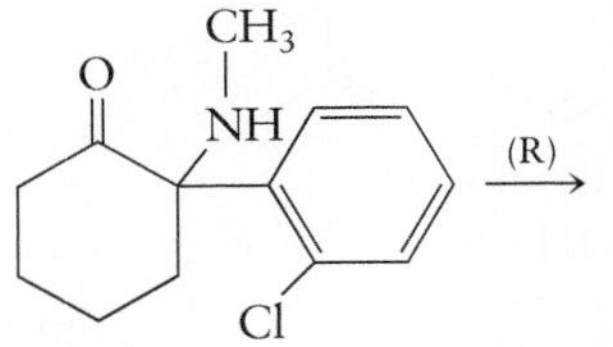

ketamine, an anesthetic

10.29 Complete the following equations. If no reaction occurs, write "no reaction."

a. [dicyclohexyl ketone: cyclohexyl–C(=O)–cyclohexyl] $+ H_2 \xrightarrow{Pt}$

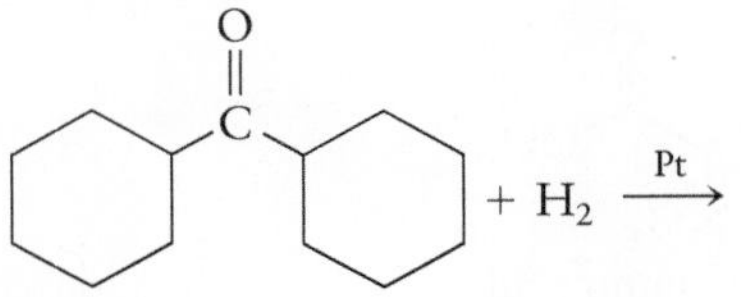

b. $CH_3CH_2-C(OH)(CH_3)-OCH_3 + CH_3CH_2CH_2-OH \xrightleftharpoons{H^+}$

c. $CH_3CH_2-C(=O)-H + 2CH_3CH(OH)CH_3 \xrightleftharpoons{H^+}$

d. [cyclopentyl]$-CH_2-C(=O)-H + (O) \longrightarrow$

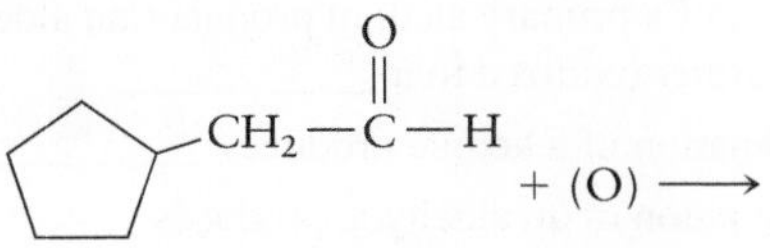

e. $CH_3CH_2CH(CH_3)-C(=O)-H + H_2 \xrightarrow{Pt}$

The Nomenclature of Carboxylic Acids and Carboxylate Salts (Section 10.4)

10.30 Write the correct IUPAC name for each of the following.

a. $CH_3CH_2CH_2CH_2-C(=O)-OH$

b. $CH_3CH_2CH(CH_3)CH_2-C(=O)-OH$

c. $CH_3CH_2CH(C_6H_5)CH_2CH_2-C(=O)-OH$ [phenyl ring on the CH]

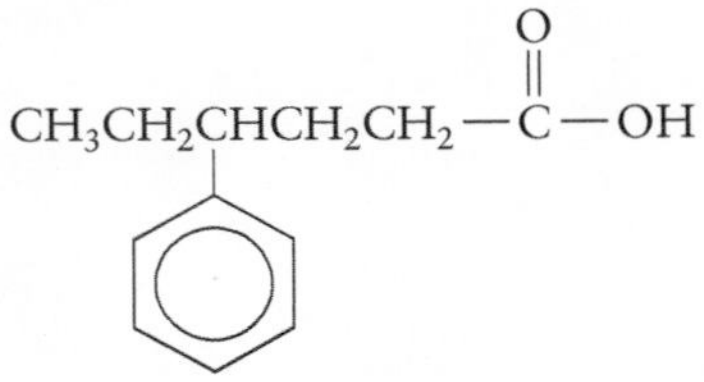

d. [benzene ring with C(=O)–OH, CH$_3$ ortho, and CH$_3$ para]

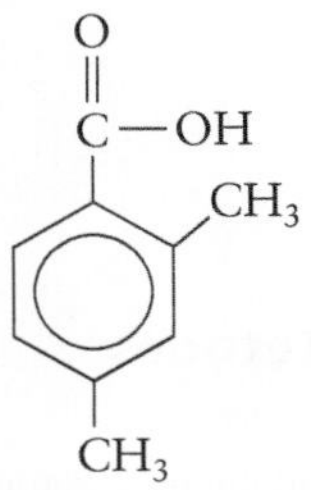

10.31 Write the correct IUPAC name for each of the following.

a. $CH_3CH_2CH_2-C(=O)-OH$

b. $CH_3-O-CH_2CH_2-C(=O)-OH$

c. $CH_3CH(CH_3)CH(Br)CH_2-C(=O)-OH$

d. [benzene ring with C(=O)–OH and CH$_2$CH$_3$ meta]

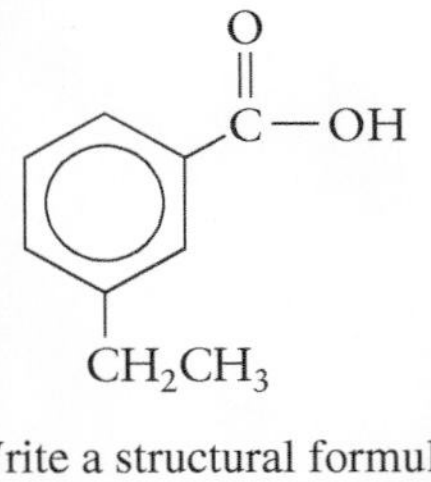

10.32 Write a structural formula for each of the following.

a. pentanoic acid

b. 2-bromo-3-methylhexanoic acid

c. 4-propylbenzoic acid

10.33 Write a structural formula for each of the following.

a. hexanoic acid

b. 4-bromo-3-methylpentanoic acid

c. *o*-ethylbenzoic acid

10.34 Give the IUPAC name for each of the following.

a. $CH_3CH_2-\overset{\displaystyle O}{\overset{\|}{C}}-O^-Na^+$

b. $CH_3\overset{\displaystyle CH_3}{\overset{|}{C}H}-\overset{\displaystyle O}{\overset{\|}{C}}-O^-K^+$

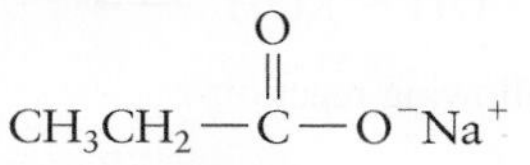

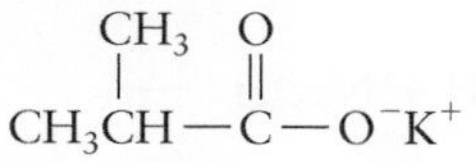

c. (benzene ring with NO_2 and $-\overset{\displaystyle \|}{\underset{\displaystyle O}{C}}-O^-Na^+$ in meta positions)

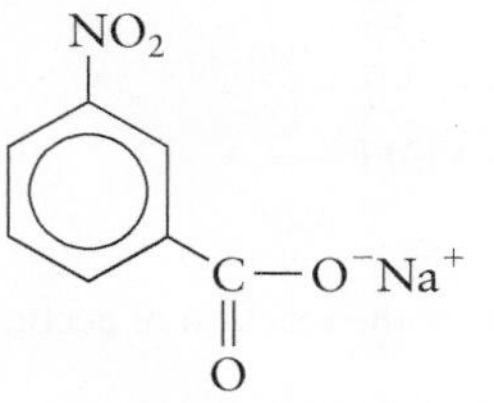

10.35 Give the IUPAC name for each of the following.

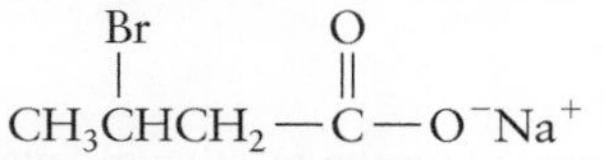

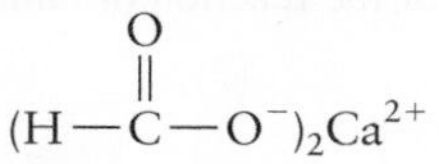

a. $CH_3\overset{\displaystyle Br}{\overset{|}{C}H}CH_2-\overset{\displaystyle O}{\overset{\|}{C}}-O^-Na^+$

b. $(H-\overset{\displaystyle O}{\overset{\|}{C}}-O^-)_2Ca^{2+}$

c. $C_6H_5-O-CH_2CH_2-\overset{\displaystyle O}{\overset{\|}{C}}-O^-K^+$

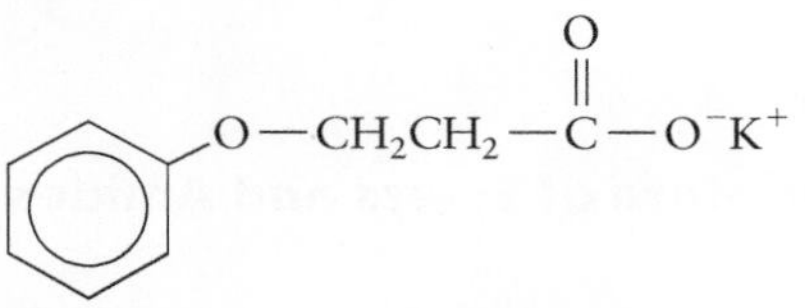

10.36 Draw structural formulas for the following.

a. sodium formate

b. calcium benzoate

c. sodium 2-methylbutanoate

10.37 Draw structural formulas for the following.

a. potassium ethanoate

b. sodium *m*-methylbenzoate

c. sodium 2-methylbutanoate

10.38 Name each of the following.

a. The sodium salt of valeric acid

b. The magnesium salt of lactic acid

c. The potassium salt of citric acid

10.39 Name each of the following.

a. The potassium salt of acetic acid

b. The sodium salt of benzoic acid

c. The magnesium salt of oxalic acid

10.40 Give the name of a carboxylic acid or carboxylate salt used in each of the following ways.

a. As a soap

b. As a mold inhibitor used in bread

c. As a food additive noted for its pH buffering ability

d. As a preservative used in soft drinks

10.41 Give the name of a carboxylic acid or carboxylate salt used in each of the following ways.

a. As a treatment for athlete's foot

b. As a general food preservative used to pickle vegetables

c. As a food preservative found naturally in many foods

d. To maintain the structure of foams and gels

10.42 Which of the following structures contains a carboxyl group?

a. $CH_3CH_2-\underset{\displaystyle O}{\underset{\|}{C}}-CH_3$

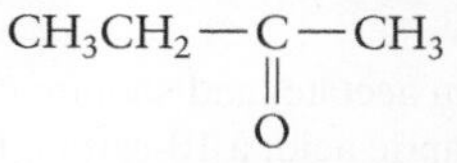

b. CH_3-CH_2-COOH

c. $CH_3-\underset{\displaystyle CH_2CH_3}{\underset{|}{C}H}-CO_2H$

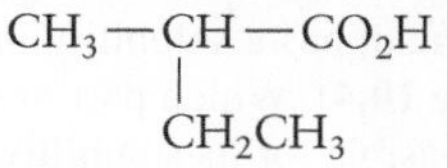

d. (cyclopentane ring)$-\overset{\displaystyle O}{\overset{\|}{C}}-OH$

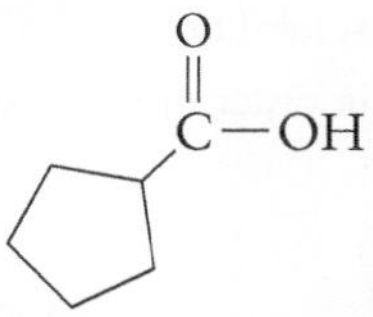

10.43 Which of the following structures contains a carboxyl group?

a. (cyclopentane ring)$-\overset{\displaystyle O}{\overset{\|}{C}}-H$

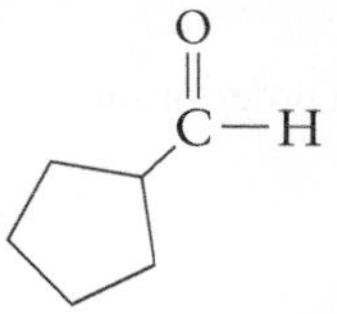

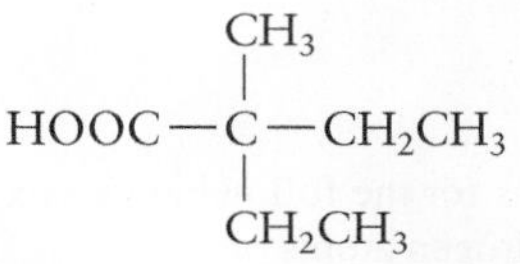

b. $HOOC-\overset{\displaystyle CH_3}{\overset{|}{\underset{\displaystyle CH_2CH_3}{\underset{|}{C}}}}-CH_2CH_3$

c. $HO-\underset{\displaystyle O}{\underset{\|}{C}}-CH_2-$(cyclohexane ring)

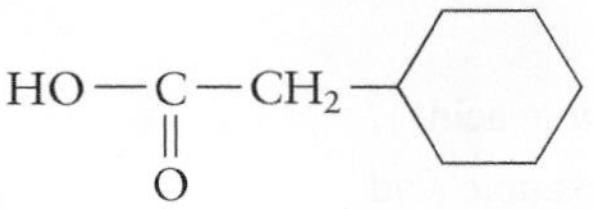

d. CH_3-CO_2H

10.44 What structural features are characteristic of fatty acids? Why are fatty acids given that name?

10.45 What compound is responsible for the sour or tart taste of Italian salad dressing (vinegar and oil)?

10.46 What carboxylic acid is present in sour milk and sauerkraut?

Properties of Carboxylic Acids (Section 10.5)

10.47 List the following compounds in order of increasing boiling point.

a. 2-butanone

b. propanoic acid

c. pentane

d. 1-butanol

10.48 List the following compounds in order of increasing boiling point.

a. pentanal

b. hexane

c. 1-pentanol

d. butanoic acid

10.49 Use a dotted line to show hydrogen bonding between an acetic acid molecule and:

a. Another acetic acid molecule

b. Water molecules

10.50 Use a dotted line to show hydrogen bonding between a propanoic acid molecule and:

a. Another propanoic acid molecule

b. Water molecules

10.51 Why are acetic acid, sodium acetate, and sodium caprate all soluble in water, whereas capric acid, a 10-carbon fatty acid, is not?

10.52 Caproic acid, a six-carbon acid, has a solubility in water of 1 g/100 mL of water (**Table 10.4**). Which part of the structure of caproic acid is responsible for its solubility in water, and which part prevents greater solubility?

10.53 List the following compounds in order of increasing water solubility.

a. hexane

b. 2-pentanol

c. 3-pentanone

d. pentanoic acid

10.54 List the following compounds in order of increasing water solubility.

a. ethoxyethane

b. propanoic acid

c. pentane

d. 1-butanol

10.55 Draw structural formulas for the following carboxylic acids and circle the acidic hydrogen atoms present.

a. pentanoic acid

b. 3,4-dichlorobutanoic acid

c. 3,3,5-trimethylhexanoic acid

10.56 Draw structural formulas for the following carboxylic acids and circle the acidic hydrogen atoms present.

a. 3-bromopropanoic acid

b. butanoic acid

c. 2,3,4-trimethylpentanoic acid

10.57 As we discuss the cellular importance of lactic acid in a later chapter, we will refer to this compound as lactate. Explain why.

$$CH_3-\underset{}{\overset{OH}{\overset{|}{C}H}}-\overset{O}{\overset{\|}{C}}-OH$$

lactic acid

10.58 Write an equation to illustrate the equilibrium that is present when propanoic acid is dissolved in water. What structure predominates when OH^- is added to raise the pH to 12? What structure predominates as acid is added to lower the pH to 2?

10.59 Complete each of the following reactions.

a. $$H-\overset{O}{\overset{\|}{C}}-OH + NaOH \longrightarrow$$

b. $$CH_3CH_2CH_2-\overset{O}{\overset{\|}{C}}-OH + KOH \longrightarrow$$

10.60 Complete each of the following reactions.

a. $$CH_3(CH_2)_7-\overset{O}{\overset{\|}{C}}-OH + NaOH \longrightarrow$$

b. 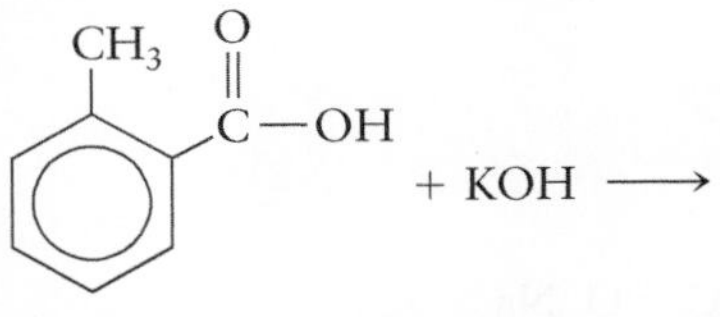 (benzene ring bearing CH_3 and, on the adjacent carbon, $\overset{O}{\overset{\|}{C}}-OH$) $+ KOH \longrightarrow$

10.61 Write a balanced equation for the reaction of acetic acid with each of the following.

a. NaOH

b. KOH

c. $Ca(OH)_2$

10.62 Write a balanced equation for the reaction of butanoic acid with each of the following.

a. KOH

b. H_2O

c. NaOH

The Nomenclature of Esters and Amides (Section 10.6)

10.63 Assign common names to the following esters. Refer to **Table 10.3** for the common names of the acids.

a. $$CH_3CH_2CH_2CH_2CH_2-\overset{O}{\overset{\|}{C}}-O-\underset{CH_3}{\underset{|}{C}HCH_3}$$

b. $$CH_3-\overset{O}{\overset{\|}{C}}-O-CH_2CH_3$$

10.64 Assign common names to the following esters. Refer to **Table 10.3** for the common names of the acids.

a. $$CH_3CH_2-\overset{O}{\overset{\|}{C}}-O-CH_3$$

b. $$CH_3-\overset{O}{\overset{\|}{C}}-O-CH_2CH_2CH_3$$

10.65 Give the IUPAC name for each of the following.

a. $$CH_3CH_2-O-\overset{O}{\overset{\|}{C}}-CH_2CH_3$$

b. $$CH_3CH_2CH_2CH_2-\overset{O}{\overset{\|}{C}}-O-CH_2CH_2CH_3$$

10.66 Give the IUPAC name for each of the following.

a. 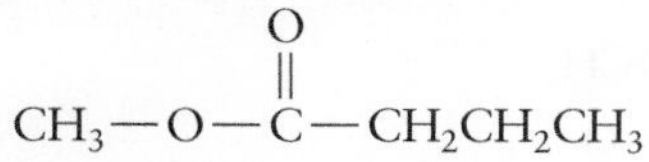 $CH_3-O-\overset{O}{\overset{\|}{C}}-CH_2CH_2CH_3$

b. 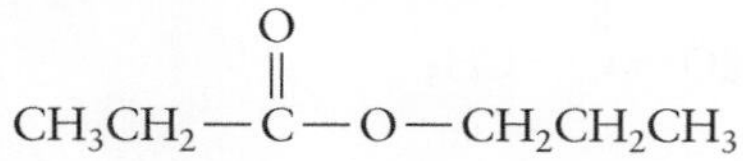 $CH_3CH_2-\overset{O}{\overset{\|}{C}}-O-CH_2CH_2CH_3$

10.67 Assign IUPAC names to the esters produced by a reaction between butanoic acid and the following.

a. 1-butanol

b. methanol

c. 1-propanol

10.68 Draw structural formulas for the following.

a. methyl ethanoate

b. propyl 2-bromobenzoate

c. ethyl 3,4-dimethylpentanoate

10.69 Draw structural formulas for the following.

a. phenyl formate

b. methyl 4-nitrobenzoate

c. ethyl 2-chloropropanoate

10.70 Assign IUPAC names to the simple esters produced by a reaction between ethanoic acid and the following.

a. ethanol

b. 1-propanol

c. 1-butanol

10.71 Identify which of the following compounds are amides.

a. 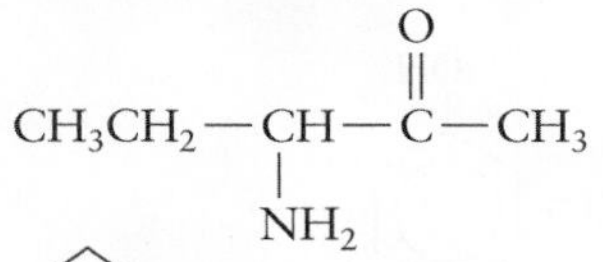 $CH_3CH_2-\underset{\underset{NH_2}{|}}{CH}-\overset{O}{\overset{\|}{C}}-CH_3$

b. (five-membered ring) $N-H$, $=O$

c. $H_2N-\overset{O}{\overset{\|}{C}}-CH_2CH_3$

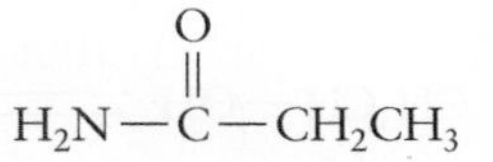

d. 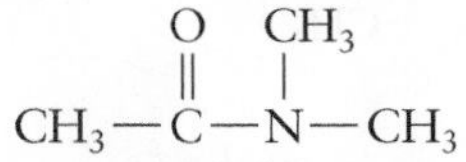 $CH_3-\overset{O}{\overset{\|}{C}}-\overset{CH_3}{\overset{|}{N}}-CH_3$

10.72 Identify which of the following compounds are amides.

a. 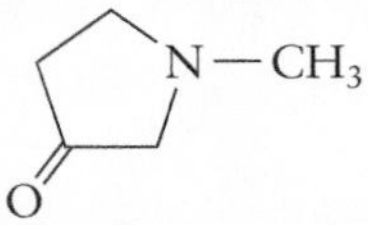 (five-membered ring) $N-CH_3$, $=O$

b. 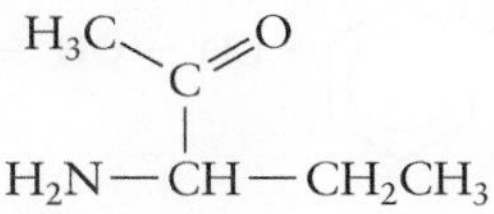 $H_3C-\overset{}{C}=O$ attached to $H_2N-CH-CH_2CH_3$

c. $CH_3-\overset{O}{\overset{\|}{C}}-NH-CH_3$

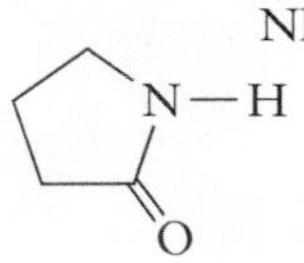

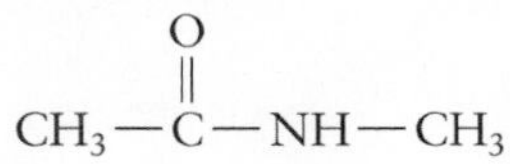

d. 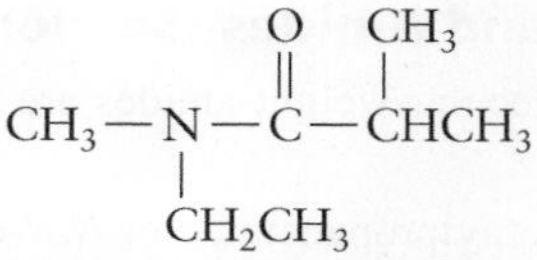 $CH_3-\underset{\underset{CH_2CH_3}{|}}{N}-\overset{O}{\overset{\|}{C}}-\overset{CH_3}{\overset{|}{C}HCH_3}$

10.73 Assign IUPAC names to the following amides.

a. 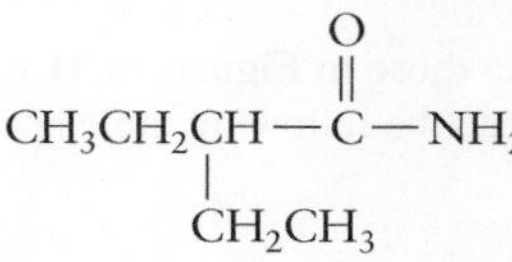 $CH_3CH_2\underset{\underset{CH_2CH_3}{|}}{CH}-\overset{O}{\overset{\|}{C}}-NH_2$

b. $CH_3CH_2CH_2-\overset{O}{\overset{\|}{C}}-NH-$ (benzene ring)

c. (benzene ring)$-\overset{O}{\overset{\|}{C}}-NH-\underset{\underset{CH_3}{|}}{CH}CH_3$

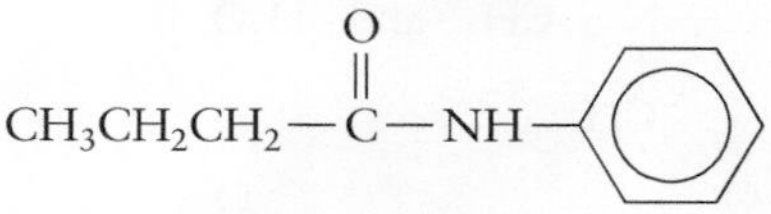

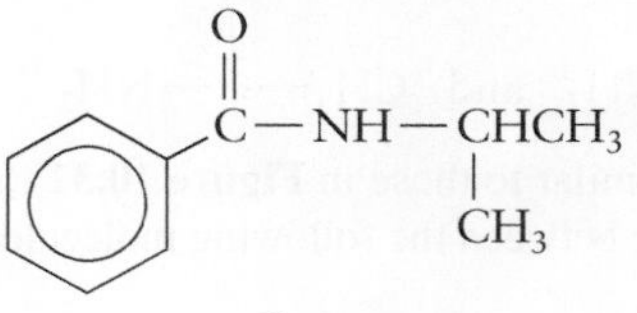

d. 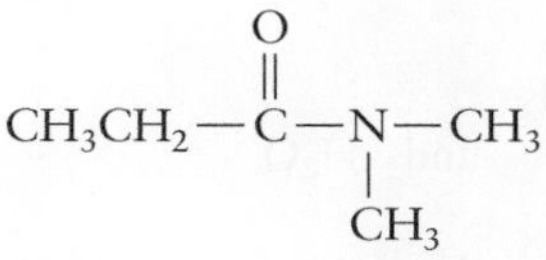 $CH_3CH_2-\overset{O}{\overset{\|}{C}}-\underset{\underset{CH_3}{|}}{N}-CH_3$

10.74 Assign IUPAC names to the following amides.

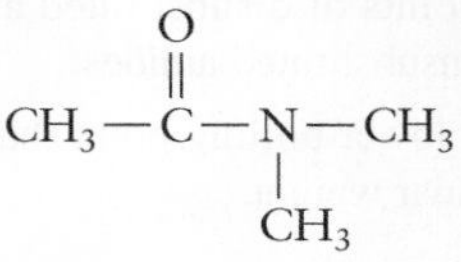

a. 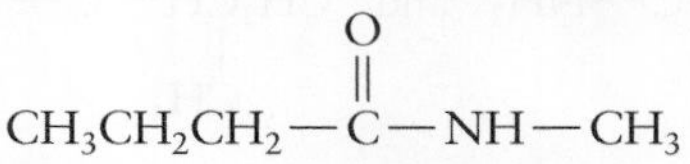 $CH_3CH_2CH_2-\overset{O}{\overset{\|}{C}}-NH-CH_3$

b. $CH_3-\overset{O}{\overset{\|}{C}}-\underset{\underset{CH_3}{|}}{N}-CH_3$

c. (benzene ring with CH_3)$-\overset{O}{\overset{\|}{C}}-NH_2$

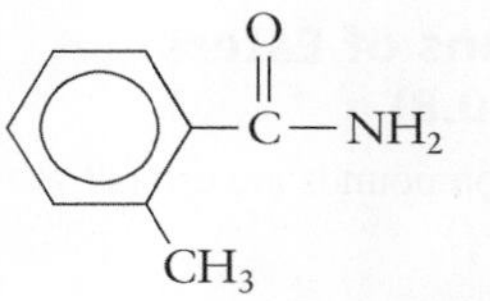

d. 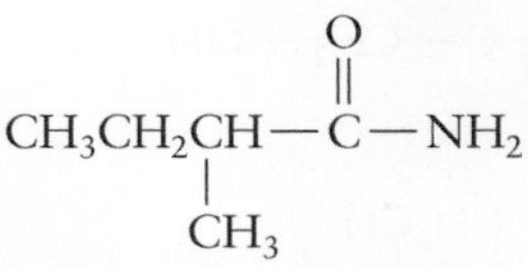 $CH_3CH_2\underset{\underset{CH_3}{|}}{CH}-\overset{O}{\overset{\|}{C}}-NH_2$

10.75 Draw structural formulas for the following amides.

a. benzamide

b. *N*-methylethanamide

c. *N*-methyl-3-phenylbutanamide

10.76 Draw structural formulas for the following amides.

a. butanamide

b. *N*-ethylbenzamide

c. *N,N*-dimethylpropanamide

Properties of Esters and Amides (Section 10.7)

10.77 Explain why low-molecular-weight amides are relatively water soluble.

10.78 Would you expect *N*-ethylpropanamide or *N,N*-dimethylpropanamide to have the higher boiling point? Explain your answer.

10.79 Draw diagrams similar to those in **Figure 10.31** to illustrate hydrogen bonding between the following molecules.

a. $CH_3-\overset{O}{\overset{\|}{C}}-\underset{CH_3}{\underset{|}{N}}-CH_3$ and H_2O

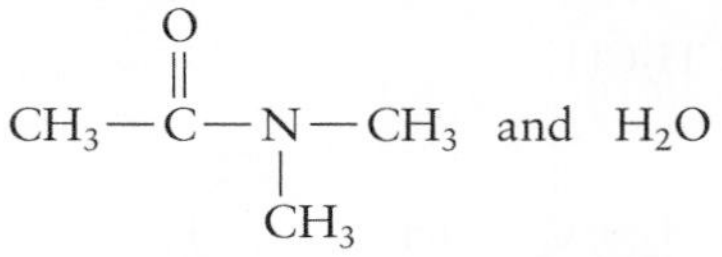

b. $CH_3-\overset{O}{\overset{\|}{C}}-NH_2$ and $CH_3-\overset{O}{\overset{\|}{C}}-NH_2$

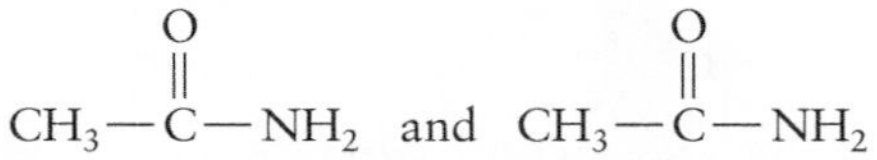

10.80 Draw diagrams similar to those in **Figure 10.31** to illustrate hydrogen bonding between the following molecules.

a. (cyclopentyl)$-\overset{O}{\overset{\|}{C}}-NH_2$ and H_2O

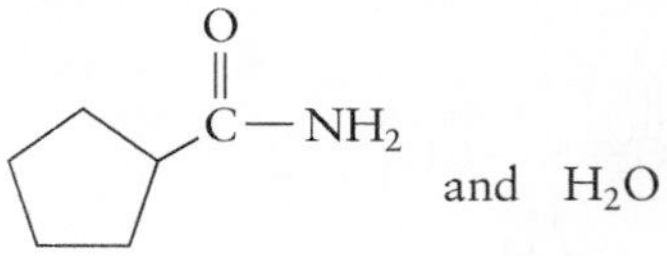

b. $CH_3\underset{CH_3}{\underset{|}{CH}}-\overset{O}{\overset{\|}{C}}-NH_2$ and $CH_3\underset{CH_3}{\underset{|}{CH}}-\overset{O}{\overset{\|}{C}}-NH_2$

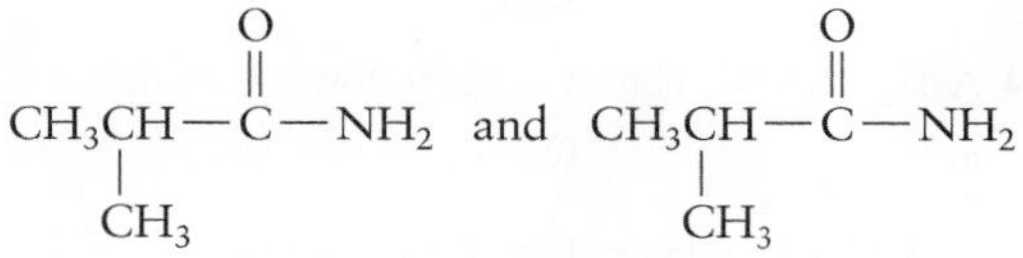

10.81 Explain why the boiling points of disubstituted amides are often lower than those of unsubstituted amides.

10.82 Explain why esters have a lower boiling point than carboxylic acids of similar molecular weight.

Formation and Reactions of Esters and Amides (Section 10.8)

10.83 Which of the following compounds are esters?

a. $CH_3-O-CH_2-\overset{O}{\overset{\|}{C}}-CH_2-OH$

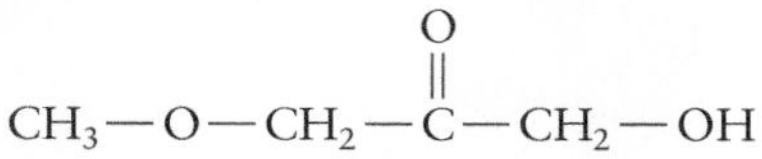

b. (phenyl)$-\overset{O}{\overset{\|}{C}}-O-CH_2CH_3$

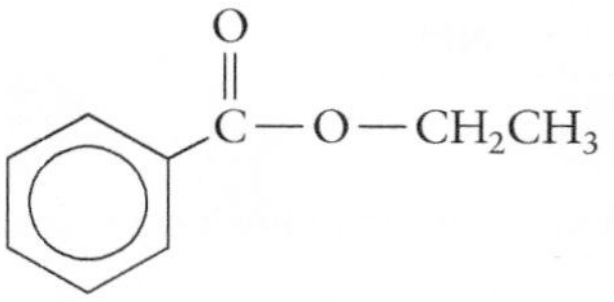

c. CH_3-O-CH_3

d. (cyclohexane ring with $=O$ and $-O-CH_3$ on the same carbon)

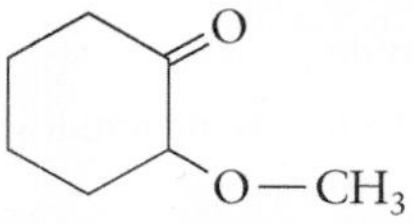

10.84 Which of the following compounds are esters?

a. (phenyl)$-CH_2-O-CH_2-OH$

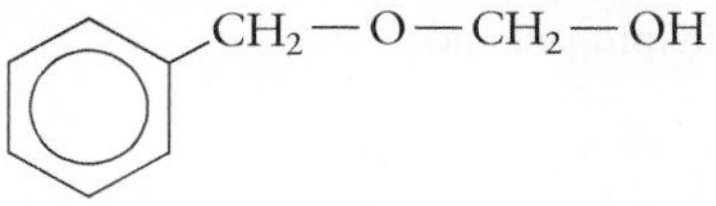

b. $CH_3-\overset{O}{\overset{\|}{C}}-O-CH_2CH_3$

c. $CH_3CH_2-\underset{OCH_3}{\underset{|}{CH}}-OCH_3$

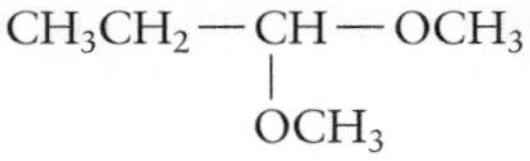

d. (phenyl)$-O-\overset{O}{\overset{\|}{C}}-CH_3$

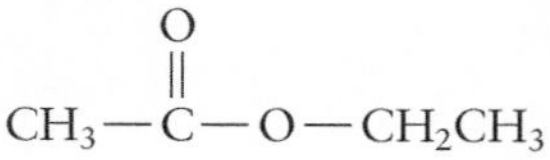

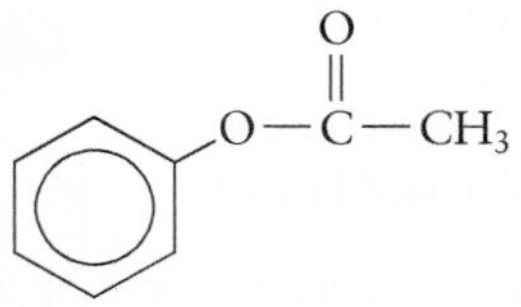

10.85 Draw the structural formula of the ester made when each of the following carboxylic acids and alcohols react.

a. propanoic acid and 2-propanol

b. ethanoic acid and 1-propanol

c. 2-methylbutanoic acid and ethanol

d. methanoic acid and 2-butanol

10.86 Draw the structural formula of the ester made when each of the following carboxylic acids and alcohols react.

a. methanoic acid and methanol

b. propanoic acid and ethanol

c. 2-methylpropanoic acid and 2-butanol

d. butanoic acid and 2-propanol

10.87 Complete the following reactions.

a. $CH_3-\overset{O}{\overset{\|}{C}}-O-\overset{O}{\overset{\|}{C}}-CH_3 + CH_3-OH \longrightarrow$

b. $CH_3\underset{CH_3}{\underset{|}{CH}}-\overset{O}{\overset{\|}{C}}-OH +$ (phenyl)$-CH_2-OH \xrightleftharpoons{H^+,\text{ heat}}$

c. $CH_3\underset{CH_3}{\underset{|}{CH}}-\overset{O}{\overset{\|}{C}}-Cl +$ (phenol, C_6H_5-OH) $\longrightarrow$

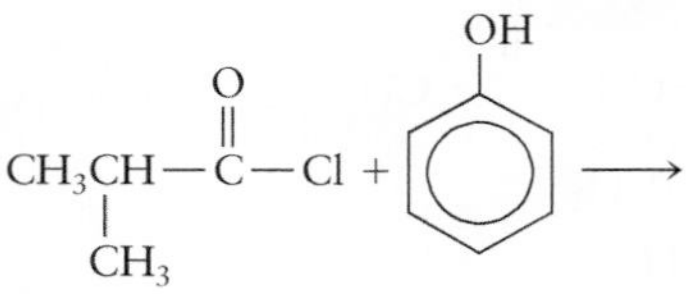

10.88 Complete the following reactions.

a. (phenyl)$-\overset{O}{\overset{\|}{C}}-OH + CH_3\underset{CH_3}{\underset{|}{CH}}-OH \xrightleftharpoons{H^+,\text{ heat}}$

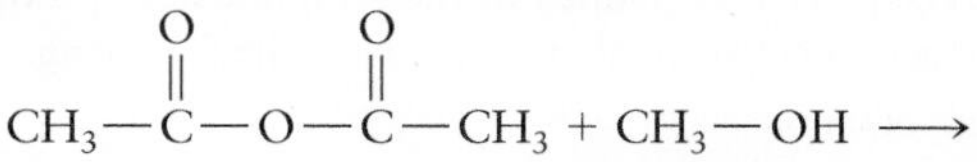

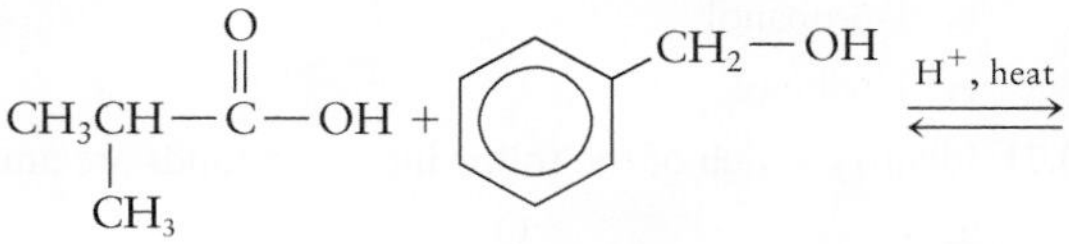

b. (phenyl)$-\overset{O}{\overset{\|}{C}}-Cl + CH_3\underset{CH_3}{\underset{|}{CH}}-OH \longrightarrow$

c. (phenyl)$-\overset{O}{\overset{\|}{C}}-O-\overset{O}{\overset{\|}{C}}-$(phenyl) $+ CH_3-OH \longrightarrow$

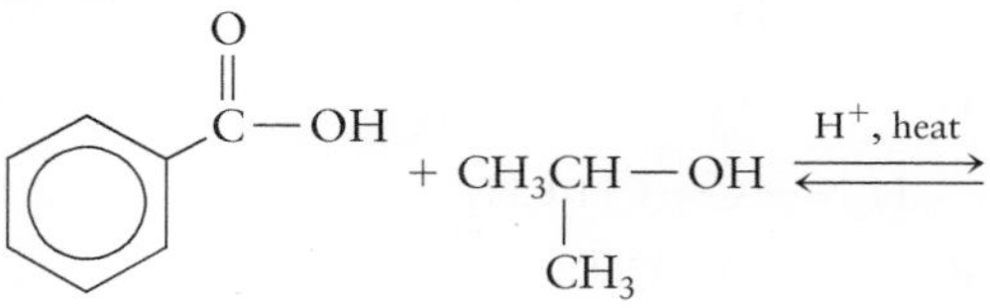

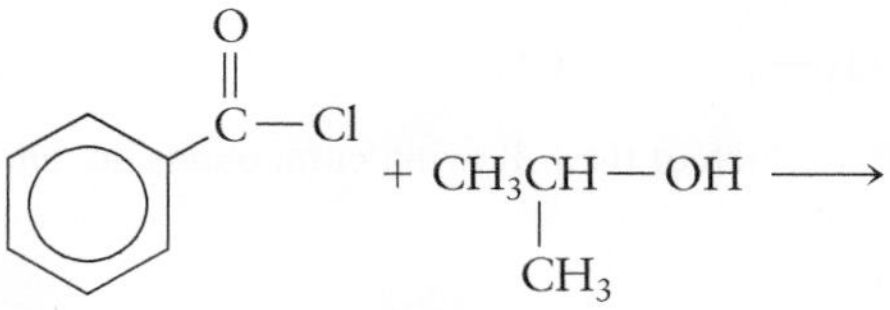

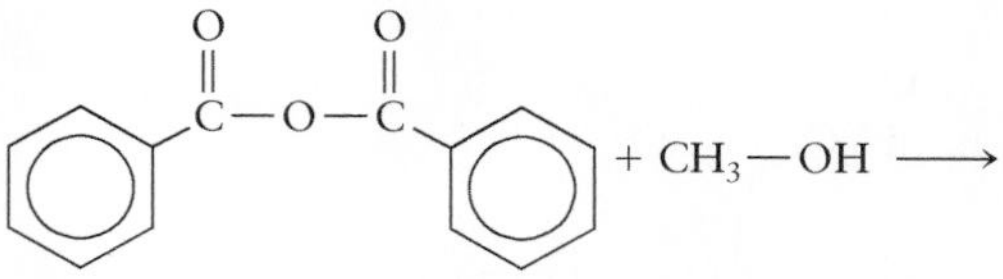

10.89 Draw the structural formulas of the "parent" carboxylic acid and the "parent" alcohol reacted to make each of the following esters.

a.

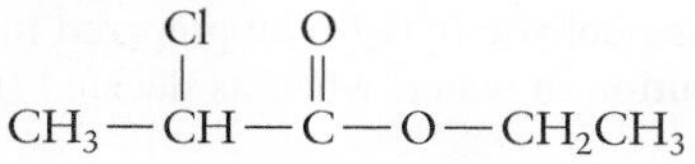

$$\begin{array}{c} \quad Cl \quad\;\; O \\ \quad | \quad\;\;\; \| \\ CH_3-CH-C-O-CH_2CH_3 \end{array}$$

b.

$$\begin{array}{c} \qquad\quad O \\ \qquad\quad \| \\ CH_3CH_2-O-C-\text{(cyclopentyl)} \end{array}$$

c.

$$\begin{array}{c} CH_3\;O \\ | \quad \| \\ CH_3C-C-O-CH_3 \\ | \\ CH_3 \end{array}$$

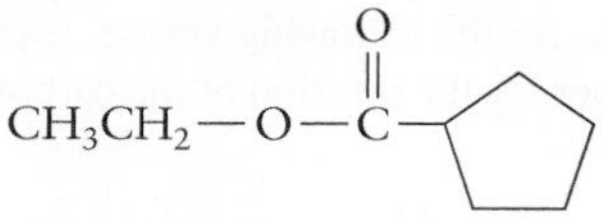

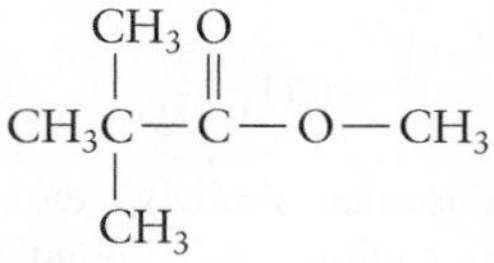

d.

$$\begin{array}{c} \qquad\qquad\qquad O \\ \qquad\qquad\qquad \| \\ CH_3CH_2CH_2CH_2-O-C-CH_3 \end{array}$$

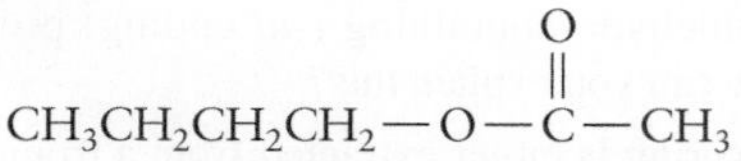

10.90 Draw the structural formulas of the "parent" carboxylic acid and the "parent" alcohol reacted to make each of the following esters.

a.

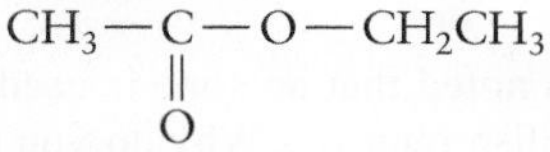

$$\begin{array}{c} CH_3-C-O-CH_2CH_3 \\ \| \\ O \end{array}$$

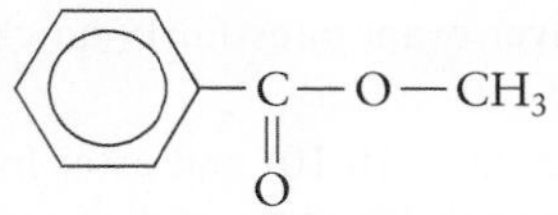

b.

$$\begin{array}{c} \text{(phenyl)}-C-O-CH_3 \\ \| \\ O \end{array}$$

c.

$$\begin{array}{c} \qquad\qquad\quad O \\ \qquad\qquad\quad \| \\ CH_3CH_2CH_2-O-C-CH_3 \end{array}$$

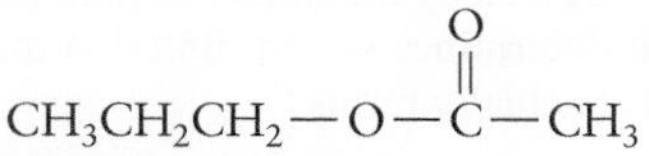

d.

$$\begin{array}{c} CH_3-C-O-CHCH_3 \\ \| \qquad\quad | \\ O \qquad\; CH_3 \end{array}$$

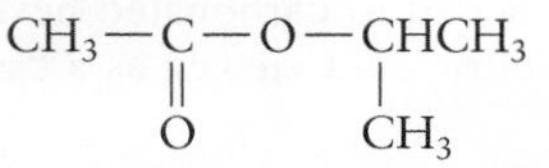

10.91 Give the structures of the ester that forms when butanoic acid is reacted with the following.

a. propanol

b. 3-methylcyclopentanol

c. 3-chloro-2-butanol

10.92 Give the structures of the ester that forms when propanoic acid is reacted with the following.

a. methyl alcohol

b. phenol

c. 2-methyl-1-propanol

10.93 The structures of two esters used as artificial flavorings are given below. Write the structure of the acid and the alcohol from which each ester could be synthesized.

a. Orange flavoring

$$\begin{array}{c} \quad O \\ \quad \| \\ CH_3-C-O-(CH_2)_7-CH_3 \end{array}$$

b. Raspberry flavoring

$$\begin{array}{c} \quad O \\ \quad \| \\ H-C-O-CH_2CH_3 \end{array}$$

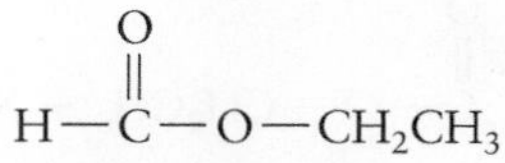

10.94 The structures of two esters used as artificial flavorings are given below. Write the structure of the acid and the alcohol from which each ester could be synthesized.

a. Pineapple flavoring

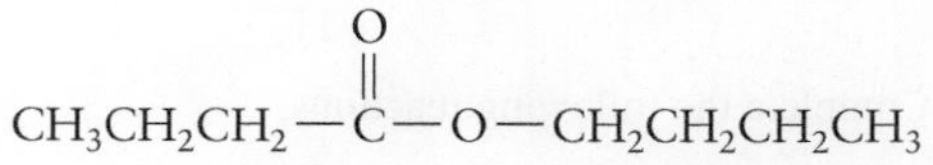

$$\begin{array}{c} \qquad\qquad\quad O \\ \qquad\qquad\quad \| \\ CH_3CH_2CH_2-C-O-CH_2CH_2CH_2CH_3 \end{array}$$

b. Apple flavoring

$$\begin{array}{c} \qquad\qquad\quad O \\ \qquad\qquad\quad \| \\ CH_3CH_2CH_2-C-O-CH_3 \end{array}$$

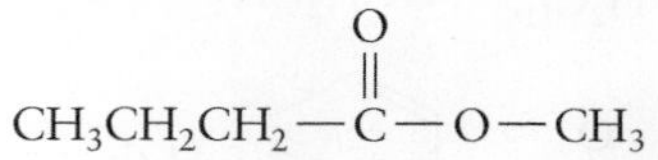

10.95 Draw the structural formulas for the carboxylic acid chloride molecule used to make each of the following amides.

a.

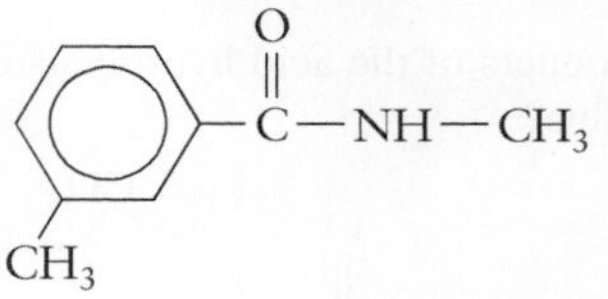

$$\begin{array}{c} \qquad\qquad O \\ \qquad\qquad \| \\ \text{(benzene ring)}-C-NH-CH_3 \\ | \\ CH_3 \end{array}$$

b.

$$\begin{array}{c} \qquad\qquad\quad O \\ \qquad\qquad\quad \| \\ CH_3CH_2CH_2-C-NH-CH_2CH_3 \end{array}$$

c.

$$\begin{array}{c} CH_3\;\; O \;\; CH_3 \\ | \qquad \| \qquad | \\ CH_3CH-C-N-CH_3 \end{array}$$

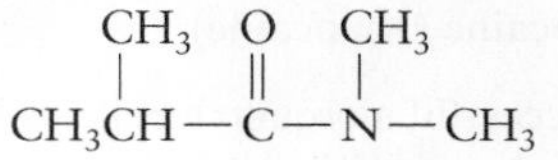

d.

$$\begin{array}{c} CH_3CH_2 \\ | \\ CH_3CH_2-N-C-CH_2CH_3 \\ \| \\ O \end{array}$$

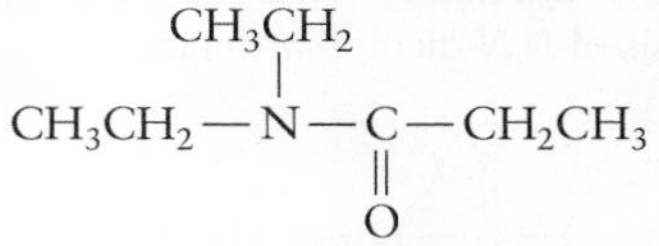

10.96 Draw the structural formulas for the acid chloride molecule that is used to make each of the following amides.

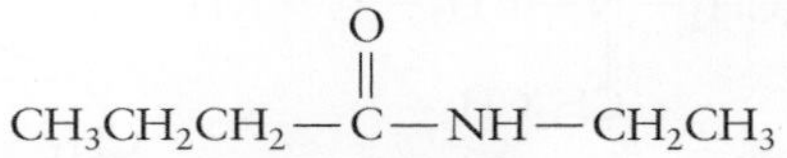

a.

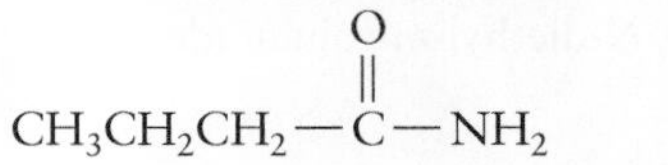

$$\begin{array}{c} \qquad\qquad\quad O \\ \qquad\qquad\quad \| \\ CH_3CH_2CH_2-C-NH_2 \end{array}$$

b.

$$\begin{array}{c} \quad O \\ \quad \| \\ H-C-NH-CH_2CH_2CH_3 \end{array}$$

c.

$$\begin{array}{c} \qquad\qquad O \\ \qquad\qquad \| \\ \text{(phenyl)}-C-NH-\text{(phenyl)} \end{array}$$

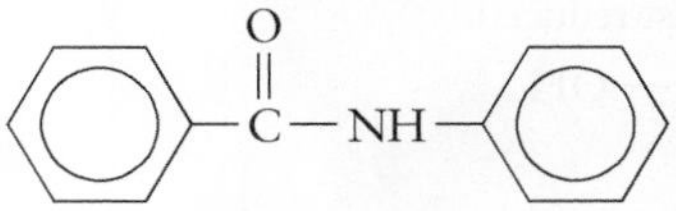

d.

$$\begin{array}{c} CH_3 \\ | \\ \text{(cyclopentyl)}-N-C-CH_3 \\ \| \\ O \end{array}$$

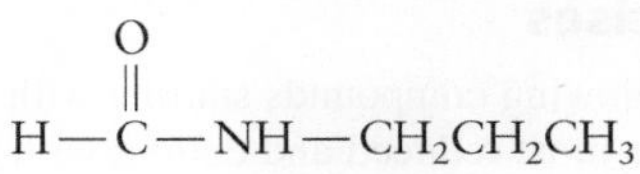

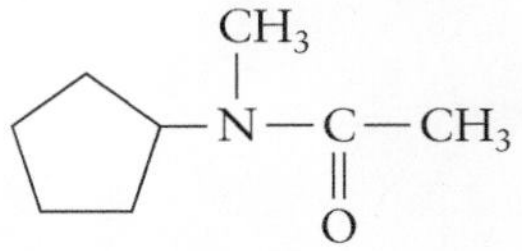

10.97 Complete the following reactions.

a. $$C_6H_5-\overset{\overset{\displaystyle O}{\|}}{C}-O-CH_2CH_3 + NaOH \longrightarrow$$

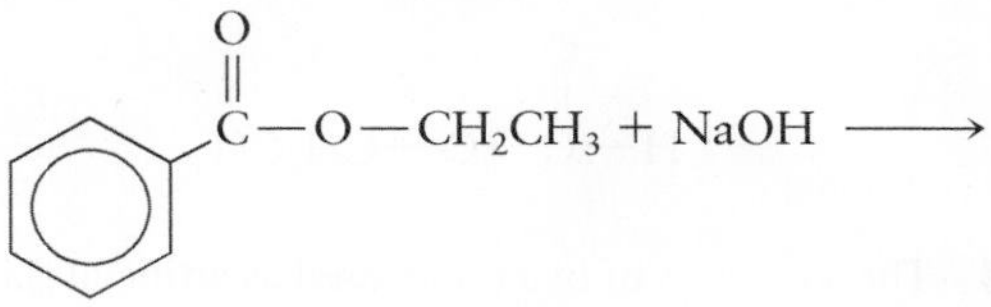

b. $$CH_3(CH_2)_4-\overset{\overset{\displaystyle O}{\|}}{C}-O-\underset{\underset{\displaystyle CH_3}{|}}{CHCH_3} + H_2O \overset{H^+}{\rightleftarrows}$$

10.98 Complete the following reactions.

a. $$H_3C(H_2C)_{16}C(=O)OCH_2CH_3 + NaOH \longrightarrow$$

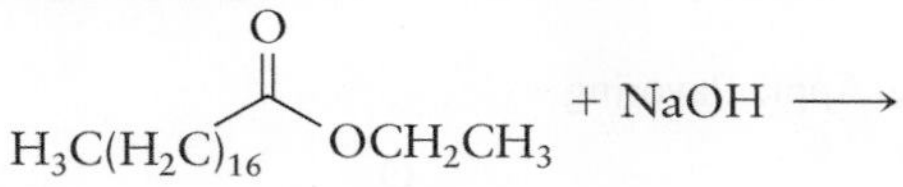

b. $$(CH_3)_2CH-C(=O)-O-C_6H_5 + H_2O \overset{H^+}{\rightleftarrows}$$

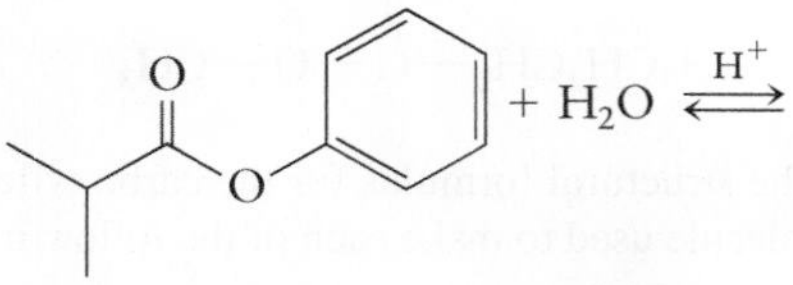

10.99 What are the products of the acid hydrolysis of the local anesthetic lidocaine?

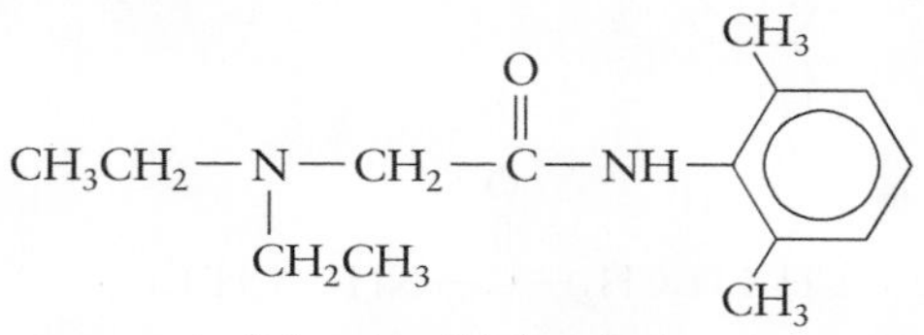

$$CH_3CH_2-\underset{\underset{\displaystyle CH_2CH_3}{|}}{N}-CH_2-\overset{\overset{\displaystyle O}{\|}}{C}-NH-C_6H_3(CH_3)_2$$

lidocaine (Xylocaine)

10.100 One of the most successful mosquito repellents has the following structure and name. What are the products of the basic hydrolysis of *N,N*-diethyl-*m*-toluamide?

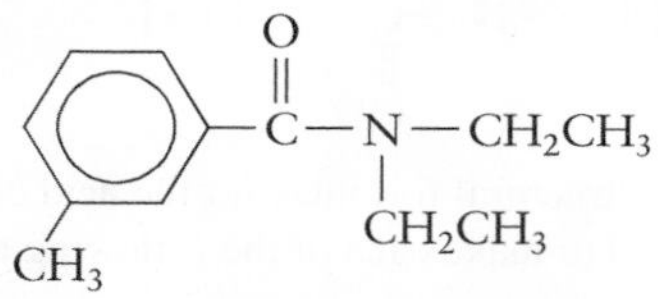

$$m\text{-}CH_3C_6H_4-\overset{\overset{\displaystyle O}{\|}}{C}-\underset{\underset{\displaystyle CH_2CH_3}{|}}{N}-CH_2CH_3$$

N, *N*-diethyl-*m*-toluamide

Additional Exercises

10.101 Arrange the following compounds starting with the carbon atom that is the most reduced and ending with the carbon that is the least reduced.

a. $R-CH_2-OH$

b. $R-\overset{\overset{\displaystyle O}{\|}}{C}H$

c. $R-CH_3$

10.102 Explain how a sodium bicarbonate ($NaHCO_3$) solution can be used to distinguish between a carboxylic acid solution and an alcohol solution.

10.103 Why can formaldehyde (CH_2O) be prepared in the form of a 37% solution in water, whereas decanal ($C_{10}H_{20}O$) cannot?

10.104 In the IUPAC name for the following ketone, it is not common to use a number for the position of the carbonyl group. Why not?

$$CH_3CH_2-\overset{\overset{\displaystyle O}{\|}}{C}-CH_3$$

10.105 In the labels of some consumer products, ketone components listed with an *-one* ending can be found. However, very few aldehyde-containing (*-al* ending) products are found. How can you explain this?

10.106 Vanilla flavoring is either extracted from a tropical orchid or synthetically produced from wood pulp by-products. What differences in chemical structure would you expect in these two commercial products: vanilla extract and imitation vanilla extract?

10.107 In **Figure 10.7**, it is noted that acetone is used as a solvent in fingernail polish remover. Why do you think fingernail polish remover evaporates fairly quickly when used?

10.108 Ester formation (Reaction 10.10) and ester hydrolysis (Reaction 10.12) are exactly the same reaction, only written in reverse. What determines which direction the reaction proceeds and what actually forms?

10.109 Citric acid is often added to carbonated beverages as a flavoring. Why is citric acid viewed as a "safe" food additive?

10.110* Write equations to show how the following conversions can be achieved. More than one reaction is required, and reactions from earlier chapters may be necessary.

a. $$CH_3CH{=}CHCH_3 \longrightarrow CH_3-\overset{\overset{\displaystyle O}{\|}}{C}-CH_2CH_3$$

b. $$CH_3CH_2CH_2-OH \longrightarrow CH_3CH_2-\overset{\overset{\displaystyle O}{\|}}{C}-OH$$

10.111* Write equations to show how the following conversions can be achieved. More than one reaction is required, and reactions from earlier chapters may be necessary.

a.

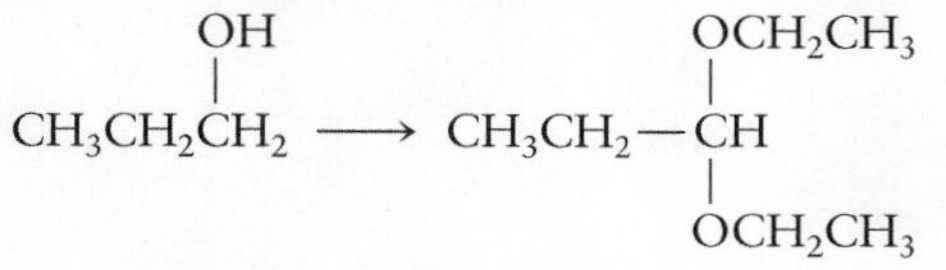

$$\underset{\displaystyle CH_3CH_2CH_2}{\overset{\displaystyle OH}{|}} \longrightarrow CH_3CH_2-\overset{\displaystyle OCH_2CH_3}{\underset{\displaystyle OCH_2CH_3}{CH}}$$

b.

$$CH_3\overset{\displaystyle CH_3}{\overset{|}{C}}H-\overset{\displaystyle O}{\overset{\|}{C}}-CH_3 \longrightarrow CH_3\overset{\displaystyle CH_3}{\overset{|}{C}}=CH-CH_3$$

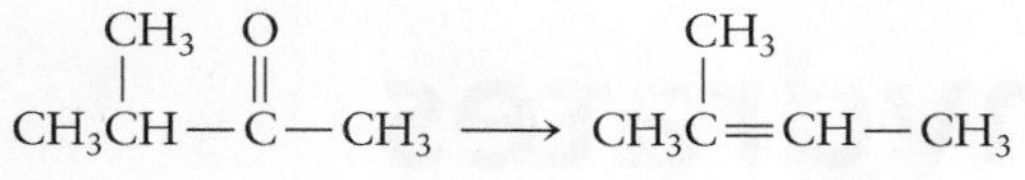

10.112 Define and compare the terms *ester hydrolysis* and *saponification.*

11 Carbohydrates

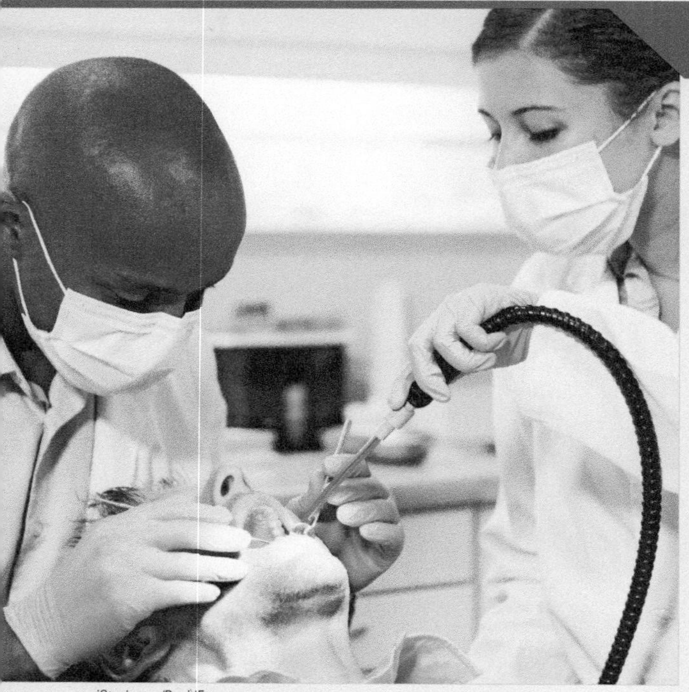

iStock.com/RapidEye

Health Career Focus

Dental Assistant

I love working with children, and I get that opportunity in my work at the dentist's office. In college, I made a brochure for my chemistry course that also earned credit in my dental program. It showed different foods and how they affect the teeth. I still use that at work to teach kids about dental health.

Most children know that sugar is bad for their teeth, but many are surprised to learn that enzymes in the mouth turn starches into sugar. So, carbohydrate-rich foods like crackers and even potatoes can be as damaging as candy.

When I give a child their gift bag—with a toothbrush, floss, and toothpaste—at the end of the appointment, I think about the difference I'm making in the health of that child. It's rewarding to know that my work is important.

Follow-up to this Career Focus appears at the end of the chapter before the *Concept Summary*.

Learning Objectives

When you have completed your study of Chapter 11, you should be able to:

1. Describe the four major functions of carbohydrates in living organisms. **(Section 11.1)**
2. Classify carbohydrates as monosaccharides, disaccharides, or polysaccharides. **(Section 11.1)**
3. Identify chiral carbon atoms in molecules. **(Section 11.2)**
4. Draw pairs of enantiomers using wedges and dashes. **(Section 11.2)**
5. Use Fischer projections to represent D- and L-enantiomers. **(Section 11.3)**
6. Identify the following functional groups: hemiacetal, hemiketal, acetal, and ketal. **(Section 11.3)**
7. Draw Haworth projections of carbohydrate anomers. **(Section 11.3)**
8. Classify monosaccharides as aldoses or ketoses. **(Section 11.4)**
9. Classify monosaccharides according to the number of carbon atoms they contain. **(Section 11.4)**
10. Describe uses of four important monosaccharides. **(Section 11.5)**
11. Describe the purpose of Benedict's test. **(Section 11.6)**
12. Identify glycosidic linkages in carbohydrate derivatives. **(Section 11.6)**
13. Describe the uses of three important disaccharides. **(Section 11.7)**
14. Describe the linkage in disaccharides. **(Section 11.7)**
15. Write reactions for the condensation and hydrolysis of disaccharides. **(Section 11.7)**
16. Describe the uses of three important polysaccharides. **(Section 11.8)**
17. Describe the linkage in polysaccharides. **(Section 11.8)**

Chapter 11 marks the start of our study of the four major types of **biomolecules**: carbohydrates (Chapter 11), proteins (Chapter 12), nucleic acids (Chapter 13), and lipids (Chapter 14). Biomolecules are organic because they are compounds of carbon, but they are also a starting point for **biochemistry**, the chemistry of living organisms. Thus, Chapters 11–14 on biomolecules represent an overlap of two areas of chemistry: organic and biochemistry. We begin Chapter 11 by exploring the general functions of sugars in biological systems. Then we examine their structures starting with the stereochemistry of monomers and expanding to Fischer projections, cyclization, and Haworth projections. Next, we examine how monomers are classified, explore specific important monosaccharides, and learn about their physical and chemical properties. We finish the chapter by expanding our discussion to disaccharides and complex sugars.

biomolecule A general term referring to organic compounds essential to life.

biochemistry A study of the compounds and processes associated with living organisms.

Billion Photos/Shutterstock.com

Figure 11.1 Foods such as pasta, bread, and rice are high in carbohydrates, which supply energy for the body and supply carbon for the synthesis of cell components.

11.1 Carbohydrate Structure and Function

Learning Objective 1 Describe the four major functions of carbohydrates in living organisms.

Learning Objective 2 Classify carbohydrates as monosaccharides, disaccharides, or polysaccharides.

Carbohydrates are compounds of tremendous biological and commercial importance. Widely distributed in nature, they include such familiar substances as cellulose, table sugar, and starch (see **Figure 11.1**). Carbohydrates have four important functions in living organisms:

- To provide energy through their oxidation
- To supply carbon for the synthesis of cell components
- To serve as a stored form of chemical energy
- To form a part of the structural elements of some cells and tissues

The most striking chemical characteristic of carbohydrates is the large number of functional groups they have. For example, ribose (**Figure 11.2**) has a functional group on every carbon atom of its five-carbon chain. This is the origin of the modern definition: **Carbohydrates** are polyhydroxy aldehydes or ketones, or substances that yield such compounds on hydrolysis.

H O
 \\ //
  C
  |
H—C—OH
  |
H—C—OH
  |
H—C—OH
  |
  CH_2OH

ribose

Figure 11.2 Condensed structural formula of ribose ($C_5H_{10}O_5$).

carbohydrate A polyhydroxy aldehyde or ketone, or substance that yields such compounds on hydrolysis.

Carbohydrates can be classified according to the size of the molecules (**Figure 11.3**). **Monosaccharides** consist of a single polyhydroxy aldehyde or ketone unit, whereas **disaccharides** are composed of two monosaccharide units covalently bonded together. **Oligosaccharides** (less common and of minor importance) contain 3 to 10 units, and **polysaccharides** consist of very long chains of linked monosaccharide units.

monosaccharide A simple carbohydrate most commonly consisting of three to six carbon atoms.

disaccharide A carbohydrate formed by the combination of two monosaccharide units.

oligosaccharide A carbohydrate formed by the combination of three to ten monosaccharide units.

polysaccharide A carbohydrate formed by the combination of many monosaccharide units.

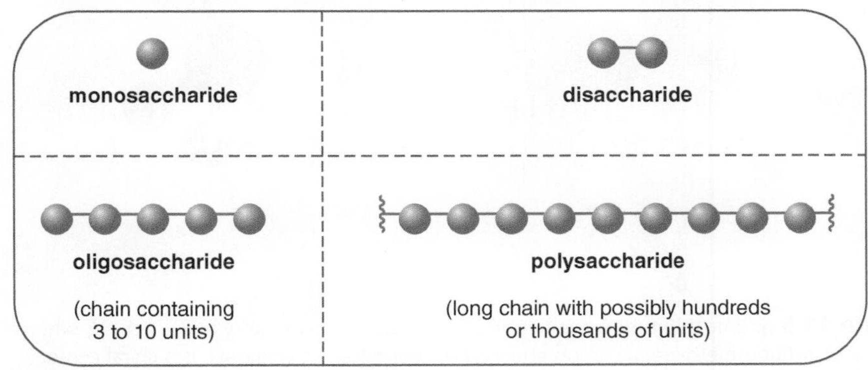

Figure 11.3 Carbohydrate classification.

11.2 Carbohydrate Stereochemistry

Learning Objective 3 Identify chiral carbon atoms in molecules.

Learning Objective 4 Draw pairs of enantiomers using wedges and dashes.

In order to appreciate the intricacies of carbohydrate structure, we must first understand the three-dimensional arrangement of atoms bonded to carbon. We have previously defined terms like *stereoisomers* (**Section 8.6**) as compounds with the same molecular formula, but a different spatial arrangement of the atoms. We must now be more specific about subsets of stereoisomers as we approach the building blocks of biochemistry, such as carbohydrates in this chapter and amino acids in Chapter 12.

When truly representing the tetrahedral nature of carbon atoms, wedges and dashes are used to help communicate exactly where bonds are located in three-dimensional space (**Figure 11.4**). The wedged bond indicates a bond coming out of the plane of the page, whereas a dashed bond indicates a bond receding back behind the plane. The other two bonds exist in the same plane as the page, so they are represented with regular-weight lines.

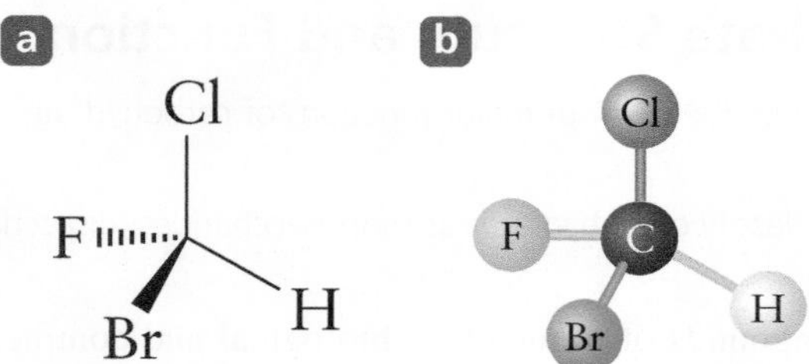

Figure 11.4 Bromochlorofluoromethane in (a) wedge-and-dash notation and (b) ball-and-stick notation.

chiral carbon A carbon atom with four different groups attached.

chiral A descriptive term for compounds or objects that cannot be superimposed on their mirror image.

A carbon atom with four different groups attached is called a **chiral carbon**. The carbon atom in bromochlorofluoromethane is chiral because it is attached to H, Cl, Br, and F (**Figures 11.4** and **11.5**). Additionally, when an object cannot be superimposed on its mirror image, the object is said to be **chiral**. Thus, a hand (**Figure 11.6**), a glove, and a shoe are chiral objects because they cannot be superimposed on their mirror images. However, a sphere, a cube, and bromochloromethane all have an internal plane of symmetry (**Figure 11.5**) and are said to be *achiral* objects because they can be superimposed on their mirror images.

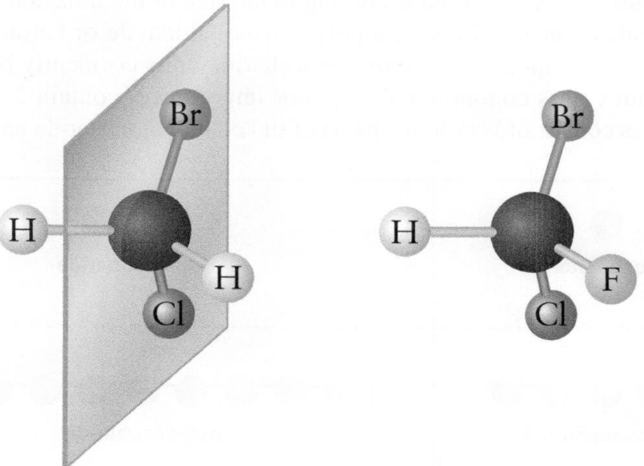

Figure 11.5 Bromochloromethane has an internal plane of symmetry and is achiral, whereas bromochlorofluoromethane, a carbon attached to four different elements, is a chiral center.

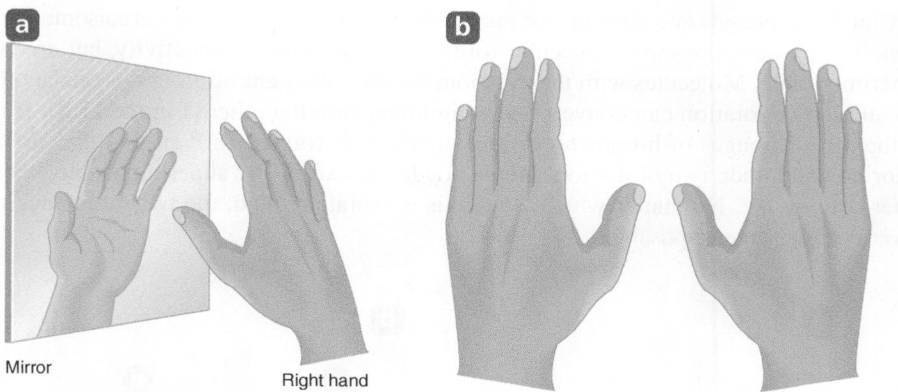

Figure 11.6 Are your hands superimposable? (a) The reflection of a right hand is a left hand. (b) A right hand and a left hand cannot be superimposed.

Example 11.1 Identifying Chiral Carbons

Which of the following compounds contain a chiral carbon?

a. $CH_3CH(OH)CH_3$ b. $CH_3CH(OH)CH_2CH_3$ c. $CH_3C(=O)CH_2CH_3$

Solution

It may help to draw the structures in a more expanded form.

a. The central carbon atom has one H, one OH, but two CH_3 groups attached. It is not chiral (achiral). Neither of the other two carbons has four different groups.

$$CH_3-\underset{OH}{\overset{H}{C}}-CH_3$$

b. The central carbon atom in 2-butanol (**Figure 11.7**) has one H, one OH, one CH_3, and one CH_2CH_3 group attached. Both CH_3 and CH_2CH_3 are bonded to the central carbon atom through a carbon atom. However, when analyzing for chiral carbons, we look at the entire alkyl group attached. CH_3 is different than CH_2CH_3, so the central carbon (identified by an asterisk *) is chiral. None of the other carbons are chiral.

$$CH_3-\underset{OH}{\overset{H}{C^*}}-CH_2CH_3$$

c. The carbonyl carbon is attached to only three groups, so it is not chiral, even though the three groups are different. None of the other carbons has four different groups attached.

$$CH_3-\underset{O}{\overset{}{C}}(=O)-CH_2-CH_3$$

Figure 11.7 Esters of 2-butanol are often used in colognes and perfumes, due to their volatile and fragrant nature.

progressman/Shutterstock.com

✔ **Learning Check 11.1** Which of the carbon atoms highlighted with a screen is chiral?

a. $CH_2OH-\boxed{CH}(OH)-CH_2OH$

b. $CHO-\boxed{CH}(OH)-CH(OH)-CH_2OH$

c. $\boxed{CH_2OH}-C(=O)-CH(OH)-CH_2OH$

d. $CH_2OH-\boxed{CH}$ in a ring: $CH-O-CH_2-CH_2-CH_2-CH_2$ (ring closed back to CH)

enantiomers Stereoisomers that are nonsuperimposable mirror images.

Chiral compounds and their mirror images display a special type of stereoisomerism—namely, they have the same molecular formula and the same connectivity, but are nonsuperimposable. Molecules with this relationship are called **enantiomers** of one another. No amount of rotation can convert one enantiomer into the other (**Figure 11.8**). If one of the mirror images of bromochlorofluoromethane is rotated 180° around the carbon–chlorine bond, then two of the four atoms bonded to carbon are superimposable, but the other two are not. No matter which bond axis we rotate around, the two molecules will never become superimposable.

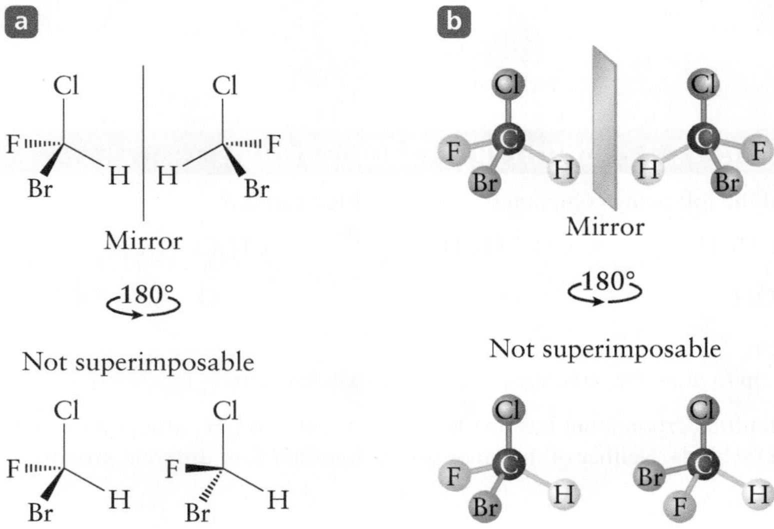

Figure 11.8 Enantiomers of bromochlorofluoromethane in (a) wedge-and-dash format and in (b) ball-and-stick form. A rotation of 180° around the carbon–chlorine bond in each shows that the original enantiomer and its mirror image are *not* superimposable.

Glyceraldehyde, the simplest carbohydrate, is a chiral molecule. Thus, there are two enantiomers of glyceraldehyde (see **Figure 11.9**). As we explain in **Section 11.3**, the two enantiomers of glyceraldehyde are called L-glyceraldehyde and

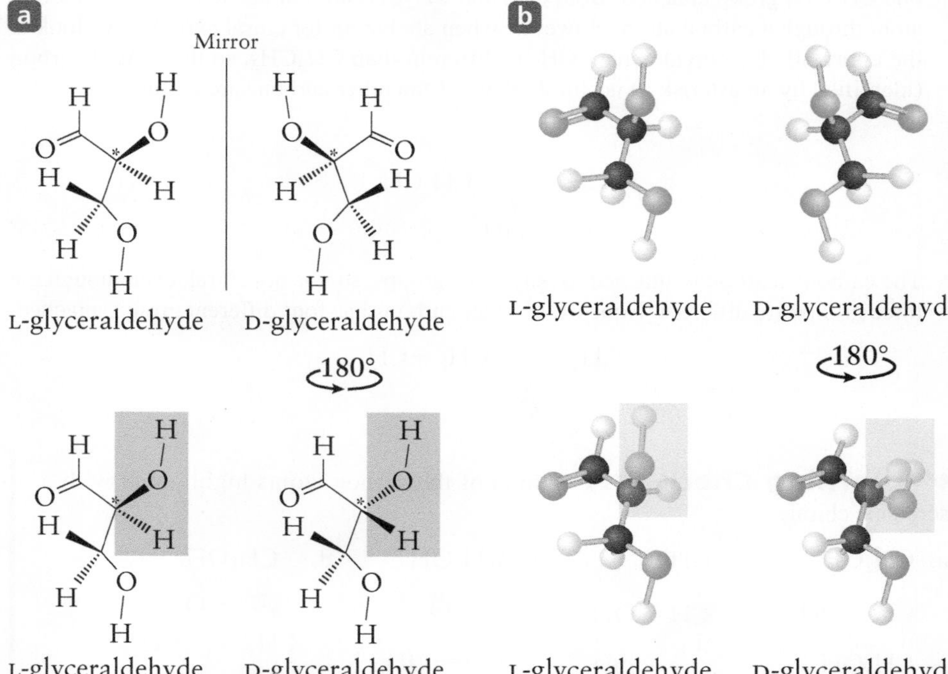

Figure 11.9 (a) The top part of the figure shows the skeletal structure of L-glyceraldehyde and its mirror image, D-glyceraldehyde. In the bottom part, D-glyceraldehyde has been rotated so that the skeletal framework matches L-glyceraldehyde. At the chiral carbon (denoted by an *), the hydroxyl group and H atom have been highlighted to denote where those substituents are placed in three dimensions, confirming they are enantiomers. (b) The same relationships between L- and D-glyceraldehyde are shown, but this time with a ball-and-stick model.

D-glyceraldehyde in order to distinguish between them. As the molecules we consider increase in size and complexity, it can become more difficult to identify where the chiral carbon resides. However, ask yourself if there is a carbon present where the four substituents differ at some point as each chain extends throughout the rest of the molecule. Unless two substituents contain the exact same atoms, in the exact same order of connectivity, throughout the entirety of the chain extending from carbon, they cannot be considered identical. In the case of glyceraldehyde, the four unique substituents bonded to the chiral carbon are an aldehyde group, a hydrogen atom, a hydroxyl (–OH) group, and a $-CH_2OH$ group. The two enantiomers of glyceraldehyde cannot be superimposed upon one another.

Organic molecules (and especially carbohydrates) may contain more than one chiral carbon (**Figure 11.10**). Erythrose, for example, has two chiral carbon atoms. When a molecule contains more than one chiral carbon, the possibility exists for two arrangements of attached groups at each chiral carbon atom. Thus, nearly all molecules with two chiral carbon atoms exist as two pairs of enantiomers, for a total of four stereoisomers. In a similar manner, when there are three chiral carbons present, there are eight stereoisomers (four pairs of enantiomers). The general formula is:

Maximum number of possible stereoisomers $= 2^n$, where n is the number of chiral carbon atoms

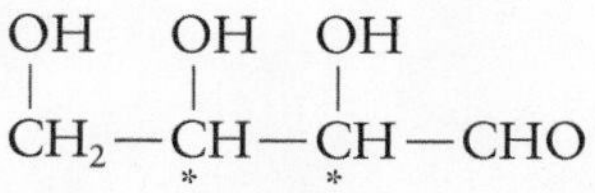

OH OH OH
$CH_2-\overset{*}{C}H-\overset{*}{C}H-CHO$
erythrose

Figure 11.10 Condensed chemical structure of erythrose with * indicating chiral centers.

Example 11.2 Drawing Enantiomers in Wedge-and-Dash Notation

The carbohydrate deoxyribose can react to form a ring, which is the way it is found in DNA (see Chapter 13); serine is an amino acid, a building block of proteins (Chapter 12).

Draw enantiomers of the following molecules in wedge-and-dash notation.

a. O=CH–CH_2–CH(OH)–CH(OH)–CH_2(OH)

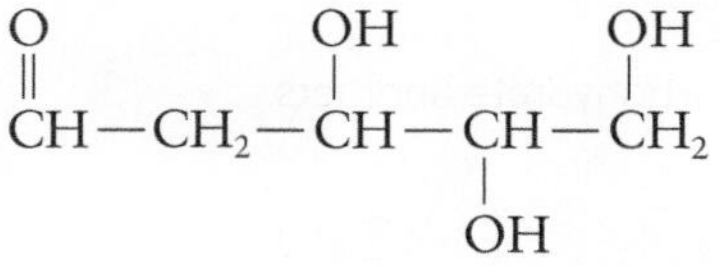

deoxyribose

b. HO–CH_2–CH(NH_2)–COOH

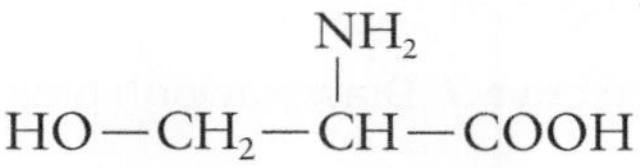

serine

Solution

a. The third and fourth carbons of deoxyribose (denoted with *) are chiral, because each is bonded to four different groups. In the left representation, the hydrocarbon chain is drawn in the plane of the page, and we arbitrarily chose to represent the –OH at position 3 with a dash and the –OH at position 4 with a wedge. Thus, the enantiomer (right) must be drawn with the hydrocarbon in the plane of the page, but rotated 180 degrees with the –OH at position 4 drawn as a wedge, and the –OH at position 3 drawn as a dash in order to make a mirror image. There would be another pair of enantiomers where the dash and wedge switch positions on carbons 3 and 4, respectively, since there are two chiral centers.

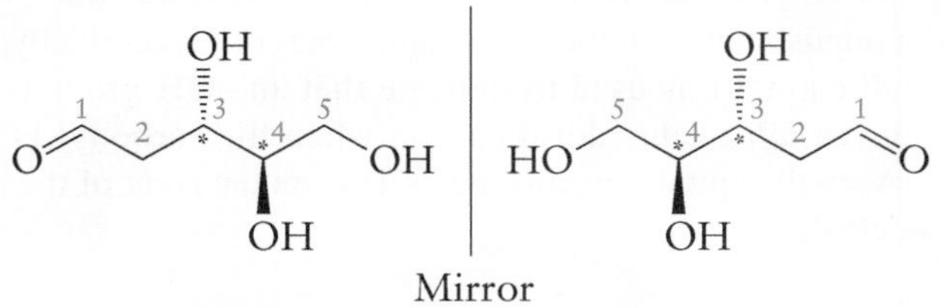

OH OH
1 2 3 4 5 | 5 4 3 2 1
O OH | HO O
OH | OH
Mirror

b. The second carbon of serine (denoted with *) is chiral, because it is bonded to four different groups: H, $-NH_2$, $-CH_2OH$, and $-COOH$. In the left representation, the hydrocarbon chain is drawn in the plane of the page, and we arbitrarily chose to represent the $-NH_2$ with a wedge. Thus, the enantiomer (right) must be drawn with the hydrocarbon in the plane of the page, but rotated 180 degrees with the $-NH_2$ on a wedge in order to make a mirror image.

HO–C3–C*2(NH$_2$)–C1(=O)–OH | HO–C1(=O)–C*2(NH$_2$)–C3–OH

Mirror

✔ Learning Check 11.2 Draw enantiomers of the amino acid threonine in wedge-and-dash notation.

$$CH_3-\underset{|}{\overset{OH}{CH}}-\underset{|}{\overset{NH_2}{CH}}-COOH$$

Source: Emil Fischer

Figure 11.11 Emil Fischer (1852–1919), a German chemist, won the Nobel Prize in Chemistry in 1902 for his contributions to the field, including Fischer projections, esterification, and an early understanding of proteins.

11.3 Fischer Projections, Cyclization, and Haworth Projections

Learning Objective 5 Use Fischer projections to represent D- and L-enantiomers.

Learning Objective 6 Identify the following functional groups: hemiacetal, hemiketal, acetal, and ketal.

Learning Objective 7 Draw Haworth projections of carbohydrate anomers.

11.3A Fischer Projections

Fischer projection A method of depicting three-dimensional shapes for chiral molecules.

It can be cumbersome to draw molecules in the three-dimensional shapes shown for the two enantiomers of larger organic molecules like carbohydrates. But there is another way to represent these mirror images in two dimensions. Emil Fischer (**Figure 11.11**), a chemist known for pioneering work in carbohydrate chemistry, introduced a method late in the nineteenth century. His two-dimensional structures, called **Fischer projections**, are illustrated in **Figure 11.12** for glyceraldehyde.

In Fischer projections, the chiral carbon is represented by the intersection of two lines. If the compound is a carbohydrate, the carbonyl group is placed at or near the top. The molecule is also positioned so that the two bonds coming toward you out of the plane of the paper are drawn horizontally in the Fischer projection. The two bonds projecting away from you into the plane of the paper are drawn vertically. Thus, when you see a Fischer projection, you must realize that the molecule has a three-dimensional shape with horizontal bonds coming toward you and vertical bonds going away from you.

We can also distinguish between the two enantiomers of glyceraldehyde using Fischer projections. A small capital L is used to indicate that an –OH group (or another functional group) is on the *left* of the chiral carbon when the carbonyl is at the top as in L-glyceraldehyde. A small capital D means the –OH is on the *right* of the chiral carbon, as it is for D-glyceraldehyde.

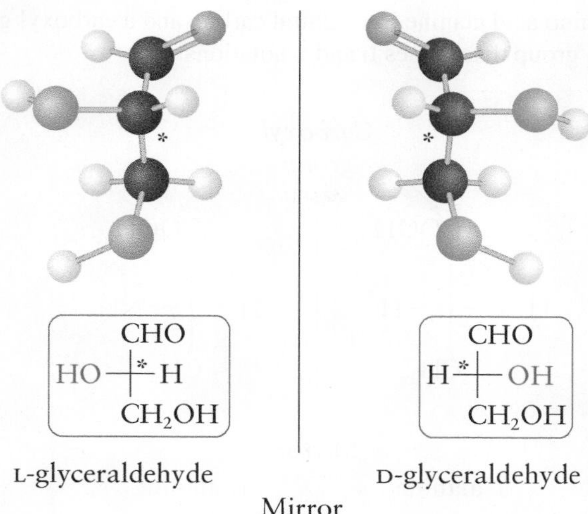

Figure 11.12 Ball-and-stick models (top) and Fischer projections (bottom) of the two enantiomers of glyceraldehyde.

Example 11.3 Drawing Enantiomers as Fischer Projections

Lactic acid is a by-product of ATP formation (**Figure 11.13** and Chapter 15), and alanine is an amino acid (Chapter 12).

Draw Fischer projections for the D and L forms of the following.

a. $CH_3—CH(OH)—COOH$ lactic acid

b. $CH_3—CH(NH_2)—COOH$ alanine

Figure 11.13 Although lactic acid accumulates in muscles during strenuous exercise, research has shown it is not the direct cause of symptoms such as muscle fatigue.

Solution

a. The second carbon of lactic acid is chiral, because it is bonded to four different groups: H, OH, CH_3, and COOH. The chiral carbon is placed at the intersection of two lines. Lactic acid is not an aldehyde, but it does contain a carbon–oxygen double bond in the carboxyl group (–COOH). This carboxyl group is placed at the top of the vertical line. The direction of the –OH groups on the chiral carbons determines the D and L notations:

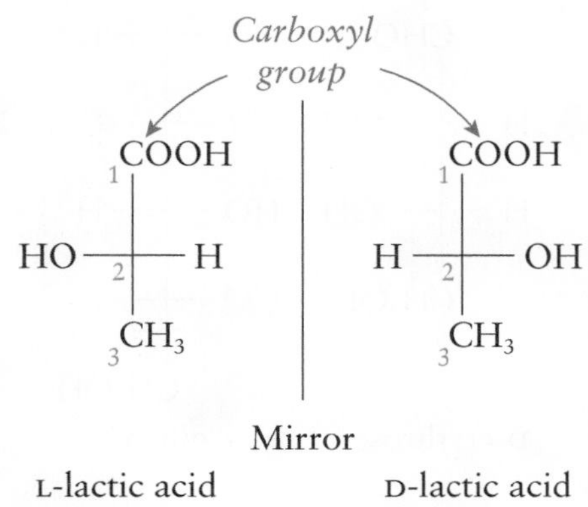

b. Similarly, the amino acid alanine has a chiral carbon and a carboxyl group. The direction of the $-NH_2$ group determines D and L notations:

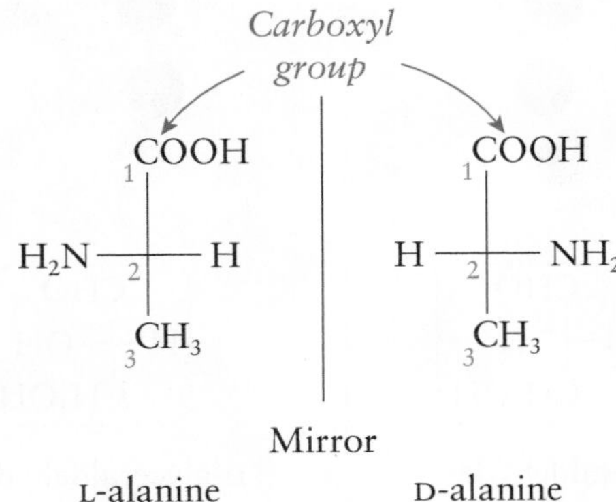

> ✔ **Learning Check 11.3** Draw Fischer projections for the D and L forms of the amino acid valine.
>
> $$CH_3-\underset{\displaystyle CH_3}{CH}-\underset{\displaystyle NH_2}{CH}-COOH$$

The existence of more than one chiral carbon in a carbohydrate molecule could lead to confusion when trying to assign D and L designations. This is avoided, however, by focusing on the hydroxy group attached to the chiral carbon *farthest* from the carbonyl group. By convention, the D family of compounds is that in which this hydroxy group projects to the right when the carbonyl is at the top of the Fischer projection. In the L family of compounds, it projects to the left:

Carbonyl group (arrows point to each CHO)

L-erythrose	D-erythrose	L-glucose	D-glucose
CHO	CHO	CHO	CHO
HO—H	H—OH	HO—H	H—OH
HO—H	H—**OH**	H—OH	HO—H
CH_2OH	CH_2OH	HO—H	H—OH
		HO—H	H—**OH**
		CH_2OH	CH_2OH

Example 11.4 Identifying D and L Isomers

Ribose is a component of ribonucleic acid (RNA), which plays a central role in protein synthesis (see **Figure 11.14** and Chapter 13). Is the following ribose structure D or L?

```
        CHO
         |
  HO ----+---- H
         |
  HO ----+---- H
         |
  HO ----+---- H
         |
       CH2OH
```

Solution

CHO is an abbreviation for the carbonyl group of an aldehyde. The intersections of lines represent chiral carbon atoms. Focus on the hydroxy group attached to the chiral carbon farthest from the carbonyl group. It is highlighted in blue in the structure in the margin. The hydroxy group is pointing to the left, so this is L-ribose.

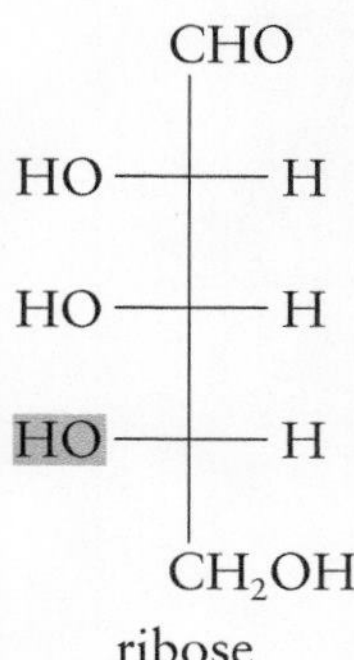

✔ Learning Check 11.4 Identify each structure as D or L.

a.

```
       CH2OH
         |
        C=O
         |
  HO ----+---- H
         |
   H ----+---- OH
         |
   H ----+---- OH
         |
       CH2OH
```

b.

```
        CHO
         |
   H ----+---- OH
         |
  HO ----+---- H
         |
       CH2OH
```

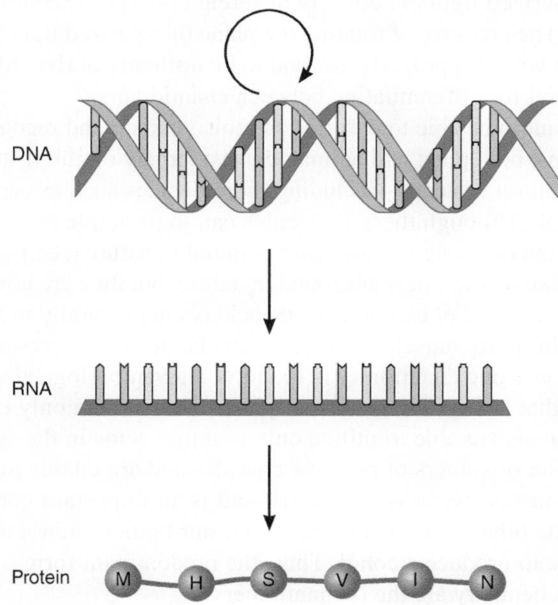

Figure 11.14 According to the central dogma of biology, genetic information flows from DNA to RNA to proteins. Thus, RNA provides the link between DNA and the synthesis of proteins.

The physical properties of the two compounds that make up a pair of D and L isomers (enantiomers) are generally the same. The only exception is the way solutions of the two compounds affect polarized light that is passed through them. Light is polarized by passing it through a special polarizing lens, such as those found in polarized sunglasses. When polarized light is passed through a solution of one enantiomer, the plane of polarization of the light is rotated to either the right or the left when viewed by looking toward the source of the light (see **Figure 11.15**). The other enantiomer rotates it the same amount but in the opposite direction. The enantiomer that rotates it to the left is called the **levorotatory** (to the left) or (−) enantiomer. The one that rotates it to the right is the **dextrorotatory** (to the right) or (+) enantiomer.

levorotatory Rotates plane-polarized light to the left.

dextrorotatory Rotates plane-polarized light to the right.

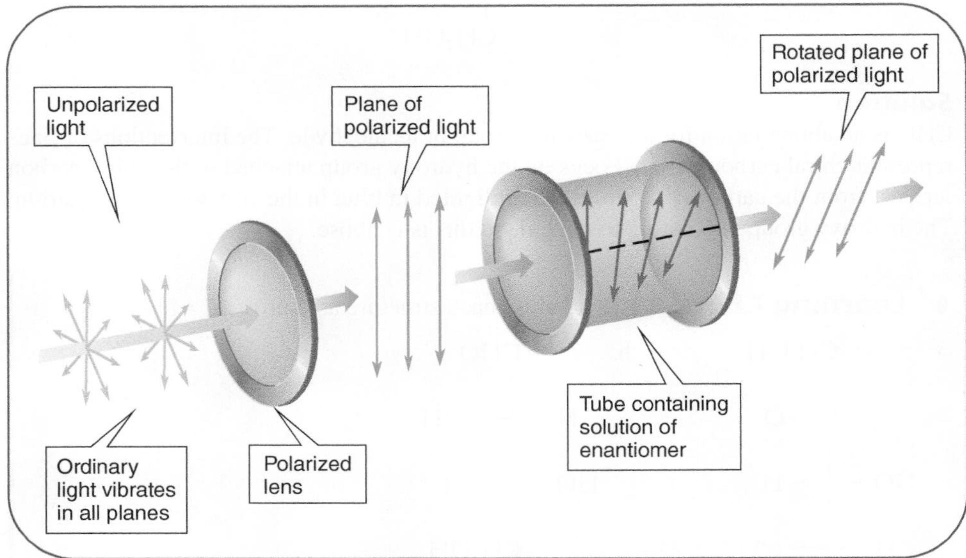

Figure 11.15 The rotation of plane-polarized light by the solution of an enantiomer.

The D and L designations do not correspond to *dextrorotatory* and *levorotatory*—instead, they represent only the spatial relationships of a Fischer projection. Thus, some D compounds rotate polarized light to the right (dextrorotatory), and some rotate it to the left (levorotatory). The same is true for L compounds. The spatial relationships D and L and the rotation of polarized light are entirely different concepts and should not be confused with each other. The property of rotating the plane of polarized light is called optical activity, and molecules with this property are said to be **optically active**. Measurements of optical activity are useful for differentiating between enantiomers.

optically active molecule A molecule that rotates the plane of polarized light.

Why is it so important to be able to describe stereoisomerism and recognize it in molecules? Living organisms, both plant and animal, consist largely of chiral substances. Most of the compounds in natural systems—including biomolecules such as carbohydrates and amino acids—are chiral. Although these molecules can in principle exist as a mixture of stereoisomers, quite often only one stereoisomer is found in nature (see **Figure 11.16**).

In some instances, two enantiomers are found in nature, but they are not found together in the same biological system. For example, lactic acid occurs naturally in both forms. The L-lactic acid is found in living muscle, whereas the D-lactic acid is present in sour milk. When we realize that only one enantiomer is found in a given biological system, it is not too surprising to find that the system can usually use or assimilate only one enantiomer. For instance, most animals are able to utilize only L-amino acids in the synthesis of proteins. Humans utilize the D-isomers of monosaccharides and are unable to metabolize the L-isomers. D-Glucose tastes sweet, is nutritious, and is an important component of our diets. L-Glucose, on the other hand, is tasteless, and our bodies cannot use it. Yeast can ferment only D-glucose to produce alcohol. Thus, the predominant form of carbohydrates that are relevant to biochemistry are the D-enantiomers.

Figure 11.16 The pleasant natural flavor of spearmint is due to the L-enantiomer of carvone, a cyclic ketone. The D-enantiomer smells herbaceous and woody.

11.3B Acetal Formation and Haworth Projections

So far, we have represented the monosaccharides as open-chain polyhydroxy aldehydes and ketones. These representations are useful in discussing the structural features and stereochemistry of monosaccharides. Aldehydes react with alcohols to form **hemiacetals**, a functional group that is very important in carbohydrate structure. Hemiacetals are usually not very stable and are difficult to isolate. With excess alcohol present and with an acid catalyst, however, a stable product called an **acetal** is formed. Similar functional groups, called *hemiketals* and *ketals*, are formed in the same way from ketones with another alkyl group replacing the H atom bonded to the central carbon.

hemiacetal A compound that contains the functional group

$$-\underset{\displaystyle OR}{\overset{\displaystyle OH}{\underset{|}{\overset{|}{C}}}}-H$$

acetal A compound that contains the functional group

$$-\underset{\displaystyle OR}{\overset{\displaystyle OR}{\underset{|}{\overset{|}{C}}}}-H$$

General reaction:

$$R-\overset{\overset{\displaystyle O}{\|}}{C}-H + R'-OH \rightleftharpoons R-\underset{\underset{\displaystyle OR'}{|}}{\overset{\overset{\displaystyle OH}{|}}{C}}-H \xrightleftharpoons[]{H^+ \text{ and } R'-OH} R-\underset{\underset{\displaystyle OR'}{|}}{\overset{\overset{\displaystyle OR'}{|}}{C}}-H + H_2O \quad (11.1)$$

aldehyde — alcohol — hemiacetal intermediate — acetal

Specific example:

$$CH_3-\overset{\overset{\displaystyle O}{\|}}{C}-H + CH_3-OH \rightleftharpoons CH_3-\underset{\underset{\displaystyle OCH_3}{|}}{\overset{\overset{\displaystyle OH}{|}}{C}}-H \xrightleftharpoons[]{H^+ \text{ and } CH_3-OH} CH_3-\underset{\underset{\displaystyle OCH_3}{|}}{\overset{\overset{\displaystyle OCH_3}{|}}{C}}-H + H_2O$$

acetaldehyde — methanol — hemiacetal intermediate — acetal

Hemiacetals and hemiketals contain an –OH group, hydrogen, and an –OR group on the same carbon. Thus, they appear to have alcohol and ether functional groups. Acetals and ketals contain a carbon that has a hydrogen and two –OR groups attached. They look like diethers:

$$CH_3-\underset{\underset{\displaystyle OCH_3}{|}}{\overset{\overset{\displaystyle OH}{|}}{C}}-H \qquad CH_3-\underset{\underset{\displaystyle OCH_3}{|}}{\overset{\overset{\displaystyle OCH_3}{|}}{C}}-H$$

Hemiacetal carbon — *Acetal carbon*

Some molecules, such as open-chain carbohydrates, contain both an –OH and a C=O group on different carbon atoms. In such cases, an intramolecular (within the molecule) reaction can occur, and a cyclic hemiacetal or hemiketal can be formed:

open-chain molecule containing an aldehyde and alcohol group (5 CH_2–O–H, 4 H_2C, 3 CH_2, 2 CH_2, 1 C=O, H) $\rightleftharpoons$ cyclic structure containing a hemiacetal group (ring of carbons 1–5 and O; C1 bears OH and H)

Cyclic hemiacetals and hemiketals are much more stable than the open-chain compounds discussed previously. All monosaccharides with at least five carbon atoms exist predominantly as cyclic hemiacetals. To help depict the cyclization of glucose (**Figure 11.17**), the open-chain structure has been rotated around to position the functional groups in closer

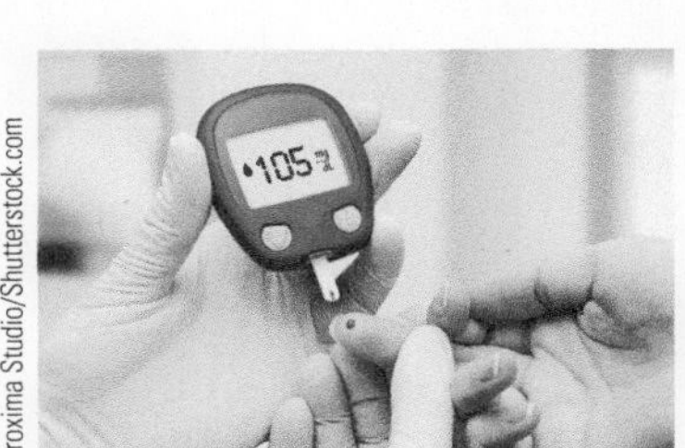

Proxima Studio/Shutterstock.com

Figure 11.17 Patients with diabetes must test their blood sugar (blood glucose levels) multiple times a day.

proximity. Notice the numbering of the carbon atoms that begins at the end of the chain, giving the lowest number to the carbonyl group carbon:

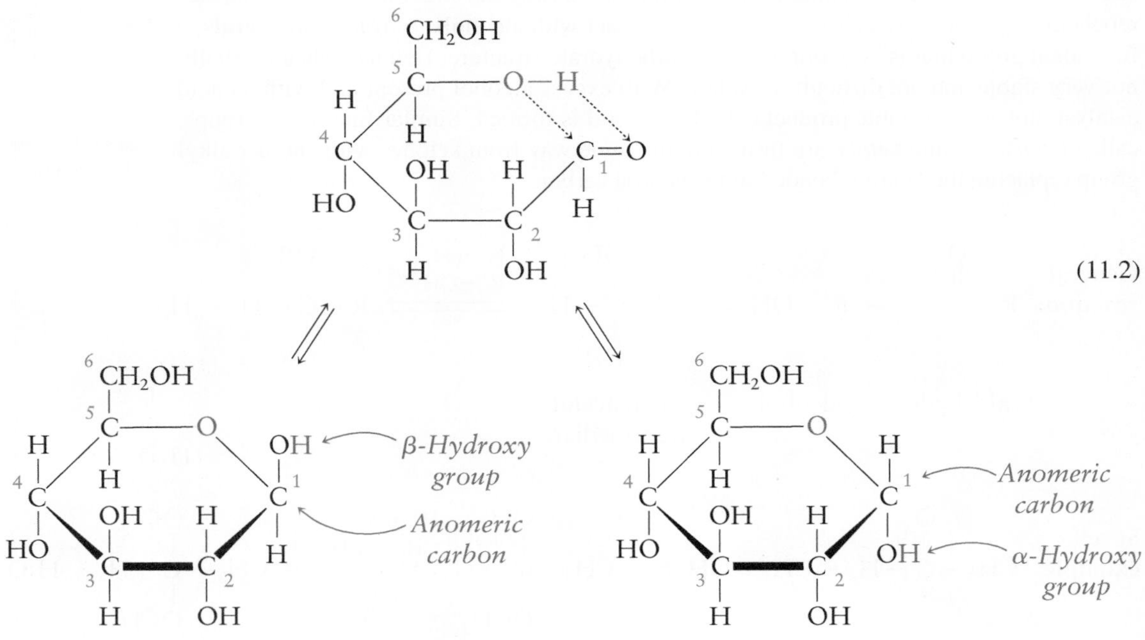

(11.2)

β-Hydroxy group

Anomeric carbon

β-D-glucose

Anomeric carbon

α-Hydroxy group

α-D-glucose

pyranose ring A six-membered sugar ring system containing an oxygen atom.

Haworth structure A method of depicting three-dimensional carbohydrate structures.

anomeric carbon An acetal, ketal, hemiacetal, or hemiketal carbon atom that gives rise to two stereoisomers.

anomers Stereoisomers that differ in the three-dimensional arrangement of groups at the carbon of an acetal, ketal, hemiacetal, or hemiketal group.

In the reaction, the alcohol group on carbon 5 adds to the aldehyde group on carbon 1. The result is a **pyranose ring**, a six-membered ring containing an oxygen atom. The attached groups have been drawn above or below the plane of the ring. This kind of drawing, called a **Haworth structure**, is used extensively in carbohydrate chemistry.

As the reaction occurs, a new chiral carbon is produced at position 1. Thus, two stereoisomers are possible: one with the –OH group pointing down (the α form), and the other with the –OH group pointing up (the β form). The C-1 carbon is called an **anomeric carbon**, and the α and β forms are called **anomers**. The following condensed structures for the cyclic compounds omit the carbon atoms in the ring and the hydrogen atoms attached to the ring carbons:

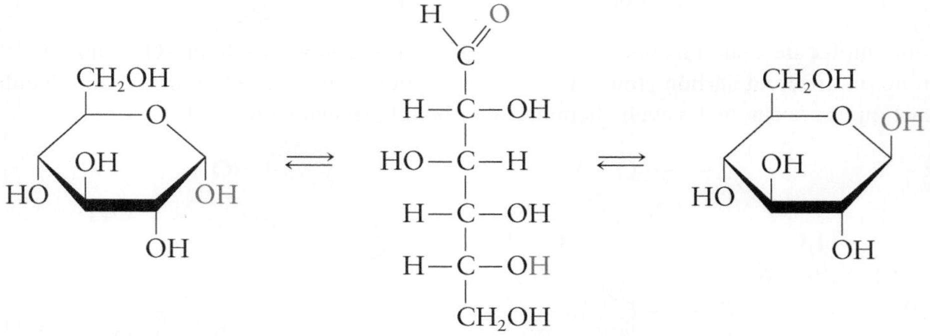

α-D-glucose (36%) ⇌ D-glucose (0.02%) ⇌ β-D-glucose (64%)

Thus, there are three forms of D-glucose: an open-chain structure and two cyclic forms. Because the reaction to form a cyclic hemiacetal is reversible, the three isomers of D-glucose are interconvertible. Studies indicate that the equilibrium distribution in aqueous solutions is approximately 36% α-D-glucose, 0.02% open-chain form, and 64% β-D-glucose.

furanose ring A five-membered sugar ring system containing an oxygen atom.

Other monosaccharides also form cyclic structures. However, some form five-membered rings, called **furanose rings**, rather than the six-membered kind. An example is

D-fructose (**Figure 11.18**). The five-membered ring cyclization (see Equation 11.3) occurs like the ring formation in glucose (see Equation 11.2). That is, an alcohol adds across the carbonyl double bond (in this case, a ketone). The orientation of the –OH group at position 2 determines whether fructose is in the α or β form:

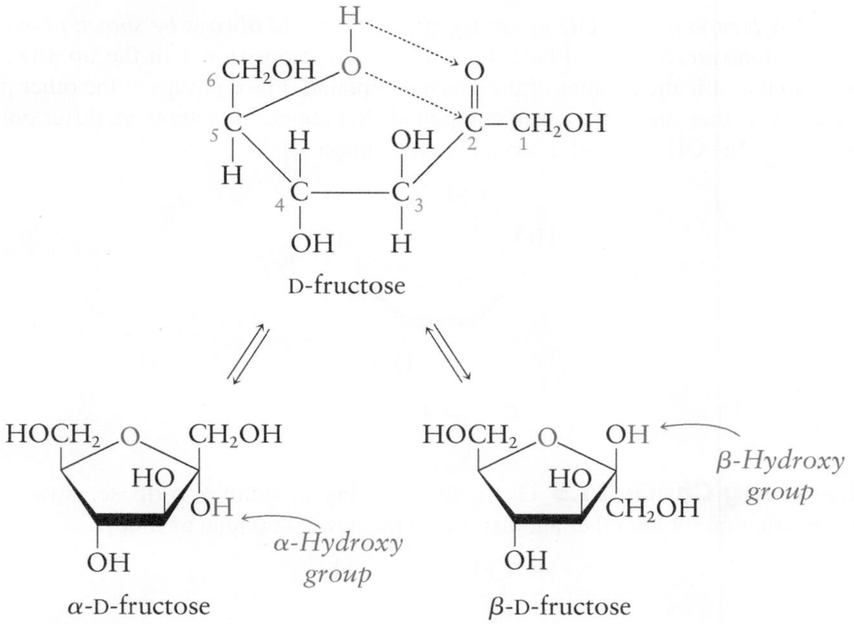

D-fructose (11.3)

α-D-fructose (α-Hydroxy group) ⇌ D-fructose ⇌ β-D-fructose (β-Hydroxy group)

Figure 11.18 Fructose is the naturally occurring sugar in honey and many fruits. It is also one of the components of sucrose (table sugar).

In drawing cyclic Haworth structures of monosaccharides, certain rules should be followed so that all structures are represented consistently.

Step 1: *Draw the ring with its oxygen to the back.*

Step 2: *Put the anomeric carbon on the right side of the ring.*

Step 3: *The terminal –CH_2OH group (position 6) should always be shown above the ring for D-monosaccharides.* In Haworth structures, remember that any groups attached to position 1 (or anomeric) carbon can be drawn above or below the ring depending on whether you are drawing the α or β anomer.

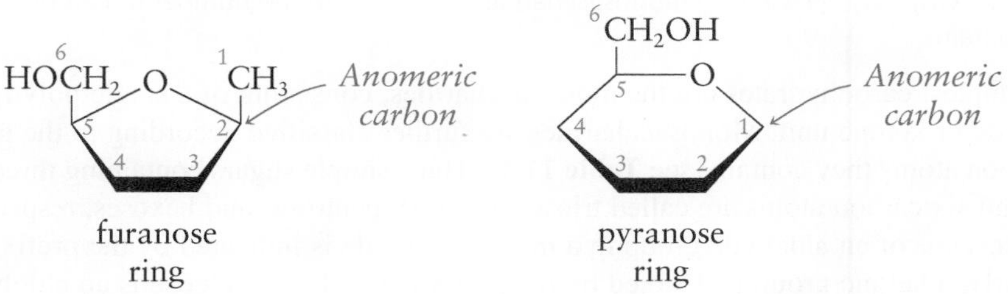

furanose ring (Anomeric carbon)   pyranose ring (Anomeric carbon)

Example 11.5 Drawing Haworth Structures

D-Galactose (**Figure 11.19**) exists predominantly in cyclic forms. Given the following structure, draw the Haworth structure for the other anomer. Label the new compound as α or β.

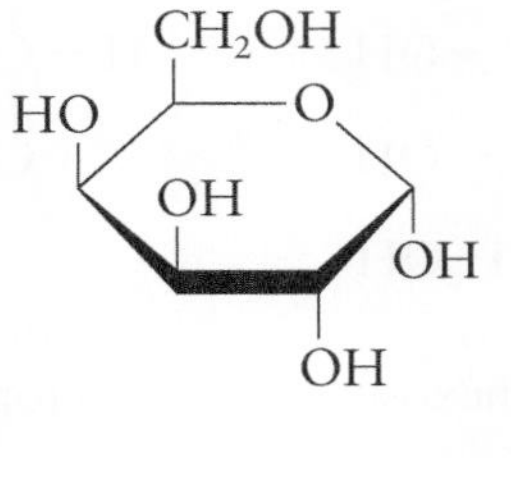

Figure 11.19 Foods high in galactose include various dairy products (milk, cheese, and yogurt). Lactose is a disaccharide consisting of galactose and glucose.

Solution

Step 1: *Draw the pyranose ring with the oxygen atom to the back.*

Step 2: *Put the anomeric carbon on the right side of the ring*. Number the ring starting at the right side. Position number 1 will be the anomeric carbon.

Step 3: *The terminal –CH_2OH group (position 6) should always be shown above the ring for* D*-monosaccharides*. Place the –OH group at position 1 in the up direction (β form) so that it is the anomer of the given compound. Place groups at the other positions exactly as they are in the given compound. Remember that anomers differ only in the position of the OH attached to the anomeric carbon.

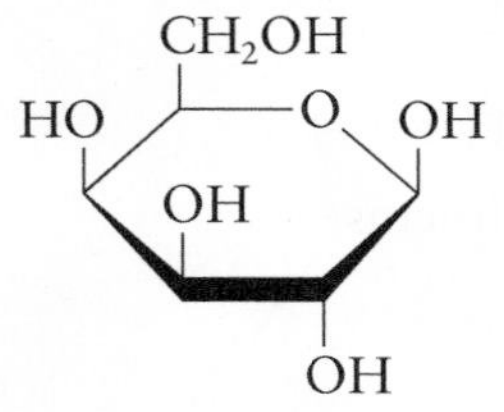

CH_2OH
HO O OH
OH
OH

β-D-galactose

✔ **Learning Check 11.5** Given the following structure of D-ribose, draw the Haworth structure for the other anomer. Label the new compound as α or β.

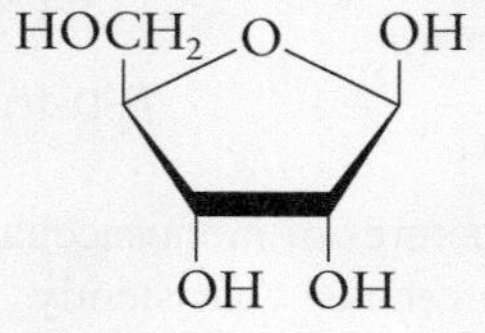

$HOCH_2$ O OH
OH OH

D-ribose

Table 11.1 Monosaccharide Classification Based on the Number of Carbons in Their Chains

Number of Carbon Atoms	Sugar Class
3	Triose
4	Tetrose
5	Pentose
6	Hexose

11.4 Classification of Monosaccharides

Learning Objective 8 Classify monosaccharides as aldoses or ketoses.

Learning Objective 9 Classify monosaccharides according to the number of carbon atoms they contain.

The simplest carbohydrates are the monosaccharides, consisting of a single polyhydroxy aldehyde or ketone unit. Monosaccharides are further classified according to the number of carbon atoms they contain (see **Table 11.1**). Thus, simple sugars containing three, four, five, and six carbon atoms are called trioses, tetroses, pentoses, and hexoses, respectively. The presence of an aldehyde group in a monosaccharide is indicated by the prefix *aldo-*. Similarly, a ketone group is denoted by the prefix *keto-*. Thus, glucose is an aldohexose, whereas ribulose is a ketopentose:

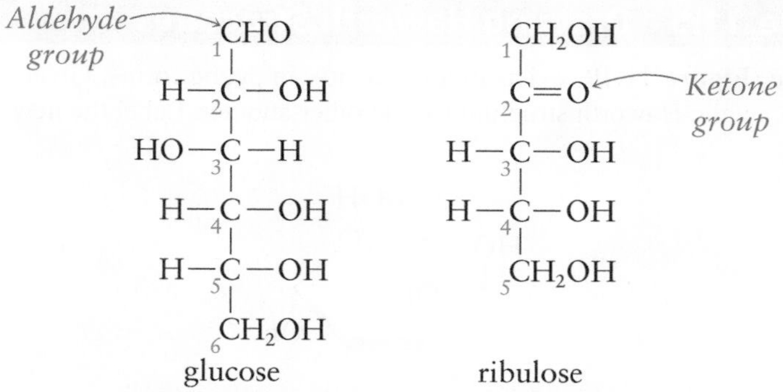

Aldehyde group → ^{1}CHO
H–C_2–OH
HO–C_3–H
H–C_4–OH
H–C_5–OH
$_6CH_2OH$

glucose
an aldohexose

$_1CH_2OH$
$_2$C=O ← *Ketone group*
H–C_3–OH
H–C_4–OH
$_5CH_2OH$

ribulose
a ketopentose

Example 11.6 Classification of Carbohydrates by Number of Carbons

Erythrulose (see **Figure 11.20**) is a natural sugar present in red berries such as raspberries. Classify erythrulose by combining the aldehyde–ketone designation with terminology indicating the number of carbon atoms as just described.

$$\begin{array}{c} CH_2OH \\ | \\ C{=}O \\ | \\ H{-}C{-}OH \\ | \\ CH_2OH \end{array}$$

Irina Bg/Shutterstock.com

Figure 11.20 Erythrulose is a monosaccharide used in many self-tanning products because of its natural ability to dye the skin.

Solution

Erythrulose has four carbons (tetrose) and is a ketone (ketose), which makes it a ketotetrose.

$$\begin{array}{c} {}^{1}CH_2OH \\ | \\ {}^{2}C{=}O \leftarrow \textit{Ketone group} \\ | \\ H{-}{}^{3}C{-}OH \\ | \\ {}^{4}CH_2OH \end{array}$$

✔ Learning Check 11.6 Classify each of the following monosaccharides by combining the aldehyde–ketone designation with terminology indicating the number of carbon atoms.

a. $$\begin{array}{c} CHO \\ | \\ H{-}C{-}OH \\ | \\ HO{-}C{-}H \\ | \\ H{-}C{-}OH \\ | \\ CH_2OH \end{array}$$

b. $$\begin{array}{c} CHO \\ | \\ HO{-}C{-}H \\ | \\ H{-}C{-}OH \\ | \\ CH_2OH \end{array}$$

Most monosaccharides are aldoses, and almost all natural monosaccharides belong to the D series. The family of D-aldoses is shown in **Figure 11.21**. D-Glyceraldehyde, the smallest monosaccharide with a chiral carbon, is the standard on which the whole series is based. Notice that the bottom chiral carbon in each compound is directed to the right. The 2^n formula tells us there must be 2 aldo trioses, 4 aldo tetroses, 8 aldo pentoses, and 16 aldo hexoses. Half of those are the D-compounds shown in **Figure 11.21**. The other half (not shown) are the corresponding L-monosaccharides.

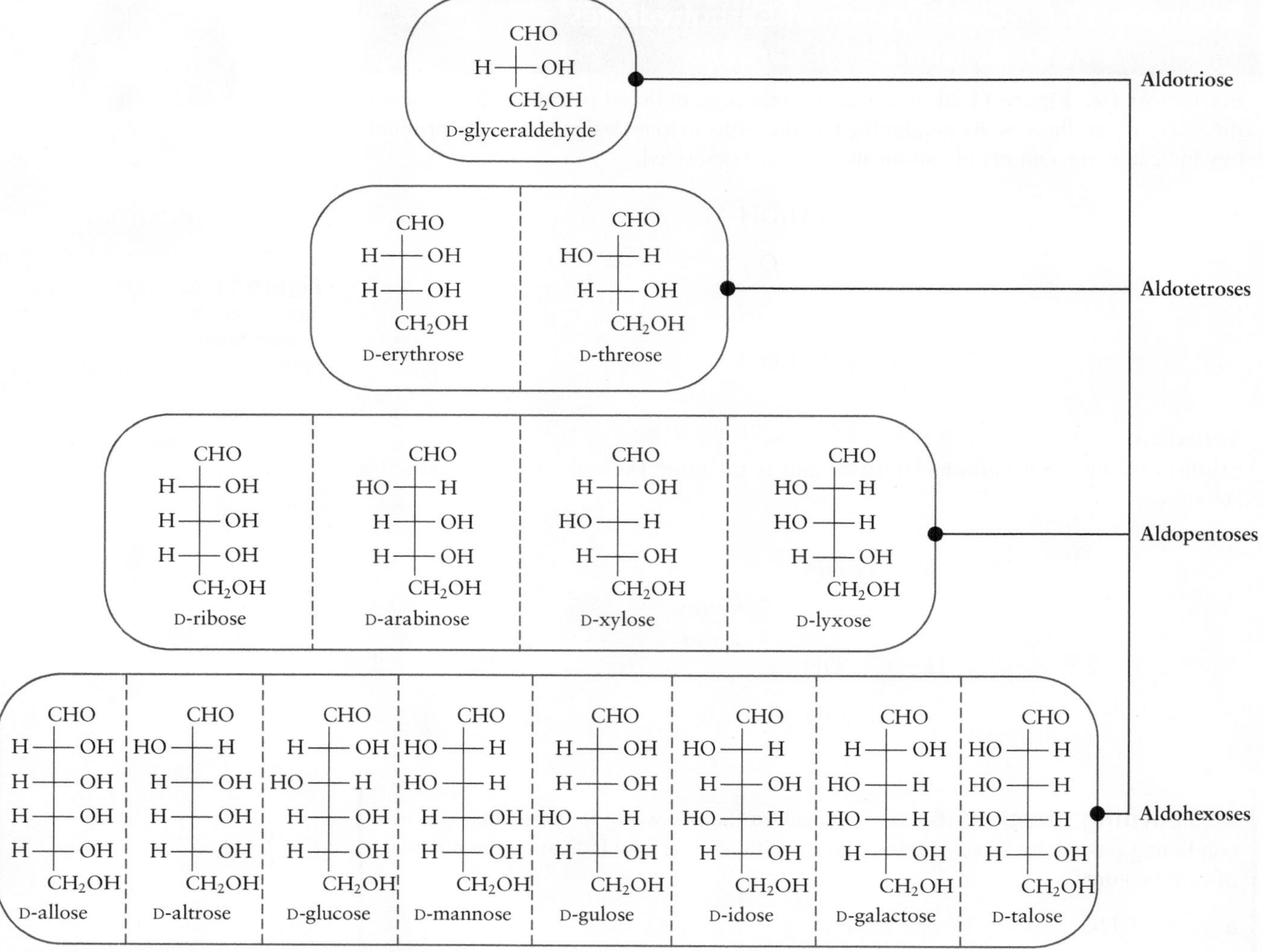

Figure 11.21 The family of D-aldoses, shown in Fischer projections.

11.5 Important Monosaccharides

Learning Objective 10 Describe uses of four important monosaccharides.

Two pentoses, ribose and deoxyribose, are extremely important because they are used in the synthesis of nucleic acids (DNA and RNA), substances that are essential in protein synthesis and the transfer of genetic material (see **Figure 11.22** and Chapter 13). Ribose (along with phosphate groups) forms the long chains that make up the backbone of ribonucleic acid (RNA). Deoxyribose (along with phosphate groups) forms the long chains that make up the backbone of deoxyribonucleic acid (DNA). Deoxyribose differs from ribose in that the OH group on carbon 2 has been replaced by a hydrogen atom (deoxy form).

$HOCH_2$ O OH (5, 4, 1, 3, 2) OH OH

β-D-ribose
found in RNA

$HOCH_2$ O OH (5, 4, 1, 3, 2) OH

Two hydrogens at this position

β-D-deoxyribose
found in DNA

Figure 11.22 Deoxyribose makes up the structural backbone of each DNA molecule.

Of the monosaccharides, the hexose glucose is the most important nutritionally and the most abundant in nature. Glucose is present in honey and fruits such as grapes, figs, and dates. Ripe grapes, for example, contain 20% to 30% glucose. Glucose is sometimes called dextrose (**Figure 11.23**). Glucose is also known as blood sugar, because it is the sugar transported by the blood to body tissues to satisfy energy requirements. Other sugars absorbed into the body must be converted to glucose by the liver. Glucose is commonly used as a sweetener in confections and other foods, including some baby foods.

β-D-glucose

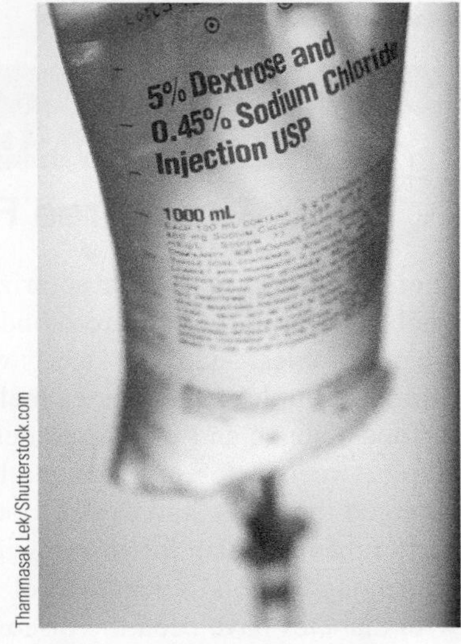

Figure 11.23 A dextrose (glucose) solution prepared for intravenous use.

Galactose is a hexose with a structure very similar to that of glucose. The only difference between the two is the orientation of the hydroxy group attached to carbon 4. Like glucose, galactose can exist in α, β, or open-chain forms; the β form is as follows:

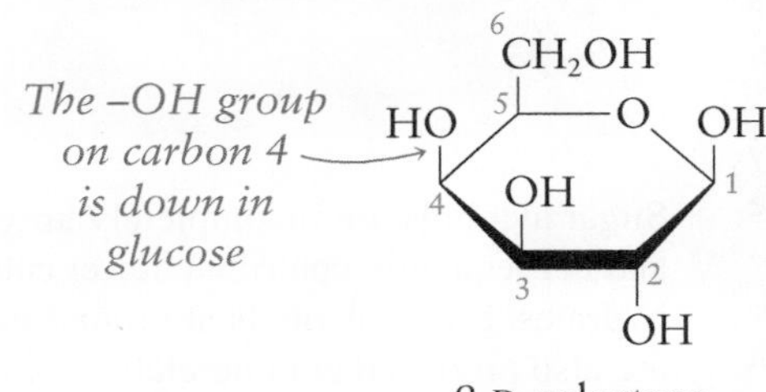

β-D-galactose

Galactose is synthesized in the mammary glands, where it is incorporated into lactose, the sugar found in milk. It is also a component of substances present in nerve tissue. Fructose

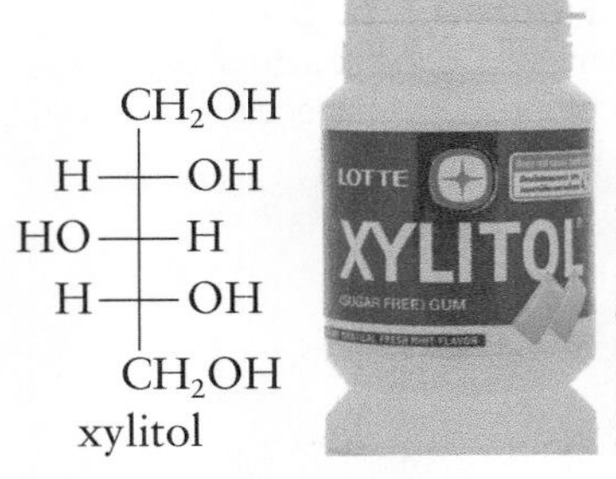

Keith Homan/Shutterstock.com

Figure 11.24 Most chewing gums are sweetened with sugars that can increase the chance of cavities. Xylitol is the only sweetener that has been shown to reduce the potential for tooth decay.

is the most important ketohexose. It is also known as levulose and fruit sugar because of its presence in many fruits. It is present in honey in a 1:1 ratio with glucose and is abundant in corn syrup. This sweetest of the common monosaccharides is important as a food sweetener because less fructose is needed than other sugars to achieve the same degree of sweetness.

11.6 Properties of Monosaccharides

Learning Objective 11 Describe the purpose of Benedict's test.

Learning Objective 12 Identify glycosidic linkages in carbohydrate derivatives.

11.6A Physical Properties

Carbohydrates are called sugars because they taste sweet (see **Figure 11.24**). The degree of sweetness varies, as shown in **Table 11.2**. For instance, fructose is about 73% sweeter than sucrose (ordinary table sugar). All carbohydrates are solids at room temperature, and because of the many –OH groups present, monosaccharide carbohydrates are extremely soluble in water. The –OH groups form numerous hydrogen bonds with the surrounding water molecules. In the body, this solubility allows carbohydrates to be transported rapidly in the circulatory system.

Table 11.2 The Relative Sweetness of Sugars (Sucrose = 1.00)

Sugar	Relative Sweetness
Lactose	0.16
Galactose	0.22
Maltose	0.32
Xylose	0.40
Glucose	0.74
Sucrose	1.00
Fructose	1.73

11.6B Oxidation

Open-chain forms of monosaccharides exist as aldehydes or hydroxyketones and are readily oxidized by weak oxidizing agents due to the presence of adjacent alcohols in the carbohydrate structure. Sugars that can be oxidized by weak oxidizing agents are

Health Connections 11.1

Sugar-Free Foods and Diabetes

Diabetes is a disease characterized by an impairment of the body's ability to utilize blood glucose. In the past, ice cream, cookies, and other sweets were considered to be off-limits to people with diabetes mellitus. However, advances in food science and a better understanding of the relationship between different foods and blood glucose levels have changed the rules, and sweet foods are now acceptable as part of the meal plan for people with diabetes.

Among the food science developments helpful to people with diabetes have been the discovery and commercial distribution of several synthetic noncarbohydrate sweetening agents. The U.S. Food and Drug Administration (FDA) has approved six products or recognized them as safe for the public, including people with diabetes: saccharin, neotame, acesulfame-K, aspartame, sucralose, and advantame.

Each of these sweeteners is noncaloric or provides very few calories when used in the quantities necessary to provide sweetness. Thus, these substances are useful dietary substitutes for sucrose or other calorie-laden sweeteners for people who wish to control calorie intake or for those who must avoid sugar, such as people with diabetes.

Other sweeteners that have expanded the food options for people with diabetes are known as sugar alcohols. These are carbohydrate derivatives, such as sorbitol, in which the carbon–oxygen double bond of the aldehyde or ketone has been converted to an alcohol:

J. Schuyler

CH_2OH
H—C—OH
HO—C—H
H—C—OH
H—C—OH
CH_2OH
sorbitol

Sugar alcohols are incompletely absorbed during digestion and consequently contribute fewer calories than other carbohydrates. Sugar alcohols are found naturally in fruits. They are also produced commercially from carbohydrates for use in sugar-free candies, cookies, and chewing gum.

called **reducing sugars**. Thus, all monosaccharides are reducing sugars. We can test for the presence of reducing sugars in the laboratory using **Benedict's reagent**, a mild oxidizing solution containing Cu^{2+} ions. As the open-chain form of a monosaccharide is oxidized, Cu^{2+} is reduced and precipitated as Cu_2O, a red-orange solid (**Figure 11.25**).

reducing sugar A sugar that can be oxidized by Cu^{2+} solutions.

Benedict's reagent A mild oxidizing solution containing Cu^{2+} ions used to test for the presence of aldehydes.

General reaction:

$$\underset{\text{hydroxyketone)}}{\underset{\text{(aldehyde or}}{\text{Reducing sugar}}} + \underset{\text{deep blue solution}}{\underset{\text{(complex)}}{Cu^{2+}}} \longrightarrow \text{oxidized compound} + \underset{\text{precipitate}}{\underset{\text{red-orange}}{Cu_2O}} \quad (11.4)$$

Specific example:

$$\begin{array}{c} \text{CHO} \\ | \\ \text{H}-\text{C}-\text{OH} \\ | \\ \text{HO}-\text{C}-\text{H} \\ | \\ \text{H}-\text{C}-\text{OH} \\ | \\ \text{H}-\text{C}-\text{OH} \\ | \\ \text{CH}_2\text{OH} \\ \text{D-glucose} \end{array} + Cu^{2+} \longrightarrow \begin{array}{c} \text{COOH} \\ | \\ \text{H}-\text{C}-\text{OH} \\ | \\ \text{HO}-\text{C}-\text{H} \\ | \\ \text{H}-\text{C}-\text{OH} \\ | \\ \text{H}-\text{C}-\text{OH} \\ | \\ \text{CH}_2\text{OH} \\ \text{D-gluconic} \\ \text{acid} \end{array} + Cu_2O$$

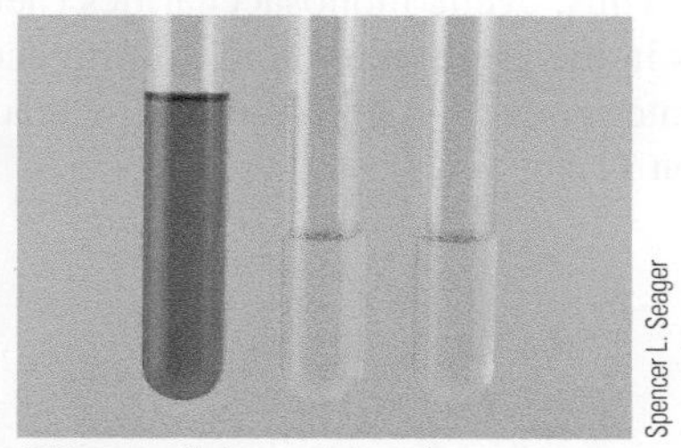

1 From left to right, three test tubes containing Benedict's reagent, 0.5% glucose solution, and 2.0% glucose solution.

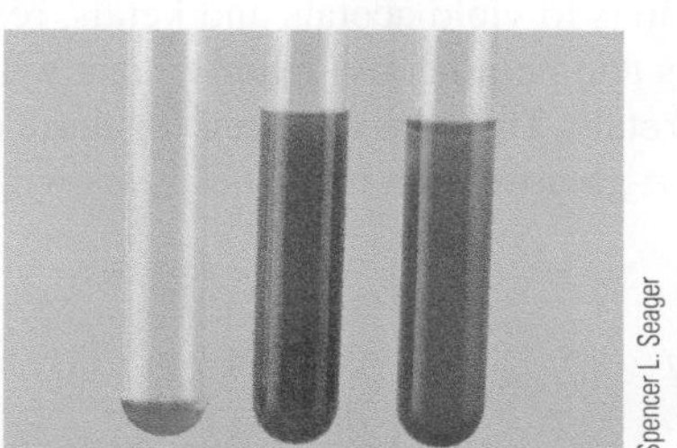

2 The addition of Benedict's reagent from the first tube produces colors (due to the red Cu_2O) that indicate the amount of glucose present.

Figure 11.25 An example of using Benedict's reagent with glucose.

11.6C Phosphate Esters

The hydroxy groups of monosaccharides can behave as alcohols and react with acids to form esters. When the central carbon atom of an ester is replaced by a phosphorus atom, a new functional group called a **phosphate ester** is formed. Esters formed from phosphoric acid and various monosaccharides are found in all cells, and some serve as important intermediates in carbohydrate metabolism and as backbones to other biomolecules (**Figure 11.26**). The structures of two representative phosphate esters are as follows:

phosphate ester An organic compound having the following general form:

$$\begin{array}{c} \text{O} \\ \| \\ -\text{O}-\text{P}-\text{OH} \\ | \\ \text{OH} \end{array}$$

$CH_2-O-P(=O)(O^-)_2$ attached to a pyranose ring with OH, HO, OH, OH substituents

glucose 6-phosphate

$^-O-P(=O)(O^-)-O-CH_2$ attached to a furanose ring with OH, HO, CH_2OH, OH substituents

fructose 6-phosphate

Figure 11.26 The phosphodiester bond connects deoxyribose and the phosphate groups that make up the backbone of DNA.

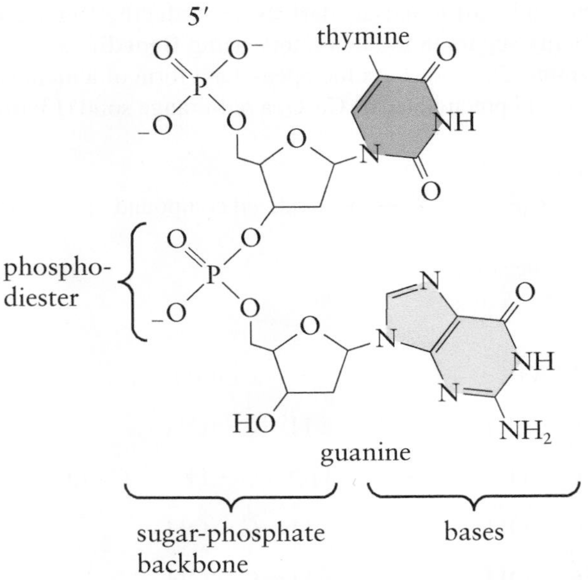

11.6D Glycoside Formation

glycoside Another name for a carbohydrate containing an acetal or ketal group.

Recall from **Section 11.3** that hemiacetals and hemiketals can react with alcohols in acid solutions to yield acetals and ketals, respectively. Thus, cyclic monosaccharides (hemiacetals and hemiketals) readily react with alcohols in the presence of acid to form acetals and ketals. The general name for these carbohydrate products is **glycosides**. The reaction of α-D-glucose with methanol is shown in Equation 11.5:

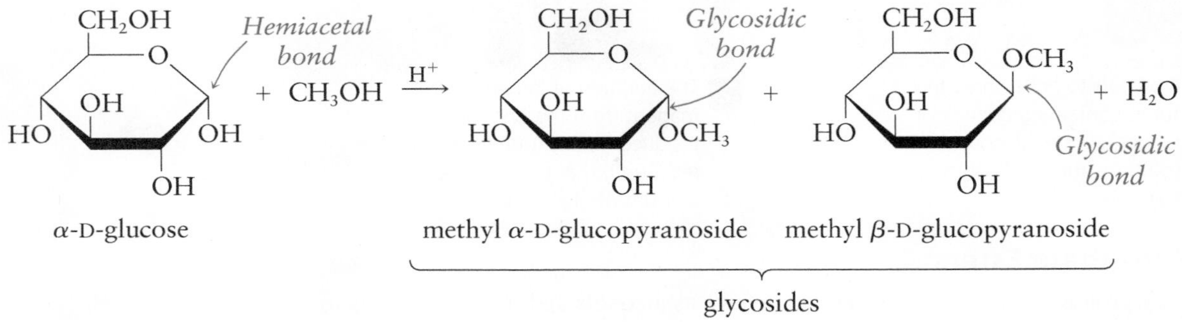

The hemiacetal in glucose is located at position 1, so the acetal forms at that same position. All the other –OH groups in glucose are ordinary alcohol groups and do not react under these conditions. As the glycoside reaction takes place, a new bond is established between the pyranose ring and the –OCH_3 groups. The new group may point up or down from the ring. Thus, a mixture of α- and β-glycosides is formed. The new covalent bond between carbon and oxygen that joins the –OR group to the carbohydrate ring is called the **glycosidic linkage**, or **glycosidic bond**.

glycosidic linkage or **glycosidic bond** A covalent bond between carbon and oxygen that joins the –OR group to the carbohydrate ring.

Although carbohydrate cyclic hemiacetals and hemiketals are in equilibrium with open-chain forms of the monosaccharides, the glycosides (acetals and ketals) are much more stable and do not exhibit open-chain forms. Therefore, glycosides of monosaccharides are not reducing sugars. As we explain in **Sections 11.7** and **11.8**, both disaccharides and polysaccharides are examples of glycosides in which monosaccharide units are joined together by acetal (glycosidic) bonds.

Example 11.7 Identifying Glycosidic Linkages

Deoxyribose is a component of deoxyribonucleic acid (DNA). A modification of deoxyribose is shown here. Highlight any acetal or ketal group and use an arrow to identify the glycosidic bond.

$HOCH_2$ O $OCH_2CH_2CH_3$

OH

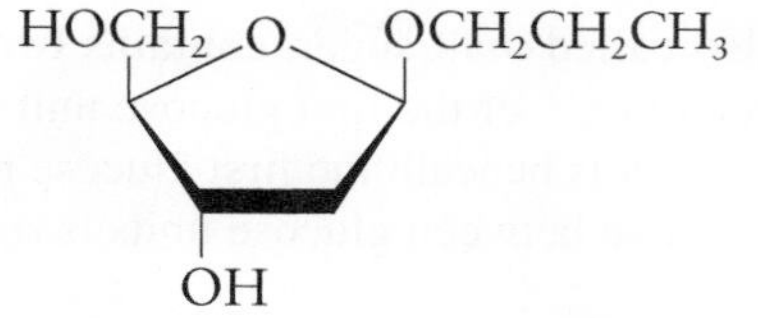

Solution

Step 1: *Identify the acetal or ketal groups.* Notice that the group highlighted in the following structure shows a carbon attached to two –OR groups and a hydrogen (attached but not shown,) making it an acetal.

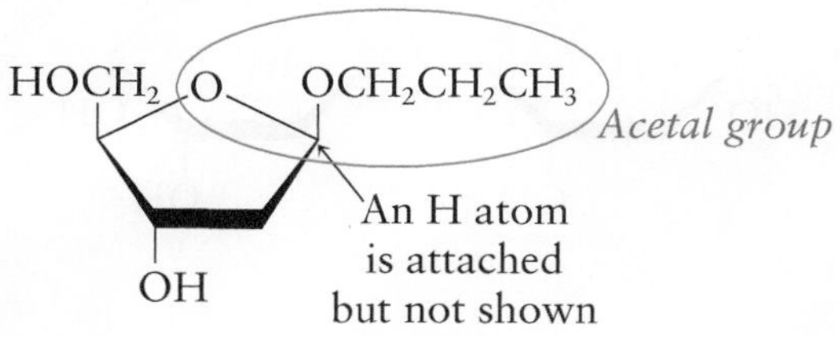

$HOCH_2$ O $OCH_2CH_2CH_3$ *Acetal group*

An H atom is attached but not shown

OH

Step 2: *Identify the glycosidic bond.* The bond between the furanose ring and the $-OCH_2CH_2CH_3$ group is the glycosidic bond.

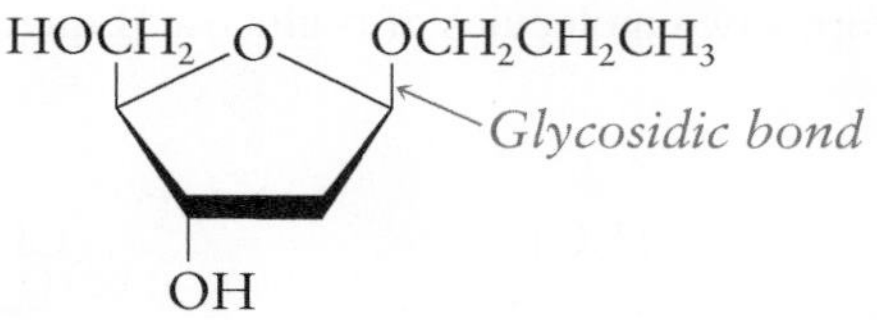

$HOCH_2$ O $OCH_2CH_2CH_3$ *Glycosidic bond*

OH

✔ Learning Check 11.7 Two glycosides are shown here. Circle any acetal or ketal groups and use an arrow to identify the glycosidic linkages.

a. CH_2OH, HO, O, OH, OCH_2CH_3, OH

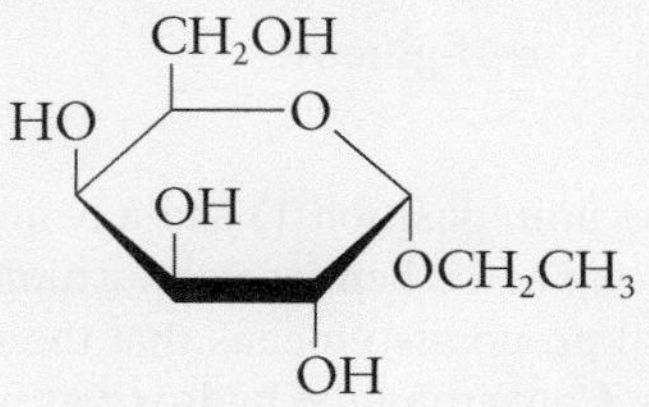

b. $HOCH_2$, O, OCH_3, CH_2OH, HO, OH

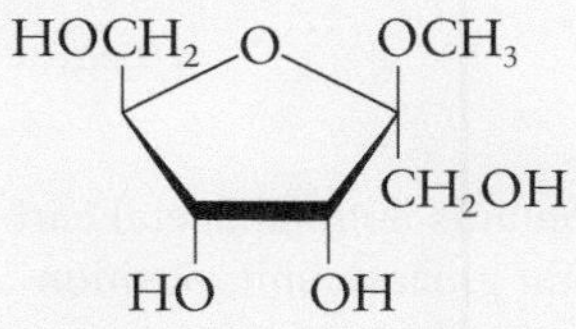

11.7 Disaccharides

Learning Objective 13 Describe the uses of three important disaccharides.

Learning Objective 14 Describe the linkage in disaccharides.

Learning Objective 15 Write reactions for the condensation and hydrolysis of disaccharides.

Disaccharides are sugars composed of two monosaccharide units linked together through a condensation reaction. The two individual units are thus bonded together through a glycosidic linkage. They can likewise undergo hydrolysis (the reverse reaction) to yield

Sokor Space/Shutterstock.com

Figure 11.27 Naturally occurring sources of maltose include grains like wheat and barley.

their monosaccharide building blocks by boiling with dilute acid or reacting them with appropriate enzymes. Nutritionally, the most important members of this group are maltose, lactose, and sucrose.

11.7A Maltose

Maltose (**Figure 11.27**), also called malt sugar, contains two glucose units joined by a glycosidic linkage between carbon 1 of the first glucose unit and carbon 4 of the second unit (Equation 11.6). The oxygen is beneath the first glucose ring, so the configuration of carbon 1 in the glycosidic linkage between glucose units is α. The linkage is symbolized as $\alpha(1 \rightarrow 4)$.

$\alpha(1 \rightarrow 4)$ Glycosidic linkage

α-D-glucose + α-D-glucose $\longrightarrow$ maltose + H_2O (11.6)

Maltose, which is found in germinating grain, is formed during the digestion (hydrolysis) of starch, a polysaccharide, to glucose. Its name is derived from the fact that during the germination (or malting) of barley, starch is hydrolyzed, and the disaccharide is formed. On hydrolysis, maltose forms two molecules of D-glucose (Equation 11.7):

An acetal *A cyclic hemiacetal*

maltose + H_2O $\xrightarrow{H^+}$ α-D-glucose + α-D-glucose (11.7)

Maltose contains both an acetal carbon (left glucose unit, position 1) and a hemiacetal carbon (right glucose unit, position 1). The –OH group at the hemiacetal position can point either up or down. The presence of a hemiacetal group also means that the right glucose ring can open up to expose an aldehyde group. Converting the hydroxy group to an aldehyde is an oxidation, so something else must be reduced because oxidation can not happen without an accompanying reduction. Thus, maltose is a reducing sugar as discussed in **Section 11.6B**.

Hemiacetal group

maltose $\rightleftharpoons$

Aldehyde group

11.7B Lactose

Lactose (milk sugar) constitutes 5% cow's milk and 7% human milk by weight (see **Figure 11.28**). Pure lactose is obtained from whey, the watery by-product of cheese production. Lactose is composed of one molecule of D-galactose and one of D-glucose. The linkage between the two sugar units is $\beta(1 \rightarrow 4)$ because the oxygen of the glycosidic linkage points up from the galactose ring:

Figure 11.28 Cow's milk is about 5% lactose. What two products form when the lactose is hydrolyzed in the child's digestive system?

The presence of a hemiacetal group in the glucose unit makes lactose a reducing sugar.

Example 11.8 Describing Disaccharide Linkages

Gentiobiose is a disaccharide formed when glucose is caramelized by heating. Describe the linkage between the two components.

Solution

At position 1 in the glucose unit on the left, the oxygen is directed upward, so it is β. The glucose unit on the right is connected through position 6, so the linkage is $\beta(1 \rightarrow 6)$.

✔ Learning Check 11.8 The disaccharide cellobiose is formed upon the hydrolysis of cellulose. Describe the linkage between the two units.

Figure 11.29 Hummingbirds depend on sucrose and other carbohydrates of nectar for their energy.

11.7C Sucrose

The disaccharide sucrose (common household sugar) is extremely abundant in the plant world. It occurs in many fruits; in the nectar of flowers (see **Figure 11.29**); and in the juices of many plants, especially sugar cane and sugar beets. Sucrose contains two monosaccharides, glucose and fructose, joined together by a linkage that is α from carbon 1 of glucose and β from carbon 2 of fructose:

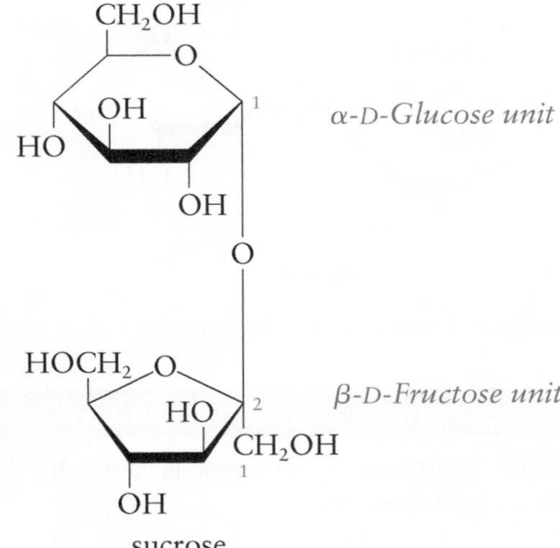

Neither ring in the sucrose molecule contains a hemiacetal or hemiketal group that is necessary for ring opening because the anomeric positions in both glucose and fructose are part of the glycosidic linkage that connects the rings to each other. Thus, in contrast to maltose and lactose, both rings of sucrose are locked in the cyclic form. Sucrose is therefore not a reducing sugar. **Table 11.3** summarizes the features of the disaccharides maltose, lactose, and sucrose.

Table 11.3 Some Important Disaccharides

Name	Monosaccharide Constituents	Glycoside Linkage	Source
Maltose	Two glucose units	$\alpha(1 \rightarrow 4)$	Hydrolysis of starch
Lactose	Galactose and glucose	$\beta(1 \rightarrow 4)$	Mammalian milk
Sucrose	Glucose and fructose	$\alpha\text{-}1 \rightarrow \beta\text{-}2$	Sugar cane and sugar beet juices

11.8 Polysaccharides

Learning Objective 16 Describe the uses of three important polysaccharides.

Learning Objective 17 Describe the linkage in polysaccharides.

Just as two sugar units are linked together to form a disaccharide, additional units can be added to form larger saccharides. Nature does just this in forming polysaccharides, which are condensation polymers containing thousands of units. Because of their size, polysaccharides are not water soluble, but some polysaccharides become viscous when heated in water because of their many hydroxyl groups. Thus, the polysaccharide known as starch can be used as a thickener in sauces, gravies, pie fillings, and other food preparations. As shown in **Table 11.4**, the properties of polysaccharides differ markedly from those of monosaccharides and disaccharides.

Table 11.4 Properties of Polysaccharides Compared with Those of Monosaccharides and Disaccharides

Property	Monosaccharides and Disaccharides	Polysaccharides
Molecular weight	Low	Very high
Taste	Sweet	Tasteless
Solubility in water	Soluble	Insoluble or viscous
Size of particles	Pass through a membrane	Do not pass through a membrane

11.8A Starch

Starch is a polymer consisting entirely of D-glucose units. It is the major storage form of D-glucose in plants. Two starch fractions, amylose (10% to 20%) and amylopectin (80% to 90%), can usually be isolated from plants. Amylose is made up of long unbranched chains of glucose units connected by $\alpha(1 \rightarrow 4)$ linkages (see **Figure 11.30**). This is the same type of linkage found in maltose (see Equation 11.6). The long chain, often containing between 1000 and 2000 glucose units, is flexible enough to allow the molecules to twist into the shape of a helix (see **Figure 11.31**).

$\alpha(1 \rightarrow 4)$ *Linkage*

Figure 11.30 The structure of amylose.

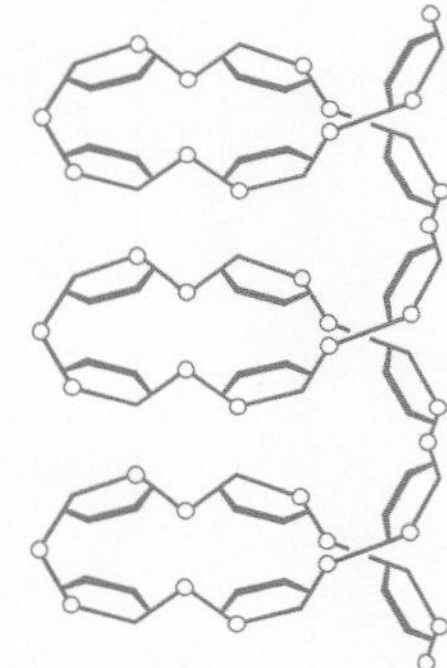

Figure 11.31 The helical conformation of starch.

An important test for the presence of starch is the reaction that occurs between iodine (I_2) and the coiled form of amylose. The product of the reaction is deep blue in color (see **Figure 11.32a**) and is thought to consist of the amylose helix filled with iodine molecules (see **Figure 11.32b**). This same iodine reaction is also widely used to monitor the hydrolysis of starch. The color gradually fades and finally disappears as starch is hydrolyzed by either acid or enzymes to form dextrins (smaller polysaccharides), then maltose, and finally glucose. The disappearance of the deep blue iodine color is thought to be the result of the breakdown of the starch helix.

a

© Michael C. Slabaugh

b

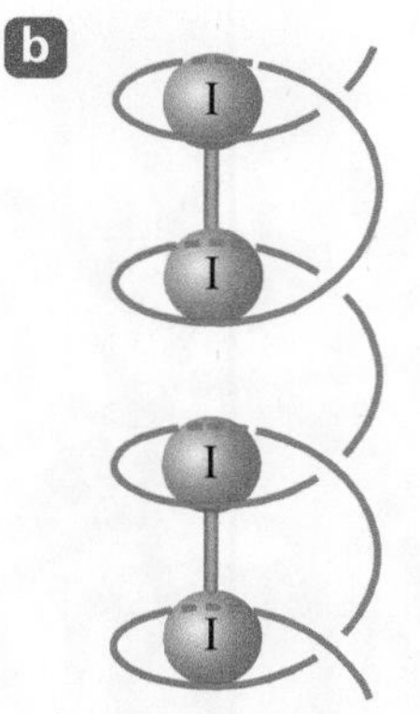

The starch–iodine complex

Figure 11.32 (a) A deep blue color is the characteristic result when a solution of iodine encounters the starch in potatoes. What other foods would you expect to give a positive starch test? (b) The proposed structure of the starch–iodine complex.

Amylopectin, the second component of starch, is not a straight-chain molecule like amylose but contains random branches. The branching point is the $\alpha(1 \rightarrow 6)$ glycosidic linkage (see **Figure 11.33**). There are usually 24 to 30 D-glucose units, all connected by $\alpha(1 \rightarrow 4)$ linkages, between each branch point of amylopectin. These branch points give amylopectin the appearance of a bushy molecule. Amylopectin contains as many as 100,000 glucose units in one gigantic molecule.

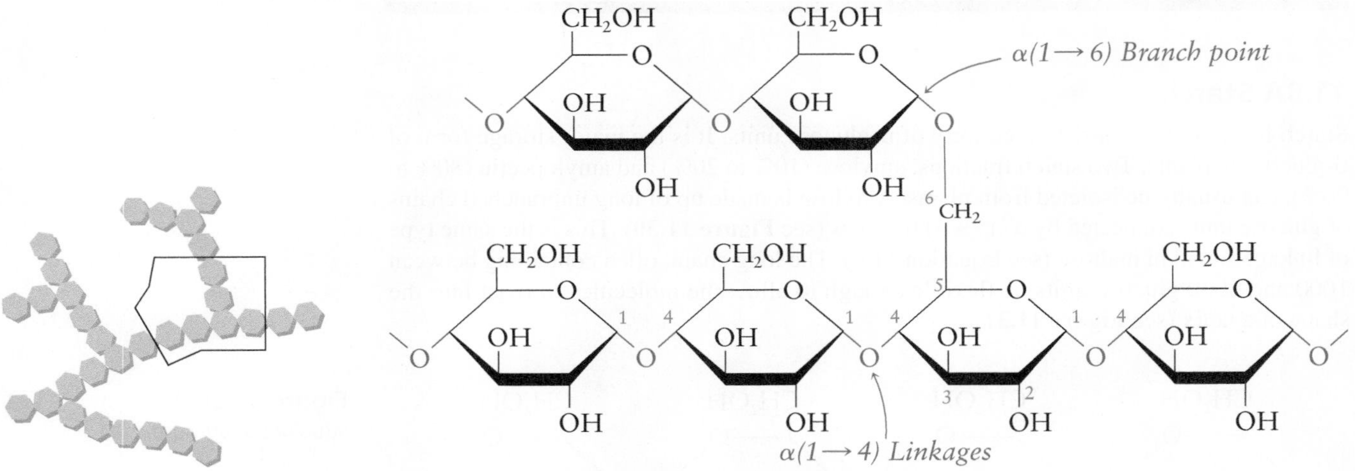

Figure 11.33 The partial structure of an amylopectin molecule. Glycogen (**Section 11.8B**) has a similar structure.

11.8B Glycogen

The polysaccharide glycogen is sometimes called animal starch because it is a storage carbohydrate for animals that is analogous to the starch of plants. It is especially abundant in the liver and muscles, where excess glucose taken in by an animal is stored for future use. On hydrolysis, glycogen forms D-glucose, which helps maintain normal blood sugar levels and provides the muscles with energy. Structurally, glycogen is very similar to amylopectin, containing both $\alpha(1 \rightarrow 4)$ and $\alpha(1 \rightarrow 6)$ linkages between glucose units (see **Figure 11.34**). The main difference between amylopectin and glycogen is that glycogen is even more highly branched. There are only 8 to 12 D-glucose units between branch points.

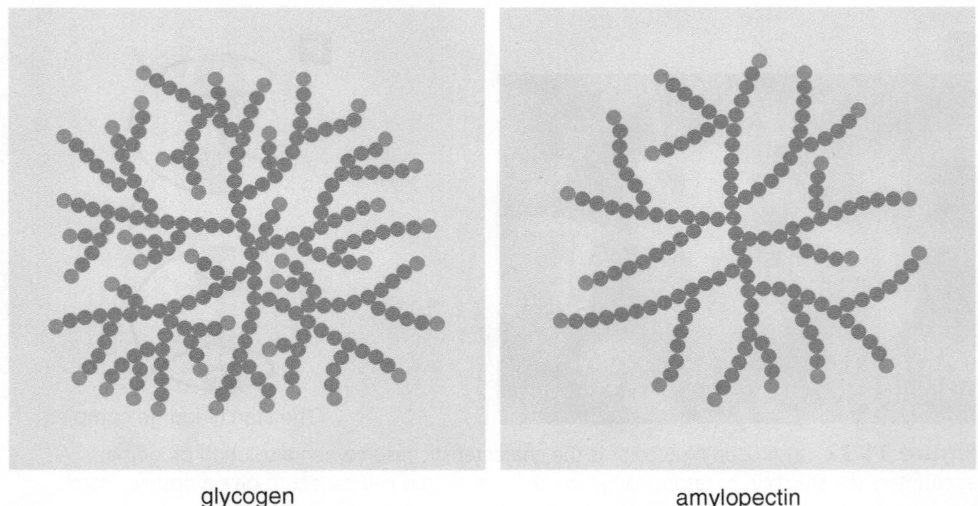

Figure 11.34 A simplified representation of the branched polysaccharide glycogen (branches every 8 to 12 glucose units). Amylopectin is much less densely branched (branches every 24 to 30 glucose units). Each small circle represents a single glucose unit.

11.8C Cellulose

Cellulose is the most important structural polysaccharide (see **Figure 11.35**). It is the material in plant cell walls that provides strength and rigidity. Wood is about 50% cellulose. Like amylose, cellulose is a linear polymer consisting of D-glucose units joined by $1 \rightarrow 4$ linkages. It may contain 300 to 3000 glucose units in one molecule. The main structural difference between amylose and cellulose is that all the $1 \rightarrow 4$ glycosidic linkages in cellulose are β instead of α (see **Figure 11.36**). This small difference causes the shapes of the molecules to be quite different. The $\alpha(1 \rightarrow 4)$-linked amylose tends to form loose spiral structures (see **Figure 11.30**), whereas $\beta(1 \rightarrow 4)$-linked cellulose tends to form extended straight chains. These chains become aligned side by side to form well-organized, water-insoluble fibers in which the hydroxy groups form numerous hydrogen bonds with the neighboring chains. These parallel chains of cellulose confer rigidity and strength to the molecules.

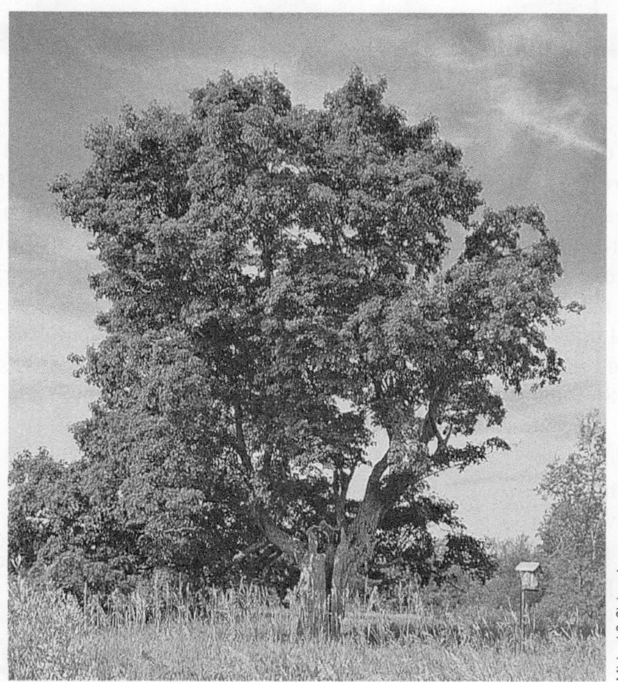

Figure 11.35 Cellulose and the plant kingdom are responsible for much of the world's natural beauty.

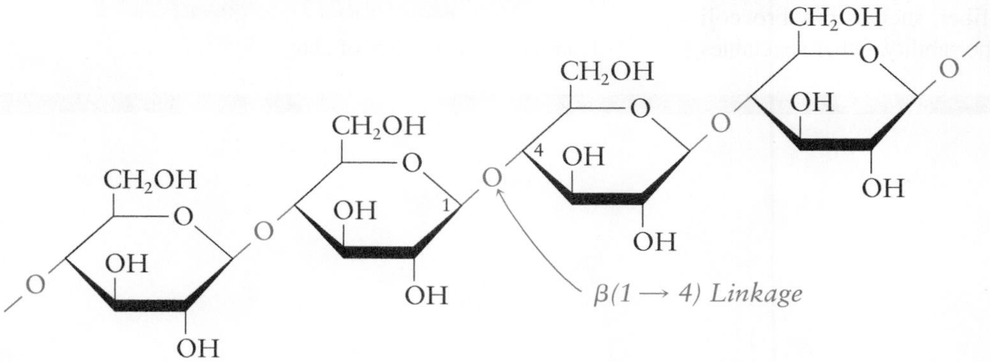

Figure 11.36 The structure of cellulose.

Helen Camacaro/Getty Images

Figure 11.37 Sources of dietary fiber.

Although starch with its α linkages is readily digestible, humans and other animals lack the enzymes necessary to hydrolyze the β linkages of cellulose. Thus, cellulose passes unchanged through the digestive tract and does not contribute to the caloric value of food. However, it still serves a useful purpose in digestion. Cellulose, a common constituent of dietary fiber (**Figure 11.37**), is the roughage that provides bulk, stimulates contraction of the intestines, and aids the passage of food through the digestive system.

Animals that use cellulose as a food do so only with the help of bacteria that possess the enzymes that are necessary for breaking it down. Herbivores such as cows, sheep, and horses are good examples. Each has a colony of such bacteria somewhere in their digestive system. The simple carbohydrates resulting from the bacterial hydrolysis of cellulose can then be used. Fortunately, the soil also contains organisms with appropriate enzymes. Otherwise, debris from dead plants would accumulate rather than decompose.

Health Connections 11.2

Put Fiber into Snacks and Meals

Fiber is a type of carbohydrate found only in food derived from plants. It is not found in any meat or dairy products. While not considered to be a "nutrient," fiber still plays a vital role in maintaining good digestive and cardiovascular health, as well as weight loss and weight maintenance. Research indicates that eating a diet high in fiber also reduces the risk of developing cancer. Fiber, along with adequate fluid intake, plays a cleansing role in the body. The recommended daily fiber intake is 21 to 25 grams for females and 30 to 38 grams for males. Nutrition labels on foods list fiber content. This is usually given in grams per serving, or in percentage of daily recommended amounts per serving. Most foods that contain fiber offer small amounts (2 to 5 grams) per serving, so including fiber throughout the day in snacks as well as meals is important in order to reach a goal of 25 to 30 grams per day.

Knowing which foods contain fiber assists in making healthy choices and reaching daily fiber intake goals. Fruits are a good place to start when counting daily fiber intake. Fresh raspberries contain 8 grams of fiber per cup, which is more than 30% of the daily fiber intake recommendation. Apples, oranges, and bananas each contain 3 to 5 grams per serving. Vegetable snacks that contribute fiber, such as raw broccoli and carrots, offer convenience and portability. Other vegetables might not make suitable raw snacks, but can add significant fiber to the diet when added to daily meals. Good vegetable choices are cooked peas, beans, corn, and artichokes. Grains and grain products are significant sources of fiber.

A great way to boost fiber intake is to sprinkle high-fiber foods on the foods you normally eat. Oat bran, wheat bran, ground flax seeds, or berries can easily be added to sweet or savory foods such as oatmeal, pancakes, soups, baked potatoes, casseroles, or salads. Any meal can become a high-fiber meal with a sprinkling of bran.

Nitr/Shutterstock.com

Fruits are a good source of fiber.

Health Career Description

Dental Assistant

A wide range of duties exists within the dental assisting field. Some assistants perform mostly office support and help with making patients comfortable, as well as collecting insurance information and payments. Other assistants may perform treatments such as applying topical anesthetics, performing coronal polishing, or applying sealants or fluoride treatments.

Assistants that provide only office assistance and help with patients are not required to have any certification, but those who practice dental treatments typically require certification or licensure. Some programs are 1 year long and end in certification; others span 2 years and culminate in an associate's degree.

The American Dental Association lists the duties of a dental assistant as:

- Assisting the dentist during a variety of treatment procedures
- Taking and developing dental radiographs (X rays)
- Asking about the patient's medical history and taking blood pressure and pulse
- Serving as an infection control officer, developing infection control protocol, and preparing and sterilizing instruments and equipment
- Helping patients feel comfortable before, during, and after dental treatment
- Providing patients with instructions for oral care following surgery or other dental treatment procedures, such as the placement of a restoration (filling)
- Teaching patients appropriate oral hygiene strategies to maintain oral health (e.g., tooth brushing, flossing, and nutritional counseling)
- Taking impressions of patients' teeth for study casts (models of teeth)
- Performing office management tasks that require the use of a computer
- Communicating with patients and suppliers (e.g., scheduling appointments, answering the telephone, billing, and ordering supplies)
- Helping to provide direct patient care in all dental specialties, including orthodontics, pediatric dentistry, periodontics, and oral surgery

A dental assistant is not a dental hygienist. Hygienists' work focuses on cleaning the teeth. Dental hygienists typically hold at least an associate's degree and often a bachelor's degree. According to the Bureau of Labor Statistics, the average salary for a dental assistant is \$38,000 annually and is "growing at 11% (much faster than average)." Demand for dental assistant jobs is expected to continue to grow and is a secure bet for a career that will be in demand for the foreseeable future.

Sources: American Dental Association. Dental Assistants. Retrieved Apr. 21, 2020 from: https://www.ada.org/en/education-careers/careers-in-dentistry/dental-team-careers/dental-assistant; Bureau of Labor Statistics. Dental Assistants. Retrieved April 21, 2020 from: https://www.bls.gov/ooh/ healthcare/dental-assistants.htm

Concept Summary

11.1 Carbohydrate Structure and Function

Learning Objectives: **Describe the four major functions of carbohydrates in living organisms. Classify carbohydrates as monosaccharides, disaccharides, or polysaccharides.**

- Carbohydrates are polyhydroxy aldehydes or ketones, or substances that yield such compounds on hydrolysis.
- Carbohydrates are used as energy sources, as biosynthetic intermediates, for energy storage, and as structural elements in organisms.
- Carbohydrates can exist either as single units (monosaccharides) or joined together in molecules ranging from two units (disaccharides) to hundreds of units (polysaccharides).

11.2 Carbohydrate Stereochemistry

Learning Objectives: **Identify chiral carbon atoms in molecules. Draw pairs of enantiomers using wedges and dashes.**

- Wedge-and-dash notation is used to visualize a molecule in three-dimensional space.
- A carbon with four different ligands attached to it is defined as a chiral carbon.
- Carbohydrates, along with many other natural substances, exhibit a type of isomerism in which two isomers are mirror images of each other (enantiomers).
- When a molecule has more than one chiral carbon, the maximum number of stereoisomers possible is 2^n, where n is the number of chiral carbons.

11.3 Fischer Projections, Cyclization, and Haworth Projections

Learning Objectives: **Use Fischer projections to represent D- and L-enantiomers. Identify the following functional groups: hemiacetal, hemiketal, acetal, and ketal. Draw Haworth projections of carbohydrate anomers.**

- A useful way of depicting the structure of chiral molecules employs crossed lines (Fischer projections) to represent chiral carbon atoms.
- The prefixes D- and L- are used to distinguish between enantiomers.
- Signs indicating the rotation of plane-polarized light to the right (+) or to the left (−) may also be used to designate enantiomers.
- In a reaction that is very important in sugar chemistry, an alcohol can add across the carbonyl group of an aldehyde to produce a hemiacetal.
- The substitution reaction of a second alcohol molecule with the hemiacetal produces an acetal.
- Ketones can undergo similar reactions to form hemiketals and ketals.
- Pentoses and hexoses form cyclic hemiacetals or hemiketals whose structures can be represented by Haworth structures.
- Two isomers referred to as anomers (the α and β forms) are produced in the cyclization reaction.

11.4 Classification of Monosaccharides

Learning Objectives: **Classify monosaccharides as aldoses or ketoses. Classify monosaccharides according to the number of carbon atoms they contain.**

- Monosaccharides that contain an aldehyde group are called aldoses.
- Those containing a ketone group are ketoses.
- Monosaccharides are also classified by the number of carbon atoms as trioses, tetroses, and so on.
- Most natural monosaccharides belong to the D family.

11.5 Important Monosaccharides

Learning Objective: **Describe uses of four important monosaccharides.**

- Ribose and deoxyribose are important as components of nucleic acids.
- The hexoses glucose, galactose, and fructose are the most important nutritionally and the most abundant in nature.
- Glucose, also known as blood sugar, is transported within the bloodstream to body tissues, where it supplies energy.

11.6 Properties of Monosaccharides

Learning Objectives: **Describe the purpose of Benedict's test. Identify glycosidic linkages in carbohydrate derivatives.**

- Monosaccharides are sweet-tasting solids that are very soluble in water.
- Noncarbohydrate low-calorie sweeteners such as aspartame have been developed as sugar substitutes.
- All monosaccharides are oxidized by Benedict's reagent and are called reducing sugars.
- Esters formed from phosphoric acid and various monosaccharides are found in biomolecules (e.g., DNA), and are called phosphate esters.
- Monosaccharides can react with alcohols to produce acetals or ketals that are called glycosides.

11.7 Disaccharides

Learning Objectives: **Describe the uses of three important disaccharides. Describe the linkage in disaccharides. Write reactions for the condensation and hydrolysis of disaccharides.**

- Monosaccharides combine to form disaccharides via condensation reactions, and disaccharides break down to monosaccharides via hydrolysis reactions.
- Glycosidic linkages join monosaccharide units together to form disaccharides.
- Three important disaccharides are maltose [two glucose units $\alpha(1 \rightarrow 4)$-linked], lactose [a galactose linked to glucose by a $\beta(1 \rightarrow 4)$ glycosidic linkage], and sucrose (α-glucose joined to β-fructose).

11.8 Polysaccharides

Learning Objectives: **Describe the uses of three important polysaccharides. Describe the linkage in polysaccharides.**

- Cellulose, starch, and glycogen are three important polysaccharides.
- Starch is the major storage form of glucose in plants.
- Glycogen is the storage form of glucose in animals.
- Cellulose is the structural material of plants.

Key Terms and Concepts

Acetal (11.3)
Anomeric carbon (11.3)
Anomers (11.3)
Benedict's reagent (11.6)
Biochemistry (Introduction)
Biomolecule (Introduction)
Carbohydrate (11.1)
Chiral (11.2)
Chiral carbon (11.2)
Dextrorotatory (11.3)
Disaccharide (11.1)
Enantiomers (11.2)
Fischer projection (11.3)
Furanose ring (11.3)
Glycoside (11.6)
Glycosidic linkage or Glycosidic bond (11.6)
Haworth structure (11.3)
Hemiacetal (11.3)
Levorotatory (11.3)
Monosaccharide (11.1)
Oligosaccharide (11.1)
Optically active molecule (11.3)
Phosphate ester (11.6)
Polysaccharide (11.1)
Pyranose ring (11.3)
Reducing sugar (11.6)

Key Equations

1. Hemiacetal and acetal formation **(Section 11.3)**	$R-\overset{O}{\overset{\|\|}{C}}-H + R'-OH \rightleftharpoons R-\underset{OR'}{\overset{OH}{C}}-H \xrightleftharpoons[R'-OH]{H^+ \text{ and}} R-\underset{OR'}{\overset{OR'}{C}}-H + H_2O$ aldehyde, alcohol, hemiacetal intermediate, acetal	Equation 11.1
2. Pyranose ring formation **(Section 11.3)**	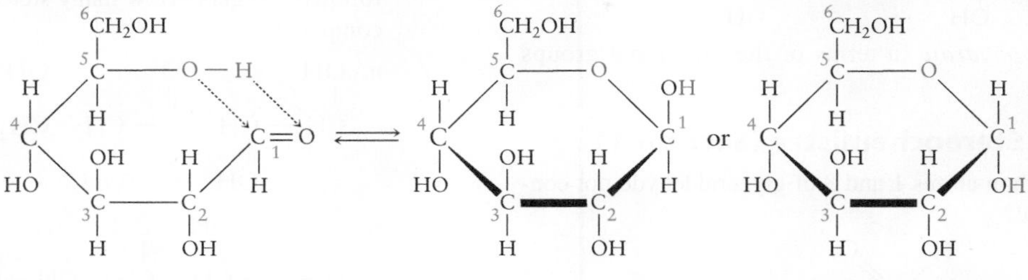 (open-chain D-glucose ⇌ ring forms) β-D-glucose or α-D-glucose	Equation 11.2
3. Furanose ring formation **(Section 11.3)**	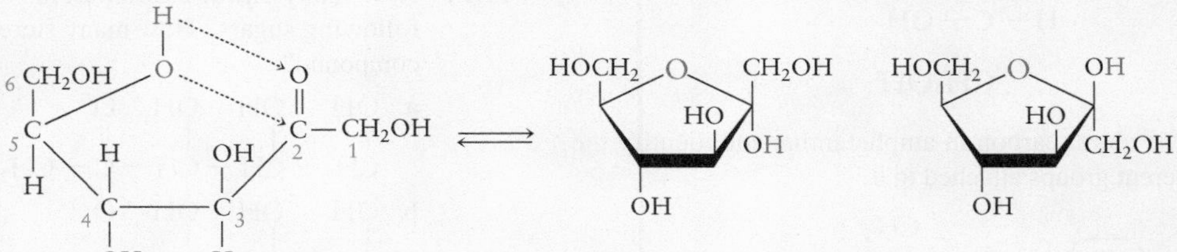 D-fructose ⇌ α-D-fructose, β-D-fructose	Equation 11.3
4. Oxidation of a sugar **(Section 11.6)**	reducing sugar + $Cu^{2+} \longrightarrow$ oxidized compound + Cu_2O	Equation 11.4
5. Glycoside formation **(Section 11.6)**	monosaccharide + alcohol $\xrightarrow{H^+}$ acetals or ketals	Equation 11.5
6. Condensation of monosaccharides to form disaccharides **(Section 11.7)**	two monosaccharides $\xrightarrow{\text{Enzymes}}$ disaccharide + H_2O	Equation 11.6
7. Hydrolysis of disaccharides **(Section 11.7)**	disaccharide + $H_2O \xrightarrow[\text{Enzymes}]{H^+ \text{ or}}$ two monosaccharides	Equation 11.7

Exercises

Even-numbered exercises are answered in Appendix B.

Exercises with an asterisk (*) are more challenging.

Carbohydrate Structure and Function (Section 11.1)

11.1 What are the four important roles of carbohydrates in living organisms?

11.2 Describe whether each of the following substances serves primarily as an energy source, a form of stored energy, or a structural material (some serve as more than one).

a. cellulose **b.** sucrose, table sugar

c. glycogen **d.** starch

11.3 What are the structural differences among monosaccharides, disaccharides, and polysaccharides?

11.4 Match the terms *carbohydrate*, *monosaccharide*, *disaccharide*, and *polysaccharide* to each of the following (more than one term may fit).

a. table sugar **b.** (ring structure with CH_2OH, O, OH, OH, HO, OH)

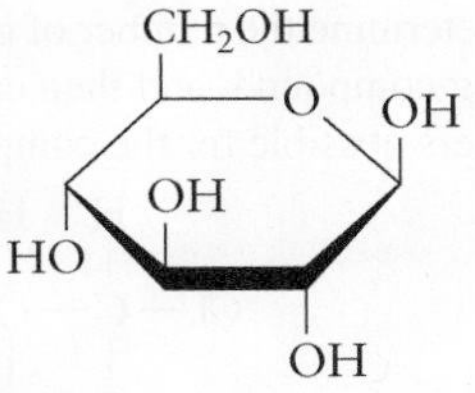

c. starch **d.** glycogen

11.5 Match the terms *carbohydrate*, *monosaccharide*, *disaccharide*, and *polysaccharide* to each of the following (more than one term may fit).

a. fructose

b. cellulose

c.

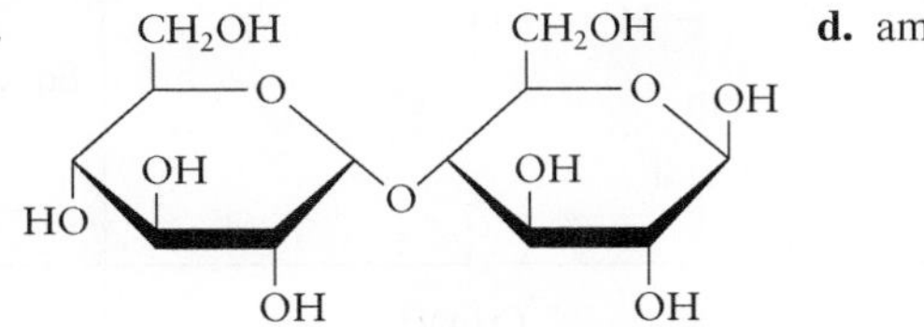

d. amylose

11.6 Define *carbohydrate* in terms of the functional groups present.

Carbohydrate Stereochemistry (Section 11.2)

11.7 Why are carbon atoms 1 and 3 of glyceraldehyde not considered chiral?

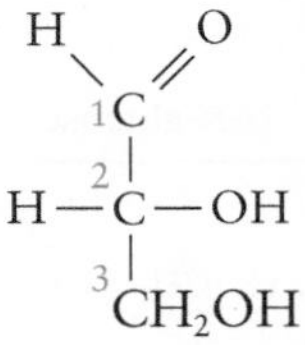

$$\mathrm{H-\overset{1}{C}(=O)-\overset{2}{C}H(OH)-\overset{3}{C}H_2OH}$$

11.8 Locate the chiral carbon in amphetamine and identify the four different groups attached to it.

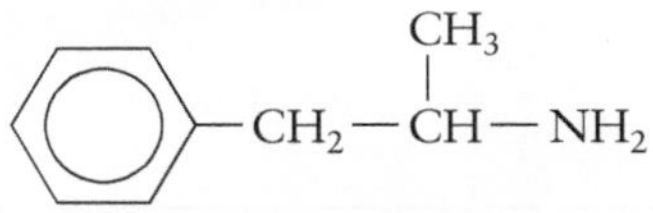

$$\mathrm{C_6H_5-CH_2-CH(CH_3)-NH_2}$$

11.9 Determine whether the circled carbon atom in each of the following compounds is a chiral center.

a.  $\mathrm{CH_3-\textcircled{C}H_2-OH}$

b. $\mathrm{CH_3-\textcircled{C}H(CH_3)-NH_2}$

c. $\mathrm{CH_3-\textcircled{C}H(Cl)-OH}$

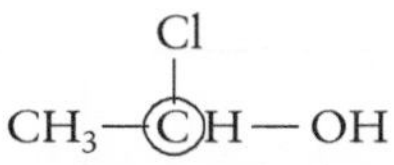

d. 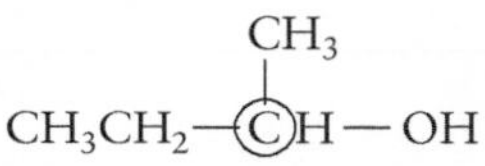 $\mathrm{CH_3CH_2-\textcircled{C}H(CH_3)-OH}$

11.10 Determine whether the circled carbon atom in each of the following compounds is a chiral center.

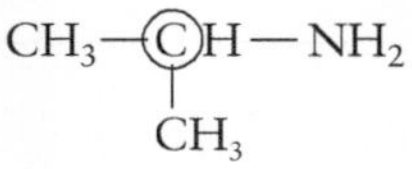

a. 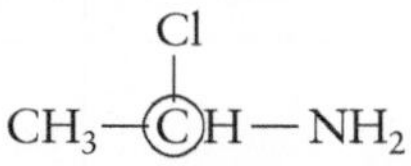 $\mathrm{CH_3-\textcircled{C}H(Cl)-NH_2}$

b. $\mathrm{CH_3CH_2-\textcircled{C}H(CH_2CH_2CH_3)-CH_3}$

c. $\mathrm{CH_3CH_2-\textcircled{C}H(OH)-CH_2CH_3}$

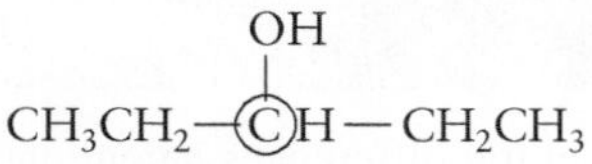

d. 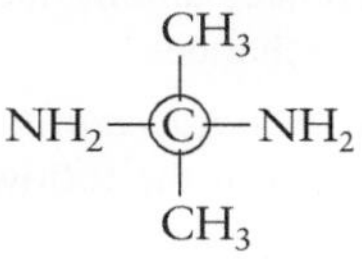 $\mathrm{NH_2-\textcircled{C}(CH_3)_2-NH_2}$

11.11 Determine the number of chiral carbon atoms in the following compound, and then calculate the number of stereoisomers possible for the compound.

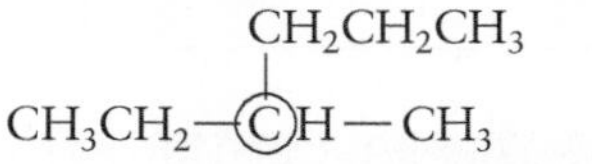

$$\mathrm{Cl-CH(Br)-CH(OH)-CH(Br)-Cl}$$

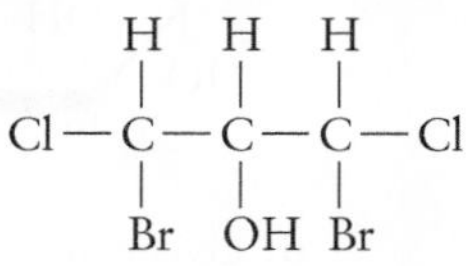

11.12 Determine the number of chiral carbon atoms in the following compound, and then calculate the number of stereoisomers possible for the compound.

$$\mathrm{O{=}CH-CH(Cl)-CH(Cl)-NH_2}$$

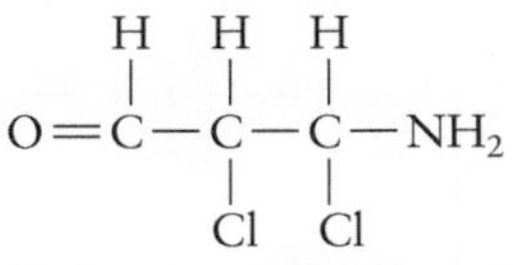

11.13 How many chiral carbon atoms are there in each of the following sugars? How many stereoisomers exist for each compound?

a. 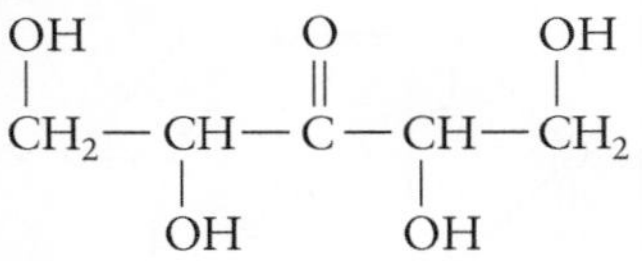 $\mathrm{CH_2(OH)-CH(OH)-C(=O)-CH(OH)-CH_2(OH)}$

b. 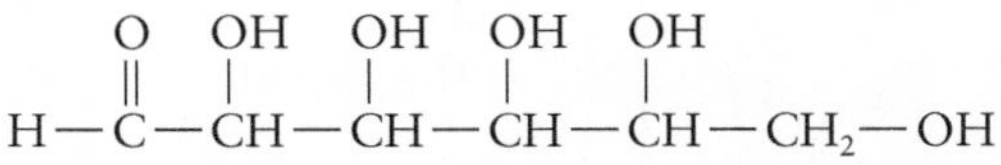 $\mathrm{H-C(=O)-CH(OH)-CH(OH)-CH(OH)-CH(OH)-CH_2-OH}$

11.14 How many chiral carbon atoms are there in each of the following sugars? How many stereoisomers exist for each compound?

a. 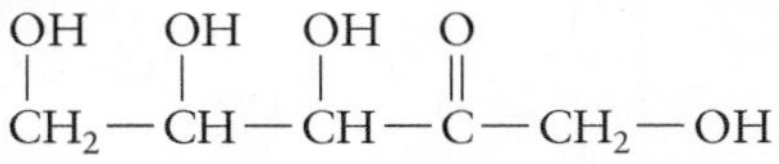 $\mathrm{CH_2(OH)-CH(OH)-CH(OH)-C(=O)-CH_2-OH}$

b. 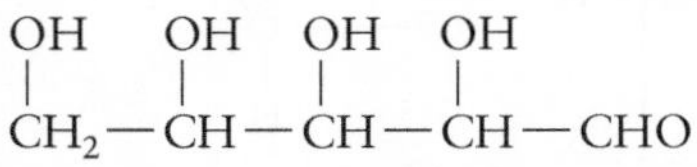 $\mathrm{CH_2(OH)-CH(OH)-CH(OH)-CH(OH)-CHO}$

11.15 Which of the following molecules can have enantiomers? Identify any chiral carbon atoms.

a. 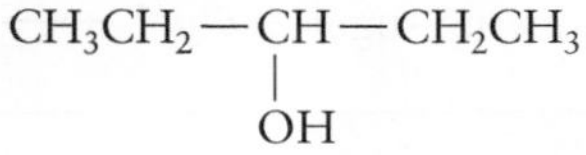 $\mathrm{CH_3CH_2-CH(OH)-CH_2CH_3}$

b. 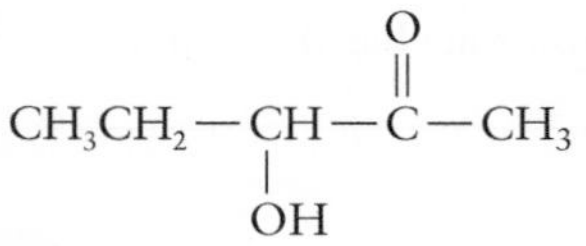 $\mathrm{CH_3CH_2-CH(OH)-C(=O)-CH_3}$

c. $\mathrm{C_5H_9-C(OH)(CH_3)-CH_2CH_3}$ (cyclopentyl)

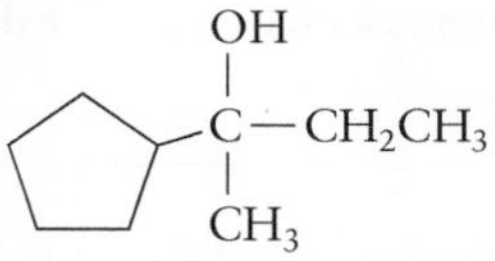

11.16* Which of the following molecules can have enantiomers? Identify any chiral carbon atoms.

a. 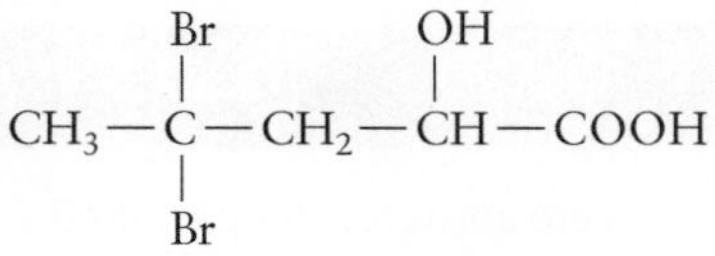 $\mathrm{CH_3-CBr_2-CH_2-CH(OH)-COOH}$

b. $\mathrm{C_6H_5-CH(OH)-COOH}$

c. $\mathrm{CH_2OH}$ attached to a six-membered ring containing O

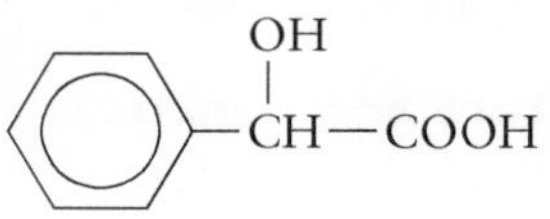

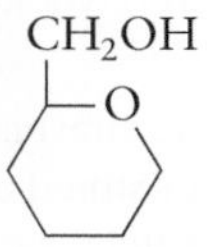

11.17 Draw the pairs of enantiomers of the following molecules in skeletal form using wedges and dashes.

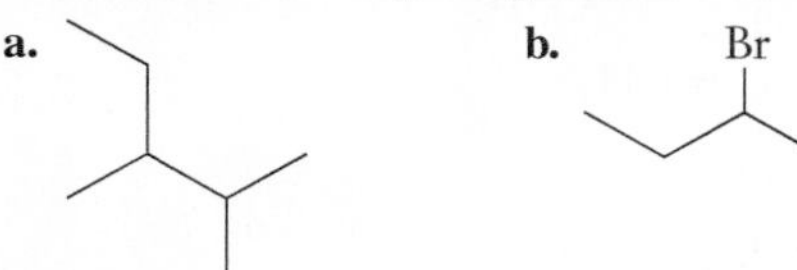

a. **b.** Br

11.18 Draw the pairs of enantiomers of the following molecules in skeletal form using wedges and dashes.

a. 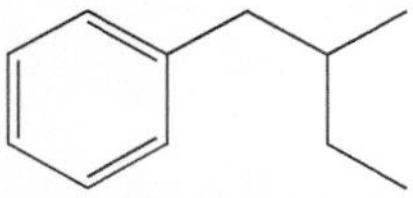 **b.** 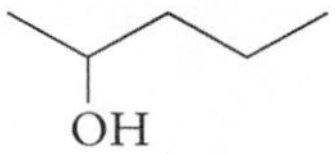 OH

11.19 Draw the enantiomer of the following molecules with multiple chiral centers in skeletal form using wedges and dashes.

a. 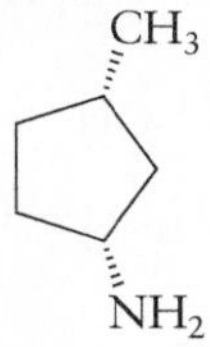 CH_3, NH_2 **b.** 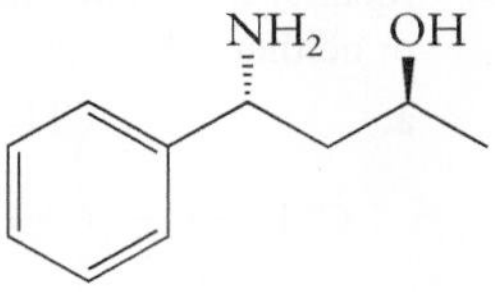 NH_2 OH

11.20 Draw the enantiomer of the following molecules with multiple chiral centers in skeletal form using wedges and dashes.

a. 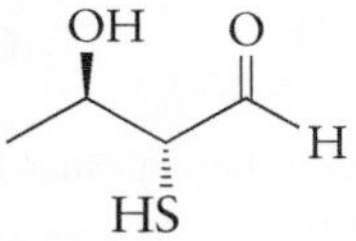 OH O H HS **b.** 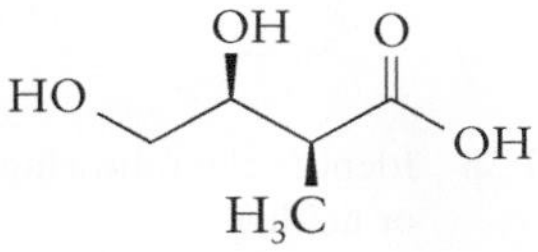 OH O HO OH H_3C

Fischer Projections, Cyclization, and Haworth Projections (Section 11.3)

11.21 Draw the Fischer projection of the enantiomer for each of the following compounds.

a.

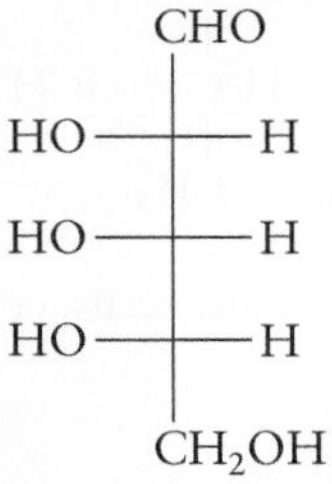

CHO
HO—H
HO—H
HO—H
CH_2OH

b.

CH_2OH
C=O
HO—H
HO—H
CH_2OH

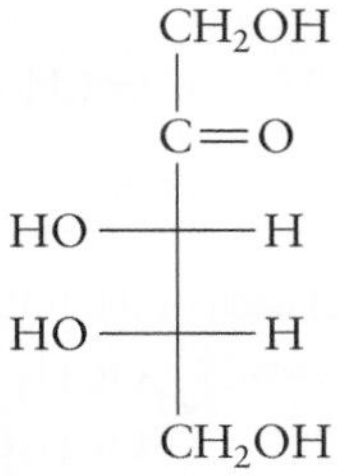

11.22 Draw the Fischer projection of the enantiomer for each of the following compounds.

a.

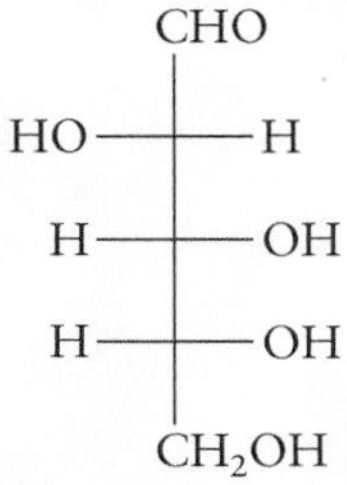

CHO
HO—H
H—OH
H—OH
CH_2OH

b.

CH_2OH
C=O
H—OH
H—OH
CH_2OH

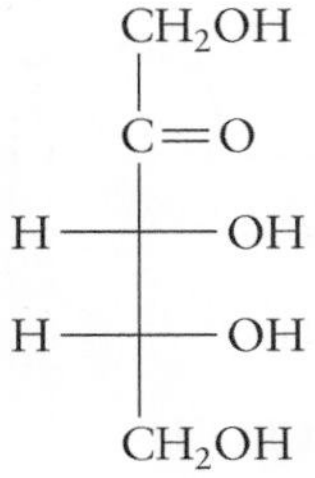

11.23 Explain what the following Fischer projection denotes about the three-dimensional structure of the compound.

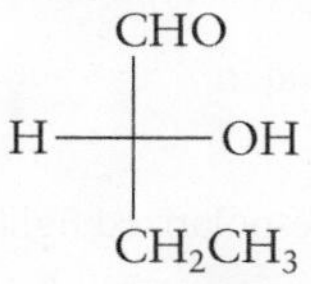

CHO
H—OH
CH_2CH_3

11.24 Identify each of the following as a D or an L form and draw the structural formula of the enantiomer.

a.

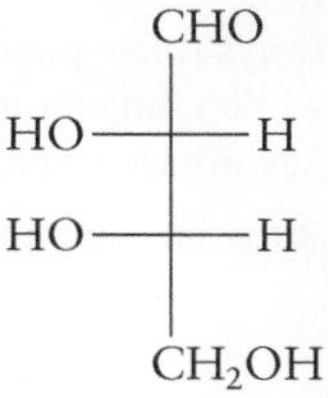

CHO
HO—H
HO—H
CH_2OH

b.

CH_2OH
C=O
HO—H
H—OH
CH_2OH

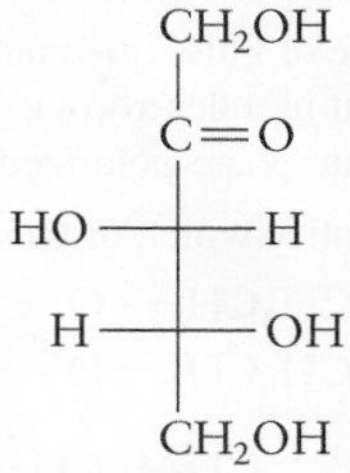

11.25 Identify each of the following as a D or an L form and draw the structural formula of the enantiomer.

a.

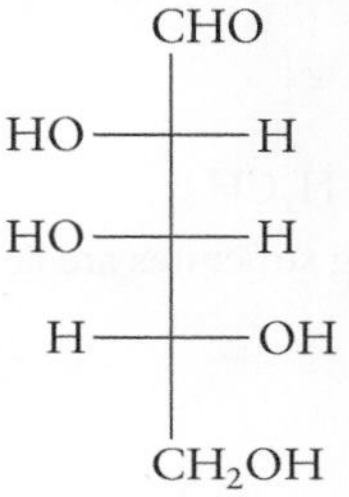

CHO
HO—H
HO—H
H—OH
CH_2OH

b.

CH_2OH
C=O
H—OH
HO—H
HO—H
CH_2OH

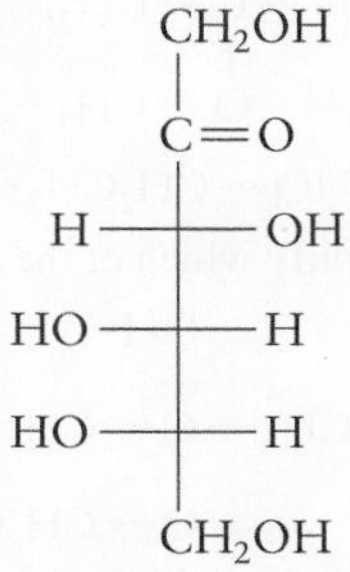

11.26* Draw Fischer projections for both the D and L isomers of the following.

a. 2,3-dihydroxypropanoic acid

b. 2-chloro-3-hydroxypropanoic acid

11.27* Draw Fischer projections for both the D and L isomers of the following amino acids.

a. alanine, $H_2N—CH—COOH$ (with CH_3 on CH)

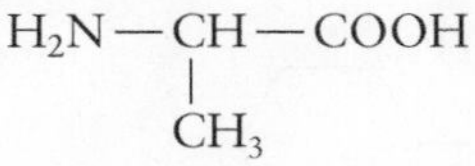

b. leucine, $H_2N—CH—COOH$ (with CH_2—CH(H_3C)(CH_3) on CH)

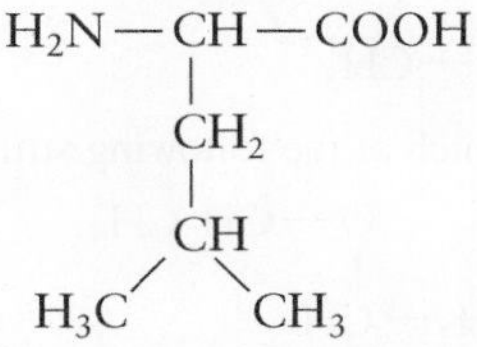

11.28 Why is the study of chiral molecules important in biochemistry?

11.29 What differences would you expect to see in the following physical properties between (+)-alanine and (−)-alanine?

a. Freezing point

b. Solubility in an achiral solvent

c. Density

d. Optical activity

11.30 What differences would you expect to see in the following physical properties between (+)-erythrose and (−)-erythrose?

a. Solubility in water

b. Melting point

c. Effect on plane-polarized light

d. Boiling point

11.31 What physical property is characteristic of optically active molecules?

11.32 True or False? D-Enantiomers always rotate plane-polarized light in a dextrorotary fashion, while L-enantiomers always rotate plane-polarized light in a levorotary fashion.

11.33 Identify which of the following structures are hemiacetals.

a. $CH_3CH_2{-}O{-}CH_2CH_3$

b. 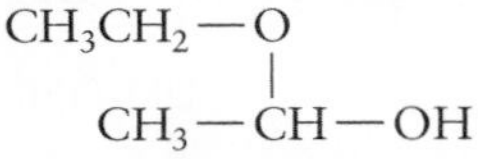 $CH_3CH_2{-}O$
$CH_3{-}CH{-}OH$

c. 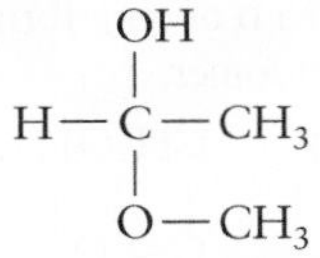 OH
$H{-}C{-}CH_3$
$O{-}CH_3$

d. $HO{-}CH_2CH_2{-}O{-}CH_2CH_3$

11.34 Identify which of the following structures are hemiacetals.

a. 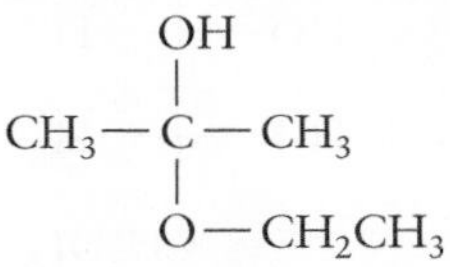 OH
$CH_3{-}C{-}CH_3$
$O{-}CH_2CH_3$

b. OH
$CH_3{-}C{-}CH_2CH_3$
OH

c. $CH_3CH_2{-}CH{-}CH_2CH_3$
$O{-}CH_2CH_3$

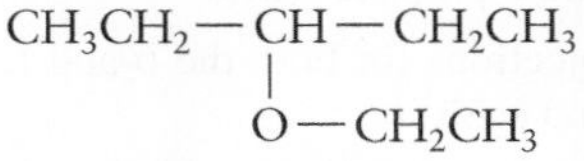

d. 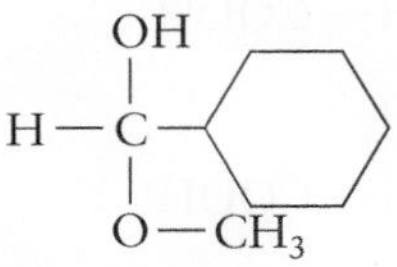 OH
$H{-}C{-}$(cyclohexane ring)
$O{-}CH_3$

11.35 Identify which of the following structures are acetals.

a. 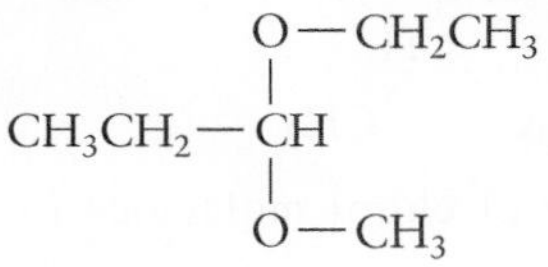 $O{-}CH_2CH_3$
$CH_3CH_2{-}CH$
$O{-}CH_3$

b. 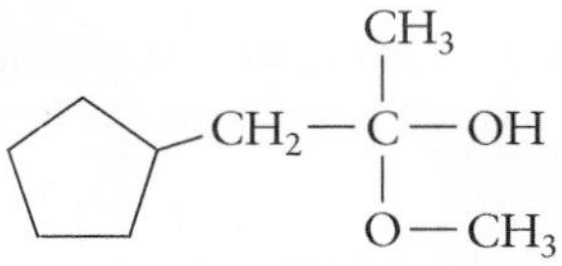 CH_3
(cyclopentane ring)$CH_2{-}C{-}OH$
$O{-}CH_3$

c. $O{-}CH_3$
$CH_3{-}C{-}O{-}CH_3$
CH_3

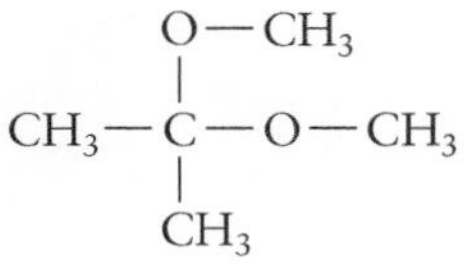

11.36 Identify which of the following structures are acetals.

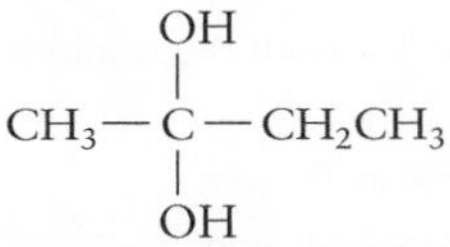

a. 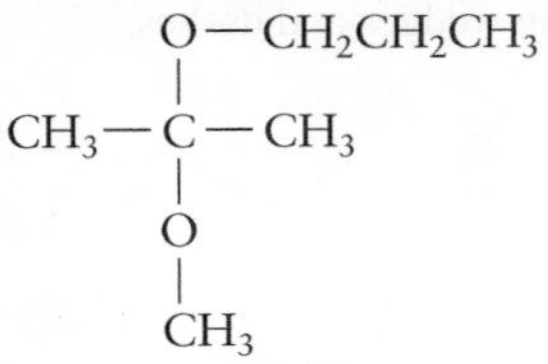 $O{-}CH_2CH_2CH_3$
$CH_3{-}C{-}CH_3$
O
CH_3

b. CH_2CH_3
$HC{-}O{-}CH_2CH_3$
$O{-}CH_3$

c. $CH_3{-}O{-}CHCH_2CH_3$
$O{-}CH_2CH_2CH_3$

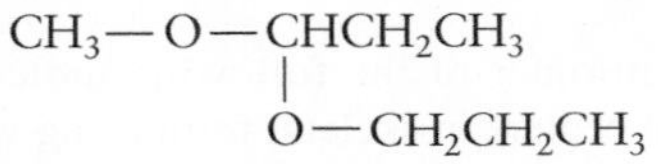

11.37 Identify the following structures as hemiacetals, hemiketals, or neither.

a. 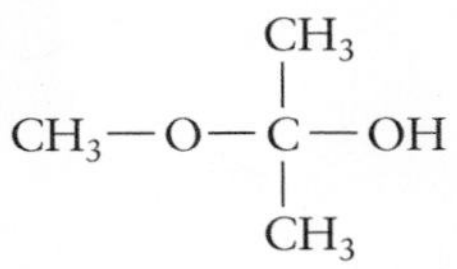 CH_3
$CH_3{-}O{-}C{-}OH$
CH_3

b. OH
$CH_3CH_2CH{-}OCH_2CH_3$

c. OH (cyclopentane ring) OCH_3

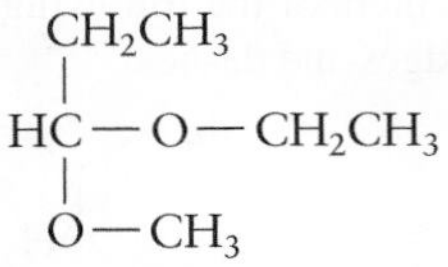

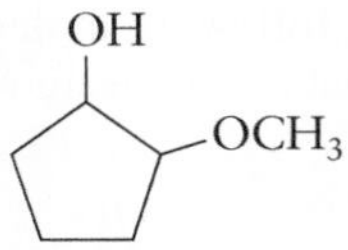

d. 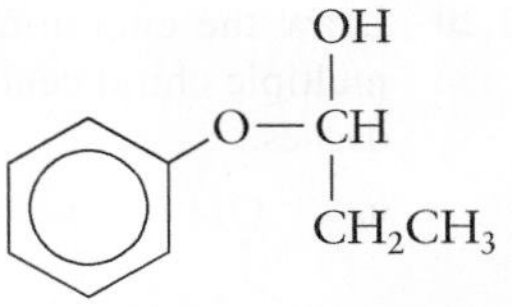 OH
(benzene ring)$O{-}CH$
CH_2CH_3

11.38 Identify the following structures as hemiacetals, hemiketals, or neither.

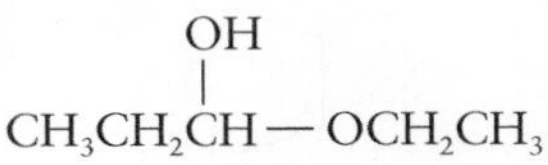

a. 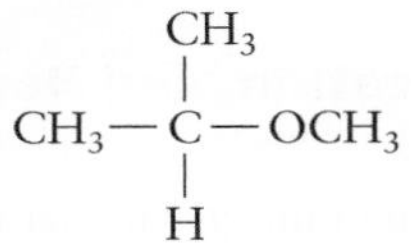 CH_3
$CH_3{-}C{-}OCH_3$
H

b. OCH_3
$CH_3CH_2CH{-}OH$

c. OH
$CH_3{-}C{-}CH_2CH_3$
OH

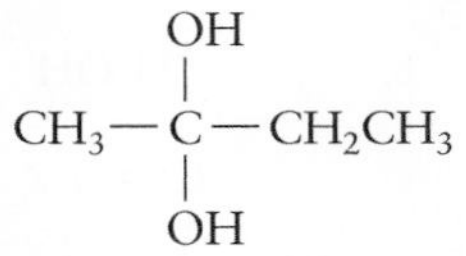

d. 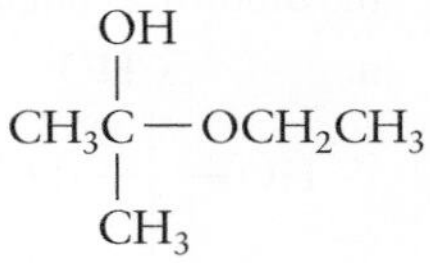 OH
$CH_3C{-}OCH_2CH_3$
CH_3

11.39 Label each of the following as acetals, ketals, or neither.

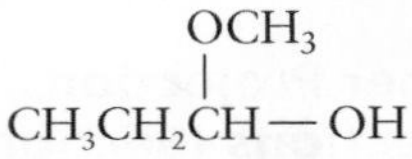

a. 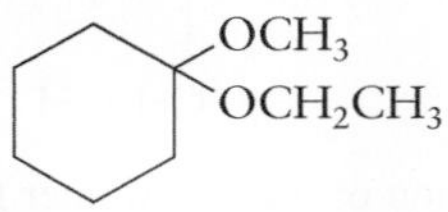 (cyclohexane ring) OCH_3
OCH_2CH_3

b. 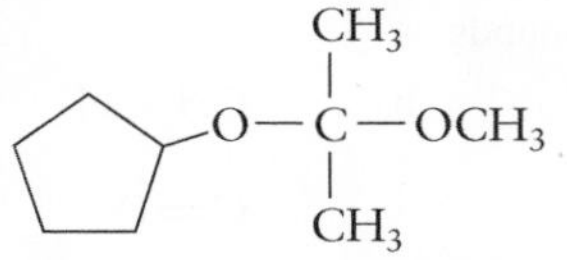 CH_3
(cyclopentane ring)$O{-}C{-}OCH_3$
CH_3

c. OCH_3
CH_2
OCH_2CH_3

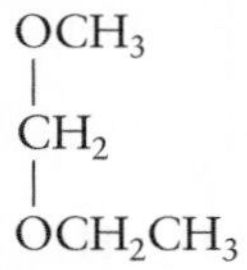

11.40 Label each of the following as acetals, ketals, or neither.

a. 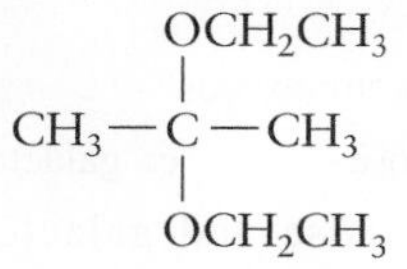 $CH_3-C(OCH_2CH_3)_2-CH_3$ b. (cyclopentyl)$-O-CHCH_3$ with O-cyclopentyl

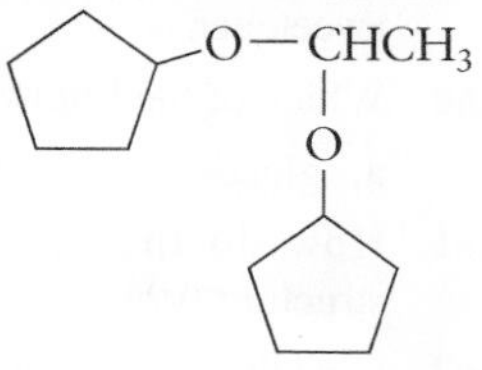

c. $CH_3CH_2-C(OH)(OCH_3)-CH_3$

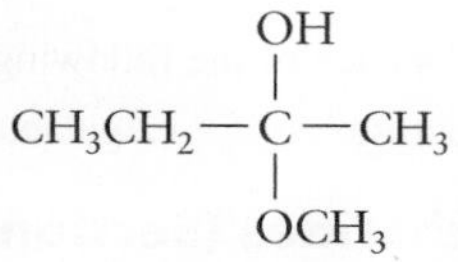

11.41 Label each of the following structures as a cyclic hemiacetal, a hemiketal, an acetal, a ketal, or none of these.

a. 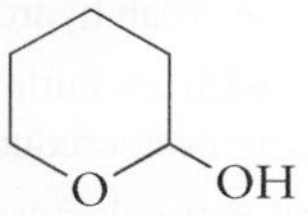 OH b. 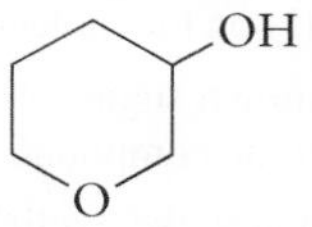 OH c. 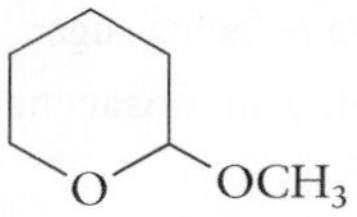 OCH_3

11.42 Label each of the following structures as a hemiacetal, a hemiketal, an acetal, a ketal, or none of these.

a. 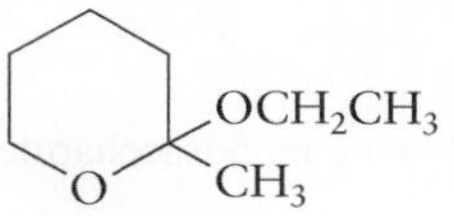 OCH_2CH_3, CH_3 b. 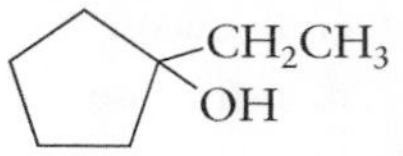 CH_2CH_3, OH c. 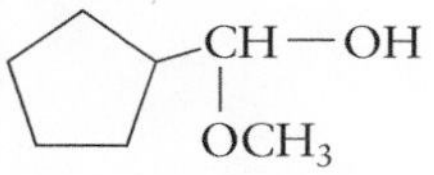 $CH-OH$, OCH_3

11.43 Label each of the following structures as a cyclic hemiacetal, a hemiketal, an acetal, a ketal, or none of these.

a. 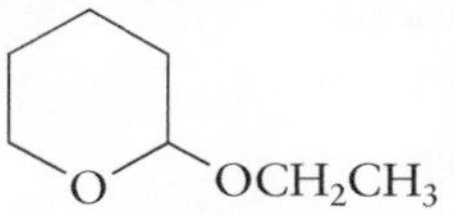 OCH_2CH_3 b. 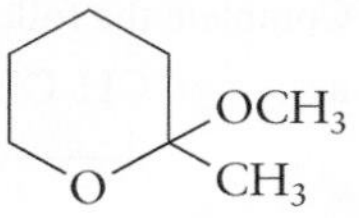 OCH_3, CH_3 c. 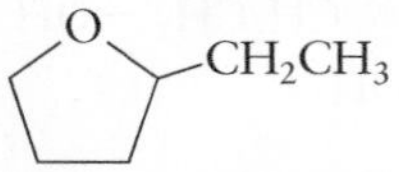 CH_2CH_3

11.44 Circle and identify the hemiacetal/hemiketal functional groups within the following cyclic carbohydrates.

a. 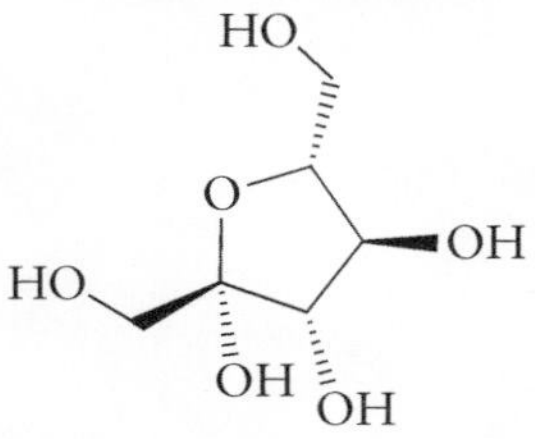 HO, HO, OH, OH, OH

b. 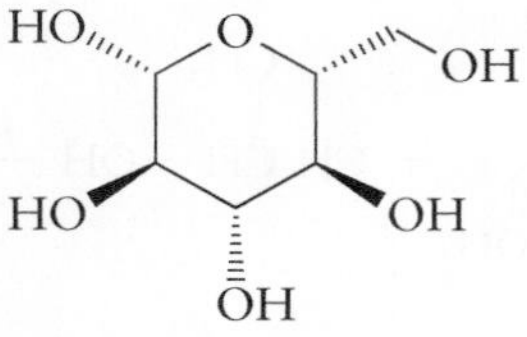 HO, OH, HO, OH, OH

11.45 Identify each of the following as an α or a β form and draw the structural formula of the other anomer.

a. 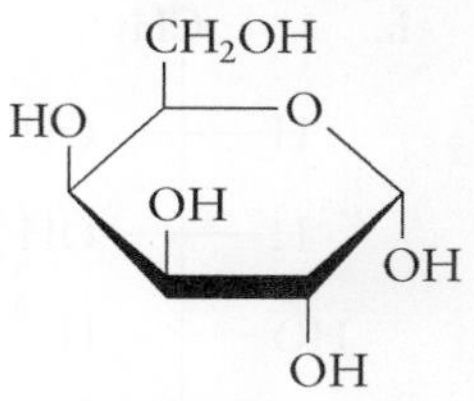 CH_2OH, HO, OH, OH, OH b. 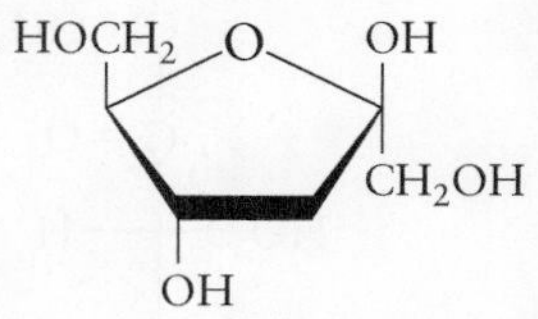 $HOCH_2$, OH, CH_2OH, OH

11.46 Identify each of the following as an α or a β form and draw the structural formula of the other anomer.

a. 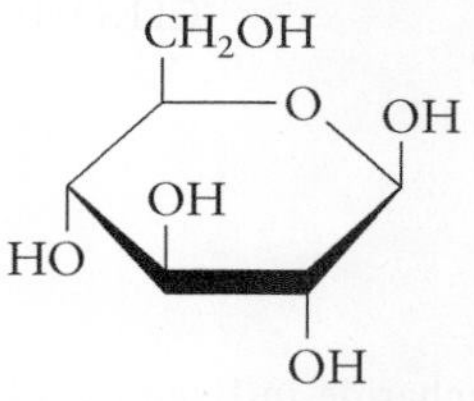 CH_2OH, OH, OH, HO, OH b. 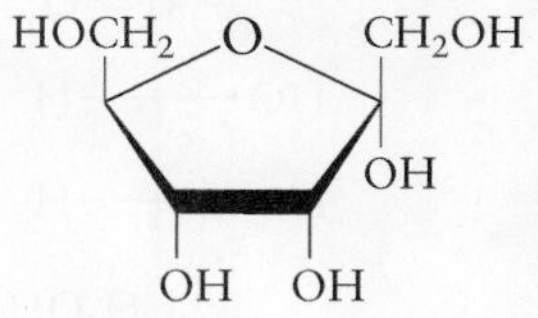 $HOCH_2$, CH_2OH, OH, OH OH

11.47 The structure of mannose differs from glucose only in the direction of the –OH group at position 2. Draw and label Haworth structures for α- and β-mannose.

11.48 The structure of talose differs from galactose only in the direction of the –OH group at position 2. Draw and label Haworth structures for α- and β-talose.

11.49 What is the difference between pyranose and furanose rings?

Classification of Monosaccharides (Section 11.4)

11.50 Classify each of the following monosaccharides as an aldose or a ketose.

a. 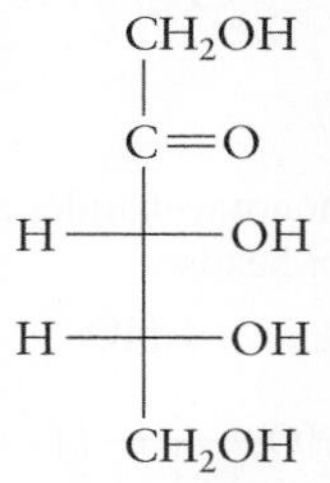 CH_2OH; $C=O$; H–OH; H–OH; CH_2OH

b. CHO; HO–H; H–OH; HO–H; H–OH; CH_2OH

c. CHO; HO–H; H–OH; CH_2OH

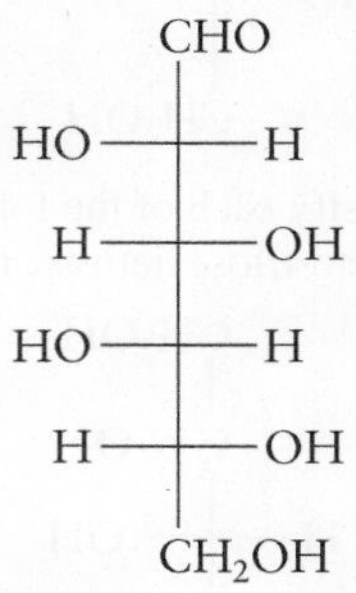

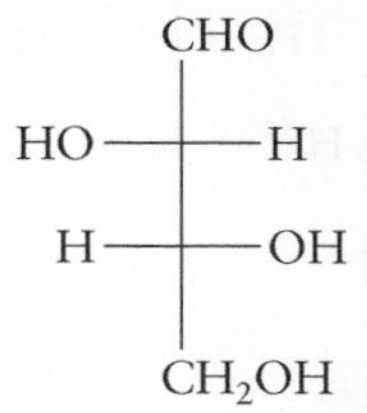

11.51 Classify each of the following monosaccharides as an aldose or a ketose.

a.

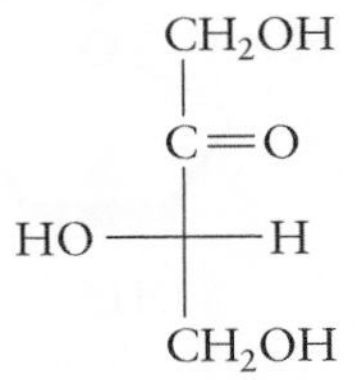

CH_2OH
C=O
HO—H
CH_2OH

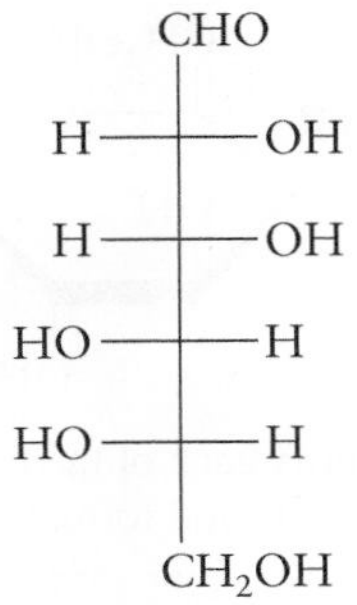

b.

CHO
H—OH
H—OH
HO—H
HO—H
CH_2OH

c.

CH_2OH
C=O
HO—H
HO—H
CH_2OH

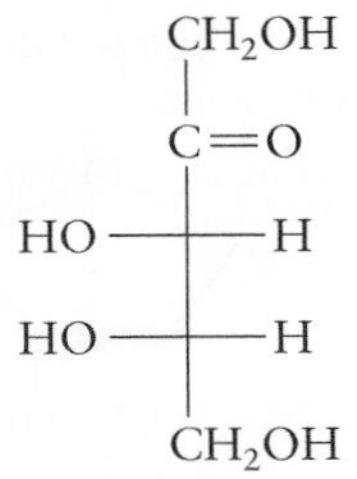

11.52 Classify each monosaccharide in Exercise 11.50 by the number of carbon atoms it has (e.g., tetrose).

11.53 Classify each monosaccharide in Exercise 11.51 by the number of carbon atoms it has (e.g., tetrose).

11.54 Classify each of the following monosaccharides as an aldo- or keto-triose, tetrose, pentose, or hexose.

a.

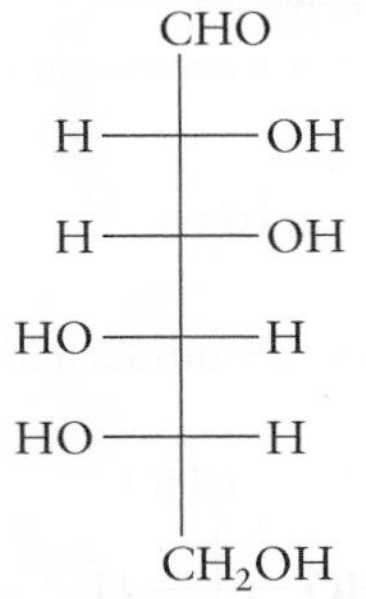

CHO
H—OH
H—OH
HO—H
HO—H
CH_2OH

b.

CH_2OH
C=O
H—OH
H—OH
CH_2OH

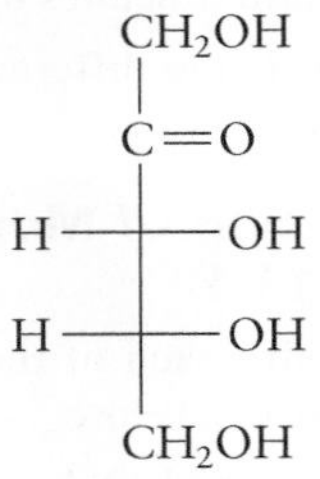

11.55 Classify each of the following monosaccharides as an aldo- or keto-triose, tetrose, pentose, or hexose.

a.

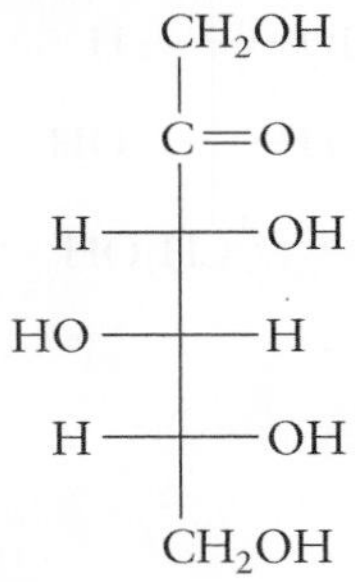

CH_2OH
C=O
H—OH
HO—H
H—OH
CH_2OH

b.

CHO
HO—H
H—OH
HO—H
CH_2OH

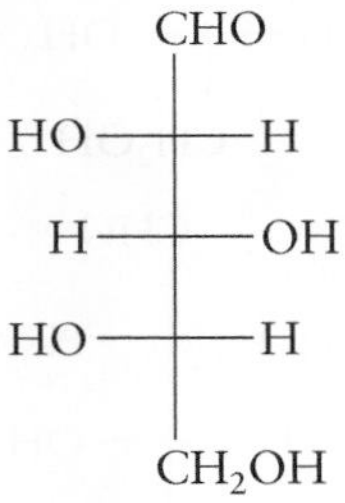

11.56 Draw Fischer projections of any aldotetrose and any ketopentose.

Important Monosaccharides (Section 11.5)

11.57 What monosaccharides are used in the synthesis of nucleic acids?

11.58 Explain why D-glucose can be injected directly into the bloodstream to serve as an energy source.

11.59 Explain why fructose can be used as a low-calorie sweetener.

11.60 Which of the following are ketohexoses?

a. glucose **b.** fructose **c.** galactose

11.61 How do the hexoses glucose and galactose differ structurally?

11.62 Give the natural sources for each of the following.

a. glucose **b.** fructose **c.** galactose

Properties of Monosaccharides (Section 11.6)

11.63 Explain why certain carbohydrates are called sugars.

11.64 Using a Fischer projection of a glucose molecule, identify the site(s) on the molecule where water can hydrogen-bond.

11.65 An unknown sugar failed to react with a solution of Cu^{2+}. Classify the compound as a reducing or nonreducing sugar.

11.66 Explain how the cyclic compound β-D-galactose can react with Cu^{2+} and be classified as a reducing sugar.

11.67 Classify each of the following monosaccharides as a *reducing* or *nonreducing* sugar.

a. D-deoxyribose

b. D-galactose

c. D-glucose

d. D-fructose

11.68 Identify which of the following monosaccharides will give a positive Benedict's test.

a. D-deoxyribose

b. D-galactose

c. D-glucose

d. D-fructose

11.69 Complete the following reactions.

a. 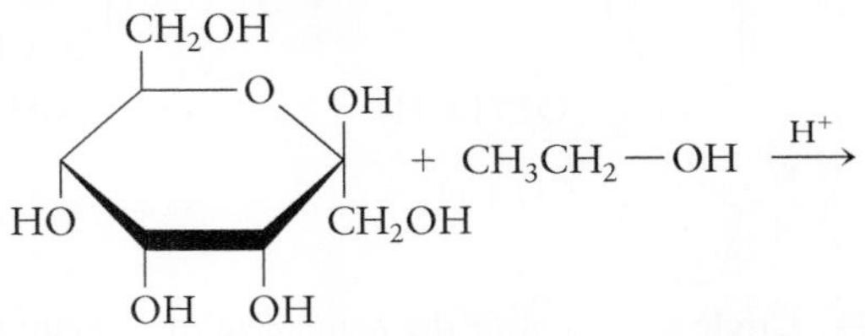 (ring structure with CH_2OH, O, OH, HO, CH_2OH, OH, OH) $+ CH_3CH_2—OH \xrightarrow{H^+}$

b.

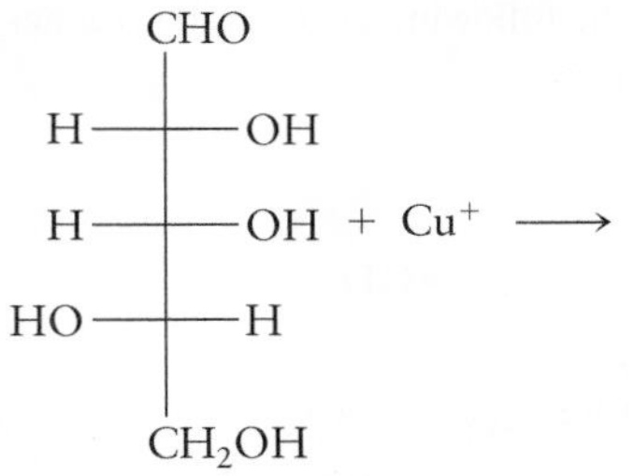

CHO
H—OH
H—OH $+ Cu^+ \longrightarrow$
HO—H
CH_2OH

c. 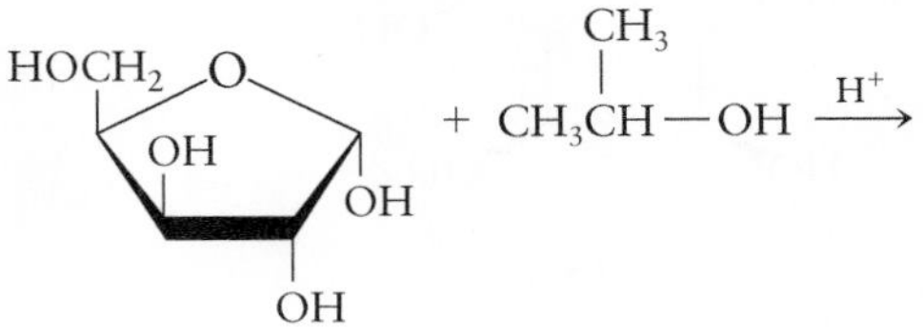 (ring structure with $HOCH_2$, O, OH, OH, OH) $+ CH_3CH(CH_3)—OH \xrightarrow{H^+}$

11.70 Complete the following reactions.

a.

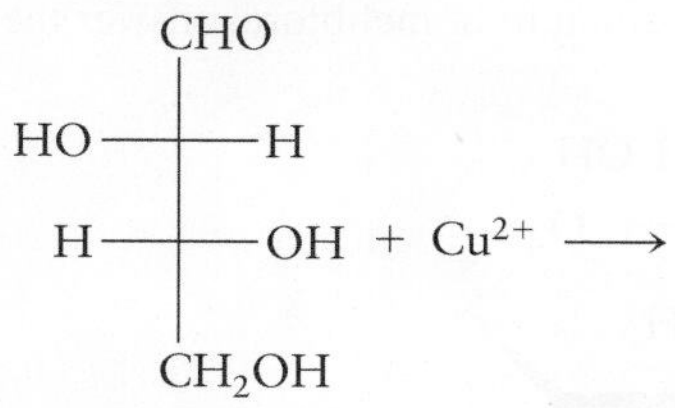

CHO
HO—H
H—OH + Cu^{2+} ⟶
CH_2OH

b.

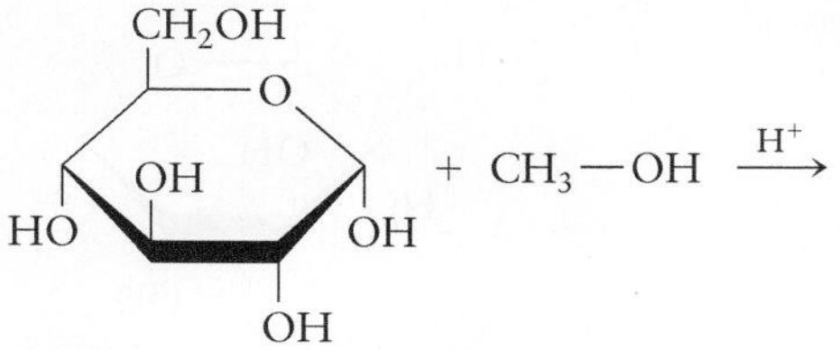

CH_2OH
O
OH
HO OH
OH
$+ CH_3—OH \xrightarrow{H^+}$

c.

$HOCH_2$ O OH
HO
CH_2OH
OH
$+ CH_3CH_2CH_2—OH \xrightarrow{H^+}$

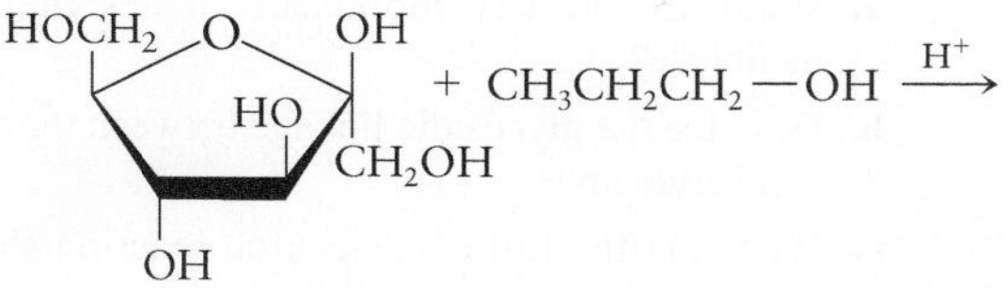

11.71 Use an arrow to identify the glycosidic linkage in each of the following.

a.

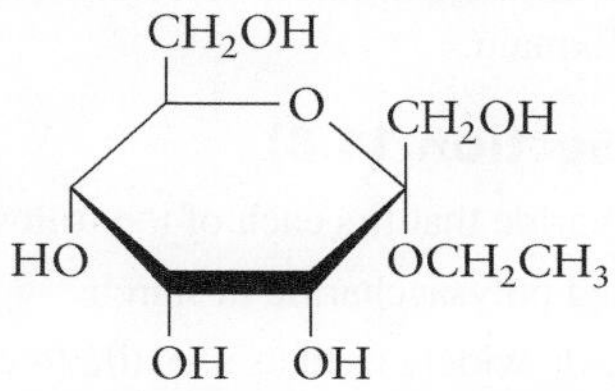

CH_2OH
O CH_2OH
HO OCH_2CH_3
OH OH

b.

$HOCH_2$ O $OCH_2CH_2CH_3$
OH OH

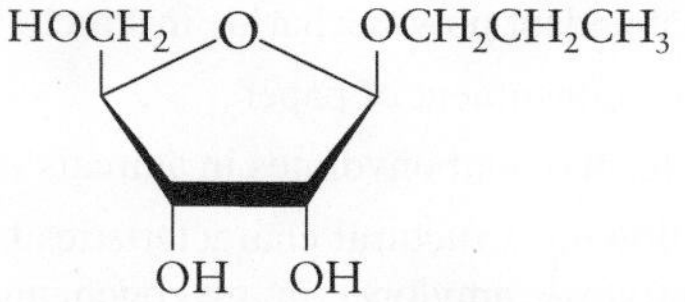

11.72 Use an arrow to identify the glycosidic linkage in each of the following.

a.

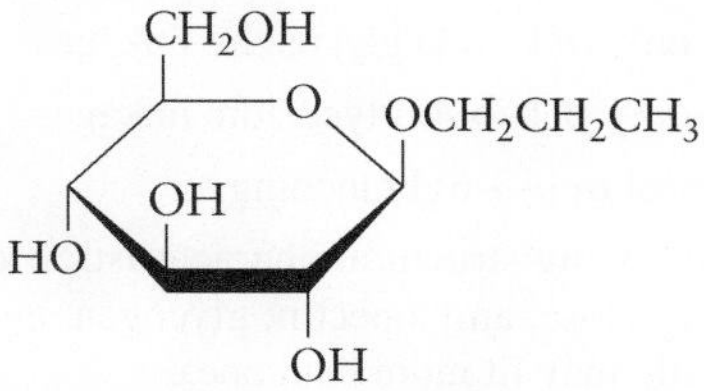

CH_2OH
O $OCH_2CH_2CH_3$
OH
HO
OH

b.

$HOCH_2$ O CH_2OH
HO
OCH_2CH_3
OH

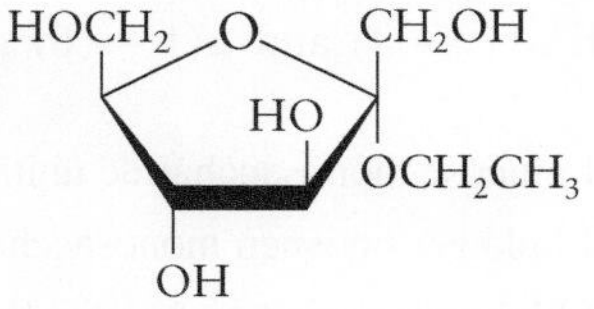

11.73 Draw the structural formula of the alcohol used to make each of the glycosides in Exercise 11.71.

11.74 Draw the structural formula of the alcohol used to make each of the glycosides in Exercise 11.72.

Disaccharides (Section 11.7)

11.75 What monosaccharides are used to make the following disaccharides?

a. sucrose **b.** lactose **c.** maltose

11.76 Identify a disaccharide that fits each of the following.

a. The most common household sugar

b. Formed during the digestion of starch

c. An ingredient of human milk

d. Found in germinating grain

e. Hydrolyzes when cooked with acidic foods to give invert sugar

f. Found in high concentrations in sugar cane

11.77 What type of linkage is broken when disaccharides are hydrolyzed to monosaccharides?

11.78 Explain the process of how the hemiacetal group of a lactose molecule is able to react with Benedict's reagent.

11.79 Identify the monosaccharides present in each of the following disaccharides as glucose, galactose, or fructose.

a.

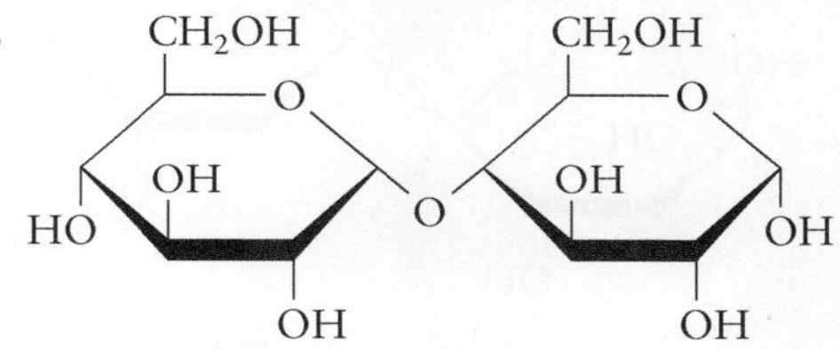

b.

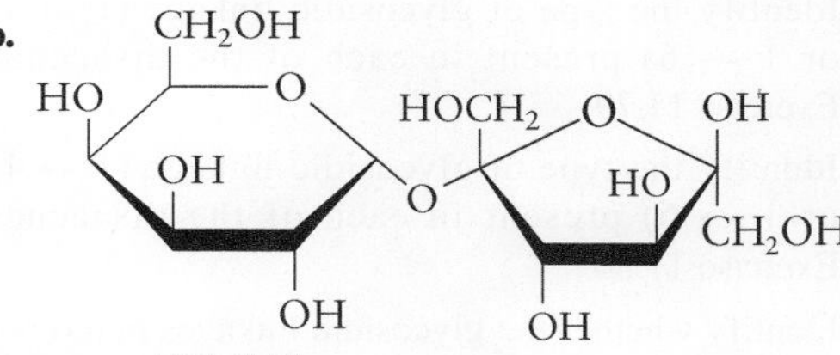

c.

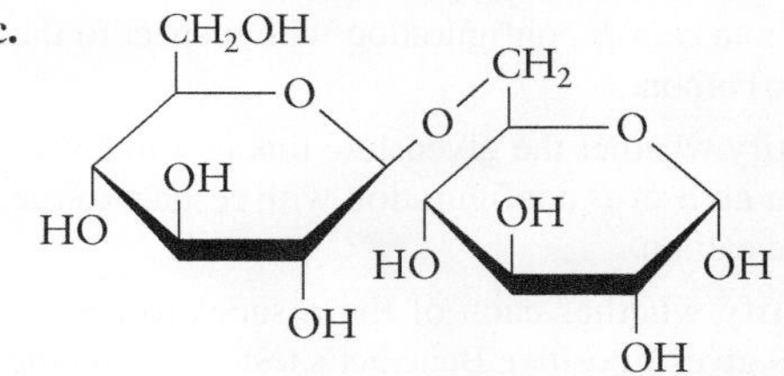

11.80 Identify the monosaccharides present in each of the following disaccharides as glucose, galactose, or fructose.

a.

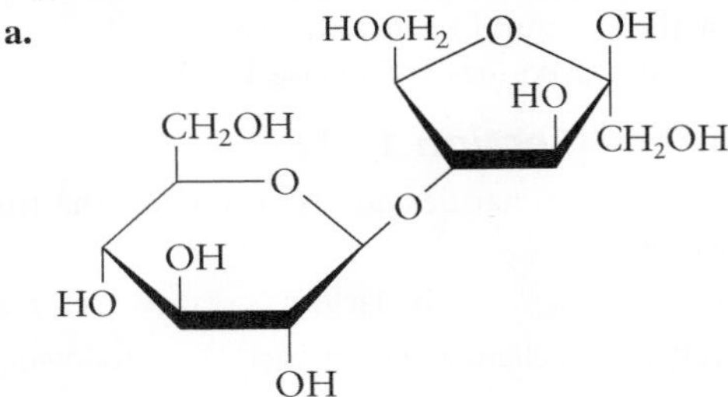

b.

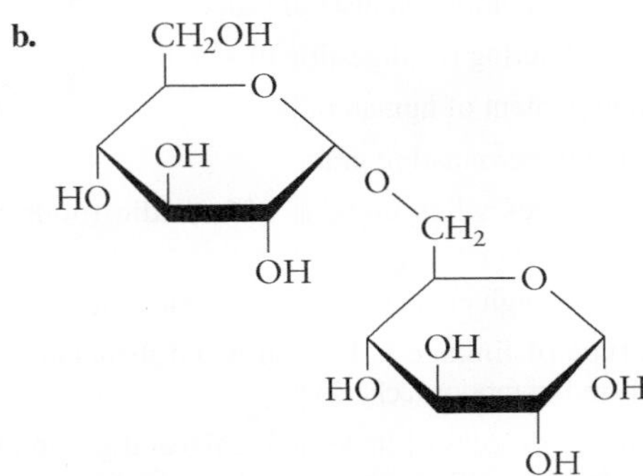

c.

11.81 Identify the type of glycosidic linkage ($1 \rightarrow 4$, $1 \rightarrow 5$, or $1 \rightarrow 6$) present in each of the disaccharides in Exercise 11.79.

11.82 Identify the type of glycosidic linkage ($1 \rightarrow 4$, $1 \rightarrow 5$, or $1 \rightarrow 6$) present in each of the disaccharides in Exercise 11.80.

11.83 Identify whether the glycosidic linkages in Exercise 11.79 are in an α or β configuration with respect to the C-1 anomeric carbon.

11.84 Identify whether the glycosidic linkages in Exercise 11.80 are in an α or β configuration with respect to the C-1 anomeric carbon.

11.85 Identify whether each of the disaccharides in Exercise 11.79 gives a positive Benedict's test as a reducing sugar.

11.86 Identify whether each of the disaccharides in Exercise 11.80 gives a positive Benedict's test as a reducing sugar.

11.87 Draw Haworth projection formulas of the disaccharides maltose and sucrose. Label the hemiacetal, hemiketal, acetal, or ketal carbons.

11.88 Using structural characteristics, show why:

a. lactose is a reducing sugar

b. sucrose is not a reducing sugar

11.89 The disaccharide melibiose is present in some plant juices. Based on the structure of melibiose, answer the following questions.

a. What are the two monosaccharides that make up melibiose?

b. Describe the glycosidic linkage between the two monosaccharide units.

c. Write out the full condensation reaction showing the individual monosaccharides combining to form the disaccharide.

11.90 Based on the your answers in Exercise 11.89, is melibiose a reducing sugar? Explain.

Polysaccharides (Section 11.8)

11.91 Name a polysaccharide that fits each of the following.

a. The unbranched polysaccharide in starch

b. A polysaccharide widely used as a textile fiber

c. The most abundant polysaccharide in starch

d. The primary constituent of paper

e. A storage form of carbohydrates in animals

11.92 Match the following structural characteristics to the polysaccharides amylose, amylopectin, glycogen, and cellulose (a characteristic may fit more than one).

a. Is composed of unbranched molecular chains

b. Contains only $\alpha(1 \rightarrow 4)$ glycosidic linkages

c. Contains only $\beta(1 \rightarrow 4)$ glycosidic linkages

d. Has the most $\alpha(1 \rightarrow 6)$ branching

11.93 Match the following structural characteristics to the polysaccharides amylose, amylopectin, glycogen, and cellulose (a characteristic may fit more than one).

a. Contains both $\alpha(1 \rightarrow 4)$ and $\alpha(1 \rightarrow 6)$ glycosidic linkages

b. Is composed of glucose monosaccharide units

c. Contains acetal linkages between monosaccharide units

d. Is composed of highly branched molecular chains

11.94 List the structural differences and similarities between:

a. glycogen and amylopectin

b. amylose and amylopectin

c. amylose and cellulose

11.95 Polysaccharides are abundant in celery, and yet celery is a good snack for people on a diet. Explain why.

Additional Exercises

11.96 There are two solutions:

1. A 10% (w/v) water–glucose solution
2. A 10% (w/v) water–maltose solution

Can you differentiate between the two solutions based on boiling points? Explain your answer.

11.97 Why does one mole of D-glucose react with only one mole of methanol (instead of two) to form an acetal?

11.98 If 100 g of sucrose is completely hydrolyzed to invert sugar, how many grams of glucose will be present?

11.99 Hexanal and L-hexanol are liquids at room temperature, but glucose (a hexose) is a solid. Give two reasons that may explain this phenomenon.

Chemistry for Thought

11.100 A sample of starch is found to have a molecular weight of 2.80×10^5 amu. How many glucose units are present in a molecule of the starch?

11.101 Suppose human intestinal bacteria were genetically altered so they could hydrolyze the β linkages of cellulose. Would the results be beneficial? Explain.

11.102 Aspartame (Nutrasweet) contains calories, yet it is used in diet drinks. Explain how a drink can contain aspartame as a sweetener and still be low in calories.

11.103 The open-chain form of glucose constitutes only a small fraction of any glucose sample. Yet, when Cu^{2+} is used to oxidize the open-chain form, nearly all the glucose in a sample reacts. Use Le Châtelier's principle to explain this observation.

11.104 Answer the question raised in **Figure 11.32**. What foods in addition to potatoes would you expect to give a positive starch test?

11.105 From **Figure 11.28**, what two products form when the lactose in cow's milk is hydrolyzed in a child's digestive system?

11.106 Amylose is a straight-chain glucose polymer similar to cellulose. What would happen to the longevity of paper manufactured with amylose instead of cellulose?

11.107 Amylopectin is a component of starch, yet it does not give a positive iodine test (turn deep blue) for the presence of starch when it is tested in its pure form. Why?

12 Amino Acids, Proteins, and Enzymes

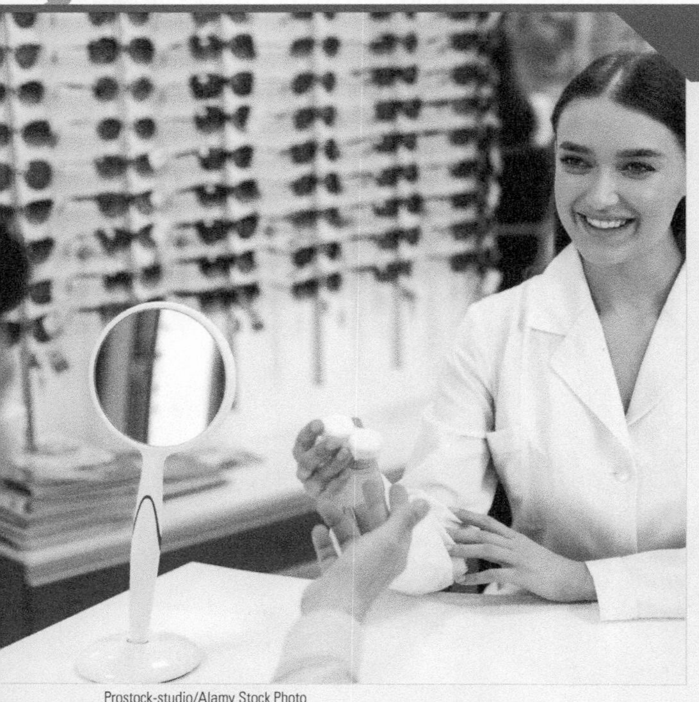
Prostock-studio/Alamy Stock Photo

Health Career Focus

Optician

"These contacts are comfortable, but avoid the temptation to sleep in them. And it's important to use this solution that removes protein buildup," I explain to my client. "Your eye adds a little protein every day. At first, it helps the lens feel comfortable, but over time, it can cause problems. You want to rub the lens in your hand—the friction helps loosen the protein, and an enzyme in the solution dissolves it."

I feel that explaining "how things work" helps my clients comply with instructions. If you just give someone the "rules," with no reason why, they're likely to do their own thing—like sleep in their contacts.

I enjoy teaching people about vision and helping them pick out eyeglass frames. I also like the scientific and technical side of my job that incorporates math and chemistry. I enjoy helping people with one of the most important aspects of their health—eyesight.

Follow-up to this Career Focus appears at the end of the chapter before the *Concept Summary*.

Learning Objectives

When you have completed your study of Chapter 12, you should be able to:

1. Identify the characteristic parts and stereochemistry of alpha-amino acids. **(Section 12.1)**
2. Draw structural formulas to illustrate the various ionic forms assumed by amino acids. **(Section 12.2)**
3. Write reactions to represent the formation of peptides and the oxidation of cysteine. **(Section 12.2)**
4. Explain what is meant by the primary structure of proteins. **(Section 12.3)**
5. Describe the role of hydrogen bonding in the secondary structure of proteins. **(Section 12.3)**
6. Describe the role of side-chain interactions in the tertiary structure of proteins. **(Section 12.3)**
7. Explain what is meant by the quaternary structure of proteins. **(Section 12.3)**
8. Describe four biological functions of proteins. **(Section 12.4)**
9. Describe three reactions that enzymes catalyze. **(Section 12.5)**
10. Describe enzyme activity in terms of efficacy and mechanism of action. **(Section 12.6)**
11. Identify four factors that affect enzyme activity. **(Section 12.7)**
12. Compare the mechanisms of competitive and noncompetitive enzyme inhibition and allosteric regulation. **(Section 12.8)**
13. Name three enzymes that can be used in diagnostic testing to identify the presence of disease. **(Section 12.9)**

Proteins are indispensable components of all living things, where they play crucial roles in all biological processes. In this chapter, we begin with a discussion of the structures and chemistry of the building blocks of proteins. Then we show how amino acids covalently bond together to form peptide chains, and learn how those chains fold on themselves and interact with one another to form complex proteins. Protein structure naturally leads us into a discussion of protein function and classification, with a special focus on enzymes. We conclude the chapter with a discussion of enzyme activity, inhibition, and the role of enzymes in medicine and disease.

12.1 Amino Acid Structure and Classification

Learning Objective 1 Identify the characteristic parts and stereochemistry of alpha-amino acids.

12.1A Structure and Stereochemistry

All proteins, regardless of their structure and function, are biopolymers composed of amino acid building blocks. Every amino acid contains both an amine and a carboxyl group.

There are 20 common, naturally occurring amino acids that make up proteins. They are known more specifically as **alpha (α) amino acids** because the amino group is attached to the alpha carbon (the carbon next to the carbonyl carbon of the carboxylate), as shown in **Figure 12.1**.

alpha (α) amino acid An organic compound containing both an amino group and a carboxylate group, with the amino group attached to the carbon next to the carboxylate group.

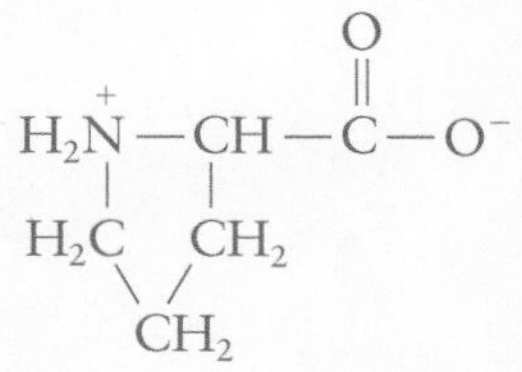

With all of these structural features in common, each amino acid is differentiated, then, by its characteristic side chain, abbreviated generically with an R. These side chains contain their own functional groups (alcohol, amine, aromatic, etc.) that confer individual characteristics to each amino acid. For instance, the amino acid proline is unique as the only amino acid containing a cyclic functional group (see margin right).

Note in **Figure 12.1b** the presence of the bolded, wedged bond between the alpha carbon and the R (variable) group when represented as a skeletal structure. This indicates that the bond is coming out of the plane of the page and is the favored stereoisomer for the vast majority of the naturally occurring amino acids.

$$H_2\overset{+}{N}-CH-\overset{\overset{O}{\|}}{C}-O^-$$
(H_2C and CH_2 below, joined by CH_2, forming a ring)

The structure of proline differs slightly from the general formula because the amino group and R group are part of a ring.

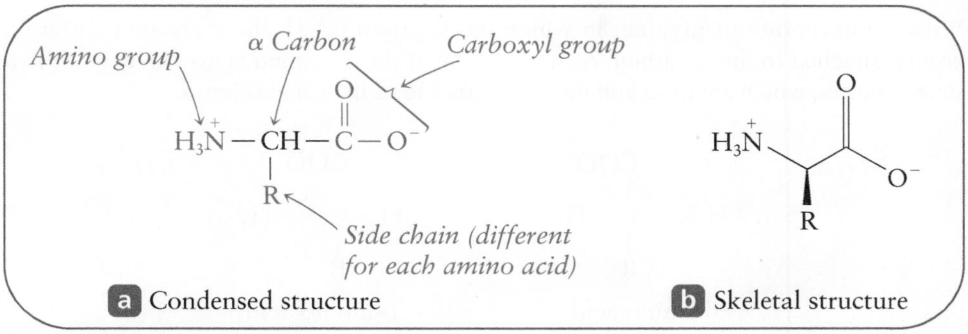

Amino group; α Carbon; Carboxyl group

$$H_3\overset{+}{N}-\underset{\underset{R}{|}}{CH}-\overset{\overset{O}{\|}}{C}-O^-$$

Side chain (different for each amino acid)

a Condensed structure

b Skeletal structure

Figure 12.1 The general structure of α-amino acids shown in ionic form.

Example 12.1 Drawing α-Amino Acids

Draw the skeletal structure of the amino acid valine, shown here in its condensed form.

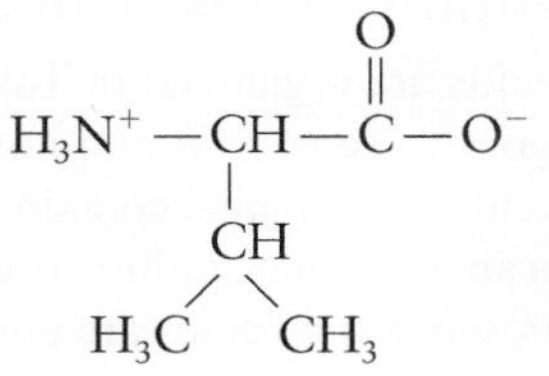

$$H_3N^+-\underset{\underset{\underset{H_3C\quad CH_3}{CH}}{|}}{CH}-\overset{\overset{O}{\|}}{C}-O^-$$

Figure 12.2 Lysine is one of the essential amino acids, meaning the human body cannot synthesize it on its own. It must be consumed through eating foods like meat, dairy, or legumes.

Solution

Use **Figure 12.1b** as a guide.

$H_3\overset{+}{N}$ O O^-

✔ Learning Check 12.1 Draw the skeletal structure of the amino acid lysine (**Figure 12.2**), shown here in its condensed form.

$$\begin{array}{c} H_3\overset{+}{N}-CH-COO^- \\ | \\ CH_2 \\ | \\ CH_2 \\ | \\ CH_2 \\ | \\ CH_2 \\ | \\ NH_3^+ \end{array}$$

In **Section 11.3**, we represented carbohydrates as Fischer projections. The general formula for the 20 amino acids can also be represented as a Fischer projection, with the alpha carbon at the intersection of the Fischer projection shown explicitly:

$$\begin{array}{c} H \\ | \\ H_3\overset{+}{N}-C-COO^- \\ | \\ R \end{array}$$

With the exception of glycine, in which the R group is –H, there are four different groups attached to the α carbon. As a result, 19 of the 20 amino acids can exist as two stereoisomers, which are most commonly referred to as the L and D forms:

$\overset{+}{N}H_3$ on the left →

$$\begin{array}{c} COO^- \\ | \\ H_3\overset{+}{N}-C-H \\ | \\ R \end{array} \qquad \begin{array}{c} COO^- \\ | \\ H-C-\overset{+}{N}H_3 \\ | \\ R \end{array}$$

← *$\overset{+}{N}H_3$ on the right*

L-amino acid D-amino acid

As we saw when we studied carbohydrates, one of the two possible forms usually predominates in natural systems. With few exceptions, the amino acids found in living systems exist in the L form.

12.1B Classification of Amino Acids: The Side Chains

The structures of the 20 amino acids are organized in **Table 12.1** based on the polarity of their side chains. The four categories are neutral and nonpolar, neutral and polar, acidic and polar, and basic and polar. Acidic side chains contain an additional carboxylate group, whereas basic side chains contain an additional amino group. The three-letter and one-letter abbreviations that are used to represent amino acids are given after the names in **Table 12.1**.

Table 12.1 Structures of the Common Amino Acids,* Their Three-Letter Abbreviations, and Their One-Letter Abbreviations

NEUTRAL, NONPOLAR SIDE CHAINS

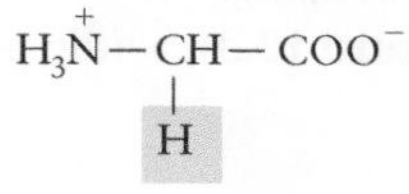 $H_3\overset{+}{N}-CH-COO^-$ $\vert$ H **Glycine (Gly) G**	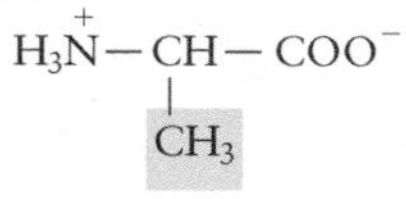 $H_3\overset{+}{N}-CH-COO^-$ $\vert$ CH_3 **Alanine (Ala) A**	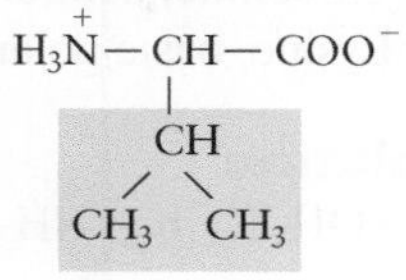 $H_3\overset{+}{N}-CH-COO^-$ $\vert$ CH $/ \quad \backslash$ $CH_3 \quad CH_3$ **Valine (Val) V**	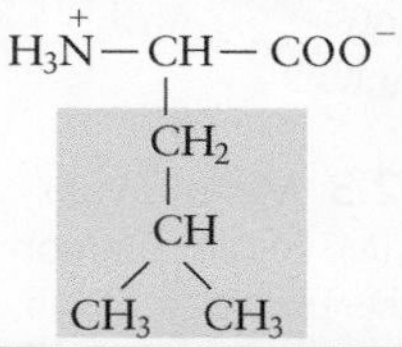 $H_3\overset{+}{N}-CH-COO^-$ $\vert$ CH_2 $\vert$ CH $/ \quad \backslash$ $CH_3 \quad CH_3$ **Leucine (Leu) L**
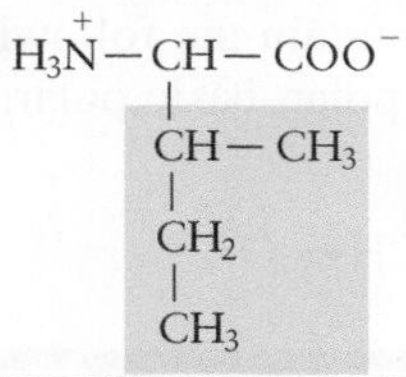 $H_3\overset{+}{N}-CH-COO^-$ $\vert$ $CH-CH_3$ $\vert$ CH_2 $\vert$ CH_3 **Isoleucine (Ile) I**	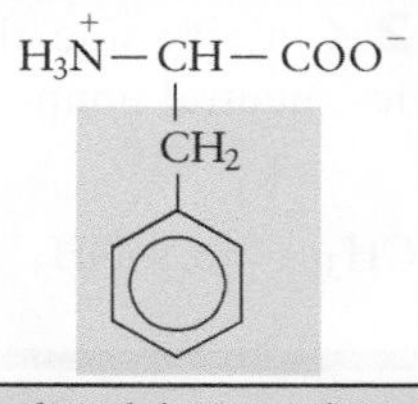 $H_3\overset{+}{N}-CH-COO^-$ $\vert$ CH_2 $\vert$ C_6H_5 (benzene ring) **Phenylalanine (Phe) F**	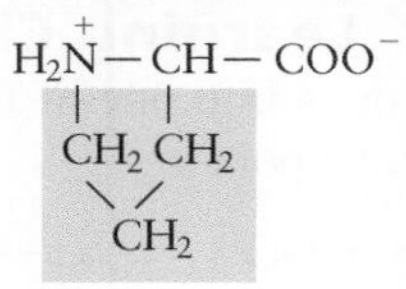 $H_2\overset{+}{N}-CH-COO^-$ $\vert \qquad \vert$ $CH_2 \; CH_2$ $\backslash \; /$ CH_2 **Proline (Pro) P**	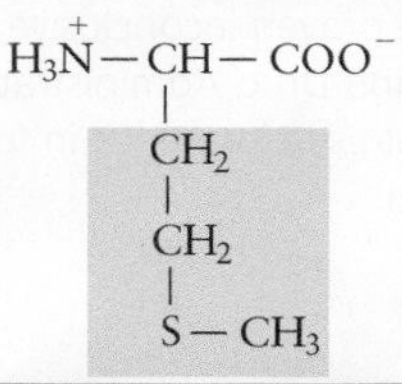 $H_3\overset{+}{N}-CH-COO^-$ $\vert$ CH_2 $\vert$ CH_2 $\vert$ $S-CH_3$ **Methionine (Met) M**

NEUTRAL, POLAR SIDE CHAINS

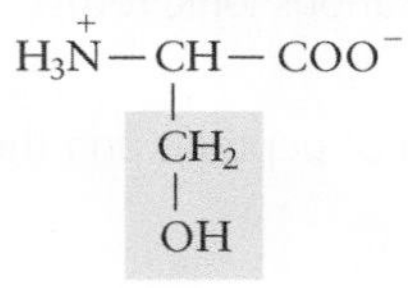 $H_3\overset{+}{N}-CH-COO^-$ $\vert$ CH_2 $\vert$ OH **Serine (Ser) S**	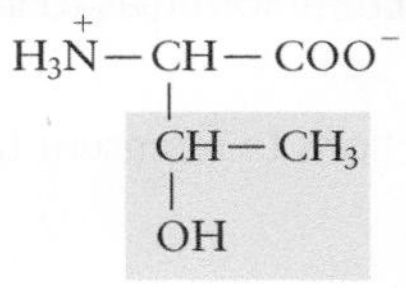 $H_3\overset{+}{N}-CH-COO^-$ $\vert$ $CH-CH_3$ $\vert$ OH **Threonine (Thr) T**	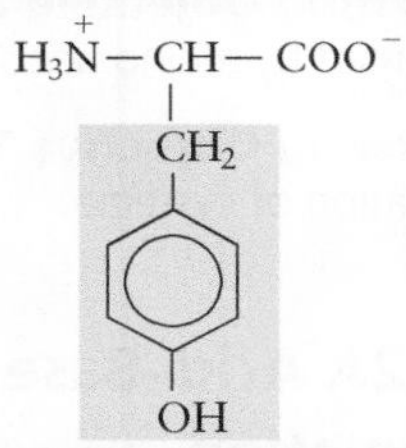 $H_3\overset{+}{N}-CH-COO^-$ $\vert$ CH_2 $\vert$ (benzene ring) $\vert$ OH **Tyrosine (Tyr) Y**	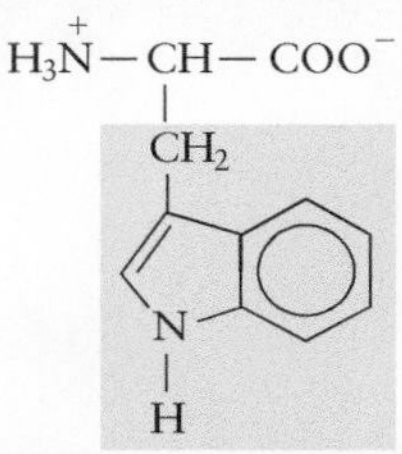 $H_3\overset{+}{N}-CH-COO^-$ $\vert$ CH_2 $\vert$ (indole ring) N $\vert$ H **Tryptophan (Trp) W**
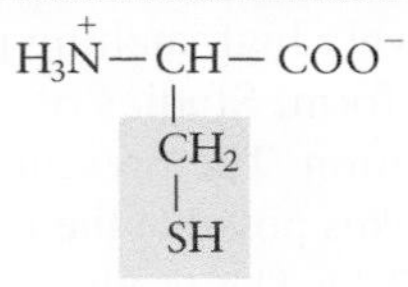 $H_3\overset{+}{N}-CH-COO^-$ $\vert$ CH_2 $\vert$ SH **Cysteine (Cys) C**	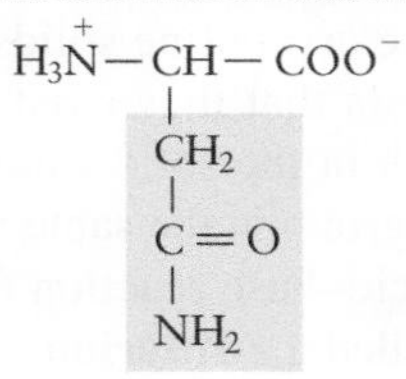 $H_3\overset{+}{N}-CH-COO^-$ $\vert$ CH_2 $\vert$ $C=O$ $\vert$ NH_2 **Asparagine (Asn) N**	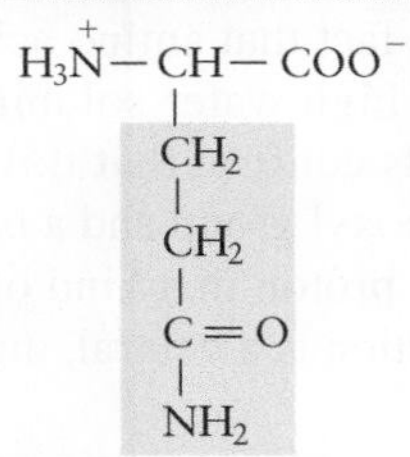 $H_3\overset{+}{N}-CH-COO^-$ $\vert$ CH_2 $\vert$ CH_2 $\vert$ $C=O$ $\vert$ NH_2 **Glutamine (Gln) Q**	

BASIC, POLAR SIDE CHAINS

$H_3\overset{+}{N}-CH-COO^-$ $\vert$ CH_2 (imidazole ring: $H\overset{+}{N}$, NH) **Histidine (His) H**	$H_3\overset{+}{N}-CH-COO^-$ $\vert$ CH_2 $\vert$ CH_2 $\vert$ CH_2 $\vert$ CH_2 $\vert$ NH_3^+ **Lysine (Lys) K**	$H_3\overset{+}{N}-CH-COO^-$ $\vert$ CH_2 $\vert$ CH_2 $\vert$ CH_2 $\vert$ NH $\vert$ $C=NH_2^+$ $\vert$ NH_2 **Arginine (Arg) R**

ACIDIC, POLAR SIDE CHAINS

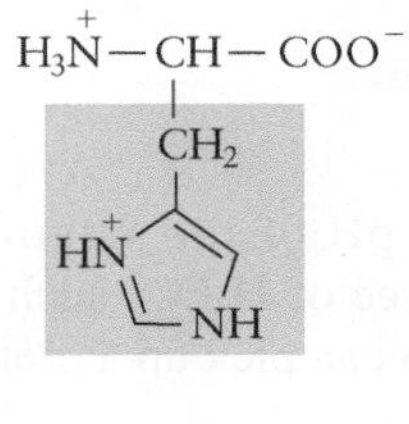 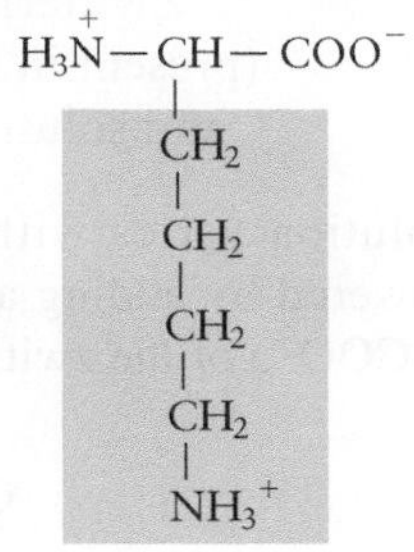 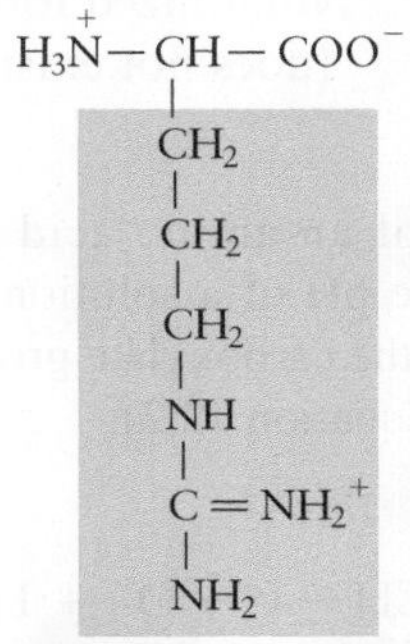 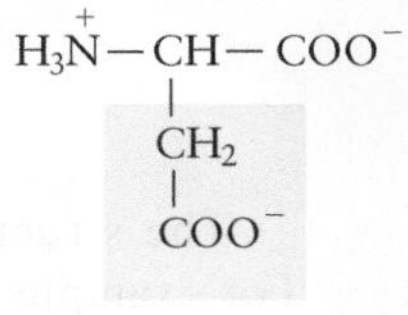 $H_3\overset{+}{N}-CH-COO^-$ $\vert$ CH_2 $\vert$ COO^- **Aspartate (Asp) D**	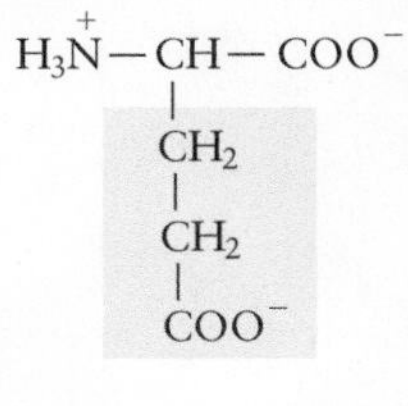 $H_3\overset{+}{N}-CH-COO^-$ $\vert$ CH_2 $\vert$ CH_2 $\vert$ COO^- **Glutamate (Glu) E**

*Side chains are highlighted using boxed color screens to help identify where the amino acid backbone ends and the R group begins.

JustStockPhotos/Alamy Stock Photo

Figure 12.3 Monosodium glutamate (MSG) is a common food additive used as a flavor enhancer in many processed foods. Though studies of its potential harm have proven inconclusive, the Food and Drug Administration (FDA) requires its presence in foods to be listed.

Example 12.2 Classifying Amino Acid Side Chains

For each of the following amino acids, use **Table 12.1** to identify a structural feature of the side chain that places the amino acid in that category.

a. Threonine, neutral polar
b. Lysine, basic polar
c. Valine, neutral nonpolar
d. Glutamate (**Figure 12.3**), acidic polar

Solution

a. –OH b. $-NH_3^+$ c. $\underset{\displaystyle CH_3}{\underset{|}{-CHCH_3}}$ d. $-COO^-$

✔ **Learning Check 12.2** Classify side chains that contain the following groups into one of four categories: neutral nonpolar, neutral polar, basic polar, or acidic polar.

a. $-COO^-$ b. $-CH_2-S-CH_3$ c. $-NH_3^+$ d. $-CH_2-OH$

12.2 Reactions of Amino Acids

Learning Objective 2 Draw structural formulas to illustrate the various ionic forms assumed by amino acids.

Learning Objective 3 Write reactions to represent the formation of peptides and the oxidation of cysteine.

12.2A Acid–Base Chemistry of Amino Acids: Zwitterion Formation

The structural formulas used to this point have shown amino acids in an ionized form. The fact that amino acids are white crystalline solids with relatively high melting points and high water solubilities suggests that they exist in an ionic form. Studies of amino acids confirm that this is true both in the solid state and in solution. The presence of a carboxyl group and a basic amino group in the same molecule makes possible the transfer of a proton in a kind of internal acid–base reaction (Equation 12.1). The product of this reaction is a neutral, dipolar ion called a **zwitterion**.

zwitterion A dipolar ion that carries both a positive and a negative charge as a result of an internal acid–base reaction in an amino acid molecule. The overall net charge on the zwitterion is neutral.

$$H_2N-\underset{\displaystyle R}{\underset{|}{CH}}-\overset{\displaystyle O}{\overset{\|}{C}}-OH \rightarrow H_3\overset{+}{N}-\underset{\displaystyle R}{\underset{|}{CH}}-\overset{\displaystyle O}{\overset{\|}{C}}-O^- \quad (12.1)$$

Nonionized form (does not exist) — Zwitterion (present in solids and solutions)

The structure of an amino acid in solution varies with the pH of the solution. For example, if the pH of a solution is lowered by adding a source of H_3O^+, such as hydrochloric acid, the carboxylate group ($-COO^-$) of the zwitterion can pick up a proton to form –COOH (Equation 12.2):

$$H_3\overset{+}{N}-\underset{\displaystyle R}{\underset{|}{CH}}-\overset{\displaystyle O}{\overset{\|}{C}}-O^- + H_3O^+ \rightarrow H_3\overset{+}{N}-\underset{\displaystyle R}{\underset{|}{CH}}-\overset{\displaystyle O}{\overset{\|}{C}}-OH + H_2O \quad (12.2)$$

Zwitterion (0 net charge) — (Positive net charge)

The zwitterion form has a net charge of 0, but the form in acidic solution has a net positive charge.

When the pH of the solution is increased by adding OH^-, the $-NH_3^+$ of the zwitterion can lose a proton, and the zwitterion is converted into a negatively charged form (Equation 12.3):

$$\underset{\text{Zwitterion (0 net charge)}}{H_3\overset{+}{N}-\underset{\underset{R}{|}}{CH}-\overset{\overset{O}{\|}}{C}-O^-} + {}^-OH \rightarrow \underset{\text{(Negative net charge)}}{H_2N-\underset{\underset{R}{|}}{CH}-\overset{\overset{O}{\|}}{C}-O^-} + H_2O \qquad (12.3)$$

Changes in pH also affect acidic and basic side chains and those proteins in which they occur. Thus, all amino acids exist in solution in ionic forms, but the actual form (and charge) is determined by the solution pH.

Because amino acids assume a positively charged form in acidic solutions and a negatively charged form in basic solutions, there is some pH between the acidic and basic extremes at which the amino acid has no net charge. This pH is called the **isoelectric point**. Each amino acid has a unique and characteristic isoelectric point. Those with neutral R groups are all near a pH of 6 (e.g., pH = 6.06 for glycine), those with basic R groups have higher values (e.g., pH = 9.47 for lysine), and the two acidic amino acids have lower values (e.g., pH = 2.98 for aspartate). Proteins, which are composed of amino acids, also have characteristic isoelectric points (**Figure 12.4**).

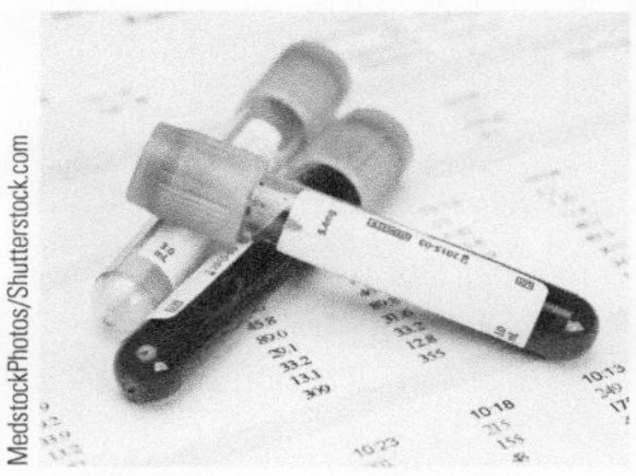

MedstockPhotos/Shutterstock.com

Figure 12.4 Human albumin, a major component of plasma, has an isoelectric point of around 5, lower than the buffered pH of blood (~7.4). This lower isoelectric point prevents the protein from becoming neutral and aggregating unnecessarily in the blood.

isoelectric point The characteristic solution pH at which an amino acid has a net charge of 0.

Example 12.3 Drawing Zwitterion Structures

Draw the structure of the amino acid leucine:

$$H_3\overset{+}{N}-\underset{\underset{\underset{\underset{CH_3\ \ CH_3}{/\ \ \backslash}}{CH}}{\underset{|}{CH_2}}}{\underset{|}{CH}}-\overset{\overset{O}{\|}}{C}-O^-$$

a. In acidic solution at a pH below the isoelectric point
b. In basic solution at a pH above the isoelectric point

Solution

a. In acidic solution, the carboxylate group of the zwitterion picks up a proton (see Equation 12.2).

$$H_3\overset{+}{N}-\underset{\underset{\underset{\underset{CH_3\ \ CH_3}{/\ \ \backslash}}{CH}}{\underset{|}{CH_2}}}{\underset{|}{CH}}-\overset{\overset{O}{\|}}{C}-O^- + H_3O^+ \rightarrow H_3\overset{+}{N}-\underset{\underset{\underset{\underset{CH_3\ \ CH_3}{/\ \ \backslash}}{CH}}{\underset{|}{CH_2}}}{\underset{|}{CH}}-\overset{\overset{O}{\|}}{C}-OH + H_2O$$

b. In basic solution, the $H_3\overset{+}{N}-$ group of the zwitterion loses a proton (see Equation 12.3).

$$H_3\overset{+}{N}-\underset{\underset{\underset{\displaystyle CH_3 \quad CH_3}{\displaystyle CH}}{\displaystyle CH_2}}{CH}-\overset{\overset{\displaystyle O}{\|}}{C}-O^- + OH^- \rightarrow H_2N-\underset{\underset{\underset{\displaystyle CH_3 \quad CH_3}{\displaystyle CH}}{\displaystyle CH_2}}{CH}-\overset{\overset{\displaystyle O}{\|}}{C}-O^- + H_2O$$

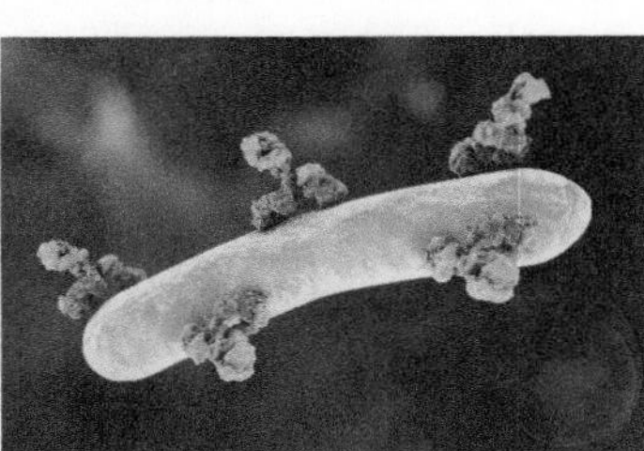
KATERYNA KON/SCIENCE PHOTO LIBRARY/Getty Images

Figure 12.5 Serine is used in the production of antibodies. The image shows antibodies attacking bacterium.

✔ **Learning Check 12.3** Draw the structure of the amino acid serine (**Figure 12.5**):

$$H_3\overset{+}{N}-\underset{\displaystyle CH_2-OH}{CH}-\overset{\overset{\displaystyle O}{\|}}{C}-O^-$$

a. In acidic solution at a pH below the isoelectric point
b. In basic solution at a pH above the isoelectric point

12.2B Peptide Formation

Recall from **Section 10.8** that amides can be thought of as being derived from a carboxylic acid and an amine (Equation 12.4), although in actual practice acid chlorides are generally used because they are more reactive than carboxylic acids.

dipeptide A compound formed when two amino acids are bonded by an amide linkage.

peptide linkage or **peptide bond** The amide linkage between amino acids that results when the amino group of one amino acid reacts with the carboxylate group of another.

General reaction:

$$\underset{\text{carboxylic acid}}{R-\overset{\overset{\displaystyle O}{\|}}{C}-OH} + \underset{\text{amine}}{H_2N-R'} \xrightarrow{\text{Heat}} \underset{\text{amide}}{R-\overset{\overset{\displaystyle O}{\|}}{C}-NH-R'} + H_2O \qquad (12.4)$$

Amide linkage (C–NH bond)

In the same hypothetical way, it is possible to envision the carboxylate group of one amino acid reacting with the amino group of a second amino acid. For example, glycine could combine with alanine:

Specific reaction:

$$\underset{\text{glycine}}{H_3\overset{+}{N}-CH_2-\overset{\overset{\displaystyle O}{\|}}{C}-O^-} + \underset{\text{alanine}}{H_3\overset{+}{N}-\underset{\displaystyle CH_3}{CH}-\overset{\overset{\displaystyle O}{\|}}{C}-O^-} \longrightarrow \underset{\text{Gly–Ala (a dipeptide)}}{H_3\overset{+}{N}-CH_2-\overset{\overset{\displaystyle O}{\|}}{C}-NH-\underset{\displaystyle CH_3}{CH}-\overset{\overset{\displaystyle O}{\|}}{C}-O^-} + H_2O$$

Peptide linkage (C–NH bond); N-terminus (Gly); C-terminus (Ala)

N-terminal residue The amino acid on the end of a chain that has an unreacted or free amino group.

C-terminal residue The amino acid on the end of a chain that has an unreacted or free carboxylate group.

Compounds of this type made up of two amino acids are called **dipeptides**, and the amide linkage that holds them together is called a **peptide linkage** or **peptide bond**. In Gly–Ala, Gly is the **N-terminal residue** (or N-terminus) because it has the free (unreacted) amino group, whereas Ala is the **C-terminal residue** (or C-terminus) because it has the free (unreacted) carboxyl group.

The dipeptide, Ala–Gly, can also be formed by linking glycine and alanine. Ala–Gly is different than Gly–Ala, however, because Ala is now the N-terminus and Gly is now the C-terminus. These dipeptides are structural isomers of one another, and each has a unique set of properties.

Specific reaction:

$$\underset{\text{alanine}}{H_3\overset{+}{N}-\underset{\underset{CH_3}{|}}{CH}-\overset{\overset{O}{\|}}{C}-O^-} + \underset{\text{glycine}}{H_3\overset{+}{N}-CH_2-\overset{\overset{O}{\|}}{C}-O^-} \longrightarrow \underset{\text{N-terminus}\quad\text{Ala–Gly}\quad\text{C-terminus}}{H_3\overset{+}{N}-\underset{\underset{CH_3}{|}}{CH}-\overset{\overset{O}{\|}}{C}-NH-CH_2-\overset{\overset{O}{\|}}{C}-O^-} + H_2O$$

Peptide linkage (arrow pointing to the C–NH bond)

In nature (including in our bodies), the reactions that form peptide bonds are more complex than the equations for forming dipeptides suggest. In fact, our DNA encodes the instructions required for the cellular machinery to build peptide chains via a complex multistep process.

Example 12.4 Identifying Peptide Features

In the tripeptide shown, identify the N-terminal and C-terminal amino acids. Use arrows to point to any peptide linkages.

$$H_3\overset{+}{N}-\underset{\underset{\underset{CH_3}{|}}{\underset{CHCH_3}{|}}}{CH}-\overset{\overset{O}{\|}}{C}-NH-\underset{\underset{CH_2SH}{|}}{CH}-\overset{\overset{O}{\|}}{C}-NH-\underset{\underset{\underset{OH}{|}}{\underset{CHCH_3}{|}}}{CH}-\overset{\overset{O}{\|}}{C}-O^-$$

Solution

valine cysteine threonine

$$H_3\overset{+}{N}-\underset{\underset{\underset{CH_3}{|}}{\underset{CHCH_3}{|}}}{CH}-\overset{\overset{O}{\|}}{C}\swarrow-NH-\underset{\underset{CH_2SH}{|}}{CH}-\overset{\overset{O}{\|}}{C}\swarrow-NH-\underset{\underset{\underset{OH}{|}}{\underset{CHCH_3}{|}}}{CH}-\overset{\overset{O}{\|}}{C}-O^-$$

N-terminal amino acid (valine) *C-terminal amino acid* (threonine)

✔ Learning Check 12.4 The artificial sweetener aspartame (**Figure 12.6**) is the methyl ester of the dipeptide shown. Identify the N-terminal and C-terminal amino acids. Use an arrow to indicate the peptide linkage.

$$H_3\overset{+}{N}-\underset{\underset{CH_2-COO^-}{|}}{CH}-\overset{\overset{O}{\|}}{C}-NH-\underset{\underset{\underset{C_6H_5}{|}}{\underset{CH_2}{|}}}{CH}-\overset{\overset{O}{\|}}{C}-O^-$$

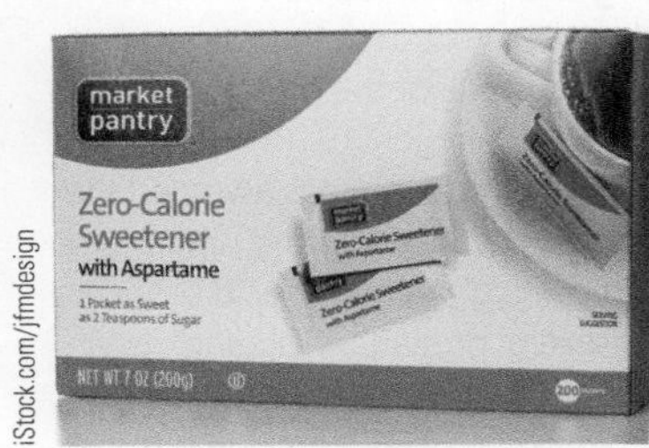

Figure 12.6 As we saw in Health Connection 8.1, aspartame is one of the most common artificial sweeteners. This dipeptide tricks our taste buds into thinking it is the disaccharide sucrose!

The presence of amino and carboxylate groups on the ends of dipeptides allows for the attachment of a third amino acid to form a tripeptide. For example, valine could be attached to Ala–Gly to form Ala–Gly–Val:

Two peptide linkages

$$H_3\overset{+}{N}-\underset{\underset{CH_3}{|}}{CH}-\overset{\overset{O}{\parallel}}{C}-NH-CH_2-\overset{\overset{O}{\parallel}}{C}-NH-\underset{\underset{\underset{\underset{CH_3}{|}}{CH-CH_3}}{|}}{CH}-\overset{\overset{O}{\parallel}}{C}-O^- + H_2O$$

Ala–Gly–Val

peptide An amino acid polymer of short chain length.

polypeptide An amino acid polymer of intermediate chain length containing up to 50 amino acid residues.

protein An amino acid polymer made up of more than 50 amino acids.

amino acid residue The parts of amino acids involved in peptide, polypeptide, or protein chain formation by the removal of water.

More amino acids can react in the same way to form a tetrapeptide, a pentapeptide, and so on, until a chain of hundreds or even thousands of amino acids has formed. The compounds with the shortest chains are often simply called **peptides**, those with longer chains are called **polypeptides**, and those with still longer chains are called **proteins**. Chemists differ about where to draw the lines in the use of these names, but generally, polypeptide chains with more than 50 amino acids are called proteins. However, the terms *protein* and *polypeptide* are often used interchangeably.

Amino acids that have been incorporated into chains are called **amino acid residues**. According to convention, the N-terminal residue is written on the left end of the peptide chain, whereas the C-terminal residue is written on the right end of the chain. Thus, alanine is the N-terminal residue and valine is the C-terminal residue in Ala–Gly–Val:

Alanine, the N-terminal residue

$$H_3\overset{+}{N}-\underset{\underset{CH_3}{|}}{CH}-\overset{\overset{O}{\parallel}}{C}-NH-CH_2-\overset{\overset{O}{\parallel}}{C}-NH-\underset{\underset{\underset{\underset{CH_3}{|}}{CH-CH_3}}{|}}{CH}-\overset{\overset{O}{\parallel}}{C}-O^-$$

Valine, the C-terminal residue

Ala–Gly–Val

Keystone Press/Alamy Stock Photo

Figure 12.7 People with type I diabetes must treat themselves with regular shots of the protein insulin. Frederick Sanger (1918–2013) was a biochemist who was awarded the Nobel Prize in Chemistry in 1958 for discovering the amino acid sequence of several proteins, including insulin. Sanger also won the Nobel Prize in Chemistry in 1980, making him one of four persons who have won two science Nobels.

Peptides are named by starting at the N-terminal end of the chain and listing the amino acid residues in order from left to right, ending with the C-terminal end. Structural formulas and even full names for large peptides quickly become unwieldy and time-consuming to write. This is simplified by representing peptide and protein structures with three-letter abbreviations for the amino acid residues and dashes for the peptide linkages, as in Ala–Gly–Val.

As peptide chains grow, the number of possible combinations increases exponentially. For instance, the tripeptide Ala–Gly–Val has five other possible arrangements: Ala–Val–Gly, Val–Ala–Gly, Val–Gly–Ala, Gly–Val–Ala, and Gly–Ala–Val. Insulin (**Figure 12.7**), on the other hand, with 51 amino acids, has 1.55×10^{66} different possible sequences! Out of all of them, it is remarkable that our bodies reliably produce the only one that will contribute to the life process.

Example 12.5 Drawing Peptide Structures

a. Use Table 12.1 to draw the condensed structure of the tetrapeptide Val–Asn–Cys–Asp.
b. Next, draw the expanded skeletal structure of the tetrapeptide Val–Asn–Cys–Asp.

Solution

a.

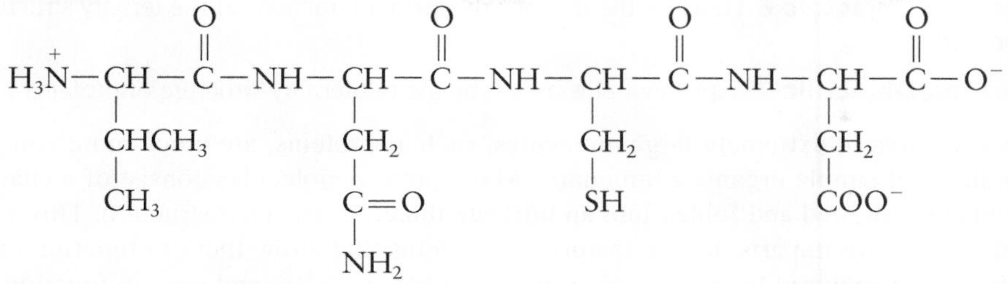

b.

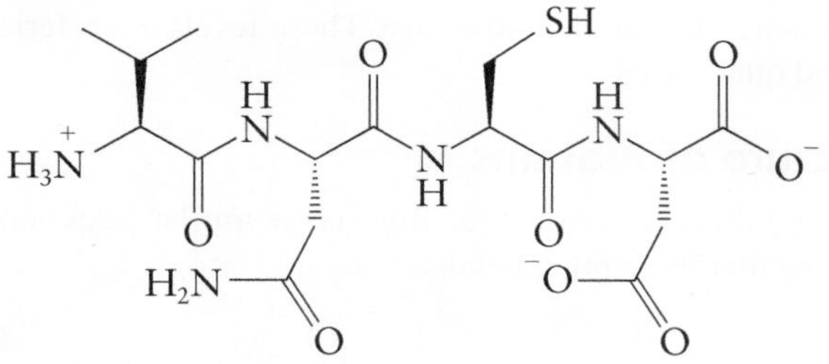

✔ Learning Check 12.5

a. Use **Table 12.1** to draw the condensed structure of the following tetrapeptide.
b. Next, draw the expanded skeletal structure of the tetrapeptide.

Phe–Cys–Ser–Ile

12.2C The Oxidation of Cysteine

Cysteine, the only sulfhydryl (–SH)-containing amino acid of the 20 commonly found in proteins, has a chemical property not shared by the other 19. Cysteine is a thiol, and thiols can easily be oxidized to form a disulfide bond (–S–S–). Thus, two cysteine molecules react readily to form a disulfide compound called cystine (see **Figure 12.8** and Equation 12.5). The disulfide, in turn, is easily converted back to –SH groups by the action of reducing agents:

iStock.com/piotr_malczyk

Figure 12.8 Cystine can sometimes crystallize in the urine, leading to cystine stones (cystinuria).

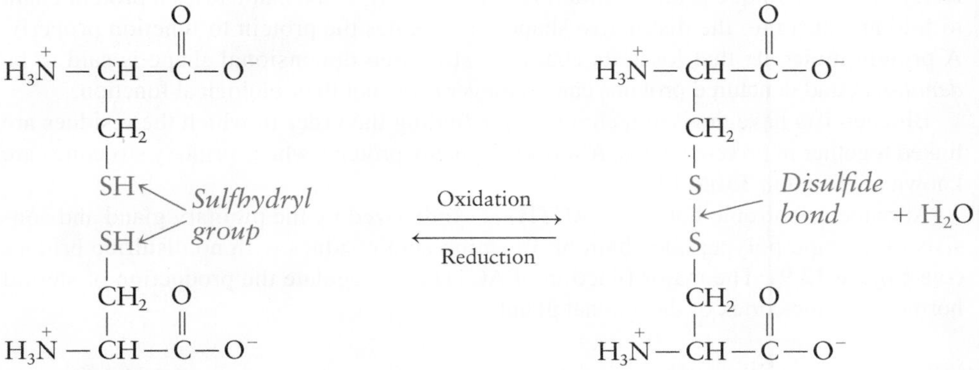

cysteine   cystine

(12.5)

This reaction is important in establishing and maintaining the structure of some proteins (**Section 12.3C**).

12.3 Protein Structure

Learning Objective 4 Explain what is meant by the primary structure of proteins.

Learning Objective 5 Describe the role of hydrogen bonding in the secondary structure of proteins.

Learning Objective 6 Describe the role of side-chain interactions in the tertiary structure of proteins.

Learning Objective 7 Explain what is meant by the quaternary structure of proteins.

The structures of extremely large molecules, such as proteins, are much more complex than those of simple organic compounds. Many protein molecules consist of a chain of amino acids twisted and folded into an intricate three-dimensional structure. This structural complexity imparts unique features to proteins that allow them to function in the diverse ways required by biological systems. To better understand protein function, we must look at the four levels of organization in their structure. These levels are referred to as primary, secondary, tertiary, and quaternary.

12.3A The Primary Structure of Proteins

Every protein, whether it's hemoglobin, keratin, or insulin, has a similar backbone of carbon and nitrogen atoms held together by peptide bonds:

$$-NH-\underset{\displaystyle R}{\underset{|}{CH}}-\overset{\displaystyle O}{\overset{\|}{C}}-NH-\underset{\displaystyle R'}{\underset{|}{CH}}-\overset{\displaystyle O}{\overset{\|}{C}}-NH-\underset{\displaystyle R''}{\underset{|}{CH}}-\overset{\displaystyle O}{\overset{\|}{C}}-NH-\underset{\displaystyle R'''}{\underset{|}{CH}}-\overset{\displaystyle O}{\overset{\|}{C}}-NH-\underset{\displaystyle R''''}{\underset{|}{CH}}-\overset{\displaystyle O}{\overset{\|}{C}}-$$

Protein backbone (in color)

The difference in proteins is the variation in the length of the backbone and the sequence of the side chains (R groups) that are attached to the backbone. The order in which amino acid residues are linked together in a protein is called the **primary structure**.

primary protein structure The linear sequence of amino acid residues in a protein chain.

Each different protein in a biological organism has a unique sequence of amino acid residues (i.e., a unique primary structure). It is this sequence that causes a protein chain to fold and curl into the distinctive shape that enables the protein to function properly. A protein molecule that loses its characteristic three-dimensional shape is said to be *denatured*, and denatured proteins can no longer carry out their biological function.

Biochemists have devised techniques for finding the order in which the residues are linked together in protein chains. A few of the many proteins whose primary structures are known are listed in **Table 12.2**.

Adrenocorticotropic hormone (ACTH) is synthesized by the pituitary gland and consists of a single polypeptide chain of 39 amino acid residues with no disulfide bridges (see **Figure 12.9**). The major function of ACTH is to regulate the production of steroid hormones in the cortex of the adrenal gland.

Figure 12.9 The amino acid sequence of ACTH.

Health Connections 12.1

Higher-Protein Diet

There is extensive debate among experts over which type of diet is the most effective for weight loss. Protein has several potential benefits due to its effect on hunger, metabolism, and body composition. Of the three macronutrients—protein, carbohydrates, and fat—protein has the greatest ability to satisfy hunger, which is critical for dieters. Studies have found that eating larger amounts of protein, especially at breakfast, helps improve hunger control throughout the day.

Protein can also boost metabolism slightly, which is particularly important during dieting, when metabolism can drop significantly. Protein may also benefit metabolism through its impact on body composition. Research has shown that protein in the diet can help maintain lean body mass during weight loss, particularly when combined with resistance training.

Finally, eating slightly more protein during weight loss may help preserve bone mineral content, which is particularly important in women as they get older and their risk of osteoporosis increases.

The bottom line is that calories count the most when it comes to losing weight. However, consuming a slightly higher percentage of protein (25% to 30%) while consuming fewer calories may make weight loss slightly easier and more sustainable.

TheFarAwayKingdom/Shutterstock.com

Protein satisfies hunger.

Table 12.2 Some Proteins Whose Sequence of Amino Acids Is Known

Name	Origin	Action	Number of Amino Acids
Adrenocorticotropic hormone (ACTH)	Pituitary	Stimulates production of adrenal hormones	39
Insulin	Pancreas	Controls metabolism of carbohydrates	51
Gastrin	Stomach	Stimulates stomach to secrete acid	101
Myoglobin	Heart and skeletal muscles	Stores oxygen in cells	153
Human growth hormone	Pituitary	Stimulates growth and cell production	191
Carboxypeptidase A	Pancreas	An enzyme that removes the amino acid residue from the C-terminal of a peptide chain; aids in digestion, blood clotting, and reproduction	307
Gamma globulin	Lymphocytes and plasma cells	Make up antibodies in the immune system	1320

12.3B The Secondary Structure of Proteins

If the only structural characteristics of proteins were their amino acid sequences (primary structures), then all protein molecules would consist of long chains arranged in random fashion. However, protein chains fold and become aligned in such a way that certain orderly patterns result. These orderly patterns, referred to as **secondary structures**, result from hydrogen bonding and include the α-helix (alpha-helix) and the β-pleated (beta-pleated) sheet.

In 1951, chemists Linus Pauling (1901–1994) and Robert Corey (1897–1971) suggested that proteins could exist in the shape of an **α-helix**, a form in which a single protein

secondary protein structure The arrangement of protein chains into patterns as a result of hydrogen bonds between amide groups of amino acid residues in the chain. The common secondary structures are the α-helix and the β-pleated sheet.

α-helix The helical structure in proteins that is maintained by hydrogen bonds.

chain twists so that it resembles a coiled helical spring (see **Figure 12.10**). The chain is held in the helical shape by numerous intramolecular hydrogen bonds between carbonyl oxygens and amide hydrogens in adjacent turns of the helical backbone:

$$\overset{\delta^-}{\underset{|}{-\mathrm{C}}=\mathrm{O}} \cdots\cdot \overset{\delta^+}{\mathrm{H}-\underset{|}{\mathrm{N}}-}$$

The carbonyl group of each amino acid residue is hydrogen-bonded to the amide hydrogen of the amino acid that is four residues away in the chain; thus, all amide groups in the helix are hydrogen-bonded (see **Figure 12.10**). The protein backbone forms the coil, and the side chains (R groups) extend outward from the coil.

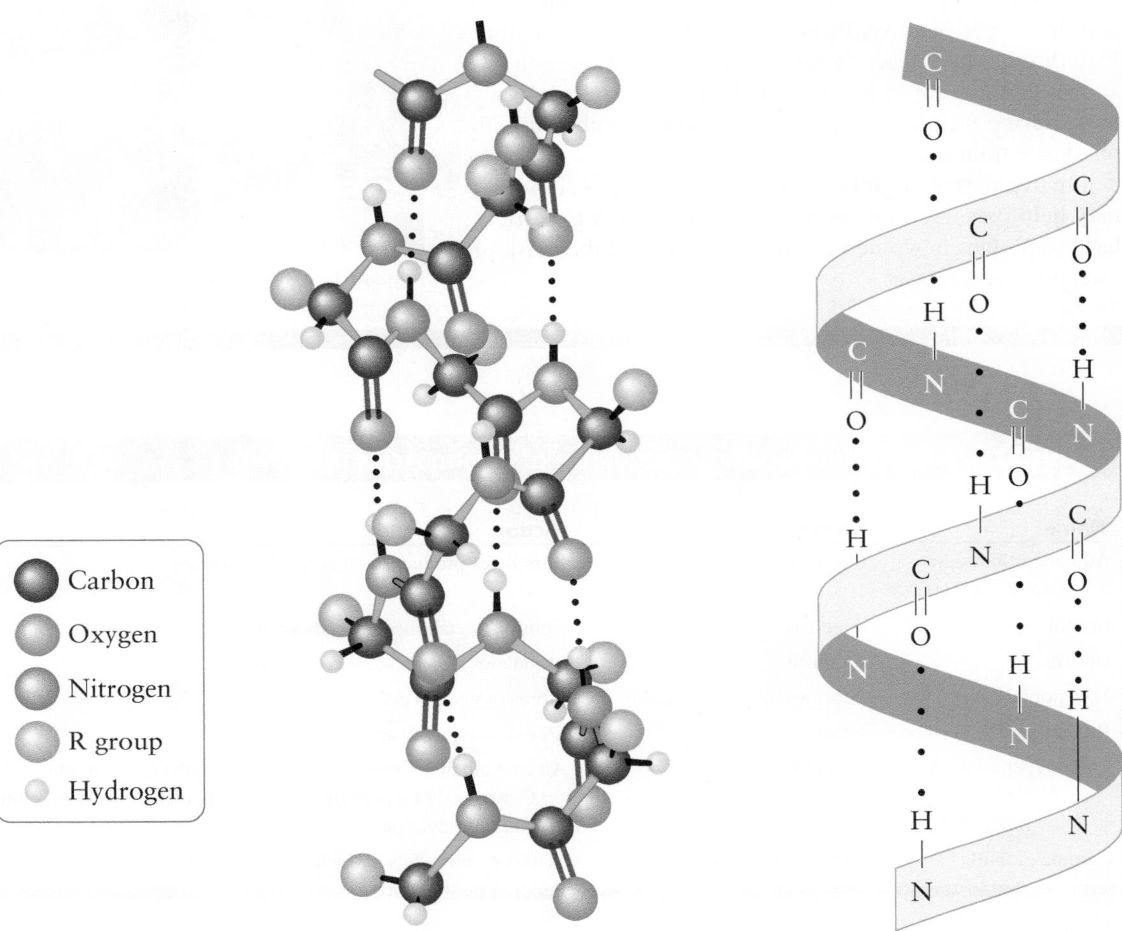

Figure 12.10 Two representations of the α-helix, showing hydrogen bonds (dotted lines) between amide groups.

Six years after Pauling and Corey proposed the α-helix, its presence in proteins was detected. Since that time, it has been found that the amount of α-helical content is quite variable in proteins. In some, it is the major structural component, whereas in others, a random coil predominates and there is little or no α-helix coiling.

fibrous proteins Proteins whose secondary structure is made up of cable-like coiled coils.

The proteins keratin (found in hair), myosin (found in muscle), epidermin (found in skin), and fibrin (found in blood clots) are all common examples of **fibrous proteins**. In these proteins, two or more helices interact to create a coiled-coil, a cable-like structure (see **Figure 12.11**). These cables make up bundles of fibers that lend strength to the tissue in which they are found.

Figure 12.11 The coiled-coil of three α-helices in the keratins of hair and wool.

Another common type of secondary structure found in proteins is called a ***β*-pleated sheet**. It results when several protein chains lie side by side and are held in position by hydrogen bonds between the carbonyl oxygens of one chain and the amide hydrogens of an adjacent chain (see **Figure 12.12**).

***β*-pleated sheet** A secondary protein structure in which protein chains are aligned side by side in a sheetlike array held together by hydrogen bonds.

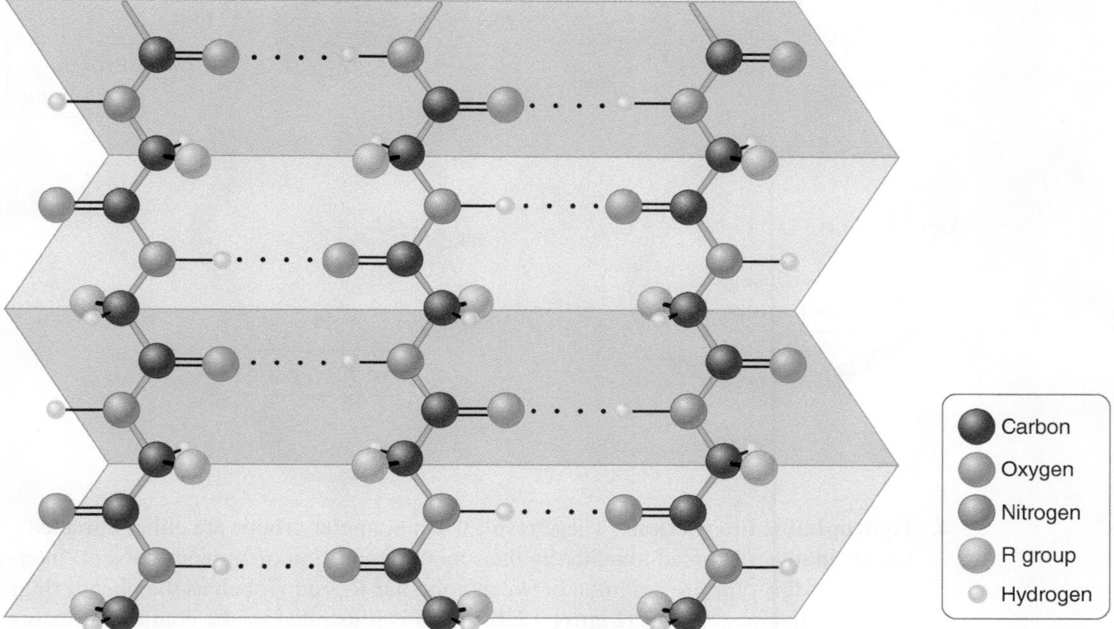

Figure 12.12 The β-pleated sheet. Each group of four dots represents a hydrogen bond between adjacent protein chains.

12.3C The Tertiary Structure of Proteins

Tertiary structure, the next higher level of complexity in protein structure, refers to the bending and folding of the protein into a specific three-dimensional shape. This bending and folding may seem rather disorganized, but nevertheless it results in a favored arrangement for a given protein. The tertiary structure results from interactions between the side chains of the amino acid residues. These R-group interactions are of four types:

tertiary protein structure A specific three-dimensional shape of a protein resulting from interactions between R groups of the amino acid residues in the protein.

1. **Disulfide bridges**: A disulfide linkage can form between two cysteine residues that are close to each other in the same chain or between cysteine residues in different chains. The existence and location of these disulfide bridges is a part of the tertiary structure because these interactions hold the protein chain in a loop or some other three-dimensional shape.
2. **Salt bridges**: These interactions are a result of ionic bonds that form between the ionized side chain of an acidic amino acid ($-COO^-$) and the ionized side chain of a basic amino acid ($-NH_3^+$).
3. **Hydrogen bonds**: Hydrogen bonds can form between a variety of side chains, especially those that possess the following functional groups:

$$-OH \qquad -NH_2 \qquad -\overset{\overset{\displaystyle O}{\|}}{C}-NH_2$$

One possible interaction, between the –OH groups of two serine residues, is shown in **Figure 12.13**. We saw in **Section 12.3B** that hydrogen bonding also determines the secondary structure of proteins. The distinction is that the tertiary hydrogen bonding occurs between R groups in tertiary structure, whereas it occurs between backbone —C=O and –N groups in secondary structure.

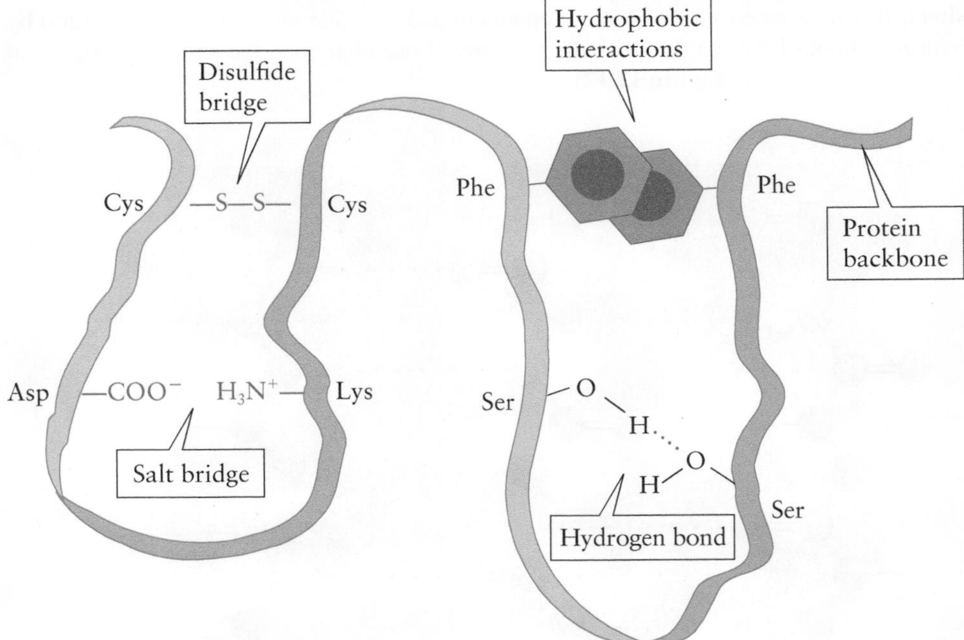

Figure 12.13 R-group interactions lead to tertiary protein structure.

4. **Hydrophobic interactions**: These result when nonpolar groups are either attracted to one another or forced together by their mutual repulsion of aqueous solvent. Interactions of this type are common between nonpolar R groups such as the phenyl rings of phenylalanine residues (**Figure 12.13**). This is illustrated by the compact structure of some proteins in aqueous solutions (see **Figure 12.14**). This shape results because polar groups are pointed outward toward the aqueous solvent, whereas nonpolar groups are pointed inward, away from the water molecules. This type of interaction is weaker than the other three, but it usually acts over large surface areas so that the net effect is an interaction strong enough to stabilize the tertiary structure.

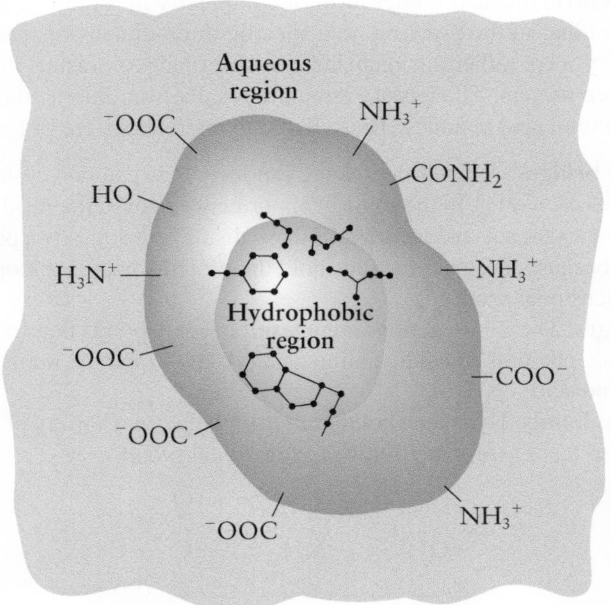

Figure 12.14 A protein with a hydrophobic region on the inside and polar groups on the outside extending into the aqueous surroundings.

Example 12.6 Identifying R-Group Interactions

What kind of R-group interaction might be expected if the following side chains were in close proximity?

a. $-\underset{|}{\overset{CH_3}{}}CHCH_3$ and $-CH_2-C_6H_5$ (benzene ring)

b. $-CH_2CH_2-COO^-$ and $-CH_2-$(imidazolium ring: $\overset{+}{H}N$ … NH)

c. $-CH_2-OH$ and $-CH_2-\overset{O}{\overset{\|}{C}}-NH_2$

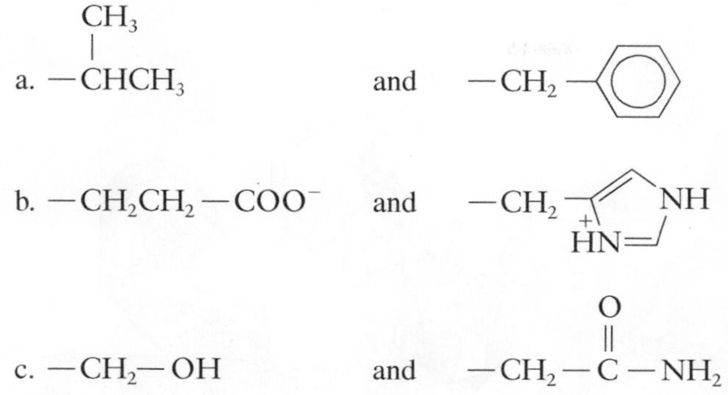

Solution

a. Hydrophobic interactions (both side chains are nonpolar)
b. Salt bridge (one side chain is positively charged, whereas the other is negatively charged)
c. Hydrogen bonding (the hydrogen of the –OH group can hydrogen-bond with the oxygen or the nitrogen of the amide group)

✔ Learning Check 12.6 What kind of R-group interaction might be expected if the following side chains were in close proximity?

a. $-\overset{CH_3}{\overset{|}{C}}HCH_2CH_3$ and $-\overset{CH_3}{\overset{|}{C}}HCH_2CH_3$

b. $-\overset{O}{\overset{\|}{C}}-NH_2$ and $-C_6H_4-OH$ (benzene ring)

c. $-CH_2-\overset{O}{\overset{\|}{C}}-O^-$ and $-CH_2CH_2CH_2CH_2-NH_3^+$

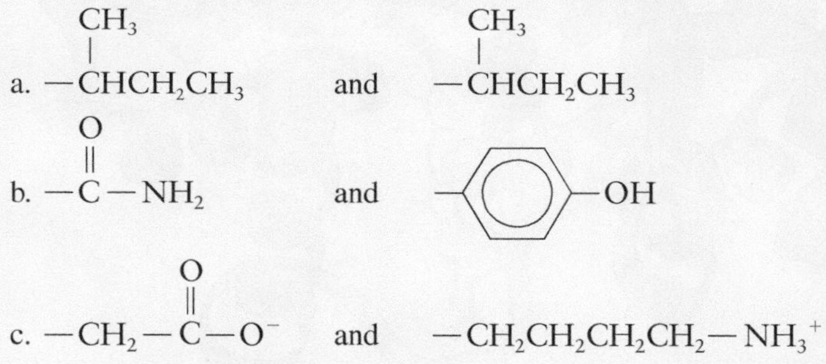

12.3D The Quaternary Structure of Proteins

quaternary protein structure The arrangement of subunits that form a larger protein.

subunit A polypeptide chain having primary, secondary, and tertiary structural features that is a part of a larger protein.

The final level of protein structure, called **quaternary structure**, occurs when multiple tertiary structures (also known as **subunits**) combine to form a larger protein. These subunits are held together by various attractive forces like ionic attractions, disulfide bridges, hydrogen bonds, and hydrophobic effects. Each individual subunit has its own primary, secondary, and tertiary structures.

Hemoglobin is a protein that exhibits quaternary structure. Hemoglobin is made up of four individual subunits (see **Figure 12.15a**): two identical alpha chains (141 amino acid residues each) and two identical beta chains (146 amino acid residues each). Each of the four subunits contains a heme group (an iron atom sandwiched between two hydrocarbon rings; see **Figure 12.15b**) located in crevices near the exterior of the structure. The four subunits are held together through hydrophobic forces, creating an overall structure that is almost spherical.

Although hemoglobin's quaternary structure consists of four individual subunits, it is not considered to be a "large" protein. With a molecular weight of 65,000 g/mol and 574 individual amino acid residues, there are many proteins that are either smaller in size (e.g., insulin at 51 amino acids) or larger in size (e.g., myosin at 6100 amino acids). A greater appreciation for the immense size of these proteins (even the small ones) can be achieved by comparing them to a common carbohydrate like glucose. Whereas glucose (180 g/mol) has a molecular formula $C_6H_{12}O_2$, the molecular formula of hemoglobin is $C_{2952}H_{4664}O_{832}N_{812}S_8Fe_4$.

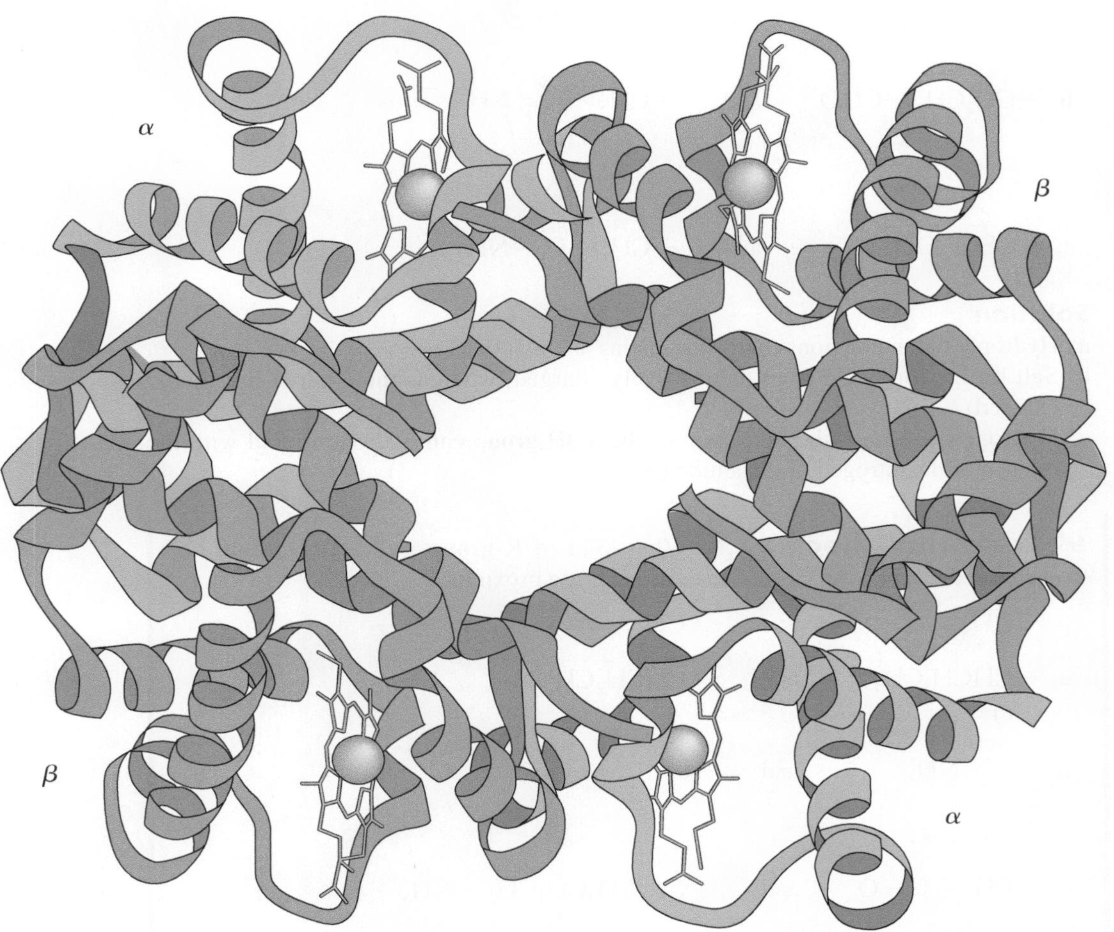

a Hemoglobin exhibits quaternary structure with two α-chains and two β-chains. The purple disks are heme groups, and the purple spheres are iron atoms.

CH_3 $CH{=}CH_2$
CH_3 CH_3
$N{-}Fe^{2+}{-}N$
$HOOCCH_2CH_2$ $CH{=}CH_2$
$HOOCCH_2CH_2$ CH_3

b The heme group uses four nitrogen atoms to bind iron.

Figure 12.15 Hemoglobin: (a) its quaternary structure and (b) the structure of its heme group.

Sorbitol dehydrogenase is a tetrameric protein that is responsible for converting sorbitol (an alcohol form of glucose) to fructose. The primary structure of each of the four subunits consists of 356 amino acid residues. The secondary structure consists of 14 alpha helices and 23 beta sheets. A bulky and basically spherical tertiary structure forms when the alpha helices and beta sheets fold back on themselves, which allows the side chains (i.e., the R groups) to interact through attractive forces (e.g., ionic bridges and disulfide bonds). The complete sorbitol dehydrogenase enzyme structure is formed when four identical tertiary subunits interact with each other through hydrogen bonding, ionic bonds, and hydrophobic attractions (**Figure 12.16**).

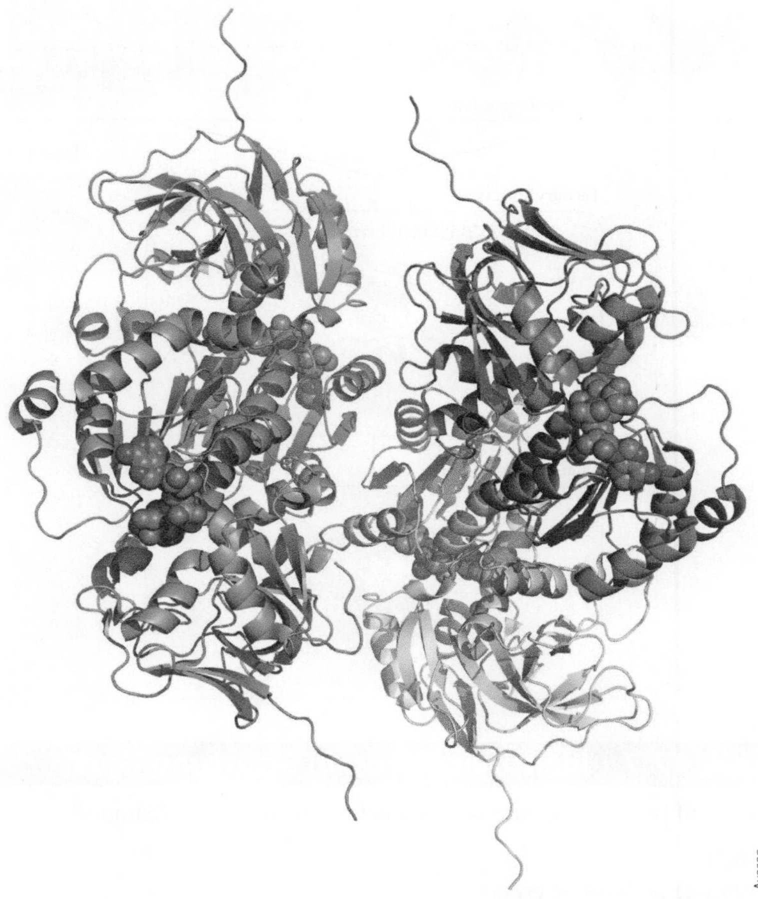

Ayacop

Figure 12.16 The structure of the human sorbitol dehydrogenase tetramer.

Sorbitol dehydrogenase is classified as a **globular protein** because of its overall spherical tertiary structure. Globular proteins tend to form stable suspensions in water or to be soluble in water because their nonpolar side chains are concentrated in the center of the spherical structure, whereas their polar side chains are concentrated on the surface.

globular protein A spherical protein that usually forms stable suspensions in water or dissolves in water.

Figure 12.17 summarizes the key points about the four levels of protein structure. That is, primary structure consists of the linear chain of amino acid residues. Those linear chains coil to form α helices or align to make β sheets, each of which is a secondary structure stabilized by hydrogen bonding. Once these secondary structural motifs are established, the protein folds even further to produce tertiary structures stabilized by interactions between R groups—mainly hydrophobic interactions, salt bridges, disulfide bonds, and to a lesser extent more hydrogen bonds.

If the protein is a monomer (i.e., a single subunit), then its folding story ends here. If it consists of multiple subunits, however, then the subunits interact via hydrophobic forces, salt bridges, and hydrogen bonding to form a unique quaternary structure.

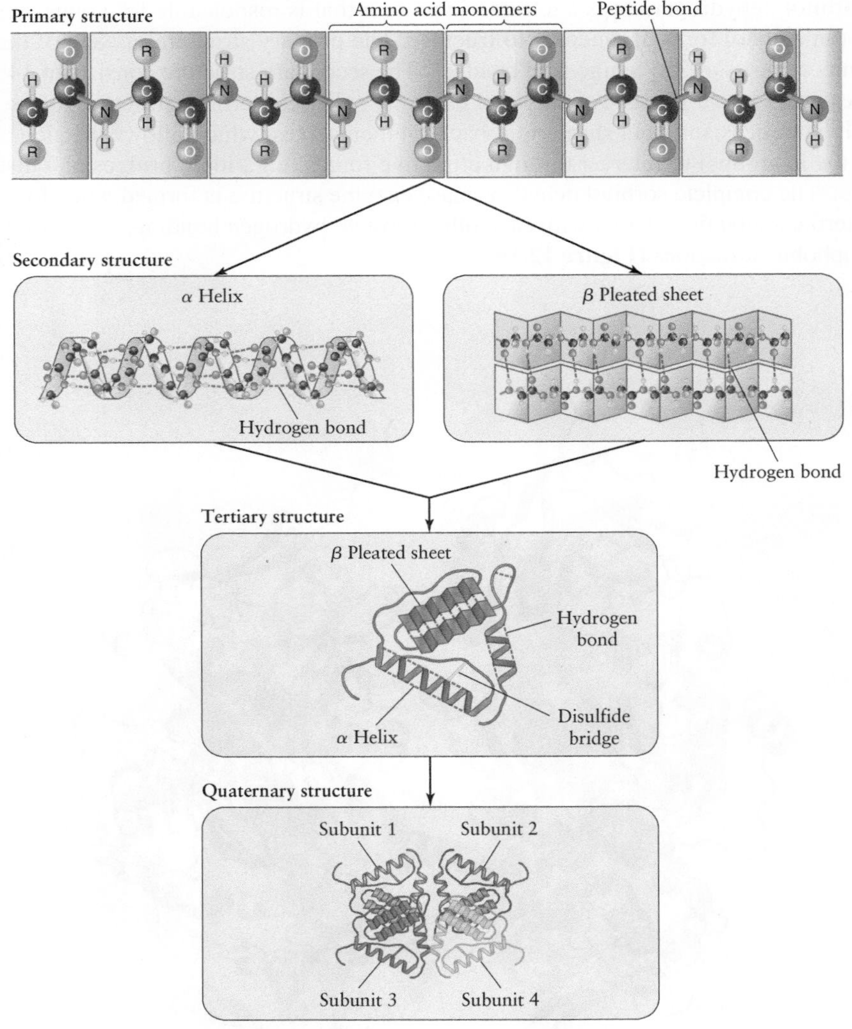

Figure 12.17 The four levels of protein structure.

Example 12.7 Identifying Protein Structural Level

In what level of protein structure would you find the following features?

a. An α-helix
b. A salt bridge between R groups
c. A peptide bond

Solution

a. Secondary: The two main types of secondary structure are the α-helix and the β-sheet.
b. Tertiary and quaternary: Tertiary and quaternary structure results from four types of interactions between R groups—namely, disulfide bridges, salt bridges, hydrogen bonds, and hydrophobic interactions.
c. Primary: Peptide bonds are what link individual amino acids together in the protein chain.

✔ Learning Check 12.7 In what level of protein structure would you find the following features?

a. Intertwining of two polypeptide chains
b. β-pleated sheet
c. Hydrophobic interactions between R groups

12.4 Protein Functions

Learning Objective 8 Describe four biological functions of proteins.

Proteins have crucial roles in all biological processes. The various protein functions discussed here are summarized in **Table 12.3**.

1. **Catalytic function:** Nearly all the reactions that take place in living organisms are catalyzed by proteins functioning as enzymes. Without these catalytic proteins, biological reactions would take place too slowly to support life. Enzyme-catalyzed reactions range from relatively simple processes to the very complex, such as the duplication of hereditary material in cell nuclei.
2. **Structural function:** Recall from **Section 11.8C** that the main structural material for plants is cellulose. In the animal kingdom, structural materials other than the inorganic components of the skeleton are composed of protein. Collagen, a fiberlike protein, is responsible for the mechanical strength of skin (see **Figure 12.18**) and bone. Keratin, the chief constituent of hair, skin, and fingernails, is another example.
3. **Storage function:** Some proteins provide a way to store small molecules or ions. Ovalbumin, for example, is a stored form of amino acids that is used by embryos developing in bird eggs. Casein, a milk protein, and gliadin, in wheat seeds, are also stored forms of protein intended to nourish animals and plants, respectively. Ferritin, which is found in all cells of the body, attaches to iron ions and forms a storage complex in humans and other animals.

Figure 12.18 The elasticity of skin is due to collagen. As we age, collagen begins to break down, giving rise to wrinkles.

Table 12.3 Biological Functions of Proteins

Function	Examples	Occurrence or Role
Catalysis	Lactate dehydrogenase	Oxidizes lactic acid
	Cytochrome c	Transfers electrons
	DNA polymerase	Replicates and repairs DNA
Structure	Viral-coat proteins	Sheath around nucleic acid of viruses
	Glycoproteins	Cell coats and walls
	α-Keratin	Skin, hair, feathers, nails, and hooves
	β-Keratin	Silk of cocoons and spiderwebs
	Collagen	Fibrous connective tissue
	Elastin	Elastic connective tissue
Storage	Ovalbumin	Egg-white protein
	Casein	A milk protein
	Ferritin	Stores iron in the spleen
	Gliadin	Stores amino acids in wheat
	Zein	Stores amino acids in corn
Protection	Antibodies	Form complexes with foreign proteins
	Fibrinogen	Involved in blood clotting
	Thrombin	Involved in blood clotting
Regulation	Insulin	Regulates glucose metabolism
	Growth hormone	Stimulates growth of bone
Nerve impulse transmission	Rhodopsin	Involved in vision
	Acetylcholine receptor protein	Impulse transmission in nerve cells
Movement	Myosin	Thick filaments in muscle fiber
	Actin	Thin filaments in muscle fiber
	Dynein	Movement of cilia and flagella
Transport	Hemoglobin	Transports O_2 in blood
	Myoglobin	Transports O_2 in muscle cells
	Serum albumin	Transports fatty acids in blood
	Transferrin	Transports iron in blood
	Ceruloplasmin	Transports copper in blood

antibody A substance that helps protect the body from invasion by viruses, bacteria, and other foreign substances.

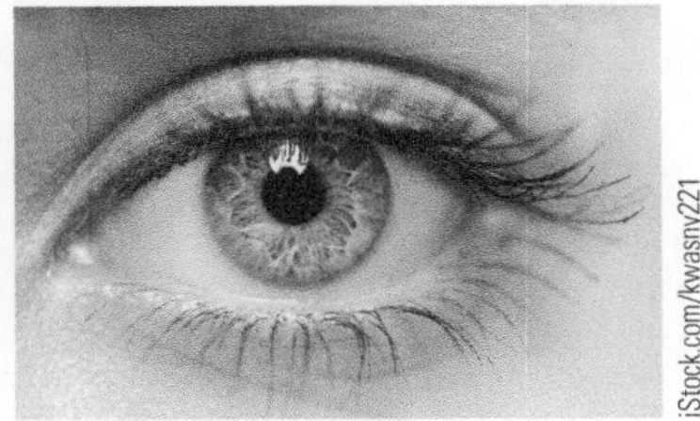

iStock.com/kwasny221

Figure 12.19 Rhodopsin is extremely sensitive to light, and thus enables vision in low-light conditions.

© Michael Slabaugh

Figure 12.20 One miracle of life is that hundreds of chemical reactions take place simultaneously in each living cell; the activity of enzymes in the cells makes it all possible.

enzyme A biomolecule that catalyzes chemical reactions.

substrate The substance that undergoes a chemical change catalyzed by an enzyme.

absolute specificity The characteristic of an enzyme that it acts on one and only one substance.

relative specificity The characteristic of an enzyme that it acts on several structurally related substances.

4. **Protective function: Antibodies** are tremendously important proteins that protect the body from disease. These highly specific proteins combine with and help destroy viruses, bacteria, and other foreign substances that get into the blood or tissue of the body. Blood clotting, another protective process, is carried out by the proteins thrombin and fibrinogen. Without this process, even small wounds would result in life-threatening bleeding.
5. **Regulatory function:** Numerous body processes are regulated by hormones, many of which are proteins. Examples are growth hormone, which regulates the growth rate of young animals, and thyrotropin, which stimulates the activity of the thyroid gland.
6. **Nerve impulse transmission:** Some proteins behave as receptors of small molecules that pass between gaps (synapses) separating nerve cells. In this way, they transmit nerve impulses from one nerve to another. Rhodopsin, a protein found in the rod cells of the retina of the eye, functions this way in the vision process (see **Figure 12.19**).
7. **Movement function:** Every time we climb stairs, push a button, or blink an eye, we use muscles that have proteins as their major components. The proteins actin and myosin are particularly important in processes involving movement. They are long-filament proteins that slide along each other during muscle contraction.
8. **Transport function:** Numerous small molecules and ions are transported effectively through the body only after binding to proteins. For example, fatty acids are carried between fat (adipose) tissue and other tissues or organs by serum albumin, a blood protein. Hemoglobin carries oxygen from the lungs to other body tissues, and transferrin is a carrier of iron in blood plasma.

Proteins perform other functions, but these are among the most important. They demonstrate that properly functioning cells of living organisms must contain many proteins. It has been estimated that a typical human cell contains 9000 different proteins, and that a human body contains about 100,000 different proteins!

12.5 Classes of Enzymes

Learning Objective 9 Describe three reactions that enzymes catalyze.

The catalytic behavior of proteins acting as **enzymes** is one of the most important functions performed by cellular proteins. Without catalysts, most cellular reactions would take place much too slowly to support life (see **Figure 12.20**). With the exception of a few recently discovered RNA molecules that catalyze their own reactions, all enzymes are globular proteins. Enzymes are the most efficient of all known catalysts; some can increase reaction rates by 1020 times that of uncatalyzed reactions.

Enzymes are well suited to their essential roles in living organisms in three major ways:

- They have enormous catalytic power.
- They are highly specific in the reactions they catalyze.
- Their activity as catalysts can be regulated.

Part of the reasons enzymes are highly specific in the reactions they catalyze is that they have evolved to be specific for certain **substrates** (i.e., the substance the enzyme acts upon). Enzymes act with different levels of specificity, from acting on a single, specific substrate (**absolute specificity**) to reacting with structurally related substrates (**relative specificity**).

Enzymes are also categorized based on the type of reaction they catalyze, many of which are fundamental reactions from organic chemistry that happen to take place in the body (**Table 12.4**). For example, reactions from previous chapters like oxidation (**Figure 12.21**) and reduction, alkene addition, and hydrolysis are among those found in both organic chemistry and within many enzymes.

Figure 12.21 The enzyme phenolase acts on the phenols present in apples to react with oxygen and produce the familiar brownish coloration. The left half of the apple has been freshly sliced, while the right half has been exposed to air (oxygen) for a few minutes.

Table 12.4 The Classification of Enzymes

Group Name	Type of Reaction Catalyzed
Oxidoreductases	Oxidation–reduction reactions
Transferases	Transfer of functional groups
Hydrolases	Hydrolysis reactions
Lyases	Addition to double bonds or the reverse of that reaction
Isomerases	Isomerization reactions
Ligases	Formation of bonds with ATP cleavage[a]

[a] ATP is discussed in **Chapter 15**.

12.6 Enzyme Activity

Learning Objective 10 Describe enzyme activity in terms of efficacy and mechanism of action.

12.6A Catalytic Efficiency

Catalysts increase the rate of chemical reactions but are not used up in the process. Although a catalyst actually participates in a chemical reaction, it is not permanently modified and may be used again and again. Enzymes are true catalysts that lower activation energies (see **Figure 12.22**), thereby allowing chemical reactions to achieve equilibrium more rapidly.

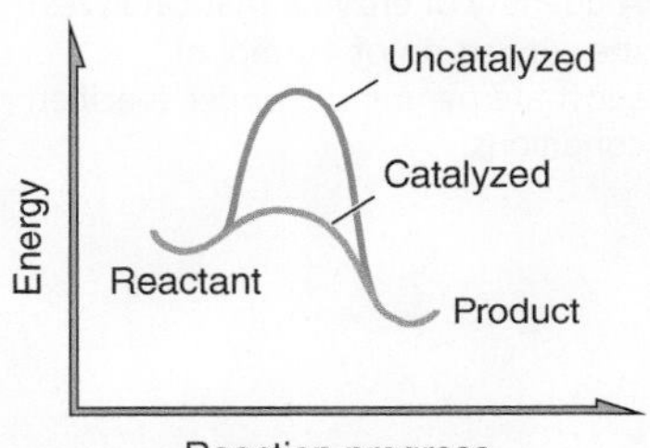

Figure 12.22 An energy diagram for a reaction when uncatalyzed and catalyzed.

Many of the important enzyme-catalyzed reactions are similar to the organic reactions we studied in Chapters 8–10, such as ester hydrolysis and alcohol oxidation. However, laboratory conditions cannot match what happens when these reactions are carried out in the body, where enzymes cause them to proceed under mild pH and temperature conditions. In addition, enzyme catalysis within the body can accomplish in seconds the reactions that ordinarily take weeks or even months under laboratory conditions.

A good example of the influence of enzymes on the rates of reactions essential to life is the process that moves carbon dioxide, a waste product of cellular respiration, out of the body. Before this can be accomplished, the carbon dioxide must be combined with water to form carbonic acid:

$$CO_2 + H_2O \xrightarrow{\text{Carbonic anhydrase}} H_2CO_3$$

In the absence of the appropriate enzyme, carbonic anhydrase, the formation of carbonic acid takes place much too slowly to support the required exchange of carbon dioxide between the blood and the lungs. But in the presence of carbonic anhydrase, this vital reaction proceeds rapidly. Each molecule of the enzyme can catalyze the formation of carbonic acid at the rate of 36 million molecules per minute.

Enzyme activity refers in general to the catalytic ability of an enzyme to increase the rate of a reaction. The amazing rate (36 million molecules per minute) at which one molecule of carbonic anhydrase converts carbon dioxide to carbonic acid is called the **turnover number**, and is one of the highest known for enzyme systems. More common turnover numbers for enzymes are closer to 1000 reactions per minute. Nevertheless, even such low numbers dramatize the speed with which a small number of enzyme molecules can transform a large number of substrate molecules. **Table 12.5** lists the turnover numbers of several enzymes.

enzyme activity The rate at which an enzyme catalyzes a reaction.

turnover number The number of molecules of substrate acted on by one molecule of enzyme per minute.

Experiments that measure enzyme activity are called *enzyme assays*. Assays for blood enzymes are routinely performed in clinical laboratories. Such assays are often done by determining how fast the characteristic color of a product forms or the color of a substrate decreases. Reactions in which protons (H^+) are produced or used up can be followed by measuring how fast the pH of the reacting mixture changes with time.

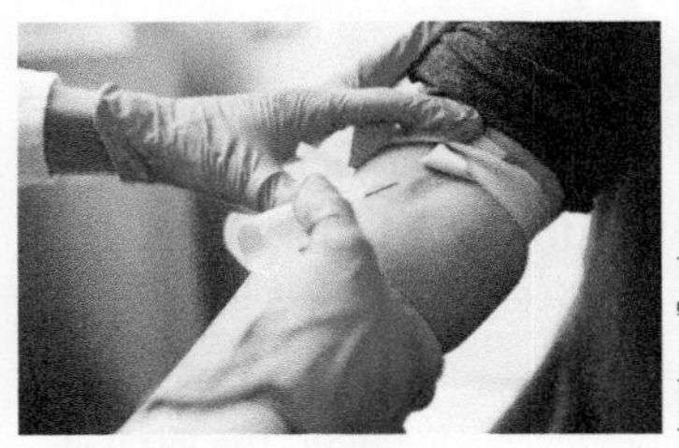
istock.com/Renphoto

Figure 12.23 Elevated levels of lactate dehydrogenase (LDH) can indicate either acute or chronic tissue damage.

Table 12.5 Examples of Enzyme Turnover Numbers

Enzyme	Turnover Number (per minute)	Reaction Catalyzed
Carbonic anhydrase	36,000,000	$CO_2 + H_2O \rightleftharpoons H_2CO_3$
Catalase	5,600,000	$2H_2O_2 \rightleftharpoons 2H_2O + O_2$
Acetylcholinesterase	1,500,000	Hydrolysis of acetylcholine
Chymotrypsin	6000	Hydrolysis of specific peptide bonds
DNA polymerase I	900	Addition of nucleotide monomers to DNA chains
Lactate dehydrogenase (**Figure 12.23**)	60,000	Pyruvate + NADH + $H^+ \rightleftharpoons$ lactate + NAD^+
Penicillinase	120,000	Hydrolysis of penicillin

enzyme international unit (IU) A quantity of enzyme that catalyzes the conversion of 1 μmol of substrate per minute under specified conditions.

Because some clinical assays are done many times a day, the procedures for running the assays have been automated and computerized. The determined enzyme activity levels are usually reported in terms of **enzyme international units (IU)**, which define enzyme activity as the amount of enzyme that will convert a specified amount of substrate to a product within a certain time.

One standard enzyme international unit (1 IU) is the quantity of enzyme that catalyzes the conversion of 1 micromole (μmol) of substrate per minute under specified reaction conditions. Thus, unlike the turnover number, which is a constant characteristic for one molecule of a particular enzyme, international units measure how much enzyme is present. For example, an enzyme preparation with an activity corresponding to 40 IU contains a concentration of enzyme 40 times greater than the standard. This is a useful way to measure enzyme activity because the level of enzyme activity compared with normal activity is significant in the diagnosis of many diseases (see **Section 12.9**).

12.6B The Mechanism of Enzyme Action

active site The location on an enzyme where a substrate is bound and catalysis occurs.

Although enzymes differ widely in structure and specificity, a general theory has been proposed to account for their catalytic behavior. Because enzyme molecules are large compared with the molecules whose reactions they catalyze (substrate molecules), it has been proposed that substrate and enzyme molecules come in contact and interact over only a small region of the enzyme surface. This region of interaction is called the **active site**.

The binding of a substrate molecule to the active site of an enzyme may occur through hydrophobic attraction, hydrogen-bonding, and/or ionic bonding. The complex formed when a substrate and an enzyme bind is called the enzyme–substrate (ES) complex. Once this complex is formed, the conversion of substrate (S) to product (P) may take place:

General reaction:

$$\underset{\text{enzyme}}{E} + \underset{\text{substrate}}{S} \rightleftharpoons \underset{\text{enzyme–substrate complex}}{ES} \longrightarrow \underset{\text{enzyme}}{E} + \underset{\text{product}}{P} \qquad (12.6)$$

Specific example:

$$\underset{\text{Enzyme}}{\text{sucrase}} + \underset{\text{Substrate}}{\text{sucrose}} \rightleftharpoons \begin{bmatrix}\text{sucrase–}\\\text{sucrose}\\\text{complex}\end{bmatrix} \longrightarrow \underset{\text{Enzyme}}{\text{sucrase}} + \underset{\text{Products}}{\text{glucose + fructose}}$$

The chemical transformation of the substrate occurs at the active site, usually aided by enzyme functional groups that participate directly in the making and breaking of chemical bonds. After chemical conversion has occurred, the product is released from the active site, and the enzyme is free for another round of catalysis.

According to the **lock-and-key theory**, enzyme surfaces will accommodate only those substrates having specific shapes and sizes. Thus, only specific substrates "fit" a given enzyme and can form complexes with it (see **Figure 12.24a**).

lock-and-key theory A theory of enzyme specificity proposing that a substrate has a shape fitting that of the enzyme's active site, as a key fits a lock.

A limitation of the lock-and-key theory is the implication that enzyme conformations are fixed or rigid. Research results suggest, however, that the active sites of some enzymes are not rigid. The **induced-fit theory**, which is a modification of the lock-and-key theory, proposes that enzymes have somewhat flexible conformations that may adapt to incoming substrates. The active site has a shape that becomes complementary to that of the substrate as the substrate binds to the enzyme active site (see **Figure 12.24b**).

induced-fit theory A theory of enzyme action proposing that the conformation of an enzyme changes to accommodate an incoming substrate.

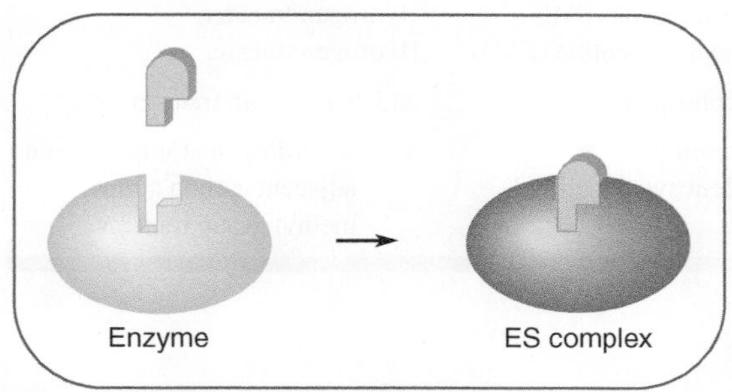

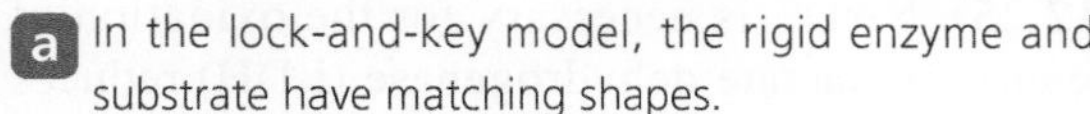

a In the lock-and-key model, the rigid enzyme and substrate have matching shapes.

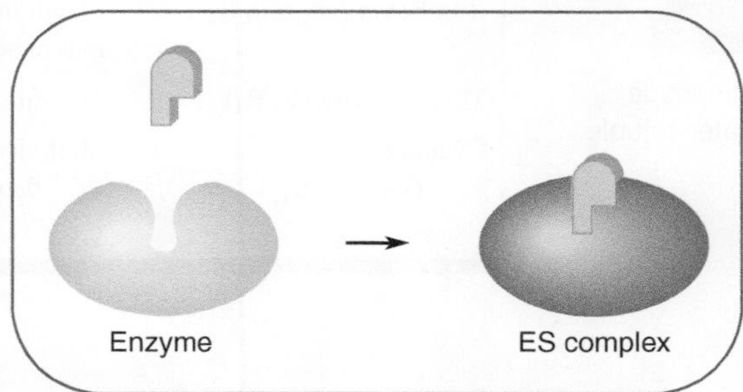

b In the induced-fit model, the flexible enzyme changes shape to match the substrate.

Figure 12.24 Two models representing enzyme action.

Recall from **Section 12.3** that proteins may be simple (containing only amino acid residues) or conjugated with a prosthetic group present (e.g., the heme group in hemoglobin; see **Figure 12.15**). Many enzymes are simple proteins, whereas many others function only in the presence of specific nonprotein molecules or metal ions. If these nonprotein components are tightly bound to and form an integral part of the enzyme structure, they are true prosthetic groups. Often, however, a nonprotein component is only weakly bound to the enzyme and is easily separated from the protein structure. This type of nonprotein component is referred to as a **cofactor**. When the cofactor is an organic substance, it is called a **coenzyme**. The cofactor may also be an inorganic ion (usually a metal ion). The protein portion of enzymes requiring a cofactor is called the **apoenzyme**. Thus, the combination of an apoenzyme and a cofactor produces an active enzyme:

cofactor A nonprotein molecule or ion required by an enzyme for catalytic activity.

coenzyme An organic molecule required by an enzyme for catalytic activity.

apoenzyme A catalytically inactive protein formed by removal of the cofactor from an active enzyme.

$$\text{apoenzyme} + \text{cofactor (coenzyme or inorganic ion)} \longrightarrow \text{active enzyme} \quad (12.7)$$

Typical inorganic ions are metal ions such as Mg^{2+}, Zn^{2+}, and Fe^{2+}. For example, the enzyme carbonic anhydrase functions only when Zn^{2+} is present, and rennin needs Ca^{2+} in order to curdle milk. Numerous other metal ions are essential for proper enzyme function in humans; hence, they are required for good health.

Like metal ions, the small organic molecules that act as coenzymes bind reversibly to an enzyme and are essential for its activity. An interesting feature of coenzymes is that many of them are formed in the body from vitamins (see **Table 12.6**), which explains why it is necessary to have certain vitamins in the diet for good health. For example, the coenzyme nicotinamide adenine dinucleotide (NAD^+), which is a necessary part of some enzyme-catalyzed oxidation–reduction reactions, is formed from the vitamin

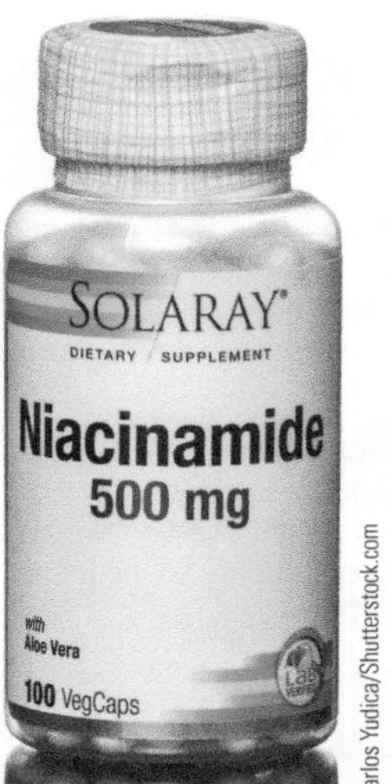

Carlos Yudica/Shutterstock.com

Figure 12.25 Nicotinamide, or niacinamide, is a water-soluble form of vitamin B_3.

Table 12.6 Vitamins and Their Coenzyme Forms

Vitamin	Coenzyme Form	Function
Biotin	Biocytin	Carboxyl group removal or transfer
Folacin	Tetrahydrofolic acid	One-carbon group transfer
Lipoic acid	Lipoamide	Acyl group transfer
Niacin (vitamin B_3)	Nicotinamide adenine dinucleotide (NAD^+)	Hydrogen transfer
	Nicotinamide adenine dinucleotide phosphate ($NADP^+$)	Hydrogen transfer
Pantothenic acid	Coenzyme A (CoA)	Acyl group carrier
Pyridoxal, pyridoxamine, pyridoxine (B_6 group)	Pyridoxal phosphate	Amino group transfer
Riboflavin (vitamin B_2)	Flavin mononucleotide (FMN)	Hydrogen transfer
	Flavin adenine dinucleotide (FAD)	Hydrogen transfer
Thiamin (vitamin B_1)	Thiamin pyrophosphate (TPP)	Aldehyde group transfer
Cyanocobalamin (vitamin B_{12})	Methylcobalamin or deoxyadenosylcobalamin	Shift of hydrogen atoms between adjacent carbon atoms; methyl group transfer

precursor nicotinamide (see **Figure 12.25**). NAD^+ is necessary for the oxidation of lactate to pyruvate. In the process, the enzyme lactate dehydrogenase (LDH) reduces NAD^+ to NADH:

$$\underset{\text{lactate}}{CH_3-\underset{|}{\overset{OH}{CH}}-COO^-} + NAD^+ \xrightleftharpoons{\text{Lactate dehydrogenase}} \underset{\text{pyruvate}}{CH_3-\overset{O}{\overset{\|}{C}}-COO^-} + NADH + H^+$$

Like other cofactors, NAD^+ is written separately from the enzyme so that the change in its structure may be shown and to emphasize that the enzyme and cofactor are easily separated.

We can see from the reaction that the coenzyme NAD^+ is essential because it is the actual oxidizing agent. NAD^+ is typical of coenzymes that often aid in the transfer of chemical groups from one compound to another. In the reaction shown, NAD^+ accepts hydrogen from lactate and will transfer it to other compounds in subsequent reactions. In **Chapter 15**, we discuss in greater detail several coenzymes that act as shuttle systems in the exchange of chemical substances among various biochemical pathways.

PF-(bygone1)/Alamy Stock Photo

Figure 12.26 Maud Menten (1879–1960), a biomedical researcher, made significant contributions to the field of mathematics dealing with the rate of enzyme-catalyzed reactions.

12.7 Factors Affecting Enzyme Activity

Learning Objective 11 Identify four factors that affect enzyme activity.

Several factors affect the rate of enzyme-catalyzed reactions, a field advanced in part by the pioneering biomedical researcher Maud Menten (**Figure 12.26**), who made several contributions to quantifying these processes. The most important factors are enzyme concentration, substrate concentration, temperature, and pH. In this section, we look at each of these factors in some detail. In **Section 12.8**, we consider another very important factor, the presence of enzyme inhibitors.

12.7A Enzyme Concentration

In an enzyme-catalyzed reaction, the concentration of enzyme is normally very low compared with the concentration of substrate. When the enzyme concentration is increased, the concentration of ES also increases in compliance with reaction rate theory.

$$E + S \longrightarrow ES$$

Increased [E] produced more [ES]

Thus, the availability of more enzyme molecules to catalyze a reaction leads to the formation of more ES and a higher reaction rate. The effect of enzyme concentration on the rate of a reaction is shown in **Figure 12.27**. According to the graph, the rate of a reaction is directly proportional to the concentration of the enzyme—that is, if the enzyme concentration is doubled, the rate of conversion of substrate to product is also doubled.

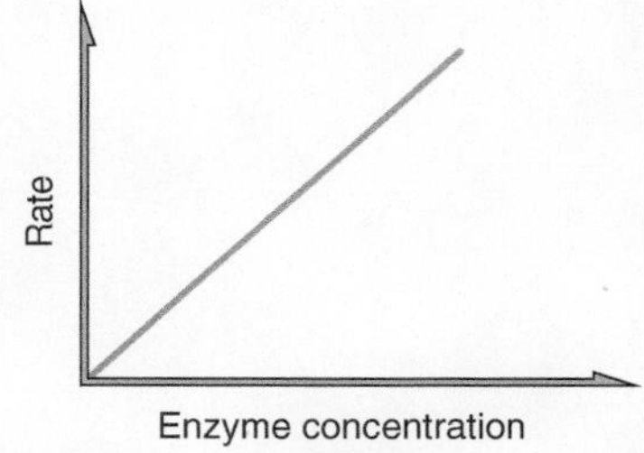

Figure 12.27 The dependence of reaction rate (initial velocity) on enzyme concentration.

12.7B Substrate Concentration

The graph in **Figure 12.28** shows that the concentration of substrate significantly influences the rate of an enzyme-catalyzed reaction. Initially, the rate is responsive to increases in substrate concentration. However, at a certain concentration, the rate levels out and remains constant. This maximum rate (symbolized by V_{max}) occurs because the enzyme is saturated with substrate and cannot work any faster under the conditions imposed.

12.7C Temperature

Enzyme-catalyzed reactions, like all chemical reactions, have rates that increase with temperature (see **Figure 12.29**). However, because enzymes are proteins, there is a temperature limit beyond which the enzyme becomes vulnerable to denaturation. Thus, every enzyme-catalyzed reaction has an **optimum temperature** (indicated by the dashed line in **Figure 12.29**). It is usually in the range of 25 °C–40 °C. Above or below that value, the reaction rate will be lower.

optimum temperature The temperature at which enzyme activity is highest.

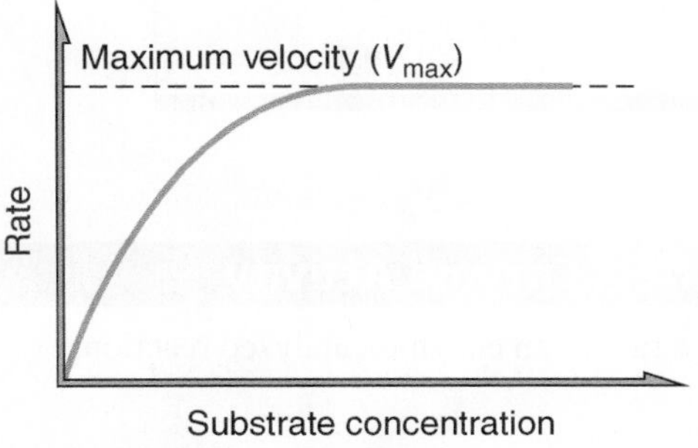

Figure 12.28 The dependence of reaction rate (initial velocity) on substrate concentration.

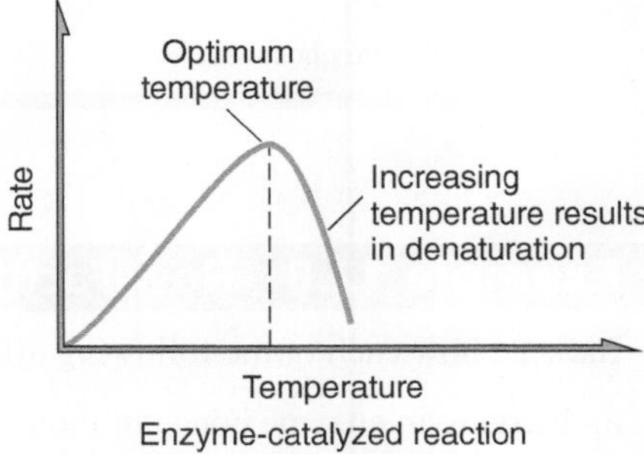

Figure 12.29 The effect of temperature on enzyme-catalyzed reaction rates.

12.7D pH

The graph of enzyme activity as a function of pH is somewhat similar to the behavior as a function of temperature (see **Figure 12.30**). Notice in **Figure 12.30** that an enzyme is most effective in a narrow pH range and is less active at pH values lower or higher than this optimum. This variation in enzyme activity with changing pH may be due to the influence of pH on acidic and basic side chains within the active site. In addition, most enzymes are denatured by pH extremes. Pickled foods (see **Figure 12.31**) resist spoilage because the enzymes of microorganisms are less active under acidic conditions.

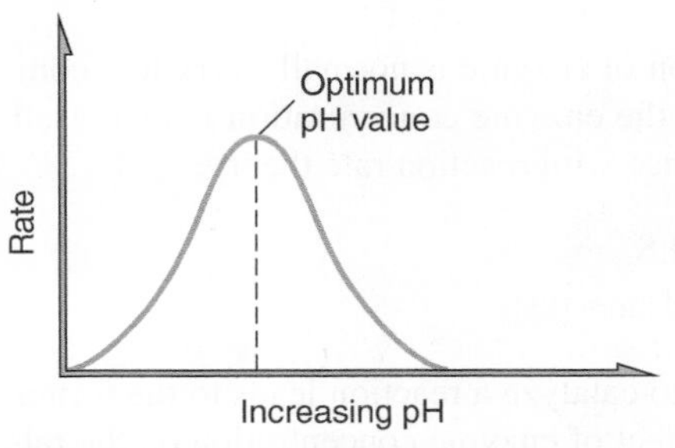

Figure 12.30 A typical plot of the effect of pH on reaction rate.

skyradar/20 images/Pixabay

Figure 12.31 Pickles are prepared under acidic conditions. Why do they resist spoilage?

optimum pH The pH at which enzyme activity is highest.

Many enzymes have an **optimum pH** near 7, the pH of most biological fluids. However, the optimum pH of a few is considerably higher or lower than 7. For example, pepsin, a digestive enzyme of the stomach (**Figure 12.32**), shows maximum activity at a pH of about 1.5, the pH of gastric fluids. **Table 12.7** lists optimum pH values for several enzymes.

Table 12.7 Examples of Optimum pH for Enzyme Activity

Enzyme	Source	Optimum pH
Pepsin	Gastric mucosa	1.5
β-Glucosidase	Almond	4.5
Sucrase	Intestine	6.2
Urease	Soybean	6.8
Catalase	Liver	7.0
Succinate dehydrogenase	Beef heart	7.6
Arginase	Beef liver	9.0
Alkaline phosphatase	Bone	9.5

a

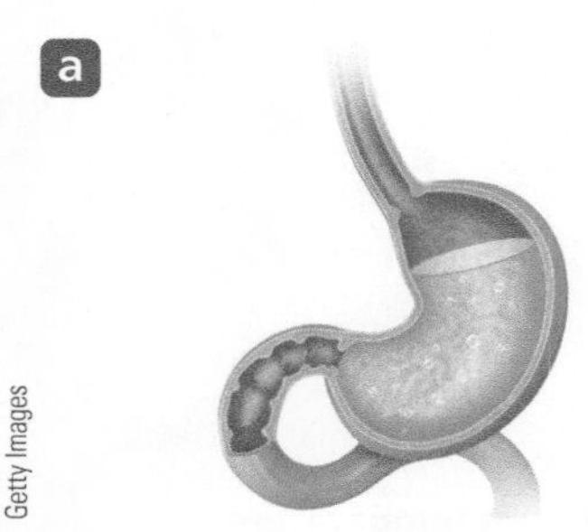

PIXOLOGICSTUDIO/SCIENCE PHOTO LIBRARY/ Getty Images

b

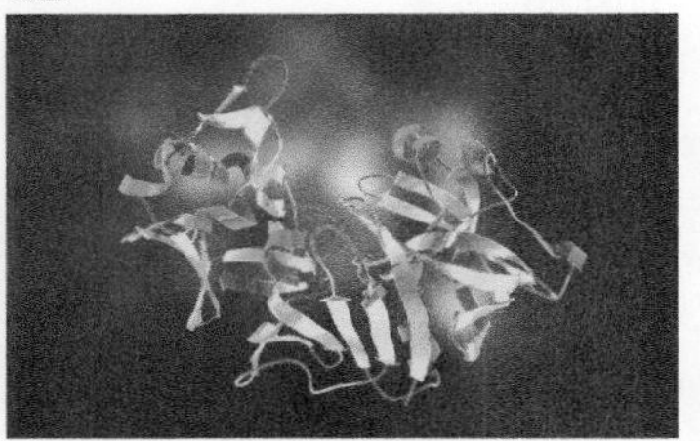

KATERYNA KON/SCIENCE PHOTO LIBRARY/Getty Images

Figure 12.32 (a) The stomach has a pH of around 2.0, significantly lower than blood pH (typically 7.35–7.45; see **Section 7.7**). (b) Enzymes like pepsin thrive in this much more acidic environment.

Example 12.8 How Factors Affect Enzyme Activity

Indicate how each of the following affects the rate of an enzyme-catalyzed reaction.

a. Increase in enzyme concentration
b. Increase in substrate concentration

Solution

a. Enzymes speed up reactions, so an increase in enzyme concentration will increase the rate of the reaction (see **Figure 12.27**).
b. Increasing the concentration of substrate will allow more effective collisions, and thus increase the rate of the reaction until the enzyme is saturated with substrate (reaching maximum velocity; see **Figure 12.28**).

✔ **Learning Check 12.8** Indicate how each of the following affects the rate of an enzyme-catalyzed reaction.

a. Increase in temperature (not exceeding the optimum temperature)
b. Decrease in pH from the optimum pH

Health Connections 12.2

No Milk, Please

Lactose intolerance is a condition that causes a person to be unable to digest lactose, the natural sugar found in milk and dairy products. An estimated 65% to 70% of the world's population live with this condition today.

Lactase is an enzyme in our bodies that catalyzes the breakdown of the disaccharide lactose into the monosaccharides glucose and galactose. When people are lactose intolerant, they do not produce enough lactase enzyme to catalyze the total breakdown of all the lactose that has been ingested.

In order to overcome the inability of the body to create enough lactase enzymes, lactose-intolerant people can take lactase supplements before they eat or drink milk or dairy products. Some milk and dairy products contain the enzyme as an additive. The milk product called "Lactaid" is milk that contains added lactase enzyme to help those who are lactose intolerant.

When lactose-intolerant people drink milk without adding lactase, the lactose disaccharide eventually makes its way into the large intestine and begins to cause pain and other discomfort. The bacteria in the large intestine begin to break down the lactose. When they do this, they also release gas, causing uncomfortable bloating and cramps.

When lactose enters the large intestine, it osmotically draws in water and causes diarrhea. The lactose also undergoes fermentation in the large intestine and produces lactic acid, which can be detected in the stool. That is one way to test for lactose intolerance in a person. Lactose intolerance is often passed on genetically, and unfortunately, there is no known cure for the condition.

Prostock-studio/Shutterstock.com

Drinking milk can be painful for individuals who are lactose intolerant.

12.8 Enzyme Inhibition and Regulation

Learning Objective 12 Compare the mechanisms of competitive and noncompetitive enzyme inhibition and allosteric regulation.

An **enzyme inhibitor** is any substance that can decrease the rate of an enzyme-catalyzed reaction. An understanding of enzyme inhibition is important for several reasons. First, the characteristic function of many poisons and some medicines is to inhibit one or more enzymes and to decrease the rates of the reactions they catalyze. Second, some substances normally found in cells inhibit specific enzyme-catalyzed reactions and thereby provide a means for the internal regulation of cellular metabolism. Enzyme inhibitors are classified as reversible or irreversible, depending on how they behave at the molecular level.

enzyme inhibitor A substance that decreases the activity of an enzyme.

12.8A Irreversible Inhibition

An irreversible inhibitor forms a covalent bond with a specific functional group of the enzyme and as a result renders the enzyme inactive. A number of very deadly poisons act as irreversible inhibitors. The cyanide ion (CN^-) is an irreversible enzyme inhibitor. It is extremely toxic and acts very rapidly. The cyanide ion binds the iron cofactor of cytochrome oxidase, forming a stable complex, and prevents it from functioning properly. As a result, cell respiration stops, causing death in a matter of minutes (see **Figure 12.33**).

$$\underset{\text{cytochrome oxidase}}{\text{Cyt—Fe}^{3+}} + \underset{\text{cyanide ion}}{\text{CN}^-} \rightarrow \underset{\text{stable complex}}{\text{Cyt—Fe—CN}^{2+}}$$

ExplorerBob/287 images/Pixabay

Figure 12.33 The seeds (also known as stones or pits) of fruits like apricots, cherries, plums, and peaches contain small amounts of a compound that releases hydrogen cyanide when ingested.

Any antidote for cyanide poisoning must be administered quickly. One antidote is sodium thiosulfate. This substance converts the cyanide ion to a thiocyanate ion that does not bind to the iron of cytochrome oxidase.

$$\underset{\text{cyanide}}{CN^-} + \underset{\text{thiosulfate}}{S_2O_3^{2-}} \longrightarrow \underset{\text{thiocyanate}}{SCN^-} + \underset{\text{sulfite}}{SO_3^{2-}}$$

The toxicity of heavy metals such as mercury and lead is due to their ability to render the protein part of enzymes ineffective. These metals act by combining with the –SH groups found on many enzymes:

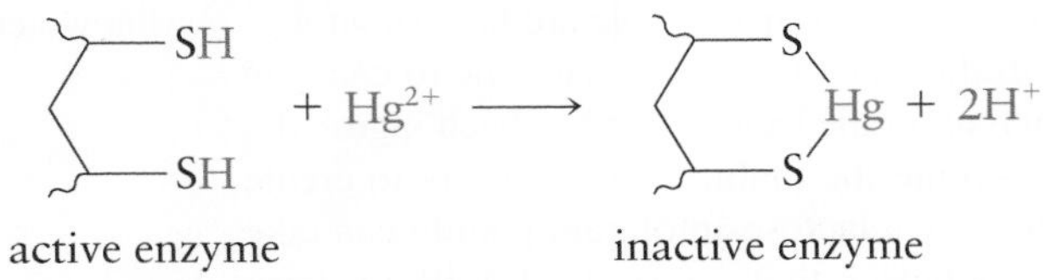

$$\underset{\text{active enzyme}}{\text{Enz}(SH)_2} + Hg^{2+} \longrightarrow \underset{\text{inactive enzyme}}{\text{Enz}(S_2Hg)} + 2H^+$$

Both mercury and lead poisoning can cause permanent neurological damage if untreated. Heavy-metal poisoning is treated by administering chelating agents—substances that combine with the metal ions and hold them very tightly. Ethylenediaminetetraacetic acid (EDTA) effectively chelates all heavy metals except mercury. The calcium salt of EDTA is administered intravenously. In the body, calcium ions of the salt are displaced by heavy-metal ions, such as lead, that bind to the chelate more tightly. The lead–EDTA complex is soluble in body fluids and is excreted in the urine.

$$CaEDTA^{2-} + Pb^{2+} \longrightarrow PbEDTA^{2-} + Ca^{2+}$$

antibiotic A substance produced by one microorganism that kills or inhibits the growth of other microorganisms.

Not all enzyme inhibitors act as poisons toward the body; some, in fact, are useful therapeutic agents. Sulfa drugs and the group of compounds known as penicillins are two well-known families of **antibiotics** that inhibit specific enzymes essential to the life processes of bacteria. The first penicillin was discovered in 1928 by Alexander Fleming (1881–1955). He noticed that the bacteria on a culture plate did not grow in the vicinity of a mold that had contaminated the culture; the mold was producing penicillin. The structure of ampicillin, the most common penicillin used, is shown in **Figure 12.34**. Penicillins interfere with transpeptidase, an enzyme that is important in bacterial cell wall construction. The inability to form strong cell walls prevents the bacteria from surviving.

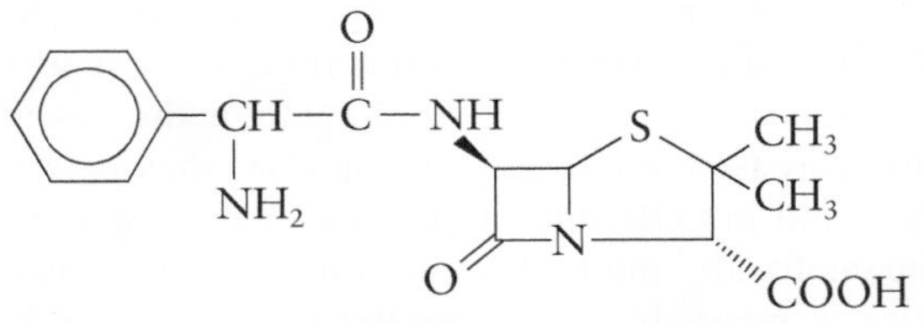

Figure 12.34 Ampicillin is a specific penicillin. Penicillins are used to treat certain infections caused by bacteria such as pneumonia; strep throat; scarlet fever; and ear, skin, and mouth infections.

ampicillin

12.8B Reversible Inhibition

A reversible inhibitor (in contrast to one that is irreversible) reversibly binds to an enzyme. An equilibrium is established, so the inhibitor (I) can be removed from the enzyme by shifting the equilibrium:

$$E + I \rightleftarrows EI$$

Reversible inhibitors can be competitive or noncompetitive.

competitive inhibitor An inhibitor that competes with substrate for binding at the active site of the enzyme.

A **competitive inhibitor** binds to the active site of an enzyme and thus "competes" with substrate molecules for the active site. Competitive inhibitors often have molecular structures that are similar to the normal substrate of the enzyme. The nature of competitive inhibition is represented in **Figure 12.35**. There is competition between the substrate and

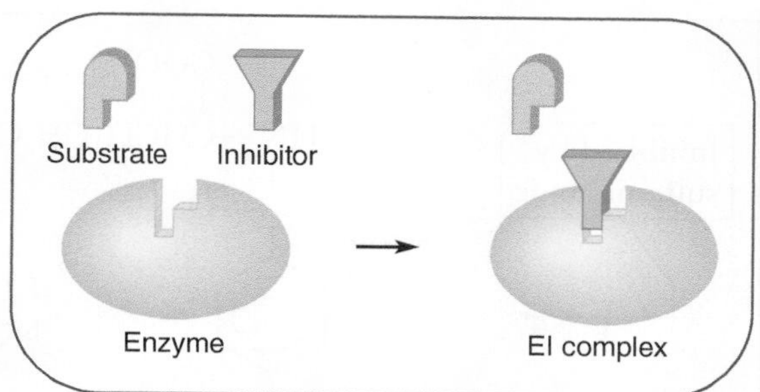

Figure 12.35 The behavior of competitive inhibitors. When the active site is occupied by the inhibitor (I), the substrate is prevented from binding.

the inhibitor for the active site. Once the inhibitor combines with the enzyme, the active site is blocked, preventing further catalytic action.

Competitive inhibition can be reversed by increasing the concentration of substrate. If enough substrate is added, it eventually displaces the bound inhibitor. Thus, the competition between inhibitor and substrate is won by whichever molecular species is in greater concentration.

The competitive inhibition of succinate dehydrogenase by malonate is a classic example. Succinate dehydrogenase catalyzes the oxidation of the substrate succinate to form fumarate by transferring two hydrogens to the coenzyme FAD:

$$\begin{array}{c} COO^- \\ | \\ CH_2 \\ | \\ CH_2 \\ | \\ COO^- \\ \text{succinate} \end{array} + FAD \xrightarrow{\text{Succinate dehydrogenase}} \begin{array}{c} COO^- \\ | \\ CH \\ \| \\ HC \\ | \\ COO^- \\ \text{fumarate} \end{array} + FADH_2$$

Malonate, which has a structure similar to succinate, competes for the active site of succinate dehydrogenase and thus inhibits the enzyme:

$$\begin{array}{c} COO^- \\ | \\ CH_2 \\ | \\ CH_2 \\ | \\ COO^- \\ \text{succinate} \end{array} \qquad \begin{array}{c} COO^- \\ | \\ CH_2 \\ | \\ COO^- \\ \\ \\ \text{malonate} \end{array}$$

The action of sulfa drugs, such as sulfanilamide (**Figure 12.36**), on bacteria is another example of competitive enzyme inhibition. Sulfanilamide resembles the compound *p*-aminobenzoic acid (**Figure 12.37**), a chemical that is essential for folic acid production and bacterial growth. Sulfanilamide competes with *p*-aminobenzoic acid for binding to the active site of the bacterial enzyme involved and prevents it from functioning properly. Because the enzyme binds readily to either of these molecules, the introduction of large quantities of sulfanilamide into a patient's body causes most of the active sites to be bound to the inhibitor rather than the *p*-aminobenzoic acid substrate. Thus, the synthesis of folic acid is stopped or slowed, and the bacteria are prevented from multiplying. Meanwhile, the normal body defenses work to destroy them.

Human beings also require folic acid, but they get it from their diet. Consequently, sulfa drugs exert no toxic effect on humans. However, one danger of sulfa drugs, or of

Food Collection/Alamy Stock Photo

Figure 12.36 The improper use of sulfanilamide elixir (sulfanilamide dissolved in the organic solvent diethylene glycol) poisoned more than 100 people in the late 1930s and led to new oversight powers of the Food and Drug Administration (FDA).

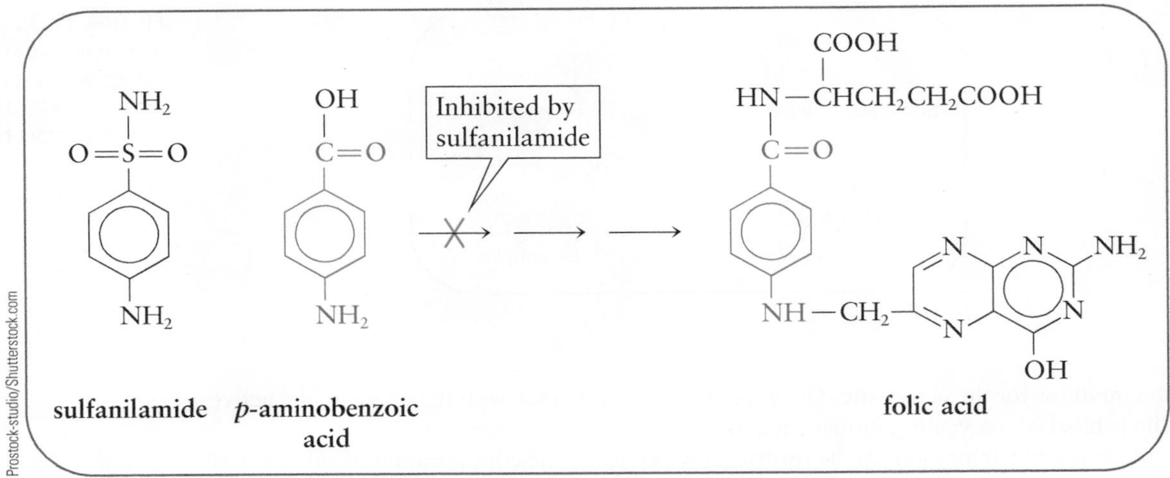

Figure 12.37 Structural relationships between sulfanilamide, *p*-aminobenzoic acid, and folic acid.

any antibiotic, is that excessive amounts can destroy important intestinal bacteria that perform many symbiotic, life-sustaining functions. Although sulfa drugs have largely been replaced by antibiotics such as the penicillins and tetracyclines, some are still used to treat certain bacterial infections, such as those occurring in the urinary tract.

noncompetitive inhibitor An inhibitor that binds to the enzyme at a location other than the active site.

A **noncompetitive inhibitor** bears no resemblance to the normal enzyme substrate and binds reversibly to the surface of an enzyme at a site other than the catalytically active site (see **Figure 12.38**). The interaction between the enzyme and the noncompetitive inhibitor causes the three-dimensional shape of the enzyme and its active site to change. The enzyme is no longer able to bind the normal substrate, or the substrate is improperly bound in a way that prevents the catalytic groups of the active site from participating in catalyzing the reaction.

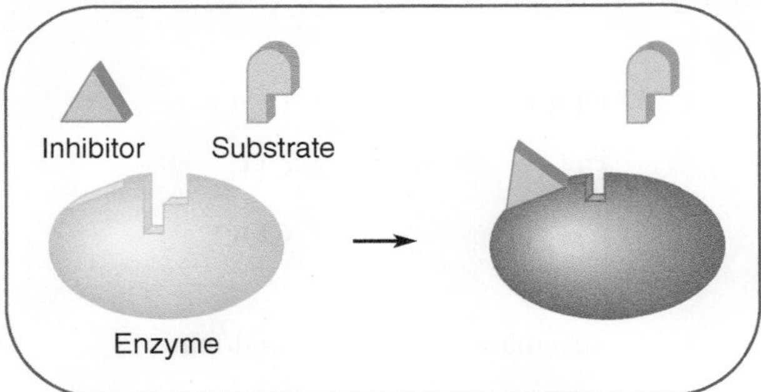

Figure 12.38 The behavior of noncompetitive inhibitors. The inhibitor binds to the enzyme at a site other than the active site and changes the shape of the active site.

Unlike competitive inhibition, noncompetitive inhibition cannot be reversed by the addition of more substrate because additional substrate has no effect on the enzyme-bound inhibitor—that is, it can't displace the inhibitor because it can't bond to the site occupied by the inhibitor.

An excellent example of noncompetitive inhibition is the five-step synthesis of the amino acid isoleucine (see **Figure 12.39**). Threonine deaminase, the enzyme that catalyzes the first step in the conversion of threonine to isoleucine, is subject to inhibition by the final product, isoleucine.

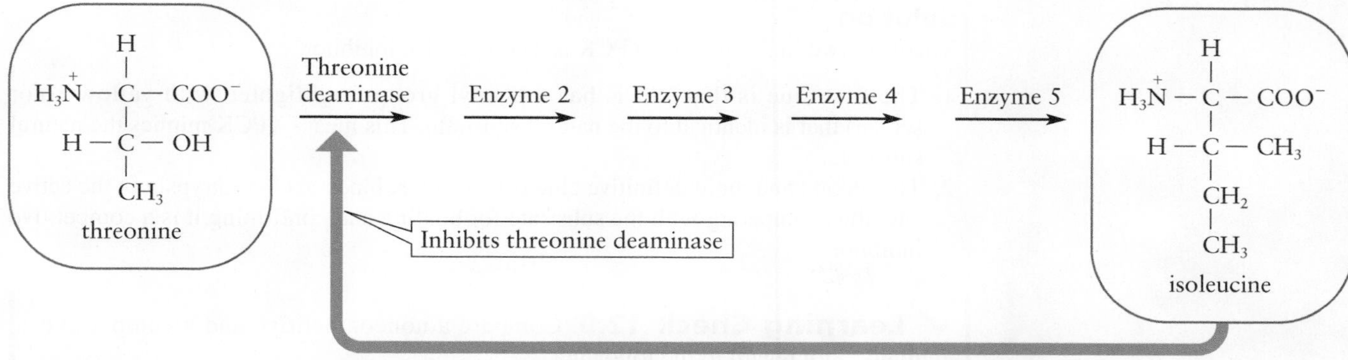

Figure 12.39 The allosteric regulation of threonine deaminase by isoleucine, an example of feedback inhibition.

The structures of isoleucine and threonine are quite different, so isoleucine is not a competitive inhibitor. Also, the site to which isoleucine binds to the enzyme is different from the enzyme active site that binds to threonine. This second site, called the allosteric site, specifically recognizes isoleucine, whose presence there induces a change in the conformation of the enzyme such that threonine binds poorly to the active site. Thus, isoleucine exerts an inhibiting effect on the enzyme activity. As a result, the reaction slows as the concentration of isoleucine increases, and no excess isoleucine is produced. When the concentration of isoleucine falls to a low enough level, the enzyme becomes more active, and more isoleucine is synthesized.

This type of **allosteric regulation** in which the enzyme that catalyzes the first step of a series of reactions is inhibited by the final product is called **feedback inhibition**. The control of enzyme activity by allosteric regulation of key enzymes is of immense benefit to an organism because it allows the concentration of cellular products to be maintained within very narrow limits.

allosteric regulation When a substance binds to an enzyme at a location other than the active site and alters the catalytic activity.

feedback inhibition A process in which the end product of a sequence of enzyme-catalyzed reactions inhibits an earlier step in the process.

Example 12.9 Classifying Enzyme Inhibitors

Chymotrypsin is an enzyme that helps digest proteins. The chymotrypsin inhibitor tosyl phenylalanyl chloromethyl ketone (TPCK) is shown below side by side with chymotrypsin's proper substrate. Given that the inhibitor permanently covalently bonds to a histidine residue in the enzyme's active site, TPCK can be classified as an irreversible inhibitor. Further classify TPCK as a competitive or noncompetitive inhibitor. Provide your reasoning for your decision.

Natural substrate for chymotrypsin

tosyl-L-phenylalanine chloromethyl ketone (TPCK)

Solution

There are two major reasons TPCK is a competitive inhibitor.

1. The first clue is that TPCK has a phenyl group (highlighted with yellow color screen) that is identical to the natural substrate. This means TPCK mimics the natural substrate.
2. The second and more definitive clue is that TPCK binds to chymotrypsin in the active site, thus competing with the substrate for binding and confirming it is a competitive inhibitor.

✔ **Learning Check 12.9** Compare a noncompetitive and a competitive inhibitor with regard to the following.

a. Resemblance to substrate
b. Binding site on the enzyme
c. The effect of increasing substrate concentration

12.9 Enzymes in Medical Applications and Disease

Learning Objective 13 Name three enzymes that can be used in diagnostic testing to identify the presence of disease.

Certain enzymes are normally found almost exclusively inside tissue cells and are released into the blood and other biological fluids only when these cells are damaged or destroyed. Because of the normal breakdown and replacement of tissue cells that go on constantly,

Study Tools 12.1

A Summary Chart of Enzyme Inhibitors

The following summary chart shows how the various types of enzyme inhibitors are related.

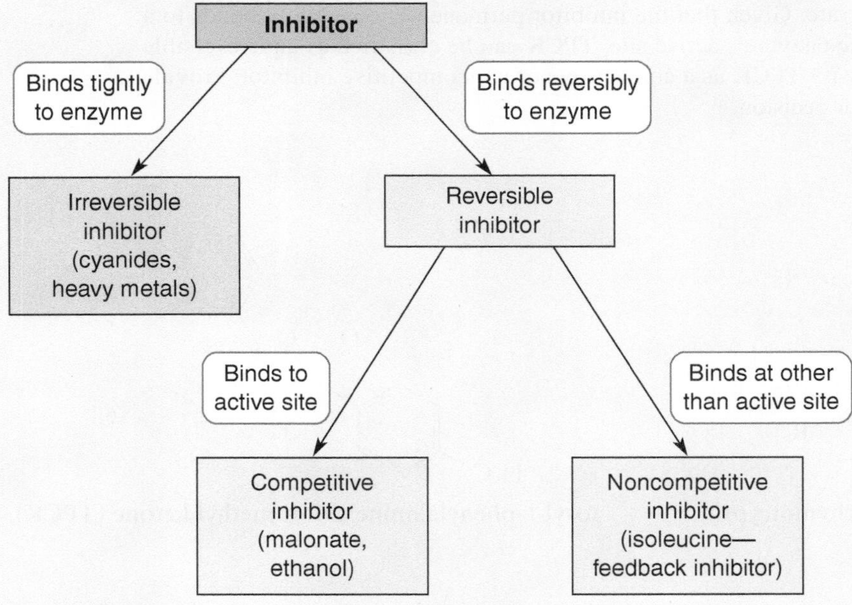

the blood serum contains these enzymes but at very low concentrations. However, blood serum levels of cellular enzymes increase significantly when excessive cell injury or destruction occurs or when cells grow rapidly as a result of cancer. Changes in blood serum concentrations of specific enzymes can be used clinically to detect cell damage or uncontrolled growth, and even to suggest the site of the damage or cancer. Also, the extent of cell damage can often be estimated by the magnitude of the serum concentration increase above normal levels. For these reasons, the measurement of enzyme concentrations in blood serum and other biological fluids has become a major diagnostic tool, particularly in diagnosing diseases of the heart (see **Figure 12.40**), liver, pancreas, prostate, and bones. In fact, certain enzyme determinations are performed so often that they have become routine procedures in the clinical chemistry laboratory. **Table 12.8** lists some enzymes used in medical diagnosis.

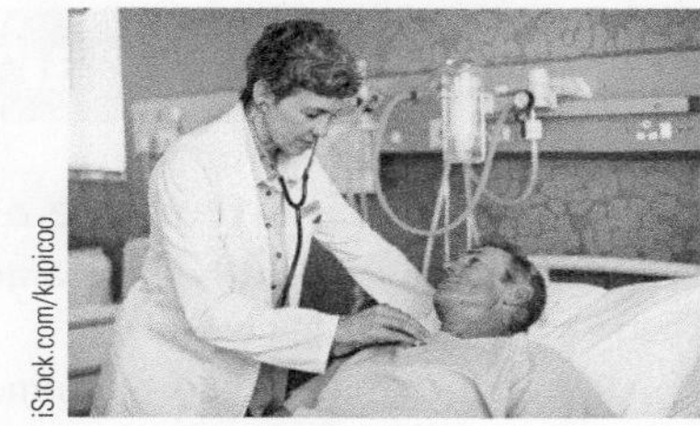

Figure 12.40 Enzyme assays are extremely important in the diagnosis of heart attacks.

Table 12.8 Diagnostically Useful Assays of Blood Serum Enzymes

Enzyme	Pathological Condition
Acid phosphatase	Prostate cancer
Alanine transaminase (ALT)	Hepatitis
Alkaline phosphatase (ALP)	Liver or bone disease
Amylase	Diseases of the pancreas
Aspartate transaminase (AST)	Heart attack or hepatitis
Creatine kinase (CK)	Heart attack
Lactate dehydrogenase (LDH)	Heart attack, liver damage
Lipase	Acute pancreatitis
Lysozyme	Monocytic leukemia

Health Career Description

Optician

Opticians assist clients with eyeglasses, contact lenses, or other visual aids prescribed by ophthalmologists or optometrists. They work in either stand-alone clinics or in in-store clinics that sell optical products. Opticians need at least a high school diploma, but many enter the field with an associate's degree. Many states require a license to work as an optician.

According to the Bureau of Labor Statistics, "Employment of opticians is projected to grow 7 percent from 2018 to 2028, faster than the average for all occupations. An aging population and increasing rates of chronic disease are expected to lead to greater demand for corrective eyewear." The average annual income for an optician is listed as $37,840.

Opticians may be called upon to offer opinions as to how certain frames look on a client and whether they are an attractive choice. Tact, kindness, and attention to detail are important skills for the optician to possess. Ability to discern minute details and a strong understanding of fashion trends is helpful in this line of work.

Opticians work with clients of all ages, from infants to the elderly. Some eyewear is for special circumstances such as sports, computer work, reading magnification, or protection in a hazardous job. Technicians must understand specific uses for different eyewear and be able to help clients find the product that will best meet their needs. Opticians must take precise measurements of such things as interpupillary distance and determine whether bifocal or trifocal lenses are sitting at the appropriate level on the individual's face. Work as an optician demands attention to detail and precision. This is a career that is a good fit for someone who enjoys helping people.

Sources: Bergenske, P. (2002). Protein Lens Care Update Removal without Enzymes. Retrieved Apr. 26, 2020 from https://www.reviewofoptometry.com/article/lens-care-update-protein-removal-without-enzymes; Bureau of Labor Statistics. (n.d.) Opticians. Retrieved Apr. 26, from https://www.bls.gov/ooh/healthcare/opticians-dispensing.htm; Esposito, L. (2018, Sept.14).U.S News and World Report: When Contact Lenses Cause Vision Problems. Retrieved Apr. 26, 2020 from https://health.usnews.com/health-care/patient-advice/articles/2018-09-14/when-contact-lenses-cause-vision-problems.

Concept Summary

12.1 Amino Acid Structure and Classification

Learning Objective: **Identify the characteristic parts and stereochemistry of alpha-amino acids.**

- All proteins are polymers of amino acids, which are bifunctional organic compounds that contain both an amino group and a carboxyl group.
- Differences in the R groups of amino acids result in differences in the properties of amino acids and proteins.
- The acidic and basic properties of proteins are determined by the acidic or basic character of the R groups of the amino acids constituting the protein.
- Protein side chains can be neutral and nonpolar, neutral and polar, basic, or acidic.

12.2 Reactions of Amino Acids

Learning Objectives: **Draw structural formulas to illustrate the various ionic forms assumed by amino acids. Write reactions to represent the formation of peptides and the oxidation of cysteine.**

- The presence of both amino groups and carboxyl groups in amino acids makes it possible for amino acids to exist in several ionic forms, including the form of a zwitterion.
- The zwitterion is a dipolar form in which the net charge on the ion is zero.
- Amino acids can undergo reactions characteristic of any functional group in the molecule.
- Two important reactions are:
  - The reaction of two cysteine molecules to form a disulfide
  - The reaction of amino groups and carboxylate groups of different molecules to form peptide (amide) linkages

12.3 Protein Structure

Learning Objectives: **Explain what is meant by the primary structure of proteins. Describe the role of hydrogen-bonding in the secondary structure of proteins. Describe the role of side-chain interactions in the tertiary structure of proteins. Explain what is meant by the quaternary structure of proteins.**

- The primary structure of a protein is the sequence of amino acids in the polymeric chain.
- This gives all proteins an identical backbone of carbon and nitrogen atoms held together by peptide linkages.
- The difference in proteins is the sequence of R groups attached to the backbone.
- Protein chains are held in characteristic shapes called secondary structures by hydrogen bonds.
- Two specific structures that have been identified are the α-helix and the β-pleated sheet.
- Tertiary protein structure results from interactions between the R groups of protein chains.
- These interactions include disulfide bridges, salt bridges, hydrogen bonds, and hydrophobic attractions.
- Some functional proteins consist of two or more polypeptide chains held together by forces such as ionic attractions, disulfide bridges, hydrogen bonds, and hydrophobic forces.
- The arrangement of these polypeptides to form the functional protein is called the quaternary structure of the protein.

12.4 Protein Functions

Learning Objective: **Describe four biological functions of proteins.**

- Proteins have many different biological functions.
- Proteins catalyze chemical reactions, provide storage and protection, regulate biochemical processes, and more.
- Some proteins are also responsible for the movement of biological systems, while others transport various chemical species.

12.5 Classes of Enzymes

Learning Objective: **Describe three reactions that enzymes catalyze.**

- Enzymes are grouped into six major classes—namely, oxidoreductases, transferases, hydrolases, lyases, isomerases, and ligases—on the basis of the type of reaction catalyzed.

12.6 Enzyme Activity

Learning Objective: **Describe enzyme activity in terms of efficacy and mechanism of action.**

- The catalytic ability of enzymes is described by turnover number and enzyme international units.
- Experiments that measure enzyme activity are referred to as enzyme assays.
- The behavior of enzymes is explained by a theory in which the formation of an enzyme–substrate complex is assumed to occur.
- The specificity of enzymes is explained by the lock-and-key theory and the induced-fit theory.

12.7 Factors Affecting Enzyme Activity

Learning Objective: **Identify four factors that affect enzyme activity.**

- The catalytic activity of enzymes is influenced by numerous factors.
- The most important are substrate concentration, enzyme concentration, temperature, and pH.

12.8 Enzyme Inhibition and Regulation

Learning Objective: **Compare the mechanisms of competitive and noncompetitive enzyme inhibition and allosteric regulation.**

- Chemical substances called inhibitors decrease the rates of enzyme-catalyzed reactions.
- Irreversible inhibitors render enzymes permanently inactive and include several very toxic substances, such as the cyanide ion and heavy-metal ions.
- Reversible inhibitors are of two types: competitive and noncompetitive.

12.9 Enzymes in Medical Applications and Disease

Learning Objective: **Name three enzymes that can be used in diagnostic testing to identify the presence of disease.**

- Numerous enzymes have become useful as aids in diagnostic medicine.
- The presence of specific enzymes in body fluids, such as blood, has been related to certain pathological conditions.

Key Terms and Concepts

Absolute specificity (12.5)
Active site (12.6)
α-helix (12.3)
Allosteric regulation (12.8)
Alpha (α) amino acid (12.1)
Amino acid residue (12.2)
Antibiotic (12.8)
Antibody (12.4)
Apoenzyme (12.6)
β-pleated sheet (12.3)
Coenzyme (12.6)
Cofactor (12.6)
Competitive inhibitor (12.8)
C-terminal residue (12.2)
Dipeptide (12.2)
Enzyme (12.5)
Enzyme activity (12.6)
Enzyme inhibitor (12.8)
Enzyme international unit (IU) (12.6)
Feedback inhibition (12.8)
Fibrous protein (12.3)
Globular protein (12.3)
Induced-fit theory (12.6)
Isoelectric point (12.2)
Lock-and-key theory (12.6)
Noncompetitive inhibitor (12.8)
N-terminal residue (12.2)
Optimum pH (12.7)
Optimum temperature (12.7)
Peptide (12.2)
Peptide linkage (peptide bond) (12.2)
Polypeptide (12.2)
Primary protein structure (12.3)
Protein (12.2)
Quaternary protein structure (12.3)
Relative specificity (12.5)
Secondary protein structure (12.3)
Substrate (12.5)
Subunit (12.3)
Tertiary protein structure (12.3)
Turnover number (12.6)
Zwitterion (12.2)

Key Equations

1. Formation of a zwitterion (**Section 12.2**)	$H_2N-CH(R)-C(=O)-OH \rightarrow H_3\overset{+}{N}-CH(R)-C(=O)-O^-$	Equation 12.1
2. Conversion of a zwitterion to a cation in an acidic solution (**Section 12.2**)	$H_3\overset{+}{N}-CH(R)-C(=O)-O^- + H_3O^+ \rightarrow H_3\overset{+}{N}-CH(R)-C(=O)-OH + H_2O$	Equation 12.2
3. Conversion of a zwitterion to an anion in a basic solution (**Section 12.2**)	$H_3\overset{+}{N}-CH(R)-C(=O)-O^- + OH^- \rightarrow H_2N-CH(R)-C(=O)-O^- + H_2O$	Equation 12.3
4. Dehydration of a carboxylic acid and an amine to form a amide (peptide) bond (**Section 12.2**)	$R-C(=O)-OH + R'-NH_2 \xrightarrow{\text{Heat}} R-C(=O)-NH-R' + H_2O$ A carboxylic acid; An amine; An amide (*Amide linkage* indicated at C–NH bond)	Equation 12.4
5. Oxidation of cysteine to cystine (**Section 12.2**)	$2\ H_3\overset{+}{N}-CH(CH_2-SH)-C(=O)-O^- \underset{\text{Reduction}}{\overset{\text{Oxidation}}{\rightleftarrows}} H_3\overset{+}{N}-CH(C(=O)O^-)-CH_2-S-S-CH_2-CH(C(=O)O^-)-\overset{+}{N}H_3 + H_2O$	Equation 12.5
6. Mechanism of enzyme action (**Section 12.6**)	$E + S \rightleftarrows ES \rightarrow E + P$	Equation 12.6
7. Formation of an active enzyme (**Section 12.6**)	apoenzyme + cofactor (coenzyme or inorganic ion) $\rightarrow$ active enzyme	Equation 12.7

Exercises

Even-numbered exercises are answered in Appendix B.

Exercises with an asterisk (*) are more challenging.

Amino Acid Structure and Classification (Section 12.1)

12.1 What functional groups are found in all amino acids?

12.2 Draw structural formulas for the following amino acids, identify the chiral carbon atom in each one, and circle the four different groups attached to the chiral carbon.

a. threonine

b. aspartate

c. serine

d. phenylalanine

12.3 Isoleucine contains two chiral carbon atoms. Draw the structural formula for isoleucine twice. In the first, identify one of the chiral carbons and circle the four groups attached to it. In the second, identify the other chiral carbon and circle the four groups attached to it.

12.4 Classify each of the amino acids in Exercise 12.2 as neutral and nonpolar, neutral and polar, basic and polar, or acidic and polar.

12.5* Draw Fischer projections representing the D and L forms of the following.

a. aspartate

b. phenylalanine

Reactions of Amino Acids (Section 12.2)

12.6 What is meant by the term *zwitterion*?

12.7 Draw the zwitterion structure for each of the following amino acids.

a. phenylalanine

b. glycine

c. leucine

12.8 Draw the zwitterion structure for each of the following amino acids.

a. methionine

b. threonine

c. cysteine

12.9 What is meant by the term *isoelectric point*?

12.10* Write structural formulas to show the form the following amino acids would have in a solution with a pH higher than the amino acid isoelectric point.

a. cysteine

b. alanine

12.11 Write structural formulas to show the form the following amino acids would have in a solution with a pH higher than the amino acid isoelectric point.

a. glycine

b. isoleucine

12.12* Write structural formulas to show the form the following amino acids would have in a solution with a pH lower than the amino acid isoelectric point.

a. valine

b. threonine

12.13 Write structural formulas to show the form the following amino acids would have in a solution with a pH lower than the amino acid isoelectric point.

a. methionine

b. serine

12.14 Write structural formulas to represent the formation of the two dipeptides that form when leucine and threonine react.

12.15 Write structural formulas to represent the formation of the two dipeptides that form when asparagine and glutamine react.

12.16 Write a complete structural formula and an abbreviated formula for the tripeptide formed from aspartate, cysteine, and valine in which the C-terminal residue is cysteine and the N-terminal residue is valine.

12.17 Identify the amino acids contained in the following tripeptide.

$$H_3\overset{+}{N}-CH_2-\overset{O}{\overset{\|}{C}}-\overset{H}{\overset{|}{N}}-\underset{CH_2-OH}{\underset{|}{CH}}-\overset{O}{\overset{\|}{C}}-\overset{H}{\overset{|}{N}}-\underset{CH_3}{\underset{|}{\underset{CH-OH}{\underset{|}{CH}}}}-COO^-$$

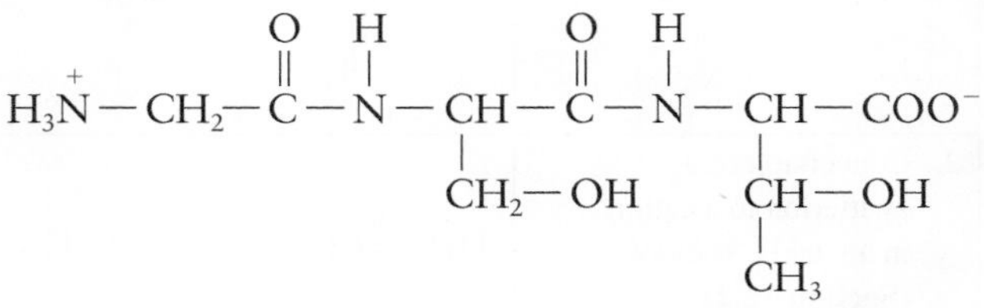

12.18 Identify the amino acids contained in the following tripeptide.

$$H_3\overset{+}{N}-\underset{CH_3}{\underset{|}{CH}}-\overset{O}{\overset{\|}{C}}-\overset{H}{\overset{|}{N}}-\underset{CH_2-SH}{\underset{|}{CH}}-\overset{O}{\overset{\|}{C}}-\overset{H}{\overset{|}{N}}-\underset{CH_3}{\underset{|}{\underset{CH-CH_3}{\underset{|}{CH}}}}-COO^-$$

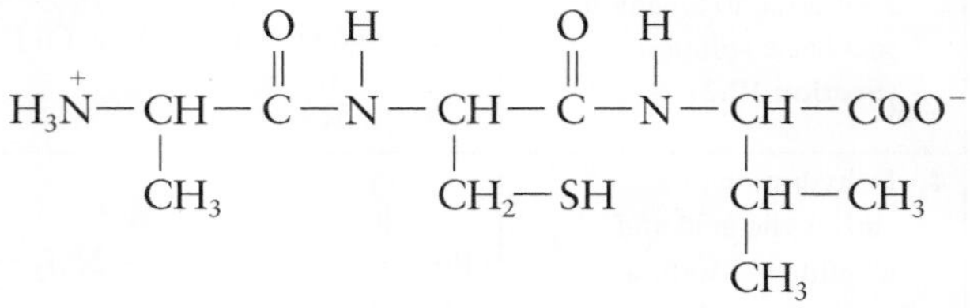

12.19 Draw the tripeptide in Exercise 12.17 and label the N-terminal and C-terminal residues. Then circle and label any peptide bonds.

12.20 Draw the tripeptide in Exercise 12.18 and label the N-terminal and C-terminal residues. Then circle and label any peptide bonds.

Protein Structure (Section 12.3)

12.21 Describe what is meant by the term *primary structure of proteins*.

12.22 Explain why two polypeptides with the same amino acid composition perform different biochemical functions.

12.23 What type of bonding is present to account for the primary structure of proteins?

12.24 Write the structure for a protein backbone. Make the backbone long enough to attach four R groups symbolizing amino acid side chains.

12.25 Describe the differences between alpha and beta secondary structures of proteins.

12.26 Explain why two proteins that have different primary structures can both form alpha or beta secondary structures.

12.27 What type of bonding between amino acid residues is most important in holding a protein or polypeptide in a specific secondary configuration?

12.28 Which amino acids have side-chain groups that can form salt bridges?

12.29 How do hydrogen bonds involved in tertiary protein structures differ from those involved in secondary structures?

12.30 Refer to **Table 12.1** and identify which of the following amino acid pairs will *not* form R-group interactions necessary for the tertiary structure of a protein.

a. lysine and glutamate

b. tryptophan and glutamine

c. histidine and proline

d. cysteine and methionine

12.31* Refer to **Table 12.1** and list the type of side-chain interaction expected between the side chains of the following pairs of amino acid residues.

a. tyrosine and glutamine

b. aspartate and lysine

c. leucine and isoleucine

d. phenylalanine and valine

12.32 What is meant by the term *quaternary protein structure*?

12.33 What types of forces give rise to quaternary structure?

12.34 Describe the quaternary protein structure of hemoglobin.

12.35 What is meant by the term *subunit*?

Protein Functions (Section 12.4)

12.36 What is the role of enzymes in the body?

12.37 Classify each of the following proteins into one of the eight functional categories of proteins.

a. insulin

b. rhodopsin

c. antibodies

12.38 Classify each of the following proteins into one of the eight functional categories of proteins.

a. collagen

b. hemoglobin

c. cytochrome c

12.39 Classify each of the following proteins into one of the eight functional categories of proteins.

a. myosin

b. serum albumin

c. thyrotropin

Classes of Enzymes (Section 12.5)

12.40 Refer to **Table 12.4**. Identify the enzyme classification based on the reaction each protein catalyzes.

a. Hydrolyzes ether linkages

b. Transfers –OH groups between molecules

c. Dehydrates alcohols, forming carbon–carbon double bonds

d. Uses ATP to join two molecules

12.41 Refer to **Table 12.4**. Identify the enzyme classification based on the reaction each protein catalyzes.

a. Converts an L-isomer to a D-isomer

b. Oxidizes nitrite

c. Hydrolyzes glycosidic linkages

d. Transfers amine groups from urea to other molecules

12.42 Match the following general enzyme names and reactions catalyzed.

Enzyme	Reaction Catalyzed
a. Decarboxylase	Formation of ester linkages
b. Phosphatase	Removal of carboxyl groups from compounds
c. Peptidase	Hydrolysis of peptide linkages
d. Esterase	Hydrolysis of phosphate ester linkages

12.43 Match the following general enzyme names and reactions catalyzed.

Enzyme	Reaction Catalyzed
a. Dehydrolase	Adds oxygen to molecules
b. Dehydrogenase	Adds carboxyl groups to molecules
c. Oxygenase	Removes water molecules
d. Carboxylase	Removes hydrogens from molecules

Enzyme Activity (Section 12.6)

12.44 List two ways that enzyme catalysis of a reaction is superior to normal laboratory conditions.

12.45 What is the relationship between an enzyme and the energy of activation for a reaction?

12.46 Why are so many different enzymes needed?

12.47 Define what is meant by the term *enzyme specificity*.

12.48 What observations may be used in experiments to determine enzyme activity?

12.49 What is an enzyme international unit? Why is the international unit a useful method of expressing enzyme activity in medical diagnoses?

12.50 Explain what is meant by the following equation.

$$E + S \rightleftarrows ES \longrightarrow E + P$$

12.51 In what way are the substrate and active site of an enzyme related?

12.52 Match the following substrates with the groups found in different enzyme active sites that would allow an enzyme–substrate complex to form.

Substrate	Enzyme Active Site
a. $^{-}OOC—CH_2—OH$	Fe^{2+}
b. $CH_3CH_2—COO^-$	
c. $CH_3CH_2CH_2CH_2CH_3$	HO–

12.53 Match the following substrates with the groups found in different enzyme active sites that would allow an enzyme–substrate complex to form.

Substrate	Enzyme Active Site
a. $^{-}OOC—CH_2—NH_3^+$	
b. $CH_3CH_2—NH_2$	$-COO^-$
c. $CH_3CH=CHCH_3$	HO–

12.54 Explain why the forces that hold a substrate in an enzyme active site are not covalent bonds.

12.55 Compare the lock-and-key theory with the induced-fit theory.

Factors Affecting Enzyme Activity (Section 12.7)

12.56 What happens to the rate of an enzyme-catalyzed reaction as substrate concentration is raised beyond the saturation point?

12.57 How would you expect hypothermia to affect enzyme activity in the body?

12.58 A lab tech is determining the V_{max} for an enzyme by periodically adding more substrate to the assay over time. Instead of adding substrate to the assay, one time he mistakenly added enzyme. Explain what will happen to the V_{max} value.

12.59 Explain why temperature can affect enzyme reaction rates in two ways: An increase in temperature can increase the rate of the reaction, or it can slow down and stop the reaction.

12.60 Explain why all enzymes do not have the same optimum pH.

12.61 When handling or storing solutions of enzymes, the pH is usually kept near 7.0. Explain why.

Enzyme Inhibition and Regulation (Section 12.8)

12.62 Distinguish between irreversible and reversible enzyme inhibition.

12.63 Distinguish between competitive and noncompetitive enzyme inhibition.

12.64 An enzyme assay had a competitive inhibitor added to it at point A on the graph below. Complete the graph, illustrating what would happen to enzyme activity if more substrate were added at point B.

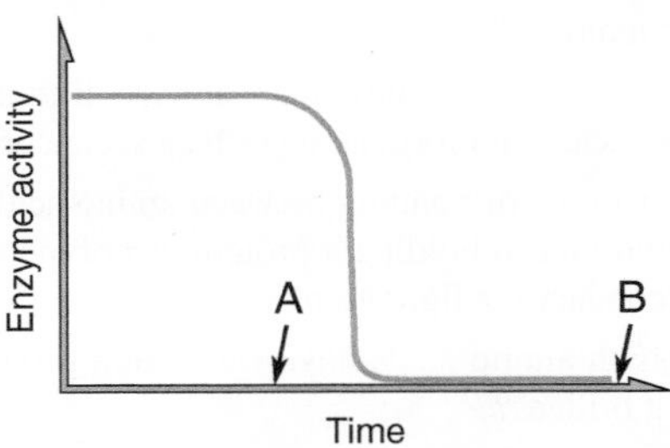

12.65 An enzyme assay had a noncompetitive inhibitor mixture added to it at point A on the following graph. Complete the graph, illustrating what would happen to enzyme activity if more substrate were added at point B.

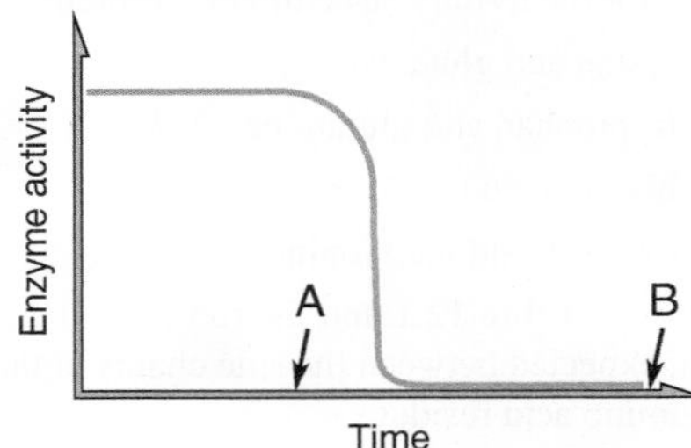

12.66 Determine what type(s) of enzyme inhibitor(s) each of the following statements characterizes. Choices are irreversible, reversible-competitive, reversible-noncompetitive, or more than one of these (list which ones).

a. The enzyme active site can be occupied by substrate or inhibitor, but not both.

b. Adding more substrate does not reverse the inhibitor effect.

c. The substrate and inhibitor structure do not have to resemble each other

d. The substrate and inhibitor can bind to the enzyme at the same time.

12.67 Determine what type(s) of enzyme inhibitor(s) each of the following statements characterizes. Choices are irreversible, reversible-competitive, reversible-noncompetitive, or more than one of these (list which ones).

a. Adding more substrate reverses the inhibitor effect.

b. The substrate and inhibitor structure resemble each other.

c. The substrate and inhibitor cannot bind to the enzyme at the same time.

d. It may cause enzyme denaturation.

Enzymes in Medical Applications and Disease (Section 12.9)

12.68 Provide three reasons a medical professional might run a test to check specific enzyme concentrations in a patient's blood serum.

12.69 List four diseases and the corresponding enzyme detected in blood serum used to diagnose the disease.

Additional Exercises

12.70 What is the conjugate acid for the following zwitterion? What is the conjugate base for it?

$$H_3\overset{+}{N}-\underset{\displaystyle R}{\overset{\displaystyle H}{\underset{|}{\overset{|}{C}}}}-COO^-$$

12.71 The K_a values recorded for alanine are 5.0×10^{-3} and 2.0×10^{-10}. Why are two values recorded? What functional groups do the values correspond to? How many K_a values would be recorded for glutamate?

Chemistry for Thought

12.72 In **Section 12.2**, you learned that amino acids like alanine are crystalline solids with high melting points. However, the ethyl ester of alanine, which has a free NH_2 group, melts at only 87 °C. Explain why the ethyl ester melts 200 degrees below the melting point of alanine.

13 Nucleic Acids and Protein Synthesis

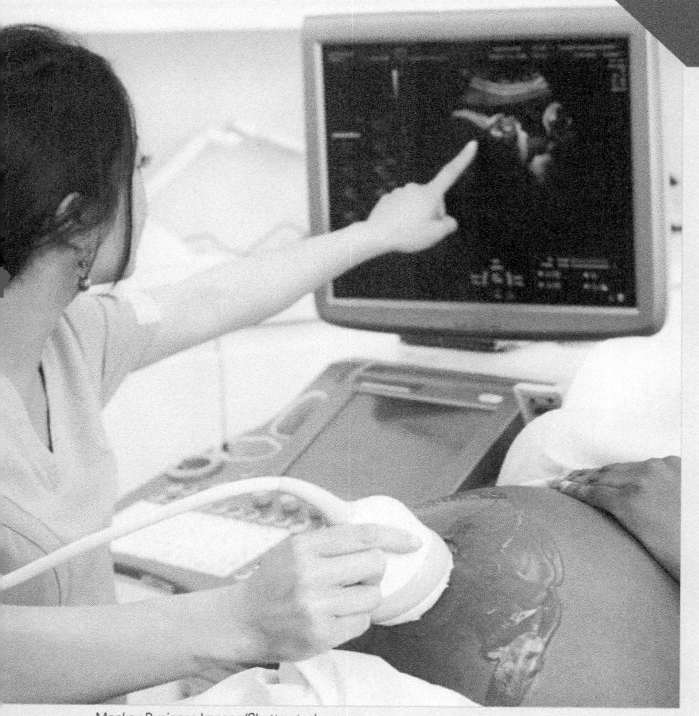
Monkey Business Images/Shutterstock.com

Health Career Focus

Sonographer

When I was pregnant I would count the days left until the appointments with the ultrasound technicians, because I knew I would get to see my baby and listen to her heartbeat.

Looking at her face on the computer screen made me think about the traits she would inherit through our DNA. Would she get her father's eyes or my shiny long hair? Even though it happens every day, I still marveled at how a single embryo could develop into a complex tiny human being!

Our ultrasound technicians were with us from the start. They were always very friendly and professional. They would describe the various developmental stages we were seeing early in the pregnancy. And later, they worked hard to obtain the needed measurements and get good pictures of the baby's face. They were patient explaining any confusing images—and took plenty of time to answer our questions.

Follow-up to this Career Focus appears at the end of the chapter before the *Concept Summary*.

Learning Objectives

When you have completed your study of Chapter 13, you should be able to:

1. Describe the central dogma of molecular biology. **(Section 13.1)**
2. Identify the components of nucleotides. **(Section 13.2)**
3. Classify the sugars and bases of nucleotides. **(Section 13.2)**
4. Describe the structure of DNA. **(Section 13.3)**
5. Write the complementary base strand of a DNA sequence. **(Section 13.3)**
6. Outline the process of DNA replication. **(Section 13.4)**
7. Describe three differences between DNA and RNA. **(Section 13.5)**
8. Describe the function of the three main types of cellular RNA. **(Section 13.5)**
9. Describe the process by which RNA is synthesized from DNA in cells. **(Section 13.6)**
10. Explain how the genetic code functions in the flow of genetic information. **(Section 13.7)**
11. Outline the process by which proteins are synthesized in cells. **(Section 13.7)**
12. Describe the downstream effects of genetic mutations in a living organism. **(Section 13.7)**
13. Describe three viruses and the diseases they cause in human patients. **(Section 13.8)**
14. Explain the process of immunization using vaccines. **(Section 13.8)**
15. Describe the technology used in molecular cloning. **(Section 13.9)**
16. Describe one biomedical application of CRISPR. **(Section 13.9)**

We begin Chapter 13 with the central dogma of molecular biology—namely, how genetic information flows from DNA to RNA to proteins. Once this big picture has been established, we pivot to the building blocks of DNA and RNA and how they combine to form the polymers that carry our genetic information. This section also explores in more detail the processes of DNA replication and RNA transcription. Next, we unravel the mystery of how ribosomes read RNA and translate it into proteins. This includes an in-depth discussion of codons and how genetic mutations lead to protein-related diseases. We then explore the role of DNA and RNA in the battle against viruses as we learn about vaccines and recombinant DNA techniques such as cloning and CRISPR.

13.1 The Flow of Genetic Information

Learning Objective 1 Describe the central dogma of molecular biology.

One of the most remarkable properties of living cells is their ability to produce nearly exact replicas of themselves through hundreds of generations. Such a process requires that certain types of information be passed unchanged from one generation to the next. The transfer of the necessary genetic information to new cells is accomplished by means of biomolecules called **nucleic acids**.

nucleic acid A biomolecule involved in the transfer of genetic information from existing cells to new cells.

These high-molecular-weight compounds represent coded information, much as words represent information in a book. It is the nearly infinite variety of possible structures that enables nucleic acids to represent the huge amount of information that must be transmitted sexually or asexually to reproduce a living organism. This information controls the inherited characteristics of the individuals in the new generation and determines the life processes as well (see **Figure 13.1**).

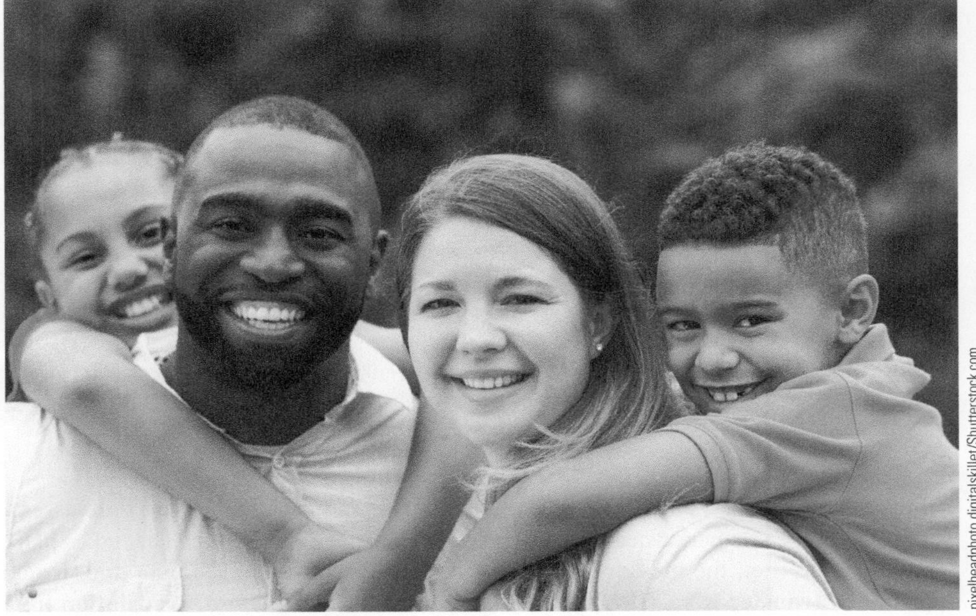

pixelheadphoto digitalskillet/Shutterstock.com

Figure 13.1 Nucleic acids passed from the parents to offspring determine the inherited characteristics.

ribonucleic acid (RNA) A nucleic acid found mainly in the cytoplasm of cells.

deoxyribonucleic acid (DNA) A nucleic acid found primarily in the nuclei of cells.

central dogma of molecular biology The well-established process by which genetic information stored in DNA molecules is expressed in the structure of synthesized proteins.

gene An individual section of a DNA molecule that is the fundamental unit of heredity.

transcription The transfer of genetic information from a DNA molecule to a molecule of messenger RNA.

messenger RNA (mRNA) RNA that carries genetic information from the DNA in the cell nucleus to the site of protein synthesis in the cytoplasm.

There are two types of nucleic acids—namely, **ribonucleic acid (RNA)** and **deoxyribonucleic acid (DNA)**. Here in **Section 13.1**, our focus is on how the genetic information stored in DNA is expressed within the cell. This process of expression is so well established that it is called the **central dogma of molecular biology**.

According to this principle, genetic information contained in DNA molecules is transferred to RNA molecules. The transferred information stored in RNA molecules is then expressed in the structure of synthesized proteins. In other words, the genetic information in DNA (genes) directs the synthesis of certain proteins.

There is a specific DNA **gene** for every protein in the body. DNA does not direct the synthesis of carbohydrates, lipids, or the other nonprotein substances essential for life. However, these other materials are manufactured by the cell through reactions made possible by enzymes (proteins) produced under the direction of DNA. Thus, in this respect, the information stored in DNA really does determine every characteristic of the living organism.

Two steps are involved in the flow of genetic information: transcription and translation. In higher organisms, the DNA containing the stored information is located in the nucleus of the cell, and protein synthesis occurs in the cytoplasm (**Figure 13.2**). Thus, the stored information must first be carried out of the nucleus. This is accomplished by **transcription**, the process by which the necessary information from a DNA molecule is transferred onto a molecule of **messenger RNA (mRNA)**. The mRNA carries the information (the message) from the nucleus to the site of protein synthesis in the cellular cytoplasm.

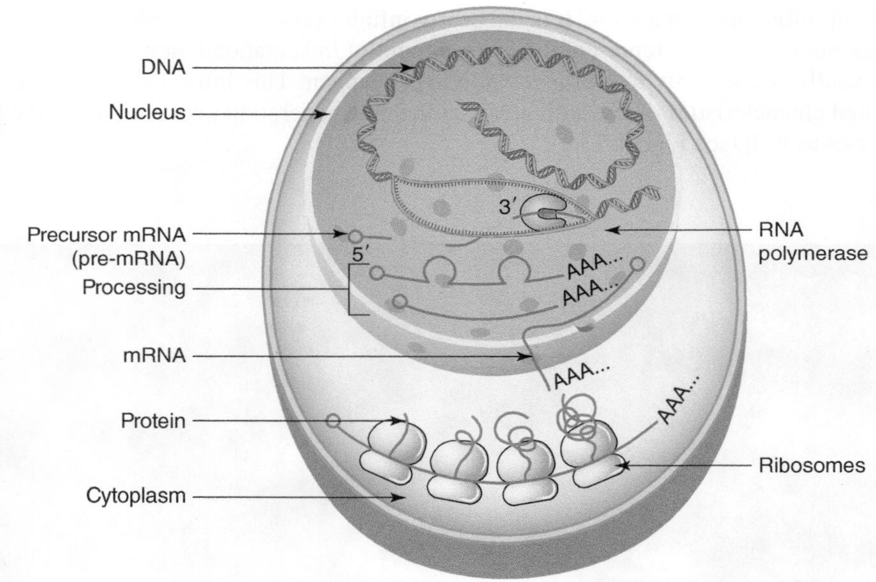

Figure 13.2 In human cells, DNA is transcribed into pre-mRNA and then edited, resulting in mRNA. The mRNA is then exported into the cytoplasm, where tRNA molecules and ribosomes translate it into a protein.

transfer RNA (tRNA) RNA that delivers individual amino acid molecules to the site of protein synthesis.

translation The conversion of the code carried by messenger RNA into an amino acid sequence of a protein.

ribosome A subcellular particle that serves as the site of protein synthesis in all organisms.

In the second step, the mRNA serves as a template on which amino acids are assembled in the proper sequence necessary to produce the specified protein. This takes place when the code (i.e., the message) carried by mRNA is translated into an amino acid sequence by **transfer RNA (tRNA)**. There is an exact "word-to-word" **translation** from mRNA to tRNA. Thus, each "word" in the mRNA language has a corresponding "word" in the amino acid language that tRNAs and **ribosomes** must read to build a growing peptide chain. This communicative relationship between mRNA and amino acids is called the *genetic code* (see **Section 13.7**). **Figure 13.2** summarizes the mechanisms for the flow of genetic information in the cell.

13.2 The Components of Nucleic Acids

Learning Objective 2 Identify the components of nucleotides.

Learning Objective 3 Classify the sugars and bases of nucleotides.

DNA is found primarily in the nuclei of living cells. Meanwhile, RNA is found in the nucleus and mitochondria, but it is mainly located in the cytoplasm. Both DNA and RNA are biopolymers, consisting of long, linear molecules. The repeating structural units (i.e., the monomers) of DNA and RNA are called **nucleotides**. Nucleotides, in turn, are composed of three simpler components: a nitrogenous base, a sugar, and a phosphate (see **Figure 13.3**).

nucleotide The repeating structural unit (the monomer) of polymeric nucleic acids DNA and RNA.

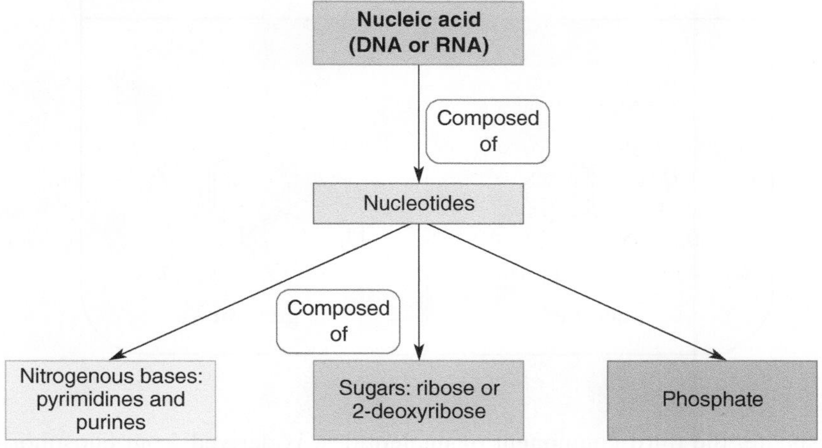

Figure 13.3 The composition of nucleotides.

Each of the five bases commonly found in nucleic acids are heterocyclic compounds that can be classified as either a pyrimidine or a purine, the parent compounds from which the bases are derived. Purines, one of the classes of nitrogenous bases that make up nucleic acids, are broken down in the body into uric acid. When too much uric acid accumulates within joints, a condition known as gout can arise (**Figure 13.4**).

pyrimidine   purine

The three pyrimidine bases are uracil (U), thymine (T), and cytosine (C), whereas adenine (A) and guanine (G) are the two purine bases. Adenine, guanine, and cytosine are found in both DNA and RNA, but uracil is ordinarily found only in RNA, and thymine only in DNA. Structural formulas of the five bases are given in **Figure 13.5**.

The sugar component of RNA is D-ribose, as the name ribonucleic acid implies. In deoxyribonucleic acid (DNA), the sugar is D-deoxyribose, which lacks a hydroxy group at the second position of the heterocyclic ring. The carbon atoms in the sugar are designated with a number followed by a prime to distinguish them from the atoms in the bases. Both sugars are present in the β configuration in nucleotides.

D-ribose   D-deoxyribose

β configuration

no –OH group

Figure 13.4 A by-product of purine digestion is uric acid. The build up of uric acid can result in crystals that cause an inflammation in the joints called gout.

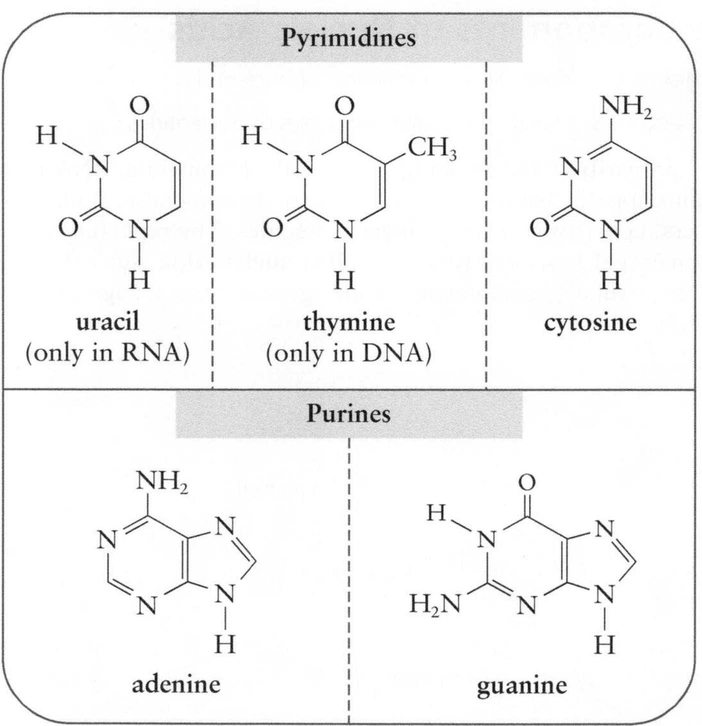

Figure 13.5 The pyrimidine and purine bases of nucleic acids.

Phosphate, the third component of nucleotides, is derived from phosphoric acid (H_3PO_4), which under cellular pH conditions exists in its ionic form:

$$^{-}O-\overset{\overset{\displaystyle O}{\|}}{\underset{\underset{\displaystyle O^{-}}{|}}{P}}-OH$$

Example 13.1 Identifying Bases

Indicate whether each of the following bases is a purine or a pyrimidine. Also, state whether it is found in DNA, RNA, or both.

a. Cytosine
b. Uracil
c. Guanine

Solution

a. Pyrimidine, both DNA and RNA
b. Pyrimidine, RNA
c. Purine, both DNA and RNA

✔ Learning Check 13.1 Indicate whether each of the following bases is a purine or pyrimidine. Also, state whether it is found in DNA, RNA, or both.

a. Adenine
b. Thymine

The formation of a nucleotide from these three components is represented in Equation 13.1:

adenine

phosphate + ribose → adenosine 5′-monophosphate (AMP) + $2H_2O$ (13.1)

phosphate

ribose

adenosine 5′-monophosphate
(AMP)

The adenosine nucleotide in Equation 13.1 is found in RNA (the sugar is ribose), but a build up of adenosine can lead to complications (**Figure 13.6**). Equation 13.1 shows that the base in a nucleotide is always attached to the 1′ position of the sugar, and the phosphate is generally located at the 5′ position. Thus, the general structure of a nucleotide can be represented as follows:

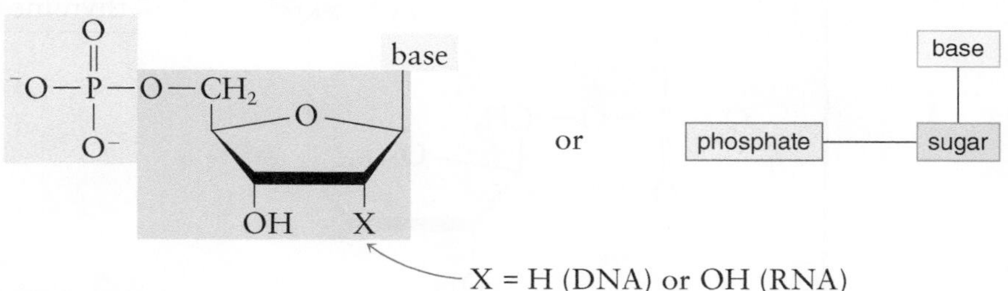

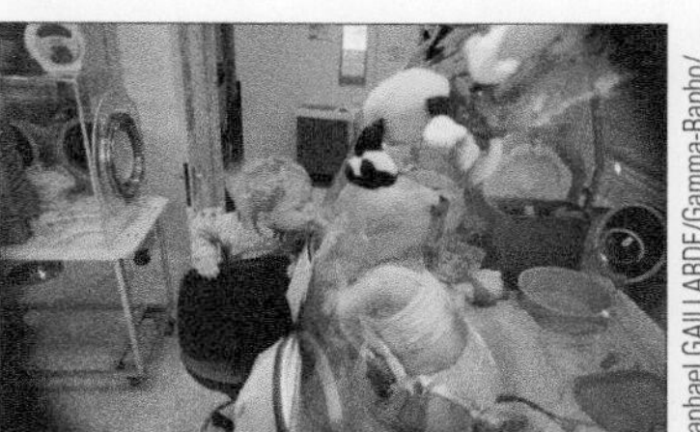

Raphael GAILLARDE/Gamma-Rapho/Getty Images

Figure 13.6 People with severe combined immunodeficiency (SCID) experience a build up of adenosine as a result of mutated enzymes. The accumulation of adenosine interferes with lymphocytes and causes irreparable damage to the immune system.

13.3 The Structure of DNA

Learning Objective 4 Describe the structure of DNA.

Learning Objective 5 Write the complementary base strand of a DNA sequence.

13.3A The Primary Structure of DNA

DNA molecules are among the largest known, containing between 1 million and 100 million nucleotide units. The nucleotides are joined together in nucleic acids by phosphate groups that connect the 5′ carbon of one nucleotide to the 3′ carbon of the next. These linkages are referred to as phosphodiester bonds (**Section 11.6** and **Figure 13.7**). The result is a chain of alternating phosphate and sugar units to which the bases are attached. The sugar–phosphate chain is referred to as the **nucleic acid backbone** (shown in blue in **Figure 13.7**), and it is constant throughout the entire DNA molecule. Thus, one DNA molecule differs from another only in the sequence, or order of attachment, of the bases along the backbone. Just as the amino acid sequence of a protein determines the primary structure of the protein, the order of the bases provides the primary structure of DNA.

nucleic acid backbone The sugar–phosphate chain that is common to all nucleic acids.

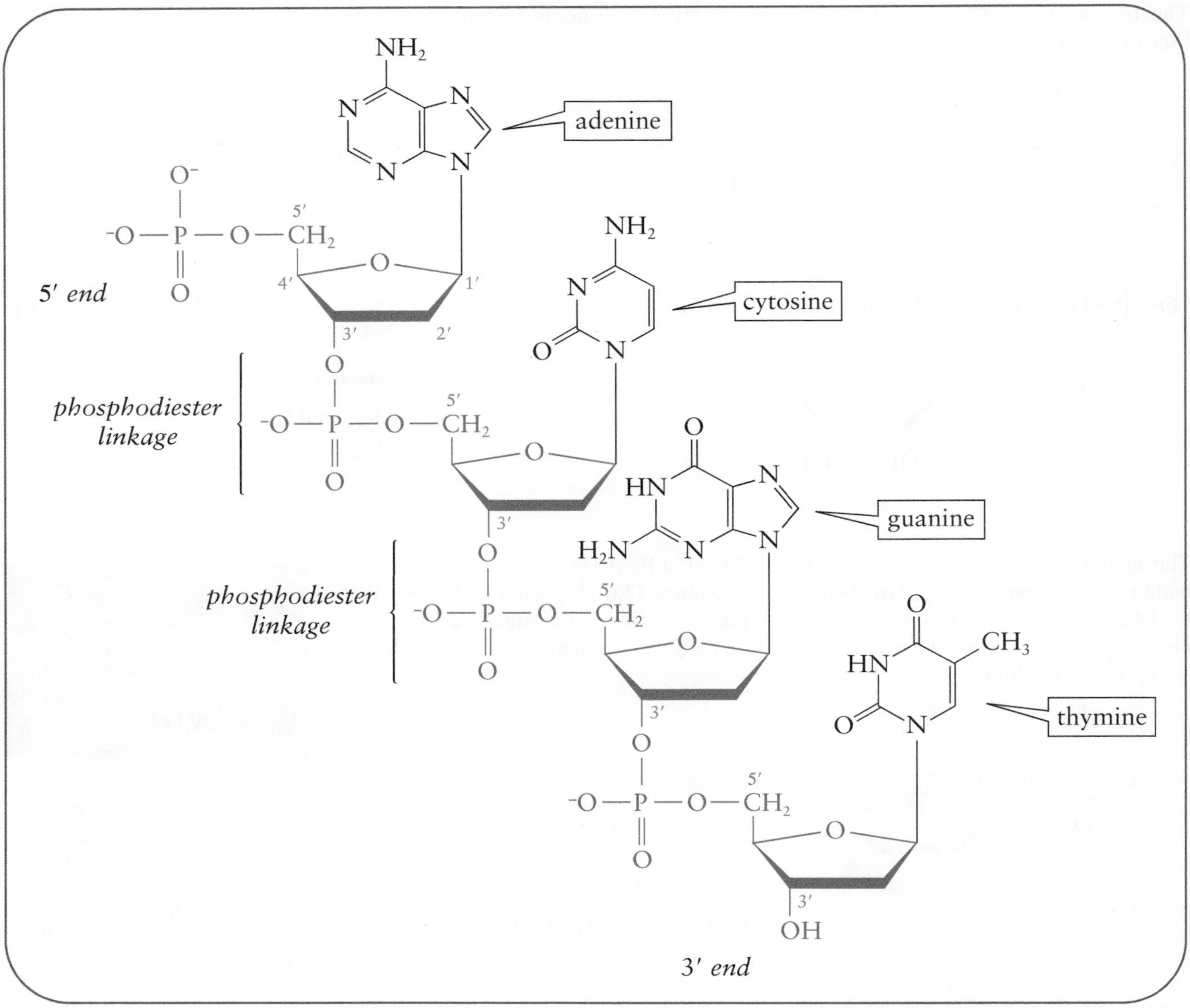

Figure 13.7 The structure of ACGT, a tetranucleotide segment of DNA. The nucleic acid backbone is colored blue.

The end of the polynucleotide segment of DNA in **Figure 13.7** that has no nucleotide attached to the 5′ CH_2 is called the 5′ end of the segment. The other end of the chain is the 3′ end. By convention, the sequence of bases along the backbone is read from the 5′ end to the 3′ end, so the tetranucleotide in **Figure 13.7** is ACGT, not TGCA. The sequence of bases can be abbreviated in this way because the backbone structure along the chain does not vary.

Example 13.2 Drawing Trinucleotide Structures

Draw the structural formula for a trinucleotide portion of DNA with the sequence AGT. Abbreviate the structures of the bases, using the letters AGT. Point out the 5′ and 3′ ends of the molecule. Indicate with arrows the phosphodiester linkages.

Solution

5′ *end* $^{-}O-\overset{O^-}{\underset{O}{P}}-O-CH_2$ — A

O

phosphodiester linkage $^{-}O-\underset{O}{P}-O-CH_2$ — G

O

phosphodiester linkage $^{-}O-\underset{O}{P}-O-CH_2$ — T

OH

3′ *end*

Pictorial Press Ltd/Alamy Stock Photo

Figure 13.8 Rosalind Franklin (1920–1958), a research scientist at Kings College London, provided the X-ray crystallography data that led to the discovery of the DNA double helix. Franklin was a trailblazer in X-ray crystallography and virus research. Although many believe she should have been included in the 1962 Nobel Prize, she passed away due to ovarian cancer at only 37 years old, 4 years before the honor was awarded.

✔ **Learning Check 13.2** Draw the structural formula for a trinucleotide portion of DNA with the sequence CTG. Abbreviate the structures of the bases, using the letters CTG. Point out the 5′ and 3′ ends of the molecule. Indicate with arrows the phosphodiester linkages.

13.3B The DNA Double Helix

The discovery of the three-dimensional structure of DNA was among the great scientific quests of the first half of the twentieth century. Scientists had gathered data that had given clues to some of the properties of DNA, but the ultimate three-dimensional structure remained elusive. For example, scientists knew that DNA contained equal amounts of adenine and thymine and also, equal amounts of guanine and cytosine. In fact, human DNA contains 20% guanine, 20% cytosine, 30% adenine, and 30% thymine.

Molecular biologist James Watson (1928–) and biologist Francis Crick (1916–2004) were two of the scientists seeking to put the many pieces of the puzzle together. Watson and Crick had been working on many different models of DNA's structure in collaboration with other scientists. However, critical information from Rosalind Franklin (**Figure 13.8**), a pioneer in X-ray crystallography, proved to be crucial in unlocking the mystery of DNA's true form. Based on these data, Watson and Crick made the breakthrough revelation that the DNA bases, the purines and pyrimidines, were on the inside of a helical structure. This led to the realization that the actual three-dimensional structure of DNA was, in fact, a double helix (**Figure 13.9**). Watson, Crick, and colleague Maurice Wilkins shared the 1962 Nobel Prize for Physiology or Medicine for this work. Franklin deserved to be recognized as well, but no more than three living persons can share one award, and she had passed 4 years earlier.

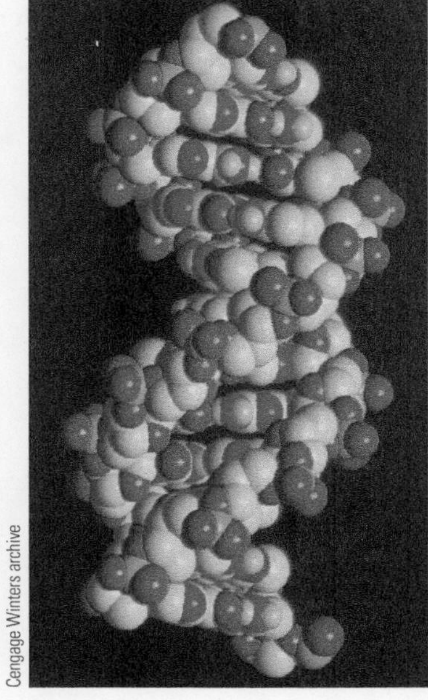

Cengage Winters archive

Figure 13.9 A three-dimensional molecular model of the DNA double helix.

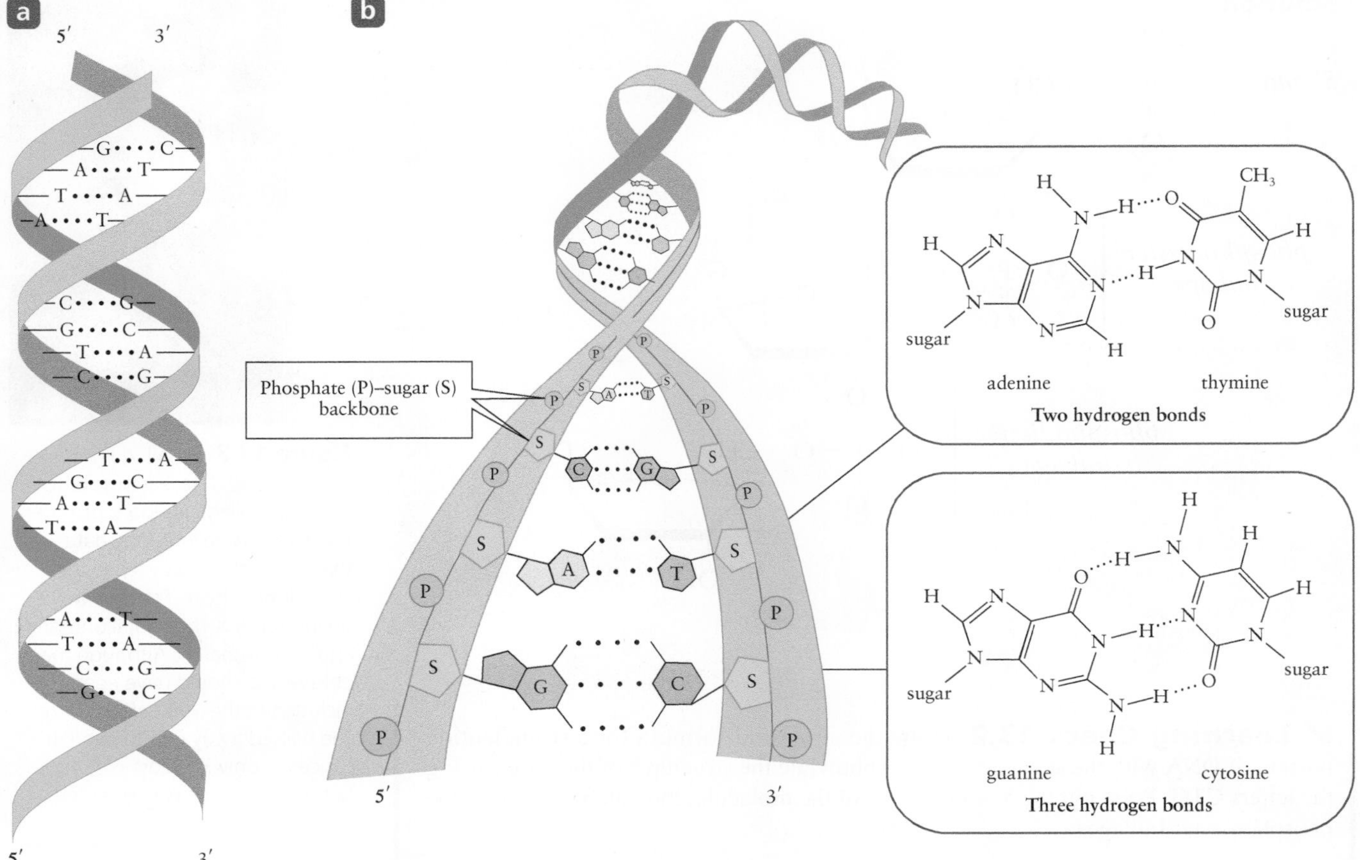

Figure 13.10 Schematic drawing of (a) the double helix of DNA and (b) base-pairing. Hydrogen-bonding between complementary base pairs holds the two strands of a DNA molecule together.

complementary DNA strands The two strands in double-helical DNA align such that adenine in one strand hydrogen-bonds to thymine in the other, and guanine in one strand hydrogen-bonds to cytosine in the other.

In the double helix, two strands of DNA are oriented in opposite directions—that is, one DNA strand runs from 3′ to 5′ while the **complementary DNA strand** runs 5′ to 3′ (**Figure 13.10a**). The two strands are said to be antiparallel. Also, because the base pairs are on the inside of the sugar–phosphate backbone, they hydrogen-bond with one another. Adenine forms two hydrogen bonds with thymine, whereas guanine forms three hydrogen bonds with cytosine (**Figure 13.10b**). So, for every adenine on one strand, a thymine appears opposite it on the other strand, and vice versa. The same complementary relationship holds true for cytosine and guanine. Thus, the base pairs resemble rungs on a ladder, with hydrogen bonds stabilizing the two DNA strands to form the double helix.

By understanding how strands of DNA are themselves arranged, the genetic code could finally begin to be unlocked. In fact, the original article published by Watson and Crick in 1953 in the scientific journal *Nature* states, "It has not escaped our notice that the specific pairing we have postulated immediately suggests a possible copying mechanism for the genetic material." The discovery of the DNA double helix was truly among the most significant scientific contributions of the last century.

Example 13.3 Writing Base Sequences

One strand of a DNA molecule has the base sequence CCATTG. What is the base sequence for the complementary strand?

Solution

Three things must be remembered and used to solve this problem:

1. The base sequence of a DNA strand is always written from the 5′ end to the 3′ end.
2. Adenine (A) is always paired with its complementary base thymine (T), and guanine (G) is always paired with its complement cytosine (C).
3. In double-stranded DNA, the two strands run in opposite directions, so that the 5′ end of one strand is associated with the 3′ end of the other strand.

Original strand:	5′ end C–C–A–T–T–G 3′ end
COMPLEMENTARY STRAND:	3′ end G–G–T–A–A–C 5′ end

When we follow convention and write the sequence of the complementary strand in the 5′ to 3′ direction, it becomes

5′ C–A–A–T–G–G 3′

✔ **Learning Check 13.3** Write the complementary base sequence for the DNA strand TTACG.

13.4 DNA Replication

Learning Objective 6 Outline the process of DNA replication.

DNA is responsible for one of the most essential functions of living organisms, the storage and transmission of hereditary information. A human cell normally contains 46 structural units called **chromosomes** (see **Figure 13.11**). Each chromosome contains one molecule of DNA coiled tightly about a group of small, basic proteins called histones. Individual sections of DNA molecules make up genes, the fundamental units of heredity.

chromosome A tightly packed bundle of DNA and protein that is involved in cell division.

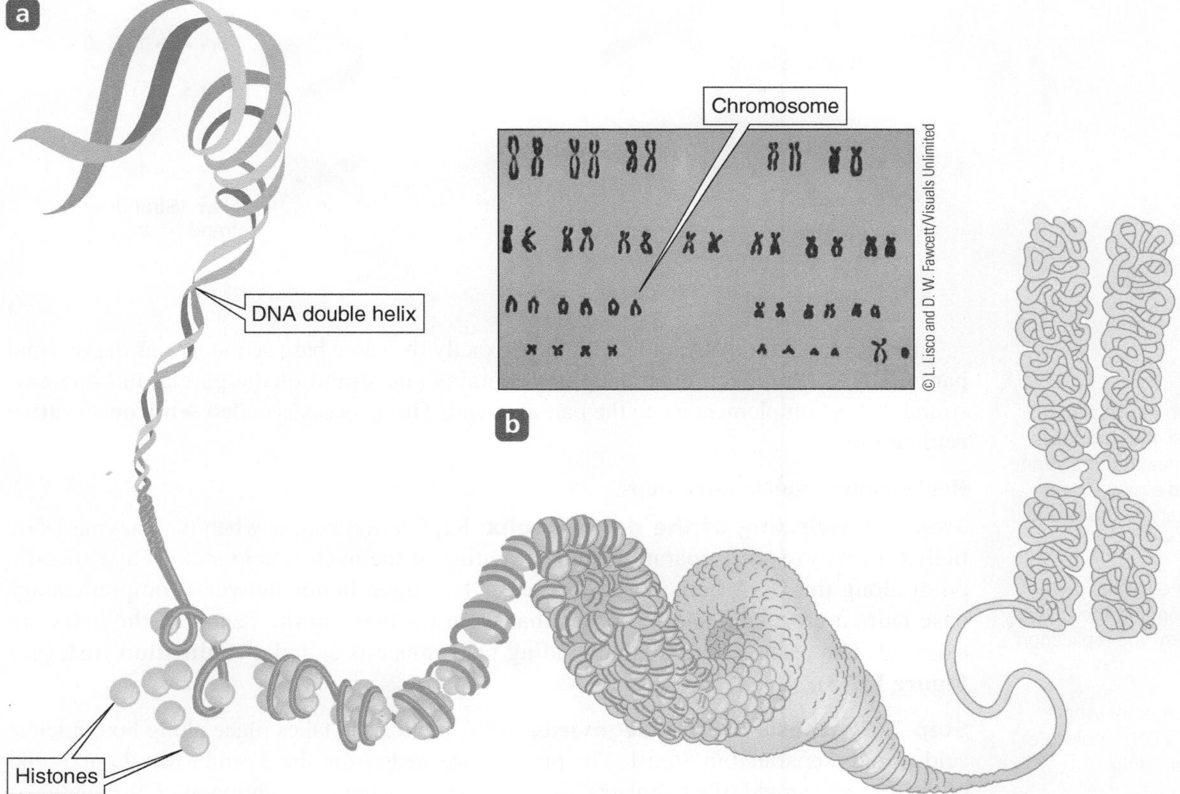

Figure 13.11 Chromosomes. (a) Each chromosome is a protein-coated strand of multicoiled DNA. (b) The 46 chromosomes of a human cell.

Recall from **Section 13.1** that each gene directs the synthesis of a specific protein. The number of genes contained in the structural unit of an organism varies with the type of organism. For example, a virus, the smallest structure known to carry genetic information, is thought to contain from a few to several hundred genes. A bacterial cell, such as *Escherichia coli* (*E. coli*), contains about 1000 genes, whereas a human cell contains approximately 25,000.

replication The process by which an exact copy of a DNA molecule is produced.

The process by which an exact copy of DNA is produced is called **replication** (**Figure 13.12**). It occurs when two strands of DNA separate and each serves as a template (pattern) for the construction of its own complement. The process generates new DNA double-stranded molecules that are exact replicas of the original DNA molecule.

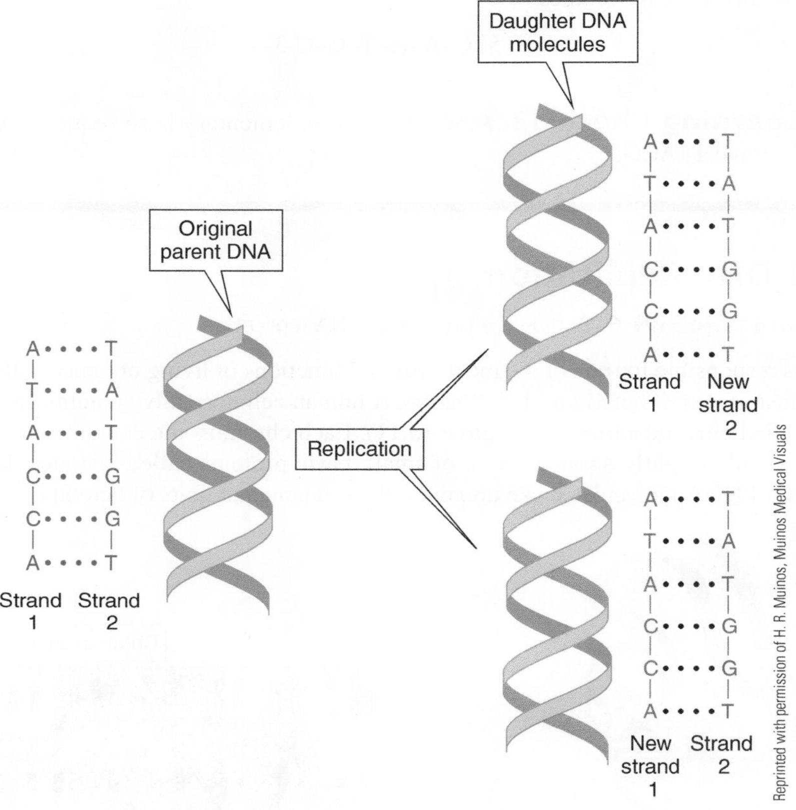

Figure 13.12 A schematic diagram of the replication of DNA.

The two daughter DNA molecules have exactly the same base sequences as the original parent DNA. Moreover, each daughter contains one strand of the parent and one new strand that is complementary to the parent strand. This process is called **semiconservative replication**.

semiconservative replication A replication process that produces DNA molecules containing one strand from the parent and a new strand that is complementary to the strand from the parent.

Replication occurs in three steps.

Step 1. Unwinding of the double helix: Replication begins when the enzyme **DNA helicase** catalyzes the separation and unwinding of the nucleic acid strands at a specific point along the DNA helix. In this process, hydrogen bonds between complementary base pairs are broken, and the bases that were formerly in the center of the helix are exposed. The point where this unwinding takes place is called a **replication fork** (see **Figure 13.13**).

DNA helicase An enzyme responsible for unwinding the DNA double strand to form the replication fork.

replication fork A point where the double helix of a DNA molecule unwinds during replication.

Step 2. Synthesis of DNA segments: DNA replication takes place along both nucleic acid strands separated in Step 1. The process proceeds from the 3′ end toward the 5′ end of the exposed strand (the template). Because the two strands are antiparallel, the synthesis of new nucleic acid strands proceeds toward the replication fork on one exposed strand and away from the replication fork on the other strand (see **Figure 13.13**). New (daughter)

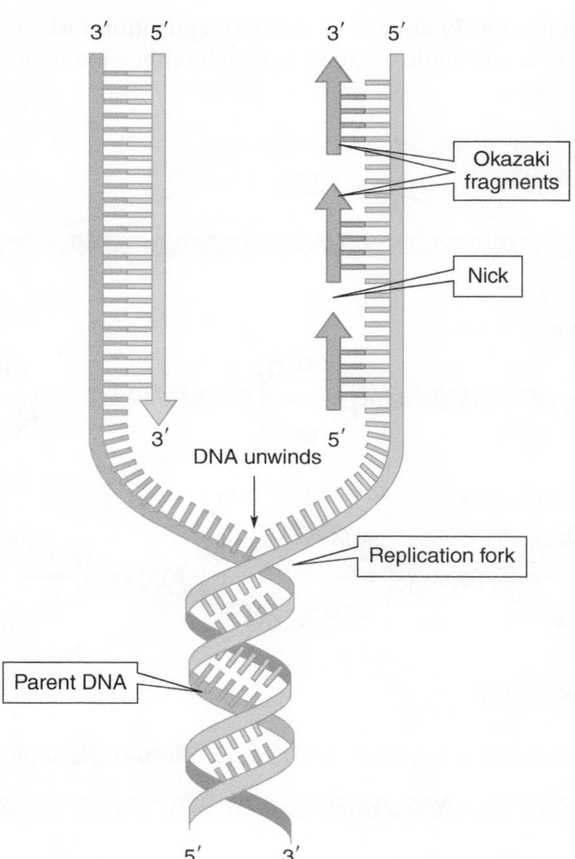

Figure 13.13 The replication of DNA. Both new strands are growing in the 5′ to 3′ direction.

DNA strands form as nucleotides, complementary to those on the exposed strands, and are linked together under the influence of the enzyme **DNA polymerase**. The daughter chain grows from the 5′ end toward the 3′ end as the process moves from the 3′ end of the exposed (template) strand toward the 5′ end. As the synthesis proceeds, a second replication fork is created when a new section of DNA unwinds. The daughter strand that was growing toward the first replication fork continues growing smoothly toward the new fork. However, the other daughter strand was growing away from the first fork. A new segment of this strand begins growing from the new fork but is not initially bonded to the segment that grew from the first fork. Thus, as the parent DNA progressively unwinds, this daughter strand is synthesized as a series of fragments that are joined together in Step 3. The gaps or breaks between segments in this daughter strand are called nicks, and the DNA fragments separated by the nicks are called **Okazaki fragments** after their discoverer, Reiji Okazaki (1930–1975) (**Figure 13.13**) and his wife and collaborator Tsuneko Okazki (1933–).

DNA polymerase A family of enzymes responsible for building daughter DNA strands complementary to parent template strands during replication.

Okazaki fragment A DNA fragment produced during replication as a result of strand growth in a direction away from the replication fork.

DNA ligase An enzyme that joins together Okazaki fragments on a daughter DNA strand.

Step 3. Closing the nicks: One daughter DNA strand is synthesized without any nicks, but the Okazaki fragments of the other strand must be joined together. An enzyme called **DNA ligase** catalyzes this final step of DNA replication. The result is two DNA double-helical molecules that are identical, each of which consists of one original (parent) strand and one new (daughter) strand.

Observations with an electron microscope show that the replication of DNA molecules in eukaryotic cells occurs simultaneously at many points along the original DNA molecule. These replication zones blend together as the process of replication continues. This is necessary if long molecules are to be replicated rapidly. For example, it is estimated that the replication of the largest chromosome in *Drosophila* (the fruit fly) (**Figure 13.14**) would take more than 16 days if there were only one origin for replication. Research results indicate that the actual replication is accomplished in less than 3 minutes because

Tomasz Klejdysz/Shutterstock.com

Figure 13.14 *Drosophila*, commonly known as a fruit fly.

it takes place simultaneously at more than 6000 replication forks per DNA molecule. **Figure 13.15** shows a schematic diagram in which replication is proceeding at several points in the DNA chain.

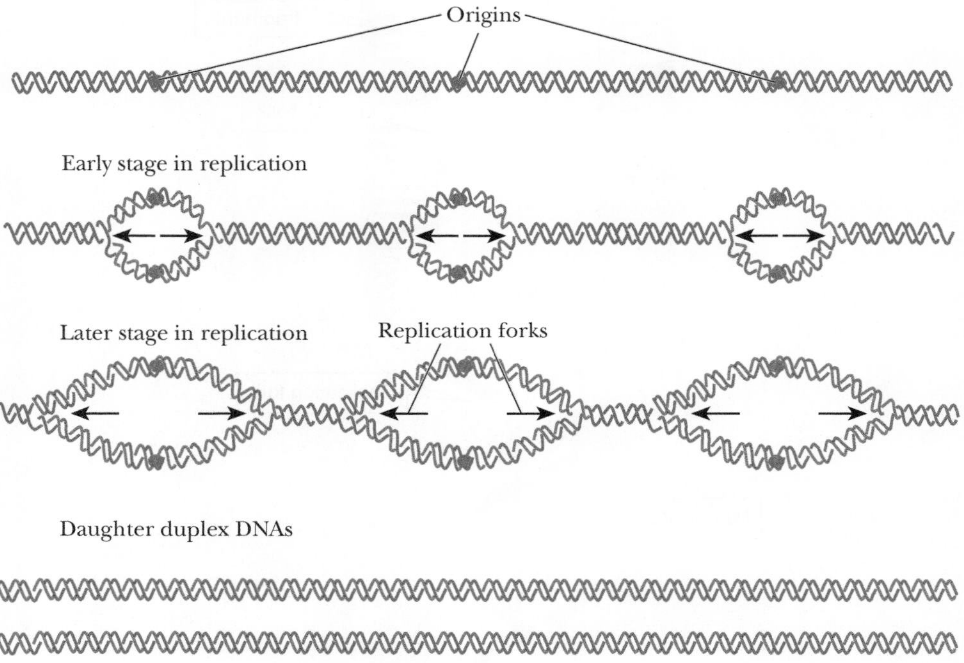

Figure 13.15 A schematic representation of eukaryotic chromosome replication, showing several replication forks. From Campbell, *Biochemistry*, 8E. © 2018 Cengage Learning.

polymerase chain reaction (PCR) A laboratory technique used to amplify small amounts of DNA.

Knowledge of the DNA replication process led scientist Kary Mullis (**Figure 13.16**) in 1983 to a revolutionary laboratory technique called the **polymerase chain reaction (PCR)** (**Figure 13.17**). The PCR technique mimics the natural process of replication, in which the DNA double helix unwinds. As the two strands separate, DNA polymerase makes a copy using each strand as a template. To perform PCR, a small quantity of the target DNA is added to a test tube along with a buffered solution containing DNA polymerase; the cofactor $MgCl_2$, which is needed by the polymerase to function; the four nucleotides (the four building blocks); and primers. The primers are short polynucleotide segments that will bind to the separated DNA strands and serve as starting points for new chain growth. The PCR mixture is taken through three-step replication cycles:

1. **Denaturing:** Heat (94 °C to 96 °C) is used for one to several minutes to unravel (denature) the DNA into single strands. This replaces the DNA helicase used to unwind the DNA in nature.
2. **Annealing:** The tube is cooled to 50 °C to 65 °C for one to several minutes, during which the primers, which are complementary to the target DNA sequence, hydrogen-bond to the separated strands of target DNA.
3. **Extension:** The tube is heated to 72 °C for one to several minutes, during which time the DNA polymerase synthesizes new strands of DNA.

Each cycle doubles the amount of DNA. Following 30 such cycles, a theoretical amplification factor of 1 billion is attained.

Agence Opale/Alamy Stock Photo

Figure 13.16 Kary Mullis (1944–2019) developed the polymerase chain reaction (PCR). It can be used to amplify minute amounts of DNA into millions of copies!

Almost overnight, PCR became a standard research technique for detecting all manner of mutations associated with genetic disease. PCR can also be used to detect the presence of unwanted DNA, as in the case of a bacterial or viral infection. Conventional tests that involve the culture of microorganisms or the use of antibodies can take weeks to perform. PCR offers a fast and simple alternative. The ability of PCR to utilize degraded DNA samples and sometimes the DNA from single cells is of great interest to forensic scientists. PCR has also permitted DNA to be amplified from some unusual sources, such as extinct mammals, Egyptian mummies, and ancient insects trapped in amber.

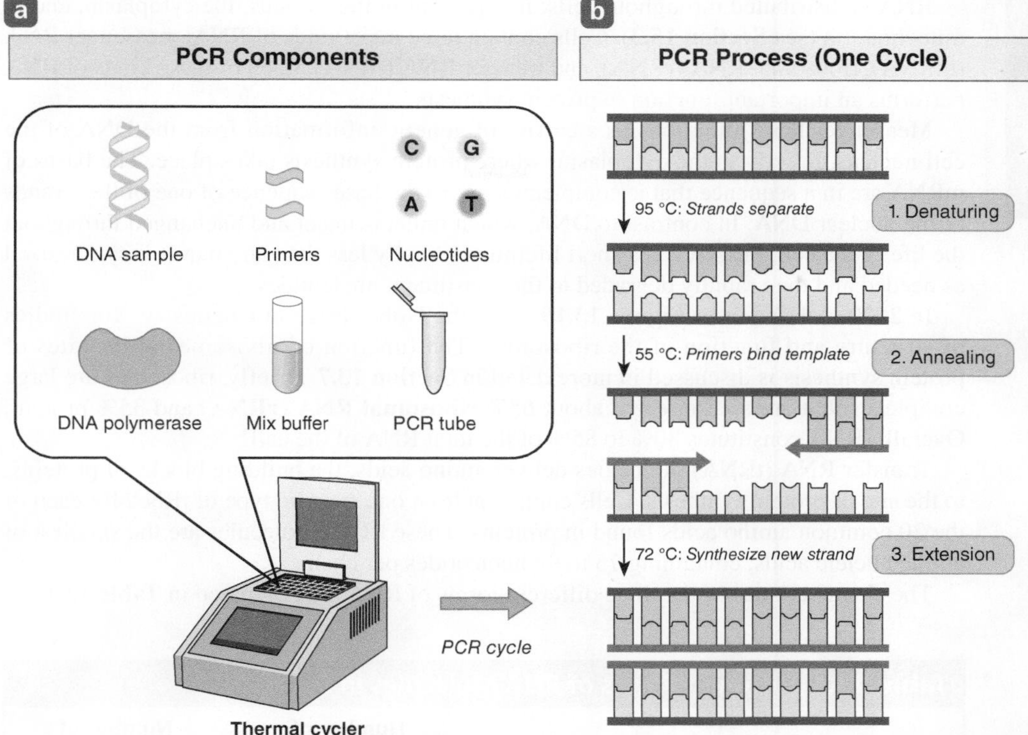

Figure 13.17 Polymerase chain reaction (PCR). (a) The components required for the PCR reaction. (b) The three-step process of DNA amplification using PCR.

13.5 Ribonucleic Acid (RNA)

Learning Objective 7 Describe three differences between DNA and RNA.

Learning Objective 8 Describe the function of the three main types of cellular RNA.

RNA, like DNA, is a long, unbranched polymer consisting of nucleotides joined by phosphodiester bonds. The number of nucleotides in an RNA molecule ranges from as few as 73 to as many as thousands. The primary structure of RNA differs from that of DNA in two ways. As mentioned in **Section 13.2**, the sugar unit in RNA is ribose rather than deoxyribose. The other difference is that RNA contains the base uracil (U) instead of thymine (T).

The secondary structure of RNA is also different from that of DNA. RNA molecules are single-stranded, except in some viruses. Consequently, an RNA molecule need not contain complementary base ratios of 1:1. In general, however, RNA molecules, are highly structured and contain stem–loop regions where the "stem" portion consists of bases that hydrogen bond with corresponding bases in the same RNA strand (see **Figure 13.18**). In these regions, adenine (A) pairs with uracil (U), and guanine (G) pairs with cytosine (C). The proportion of helical regions varies over a wide range depending on the kind of RNA studied, but a value of 50% is typical.

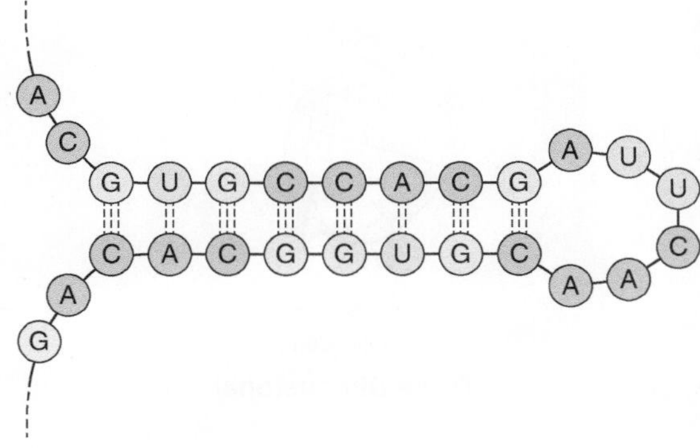

Figure 13.18 A portion of RNA that has folded back on itself and formed a double-helical region.

Figure 13.19 Ada Yonath (1939–), a crystallographer, received the 2009 Nobel Prize in Chemistry for her pioneering work on the structure of ribosomes.

ribosomal RNA (rRNA) RNA that constitutes about 65% of the material in ribosomes, the sites of protein synthesis.

RNA is distributed throughout cells; it is present in the nucleus, the cytoplasm, and in mitochondria (see **Section 15.3**). Cells contain three main kinds of RNA: messenger RNA (mRNA), ribosomal RNA (rRNA), and transfer RNA (tRNA). Each of these kinds of RNA performs an important function in protein synthesis.

Messenger RNA functions as a carrier of genetic information from the DNA of the cell nucleus directly to the cytoplasm, where protein synthesis takes place. The bases of mRNA are in a sequence that is complementary to the base sequence of one of the strands of the nuclear DNA. In contrast to DNA, which remains intact and unchanged throughout the life of the cell, mRNA has a short lifetime—usually less than an hour. It is synthesized as needed and then rapidly degraded to the constituent nucleotides.

In 2009, Ada Yonath (**Figure 13.19**) won the Nobel Prize in Chemistry "for studies of structure and function of the ribosome." The function of ribosomes at the sites of protein synthesis is discussed in more detail in **Section 13.7**. Briefly, ribosomes are large complexes that are composed of about 65% **ribosomal RNA (rRNA)** and 35% protein. Overall, rRNA constitutes 80% to 85% of the total RNA of the cell!

Transfer RNA (tRNA) molecules deliver amino acids, the building blocks of proteins, to the site of protein synthesis. Cells contain at least one specific type of tRNA for each of the 20 common amino acids found in proteins. These tRNA molecules are the smallest of all the nucleic acids, containing 73 to 93 nucleotides per chain.

The characteristics of the three different forms of RNA are compared in **Table 13.1**.

Table 13.1 Different Forms of RNA Molecules in *E. coli*

Class of RNA	% in Cells	Number of RNA Subtypes	Number of Nucleotides
Ribosomal RNA (rRNA)	80	3	*120, 1700, 3700
Transfer RNA (tRNA)	15	46	73 to 93
Messenger RNA (mRNA)	5	Many	75 to 3000

* The ribosome consists of a protein–RNA complex with three strands of varying length instead of a single range of values.

Because of their small size, a number of tRNAs have been studied extensively. **Figure 13.20** shows representations of the secondary and tertiary structures of a typical tRNA. This tRNA molecule, like all others, has regions where there is hydrogen-bonding

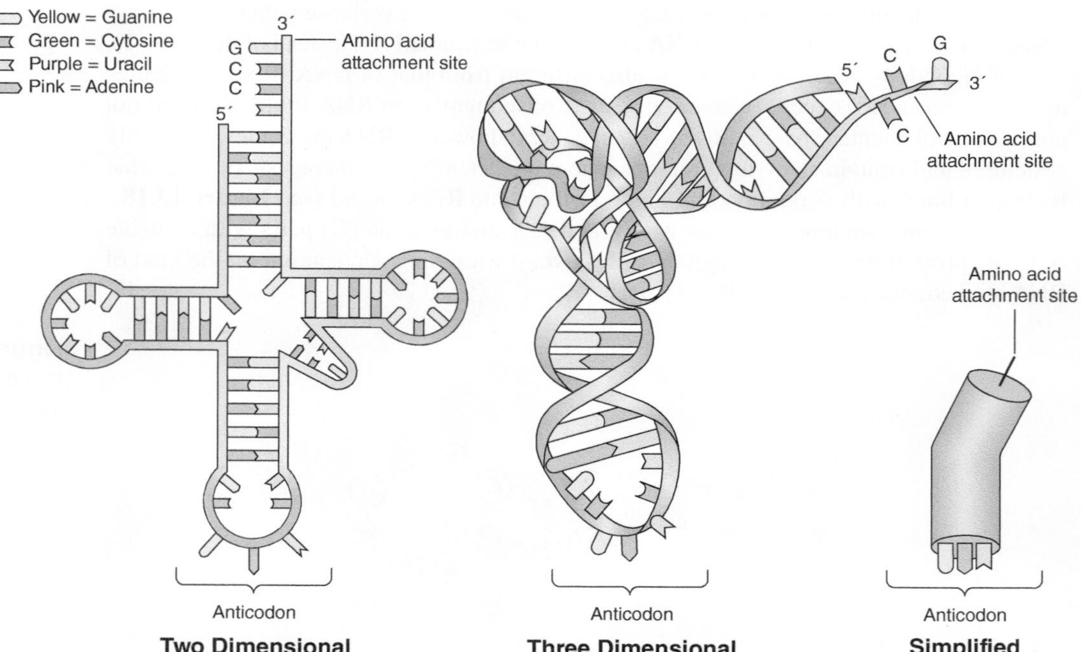

Figure 13.20 Transfer RNA (tRNA) structure. (a) The secondary structure of a tRNA is typically a cloverleaf. (b) The tertiary structure of a tRNA shows the cloverleaf folded over into a three-dimensional L shape. (c) For more complicated translation schemes, tRNAs are often simplified to cartoons with only the amino acid attachment site and anticodon indicated.

between complementary bases, and regions (loops) where there is no hydrogen-bonding. Two regions of tRNA molecules have important functions during protein synthesis. One of these regions, designated the **anticodon**, enables the tRNA to bind to mRNA during protein synthesis. The second important site is the 3′ end of the molecule, which binds to an amino acid and transports it to the site of protein synthesis.

anticodon A three-base sequence in tRNA that is complementary to one of the codons in mRNA.

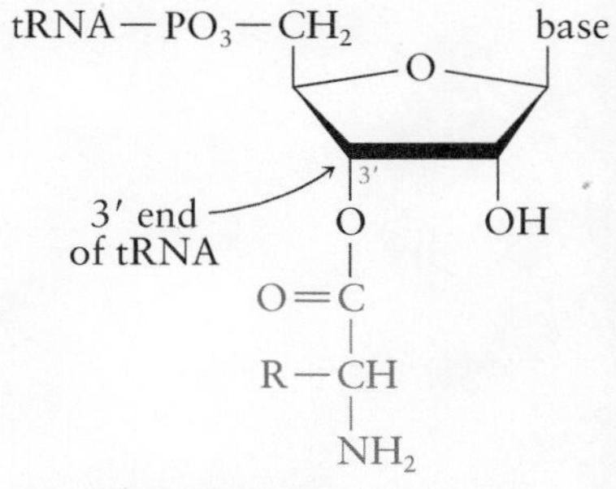

Each amino acid is joined to the 3′ end of its specific tRNA by an ester bond that forms between the carboxyl group of the amino acid and the 3′ hydroxy group of ribose. The reaction is catalyzed by an enzyme that matches tRNA molecules to their proper amino acids. These enzymes, which are very specific for both the structure of the amino acid and the tRNA, rarely cause a bond to form between an amino acid and the wrong tRNA. When a tRNA molecule is attached to its specific amino acid, it is said to be "activated" because it is ready to participate in protein synthesis. **Figure 13.21** shows the structure of an activated tRNA. In **Section 13.7**, a simplified schematic representation is used to describe protein synthesis.

Figure 13.21 General structure of an activated tRNA. The amino acid component is highlighted in blue.

13.6 Transcription: RNA Synthesis from DNA

Learning Objective 9 Describe the process by which RNA is synthesized from DNA in cells.

An enzyme called **RNA polymerase** catalyzes the synthesis of RNA. During the first step of the process, the DNA double helix begins to unwind at a point near the gene that is to be transcribed. Because the end product will be single-stranded, only one strand of the DNA molecule is transcribed. Ribonucleotides are linked along the unwound DNA strand in a sequence determined by complementary base-pairing of the DNA strand bases and ribonucleotide bases.

RNA polymerase A family of enzymes responsible for building nascent RNA strands from complementary DNA templates during transcription.

The DNA strand always has one sequence of bases recognized by RNA polymerase as the initiation or starting point. Starting at this point, the enzyme catalyzes the synthesis of mRNA in the 5′ to 3′ direction until it reaches another sequence of bases in the DNA template that designates the termination point. Because the complementary chains of RNA and DNA run in opposite directions, the enzyme must move along the DNA template in the 3′ to 5′ direction (see **Figure 13.22**). Once the RNA molecule has been synthesized, it moves away from the DNA template, which then rewinds to form the original double-helical structure. Messenger RNA, transfer RNA, and ribosomal RNA are all synthesized in the same way, with DNA serving as a template.

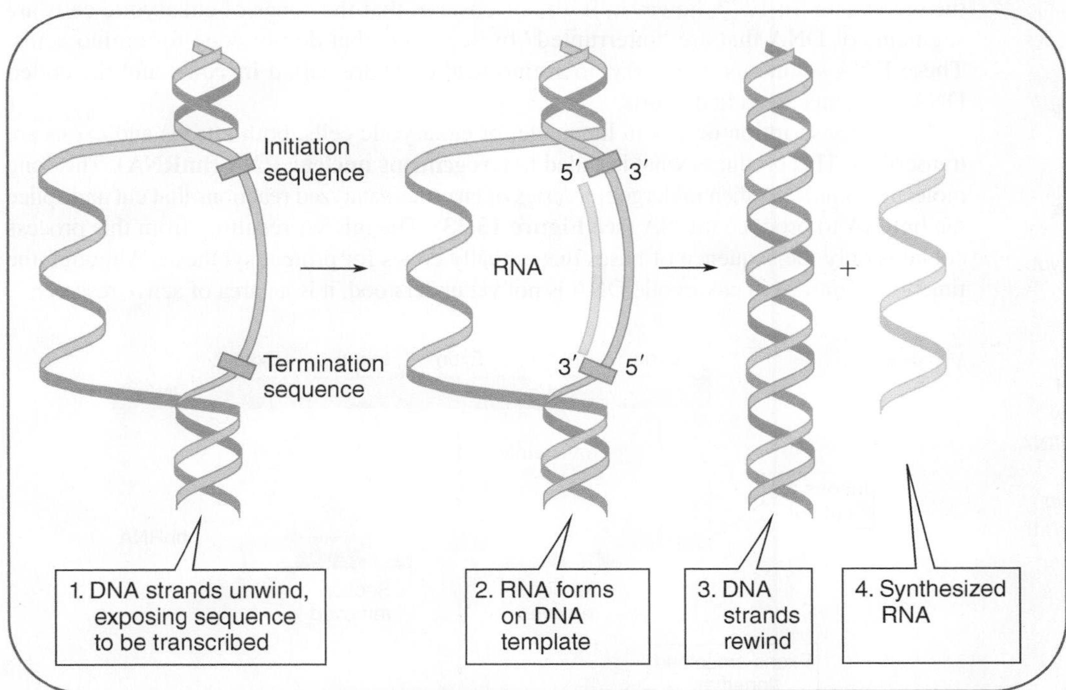

Figure 13.22 Transcription: The synthesis of RNA.

Example 13.4 Writing RNA Sequences

Write the sequence for the RNA that could be synthesized using the following DNA base sequence as a template.

5′ G–C–A–A–C–T–T–G 3′

Solution

RNA synthesis begins at the 3′ end of the DNA template and proceeds toward the 5′ end. The complementary RNA strand is formed from the bases C, G, A, and U. Uracil (U) is the complement of adenine (A) on the DNA template.

		Direction of strand $\longrightarrow$	
DNA template:	5′	G–C–A–A–C–T–T–G	3′
New mRNA:	3′	C–G–U–U–G–A–A–C	5′
		$\longleftarrow$ Direction of strand	

Writing the sequence of the new mRNA in the 5′ to 3′ direction, it becomes:

5′ C–A–A–G–U–U–G–C 3′

✔ Learning Check 13.4 Write the sequence for the mRNA that could be synthesized on the following DNA template.

5′ A–T–T–A–G–C–C–G 3′

Although the general process of transcription operates universally, there are differences in detail between the processes in prokaryotes. single-celled organisms without a nucleus, and eukaryotes, multicelled organisms with a nucleus (see **Section 14.8**). The genes found in prokaryotic cells exist as a continuous segment of a DNA molecule. Transcription of this gene segment produces mRNA that undergoes translation into a protein almost immediately because there is no nuclear membrane in a prokaryotic cell separating DNA from the cytoplasm. In 1977, however, it was discovered that the genes of eukaryotic cells are segments of DNA that are "interrupted" by segments that do not code for amino acids. These DNA segments that carry no amino acid code are called **introns**, and the coded DNA segments are called **exons**.

intron A segment of a eukaryotic DNA molecule that carries no codes for amino acids.

exon A segment of a eukaryotic DNA molecule that is coded for amino acids.

heterogeneous nuclear RNA (hnRNA) RNA produced when both introns and exons of eukaryotic cellular DNA are transcribed.

When transcription occurs in the nuclei of eukaryotic cells, both introns and exons are transcribed. This produces what is called **heterogeneous nuclear RNA (hnRNA)**. This long molecule of hnRNA then undergoes a series of enzyme-catalyzed reactions that cut and splice the hnRNA to produce mRNA (see **Figure 13.23**). The mRNA resulting from this process contains only the sequence of bases that actually codes for protein synthesis. Although the function of introns in eukaryotic DNA is not yet understood, it is an area of active research.

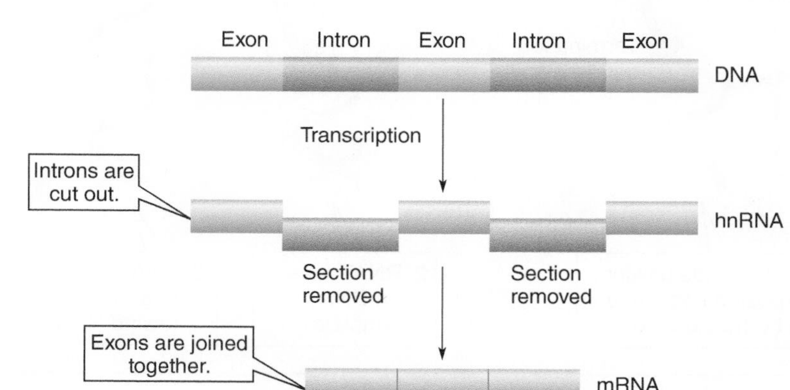

Figure 13.23 Segments of hnRNA formed from introns of eukaryotic cells are removed by special enzymes to produce mRNA. One way to remember these definitions is that introns stay *in* the nucleus, whereas exons *exit* the nucleus as mRNA.

13.7 The Genetic Code and Protein Synthesis

Learning Objective 10 Explain how the genetic code functions in the flow of genetic information.

Learning Objective 11 Outline the process by which proteins are synthesized in cells.

Learning Objective 12 Describe the downstream effects of genetic mutations in a living organism.

13.7A Discovery and Characteristics of Codons

By 1961, it was clear that the sequence of bases in DNA serves to direct the synthesis of mRNA, and that the sequence of bases in mRNA corresponded to the order of amino acids in a particular protein. However, the genetic code, the exact relationship between mRNA sequences and amino acids, was unknown.

At least 20 mRNA "words" are needed to represent uniquely each of the 20 common amino acids found in proteins. If the mRNA words consisted of a single letter represented by a base (A, C, G, or U), only four amino acids could be uniquely represented. Thus, it was proposed that it is not one mRNA base but a combination of bases that codes for each amino acid. For example, if a code "word" consists of a sequence of two mRNA bases, there are $4^2 = 16$ possible combinations, so only 16 amino acids could be represented uniquely.

This is a more extensive code, but it still contains too few "words" to do the job for 20 amino acids. If the code consists of a sequence of three bases, there are $4^3 = 64$ possible combinations, more than enough to specify uniquely each amino acid in the primary sequence of a protein. Research has confirmed that nature does indeed use three-letter code words (a triplet code) to store and express genetic information. Each sequence of three nucleotide bases that represents code words on mRNA molecules is called a **codon** (**Figure 13.24**).

codon A sequence of three nucleotide bases that represents a code word on mRNA molecules.

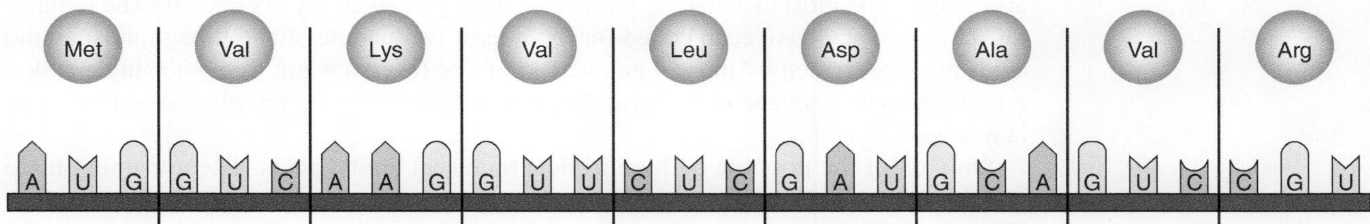

Figure 13.24 An example mRNA sequence separated into the codon with their corresponding amino acid three-letter abbreviation depicted in the circle above them.

In 1961, Marshall Nirenberg (1927–2010) and his coworkers at the National Institutes of Health (NIH) in Bethesda, Maryland, attempted to determine which three-letter codons correspond to each amino acid. They made a synthetic molecule of mRNA consisting of uracil bases only. Thus, this mRNA contained only one codon, the triplet UUU. They incubated this synthetic mRNA with ribosomes, amino acids, tRNAs, and the appropriate enzymes for protein synthesis. They obtained a polypeptide that consisted only of phenylalanine (Phe). Thus, the first word of the genetic code had been deciphered; UUU codes for phenylalanine.

A series of similar experiments by Nirenberg and other researchers followed, and by 1967, the entire genetic code had been broken. The complete code is shown in **Figure 13.25**.

The genetic code has several important characteristics. First, it applies almost universally. With very minor exceptions, the same amino acid is represented by the same three-base codon (or codons) in every organism. Second, most of the amino acids are represented by more than one codon—and are said to be **degenerate**. This is particularly apparent in the third (light red) ring of **Figure 13.25**.

degenerate A code in which different codons specify the same amino acid.

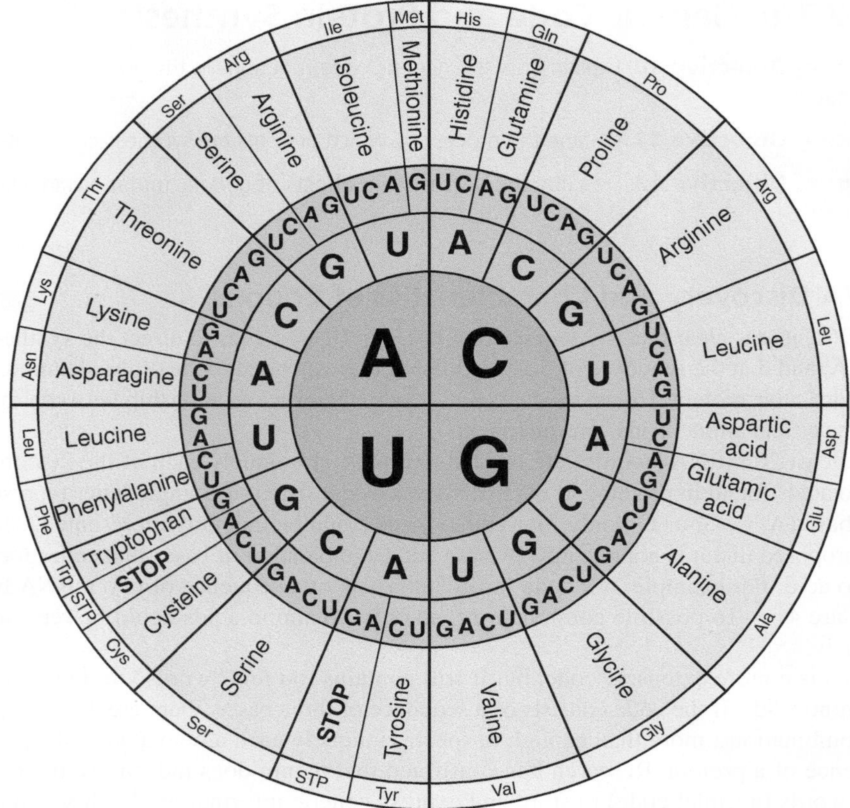

Figure 13.25 The genetic code that relates codons to amino acids. The first letter of the three-letter codon is at the center circle in blue. It is followed by the second letter in the first ring (shown in green) and the third letter in the third ring (light red). The amino acids coded by the codon are shown in the outermost rings in purple (full name) and beige (three-letter abbreviation).

Only methionine and tryptophan are represented by single codons. Leucine, serine, and arginine are the most degenerate, with each one represented by six codons. The remaining 15 amino acids are each coded for by at least two codons. Even though most amino acids are represented by more than one codon, the reverse is not true—no single codon represents more than one amino acid. Each three-base codon represents one and only one amino acid.

Only 61 of the possible 64 base triplets represent amino acids. The remaining three (UAA, UAG, and UGA) are signals for chain terminations and are called STOP codons (**Figure 13.25**). They tell the ribosome when the primary structure of the synthesized protein is complete and it is time to stop adding amino acids to the chain. The presence of stop signals in the code implies that START codons must also exist. There is only one initiation (start) codon, AUG, which also is the codon for the amino acid methionine. AUG functions as an initiation codon only when it occurs as the first codon of a sequence. As we explain next in **Section 13.7B**, protein synthesis begins when this happens. The general characteristics of the genetic code are summarized in **Table 13.2**.

Table 13.2 General Characteristics of the Genetic Code

Characteristic	Example
Codons are three-letter words.	GCA = alanine
The code is degenerate.	GCA, GCC, GCG, GCU all represent alanine
The code is precise.	GCC represents only alanine
Chain initiation is coded.	AUG = methionine
Chain termination is coded.	UAA, UAG, and UGA
The code is almost universal.	GCA = alanine, perhaps, in all organisms

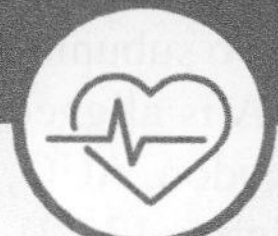

Health Connections 13.1

Is There a DNA Checkup in Your Future?

Today, the work of reading a human genome—which initially required thousands of scientists much longer than 10 years to accomplish at a cost of $2.7 billion—can be done in one lab in a matter of hours at a cost of $100.

What is the benefit of knowing the genome of a human? In 2000, only a handful of genes had been identified that were known to influence the development of common diseases such as heart disease, diabetes, and some cancers. Today, as a result of research in the Human Genome Project, the number of such predictive genes has increased to 600 to 700 and involves more than 100 diseases.

While it is probably a number of years in the future, one clinical expectation from the Human Genome Project is that physicians will one day be able to prescribe personal genome readings for patients. The results of such DNA check-ups might allow physicians to precisely estimate individual risks for a significant number of common diseases and then provide advice on lifestyle changes, or even prescribe preventive medications.

Continuing efforts are underway to reduce the costs associated with DNA-sequencing equipment and procedures to bring them in line with the costs of relatively routine diagnostic procedures. So, it appears probable that DNA checkups are likely to be in your future, especially if you are relatively young.

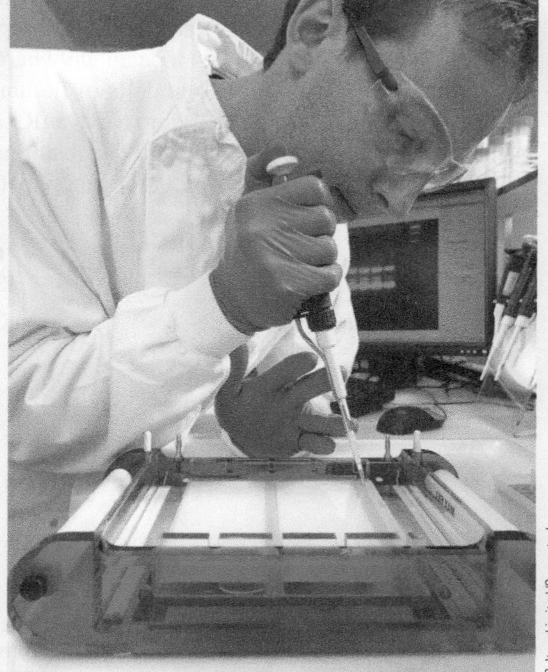

Cultura Limited/Superstock

A researcher prepares a sample for DNA sequencing.

13.7B Translation Process

To this point, you have become familiar with the molecules that participate in protein synthesis and the genetic code, the language that directs the synthesis. We now investigate the actual process by which polypeptide chains are assembled. There are three major stages in protein synthesis (see **Figure 13.26**): initiation of the polypeptide chain, elongation of the chain, and termination of the completed polypeptide chain.

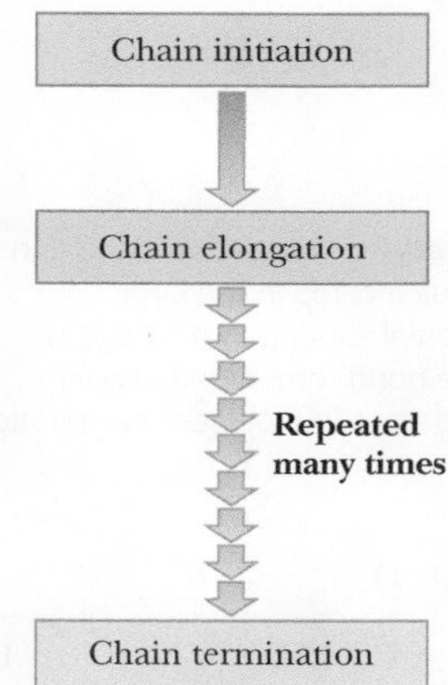

Figure 13.26 The stages of protein synthesis. [From Campbell, *Biochemistry*, 8E. © 2018 Cengage Learning.]

Initiation of the Polypeptide Chain

The first amino acid that starts the process of protein synthesis in prokaryotic (bacterial) cells is a derivative of methionine. This compound, *N*-formylmethionine, initiates the growing polypeptide chain as the *N*-terminal amino acid. Most proteins do not have *N*-formylmethionine as their *N*-terminal amino acid, however, which indicates that the *N*-formylmethionine is cleaved from the finished protein when protein synthesis is completed.

$$\underbrace{H-\overset{\overset{\displaystyle O}{\|}}{C}}_{\textit{Formyl group}}-NH-\underset{\underset{\displaystyle CH_2-S-CH_3}{|}}{\underset{\displaystyle CH_2}{\underset{|}{CH}}}-\overset{\overset{\displaystyle O}{\|}}{C}-O^-$$

N-formylmethionine
(fMet)

A ribosome is a large complex consisting of both protein and rRNA. It has two subunits, a large subunit and a small subunit. The initiation process begins when mRNA is aligned on the surface of a small ribosomal subunit in such a way that the initiating codon, AUG, occupies a specific site on the ribosome called the P site (the peptidyl site) (**Figure 13.27**). When the AUG codon is used this way to initiate synthesis of a polypeptide chain in prokaryotic cells, it represents *N*-formylmethionine (fMet) instead of methionine. When AUG is located anywhere else on the mRNA, it simply represents methionine. For the eukaryotic cells of humans, AUG always specifies methionine, even when it is the initiating codon. Next, a tRNA molecule with its attached fMet binds to the codon through hydrogen bonds. The resulting complex binds to the large ribosomal subunit to form a unit called an initiation complex (see **Figure 13.27**).

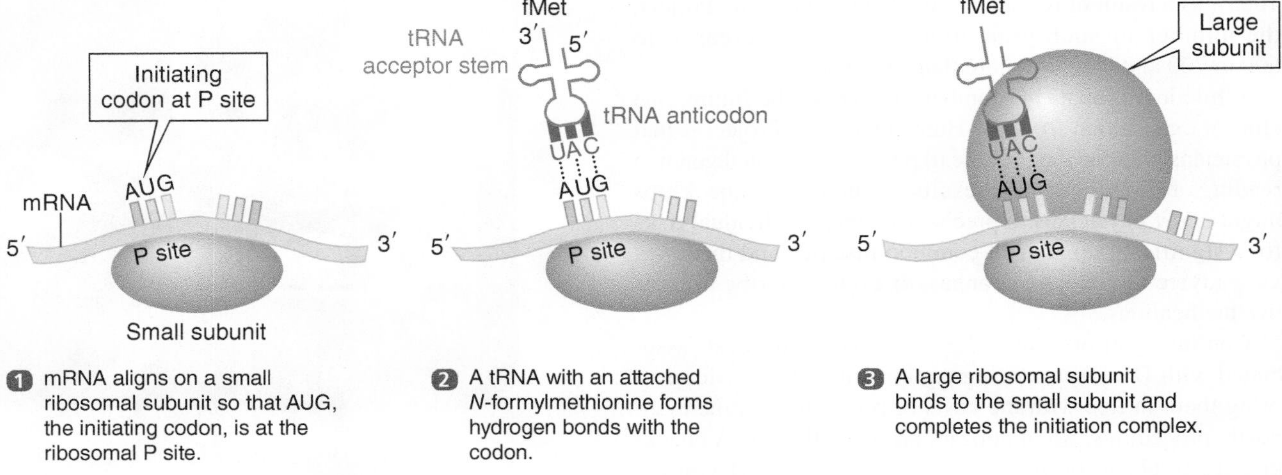

Figure 13.27 Initiation complex formation.

Elongation of the Chain

peptidyl transferase A protein enzyme located in the large ribosomal subunit that catalyzes amide bond formation between amino acids in a nascent polypeptide.

A second site, called the A site (the aminoacyl site), is located on the mRNA–ribosome complex next to the P site. The A site is where an incoming tRNA carrying the next amino acid will bond. Each of the tRNA molecules representing the 20 amino acids can try to fit the A site, but only the one with the correct anticodon that is complementary to the next codon on the mRNA will fit properly. Once at the A site, the second amino acid (in this case, phenylalanine) is linked to *N*-formylmethionine by a peptide bond whose formation is catalyzed by the enzyme **peptidyl transferase** in the ribosome (see **Figure 13.28**). After the peptide bond forms, the tRNA bound to the P site is "empty," and the growing polypeptide chain is now attached to the tRNA bound to the A site.

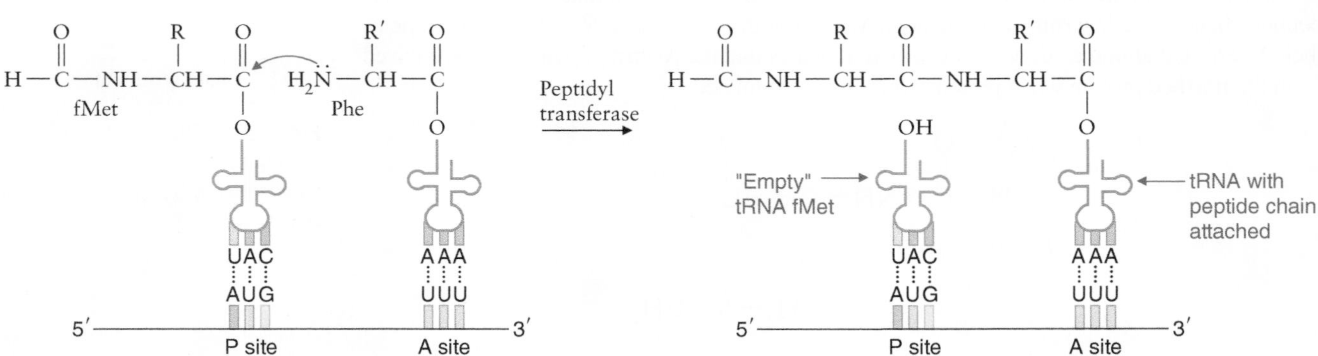

Figure 13.28 Peptidyl transferase links the amino acid at the P site through a peptide bond to the amino acid at the A site.

In the next phase of elongation, the whole ribosome moves one codon along the mRNA toward the 3′ end. As the ribosome moves, the empty tRNA is released from the P site, and tRNA attached to the peptide chain moves from the A site to the P site. This movement of the ribosome along the mRNA is called **translocation** and makes the A site available to receive the next tRNA with the proper anticodon. The amino acid carried by this tRNA bonds to the peptide chain, and the elongation process is repeated. This occurs over and over until the entire polypeptide chain is synthesized. The elongation process is represented in **Figure 13.29** for the synthesis of the tripeptide fMet–Phe–Val.

translocation The movement of the ribosome from one codon to the next along the mRNA in the 5′ to 3′ direction during the elongation phase of protein translation.

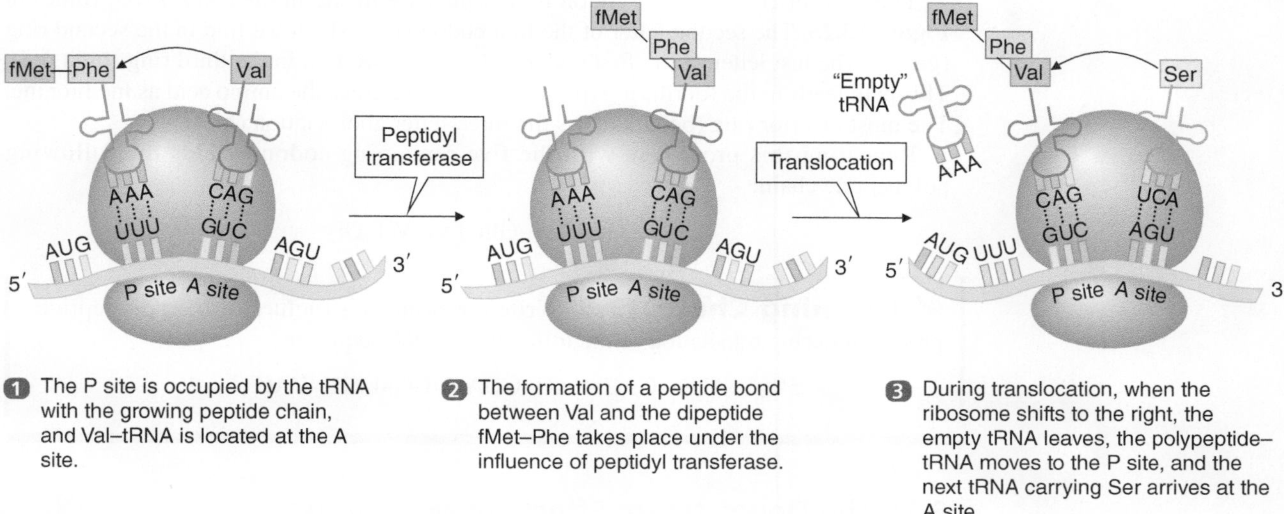

fMet–Phe
Val
AAA
CAG
UUU
GUC
AUG
AGU
5′
3′
P site
A site
Peptidyl transferase
fMet
Phe
Val
"Empty" tRNA
Translocation
AAA
Ser
CAG
UCA
GUC
AGU
AUG UUU
1 The P site is occupied by the tRNA with the growing peptide chain, and Val–tRNA is located at the A site.
2 The formation of a peptide bond between Val and the dipeptide fMet–Phe takes place under the influence of peptidyl transferase.
3 During translocation, when the ribosome shifts to the right, the empty tRNA leaves, the polypeptide–tRNA moves to the P site, and the next tRNA carrying Ser arrives at the A site.

Figure 13.29 Polypeptide chain elongation.

The Termination of Polypeptide Synthesis

The chain elongation process continues, and polypeptide synthesis continues until the ribosome complex reaches a stop codon (UAA, UAG, or UGA) on the mRNA. At that point, a specific protein known as a termination factor binds to the stop codon and catalyzes the hydrolysis of the completed polypeptide chain from the final tRNA. The "empty" ribosome dissociates and can then bind to another strand of mRNA to once again begin the process of protein synthesis.

Several ribosomes can move along a single strand of mRNA one after another (see **Figure 13.30**). Thus, several identical polypeptide chains can be synthesized almost simultaneously from a single mRNA molecule. This markedly increases the efficiency of utilization of the mRNA. Such complexes of several ribosomes and mRNA are called **polyribosomes** or **polysomes**. Growing polypeptide chains extend from the ribosomes into the cellular cytoplasm and spontaneously fold to assume characteristic three-dimensional secondary and tertiary configurations.

polyribosome or **polysome** A complex of mRNA and several ribosomes.

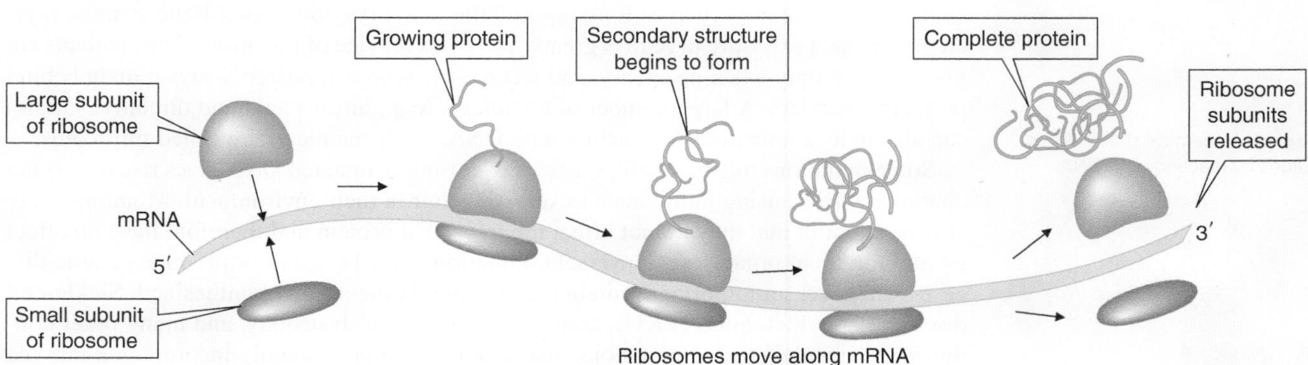

Growing protein
Secondary structure begins to form
Complete protein
Large subunit of ribosome
Ribosome subunits released
mRNA
5′
3′
Small subunit of ribosome
Ribosomes move along mRNA

Figure 13.30 A polyribosome consists of several ribosomes proceeding simultaneously along mRNA.

Example 13.5 Translating Polypeptide Structures

Write the primary structure of the polypeptide produced during translation of the following mRNA sequence.

5′ AUG-UUG-GAG-AAA-GUA-GGA 3′

Solution

To solve this problem, we must refer to the genetic code in **Figure 13.25**.

The first letter of the first codon is A, which we locate in the central ring (blue) in **Figure 13.25**. The second letter of the first codon is U, which we find in the second ring (green). The last letter of the first codon is G, which we find in the third ring (light red). This path leads to the fourth ring (purple), which identifies the amino acid as methionine. The most exterior ring (beige) shows the three-letter abbreviation (Met).

Repeating this procedure with the five remaining codons yields the following polypeptide chain:

Met–Leu–Glu–Lys–Val–Gly

✔ **Learning Check 13.5** Write the primary structure of the polypeptide produced during translation of the following mRNA sequence.

5′ AUG–CAC–CAU–GUA–UUG–UGU–UAG 3′

13.7C The Downstream Effects of Mutations

The base-pairing mechanism introduced in **Section 13.4** provides a nearly perfect way to copy a DNA molecule during replication. However, not even the copying mechanism involved in DNA replication is totally error free. It is estimated that, on average, one in every 10^{10} bases of DNA (i.e., 1 in 10 billion) is replicated incorrectly. Any change resulting in an incorrect sequence of bases on DNA is called a **mutation**. The faithful transcription of mutated DNA leads to an incorrect base sequence in mRNA. This can lead to an incorrect amino acid sequence for a protein, or possibly the failure of a protein to be synthesized at all.

mutation Any change resulting in an incorrect base sequence on DNA.

The three main types of mutations that result in translatable damage to DNA are called substitution, insertion, and deletion (**Figure 13.31**). In base substitution, a single base is substituted for another base. For example, if a CTT is mutated to TTT in the coding region of DNA, then the amino acid in the resulting protein changes from glutamic acid to lysine. In a base insertion mutation, extra nucleotides are added to a DNA region. This changes the reading frame such that the protein product made is truncated or nonfunctional. In a deletion mutation, one or more nucleotides are deleted from DNA. This also changes the reading frame, which often leads to a nonfunctional protein product.

Some mutations occur naturally during DNA replication; others can be induced by environmental factors such as ionizing radiation (X-rays, ultraviolet light, gamma rays, etc.). Repeated exposure to X-rays greatly increases the rate of mutation. Thus, patients are given X-rays only when necessary, and technicians who administer X-rays remain behind protective barriers. A large number of chemicals (e.g., nitrous acid and dimethyl sulfate) can also induce mutations by reacting with DNA. Such chemicals are called **mutagens**.

mutagen A chemical that induces mutations by reacting with DNA.

Some mutations might benefit a species over time as mutated DNA gives rise to varying characteristics causing individuals to survive better in their environment. Mutations may also be silent in that they do not affect the translated protein and therefore have no affect on an individual organism. Conversely, mutations may be lethal or produce genetic diseases whenever an important protein (or enzyme) is incorrectly synthesized. Sickle-cell disease, phenylketonuria (PKU), hemophilia, muscular dystrophy, and many other conditions are results of such mutations that have become permanently incorporated into the genetic makeup of certain individuals.

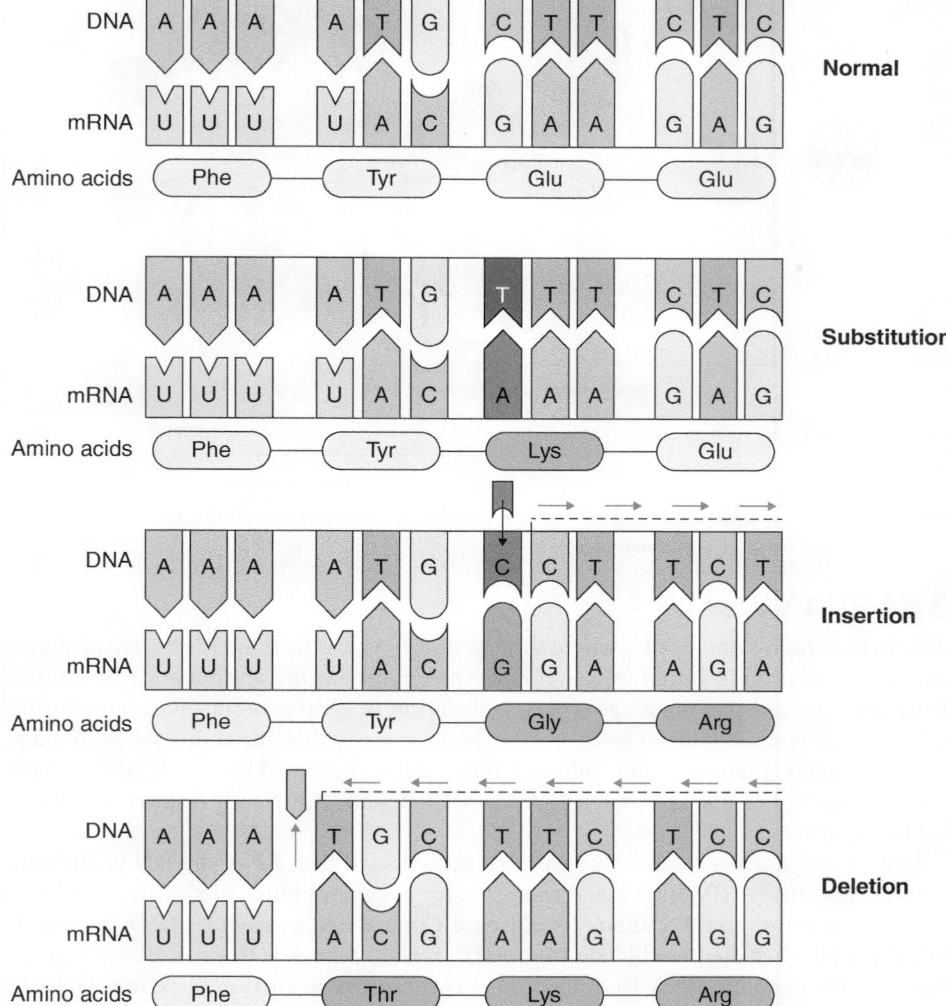

Figure 13.31 The types of mutations resulting in translatable damage are base substitution, insertion, and deletion.

In sickle cell disease, a DNA base mutation causes a hydrophobic valine to replace a hydrophilic glutamic acid in hemoglobin. This single mutation makes the protein fold differently. The adjacent hemoglobin peptide chains try to pack together to prevent exposing the hydrophobic valine to the aqueous environment (our blood!). As the hemoglobin chains stack together, they form long rod-like structures inside the red blood cells that warp the shape to resemble that of a "sickle" (**Figure 13.32**). The sickled blood cells have a tendency to clog blood capillaries, which has many catastrophic side effects, such as anemia, vision problems, and even strokes.

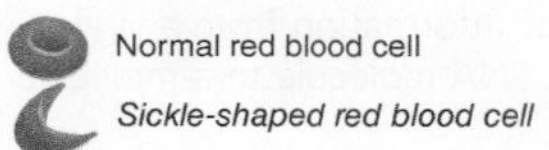

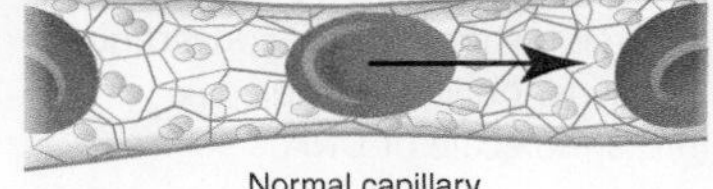

Normal capillary

Sickle cell disease

Figure 13.32 In sickle cell disease, blood cells take on a "sickled" shape and tend to clump together, causing painful blockages in the blood capillaries.

13.8 Viruses

Learning Objective 13 Describe three viruses and the diseases they cause in human patients.

Learning Objective 14 Explain the process of immunization using vaccines.

Viruses are microscopic agents, ~50-fold smaller than standard animal cells (**Figure 13.33**), that infect all forms of life. Generally they consist of strands of nucleic acids, either DNA or RNA, enclosed in a protein shell. Once viruses infect a host, they force the host cell machinery to replicate their genetic material and translate viral proteins that make up a nascent virus, called a budding virion.

virus An infectious agent made of nucleic acids (RNA or DNA) that can affect all forms of life and force the host organism to replicate its genetic material.

Figure 13.33 A comparison of the relative sizes of cells, virions, bacteria, and eggs on a logarithmic scale.

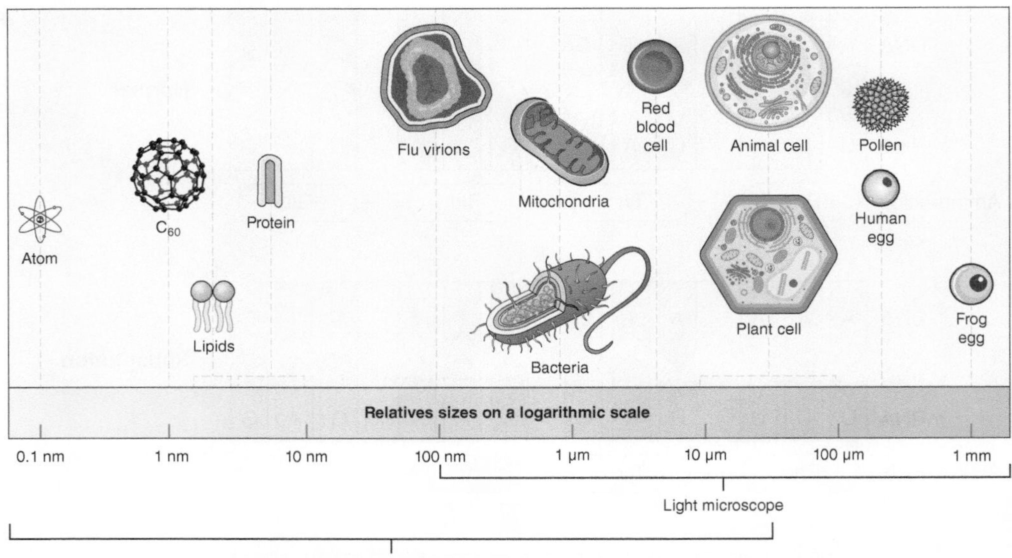

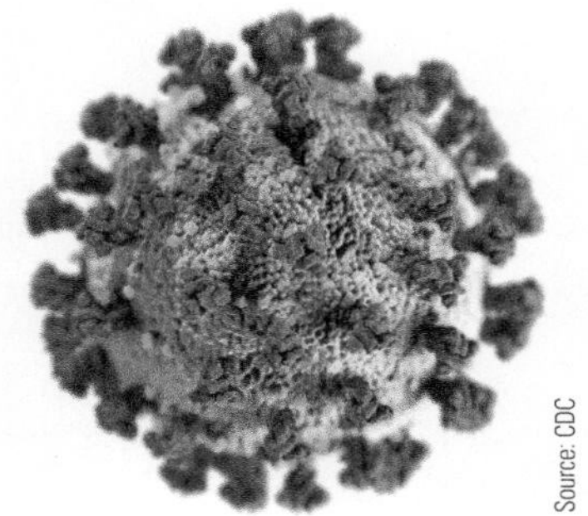

Figure 13.34 The COVID-19 virion viewed through an electron microscope.

reverse transcription The transfer of genetic information from a viral genomic RNA molecule to a molecule of DNA.

reverse transcriptase The viral enzyme responsible for converting a viral genomic RNA molecule into a molecule of DNA.

integrase The viral enzyme responsible for splicing retroviral DNA into the host cell DNA.

13.8A RNA Viruses

RNA viruses house one or two whole viral genomes in the form of RNA. Because their genetic information is stored as RNA, it serves directly as the template for replication of the viral genome (in most cases) and translation of viral proteins necessary for the budding virions. Examples of RNA viruses include many that cause disease in humans, such as measles, mumps, polio, rabies, Ebola, and influenza. The family of coronaviruses that has led to disease such as SARS, MERS, and COVID-19 (**Figure 13.34**) are also RNA viruses.

Retroviruses are a subclass of viruses that make use of both RNA and DNA. The most notable retrovirus is HIV, the virus responsible for acquired immunodeficiency syndrome (AIDS). According to the Centers for Disease Control and Prevention (CDC), over 35 million people have died worldwide from AIDS-related illnesses. Retroviruses are unique because their genomic RNA first undergoes **reverse transcription** back into DNA by the viral enzyme **reverse transcriptase** (**Figure 13.35**), before the DNA is integrated into the host cell genome in the nucleus by the enzyme **integrase** (**Figure 13.36**).

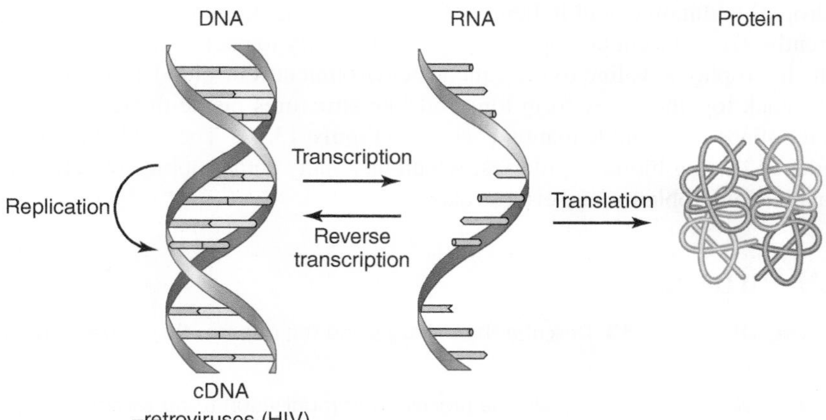

Figure 13.35 The central dogma of molecular biology with the addition of reverse transcription.

Once the retroviral DNA is spliced into the host DNA, the host cell unknowingly transcribes the viral genome and translates the mRNA into viral proteins necessary for viral replication and virion formation. Retroviruses are difficult for the immune system to combat because they exhibit extensive genomic mutations, preventing the immune system from

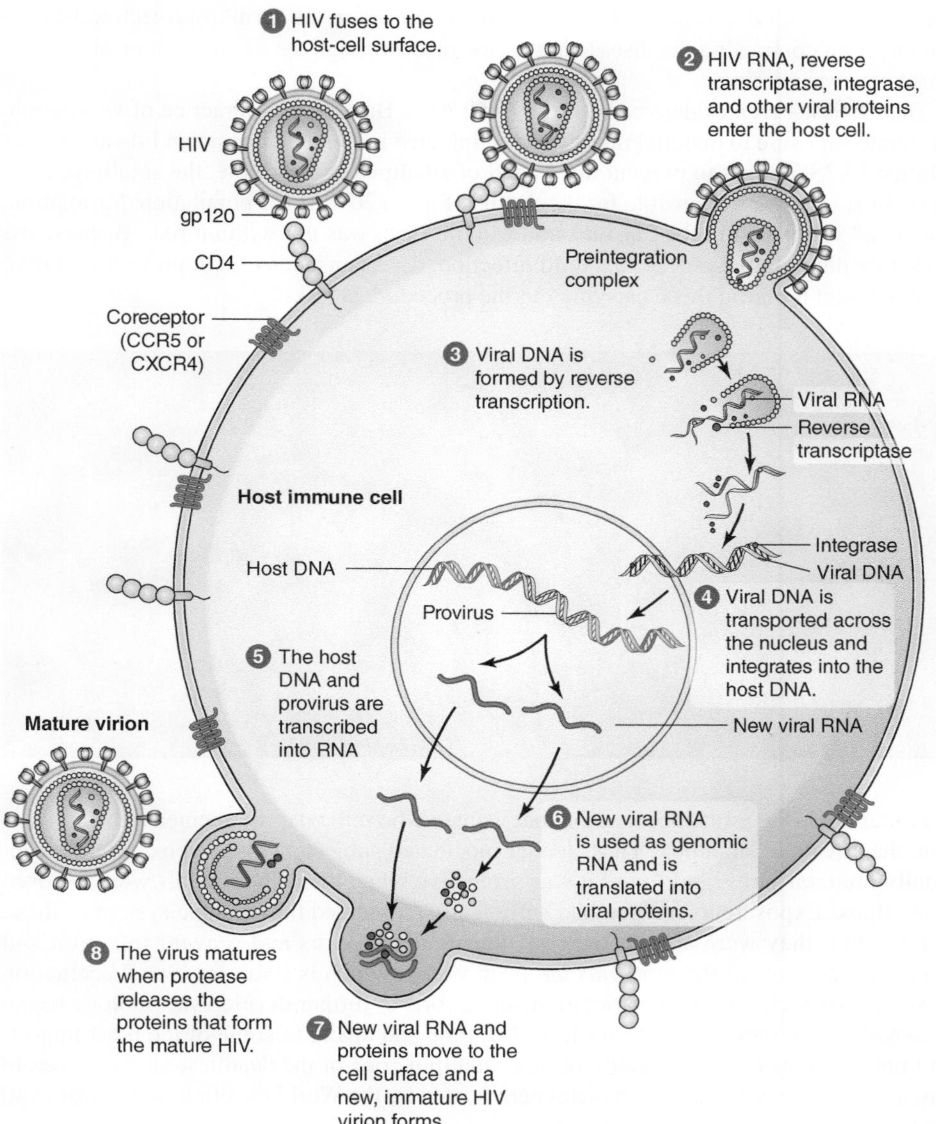

1 HIV fuses to the host-cell surface.
2 HIV RNA, reverse transcriptase, integrase, and other viral proteins enter the host cell.
HIV
gp120
CD4
Coreceptor (CCR5 or CXCR4)
Preintegration complex
3 Viral DNA is formed by reverse transcription.
Viral RNA
Reverse transcriptase
Host immune cell
Integrase
Viral DNA
Host DNA
Provirus
4 Viral DNA is transported across the nucleus and integrates into the host DNA.
5 The host DNA and provirus are transcribed into RNA
Mature virion
New viral RNA
6 New viral RNA is used as genomic RNA and is translated into viral proteins.
8 The virus matures when protease releases the proteins that form the mature HIV.
7 New viral RNA and proteins move to the cell surface and a new, immature HIV virion forms.

Figure 13.36 The retroviral life cycle.

forming effective antibodies to the virus. The high error rate of reverse transcriptase is in part responsible for the increased genetic variability and resulting rapid evolution of retroviruses.

13.8B DNA Viruses

DNA viruses can either be single-stranded or double-stranded. DNA viruses typically replicate within the host cell's nucleus, but unlike retroviruses, their DNA does not need to integrate into the host cell genome for viral replication to take place. Because DNA is inherently more stable than RNA, DNA viruses are often more stable than RNA viruses, which results in fewer mutations. Notable DNA viruses include herpes; smallpox; and the adenovirus, a type of virus that typically causes respiratory or common cold symptoms, especially in children (**Figure 13.37**). In fact, the devastation due to the smallpox pandemic was the driving force for the invention of the first methods of **immunization** (see **Section 13.8C**).

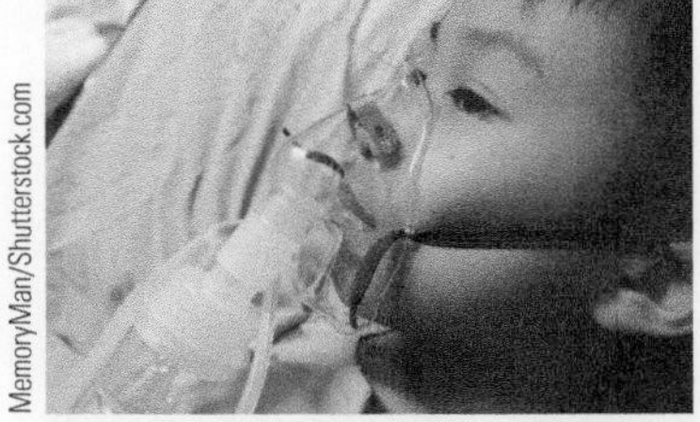

MemoryMan/Shutterstock.com

Figure 13.37 Pediatric patient exhibiting respiratory distress due to adenovirus.

immunization A medical treatment to prevent an individual from developing a disease upon exposure to its source.

13.8C Vaccines

Because viruses hijack the cell machinery of their host organism, cures to completely eradicate the virus are difficult to administer without risk of harming the host itself. The infectious damage viruses cause has prompted advances in medicine to combat their spread through human populations—efforts that continue to evolve into the present day.

vaccine A medical treatment utilizing parts of a virus or bacteria in order to instigate an immune response (antibody production) against a disease to prevent future illness during secondary exposure to the virus or bacteria.

The most successful treatment of viruses actually comes in the form of protecting healthy humans from contracting the disease in the first place through the use of biological applications known as **vaccines**.

The earliest vaccines date back to the late 1700s. Based on the practice of variolation, deliberate exposure to material taken from an infected individual, physician Edward Jenner (**Figure 13.38**) sought to prevent the spread of smallpox. At the time, the smallpox virus was estimated to be responsible for the deaths of up to 20% of the population. Variolation, which had originated in the Far East and Middle East, was not without risk. Because the procedure did not always lead to a mild infection, it was unsuccessful at preventing future infection, and it placed those carrying out the procedure at risk.

History_docu_photo/Alamy Stock Photo

Figure 13.38 Edward Jenner (1749–1823) inoculated subjects with cowpox, and discovered it provided patients immunity against smallpox.

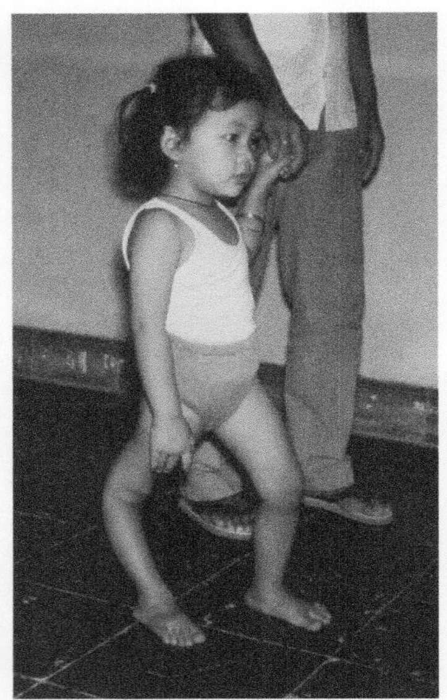

Source: Center for Disease Control (CDC) - PHIL

Figure 13.39 Polio infects motor neutrons in the brain, often causing paralysis. Post-polio syndrome is a condition where the symptoms remain long-term, causing complications such as the leg muscle weakness depicted in this image.

Jenner made the astute observation that women who milked cows seemed to be immune from the effects of smallpox. Thus, Jenner inoculated subjects with cowpox, a related but usually nonfatal virus, and found it was protective when those individuals were exposed to smallpox. Exposure to cowpox had sufficiently stimulated the immune system of those affected that they were able to fight off the smallpox virus and prevent infection and spread. Jenner coined the term *vaccine* from *vacca*, which is Latin for *cow*. Vaccination proved to be much safer than variolation, and efforts to further develop vaccinology began in earnest. Since then, vaccinations have led, time and again, to some of the most important improvements in public health. In fact, smallpox, one of the deadliest viral diseases in human history, was declared completely eradicated by the World Health Assembly in 1980 due to a worldwide vaccination campaign.

Landmark successes in the field of vaccinology include vaccines by Louis Pasteur (1822–1895) against rabies and Jonas Salk (1914–1995) against polio (**Figure 13.39**). Successful childhood immunization campaigns have led to dramatic reductions in cases of viral diseases such as measles, mumps, rubella, and polio. Unfortunately, vaccine uptake has slowed in recent years. Some vaccines may provide life-long protection, while others, like vaccines targeting influenza, are given on a yearly basis. This is due, in part, to the rapid evolution of influenza viruses and the number of different types of influenza that may spread during a given flu season. Scientists actually gather on an annual basis as part of the World Health Organization's Global Influenza Surveillance and Response System in order to predict the strains of influenza that will most likely spread during the next flu season and thus optimize the upcoming flu vaccine.

The most rapid example of vaccine development in human history came in response to the COVID-19 pandemic (**Figure 13.40**) that began toward the end of 2019. COVID-19 is the disease caused by exposure to the SARS-CoV-2 virus that had caused approximately 7 million deaths as of July 2023 according to the World Health Organization. The race for a COVID vaccine quickly became the most urgent scientific quest of the twenty-first century in order to reverse the global pandemic. Ultimately, new vaccine technology, along with manufacturing the vaccine at the same time as the vaccine was being tested in clinical trials, allowed an mRNA vaccine against COVID-19 to be developed and approved for use with patients beginning in late December 2021.

MR.Cole_Photographer/Moment/Getty Images

Figure 13.40 A busy street lined with commuters in medical masks protecting themselves from infection during the height of the COVID-19 pandemic.

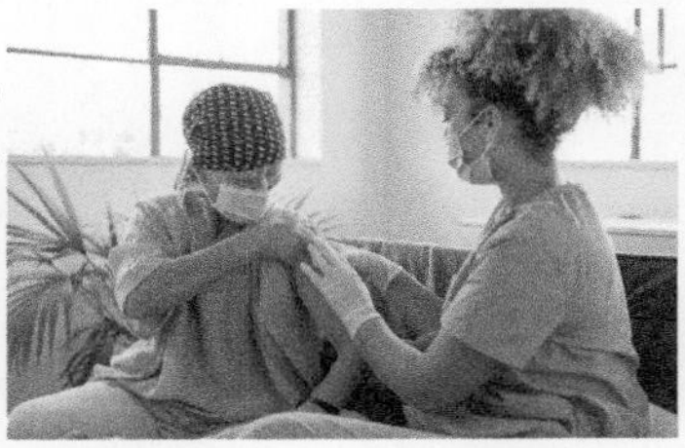

While previous generations of vaccines used attenuated viruses (weakened or dead forms of the virus), these mRNA vaccines, which had been researched and developed over the last two decades, were generated based on the genetic code of the virus itself. By inoculating with strategic bits of viral nucleic acid, rather than weakened or dead viruses, patients' cell machinery is able to generate antibodies in order to fight off infection with no risk of the vaccine itself causing disease. While the rapid deployment of these COVID mRNA vaccines has caused some hesitancy in the population, current regulatory studies have found the vaccines to be relatively safe and significantly effective at preventing the development of severe disease from COVID-19. These preventative measures are especially helpful for immunocompromised individuals and others who are at higher risk of developing complications upon COVID-19 infection (**Figure 13.41**).

Fly View Productions/Getty Images

Figure 13.41 A cancer patient receiving a COVID-19 vaccine.

Despite the success of these mRNA vaccines as of late 2022, mutated forms of SARS-CoV-2 have continued to evolve through persistent population spread. This development has meant the eradication of COVID-19 remains elusive. Next-generation vaccines remain in development through alternate admission pathways (e.g., nasal vaccines), which may provide more sterilizing immunity, or through so-called pan-coronavirus vaccines, which would be effective against all possible mutations of the virus. Furthermore, the rapidly accelerating mRNA vaccine technology is also being studied for use against other viruses, such as influenza, and even as a possible tool in the fight against cancer.

13.9 Recombinant DNA

Learning Objective 15 Describe the technology used in molecular cloning.

Learning Objective 16 Describe one biomedical application of CRISPR.

Remarkable technology is available that allows segments of DNA from one organism to be introduced into the genetic material of another organism. The resulting new DNA (containing the foreign segment) is referred to as **recombinant DNA**. The application of this technology, commonly called genetic engineering, has produced major advances in human health care and holds the promise for a future of exciting advances in biology, agriculture, and many other areas of study.

recombinant DNA DNA of an organism that contains genetic material from another organism.

An early success of genetic engineering was the introduction of the gene for human insulin into the DNA of the common bacterium *E. coli*. The bacterium then transcribed and translated the information carried by the gene and produced the protein hormone. By culturing such *E. coli*, it has become possible to produce and market large quantities of human insulin. The availability of human insulin is very important for people who have diabetes and are allergic to the insulin traditionally used, which is isolated from the pancreatic tissue of slaughtered pigs or cattle. **Table 13.3** lists several other medically important materials that have been produced through genetic engineering.

Table 13.3 Some Substances Produced by Genetic Engineering

Substance	Use
Human insulin	Treats diabetes mellitus
Human growth hormone	Treats dwarfism
Interferon	Fights viral infections
Hepatitis B vaccine	Protects against hepatitis
Malaria vaccine	Protects against malaria
Tissue plasminogen activator	Promotes the dissolution of blood clots

13.9A Molecular Cloning

restriction enzyme A protective enzyme found in some bacteria that catalyzes the cleaving of all but a few specific types of DNA.

The discovery of restriction enzymes in the 1960s and 1970s made genetic engineering possible. **Restriction enzymes**, which are found in a wide variety of bacterial cells, catalyze the cleaving of DNA molecules. These enzymes are normally part of a mechanism that protects certain bacteria from invasion by foreign DNA, such as that of a virus. In these bacteria, some of the bases in their DNA have methyl groups attached:

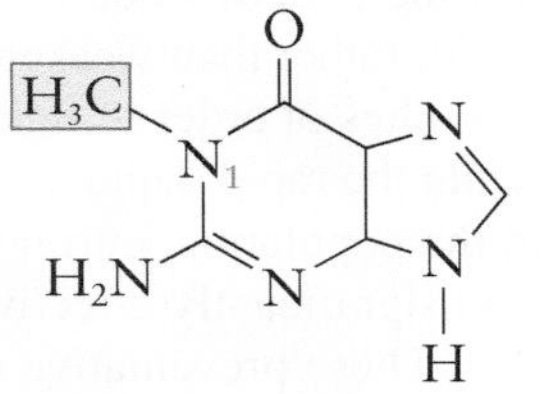

1-methylguanine

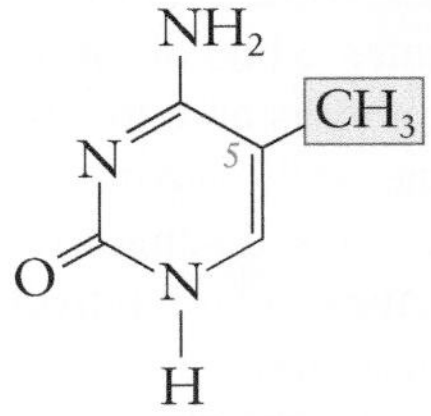

5-methylcytosine

The methylated DNA of these bacteria is left untouched by the restriction enzymes, but foreign DNA that lacks the methylated bases undergoes rapid cleavage of both strands and thus becomes nonfunctional. Because there is a "restriction" on the type of DNA allowed in the bacterial cell, the protective enzymes are called restriction enzymes.

palindrome A section of DNA in which the two strands have the same sequence but run in opposite directions.

Restriction enzymes act at sites on DNA called palindromes. In language, a palindrome is any word or phrase that reads the same in either direction, such as "racecar" or "Hannah." For double-stranded DNA, a **palindrome** is a section in which the two strands have the same sequence but run in opposite directions. In the following examples of DNA palindromes, the arrows indicate the points of attack by restriction enzymes:

```
     ↓                      ↓
5′ C-C-G-C-G-G 3′     5′ G-G-A-T-C-C 3′
   : : : : : :           : : : : : :
3′ G-G-C-G-C-C 5′     3′ C-C-T-A-G-G 5′
       ↑                          ↑
```

At least 100 bacterial restriction enzymes are known, and each catalyzes DNA cleavage in a specific and predictable way. These enzymes are the tools used to take DNA apart and reduce it to fragments of known size and nucleotide sequence.

Another set of enzymes that is important in genetic engineering, called DNA ligases (**Section 13.4**), has been known since 1967. Recall that these enzymes normally function to connect DNA fragments during replication. In genetic engineering, however, they are used to put together pieces of DNA produced by restriction enzymes.

The introduction of a new DNA segment (gene) into a bacterial cell requires the assistance of a DNA carrier called a **vector**. The vector is often a circular piece of double-stranded DNA called a **plasmid**. Plasmids range in size from 2000 to several hundred thousand nucleotides and are found in the cytoplasm of bacterial cells. Plasmids function as accessories to chromosomes by carrying genes for the inactivation of antibiotics and the production of toxins. They have the unusual ability to replicate independently of chromosomal DNA. A typical bacterial cell contains about 20 plasmids and one or two chromosomes.

vector A carrier of foreign DNA into a cell.

plasmid Circular, double-stranded DNA found in the cytoplasm of bacterial cells.

The recombinant DNA technique begins with the isolation of a plasmid from a bacterium. A restriction enzyme is added to the plasmid, which is cleaved at a specific site:

```
~G–G–A–T–C–C~   Restriction    ~G                  G–A–T–C–C~
 ⋮  ⋮  ⋮  ⋮  ⋮  ⋮      enzyme       ⋮                           ⋮
~C–C–T–A–G–G~  ———————→   ~C–C–T–A–G                   G~
                                    Sticky ends
```

Because a plasmid is circular, cleaving it this way produces a double-stranded chain with two ends (see **Figure 13.42**). These are called sticky ends because each has one strand in which several bases are unpaired and ready to pair up with a complementary section if available.

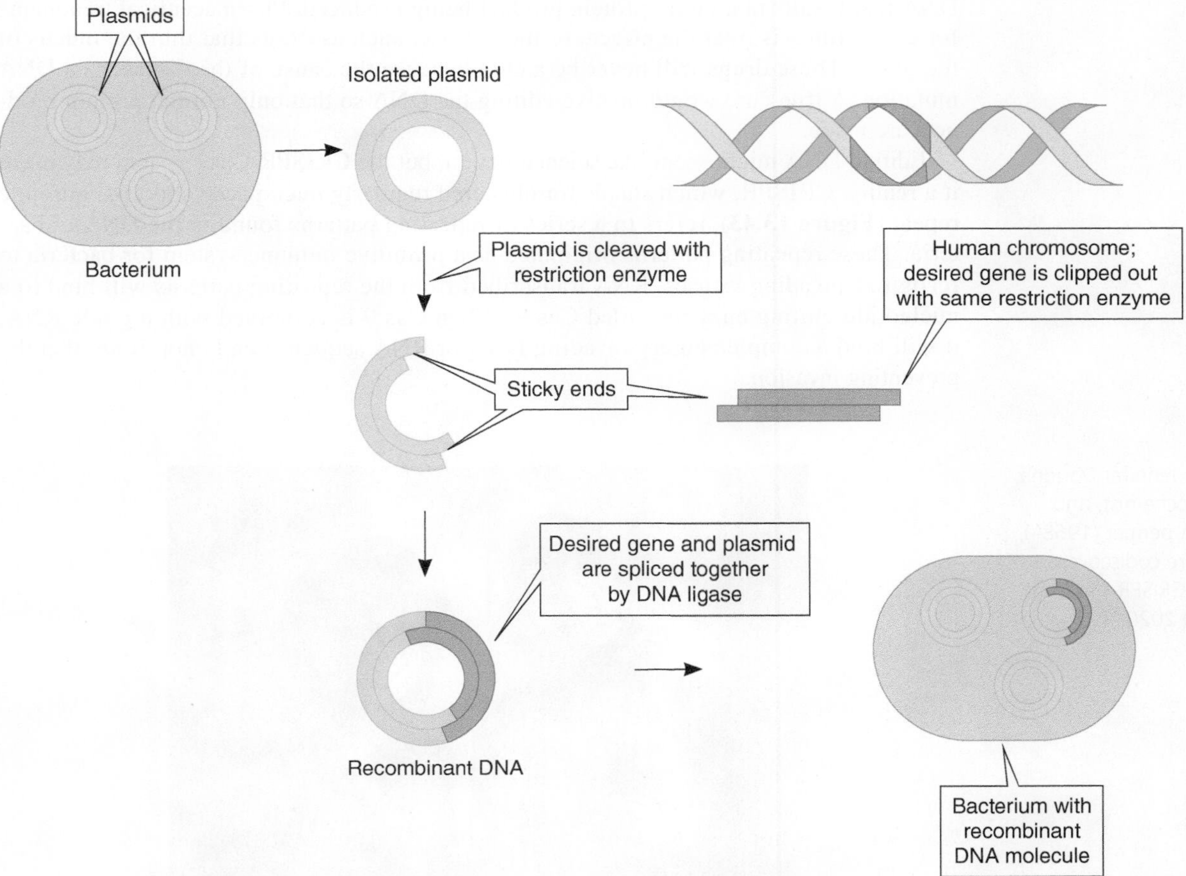

Figure 13.42 The formation of recombinant DNA.

The next step is to provide the sticky ends with complementary sections for pairing. A human chromosome is cleaved into several fragments using the same restriction enzyme. Because the same enzyme is used to cleave the human DNA, the same sticky ends result:

```
G-A-T-C-C~~~~~~~~~~~~~~~~~~G
        :  Human DNA segment :
        G~~~~~~~~~~~~~~~~~~C-C-T-A-G
```

To splice the human gene into the plasmid, the two are brought together under conditions suitable for the formation of hydrogen bonds between the complementary bases of the sticky ends. The breaks in the strands are then joined by using DNA ligase, and the plasmid once again becomes a circular piece of double-stranded DNA that now incorporates both human and bacterial DNA (see **Figure 13.42**); recombinant DNA has been formed.

Bacterial cells are bathed in a solution containing recombinant DNA plasmids, which diffuse into the cells. When the bacteria reproduce, they replicate all the genes, including the recombinant DNA plasmid. Bacteria multiply quickly, and soon a large number of bacteria, all containing the modified plasmid, are manufacturing the new protein as directed by the recombinant DNA. The use of this technology makes possible (in principle) the large-scale production of virtually any polypeptide or protein. Because these substances play so many essential roles in the body, genetic engineering has the potential to make tremendous contributions to improving health care in the future.

13.9B CRISPR

Cystic fibrosis is one of many diseases that are caused by an error (a mutation) in our DNA that results in a faulty protein product being produced. Pharmaceutical treatments for cystic fibrosis treat the effects of the disease, such as drugs that thin the mucus of the lungs. These drugs will never be a cure because the cause of the disease is a DNA mutation. A true cure would involve editing the DNA so that only normal protein products are made.

Editing DNA might seem like science fiction, but the CRISPR-Cas9 system may make it a reality. CRISPR, which stands for clustered regularly interspaced short palindromic repeat (**Figure 13.43**), refers to a series of repeating patterns found in the DNA of bacteria. These repeating patterns originated as a primitive immune system for bacteria to recognize invading viruses. RNA transcribed from the repeating patterns will bind to a nucleotide-cutting enzyme called Cas 9. When Cas 9 is combined with a guide RNA, it will bind a complementary invading DNA or RNA sequence and chop it up, thereby preventing invasion.

Alexander Heinl/picture alliance/Getty Images

Figure 13.43 Jennifer Doudna (1964–) (left), biochemist, and Emmanuelle Charpentier (1968–), microbiologist, are codiscoverers of the genetic tool CRISPR-Cas9 and cowinners of the 2020 Nobel Prize in Chemistry.

Although cutting viral DNA or RNA will inactivate it, cutting human DNA results in the activation of DNA repair mechanisms. Because the guide RNA–Cas 9 complex is recruited to certain regions of DNA, it is possible to edit very specific regions of DNA. CRISPR–Cas 9 technology has been used to change a single base of DNA, replace a faulty gene, splice in new genes, and inactivate genes. With this in mind, is a cure for cystic fibrosis around the corner?

Some biomedical treatments involving CRISPR are already underway. In the spring of 2021, the U.S. Food and Drug Administration (FDA) approved a novel approach involving CRISPR to treat humans with severe sickle cell disease (**Figure 13.44**). The clinical trial will take place over 4 years and the results of the trial, if successful, could forever change the future of how we treat diseases mapped to genetic mutations. Among the many challenges to CRISPR-based treatments is how to deliver CRISPR–Cas 9 gene-editing technology to every cell. Challenges aside, gene-editing technology is being used in new and inventive ways every day!

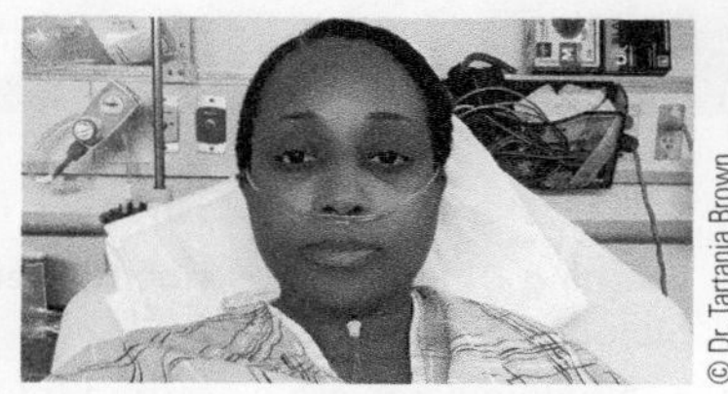

Figure 13.44 The U.S. FDA has approved the first test of CRISPR to correct the genetic defect causing sickle cell disease.

Health Career Description

Sonographer

According to the American Registry for Diagnostic Medical Sonography (ARDMS), "Sonography is a painless non-invasive procedure that uses high-frequency sound waves to produce visual images of organs, tissues, or blood flow inside the body."

Sonography encompasses several specialties, including abdominal sonographer, breast sonographer, neurosonographer, gynecological sonographer, echocardiographer, vascular sonographer, and musculoskeletal sonographer. Each of these specialties requires separate certifications found under ARDMS sonography assessments.

Sonographers use a transducer to introduce sound waves into the body that may then be converted to visual imagery. Sonographers create videos of the examinations performed and write reports interpreting the results for physicians.

Students may become sonographers by pursuing a 2-year program at a technical college or a 4-year bachelor's degree. They may also pursue a 1-year certificate if they are already trained in a health profession. Weber State University in Ogden, Utah, offers a multispecialty bachelor's degree called diagnostic medical sonography (DMS). This bachelors-degree program builds on an associate of radiologic sciences degree. At the end of the 4-year degree program, students are ready to test in cardiac, medical, and vascular sonography as well as a variety of subspecialties.

Because sonographers work with patients in very intimate settings, excellent communication skills and strict adherence to privacy standards are essential. Sonographers must be able to treat patients with respect and kindness, while also completing the job of obtaining the images ordered by the physician.

Sonographers may work in a variety of settings depending on specialty. A wide range of possibilities exist for sonographers, from high-demand jobs in hospitals, to more predictable regular shift work in clinics. The need for sonographers is expected to increase in the future, making it an excellent career choice.

Sources: American Registry for Diagnostic Medical Sonographers (ARDMS). (n.d.). How to Become a Sonographer. Retrieved May 17, 2020 from https://www.ardms.org/how-to-become-a-sonographer/; Weber State University. (n.d.). Diagnostic Medical Sonography. Retrieved May 17, from 2020 from https://www.weber.edu/RadSci/diagnosticemphases.html.

Concept Summary

13.1 The Flow of Genetic Information

Learning Objective: **Describe the central dogma of molecular biology.**

- Nucleic acids are classified into two categories: ribonucleic acids (RNA) and deoxyribonucleic acids (DNA).
- The flow of genetic information occurs in two steps called transcription and translation.
- In transcription, information stored in DNA molecules is passed to molecules of messenger RNA (mRNA).
- In translation, the mRNA serves as a template that directs the assembly of amino acids into proteins.
- Transcription occurs in the cell's nucleus, whereas translation occurs at the ribosome in the cytoplasm.

13.2 The Components of Nucleic Acids

Learning Objectives: **Identify the components of nucleotides. Classify the sugars and bases of nucleotides.**

- DNA and RNA are polymers made up of monomers called nucleotides.
- All nucleotides are composed of a pyrimidine or purine base, a sugar, and phosphate.
- The sugar component of RNA is ribose, whereas that of DNA is deoxyribose.
- The bases adenine, guanine, and cytosine are found in all nucleic acids.
- Uracil is found only in RNA.
- Thymine is found only in DNA.

13.3 The Structure of DNA

Learning Objectives: **Describe the structure of DNA. Write the complementary base strand of a DNA sequence.**

- The nucleotides of DNA are joined by phosphodiester linkages between phosphate groups and sugars.
- The resulting sugar–phosphate backbone is the same for all DNA molecules, but the order of attached bases along the backbone varies.
- This order of nucleotides with attached bases is the primary structure of nucleic acids.
- The three-dimensional structure is a double-stranded helix held together by hydrogen bonds between complementary base pairs on the strands.

13.4 DNA Replication

Learning Objective: **Outline the process of DNA replication.**

- The replication of DNA occurs when a double strand of DNA unwinds at specific points.
- The exposed bases match up with complementary bases of nucleotides.
- The nucleotides bind together to form two new strands that are complementary to the strands that separated.
- Thus, the two new DNA molecules each contain one old strand and one new strand (semiconservative replication).

13.5 Ribonucleic Acid (RNA)

Learning Objectives: **Describe three differences between DNA and RNA. Describe the function of the three main types of cellular RNA.**

- Three main forms of ribonucleic acid are found in cells:
  - Messenger RNA (mRNA)
  - Ribosomal RNA (rRNA)
  - Transfer RNA (tRNA)
- mRNA carries genetic information from the DNA in the cell nucleus to the site of protein synthesis in the cytoplasm.
- rRNA is the RNA portion of the protein–RNA complex that is the ribosome.
- tRNAs bind and deliver individual amino acid molecules to the site of protein synthesis.
- All RNA molecules are single-stranded, but the single strands can hydrogen-bond with themselves, resulting in stem–loops or folds.

13.6 Transcription: RNA Synthesis from DNA

Learning Objective: **Describe the process by which RNA is synthesized from DNA in cells.**

- The various RNAs are synthesized in a way that is similar to how DNA is replicated.
- Nucleotides with complementary bases align themselves against one strand of a partially unwound DNA segment containing the genetic information that is to be transcribed.
- The aligned nucleotides bond to form the RNA.
- In eukaryotic cells, the produced RNA is heterogeneous and is cut and spliced after being synthesized to produce the functional RNA.

13.7 The Genetic Code and Protein Synthesis

Learning Objectives: **Explain how the genetic code functions in the flow of genetic information. Outline the process by which proteins are synthesized in cells. Describe the downstream effects of genetic mutations in a living organism.**

- The genetic code is a series of three-letter "words" that represent the amino acids of proteins as well as start and stop signals for protein synthesis.
- The letters of the "words" are the bases found on mRNA, and the words on mRNA are called codons.
- The genetic code is the same for all organisms and is degenerate for most amino acids.
- The translation step in the flow of genetic information results in the synthesis of proteins.
- The synthesis takes place when properly coded mRNA forms a complex with the component of a ribosome.
- Transfer RNA molecules carrying amino acids align themselves along the mRNA in an order representing the correct primary structure of the protein.
- The order is determined by the matching of complementary codons on the mRNA to anticodons on the tRNA.
- The amino acids sequentially bond together to form the protein, which then spontaneously forms characteristic secondary and tertiary structures.

- Any change that results in an incorrect sequence of bases on DNA is called a mutation.
- The three main types of DNA mutation are substitution, insertion, and deletion.
- Substitution mutations occur when a single base is substituted for another base.
- Base substitutions can be silent, resulting in no change in translation, or they can cause an amino acid swap.
- Insertion mutations occur when extra nucleotides are added to a DNA region.
- Insertion changes the reading frame such that the protein product made is truncated or nonfunctional.
- In a deletion mutation, one or more nucleotides are deleted from DNA. This also changes the reading frame, which often leads to a nonfunctional protein product.
- Some mutations occur naturally during DNA replication, whereas others are induced by environmental factors.
- Some mutations are beneficial to organisms.
- Other mutations may be lethal or result in genetic diseases.

13.8 Viruses

Learning Objective: **Describe three viruses and the diseases they cause in human patients. Explain the process of immunization using vaccines.**

- Viruses are microscopic agents that infect all forms of life.
- They consist of strands of nucleic acids, either DNA or RNA, enclosed in a protein shell.
- Viruses infect host cells and force them to replicate viral genetic material and produce new viruses that bud from the host cell.
- Viruses can be broadly classified as RNA or DNA.
- Retroviruses are a subclass of RNA viruses that must reverse transcribe their genomic RNA into DNA prior to integration into the host cell genome.
- Reverse transcriptase, the enzyme that converts retroviral RNA into DNA, increases genetic variability because it is highly error prone.
- Increased genetic variability leads to virus evolution and helps the virus evade the host's immune system.
- Variolation was a crude form of immunization in which a patient was deliberately exposed to material taken from an infected individual to produce antibodies to a virus.
- Post-variolation, traditional vaccines use weakened or dead virus to instigate an immune response that will fight off a virus in future exposures.
- Novel mRNA vaccines use nucleic acids to encode bits of viral genome such that the cell machinery of the patient generates antibodies.

13.9 Recombinant DNA

Learning Objective: **Describe the technology used in molecular cloning. Describe one biomedical application of CRISPR.**

- The discovery and application of restriction enzymes and DNA ligases have resulted in a technology called molecular cloning.
- Molecular cloning involves the isolation of genes (DNA) that code for specific useful proteins and the introduction of these genes into the DNA of bacteria.
- The new (recombinant) DNA in the rapidly reproducing bacteria mediates production of the useful protein, which is then isolated for use.
- CRISPR stands for clustered regularly interspaced short palindromic repeat.
- The CRISPR technique involves a nucleotide-cutting enzyme called Cas 9, which when combined with a guide RNA, will bind a complementary invading DNA or RNA sequence and chop it up.
- Because the guide RNA–Cas 9 complex is recruited to certain regions of DNA, it is possible to edit very specific regions of DNA.
- CRISPR–Cas 9 technology has been used to change a single base of DNA, replace a faulty gene, splice in new genes, and inactivate genes.
- CRISPR-based treatments are being developed to treat sickle cell disease and potentially other diseases in the future.

Key Terms and Concepts

Anticodon (13.5)
Central dogma of molecular biology (13.1)
Chromosome (13.4)
Codon (13.7)
Complementary DNA strands (13.3)
Degenerate (13.7)
Deoxyribonucleic acid (DNA) (13.1)
DNA helicase (13.4)
DNA ligase (13.4)
DNA polymerase (13.4)
Exon (13.6)
Gene (13.1)
Heterogeneous nuclear RNA (hnRNA) (13.6)
Immunization (13.8)
Integrase (13.8)
Intron (13.6)
Messenger RNA (mRNA) (13.1)
Mutagen (13.7)
Mutation (13.7)
Nucleic acid (13.1)
Nucleic acid backbone (13.3)
Nucleotide (13.2)
Okazaki fragment (13.4)
Palindrome (13.9)
Peptidyl transferase (13.7)
Plasmid (13.9)
Polymerase chain reaction (PCR) (13.4)
Polyribosome (polysome) (13.7)
Recombinant DNA (13.9)
Replication (13.4)
Replication fork (13.4)
Restriction enzyme (13.9)
Reverse transcriptase (13.8)
Reverse transcription (13.8)
Ribonucleic acid (RNA) (13.1)
Ribosomal RNA (rRNA) (13.5)
Ribosome (13.1)
RNA polymerase (13.6)
Semiconservative replication (13.4)
Transcription (13.1)
Transfer RNA (tRNA) (13.1)
Translation (13.1)
Translocation (13.7)
Vaccine (13.8)
Vector (13.9)
Virus (13.8)

Key Equation

1. Formation a nucleotide (Section 13.2)	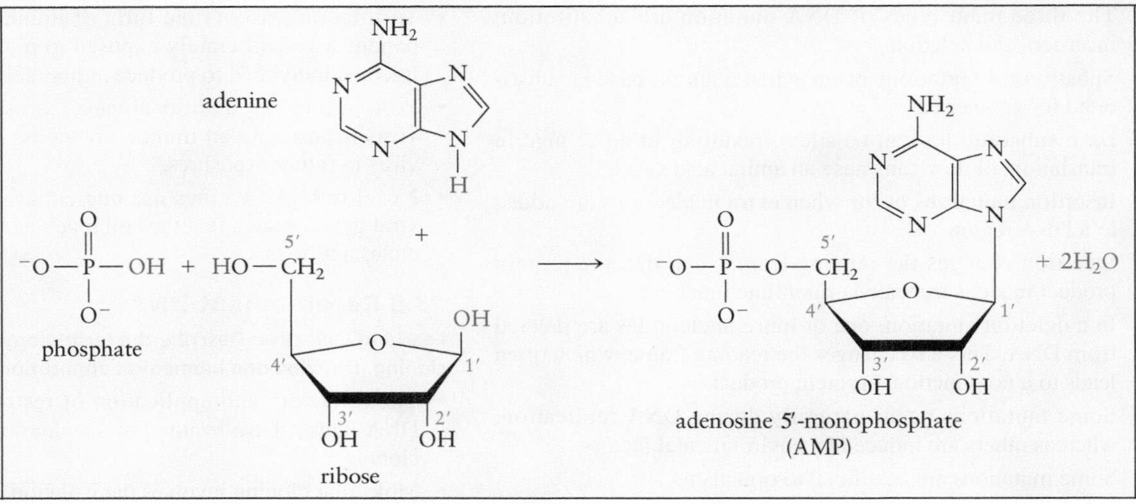 $^{-}O-P(=O)(O^-)-OH$ (phosphate) + $HO-CH_2$ (5′) ribose (with adenine attached at 1′; 3′ OH, 2′ OH) $\longrightarrow$ $^{-}O-P(=O)(O^-)-O-CH_2$ (5′) ribose–adenine (3′ OH, 2′ OH) $+ 2H_2O$ adenosine 5′-monophosphate (AMP)	Equation 13.1

Exercises

Even-numbered exercises are answered in Appendix B.

Exercises with an asterisk (*) are more challenging.

The Flow of Genetic Information (Section 13.1)

13.1 What is the principal location of DNA within the eukaryotic cell?

13.2 What is the principal location of RNA within the eukaryotic cell?

13.3 What is the central dogma of molecular biology?

13.4 In the flow of genetic information, what is meant by the terms *transcription* and *translation*?

13.5 In simple terms, what is a gene?

The Components of Nucleic Acids (Section 13.2)

13.6 Which pentose sugar is present in DNA? In RNA?

13.7 Name the three components of nucleotides.

13.8 Indicate whether each of the following is a pyrimidine or a purine.

a. guanine
b. thymine
c. uracil
d. cytosine
e. adenine

13.9 Which bases are found in DNA? In RNA?

13.10 List the different choices available for each of the following subunits in the given type of nucleotide.

a. Pentose sugar subunit in DNA nucleotides
b. Base subunit in RNA nucleotides
c. Phosphate subunit in RNA nucleotides

13.11 List the different choices available for each of the following subunits in the given type of nucleotide.

a. Pentose sugar subunit in RNA nucleotides
b. Base subunit in DNA nucleotides
c. Phosphate subunit in DNA nucleotides

13.12 Write the structural formula for the nucleotide thymidine 5′-monophosphate. The base component is thymine. (See **Equation 13.1** for an example nucleotide structure.)

13.13 Write the structural formula for the nucleotide guanosine 5′-monophosphate. The base component is guanine; the sugar is ribose. (See **Equation 13.1** for an example nucleotide structure.)

13.14 Indicate whether each of the following statements about the nucleotide shown is true or false.

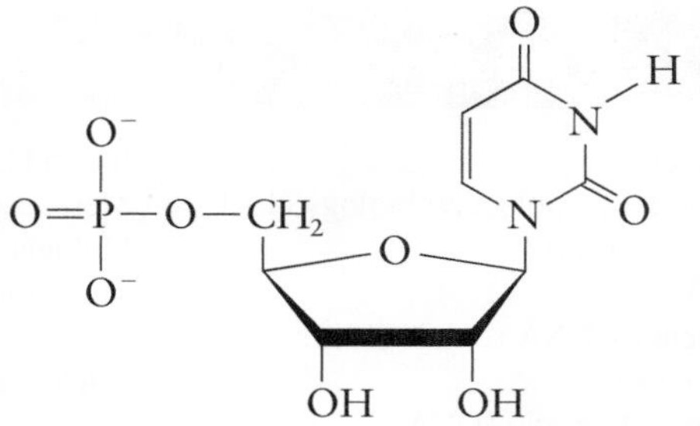

$O=P(O^-)(O^-)-O-CH_2$–ribose (OH, OH)–N (uracil ring: C=O, N–H, C=O)

a. The base is a purine derivative.
b. The phosphate group is attached to the sugar unit at carbon 3′.
c. The sugar unit is ribose.
d. The nucleotide can only be a component of DNA.

13.15 Indicate whether each of the following statements about the nucleotide shown is true or false.

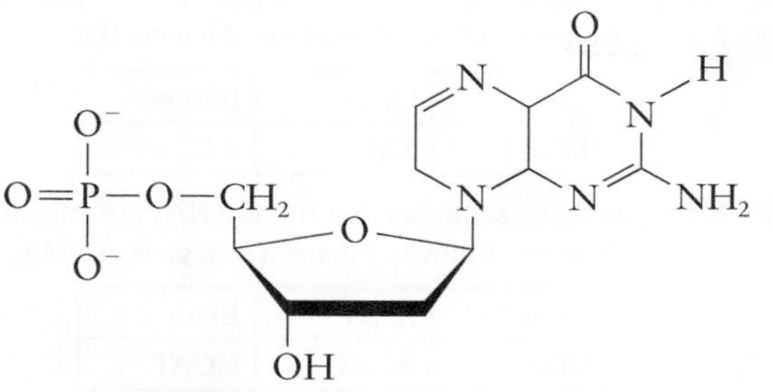

a. The sugar unit is deoxyribose.

b. The sugar unit is attached to the base unit at carbon 1′.

c. The base is a pyrimidine derivative.

d. The nucleotide could be a component of both DNA and RNA.

The Structure of DNA (Section 13.3)

13.16 In what way might two DNA molecules that contain the same number of nucleotides differ?

13.17 Identify the 3′ and 5′ ends of the DNA segment AGTCAT.

13.18 What data obtained from the chemical analysis of DNA supported the idea of complementary base-pairing in DNA proposed by Watson and Crick?

13.19 Describe the three-dimensional structure of DNA as proposed by Watson and Crick.

13.20 Describe the role of hydrogen bonding in the three-dimensional structure of DNA.

13.21 How many total hydrogen bonds would exist between the following strands of DNA and their complementary strands?

a. CAGTAG

b. TTGACA

13.22 How many total hydrogen bonds would exist between the following strands of DNA and their complementary strands?

a. GCATGC

b. TATGGC

13.23 Of the two DNA strands in **Exercise 13.21**, determine which strand of DNA and its complementary strand would require the most energy to break apart. Explain your answer.

13.24 Of the two DNA strands in **Exercise 13.22**, determine which strand of DNA and its complementary strand would require the most energy to break apart. Explain your answer.

13.25 The base content of a particular double-stranded DNA segment is 43% adenine. What is the percentage of each of the following bases in the DNA segment?

a. thymine

b. guanine

c. cytosine

13.26 The base content of a particular double-stranded DNA segment is 36% cytosine. What is the percentage of each of the following bases in the DNA segment?

a. adenine

b. guanine

c. thymine

13.27 A strand of DNA has the base sequence ATGCATC. Write the base sequence for the complementary strand.

13.28 A strand of DNA has the base sequence GATTCA. Write the base sequence for the complementary strand.

13.29 A single strand of DNA has the base composition of 19% A, 34% C, 28% G, and 19% T. What is the percent base composition for the complementary DNA strand?

13.30 A single strand of DNA has the base composition of 21% A, 29% C, 36% G, and 14% T. What is the percent base composition for the complementary DNA strand?

DNA Replication (Section 13.4)

13.31 What is a chromosome? How many chromosomes are in a human cell? What is the approximate number of genes in a human cell?

13.32 What is semiconservative replication?

13.33 What is a replication fork?

13.34 Describe the function of the enzyme helicase in the replication of DNA.

13.35 List the steps involved in DNA replication.

13.36 What enzymes are involved in DNA replication?

13.37 In what direction is a new DNA strand formed?

13.38 Explain how the synthesis of a DNA daughter strand growing toward a replication fork differs from the synthesis of a daughter strand growing away from a replication fork.

13.39 The base sequence ACGTCT represents a portion of a single strand of DNA. (a) Draw the complete double-stranded molecule for this portion of the strand. (b) Use the representation to illustrate the replication of the DNA strand. Be sure to clearly identify the nucleotide bases involved, the new strands formed, and the daughter DNA molecules.

13.40 The base sequence GTACGT represents a portion of a single strand of DNA. (a) Draw the complete double-stranded molecule for this portion of the strand. (b) Use the representation to illustrate the replication of the DNA strand. Be sure to clearly identify the nucleotide bases involved, the new strands formed, and the daughter DNA molecules.

13.41 Explain the origin of Okazaki fragments.

13.42 Two daughter molecules, Q and R, are formed in the replication of a DNA molecule. The base sequence below is part of the newly formed daughter molecule Q.

5′ AACTGG 3′

With this information, write the base sequence (5′ → 3′) for each of the following.

a. The newly formed strand in daughter molecule R

b. The "parent" strand for daughter molecule Q

c. The "parent" strand for daughter molecule R

13.43 Two daughter molecules, S and T, are formed in the replication of a DNA molecule. The base sequence below is part of the newly formed daughter molecule T:

5′ ACGTAT 3′

With this information, write the base sequence (5′ → 3′) for each of the following.

a. The newly formed strand in daughter molecule T

b. The "parent" strand for daughter molecule S

c. The "parent" strand for daughter molecule T

13.44 DNA ligase closes nicks between Okazaki fragments during DNA replication. Three proposed mechanisms for how the enzyme works are provided. Identify the correct mechanism of ligase function. Explain your choice.

a. Ligase adds a base unit to the deoxyribose sugar unit carbon 1′, thereby closing the nick.

b. Ligase adds new base units to the 3′ end of the newly replicated Okazaki fragment.

c. Ligase forms a phosphodiester bond between the 5′ phosphate of one Okazaki fragment and the 3′ OH of the deoxyribose sugar on the next Okazaki fragment, thereby closing the nick.

Ribonucleic Acid (RNA) (Section 13.5)

13.45 How does the sugar–phosphate backbone of RNA differ from the backbone of DNA?

13.46 Compare the secondary structures of RNA and DNA.

13.47 Briefly describe the characteristics and functions of the three types of cellular RNA.

13.48 Identify whether each of the following describes mRNA, rRNA, or tRNA.

a. Associated with a series of proteins in a complex structure

b. Contains genetic information needed for protein synthesis

c. Delivers amino acids to protein synthesis sites

13.49 What are the two important regions of a tRNA molecule?

13.50

Use arrows and label the following on the two-dimensional representation of a tRNA molecule provided.

a. The 3′ end and 5′ end of the tRNA molecule

b. The anticodon on the tRNA molecule

c. Helical areas caused by hydrogen bonding between complementary bases along the tRNA molecule

d. Areas where no hydrogen bonding occurs between bases along the tRNA molecule

Transcription: RNA Synthesis from DNA (Section 13.6)

13.51 Briefly describe the synthesis of mRNA in the transcription process.

13.52 Write the base sequence for the RNA that would be formed during transcription from the DNA template strand with the base sequence GCCATATTG.

13.53 Write the base sequence for the RNA that would be formed during transcription from the DNA template strand with the base sequence ATTGACTCG.

13.54 Write the base sequence for the DNA template that coded for the RNA strand with the base sequence UUGCGAAUC.

13.55 Write the base sequence for the DNA template that coded for the RNA strand with the base sequence AUGCGUACG.

13.56 Write the base sequence for the mRNA, not pre-mRNA, formed from the following portion of a gene (DNA).

Intron	Exon	Intron
TCAG	TACG	TTCA

13.57 Write the base sequence for the mRNA, not pre-mRNA, formed from the following portion of a gene (DNA).

Exon	Intron	Exon
TTAC	AACG	GCAT

The Genetic Code and Protein Synthesis (Section 13.7)

13.58 What is a codon?

13.59 For each of the following mRNA codons, give the tRNA anticodon and use the genetic code in **Figure 13.25** to determine the amino acid being coded for by the codon.

a. UAU

b. CAU

c. UCA

d. UCU

13.60 For each of the following mRNA codons, give the tRNA anticodon and use the genetic code in **Figure 13.25** to determine the amino acid being coded for by the codon.

a. GUA

b. CCC

c. CAC

d. CCA

13.61 Write the sequence of amino acids coded for by the following mRNA sequence.

5′ UGA–GAC–CUA–GGA–GGC 3′

13.62 Write the sequence of amino acids coded for by the following mRNA sequence.

5′ GAA–GAC–CUA–GGA–GGC 3′

13.63 Explain why the base sequence ATC cannot be a codon.

13.64 Explain why the base sequence AGAC cannot be a codon.

13.65 Describe the experiment that allowed researchers to first identify the codon for a specific amino acid.

13.66 Which of the following statements about the genetic code are true and which are false? Correct each false statement.

a. Each codon is composed of four bases.

b. Some amino acids are represented by more than one codon.

c. All codons represent an amino acid.

d. Each living species is thought to have its own unique genetic code.

e. The codon AUG at the beginning of a sequence is a signal for protein synthesis to begin at that codon.

f. It is not known if the code contains stop signals for protein synthesis.

13.67 The β-chain of hemoglobin is a protein that contains 146 amino acid residues. What minimum number of nucleotides must be present in a strand of mRNA that is coded for this protein?

13.68 What is a polysome?

13.69 Beginning with DNA, describe in simple terms (no specific codons, etc.) how proteins are coded and synthesized.

13.70 List the three major stages in protein synthesis.

13.71 Does protein synthesis begin with the *N*-terminal or with the *C*-terminal amino acid?

13.72 What is the A site and what is the P site?

13.73* Beginning with DNA, describe specifically the coding and synthesis of the following tetrapeptide that represents the first four amino acid residues of the hormone oxytocin: Gly–Leu–Pro–Cys. Be sure to include processes such as formation of mRNA (use correct codons, etc.), attachment of mRNA to a ribosome, attachment of tRNA to mRNA–ribosome complex, and so on.

13.74 What is a genetic mutation?

13.75 Briefly explain how genetic mutations can do the following.

a. Harm an organism

b. Help an organism

13.76 What is the result of a genetic mutation that causes the mRNA sequence GCC to be replaced by CCC?

13.77 Write the amino acid sequence made in each of the following ways for the codon sequence GGC–UAU–AGU–AGA.

a. Translation proceeds in a normal manner.

b. A mutation changes AGA to AGG.

c. A mutation changes AGA to AUA.

13.78 Write the amino acid sequence made in each of the following ways for the codon sequence UCG–GGA–AUA–UGG.

a. Translation proceeds in a normal manner.

b. A mutation changes UCG to ACG.

c. A mutation changes UCG to UCU.

Viruses (Section 13.8)

13.79 Describe in simple terms what a virus consists of.

13.80 It is said that viruses "hijack" host cell machinery to replicate. Briefly explain what this phrase means.

13.81 Describe one example from the three groups of viruses—namely, RNA, DNA, and retroviruses—and explain the disease each virus causes in human patients.

13.82 Contrast how an RNA virus, a retrovirus, and a DNA virus are different.

13.83 Describe the roles of the enzymes reverse transcriptase and integrase in the retroviral life cycle.

13.84 Explain one reason that retroviruses rapidly evolve.

13.85 In simple detail, describe the process of immunization using vaccines.

13.86 Contrast the methods used to create traditional vaccines and mRNA vaccines.

Recombinant DNA (Section 13.9)

13.87 Explain the function and importance of restriction enzymes and DNA ligase in the formation of recombinant DNA.

13.88 Explain how recombinant DNA differs from normal DNA.

13.89 Explain what plasmids are and how they are used to get bacteria to synthesize a new protein that they normally do not synthesize.

13.90 List three substances likely to be produced on a large scale by genetic engineering, and give an important use for each.

13.91 What does CRISPR stand for?

13.92 What organism was CRISPR discovered in, and what is the role of CRISPR in nature?

13.93 Explain in simple detail the CRISPR technique.

13.94 Identify two potential biomedical applications of CRISPR.

Additional Exercises

13.95 Explain how heating a double-stranded DNA fragment to 94 °C to 96 °C when performing PCR causes the DNA to unravel into single strands.

13.96 Review **Figure 13.7** to recall the details of nucleic acid backbones and suggest a reaction that would be used to eliminate introns from pre-mRNA. Also suggest a reaction that would be used to join the exon segments together.

13.97 A segment of DNA has an original code of –ACA–. This segment was mutated to –AAA–. After the mutation occurred, the protein that was coded for was no longer able to maintain its tertiary structure. Explain why.

Chemistry for Thought

13.98 Genetic engineering shows great promise for the future but has been controversial at times. Discuss with some classmates the pros and cons of genetic engineering. List two benefits and two concerns that come from your discussion.

13.99 Two samples of DNA are compared, and one has a greater percentage of guanine–cytosine base pairs. How should that greater percentage affect the attractive forces holding the double helix together?

13.100 Azidothymidine (AZT), a drug used to fight HIV, is believed to act as an enzyme inhibitor. What type of enzyme inhibition is most likely caused by this drug?

13.101 The genetic code contains three stop signals and 61 codons that code for 20 amino acids. From the standpoint of mutations, why are we fortunate that the genetic code does not have only the required 20 amino acid codons and 44 stop signals?

13.102 If DNA specifies only the primary structure of a protein, how does the correct three-dimensional protein structure develop?

13.103 During DNA replication, about 1000 nucleotides are added per minute per molecule of DNA polymerase. The genetic material in a single human cell consists of about 3 billion nucleotides.

a. How many years would it take one DNA polymerase enzyme to replicate the genetic material in one cell?

b. How many DNA polymerase molecules would be needed to replicate a cell's genetic material in 10 minutes (assuming equidistance between replication forks)?

14 Lipids

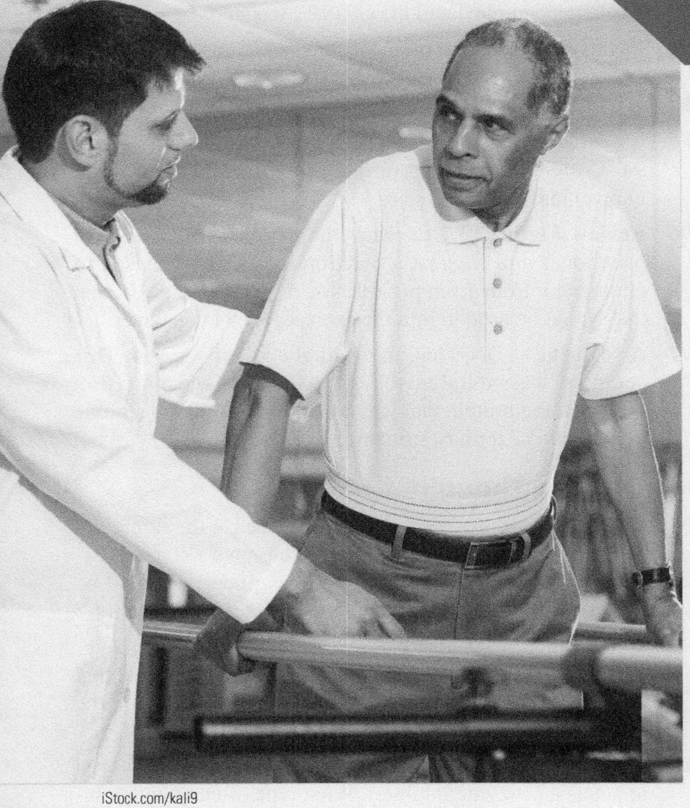
iStock.com/kali9

Health Career Focus

Physical Therapy Assistant

My cholesterol was high, and my knee pain excruciating. My doctor said the best way to improve my health was to lose weight. I tried a low-fat diet, but what I really needed was exercise—impossible because my knees hurt so much! Then, my doctor suggested physical therapy.

"I'll give you personalized exercises to build up tiny muscles in your knee joint," explained the physical therapy assistant. By strengthening the knees in this way, I would eventually be able return to walking and hiking—and my weight should improve.

The therapist took all sorts of measurements of my range of motion and strength, then wrote an individualized plan, consisting of surprisingly simple exercises. I "worked out" sitting in a chair or lying on a table, or I rode a recumbent bike. It wasn't painful, like I expected. Over time, my strength increased and I could put weight on my knees without pain. The more I moved, the better I felt. Physical therapy changed my life.

Follow-up to this Career Focus appears at the end of the chapter before the *Concept Summary*.

Learning Objectives

When you have completed your study of Chapter 14, you should be able to:

1. List two of the major functions of lipids. **(Section 14.1)**
2. Classify lipids as saponifiable or nonsaponifiable. **(Section 14.1)**
3. Describe four general characteristics of fatty acids. **(Section 14.2)**
4. Draw structural formulas of triglycerides given the formulas of the component parts. **(Section 14.3)**
5. Describe the structural similarities and differences of fats and oils. **(Section 14.3)**
6. Write key reactions for fats and oils. **(Section 14.4)**
7. Compare and contrast the structures of fats and waxes. **(Section 14.5)**
8. Draw structural formulas for phosphoglycerides. **(Section 14.6)**
9. Describe two uses for phosphoglycerides. **(Section 14.6)**
10. Draw structural formulas for sphingolipids. **(Section 14.7)**
11. Describe two uses for sphingolipids. **(Section 14.7)**
12. Describe the major features of cell membrane structure. **(Section 14.8)**
13. Identify the structural characteristic that is typical of steroids. **(Section 14.9)**
14. List five important groups of steroids in the body. **(Section 14.9)**

We begin Chapter 14 by explaining how lipids are classified. We next examine the structure of fatty acids, one of the major components of both simple and complex lipids. Then we discuss the structure and resulting chemical properties of fats, the first group of simple lipids, followed by the structure of waxes, the second group of simple lipids. We then explore the structures of the complex lipids: phosphoglycerides and sphingolipids. This leads to a discussion of their roles in lipid bilayers of cell membranes and drug delivery in biological systems. We then wrap up the chapter with steroids and their biological functions as cell membrane components and hormones.

Itchy Feet/Shutterstock.com

Figure 14.1 Waxes form a protective coating on these leaves.

14.1 The Classification of Lipids

Learning Objective 1 List two of the major functions of lipids.

Learning Objective 2 Classify lipids as saponifiable or nonsaponifiable.

The group of compounds called lipids is made up of substances with widely different compositions and structures. Unlike the other macromolecules, which are defined in terms of their polymer structures, lipids are defined in terms of a physical property—their solubility. **Lipids** are biological molecules that are insoluble in water but soluble in nonpolar solvents. Lipids are the waxy, greasy, or oily compounds found in plants and animals (see **Figures 14.1** and **14.2**). Lipids repel water, which is a useful characteristic of protective wax coatings found on some plants. Fats and oils are energy-rich and have relatively low densities. These properties account for their use as storage forms of energy in plants and animals. Still other lipids are used as structural components, especially in the formation of cellular membranes.

Lipids are classified as saponifiable or nonsaponifiable. Saponification is the process in which esters are hydrolyzed under basic conditions (see **Section 10.8**). Triglycerides, waxes, phospholipids, and sphingolipids are esters, so they are saponifiable lipids. Nonsaponifiable lipids are not esters and cannot be hydrolyzed. Steroids and prostaglandins belong to this class.

Figure 14.3 shows how these different types of lipids are classified. Notice that saponifiable lipids are further classified as simple and complex, based on the number of components in their structure. **Simple lipids** contain just two types of components (fatty acids and an alcohol), whereas **complex lipids** contain more than two (fatty acids, an alcohol, plus other components).

iStock.com/Mike Matas

Figure 14.2 Thick layers of fat help insulate penguins against low temperatures.

lipid A biological compound that is soluble only in nonpolar solvents.

simple lipid An ester-containing lipid with just two types of components: an alcohol and one or more fatty acids.

complex lipid An ester-containing lipid with more than two types of components: an alcohol, fatty acids, and other components.

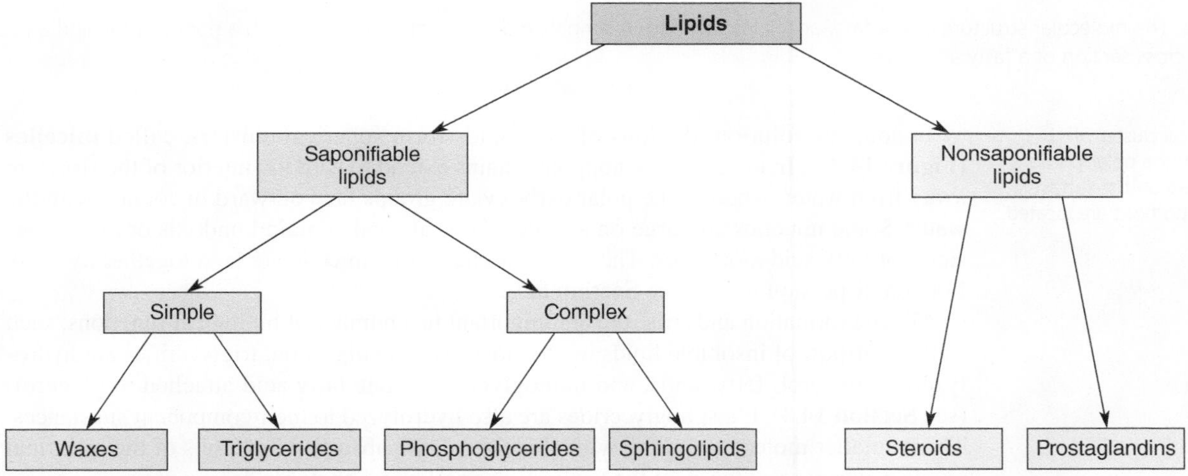

Figure 14.3 The classification of the major types of lipids.

14.2 Fatty Acids

Learning Objective 3 Describe four general characteristics of fatty acids.

Prior to discussing the chemical structures and properties of the saponifiable lipids, it is useful to first describe the chemistry of fatty acids, the fundamental building blocks of many lipids. Fatty acids (defined in **Section 10.4**) are long-chain carboxylic acids, as shown in **Figure 14.4a**. It is the long nonpolar tails of fatty acids that are responsible for most of the fatty or oily characteristics of fats. The carboxyl group, or polar head of fatty acids, is very hydrophilic under conditions of physiological pH, where it exists as the carboxylate anion $-COO^-$ (**Figure 14.4b**).

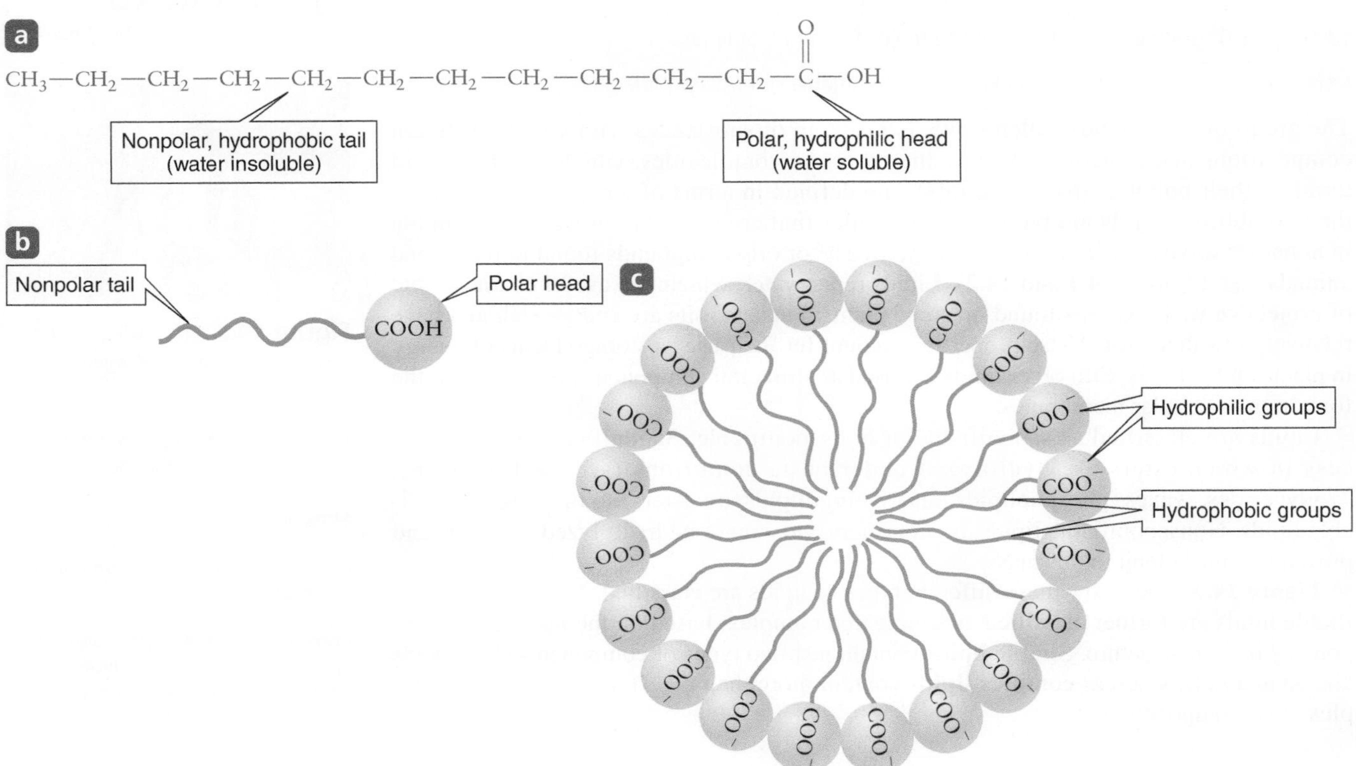

Figure 14.4 (a) The molecular structure of the fatty acid lauric acid. (b) A simplified diagram of a fatty acid with a nonpolar tail and a polar head. (c) A cross section of a fatty acid micelle.

micelle A spherical cluster of molecules in which the polar portions of the molecules are on the surface and the nonpolar portions are located in the interior.

In aqueous solution, the ions of fatty acids form spherical clusters, called **micelles** (**Figure 14.4c**). In micelles, the nonpolar chains extend toward the interior of the structure away from water, whereas the polar carboxylate groups face outward in contact with the water. Some micelles are large on a molecular scale and contain hundreds or even thousands of fatty acid molecules. The nonpolar chains in a micelle are held together by weak London dispersion forces (see **Section 5.3**).

Micelle formation and structure are important in a number of biological functions, such as the transport of insoluble lipids in the blood. During digestion, triglycerides are hydrolyzed to glycerol, fatty acids, and monoglycerides (one fatty acid attached to glycerol) (see **Section 14.3**). Phosphoglycerides are also hydrolyzed to their component substances. These smaller molecules, along with cholesterol, are absorbed into cells of the intestinal mucosa, where resynthesis of the triglycerides and phosphoglycerides occurs.

chylomicron A lipoprotein aggregate found in the lymph and the bloodstream.

For transport within the aqueous environment of the lymph and blood, these insoluble lipids are complexed with proteins to form lipoprotein aggregates called **chylomicrons**. The chylomicrons (which contain triglycerides, phosphoglycerides, and cholesterol) pass into the lymph system and then into the bloodstream. Chylomicrons are modified by the liver into smaller lipoprotein particles, the form in which most lipids are transported to various parts of the body by the bloodstream (see **Figure 14.5**).

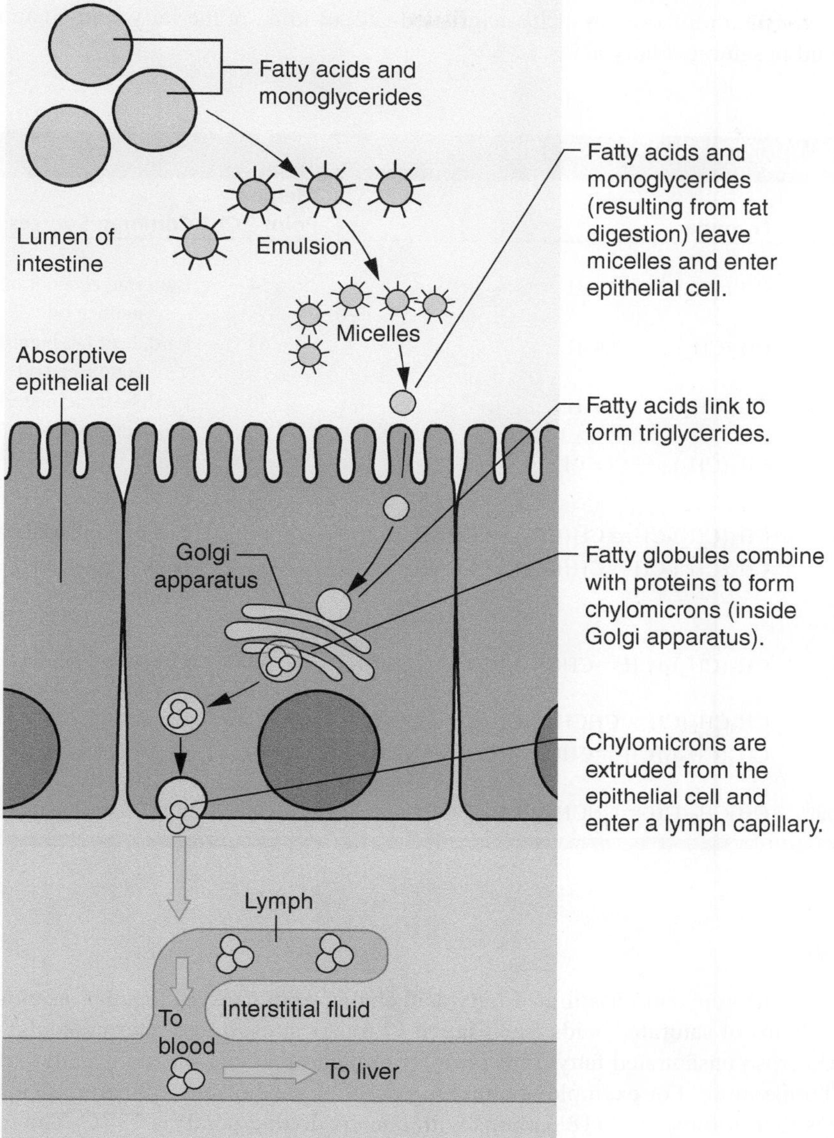

Fatty acids and monoglycerides
Lumen of intestine
Emulsion
Micelles
Fatty acids and monoglycerides (resulting from fat digestion) leave micelles and enter epithelial cell.
Absorptive epithelial cell
Fatty acids link to form triglycerides.
Golgi apparatus
Fatty globules combine with proteins to form chylomicrons (inside Golgi apparatus).
Chylomicrons are extruded from the epithelial cell and enter a lymph capillary.
Lymph
Interstitial fluid
To blood
To liver

Figure 14.5 The digestion and absorption of triglycerides.

The fatty acids found in natural lipids have the following characteristics in common:

1. They are usually straight-chain carboxylic acids (no branching).
2. The sizes of most common fatty acids range from 10 to 20 carbons.
3. Fatty acids usually have an even number of carbon atoms (including the carboxyl group carbon).
4. Fatty acids can be saturated (containing no double bonds between carbons) or unsaturated (containing one or more double bonds between carbons). Apart from the carboxyl group and double bonds, there are usually no other functional groups present.

Table 14.1 lists some important fatty acids, along with their formulas and melting points. Unsaturated fatty acids usually contain double bonds in the *cis* configuration (i.e., both carbon chains on the same side of the double bond) rather than the *trans* configuration (i.e., both carbon chains on opposite sides of the double bond).

$$\underset{\text{cis}}{\overset{\text{H}\diagdown\quad\diagup\text{H}}{\text{long chain}\diagup\text{C}=\text{C}\diagdown\text{long chain}}} \qquad \underset{\text{trans}}{\overset{\text{H}\diagdown\quad\diagup\text{long chain}}{\text{long chain}\diagup\text{C}=\text{C}\diagdown\text{H}}}$$

The cis configuration creates a characteristic bend, or kink, in the fatty acid chain that is not found in saturated fatty acids.

Table 14.1 Some Important Fatty Acids

Compound Type and Number of Carbons	Name	Formula	Melting Point (°C)	Common Sources
Saturated				
14	Myristic acid	$CH_3(CH_2)_{12}—COOH$	54	Butterfat, coconut oil, nutmeg oil
16	Palmitic acid	$CH_3(CH_2)_{14}—COOH$	63	Lard, beef fat, butterfat, cottonseed oil
18	Stearic acid	$CH_3(CH_2)_{16}—COOH$	70	Lard, beef fat, butterfat, cottonseed oil
20	Arachidic acid	$CH_3(CH_2)_{18}—COOH$	76	Peanut oil
Monounsaturated				
16	Palmitoleic acid	$CH_3(CH_2)_5CH=CH(CH_2)_7—COOH$	−1	Cod liver oil, butterfat
18	Oleic acid	$CH_3(CH_2)_7CH=CH(CH_2)_7—COOH$	13	Lard, beef fat, olive oil, peanut oil
Polyunsaturated				
18	Linoleic acid[a]	$CH_3(CH_2)_4(CH=CHCH_2)_2(CH_2)_6—COOH$	−5	Cottonseed oil, soybean oil, corn oil, linseed oil
18	Linolenic acid[b]	$CH_3CH_2(CH=CHCH_2)_3(CH_2)_6—COOH$	−11	Linseed oil, corn oil
20	Arachidonic acid[a]	$CH_3(CH_2)_4(CH=CHCH_2)_4(CH_2)_2—COOH$	−50	Corn oil, linseed oil, animal tissues
20	Eicosapentaenoic acid[b]	$CH_3CH_2(CH=CHCH_2)_5(CH_2)_2—COOH$	−54	Fish oil, seafoods

[a] Omega-6 fatty acid.

[b] Omega-3 fatty acid.

These kinks prevent unsaturated fatty acid chains from packing together as closely as do the chains of saturated acids (see **Figure 14.6**). As a result, the intermolecular forces are weaker, so unsaturated fatty acids have lower melting points and are usually liquids at room temperature. For example, the melting point of stearic acid (18 carbons) is 71 °C, whereas that of oleic acid (18 carbons with one cis double bond) is 13 °C. The melting point of linoleic acid (18 carbons and two double bonds) is even lower (−5 °C). Chain length also affects the melting point: Palmitic acid (16 carbons) melts at 7 °C lower than stearic acid (18 carbons). The presence of double bonds and the length of fatty acid chains in membrane lipids partly explain the fluidity of biological membranes, an important feature discussed in **Section 14.8**.

essential fatty acid A fatty acid needed by the body but not synthesized within the body.

The human body can synthesize all except two of the fatty acids it needs. These two, linoleic acid and linolenic acid, are polyunsaturated fatty acids, and each contains 18 carbon atoms (see **Table 14.1**). Because they are not synthesized within the body and must be obtained from the diet, they are called **essential fatty acids**. Both are widely distributed in plant and fish oils. In the body, both acids are used to produce hormone-like substances that regulate a wide range of functions and characteristics, including blood pressure, blood clotting, blood lipid levels, the immune response, and the inflammation response to injury and infection.

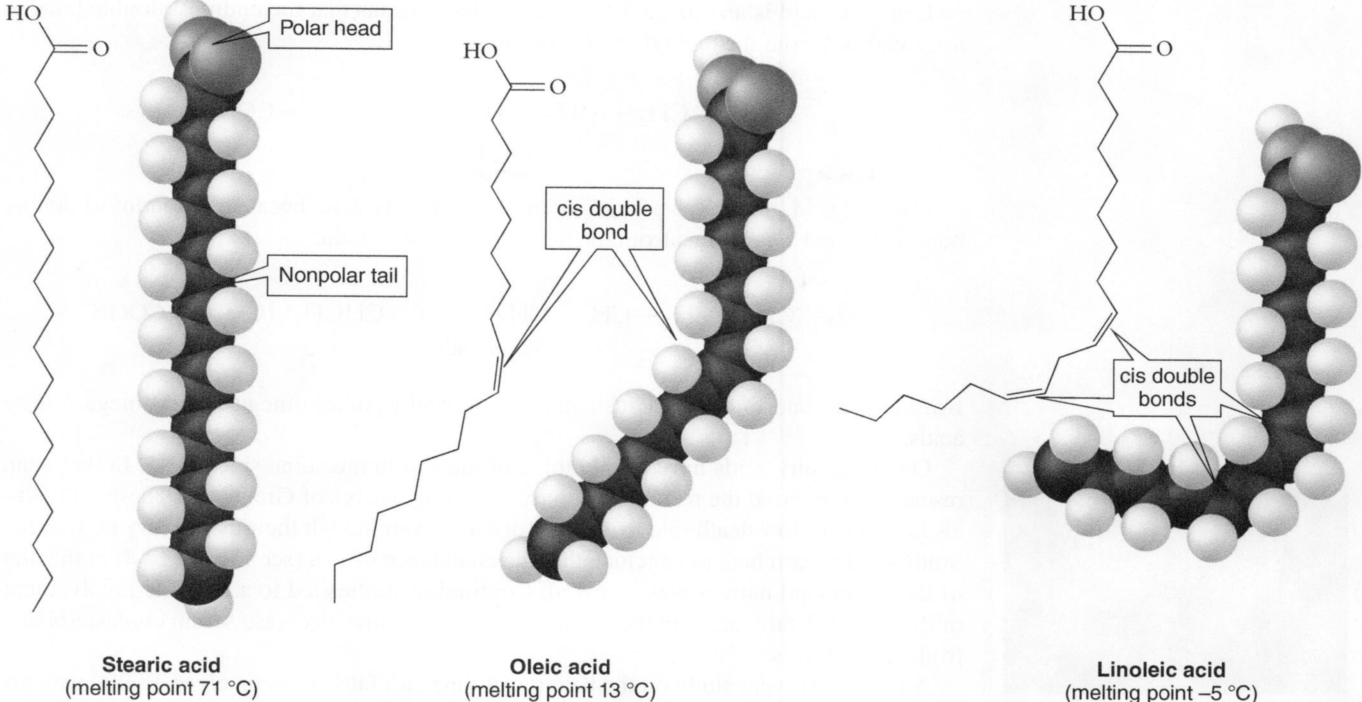

Figure 14.6 Skeletal structures and space-filling models of fatty acids. Unsaturated fatty acids (e.g., oleic acid and linoleic acid) do not pack as tightly as saturated fatty acids (e.g., stearic acid), and their melting points are lower than those of saturated acids.

Example 14.1 Fatty Acids

Myristic acid is saturated and found in solid butter and coconut oil, whereas oleic acid is monounsaturated and found in olive oil (**Figure 14.7**). Using your knowledge of saturation, place each of the following fatty acids into one of three categories: saturated, monounsaturated, or polyunsaturated. State whether each fatty acid is expected to be a solid or a liquid at room temperature.

a. $CH_3(CH_2)_3CH{=}CHCH_2CH{=}CH(CH_2)_6{-}COOH$
b. $CH_3(CH_2)_8{-}COOH$
c. $CH_3(CH_2)_6CH{=}CH(CH_2)_4{-}COOH$

Solution

a. The presence of two double bonds means that this is a polyunsaturated fatty acid. Polyunsaturated fatty acids are liquid at room temperature.
b. With no carbon–carbon double bonds, this is a saturated fatty acid and is solid at room temperature.
c. One carbon–carbon double bond is present. It is a monounsaturated fatty acid and is liquid at room temperature.

Figure 14.7 Coconut oil consists of saturated myristic acid and occurs as a solid at room temperature, whereas olive oil consists of monounsaturated oleic acid and occurs as a liquid at room temperature.

✔ Learning Check 14.1 Unusual fatty acids are sometimes found in seed oils. Classify vernolic acid as saturated, monounsaturated, or polyunsaturated. Further classify it as a solid or liquid at room temperature.

$$CH_3(CH_2)_4\overset{\overset{O}{/\ \backslash}}{CHCH}CH_2CH{=}CH(CH_2)_7{-}COOH$$

Linolenic acid is an omega-3 fatty acid, which means that the endmost double bond is *three* carbons from the methyl end of the chain:

$$\overset{1}{CH_3}-\overset{2}{CH_2}-(\overset{3}{CH}{=}CHCH_2)_3-(CH_2)_6-COOH$$

linolenic acid

Linoleic acid, on the other hand, is an omega-6 fatty acid, because the endmost double bond is located *six* carbons from the methyl end of the chain.

$$\overset{1}{CH_3}-\overset{2}{CH_2}-\overset{3}{CH_2}-\overset{4}{CH_2}-\overset{5}{CH_2}-(\overset{6}{CH}{=}CHCH_2)_2(CH_2)_6-COOH$$

linoleic acid

Both linoleic and linolenic acids can be converted to other omega-3 and omega-6 fatty acids.

Omega-3 fatty acids have been a topic of interest in medicine since 1985. In that year, researchers reported the results of a study involving natives of Greenland. These individuals have a very low death rate from heart disease, even though their diet is very high in fat. Studies led researchers to conclude that the abundance of fish (see **Figure 14.8**) in the diet of the Greenland natives was involved. Continuing studies led to a possible involvement of the omega-3 fatty acids in the oil of the fish, which may decrease serum cholesterol and triglyceride levels.

A recent five-year study on the benefits of omega-3 fatty acids involved 18,000 patients with unhealthy cholesterol levels. At the end of the study, patients receiving omega-3 fatty acids had superior cardiovascular function, and nonfatal coronary events were significantly reduced.

Beat J Korner/Shutterstock.com

Figure 14.8 Salmon and other cold water fish are excellent sources of omega-3 fatty acids.

Health Connections 14.1

Consider the Mediterranean Diet

If you are looking for a heart-healthy eating plan, the Mediterranean diet might be right for you. The Mediterranean diet incorporates the basics of healthy eating—plus a splash of flavorful olive oil and perhaps even an occasional glass of red wine. These components characterize the traditional cooking style of countries that border the Mediterranean Sea.

Most healthful diets include fruits, vegetables, fish, and whole grains, and they limit unhealthy fats. While these parts of a healthful diet remain tried and true, subtle variations or differences in the proportions of certain foods might make a difference in your risk for heart disease.

Research has shown that the traditional Mediterranean diet reduces the risk of heart disease. In fact, an analysis of more than 1.5 million healthy adults demonstrated that following a Mediterranean diet was associated with a reduced risk of death from heart disease and cancer, as well as a reduced incidence of Parkinson's and Alzheimer's diseases.

The Mediterranean diet emphasizes the following dietary practices:

1. Eating primarily plant-based foods, such as fruits, vegetables, whole grains, legumes, and nuts
2. Replacing butter with healthy fats such as olive oil
3. Using herbs and spices instead of salt to flavor foods
4. Limiting red meat to no more than a few times a month
5. Eating fish and poultry at least twice a week
6. Drinking red wine in moderation (optional)

alvarez/Getty Images

The Mediterranean diet incorporates a variety of healthy foods.

14.3 The Structure of Fats and Oils

Learning Objective 4 Draw structural formulas of triglycerides given the formulas of the component parts.

Learning Objective 5 Describe the structural similarities and differences of fats and oils.

Animal fats and vegetable oils are the most widely occurring lipids. Chemically, fats and oils are both esters, which consist of an alcohol portion and an acid portion (**Section 10.6**). In fats and oils, the alcohol portion is always derived from glycerol, and the acid portion is furnished by fatty acids. Because glycerol has three –OH groups, a single molecule of glycerol can be attached to three different acid molecules. An example is the esterification reaction of stearic acid and glycerol shown in Equation 14.1:

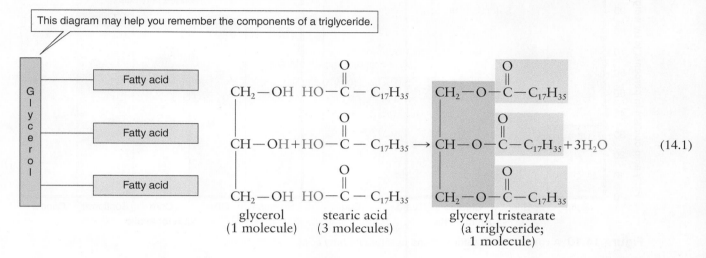

This diagram may help you remember the components of a triglyceride.

Glycerol	
	Fatty acid
	Fatty acid
	Fatty acid

$$\begin{array}{l} CH_2-OH \quad HO-\overset{\overset{\large O}{\|}}{C}-C_{17}H_{35} \\ | \\ CH-OH + HO-\overset{\overset{\large O}{\|}}{C}-C_{17}H_{35} \\ | \\ CH_2-OH \quad HO-\overset{\overset{\large O}{\|}}{C}-C_{17}H_{35} \end{array} \longrightarrow \begin{array}{l} CH_2-O-\overset{\overset{\large O}{\|}}{C}-C_{17}H_{35} \\ | \\ CH-O-\overset{\overset{\large O}{\|}}{C}-C_{17}H_{35} + 3H_2O \\ | \\ CH_2-O-\overset{\overset{\large O}{\|}}{C}-C_{17}H_{35} \end{array} \qquad (14.1)$$

glycerol (1 molecule) stearic acid (3 molecules) glyceryl tristearate (a triglyceride; 1 molecule)

The resulting esters are called **triglycerides** or **triacylglycerols**.

triglyceride or **triacylglycerol** A triester of glycerol in which all three alcohol groups are esterified.

The fatty acid components in naturally occurring triglyceride molecules are rarely identical (as in the case of glyceryl tristearate). In addition, natural triglycerides (fats and oils) are usually mixtures of different triglyceride molecules. Butterfat, for example, contains at least 14 different fatty acid components (**Figure 14.9**).

Figure 14.9 The fat in butter contains at least 14 different fatty acids!

Example 14.2 Writing Structures for Triglycerides

Write the structure for the triglyceride that is formed from glycerol and three molecules of myristic acid.

Solution

Using Equation 14.1 as a guide, position a molecule of glycerol to the left. Align three molecules of myristic acid on the right (see **Table 14.1**). Finally, eliminate water molecules to form the three ester linkages.

$$\begin{array}{l} CH_2-OH \quad HO-\overset{\overset{\large O}{\|}}{C}-(CH_2)_{12}CH_3 \\ | \\ CH-OH + HO-\overset{\overset{\large O}{\|}}{C}-(CH_2)_{12}CH_3 \\ | \\ CH_2-OH \quad HO-\overset{\overset{\large O}{\|}}{C}-(CH_2)_{12}CH_3 \end{array} \longrightarrow \begin{array}{l} CH_2-O-\overset{\overset{\large O}{\|}}{C}-(CH_2)_{12}CH_3 \\ | \\ CH-O-\overset{\overset{\large O}{\|}}{C}-(CH_2)_{12}CH_3 + 3H_2O \\ | \\ CH_2-O-\overset{\overset{\large O}{\|}}{C}-(CH_2)_{12}CH_3 \end{array}$$

✔ Learning Check 14.2 Write one structure (several possibilities exist) for a triglyceride derived from stearic acid, oleic acid, and palmitic acid.

fat A triglyceride that is solid at room temperature.

oil A triglyceride that is liquid at room temperature.

With some exceptions, triglycerides that come from animals other than fish are solids at room temperature and are called **fats**. Triglycerides from plants or fish are liquids at room temperature and are usually referred to as **oils**. Although both fats and oils are triglycerides, they differ structurally in one important respect—the degree of unsaturation. As shown in **Figure 14.10**, animal fats contain primarily triglycerides of long-chain saturated fatty acids (higher melting points). In contrast, vegetable oils, such as corn oil and soybean oil, consist of triglycerides containing unsaturated fatty acids (lower melting points). Thus, the structural and physical properties of fatty acids determine the properties of the triglycerides derived from them.

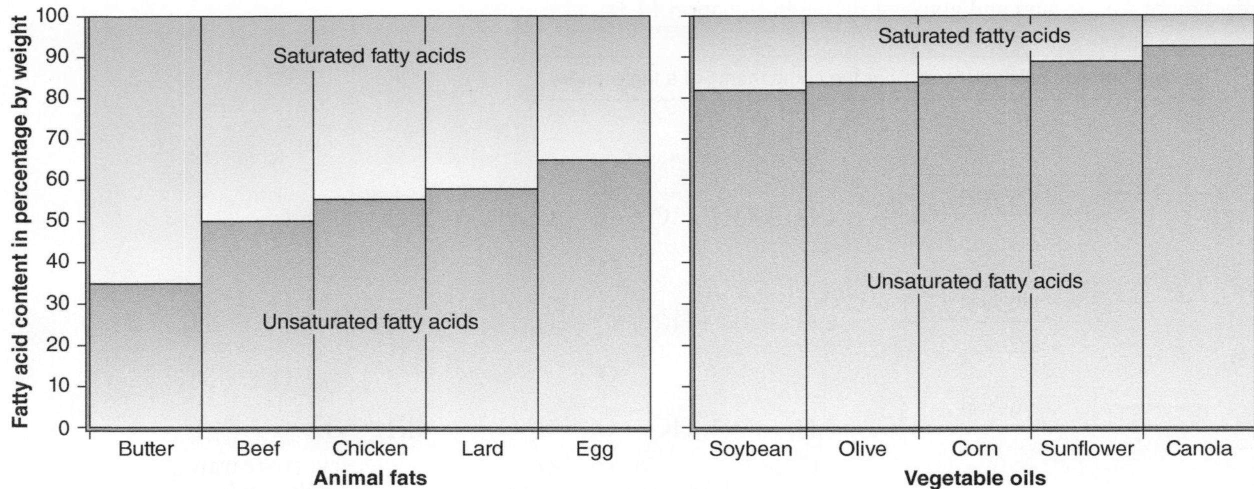

Figure 14.10 A comparison of saturated and unsaturated fatty acids in some foods.

Excessive fat in the diet is recognized as one risk factor influencing the development of chronic disease. The main concern about excessive dietary fat, especially saturated fat, centers on its role in raising blood cholesterol levels (see **Figure 14.11**). Saturated fats are considered more harmful to heart health because the lack of a double bond "kink" allows the fatty acids to stack more effectively in arteries. High blood cholesterol is a recognized risk factor in the development of coronary heart disease, the leading cause of death in Americans every year.

Figure 14.11 Steak contains a substantial amount of solid fat, consisting of saturated fatty acids. Consuming an excessive amount of saturated fats can lead to high blood cholesterol and heart disease.

14.4 Chemical Properties and Reactions of Fats and Oils

Learning Objective 6 Write key reactions for fats and oils.

The chemical properties of triglycerides are typical of esters and alkenes because these are the two functional groups present.

14.4A Hydrolysis

Hydrolysis is one of the most important reactions of fats and oils, just as it was for carbohydrates (see **Section 11.7**). The treatment of fats or oils with water and an acid catalyst causes them to hydrolyze to form glycerol and fatty acids. This reaction (see Equation 14.2) is simply the reverse of ester formation:

$$\begin{array}{l} CH_2-O-\overset{O}{\overset{\|}{C}}-(CH_2)_{14}CH_3 \\ \;| \\ CH-O-\overset{O}{\overset{\|}{C}}-(CH_2)_7CH{=}CH(CH_2)_7CH_3 \\ \;| \\ CH_2-O-\overset{O}{\overset{\|}{C}}-(CH_2)_6(CH_2CH{=}CH)_2(CH_2)_4CH_3 \end{array} + 3H_2O \xrightarrow[\text{lipase}]{H^+ \text{ or}} \quad (14.2)$$

$$\underset{\text{glycerol}}{\begin{array}{l} CH_2-OH \\ \;| \\ CH-OH \\ \;| \\ CH_2-OH \end{array}} + \left\{ \begin{array}{l} \underset{\text{palmitic acid}}{CH_3(CH_2)_{14}-\overset{O}{\overset{\|}{C}}-OH} \\ \underset{\text{oleic acid}}{CH_3(CH_2)_7CH{=}CH(CH_2)_7-\overset{O}{\overset{\|}{C}}-OH} \\ \underset{\text{linoleic acid}}{CH_3(CH_2)_4(CH{=}CHCH_2)_2(CH_2)_6-\overset{O}{\overset{\|}{C}}-OH} \end{array} \right.$$

Enzymes (lipases) of the digestive system also catalyze hydrolysis (**Figure 14.12**). This reaction represents the only important change that takes place in fats and oils during digestion. The breakdown of cellular fat deposits to supply energy also begins with the lipase-catalyzed hydrolysis reaction.

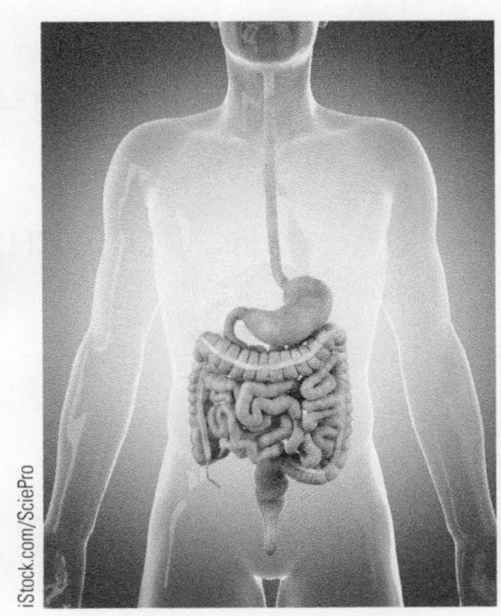

Figure 14.12 Lipases are enzymes that break down fats and oils into glycerol and fatty acids via hydrolysis reactions.

Example 14.3 Writing Hydrolysis Reactions

Write the structure of the hydrolysis products of the following reaction.

$$\begin{array}{l} CH_2-O-\overset{O}{\overset{\|}{C}}-(CH_2)_{16}CH_3 \\ \;| \\ CH-O-\overset{O}{\overset{\|}{C}}-(CH_2)_6(CH_2CH{=}\,CH)_2(CH_2)_4CH_3 + 3H_2O \xrightarrow{H^+} \\ \;| \\ CH_2-O-\overset{O}{\overset{\|}{C}}-(CH_2)_{16}CH_3 \end{array}$$

Solution

Using Equation 14.2 as a guide, the products will be glycerol and three molecules of fatty acid. In this case, two of the three fatty acid molecules are identical.

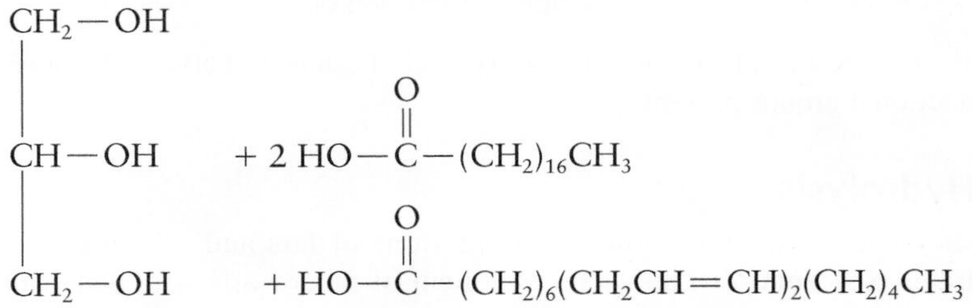

$$\begin{array}{l} CH_2-OH \\ | \\ CH-OH \\ | \\ CH_2-OH \end{array} \begin{array}{l} \\ + 2\,HO-\overset{\overset{\large O}{\|}}{C}-(CH_2)_{16}CH_3 \\ \\ + HO-\overset{\overset{\large O}{\|}}{C}-(CH_2)_6(CH_2CH=CH)_2(CH_2)_4CH_3 \end{array}$$

✔ **Learning Check 14.3** Write the structures of the hydrolysis products of the following reaction.

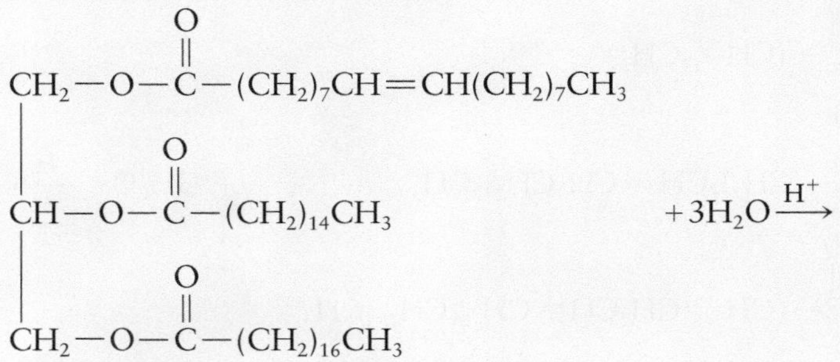

$$\begin{array}{l} CH_2-O-\overset{\overset{\large O}{\|}}{C}-(CH_2)_7CH=CH(CH_2)_7CH_3 \\ | \\ CH-O-\overset{\overset{\large O}{\|}}{C}-(CH_2)_{14}CH_3 \\ | \\ CH_2-O-\overset{\overset{\large O}{\|}}{C}-(CH_2)_{16}CH_3 \end{array} + 3H_2O \xrightarrow{H^+}$$

14.4B Saponification

soap A salt of a fatty acid often used as a cleaning agent.

When triglycerides are reacted with a strong base, the process of saponification (soap making) occurs. In this commercially important reaction, the products are glycerol and the salts of fatty acids, which are also called **soaps** (Equation 14.3).

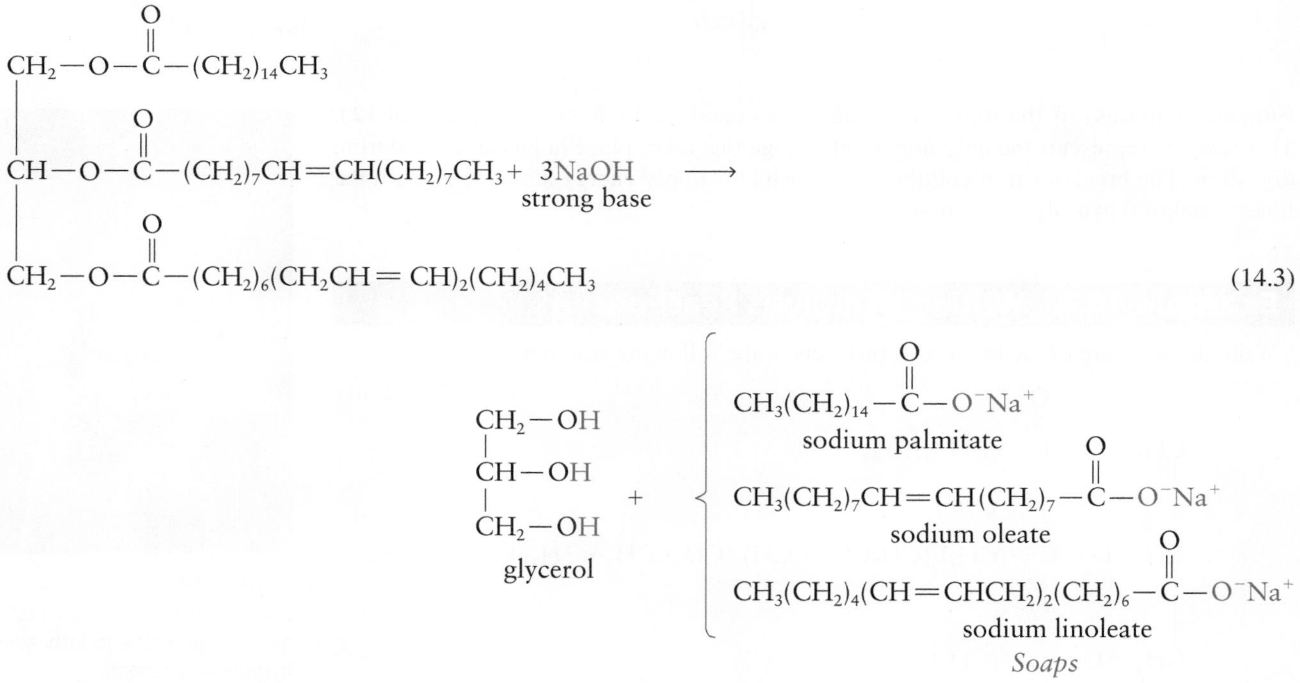

$$\begin{array}{l} CH_2-O-\overset{\overset{\large O}{\|}}{C}-(CH_2)_{14}CH_3 \\ | \\ CH-O-\overset{\overset{\large O}{\|}}{C}-(CH_2)_7CH=CH(CH_2)_7CH_3 + \underset{\text{strong base}}{3NaOH} \longrightarrow \\ | \\ CH_2-O-\overset{\overset{\large O}{\|}}{C}-(CH_2)_6(CH_2CH=CH)_2(CH_2)_4CH_3 \end{array} \quad (14.3)$$

$$\underset{\text{glycerol}}{\begin{array}{l} CH_2-OH \\ | \\ CH-OH \\ | \\ CH_2-OH \end{array}} + \left\{ \begin{array}{c} \underset{\text{sodium palmitate}}{CH_3(CH_2)_{14}-\overset{\overset{\large O}{\|}}{C}-O^-Na^+} \\ \underset{\text{sodium oleate}}{CH_3(CH_2)_7CH=CH(CH_2)_7-\overset{\overset{\large O}{\|}}{C}-O^-Na^+} \\ \underset{\text{sodium linoleate}}{CH_3(CH_2)_4(CH=CHCH_2)_2(CH_2)_6-\overset{\overset{\large O}{\|}}{C}-O^-Na^+} \end{array} \right.$$

Soaps

The salts obtained from saponification depend on the base used. Sodium salts, known as hard soaps (see **Figure 14.13**), are found in most bar soaps, whereas potassium salts, known as soft soaps, are used in some shaving creams and liquid soap preparations. In traditional soap making, animal fat is the source of triglycerides, and lye (crude NaOH) or an aqueous extract of wood ashes is the source of the base.

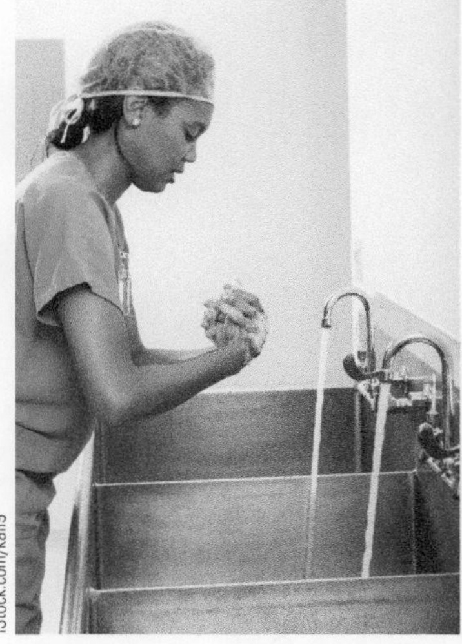

Figure 14.13 The use of soaps by health professionals is vital in controlling the spread of infectious microorganisms.

Example 14.4 Writing Saponification Reactions

Write structures for the products formed when the following triglyceride is saponified using NaOH.

$$\begin{array}{l} CH_2-O-\overset{\overset{\large O}{\|}}{C}-(CH_2)_{16}CH_3 \\ | \\ CH-O-\overset{\overset{\large O}{\|}}{C}-(CH_2)_{16}CH_3 \qquad + 3NaOH \longrightarrow \\ | \\ CH_2-O-\overset{\overset{\large O}{\|}}{C}-(CH_2)_{16}CH_3 \end{array}$$

Solution

Based on Equation 14.3, the products will be glycerol and three molecules of fatty acid salt. In this case, the three molecules of fatty acid salt are identical.

$$\begin{array}{l} CH_2-OH \\ | \\ CH-OH \qquad + 3Na^{+-}O-\overset{\overset{\large O}{\|}}{C}-(CH_2)_{16}CH_3 \\ | \\ CH_2-OH \end{array}$$

✔ **Learning Check 14.4** Write structures for the products formed when the following triglyceride is saponified using KOH.

$$\begin{array}{l} CH_2-O-\overset{\overset{\large O}{\|}}{C}-(CH_2)_{16}CH_3 \\ | \\ CH-O-\overset{\overset{\large O}{\|}}{C}-(CH_2)_7CH=CH(CH_2)_7CH_3 \\ | \\ CH_2-O-\overset{\overset{\large O}{\|}}{C}-(CH_2)_7CH=CH(CH_2)_7CH_3 \end{array}$$

14.4C Hydrogenation

Recall from **Section 8.8** that double bonds can be reduced to single bonds by treatment with hydrogen (H_2) in the presence of a catalyst. An important commercial reaction of fats and oils uses this same hydrogenation process, in which some of the fatty acid double bonds are converted to single bonds. The result is a decrease in the degree of unsaturation and a corresponding increase in the melting point of the fat or oil (see **Figure 14.10**). The peanut oil in many popular brands of peanut butter has been partially hydrogenated to convert the oil to a semisolid that does not separate. Hydrogenation is most often used in the production of semisolid cooking shortenings (such as Crisco) or margarines from liquid vegetable oils (see **Figure 14.14**).

Partially hydrogenated vegetable oils were developed in part to help displace highly saturated animal fats used in frying, baking, and spreads. It is important to stop the reaction before all the double bonds are saturated. If not, the completely saturated product is hard and waxy—not the smooth, creamy product desired by consumers. Equation 14.4 gives an example in which some of the double bonds react, showing one of the possible products.

$$\begin{array}{l} CH_2-O-\overset{\overset{\displaystyle O}{\|}}{C}-(CH_2)_{14}CH_3 \\ | \\ CH-O-\overset{\overset{\displaystyle O}{\|}}{C}-(CH_2)_7CH=CH(CH_2)_7CH_3 \\ | \\ CH_2-O-\overset{\overset{\displaystyle O}{\|}}{C}-(CH_2)_6(CH_2CH=CH)_2(CH_2)_4CH_3 \end{array} + 2H_2 \xrightarrow{Ni} \begin{array}{l} CH_2-O-\overset{\overset{\displaystyle O}{\|}}{C}-(CH_2)_{14}CH_3 \\ | \\ CH-O-\overset{\overset{\displaystyle O}{\|}}{C}-(CH_2)_7CH=CH(CH_2)_7CH_3 \\ | \\ CH_2-O-\overset{\overset{\displaystyle O}{\|}}{C}-(CH_2)_6(CH_2CH_2CH_2)_2(CH_2)_4CH_3 \end{array} \quad (14.4)$$

© Mike Slabaugh

Figure 14.14 Hydrogenation of vegetable oils is used to prepare both shortenings and margarines. How do oils used in automobiles differ chemically from these vegetable oils?

Example 14.5 Writing Hydrogenation Reactions

Write the structure of the product formed when the following triglyceride is completely hydrogenated in the presence of a nickel catalyst.

$$\begin{array}{l} CH_2-O-\overset{\overset{\displaystyle O}{\|}}{C}-(CH_2)_{14}CH_3 \\ | \\ CH-O-\overset{\overset{\displaystyle O}{\|}}{C}-(CH_2)_7CH=CHCH_2CH=CH(CH_2)_4CH_3 \\ | \\ CH_2-O-\overset{\overset{\displaystyle O}{\|}}{C}-(CH_2)_7CH=CHCH_2CH=CH(CH_2)_4CH_3 \end{array}$$

Solution

Using Equation 14.4 as a guide, the four carbon–carbon double bonds would each react with a molecule of hydrogen, eliminating the double bonds.

$$\begin{array}{l} CH_2-O-\overset{\overset{\displaystyle O}{\|}}{C}-(CH_2)_{14}CH_3 \\ | \\ CH-O-\overset{\overset{\displaystyle O}{\|}}{C}-(CH_2)_{16}CH_3 \\ | \\ CH_2-O-\overset{\overset{\displaystyle O}{\|}}{C}-(CH_2)_{16}CH_3 \end{array}$$

✔ **Learning Check 14.5** If an oil with the following structure is completely hydrogenated, what is the structure of the product?

$$CH_2-O-\overset{O}{\overset{||}{C}}-(CH_2)_7CH=CH(CH_2)_7CH_3$$

$$CH-O-\overset{O}{\overset{||}{C}}-(CH_2)_7CH=CHCH_2CH=CHCH_2CH=CHCH_2CH_3$$

$$CH_2-O-\overset{O}{\overset{||}{C}}-(CH_2)_7CH=CHCH_2CH=CH(CH_2)_4CH_3$$

During hydrogenation, some fatty acid molecules in the common cis configuration (see **Section 14.2**) are isomerized into trans fatty acids (also called trans fats). The main sources of trans fatty acids in the diet are stick margarine, shortening, and high-fat baked goods (**Figure 14.14**). Recent clinical studies have shown that blood cholesterol levels may be raised by the consumption of trans fatty acids. Current dietary advice is to reduce the consumption of saturated fatty acids and products that contain significant amounts of trans fatty acids. In 2015, the Food and Drug Administration (FDA) took action to eliminate the use of partially hydrogenated oils, the major source of trans fats in the food supply. This action is expected to reduce coronary heart disease and prevent thousands of fatal heart attacks each year in the United States. In 2018, the World Health Organization (WHO) called for a worldwide ban of artificial trans fats by 2023. And while that goal may not be reached, many countries have passed legislation banning, or greatly reducing the amount of trans fats allowed in food.

Study Tools 14.1

A Reaction Map for Triglycerides

We've discussed three different reactions of triglycerides. Remember, when you're trying to decide what happens in a reaction, focus your attention on the functional groups.

After identifying a triglyceride as a starting material, decide what other reagent is involved and which of the three pathways is appropriate.

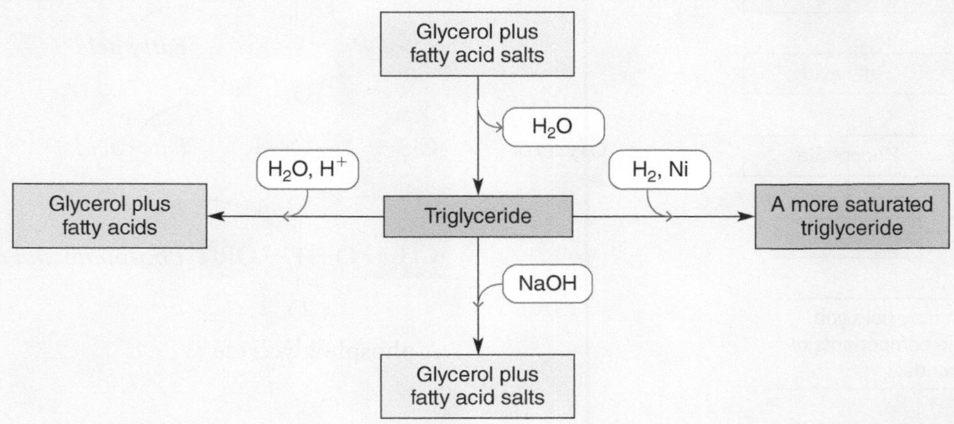

14.5 Waxes

Learning Objective 7 Compare and contrast the structures of fats and waxes.

wax An ester of a long-chain fatty acid and a long-chain alcohol.

Waxes represent a second group of simple lipids that, like fats and oils, are esters of fatty acids. However, the alcohol portion of waxes is derived from long-chain alcohols (12 to 32 carbons) rather than glycerol. The following diagram may help you remember the components of a wax.

Beeswax, for example, contains a wax with the following structure (**Figure 14.15**):

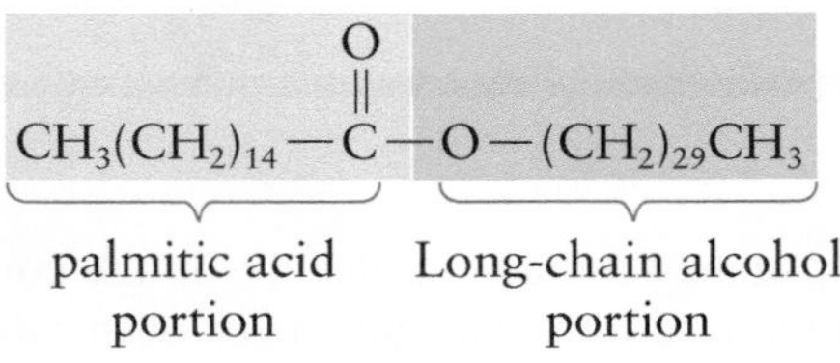

Figure 14.15 Beeswax is used in lip balm, candles, hand creams, salves, moisturizers, and cosmetics such as eye shadow.

Waxes are water insoluble and not as easily hydrolyzed as fats and oils. Consequently, they often occur in nature as protective coatings on feathers, fur, skin, leaves, and fruits. Sebum, a secretion of the sebaceous glands of the skin, contains many different waxes. It keeps the skin soft and prevents dehydration. Waxes are used commercially to make cosmetics, candles, ointments, and protective polishes.

14.6 Phosphoglycerides

Learning Objective 8 Draw structural formulas for phosphoglycerides.

Learning Objective 9 Describe two uses for phosphoglycerides.

phosphoglyceride A complex lipid containing glycerol, fatty acids, phosphoric acid, and an aminoalcohol component.

phospholipid A phosphorus-containing lipid.

Phosphoglycerides are complex lipids that serve as the major components of cell membranes. Phosphoglycerides and related compounds are also referred to more generally as **phospholipids**.

In these compounds, one of the –OH groups of glycerol is joined by an ester linkage to phosphoric acid, which is linked, in turn, to another alcohol (usually an aminoalcohol). The other two –OH groups of glycerol are linked to fatty acids, resulting in the following general structure:

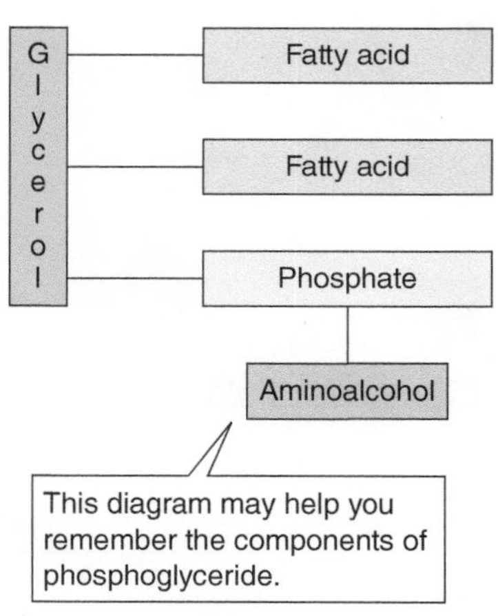

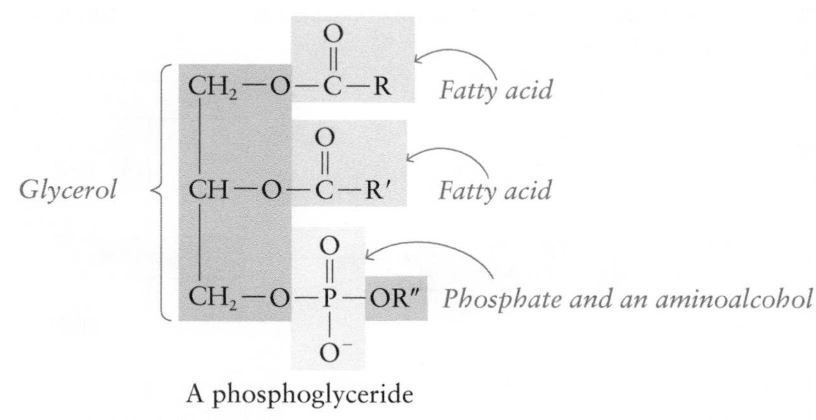

A phosphoglyceride

The most abundant phosphoglycerides have one of the alcohols choline, ethanolamine, or serine attached to the phosphate group. These aminoalcohols are shown here in their charged forms:

$$\mathrm{HO{-}CH_2CH_2{-}\overset{+}{N}(CH_3)_3} \qquad \mathrm{HO{-}CH_2CH_2{-}\overset{+}{N}H_3} \qquad \mathrm{HO{-}CH_2\underset{\displaystyle COO^-}{\underset{|}{CH}}{-}\overset{+}{N}H_3}$$

choline (a quaternary ammonium cation) | ethanolamine (cation form) | serine (two ionic groups present)

A typical phosphoglyceride is phosphatidylcholine, which is commonly called **lecithin**:

lecithin A phosphoglyceride containing choline.

$$
\begin{array}{l}
1\ \mathrm{CH_2{-}O{-}\overset{\displaystyle O}{\overset{\|}{C}}{-}(CH_2)_{16}CH_3} \\
\quad | \\
2\ \mathrm{CH{-}O{-}\overset{\displaystyle O}{\overset{\|}{C}}{-}(CH_2)_7CH{=}CHCH_2CH{=}CH(CH_2)_4CH_3} \\
\quad | \\
3\ \mathrm{CH_2{-}O{-}\underset{\displaystyle O^-}{\underset{|}{\overset{\displaystyle O}{\overset{\|}{P}}}}{-}O{-}CH_2CH_2{-}\overset{+}{N}(CH_3)_3}
\end{array}
$$

phosphatidylcholine

All phosphoglycerides that contain choline are classified as lecithins. Because different fatty acids may be bonded at positions 1 and 2 of the glycerol, a number of different lecithins are possible. The lecithin shown above contains a negatively charged phosphate group and a positively charged quaternary nitrogen. These charges make that end of the molecule strongly hydrophilic, whereas the rest of the molecule is hydrophobic. This structure of lecithins enables them to function as very important structural components of most cell membranes (see **Section 14.8**).

Panther Media GmbH/Alamy Stock Photo

Figure 14.16 Chocolate candy often contains lecithins as emulsifying agents.

This structure also allows lecithins to function as emulsifying and micelle-forming agents. Such micelles play an important role in the transport of lipids (**Section 14.2**). Lecithin (phosphatidylcholine) is commercially extracted from soybeans for use as an emulsifying agent in food. It gives a smooth texture to such products as margarine and chocolate candies (see **Figure 14.16**).

Phosphoglycerides in which the alcohol is ethanolamine or serine, rather than choline, are called **cephalins**, which like lecithins can act as emulsifiers. They are found in most cell membranes and are particularly abundant in brain tissue. Cephalins are also found in blood platelets, where they play an important role in the blood-clotting process (**Figure 14.17**). A typical cephalin is represented as follows:

cephalin A phosphoglyceride containing ethanolamine or serine.

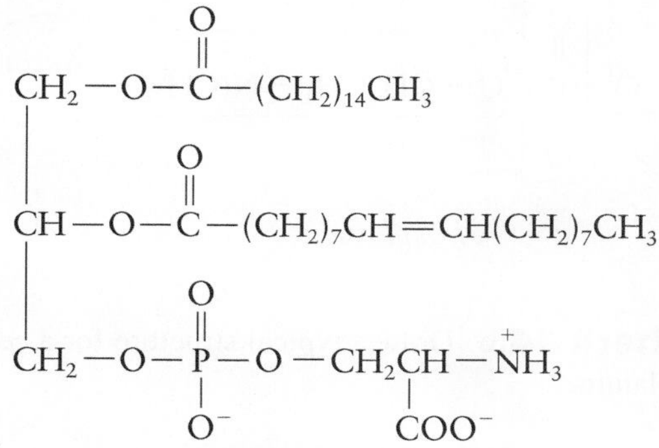

Figure 14.17 Depiction of a blood clot forming inside a blood vessel.

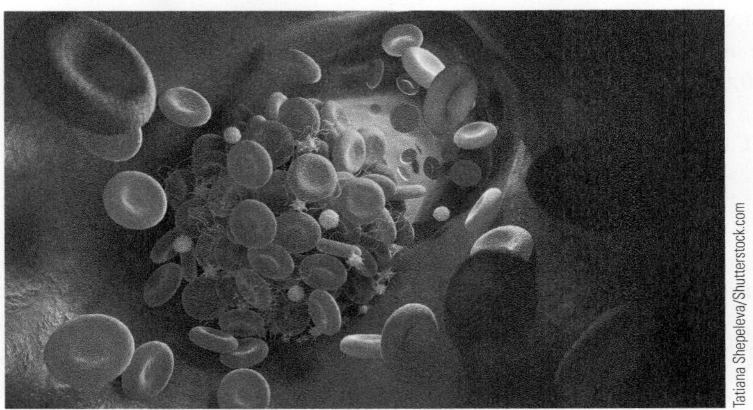

Example 14.6 Writing Phosphoglyceride Structures

In the following phosphoglyceride structure, circle the component parts and label each circle with one of these terms: *glycerol*, *fatty acid*, *phosphate* or *aminoalcohol*.

$$
\begin{array}{l}
CH_2-O-\overset{\overset{\displaystyle O}{\|}}{C}-(CH_2)_{16}CH_3 \\
| \\
CH-O-\overset{\overset{\displaystyle O}{\|}}{C}-(CH_2)_7CH=CHCH_2CH=CH(CH_2)_4CH_3 \\
| \\
CH_2-O-\underset{\underset{\displaystyle O^-}{|}}{\overset{\overset{\displaystyle O}{\|}}{P}}-O-CH_2CH_2-\overset{+}{N}(CH_3)_3
\end{array}
$$

Solution

Following the example at the beginning of **Section 14.6**:

$$
\begin{array}{l}
\boxed{CH_2-O}\boxed{-\overset{\overset{\displaystyle O}{\|}}{C}-(CH_2)_{16}CH_3} \\
| \\
CH-O\boxed{-\overset{\overset{\displaystyle O}{\|}}{C}-(CH_2)_7CH=CHCH_2CH=CH(CH_2)_4CH_3} \\
| \\
CH_2-O-\boxed{\underset{\underset{\displaystyle O^-}{|}}{\overset{\overset{\displaystyle O}{\|}}{P}}}-\boxed{O-CH_2CH_2-\overset{+}{N}(CH_3)_3}
\end{array}
$$

Fatty acid

Aminoalcohol

Glycerol

Phosphate

✔ **Learning Check 14.6** Draw a typical structure for a cephalin containing the cation of ethanolamine.

$$HO-CH_2CH_2-\overset{+}{N}H_3$$

14.7 Sphingolipids

Learning Objective 10 Draw structural formulas for sphingolipids.

Learning Objective 11 Describe two uses for sphingolipids.

sphingolipid A complex lipid containing the aminoalcohol sphingosine.

Sphingolipids, a second type of complex lipid found in cell membranes, do not contain glycerol. Instead, they contain sphingosine, a long-chain unsaturated aminoalcohol.

$$\begin{array}{l} CH_3(CH_2)_{12}CH{=}CH-CH-OH \\ \qquad\qquad\qquad\qquad\quad\;\; | \\ \qquad\qquad\qquad\qquad\quad CH-NH_2 \\ \qquad\qquad\qquad\qquad\quad\;\; | \\ \qquad\qquad\qquad\qquad\quad CH_2OH \end{array}$$

sphingosine

This substance and a typical sphingolipid, called a sphingomyelin, are represented as follows:

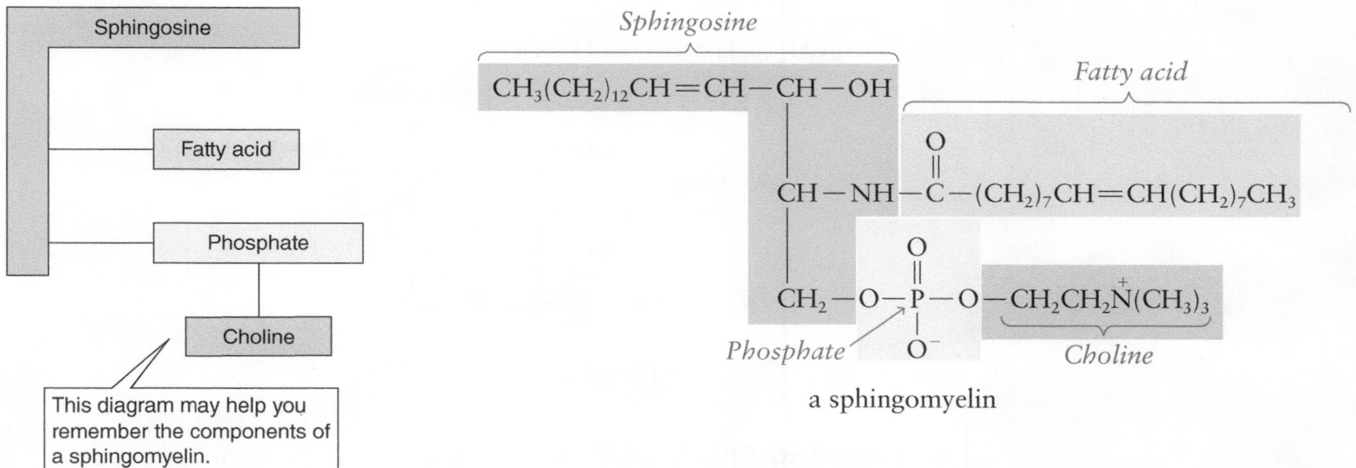

In sphingomyelin, choline is attached to sphingosine through a phosphate group. A single fatty acid is also attached to the sphingosine, but an amide linkage is involved instead of an ester linkage. A number of sphingomyelins are known, and they differ only in the fatty acid component. Large amounts of sphingomyelins are found in brain and nerve tissue and in the protective myelin sheath that surrounds nerves (**Figure 14.18**).

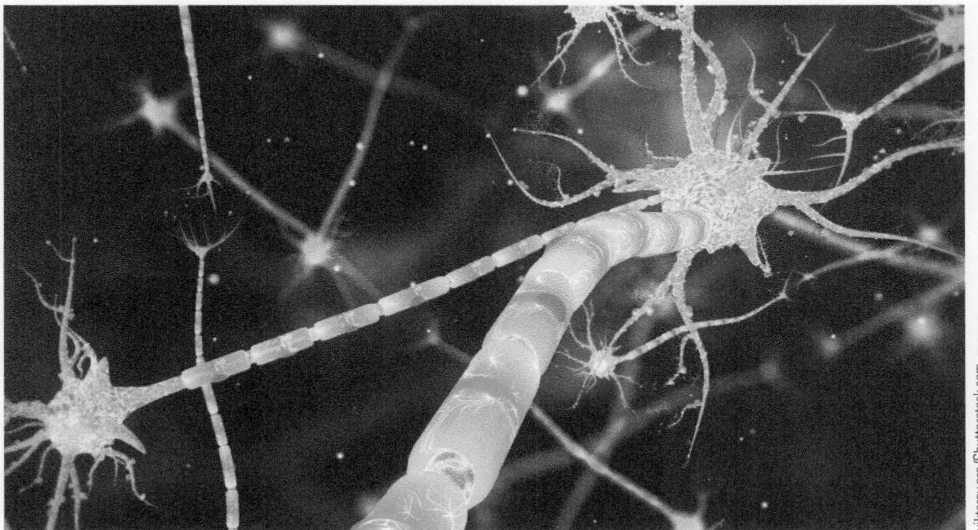

Figure 14.18 Depiction of the mylein sheath covering a nerve cell.

Example 14.7 Writing Sphingomyelin Structures

In the following sphingomyelin structure, circle the component parts and label each circle with one of these terms: *sphingosine*, *fatty acid*, *phosphoric acid*, or *choline*.

$$
\begin{array}{l}
CH_3(CH_2)_{12}CH{=}CH{-}CH{-}OH \\
\qquad\qquad\qquad\qquad\quad | \\
\qquad\qquad\qquad\qquad\quad CH{-}NH{-}\overset{O}{\overset{\|}{C}}{-}(CH_2)_5CH{=}CH(CH_2)_7CH_3 \\
\qquad\qquad\qquad\qquad\quad | \\
\qquad\qquad\qquad\qquad\quad CH_2{-}O{-}\underset{O^-}{\underset{|}{\overset{O}{\overset{\|}{P}}}}{-}O{-}CH_2CH_2\overset{+}{N}(CH_3)_3
\end{array}
$$

Solution

Using the structural guide provided at the beginning of this section:

$$
\begin{array}{l}
\boxed{CH_3(CH_2)_{12}CH{=}CH{-}CH{-}OH} \\
\qquad\qquad\qquad\qquad\quad | \\
\qquad\qquad\qquad\qquad\quad CH{-}NH{-}\boxed{\overset{O}{\overset{\|}{C}}{-}(CH_2)_5CH{=}CH(CH_2)_7CH_3} \\
\qquad\qquad\qquad\qquad\quad | \\
\qquad\qquad\qquad\qquad\quad CH_2{-}\boxed{O{-}\underset{O^-}{\underset{|}{\overset{O}{\overset{\|}{P}}}}{-}O}{-}\boxed{CH_2CH_2\overset{+}{N}(CH_3)_3}
\end{array}
$$

Fatty acid

Sphingosine

Choline

Phosphate

✔ Learning Check 14.7 Write the structure of a sphingomyelin containing the following fatty acid (linoleic acid).

$$CH_3(CH_2)_4(CH{=}CHCH_2)_2(CH_2)_6{-}COOH$$

---

glycolipid A complex lipid containing a sphingosine, a fatty acid, and a carbohydrate.

Glycolipids are another type of sphingolipid. Unlike sphingomyelins, however, these complex lipids contain carbohydrates (usually monosaccharides such as glucose or galactose). Glycolipids are often called cerebrosides because of their abundance in brain tissue. The following structure is of a typical cerebroside:

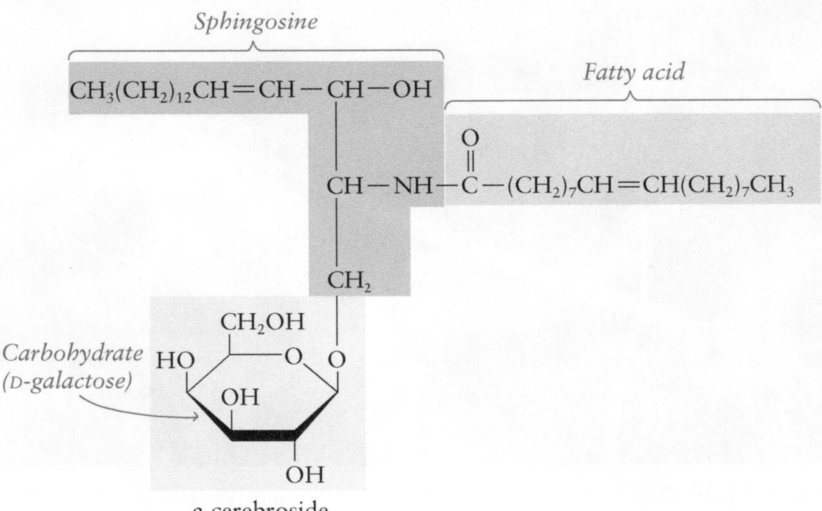

a cerebroside

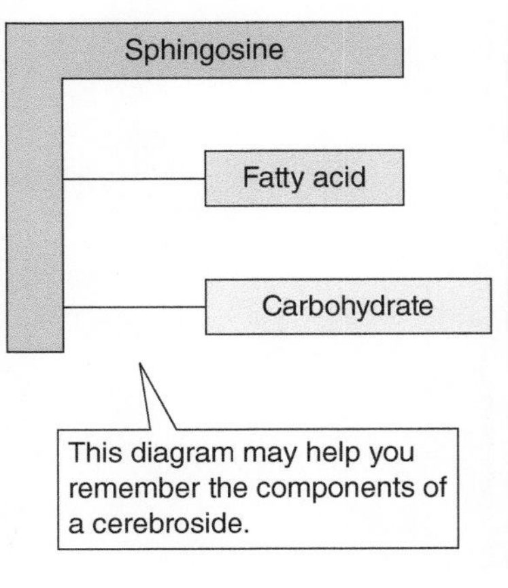

Glycolipids are also found on the surface of red blood cells, where they are referred to as antigens. The antigens expressed on your red blood cells determine your blood type (**Figures 14.19**). For instance, if an individual has blood type B, then their red blood cells will express the glycolipid antigen B, which has an extra galactose (Gal) relative to antigen O and a galactose in place of *N*-acetylgalactosamine (GalNAc) when compared to antigen A.

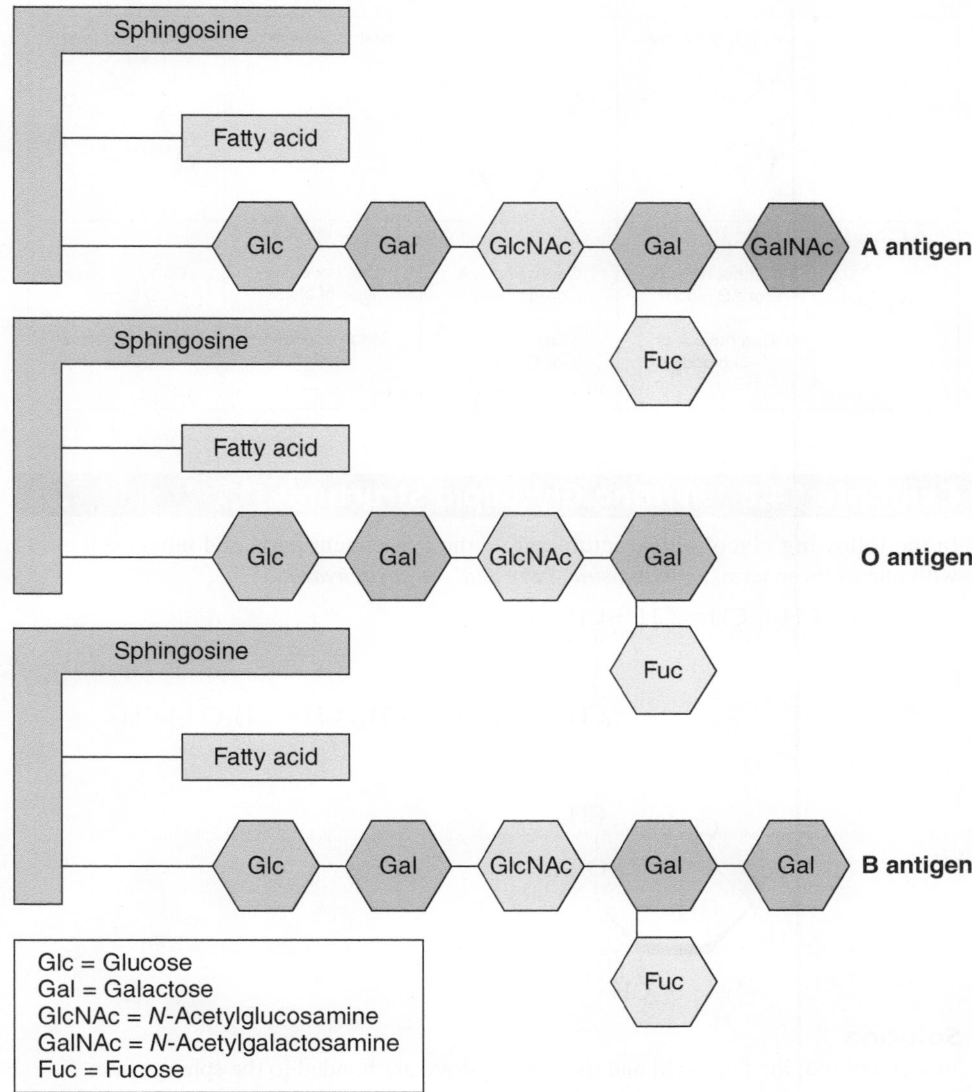

Figure 14.19 Blood type antigens A, B, and O.

Because an individual with blood type B expresses only the B antigen, they will develop antibodies to the A antigen and cannot receive blood from an individual with blood type A or AB. If they do, their immune system will detect those cells as foreign and attack them (**Figure 14.20**). Instead, if a blood transfusion is necessary for any medical procedure, they must receive only blood type B or alternatively blood type O. Why can an individual with blood type B accept blood type O? The answer lies in the structure of the glycolipid! If you look closely at **Figure 14.19**, all blood antigens consist of the same structure through the sugar fucose (Fuc), and it is only the addition of GalNAc or Gal that results in antigen A or B, respectively.

Figure 14.20 Blood type. (a) Depiction of blood type antigens expressed on red blood cells (RBC). (b) Antibodies formed in plasma to incompatible blood types. (c) Explanation of allowed donors based on patient blood type.

	Type A	Type B	Type AB	Type O
Antigen (on RBC)	Antigen A 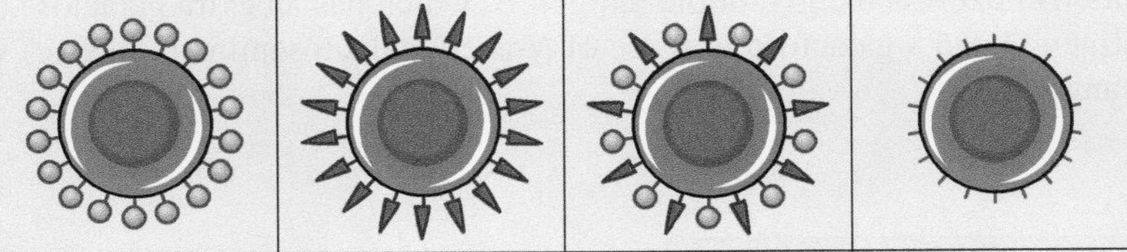	Antigen B	Antigens A + B	Neither A or B
Antibody (in plasma)	Anti-B Antibody 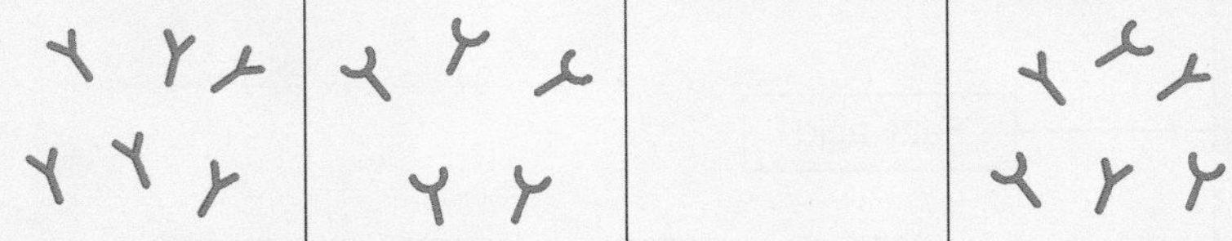	Anti-A Antibody	Neither Antibody	Both Antibodies
Blood Donors	Cannot have B or AB blood Can have A or O blood	Cannot have A or AB blood Can have B or O blood	Can have any type of blood Is the universal recipient	Can only have O blood Is the universal donor

Example 14.8 Writing Glycolipid Structures

In the following glycolipid structure, circle the component parts and label each circle with one of these terms: *sphingosine*, *fatty acid*, or *carbohydrate*.

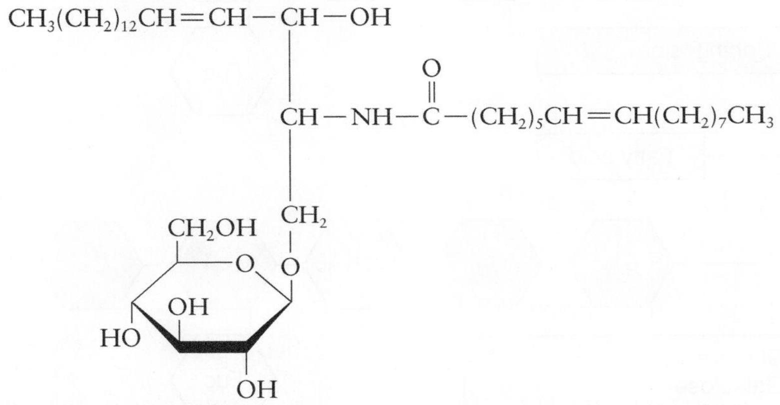

$CH_3(CH_2)_{12}CH{=}CH-CH-OH$

$CH-NH-C(=O)-(CH_2)_5CH{=}CH(CH_2)_7CH_3$

CH_2

CH_2OH O O OH HO OH

Solution

In a glycolipid, the fatty acid and the carbohydrate are bonded to the sphingosine:

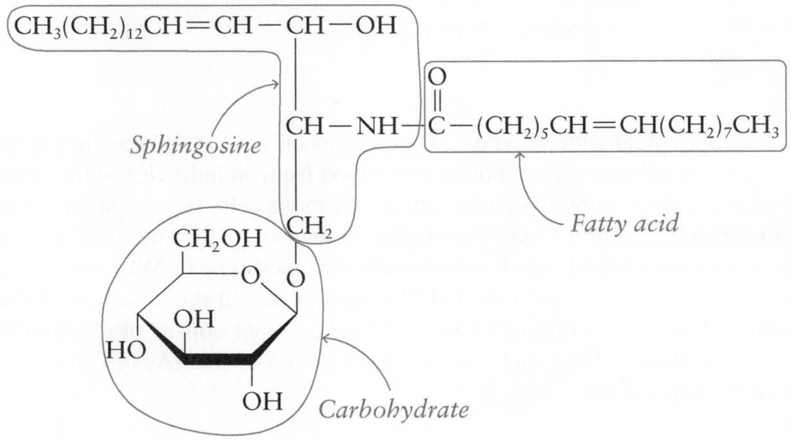

$CH_3(CH_2)_{12}CH{=}CH-CH-OH$

Sphingosine

$CH-NH-C(=O)-(CH_2)_5CH{=}CH(CH_2)_7CH_3$

Fatty acid

CH_2

CH_2OH O O OH HO OH

Carbohydrate

✔ **Learning Check 14.8** Write the structure of the glycolipid containing the carbohydrate glucose and the fatty acid linoleic acid.

CH_2OH

O OH

OH

HO

OH

glucose

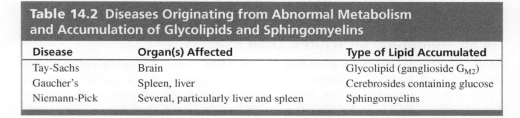

Several human diseases are known to result from an abnormal accumulation of sphingomyelins and glycolipids in the body (see **Table 14.2**). Research has shown that each of these diseases is the result of an inherited absence of an enzyme needed to break down these complex lipids. For instance, the buildup of glycolipids in Tay-Sachs disease eventually becomes toxic, damaging the nerve cells in the brain and spine (**Figure 14.21**). Due to the severity of the nerve cell damage the life expectancy of an individual with Tay-Sachs disease is only 3 to 5 years.

Richard Swingler/Mirrorpix/NEWSCOM

Figure 14.21 Tay-Sachs diseases results from an excessive buildup of glycolipids in the brain and spine of affected individuals.

Table 14.2 Diseases Originating from Abnormal Metabolism and Accumulation of Glycolipids and Sphingomyelins

Disease	Organ(s) Affected	Type of Lipid Accumulated
Tay-Sachs	Brain	Glycolipid (ganglioside G_{M2})
Gaucher's	Spleen, liver	Cerebrosides containing glucose
Niemann-Pick	Several, particularly liver and spleen	Sphingomyelins

14.8 Biological Membranes

Learning Objective 12 Describe the major features of cell membrane structure.

14.8A Cell Structure

Two cell types are found in living organisms: prokaryotic and eukaryotic (**Figure 14.22**). **Prokaryotic cells**, comprising bacteria and cyanobacteria (blue-green algae), are smaller and less complex than eukaryotic cells. The more complex **eukaryotic cells** make up the tissues of all other organisms, including humans. Both types of cells are essentially tiny membrane-enclosed units of fluid containing the chemicals involved in life processes.

In addition, eukaryotic cells contain small bodies called **organelles**, which are suspended in the cellular fluid, or cytoplasm. Organelles are the sites of many specialized functions in eukaryotes. The most prominent organelles and their functions are listed in **Table 14.3**. Membranes themselves perform two vital functions in living organisms: The

prokaryotic cell A simple unicellular organism that contains no nucleus and no membrane-enclosed organelles.

eukaryotic cell A cell containing membrane-enclosed organelles, particularly a nucleus.

organelle A specialized structure within a cell that performs a specific function.

Table 14.3 The Functions of Some Cellular Organelles

Organelle	Function
Endoplasmic reticulum	Synthesis of proteins, lipids, and other substances
Lysosome	Digestion of substances taken into cells
Microfilaments and microtubules	Cellular movements
Mitochondrion	Cellular respiration and energy production
Nucleus	Contains hereditary material (DNA), which directs protein synthesis
Plastids (plants only)	Contains plant pigments such as chlorophyll (photosynthesis)
Ribosome	Protein synthesis

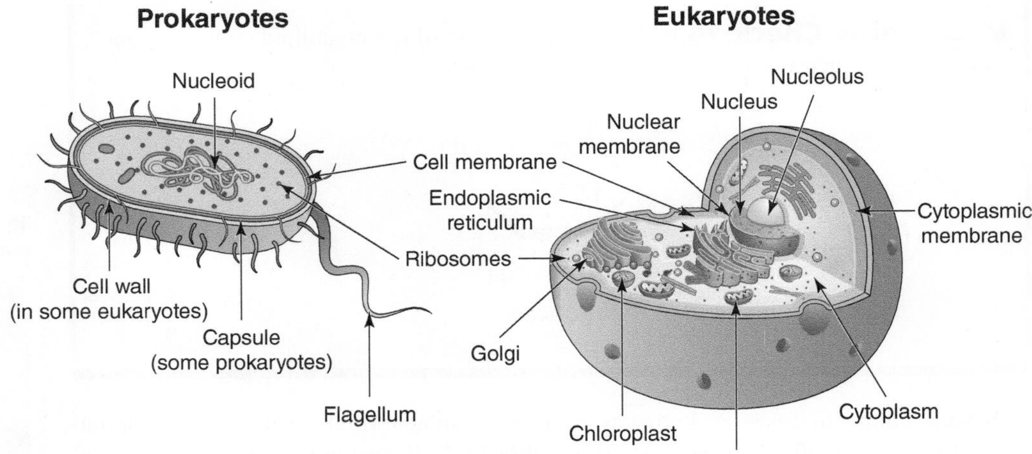

Figure 14.22 A comparison of the basic characteristics of prokaryotic and eukaryotic cells.

external cell membrane acts as a selective barrier between the living cell and its environment, and internal membranes surround some organelles, creating cellular compartments that have separate organization and functions.

14.8B Membrane Structures: Lipid Bilayers

fluid-mosaic model A model of membrane structure in which proteins are embedded in a flexible lipid bilayer.

Most cell membranes contain about 60% lipid and 40% protein. Phosphoglycerides (such as lecithin and cephalin), sphingomyelin, and cholesterol are the predominant types of lipids found in most membranes. Precisely how the lipids and proteins are organized to form membranes has been the subject of a great deal of research. A widely accepted model called the **fluid-mosaic model** is diagrammed in **Figure 14.23**.

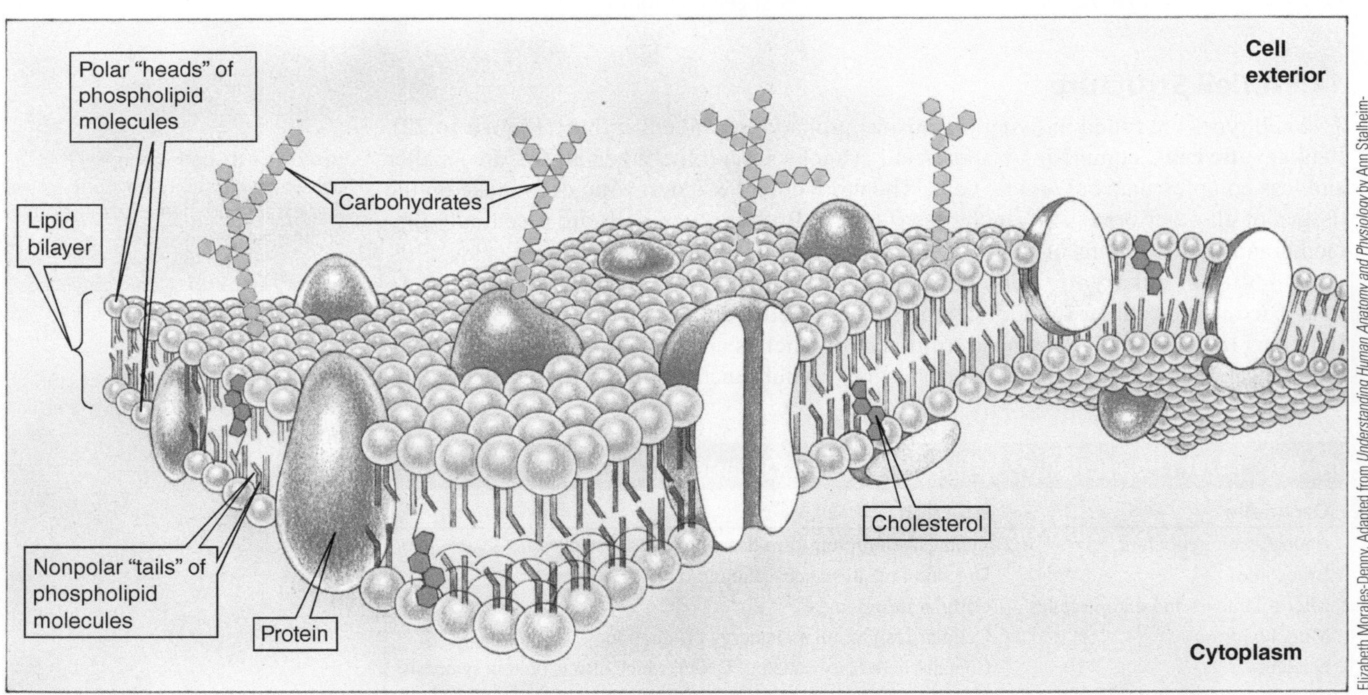

Elizabeth Morales-Denny. Adapted from *Understanding Human Anatomy and Physiology* by Ann Stalheim-Smith and Greg K. Fitch (Copyright © West Publishing Company, 1993). Used with permission.

Figure 14.23 The fluid-mosaic model of membrane structure. Phosphoglycerides are the chief lipid component. They are arranged in a bilayer. Proteins float like icebergs in a sea of lipids.

The lipids are organized in a **lipid bilayer** in which the hydrophobic chains extend toward the inside of the bilayer, and the hydrophilic groups (the phosphate groups and other polar groups) are oriented toward the outside, where they come in contact with water (see **Figure 14.24**). Like the micelle structure discussed in **Section 14.2**, a lipid bilayer is a very stable arrangement where "like" associates with "like." The hydrophobic tails of the lipids are protected from water, and the hydrophilic heads are in a position to interact with water. When a lipid bilayer is broken and the interior hydrocarbon tails are exposed to water, the resulting repulsion causes the bilayer to re-form, and the break seals spontaneously.

lipid bilayer A structure found in membranes, consisting of two sheets of lipid molecules arranged so that the hydrophobic portions are facing each other, and the hydrophilic portions are positioned to interact with water.

Membrane lipids usually contain unsaturated fatty acid chains that fit into bilayers more loosely than do saturated fatty acids (see **Section 14.2**). This increases the flexibility or fluidity of the membrane. Some of the proteins in the membrane float in the lipid bilayer like icebergs in the sea, whereas others extend completely through the bilayer. The lipid molecules are free to move laterally within the bilayer like dancers on a crowded dance floor—hence the term *fluid mosaic*.

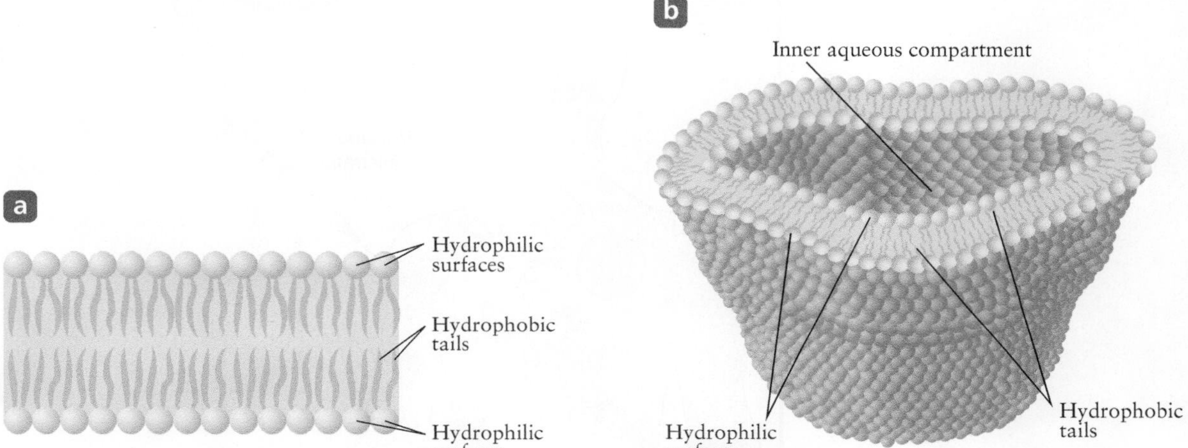

Figure 14.24 Lipid bilayers. (a) Portion of a bilayer. Circles represent the charged polar portion of phosphoglycerides. The nonpolar hydrocarbon tails are within the interior of the bilayer. (b) Cutaway view of a lipid bilayer vesicle, showing an inner aqueous compartment.

14.8C The Role of Lipids in Drug Delivery

Lipids are integral to the delivery of many therapeutics within the human body. For drug molecules to be beneficial, they must traverse through many systems in the body that may wish to reroute them from their intended target. One crucial feature of drugs is their bioavailability—that is, the amount of drug that can actually enter the body's circulation, especially for drugs that are taken orally and must be absorbed through the gastrointestinal (GI) system. A problem may occur, if a potentially potent drug molecule has low solubility in the aqueous environment of the body. One strategy to deliver these kinds of therapeutics is to use lipids to enhance the solubility, and thus the bioavailability, of these drugs. These lipid delivery systems come in many types and forms.

One such strategy involves the use of lipid nanoparticles. These tiny spheres surround the drug molecule to be delivered, shielding it through its journey to its final destination in the body. A recent application of these lipid nanoparticles has been their use in aiding the delivery of mRNA COVID-19 vaccines (**Figure 14.25**), which are discussed in Chapter 13. The fragments of mRNA within the vaccines have been designed to code for the spike protein of the SARS-CoV-2 virus so that the immune system can generate matched antibodies that will recognize and destroy the virus when it enters the body. On their own, however, these small fragments of genetic material are very unstable. They would be easily consumed by a variety of enzymes in the body, and thus unable to reach their final destination—namely, our own cell machinery. But by enveloping the mRNA within lipid nanoparticles, these nucleic acid strands are sufficiently protected to reach their target.

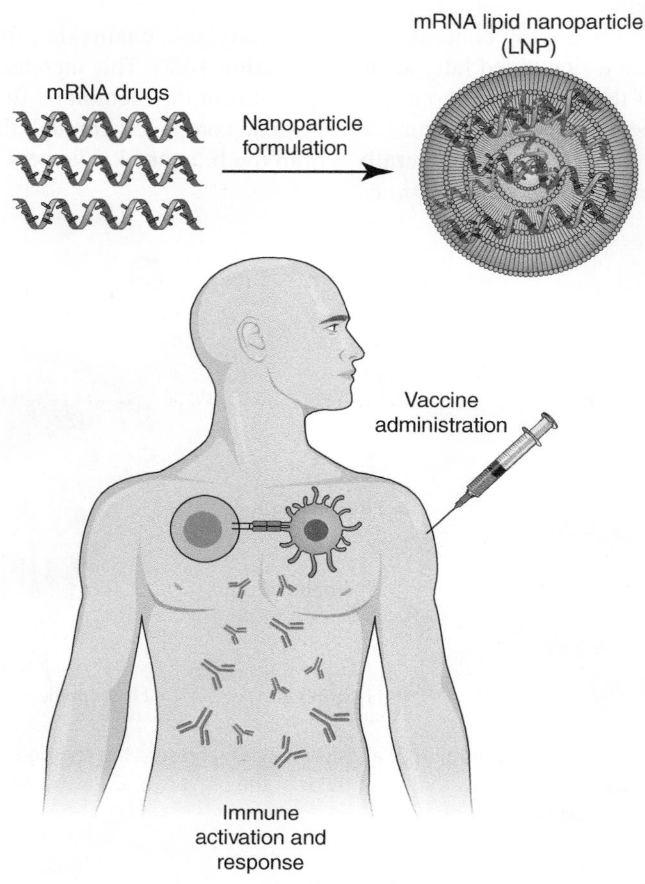

Figure 14.25 Lipid nanoparticle delivery of mRNA in COVID-19 vaccine.

14.9 Steroids and Hormones

Learning Objective 13 Identify the structural characteristic that is typical of steroids.

Learning Objective 14 List five important groups of steroids in the body.

steroid A compound containing four rings fused in a particular pattern.

Steroids exhibit the distinguishing feature of other lipids—that is, they are soluble in nonpolar solvents. Structurally, however, they are completely different from the lipids already discussed. **Steroids** have as their basic structure a set of three 6-membered rings and a single 5-membered ring fused together:

steroid ring system

14.9A Cholesterol

The most abundant steroid in the human body, and the most important, is cholesterol:

CH_3 CH_3
$CH(CH_2)_3CHCH_3$
H_3C
H_3C
HO

cholesterol

Cholesterol is an essential component of cell membranes and is a precursor for other important steroids, including bile salts, male and female sex hormones, vitamin D, and the adrenocorticoid hormones. Cholesterol for use in these important functions is synthesized in the liver. Additional amounts of cholesterol are present in the foods we eat.

Cholesterol has received considerable attention because of a correlation between its levels in the blood and the disease known as atherosclerosis, or hardening of the arteries (**Figure 14.26**). Although our knowledge of the role played by cholesterol in atherosclerosis is incomplete, it is now considered advisable to reduce the amount of cholesterol in the foods we eat. In addition, reducing the amount of saturated fatty acids in the diet appears to lower cholesterol production by the body.

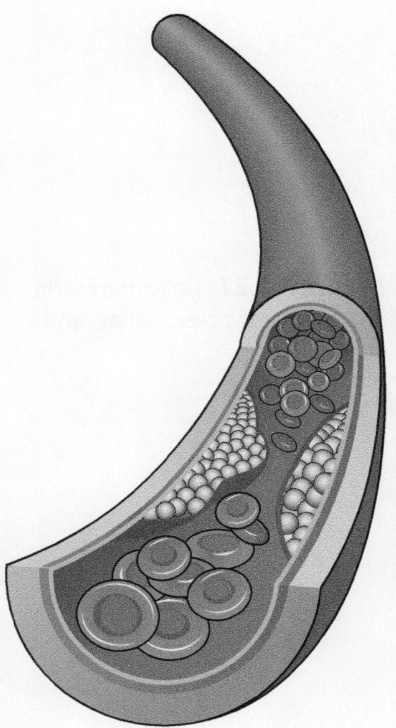

Figure 14.26 Buildup of cholesterol in the arteries results in atherosclerosis.

14.9B Bile Salts

Bile is a yellowish-brown or green liver secretion that is stored and concentrated in the gallbladder. The entry of partially digested fatty food into the small intestine causes the gallbladder to contract and empty bile into the intestine. Bile does not catalyze any digestive reactions, but it is important in lipid digestion. Fats are more difficult to hydrolyze than starch and proteins because fats are insoluble in water. In the watery medium of the digestive system, fats tend to form large globules with a limited surface area exposed to attack by the digestive juices. The chief constituents of bile are the bile salts and two waste components: cholesterol and bile pigments. One of the principal bile salts is sodium glycocholate:

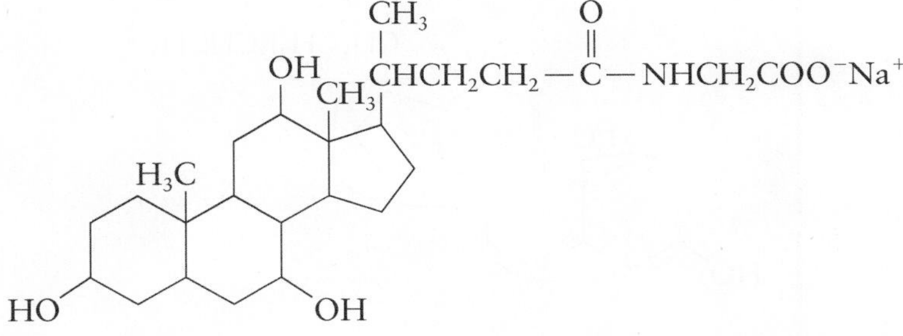

sodium glycocholate

The bile salts act much like soaps: They emulsify the lipids and break the large globules into many smaller droplets. After such action, the total amount of lipids is still the same, but a much larger surface area is available for the hydrolysis reactions. Some researchers feel that bile salts also remove fatty coatings from particles of other types of food and thus aid in their digestion.

A second function of bile salts is to emulsify cholesterol found in the bile. Excess body cholesterol is concentrated in bile and is passed into the small intestine for excretion. If cholesterol levels become too high in the bile, or if the concentration of bile salts is too low, the cholesterol precipitates and forms gallstones (see **Figure 14.27**). The most common gallstones contain about 80% cholesterol and are colored by entrapped bile pigments.

The passage of a gallstone from the gallbladder down the common bile duct to the intestine causes excruciating pain. Sometimes one or more stones become lodged in the duct and prevent bile from passing into the duodenum, and fats can no longer be digested normally. A person with this condition feels great pain and becomes quite nauseated and ill. Bile pigments are absorbed into the blood, the skin takes on a yellow coloration, and the stool becomes gray-colored because of lack of excreted bile pigments. If the condition becomes serious, both the gallbladder and the stones can be surgically removed.

iStock.com/umdash9

Figure 14.27 Gallstones form in a variety of sizes, shapes, and colors.

Example 14.9 Identifying Steroids

All steroids have a unique ring system. In the following structure of cholesterol, circle the ring system that makes this compound a steroid.

CH_3 CH_3
$CH(CH_2)_3CHCH_3$
H_3C
H_3C
HO

Solution

CH_3 CH_3
$CH(CH_2)_3CHCH_3$
H_3C
H_3C
HO

✔ Learning Check 14.9 In the structure of progesterone, a female sex hormone, circle the ring system that makes this compound a steroid.

CH_3
$C{=}O$
H_3C
H_3C
O

A number of steroids in the body serve important roles as **hormones**. The two major categories of steroid hormones are the adrenocorticoid hormones and the sex hormones.

hormone A chemical messenger secreted by specific glands and carried by the blood to a target tissue, where it triggers a particular response.

14.9C Adrenocorticoid Hormones

Adrenal glands are small mounds of tissue located at the top of each kidney. The outer layer of the gland, the adrenal cortex, produces a number of potent steroid hormones, the adrenocorticoids. They are classified into two groups according to function: Mineralocorticoids regulate the concentration of ions (mainly Na^+) in body fluids, and glucocorticoids enhance carbohydrate metabolism.

Cortisol, the major glucocorticoid, functions to increase the glucose and glycogen concentrations in the body. This is accomplished by the conversion of lactate and amino acids from body proteins into glucose and glycogen. The reactions take place in the liver under the influence of cortisol.

cortisol

Cortisol and its ketone derivative, cortisone (see **Figure 14.28**), exert powerful anti-inflammatory effects in the body. These, or similar synthetic derivatives such as prednisolone, are used to treat inflammatory diseases such as rheumatoid arthritis and bronchial asthma.

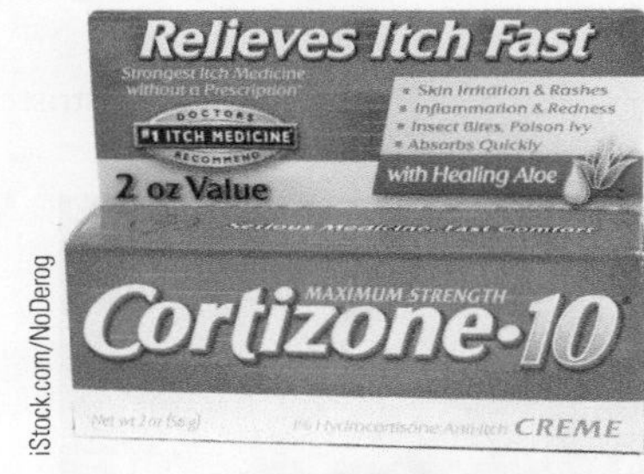

Figure 14.28 A valuable tool in treating inflammation, cortisone is used to treat joint swelling and pain.

cortisone

prednisolone

By far the most important mineralocorticoid is aldosterone, which influences the absorption of Na^+ and Cl^- in kidney tubules (**Figure 14.29**). An increase in the level of the hormone results in a corresponding increase in absorption of Na^+ and Cl^-. Because the concentration of Na^+ in tissues influences the retention of water, aldosterone is involved in water balance in the body.

aldosterone

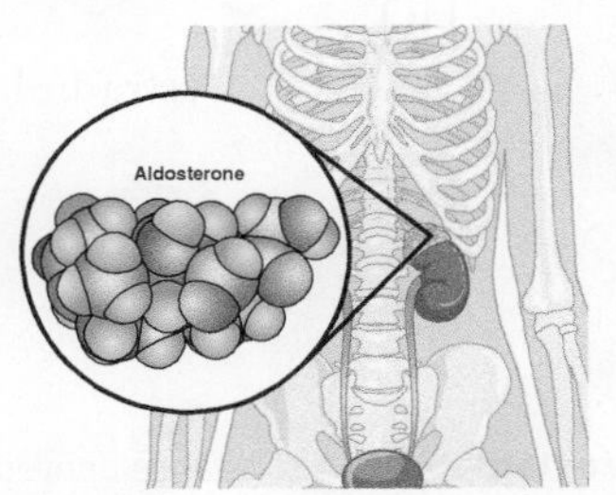

Figure 14.29 The steroid hormone aldosterone is produced in the adrenal glands located at the top of each kidney.

14.9D Sex Hormones

The testes and ovaries produce steroids that function as sex hormones. Secondary sex characteristics that appear at puberty—including a deep voice, beard, and increased muscular development in males and a higher voice, increased breast size, and lack of facial hair in females—develop under the influence of these sex hormones.

The testes in males perform two functions: One is to produce sperm; the other is to produce male sex hormones (androgens), the most important of which is testosterone. Testosterone promotes the normal growth of the male genital organs and aids in the development of secondary sex characteristics.

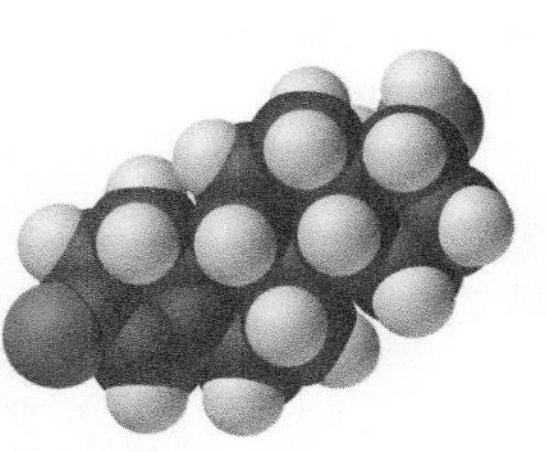

testosterone

testosterone

methandrostenolone

ALL best fitness is HERE/Shutterstock.com

Figure 14.30 The use of anabolic steroids is banned by sports organizations due to their detrimental side effects that affect one's health.

Growth-promoting (anabolic) steroids, including the male hormone testosterone and its synthetic derivatives such as methandrostenolone (Dianabol), are among the most widely used drugs banned by sports-governing organizations because they are alleged to be dangerous and confer an unfair advantage to users. These drugs, used by both male and female athletes, promote muscular development without excessive masculinization. Their use is particularly prevalent in sports where strength and muscle mass are advantageous, such as weight lifting (**Figure 14.30**), track and field events like the shot put and hammer throw, and body building. The side effects of anabolic steroids range from acne to deadly liver tumors. Effects on the male reproductive system include testicular atrophy, a decrease in sperm count, and, occasionally, temporary infertility.

The primary female sex hormones are the estrogens (estradiol and estrone) as well as progesterone. Estrogens and progesterone are important in the reproductive process. Estrogens are involved in egg (ovum) development in the ovaries, and progesterone causes changes in the wall of the uterus to prepare it to accept a fertilized egg and maintain the resulting pregnancy.

estradiol

estrone

progesterone

14.9E Prostaglandins

This group of compounds was given its name because the first prostaglandins were identified from the secretions of the male prostate gland. Recent research has identified as many as 20 prostaglandins in a variety of tissues within both males and females. **Prostaglandins** are cyclic compounds synthesized in the body from the 20-carbon unsaturated fatty acid arachidonic acid. Prostaglandins are designated by codes that refer to the ring substituents and the number of side-chain double bonds. For example, prostaglandin E_2 (PGE_2) has a carbonyl group on the ring and two chains extending off the ring, each of which contains a double bond among other functional groups:

prostaglandin A substance derived from unsaturated fatty acids with hormone-like effects on a number of body tissues.

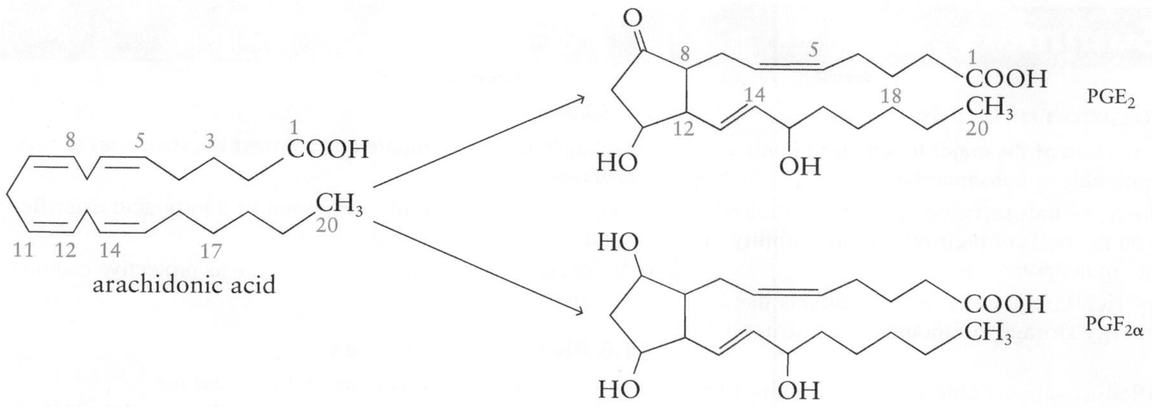

arachidonic acid

PGE$_2$

PGF$_{2\alpha}$

Prostaglandins are similar to hormones in the sense that they are intimately involved in a host of body processes. Clinically, it has been found that prostaglandins are involved in almost every phase of the reproductive process; for example, they can act to regulate menstruation, prevent conception, and induce uterine contractions during childbirth. Certain prostaglandins stimulate blood clotting. They also lead to inflammation and fever. It has been found that aspirin inhibits prostaglandin production. This explains in part why aspirin is such a powerful drug in the treatment of inflammatory diseases such as arthritis.

From a medical viewpoint, prostaglandins have enormous therapeutic potential. For example, PGE$_2$ and PGF$_2$ induce labor (see **Figure 14.31**) and are used for therapeutic abortion in early pregnancy. PGE$_2$ in aerosol form is used to treat asthma. It opens up the bronchial tubes by relaxing the surrounding muscles. Other prostaglandins inhibit gastric secretions and are used to treat peptic ulcers. Many researchers believe that when prostaglandins are fully understood, they will be useful for treating a much wider variety of ailments.

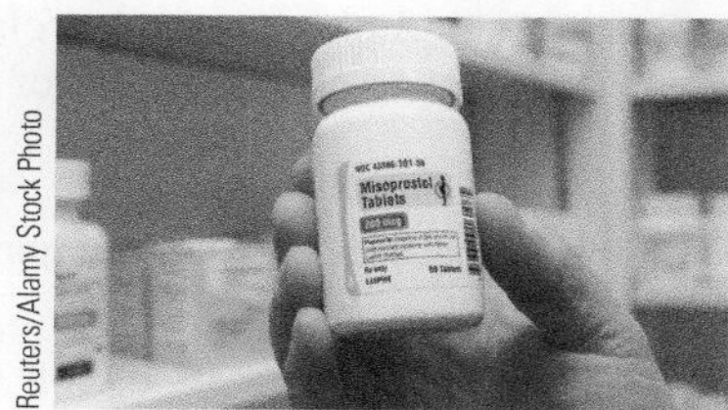

Reuters/Alamy Stock Photo

Figure 14.31 Pharmaceutical companies manufacture a number of medications containing prostaglandins. For example, the misoprostol shown here is used to induce contractions to begin labor.

Health Career Description

Physical Therapy Assistant

Physical therapy assistants (PTAs) work under the direction of physical therapists who hold a doctorate in physical therapy (DPT). Physical therapy assistants work with patients of all ages who experience difficulty with functional mobility due to diseases, accidents, or aging. PTAs teach exercises, record patient progress, take measurements, and support patient wellness through education and empathetic care. According to the American Association of Physical Therapists, "Care provided by a PTA may include teaching patients/clients exercise for mobility; strength and coordination; training for activities such as walking with crutches, canes, or walkers, massage; and the use of physical agents and electrotherapy such as ultrasound and electrical stimulation."

Physical therapy assisting is a growing career field. The Bureau of Labor Statistics predicts the demand for physical therapy assistants to grow at a rate of 26% per year and lists the average annual income for PTAs at $48,990. This field is growing quickly because of a rapid increase in the elderly population and an increase in access to care due to the Affordable Care Act. PTAs require an associate's degree from a technical college or university offering a PTA program and must pass a national licensing exam.

PTAs spend the majority of their day on their feet moving about between patients and demonstrating various exercises. They must have a minimum level of physical fitness to perform their job. Because they work closely with patients, including touching patients for treatments, they must have outstanding communication skills, and a friendly, professional demeanor, as well as the ability to protect patient confidentiality.

PTAs have the opportunity to make lasting improvements in the quality of life of patients experiencing pain and limitations in their daily lives. They work with patients of all abilities, including athletes, pregnant patients, patients preparing for or recovering from surgery, the elderly, and those experiencing unexplained pain. PTAs play an important role in helping people across a wide spectrum of circumstances to realize their full physical potential.

Sources: American Physical Therapy Association (APTA). (n.d.) Who Are Physical Therapist Assistants? Retrieved Apr. 26, 2020 from https:// www.apta.org /AboutPTAs/Bureau of Labor Statistics. (n.d.) Physical Therapist Assistants and Aides. Retrieved Apr. 26, 2020 from https:// www.bls.gov/ooh/healthcare /physical-therapist-assistants-and-aides.htm

Concept Summary

14.1 The Classification of Lipids

Learning Objectives: List two of the major functions of lipids. Classify lipids as saponifiable or nonsaponifiable.

- Lipids are a family of naturally occurring compounds grouped together on the basis of their relative insolubility in water and solubility in nonpolar solvents.
- Lipids are energy-rich compounds that organisms use as waxy coatings, energy storage compounds, and structural components.
- Lipids are classified as saponifiable (ester-containing) or nonsaponifiable.
- Saponifiable lipids are further classified as simple or complex, depending on the number of structural components.

14.2 Fatty Acids

Learning Objective: Describe four general characteristics of fatty acids.

- A fatty acid consists of a long nonpolar chain of carbon atoms, with a polar carboxylic acid group at one end.
- Most natural fatty acids contain an even number of carbon atoms.
- Fatty acids may be saturated, unsaturated, or polyunsaturated (i.e., containing two or more double bonds).
- In aqueous solutions, the ions of fatty acids form spherical clusters called micelles.
- During digestion, fats and phosphoglycerides are hydrolyzed to their component substances and are absorbed into intestinal cells.
- Once lipids are absorbed by the intestinal mucosa, triglycerides and phosphoglycerides are resynthesized and packaged for transport through lymph and blood.
- Lipids travel through aqueous lymph and blood in complex with proteins. The resulting aggregates are called chylomicrons.

14.3 The Structure of Fats and Oils

Learning Objectives: Draw structural formulas of triglycerides given the formulas of the component parts. Describe the structural similarities and differences of fats and oils.

- Triglycerides or triacylglycerols in the form of fats and oils are the most abundant lipids.
- Fats and oils are simple lipids that are esters of glycerol and fatty acids.
- The difference between fats and oils is the melting point, which is essentially a function of the fatty acids in the compound.

14.4 Chemical Properties and Reactions of Fats and Oils

Learning Objective: Write key reactions for fats and oils.

- Fats and oils can be hydrolyzed in the presence of acid to produce glycerol and fatty acids.
- When the hydrolysis reaction is carried out in the presence of a strong base, salts of the fatty acids (soaps) are produced.
- During hydrogenation, some multiple bonds of unsaturated fatty acids contained in fats or oils react with hydrogen and are converted to single bonds.

14.5 Waxes

Learning Objective: Compare and contrast the structures of fats and waxes.

- Waxes are simple lipids composed of a fatty acid esterified with a long-chain alcohol.
- Waxes are insoluble in water and serve as protective coatings in nature.

14.6 Phosphoglycerides

Learning Objectives: Draw structural formulas for phosphoglycerides. Describe two uses for phosphoglycerides.

- Phosphoglycerides consist of glycerol esterified to two fatty acids and phosphoric acid.
- The phosphoric acid is further esterified to choline (in the lecithins) and to ethanolamine or serine (in the cephalins).
- The phosphoglycerides are particularly important in membrane formation.

14.7 Sphingolipids

Learning Objectives: Draw structural formulas for sphingolipids. Describe two uses for sphingolipids.

- Sphingolipids are complex lipids that contain a backbone of sphingosine rather than glycerol and only one fatty acid component.
- Sphingolipids are abundant in brain and nerve tissue.
- When expressed on the surface of blood cells, sphingolipids are called antigens, and they determine blood type: A, B, AB, or O.
- Individuals lacking enzymes that digest sphingolipids develop diseases such as Tay-Sachs, Gaucher's, and Niemann-Pick.

14.8 Biological Membranes

Learning Objective: Describe the major features of cell membrane structure.

- Membranes surround tissue cells as a selective barrier, and they encase the organelles found in eukaryotic cells.
- Membranes contain both proteins and lipids.
- According to the fluid-mosaic model, the lipids are arranged in a bilayer fashion, with the hydrophobic portions on the inside of the bilayer.
- Proteins float in the bilayer.
- Lipids can be utilized in nanoparticles for the delivery of many therapeutics within the human body.

14.9 Steroids and Hormones

Learning Objectives: Identify the structural characteristic that is typical of steroids. List five important groups of steroids in the body.

- Steroids are compounds that have four rings fused together in a specific way.
- The most abundant steroid in humans is cholesterol.
- Cholesterol serves as a starting material for other important steroids such as bile salts, adrenocorticoid hormones, and sex hormones.

- Hormones are chemical messengers that are synthesized by specific glands and that affect various target tissues in the body.
- The adrenal cortex produces a number of steroid hormones that regulate carbohydrate utilization (the glucocorticoids) and electrolyte balance (the mineralocorticoids).
- The testes and ovaries produce steroid hormones that determine secondary sex characteristics and regulate the reproductive cycle in females.
- These compounds are synthesized from the 20-carbon fatty acid arachidonic acid.
- They exert many hormone-like effects on the body.
- They are used therapeutically to induce labor, treat asthma, and control gastric secretions.

Key Terms and Concepts

Cephalin (14.6)
Chylomicron (14.2)
Complex lipid (14.1)
Essential fatty acid (14.2)
Eukaryotic cell (14.8)
Fat (14.3)
Fluid-mosaic model (14.8)
Glycolipid (14.7)
Hormone (14.9)
Lecithin (14.6)
Lipid (14.1)
Lipid bilayer (14.8)
Micelle (14.2)
Oil (14.3)
Organelle (14.8)
Phosphoglyceride (14.6)
Phospholipid (14.6)
Prokaryotic cell (14.8)
Prostaglandin (14.9)
Simple lipid (14.1)
Soap (14.4)
Sphingolipid (14.7)
Steroid (14.9)
Triglyceride or triacylglycerol (14.3)
Wax (14.5)

Key Equations

1. Esterification reaction producing a triglyceride from glycerol and fatty acids—general reaction **(Section 14.3)**	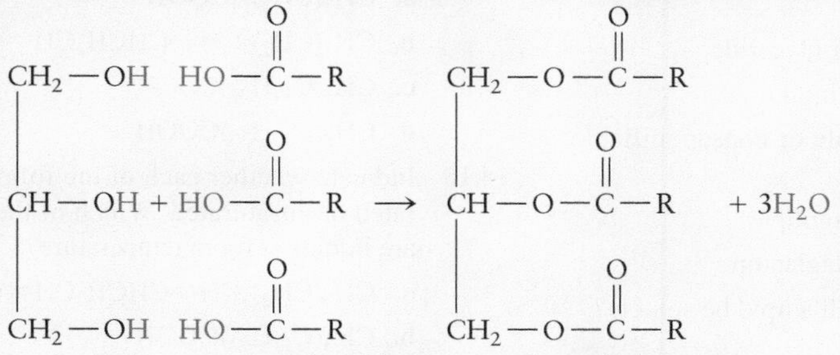 $\begin{array}{lcl} CH_2-OH \quad HO-\overset{\overset{O}{\parallel}}{C}-R & & CH_2-O-\overset{\overset{O}{\parallel}}{C}-R \\ CH-OH + HO-\overset{\overset{O}{\parallel}}{C}-R & \longrightarrow & CH-O-\overset{\overset{O}{\parallel}}{C}-R \quad + 3H_2O \\ CH_2-OH \quad HO-\overset{\overset{O}{\parallel}}{C}-R & & CH_2-O-\overset{\overset{O}{\parallel}}{C}-R \end{array}$ glycerol (1 molecule) stearic acid (3 molecules) glyceryl tristearate (a triglyceride; 1 molecule)	Equation 14.1
2. Hydrolysis of a triglyceride to glycerol and fatty acids—general reaction **(Section 14.4)**	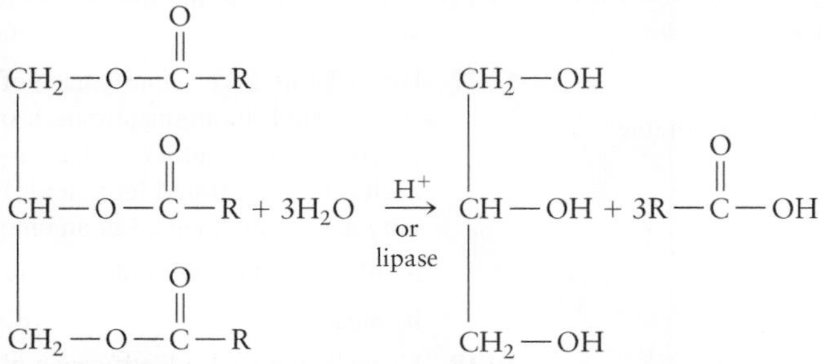 $\begin{array}{lcl} CH_2-O-\overset{\overset{O}{\parallel}}{C}-R & & CH_2-OH \\ CH-O-\overset{\overset{O}{\parallel}}{C}-R + 3H_2O & \xrightarrow[\text{or lipase}]{H^+} & CH-OH + 3R-\overset{\overset{O}{\parallel}}{C}-OH \\ CH_2-O-\overset{\overset{O}{\parallel}}{C}-R & & CH_2-OH \end{array}$	Equation 14.2
3. Saponification of a triglyceride to glycerol and fatty acid salts—general reaction **(Section 14.4)**	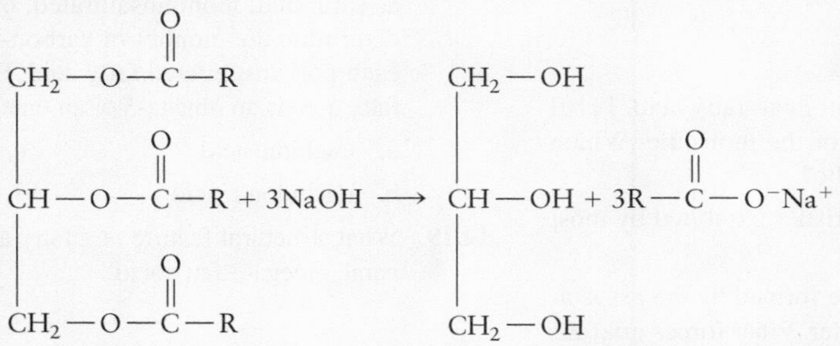 $\begin{array}{lcl} CH_2-O-\overset{\overset{O}{\parallel}}{C}-R & & CH_2-OH \\ CH-O-\overset{\overset{O}{\parallel}}{C}-R + 3NaOH & \longrightarrow & CH-OH + 3R-\overset{\overset{O}{\parallel}}{C}-O^-Na^+ \\ CH_2-O-\overset{\overset{O}{\parallel}}{C}-R & & CH_2-OH \end{array}$	Equation 14.3

4. Hydrogenation of a triglyceride—general reaction **(Section 14.4)**	$$\begin{array}{l} CH_2-O-\overset{O}{\overset{\|}{C}}-R \\ \vert \\ CH-O-\overset{O}{\overset{\|}{C}}-(CH_2)_7CH{=}CH(CH_2)_7CH_3 + H_2 \xrightarrow{Ni} \\ \vert \\ CH_2-O-\overset{O}{\overset{\|}{C}}-R \end{array} \begin{array}{l} CH_2-O-\overset{O}{\overset{\|}{C}}-R \\ \vert \\ CH-O-\overset{O}{\overset{\|}{C}}-(CH_2)_{16}CH_3 \\ \vert \\ CH_2-O-\overset{O}{\overset{\|}{C}}-R \end{array}$$	Equation 14.4

Exercises

Even-numbered exercises are answered in Appendix B.

Exercises with an asterisk (*) are more challenging.

The Classification of Lipids (Section 14.1)

14.1 What is the basis for deciding if a substance is a lipid?

14.2 List two major functions of lipids in the human body.

14.3 What functional group is common to all saponifiable lipids?

14.4 Classify the following as saponifiable or nonsaponifiable lipids.

a. A steroid **c.** A phosphoglyceride

b. A wax **d.** A simple lipid

14.5 Classify the following as saponifiable or nonsaponifiable lipids.

a. A triglyceride **c.** A sphingolipid

b. A complex lipid **d.** A prostaglandin

14.6 In which of the following solvents will a lipid be soluble?

a. $CH_3(CH_2)_4CH_3$ (nonpolar)

b. $CH_3-\underset{}{\overset{CH_3}{\overset{|}{CH}}}-OH$ (polar)

c. $CH_3CH_2-O-CH_2CH_2CH_3$ (nonpolar)

d. H_2O (polar)

14.7 In which of the following solvents will a lipid be soluble?

a. CCl_4 (nonpolar)

b. $CH_3(CH_2)_3CH_3$ (nonpolar)

c. CH_3CH_2-OH (polar)

d. 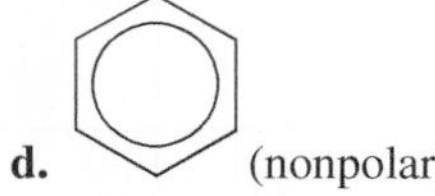 (nonpolar)

Fatty Acids (Section 14.2)

14.8 Draw the structure of a typical saturated fatty acid. Label the polar and nonpolar portions of the molecule. Which portion is hydrophobic? Hydrophilic?

14.9 Describe four structural characteristics exhibited by most fatty acids.

14.10 Describe the structure of a micelle formed by the association of fatty acid molecules in water. What forces hold the micelle together?

14.11 What is a chylomicron, and what is its function?

14.12 What mechanism exists in the body for transporting insoluble lipids?

14.13 Name two essential fatty acids, and explain why they are called essential.

14.14 Indicate whether each of the following fatty acids is saturated or unsaturated. Which of them are solids, and which are liquids at room temperature?

a. $CH_3(CH_2)_{14}COOH$

b. $CH_3(CH_2)_4CH{=}CHCH_2CH{=}CH(CH_2)_7COOH$

c. $CH_3(C_{14}H_{24})COOH$

d. $CH_3(C_{10}H_{20})COOH$

14.15 Indicate whether each of the following fatty acids is saturated or unsaturated. Which of them are solids, and which are liquids at room temperature?

a. $CH_3(CH_2)_3CH{=}CHCH_2CH{=}CHCH_2COOH$

b. $CH_3(C_{12}H_{18})COOH$

c. $CH_3(C_8H_{16})COOH$

d. $CH_3(CH_2)_{11}COOH$

14.16* Explain why the melting points of unsaturated fatty acids are lower than those of saturated fatty acids.

14.17 Using **Table 14.1**, identify each of the following fatty acids as saturated, monounsaturated, or polyunsaturated. Then determine the number of carbon–carbon double bonds in each polyunsaturated fatty acid. Finally, state whether the fatty acid is an omega-3 or an omega-6 fatty acid.

a. eicosapentaenoic acid **c.** linoleic acid

b. stearic acid **d.** oleic acid

14.18 Using **Table 14.1**, identify each of the following fatty acids as saturated, monounsaturated, or polyunsaturated. Then determine the number of carbon–carbon double bonds in each polyunsaturated fatty acid. Finally, state whether the fatty acid is an omega-3 or an omega-6 fatty acid.

a. arachidic acid **c.** palmitoleic acid

b. arachidonic acid **d.** myristic acid

14.19 What structural feature of a fatty acid is responsible for the name omega-3 fatty acid?

The Structure of Fats and Oils (Section 14.3)

14.20

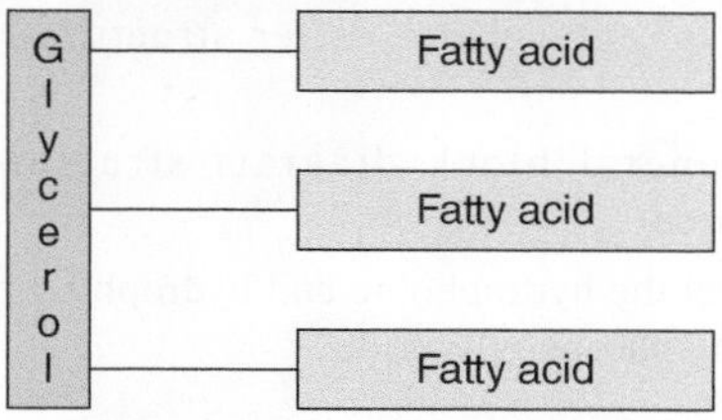

Glycerol	
G–l–y–c–e–r–o–l	Fatty acid
	Fatty acid
	Fatty acid

Draw block diagrams similar to the one provided to show the four different triglycerides that can be made from myristic acid, oleic acid, and glycerol.

14.21

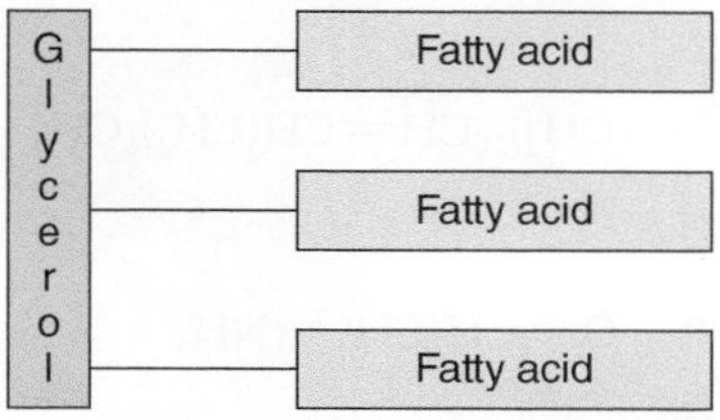

Glycerol	
G–l–y–c–e–r–o–l	Fatty acid
	Fatty acid
	Fatty acid

Draw block diagrams similar to the one provided to show the three different triglycerides that can be made from stearic acid, arachidonic acid, linolenic acid, and glycerol.

14.22 Draw the structure of a triglyceride that contains one myristic acid, one palmitoleic acid, and one linoleic acid. Identify the ester bonds. (See **Table 14.1**.)

14.23 Draw the structure of a triglyceride that contains one palmitic acid, one arachidonic acid, and one oleic acid. Identify the ester bonds. (See **Table 14.1**.)

14.24 How are fats and oils structurally similar? How are they different?

14.25 From **Figure 14.10**, arrange the following substances in order of increasing percentage of unsaturated fatty acids: chicken fat, beef fat, corn oil, butter, and sunflower oil.

14.26 From what general source do triglycerides tend to have more saturated fatty acids? More unsaturated fatty acids?

14.27 The percentage of fatty acid composition of triglycerides A and B are reported below. Predict which triglyceride has the lower melting point.

	Palmitic Acid	Stearic Acid	Oleic Acid
Triglyceride A	30.4	47.8	21.8
Triglyceride B	9.6	7.2	83.2

14.28 The percentage of fatty acid composition of triglycerides A and B are reported below. Predict which triglyceride has the lower melting point.

	Arachidonic Acid	Myristic Acid	Palmitoleic Acid
Triglyceride A	18.9	73.5	7.6
Triglyceride B	39.6	16.1	44.3

14.29 Why is the amount of saturated fat in the diet a health concern?

Chemical Properties and Reactions of Fats and Oils (Section 14.4)

14.30 What process is used to prepare a number of useful products such as margarines and cooking shortenings from vegetable oils?

14.31 Draw structural formulas for all the products you would get from the complete hydrolysis of the following triglyceride.

$$CH_2-O-\overset{\overset{O}{\|}}{C}-(CH_2)_5CH{=}CH(CH_2)_3CH{=}CH(CH_2)_4CH_3$$
$$CH-O-\overset{\overset{O}{\|}}{C}-(CH_2)_{18}CH_3$$
$$CH_2-O-\overset{\overset{O}{\|}}{C}-(CH_2)_2CH{=}CH_2(CH_2)_4CH{=}CH(CH_2)_2CH_3$$

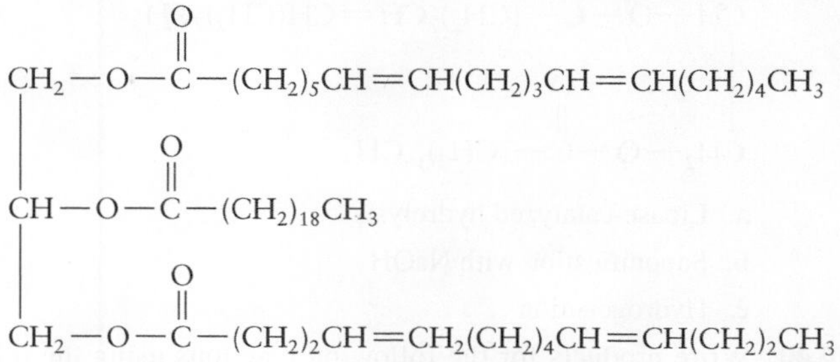

14.32 Draw structural formulas for all the products you would get from the complete hydrolysis of the following triglyceride.

$$CH_2-O-\overset{\overset{O}{\|}}{C}-(CH_2)_3CH{=}(CH_2)_3CH{=}CH(CH_2)_4CH_3$$
$$CH-O-\overset{\overset{O}{\|}}{C}-(CH_2)_6CH{=}CH(CH_2)_6CH_3$$
$$CH_2-O-\overset{\overset{O}{\|}}{C}-(CH_2)_{16}CH_3$$

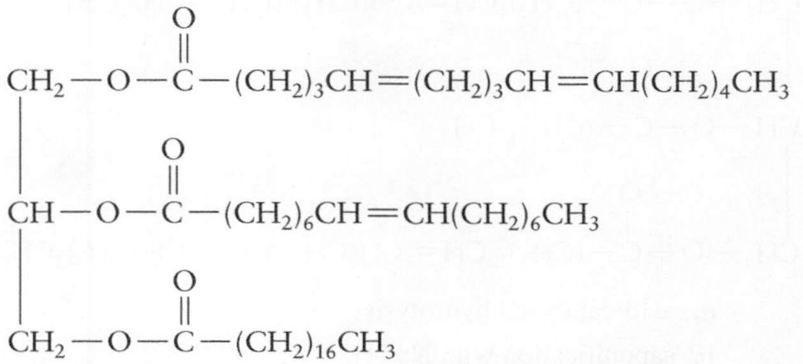

14.33 Draw structural formulas for all the products you would get from the saponification of the triglyceride in Exercise 14.31 with NaOH.

14.34 Draw structural formulas for all the products you would get from the saponification of the triglyceride in Exercise 14.32 with KOH.

14.35 Draw a structural formula for the product you would get from the complete hydrogenation of the triglyceride in Exercise 14.31.

14.36 Draw a structural formula for the product you would get from the complete hydrogenation of the triglyceride in Exercise 14.32.

14.37 How many H_2 molecules are needed to react with the triglyceride molecule in Exercise 14.31 for its complete hydrogenation?

14.38 How many H_2 molecules are needed to react with the triglyceride molecule in Exercise 14.32 for its complete hydrogenation?

14.39 Write products for the following reactions using the oil provided.

$$\begin{array}{l} CH_2-O-\overset{\overset{\displaystyle O}{\|}}{C}-(CH_2)_7CH=CH(CH_2)_7CH_3 \\ | \\ CH-O-\overset{\overset{\displaystyle O}{\|}}{C}-(CH_2)_7CH=CH(CH_2)_7CH_3 \\ | \\ CH_2-O-\overset{\overset{\displaystyle O}{\|}}{C}-(CH_2)_{16}CH_3 \end{array}$$

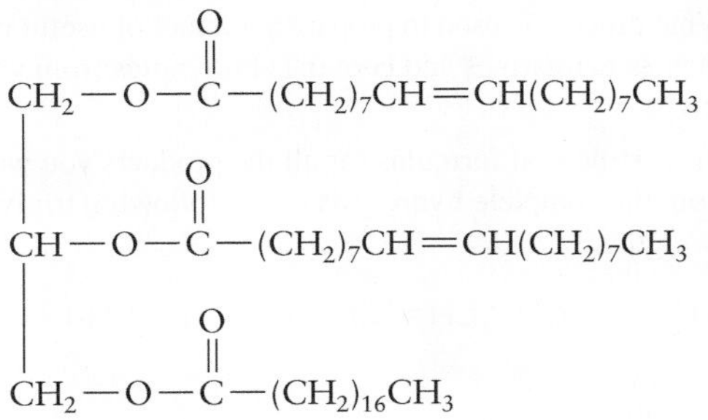

a. Lipase-catalyzed hydrolysis

b. Saponification with NaOH

c. Hydrogenation

14.40 Write products for the following reactions using the oil provided.

$$\begin{array}{l} CH_2-O-\overset{\overset{\displaystyle O}{\|}}{C}-(CH_2)_4CH=CH(CH_2)_2(CH=CH)_2CH_3 \\ | \\ CH-O-\overset{\overset{\displaystyle O}{\|}}{C}-(CH_2)_{14}CH_3 \\ | \\ CH_2-O-\overset{\overset{\displaystyle O}{\|}}{C}-(CH_2)_{12}CH=CH(CH_2)_2CH=CH(CH_2)_2CH_3 \end{array}$$

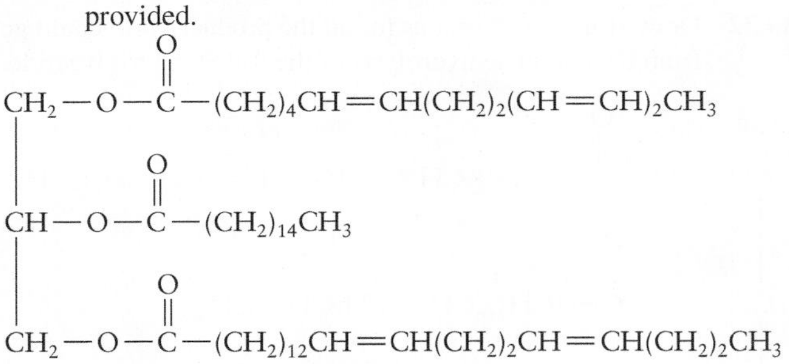

a. acid-catalyzed hydrolysis

b. saponification with NaOH

c. hydrogenation

14.41 Why is the hydrogenation of vegetable oils of great commercial importance?

14.42 In general terms, name the products of the following reactions.

a. acid hydrolysis of a fat

b. acid hydrolysis of an oil

c. saponification of a fat

d. saponification of an oil

Waxes (Section 14.5)

14.43 Draw the structure of a wax formed from oleic acid and cetyl alcohol ($CH_3(CH_2)_{14}CH_2—OH$).

14.44 Like fats, waxes are esters of long-chain fatty acids. What structural difference exists between them that warrants placing waxes in a separate category?

14.45 Draw the structure of a wax formed from stearic acid and 1-dodecanol ($CH_3(CH_2)_{10}CH_2—OH$).

14.46 Draw the structural formulas of the fatty acid and the long-chain alcohol used to make the following wax.

$$CH_3(CH_2)_4(CH=CHCH_2)_2(CH_2)_6-\overset{\overset{\displaystyle O}{\|}}{C}-O-(CH_2)_{16}CH_3$$

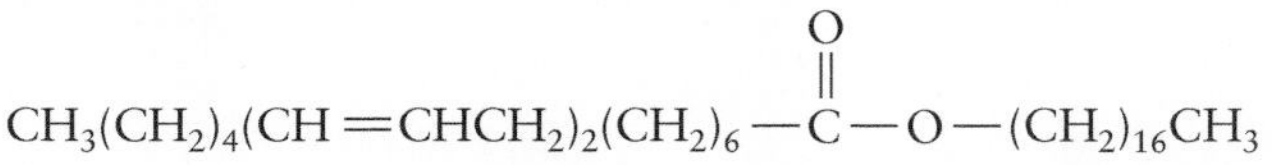

14.47 Draw the structural formulas of the fatty acid and the long-chain alcohol used to make the following wax.

$$CH_3(CH_2)_7CH=CH(CH_2)_7-\overset{\overset{\displaystyle O}{\|}}{C}-O-(CH_2)_{21}\underset{}{\overset{\overset{\displaystyle CH_3}{|}}{C}H}-CH_3$$

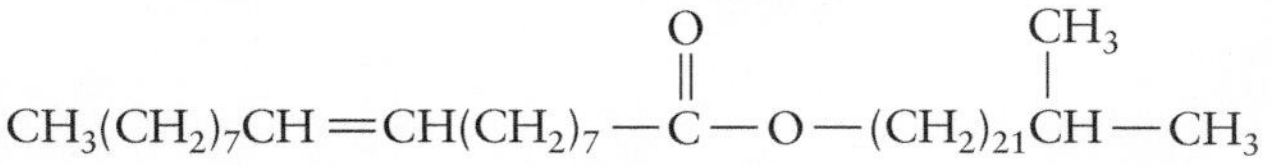

14.48 What role do waxes play in nature?

Phosphoglycerides (Section 14.6)

14.49 How do phosphoglycerides differ structurally from triglycerides?

14.50 Draw the general block diagram structure of a phosphoglyceride.

14.51 Circle and label the hydrophobic and hydrophilic portions of the following phosphoglyceride.

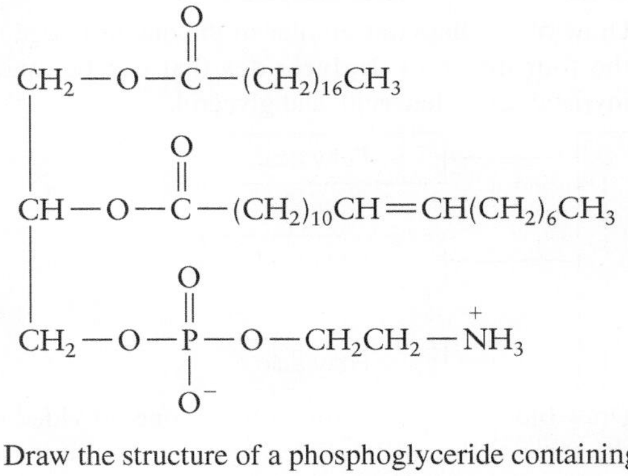

$$\begin{array}{l} CH_2-O-\overset{\overset{\displaystyle O}{\|}}{C}-(CH_2)_{16}CH_3 \\ | \\ CH-O-\overset{\overset{\displaystyle O}{\|}}{C}-(CH_2)_{10}CH=CH(CH_2)_6CH_3 \\ | \\ CH_2-O-\overset{\overset{\displaystyle O}{\|}}{\underset{\underset{\displaystyle O^-}{|}}{P}}-O-CH_2CH_2-\overset{+}{N}H_3 \end{array}$$

14.52 Draw the structure of a phosphoglyceride containing serine.

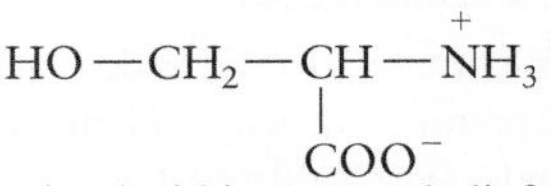

$$HO-CH_2-\underset{\underset{\displaystyle COO^-}{|}}{CH}-\overset{+}{N}H_3$$

Is it a lecithin or a cephalin?

14.53 Draw the structure of a phosphoglyceride containing choline.

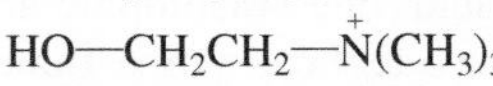

$$HO-CH_2CH_2-\overset{+}{N}(CH_3)_3$$

14.54 What is the structural difference between a lecithin and a cephalin?

14.55 Draw the structure of a lecithin. What structural features make lecithin an important commercial emulsifying agent in certain food products?

14.56 Describe two biological roles served by the lecithins.

14.57 Where are cephalins found in the human body?

Sphingolipids (Section 14.7)

14.58 Draw the general block diagram structure of a sphingolipid.

14.59 List two structural differences between sphingolipids and phosphoglycerides.

14.60 Circle and label the hydrophobic and hydrophilic portions of the following sphingolipid.

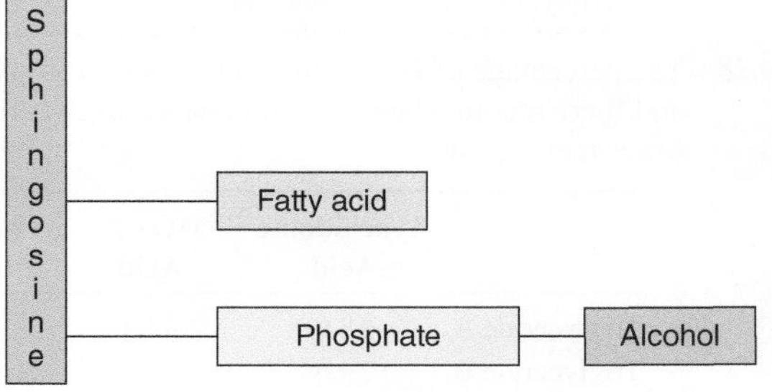

14.61 Circle and label the hydrophobic and hydrophilic portions of the following glycolipid.

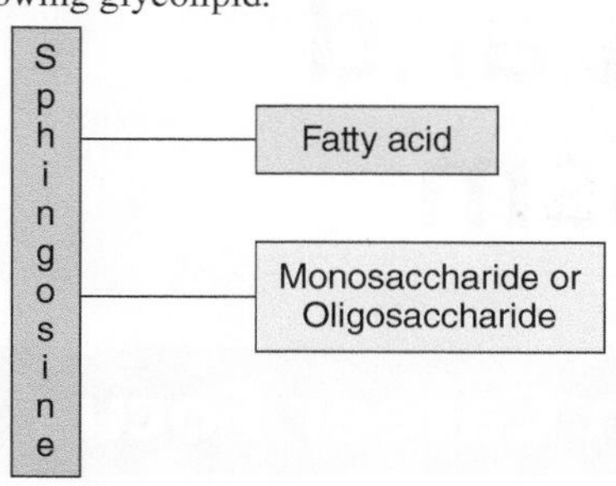

14.62 Describe the structural similarities and differences between the sphingomyelins and the glycolipids.

14.63 Give another name for glycolipids. In what tissues are they found?

14.64 Draw the general block diagram structure of the glycolipid blood antigen A.

14.65 Draw the general block diagram structure of the glycolipid blood antigen B.

14.66 Explain why a person with blood type A cannot receive blood from a person with blood type B.

14.67 Explain why persons with blood type O are universal donors, meaning they can donate to patients of blood type A, B, AB, and O.

14.68 List three diseases caused by abnormal metabolism and accumulation of sphingolipids.

Biological Membranes (Section 14.8)

14.69 Where would you find membranes in a prokaryotic cell? Where would you find membranes in a eukaryotic cell? What functions do the membranes serve?

14.70 What three classes of lipids are found in membranes?

14.71 Cholesterol, a hydrophobic steroid compound, is found in cell membranes. Would you expect to find cholesterol internally or on the surfaces of the cell membrane lipid bilayer? Explain why.

14.72 How does the polarity of the phosphoglycerides contribute to their function of forming cell membranes?

14.73 Describe the major features of the fluid-mosaic model of cell membrane structure.

14.74 Explain how lipid nanoparticles can help in the delivery of medicines to target tissues in human patients.

14.75 Give one example of a recent medical treatment that utilizes lipid nanoparticles.

Steroids and Hormones (Section 14.9)

14.76 Draw the characteristic chemical structure that applies to all steroid molecules.

14.77 How are testosterone and progesterone structurally similar? How are they different?

14.78 What structural features characterize the prostaglandins?

14.79 Compare the structures of testosterone, methandrostenolone, and progesterone. What structural feature makes the synthetic methandrostenolone a male sex hormone and not a female sex hormone?

14.80 Explain how bile salts aid in the digestion of lipids.

14.81 List three important groups of compounds the body synthesizes from cholesterol. Give the location in bodily tissues where cholesterol is an essential component.

14.82 Cholesterol and bile salts, such as glycocholate, are steroids. Explain how structural differences between cholesterol and glycocholate allow glycocholate to emulsify cholesterol for excretion from the gall bladder.

14.83 What symptoms might indicate the presence of gallstones in the gallbladder?

14.84 What is the major component in gallstones?

14.85 What is a hormone? What are the two major categories of steroid hormones?

14.86 Name the two groups of adrenocorticoid hormones, give a specific example of each group, and explain the function of those compounds in the body.

14.87 Name the primary male sex hormone and the three principal female sex hormones.

14.88 What advantages do anabolic steroids provide athletes? What side effects are associated with their use?

14.89 What role do the estrogens and progesterone serve in preparation for pregnancy?

14.90 Why is it suggested that some people restrict cholesterol intake in their diet?

14.91 What compound serves as a starting material for prostaglandin synthesis?

14.92 What body processes appear to be regulated in part by prostaglandins?

14.93 Prostaglandins perform functions similar to those of hormones. Explain why prostaglandins are not considered hormones. (*Hint:* Think of their points of origin.)

14.94 Name three therapeutic uses of prostaglandins.

Additional Exercises

14.95 Unsaturated fatty acids are susceptible to oxidation, which causes rancidity in food products. Suggest a way to decrease the amount of oxidative rancidity occurring in a food product.

14.96 Suggest a reason why complex lipids, such as phosphoglycerides and sphingomyelin, are more predominant in cell membranes than simple lipids.

14.97 Which of the following waxes would have the lowest melting point? Explain your answer.

a. $CH_3(CH_2)_{14}-\overset{\overset{\large O}{\|}}{C}-O-(CH_2)_{29}CH_3$

b. $CH_3(CH_2)_7CH{=}CH(CH_2)_5-\overset{\overset{\large O}{\|}}{C}-O-(CH_2)_{29}CH_3$

Chemistry for Thought

14.98 Gasoline is soluble in nonpolar solvents. Why is gasoline not classified as a lipid?

14.99 The structure of cellular membranes is such that ruptures are closed naturally. Describe the molecular forces that cause the closing to occur.

14.100 The five vegetable oils listed in **Figure 14.10** are derived from seeds. Why do you think seeds are rich in oils?

14.101 In what ways are the structural features of lecithin similar to those of a soap?

15 Nutrition and Metabolism

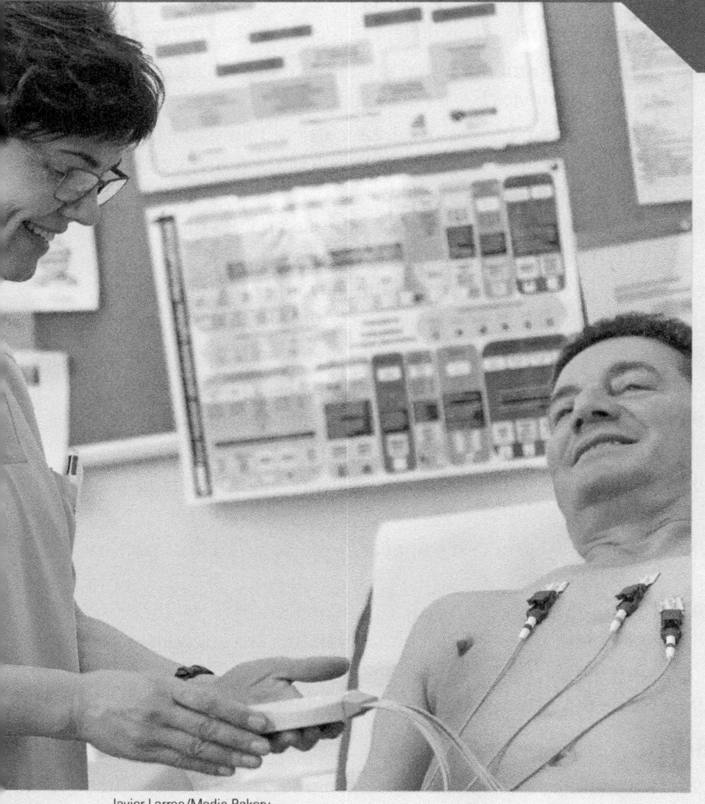
Javier Larrea/Media Bakery

Health Career Focus

Physician Assistant

The hardest part is getting your patient experience. I'm lucky that I went to a high school that allowed students release time to attend a local technical college. I took advantage of that and earned a certificate in medical assisting by the time I graduated, so I worked at a doctor's office throughout college. The more intensive physician assistant program I completed later required 2000 hours of patient experience, but some require twice that!

I didn't decide to become a PA right away. I didn't even decide to go to college right away. I got married and started a family. I liked the medical assisting job, but working with PAs helped me see more possibilities. I started dreaming about it, then I started planning for it.

My physician assistant program accepted online credit, and since I have kids, that made it much easier for me to meet the requirements. I could do my schoolwork for an hour or two after my children went to bed. It wasn't easy, but I'm so glad I did it. It's a lot of hard work, but if you want it, you'll find a way.

Follow-up to this Career Focus appears at the end of the chapter before the *Concept Summary*.

Learning Objectives

When you have completed your study of Chapter 15, you should be able to:

1. Describe the primary functions in the body of carbohydrates, lipids, proteins, vitamins, and minerals. **(Section 15.1)**
2. Define metabolism, anabolism, and catabolism. **(Section 15.2)**
3. Outline the three stages in the extraction of energy from food. **(Section 15.2)**
4. Explain how ATP plays a central role in the production and use of cellular energy. **(Section 15.3)**
5. Identify the monosaccharide products of the digestion of carbohydrates. **(Section 15.4)**
6. Explain the importance to the body of maintaining proper blood sugar levels. **(Section 15.4)**
7. Describe the process and cellular regulation of glycolysis and the citric acid cycle. **(Section 15.5)**
8. Describe in detail the process of oxidative phosphorylation and how it results in ATP production. **(Section 15.5)**
9. Explain the importance of glycogenesis, glycogenolysis, and gluconeogenesis. **(Section 15.6)**
10. Describe how hormones regulate carbohydrate metabolism. **(Section 15.6)**
11. Describe the metabolic pathways by which glycerol and fatty acids are catabolized. **(Section 15.7)**
12. Name the three ketone bodies. **(Section 15.7)**
13. List the conditions that cause the overproduction of ketone bodies. **(Section 15.7)**
14. Describe the pathway for fatty acid synthesis. **(Section 15.7)**
15. Describe the source and function of the body's amino acid pool. **(Section 15.8)**
16. Explain the chemical fate of the nitrogen atoms and carbon skeleton of amino acids. **(Section 15.8)**
17. Explain the relationship between intermediates of carbohydrate metabolism and the synthesis of nonessential amino acids. **(Section 15.8)**

The Science of nutrition is an applied field that focuses on the study of food, water, and other nutrients and the ways in which living organisms utilize them. Nutrition scientists use the principles of nutrition to determine the proper components of a sound diet, the best way to maintain proper body weight, or the foods and other nutrients needed by people with specific illnesses or injuries. The fuels of the human body are the sugars, lipids, and proteins derived from food (see **Figure 15.1**). The reactions that release energy from these substances are among the body's most important biochemical processes.

In this chapter, we present an introduction to nutrition and how cells extract energy from food. We also explain how the chemical degradation of glucose provides energy in the cell, how glucose is stored, how the blood glucose level is regulated, and how diabetes results from changes in blood glucose regulation mechanisms. We then examine the storage, degradation, and synthesis of lipids followed by the key highlights of the degradation and synthesis of amino acids.

nutrition An applied science that studies food, water, and other nutrients and the ways living organisms use them.

macronutrient A substance needed by the body in relatively large amounts.

micronutrient A substance needed by the body only in small amounts.

Reference Daily Intakes (RDIs) A set of standards for protein, vitamins, and minerals; used on food labels as part of the Daily Values.

Daily Reference Values (DRVs) A set of standards for nutrients and food components (such as fat and fiber) that have important relationships with health; used on food labels as part of the Daily Values.

Daily Values (DVs) Reference values developed by the FDA specifically for use on food labels. The Daily Values represent two sets of standards: Reference Daily Intakes (RDIs) and Daily Reference Values (DRVs).

15.1 The Basics of Nutrition

Learning Objective 1 Describe the primary functions in the body of carbohydrates, lipids, proteins, vitamins, and minerals.

A human body must be supplied with appropriate nutrients if it is to function properly and remain healthy. Some nutrients are required in relatively large amounts (more than one gram per day) because they provide energy and materials required to repair damaged tissue or build new tissue. These **macronutrients** are the carbohydrates, lipids, and proteins contained in food. **Micronutrients** are required by the body in only small amounts (milligrams or micrograms daily). Micronutrients are classified as either vitamins or minerals.

In addition to macronutrients and micronutrients, the human body must receive appropriate amounts of water and fiber. Water is important because 45% to 75% of the human body mass is water. Fiber is an indigestible plant material composed primarily of cellulose. It is neither a macronutrient nor a micronutrient, but it prevents or relieves constipation by absorbing water and softening the stool for easier elimination.

A number of countries of the world have established nutritional guidelines for their citizens in an attempt to improve and maintain good health nationally. In the United States, the Nutrition Labeling and Education Act of 1990 brought sweeping changes to the regulations that define what is required on a food label. The official guidelines are called **Reference Daily Intakes (RDIs)** for proteins and 19 vitamins and minerals, and **Daily Reference Values (DRVs)** for other nutrients of public health importance. For simplicity, all reference values on food labels are referred to as **Daily Values (DVs)** (see **Figure 15.1**). In 2016, the Food and Drug Administration (FDA) finalized the new Nutrition Facts label for packaged foods. The new label makes it easier for consumers to make more informed food choices. The FDA decided to use 2000 calories as a standard for energy intake in calculating the DRVs.

A 2000-calorie diet is considered about right for many adults. Recall from **Chapter 1** that the nutritional calorie is equal to 1 kilocalorie (kcal) of energy. The U.S. Department of Agriculture (USDA) has issued the food guide MyPlate, which was designed to replace the old MyPyramid posters with new recommendations (see **Figure 15.2**).

Nutrition Facts

8 Servings per container	
Serving size	**2/3 cup (55 g)**
Amount per serving	
Calories	**230**
	% Daily Value*
Total Fat 8 g	**10%**
Saturated Fat 1 g	**5%**
Trans Fat 0 g	
Cholesterol 0 mg	**0%**
Sodium 160 mg	**7%**
Total Carbohydrate 37 g	**13%**
Dietary Fiber 4 g	**14%**
Total sugars 12 g	
Includes 10 g Added Sugars	**20%**
Protein 3 g	
Vitamin D 2 mcg	10%
Calcium 260 mg	20%
Iron 8 mg	45%
Potassium 235 mg	5%

*The % Daily Value (DV) tells you how much a nutrient in a serving of food contributes to a daily diet. 2000 calories a day is used for general nutrition advice.

Figure 15.1 An example of a food label.

15.1A The Macronutrients: Carbohydrates, Lipids, and Proteins

Carbohydrates

Carbohydrates are ideal energy sources for most bodily functions. They also provide useful materials for the synthesis of cell and tissue components. Carbohydrates, moreover, are relatively inexpensive and readily available, so they are used worldwide as the main dietary source of energy.

Nevertheless, many people try to avoid foods rich in carbohydrates because they are thought to be fattening. It is now recognized that this reputation is generally undeserved. Most of the excess calories associated with eating carbohydrates are actually due to the high-calorie foods eaten *with* the carbohydrates—for example, the butter added to potatoes or bread.

Dietary carbohydrates can be classified as simple or complex. **Simple carbohydrates** are the sugars we classified in **Chapter 11** as monosaccharides and disaccharides. **Complex carbohydrates** consist essentially of the polysaccharides amylose and amylopectin, which are collectively called starch. Cellulose, another polysaccharide, is also a complex carbohydrate; however, it cannot be digested by humans, so it serves a nonnutritive role as fiber instead.

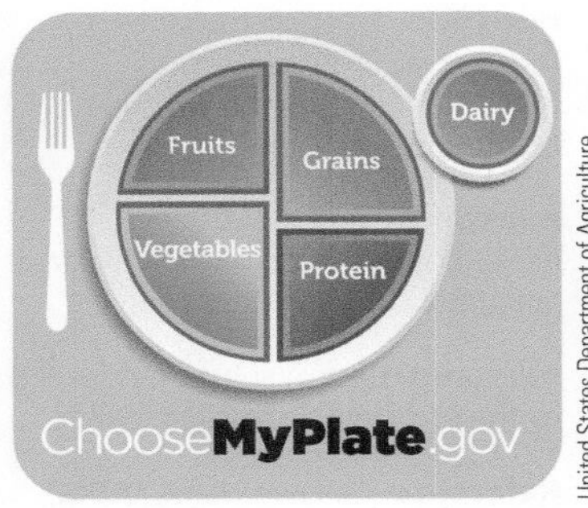

Figure 15.2 The food guide MyPlate developed by the U.S. Department of Agriculture. ChooseMyPlate.gov will help you select foods and amounts that are right for you.

simple carbohydrates Monosaccharides and disaccharides; commonly called sugars.

complex carbohydrates The polysaccharides amylose and amylopectin; collectively called starch.

Lipids

Lipids (fats), like carbohydrates, have a negative dietary image. In part, this is because the word *fat* has several meanings, and its relationship to health is sometimes misunderstood. Some people think of fat as something to be avoided at all costs, regardless of body size. As we explained in **Chapter 14**, a number of different substances are classified as lipids. However, about 95% of the lipids in foods and in our bodies are triglycerides (**Figure 15.3**), in which three fatty acid molecules are bonded to a molecule of glycerol by ester linkages (see Equation 14.1).

Lipids are important dietary constituents for a number of reasons: They are a concentrated source of energy and provide more than twice the energy of an equal mass of carbohydrate; they contain some fat-soluble vitamins and help carry them through the body; and some fatty acids needed by the body cannot be synthesized and must come from the diet—they are the essential fatty acids. The essential fatty acids are the polyunsaturated linoleic and linolenic acids. Generally, vegetable oils are good sources of unsaturated fatty acids. Some oils that are especially rich in linoleic acid come from corn, cottonseed, soybean, and wheat germ. Infants especially need linoleic acid for good health and growth; human breast milk contains a higher percentage of it than does cow's milk.

Figure 15.3 Foods fried in oil are rich in triglycerides.

We still do not have a complete understanding of the relationship between dietary lipids and health. A moderate amount of fat is needed in everyone's diet, but many people consume much more than required. Research results indicate a correlation exists between the consumption of too much fat (and fat of the wrong type—namely, saturated fatty acids) and two of our greatest health problems: obesity and cardiovascular disease. Because of these results, the recommended percentage of calories obtained from fats (total, both saturated and unsaturated) has been reduced from the national average of almost 42% to no more than 30%. The 2015–2020 Dietary Guidelines issued by the HHS and the USDA recommend the consumption of less than 10% of calories per day from saturated fats (**Figure 15.4**).

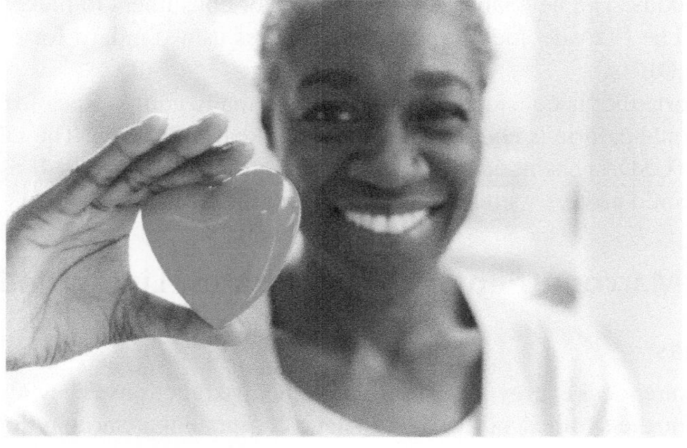

Figure 15.4 Reducing saturated fat consumption to less than 10% of calories per day can help prevent heart disease.

Proteins

Proteins are the only macronutrients for which a Reference Daily Intake (RDI) has been established. The RDI varies for different groups of people. For example, **Table 15.1** shows that nursing mothers have a greater need for protein than do other adults. This reflects one of the important uses of protein in the body—namely, the production of new tissue as the body grows.

In addition, proteins are needed for the maintenance and repair of cells and for the production of enzymes, hormones, and other important nitrogen-containing compounds of the body. Proteins can also be used to supply energy; they provide about 4 cal/g, the same as carbohydrates. It is recommended that 12% of the calories obtained from food be obtained from dietary proteins (see **Figure 15.5**).

As discussed in **Chapter 12**, proteins are natural polymers of amino acids joined by amide (or peptide) linkages (see Equation 12.4). On digestion, proteins are broken down to individual amino acids that are absorbed into the body's amino acid pool and used as needed. The **essential amino acids** listed in **Table 15.2** must be obtained from the diet because they cannot be synthesized in the body in sufficient amounts to satisfy the body's needs.

The minimum quantity of each essential amino acid needed per day by an individual will vary, depending on the use in the body. For example, results of one study showed that young men need a daily average of only 7 g (0.034 mol) of tryptophan but 31 g (0.21 mol) of methionine. According to this study, then, young men require about seven times as many moles of methionine as they do of tryptophan to carry out the processes involving amino acids.

Proteins in foods are classified as **complete proteins** if they contain all the essential amino acids in the proper proportions needed by the body. Protein foods that do not meet this requirement are classified as incomplete. Several protein-containing foods are depicted in **Figure 15.5**.

Table 15.1 The RDI for Protein

Group	RDI (g)
Pregnant women	60
Nursing mothers	65
Infants under age 1	14
Children age 1 to 4	16
Adults	50

essential amino acid An amino acid that cannot be synthesized within the body at a rate adequate to meet metabolic needs.

complete protein A protein that contains all the essential amino acids in a sufficient quantity needed for the body.

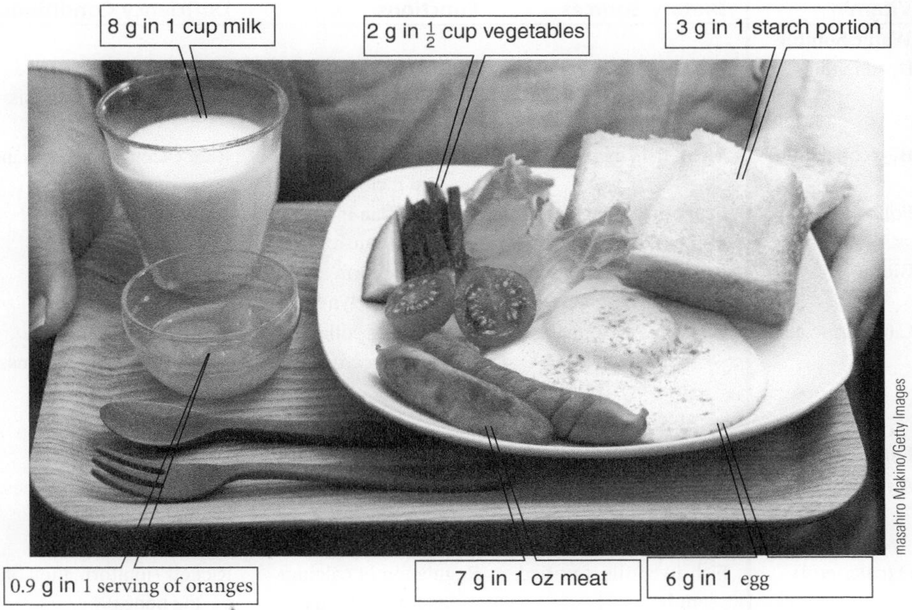

Figure 15.5 The protein content of several foods.

Table 15.2 The Essential Amino Acids*

Histidine
Isoleucine
Leucine
Lysine
Methionine
Phenylalanine
Threonine
Tryptophan
Valine

*For the structures of all 20 common amino acids, see Table 12.1.

Example 15.1 Describing Functions of Macronutrients

List a primary function for each of the macronutrients.

a. Carbohydrates b. Lipids c. Proteins

Solution

a. Energy source for body functions
b. Concentrated energy source for body functions
c. Production of new tissue

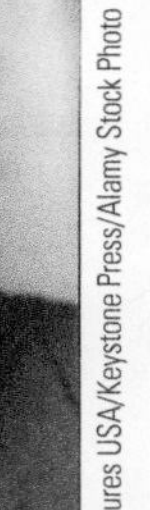

Figure 15.6 Dorothy Crowfoot Hodgkin was awarded the Nobel Prize for Chemistry in 1964 for determining the complicated structure of vitamin B_{12}.

> ✔ **Learning Check 15.1** List another primary function for each macronutrient.
>
> a. Carbohydrates b. Lipids c. Proteins

15.1B The Micronutrients: Vitamins and Minerals

Vitamins are organic micronutrients that the body cannot produce in amounts needed for good health. The highly polar nature of some vitamins renders them water soluble. Nine water-soluble vitamins have been identified: folic acid, thiamine, riboflavin, niacin, pantothenic acid, biotin, vitamin B_6, vitamin B_{12} (**Figure 15.6**), and vitamin C. **Table 15.3** presents some of these essential vitamins, their functions, and the conditions caused when they are deficient. With the exception of vitamin C (see **Figure 15.7**), all water-soluble vitamins have been shown to function as coenzymes.

Some vitamins (A, D, E, and K) have very nonpolar molecular structures and therefore dissolve only in nonpolar solvents. In the body, the nonpolar solvents are the lipids we have classified as fats, so these vitamins are called fat soluble. The fat-soluble vitamins have diverse functions in the body, and they act somewhat like hormones (**Table 15.3**). Care must be taken to avoid overdoses of the fat-soluble vitamins. Toxic effects are known to occur, especially with vitamin A, when excess amounts of these vitamins accumulate in body tissue. Excesses of water-soluble vitamins, on the other hand, are excreted readily through the kidneys and are not normally a problem.

Figure 15.7 Although oranges are recognized as a good source of vitamin C (50 mg per 100 g), strawberries (60 mg per 100 g) and cauliflower (70 mg per 100 g) are even better.

Table 15.3 Vitamin Sources, Functions, and Deficiency Conditions

Vitamin	Dietary Sources	Functions	Deficiency Conditions
Water Soluble			
B_1 (thiamine)	Bread, beans, nuts, milk, peas, pork, rice bran	Coenzyme in decarboxylation reactions	Beriberi: nausea, severe exhaustion, paralysis
B_{12} (cobalamin)	Meat, fish, eggs, milk	Coenzyme in amino acid metabolism	Rare except in vegetarians; pernicious anemia
Folic acid	Leafy green vegetables, peas, beans	Coenzyme in methyl group transfers	Anemia
Biotin	Found widely; egg yolk, liver, yeast, nuts	Coenzyme form used in fatty acid synthesis	Dermatitis, muscle weakness
C (ascorbic acid)	Citrus fruits, tomatoes, green pepper, strawberries, leafy green vegetables	Synthesis of collagen for connective tissue	Scurvy: tender tissues; weak, bleeding gums; swollen joints
Fat Soluble			
A (retinol)	Eggs, butter, cheese, dark green and deep orange vegetables	Synthesis of visual pigments	Inflamed eye membranes, night blindness, scaliness of skin
D (calciferol)	Fish-liver oils, fortified milk	Regulation of calcium and phosphorus metabolism	Rickets (malformation of the bones)
K	Cabbage, potatoes, peas, leafy green vegetables	Synthesis of blood-clotting substances	Blood-clotting disorders
E (tocopherol)	Whole-grain cereals, margarine, vegetable oil	Prevention of oxidation of vitamin A and fatty acids	Breakage of red blood cells

mineral A metal or nonmetal used in the body in the form of ions or compounds.

Minerals are generally inorganic substances. For example, a food that is a good source of phosphorus does not contain elemental phosphorus, a caustic nonmetal, but probably contains phosphate ions such as HPO_4^{2-} and $H_2PO_4^-$ or, more likely, organic substances that contain phosphorus, such as nucleic acids and phospholipids.

In the body, the elements classified as minerals are never used in elemental form, but rather in the form of ions or compounds. They are classified as **major minerals** or **trace minerals** on the basis of the amount present in the body, as shown in **Table 15.4** and **Figure 15.8**. In general, the major minerals are required in larger daily amounts than are the trace minerals. For example, the RDIs for calcium and phosphorus, the major minerals

major mineral A mineral found in the body in quantities greater than 5 g.

trace mineral A mineral found in the body in quantities less than 5 g.

Table 15.4 Major and Trace Mineral Sources, Functions, and Deficiency Conditions

Mineral	Dietary Sources	Functions	Deficiency Conditions
Major Minerals			
Calcium	Dairy foods, dark green vegetables	Bone and teeth formation, blood clotting, nerve impulse transmission	Stunted growth, rickets, weak and brittle bones
Chlorine	Table salt, seafood, meat	HCl in gastric juice, acid–base balance	Muscle cramps, apathy, excessive bleeding, reduced appetite
Magnesium	Whole-grain cereals, meat, nuts, milk, legumes	Activation of enzymes, protein synthesis	Inhibited growth, weakness, spasms
Phosphorus	Milk, cheese, meat, fish, grains, legumes, nuts	Enzyme component, acid–base balance, bone and tooth formation	Weakness, calcium loss, weak bones
Potassium	Meat, milk, many fruits, cereals, legumes	Acid–base and water balance, nerve function	Muscle weakness, paralysis
Sodium	Most foods except fruit	Acid–base and water balance, nerve function	Muscle cramps, apathy, reduced appetite
Sulfur	Protein foods	Component of proteins	Deficiencies are very rare
Trace Minerals			
Copper	Drinking water, liver, grains, legumes, nuts	Component of numerous enzymes, hemoglobin formation	Anemia, fragility of arteries, low appetite, and inhibited growth
Iodine	Iodized salt, fish, dairy products	Component of thyroid hormones	Hypothyroidism, goiter
Iron	Liver, lean meat, whole grains, dark green vegetables	Component of enzymes and hemoglobin	Anemia
Manganese	Grains, beet greens, legumes, fruit	Component of enzymes	Deficiencies are rare

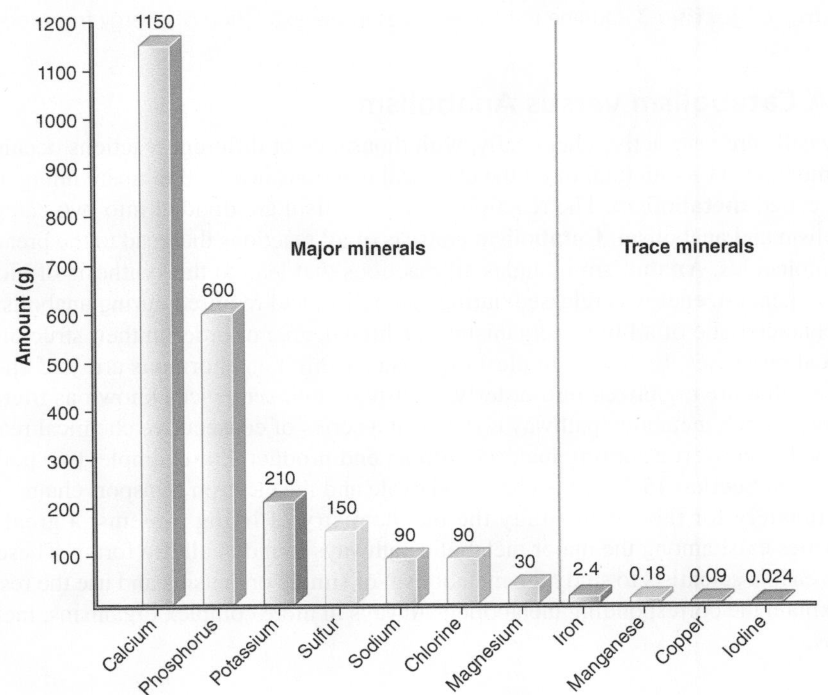

Figure 15.8 Minerals in a 130 lb person. The major minerals are those present in amounts larger than 5 g (a teaspoon); only four of the numerous trace minerals are shown. Only calcium and phosphorus appear in amounts larger than a pound (454 g).

found in the body in largest amounts, are 800 mg. The RDIs for iron and iodine, two of the trace minerals, are 10 to 18 mg and 0.12 mg, respectively. By way of comparison, the RDIs for vitamin C and niacin are 55 to 60 mg and 18 mg, respectively. Compounds of some major minerals (Ca and P) are the primary inorganic structural components of bones and teeth. Other major minerals (Na, K, Cl, and Mg) form principal ions that are distributed throughout the body's various fluids. Some trace minerals are components of vitamins (Co), enzymes (Zn and Se), hormones (I), or specialized proteins (Fe and Cu). Thus, even though trace minerals are required in small quantities, their involvement in critical enzymes, hormones, and the like makes them equally as important for good health as the major minerals.

Example 15.2 Identifying Mineral Functions

Identify a mineral that fits each of the following descriptions.

a. A component of thyroid hormones
b. A component of enzymes and hemoglobin
c. Present in gastric juice

Solution

a. Iodine b. Iron c. Chlorine

✔ **Learning Check 15.2** Identify a mineral associated with each of the following deficiency conditions.

a. Muscle weakness, paralysis
b. Weak and brittle bones
c. Goiter

15.2 An Overview of Metabolism

Learning Objective 2 Define metabolism, anabolism, and catabolism.

Learning Objective 3 Outline the three stages in the extraction of energy from food.

15.2A Catabolism versus Anabolism

metabolism The sum of all reactions occurring in an organism.

catabolism All reactions involved in the breakdown of biomolecules.

anabolism All reactions involved in the synthesis of biomolecules.

metabolic pathway A sequence of reactions used to produce one product or accomplish one process.

Living cells are very active chemically, with thousands of different reactions occurring at the same time. The sum total of all the chemical reactions involved in maintaining a living cell is called **metabolism**. The reactions of metabolism are divided into two categories, catabolism and anabolism. **Catabolism** consists of all reactions that lead to the breakdown of biomolecules. **Anabolism** includes all reactions that lead to the synthesis of biomolecules. In general, energy is released during catabolism and required during anabolism.

A characteristic of all living organisms is a high degree of order in their structures and chemical processes. Even the simplest organisms exhibit an enormous array of chemical reactions that are organized into orderly, well-regulated sequences known as **metabolic pathways**. Each metabolic pathway consists of a series of consecutive chemical reactions or steps that convert a starting material into an end product. For example, two pathways discussed in **Section 15.5** are the citric acid cycle and the electron transport chain.

Fortunately for those who study the biochemistry of living systems, a great many similarities exist among the major metabolic pathways found in all life forms. These similarities enable scientists to study the metabolism of simple organisms and use the results to help explain the corresponding metabolic pathways in more complex organisms, including humans.

15.2B The Catabolism of Food

The three stages involved in the catabolic extraction of energy from food are illustrated in **Figure 15.9**.

Stage 1 is *digestion*. Digestion occurs when large, complex molecules are chemically broken into relatively small, simple ones. The most common reaction in digestion is hydrolysis, in which proteins are converted to amino acids, carbohydrates are converted to monosaccharides (primarily glucose, galactose, and fructose), and fats are converted to fatty acids and glycerol. The smaller molecules are then absorbed into the body through the lining of the small intestine.

Stage 2 is *acetyl CoA production*. The small molecules from digestion are degraded to even simpler units, primarily the two-carbon acetyl portion of acetyl coenzyme A (acetyl CoA):

$$\underbrace{CH_3-\overset{\overset{\displaystyle O}{\|}}{C}}_{\text{acetyl}}\underbrace{-S-CoA}_{\text{coenzyme A}}$$

Some energy is released in the second stage, but much more is produced during the oxidation of the acetyl units of acetyl CoA in the third stage of catabolism.

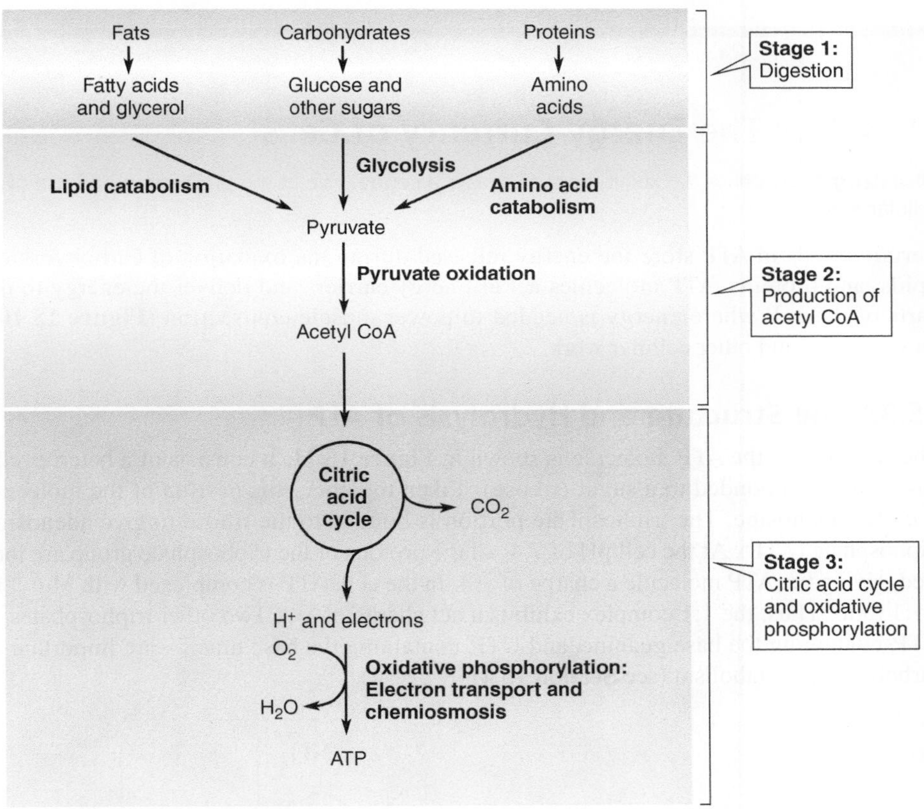

Figure 15.9 The three stages in the extraction of energy from food.

Stage 3 is *the citric acid cycle and oxidative phosphorylation* (the electron transport chain coupled with **chemiosmosis**). Because the reactions of Stage 3 are the same regardless of the type of food being degraded, Stage 3 is sometimes referred to as the **common catabolic pathway**. Energy released during Stage 3 appears in the form of energy-rich molecules of ATP. The whole purpose of the catabolic pathway is to convert the chemical energy in foods to molecules of ATP (**Section 15.3**). The ways in which carbohydrates, lipids, and proteins provide the molecules that are degraded by reactions of the common catabolic pathway are discussed in **Sections 15.4**, **15.7**, and **15.8**.

chemiosmosis The mechanism wherein protons flow across the inner mitochondrial membrane during operation of the electron transport chain, thus providing the energy for ATP synthesis.

common catabolic pathway The reactions of the citric acid cycle plus those of oxidative phosphorylation.

Example 15.3 Identify Stages of Energy Extraction

Use **Figure 15.9** to identify the stage of energy extraction associated with each of the following.

a. Oxidative phosphorylation b. Digestion c. Pyruvate oxidation

Solution

a. The third state of the catabolic extraction of energy consists of the citric acid cycle and oxidative phosphorylation, so the answer is Stage 3.
b. The first stage of energy extraction involves breaking down molecules via digestion, so the answer is Stage 1.
c. The second stage of catabolic extraction of energy involves the two processes of glycolysis and pyruvate oxidation to produce the final product of acetyl CoA, so the answer is Stage 2.

✔ Learning Check 15.3 Identify the stage of energy extraction associated with each of the following.

a. Electron transport
b. Conversion of fats to fatty acids and glycerol
c. Production of acetyl CoA

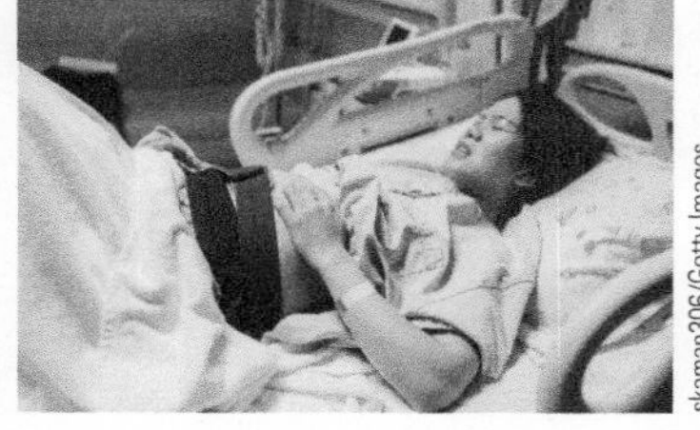

Figure 15.10 All organisms use metabolic pathways to produce ATP to deliver energy to our cells. Some of that energy is used to power muscle contractions from your heart beating and labor pain experienced during childbirth!

15.3 ATP: The Energy Currency of Cells

Learning Objective 4 Explain how ATP plays a central role in the production and use of cellular energy.

Certain bonds in ATP store the energy released during the oxidation of carbohydrates, lipids, and proteins. ATP molecules act as energy carriers and deliver the energy to the parts of the cell where energy is needed to power muscle contraction (**Figure 15.10**), biosynthesis, and other cellular work.

15.3A The Structure and Hydrolysis of ATP

The structure of the ATP molecule is shown in **Figure 15.11**. It consists of a heterocyclic base (adenine) bonded to a sugar (ribose). Taken together, this portion of the molecule is called adenosine. The triphosphate portion is bonded to the ribose to give adenosine triphosphate (ATP). At the cell pH of 7.4, all the protons of the triphosphate group are ionized, giving the ATP molecule a charge of −4. In the cell, ATP is complexed with Mg^{2+} in a 1:1 ratio. Thus, the 1:1 complex exhibits a net charge of −2. Two other triphosphates—GTP, containing the base guanine, and UTP, containing the base uracil—are important in carbohydrate metabolism (see **Section 15.4**).

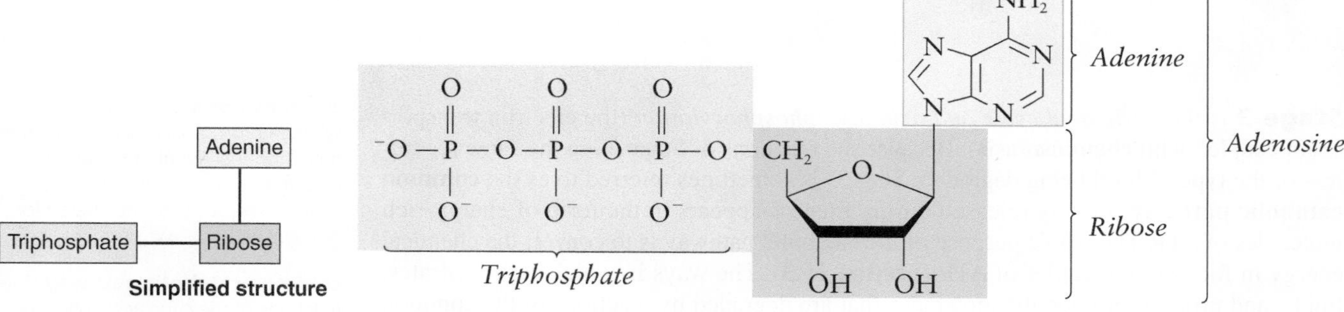

Figure 15.11 The structure of ATP.

The triphosphate end of ATP is the part of the molecule that is important in the transfer of biochemical energy. The key reaction is the transfer of a phosphoryl group, $-PO_3^{2-}$, from ATP to another molecule. For example, during the hydrolysis of ATP in water, a phosphoryl group is removed from ATP. The products of this hydrolysis are adenosine diphosphate (ADP) and a phosphate ion, which is often referred to as an inorganic phosphate, P_i, or simply as *phosphate*.

$$\text{adenosine}-O-\underset{O^-}{\overset{O}{\overset{\|}{\underset{|}{P}}}}-O-\underset{O^-}{\overset{O}{\overset{\|}{\underset{|}{P}}}}-O-\underset{O^-}{\overset{O}{\overset{\|}{\underset{|}{P}}}}-O^- + H-OH \longrightarrow$$

ATP

$$\text{adenosine}-O-\underset{O^-}{\overset{O}{\overset{\|}{\underset{|}{P}}}}-O-\underset{O^-}{\overset{O}{\overset{\|}{\underset{|}{P}}}}-O^- + HO-\underset{O^-}{\overset{O}{\overset{\|}{\underset{|}{P}}}}-O^- + H^+ \qquad \Delta G^{\circ\prime} = -7.3 \text{ kcal/mol}$$

ADP inorganic phosphate (P_i)

or

$$ATP + H_2O \rightarrow ADP + P_i + H^+ \qquad \Delta G^{\circ\prime} = -7.3 \text{ kcal/mol} \qquad (15.1)$$

The removal of a phosphoryl group from ATP by water is accompanied by a release of energy. This energy is called **Gibbs free energy** and is represented by the symbol ΔG. When energy is released by a reaction or process, ΔG will have a negative value. When energy is absorbed, ΔG will have a positive value. When ΔG is measured under standard conditions (temperature, concentration, etc.), it is represented by ΔG°, and a superscript prime ($\Delta G^{\circ\prime}$) is added if ΔG° is measured at body conditions. The liberated free energy is available for use by the cell to carry out processes requiring an input of energy. This energy-releasing hydrolysis reaction is represented in **Figure 15.12**.

change in Gibbs free energy (ΔG) The difference in the product and reactant energies in a chemical reaction.

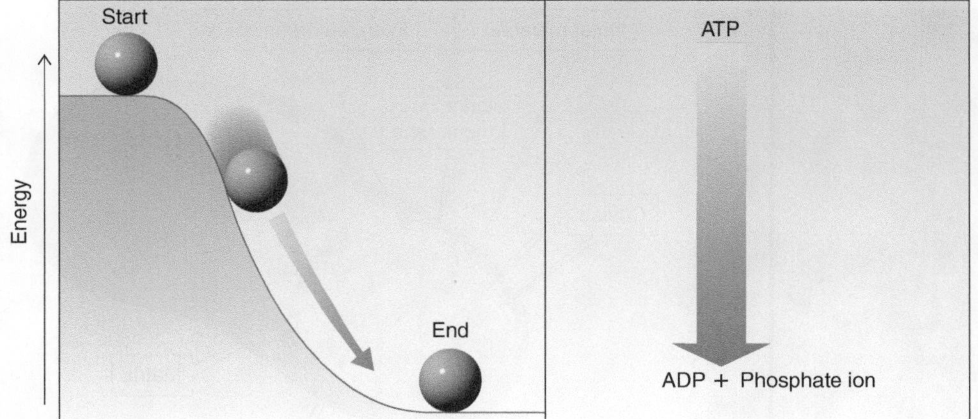

a A ball rolls down a hill, releasing potential energy.

b ATP is hydrolyzed to produce ADP and phosphate ion, releasing energy. The release of energy when a ball rolls down a hill is analogous to the release of energy in a chemical reaction.

Figure 15.12 Schematic representation of the lowering of energy when (a) a ball rolls down hill and (b) ATP is hydrolyzed to ADP + P_i. From Campbell, *Biochemistry*, 9E. © 2018 Cengage Learning.

15.3B Mitochondria: The Powerhouse of the Cell

In biological systems, ATP functions as an immediate donor of free energy rather than as a storage form of free energy. In a typical cell, an ATP molecule is hydrolyzed within one minute following its formation. Thus, the turnover rate of ATP is very high. For example, a resting human body hydrolyzes ATP at the rate of about 40 kg every 24 hours. During strenuous exertion, this rate of ATP utilization may be as high as 0.5 kg per minute. Motion, biosynthesis, and other forms of cellular work can occur only if ATP is continuously regenerated from ADP. **Figure 15.13** depicts the central role served by the ATP–ADP cycle in linking energy production with energy utilization.

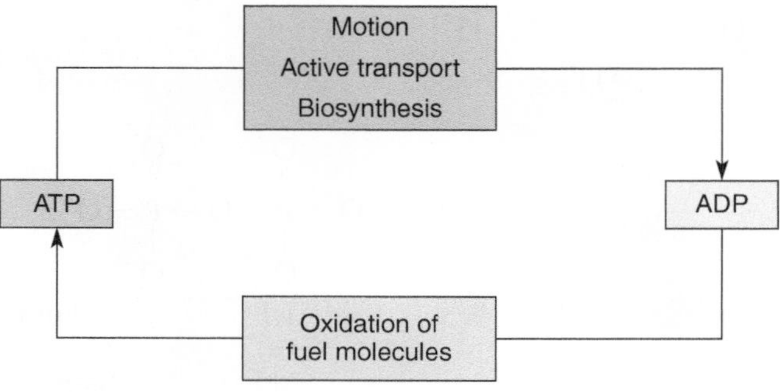

Figure 15.13 The ATP–ADP cycle. ATP, which supplies the energy for cellular work, is continuously regenerated from ADP during the oxidation of fuel molecules.

mitochondrion A cellular organelle in which reactions of the common catabolic pathway occur.

Enzymes that catalyze the formation of ATP, as well as the other processes in the common catabolic pathway, are all located within the cell in organelles called **mitochondria** (see **Figure 15.14**). Mitochondria are often called cellular "power stations" because they are the sites for most ATP synthesis in the cells.

A mitochondrion contains both inner and outer membranes. As a result of extensive folding, the inner membrane has a surface area that is many times that of the outer membrane. The folds of the inner membrane are called *cristae*, and the gel-filled space that surrounds them is called the *matrix*. The enzymes for ATP synthesis (electron transport and oxidative phosphorylation) are located on the matrix side of the cristae. The enzymes for the citric acid cycle are found within the matrix, attached to or near the surface of the inner membrane. Thus, all the enzymes involved in the common catabolic pathway are in close proximity.

a

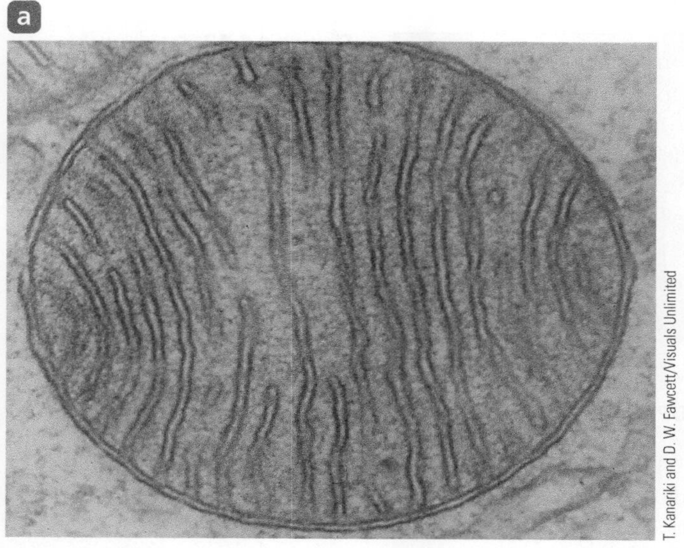

T. Kanariki and D. W. Fawcett/Visuals Unlimited

b

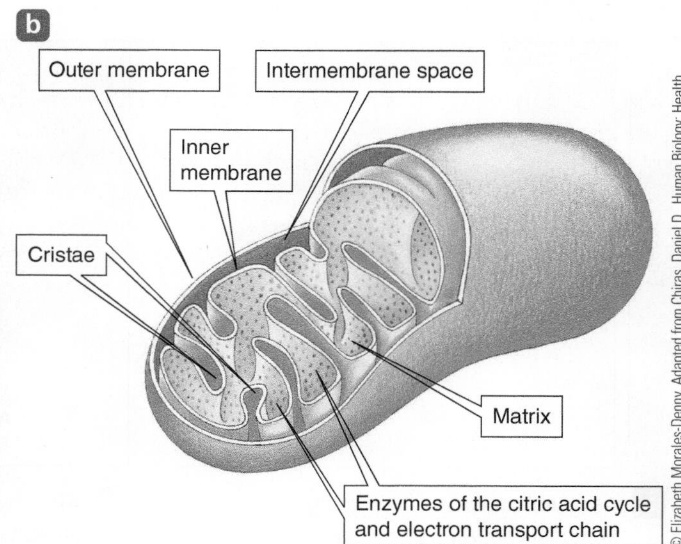

Figure 15.14 The mitochondrion: (a) a photomicrograph; (b) a schematic drawing, cut away to reveal the internal organization.

15.4 Carbohydrate Metabolism and Blood Glucose

Learning Objective 5 Identify the monosaccharide products of the digestion of carbohydrates.

Learning Objective 6 Explain the importance to the body of maintaining proper blood sugar levels.

Figure 15.15 Carbohydrates are a major source of energy for human activity.

Carbohydrates play major roles in cell metabolism. The central substance in carbohydrate metabolism is glucose; it is the key food molecule for most organisms. As the major carbohydrate metabolic pathways are discussed, notice that each one involves either the use or the synthesis of glucose.

The major function of dietary carbohydrates is to serve as an energy source (see **Figure 15.15**). In a typical American diet, carbohydrates meet about 45% to 55% of the daily energy needs. Fats and proteins furnish the remaining energy. During carbohydrate digestion, di- and polysaccharides are hydrolyzed to monosaccharides, primarily glucose, fructose, and galactose:

$$\text{polysaccharides} + H_2O \xrightarrow{\text{digestion}} \text{glucose}$$

$$\text{sucrose} + H_2O \xrightarrow{\text{digestion}} \text{glucose} + \text{fructose}$$

$$\text{lactose} + H_2O \xrightarrow{\text{digestion}} \text{glucose} + \text{galactose}$$

$$\text{maltose} + H_2O \xrightarrow{\text{digestion}} \text{glucose}$$

After digestion is completed, glucose, fructose, and galactose are absorbed into the bloodstream through the lining of the small intestine and transported to the liver. In the liver, fructose and galactose are rapidly converted to glucose or to compounds that are metabolized by the same pathway as glucose. Glucose is by far the most plentiful monosaccharide found in blood, and the term *blood sugar* usually refers to glucose. In adults, the normal **blood sugar level** measured after a fast of 8 to 12 hours is in the range of 70 to 110 mg/100 mL (**Figure 15.16**). (In clinical reports, such levels are often expressed in units of milligrams per deciliter, mg/dL, where 100 mL = 1 dL.) The blood sugar level reaches a maximum of approximately 140 to 160 mg/100 mL about 1 hour after a carbohydrate-containing meal. It returns to normal after 2 to 2½ hours.

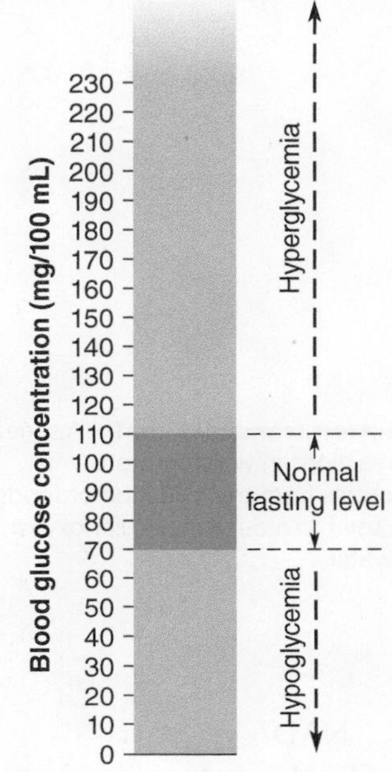

Figure 15.16 Blood glucose levels.

If a blood sugar level is below the normal fasting level, a condition called **hypoglycemia** exists. Because glucose is the only nutrient normally used by the brain for energy, mild hypoglycemia leads to dizziness and fainting as brain cells are deprived of energy. Severe hypoglycemia can cause convulsions and coma. When the blood glucose concentration is above normal, the condition is referred to as **hyperglycemia**. If blood glucose levels exceed approximately 180 mg/100 mL, the sugar is not completely reabsorbed by the kidneys, and glucose is excreted in the urine. The blood glucose level at which this occurs is called the **renal threshold**, and the condition in which glucose appears in the urine is called **glucosuria**. Prolonged hyperglycemia at a glucosuric level is serious because it indicates that something is wrong with the body's normal ability to control the blood sugar level.

The liver is the key organ involved in regulating the blood glucose level. The liver responds to the increase in blood glucose that follows a meal by removing glucose from the bloodstream. The removed glucose is primarily converted to glycogen or triglycerides for storage. Similarly, when blood glucose levels are low, the liver responds by converting stored glycogen to glucose and by synthesizing new glucose from noncarbohydrate substances.

blood sugar level The amount of glucose present in blood, normally expressed as milligrams per 100 mL of blood.

hypoglycemia A lower-than-normal blood sugar level.

hyperglycemia A higher-than-normal blood sugar level.

renal threshold The blood glucose level at which glucose begins to be excreted in the urine.

glucosuria A condition in which elevated blood sugar levels result in the excretion of glucose in the urine.

15.5 An Overview of Cellular Respiration

Learning Objective 7 Describe the process and cellular regulation of glycolysis and the citric acid cycle.

Learning Objective 8 Describe in detail the process of oxidative phosphorylation and how it results in ATP production.

cellular respiration The entire process involved in the use of oxygen by cells.

During **cellular respiration**, both plants and animals combine glucose and other energy-rich compounds with oxygen from the air to produce carbon dioxide and water in an exothermic reaction. Cellular respiration, which must not be confused with pulmonary respiration (breathing), is represented as:

$$\text{glucose and other storage forms of energy} + O_2 \xrightarrow[\text{respiration}]{\text{cellular}} CO_2 + H_2O + \text{energy released} \quad (15.2)$$

A portion of the energy released during cellular respiration is captured within the cells in the form of adenosine triphosphate (ATP), from which energy can be obtained and used directly for the performance of biological work as discussed in **Section 15.3**. The remainder of the energy from cellular respiration is liberated as heat.

During cellular respiration, oxygen is reduced when it accepts electrons and hydrogen ions. The electrons come from the oxidation of fuel molecules, such as glucose, but the electrons are not transferred directly to the oxygen from the fuel molecules or their breakdown products. Instead, glucose and other fuel molecules first transfer the electrons to special coenzyme carriers. The reduced forms of these coenzymes then transfer the electrons to oxygen through the reactions of the **electron transport chain** (see **Section 15.5D**). ATP is formed from ADP and P_i as a result of this flow of electrons.

electron transport chain A series of reactions in which protons and electrons from the oxidation of foods are used to reduce molecular oxygen to water.

The first of these coenzymes, nicotinamide adenine dinucleotide (NAD^+), has a ring that gets reduced when it accepts two electrons and one proton, forming coenzyme NADH (Equation 15.3). The reduction of NAD^+ is discussed further in **Section 15.5A** when we describe how pyruvate forms via glycolysis.

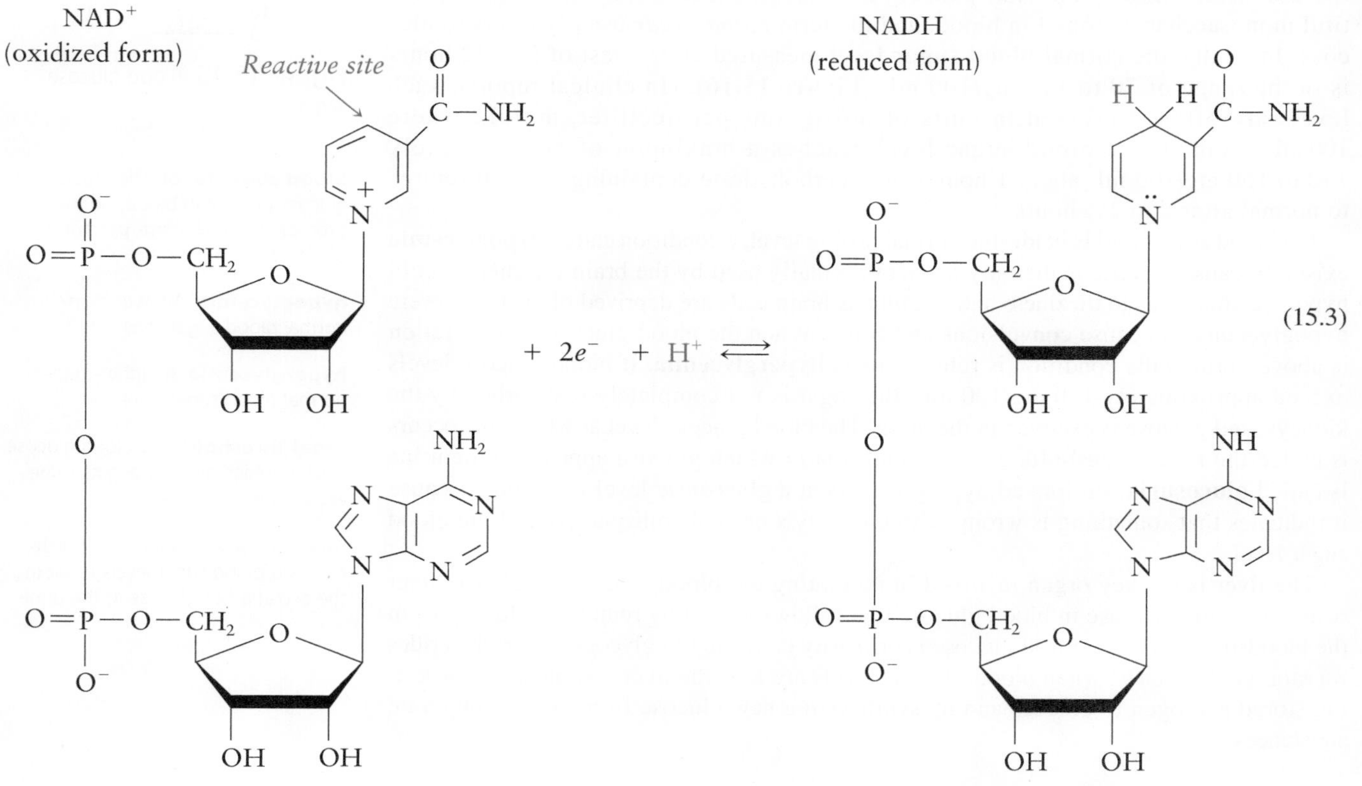

NAD⁺ (oxidized form) Reactive site $+ 2e^- + H^+ \rightleftarrows$ NADH (reduced form) (15.3)

15.5A Glycolysis

Glycolysis is Stage 1 of cellular respiration. It consists of the series of 10 reactions shown in **Figure 15.17**, with the net result being the conversion of one glucose molecule into two molecules of pyruvate. All enzymes for glycolysis are present in the cellular cytoplasm.

glycolysis A series of reactions by which glucose is oxidized to pyruvate.

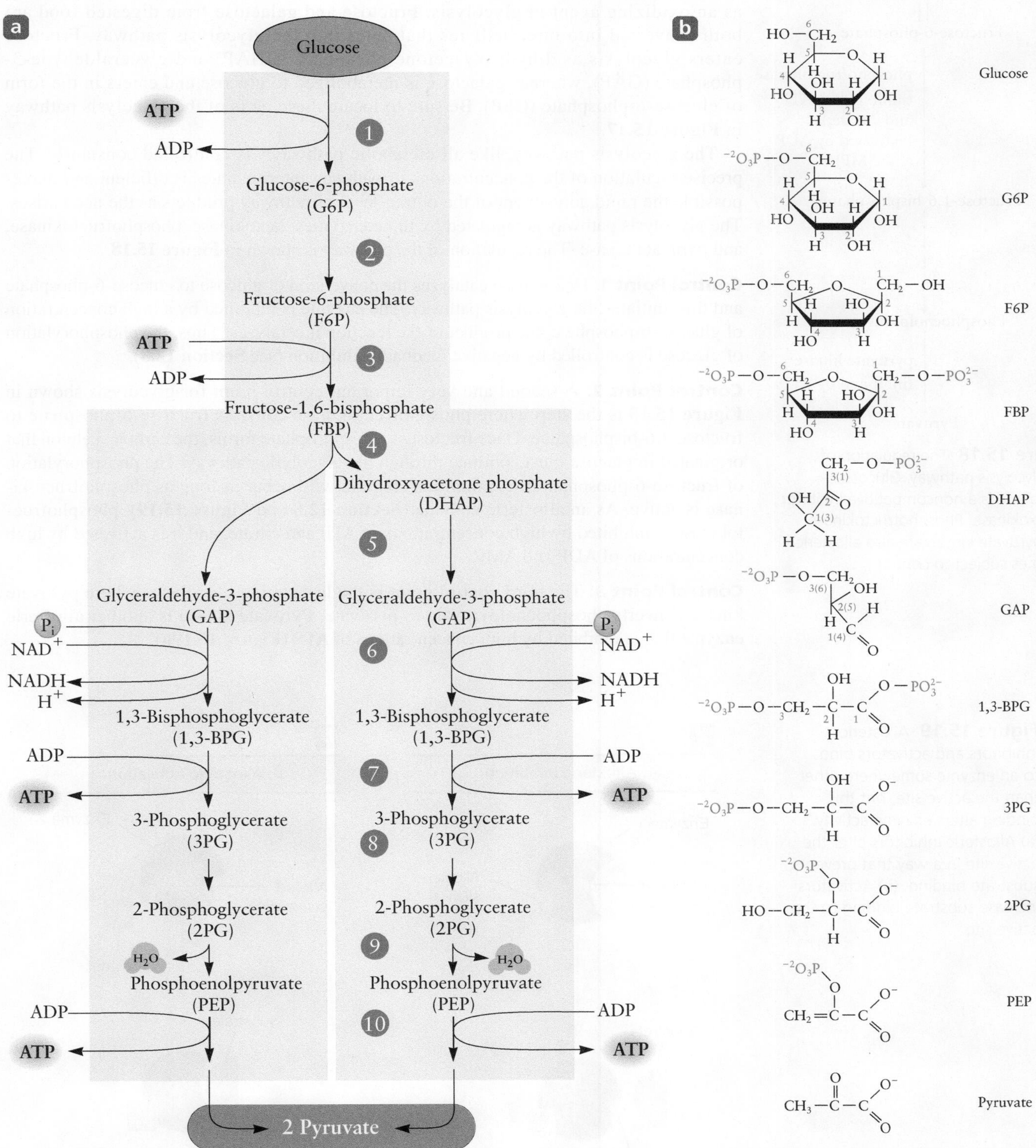

Figure 15.17 A summary of glycolysis. (a) Schematic of the 10 steps of glycolysis. (b) Chemical structures of the intermediates of glycolysis.

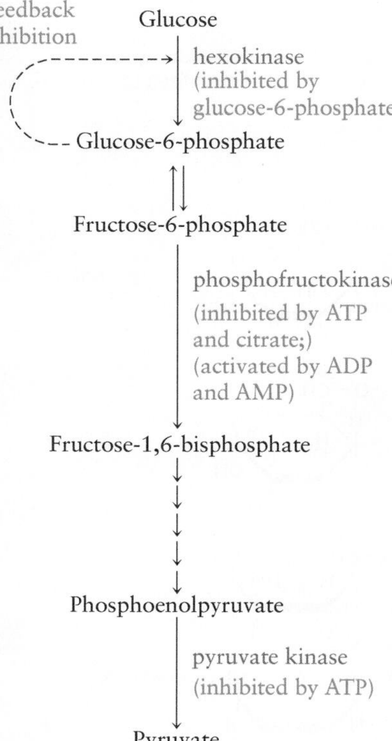

Figure 15.18 The regulation of the glycolysis pathway. Glucose-6-phosphate is a noncompetitive inhibitor of hexokinase. Phosphofructokinase and pyruvate kinase are also allosteric enzymes subject to control.

The addition of all these reactions gives the following net reaction for glycolysis:

$$\text{glucose} + 2\text{P}_\text{i} + 2\text{ADP} + 2\text{NAD}^+ \rightarrow 2\ \text{pyruvate} + 2\text{ATP} + 2\text{NADH} + 2\text{H}^+ + 2\text{H}_2\text{O} \qquad (15.4)$$

Equation 15.4 indicates a net gain of 2 mol of ATP for every 1 mol of glucose converted to pyruvate. In addition, 2 mol of the coenzyme NADH is formed when NAD^+ serves as an oxidizing agent in glycolysis. Fructose and galactose from digested food are both converted into intermediates that enter into the glycolysis pathway. Fructose enters glycolysis as dihydroxyacetone phosphate (DHAP) and glyceraldehyde-3-phosphate (GAP), whereas galactose is metabolized to glucose and enters in the form of glucose-6-phosphate (G6P). Be sure to locate these parts of the glycolysis pathway in **Figure 15.17**.

The glycolysis pathway, like all metabolic pathways, is controlled constantly. The precise regulation of the concentrations of pathway intermediates is efficient and makes possible the rapid adjustment of the output level of pathway products as the need arises. The glycolysis pathway is regulated by three enzymes: hexokinase, phosphofructokinase, and pyruvate kinase. The regulation of the pathway is shown in **Figure 15.18**.

Control Point 1: Hexokinase catalyzes the conversion of glucose to glucose-6-phosphate and thus initiates the glycolysis pathway. The enzyme is inhibited by a high concentration of glucose-6-phosphate, the product of the reaction it catalyzes. Thus, the phosphorylation of glucose is controlled by negative feedback inhibition (see **Section 12.8**).

Control Point 2: A second and very important control point for glycolysis shown in **Figure 15.18** is the step where phosphofructokinase converts fructose-6-phosphate to fructose-1,6-bisphosphate. Once fructose-1,6-bisphosphate forms, the carbon skeleton that originated in glucose must continue through the glycolytic pathway. The phosphorylation of fructose-6-phosphate is a committed step that will occur as long as phosphofructokinase is active. As an allosteric enzyme (**Section 12.8** and **Figure 15.19**), phosphofructokinase is inhibited by high concentrations of ATP and citrate, and it is activated by high concentrations of ADP and AMP.

Control Point 3: The third control point is the last step of glycolysis, where pyruvate kinase converts phosphoenolpyruvate to pyruvate. Pyruvate kinase is another allosteric enzyme that is inhibited by high concentrations of ATP (**Figure 15.19a**).

Figure 15.19 Allosteric inhibitors and activators bind to an enzyme somewhere other than the active site, but this binding alters enzyme activity. (a) Allosteric inhibitors alter the active site in a way that prevents substrate binding. (b) Activators increase substrate binding to the active site.

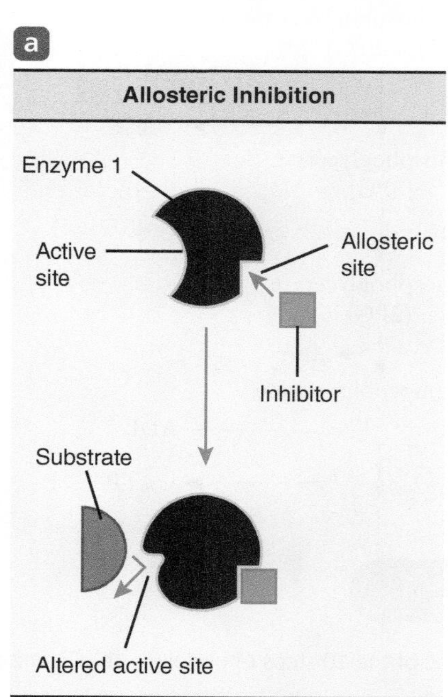

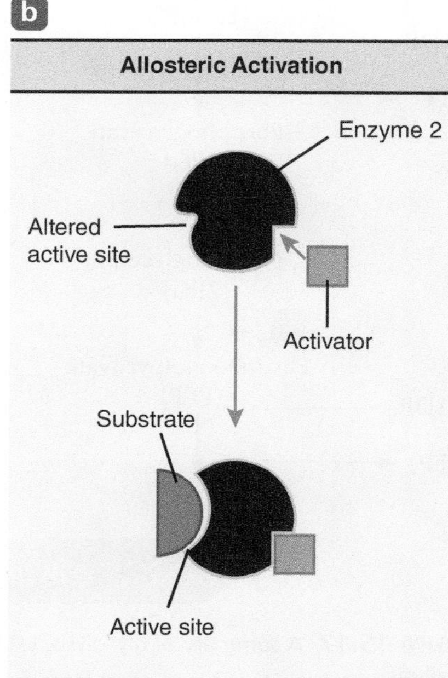

To understand how these controls interact, remember that when the glycolysis pathway is operating at a maximum rate, so also are the citric acid cycle and the electron transport chain. As a result of their activity, large quantities of ATP are produced. The entire pathway continues to operate at a high rate, as long as the ATP being produced is used up by processes within the cell. However, if ATP use decreases, the concentration of ATP increases, and more diffuses to the binding sites of the allosteric enzymes sensitive to ATP. Thus, the negative feedback loop causes the activity of these enzymes to decrease. When phosphofructokinase is inhibited in this way, the entire glycolysis pathway slows or stops. As a result, the concentration of glucose-6-phosphate increases, thus inhibiting hexokinase and the initial step of the glycolysis pathway. Conversely, when ATP concentrations are low, the concentrations of AMP and ADP are high. The AMP and ADP function as activators (**Figure 15.19**) that influence phosphofructokinase and thereby speed up the entire glycolysis pathway.

Example 15.4 Regulating Glycolysis

Describe the three control points in the glycolysis pathway by listing the reactions involved and explaining the type of regulation at each point.

Solution

The glycolysis pathway is regulated at three different steps by inhibitors and one step by an activator. The control points are:

1. The reaction of glucose → glucose-6-phosphate is inhibited by glucose-6-phosphate.
2. The reaction of fructose-6-phosphate → fructose-1,6-bisphosphate is inhibited by ATP and citrate and activated by ADP and AMP.
3. Phosphoenolpyruvate → pyruvate is inhibited by ATP.

✔ Learning Check 15.4 ATP is an allosteric regulator of phosphofructokinase and pyruvate kinase in the glycolysis pathway. Explain in your own words why it makes sense that ATP is an inhibitor of these enzymes.

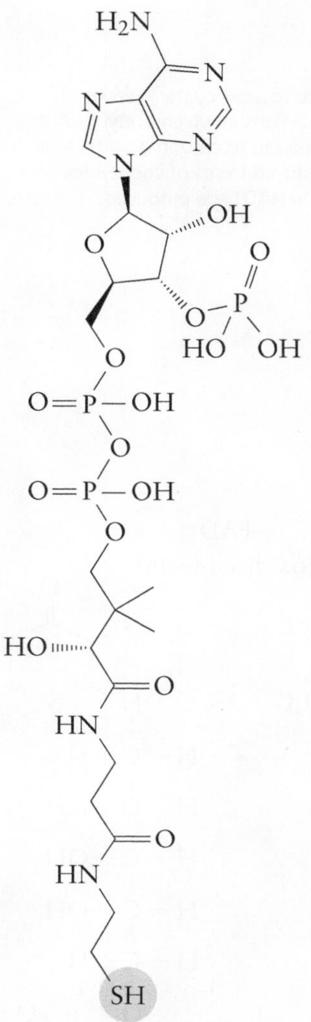

Figure 15.20 The structure of coenzyme A (CoA). The sulfhydryl group (–SH) is the reactive site.

15.5B Pyruvate Oxidation

The sequence of reactions that converts glucose to pyruvate is very similar in all organisms and in all kinds of cells. As glucose is oxidized to pyruvate in glycolysis, NAD^+ is reduced to NADH. Pyruvate, in turn, is metabolized in ways that regenerate NAD^+ so that glycolysis can continue. The mitochondrial membrane is permeable to pyruvate formed by glycolysis in the cytoplasm, and under **aerobic** conditions (a plentiful oxygen supply), pyruvate reacts with coenzyme A (CoA; **Figure 15.20**) and NAD^+ inside the mitochondria to produce acetyl CoA, NADH, and CO_2 during Stage 2 of cellular respiration.

aerobic In the presence of oxygen.

$$\underset{\text{pyruvate}}{CH_3-\overset{\overset{\displaystyle O}{\|}}{C}-COO^-} + CoA-S-H + NAD^+ \xrightarrow[\text{complex}]{\substack{\text{pyruvate}\\ \text{dehydrogenase}}} \underset{\text{acetyl CoA}}{CH_3-\overset{\overset{\displaystyle O}{\|}}{C}-S-CoA} + NADH + CO_2 \qquad (15.5)$$

The NAD^+ is an electron carrier that acts as an oxidizing agent in Equation 15.5 and in Step 6 of the glycolysis pathway. NAD^+ is regenerated when NADH transfers its electrons to O_2 in the electron transport chain (see **Section 15.5D**). Acetyl CoA formed in Equation 15.5 can enter the citric acid cycle; however, not all acetyl CoA generated metabolically does. Instead, some also serves as a starting material for fatty acid biosynthesis (**Section 15.7**). The fatty acid components of triglycerides, for example, are synthesized using acetyl CoA. This is why the intake of excess dietary carbohydrates can lead to an increase in body fat storage.

15.5C The Citric Acid Cycle

citric acid cycle A series of reactions in which acetyl CoA is oxidized to carbon dioxide, and reduced forms of coenzymes FAD and NAD^+ are produced.

Stage 3 in the oxidation of fuel molecules begins when the two-carbon acetyl units (of acetyl CoA) enter the **citric acid cycle**. This process is called the citric acid cycle because one of the key intermediates is citric acid. However, it is also called the tricarboxylic acid cycle (TCA) in reference to the three carboxylic acid groups in citric acid, and the Krebs cycle in honor of Sir Hans A. Krebs, who deduced its reaction sequence in 1937.

Flavin adenine dinucleotide (FAD) is the other major electron carrier in the oxidation of fuel molecules during the citric acid cycle of cellular respiration. This coenzyme is a derivative of ADP and the vitamin riboflavin. The reactive site is located within the riboflavin ring system (see Equation 15.6). As in the reduction of NAD^+, substrates of enzymes that use FAD as the coenzyme give up two electrons to the coenzyme. However, unlike NAD^+, the FAD coenzyme accepts both of the hydrogen atoms lost by the substrate. Thus, the abbreviation for the reduced form of FAD is $FADH_2$:

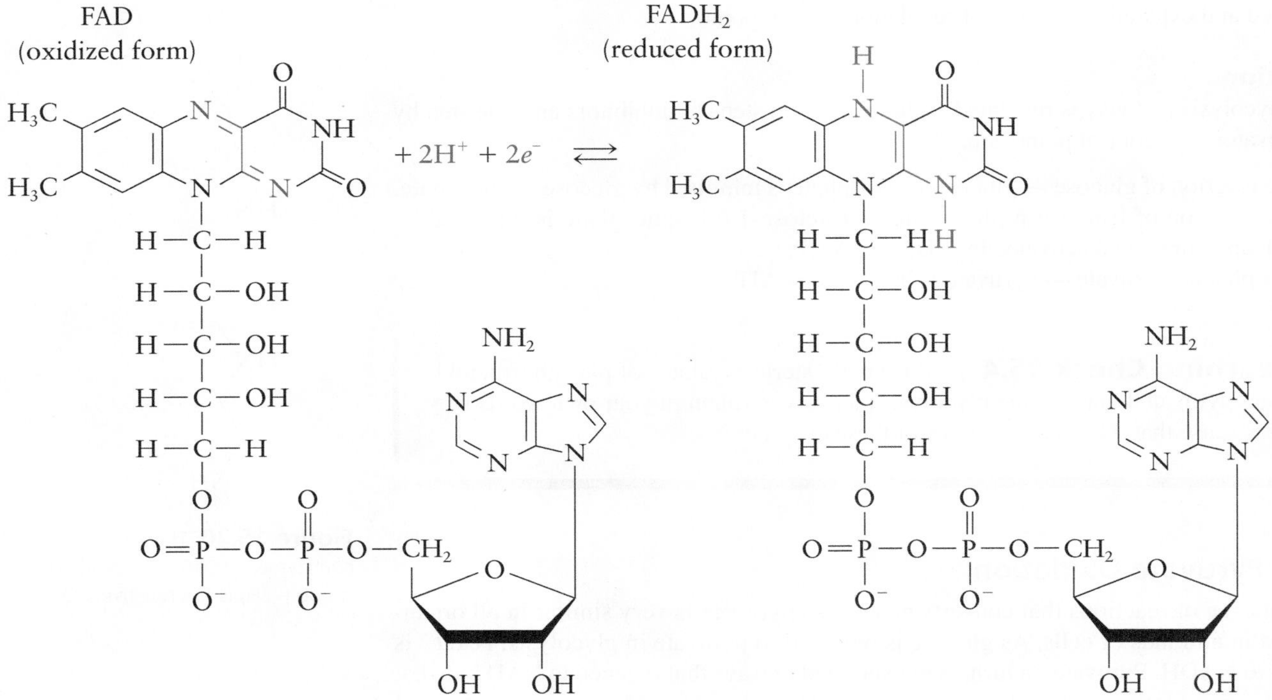

FAD (oxidized form) $+ 2H^+ + 2e^- \rightleftarrows$ $FADH_2$ (reduced form) (15.6)

The reactions of the citric acid cycle occur within the matrix of the mitochondrion. **Figure 15.21** summarizes the details of the eight reactions that make up the citric acid cycle.

Of particular importance are those reactions in which molecules of NADH and $FADH_2$ are produced—namely Steps 3, 4, 6, and 8. Notice that a two-carbon unit (the acetyl group) enters the cycle and that two carbon atoms are liberated in the form of two CO_2 molecules, one each in Steps 3 and 4. Thus, this series of reactions both begins and ends with a C_4 compound, oxaloacetate. This is why the pathway is called a cycle. In each trip around the cycle, the starting material is regenerated and the reactions can proceed again as long as more acetyl CoA is available as fuel.

The individual reactions of the cycle can be added to give the following overall net equation:

$$\text{acetyl CoA} + 3NAD^+ + FAD + GDP + P_i + 2H_2O \rightarrow 2CO_2 + \text{CoA—S—H} + 3NADH + 2H^+ + FADH_2 + GTP \qquad (15.7)$$

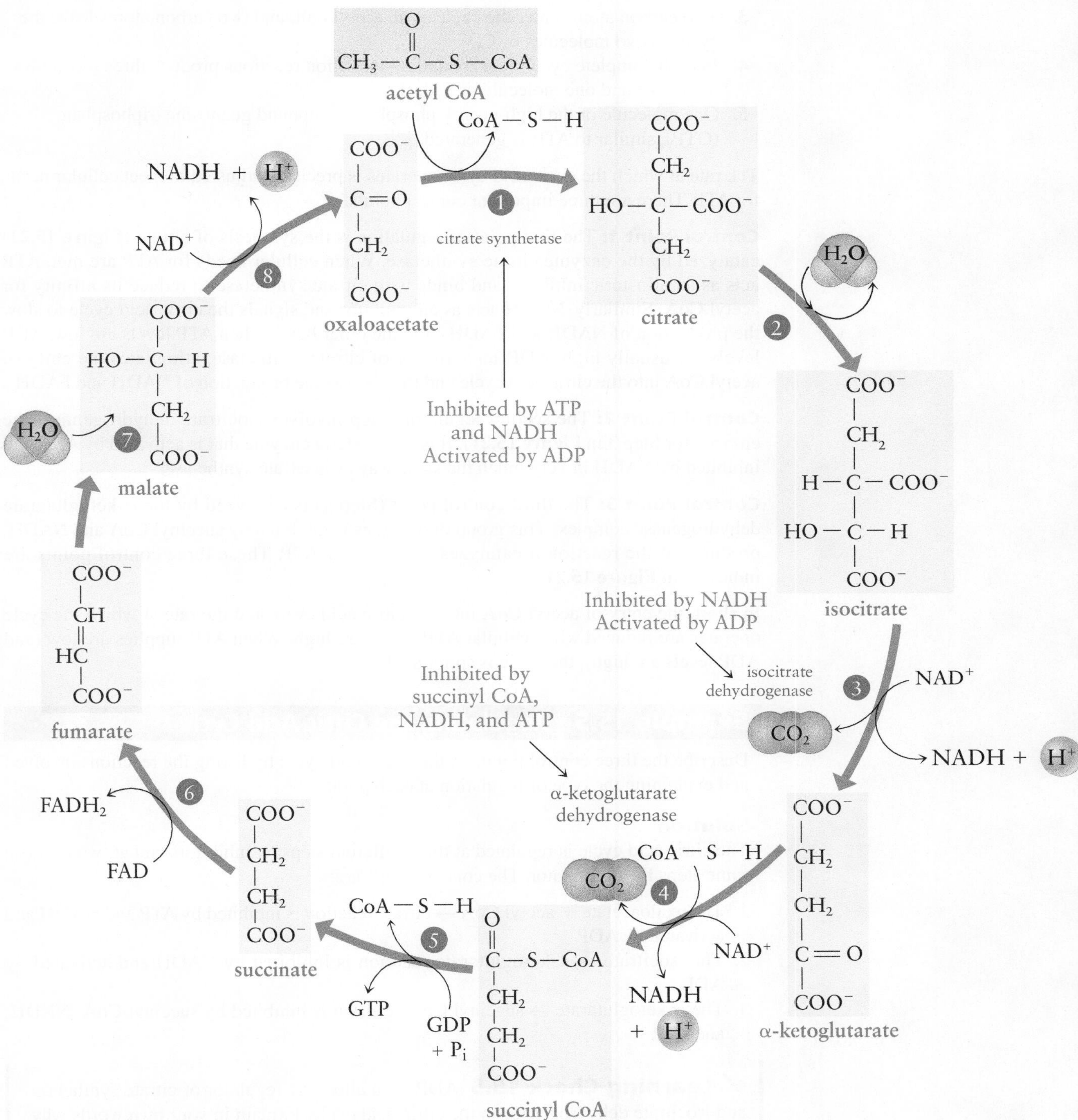

Figure 15.21 The citric acid cycle.

Some important features of the citric acid cycle are the following:

1. Acetyl CoA, available from the breakdown of carbohydrates, lipids, and amino acids, is the fuel of the citric acid cycle.
2. The operation of the cycle requires a supply of the oxidizing agents NAD^+ and FAD. The cycle depends on reactions of the electron transport chain to supply the necessary NAD^+ and FAD. Because oxygen is the final acceptor of electrons in the electron transport chain, the continued operation of the citric acid cycle depends ultimately on an adequate supply of oxygen.

3. Two carbon atoms enter the cycle as an acetyl unit, and two carbon atoms leave the cycle as two molecules of CO_2.
4. In each complete cycle, four oxidation–reduction reactions produce three molecules of NADH and one molecule of $FADH_2$.
5. One molecule of the high-energy phosphate compound guanosine triphosphate (GTP), similar to ATP, is generated.

The rate at which the citric acid cycle operates is precisely adjusted to meet cellular needs for ATP. There are three important control points.

Control Point 1: The first point of regulation is the synthesis of citrate (**Figure 15.21**) catalyzed by the enzyme citrate synthetase. When cellular needs for ATP are met, ATP acts as an allosteric inhibitor and binds with citrate synthetase to reduce its affinity for acetyl CoA. Similarly, NADH acts as an inhibitor and signals the citric acid cycle to slow the production of NADH and $FADH_2$. On the other hand, when ATP levels are low, ADP levels are usually high. ADP, an activator of citrate synthetase, stimulates the entry of acetyl CoA into the citric acid cycle and thus boosts the production of NADH and $FADH_2$.

Control Point 2: The second regulatory step involves isocitrate dehydrogenase (the enzyme for Step 3 in **Figure 15.21**). It is an allosteric enzyme that is activated by ADP and inhibited by NADH in very much the same way as is citrate synthetase.

Control Point 3: The third control point (Step 4) is catalyzed by the α-ketoglutarate dehydrogenase complex. This group of enzymes is inhibited by succinyl CoA and NADH, products of the reaction it catalyzes, and also by ATP. These three control points are indicated in **Figure 15.21**.

In short, the entry of acetyl CoA into the citric acid cycle and the rate at which the cycle operates are reduced when cellular ATP levels are high. When ATP supplies are low (and ADP levels are high), the cycle is stimulated.

Example 15.5 Regulating the Citric Acid Cycle

Describe the three control points in the citric acid cycle by listing the reactions involved and explaining the type of regulation at each point.

Solution

The citric acid cycle is regulated at three different steps by inhibitors and at two of those same steps by an activator. The control points are:

1. The oxaloacetate + acetyl CoA $\rightarrow$ citrate reaction is inhibited by ATP and NADH and activated by ADP.
2. The isocitrate $\rightarrow$ α-ketoglutarate reaction is inhibited by NADH and activated by ADP.
3. The α-ketoglutarate $\rightarrow$ succinyl CoA reaction is inhibited by succinyl CoA, NADH, and ATP.

✔ **Learning Check 15.5** ADP is an allosteric regulator of citrate synthetase and isocitrate dehydrogenase in the citric acid cycle. Explain in your own words why it makes sense that ADP is an activator of these enzymes.

15.5D Oxidative Phosphorylation: The Electron Transport Chain and Chemiosmosis

oxidative phosphorylation A process wherein the electron transport chain is coupled with chemiosmosis to form ATP from ADP and P_i.

The reduced coenzymes NADH and $FADH_2$ are end products of the citric acid cycle. In the final stage of food oxidation, known as **oxidative phosphorylation**, the hydrogen ions and electrons carried by these coenzymes combine with oxygen and form water.

$$4H^+ + 4e^- + O_2 \rightarrow 2H_2O \quad (15.8)$$

This simple exothermic (energy-releasing) process does not take place in a single step. Instead, it occurs in a series of steps that involve a number of enzymes and cofactors located within the inner membrane of the mitochondria (**Figure 15.14**). Electrons from the reduced coenzymes are passed from one electron carrier to another within the membrane in assembly-line fashion and are finally combined with molecular oxygen, the final electron acceptor. The series of reactions involved in the process is called the electron transport chain, and it represents the first of two parts of oxidative phosphorylation (the second part is called chemiosmosis). As illustrated in **Figure 15.22**, there are several intermediate electron carriers between NADH and molecular oxygen. The electron carriers are lined up in order of increasing affinity for electrons.

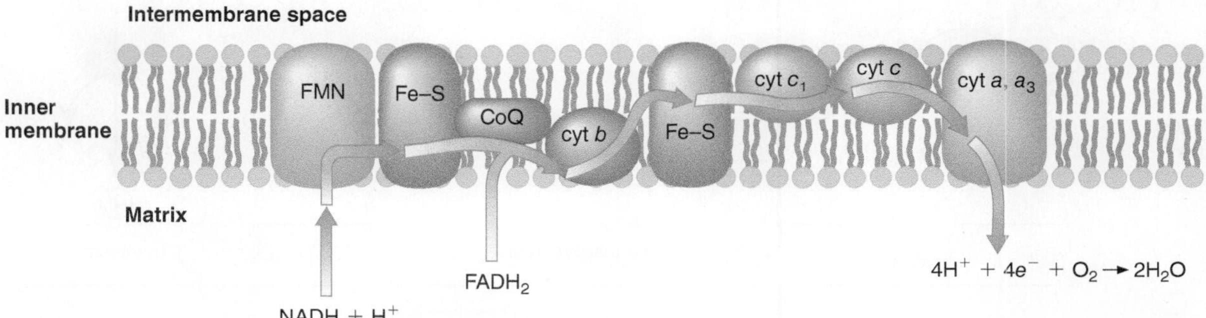

Figure 15.22 The electron transport chain. The heavy arrows show the path of the electrons as they pass from one carrier to the next.

The first electron carrier in the electron transport chain is an enzyme that contains a tightly bound coenzyme. The coenzyme has a structure similar to FAD. The enzyme formed by the combination of this coenzyme with a protein is called flavin mononucleotide (FMN). Two electrons and one H^+ ion from NADH plus another H^+ ion from a mitochondrion pass to FMN, then to an iron–sulfur (Fe–S) protein, and then to coenzyme Q (CoQ). CoQ is also the entry point into the electron transport chain for the two electrons and two H^+ ions from $FADH_2$. As NADH and $FADH_2$ release their hydrogen atoms and electrons, NAD^+ and FAD are regenerated for reuse in the citric acid cycle.

Four of the five remaining electron carriers are **cytochromes** (abbreviated cyt), which are structurally related proteins that contain an iron group. CoQ passes along the two electrons to two molecules of cytochrome *b,* and the two H^+ ions are given up to the mitochondrion. The electrons are passed along the chain, and in the final step, an oxygen atom accepts the electrons and combines with two H^+ ions to form water.

cytochrome An iron-containing enzyme located in the electron transport chain.

During chemiosmosis, the second part of oxidative phosphorylation, the cell couples the oxidations of the electron transport chain and the synthesis of ATP. The flow of electrons through the electron transport chain causes protons to be "pumped" from the mitochondrial matrix across the inner membrane and into the space between the inner and outer mitochondrial membranes (see **Figure 15.23**). This creates a difference in proton concentration across the inner mitochondrial membrane, along with an electrical potential difference.

As a result of the concentration and potential differences, protons flow back through the membrane using a channel formed by the enzyme ATP synthase. The flow of protons through this enzyme drives the phosphorylation reaction.

$$\text{ADP} + \text{P}_\text{i} \xrightarrow[\text{F}_1\text{-ATPase}]{\text{H}^+ \text{ flow through}} \text{ATP} \quad (15.9)$$

As electrons are transported along the electron transport chain, a significant amount of free energy is released (52.6 kcal/mol). Some of this energy is conserved by the synthesis

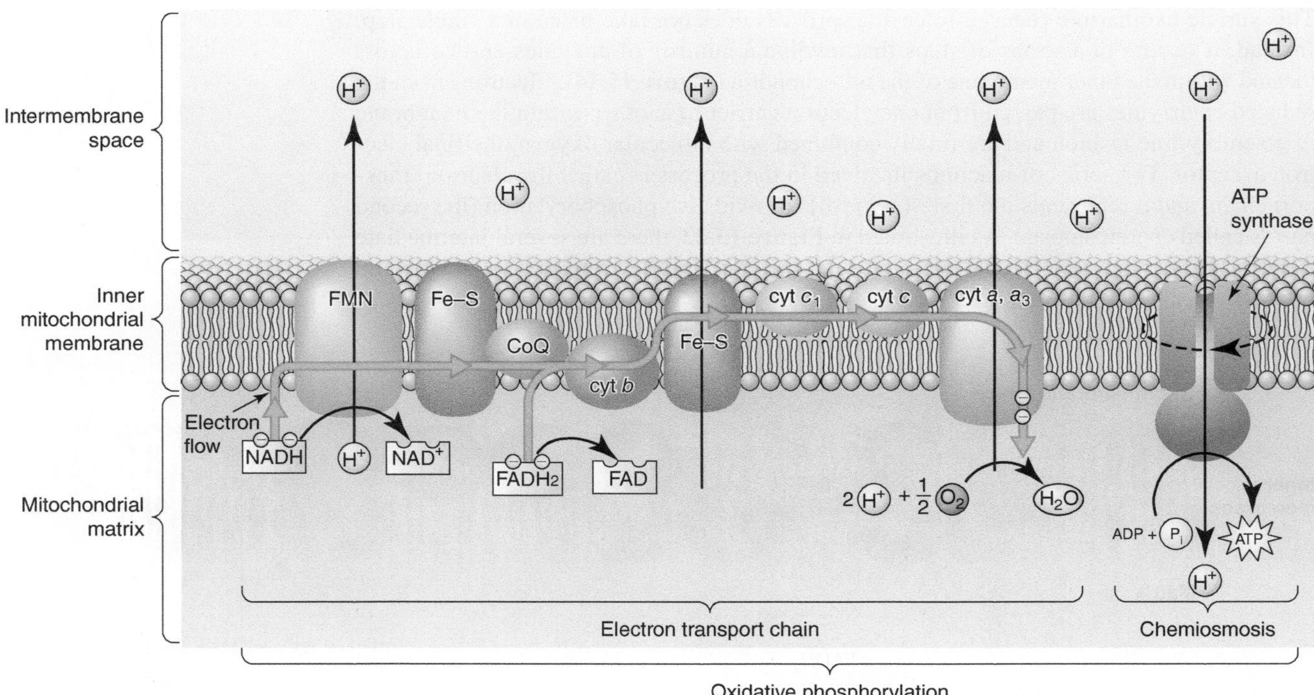

Figure 15.23 Oxidative phosphorylation. The electron transport chain pumps protons (H^+) out of the matrix. The formation of ATP accompanies the flow of protons through ATP synthase back into the matrix in a process called chemiosmosis.

of ATP from ADP and P_i during chemiosmosis. For each mole of NADH oxidized in the electron transport chain, 2.5 mol of ATP is formed. The free energy required for the phosphorylation of ADP is 7.3 kcal/mol, or 18 kcal for 2.5 mol of ATP. Because the oxidation of NADH liberates 52.6 kcal/mol, coupling the oxidation and phosphorylation reactions conserves 18 kcal/52.6 kcal, or approximately 34% of the energy released. The electron donor $FADH_2$ enters the electron transport chain later than does NADH. Thus, for every $FADH_2$ oxidized to FAD, only 1.5 ATP molecules are formed during oxidative phosphorylation, and approximately 25% of the released free energy is conserved as ATP.

Now we can calculate the energy yield for the citric acid cycle and oxidative phosphorylation combined. Recall from **Figure 15.21** that every acetyl CoA entering the citric acid cycle produces three NADH and one $FADH_2$ plus one GTP, which is equivalent in energy to one ATP. Thus, 10 ATP molecules are formed per molecule of acetyl CoA catabolized.

3 NADH ultimately produce	7.5 ATP
1 $FADH_2$ ultimately produces	1.5 ATP
1 GTP is equivalent to	1.0 ATP
	10.0 ATP total produced

15.5E ATP Produced from Glucose Oxidation

Figure 15.24 summarizes the overall process that is cellular respiration.

- **Glycolysis:** In the cellular cytoplasm, one glucose molecule is converted to two pyruvate molecules via glycolysis. Two ATP are produced and two NADH coenzymes are released and then shuttled across the mitochondrial membrane.
- **Pyruvate oxidation:** The two pyruvate molecules from glycolysis are transported into the mitochondrion, where they are oxidized to two acetyl-CoA molecules, releasing two more NADH coenzymes.
- **Citric acid cycle:** Within the mitochondrion, each of the two molecules of acetyl-CoA proceeds through the citric acid cycle, which produces a total of four CO_2, two GTP (the equivalent of two ATP), six NADH, and two $FADH_2$ molecules.

- **Oxidative phosphorylation:** It consists of the electron transport chain and chemiosmosis.
  - **Electron transport chain:** The accumulated NADH and $FADH_2$ molecules are oxidized as their electrons are passed from one electron carrier to another down the electron transport chain. They are finally combined with molecular oxygen, the last electron acceptor, and H^+, thereby producing water.
  - **Chemiosmosis:** The flow of electrons down the electron transport chain pumps protons from the mitochondrial matrix into the intermembrane space. As protons flow back down their concentration gradient, they pass through ATP synthase, resulting in the formation of approximately 28 ATP.

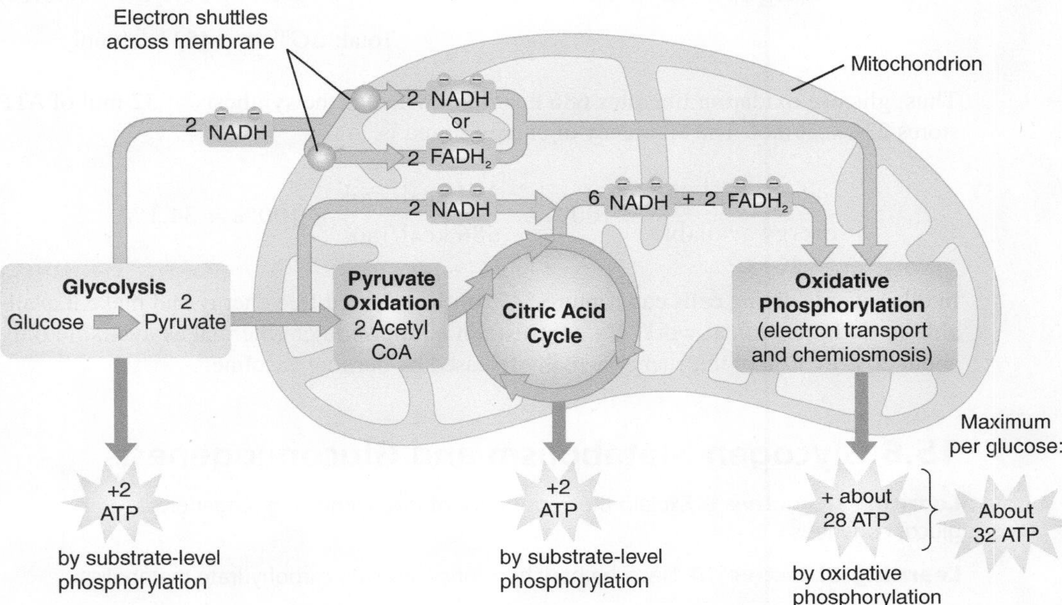

Figure 15.24 Overview of ATP production during cellular respiration.

Based on the information provided so far about glycolysis, pyruvate oxidation, the citric acid cycle, and oxidative phosphorylation, the production of ATP from the complete aerobic catabolism of 1 mol of glucose in the liver can be estimated as follows:

Catabolic step		Energy production
Glycolysis (1 glucose $\longrightarrow$ 2 pyruvate)	produces	2 NADH + 2 ATP
Oxidation of 2 pyruvates to 2 acetyl CoA	produces	2 NADH
2 pyruvates through the citric acid cycle	produces	2 $FADH_2$ + 2 GTP 6 NADH
Oxidative phosphorylation (electron transport chain and chemiosmosis)		
2 NADH formed in glycolysis		+5 ATP
2 NADH formed in the oxidation of pyruvate		+5 ATP
2 $FADH_2$ formed in the citric acid cycle		+3 ATP
6 NADH formed in the citric acid cycle		+15 ATP
Net yield of ATP		**+32 ATP**

Of the 32 mol of ATP generated, the great majority (28 mol) is formed as a result of oxidative phosphorylation.

One way to look at the efficiency of the complete oxidation of glucose is to compare the amount of free energy available in glucose and the amount of free energy stored in synthesized ATP.

Glucose oxidation:

$$C_6H_{12}O_6 + 6O_2 \rightarrow 6CO_2 + 6H_2O \qquad \Delta G^{\circ\prime} = -686 \text{ kcal/mol}$$

ATP synthesis:

$$32ADP + 32P_i \rightarrow 32ATP + 32H_2O \qquad \Delta G^{\circ\prime} = +234 \text{ kcal/mol}$$

Overall reaction:

$$C_6H_{12}O_6 + 6O_2 + 32ADP + 32P_i \rightarrow 6CO_2 + 32ATP + 38H_2O \qquad (15.10)$$

$$\text{Total: } \Delta G^{\circ\prime} = -452 \text{ kcal/mol}$$

Thus, glucose oxidation liberates 686 kcal/mol, whereas the synthesis of 32 mol of ATP stores 234 kcal/mol. The efficiency of energy stored is

$$\frac{\text{energy stored}}{\text{energy available}} \times 100\% = \frac{234 \text{ kcal/mol}}{686 \text{ kcal/mol}} \times 100\% = 34.1\%$$

In other words, living cells can capture 34% of the released free energy and make it available to do biochemical work. By contrast, an automobile engine makes available only about 20% to 30% of the energy actually released by burning gasoline.

15.6 Glycogen Metabolism and Gluconeogenesis

Learning Objective 9 Explain the importance of glycogenesis, glycogenolysis, and gluconeogenesis.

Learning Objective 10 Describe how hormones regulate carbohydrate metabolism.

15.6A Glycogen Synthesis

glycogenesis The synthesis of glycogen from glucose.

Glucose consumed in excess of immediate body requirements is converted to glycogen (see **Section 11.8**), which is stored primarily in the liver and muscle tissue. The liver of an average adult can store about 110 g of glycogen, and the muscles can store about 245 g. The process by which glucose is converted to glycogen is called **glycogenesis**. Glycogenesis can occur in all cells, but it is an especially important function of liver and muscle cells (**Figure 15.25**). The net result of this anabolic process is the bonding of glucose units to a growing glycogen chain (**Figure 15.26**), with the hydrolysis of a high-energy nucleotide, UTP (uridine triphosphate), similar to ATP, providing the energy. The enzyme glycogen synthase forms the new $\alpha(1 \rightarrow 4)$ linkages between each glucose unit and lengthens the chain. The $\alpha(1 \rightarrow 6)$ branch points on the chain are inserted by a branching enzyme called amylo-$(1,4 \rightarrow 1,6)$-transglycosylase.

The overall process of glycogenesis is summarized in Equation 15.11.

$$(\text{glucose})_n \xrightarrow[\text{UTP} \quad \text{UDP} + P_i]{} (\text{glucose})_{n+1} = \text{glycogen} \qquad (15.11)$$

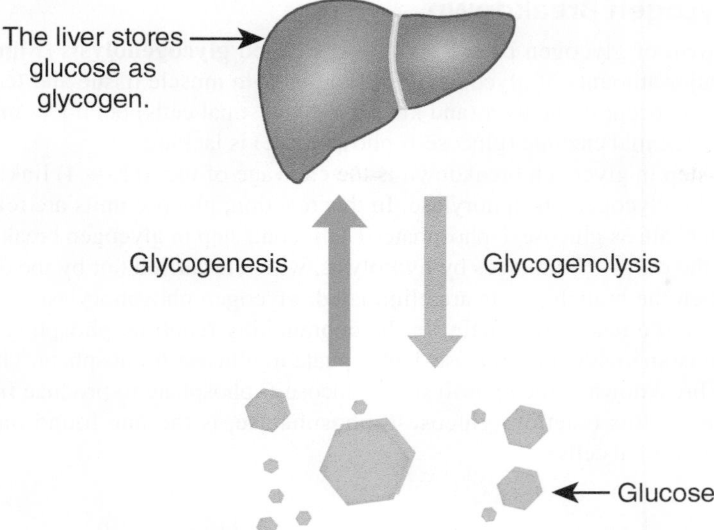

Figure 15.25 Excess glucose is converted to the polysaccharide glycogen and stored in the liver. When glucose is needed for ATP production, glycogen is broken down into glucose via glycogenolysis.

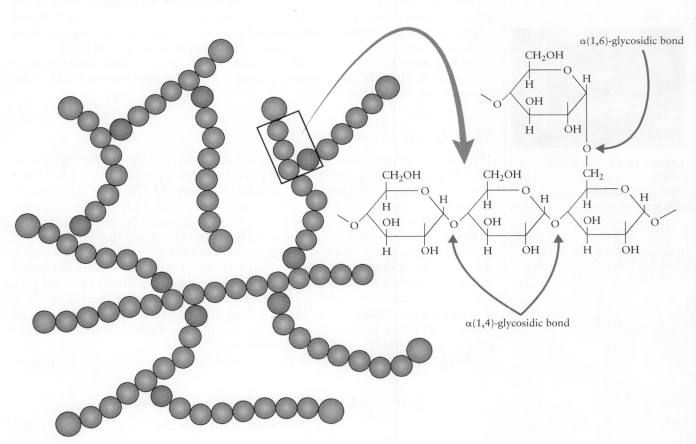

Figure 15.26 Schematic depiction and skeletal structure of glycogen.

15.6B Glycogen Breakdown

glycogenolysis The breakdown of glycogen to glucose.

The breakdown of glycogen back to glucose is called **glycogenolysis (Figure 15.25)**. Although major amounts of glycogen are stored in both muscle tissue and the liver, glycogenolysis can occur in the liver (and kidney and intestinal cells) but not in muscle tissue because one essential enzyme (glucose-6-phosphatase) is lacking.

The first step in glycogen breakdown is the cleavage of the $\alpha(1 \rightarrow 4)$ linkages, which is catalyzed by glycogen phosphorylase. In this reaction, glucose units are released from the glycogen chain as glucose-1-phosphate. The second step in glycogen breakdown is the cleavage of the $\alpha(1 \rightarrow 6)$ linkages by hydrolysis, which is carried out by the debranching enzyme. When the branch points are eliminated, glycogen phosphorylase can then continue to act on the rest of the chain. In the debranching reaction, phosphoglucomutase catalyzes the isomerization of glucose-1-phosphate to glucose-6-phosphate. The final step in glycogen breakdown is the hydrolysis of glucose-6-phosphate to produce free glucose. The enzyme for this reaction, glucose-6-phosphatase, is the one found only in liver, kidney, and intestinal cells.

$$\text{glycogen} \xrightarrow{P_i} \text{glucose-1-phosphate} \rightleftharpoons \text{glucose-6-phosphate} \xrightarrow{H_2O \;\; \curvearrowright \;\; P_i} \text{glucose} \quad (15.12)$$

Without glucose-6-phosphatase, muscle cells cannot form free glucose from glycogen. However, they can carry out the first two steps of glycogenolysis to produce glucose-6-phosphate. This form of glucose is the first intermediate in the glycolysis pathway and can still be used to produce energy even if the reaction does not proceed to the formation of glucose. In this way, muscle cells can utilize glycogen only for energy production. An important function of the liver is to maintain a relatively constant level of blood glucose, so the liver must have a broader biochemical function than muscle. Thus, it has the capacity to degrade glycogen all the way to glucose, which is released into the blood during muscular activity and between meals.

Figure 15.27 Gluconeogenesis provides a source of glucose when glycogen supplies are exhausted after vigorous exercise.

15.6C Gluconeogenesis

The supply of glucose in the form of liver and muscle glycogen can be depleted by about 12 to 18 hours of fasting. These stores of glucose are also depleted in about 1 to 2 hours as a result of heavy work or strenuous exercise (see **Figure 15.27**). Nerve tissue, including the brain, would be deprived of glucose under such conditions if the only source was glycogen. Fortunately, a metabolic pathway exists that overcomes this problem.

gluconeogenesis The synthesis of glucose from noncarbohydrate molecules.

Glucose can be synthesized from noncarbohydrate materials in a process called **gluconeogenesis**. When carbohydrate intake is low and when glycogen stores are depleted, the carbon skeletons of lactate, glycerol (derived from the hydrolysis of fats), and certain amino acids are used to synthesize pyruvate, which is then converted to glucose via a pathway of 11 chemical reactions (**Figure 15.28**). Our bodies have evolved to utilize enzymes for multiple pathways—in fact, 7 out of the 11 reactions utilize glycolysis enzymes that can catalyze the reverse reaction, leaving only 4 new enzymes necessary to produce glucose from pyruvate, highlighted below.

The addition of all these reactions gives the following net reaction for gluconeogenesis:

$$\begin{aligned} &2\ \text{pyruvate} + 4\text{ATP} + 2\text{GTP} + 2\text{NADH} + 4\text{H}^+ + 6\text{H}_2\text{O} \\ &\qquad \longrightarrow \text{glucose} + 6\text{P}_i + 4\text{ADP} + 2\text{GDP} + 2\text{NAD}^+ \quad (15.13) \end{aligned}$$

About 90% of gluconeogenesis takes place in the liver. Very little occurs in the kidneys, brain, skeletal muscle, or heart, even though these organs have a high demand for glucose. Thus, gluconeogenesis taking place in the liver helps maintain the blood glucose level so that tissues needing glucose can extract it from the blood.

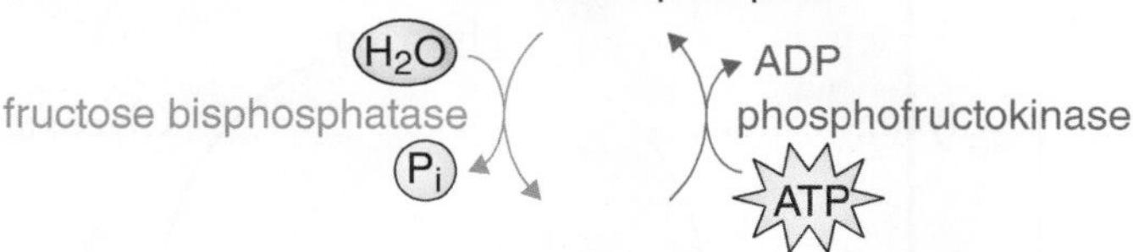

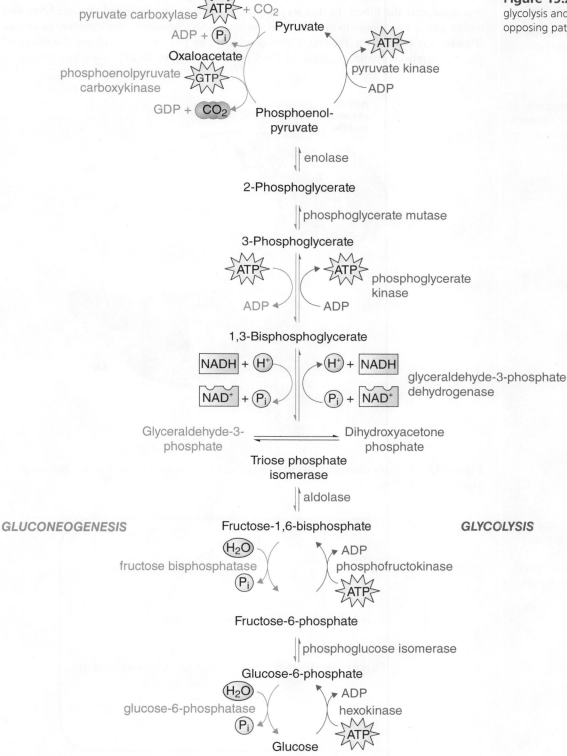

Figure 15.28 The reactions of glycolysis and gluconeogenesis are opposing pathways.

During active exercise, lactate levels increase in muscle tissue, and the compound diffuses out of the tissue into the blood. It is transported to the liver, where lactate dehydrogenase, the enzyme that catalyzes lactate formation in muscle, converts it back to pyruvate:

$$\underset{\text{lactate}}{CH_3-\overset{\overset{\displaystyle OH}{|}}{CH}-COO^-} + NAD^+ \rightleftharpoons \underset{\text{pyruvate}}{CH_3-\overset{\overset{\displaystyle O}{\|}}{C}-COO^-} + NADH + H^+ \qquad (15.14)$$

Cori cycle The process in which glucose is converted to lactate in muscle tissue, lactate is reconverted to glucose in the liver, and glucose is returned to the muscle.

The pyruvate is then converted to glucose by the gluconeogenesis pathway, and the glucose enters the blood. In this way, the liver increases a low blood glucose level and makes glucose available to muscle. This cyclic process, involving the transport of lactate from muscle to the liver, the resynthesis of glucose by gluconeogenesis, and the return of glucose to muscle tissue, is called the **Cori cycle** (see **Figure 15.29**).

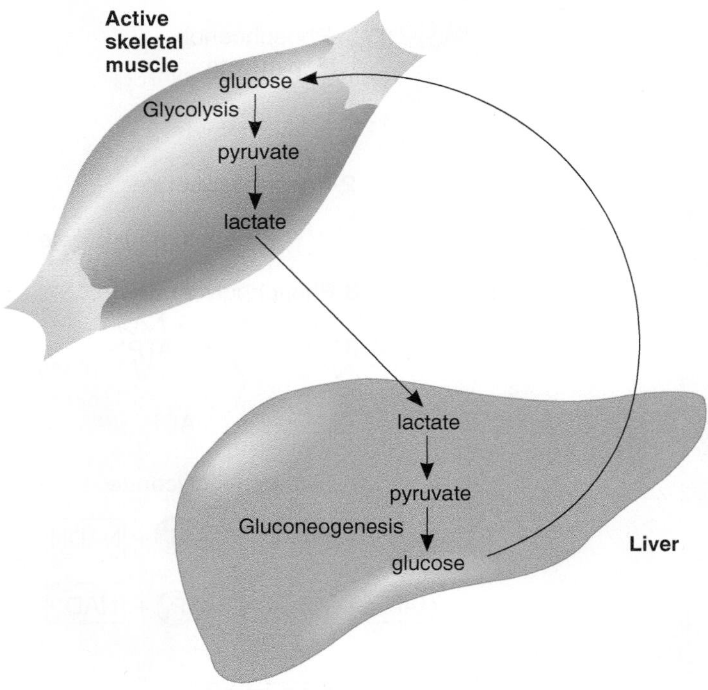

Figure 15.29 The Cori cycle, showing the relationship between glycolysis (in the muscle) and gluconeogenesis (in the liver).

Figure 15.30 shows the relationship of gluconeogenesis to the other major pathways of glucose metabolism.

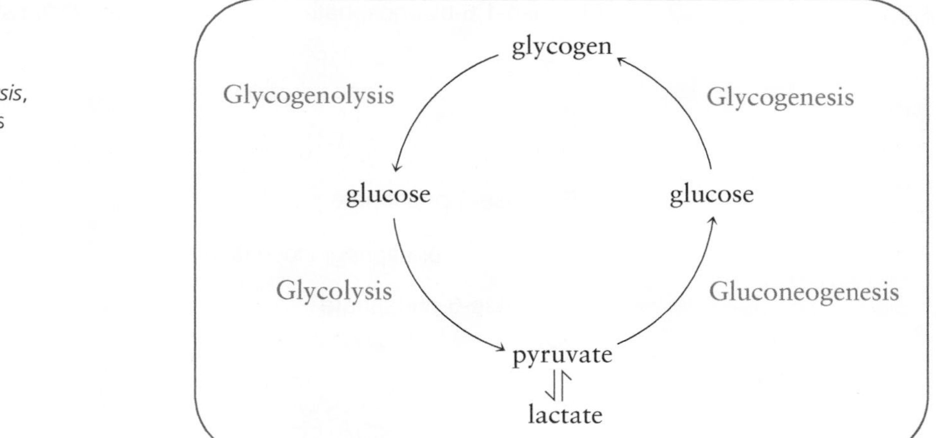

Figure 15.30 A summary of the major pathways in glucose metabolism. Note that the breakdown processes end in *-lysis*, whereas the synthesis pathways end in *-genesis*.

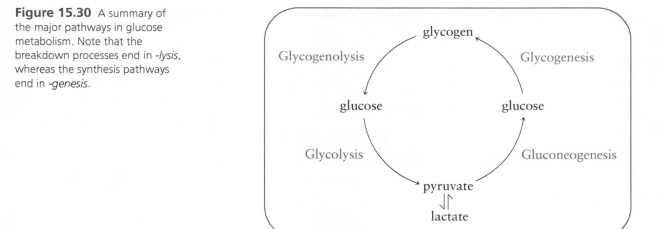

15.6D Hormone Control of Carbohydrate Metabolism

Metabolic pathways must be responsive to cellular conditions so that energy is not wasted in producing unneeded materials. Besides the regulation of enzymes at key control points (see **Section 15.5**), the body also uses three important regulatory hormones: insulin, glucagon, and epinephrine.

When carbohydrates are consumed, the blood glucose level increases (see **Section 15.4**). This increase signals the β cells of the pancreas to release a small amount of the hormone insulin into the bloodstream. The hormone enhances the absorption of glucose from the blood into cells of active tissue such as skeletal and heart muscles. In addition, insulin increases the rate of synthesis of glycogen, fatty acids, and proteins and stimulates glycolysis. Each of these activities uses up glucose. As a result of insulin released into the bloodstream, the blood glucose level begins to decrease within 1 hour after carbohydrates have been ingested and returns to normal within about 3 hours.

Glucagon, a second polypeptide hormone, is synthesized and secreted by the α cells of the pancreas. Glucagon activates the breakdown of glycogen (glycogenolysis) in the liver and thus counteracts the effect of insulin by raising blood glucose levels. Insulin and glucagon work in opposition to each other, so the blood sugar level at any given time depends, in part, on the biochemical balance between these two hormones.

Epinephrine, the third hormone affecting glucose metabolism, is synthesized by the adrenal gland. Epinephrine stimulates glycogen breakdown in muscles and, to a smaller extent, in the liver. This glycogenolysis reaction—usually a response to pain, anger, or fear—provides energy for a sudden burst of muscular activity. **Table 15.5** summarizes the hormonal control of glycogen.

Table 15.5 The Hormonal Control of Glycogen

Hormone	Source	Effect on Glycogen	Impact on Blood Glucose
Insulin	β cells of pancreas	Increases formation	Lowers blood glucose levels
Glucagon	α cells of pancreas	Activates breakdown of liver glycogen	Raises blood glucose levels
Epinephrine	Adrenal medulla	Stimulates breakdown	Raises blood glucose levels

15.7 Lipid Metabolism and Biosynthesis

Learning Objective 11 Describe the metabolic pathways by which glycerol and fatty acids are catabolized.

Learning Objective 12 Name the three ketone bodies.

Learning Objective 13 List the conditions that cause the overproduction of ketone bodies.

Learning Objective 14 Describe the pathway for fatty acid synthesis.

Glycogen and carbohydrates from dietary sources are utilized preferentially for energy production by tissues such as the brain and active skeletal muscles. When the body's stores of glycogen are depleted after only a few hours of fasting, however, fatty acids stored in triglycerides are called on as energy sources (see **Figure 15.31**).

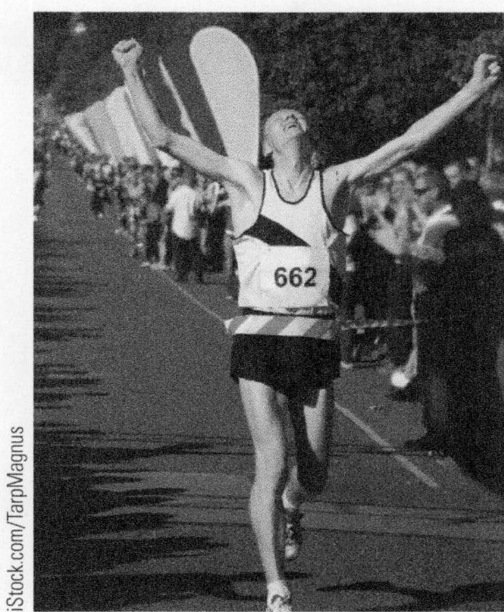

Figure 15.31 Marathon runners are initially fueled by glucose; triglycerides would become the primary fuel in later stages of the race if they didn't consume carbohydrates along the way.

Even when glycogen supplies are adequate, many body cells, such as those of resting muscle and liver, utilize energy derived from the breakdown of fatty acids. This helps to conserve the body's glycogen stores and glucose for use by brain cells and red blood cells. Red blood cells cannot oxidize fatty acids because they have no mitochondria, the site of fatty acid oxidation. Brain cells are bathed in cerebrospinal fluid and do not obtain nutrients directly from the blood because of the blood–brain barrier. Glucose and many other substances can cross the barrier into the cerebrospinal fluid, but fatty acids cannot.

When body cells need fatty acids for energy, the endocrine system is stimulated to produce several hormones, including epinephrine. In the fat cells of **adipose tissue**, epinephrine stimulates the hydrolysis of triglycerides to fatty acids and glycerol, which enter the bloodstream. This process is called **fat mobilization**. In the blood, mobilized fatty acids form a lipoprotein with the plasma protein called serum albumin. In this form, the fatty acids are transported to the tissue cells that need them. Recall that during digestion, triglycerides are hydrolyzed to glycerol, fatty acids, and monoglycerides via

adipose tissue A kind of connective tissue where triglycerides are stored.

fat mobilization The hydrolysis of stored triglycerides, followed by the entry of fatty acids and glycerol into the bloodstream.

lipases (**Figure 15.32**). The glycerol produced by the hydrolysis of the triglyceride is water soluble, so it dissolves in the blood and is also transported to cells that need it. The fatty acid tails, on the other hand, must complex with proteins forming chylomicrons for transport within the aqueous environment of the lymph and blood (**Section 14.2**, **Figure 14.5**).

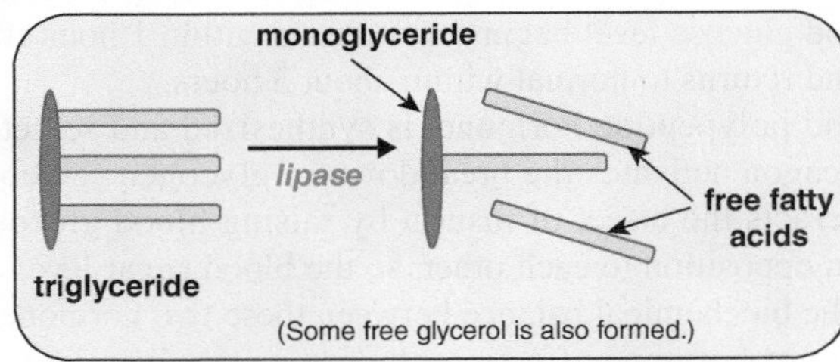

Figure 15.32 Triglycerides are hydrolyzed to glycerol, fatty acids, and monoglycerides via lipases.

15.7A Breakdown of Glycerol and Fatty Acids

The glycerol hydrolyzed from fats can provide energy to cells. Glycerol is converted in the cytoplasm of the cell to dihydroxyacetone phosphate (DHAP) in two steps:

$$\underset{\text{glycerol}}{\begin{array}{l} H_2C-OH \\ HC-OH \\ H_2C-OH \end{array}} \xrightarrow[\text{ATP} \quad \text{ADP}]{} \underset{\text{glycerol-3-phosphate}}{\begin{array}{l} H_2C-OH \\ HC-OH \\ H_2C-OPO_3^{2-} \end{array}} \xrightarrow[\text{NAD}^+ \quad \text{NADH} + \text{H}^+]{} \underset{\text{dihydroxyacetone phosphate}}{\begin{array}{l} H_2C-OH \\ C=O \\ H_2C-OPO_3^{2-} \end{array}} \qquad (15.15)$$

Dihydroxyacetone phosphate is one of the chemical intermediates of glycolysis (see **Figure 15.17**). Thus, glycerol, by entering glycolysis, can be converted to pyruvate and can contribute to cellular energy production. The pyruvate formed from glycerol can also be converted to glucose through gluconeogenesis (see **Section 15.6**).

Fatty acids that enter tissue cells cannot be oxidized to produce energy until they pass through the mitochondrial membrane. This cannot occur until the fatty acid is converted into fatty acyl CoA by reaction with coenzyme A (**Figure 15.20**). Fatty acyl refers to the $R-\overset{\overset{\large O}{\|}}{C}-$ portion of the molecule.

$$\underset{\text{fatty acid}}{R-\overset{\overset{O}{\|}}{C}-OH} + HS-CoA \xrightarrow[\text{ATP} \quad \text{AMP} + \text{PP}_i]{} \underset{\text{fatty acyl CoA}}{R-\overset{\overset{O}{\|}}{C}-S-CoA} + H_2O \qquad (15.16)$$

This reaction, which is catalyzed by acyl CoA synthetase, is referred to as activation of the fatty acid because the fatty acyl CoA is a high-energy compound. The energy needed for its synthesis is provided by the hydrolysis of ATP to AMP and PP_i and the subsequent hydrolysis of PP_i to $2P_i$. The process of forming fatty acyl CoA molecules demonstrates how the energy stored in the phosphate bonds of ATP can be used to drive chemical processes that normally could not occur. The cost to the cell of activating one molecule of fatty acid is the loss of two high-energy bonds in an ATP molecule.

The fatty acyl CoA molecules that enter mitochondria are then degraded in a catabolic process called ***β*-oxidation**. During *β*-oxidation, the second (or beta) carbon in the fatty acyl CoA is first oxidized to a ketone (Equation 15.17), and then the ketone reacts with another CoA, resulting in the release of the carbonyl carbon and *α* carbon in an acetyl unit shown in Equation 15.18.

***β*-oxidation process** A pathway in which fatty acids are broken down into molecules of acetyl CoA.

$$R-\overset{\beta}{\underset{H}{\overset{H}{C}}}-\overset{\alpha}{CH_2}-\overset{O}{\overset{\|}{C}}-S-CoA \xrightarrow{\text{oxidation}} R-\overset{\beta\ O}{\overset{\|}{C}}-\overset{\alpha}{CH_2}-\overset{O}{\overset{\|}{C}}-S-CoA \tag{15.17}$$

fatty acyl CoA → ketone

$$R-\overset{\beta\ O}{\overset{\|}{C}}-\overset{\alpha}{CH_2}-\overset{O}{\overset{\|}{C}}-S-CoA \xrightarrow{CoA-SH} R-\overset{O}{\overset{\|}{C}}-S-CoA + CH_3-\overset{O}{\overset{\|}{C}}-S-CoA \tag{15.18}$$

ketone → acetyl CoA

The β-oxidation process is repeated until the fatty acyl CoA is completely degraded to acetyl CoA. The fatty acyl CoA molecule that starts each run through the β-oxidation process is two carbons shorter than the one going through the previous run. For this reason, the β-oxidation pathway for fatty acid degradation to acetyl CoA is often called the fatty acid spiral (rather than cycle). Taken together, every run through the spiral produces one molecule each of acetyl CoA, NADH, and $FADH_2$ until the fatty acyl CoA is only four carbons long:

$$\underset{\text{fatty acyl CoA}}{R-CH_2CH_2-\overset{O}{\overset{\|}{C}}-S-CoA} + NAD^+ + FAD + H_2O + CoA-SH \longrightarrow$$

$$\underset{\substack{\text{fatty acyl CoA}\\ \text{2 carbons shorter}}}{R-\overset{O}{\overset{\|}{C}}-S-CoA} + \underset{\text{acetyl CoA}}{CH_3C-\overset{O}{\overset{\|}{}}S-CoA} + NADH + H^+ + FADH_2 \tag{15.19}$$

In other words, the complete conversion of a fatty acyl CoA to two-carbon fragments of acetyl CoA always produces one more molecule of acetyl CoA than of $FADH_2$ or NADH.

15.7B Ketone Bodies as an Indicator of Malnutrition

Nutrition scientists generally agree that a diet balanced in carbohydrates and fats as energy sources is best for good health. When such a balanced diet is followed, most of the acetyl CoA produced by fatty acid metabolism is processed through the citric acid cycle.

Under certain conditions, carbohydrate metabolism and fatty acid metabolism are no longer balanced. During a fast, for example, fatty acids from stored fats become the body's primary energy source. With a minimum amount of cellular glucose available, the level of glycolysis decreases, and a reduced amount of oxaloacetate is synthesized. In addition, oxaloacetate is used for gluconeogenesis to a greater-than-normal extent as the cells react to make their own glucose. The lack of oxaloacetate reduces the activity of the citric acid cycle. As a result of these events, more acetyl CoA is produced by fatty acid oxidation than can be processed through the citric acid cycle. As the concentration of acetyl CoA builds up, the excess is converted within the liver to three substances called **ketone bodies**—acetoacetate, β-hydroxybutyrate, and acetone.

$$CH_3-\overset{O}{\overset{\|}{C}}-CH_2-\overset{O}{\overset{\|}{C}}-O^-$$

acetoacetate

$$CH_3-\overset{OH}{\overset{|}{CH}}-CH_2-\overset{O}{\overset{\|}{C}}-O^-$$

β-hydroxybutyrate (not a ketone)

$$CH_3-\overset{O}{\overset{\|}{C}}-CH_3$$

acetone

ketone bodies Three compounds—acetoacetate, β-hydroxybutyrate, and acetone—formed from acetyl CoA.

ketonemia An elevated level of ketone bodies in the blood.

ketonuria The presence of ketone bodies in the urine.

acetone breath A condition in which acetone can be detected in the breath.

ketosis A condition in which ketonemia, ketonuria, and acetone breath exist together.

The ketone bodies are carried by the blood to body tissues—mainly the brain, heart, and skeletal muscle—where they may be oxidized to meet energy needs. β-Hydroxybutyrate does not actually contain a ketone group, but it is nevertheless included in the designation because it is always produced along with the other two. Under normal conditions, the concentration of ketone bodies in the blood averages a very low 0.5 mg/100 mL.

Diabetes mellitus, like a long fast or starvation, also produces an imbalance in carbohydrate and lipid metabolism. Even though blood glucose reaches hyperglycemic levels, a deficiency of insulin prevents the glucose from entering tissue cells in sufficient amounts to meet cellular energy needs. The resulting increase in fatty acid metabolism leads to excessive production of acetyl CoA and a substantial increase in the level of ketone bodies in the blood of poorly controlled diabetes. A concentration higher than about 20 mg/100 mL of blood is called **ketonemia** ("ketones in the blood"). At a level of about 70 mg/100 mL of blood, the renal threshold for ketone bodies is exceeded, and ketone bodies are excreted in the urine. This condition is called **ketonuria** ("ketones in the urine") (see **Figure 15.33**). A routine check of urine for ketone bodies is useful in the diagnosis of diabetes. Occasionally, the concentration of acetone in the blood reaches levels that cause it to be expelled through the lungs, and the odor of acetone can be detected on the breath. When ketonemia, ketonuria, and **acetone breath** exist simultaneously, the condition is called **ketosis**.

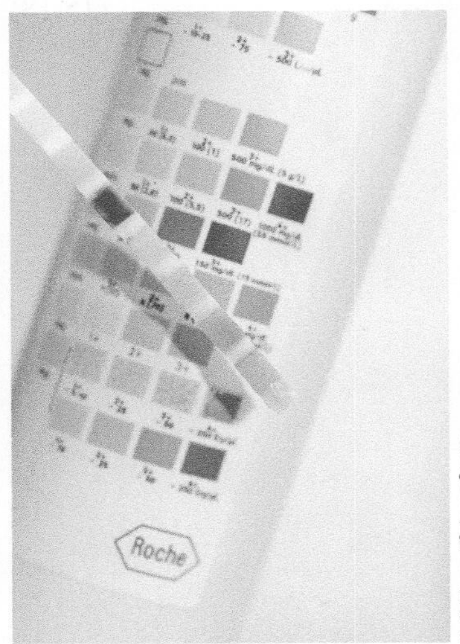

Dr. P. Marazzi/Science Source

Figure 15.33 The level of ketone bodies in urine can be measured using a test strip.

Example 15.6 Describing Ketone Bodies

Three substances are referred to as ketone bodies.

a. Which ketone bodies actually contain a ketone?
b. Which ketone body has an alcohol group?

Solution

a. Acetone and acetoacetate
b. β-Hydroxybutyrate

✔ Learning Check 15.6 Which two ketone bodies contain carboxylate groups at physiological conditions?

acidosis A low blood pH.

ketoacidosis A low blood pH due to elevated levels of ketone bodies.

The ketone bodies acetoacetate and β-hydroxybutyrate are acidic (shown as conjugate bases on the previous page), and their accumulation in the blood as ketosis worsens results in a condition of **acidosis** called **ketoacidosis** because it is caused by ketone bodies in the blood. If ketoacidosis is not controlled, the person becomes severely dehydrated because the kidneys excrete excessive amounts of water in response to low blood pH. Prolonged ketoacidosis leads to general debilitation, coma, and even death.

Patients with diabetes-related ketosis are usually given insulin as a first step in their treatment. The insulin restores normal glucose metabolism and reduces the rate of formation of ketone bodies. If the patient is also severely dehydrated, the return to fluid and acid–base balance is sped up by the intravenous administration of solutions containing sodium bicarbonate. Excessive alcohol consumption can also cause ketone bodies to be produced with acetone on the breath.

Nancy Bauer/Shutterstock.com

Figure 15.34 The results of dietary excess can be seen as this squirrel prepares for an upcoming season of scarce food.

15.7C Fatty Acid Synthesis

When organisms (including humans) take in more nutrients than are needed for their energy requirements, the excesses are not excreted; they are converted first to fatty acids and then to body fat (see **Figure 15.34**). Most of the conversion reactions take place in the liver, adipose tissue, and mammary glands. The mammary glands become especially active in the process during lactation.

The pathways for the opposing processes of fatty acid degradation and fatty acid synthesis are not simply the reverse of each other. They utilize different enzyme systems, and they take place in different cellular compartments. Degradation by the β-oxidation pathway takes place in cellular mitochondria, whereas biosynthesis of fatty acids occurs in the cytoplasm.

However, synthesis and degradation of fatty acids have one feature in common. Both processes involve units containing two carbon atoms due to the involvement of acetyl CoA. Thus, fatty acid chains are built up two carbons at a time during synthesis and broken down two carbons at a time during β-oxidation.

Acetyl CoA is generated inside mitochondria and so must be transported into the cytoplasm (**Figure 15.35**) if it is to be used to synthesize fatty acids. This transport is done by first reacting acetyl CoA with oxaloacetate to produce citrate (the first step of the citric acid cycle; see **Figure 15.21**). Mitochondrial membranes contain a citrate transport system that enables excess citrate to move out of the mitochondria into the cytoplasm. Once in the cytoplasm, the citrate reacts to regenerate acetyl CoA and oxaloacetate.

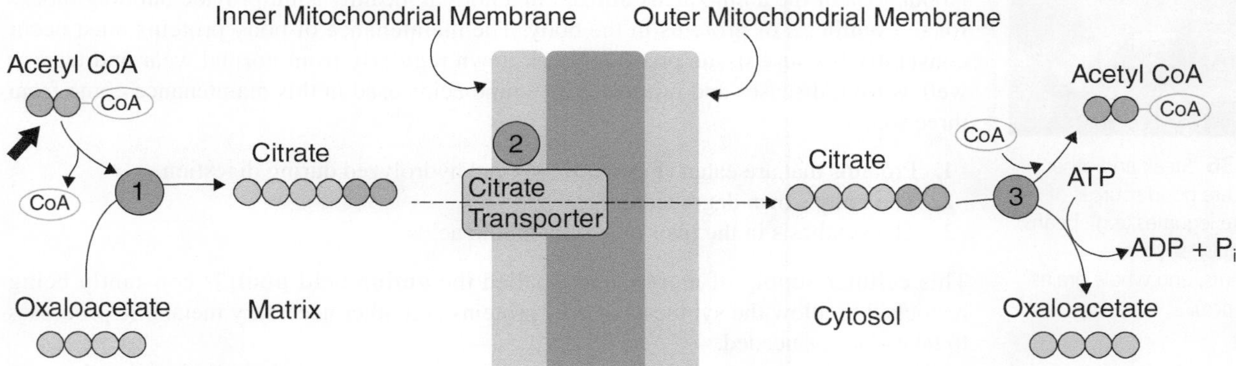

Figure 15.35 The citrate transport system.

Fatty acid synthesis occurs by way of a rather complex series of reactions catalyzed by a multienzyme complex called the *fatty acid synthetase system*. This system is made up of six enzymes and an additional protein, acyl carrier protein (ACP), to which all intermediates are attached. The summary equation for the synthesis of palmitic acid from acetyl CoA shows that a great deal of energy—namely, 7 ATP + 14 NADPH (a phosphate derivative of NADH)—is required:

$$8\underset{\text{acetyl CoA}}{CH_3-\overset{\overset{\displaystyle O}{\|}}{C}-S-CoA} + 14NADPH + 13H^+ + 7ATP \longrightarrow \tag{15.20}$$

$$\underset{\text{palmitate}}{CH_3(CH_2)_{14}COO^-} + 8CoA-SH + 6H_2O + 14NADP^+ + 7ADP + 7P_i$$

This large input of energy is stored in the synthesized fatty acids and is one of the reasons it is so difficult to lose excess weight due to fat. After synthesis, the fatty acids are incorporated into triglycerides and stored in the form of fat in adipose tissues.

The liver is the most important organ involved in fatty acid and triglyceride synthesis. It is able to modify body fats by lengthening or shortening and saturating or unsaturating the fatty acid chains. The only fatty acids that cannot be synthesized by the body are those that are polyunsaturated. However, linoleic acid (two double bonds) and linolenic acid (three double bonds) from the diet can be converted to other polyunsaturated fatty acids. This utilization of linoleic and linolenic acids as sources of other polyunsaturated fatty acids is the basis for classifying them as essential fatty acids for humans (see **Section 14.2**).

The human body can convert glucose to fatty acids, but it cannot convert fatty acids to glucose. Our cells contain no enzyme that can catalyze the conversion of acetyl CoA to pyruvate, a compound required for gluconeogenesis (**Section 15.6**). However, plants and some bacteria do possess such enzymes and convert fats to carbohydrates as part of their normal metabolism.

15.8 Amino Acid Metabolism and Biosynthesis

Learning Objective 15 Describe the source and function of the body's amino acid pool.

Learning Objective 16 Explain the chemical fate of the nitrogen atoms and carbon skeleton of amino acids.

Learning Objective 17 Explain the relationship between intermediates of carbohydrate metabolism and the synthesis of nonessential amino acids.

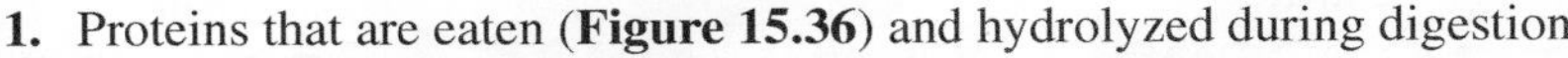

DuKai photographer/Getty Images

Figure 15.36 Steak and other animal meat are good sources of protein, as are legumes (e.g., beans, peas, and lentils), soybeans (e.g., tofu), nuts, and whole grains (e.g., brown rice).

About 75% of the amino acids utilized in a normal, healthy adult provide building blocks for the synthesis of proteins in the body. The maintenance of body proteins must occur constantly because tissue proteins break down regularly from normal wear and tear, as well as from diseases and injuries. The amino acids used in this maintenance come from three sources:

1. Proteins that are eaten (**Figure 15.36**) and hydrolyzed during digestion
2. The body's own degraded tissue
3. The synthesis in the liver of certain amino acids

This cellular supply of amino acids, called the **amino acid pool**, is constantly being restocked to allow the synthesis of new proteins and other necessary metabolic processes to take place as needed.

amino acid pool The total supply of amino acids in the body.

protein turnover The continuous process in which body proteins are hydrolyzed and resynthesized.

The dynamic process in which body proteins are continuously hydrolyzed and resynthesized is called **protein turnover**. The turnover rate, or life expectancy, of body proteins is a measure of how fast the proteins are broken down and resynthesized. The turnover rate is usually expressed as a half-life.

For example, the half-life of liver proteins is about 10 days. This means that over a 10-day period, half the proteins in the liver are hydrolyzed to amino acids and replaced by equivalent proteins. Plasma proteins also have a half-life of about 10 days, hemoglobin about 120 days, and muscle protein about 180 days. The half-life of collagen, a protein of connective tissue, is considerably longer—some estimates are as high as 1000 days. Other proteins, particularly enzyme and polypeptide hormones, have half-lives of only a few minutes. Once it is released from the pancreas, insulin has a half-life estimated to be only 7 to 10 minutes.

Because of the high turnover rate of proteins in the body, manufactured proteins do not have enough time to become a permanent part of the body structure or function. The frequent turnover of proteins is advantageous, for it allows the body to continually renew important molecules and to respond quickly to its own changing needs.

In addition to tissue protein synthesis, there is a constant draw on the amino acid pool for the synthesis of other nitrogen-containing biomolecules. These molecules include the purine and pyrimidine bases of nucleic acids; heme structures for hemoglobin and myoglobin; choline and ethanolamine, which are the building blocks of phospholipids; and neurotransmitters such as acetylcholine and dopamine. Like proteins, these compounds are also constantly being degraded and replaced. **Table 15.6** lists some of the important nitrogen-containing compounds that are synthesized from amino acids.

Unlike carbohydrates and fatty acids, amino acids in excess of immediate body requirements cannot be stored for later use. They are degraded and, depending on the organism, their nitrogen atoms are converted to ammonium ions, urea, or uric acid and excreted. Their carbon skeletons are converted to pyruvate, acetyl CoA, or one of the intermediates

Table 15.6 Compounds Derived from Amino Acids

Amino Acid	Product	Function
Tyrosine	Dopamine	Neurotransmitter
	Norepinephrine	Neurotransmitter, hormone
	Epinephrine	Hormone
	Thyroxine (T_3 and T_4)	Hormone
	Melanin	Skin pigmentation
Tryptophan	Serotonin	Neurotransmitter
Histidine	Histamine	Involved in allergic reactions
Serine	Ethanolamine	Required in cephalin synthesis
Cysteine	Taurine	A compound of bile salts

of the citric acid cycle and used for energy production, the synthesis of glucose through gluconeogenesis (see **Section 15.6**), or conversion to triglycerides. The various pathways involved in amino acid metabolism are summarized in **Figure 15.37**.

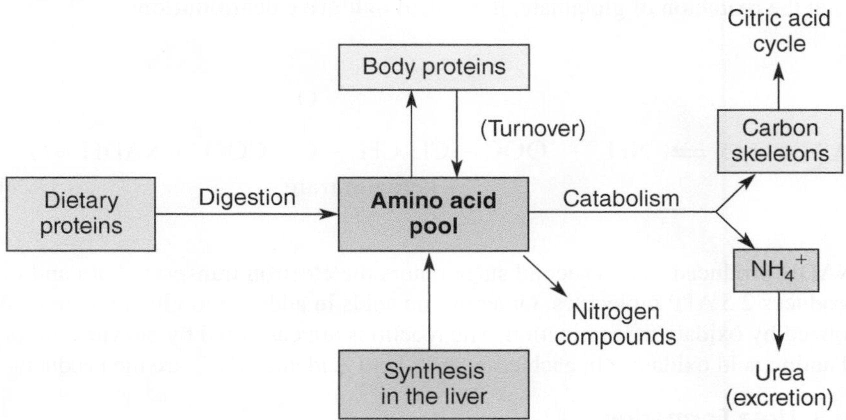

Figure 15.37 The major metabolic pathways of amino acids.

15.8A Amino Acid Catabolism: The Fate of the Nitrogen Atoms

In this section, we discuss the chemical pathways that involve the nitrogen of amino acids. In essence, there are three stages in nitrogen catabolism (see **Figure 15.38**). These processes (which occur in the liver) are transamination, deamination, and urea formation.

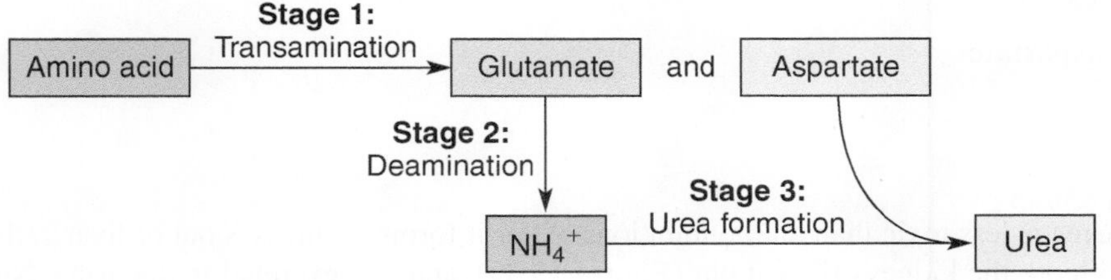

Figure 15.38 An overview of the catabolic fate of nitrogen atoms in amino acids.

transaminase An enzyme that catalyzes the transfer of an amino group.

transamination The enzyme-catalyzed transfer of an amino group to a keto acid.

Stage 1: Transamination

In the tissues, amino groups freely move from one amino acid to another under the catalytic influence of enzymes called aminotransferases, or more commonly, **transaminases**. A key reaction for amino acids undergoing catabolism is a **transamination** involving the transfer of amino groups to α-ketoglutarate. The carbon skeleton of the amino acid remains behind and is transformed into a new α-keto acid:

General reaction:

$$\underset{\substack{\text{a donor}\\ \alpha\text{-amino acid}}}{\text{R}-\overset{\overset{\displaystyle NH_3^+}{|}}{\text{CH}}-\text{COO}^-} + \underset{\substack{\alpha\text{-ketoglutarate}\\ \alpha\text{-keto acid}}}{{}^-\text{OOC}-\text{CH}_2\text{CH}_2-\overset{\overset{\displaystyle O}{\|}}{\text{C}}-\text{COO}^-} \underset{}{\overset{\text{transaminase}}{\rightleftharpoons}} \text{R}-\overset{\overset{\displaystyle O}{\|}}{\text{C}}-\text{COO}^- + \underset{\text{glutamate}}{{}^-\text{OOC}-\text{CH}_2\text{CH}_2-\overset{\overset{\displaystyle NH_3^+}{|}}{\text{CH}}-\text{COO}^-} \qquad (15.21)$$

oxidative deamination An oxidation process resulting in the removal of an amino group.

Stage 2: Deamination

This phase of amino acid catabolism uses the glutamate produced in Stage 1. The enzyme glutamate dehydrogenase catalyzes the removal of the amino group as an ammonium ion and regenerates α-ketoglutarate, which can again participate in the first stage (transamination). This reaction is the principal source of NH_4^+ in humans. Because the deamination results in the oxidation of glutamate, it is called **oxidative deamination**:

$$\underset{\text{glutamate}}{{}^-\text{OOC}-\text{CH}_2\text{CH}_2-\overset{\overset{\displaystyle NH_3^+}{|}}{\text{CH}}-\text{COO}^-} + \text{NAD}^+ + \text{H}_2\text{O} \rightleftharpoons \text{NH}_4^+ + \underset{\alpha\text{-ketoglutarate}}{{}^-\text{OOC}-\text{CH}_2\text{CH}_2-\overset{\overset{\displaystyle O}{\|}}{\text{C}}-\text{COO}^-} + \text{NADH} + \text{H}^+ \qquad (15.22)$$

The NADH produced in this second stage enters the electron transport chain and eventually produces 2.5 ATP molecules. Other amino acids in addition to glutamate may also be catabolized by oxidative deamination. The reactions are catalyzed by enzymes in the liver called amino acid oxidases. In each case, an α-keto acid and NH_4^+ are the products.

urea cycle A metabolic pathway in which ammonium ions are converted to urea.

Stage 3: Urea Formation

The ammonium ions released by the glutamate dehydrogenase reaction are toxic to the body and must be prevented from accumulating. In the **urea cycle**, which occurs only in the liver, ammonium ions are converted to urea:

$$\text{NH}_4^+ + \text{HCO}_3^- + 3\text{ATP} + 2\text{H}_2\text{O} + \underset{\text{aspartate}}{\begin{matrix}\text{COO}^-\\ |\\ \text{CH}_2\\ |\\ \text{CH}-\text{NH}_3^+\\ |\\ \text{COO}^-\end{matrix}} \longrightarrow \underset{\text{urea}}{\text{H}_2\text{N}-\overset{\overset{\displaystyle O}{\|}}{\text{C}}-\text{NH}_2} + \begin{matrix}\text{H} & & \text{COO}^-\\ & \text{C}=\text{C} & \\ {}^-\text{OOC} & & \text{H}\end{matrix} + 2\text{ADP} + 2\text{P}_i + \text{AMP} + \text{P} \qquad (15.2$$

SEBASTIAN KAULITZKI/SCIENCE PHOTO LIBRARY/Getty Images

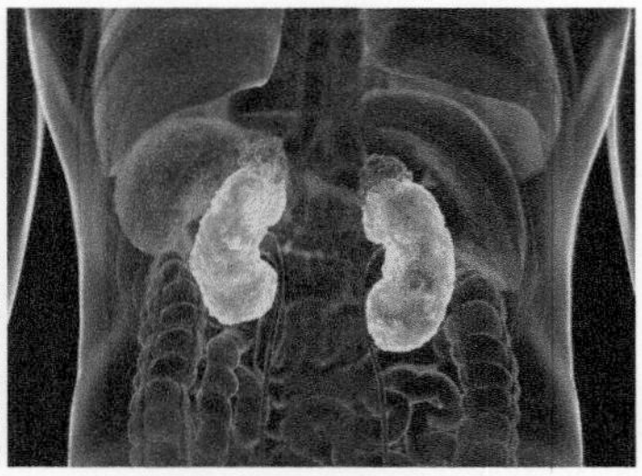

Figure 15.39 Urea is filtered from the blood through the kidneys and excreted in urine.

Urea is less toxic than ammonium ions. After it forms, it diffuses out of liver cells into the blood, the kidneys filter it out (**Figure 15.39**), and it is excreted in the urine. Normal urine from an adult usually contains about 25 to 30 g of urea daily, although the exact amount varies with the protein content of the diet.

15.8B Amino Acid Catabolism: The Fate of the Carbon Skeleton

After transamination or oxidative deamination removes the amino group of an amino acid, the remaining carbon skeleton is catabolized and converted into pyruvate, acetyl CoA, acetoacetyl CoA (which is degraded to acetyl CoA), or various substances that are intermediates in the citric acid cycle (**Figure 15.40**).

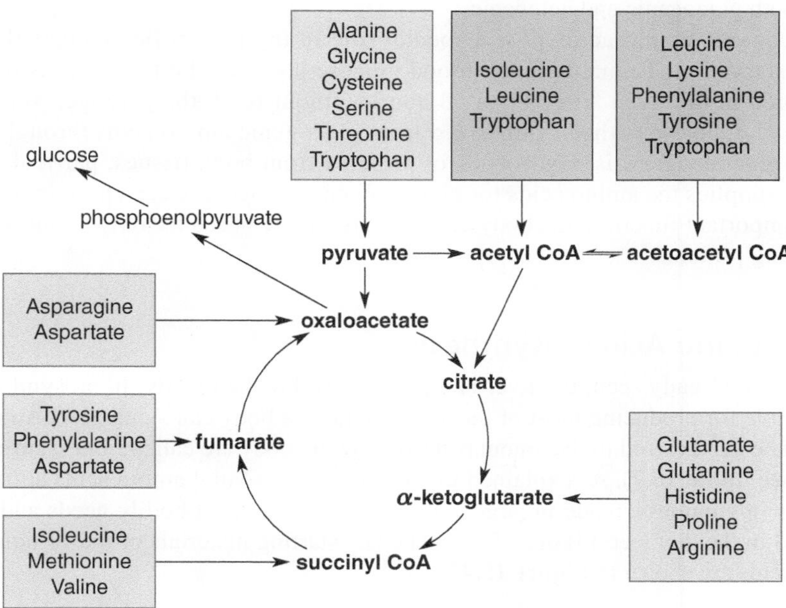

Figure 15.40 Fates of the carbon atoms of amino acids. Glucogenic amino acids are shaded purple, whereas ketogenic amino acids are shaded green.

Health Connections **15.1**

Phenylketonuria (PKU)

Phenylketonuria (PKU) is an inherited disease in which the affected person lacks the enzyme phenylalanine hydroxylase. As a result, phenylalanine cannot be converted to tyrosine, leading to the accumulation of phenylalanine and its metabolites (phenylpyruvate and phenylacetate) in the tissues and blood:

$$C_6H_5-CH_2-\overset{\overset{\displaystyle O}{\|}}{C}-COO^-$$

phenylpyruvate

$$C_6H_5-CH_2-COO^-$$

phenylacetate

Abnormally high levels of phenylpyruvate and phenylacetate damage developing brain cells, causing severe loss of cognitive ability. It is estimated that 1 out of every 20,000 newborns has PKU. A federal regulation requires that all infants be tested for this disease, which can be detected very easily by analyzing blood or urine samples for phenylalanine or phenylpyruvic acid.

The disease is treated by restricting the amount of phenylalanine in the diet of the infant starting 2 to 3 weeks after birth. Their diet is also supplemented with tyrosine because they cannot form it on their own. Strict dietary management is not as critical after the age of 6 because the brain has matured by this time and is not as susceptible to the toxic effects of phenylalanine metabolites.

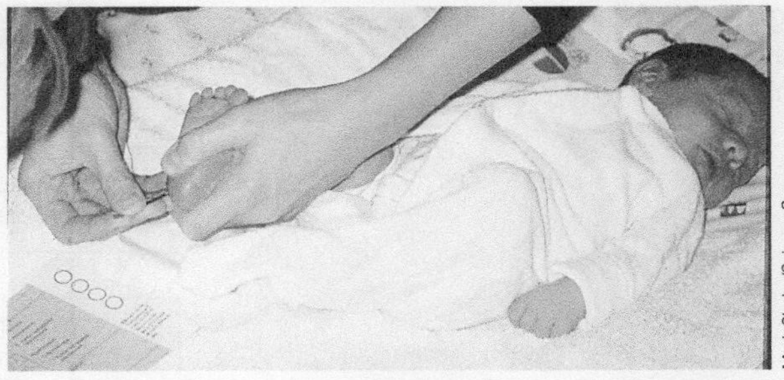

Mark Clarke/Science Source

Infants with PKU have an excellent chance for a normal healthy life, if the disease is detected and treated early. Here a nurse is taking blood samples from an infant's foot to be used in a simple chemical test for PKU.

glucogenic amino acid An amino acid whose carbon skeleton can be converted metabolically to an intermediate used in the synthesis of glucose.

ketogenic amino acid An amino acid whose carbon skeleton can be converted metabolically to acetyl CoA or acetoacetyl CoA.

The skeleton of an amino acid that is degraded into pyruvate or intermediates of the citric acid cycle can be used to make glucose and is classified as a **glucogenic amino acid**. Amino acids with skeletons that are degraded into acetyl CoA or acetoacetyl CoA cannot be converted into glucose but can, in addition to energy production, be used to make ketone bodies and fatty acids. Such amino acids are classified as **ketogenic**. Notice in **Figure 15.40** that some amino acids appear in more than one place. Tryptophan, for example, is both glucogenic and ketogenic.

Glucogenic amino acids play a special role in the metabolic changes that occur during starvation. To maintain the blood glucose level, the body first uses the glycogen stored in the liver. After 12 to 18 hours without food, the glycogen supplies are exhausted. Glucose is then synthesized from glucogenic amino acids through the process of gluconeogenesis. Hydrolysis of proteins from body tissues, particularly from muscle, supplies the amino acids for gluconeogenesis. Because each protein in the body has an important function, the body can meet its energy needs for only a limited time by sacrificing proteins.

15.8C Amino Acid Biosynthesis

nonessential amino acid An amino acid that can be synthesized within the body in adequate amounts.

As we have already seen, the liver is highly active biochemically. In biosynthesis, it is responsible for producing most of the amino acids the body can synthesize. Amino acids that can be synthesized in the amounts needed by the body are called **nonessential amino acids** (see **Table 15.7**). As explained in **Chapter 12**, essential amino acids either cannot be made or cannot be made in large enough amounts to meet bodily needs and must be included in the diet (see **Figure 15.41**). The key starting materials of the 11 nonessential amino acids are shown in **Figure 15.42**.

Figure 15.41 Good dietary sources of essential amino acids include meat, eggs, tofu, and dairy products.

Table 15.7 Essential and Nonessential Amino Acids

Essential		Nonessential	
Histidine	Threonine	Alanine	Glutamine
Isoleucine	Tryptophan	Arginine	Glycine
Leucine	Valine	Asparagine	Proline
Lysine		Aspartate	Serine
Methionine		Cysteine	Tyrosine
Phenylalanine		Glutamate	

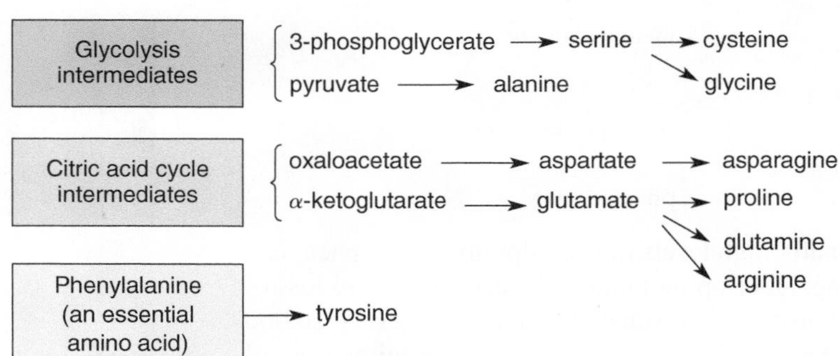

Figure 15.42 Key starting materials for the 11 nonessential amino acids.

Example 15.7 Describing Amino Acid Biosynthesis

How many amino acids may be synthesized from the following?

a. Phenylalanine
b. Intermediates of glycolysis
c. Intermediates of the citric acid cycle

Solution

a. Only tyrosine can be made from phenylalanine, so one amino acid.
b. Serine and alanine can be made directly from glycolysis intermediates and because cysteine and glycine can be made from serine, there are four amino acids that can ultimately be produced.
c. Aspartate and glutamate can be made directly from citric acid cycle intermediates. Additionally, asparagine can be made from aspartate; proline, glutamine, and arginine can be made from glutamate. Thus, six amino acids can be synthesized using intermediates from glycolysis as starting material.

✔ Learning Check 15.7 Which amino acids are

a. Essential?
b. Nonessential?

Health Career Description

Physician Assistant

As the name implies, physician assistants (PAs) "assist physicians." Their role varies by which type of physician they assist. They may work with allergists, surgeons, ob-gyn physicians, general practice physicians, or any other type of physician. PAs work in collaboration with a physician, but they enjoy a high level of autonomy. While working under the guidance and supervision of a physician, PAs maintain the ability to see patients one-on-one and to prescribe medications.

In order to become a PA, individuals must first earn a bachelor's degree and accumulate patient-contact hours. Although they may major in any subject, the most straightforward path is to major in one of the "life sciences," such as biochemistry, microbiology, biology, or zoology, because the required prerequisite coursework will count toward the requirements for these majors.

According to the University of Utah School of Medicine PA program requirements, a successful applicant will have the following items on their transcript:

- Bachelor's degree from a regionally accredited institution
- Anatomy with lab (4 credits)
- Physiology with lab (4 credits)
- Biology (4 credits)
- Chemistry (8 credits—lab credit may be included)
- Other recommended coursework:
    - Writing
    - Diversity and cultural awareness courses
    - Genetics
    - Statistics

PA applicants apply through an online service called CASPA (Centralized Application Service for Physician Assistants). Along with proof of patient-contact hours, applicants must submit a 5000-word essay explaining why they want to become a PA. The essay is an important piece of the application. It is your chance to stand out from the crowd.

After successful completion of a PA program, graduates must pass the PANCE (the Physician Assistant National Certifying Exam). This is the final hurdle in becoming a PA. Although it is common to take much longer, it's possible to complete requirements for PA certification within 6 years after beginning a bachelor's program.

Because medical school is long, expensive, and competitive, there are not enough medical doctors to keep up with the growing need for medical care. PAs fill an important need within the health care system.

Concept Summary

15.1 The Basics of Nutrition

Learning Objective: **Describe the primary functions in the body of carbohydrates, lipids, proteins, vitamins, and minerals.**

- Every human body requires certain amounts of various macronutrients and micronutrients, water, and fiber to function properly.
- The amounts needed vary with a number of factors such as body size, age, and sex.
- Various countries have established nutritional guidelines in attempts to maintain good health for their citizens.
- In the United States, the Reference Daily Intakes (RDIs) and Daily Reference Values (DRVs) are designed to represent appropriate nutrient intake for 95% of the population.
- Macronutrients are substances required by the body in relatively large amounts.
- They are used by the body to provide energy and the materials necessary to form new or replacement tissue.
- The macronutrients are carbohydrates, lipids, and proteins.
- An RDI has been established only for proteins, but other groups have recommended the amounts of carbohydrates, lipids, and proteins that should be included in the diet.
- Vitamins are organic micronutrients that the body cannot produce in the amounts needed for good health.
- A number of vitamins are highly water soluble because they are so highly polar.
- All but one of the water-soluble vitamins are known to function as coenzymes in the body and are involved in many important metabolic processes.
- Fat-soluble vitamins have nonpolar molecular structures. As a result, they are insoluble in water but soluble in fat or other nonpolar solvents.
- Fat-soluble vitamins act somewhat like hormones in the body.
- Fat-soluble vitamins do not dissolve in water-based body fluids, so they are not excreted through the kidneys.
- Amounts in excess of bodily requirements are stored in body fat.
- Thus, it is much easier to produce toxic effects by overdosing with fat-soluble vitamins than with water-soluble vitamins.
- Minerals are inorganic substances that perform many useful functions in the body.
- Those present in the body in amounts equal to or greater than 5 grams are called major minerals.
- Those present in smaller amounts are called trace minerals.

15.2 An Overview of Metabolism

Learning Objectives: **Define metabolism, anabolism, and catabolism. Outline the three stages in the extraction of energy from food.**

- Metabolism is the sum of all cellular reactions. It involves the breakdown (catabolism) and synthesis (anabolism) of molecules.
- There are three stages in the catabolism of foods to provide energy.
- In Stage 1, digestion converts foods into smaller molecules.
- In Stage 2, these smaller molecules are converted into two-carbon acetyl units that combine with coenzyme A, forming acetyl CoA.
- Stage 3 is called the common catabolic pathway. It consists of the citric acid cycle followed by oxidative phosphorylation (electron transport coupled to chemiosmosis).
- The main function of catabolism is to produce ATP molecules.

15.3 ATP: The Energy Currency of Cells

Learning Objective: **Explain how ATP plays a central role in the production and use of cellular energy.**

- Molecular ATP is the link between energy production and energy use in cells.
- ATP is called a high-energy compound because of the large amount of free energy liberated on hydrolysis.
- Cells are able to harness the free energy liberated by ATP to carry out cellular work.
- ATP is consumed immediately following its formation and is quickly regenerated from ADP as fuel molecules are oxidized.
- Mitochondria play a key role in energy production because they house the enzymes for both the citric acid cycle and oxidative phosphorylation.

15.4 Carbohydrate Metabolism and Blood Glucose

Learning Objectives: **Identify the monosaccharide products of the digestion of carbohydrates. Explain the importance to the body of maintaining proper blood sugar levels.**

- Glucose is the key food molecule for most organisms.
- It is the central substance in carbohydrate metabolism.
- During digestion, carbohydrates are hydrolyzed to the monosaccharides glucose, fructose, and galactose.
- These are absorbed into the bloodstream through the lining of the small intestine.
- Glucose is the most abundant carbohydrate in the blood.
- The liver regulates blood glucose levels so that a sufficient concentration is always available to meet the body's energy needs.
- A lower-than-normal blood glucose level is referred to as hypoglycemia.
- A higher-than-normal blood glucose level is called hyperglycemia.

15.5 An Overview of Cellular Respiration

Learning Objectives: Describe the process and cellular regulation of glycolysis and the citric acid cycle. Describe in detail the process of oxidative phosphorylation and how it results in ATP production.

- Glycolysis is a series of 10 reactions that occur in the cytoplasm of cells.
- In glycolysis, each glucose molecule is converted into two molecules of pyruvate.
- Two molecules of ATP and two molecules of NADH are produced in glycolysis.
- Both galactose and fructose are converted into intermediates that enter the glycolysis pathway.
- Under aerobic conditions, pyruvate enters mitochondria and is converted to acetyl CoA via pyruvate oxidation.
- During the oxidation of pyruvate to acetyl CoA, NAD^+ is regenerated so that glycolysis can continue.
- The acetyl CoA enters the citric acid cycle.
- The citric acid cycle consists of eight reactions that process incoming molecules of acetyl CoA.
- The carbon atoms leave the citric acid cycle as molecules of CO_2.
- The hydrogen atoms and electrons leave the cycle in the form of reduced coenzymes NADH and $FADH_2$, which are later oxidized in the electron transport chain.
- The cycle is regulated by three allosteric enzymes in response to cellular levels of ATP.
- One acetyl CoA molecule entering the citric acid cycle produces three molecules of NADH, one of $FADH_2$, and one of GTP.
- The citric acid cycle also produces intermediates used for biosynthesis.
- The electron transport chain involves a series of reactions that pass electrons from NADH and $FADH_2$ to molecular oxygen.
- During this oxidation process, protein complexes that are part of the electron transport chain are pumping protons against their concentration gradients from the mitochondrial matrix to the intermembrane space.
- Each carrier in the series has an increasing affinity for electrons.
- Four of the carriers, which are referred to as cytochromes, contain iron, which accepts and then transfers the electrons.
- As NADH and $FADH_2$ release their hydrogen atoms and electrons, NAD^+ and FAD are regenerated for return to the citric acid cycle.
- The synthesis of ATP takes place because of a flow of protons down their concentration gradient across the inner mitochondrial membrane through ATP synthase in a process called chemiosmosis.
- One molecule of NADH produces 2.5 molecules of ATP.
- $FADH_2$, which enters the chain one step later, produces 1.5 molecules of ATP.
- The complete oxidation of 1 mol of glucose produces 32 (or 30) mol of ATP.

15.6 Glycogen Metabolism and Gluconeogenesis

Learning Objectives: Explain the importance of glycogenesis, glycogenolysis, and gluconeogenesis. Describe how hormones regulate carbohydrate metabolism.

- When blood glucose levels are high, excess glucose is converted into glycogen by the process of glycogenesis.
- The glycogen is stored within the liver and muscle tissue.
- Glycogenolysis is the breakdown of glycogen into glucose.
- It occurs when muscles need energy and when the liver is restoring a low blood sugar level to normal.
- Lactate, certain amino acids, and glycerol can be converted into glucose in a series of 11 chemical reactions that opposed glycolysis.
- This process, called gluconeogenesis, takes place in the liver when glycogen supplies are being depleted and when carbohydrate intake is low.
- Gluconeogenesis from lactate is especially important during periods of high muscle activity.
- The liver converts excess lactate from the muscles into glucose, which is then cycled back to the muscles. This process is called the Cori cycle.

15.7 Lipid Metabolism and Biosynthesis

Learning Objectives: Describe the metabolic pathways by which glycerol and fatty acids are catabolized. Name the three ketone bodies. List the conditions that cause the overproduction of ketone bodies. Describe the pathway for fatty acid synthesis.

- When stored fats are needed for energy, epinephrine stimulates the hydrolysis of triglycerides to fatty acids and glycerol.
- These component substances of fat then enter the bloodstream and travel to tissues, where they are utilized.
- The glycerol available from fat mobilization is first phosphorylated and then oxidized to dihydroxyacetone phosphate (DHAP), an intermediate of glycolysis.
- By having an entry point into the glycolysis pathway, glycerol can ultimately be converted into glucose or oxidized to CO_2 and H_2O.
- The catabolism of fatty acids begins in the cytoplasm, where they are activated by combining with CoA–SH.
- After being transported into mitochondria, the degradation of fatty acids occurs by the β-oxidation pathway.
- The β-oxidation pathway produces reduced coenzymes ($FADH_2$ and NADH) and cleaves the fatty acid chain into two-carbon fragments bound to coenzyme A (acetyl CoA).
- The energy available from a fatty acid for ATP synthesis can be calculated by determining the amount of acetyl CoA, $FADH_2$, and NADH produced.
- On an equal-mass basis, fatty acids contain more than twice the energy of glucose.
- Acetoacetate, β-hydroxybutyrate, and acetone are known as ketone bodies.
- They are synthesized in the liver from acetyl CoA.

- During starvation, in unchecked diabetes, and after excessive alcohol consumption, the level of ketone bodies becomes very high, leading to ketonemia, ketonuria, acetone breath, and ketosis.
- Fatty acid synthesis occurs in the cytoplasm and uses acetyl CoA as a starting material.
- Two-carbon fragments are added to growing fatty acid molecules as energy is supplied by ATP and NADPH.

15.8 Amino Acid Metabolism and Biosynthesis

Learning Objectives: **Describe the source and function of the body's amino acid pool. Explain the chemical fate of the nitrogen atoms and carbon skeleton of amino acids. Explain the relationship between intermediates of carbohydrate metabolism and the synthesis of nonessential amino acids.**

- Amino acids and proteins are not stored within the body.
- Amino acids absorbed from digested food or synthesized in the body become part of a temporary supply or pool that may be used in metabolic processes.
- The turnover, or life expectancy, of proteins is usually expressed in half-lives.
- Protein half-lives range from several minutes to hundreds of days.
- Some amino acids in the body are converted into relatively simple but vital nitrogen-containing compounds.
- The catabolism of amino acids produces intermediates used for energy production, the synthesis of glucose, or the formation of triglycerides.
- As an amino acid undergoes catabolism, the amino group may be transferred to an α-keto acid through a process called transamination.
- In the process of deamination, an amino acid is converted into α-keto acid and ammonia.
- The urea cycle converts toxic ammonium ions into urea for excretion.
- Amino acids are classified as glucogenic or ketogenic based on their catabolic products.
- Glucogenic amino acids are degraded into pyruvate or intermediates of the citric acid cycle and can be used for glucose synthesis.
- Ketogenic amino acids are degraded into acetoacetyl CoA and acetyl CoA for energy production and triglyceride synthesis.
- The body can synthesize 11 of the 20 common amino acids, known as nonessential amino acids.
- The key starting materials are intermediates of the glycolysis pathway and the citric acid cycle and, in the case of tyrosine, phenylalanine.

Key Terms and Concepts

Acetone breath (15.7)
Acidosis (15.7)
Adipose tissue (15.7)
Aerobic (15.5)
Amino acid pool (15.8)
Anabolism (15.2)
Blood sugar level (15.4)
β-Oxidation process (15.7)
Catabolism (15.2)
Cellular respiration (15.5)
Change in Gibbs free energy (ΔG) (15.3)
Chemiosmosis (15.2)
Citric acid cycle (15.5)
Common catabolic pathway (15.2)
Complete protein (15.1)
Complex carbohydrates (15.1)
Cori cycle (15.6)
Cytochrome (15.5)
Daily Reference Values (DRVs) (15.1)
Daily Values (DVs) (15.1)
Electron transport chain (15.5)
Essential amino acid (15.1)
Fat mobilization (15.7)
Glucogenic amino acid (15.8)
Gluconeogenesis (15.6)
Glucosuria (15.4)
Glycogenesis (15.6)
Glycogenolysis (15.6)
Glycolysis (15.5)
Hyperglycemia (15.4)
Hypoglycemia (15.4)
Ketoacidosis (15.7)
Ketogenic amino acid (15.8)
Ketone bodies (15.7)
Ketonemia (15.7)
Ketonuria (15.7)
Ketosis (15.7)
Macronutrient (15.1)
Major mineral (15.1)
Metabolic pathway (15.2)
Metabolism (15.2)
Micronutrient (15.1)
Mineral (15.1)
Mitochondrion (15.3)
Nonessential amino acid (15.8)
Nutrition (Introduction)
Oxidative deamination (15.8)
Oxidative phosphorylation (15.5)
Protein turnover (15.8)
Reference Daily Intakes (RDIs) (15.1)
Renal threshold (15.4)
Simple carbohydrates (15.1)
Trace mineral (15.1)
Transaminase (15.8)
Transamination (15.8)
Urea cycle (15.8)

Key Equations

1. Hydrolysis of ATP to ADP (**Section 15.3**)	$ATP + H_2O \rightarrow ADP + P_i + H^+$	Equation 15.1
2. Cellular respiration (**Section 15.5**)	$\text{glucose} + O_2 \longrightarrow CO_2 + H_2O + \text{energy}$	Equation 15.2
3. Reduction of NAD^+ (**Section 15.5**)	$NAD^+ + 2e^- + H^+ \rightleftarrows NADH$	Equation 15.3
4. Glycolysis (**Section 15.5**)	$\text{glucose} + 2P_i + 2ADP + 2NAD^+ \longrightarrow$ $2\text{ pyruvate} + 2ATP + 2NADH + 4H^+ + 2H_2O$	Equation 15.4
5. Oxidation of pyruvate to acetyl CoA (**Section 15.5**) $CH_3{-}\overset{O}{\overset{\|}{C}}{-}COO^- + CoA{-}S{-}H + NAD^+ \longrightarrow CH_3{-}\overset{O}{\overset{\|}{C}}{-}S{-}CoA + NADH + CO_2$		Equation 15.5
6. Reduction of FAD (**Section 15.5**)	$FAD + 2e^- + H^+ \rightleftarrows FADH_2$	Equation 15.6
7. The citric acid cycle (**Section 15.5**) $\text{acetyl CoA} + 3NAD^+ + FAD + GDP + P_i + 2H_2O \longrightarrow 2CO_2 + CoA{-}S{-}H + 3NADH + 2H^+ + FADH_2 + GTP$		Equation 15.7
8. The electron transport chain (**Section 15.5**)	$4H^+ + 4e^- + O_2 \longrightarrow 2H_2O$	Equation 15.8
9. Oxidative phosphorylation (**Section 15.5**)	$ADP + P_i \longrightarrow ATP$	Equation 15.9
10. Complete oxidation of glucose (**Section 15.5**)	$C_6H_{12}O_6 + 6O_2 + 32ADP + 32P_i \longrightarrow 6CO_2 + 32ATP + 38H_2O$	Equation 15.10
11. Glycogenesis (**Section 15.6**)	$(\text{glucose})_n \xrightarrow[\text{UTP} \quad \text{UDP} + P_i]{} (\text{glucose})_{n+1} = \text{glycogen}$	Equation 15.11
12. Glycogenolysis (**Section 15.6**)	$\text{glycogen} \xrightarrow{P_i} \text{glucose-1-phosphate} \rightleftarrows \text{glucose-6-phosphate} \xrightarrow{H_2O \quad P_i} \text{glucose}$	Equation 15.12
13. Gluconeogenesis (**Section 15.6**)	$2\text{ pyruvate} + 4ATP + 2GTP + 2NADH + 2H^+ + 6H_2O$ $\longrightarrow \text{glucose} + 6P_i + 4ADP + 2GDP + 2NAD^+$	Equation 15.13
14. Conversion of lactate to pyruvate (**Section 15.6**)	$\underset{\text{lactate}}{CH_3{-}\overset{OH}{\overset{\|}{CH}}{-}COO^-} + NAD^+ \rightleftarrows \underset{\text{pyruvate}}{CH_3{-}\overset{O}{\overset{\|}{C}}{-}COO^-} + NADH + H^+$	Equation 15.14
15. Breakdown of glycerol to DHP (**Section 15.7**) glycerol ($H_2C{-}OH$ / $HC{-}OH$ / $H_2C{-}OH$) $\xrightarrow[\text{ATP} \quad \text{ADP}]{}$ glycerol-3-phosphate ($H_2C{-}OH$ / $HC{-}OH$ / $H_2C{-}OPO_3^{2-}$) $\xrightarrow[NAD^+ \quad NADH + H^+]{}$ dihydroxyacetone phosphate (DHP) ($H_2C{-}OH$ / $C{=}O$ / $H_2C{-}OPO_3^{2-}$)		Equation 15.15

Reaction	Equation	
16. Activation of fatty acids (**Section 15.7**)	$\underset{\text{fatty acid}}{R-\overset{O}{\overset{\|}{C}}-OH} + HS-CoA \xrightarrow[\text{ATP} \;\to\; \text{AMP} + PP_i]{} \underset{\text{fatty acyl CoA}}{R-\overset{O}{\overset{\|}{C}}-S-CoA} + H_2O$	Equation 15.16
17. β-Oxidation (**Section 15.7**)	$\underset{\text{fatty acyl CoA}}{R-\underset{H}{\overset{H}{\overset{\beta}{C}}}-\overset{\alpha}{CH_2}-\overset{O}{\overset{\|}{C}}-S-CoA} \xrightarrow{\text{oxidation}} \underset{\text{ketone}}{R-\overset{O}{\overset{\|}{\overset{\beta}{C}}}-\overset{\alpha}{CH_2}-\overset{O}{\overset{\|}{C}}-S-CoA}$	Equation 15.17
18. Release of acetyl unit during β-oxidation process (**Section 15.7**)	$\underset{\text{ketone}}{R-\overset{O}{\overset{\|}{\overset{\beta}{C}}}-\overset{\alpha}{CH_2}-\overset{O}{\overset{\|}{C}}-S-CoA} \xrightarrow{CoA-SH} R-\overset{O}{\overset{\|}{C}}-S-CoA + \underset{\text{acetyl CoA}}{CH_3-\overset{O}{\overset{\|}{C}}-S-CoA}$	Equation 15.18
19. Oxidation of fatty acids (**Section 15.7**)	$R-CH_2CH_2-\overset{O}{\overset{\|}{C}}-S-CoA + NAD^+ + FAD + H_2O + CoA-SH \longrightarrow R-\overset{O}{\overset{\|}{C}}-S-CoA + CH_3\overset{O}{\overset{\|}{C}}-S-CoA + NADH + H^+ + FADH_2$	Equation 15.19
20. Fatty acid synthesis (**Section 15.7**)	$8CH_3-\overset{O}{\overset{\|}{C}}-S-CoA + 14NADPH + 13H^+ + 7ATP \longrightarrow CH_3(CH_2)_{14}COO^- + 8CoA-SH + 6H_2O + 14NADP^+ + 7ADP + 7P_i$	Equation 15.20
21. Transamination (**Section 15.8**)	$R-\overset{NH_3^+}{\overset{\vert}{CH}}-COO^- + {}^-OOC-CH_2CH_2-\overset{O}{\overset{\|}{C}}-COO^- \rightleftharpoons R-\overset{O}{\overset{\|}{C}}-COO^- + {}^-OOC-CH_2CH_2-\overset{NH_3^+}{\overset{\vert}{CH}}-COO^-$	Equation 15.21
22. Oxidative deamination (**Section 15.8**)	${}^-OOC-CH_2CH_2-\overset{NH_3^+}{\overset{\vert}{CH}}-COO^- + NAD^+ + H_2O \rightleftharpoons NH_4^+ + {}^-OOC-CH_2CH_2-\overset{O}{\overset{\|}{C}}-COO^- + NADH + H^+$	Equation 15.22
23. Urea formation (**Section 15.8**)	$NH_4^+ + HCO_3^- + 3ATP + 2H_2O + \underset{\text{aspartate}}{COO^--CH_2-CH(NH_3^+)-COO^-} \longrightarrow \underset{\text{urea}}{H_2N-\overset{O}{\overset{\|}{C}}-NH_2} + \underset{\text{fumarate}}{(H)({}^-OOC)C=C(COO^-)(H)} + 2ADP + 2P_i + AMP + PP_i$	Equation 15.23

Exercises

Even-numbered exercises are answered in Appendix B.

Exercises with an asterisk (*) are more challenging.

The Basics of Nutrition (Section 15.1)

15.1 What is the principal component of dietary fiber?

15.2 What is the primary difference between a macronutrient and a micronutrient?

15.3 Calculate how many mg of sodium per day are recommended in a 2000-cal/day diet using the "Amount per serving" and "% Daily Value" information supplied in **Figure 15.1**.

15.4 Using your answer to Exercise 15.3 calculate the number of teaspoons of salt that would be needed to fulfill the daily requirements if one level teaspoon of salt contains 4250 mg of sodium.

15.5 Calculate how many grams of total carbohydrate are recommended in a 2000-cal/day diet using the "Amount per serving" and "% Daily Value" information supplied in **Figure 15.1**.

15.6 Using your answer to Exercise 15.5, calculate the number of average-sized (150 g) potatoes you would have to eat to fulfill the daily requirements if one potato contains 26 g of total carbohydrate.

15.7 Calculate how many grams of dietary fiber are recommended in a 2000-cal/day diet using the "Amount per serving" and "% Daily Value" information supplied in **Figure 15.1**.

15.8 Using your answer to Exercise 15.7, calculate the number of average-sized (150 g) potatoes you would have to eat to fulfill the daily requirements if one potato contains 2 g of dietary fiber.

15.9 Explain the importance of sufficient fiber in the diet.

15.10 List two general functions in the body for each of the macronutrients.

15.11 For each of the following macronutrients, list a single food item that would be a good source.

a. carbohydrate (simple)

b. carbohydrate (complex)

c. lipid

d. protein

15.12 Identify the class of macronutrient (carbohydrates, lipids, proteins, or more than one class) being characterized by each of the following.

a. Can be defined as simple or complex

b. Only macronutrient with RDIs for different groups of people

c. Are digested more slowly than other macronutrients

d. Are triglycerides

15.13 Identify the class of macronutrient (carbohydrates, lipids, proteins, or more than one class) being characterized by each of the following.

a. Can be defined as saturated or unsaturated

b. Are natural polymers of amino acids

c. Has components that are considered essential

d. Includes the category called "fiber"

15.14 Identify the class of macronutrient (carbohydrate, lipid, or protein) to which each of the following belongs.

a. $H_3\overset{+}{N}-CH(CH_2C_6H_5)-COO^-$

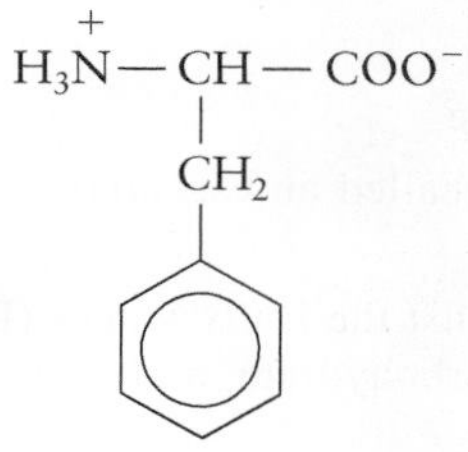

b. (disaccharide structure: two CH_2OH pyranose rings joined by O, with OH groups)

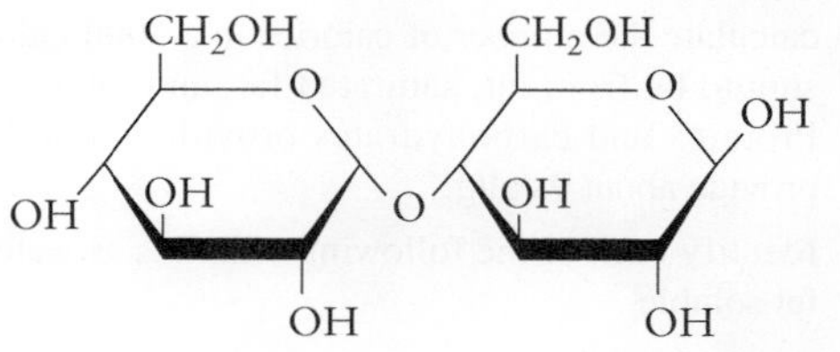

c. $HO-\overset{O}{\overset{\|}{C}}-(CH_2)_7CH{=}CHCH_2CH{=}CH(CH_2)_4CH_3$

15.15 Identify the class of macronutrient (carbohydrate, lipid, or protein) to which each of the following belongs.

a.

$$\begin{array}{l} CH_2-O-\overset{O}{\overset{\|}{C}}-(CH_2)_7CH{=}CH-(CH_2)_7CH_3 \\ | \\ CH-O-\overset{O}{\overset{\|}{C}}-(CH_2)_3CH{=}CH(CH_2)_3CH_3 \\ | \\ CH_2-O-\overset{O}{\overset{\|}{C}}-CH{=}CHCH_2CH{=}CHCH_3 \end{array}$$

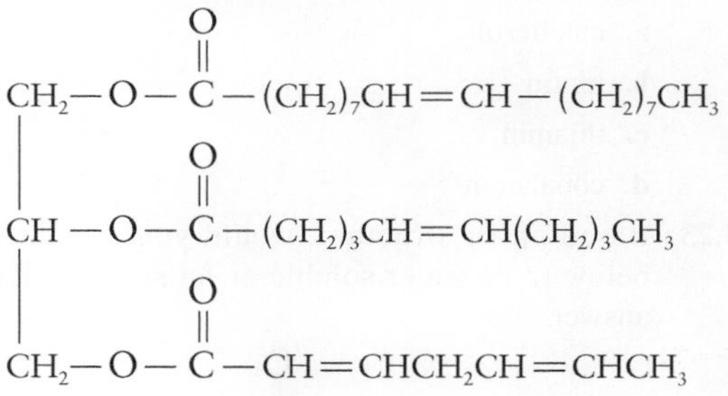

b. Ala — Trp — Arg — Ser — Cys — Gln

c. (monosaccharide ring structure: CH_2OH, O, HO, OH, OH, OH)

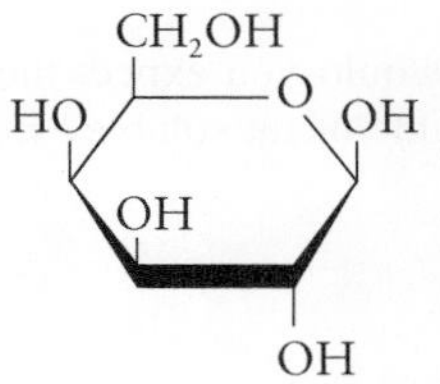

15.16 List the types of macronutrients found in each of the following food items. List the nutrients in approximate decreasing order of abundance (most abundant first, etc.).

a. Potato chips

b. Buttered toast

c. Plain toast with jam

d. Cheese sandwich

15.17 List the types of macronutrients found in each of the following food items. List the nutrients in approximate decreasing order of abundance (most abundant first, etc.).

a. A lean steak

b. A fried egg

c. Boiled broccoli

d. Chicken noodle soup

15.18 Why is linoleic acid called an essential fatty acid for humans?

15.19 Use **Figure 15.1** and list the Daily Values (DV) for fat, saturated fat, total carbohydrate, and fiber based on a 2000-calorie diet.

15.20* Use the Daily Values (DVs) given in **Figure 15.1** to calculate the number of calories in a 2000-calorie diet that should be from fat, saturated fat, and total carbohydrate. Proteins and carbohydrates provide about 4 cal/g. Fats provide about 9 cal/g.

15.21 Identify each of the following vitamins as water soluble or fat soluble.

a. tocopherol

b. niacin

c. folic acid

d. retinol

15.22 Identify each of the following vitamins as water soluble or fat soluble.

a. calciferol

b. biotin

c. thiamin

d. cobalamin

15.23 Based on its structure, would you expect the vitamin below to be water soluble or fat soluble? Explain your answer.

O
||
C—OH
N

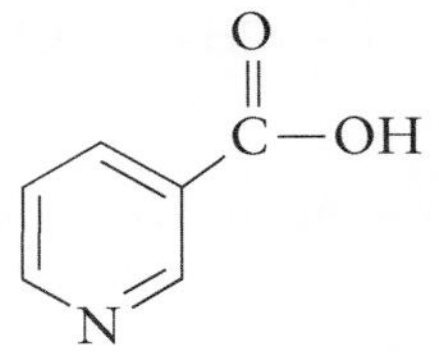

15.24 Based on its structure, would you expect the following vitamin to be water soluble or fat soluble? Explain your answer.

CH_3
CH_3
$CH{=}CH{-}\underset{CH_3}{C}{=}CH{-}CH{=}CH{-}\underset{CH_3}{C}{=}CH{-}CH_2{-}OH$
CH_3

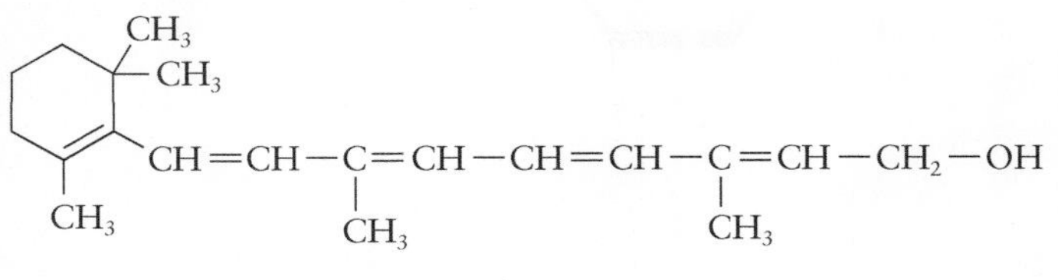

15.25 What general function is served by eight of the water-soluble vitamins?

15.26 Why is there more concern about large doses of fat-soluble vitamins than of water-soluble vitamins?

15.27 What fat-soluble vitamin deficiency is associated with the following?

a. Night blindness

b. Blood clotting

c. Calcium and phosphorus use in forming bones and teeth

d. Preventing oxidation of fatty acids

15.28 What water-soluble vitamin deficiency is associated with the following?

a. Scurvy

b. Beriberi

c. Pernicious anemia

d. Pellagra

15.29 What is the difference between a vitamin and a mineral?

15.30 What determines whether a mineral is classified as a major or trace mineral?

15.31 What are the general functions of trace minerals in the body?

An Overview of Metabolism (Section 15.2)

15.32 What is a metabolic pathway?

15.33 Classify catabolism and anabolism as synthetic or breakdown processes.

15.34 Classify catabolism and anabolism as energy-producing or energy-consuming processes.

15.35 Outline the three stages in the extraction of energy from food.

15.36 Which stage of energy production is concerned primarily with the following?

a. Formation of ATP

b. Digestion of fuel molecules

c. Consumption of O_2

d. Generation of acetyl CoA

15.37 Which stage of energy production is concerned primarily with the following?

a. Produces carbon dioxide

b. Macronutrient components degraded into two-carbon units

c. Uses oxygen to make water

d. Hydrolysis of macronutrients

15.38 In terms of energy production, what is the main purpose of the catabolic pathway?

ATP: The Energy Currency of Cells (Section 15.3)

15.39 Is ATP involved in anabolic processes or catabolic processes?

15.40 What do the abbreviations ATP, ADP, and AMP represent?

15.41 Using the partial structure provided, write the structure of ATP, ADP, and AMP.

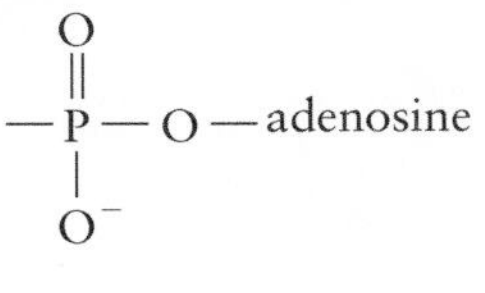

O
||
—P—O—adenosine
|
O^-

15.42 What does the symbol $\Delta G^{\circ\prime}$ represent? How does the sign for $\Delta G^{\circ\prime}$ indicate whether a reaction is exergonic or endergonic?

15.43 Which portion of the ATP molecule is particularly responsible for it being described as a high-energy compound?

15.44 In general, would you expect an anabolic process to have an overall $+\Delta G^{\circ\prime}$ or a $-\Delta G^{\circ\prime}$ value? What about a catabolic process? Explain your answers.

15.45 What is the role of mitochondria in the use of energy by living organisms?

Carbohydrate Metabolism and Blood Glucose (Section 15.4)

15.46* Name the products produced by the digestion of the following.

a. starch

b. lactose

c. sucrose

d. maltose

15.47 What type of reaction is involved as carbohydrates undergo digestion?

15.48 Describe what is meant by the terms *blood sugar level* and *normal fasting level.*

15.49 What range of concentrations for glucose in blood is considered a normal fasting level?

15.50 How do each of the following terms relate to blood sugar level?

a. Hypoglycemia

b. Hyperglycemia

c. Renal threshold

d. Glucosuria

15.51 Convert a normal blood sugar level of 100.0 mg/100.0 mL into molarity. One mol of glucose is 180.2 g.

15.52 Insulin is a hormone that acts like a key, opening doors on cell membranes that allows glucose to enter the cell so it can be metabolized. Would insulin be used to treat hyperglycemia or hypoglycemia? Explain your answer.

15.53 Explain how the synthesis and degradation of glycogen in the liver help regulate blood sugar levels.

An Overview of Cellular Respiration (Section 15.5)

15.54 What is the main purpose of glycolysis?

15.55 What is the starting material for glycolysis? The product?

15.56 In what part of the cell does glycolysis take place?

15.57 How many steps in glycolysis require ATP? How many steps produce ATP? What is the net production of ATP from one glucose molecule undergoing glycolysis?

15.58 Which coenzyme serves as an oxidizing agent in glycolysis?

15.59 Explain what is meant by feedback inhibition.

15.60 Explain how a high concentration of glucose-6-phosphate causes a decrease in the rate of glycolysis.

15.61 Explain how fructose and galactose can enter the glycolysis pathway.

15.62 Describe the cellular location and the enzyme that facilitate the production of acetyl CoA from pyruvate.

15.63 What is the primary function of the citric acid cycle in ATP production? What other vital role is served by the citric acid cycle?

15.64 Describe the citric acid cycle by identifying the following.

a. The fuel needed by the cycle

b. The form in which carbon atoms leave the cycle

c. The form in which hydrogen atoms and electrons leave the cycle

15.65 Although oxygen (O_2) is not used directly in the citric acid cycle, the cycle cannot operate under anaerobic conditions. Explain why.

15.66 The reactions in the citric acid cycle involve enzymes.

a. Identify the enzymes at the three control points.

b. Each control-point enzyme is inhibited by NADH and ATP. Explain how this enables the cell to be responsive to energy needs.

c. Two of the control-point enzymes are activated by ADP. Explain the benefit of this regulation to the cell.

15.67 What is the primary function of the electron transport chain?

15.68 Identify the cellular location for each of the following.

a. glycolysis

b. pyruvate oxidation

c. Citric acid cycle

d. Electron transport chain

15.69 Which coenzymes bring electrons to the electron transport chain?

15.70 What is the role of the cytochromes in the electron transport chain?

15.71 What is the role of oxygen in the functioning of the electron transport chain? Write a reaction that summarizes the fate of the oxygen.

15.72 Iron deficiency can reduce the number of active cytochrome centers in the electron transport chain. Use this information to explain why fatigue is a common symptom of iron deficiency.

15.73 According to the chemiosmosis, ATP synthesis results from the flow of ________ particles, across the _____ membrane through the ________ protein.

15.74 How many total moles of ATP can be produced by the complete oxidation of 1 mol of glucose in the liver?

Glycogen Metabolism and Gluconeogenesis (Section 15.6)

15.75 Compare the terms *glycogenesis* and *glycogenolysis*.

15.76 Fructose and galactose from digested food enter the glycolysis pathway. Name the intermediates of the pathway that serve as entry points for these substances.

15.77 What high-energy compound is involved in the conversion of glucose to glycogen?

15.78 Why is it important that the liver store glycogen? How does this aid in controlling blood sugar levels?

15.79 How does glycogen enter the glycolysis pathway?

15.80 What organ serves as the principal site for gluconeogenesis?

15.81 A friend has been fasting for three days. Where does the energy come from to keep his brain functioning?

15.82 What are the sources of compounds that undergo gluconeogenesis?

15.83 Explain what would happen to gluconeogenesis by the liver if your friend went on a diet totally free of carbohydrates, including all starches and sugars.

15.84 Explain the operation of the Cori cycle and its importance to muscles.

15.85 What cells produce insulin and glucagon in the body?

15.86 Describe the influence of glucagon and insulin on the following.

a. Blood sugar level

b. Glycogen formation

Lipid Metabolism and Biosynthesis (Section 15.7)

15.87 Describe the relationship between epinephrine and fat mobilization.

15.88 Describe what is meant by the term *fat mobilization*.

15.89 Which cells utilize fatty acids over glucose to satisfy their energy needs?

15.90 At what point does glycerol enter the glycolysis pathway?

15.91 In what cellular location is glycerol converted to dihydroxyacetone phosphate?

15.92 Describe the two fates of glycerol after it has been converted to an intermediate of glycolysis.

15.93 Where in the cell does fatty acid catabolism take place?

15.94 How is the fatty acid prepared for catabolism? Where in the cell does fatty acid activation take place?

15.95 Label the α- and β-positions in the following acids.

a. $CH_3CH_2CH_2CH_2CH_2-\overset{\overset{\Large O}{\|}}{C}-OH$

b. $H_3(CH_2)_6\overset{\overset{\Large OH}{|}}{CH}-CH_2-\overset{\overset{\Large O}{\|}}{C}-OH$

15.96 List the three substances known as ketone bodies.

15.97 Explain how excessive ketone bodies may form in the following.

a. During starvation

b. In patients with diabetes mellitus

15.98 Where are ketone bodies formed? What tissues utilize ketone bodies to meet energy needs?

15.99 Why does acetyl CoA build up in a low-carbohydrate diet?

15.100 Explain how ketoacidosis can result in dehydration.

15.101 Which ketone body has the highest blood level during starvation?

15.102 Why is sodium bicarbonate sometimes given to patients experiencing ketoacidosis?

15.103 Differentiate between (a) ketonemia, (b) ketonuria, (c) acetone breath, and (d) ketosis.

15.104 What substance supplies the carbons for fatty acid synthesis?

15.105 What substances furnish the energy for fatty acid synthesis?

15.106 How does citrate move from the mitochondria to the cytoplasm?

15.107 Why can the liver convert glucose to fatty acids but cannot convert fatty acids to glucose?

15.108 Which fatty acids are considered essential to the diet? Explain why.

Amino Acid Metabolism and Biosynthesis (Section 15.8)

15.109 List three vital functions served by amino acids in the body.

15.110 What are the sources of amino acids in the pool?

15.111 What happens to the nitrogen atoms in degraded amino acids?

15.112 Name four important biomolecules other than proteins that are synthesized from amino acids.

15.113 Match the terms *transamination* and *deamination* to the following.

a. An amino group is removed from an amino acid and donated to a keto acid.

b. An ammonium ion is produced.

c. New amino acids are synthesized from other amino acids.

d. A keto acid is produced from an amino acid.

15.114 In what organ of the body is urea synthesized?

15.115 What are the sources of the carbon atom and two nitrogen atoms in urea?

15.116 Write a summary equation of the formation of urea.

15.117 Differentiate between glucogenic and ketogenic amino acids.

15.118 Is any amino acid both glucogenic and ketogenic?

15.119 Identify the five locations at which the carbon skeletons of amino acids enter the citric acid cycle.

15.120 What enzyme is responsible for PKU?

15.121 Based on the following conversions, classify each amino acid as glucogenic or ketogenic.

a. aspartate $\rightarrow$ oxaloacetate

b. leucine $\rightarrow$ acetyl CoA

c. tyrosine $\rightarrow$ acetoacetyl CoA

15.122 What special role is played by glucogenic amino acids during fasting or starvation?

15.123 Differentiate between essential and nonessential amino acids.

15.124 What essential amino acid makes tyrosine nonessential?

15.125 Where do the starting materials for nonessential amino acids come from?

15.126 What amino acids are synthesized from glutamate?

Appendix A
The International System of Units

The International System of Units (SI units) was established in 1960 by the International Bureau of Weights and Measures. The system was established in an attempt to streamline the metric system, which included certain traditional units that had historical origins but that were not logically related to other metric units. The International System established fundamental units to represent seven basic physical quantities. These quantities and the fundamental units used to express them are given in Table A.1.

Table A.1 Fundamental SI Units

Physical Quantity	SI Unit	Abbreviation
Length	Meter	m
Mass	Kilogram	kg
Temperature	Kelvin	K
Amount of substance	Mole	mol
Electrical current	Ampere	A
Time	Second	s
Luminous intensity	Candela	cd

All other SI units are derived from the seven fundamental units. Prefixes are used to indicate multiples or fractions of the fundamental units (Table 1.2). In some cases, it has been found to be convenient to give derived units specific names. For example, the derived unit for force is kg m/s^2, which has been given the name newton (abbreviated N) in honor of Sir Isaac Newton (1642–1727). Some examples of derived units are given in Table A.2.

Table A.2 Examples of Derived SI Units

Physical Quantity	Definition in Fundamental Units	Specific Name	Abbreviation
Volume	m^3 (see NOTE below)	—	—
Force	kg m/s^2	newton	N
Energy	kg m^2/s^2 = N m	joule	J
Power	kg m^2/s^3 = J/s	watt	W
Pressure	kg/m s^2 = N/m^2	pascal	Pa
Electrical charge	A s	coulomb	C
Electrical potential	kg m^2/A s^3 = W/A	volt	V
Electrical resistance	kg m^2/A^2 s^3 = V/A	ohm	Ω
Frequency	1/s	hertz	Hz

NOTE: The liter (L), a popular volume unit of the metric system, has been redefined in terms of SI units as 1 dm^3 or 1/1000 m^3. It and its fractions (mL, μL, etc.) are still widely used to express volume.

Appendix B

Answers to Even-Numbered End-of-Chapter Exercises

Chapter 1
What Is Matter? (Section 1.1)

1.2 All matter occupies space and has mass. Mass is a measurement of the amount of matter in an object. Weight is a measurement of the gravitational force acting on an object.

1.4 The distance you can throw a bowling ball **(a)** will change more than the distance you can roll a bowling ball on a flat, smooth surface. The ball rolled on the ground is slowed by friction between Earth and the ball itself. Also the gravitational pull increases the closer an object is to the center of Earth, so the ball on the ground has a slightly stronger gravitational force exerted on it when compared to the ball in the air.

1.6 The attractive force of gravity for objects near Earth's surface increases as you get closer to the center of Earth (Exercise 1.5). If two people with the same mass were weighed at the equator and at the North Pole, the person at the equator would weigh less than the person at the North Pole.

Physical and Chemical Properties and Changes (Section 1.2)

1.8 **a.** This change did not involve composition; therefore, it is a **physical change**.

b. This change involves composition; therefore, it is a **chemical change**.

c. This change did not involve composition; therefore, it is a **physical change**.

d. This change involves composition; therefore, it is a **chemical change**.

1.10 **a.** The phase of matter at room temperature is a **physical property** because the composition does not change while making this observation.

b. The reaction between two substances is a **chemical property** because the composition of the products differs from the reactants.

c. Freezing point is a **physical property** because the composition does not change while making this observation.

d. The inability of a material to form new products by rusting is a **chemical property** because rust would have a different chemical composition than gold.

e. The color of a substance is a **physical property** because the composition does not change while making this observation.

Classifying Matter (Section 1.3)

1.12 Carbon dioxide is **heteroatomic**. If oxygen and carbon atoms react to form one product, then carbon dioxide must contain these two types of atoms.

1.14 Water is **heteroatomic**. If breaking water apart into its components produces both hydrogen gas and oxygen gas, then water must contain two types of atoms.

1.16 **a.** Substance A is a **compound** because it is composed of molecules that contain more than one type of atom.

b. Substance D is an **element** because it is composed of molecules that contain only one type of atom.

c. Substance E is a **compound** because it is a pure substance that can break down into at least two different materials. Substances G and J cannot be classified because no tests were performed on them.

1.18 **a.** Substance R **cannot be classified** as an element or a compound based on the information given.

b. Substance T is a **compound**.

c. The solid left in part b **cannot be classified** as an element or a compound. No tests have been performed on it.

1.20 **a.** heterogeneous

b. homogeneous

c. homogeneous

d. heterogeneous

e. homogeneous

f. homogeneous

g. heterogeneous

1.22 **b.** solution

c. solution

e. solution

f. solution

Measurement Units (Section 1.4)

1.24 In order for modern society to operate efficiently, people need a common language of measurement. Measurement, for example, is important for giving directions, keeping track of the time people work, and keeping indoor environments at a comfortable temperature and pressure.

1.26 The amount of weight that a horse could carry or drag might have been measured in stones. It could also be used to measure people or other items in the 50–500 pound range. It is likely that a large stone was picked as the standard weight for the "stone" unit. Stones may have also been used as counterweights on an old-fashioned set of balances.

The Metric System (Section 1.5)

1.28 The metric units are (a) degrees Celsius, (b) liters, (d) milligrams, and (f) seconds. The English units are (c) feet and (e) quarts.

1.30 Meters are a metric unit that could replace the English unit feet in the measurement of the ceiling height. Liters are a metric unit that could replace the English unit quarts in the measurement of the volume of a cooking pot.

1.32 **a.** $1.00\ \cancel{L}\left(\frac{1\ \mu L}{10^{-6}\ \cancel{L}}\right) = 1.00 \times 10^6\ \mu L$

b. $75\ \cancel{\text{kilowatts}}\left(\frac{1000\ \text{watts}}{1\ \cancel{\text{kilowatt}}}\right) = 7.5 \times 10^4\ \text{watts}$

c. $15\ \cancel{\text{megahertz}}\left(\frac{10^6\ \text{hertz}}{1\ \cancel{\text{megahertz}}}\right) = 1.5 \times 10^7\ \text{hertz}$

d. $200\ \cancel{\text{picometers}}\left(\frac{10^{-12}\ \text{meters}}{1\ \cancel{\text{picometer}}}\right) = 2 \times 10^{-10}\ \text{meters}$

1.34 $1\ \cancel{\text{cup}}\left(\frac{240\ \cancel{\text{mL}}}{1\ \cancel{\text{cup}}}\right)\left(\frac{1\ \text{L}}{1000\ \cancel{\text{mL}}}\right) = 0.240\ \text{L}$ or

$1\ \cancel{\text{cup}}\left(\frac{240\ \cancel{\text{mL}}}{1\ \cancel{\text{cup}}}\right)\left(\frac{1\ \text{cm}^3}{1\ \cancel{\text{mL}}}\right) = 240\ \text{cm}^3$

1.36 $4.0\ \cancel{\text{kg}}\left(\frac{2.20\ \text{lb}}{1\ \cancel{\text{kg}}}\right) = 8.8\ \text{lb}$

1.38 **a.** 1 m = 1.094 yd, so: 1 m − 1 yd = 1.094 yd − 1 yd

$= 0.094\ \cancel{\text{yd}}\left(\frac{3\ \cancel{\text{ft}}}{1\ \cancel{\text{yd}}}\right)\left(\frac{12\ \text{in}}{1\ \cancel{\text{ft}}}\right) = 3.4\ \text{in}$

b. The size of 1 °C is the same as 1 K; therefore, a change of 65 °C is also a change of 65 K.

c. $5.00\ \cancel{\text{lb}}\left(\frac{1\ \text{kg}}{2.20\ \cancel{\text{lb}}}\right) = 2.27\ \text{kg}$ with significant figures

1.40 **a.** $\frac{1\ \cancel{\text{kg}}}{1.0\ \cancel{\text{cm}^3}} \times \frac{1\ \cancel{\text{cm}^3}}{1000\ \cancel{\text{cm}^3}} \times \frac{1000\ \text{g}}{1\ \cancel{\text{kg}}} = 1.0\ \text{g/cm}^3$

b. $2.0\ \cancel{L} \times \frac{1.057\ \cancel{\text{qt}}}{1\ \text{L}} \times \frac{32\ \text{oz}}{1\ \cancel{\text{qt}}} = 68\ \text{oz}$

c. $5\ \cancel{\text{grain}} \times \frac{1\ \text{mg}}{0.015\ \cancel{\text{grain}}} = 300\ \text{mg}$

1.42 $°F = \frac{9}{5}(°C) + 32\ °F = \frac{9}{5}(36.1\ °C) + 32 = 97.0\ °F$

$°F = \frac{9}{5}(37.2\ °C) + 32 = 99.0\ °F$

Large and Small Numbers: An Introduction to Scientific Notation (Section 1.6)

1.44 **a.** 02.7×10^{-3} Improper form because no leading zero is necessary. (2.7×10^{-3})

b. 4.1×10^2 Correct.

c. 71.9×10^{-6} Improper form because only one digit should be to the left of the decimal point. (7.19×10^{-5})

d. 10^3 Improper form because a nonexponential term should be written before the exponential term. (1×10^3)

e. 0.0405×10^{-2} Improper form because one nonzero digit should be to the left of the decimal point. (4.05×10^{-4})

f. 0.119 Improper form because one nonzero digit should be to the left of the decimal point and an exponential term should be to the right of the nonexponential term. (1.19×10^{-1})

1.46 **a.** 1.4×10^4 **b.** 3.65×10^2 **c.** 2.04×10^{-3} **d.** 4.618×10^2 **e.** 1.00×10^{-3} **f.** 9.11×10^2

1.48 **a.** 1.86×10^5 mi/s **b.** 1.1×10^9 km/h

1.50 The decimal point has been moved 22 places to the left. This places 21 zeros to the right of the decimal point and before the numbers 105 g: 0.0000000000000000000000105 g.

1.52 **a.** 9.0×10^{-5} **b.** 1.4×10^7 **c.** 7.6×10^{-2} **d.** 1.9×10^1 **e.** 2.6×10^{12}

1.54 **a.** 1.26×10^1 **b.** 7.96×10^4 **c.** 5.02×10^{-5} **d.** 4.04×10^0

1.56 **a.** 2.6×10^{-5} **b.** 2.2×10^2 **c.** 6.4×10^{-4} **d.** 7.25×10^{-4} **e.** 3.1×10^{-2}

1.58 **a.** 1.7×10^{-1} **b.** 1.0×10^{-9} **c.** 2.6×10^6 **d.** 2.3×10^0 **e.** 1.5×10^7

Significant Figures (Section 1.7)

1.60 **a.** 0.01 cm **b.** 0.01 mm **c.** 0.1° **d.** 0.1 lb/in^2.

1.62 **a.** 6.0 mL **b.** 37.00 °C **c.** 9.00 s **d.** 15.5°

1.64 **a.** Measured = 5.06 lb

Exact = 16 potatoes

$\frac{5.06\ \text{lb}}{16\ \text{potatoes}} = 0.31625\ \frac{\text{lb}}{\text{potato}} = 0.316\ \frac{\text{lb}}{\text{potato}}$

b. Measured = percentages

Exact = 5 players

$\frac{71.2\% + 66.9\% + 74.1\% + 80.9\% + 63.6\%}{5\ \text{players}} = 71.3\%$

1.66 **a.** 3 ($0.04\underline{00}$) **b.** 3 **c.** 4 **d.** 2 **e.** 4 **f.** 5

1.68 **a.** 5.2 **b.** 0.104 **c.** 0.518 **d.** 1.0×10^2 **e.** 2.52×10^{-18} (assuming the 0 in 760 is significant)

1.70 **a.** 6.2 **b.** 2.30×10^2 **c.** 0.589 **d.** 0.58 **e.** 27.75 **f.** 21.64

1.72 **a.** 0.00460

b. 2.208

c. 2.65

d. −13

e. 3

f. 0.81

1.74 **a.** **Area** ($A = l \times w$) **Perimeter** [$P = 2(l) + 2(w)$]

	Area	Perimeter
Black	$A = 124.8\ cm^2$	$P = 44.80$ cm
Red	$A = 48.9\ cm^2$	$P = 45.24$ cm
Green	$A = 8.11\ cm^2$	$P = 11.46$ cm
Orange	$A = 9.0\ cm^2$	$P = 27.80$ cm

b. **Length** **Width**

Black

$$12.00\ \cancel{cm}\left(\frac{1\ m}{100\ \cancel{cm}}\right) = 0.1200\ m \quad 10.40\ \cancel{cm}\left(\frac{1\ m}{100\ \cancel{cm}}\right) = 0.1040\ m$$

Red

$$20.20\ \cancel{cm}\left(\frac{1\ m}{100\ \cancel{cm}}\right) = 0.2020\ m \quad 2.42\ \cancel{cm}\left(\frac{1\ m}{100\ \cancel{cm}}\right) = 0.0242\ m$$

Green

$$3.18\ \cancel{cm}\left(\frac{1\ m}{100\ \cancel{cm}}\right) = 0.0318\ m \quad 2.55\ \cancel{cm}\left(\frac{1\ m}{100\ \cancel{cm}}\right) = 0.0255\ m$$

Orange

$$13.22\ \cancel{cm}\left(\frac{1\ m}{100\ \cancel{cm}}\right) = 0.1322\ m \quad 0.68\ \cancel{cm}\left(\frac{1\ m}{100\ \cancel{cm}}\right) = 0.0068\ m$$

	Area ($A = l \times w$)	**Perimeter** [$P = 2(l) + 2(w)$]
Black	$A = 0.01248\ m^2$	$P = 0.4480$ m
Red	$A = 0.00489\ m^2$	$P = 0.4524$ m
Green	$A = 8.11 \times 10^{-4}\ m^2$	$P = 0.1146$ m
Orange	$A = 9.0 \times 10^{-4}\ m^2$	$P = 0.2780$ m

c. No, the number of significant figures in the answers remains constant.

Using Units in Calculations: An Introduction to Dimensional Analysis (Section 1.8)

1.76 **a.** $\frac{0.015\ grain}{1\ mg}$ **b.** $\frac{0.0338\ fl\ oz}{1\ mL}$

c. $\frac{1\ L}{1.057\ qt}$ **d.** $\frac{1\ m}{1.094\ yd}$

1.78 1.5 fl oz

1.80 1.4 mL

1.82 286 min

1.84 1.31 g/L

Calculating Percentages (Section 1.9)

1.86 5.500%

1.88 71%

1.90 Total = 987.1 mg + 213.3 mg + 99.7 mg + 14.4 mg + 0.1 mg = 1314.6 mg

$$IgG = \frac{987.1\ mg}{1314.6\ mg} \times 100 = 75.09\%;$$

$$IgA = \frac{213.3\ mg}{1314.6\ mg} \times 100 = 16.23\%;$$

$$IgM = \frac{99.7\ mg}{1314.6\ mg} \times 100 = 7.58\%$$

$$IgD = \frac{14.4\ mg}{1314.6\ mg} \times 100 = 1.10\%;$$

$$IgE = \frac{0.1\ mg}{1314.6\ mg} \times 100 = 0.008\%$$

Density and Its Applications (Section 1.10)

1.92 **a.** $\frac{39.6\ g}{50.0\ mL} = 0.792\ \frac{g}{mL}$ **b.** $\frac{243\ g}{236\ mL} = 1.03\ \frac{g}{mL}$

c. $\frac{39.54\ g}{20.0\ L} = 1.98\ \frac{g}{L}$ **d.** $\frac{222.5\ g}{25.0\ cm^3} = 8.90\ \frac{g}{cm^3}$

1.94 Volume = $(3.98\ cm)^3 = 63.0\ cm^3$

$$Density = \frac{mass}{volume} = \frac{718.3\ g}{(3.98\ cm)^3} = \frac{11.4\ g}{cm^3}$$

1.96 2.86 mL

Additional Exercises

1.98 **a.** $4.5\ km\left(\frac{1000\ m}{1\ km}\right)\left(\frac{1000\ mm}{1\ m}\right) = 4.5 \times 10^6\ mm$

b. $6.0 \times 10^6\ m\left(\frac{1\ g}{1000\ mg}\right) = 6.0 \times 10^3\ g$

c. $9.86 \times 10^{15}\ m\left(\frac{1\ km}{1000\ m}\right) = 9.86 \times 10^{12}\ km$

d. $1.91 \times 10^{-4}\ kg\left(\frac{1000\ g}{1\ kg}\right)\left(\frac{1000\ mg}{1\ g}\right) = 1.91 \times 10^2\ mg$

e. $5.0\ mg\left(\frac{1\ g}{10^9\ mg}\right)\left(\frac{1000\ mg}{1\ g}\right) = 5.0 \times 10^{-6}\ mg$

1.100 53.55 days

1.102 9.5×10^2 mg

Chemistry for Thought

1.104 **a.** To separate wood sawdust and sand, I would add water. The sawdust will float, while the sand will sink. The top layer of water and sawdust can be poured off into a filter. The water will run through the filter, leaving the sawdust in the filter. The sawdust can then be allowed to dry. The remainder of the water and sand can be poured off into a filter, and the sand can be allowed to dry.

b. To separate sugar and sand, I would add water to dissolve the sugar. I would then filter the mixture to isolate the sand. I would evaporate the water to isolate the sugar.

c. To separate iron filings and sand, I would use a magnet. The iron filings will be attracted to the magnet, while the sand will not be attracted to the magnet.

d. To separate sand soaked with oil, I would pour the mixture through a filter. The oil will go through the filter and leave the sand behind on the filter.

1.106 $44.5\ \cancel{kg}\left(\frac{2.2\ lb}{1\ \cancel{kg}}\right) = 97.9\ lb$

$\frac{44.5}{2.2} = 20.2$

This student should have used the relationship 2.2 lb = 1 kg to multiply 44.5 kg by 2.2 lb/kg to find a weight of 97.9 lb. The mistake made appears to be that of dividing 44.5 kg by 2.2 rather than multiplying by it.

1.108 9 cups

1.110 The density of the object is only 8.76 g/mL; therefore, it does not have the same density as silver and is not silver.

Chapter 2
The Periodic Table (Section 2.1)

2.2

	Group	Period
a.	VIIIB(9)	4
b.	IVA(14)	6
c.	VA(15)	4
d.	IIA(2)	6

2.4 **a.** 4 **b.** 10
c. 18

2.6 **a.** Period
b. Period

2.8 **a.** Transition metal **b.** Inner-transition metal
c. Noble gases **d.** Noble gases
e. Representative

2.10 **a.** Metal **b.** Metalloid
c. Metal **d.** Nonmetal
e. Nonmetal

Subatomic Particles (Section 2.2)

2.12

	Charge	Mass (amu)
a.	9	19
b.	20	43
c.	47	107

2.14 The number of protons and electrons are equal in a neutral atom.

a. 9 electrons **b.** 20 electrons **c.** 47 electrons

Atomic Symbols (Section 2.3)

2.16

	Symbol	Name
a.	Kr	Krypton
b.	Sn	Tin
c.	Mo	Molybdenum
d.	Nd	Neodymium

2.18

	Electrons	Protons
a.	14	14
b.	50	50
c.	74	74

Isotopes and Atomic Weights (Section 2.4)

2.20

	Protons	Neutrons	Electrons
a.	16	18	16
b.	40	51	40
c.	54	77	54

2.22 **a.** $^{28}_{14}Si$
b. $^{40}_{18}Ar$
c. $^{88}_{38}Sr$

2.24

	Mass Number	Atomic Number	Symbol
a.	19	9	$^{19}_{9}F$
b.	43	20	$^{43}_{20}Ca$
c.	107	47	$^{107}_{47}Ag$

2.26 **a.** $^{37}_{17}Cl$
b. $^{65}_{29}Cu$
c. $^{66}_{30}Zn$

2.28 **a.** $26.982 - 13 = 13.982 \approx 14$ neutrons
b. 27.0 amu

2.30 10.812 amu. The atomic weight listed for boron in the periodic table is 10.81 amu. The two values are close to one another.

2.32 63.55 amu. The atomic weight listed for copper in the periodic table is 63.55 amu. The two values are the same.

Radioactive Nuclei (Section 2.5)

2.34 **a.** Beta, gamma, positron.
b. Alpha, positron
c. Gamma, neutron

2.36 **a.** An electron
b. 2 protons and 2 neutrons
c. A positive electron

2.38

	Atomic number change	Mass number change
a.	Decrease by 2	decrease by 4
b.	Increase by 1	no change
c.	Decrease by 1	no change
d.	No change	no change
e.	Decrease by 1	no change

2.40 **a.** $^{96}_{41}Nb$
b. $^{80}_{37}Rb$
c. $^{38}_{20}Ca$

2.42 **a.** $^{204}_{82}Pb \rightarrow ? + ^{4}_{2}\alpha$ $\quad ? = ^{200}_{80}Hg$
b. $^{84}_{35}Br \rightarrow ? + ^{0}_{-1}\beta$ $\quad ? = ^{84}_{36}Kr$
c. $? + ^{0}_{-1}e \rightarrow ^{41}_{19}Br$ $\quad ? = ^{41}_{20}Ca$

d. ${}^{149}_{62}\text{Sm} \rightarrow ? + {}^{145}_{60}\text{Nd}$ $? = {}^{4}_{2}\alpha$

e. $? \rightarrow {}^{34}_{15}\text{P} + {}^{0}_{-1}\beta$ $? = {}^{34}_{14}\text{Si}$

f. ${}^{15}_{8}\text{O} + {}^{0}_{1}\beta \rightarrow ?$ $? = {}^{15}_{7}\text{N}$

2.44 **a.** ${}^{157}_{63}\text{Eu}$ (beta emission) ${}^{157}_{63}\text{EU} \rightarrow {}^{0}_{-1}\beta + {}^{157}_{64}\text{Gd}$

b. ${}^{190}_{78}\text{Pt}$ (daughter = osmium-86) ${}^{190}_{78}\text{Pt} \rightarrow {}^{4}_{2}\alpha + {}^{186}_{76}\text{Os}$

c. ${}^{138}_{62}\text{Sm}$ (electron capture) ${}^{138}_{62}\text{Sm} \rightarrow {}^{0}_{-1}\text{e} + {}^{138}_{61}\text{Pm}$

d. ${}^{188}_{80}\text{Hg}$ (daughter = Au-188) ${}^{188}_{80}\text{Hg} \rightarrow {}^{0}_{-1}\beta + {}^{188}_{79}\text{Au}$

e. ${}^{234}_{90}\text{Th}$ (beta emission) ${}^{234}_{90}\text{Th} \rightarrow {}^{0}_{-1}\beta + {}^{234}_{91}\text{Pa}$

f. ${}^{218}_{85}\text{At}$ (alpha emission) ${}^{218}_{85}\text{At} \rightarrow {}^{4}_{2}\alpha + {}^{214}_{83}\text{Bi}$

2.46 Half-life is the amount of time required for half of a sample to undergo a specific process. For example, if the half-life of a cake is one day, then half of the cake will be eaten by the first day, half of the remaining cake (1/4 of the original cake) will be eaten by the second day, half of the remaining cake (1/8 of the original cake) will be eaten by the third day, and so on.

2.48 24 hours is 6 half-lives, so 0.56 ng remains.

2.50 2.24×10^4 years

2.52 1.68×10^4 years

2.54 Long-term, low-level exposure to radiation may lead to genetic mutations because ionizing radiation can produce free radicals in exposed tissues. Short-term exposure to intense radiation destroys tissue rapidly and causes radiation sickness. Both forms of exposure have negative effects on health.

2.56 Physical units of radiation indicate the activity of a source of radiation, whereas biological units of radiation indicate the damage caused by radiation in living tissue. Examples of physical units of radiation include the curie and the becquerel. Examples of biological units of radiation include the sievert, the rad, the gray, and the rem.

2.58 0.031 grays

2.60 Radioactive isotopes can be used for diagnostic work. When the radioactive isotope concentrates in a tissue under observation, the location is called a hot spot. When the radioactive isotope is excluded or rejected by a tissue under observation, the location is called a cold spot. Both hot spots and cold spots can be used for diagnostic work.

2.62 ${}^{51}_{24}\text{Cr} + {}^{0}_{-1}\text{e} \rightarrow {}^{51}_{23}\text{V}$; daughter nucleus = vanadium-51

2.64 By using water that contains a radioactive isotope of oxygen, the oxygen gas produced could be analyzed to see if it contains the radioactive isotope of oxygen from the water or a nonradioactive isotope of oxygen from the hydrogen peroxide.

Where Are the Electrons? (Section 2.6)

2.66. Protons

2.68 **a.** X rays

b. Red light

2.70 **a.** 2 electrons

b. 6 electrons

c. 8 electrons

2.72 Four (4) orbitals are found in the second shell: one 2*s* orbital and three 2*p* orbitals.

2.74 Three (3) orbitals are found in a 4*p* subshell. The maximum number of electrons that can be located in this subshell is 6.

2.76

	Electron Configuration	Unpaired Electrons
a.	$1s^22s^22p^63s^23p^64s^23d^{10}4p^65s^1$	1
b.	$1s^22s^22p^63s^23p^2$	2
c.	$1s^22s^22p^63s^23p^64s^23d^2$	2
d.	$1s^22s^22p^63s^23p^6$	0

2.78

	Electron Configuration	Solutions
a. *s* electrons in magnesium	$1s^22s^22p^63s^2$	6
b. unpaired electrons in nitrogen	$1s^22s^22p^3$	3
c. filled subshells in Al	$1s^22s^22p^63s^23p^1$	4

2.80

	Symbol	Name
a.	C	carbon
b.	Na	sodium

2.82 **a.** $[\text{Ar}]4s^23d^{10}4p^3$

b. $[\text{Ar}]4s^23d^5$

c. $[\text{Ne}]3s^23p^2$

d. $[\text{Kr}]5s^24d^{10}5p^5$

2.84 **a.** sodium $[\text{Ne}]3s^1$ **b.** magnesium $[\text{Ne}]3s^2$

c. aluminum $[\text{Ne}]3s^23p^1$ **d.** silicon $[\text{Ne}]3s^23p^2$

e. phosphorus $[\text{Ne}]3s^23p^3$ **f.** sulfur $[\text{Ne}]3s^23p^4$

g. chlorine $[\text{Ne}]3s^23p^5$ **h.** argon $[\text{Ne}]3s^23p^6$

2.86 Cesium because it has 1 valence-shell electrons like sodium.

2.88 **a.** 8 electrons **b.** 5 electrons

c. 4 electrons **d.** 4 electrons

2.90 **a.** *p* area **b.** *d* area **c.** *s* area

Trends within the Periodic Table (Section 2.7)

2.92 Silver and gold because these elements are all in the same group on the periodic table as copper.

2.94 **a.** Ga **b.** Sb

c. C **d.** Te

2.96 **a.** K **b.** Sn

c. Mg **d.** Li

2.98 **a.** Cl **b.** Cl

c. N **d.** B

Additional Exercises

2.100 2.32×10^{-23} g

2.102 Chemical properties depend on the number of valence electrons an atom contains, not the number of neutrons an atom contains; therefore, the chemical properties of isotopes of the same element are the same because all isotopes of the same element contain the same number of electrons, including valence electrons.

Chemistry for Thought

2.104 Aluminum exists as one isotope; therefore, all aluminum atoms have the same number of protons and neutrons, which gives them all the same mass. Nickel exists as several isotopes; therefore, the individual atoms do not have the weighted average atomic mass of 58.69 amu.

2.106 While radioactive decay does occur naturally, it is unlikely that lead changes into gold naturally because lead has an atomic number of 82 and gold has an atomic number of 79, which makes a difference of 3. None of the common nuclear decay processes change the atomic number of the parent nucleus by 3.

2.108 In principle, a radioactive isotope never completely disappears by radioactive decay because only half of the sample decays per half-life, so half of the initial sample remains. In reality, however, all of a sample will decay because eventually one atom will remain in a sample. When that one atom undergoes decay, the entire atom will undergo decay, not half of the atom.

2.110 **a.** ${}^{239}_{94}\text{Pu} \rightarrow {}^{4}_{2}\text{He} + {}^{235}_{92}\text{U}$ and ${}^{131}_{53}\text{I} \rightarrow {}^{0}_{-1}\text{e} + {}^{0}_{0}\gamma + {}^{131}_{54}\text{Xe}$

b. When iodine-131 is ingested, it is readily absorbed by the body's thyroid gland. The accumulation of iodine-131 in the thyroid can lead to the irradiation of thyroid tissue and the emission of ionizing radiation, which can increase the risk of developing thyroid cancer or other thyroid disorders.

Chapter 3
An Introduction to Lewis Structures (Section 3.1)

3.2 **a.** iodine ·Ï: **b.** strontium ·Sr·
c. tin ·Sṅ· (with a dot below) **d.** sulfur ·S̈:

3.4 **a.** germanium ·Ġe· **b.** cesium ·Cs
c. indium ·İn· **d.** calcium ·Ca·

3.6
a. lithium $[\text{He}]2s^1$ ·Li
b. chlorine $[\text{Ne}]3s^23p^5$:C̤̈l·
c. titanium $[\text{Ar}]4s^23d^2$ ·Ti·
d. $[\text{Ar}]4s^23d^{10}4p^3$ ·Äs·

3.8 **a.** Any Group IIA-(2) element Ė·
b. Any Group VA-(15) element ·Ë· (with a dot below)

The Formation of Ions (Section 3.2)

3.10

	Added Electrons	Removed Electrons
a. germanium	4	14 (4 not including *d* electrons)
b. Cs	31 (7 not including *d* or *f* electrons)	1
c. element number 49	5	13 (3 not including *d* electrons)
d. calcium	16 (6 not including *d* electrons)	2

3.12

	Number of Electrons Lost/Gained	Equation
a. Cs	1 electron lost	$\text{Cs} \rightarrow \text{Cs}^+ + e^-$
b. oxygen	2 electrons gained	$\text{O} + 2e^- \rightarrow \text{O}^{2-}$
c. element number 7	3 electrons gained	$\text{N} + 3e^- \rightarrow \text{N}^{3-}$
d. iodine	1 electron gained	$\text{I} + e^- \rightarrow \text{I}^-$

3.14

	Equation	Ion Symbol
a.	$\text{Se} + 2e^- \rightarrow \text{Se}^{2-}$	Se^{2-}
b.	$\text{Rb} \rightarrow \text{Rb}^+ + e^-$	Rb^+
c.	$\text{Al} \rightarrow \text{Al}^{3+} + 3e^-$	Al^{3+}

3.16 **a.** sulfur **b.** aluminum
c. sodium **d.** chlorine

3.18 **a.** helium **b.** xenon
c. argon **d.** krypton

3.20 **a.** 29 protons, 28 electrons
b. 16 protons, 18 electrons
c. 21 protons, 18 electrons

Ionic Compounds (Section 3.3)

3.22

	Cation Formula	Anion Formation	Formula
a. Ca and Cl	$\text{Ca} \rightarrow \text{Ca}^{2+} + 2e^-$	$\text{Cl} + e^- \rightarrow \text{Cl}^-$	CaCl_2
b. lithium and bromine	$\text{Li} \rightarrow \text{Li}^+ + e^-$	$\text{Br} + e^- \rightarrow \text{Br}^-$	LiBr
c. Elements 12 and 16	$\text{Mg} \rightarrow \text{Mg}^{2+} + 2e^-$	$\text{S} + 2e^- \rightarrow \text{S}^{2-}$	MgS

3.24 **a.** BaSe **b.** Ba_3P_2
c. BaI_2 **d.** Ba_3As_2

3.26 **a.** X = −2 **b.** X = −1
c. X = −2 **d.** X = −2

3.28 **a.** NaBr = 102.89 amu **b.** CaF_2 = 78.08 amu
c. Cu_2S = 159.16 amu **d.** Li_3N = 34.83 amu

Naming Binary Ionic Compounds (Section 3.4)

3.30 **a.** binary **b.** binary
c. not binary **d.** binary
e. not binary

3.32 **a.** lithium ion **b.** magnesium ion
c. barium ion **d.** cesium ion

3.34 **a.** bromide ion **b.** oxide ion
c. phosphide ion **d.** fluoride ion

3.36 **a.** strontium sulfide **b.** calcium fluoride
c. barium chloride **d.** lithium oxide
e. magnesium oxide

3.38 **a.** CsI **b.** CaS
c. $AlCl_3$ **d.** Be_3P_2

3.40 **a.** lead(II) oxide and lead(IV) oxide
b. copper(I) chloride and copper(II) chloride
c. gold(I) sulfide and gold(III) sulfide
d. cobalt(II) oxide and cobalt(III) oxide

3.42 **a.** plumbous oxide and plumbic oxide
b. cuprous chloride and cupric chloride
c. aurous sulfide and auric sulfide
d. cobaltous oxide and cobaltic oxide

3.44 **a.** Hg_2O **b.** PbO
c. PtI_4 **d.** Cu_3N
e. CoS

Covalent Bonding (Section 3.5)

3.46 8·S: ⟶ (S₈ ring: :S—S: bonded in an eight-membered ring)

3.48 **a.** CH_4 (each H atom is bonded to the C atom)

H—C—H (with H above and below C)

b. CO_2 (each O atom is bonded to the C atom)

Ö=C=Ö

c. H_2Se (each H atom is bonded to the Se atom)

H—S̈e—H

d. NH_3 (each H atom is bonded to the N atom)

H—N̈—H (with H below N)

3.50 **a.** Boron does not have a full octet. (:C̈l: bonded to B, with :C̈l and C̈l: on B)

b. Sulfur has an expanded octet with 12 electrons around the central atom. (S bonded to six :F̈: atoms)

c. Sulfur has an expanded octet with 10 electrons around the central atom. (:O:=S:=Ö)

d. In SeO_3, selenium has 12 electrons around the central atom. (:O:=Se, with Ö= and =Ö)

3.52 **a.** ClO_2 = 67.45 amu **b.** N_2O = 44.02 amu
c. SO_2 = 64.06 amu **d.** CCl_4 = 153.81 amu

3.54 **a.** C_3H_8 = 44.09 amu
b. CO = 28.01 amu

Naming Binary Covalent Compounds (Section 3.6)

3.56 **a.** phosphorus trichloride **b.** dinitrogen monoxide
c. carbon tetrachloride **d.** boron trifluoride
e. carbon disulfide

3.58. **a.** SO_2, sulfur dioxide, covalent
b. $MgBr_2$, magnesium bromide, ionic
c. BaS, barium sulfide, ionic
d. N_2O_4, dinitrogen tetroxide, covalent

Lewis Structures of Polyatomic Ions (Section 3.7)

3.60 **a.** SO_3^{2-} (each O atom is bonded to the S atom), sulfite

[:O:=S̈(—:Ö:)(—Ö:)]$^{2-}$

b. SO_4^{2-} (each O atom is bonded to the S atom), sulfate

[:O:=S(—:Ö:)(—Ö:)=:O:]$^{2-}$

3.62 **a.** NH_4^+ (each H atom is bonded to the N atom)

```
[    H    ]+
[    |    ]
[ H—N—H   ]
[    |    ]
[    H    ]
```

b. PO_4^{3-} (each O atom is bonded to the P atom)

```
[     :Ö:     ]3−
[      |      ]
[ :Ö—P=Ö      ]
[      |      ]
[     :Ö:     ]
```

c. SO_3^{2-} (each O atom is bonded to the S atom)

```
[      :O:      ]2−
[       ||      ]
[       S       ]
[ :Ö/  ..  \Ö:  ]
```

Compounds Containing Polyatomic Ions (Section 3.8)

3.64 **a.** M_2SO_3

b. $MC_2H_3O_2$

c. MCr_2O_7

d. MPO_4

e. $M(NO_3)_3$

3.66 **a.** $Ca(ClO)_2$ calcium hypochlorite

b. $CsNO_2$ cesium nitrite

c. $MgSO_3$ magnesium sulfite

d. $K_2Cr_2O_7$ potassium dichromate

3.68 **a.** $MgSO_3$

b. $Ba(OH)_2$

c. $CaCO_3$

d. $(NH_4)_2SO_4$

e. $LiHCO_3$

3.70 **a.** $NaNO_3$ **b.** $Ca(OH)_2$

c. $MgBr_2$ **d.** K_2CO_3

3.72 **a.** $FeSO_4$, ionic, iron(II) sulfate

b. NO, covalent, nitrogen monoxide

c. CF_4, covalent, carbon tetrafluoride

Chemistry for Thought

3.74 NH_3

3.76 Potassium dichromate is $K_2Cr_2O_7$. Potassium chromate is K_2CrO_4.

Potassium phosphate is K_3PO_4. Potassium permanganate is $KMnO_4$.

The colored compounds of potassium ($K_2Cr_2O_7$, K_2CrO_4, and $KMnO_4$) have a transition metal as part of the polyatomic ion.

3.78 A negatively charged ion will be larger than a nonmetal atom of the same element. As electrons are added to the atom to form the anion, the electron cloud increases in size because the nucleus is unable to hold onto the electrons as well as in the neutral atom.

3.80 There is one pair of nonbonding electrons present on the nitrogen atom in methamphetamine.

methamphetamine

3.82 There are five nonbonding electron pairs on glycine that are located on the nitrogen and oxygen atoms: Two are on the C=O oxygen atom, two are on the OH oxygen atom, and one pair is on the NH_2 nitrogen atom.

glycine

3.84 There are four nonbonding electron pairs on propanoic acid that are located on the two oxygen atoms: Two are on the C=O oxygen, and two are on the OH oxygen.

propanoic acid

Chapter 4
Avogadro's Number: The Mole (Section 4.1)

4.2 **a.** 6.022×10^{22} atoms are in 1.60 grams of oxygen.

b. 1.90 grams of fluorine would contain the same number of atoms as 1.60 grams of oxygen.

4.4 **a.** 5.14×10^{-23} g P

b. 44.5 g Al

c. 20.95 g Kr

The Mole and Chemical Formulas (Section 4.2)

4.6 131.1 g/mol

4.8 273.3 g/mol

4.10 0.68 g BF_3

4.12 **a.** 3.01×10^{23} C atoms $= 0.25 \text{ mol } C_2H_6O \times \dfrac{2 \text{ mol C}}{1 \text{ mol } C_2H_6O} \times \dfrac{6.02 \times 10^{23} \text{ atoms}}{1 \text{ mol C}}$

b. 5.0 mol H atoms $= 0.50 \text{ mol } C_2H_{10}O \times \dfrac{10 \text{ mol H}}{1 \text{ mol } C_2H_{10}O}$

c. 2.0 g of H atoms $= 2.00 \text{ mol } C_2HBrClF_3 \times \dfrac{1.0 \text{ gH}}{1 \text{ mol } C_2HBrClF_3}$

4.14 0.0185 mol Fe

4.16 25.1% H in CH_4 and 20.1% H in C_2H_6

4.18 **a.** 23.2 g O **b.** 18.0 mol H atoms

c. 4.50×10^{23} C atoms

4.20 1.17 g

4.22 1760 g

4.24 **a.** $C_5H_8NO_4$

b. Molar mass of $C_5H_8NO_4 = (5 \times 12.01 \text{ g/mol C}) + (8 \times 1.008 \text{ g/mol H}) + (1 \times 14.01 \text{ g/mol N}) + (4 \times 16.00 \text{ g/mol O}) = 146.1 \text{ g/mol}$

c. 43.8%

$$\frac{(4 \times 16.00 \text{ g})}{146.1 \text{ g}} \times 100\% = 43.8\% \text{ O}$$

d. Yes, the risk value is 80 mg/kg, and if a 9 kg the toddler ingested 5.4 g of glutamate, they will have ingested over the risk value.

$$5.4 \text{ g } C_5H_7NO_4\left(\frac{1000 \text{ mg}}{1 \text{ g}}\right) = 5400 \text{ mg } C_5H_7NO_4$$

$$\frac{5400 \text{ mg } C_5H_7NO_4}{9 \text{ kg}} = \frac{600 \text{ mg}}{\text{kg}} > \frac{80 \text{ mg}}{\text{kg}}$$

Chemical Equations (Section 4.3)

4.26

	Reactants	Products
a.	H_2, Cl_2	HCl
b.	$KClO_3$	KCl, O_2
c.	magnesium oxide, carbon	magnesium carbon monoxide
d.	ethane, oxygen	carbon dioxide, water

4.28 **a.** $ZnS(s) + O_2(g) \rightarrow ZnO(s) + SO_2(g)$ is *not* consistent with the law of conservation of matter because the reactant (left) side of the equation has two moles of oxygen atoms, while the product (right) side of the equation has three moles of oxygen atoms.

b. $Cl_2(aq) + 2I^-(aq) \rightarrow I_2(aq) + 2Cl^-(aq)$ is consistent with the law of conservation of matter.

c. 1.50 g oxygen + 1.50 g carbon → 2.80 g carbon monoxide is *not* consistent with the law of conservation of matter because the mass of the reactants is 3.00 g, while the mass of the products is only 2.80 g.

d. $2C_2H_6(g) + 7O_2 \rightarrow 4CO_2(g) + 6H_2O(g)$ is consistent with the law of conversation of matter.

4.30 **a.** $Ag(s) + Cu(NO_3)_2(aq) \rightarrow Cu(s) + AgNO_3(aq)$

Elements	Ag	Cu	N	O
Reactant	1	1	2	6
Product	1	1	1	3

This equation is not balanced because the numbers of moles of nitrogen and oxygen are not balanced.

b. $2N_2O(g) + 3O_2(g) \rightarrow 4NO_2(g)$

Elements	N	O
Reactant	4	8
Product	4	8

This equation is balanced.

c. $Mg(s) + O_2(g) \rightarrow 2MgO(s)$

Elements	Mg	O
Reactant	1	2
Product	2	2

This equation is not balanced because the number of moles of magnesium is not balanced.

d. $H_2SO_4(aq) + Ca(OH)_2(aq) \rightarrow CaSO_4(s) + 2H_2O(\ell)$

Elements	H	S	O	Ca
Reactant	4	1	6	1
Product	4	1	6	1

This equation is balanced.

4.32 **a.** $2KClO_3(s) \rightarrow 2KCl(s) + 3O_2(g)$

b. $2C_2H_6(g) + 7O_2(g) \rightarrow 4CO_2(g) + 6H_2O(\ell)$

c. $2N_2(g) + 5O_2(g) \rightarrow 2N_2O_5(g)$

d. $MgCl_2(s) + H_2O(g) \rightarrow MgO(s) + 2HCl(g)$

e. $CaH_2(s) + H_2O(\ell) \rightarrow Ca(OH)_2(s) + 2H_2(g)$

f. $2Al(s) + Fe_2O_3(s) \rightarrow Al_2O_3(s) + 2Fe(s)$

g. $2Al(s) + 3Br_2(\ell) \rightarrow 2AlBr_3(s)$

h. $Hg_2(NO_3)_2(aq) + 2NaCl(aq) \rightarrow Hg_2Cl_2(s) + 2NaNO_3(aq)$

The Mole and Chemical Equations (Section 4.4)

4.34 **a.** $S(s) + O_2(g) \rightarrow SO_2(g)$

Statement 1: 1S atom + $1O_2$ molecule → $1SO_2$ molecule

Statement 2: 1 mole S + 1 mole O_2 → 1 mole SO_2

Statement 3: 6.02×10^{23} S atoms + 6.02×10^{23} O_2 molecules → 6.02×10^{23} SO_2 molecules

Statement 4: 32.1 g S + 32.0 g O_2 → 64.1 g SO_2

b. $Sr(s) + 2H_2O(\ell) \rightarrow Sr(OH)_2(s) + H_2(g)$

Statement 1: 1 Sr atom + 2 H_2O molecules → 1 $Sr(OH)_2$ formula unit + 1 H_2 molecule

Statement 2: 1 mole Sr + 2 moles H_2O → 1 mole $Sr(OH)_2$ + 1 mole H_2

Statement 3: 6.02×10^{23} Sr atoms + 12.0×10^{23} H_2O molecules → 6.02×10^{23} $Sr(OH)_2$ formula units + 6.02×10^{23} H_2 molecules

Statement 4: 87.6 g Sr + 36.0 g H_2O → 121.6 g $Sr(OH)_2$ + 2.0 g H_2

c. $2H_2S(g) + 3O_2(g) \rightarrow 2H_2O(g) + 2SO_2(g)$

Statement 1: 2 H_2S molecules + 3 O_2 molecules → 2 H_2O molecules + 2 SO_2 molecules

Statement 2: 2 moles H_2S + 3 moles $O_2 \rightarrow$ 2 moles H_2O + 2 moles SO_2

Statement 3: 12.0×10^{23} H_2S molecules + 18.1×10^{23} O_2 molecules $\rightarrow$ 12.0×10^{23} H_2O molecules + 12.0×10^{23} SO_2 molecules

Statement 4: 68.2 g H_2S + 96.0 g $O_2 \rightarrow$ 36.0 g H_2O + 128.2 g SO_2

d. $4NH_3(g) + 5O_2(g) \rightarrow 4NO(g) + 6H_2O(g)$

Statement 1: 4 NH_3 molecules + 5 O_2 molecules $\rightarrow$ 4 NO molecules + 6 H_2O molecules

Statement 2: 4 moles NH_3 + 5 moles $O_2 \rightarrow$ 4 moles NO + 6 moles H_2O

Statement 3: 24.1×10^{23} NH_3 molecules + 30.1×10^{23} O_2 molecules $\rightarrow$

24.1×10^{23} NO molecules + 36.1×10^{23} H_2O molecules

Statement 4: 68.0 g NH_3 + 160.0 g $O_2 \rightarrow$ 120.0 g NO + 108.0 g H_2O

e. $CaO(s) + 3C(s) \rightarrow CaC_2(s) + CO(g)$

Statement 1: 1 CaO formula unit + 3 C atoms $\rightarrow$ 1 CaC_2 formula unit + 1 CO molecule

Statement 2: 1 mole CaO + 3 moles C $\rightarrow$ 1 mole CaC_2 + 1 mole CO

Statement 3: 6.02×10^{23} CaO formula units + 18.1×10^{23} C atoms $\rightarrow$ 6.02×10^{23} CaC_2 formula units + 6.02×10^{23} CO molecules

Statement 4: 56.1 g CaO + 36.0 g C $\rightarrow$ 64.1 g CaC_2 + 28.0 g CO

4.36 $2SO_2 + O_2 \rightarrow 2SO_3(g)$

Statement 1: 2 SO_2 molecules + 1 O_2 molecule $\rightarrow$ 2 SO_3 molecules

Statement 2: 2 moles SO_2 + 1 mole $O_2 \rightarrow$ 2 moles SO_3

Statement 3: 12.0×10^{23} SO_2 molecules + 6.02×10^{23} O molecules $\rightarrow$ 12.0×10^{23} SO_3 molecules

Statement 4: 128 g SO_2 + 32.0 g $O_2 \rightarrow$ 160.g SO_3

Factors:

$\frac{12.0 \times 10^{23}\ SO_2\ \text{molecules}}{6.02 \times 10^{23}\ O_2\ \text{molecules}}$; $\frac{6.02 \times 10^{23}\ O_2\ \text{molecules}}{12.0 \times 10^{23}\ SO_2\ \text{molecules}}$; $\frac{12.0 \times 10^{23}\ SO_2\ \text{molecules}}{12.0 \times 10^{23}\ O_2\ \text{molecules}}$;

$\frac{12.0 \times 10^{23}\ SO_3\ \text{molecules}}{12.0 \times 10^{23}\ SO_2\ \text{molecules}}$; $\frac{6.02 \times 10^{23}\ O_2\ \text{molecules}}{12.0 \times 10^{23}\ SO_3\ \text{molecules}}$; $\frac{12.0 \times 10^{23}\ SO_3\ \text{molecules}}{6.02 \times 10^{23}\ O_2\ \text{molecules}}$;

$\frac{2\ \text{moles}\ SO_2}{1\ \text{mole}\ O_2}$; $\frac{1\ \text{mole}\ O_2}{2\ \text{moles}\ SO_2}$; $\frac{2\ \text{moles}\ SO_2}{2\ \text{moles}\ SO_3}$; $\frac{2\ \text{moles}\ SO_3}{2\ \text{moles}\ SO_2}$; $\frac{1\ \text{mole}\ O_2}{2\ \text{moles}\ SO_3}$; $\frac{2\ \text{moles}\ SO_3}{1\ \text{mole}\ O_2}$

$\frac{128\ g\ SO_2}{32.0\ O_2}$; $\frac{32.0\ g\ O_2}{128\ g\ SO_2}$; $\frac{128\ g\ SO_2}{160.0\ g\ SO_3}$; $\frac{160.0\ g\ SO_3}{128\ g\ SO_2}$; $\frac{32.0\ O_2}{160.0\ g\ SO_3}$; $\frac{160.0\ g\ SO_3}{32.0\ g\ O_2}$

This list does not include all possible factors.

4.38 892 g $CaCO_3$

4.40 445 g Br_2

4.42 1.02×10^3 g Mg

4.44 256 g O_2

Reaction Yields (Section 4.5)

4.46 89.0%

Types of Reactions (Section 4.6)

4.48

	O.N.	Change	Classification
a. $4\underline{Al}(s) + 3O_2(aq) \rightarrow 2Al_2O_3(s)$	$0 \rightarrow +3$	lost e^-	oxidized
b. $SO_2(g) + \underline{H}_2O(\ell) \rightarrow H_2SO_3(aq)$	$+1 \rightarrow +1$	no change	neither
c. $2K\underline{Cl}O_3(s) \rightarrow 2KCl(s) + 3O_2(g)$	$+5 \rightarrow -1$	gained e^-	reduced
d. $2\underline{C}O(g) + O_2(g) \rightarrow 2CO_2(g)$	$+2 \rightarrow +4$	lost e^-	oxidized
e. $2\underline{Na}(s) + 2H_2O(\ell) \rightarrow 2NaOH(aq) + H_2(g)$	$0 \rightarrow +1$	lost e^-	oxidized

4.50 **a.** $2Cu(s) + O_2(g) \rightarrow 2CuO(s)$

Elements	Cu	O
Reactant	0	0
Product	+2	−2

The oxidizing agent is O_2.
The reducing agent is Cu.

b. $Cl_2(aq) + 2KI(aq) \rightarrow 2KCl(aq) + I_2(aq)$

Elements	Cl	K	I
Reactant	0	+1	−1
Product	−1	+1	0

The oxidizing agent is Cl_2.
The reducing agent is KI.

c. $3MnO_2(s) + 4Al(s) \rightarrow 2Al_2O_3(s) + 3Mn(s)$

Elements	Mn	O	Al
Reactant	+4	−2	0
Product	0	−2	+3

The oxidizing agent is MnO_2.
The reducing agent is Al.

d. $2H^+(aq) + 3SO_3^{2-}(aq) \rightarrow 2NO_3^-(g) + H_2O(\ell) + 3SO_4^{2-}(aq)$

Elements	H	S	O	N
Reactant	+1	+4	−2	+5
Product	+1	+6	−2	+2

The oxidizing agent is NO_3^-.
The reducing agent is SO_3^{2-}.

e. $Mg(s) + 2\,HCl(aq) \rightarrow MgCl_2(aq) + H_2(g)$

Elements	Mg	H	Cl
Reactant	0	+1	−1
Product	+2	0	−1

The oxidizing agent is HCl.
The reducing agent is Mg.

f. $4NO_2(g) + O_2(g) \rightarrow 2N_2O_5(g)$

Elements	N	O	O
Reactant	+4	−2	0
Product	+5	−2	−2

The oxidizing agent is O_2.
The reducing agent is NO_2.

4.52 $6NaOH(aq) + 2Al(s) \rightarrow 3H_2(g) + Na_3AlO_3(aq) + \text{heat}$

The oxidizing agent is NaOH because the oxidation number for the hydrogen changes from +1 to 0.

The reducing agent is Al because the oxidation number for the aluminum changes from 0 to +3.

4.54 **a.** nonredox: decomposition

b. redox: single-replacement

c. nonredox: double-replacement

d. nonredox: combination

e. redox: combination

f. redox: combination

g. redox: combustion

4.56 Nonredox decomposition

4.58 Redox.

4.60 $\underset{0}{Cl_2(aq)} + \underset{+1-2}{H_2O(\ell)} \rightarrow \underset{+1-2+2}{HOCl(aq)} + \underset{+1-1}{HCl(aq)}$;

This is a redox reaction.

4.62 Oxidizing agent: N_2 Reducing agent: H_2

Additional Exercises

4.64 $$\frac{1.99 \times 10^{-23}\ \text{g}}{1\ \cancel{\text{C} - 12\ \text{atom}}}\left(\frac{1\ \cancel{\text{C} - 12\ \text{atom}}}{12\ \cancel{\text{protons} + \text{neutrons}}}\right)\left(\frac{14\ \cancel{\text{protons} + \text{neutrons}}}{1\ \text{C} - 14\ \text{atom}}\right) = \frac{2.32 \times 10^{-23}\ \text{g}}{1\ \text{C} - 14\ \text{atom}}$$

4.66 Due to its highly reactive nature and tendency to donate electrons (i.e., its tendency to be oxidized), lithium can participate in redox reactions as a reducing agent.

4.68 6.983 g naturally occurring elemental iron

$$\left(\frac{60.0\ \text{g}\ ^{60}\text{Fe}}{55.9\ \text{g naturally occurring elemental iron}}\right) = 7.5\ \text{g}\ ^{60}\text{Fe}$$

4.70 **a.** There are approximately 7.52×10^{-6} moles of tetrodotoxin in 2.4 mg of $C_{11}H_{17}N_3O_8$.

$$2.4\ \text{mg}\ C_{11}H_{17}N_3O_8\left(\frac{\text{g}}{1000\ \text{mg}}\right)\left(\frac{1\ \text{mol}\ C_{11}H_{17}N_3O_8}{319.3\ \text{g}\ C_{11}H_{17}N_3O_8}\right)$$
$$= 7.52 \times 10^{-6}\ \text{mol}\ C_{11}H_{17}N_3O_8$$

b. There are 4.53×10^{18} molecules of tetrodotoxin in 2.4 mg of $C_{11}H_{17}N_3O_8$.

$$7.52 \times 10^{-6}\ \text{mol}\ C_{11}H_{17}N_3O_8\left(\frac{6.022 \times 10^{23}}{1\ \text{mol}\ C_{11}H_{17}N_3O_8}\right)$$
$$= 4.53 \times 10^{-18}\ \text{molecules}\ C_{11}H_{17}N_3O_8$$

Chemistry for Thought

4.72 When a yield of more than 100% occurs for a compound prepared by precipitation from water solutions, it is likely that the "dry" compound is still moist and contains extra water mass.

4.74 Zinc changing from zinc metal into zinc ions is an oxidation reaction because the zinc atoms are losing electrons to become cations. This reaction is the source of electrons as shown in the following chemical equation:

$$Zn(s) \rightarrow Zn^{2+}(aq) + 2\,e^-$$

Chapter 5
Geometries of Molecules and Polyatomic Ions (Section 5.1)

5.2 **a.** NH_3 (each H atom is bonded to the N atom) has three bonding domains and one nonbonding domain.

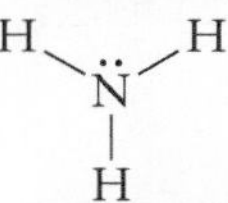

b. $BeCl_2$ (each Cl atom is bonded to the Be atom) has two bonding domains.

:C̤̈l—Be—C̤̈l:

c. ClCN (the Cl and N atoms are bonded to the C atom) has two bonding domains.

:C̤̈l—C≡N:

5.4 **a.** H_2S_2 (each H atom is bonded to the S atom)

Lewis Structure	3D Structure	VSEPR
H—S̤̈—H	H, S:, H	bent

b. PCl_3 (each Cl atom is bonded to the P atom)

Lewis Structure	3D Structure	VSEPR
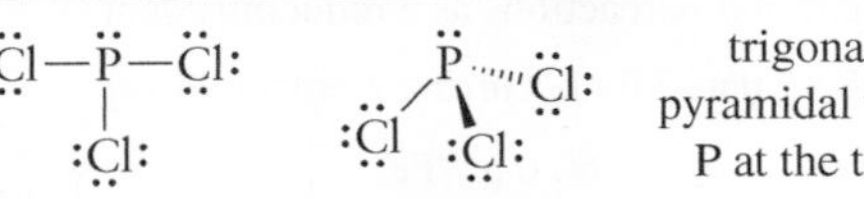		trigonal pyramidal with P at the top

c. OF_2 (each F atom is bonded to the O atom)

Lewis Structure	3D Structure	VSEPR
:F–O–F:	:F:, O:, :F	bent

5.6 **a.** NO_2^- (each O is bonded to N)

Lewis Structure	3D Structure	VSEPR
$[:O=N-O:]^-$	$[:O=N, O:]^-$	bent

b. ClO_3^- (each O is bonded to Cl)

Lewis Structure	3D Structure	VSEPR
$[:O-Cl-O:, :O:]^-$	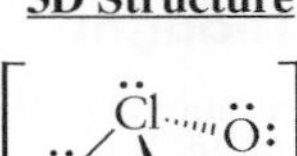	trigonal pyramidal with Cl at the top

c. CO_3^{2-} (each O is bonded to C)

Lewis Structure	3D Structure	VSEPR
$[:O=C-O:, :O:]^{2-}$	$[:O=C, O:, O:]^{2-}$	trigonal planar with C in the middle

d. $H_3-O_3^+$ (each H is bonded to O; note the positive charge; compare with $NH_4{}^+$)

Lewis Structure	3D Structure	VSEPR
$[H-O-H, H]^+$	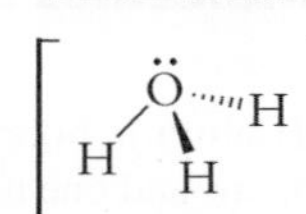	trigonal pyramidal with O at the top

5.8 Trigonal pyramidal

5.10 Bent

The Polarity of Covalent Molecules (Section 5.2)

5.12 **a.** O **b.** F

5.14 **a.** $\overrightarrow{N—F}$ **b.** $\overleftarrow{Si—Cl}$

5.16 **a.** $\overrightarrow{H—N}$ **b.** $\overleftarrow{F—Cl}$

c. $\overrightarrow{C—F}$

5.18 **a.** H—I $\quad \Delta EN = 2.5 - 2.1 = 0.4 \quad \overset{\delta^+}{H}\!\rightleftarrows\!\overset{\delta^-}{I}$

b. O–S=O $\quad \Delta EN = 3.5 - 2.5 = 1.0$

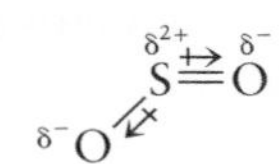

c. O–O=O $\quad \Delta EN = 3.5 - 3.5 = 0 \quad$ The bonds in this molecule are nonpolar.

5.20 **a.** ionic **b.** polar covalent

c. nonpolar covalent **d.** polar covalent

e. ionic

5.22 **a.** polar **b.** polar

c. nonpolar

5.24 **a.** polar covalent **b.** nonpolar covalent

c. nonpolar covalent

5.26 **a.** C≡O $\quad \overset{\delta^+}{C}\!\equiv\!\overset{\delta^-}{O}$ $\quad$ polar molecule (nonsymmetric charge distribution)

b. H—Se—H $\quad \delta^+$ H, δ^{2-} Se, δ^+ H $\quad$ polar molecule (nonsymmetric charge distribution)

c. AlI_3 $\quad \delta^-$ I, δ^{3+} Al, δ^- I, δ^- I $\quad$ nonpolar molecule (symmetric charge distribution)

5.28 **a.** $SOCl_2$ (S atom in the middle)
$SOCl_2$ is polar.

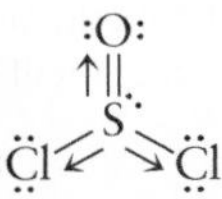

b. NF_3 (N atom in the middle)
NF_3 is polar.

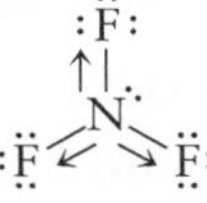

c. CH_4 (C atom in the middle)
CH_4 is nonpolar.

H–C–H (with H above and H below C)

Intermolecular Forces (Section 5.3)

5.30 **a.** London dispersion forces **b.** London dispersion forces

c. Dipole–dipole

5.32 The alcohol has higher melting and boiling points than the ether. The forces that hold the alcohol molecules together must be stronger and harder to break than the forces that hold the ether molecules together.

5.34 London dispersion forces < dipole–dipole forces < hydrogen-bonding < ion–dipole interactions.

5.36 A single water molecule can form a maximum of four hydrogen bonds.

Energy and Properties of Matter (Section 5.4)

5.38 **a.** The molecules of a liquid possess kinetic and potential energy. The kinetic energy is not great enough to overcome the attractive forces between the molecules; therefore, the molecules are able to flow together into the shape of the container, but the volume of the liquid remains constant.

b. Solid and liquids are composed of molecules with considerable attractive forces between the molecules. These attractive forces bring the molecules close together and cause them to be difficult to compress further.

c. Gases are composed of molecules with high kinetic energy and low potential energy. The molecules are small compared to the amount of space occupied by the gas. The molecules strike each other and the walls of the container; however, all of the collisions are elastic and no net energy is lost from the system. Since the molecules are in constant motion, the same average number of gas molecules will strike the walls of the container at any given time and the elastic collisions result in uniform pressure on the walls of its container.

5.40 Liquids and gases are similar in their fluidity, ability to mix, and response to temperature and pressure changes. However, liquids have stronger intermolecular forces, a fixed volume, and are denser, while gases have weaker intermolecular forces, are more compressible, and have greater diffusion rates.

5.42 Kinetic energy is the energy of motion. The particles in a solid have the lowest kinetic energy because the particles are moving the least in this phase of matter. The particles in a liquid have higher kinetic energy than a solid, as well as lower kinetic energy than a gas. The particles in a gas have the highest kinetic energy because the particles are moving the most in this phase of matter.

The potential energy of the three phases of matter is easiest to compare during a phase transition. When a solid melts into a liquid, the temperature does not change, even though energy is added to the solid; therefore, both the liquid and the solid particles have the same average kinetic energy at the melting point, and the liquid must have higher potential energy than the solid. Similarly, when a liquid evaporates into a gas, the temperature does not change, even though energy is added to the liquid; therefore, both the liquid and the gas particles have the same average kinetic energy at the boiling point and the gas must have higher potential energy than the liquid.

Changes of State (Section 5.5)

5.44 **a.** gaseous **b.** solid
c. gaseous **d.** gaseous

5.46 **a.** exothermic **b.** exothermic
c. endothermic

5.48 **a.** endothermic **b.** exothermic

5.50 **a.** endothermic **b.** exothermic

5.52 Liquid water was heated to become steam (endothermic process), and then the steam was cooled to become liquid water again (exothermic process).

5.54 Methylene chloride is a volatile liquid. When it was sprayed in the mouth, the methylene chloride absorbed heat from the tissue and evaporated. The tissue became cold and was anesthetized.

5.56 Water and ethylene glycol differ in their boiling points. Water has a boiling point of 100 °C, while ethylene glycol has a boiling point that is higher than 100 °C. To determine the identity of the two boiling liquids, use a thermometer to measure the boiling point.

5.58 Water (H_2O) has a lower vapor pressure than 1-propanol ($CH_3CHOHCH_3$). The hydrogen-bonding intermolecular forces in water are stronger than in 1-propanol, which makes it more difficult for the water molecules to escape from the liquid state and enter the gas state.

5.60 $C_5H_{12} < C_6H_{14} < C_8H_{18} < C_{10}H_{22}$

The boiling point increases as the number of carbon atoms in the molecule increases.

5.62 $Mg(OH)_2$

5.64 To obtain pure solid iodine from a mixture of solid iodine and sand, heat the mixture until the iodine sublimes and provide a cold surface above the mixture so the iodine can be deposited as a solid. The sand will not sublime, and the solid on the cold surface will be pure iodine.

5.66 When the ice and water mixture reached a constant temperature of 0.0 °C, the system shared not only the same temperature throughout, but also the same vapor pressure. The vapor pressure of both the water and the ice is 4.58 torr.

Energetics: Changes of State vs. Specific Heat (Section 5.6)

5.68 **a.** 711.3 kJ, which rounds to 710 kJ
b. 104.6 kJ, which rounds to 1.0×10^2 kJ

5.70 35.6 kcal, which rounds to 36 kcal

5.72 **a.** 3.6×10^2 cal
b. 7×10^3 cal
c. 2.40×10^3 cal

5.74 In a solar heat storage system, solids are melted and the energy is stored in the liquid form until the heat is released upon resolidification. The melting point of K_2SO_4 is 1069 °C which is much too high to be useful. The melting temperature must be low enough to easily maintain the compound in the liquid state for storage on cloudy days.

5.76 7.72×10^4 cal

Additional Exercises

5.78 N_2O. Covalent. The name of N_2O is dinitrogen monoxide.

5.80 The temperature of the boiling water would stay the same. The boiling point of water does not depend on the amount of energy added to the water. The time required to reach the boiling point of water may decrease when using two burners rather than just one.

5.82 Lead because it has the lowest specific heat

5.84 1.83 kJ

Chapter 6
Properties of Gases (Section 6.1)

6.2 N_2O kinetic energy $= 3.14 \times 10^{10} \dfrac{\text{g cm}^2}{\text{s}^2}$

Sevoflurane kinetic energy $= 3.13 \times 10^{10} \dfrac{\text{gcm}^2}{\text{s}^2}$

6.4 **a.** 0.957 atm **b.** 727 torr
c. 14.1 psi **d.** 0.966 bars

6.6 **a.** 14.3 atm **b.** 14.4 bars

c. 1.09×10^4 mm Hg **d.** 427 in. Hg

6.8 **a.** $K = 273 + °C = 273 + (-268.9\ °C) = 4.1\ K$

b. $°C = K - 273 = 14.1\ K - 273 = 258.9\ °C$

c. $K = 273 + °C = 273 + 63.7\ °C = 336.7\ K$

Pressure, Temperature, and Volume Relationships (Section 6.2)

6.10 Gas A: $P_f = 1.50$ atm Gas B: $V_f = 4.49$ L Gas C: $T_f = 221$ K

6.12 138 mL

6.14 3.70 atm

6.16 31.6 L

6.18 714 L

6.20 4.6 L

6.22 2.5 L

6.24 3.62 mL

6.26 $1.4 \times 10^4\ ft^3$

6.28 1.33 L

6.30 5.00 g/L

The Ideal Gas Law (Section 6.3)

6.32 **a.** 0.0130 moles **b.** 7.65 atm

c. 8.35 L

6.34 3.19 atm

6.36 26.3 g

6.38 2.7 moles

6.40 168 g/mol

6.42 The gas is CO_2 (MW = 44.0 g/mole).

Dalton's Law (Section 6.4)

6.44 $P_T = P_{O_2} + P_{He}$

$2100\ torr = P_{O_2} + 1680\ torr$

$P_{O_2} = 420\ torr$

$$\% = \frac{\text{part}}{\text{whole}} \times 100\%$$

$$\%He = \frac{1680\ torr}{2100\ torr} \times 100\% = 80\%\ He$$

$$\%O_2 = \frac{420\ torr}{2100\ torr} \times 100\% = 20\%\ O_2$$

Solutions and Solubility (Section 6.5)

6.46

	Solvent	Solutes
a. Liquid laundry bleach: sodium hypochlorite 5.25%, inert ingredient 94.75%	water	sodium hypochlorite
b. Rubbing alcohol: isopropyl alcohol 70%	isopropyl alcohol	water
c. Hydrogen peroxide: 3% hydrogen peroxide	water	hydrogen peroxide
d. After shave: SD alcohol, water, glycerin, fragrance, menthol, benzophenone-1, coloring	SD alcohol	water, glycerin, fragrance, menthol, benzophenone-1, coloring

6.48 **a.** Immiscible

b. Soluble

c. Soluble

6.50 **a.** Unsaturated

b. Supersaturated

c. Saturated

6.52 *Miscible* means that two liquids can be mixed in any concentration; there are no solubility limits.

6.54 This solution could become supersaturated by slowly lowering the temperature of the solution or by allowing some of the solvent to evaporate. The solution must be handled very gently because a supersaturated solution is unstable.

6.56 **a.** soluble in benzene

b. soluble in benzene

c. soluble in water

d. soluble in benzene

6.58 Insoluble. The structure indicates that halothane is a nonpolar material. Therefore, halothane would not be soluble in a polar solvent, such as water.

Electrolytes and Net Ionic Equations (Section 6.6)

6.60 **a.** $LiNO_3$ Li^+, NO_3^- **d.** KOH K^+, OH^-

b. Na_2HPO_4 $2Na^+, HPO_4^{2-}$ **e.** $MgBr_2$ $Mg^{2+}, 2Br^-$

c. $Ca(ClO_3)_2$ $Ca^{2+}, 2ClO_3^-$ **f.** $(NH_4)_2SO_4$ $2NH_4^+, SO_4^{2-}$

6.62 **a.** $SO_2(aq) + H_2O(\ell) \rightarrow H_2SO_3(aq)$

Total ionic equation: $SO_2(aq) + H_2O(\ell) \rightarrow 2H^+(aq) + SO_3^{2-}(aq)$

Spectator ions: none

Net ionic equation: $SO_2(aq) + H_2O(\ell) \rightarrow 2H^+(aq) + SO_3^{2-}(aq)$

b. $CuSO_4(aq) + Zn(s) \rightarrow ZnSO_4(aq)$

Total ionic equation:
$Cu^{2+}(aq) + SO_4^{2-}(aq) + Zn(s) \rightarrow Cu(s) + Zn^{2+}(aq) + SO_4^{2-}(aq)$

Spectator ions: $SO_4^{2-}(aq)$

Net ionic equation: $Cu^{2+}(aq) + Zn(s) \rightarrow Cu(s) + Zn^{2+}(aq)$

(Solution 6.62 continued)

c. $2KBr(aq) + 2H_2SO_4(aq) \rightarrow Br_2(aq) + SO_2(aq) + K_2SO_4(aq) + 2H_2O(\ell)$

Total ionic equation: $2K^+(aq) + 2Br^-(aq) + 4H^+(aq) + 2SO_4^{2-}(aq) \rightarrow Br_2(aq) + SO_2(aq) + 2K^+(aq) + SO_4^{2-}(aq) + 2H_2O(\ell)$

Spectator ions: $2K^+(aq), SO_4^{2-}(aq)$

Net ionic equation: $2Br(aq) + 4H^+(aq) + SO_4^{2-}(aq) \rightarrow Br_2^-(aq) + SO_2(aq) + 2H_2O(\ell)$

d. $AgNO_3(aq) + NaOH(aq) \rightarrow AgOH(s) + NaNO_3(aq)$

Total ionic equation: $Ag^+(aq) + NO_3^-(aq) + Na^+(aq) + OH^-(aq) \rightarrow AgOH(s) + Na^+(aq) + NO_3^-(aq)$

Spectator ions: $Na^+(aq), NO_3^-$

Net ionic equation: $Ag^+(aq) + OH^-(aq) \rightarrow AgOH(s)$

e. $BaCO_3(s) + 2HNO_3(aq) \rightarrow Ba(NO_3)_2(aq) + CO_2(g) + H_2O(\ell)$

Total ionic equation: $BaCO_3(s) + 2H^+(aq) + 2NO_3^-(aq) \rightarrow Ba^{2+}(aq) + 2NO_3^-(aq) + CO_2(g) + H_2O(\ell)$

Spectator ions: $2NO_3^-$

Net ionic equation: $BaCO_3(s) + 2H^+(aq) \rightarrow Ba^{2+}(aq) + CO_2(g) + H_2O(\ell)$

f. $N_2O_5(aq) + H_2O(\ell) \rightarrow + 2HNO_3(aq)$

Total ionic equation: $N_2O_5(aq) + H_2O(\ell) \rightarrow 2H^+(aq) + 2NO_3^-(aq)$

Spectator ions: none

Net ionic equation: $N_2O_5(aq) + H_2O(\ell) \rightarrow 2H^+(aq) + 2NO_3^-(aq)$

Solution Concentrations (Section 6.7)

6.64 a. 0.364 M b. 0.860 M c. 1.75 M

6.66 a. 0.500 M b. 0.200 M c. 0.26 M

6.68 a. 0.376 moles b. 0.0750 moles c. 571 mL

6.70 a. 0.820 b. 0.130 L c. 516 mL

6.72 a. 5.0% b. 5.0% c. 5.0%

6.74 a. 7.41% b. 15.3% c. 54.5% d. 44.1%

6.76 a. 11.1% b. 10.7% c. 7.48%

6.78 a. 7.5% b. 7.5% c. 3.1% d. 43.0%

6.80 25 mL

6.82 a. 5.00% b. 5.00% c. 0.900%

6.84 $\dfrac{65 \text{ g KBr}}{65 \text{ g KBr} + 100\% \cdot \text{g } H_2O} \times 100\% = 39\% \text{ (w/w)}$

Solution Preparation (Section 6.8)

6.86 a. $200. \text{ mL}\left(\dfrac{1 \text{ L}}{1000 \text{ mL}}\right)\left(\dfrac{0.150 \text{ moles } Na_2SO_4}{1 \text{ L}}\right)\left(\dfrac{142.05 \text{ g } Na_2SO_4}{1 \text{ mole } Na_2SO_4}\right) = 4.26 \text{ g } Na_2SO_4$

I would weigh out 4.26 Na_2SO_4 and add it to a 200. mL volumetric flask. I would add enough water to dissolve the Na_2SO_4, then add water up to the mark on the volumetric flask, cap, and shake to ensure the solution is homogeneous.

b. $250.0 \text{ mL}\left(\dfrac{1 \text{ L}}{1000 \text{ mL}}\right)\left(\dfrac{0.250 \text{ moles}}{1 \text{ L}}\right)\left(\dfrac{189.41 \text{ g } Zn(NO_3)_2}{1 \text{ mole } Zn(NO_3)_2}\right) = 11.8 \text{ g } Zn(NO_3)_2$

I would weigh out 11.8 g $Zn(NO_3)_2$ and add it to a 250.0 mL volumetric flask. I would add enough water to dissolve the $Zn(NO_3)_2$, then add water up to the mark on the volumetric flask cap and shake to ensure the solution is homogeneous.

c. $150.0 \text{ g} \times 2.25\% \text{ (w/w)} = 150.0 \text{ g} \times 0.0225 = 3.38 \text{ g NaCl}$

$150.0 \text{ mL} - 3.38 \text{ g} = 146.62 \text{ g } H_2O\left(\dfrac{1 \text{ mL}}{1.00 \ H_2O}\right) = 146.62 \text{ mL } H_2O \approx 147 \text{ mL } H_2O$

I would weigh out 3.38 g NaCl and add it to a 250.0 mL Erlenmeyer flask. I would measure 147 mL of water and add enough of it to the Erlenmeyer flask to dissolve the salt, then I would add the rest of the water and swirl the flask to ensure that the solution is homogeneous.

d. $125.0 \text{ mL} \times 0.75\% \text{ (w/v)} = 0.938 \text{ KCl} \approx 0.94 \text{ g KCl}$

I would add 0.94 g KCl to a 125.0 mL volumetric flask. I would add enough water to dissolve the KCl, then dilute the mixture to the mark on the volumetric flask. I would then cap the flask and shake it to ensure the solution is homogeneous.

6.88 2.34 g

6.90 a. 32.3 g b. 0.700 moles c. 31.3 mL d. 2.10 g

6.92 a. $(6.00 \text{ M})(5.00 \text{ L}) = (18.0 \text{ M})(V_c)$
$V_c = 1.67 \text{ L}$

I would add 3.00 L of water to a 5.00 L volumetric flask, then add 1.67 L of 18.0 M H_2SO_4 to the flask, let it cool, and dilute to the mark with water. (Always add acid to water, not the reverse!)

b. $(0.500 \text{ M})(250.0 \text{ mL}) = (3.00 \text{ M})(V_c)$
$V_c = 41.7 \text{ mL}$

I would add 41.7 mL of 3.00 M $CaCl_2$ to a 250.0 mL volumetric flask, then dilute to the mark with water. I would be sure to shake well.

c. $(1.50\%\ (w/v))(200.0\ mL) = (10.0\%\ (w/v))(V_c)$
$V_c = 30.0\ mL$

I would add 30.0 mL of 10.0% (w/v) KBr to a 200.0 mL volumetric flask, then dilute to the mark with water. I would be sure to shake well.

d. $(10.0\%\ (v/v))(500.0\ mL) = (50.0\%\ (v/v))(V_c)$
$V_c = 100\ mL$

I would add 100 mL of 50.0% (v/v) alcohol to a 500.0 mL volumetric flask, then dilute to the mark with water. I would be sure to shake well.

6.94 a. 0.00650 M b. 0.0488 M
c. 0.0195 M d. 0.0139 M

Osmosis and Dialysis (Section 6.9)

6.96 The water will flow from the 5.00% sugar solution into the 10.0% sugar solution because the 5.00% sugar solution contains more solvent (water) than the 10.0% sugar solution does. The 10.0% sugar solution will become diluted as osmosis takes place. Allowed enough time, the two solutions will eventually have the same concentration.

6.98 a. The hydrated sodium and chloride ions will pass through the dialyzing membrane, but the starch (colloid) will not.

b. The urea will pass through the dialyzing membrane because it is a small organic molecule, but the starch (colloid) will not.

c. The hydrated potassium and chloride ions as well as the glucose molecules will pass through the dialyzing membrane, but the albumin (colloid) will not.

Additional Exercises

6.100 $2H_2(g) + O_2(g) \rightarrow 2H_2O(\ell)$

$$2.31\ L\ H_2\left(\frac{1\ L\ O_2}{2\ L\ H_2}\right) = 1.16\ L\ O_2$$

This reaction is classified as both a combination reaction and a redox reaction.

Chapter 7
Principles of Chemical Reactions (Section 7.1)

7.2 a. Endothermic (endergonic) reaction with activation energy

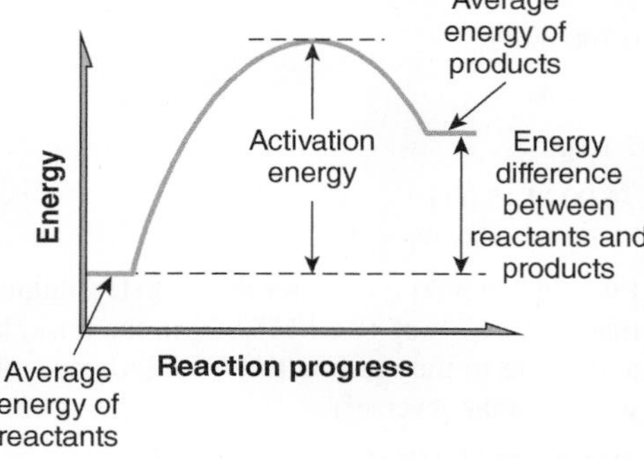

b. Endothermic (endergonic) reaction without activation energy

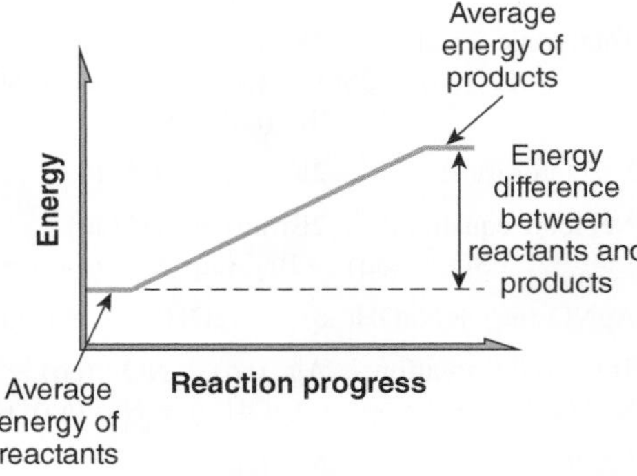

Both of these energy diagrams have the same average energy of the reactants, average energy of products, and energy difference between reactants and products. The main difference between these two energy diagrams is that the first diagram has an activation energy and the second diagram does not.

7.4 As temperature increases, the molecules move faster. The faster the molecules move, the more likely they are to collide. As the temperature increases, the molecules gain energy. The more energy the molecules have, the more they are likely to react once they collide.

7.6

a.	The space shuttle Orion leaves its pad and goes into orbit.	nonspontaneous	Rocket engines must continually operate to push the shuttle into orbit.
b.	The fuel in a booster rocket of the space shuttle burns.	spontaneous	Once the fuel is ignited, it will continue to burn. No additional energy has to be provided.
c.	Water boils at 100 °C and 1 atm pressure.	nonspontaneous	Heat must be continually supplied to maintain boiling.
d.	Water temperature increases to 100 °C at 1 atm pressure.	nonspontaneous	Increasing the temperature of water requires a continual supply of energy.
e.	Your bedroom becomes orderly.	nonspontaneous	A room will not become orderly on its own. Cleaning requires energy.

7.8 a. An automobile being pushed up a slight hill (from POV of the one pushing) — exothermic — The person pushing the car gives energy to the car.

b. Ice melting (from POV of the ice) — endothermic — Melting ice requires energy.

c. Ice melting (from POV of surroundings of the ice) — exothermic — The surroundings release heat into the ice.

d. Steam condensing to liquid water (from POV of the steam) — exothermic — Heat must be released from the steam.

e. Steam condensing to liquid water (from POV of surroundings of the steam) — endothermic — Heat must be absorbed by the surroundings.

7.10 a. the melting of a block of ice — The changing height of the block

b. the setting (hardening) of concrete — The ability of an object to penetrate

c. the burning of a candle — The changing height of the candle

7.12 (1) Heat the reactants in order to increase the energy of the reactants and the frequency of collision, (2) stir the reactants in order to increase the frequency of collision, (3) increase the surface area of the reactants to increase the number of collisions, and (4) add a catalyst to lower the activation energy for the reaction. Only three of these steps are needed for a complete answer.

Equilibrium (Section 7.2)

7.14 a. The color of the gas mixture will stop changing once equilibrium is reached.

b. The amount of solid sugar will become constant once the mixture has reached equilibrium.

c. Both the color and the pressure of the gas mixture will stop changing once equilibrium is reached.

7.16 a. $2CO + O_2 \rightleftarrows 2CO_2$ $\quad K_{eq} = \dfrac{[CO_2]^2}{[CO]^2[O_2]}$

b. $N_2O_4 \rightleftarrows 2NO_2$ $\quad K_{eq} = \dfrac{[NO_2]^2}{[N_2O_4]}$

c. $2C_2H_6 + 7O_2 \rightleftarrows 4CO_2 + 6H_2O$ $\quad K_{eq} = \dfrac{[CO_2]^4[H_2O]^6}{[C_2H_6]^2[O_2]^7}$

d. $2NO + Cl_2 \rightleftarrows 4CO_2 + 6H_2O$ $\quad K_{eq} = \dfrac{[NO]^2[Cl_2]}{[NOCl]^2}$

e. $2Cl_2O_5 \rightleftarrows O_2 + 4ClO_2$ $\quad K_{eq} = \dfrac{[O_2][ClO_2]^4}{[Cl_2O_5]^2}$

7.18 a. $Fe^{3+} + 6CN^- \rightleftarrows Fe(CN)_6^{3-}$ $\quad K_{eq} = \dfrac{[Fe(CN)_6^{3-}]}{[Fe^{3+}][CN_6^-]}$

b. $Ag^+ + 2NH_3 \rightleftarrows Ag(NH_3)_2^+$ $\quad K_{eq} = \dfrac{[Ag(NH_3)_2^+]}{[Ag^+][NH_3]^2}$

c. $Au^{3+} + 4Cl^- \rightleftarrows AuCl_4^-$ $\quad K_{eq} = \dfrac{[AuCl_4^-]}{[Au^{3+}][Cl^-]^4}$

7.20 a. $K_{eq} = \dfrac{[CO_2][H_2O]^2}{[CH_4][O_2]^2}$ $\quad CH_4 + 2O_2 \rightleftarrows CO_2 + 2H_2O$

b. $K_{eq} = \dfrac{[CH_4][H_2O]}{[H_2]^3[CO]}$ $\quad 3H_2 + CO \rightleftarrows CH_4 + 2H_2O$

c. $K_{eq} = \dfrac{[O_2]^3}{[O_3]^2}$ $\quad 2O_3 \rightleftarrows 3O_2$

d. $K_{eq} = \dfrac{[NH_3]^4[O_2]^7}{[NO_2]^4[H_2O]^6}$ $\quad 4NO_2 + 6H_2O \rightleftarrows 4NH_3 + 7O_2$

7.22 a. [reactants] smaller than [products]

b. [reactants] smaller than [products]

c. [reactants] larger than [products]

d. [reactants] larger than [products]

7.24 a. The equilibrium will shift to the right and the mixture will become less blue and more purple.

b. The equilibrium will shift to the right and more precipitate will form. Heat will also be generated and the temperature of the container will increase.

c. The equilibrium will shift to the right, the mixture will become less violet, and more heat will be produced.

d. The equilibrium will shift to the right, the mixture will become less violet, and more heat will be produced.

e. The catalyst does not have any effect on the equilibrium; however, adding NH_3 shifts the equilibrium to the left and this produces heat, lowers the pressure because there will be fewer moles of gas present, and increases the brown color of the equilibrium mixture.

7.26 a. The equilibrium will shift to the left. The amount of $LiHCO_3$ will decrease, the amount of LiOH will increase and the concentration of CO_2 will increase. Heat will be used and the container will be cooler to the touch.

b. The equilibrium will shift to the left. The concentration of CO_2 and H_2O will decrease, as will the amount of Na_2O. The amount of $NaHCO_3$ will increase. Heat will be generated and the container will feel warmer.

c. The equilibrium will shift to the left. The concentration of CO_2 will decrease, as will the amount of CaO. The amount of $CaCO_3$ will increase. Heat will be generated and the container will feel warmer.

7.28 $N_2(g) + 3H_2(g) \rightleftarrows 2NH_3(g) + \text{heat}$

a. to the right b. to the left

c. to the right d. to the left

e. no shift f. to the left

7.30 $[I_2] = 0.099$ M

7.32 0.099

Acid–Base Equilibria (Section 7.3)

7.34 **a.** $HBrO_2(aq) \rightleftarrows H^+(aq) + BrO_2^-(aq)$

b. $HS^-(aq) \rightleftarrows H^+(aq) + S^{2-}(aq)$

c. $HBr(aq) \rightleftarrows H^+(aq) + Br^-(aq)$

d. $HC_2H_3O_2(aq) \rightleftarrows H^+(aq) + C_2H_3O_2^-(aq)$

7.36

	Brønsted Acids	Brønsted Bases
a.	$HC_2O_4^-$, H_3O^+	H_2O, $C_2O_4^{2-}$
b.	HNO_2, H_3O^+	H_2O, NO_2^-
c.	H_2O, HPO_4^{2-}	PO_4^{3-}, OH^-
d.	H_2SO_3, H_3O^+	H_2O, HSO_3^-
e.	H_2O, HF	F^-, OH^-

7.38 **a.** $HF(aq) + H_2O(\ell) \rightleftarrows F^-(aq) + H_3O^+(aq)$

b. $HClO_3(aq) + H_2O(\ell) \rightleftarrows ClO_3^-(aq) + H_3O^+(aq)$

c. $HClO(aq) + H_2O(\ell) \rightleftarrows ClO^-(aq) + H_3O^+(aq)$

d. $HS^-(aq) + H_2O(\ell) \rightleftarrows S^{2-}(aq) + H_3O^+(aq)$

7.40 Conjugate Base

a. SO_4^{2-} **b.** CH_3NH_2

c. ClO_4^- **d.** NH_3

e. Cl^-

7.42 Conjugate Acid

a. H_2CO_3 **b.** HS^-

c. H_2S **d.** $H_2C_2O_4$

e. $H_2N_2O_2$

7.44 Missing Formula

a. $HAsO_4^{2-}(aq)$ **b.** $C_6H_5NH_2(aq)$

c. $H_2O(\ell)$ **d.** $(CH_3)_2NH(aq)$

e. $CH_3NH_3^+(aq)$

7.46 **a.** $H_3O^+(aq) + NH_2^-(aq) \rightleftarrows NH_3(aq) + H_2O(\ell)$

b. $H_2PO_4^-(aq) + NH_3(aq) \rightleftarrows NH_4^+(aq) + HPO_4^{2-}(aq)$

c. $HS_2O_3^-(aq) + OCl^-(aq) \rightleftarrows HOCl(aq) + S_2O_3^{2-}(aq)$

d. $H_2O(\ell) + ClO_4^-(aq) \rightleftarrows HClO_4(aq) + OH^-(aq)$

e. $H_2O(\ell) + NH_3(aq) \rightleftarrows OH^-(aq) + NH_4^+(aq)$

7.48 **a.** hydrotelluric acid **b.** hypochlorous acid

c. sulfurous acid **d.** nitrous acid

7.50 succinic acid

The pH Concept (Section 7.4)

7.52 **a.** Acidic **b.** Basic

c. Acidic **d.** Basic

7.54 **a.** $[OH^-] = \dfrac{K_w}{[H^+]} = \dfrac{1.0 \times 10^{-14}\,(mol/L)^2}{1.0 \times 10^{-7}\,mol/L} = 1.0 \times 10^{-7}\,M$

b. $[OH^-] = \dfrac{K_w}{[H^+]} = \dfrac{1.0 \times 10^{-14}\,(mol/L)^2}{1.3 \times 10^{-4}\,mol/L} = 7.7 \times 10^{-11}\,M$

c. $[OH^-] = \dfrac{K_w}{[H^+]} = \dfrac{1.0 \times 10^{-14}\,(mol/L)^2}{0.0063\,mol/L} = 1.6 \times 10^{-12}\,M$

d. $[OH^-] = \dfrac{K_w}{[H^+]} = \dfrac{1.0 \times 10^{-14}\,(mol/L)^2}{1.0 \times 10^{-5}\,mol/L} = 1.0 \times 10^{-9}\,M$

e. $[OH^-] = \dfrac{K_w}{[H^+]} = \dfrac{1.0 \times 10^{-14}\,(mol/L)^2}{5.0 \times 10^{-9}\,mol/L} = 2.0 \times 10^{-6}\,M$

7.56 **a.** $[H_3O^+] = \dfrac{K_w}{[OH^-]} = \dfrac{1.0 \times 10^{-14}\,(mol/L)^2}{1.0 \times 10^{-11}\,mol/L} = 1.0 \times 10^{-3}\,M$

b. $[H_3O^+] = \dfrac{K_w}{[OH^-]} = \dfrac{1.0 \times 10^{-14}\,(mol/L)^2}{1.3 \times 10^{-10}\,mol/L} = 7.7 \times 10^{-5}\,M$

c. $[H_3O^+] = \dfrac{K_w}{[OH^-]} = \dfrac{1.0 \times 10^{-14}\,(mol/L)^2}{1.0 \times 10^{-7}\,mol/L} = 1.0 \times 10^{-7}\,M$

d. $[H_3O^+] = \dfrac{K_w}{[OH^-]} = \dfrac{1.0 \times 10^{-14}\,(mol/L)^2}{3.2 \times 10^{-7}\,mol/L} = 3.1 \times 10^{-8}\,M$

e. $[H_3O^+] = \dfrac{K_w}{[OH^-]} = \dfrac{1.0 \times 10^{-14}\,(mol/L)^2}{5.0 \times 10^{-11}\,mol/L} = 2.0 \times 10^{-4}\,M$

7.58 **a.** Bile, pH = 8.05
$[H^+] = 10^{-8.05} = 8.9 \times 10^{-9}\,M$ basic

b. Vaginal fluid, pH = 3.93
$[H^+] = 10^{-3.93} = 1.2 \times 10^{-4}\,M$ acidic

c. Semen, pH = 7.38
$[H^+] = 10^{-7.38} = 4.2 \times 10^{-8}\,M$ basic

d. Cerebrospinal fluid, pH = 7.40
$[H^+] = 10^{-7.40} = 4.0 \times 10^{-8}\,M$ basic

e. Perspiration, pH = 6.23
$[H^+] = 10^{-6.23} = 5.9 \times 10^{-7}\,M$ acidic

7.60 **a.** $[H_3O^+] = 1.0 \times 10^{-pH} = 1.0 \times 10^{-7.2} = 6.3 \times 10^{-8}\,M$

b. $[H_3O^+] = 1.0 \times 10^{-pH} = 1.0 \times 10^{-7.8} = 1.6 \times 10^{-8}\,M$

c. $[H_3O^+] = 1.0 \times 10^{-pH} = \dfrac{6.3 \times 10^{-8}\,M\ (\text{for pH 7.2})}{1.6 \times 10^{-8}\,M\ (\text{for pH 7.8})}$

pH 7.2 is 3.9 times as acidic as pH 7.8.

7.62 **a.** pH = 3.95 $[H^+] = 10^{-3.95} = 1.1 \times 10^{-4}\,M$

$[OH^-] = \dfrac{10^{-14}}{10^{-3.95}} = 8.9 \times 10^{-11}\,M$

b. pH = 4.00 $[H^+] = 10^{-4.00} = 1.0 \times 10^{-4}\,M$

$[OH^-] = \dfrac{10^{-14}}{10^{-4.00}} = 1.0 \times 10^{-10}\,M$

c. pH = 11.86 $[H^+] = 10^{-11.86} = 1.4 \times 10^{-12}\,M$

$[OH^-] = \dfrac{10^{-14}}{10^{-11.86}} = 7.2 \times 10^{-3}\,M$

The Strengths of Acids and Bases (Section 7.5)

7.64 **a.** (weakest) acid B < acid A < acid C < acid D (strongest)

b. (weakest) base D < base C < base A < base B (strongest)

7.66 **a.** HSe^- $HSe^-(aq) \rightleftarrows H^+(aq) + Se^{2-}(aq)$ $K_a = \frac{[H^+][Se^{2-}]}{[HSe^-]}$

b. $H_2BO_3^-$ $H_2BO_3^-(aq) \rightleftarrows H^+(aq) + HBO_3^{2-}(aq)$ $K_a = \frac{[H^+][HBO_3^{2-}]}{[H_2BO_3^-]}$

c. HBO_3^{2-} $HBO_3^{2-}(aq) \rightleftarrows H^+(aq) + BO_3^{3-}(aq)$ $K_a = \frac{[H^+][BO_3^{3-}]}{[HBO_3^{2-}]}$

d. $HAsO_4^{2-}$ $HAsO_4^{2-}(aq) \rightleftarrows H^+(aq) + AsO_4^{3-}(aq)$ $K_a = \frac{[H^+][AsO_4^{3-}]}{[HAsO_4^{2-}]}$

e. $HClO$ $HClO(aq) \rightleftarrows H^+(aq) + ClO^-(aq)$ $K_a = \frac{[H^+][ClO^-]}{[HClO]}$

7.68 The 20% acetic acid solution is the weak acid solution. It is a weak acid solution because acetic acid does not completely dissociate.

Acid–Base Titrations (Section 7.6)

7.70 **a.** M*: $3RbOH(aq) + H_3PO_4(aq) \rightarrow Rb_3PO_4(aq) + 3H_2O(\ell)$

TIE**: $3Rb^+(aq) + 3OH^-(aq) + 3H^+(aq) + PO_4^{2-}(aq) \rightarrow 3Rb^+(aq) + PO_4^{3-}(aq) + 3H_2O(\ell)$

NIE***: $OH^-(aq) + H^+(aq) \rightarrow H_2O(\ell)$

b. M*: $2RbOH(aq) + H_2C_2O_4(aq) \rightarrow Rb_2C_2O_4(aq) + 2H_2O(\ell)$

TIE**: $2Rb^+(aq) + 2OH^-(aq) + 2H^+ + C_2O_4^{2-}(aq) \rightarrow 2Rb^+(aq) + C_2O_4^{2-}(aq) + 2H_2O(\ell)$

NIE***: $OH^-(aq) + H^+(aq) \rightarrow H_2O(\ell)$

c. M*: $RbOH(aq) + HC_2H_3O_2(aq) \rightarrow RbC_2H_3O_2(aq) + H_2O(\ell)$

TIE**: $Rb^+(aq) + OH^-(aq) + H^+(aq) + C_2H_3O_2^-(aq) \rightarrow Rb^+(aq) + C_2H_3O_2^-(aq) + H_2O(\ell)$

NIE***: $OH^-(aq) + H^+(aq) \rightarrow H_2O(\ell)$

7.72 $Al(OH)_3 + 3HCl \rightarrow AlCl_3 + 3H_2O$

7.74 **a.** M*: $2KOH(aq) + H_3PO_4(aq) \rightarrow K_2HPO_4(aq) + 2H_2O(\ell)$

TIE**: $2K^+(aq) + 2OH^-(aq) + 2H^+(aq) + HPO_4^{2-}(aq) \rightarrow 2K^+(aq) + HPO_4^{2-}(aq) + 2H_2O(\ell)$

NIE***: $OH^-(aq) + H^+(aq) \rightarrow H_2O(\ell)$

b. M*: $3KOH(aq) + H_3PO_4(aq) \rightarrow K_3PO_4 + 3H_2O(\ell)$

TIE**: $3K^+(aq) + 3OH^-(aq) + 3H^+(aq) + PO_4^{3-}(aq) \rightarrow 3K^+(aq) + PO_4^{3-} + 3H_2O(\ell)$

NIE***: $OH^-(aq) + H^+(aq) \rightarrow H_2O(\ell)$

c. M*: $KOH(aq) + H_2C_2O_4(aq) \rightarrow KHC_2O_4 + H_2O(\ell)$

TIE**: $K^+(aq) + OH^-(aq) + H^+(aq) + HC_2O_4^-(aq) \rightarrow K^+(aq) + HC_2O_4^- + H_2O(\ell)$

NIE***: $OH^-(aq) + H^+(aq) \rightarrow H_2O(\ell)$

7.76 **a.** 0.100 moles NaOH **b.** 0.225 moles NaOH

7.78 0.1660 M $H_2C_2O_4$

7.80 **a.** 3.50 M H_2SO_4 **b.** 0.891 M $HC_2H_3O_2$

c. 12.4 M HCl

Maintaining pH: The Balancing Act (Section 7.7)

7.82 One component reacts with acid, whereas the other component reacts with base.

7.84 $HPO_4^{2-}(aq) + H_3O^+(aq) \rightleftarrows H_2PO_4^-(aq) + H_2O(\ell)$

$H_2PO_4^-(aq) + OH^-(aq) \rightleftarrows HPO_4^{2-}(aq) + H_2O(\ell)$

7.86 **a.** $pH = 3.85 + \log\frac{(0.1)}{(0.1)} = 3.85$

b. $pH = 3.85 + \log\frac{(1)}{(1)} = 3.85$

c. The solution in part b has greater buffer capacity than the solution in part a because the higher concentration of the buffer components will allow it to react with larger added amounts of acid or base.

7.88 **a.** $pH = 4.74 + \log\frac{(0.25)}{(0.40)} = 4.54$

b. $pH = 7.21 + \log\frac{(0.40)}{(0.10)} = 7.81$

c. $pH = 7.00 + \log\frac{(0.20)}{(1.50)} = 6.12$

7.90 Three systems that cooperate to maintain blood pH in an appropriate narrow range are blood buffer, respiratory, and urinary.

7.92 No. When the blood buffering systems become taxed, the respiratory and urinary systems act to help maintain blood pH.

7.94 $Hb^- + H_3O^+ \rightleftarrows HHb + H_2O$

$HHb + OH^- \rightleftarrows Hb^- + H_2O$

7.96 Hyperventilation is a symptom of low pH in the body (acidosis).

7.98 When kidneys function to control blood pH, the concentration of H^+ in the blood decreases. As a result, the CO_2 in the blood is converted to bicarbonate ions in the blood, in order to maintain the equilibria: $H_2O + CO_2^- + H_2CO_3 \rightleftarrows H^+ + HCO_3^-$. Removal of H^+ ions causes more carbonic acid to form HCO_3^-. The source of carbonic acid is CO_2. Overall, CO_2 is converted to HCO_3^-.

7.100 The buffer system in urine that keeps the urine pH from dropping lower than 6 is the phosphate buffer system.

Additional Exercises

7.102 $HA(aq) + H_2O(\ell) \rightleftarrows H_3O^+(aq) + A^-(aq)$

$[H_3O^+] = 2.63\% \times 0.150\ M = 0.003945$

$pH = -\log[H_3O^+] = -\log(0.003945) = 2.404$

7.104 Ordinary water may be the most effective "sports drink" because the body needs to be hydrated in order to maintain fluid and electrolyte balance in the body.

7.106 Smoking is dangerous in the presence of oxygen gas. The abundance of oxygen (an oxidizing agent) would increase the reaction rate for a redox reaction occurring between any reducing agent and the oxygen gas.

7.108 $Cl^-(aq) + H_3O^+(aq) \rightleftarrows HCl(aq) + H_2O(l)$

$$[:\ddot{\underset{..}{Cl}}:]^- + \left[\begin{array}{c} H-\ddot{O}-H \\ | \\ H \end{array}\right]^+ \longrightarrow H-\ddot{\underset{..}{Cl}}: + H-\ddot{\underset{..}{O}}-H$$

The chloride ion is a Brønsted base because it is a proton acceptor. The hydronium ion is a Brønsted acid because it is a proton donor. An alternative way to look at this is reaction is that the chloride ion has a pair of electrons that it can donate to one of the hydrogen atoms in the hydronium ion in order to form a covalent bond and HCl. The electron pair donor (chloride ion) is acting as a base. The electron pair acceptor (hydronium ion) is acting as an acid.

Chapter 8
Organic Chemistry: The Story of Carbon
(Section 8.1)

8.2 Fruits and vegetables, the family pet, plastics, sugar, cotton, and wood are a few of many items composed of organic compounds.

8.4 All organic compounds contain carbon atoms.

8.6 **a.** inorganic

b. organic

c. inorganic

d. inorganic

8.8 The majority of all compounds that are insoluble in water are organic because organic compounds are often nonpolar and nonpolar compounds are insoluble in polar water.

8.10 **a.** A relatively low melting point is characteristic of organic compounds.

b. Solubility in water and lack of flammability is typical of inorganic compounds.

c. Conductivity of water solutions is typical for ionic substances (inorganic compounds).

d. Inorganics often have high melting points.

8.12 **a.**

$$\begin{array}{ccccccc} & & H & & H & & H & \\ & & | & & | & & | & \\ H & - & C & - & C & - & C & - & H \\ & & | & & | & & | & \\ & & H & & H & & H & \end{array}$$

b.

$$\begin{array}{ccccccc} & & H & & H & & H \\ & & | & & | & / & \\ H & - & C & - & C & = & C \\ & & | & & & \backslash & \\ & & H & & & & H \end{array}$$

c.

$$\begin{array}{ccccc} & & H & & O & \\ & & | & & \| & \\ H & - & C & - & C & - & H \\ & & | & & & \\ & & H & & & \end{array}$$

d.

$$\begin{array}{ccccccc} & & H & & H & & \\ & & | & & | & & \\ H & - & C & - & C & - & N & - & H \\ & & | & & | & & | & \\ & & H & & H & & H & \end{array}$$

Representations of Organic Molecules (Section 8.2)

8.14 **a.** Correct

b. Incorrect; the center carbon has five bonds.

c. Incorrect; the carbon on the right has five bonds.

d. Incorrect; the second carbon from the left has only three bonds.

8.16

$$\begin{array}{ccccc} & & H & & O & \\ & & | & & \| & \\ H_2N & - & C & - & C & - & OH \\ & & | & & & \\ & & CH_2 & & & \\ & & | & & & \\ & & C{=}O & & & \\ & & | & & & \\ & & NH_2 & & & \end{array}$$

8.18 **a.** Expanded structural formula

b. Condensed structural formula

c. Molecular formula

d. Condensed structural formula

8.20 **a.** <u>Expanded</u>

$$\begin{array}{ccccccccc} & & & & H\ \ \ H & & & & \\ & & & & \backslash\ / & & & & \\ & & H & & N & & H & & H \\ & & | & & | & & | & & | \\ H & - & C & ——— & C & - & C & = & C & - & H \\ & & | & & | & & & & \\ & & H & H - & C & - H & & & \\ & & & & | & & & & \\ & & & & H & & & & \end{array}$$

<u>Condensed</u>

$$\begin{array}{ccccc} & & NH_2 & & \\ & & | & & \\ H_3C & - & C & - & CH{=}CH_2 \\ & & | & & \\ & & CH_3 & & \end{array}$$

b. <u>Expanded</u>

$$\begin{array}{ccccccccccc} & & & & O & & H & & & & H & & O & \\ & & & & \| & & | & & & & | & & \| & \\ H & - & O & - & C & - & C & - & O & - & C & - & C & - & H \\ & & & & & & | & & & & | & & & \\ & & & & & H - & C & - H & & & H & & & \\ & & & & & & | & & & & & & & \\ & & & & & & H & & & & & & & \end{array}$$

<u>Condensed</u>

$$\begin{array}{ccccccccc} & & O & & & & & & O \\ & & \| & & & & & & \| \\ HO & - & C & - & CH & - & O & - & CH_2 & - & CH \\ & & & & | & & & & \\ & & & & CH_3 & & & & \end{array}$$

or

$$\begin{array}{ccccccc} HOOC & - & CH & - & O & - & CH_2 & - & CHO \\ & & | & & & & \\ & & CH_3 & & & & \end{array}$$

8.22 a.

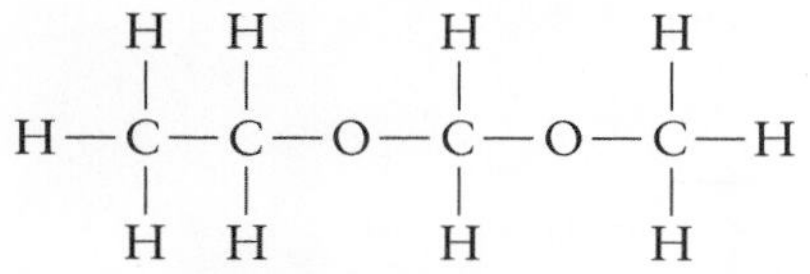

```
   H   H       H       H
   |   |       |       |
H—C—C—O—C—O—C—H
   |   |       |       |
   H   H       H       H
```

b.

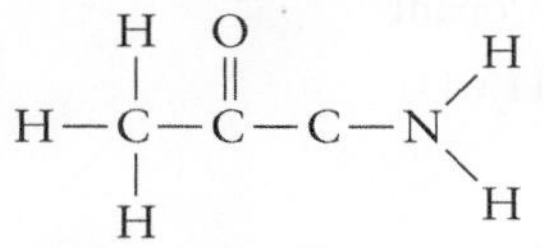

```
   H  O          H
   |  ||        /
H—C—C—C—N
   |            \
   H             H
```

8.24 a. 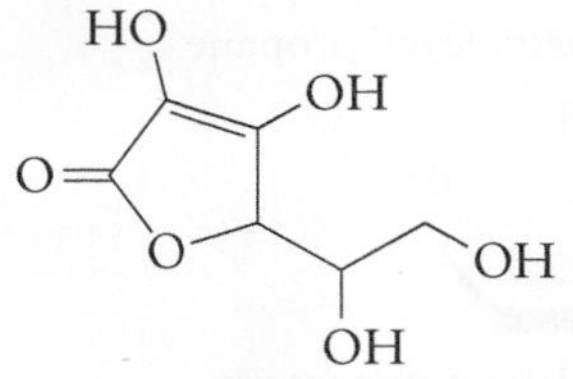 HO, OH, O, O, OH, OH

ascorbic acid

b. 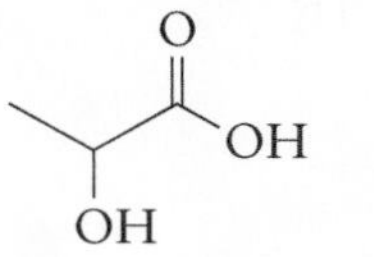 O, OH, OH

8.26 a. 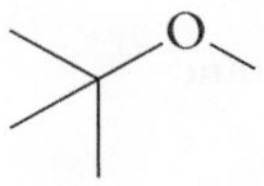 O

b. 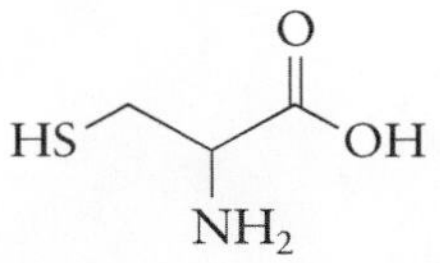 O, HS, OH, NH_2

Functional Groups (Section 8.3)

8.28 **a.** saturated **b.** unsaturated

c. unsaturated **d.** unsaturated

8.30 Name the four functional groups circled in the antibiotic cephalexin:

a. carboxylic acid **b.** amide

c. amine **d.** aromatic ring

Isomers and Conformations (Section 8.4)

8.32 Numerous organic compounds exist because each carbon atom can form four covalent bonds, including bonds to other carbon atoms, and carbon-containing molecules can exhibit isomerism. Carbon has the unique capacity to form extremely long chains that are stable. Because the same number of carbon atoms can link to form both straight and branched chains, the number of possibilities rapidly increases as the number of carbon atoms increases.

8.34 When a central carbon atom bonds to three other atoms, a trigonal planar molecular geometry exists.

8.36 **a.** not isomers (different molecular formulas —C_4H_8 and C_4H_{10})

b. isomers (same molecular formula —C_5H_{12}, but different structures)

c. not isomers (different molecular formulas —$C_4H_{10}O$ and C_4H_8O)

d. isomers (same molecular formula —C_3H_6O, but different structures)

8.38 The isomers are (a), (b), (e), and (f)—all have the molecular formula ($C_3H_6O_2$).

a. $C_3H_6O_2$ **b.** $C_3H_6O_2$

c. C_3H_8O **d.** C_4H_8O

e. $C_3H_6O_2$ **f.** $C_3H_6O_2$

g. $C_3H_8O_2$

8.40 **a.** branched (3 C longest chain)

b. normal (5 C longest chain)

c. normal (4 C longest chain)

d. normal (5 C longest chain)

8.42 **a.** the same compound, 2-methylbutane

b. the same compound, 2-methylbutane

c. structural isomers

d. the same compound, butane

Classification and Naming of Alkanes (Section 8.5)

8.44 **a.** propyl **b.** ethyl

c. *sec*-butyl **d.** *tert*-butyl

8.46 **a.** 2-methylbutane

b. 2,2-dimethylpropane

c. 2,3-dimethylpentane

d. 4,6,6-triethyl-2,7-dimethylnonane

e. 5-*sec*-butyl-2,4-dimethylnonane

8.48 **a.**

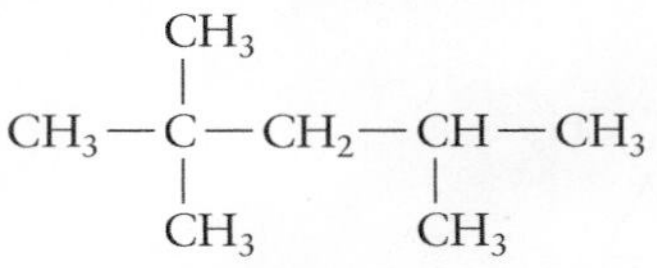

```
       CH3
       |
CH3—C—CH2—CH—CH3
       |         |
       CH3       CH3
```

2,2,4-trimethylpentane

b.

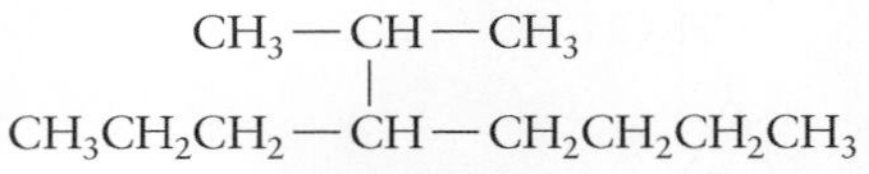

```
           CH3—CH—CH3
                |
CH3CH2CH2—CH—CH2CH2CH2CH3
```

4-isopropyloctane

c.

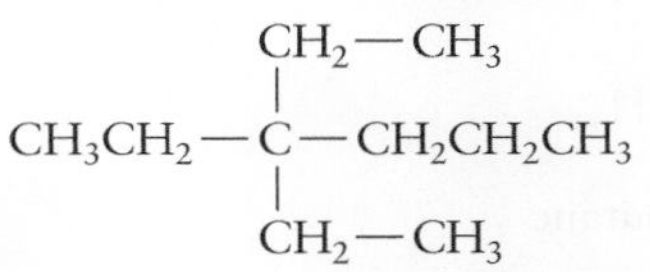

```
         CH2—CH3
         |
CH3CH2—C—CH2CH2CH3
         |
         CH2—CH3
```

3,3-diethylhexane

d.

```
                         CH3
                         |
       CH3        CH3—C—CH3
       |                 |
CH3—CH—CH2CH2—CH—CH2CH2CH2CH3
```

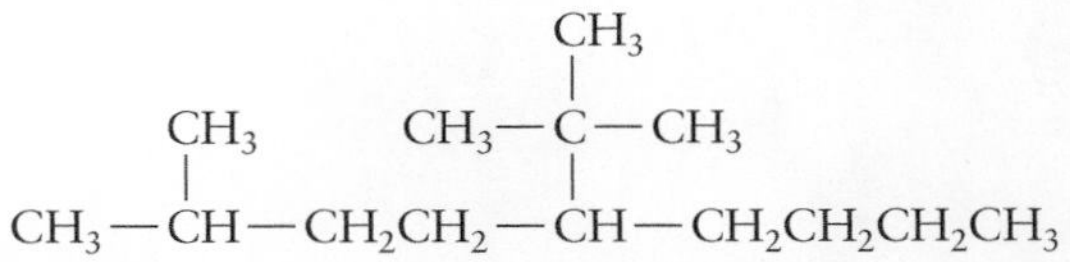

5-*tert*-butyl-2-methylnonane

8.50 Different conformations of an alkane are not considered structural isomers because the bonding order is not changed in the different conformations, which means the conformations correspond to the same molecule. Two structures would be structural isomers only if bonds were broken and remade to convert one into the other.

Cycloalkanes (Section 8.6)

8.52 **a.** two

b. one

c. zero

8.54 **a.** 1-ethyl methylcyclopentane

b. *tert*-butylcyclohexane

c. 2-chloro-1,3-dimethylcyclopentane

d. 1-ethyl-2-propylcyclopropane

8.56 **a.**

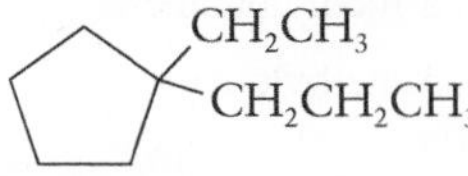

CH_2CH_3
$CH_2CH_2CH_3$

1-ethyl-1-propylcyclopentane

b.

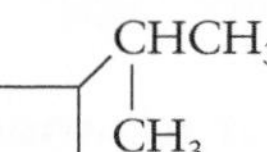

isopropylcyclobutane

c.

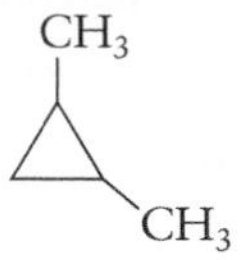

CH_3
CH_3

1,2-dimethylcyclopropane

e.

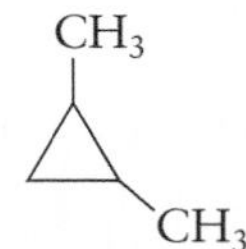

1,2-dimethylcyclopropane

8.58 **a.**

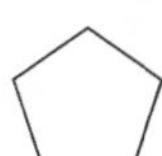

cyclopentane

b.

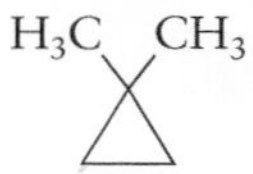

H_3C CH_3

1,1-dimethylcyclopropane

c.

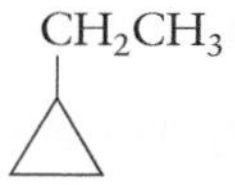

CH_2CH_3

ethylcyclopropane

d.

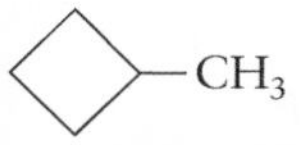

CH_3

methylcyclobutane

8.60 **a.**

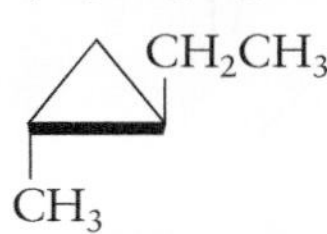

trans-1-ethyl-2-methylcyclopropane

b.

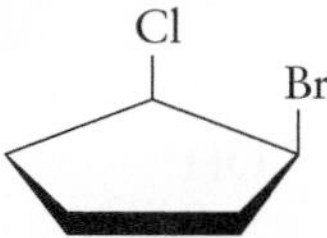

cis-1-bromo-2-chlorocyclopentane

c.

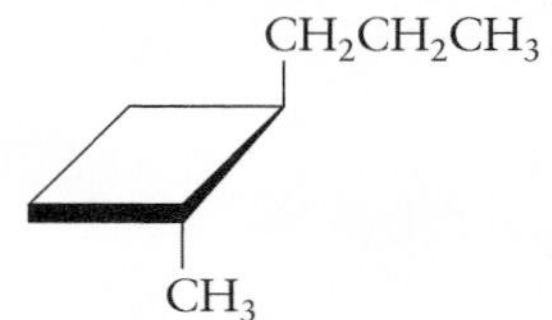

trans-1-methyl-2-propylcyclobutane

d.

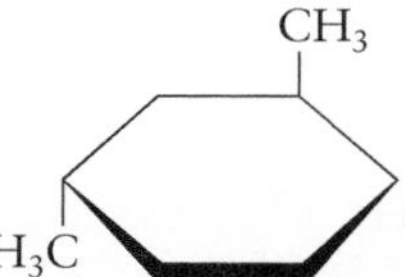

trans-1,3-dimethylcyclohexane

Alkenes and Alkynes (Section 8.7)

8.62 An alkene is a hydrocarbon that contains at least one carbon–carbon double bond.

An alkyne is a hydrocarbon that contains at least one carbon–carbon triple bond.

An aromatic hydrocarbon is a compound that contains a benzene ring or other similar feature.

8.64 **a.** unsaturated, alkene

b. unsaturated, alkene

c. unsaturated, alkene

d. not unsaturated

8.66 **a.** Br

b. H

c. $-CH_3$

d. $-CH_3$

8.68 **a.** 2-methyl-3-hexene

b. 6-methyl-2,4-heptadiene

c. cyclopentene

d. 2-pentyne

e. 2-5,dimethyl-1,5-hexadiene

f. 3-isopropyl-5-methylcyclohexene

g. 3-ethylcyclohexene

8.70 **a.** CH_3

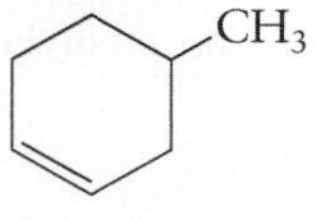

4-methylcyclohexene

b.

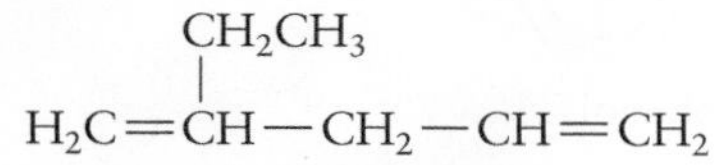

$$H_2C{=}\underset{}{\overset{CH_2CH_3}{\overset{|}{C}}}H{-}CH_2{-}CH{=}CH_2$$

2-ethyl-1,4-pentadiene

c.

$$CH_2{=}CH{-}\overset{H_3C}{\overset{|}{\underset{\underset{CH_3}{|}}{C}}}{-}\overset{\overset{CH_3}{|}}{\underset{}{CH{-}CH_3}}\!\overset{|}{CH}\quad{-}CH{=}CH{-}CH_2{-}CH_3$$

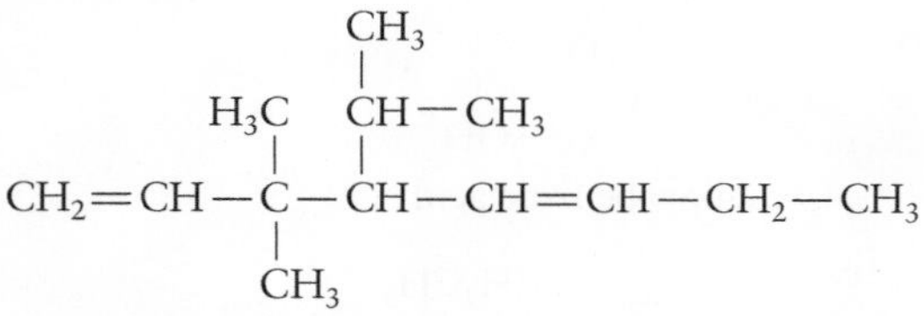

4-isopropyl-3,3-dimethyl-1,5-octadiene

8.72 **a.** C_5H_8; alkyne

$$CH{\equiv}C{-}CH_2{-}CH_2{-}CH_3$$

1-pentyne

$$CH_3{-}C{\equiv}C{-}CH_2{-}CH_3$$

2-pentyne

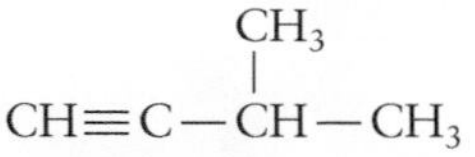

$$CH{\equiv}C{-}\overset{\overset{CH_3}{|}}{C}H{-}CH_3$$

3-methyl-1-butyne

b. C_5H_8; diene

$$CH_2{=}C{=}CH{-}CH_2{-}CH_3$$

1,2-pentadiene

$$CH_3{-}CH{=}C{=}CH{-}CH_3$$

2,3-pentadiene

$$CH_2{=}CH{-}CH{=}CH{-}CH_3$$

1,3 pentadiene

$$CH_2{=}CH{-}CH_2{-}CH{=}CH_2$$

1,4-pentadiene

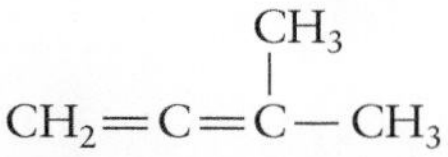

$$CH_2{=}C{=}\overset{\overset{CH_3}{|}}{C}{-}CH_3$$

3-methyl-1,2-butadiene

c. C_5H_8; cyclic alkene

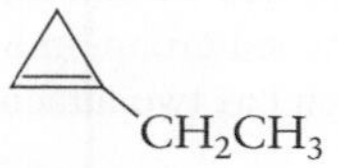

CH_2CH_3

1-ethylcyclopropene

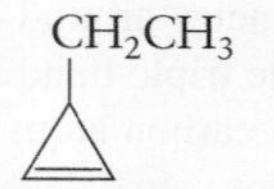

CH_2CH_3

3-ethylcycloproene

H_3C CH_3

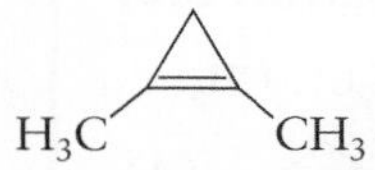

1,2-dimethylcyclopropene

CH_3 H_3C

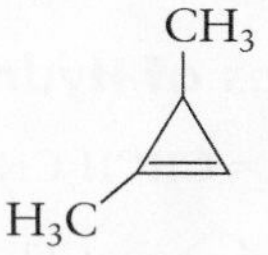

1,3-dimethylcyclopropene

H_3C CH_3

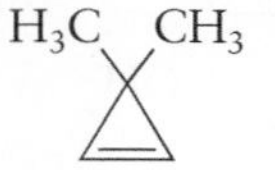

3,3-dimethylcyclopropene

CH_3

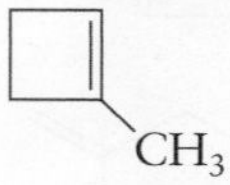

1-methylcyclobutene

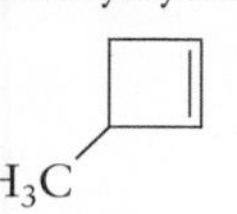

3-methylcyclobutene

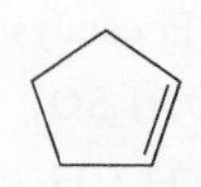

cyclopentene

8.74 Carbon atoms in a double bond have a trigonal planar geometry where the two atoms attached to the double-bonded caron atom as well as the other carbon atom in the double bond are in the same plane, separated by bond angles of 120°.

8.76 **a.** $CH_3CH_2CH_2CH_2CH{=}CH_2$; no geometric isomers

b.

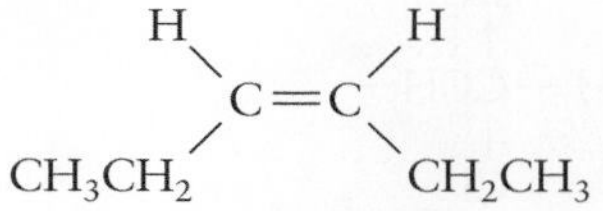

$$\begin{matrix} H & & H \\ & C{=}C & \\ CH_3CH_2 & & CH_2CH_3 \end{matrix}$$

cis-3-hexene

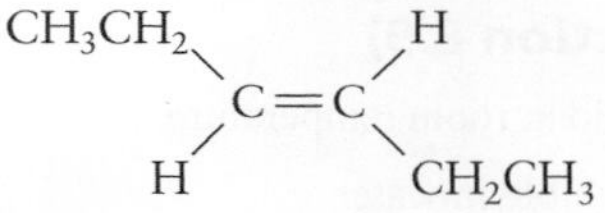

$$\begin{matrix} CH_3CH_2 & & H \\ & C{=}C & \\ H & & CH_2CH_3 \end{matrix}$$

trans-3-hexene

c.

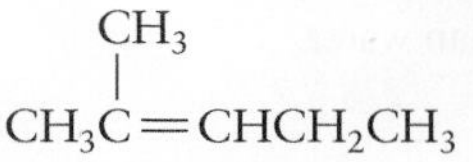

$$CH_3\overset{\overset{CH_3}{|}}{C}{=}CHCH_2CH_3$$

8.78 **a.**

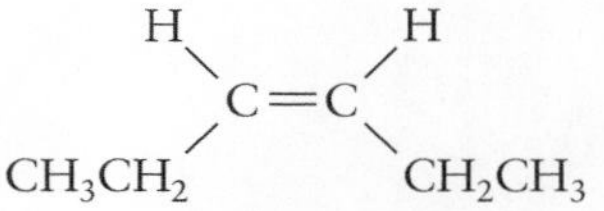

$$\begin{matrix} H & & H \\ & C{=}C & \\ CH_3CH_2 & & CH_2CH_3 \end{matrix}$$

cis-3-hexene

b.

$$\begin{matrix} CH_3H_2C & & H \\ & C{=}C & \\ H & & CH_2CH_2CH_3 \end{matrix}$$

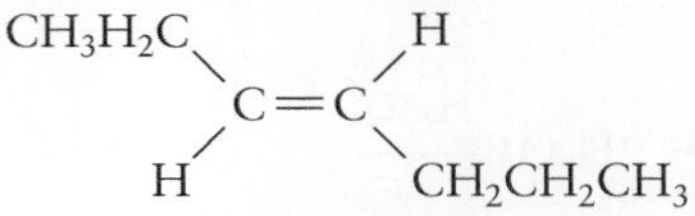

trans-3-heptene

8.80 **a.** 1,1,2,2-tetrachloroethene

b. *cis*-2-pentene

c. *trans*-1-chloropropene

d. 2-methy-2-butene

8.82 Geometric isomerism is not possible in alkynes because the geometry around a triple bond is linear. Each carbon atom in the triple bond has only one other attached group, unlike each carbon atom in a double bond, which has two attached groups.

8.84 The geometry of a triple bond is linear.

Reactions of Hydrocarbons (Section 8.8)

8.86 $CH_2 = CHCH_2CH_3 + H_2 \xrightarrow{Pt} CH_3CH_2CH_2CH_3$

8.88 **a.** 1-methylcyclopentene $+ H_2 \xrightarrow{Pt}$ methylcyclopentane

b. 2-methyl-1-butene $+ H_2O \xrightarrow{H_2SO_4}$ 2-methyl-2-butanol (OH)

8.90 **a.** H_2, Pt catalyst

b. H_2O, H_2SO_4

8.92 **a.** $-CH_2CH_2-CH_2CH_2-CH_2CH_2-$

b. $-CH(Cl)CH_2-CH(Cl)CH_2-CH(Cl)CH_2-$

c. $-CH(Cl)CH(Cl)-CH(Cl)CH(Cl)-CH(Cl)CH(Cl)-$

d. $-CCl_2CHCl-CCl_2CHCl-CCl_2CHCl-$

Physical and Chemical Properties of Hydrocarbons (Section 8.9)

8.94 **a.** Decane is a liquid at room temperature.

b. Decane is not soluble in water.

c. Decane is soluble in hexane.

d. Decane is less dense than water.

8.96 **a.** hexane

b. decane

8.98 **a.** gas

b. solid

c. liquid

8.100 (limonene: CH_3 on ring, $C(=CH_2)CH_3$ substituent) $+ 2H_2O \xrightarrow{H_2SO_4}$ (cyclohexane with H_3C and OH on one carbon, and $H_3C-C(OH)-CH_3$ on the opposite carbon)

Aromatic Compounds and Benzene (Section 8.10)

8.102 Aromatic means a molecule contains a benzene ring or one of its structural relatives.

8.104 **a.** (benzene ring with CH_3, H_3C, CH_3)
1,3,5-trimethylbenzene

b. (benzene ring with CH_2CH_3 at top and bottom)
1,4-diethylbenzene
p-diethylbenzene

8.106 **a.** *m*-bromophenol **b.** *p*-ethylaniline

8.108 **a.** (benzene ring with HO and CH_3H_2C- adjacent)
o-ethylphenol

b. (benzene ring with Cl and $-C(=O)-OH$)
m-chlorobenzoic acid

c. (benzene ring attached to C bearing CH_2CH_3, CH_3, CH_2CH_3)
3-methyl-3-phenylpentane

8.110 **a.** (benzene ring with OH and adjacent Cl)
2-chlorophenol
or
o-chlorophenol

b. (benzene ring with OH and Cl meta)
3-chlorophenol
or
m-chlorophenol

c. (benzene ring with OH and Cl para)
4-chlorophenol
or
p-chlorophenol

8.112 Benzene does not readily undergo addition reactions characteristic of other unsaturated compounds because the delocalized pi cloud of the benzene ring makes the ring so stable that addition reactions do not occur. An addition reaction would result in the loss of one of the double bonds and, consequently, would disrupt the delocalized π system. This is not favored because it results in a loss of stability. Benzene, therefore, undergoes substitution reactions instead of addition reactions.

8.114 **a.** benzene, toluene

b. riboflavin

c. phenylalanine

d. aniline

Chemistry for Thought

8.116 The study of organic compounds might be important to someone interested in the health or life sciences because living organisms are composed of mostly organic compounds and most medications are also organic compounds.

8.118 The low melting point of aspirin indicates that the molecules in aspirin have weak forces between the molecules. (*Note:* The forces are dispersion forces.)

8.120 Alkanes do not mix with water and are less dense than water; therefore, when an oil spill occurs, the oil floats on top of the water. Alkanes are relatively unreactive; therefore, short of burning the oil, a chemical reaction cannot be performed to "neutralize" the oil spill. Birds that come into contact with the oil must be cleaned with soap before they are able to fly again. Sea otters and killer whales, as well as small organisms at the bottom of the food chain, are also impacted by contact with oil.

8.122 Propene does not exhibit geometric isomerism because one of the carbon atoms in the double bond is also bonded to two hydrogen atoms. In order to exhibit geometric isomerism, both carbon atoms in the double bond must also be bonded to two unique groups.

Chapter 9
Nomenclature and Classification of Alcohols (Section 9.1)

9.2 **a.** 2-propanol

b. 3-methyl-3-pentanol

c. 2,2-dimethylcyclopentanol

d. 1,2,4-hexanentriol

9.4 **a.** methyl alcohol (methanol)

b. isopropyl alcohol (isopropanol)

c. ethyl alcohol (ethanol)

d. ethylene glycol

9.6 **a.** CH_3 above, $CH_3CCH_2CH_2CH_3$, OH below

2-methyl-2-pentanol

b. $CH_2CH_2CHCH_3$ with OH below first and third carbons

1,3-butanendiol

c. Cyclopentane ring with OH and CH_2CH_3 on the same carbon

1-ethyl-cyclopentanol

9.8 **a.** Benzene ring with CH_2CH_3 and OH on adjacent carbons

2-ethylphenol

o-ethylphenol

b. Benzene ring with OH; H_3C and CH_2CH_3 ortho to OH; CH_3 para to OH

2-ethyl-4,6-dimethylphenol

9.10 **a.** Cl—(benzene ring)—OH

p-chlorophenol

b. Benzene ring with OH; $CH(CH_3)CH_3$ ortho to OH; $H_3C{-}CH(CH_3)$ at position 5

2,5-diisopropylphenol

9.12 **a.** Tertiary

b. Secondary

c. Secondary

9.14

$CH_2(OH){-}CH_2{-}CH_2{-}CH_2{-}CH_3$
1-pentanol, primary

$H_3C{-}CH(OH){-}CH_2{-}CH_2{-}CH_3$
2-pentanol, secondary

$H_3C{-}CH_2{-}CH(OH){-}CH_2{-}CH_3$
3-pentanol, secondary

$CH_2(OH){-}CH(CH_3){-}CH_2{-}CH_3$
2-methyl-1-butanol, primary

$H_3C{-}C(OH)(CH_3){-}CH_2{-}CH_3$
2-methyl-2-butanol-tertiary

$H_3C{-}CH(OH){-}CH(CH_3){-}CH_3$
3-methyl-2-butanol, secondary

$H_3C{-}CH(CH_3){-}CH_2{-}CH_2(OH)$
3-methyl-1-butanol, primary

$H_3C{-}C(CH_3)_2{-}CH_2(OH)$
2,2-dimethyl-1-propanol, primary

9.16 Both compounds are aromatic, but only the structure on the left has the —OH attached to the benzene ring, thereby making it a phenol.

Physical Properties of Alcohols (Section 9.2)

9.18 **a.** (lowest boiling point) methanol < ethanol < 1-propanol (highest boiling point)

b. (lowest boiling point) butane < 1-propanol < ethylene glycol (highest boiling point)

9.20

a. butane or 2-butanol	2-Butanol is more soluble in water than butane because 2-butanol forms hydrogen bonds with the water molecules, while nonpolar butane cannot.
b. 2-propanol or 2-pentanol	Both compounds form hydrogen bonds with water. However, 2-propanol is more soluble in water than 2-pentanol because 2-propanol has a shorter carbon chain and lower molecular weight than 2-pentanol, which increases water solubility.
c. 2-butanol or 2,3-butanediol	2,3-Butanediol will be more soluble in water than 2-butanol because it contains two hydroxy groups, while the 2-butanol only contains one hydroxy group. The two hydroxy groups increase the hydrogen-bonding interactions with water, which increases the solubility of the compound in water.

9.22 **a.** $CH_3CH_2CHCH_2CH_3$ with O(H) hydrogen-bonded (H····O) to $CH_3CH_2CH_2$–O–H

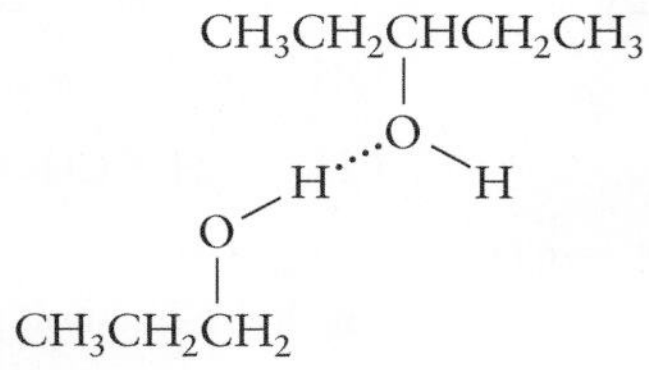

b. H–O(····H–O–H)–CH_3CHCH_3

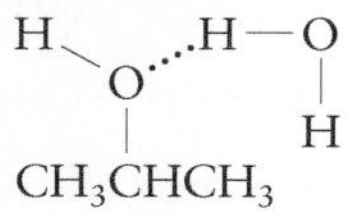

Reactions of Alcohols (Section 9.3)

9.24 **a.** H_2SO_4 + [2-methyl-1-pentanol skeletal structure, OH] ⇌ [2-methyl-1-pentene skeletal structure]

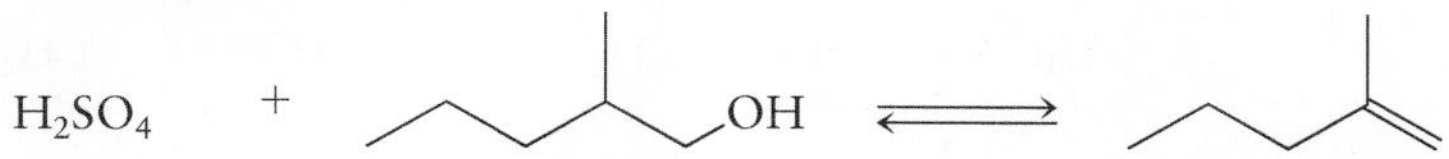

b. HNO_3 + [cyclohexanol, OH] ⇌ [cyclohexene]

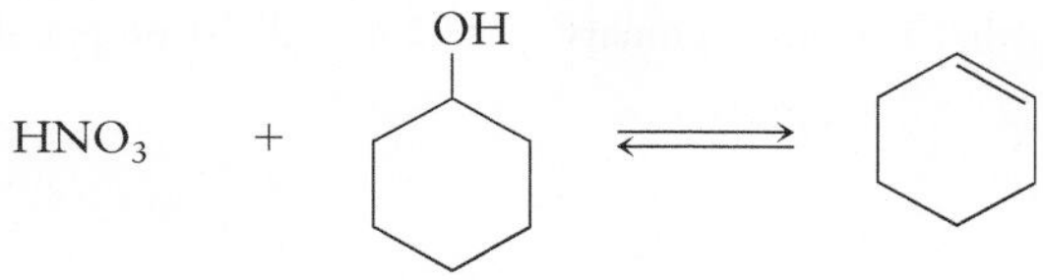

9.26

	Reactant	Product
a.	$CH_3CHCH_2CH_3$ (OH on C2)	$CH_3CH{=}CHCH_3$
b.	$CH_3CHCHCH_3$ (CH_3 on C3, OH on C2)	$CH_3CH{=}CCH_3$ (CH_3 on C3)

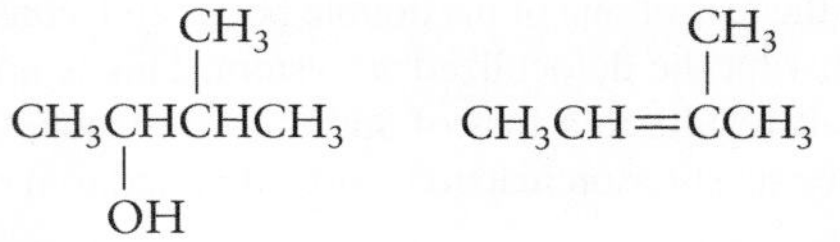

9.28 **a.** $CH_3{-}O{-}CH_3$

b. $CH_3CH_2CH_2CH_2{-}O{-}CH_2CH_2CH_2CH_3$

c. cyclohexyl–$CH_2{-}O{-}H_2C$–cyclohexyl

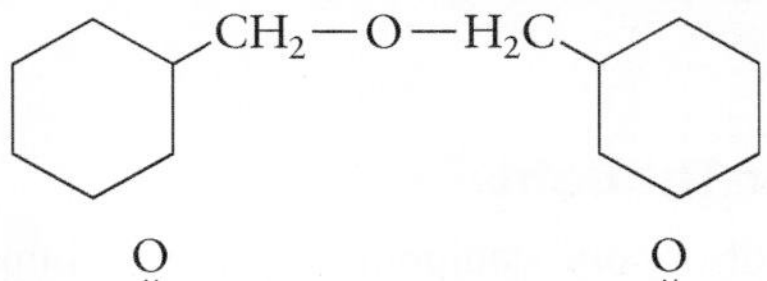

9.30 $CH_3CH_2\overset{O}{\overset{\|}{C}}{-}H$ and $CH_3CH_2\overset{O}{\overset{\|}{C}}{-}OH$

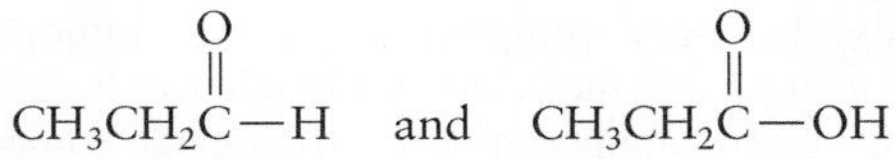

9.32 **a.** $CH_3CH{=}CHCH_3$

b. $CH_3CH_2C{=}CH_2$ (CH_3 on C2)

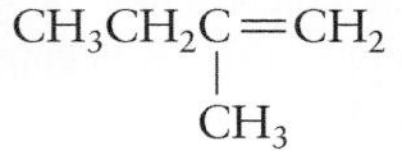

c. $CH_3CH_2CH_2{-}O{-}CH_2CH_2CH_3$

d. cyclopropyl–$CH_2{-}O{-}H_2C$–cyclopropyl

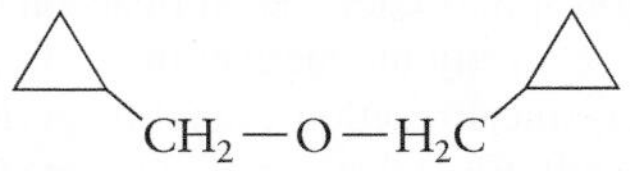

9.34 **a.** $CH_3CH_2CHCH_2{-}OH$ (CH_3 on C2)

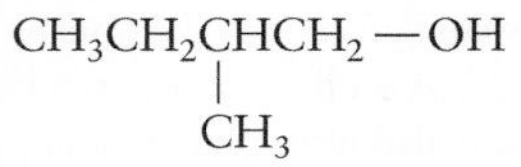

b. CH_3CHCH_3 (OH on C2) and $CH_3CH_2CH_2{-}OH$

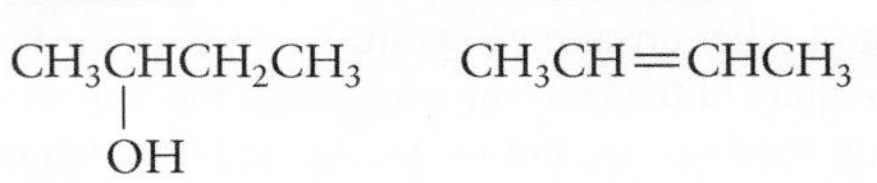

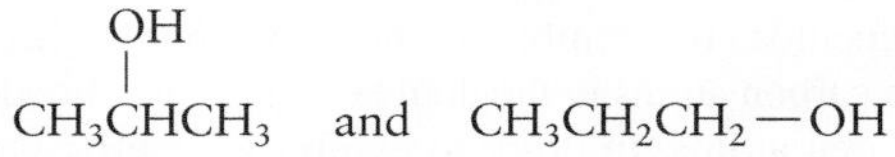

c. $CH_3{-}OH$

d. $CH_3CHCH_2{-}OH$ (CH_3 on C2)

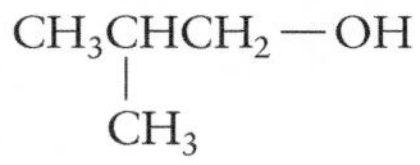

9.36 **a.** primary **b.** secondary

c. primary **d.** primary

9.38

	Product	Reactant
a.	cyclopentanone (=O)	cyclopentanol (–OH)
b.	$CH_3CH_2\overset{O}{\overset{\|}{C}}CH{-}CH_3$ (CH_3 on CH)	$CH_3CH_2CHCH{-}CH_3$ (OH on C3, CH_3 on C2)
c.	$CH_3CH_2CH_2{-}\overset{O}{\overset{\|}{C}}{-}OH$	$CH_3CH_2CH_2{-}CH_2$ (OH on terminal CH_2)

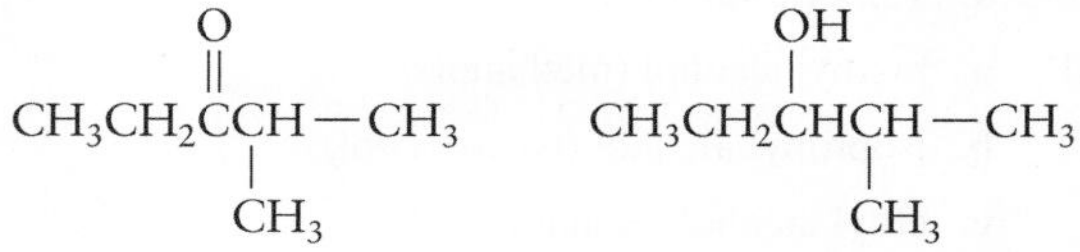

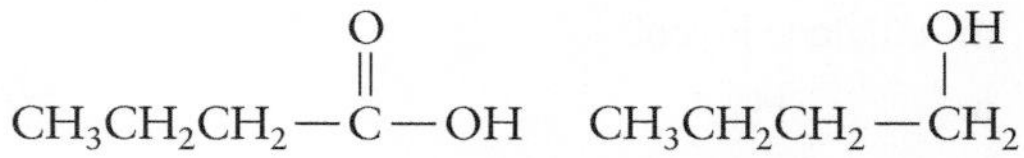

Applications of Alcohols (Section 9.4)

9.40 If in the laboratory methanol is accidentally spilled on one's clothing, it would be a serious mistake to just let it evaporate because the methanol in contact with the skin will be absorbed through the skin. Methanol is toxic for humans and can cause blindness and death. The methanol-infused clothing needs to be changed immediately.

9.42 **a.** 1,2,3-propanetriol (glycerin, glycerol)

b. ethanol (ethyl alcohol)

c. 1,2-ethanediol (ethylene glycol)

d. 2-propanol (isopropyl alcohol)

Ethers (Section 9.5)

9.44 $R-O-R'$

where the R and the R′ are hydrocarbon groups that may be the same or different.

9.46 **a.** ethyl methyl ether

b. phenyl propyl ether

c. di-*sec*-butyl-ether

9.48 **a.** methyl propyl ether

b. cyclopentyl ethyl ether

c. isopropyl phenyl ether

9.50 **a.** ethyl methyl ether

$CH_3CH_2-O-CH_3$

b. butyl phenyl

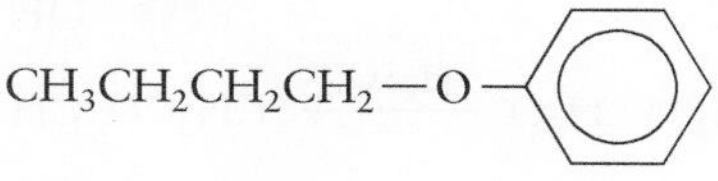

$CH_3CH_2CH_2CH_2-O-C_6H_5$ (O bonded to a benzene ring)

9.52

$$\underset{\displaystyle OH}{\underset{|}{CH_3CHCH_2CH_3}} > CH_3CH_2-O-CH_2CH_3 > \overset{\displaystyle CH_3}{\overset{|}{CH_3CHCH_2CH_3}}$$

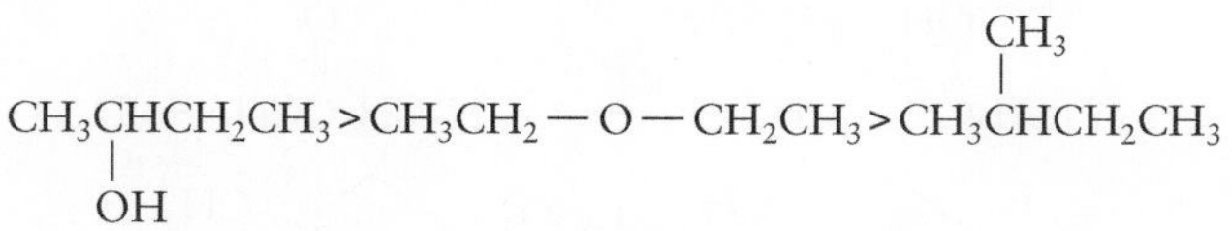

For compounds of comparable molecular weight, alcohols are more soluble than ethers, which are more soluble than alkanes.

9.54 The order is the same as in the answer to Exercise 9.52. Alcohols have the highest boiling point due to hydrogen-bonding. Pure ethers exhibit no hydrogen-bonding. Ethers have slightly higher boiling points than alkanes due to the oxygen atom: alcohol > ether > alkane.

9.56 Diethyl ether is hazardous to use as an anesthetic or as a solvent in the laboratory because it is very volatile and extremely flammable.

The Nomenclature and Classification of Amines (Section 9.6)

9.58 **a.** cyclohexylamine

b. *N*-ethylcyclohexylamine

c. *sec*-butylamine

9.60 **a.**

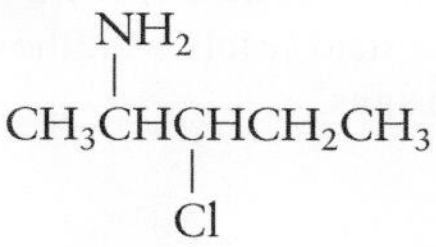

$$\underset{\displaystyle Cl}{\underset{|}{\overset{\displaystyle NH_2}{\overset{|}{CH_3CHCHCH_2CH_3}}}}$$

b.

$$\underset{\displaystyle CH_3}{\underset{|}{\overset{\displaystyle NH_2}{\overset{|}{CH_3CHCHCH_3}}}}$$

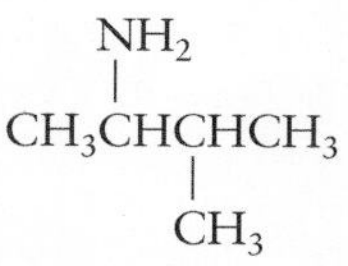

c. $CH_3CH_2-NH_2$

9.62 Primary amines have the formula $R-NH_2$

Secondary amines have the formula $\underset{\displaystyle R'}{\underset{|}{R-NH}}$

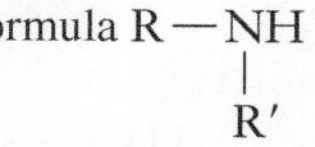

Tertiary amines have the formula $\underset{\displaystyle R'}{\underset{|}{R-N-R''}}$

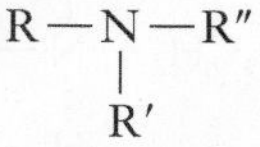

R, R′, and R″ may be the same or different.

9.64 **a.**

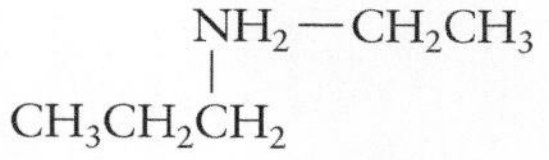

$$\underset{\displaystyle CH_3CH_2CH_2}{\underset{|}{NH_2-CH_2CH_3}}$$

secondary

b.

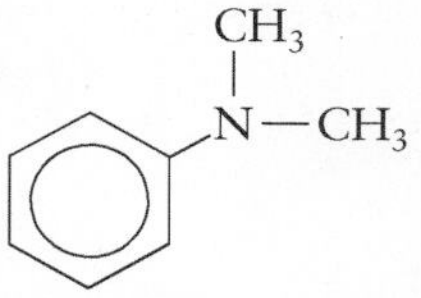

$C_6H_5-\overset{\displaystyle CH_3}{\overset{|}{N}}-CH_3$ (N bonded to a benzene ring)

tertiary

c.

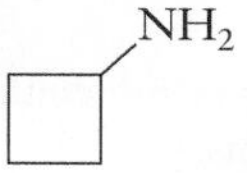

NH_2 bonded to a cyclobutane ring

primary

d. $\underset{\displaystyle CH_3}{\underset{|}{CH_3CH}}-NH-\underset{\displaystyle CH_3}{\underset{|}{CHCH_3}}$

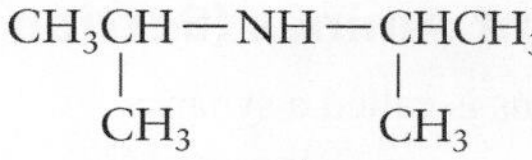

secondary

Properties of Amines (Section 9.7)

9.66 $CH_3CH_2-NH-CH_2CH_3 + H_2O \rightleftharpoons$

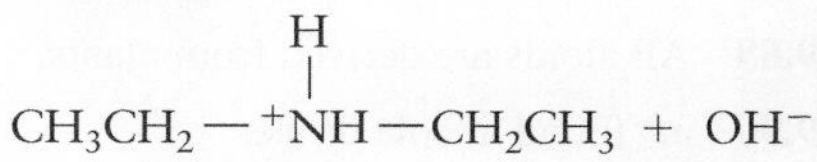

$$CH_3CH_2-{}^{+}\overset{\displaystyle H}{\overset{|}{NH}}-CH_2CH_3 + OH^-$$

9.68 **a.** OH^-

b. $H_3\overset{+}{N}-\underset{\displaystyle CH_3}{\underset{|}{CH_2CHCH_3}}$

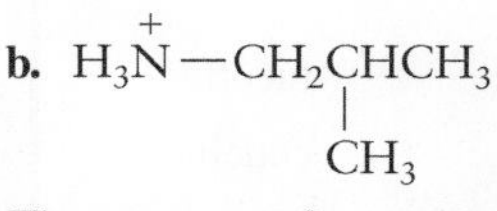

9.70 The structure of a quaternary ammonium salt differs from the structure of a salt of a tertiary amine because the nitrogen atom of a quaternary ammonium salt is bonded directly to four carbon atoms, while the nitrogen atom of a salt of a tertiary amine is bonded directly to three carbon atoms and one hydrogen atom.

9.72 The boiling points of amines are lower than those of the corresponding alcohols because the hydrogen bonds formed between nitrogen and hydrogen are weaker than the hydrogen bonds formed between oxygen and hydrogen.

9.74 [structure: tertiary amine with CH_3 on N and CH_3O on benzene ring] + HCl ⟶ [structure: amine salt with NH^+Cl^-, CH_3, CH_3O]

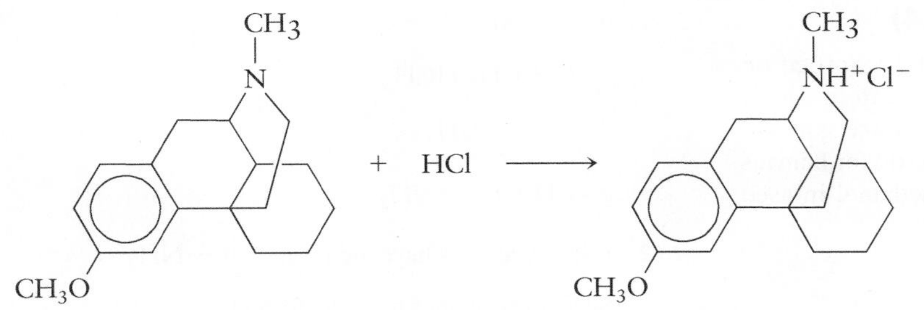

9.76 **a.**

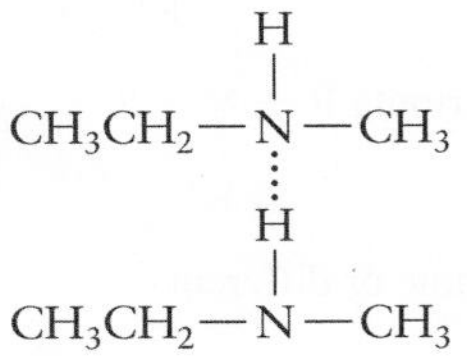

$$\begin{array}{c} H \\ | \\ CH_3CH_2-N-CH_3 \\ \vdots \\ H \\ | \\ CH_3CH_2-N-CH_3 \end{array}$$

b. [structure: pyrrolidine N hydrogen-bonded to H–O–H, and H···O of a second water molecule]

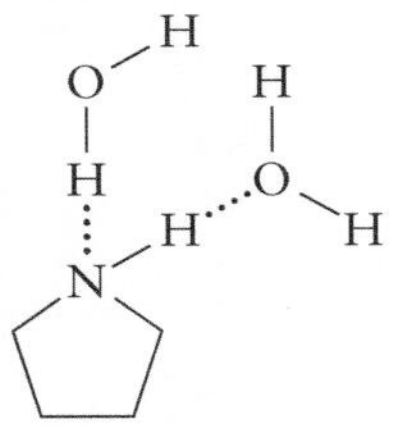

9.78 alkane < amine < alcohol

9.80 **a.** 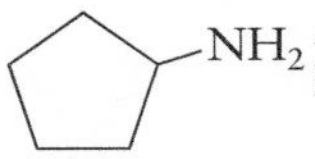 [cyclopentyl]$-NH_2$ As a primary amine, it can exhibit greater hydrogen-bonding with water molecules.

b. $CH_3CH_2-NH-CH_3$ has a lower molecular weight and fewer hydrophobic carbon atoms.

Biologically Significant Amines (Section 9.8)

9.82 The gap between neurons is called a synapse.

9.84 The two amines often associated with the biochemical theory of mental illness are norepinephrine and serotonin.

9.86 Four neurotransmitters important in the central nervous system are acetylcholine, norepinephrine, dopamine, and serotonin.

9.88 Alkaloids are derived from plants.

9.90

a. found in cola drinks	caffeine
b. used to reduce saliva flow during surgery	atropine
c. present in tobacco	nicotine
d. a cough suppressant	codeine
e. used to treat malaria	quinine
f. an effective painkiller	morphine

Additional Exercises

9.92 $R-O-H + X^- \rightarrow R-O^- + HX^-$

$R-O-H + HA \rightarrow R-OH_2^+ + A^-$

Chemistry for Thought

9.94 The three products are:

$CH_3CH_2-O-CH_2CH_3$ diethyl ether

$CH_3CH_2-O-CH_2CH_2CH_3$ ethyl propyl ether

$CH_3CH_2CH_2-O-CH_2CH_2CH_3$ dipropyl ether

Three products form because the two types of alcohol molecules can react with a molecule of the same type to produce either diethyl ether or dipropyl ether or a molecule of a different type to produce ethyl propyl ether.

9.96 Skin takes on a dry appearance after diethyl ether is spilled on the skin because diethyl ether, a volatile solvent, removes natural skin oils by dissolving them and then evaporating, taking these oils with it.

9.98 Hydrochloric acid is used to prepare an amine hydrochloride salt.

9.100 **a.**

$$CH_3CH_2CH{=}CH_2 + H_2O \xrightarrow{H_2SO_4} CH_3CH_2\overset{OH}{\overset{|}{C}H}-CH_3$$

$$CH_3CH_2\overset{OH}{\overset{|}{C}H}-CH_3 \xrightarrow{(O)} CH_3CH_2\overset{O}{\overset{\|}{C}}CH_3 + H_2O$$

b. [2-methylcyclopentanol: cyclopentane with CH_3 and OH] $\xrightarrow[180\ °C]{H_2SO_4}$ [1-methylcyclopentene with CH_3]

[1-methylcyclopentene with CH_3] $+ H_2O \xrightarrow{H_2SO_4}$ [1-methylcyclopentanol with CH_3 and OH]

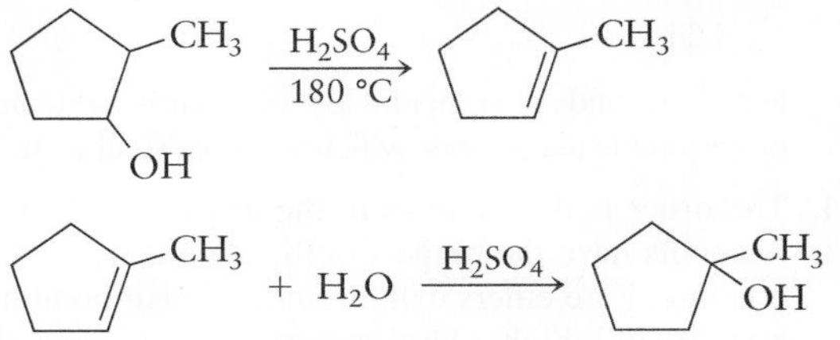

c.

$$CH_3CH_2CH_2\overset{OH}{\overset{|}{C}H_2} \xrightarrow[180\ °C]{H_2SO_4} CH_3CH_2CH{=}CH_2 + H_2O$$

$$CH_3CH_2CH{=}CH_2 + H_2O \xrightarrow{H_2SO_4} CH_3CH_2\overset{OH}{\overset{|}{C}H}-CH_3$$

$$CH_3CH_2\overset{OH}{\overset{|}{C}H}-CH_3 \xrightarrow{(O)} CH_3CH_2\overset{O}{\overset{\|}{C}}CH_3 + H_2O$$

9.102 Alkaloids can be extracted from plant tissues using dilute hydrochloric acid because the acid reacts with the amines to produce water-soluble amine salts.

Chapter 10
The Nomenclature of Aldehydes and Ketones (Section 10.1)

10.2 $\mathrm{H{-}\overset{\overset{\displaystyle O}{\|}}{C}{-}H}$ aldehyde $\mathrm{H_3C{-}\overset{\overset{\displaystyle O}{\|}}{C}{-}CH_3}$ ketone

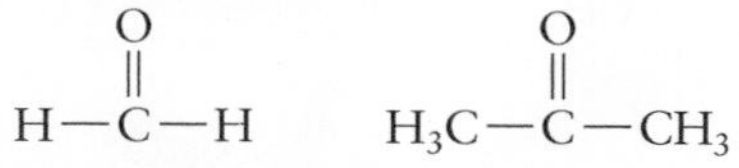

10.4 b and d

10.6 **a.** ketone **b.** neither (carboxylic acid)
c. neither (ester) **d.** ketone
e. aldehyde **f.** neither (amide)

10.8 **a.** 3-methylbutanal **b.** 4-methylheptanal
c. 2-phenylethanal **d.** butanal

10.10 **a.** 2-pentanone
b. 3-pentanone
c. 2,4-dimethyl-3-hexanone
d. 2-methyl-4-phenyl-3-pentanone

10.12 **a.** $\mathrm{CH_3CH_2CH_2\underset{\underset{\displaystyle CH_3}{|}}{C}HCH_2{-}\overset{\overset{\displaystyle O}{\|}}{C}{-}H}$

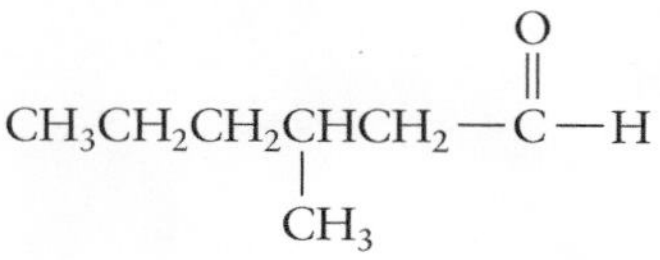

b. $\mathrm{CH_3CH_2\overset{\overset{\displaystyle CH_3}{|}}{\underset{\underset{\displaystyle CH_3}{|}}{C}}CH_2{-}\overset{\overset{\displaystyle O}{\|}}{C}{-}H}$

c. $\mathrm{CH_3\underset{\underset{\displaystyle Br}{|}}{C}H\underset{\underset{\displaystyle Br}{|}}{C}H{-}\overset{\overset{\displaystyle O}{\|}}{C}{-}H}$

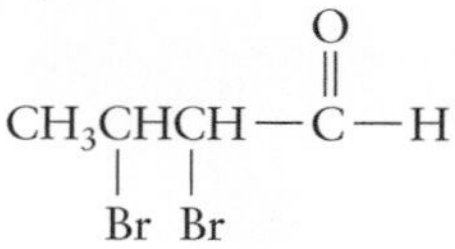

d. $\mathrm{CH_3\underset{\underset{\displaystyle C_6H_5}{|}}{C}H{-}\overset{\overset{\displaystyle O}{\|}}{C}{-}H}$

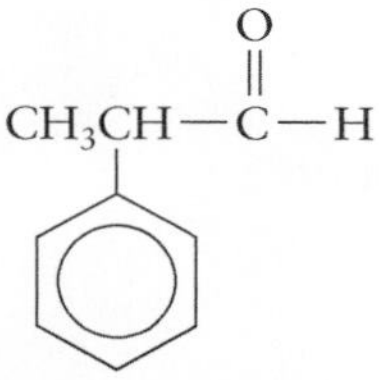

10.14 **a.** O

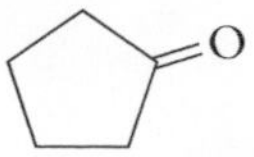

b. $\mathrm{\underset{\underset{\displaystyle Cl}{|}}{C}H_2\underset{\underset{\displaystyle Cl}{|}}{C}H{-}\overset{\overset{\displaystyle O}{\|}}{C}{-}CH_2CH_3}$

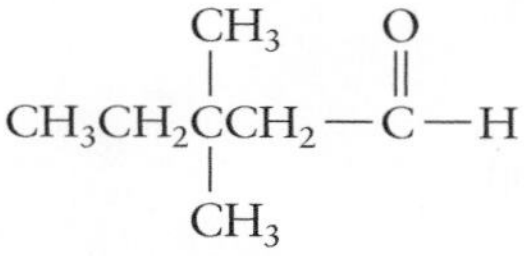

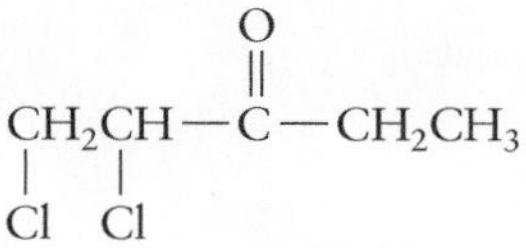

c. $\mathrm{CH_3{-}\overset{\overset{\displaystyle O}{\|}}{C}{-}CH_2CH_2CH_2CH_3}$

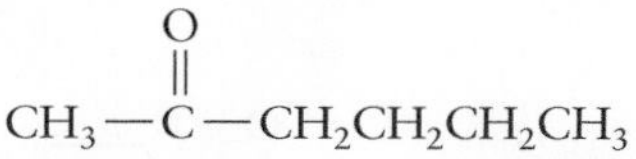

d. $\mathrm{CH_3{-}\overset{\overset{\displaystyle O}{\|}}{C}{-}\underset{\underset{\displaystyle C_6H_5}{|}}{C}HCH_2CH_3}$

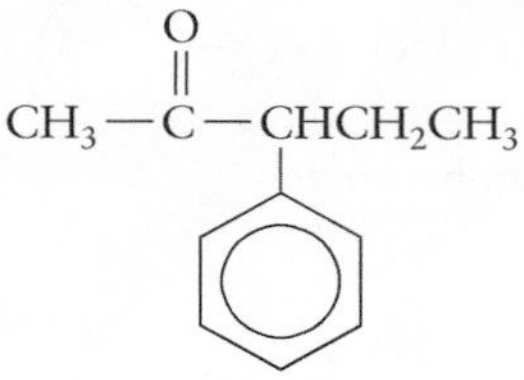

Properties of Aldehydes and Ketones (Section 10.2)

10.16 Hydrogen-bonding does not occur between molecules of aldehydes or ketones because aldehydes and ketones do not contain an oxygen atom bonded directly to a hydrogen atom.

10.18 c < b < a

10.20 **a.** H—O—H····O=C(CH$_3$)—CH$_3$ ($\mathrm{H_3C{-}\overset{\overset{\displaystyle O}{\|}}{C}{-}CH_3}$)

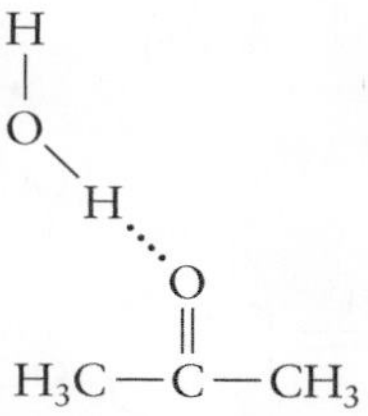

b. H—O—H····O=C—H (cyclopentyl)

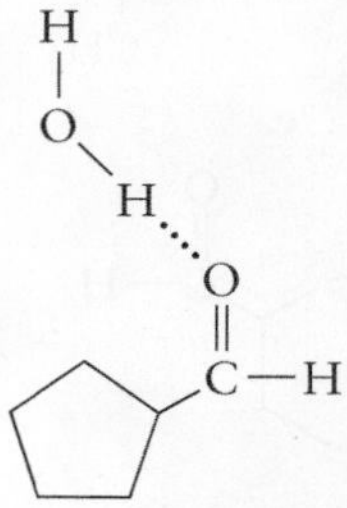

10.22 b < c < a

Reactions of Aldehydes and Ketones (Section 10.3)

10.24 **a.** O

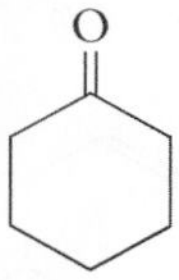

b. O H

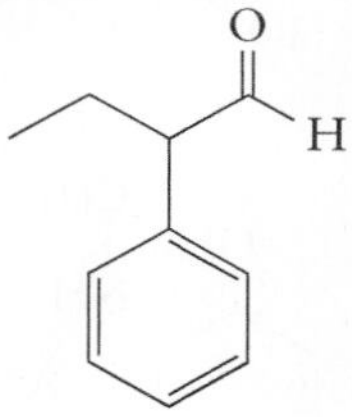

c. O H N

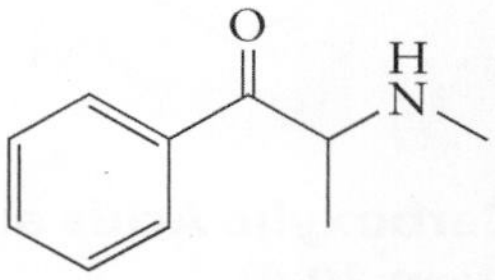

d. O

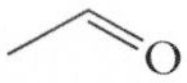

10.26 **a.** H O

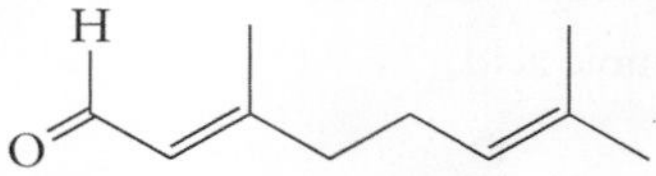

b. O

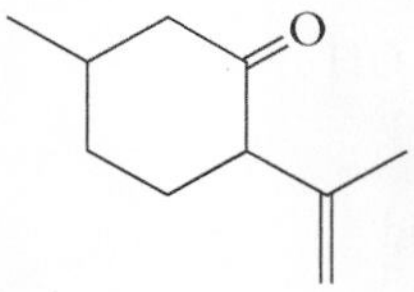

c. O O OH NH$_2$

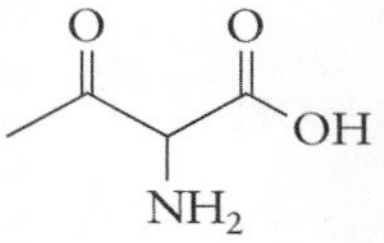

d. O

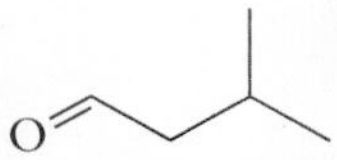

isoamyl alcohol

10.28 a. $CH_3CH_2-\overset{O}{\overset{\|}{C}}-CH_3 + H_2 \xrightarrow{Pt} CH_3CH_2-\overset{OH}{\overset{|}{C}H}-CH_3$

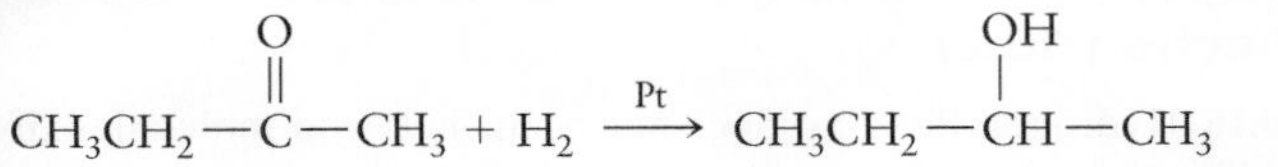

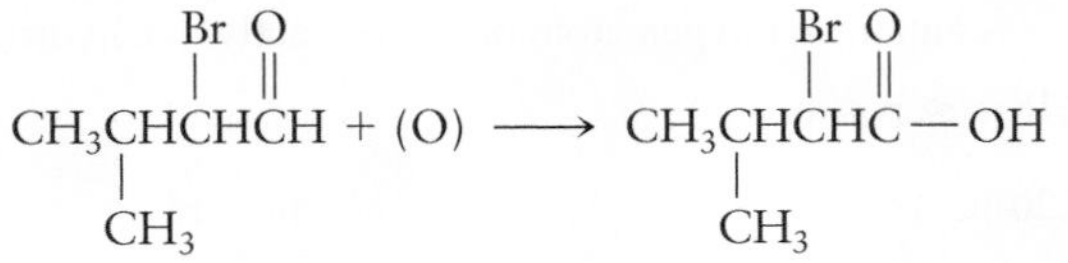

b. $CH_3\underset{CH_3}{\underset{|}{C}H}\overset{Br}{\overset{|}{C}H}\overset{O}{\overset{\|}{C}H} + (O) \longrightarrow CH_3\underset{CH_3}{\underset{|}{C}H}\overset{Br}{\overset{|}{C}H}\overset{O}{\overset{\|}{C}}-OH$

c. cyclohexyl$-\overset{O}{\overset{\|}{C}}-H + 2CH_3-OH \underset{}{\overset{H^+}{\rightleftharpoons}}$ cyclohexyl$-\underset{OCH_3}{\underset{|}{\overset{OCH_3}{\overset{|}{C}}}}-H + H_2O$

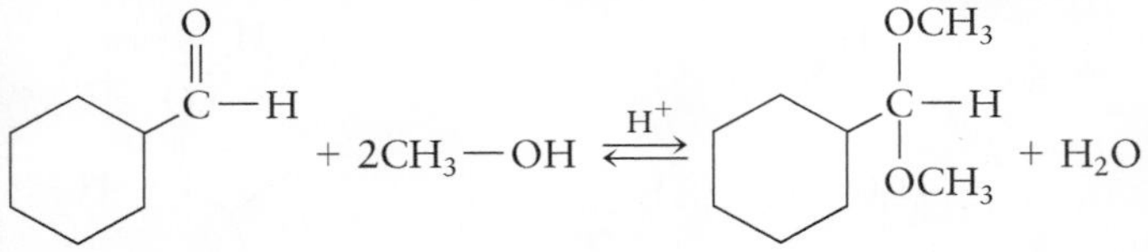

d. cyclopentanone $+ (O) \longrightarrow$ no Reaction

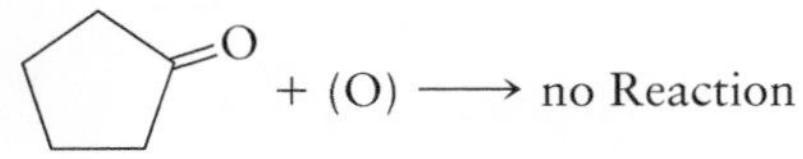

e. 2-hydroxyoxetane (OH) $+ CH_3CH_2-OH \overset{H^+}{\rightleftharpoons}$ 2-ethoxyoxetane (OCH_2CH_3) $+ H_2O$

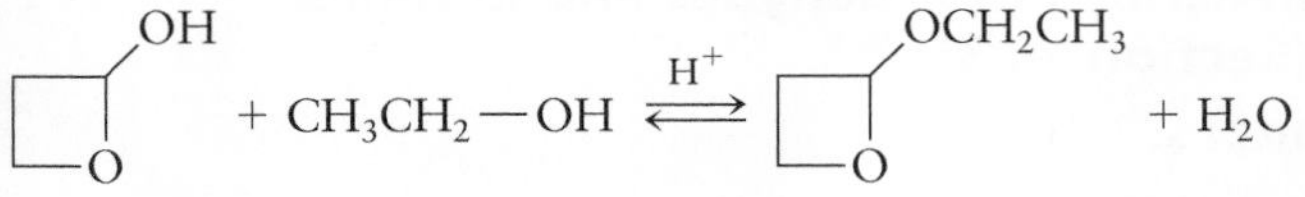

f. (ketone, O) $\xrightarrow{(R)}$ (alcohol, HO)

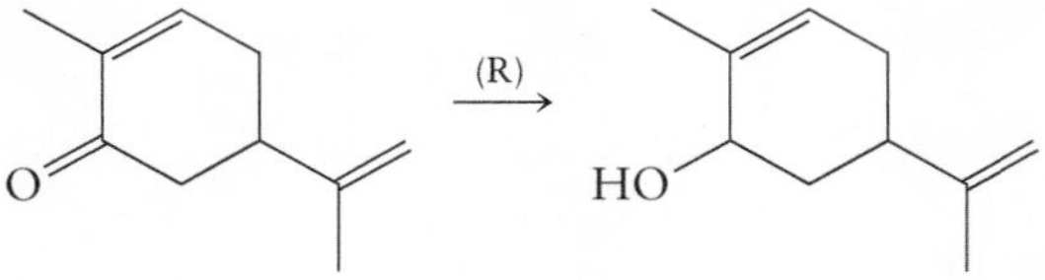

g. (ketone: O, CH_3, NH, Cl) $\xrightarrow{(R)}$ (alcohol: OH, CH_3, NH, Cl)

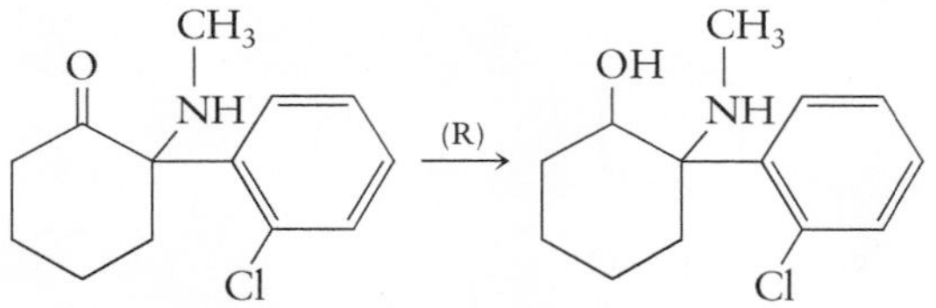

The Nomenclature of Carboxylic Acids and Carboxylate Salts (Section 10.4)

10.30 a. $CH_3CH_2CH_2CH_2-\overset{O}{\overset{\|}{C}}-OH$

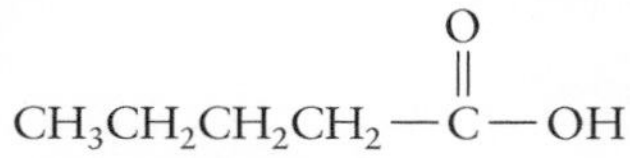

pentanoic acid

b. $CH_3CH_2\underset{CH_3}{\underset{|}{C}H}CH_2-\overset{O}{\overset{\|}{C}}-OH$

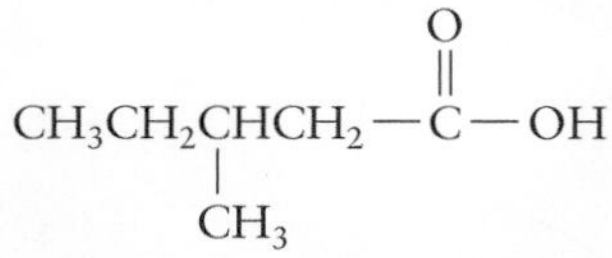

3-methylpentanoic acid

c. $CH_3CH_2\underset{C_6H_5}{\underset{|}{C}H}CH_2CH_2-\overset{O}{\overset{\|}{C}}-OH$

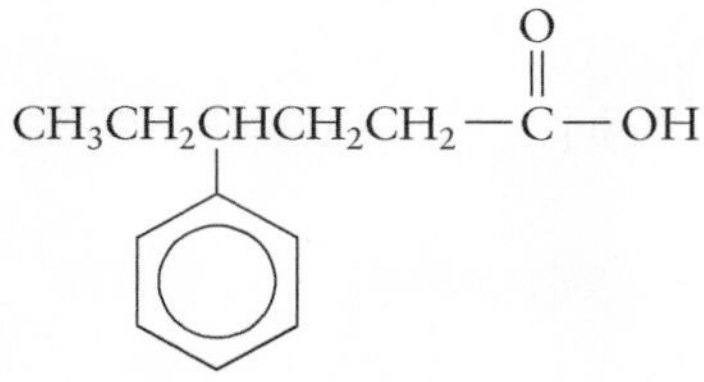

4-phenylhexanoic acid

d. (benzene ring with $\overset{O}{\overset{\|}{C}}-OH$, CH_3, CH_3)

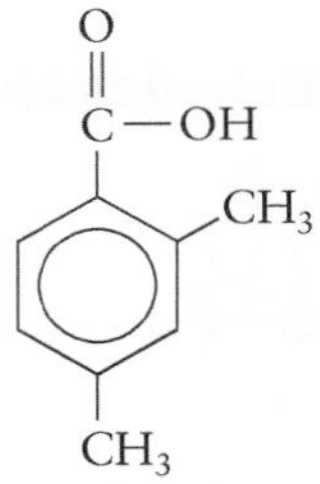

2,4-dimethylbenzoic acid

10.32 a. $CH_3CH_2CH_2CH_2-\overset{O}{\overset{\|}{C}}-OH$

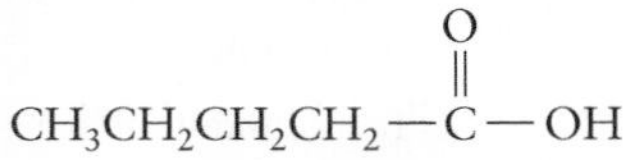

pentanoic acid

b. $CH_3CH_2CH_2\underset{CH_3}{\underset{|}{C}H}\overset{Br}{\overset{|}{C}H}-\overset{O}{\overset{\|}{C}}-OH$

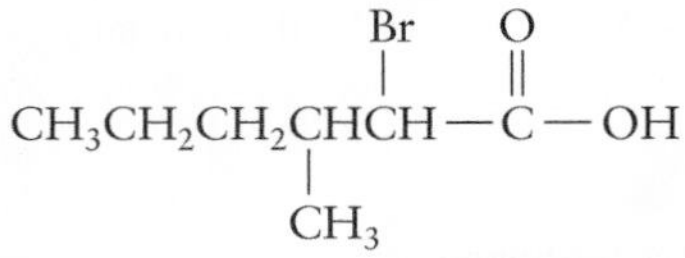

2-bromo-3-methylhexanoic acid

c. (benzene ring with $\overset{O}{\overset{\|}{C}}-OH$ and $CH_2CH_2CH_3$)

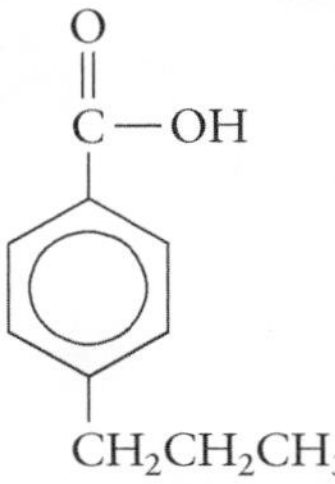

4-propylbenzoic acid

10.34 a. $CH_3CH_2-\overset{O}{\overset{\|}{C}}-O^-Na^+$

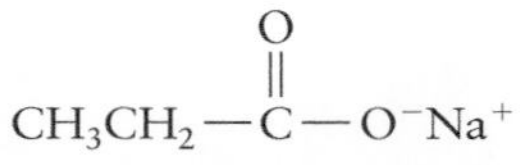

sodium propanoate

b. $CH_3\overset{CH_3}{\overset{|}{C}H}-\overset{O}{\overset{\|}{C}}-O^-K^+$

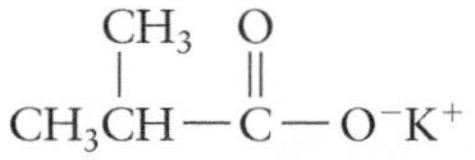

potassium 2-methylpropanoate

c. (benzene ring with NO_2 and $\underset{O}{\underset{\|}{C}}-O^-Na^+$)

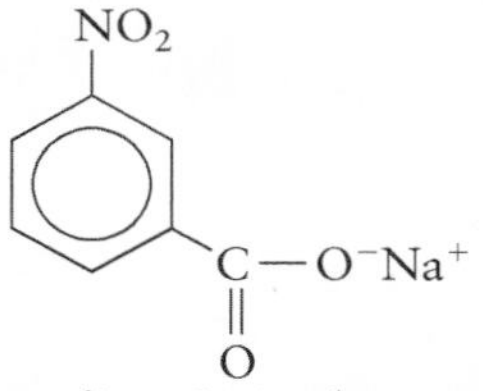

sodium 3-nitrobenzoate

10.36 a. $H-\overset{O}{\overset{\|}{C}}-O^-Na^+$

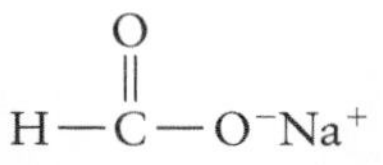

sodium formate

b. $C_6H_5-\overset{O}{\overset{\|}{C}}-O^-Ca^{2+}O^--\overset{O}{\overset{\|}{C}}-C_6H_5$

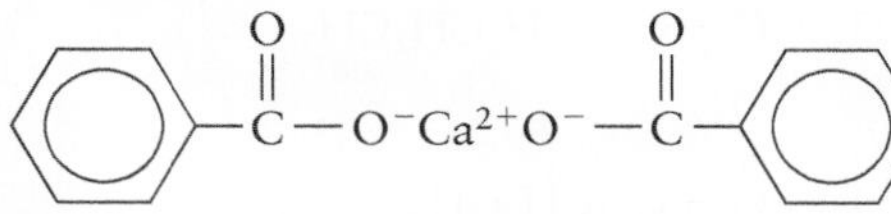

calcium benzoate

c. $CH_3CH_2\overset{CH_3}{\overset{|}{C}H}-\overset{O}{\overset{\|}{C}}-O^-Na^+$

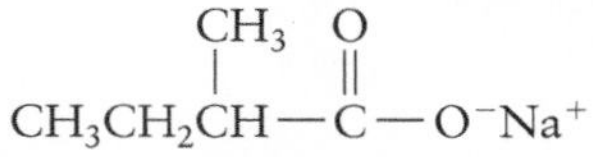

sodium 2-methylbutanoate

10.38 **a.** sodium valerate
b. magnesium lactate
c. potassium citrate

10.40 **a.** sodium stearate
b. calcium propanoate
c. citric acid and sodium citrate
d. sodium benzoate

10.42 b, c, and d

10.44 $R-\overset{O}{\overset{\|}{C}}-OH$ carboxylic acid

The structural features of a fatty acid are the carboxylic acid functional group and a long hydrocarbon tail. They are called fatty acids because they were originally isolated from fats.

10.46 The carboxylic acid present in sour milk and sauerkraut is lactic acid.

Properties of Carboxylic Acids (Section 10.5)

10.48 (b) hexane < (a) pentanal < (c) 1-pentanol < (d) butanoic acid

10.50 **a.**

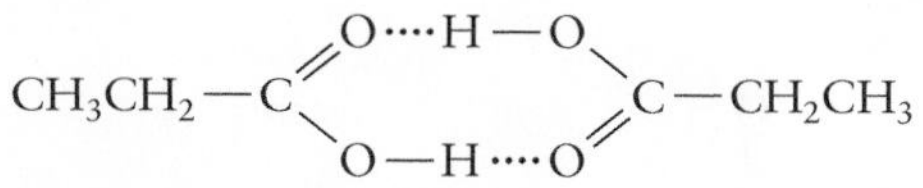

$$CH_3CH_2-C\begin{matrix}\nearrow O\cdots H-O \searrow \\ \searrow O-H\cdots O \nearrow\end{matrix}C-CH_2CH_3$$

b.

$$CH_3CH_2-\overset{O\cdots H-O-H}{\overset{\|}{C}}-O-H\cdots O(H)H \quad ;\quad H-O-H\cdots O$$

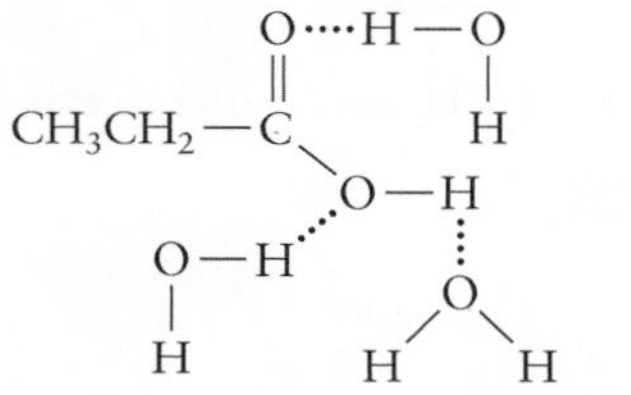

10.52 The carboxylic acid functional group allows caproic acid to be soluble in water. The solubility is limited by the aliphatic portion of the acid because it is hydrophobic.

10.54 pentane < ethoxyethane < 1-butanol < propanoic acid

10.56 **a.**

$$\underset{\displaystyle Br}{\underset{|}{CH_2}}CH_2-\overset{O}{\overset{\|}{C}}-O-\text{(H)}$$

b.

$$CH_3CH_2CH_2-\overset{O}{\overset{\|}{C}}-O-\text{(H)}$$

c.

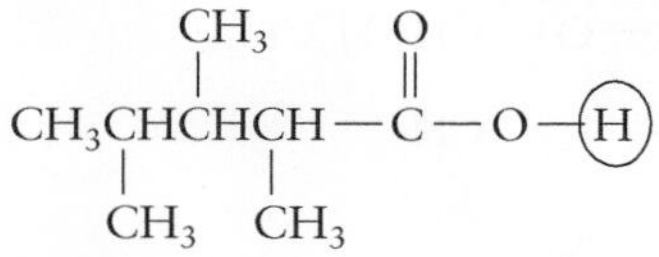

$$CH_3\underset{\displaystyle CH_3}{\underset{|}{CH}}\overset{\displaystyle CH_3}{\overset{|}{CH}}\underset{\displaystyle CH_3}{\underset{|}{CH}}-\overset{O}{\overset{\|}{C}}-O-\text{(H)}$$

10.58

$$CH_3CH_2-\overset{O}{\overset{\|}{C}}-OH + H_2O \rightleftarrows CH_3CH_2-\overset{O}{\overset{\|}{C}}-O^- + H_3O^+$$

At a pH of 12, the propanoate ion is the predominant form because the OH^- ions remove the H_3O^+ ions from solution and shift the equilibrium to the right. At a pH of 2, the propanoic acid is the predominant form because the excess H_3O^+ ions shift the equilibrium to the left.

10.60 **a.**

$$CH_3(CH_2)_7-\overset{O}{\overset{\|}{C}}-OH + NaOH \longrightarrow CH_3(CH_2)_7-\overset{O}{\overset{\|}{C}}-O^-Na^+ + H_2O$$

b.

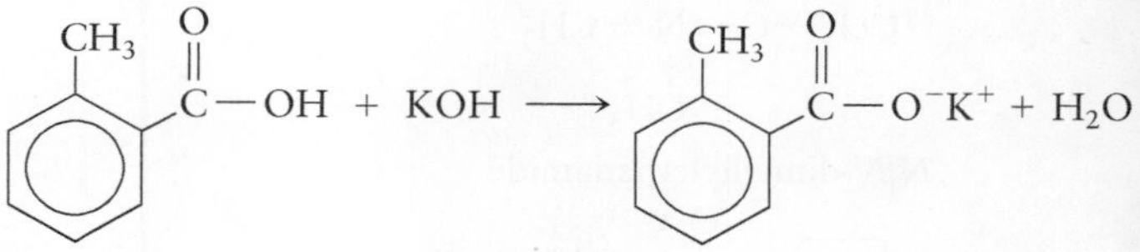

$$\text{(2-}CH_3\text{-benzene)}-\overset{O}{\overset{\|}{C}}-OH + KOH \longrightarrow \text{(2-}CH_3\text{-benzene)}-\overset{O}{\overset{\|}{C}}-O^-K^+ + H_2O$$

10.62 **a.**

$$CH_3CH_2CH_2-\overset{O}{\overset{\|}{C}}-OH + KOH \longrightarrow CH_3CH_2CH_2-\overset{O}{\overset{\|}{C}}-O^-K^+ + H_2O$$

b.

$$CH_3CH_2CH_2-\overset{O}{\overset{\|}{C}}-OH + H_2O \rightleftarrows CH_3CH_2CH_2-\overset{O}{\overset{\|}{C}}-O^- + H_3O^+$$

c.

$$CH_3CH_2CH_2-\overset{O}{\overset{\|}{C}}-OH + NaOH \longrightarrow CH_3CH_2CH_2-\overset{O}{\overset{\|}{C}}-O^-Na^+ + H_2O$$

The Nomenclature of Esters and Amides (Section 10.6)

10.64 **a.** methyl propionate
b. propyl acetate

10.66 **a.** ethyl propanonate
b. propyl pentanoate

10.68 **a.**

$$CH_3-\overset{O}{\overset{\|}{C}}-O-CH_3$$

methyl ethanoate

b.

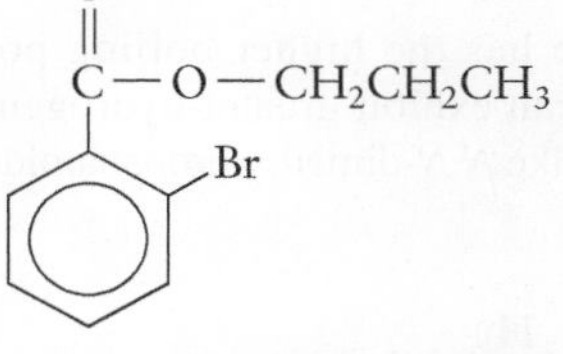

$$\text{(2-Br-benzene)}-\overset{O}{\overset{\|}{C}}-O-CH_2CH_2CH_3$$

propyl 2-bromobenzoate

c.

$$CH_3\overset{\displaystyle CH_3}{\overset{|}{CH}}\underset{\displaystyle CH_3}{\underset{|}{CH}}CH_2-\overset{O}{\overset{\|}{C}}-O-CH_2CH_3$$

ethyl 3,4-dimethylpentanoate

10.70 **a.** ethyl ethanoate
b. propyl ethanoate
c. butyl ethanoate

10.72 c and d

10.74 a. $CH_3CH_2CH_2-\overset{O}{\overset{\|}{C}}-NH-CH_3$

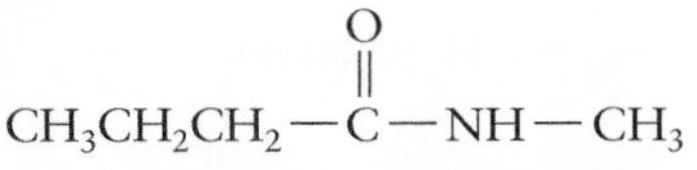

N-methylbutanamide.

b. $CH_3-\overset{O}{\overset{\|}{C}}-\underset{CH_3}{\underset{|}{N}}-CH_3$

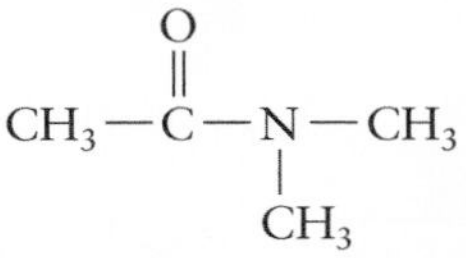

N,*N*-dimethylethanamide

c. (benzene ring with $-\overset{O}{\overset{\|}{C}}-NH_2$ and ortho CH_3)

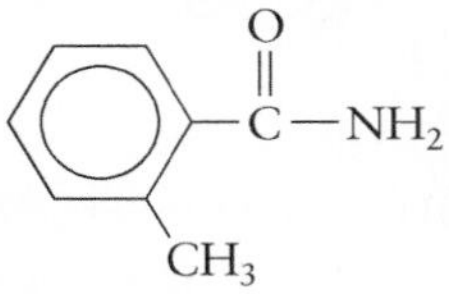

o-methylbenzamide or 2-methylbenzamide

d. $CH_3CH_2\underset{CH_3}{\underset{|}{CH}}-\overset{O}{\overset{\|}{C}}-NH_2$

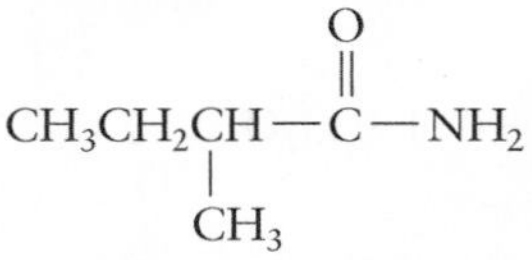

2-methylbutanamide

10.76 a. $CH_3CH_2CH_2-\overset{O}{\overset{\|}{C}}-NH_2$

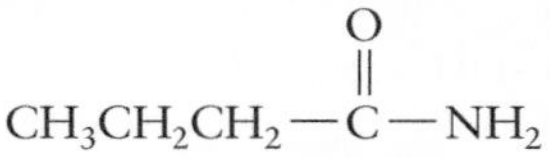

butanamide

b. (benzene ring)$-\overset{O}{\overset{\|}{C}}-NH-CH_2CH_3$

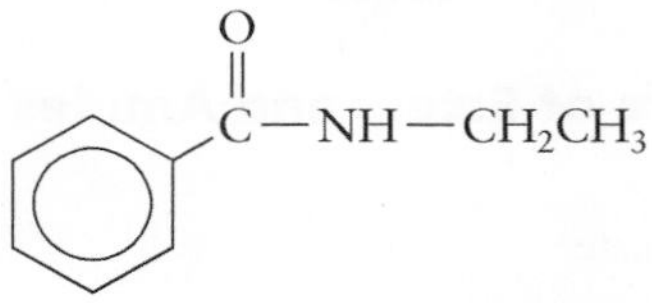

N-ethylbenzamide

c. $CH_3CH_2-\overset{O}{\overset{\|}{C}}-\underset{CH_3}{\underset{|}{N}}-CH_3$

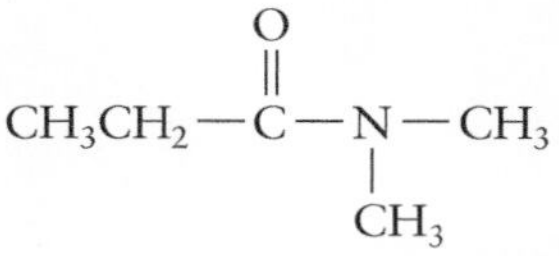

N,*N*-dimethylpropanamide

Properties of Esters and Amides (Section 10.7)

10.78 *N*-Ethylpropanamide has the higher boiling point. As a secondary amine, it can exhibit greater hydrogen-bonding than a tertiary amine like *N*,*N*-dimethylpropanamide.

10.80 a. (cyclopentyl$-\overset{O}{\overset{\|}{C}}-NH_2$ hydrogen bonded to water molecules: the carbonyl O···H–O–H; the N–H···O of water (three water molecules bonded to the N–H hydrogens and to the N))

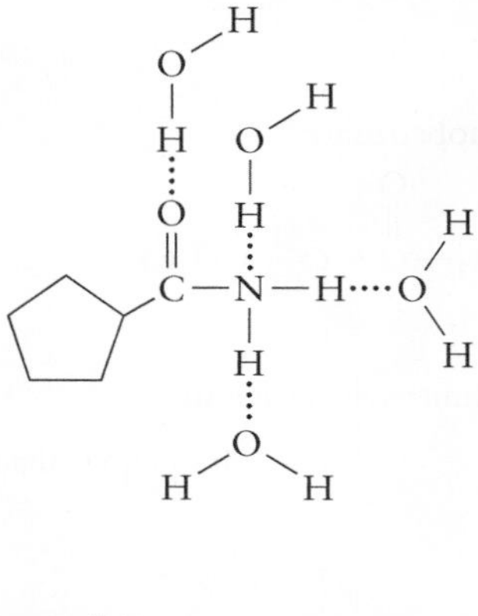

b. $H-\underset{H}{\underset{|}{N}}-\overset{O}{\overset{\|}{C}}-\underset{CH_3}{\underset{|}{CHCH_3}}$ hydrogen bonded (N–H···O=C and O···H–N) to $CH_3\underset{CH_3}{\underset{|}{CH}}-\overset{O}{\overset{\|}{C}}-\underset{H}{\underset{|}{N}}-H$

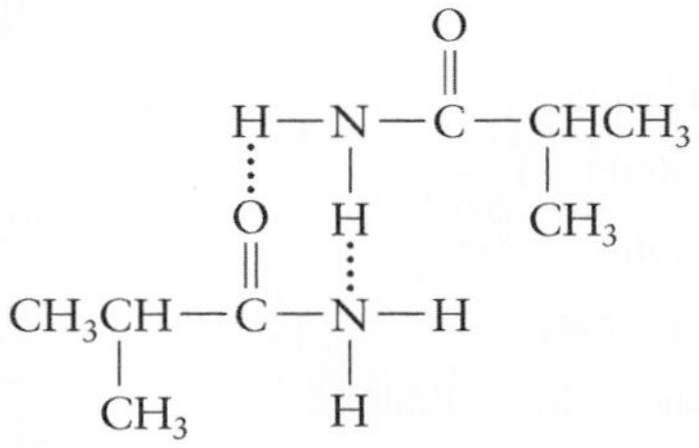

10.82 Esters have a lower boiling point than carboxylic acids of similar molecular weight due to the difference in their intermolecular forces and molecular structure. Esters can only engage in weaker London dispersion forces and dipole–dipole interactions. Carboxylic acids contain a carboxyl functional group (—COOH), which allows them to form strong intermolecular hydrogen bonds.

Formation and Reactions of Esters and Amides (Section 10.8)

10.84 a. (benzene ring)$-CH_2-O-CH_2-OH$ not an ester (ether, alcohol)

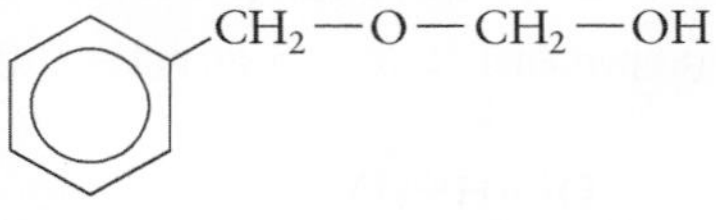

b. $CH_3-\overset{O}{\overset{\|}{C}}-O-CH_2CH_3$ ester

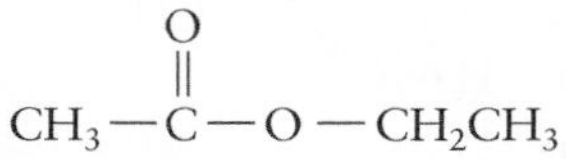

c. $CH_3CH_2-\underset{OCH_3}{\underset{|}{CH}}-OCH_3$ not an ester (acetal)

d. (benzene ring)$-O-\overset{O}{\overset{\|}{C}}-CH_3$ ester

10.86 a. $H-\overset{O}{\overset{\|}{C}}-O-CH_3$

b. $CH_3CH_2-\overset{O}{\overset{\|}{C}}-O-CH_2CH_3$

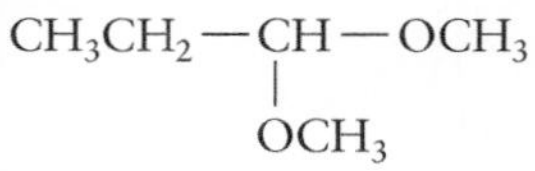

c. $CH_3\underset{CH_3}{\underset{|}{CH}}-\overset{O}{\overset{\|}{C}}-O-\underset{CH_3}{\underset{|}{CH}}CH_2CH_3$

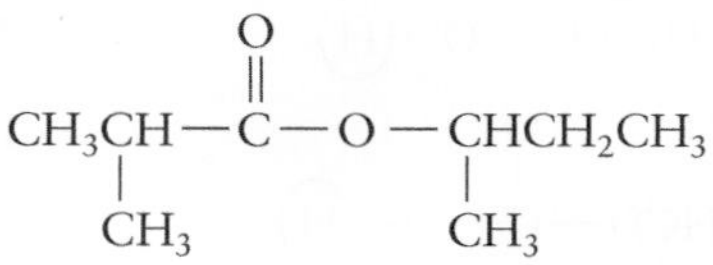

d. $CH_3CH_2CH_2-\overset{O}{\overset{\|}{C}}-O-\underset{CH_3}{\underset{|}{CH}}CH_3$

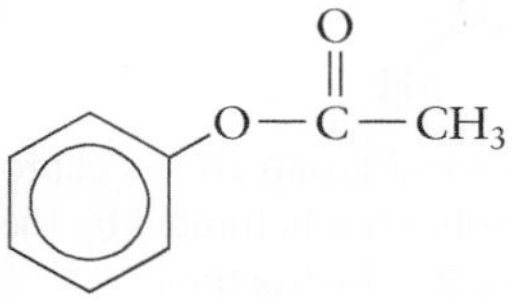

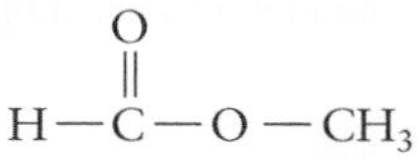

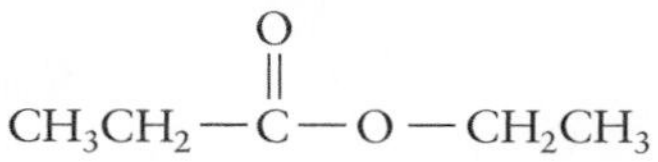

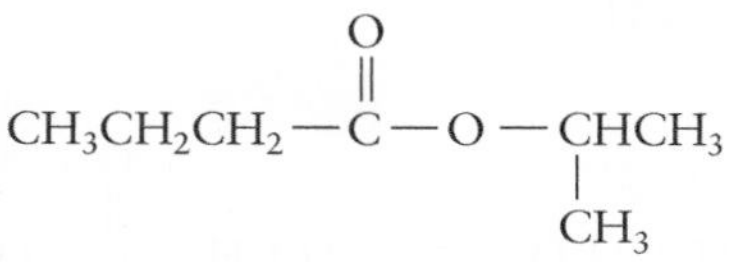

10.88 a. $C_6H_5-\overset{O}{\overset{\|}{C}}-OH + CH_3\underset{CH_3}{\underset{|}{C}H}-OH \xrightleftharpoons{H^+,\ heat} C_6H_5-\overset{O}{\overset{\|}{C}}-O-CH(CH_3)_2$ (CH_3CHCH_3) $+ H_2O$

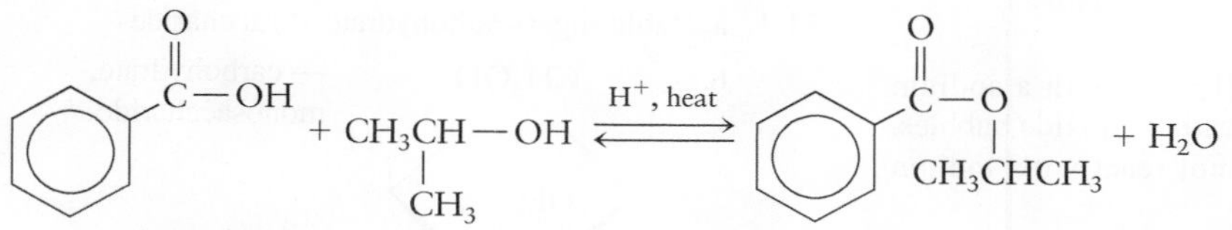

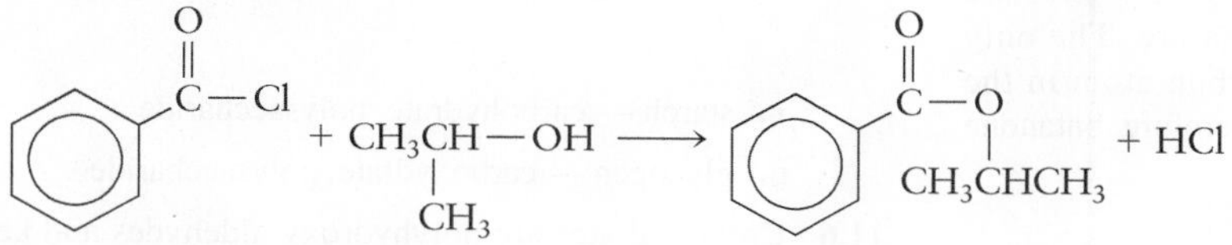

b. $C_6H_5-\overset{O}{\overset{\|}{C}}-Cl + CH_3\underset{CH_3}{\underset{|}{C}H}-OH \longrightarrow C_6H_5-\overset{O}{\overset{\|}{C}}-O-CH(CH_3)_2$ (CH_3CHCH_3) $+ HCl$

c. $C_6H_5-\overset{O}{\overset{\|}{C}}-O-\overset{O}{\overset{\|}{C}}-C_6H_5 + CH_3-OH \longrightarrow C_6H_5-\overset{O}{\overset{\|}{C}}-OH + CH_3O-\overset{O}{\overset{\|}{C}}-C_6H_5$

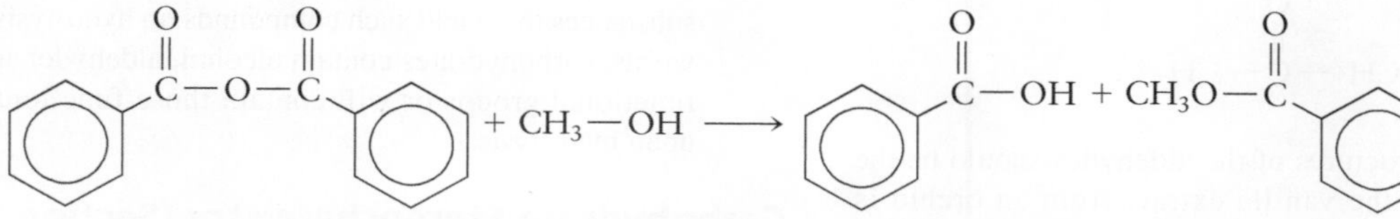

10.90 a. $CH_3-\overset{O}{\overset{\|}{C}}-OH$ and CH_3CH_2-OH

b. $C_6H_5-\overset{O}{\overset{\|}{C}}-OH$ and CH_3-OH

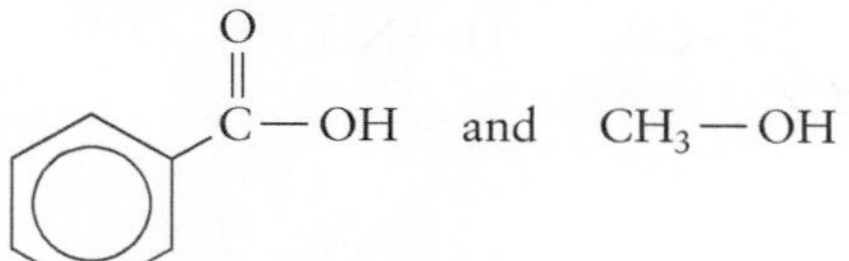

c. $CH_3-\overset{O}{\overset{\|}{C}}-OH$ and $CH_3CH_2CH_2-OH$

d. $CH_3-\overset{O}{\overset{\|}{C}}-OH$ and $CH_3\overset{OH}{\overset{|}{C}H}CH_3$

10.92 a. propanoic acid and methyl alcohol

$CH_3CH_2-\overset{O}{\overset{\|}{C}}-O-CH_3$

b. propanoic acid and phenol

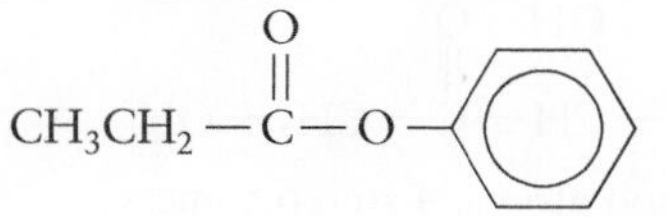

$CH_3CH_2-\overset{O}{\overset{\|}{C}}-O-C_6H_5$

(Solution 10.92 continued)

c. propanoic acid and 2-methyl-1-propanol

$CH_3CH_2-\overset{O}{\overset{\|}{C}}-O-CH_2-\overset{CH_3}{\overset{|}{C}H}-CH_3$

10.94 a. Pineapple flavoring

$CH_3CH_2CH_2-\overset{O}{\overset{\|}{C}}-OH,$

$HO-CH_2CH_2CH_2CH_3$

b. Apple flavoring

$CH_3CH_2CH_2-\overset{O}{\overset{\|}{C}}-OH,\ HO-CH_3$

10.96 a. $CH_3CH_2CH_2-\overset{O}{\overset{\|}{C}}-Cl$ **b.** $H-\overset{O}{\overset{\|}{C}}-Cl$

c. $C_6H_5-\overset{O}{\overset{\|}{C}}-Cl$ **d.** $CH_3-\overset{O}{\overset{\|}{C}}-Cl$

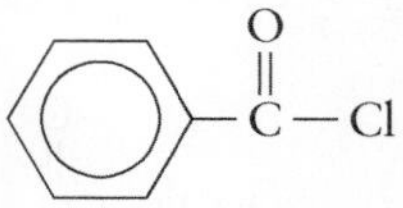

10.98 a. $CH_3(CH_2)_{16}-\overset{O}{\overset{\|}{C}}-OCH_2CH_3 + NaOH \longrightarrow CH_3(CH_2)_{16}-\overset{O}{\overset{\|}{C}}-O^-Na^+ + HOCH_2CH_3$

b. $CH_3\overset{CH_3}{\overset{|}{C}H}-\overset{O}{\overset{\|}{C}}-O-C_6H_5 + H_2O \xrightleftharpoons{H^+} CH_3\overset{CH_3}{\overset{|}{C}H}-\overset{O}{\overset{\|}{C}}-OH + HO-C_6H_5$

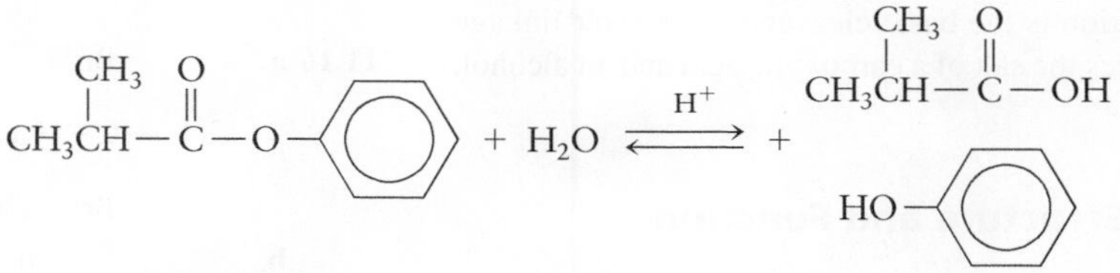

10.100 $(m\text{-}H_3C-C_6H_4)-\overset{O}{\overset{\|}{C}}-\underset{CH_2CH_3}{\underset{|}{N}}-CH_2CH_3 + NaOH \xrightarrow{heat} (m\text{-}H_3C-C_6H_4)-\overset{O}{\overset{\|}{C}}-O^-Na^+ + H-\underset{CH_2CH_3}{\underset{|}{N}}-CH_2CH_3$

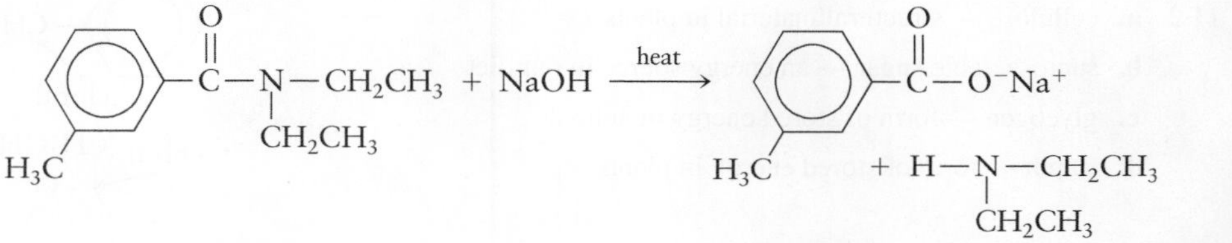

Additional Exercises

10.102 A carboxylic acid solution will react with a sodium bicarbonate solution to produce carbon dioxide bubbles, while an alcohol solution will not react with sodium bicarbonate.

10.104 Butanone is the preferable IUPAC name for this structure because 2-butanone is repetitive in nature. The only possible location of the carbonyl carbon atom in the four-carbon parent chain is carbon 2; therefore, butanone is a sufficient name for this structure.

$$CH_3CH_2-\overset{\overset{\displaystyle O}{\|}}{C}-CH_3$$

10.106 The chemical structures of the aldehydes should be the same; however, the vanilla extract from an orchid is likely to contain other natural compounds in addition to the vanillin, which may enhance the flavor of the vanilla extract. Imitation vanilla extract is typically less expensive.

10.108 Ester formation and ester hydrolysis are exactly the same reaction, only written in reverse order. The presence (or absence) of heat as well as the concentration of reactants and products determines which direction the reaction proceeds and what actually forms.

10.110 **a.**

$$CH_3CH{=}CHCH_3 + H_2O \xrightarrow{H_2SO_4} H_3C-\overset{\overset{\displaystyle OH}{|}}{CH}-CH_2CH_3$$

$$H_3C-\overset{\overset{\displaystyle OH}{|}}{CH}-CH_2CH_3 + (O) \longrightarrow H_3C-\overset{\overset{\displaystyle O}{\|}}{C}-CH_2CH_3$$

b.

$$CH_3CH_2CH_2-OH + (O) \longrightarrow CH_3CH_2-\overset{\overset{\displaystyle O}{\|}}{C}-H$$

$$CH_3CH_2-\overset{\overset{\displaystyle O}{\|}}{C}-H + (O) \longrightarrow CH_3CH_2-\overset{\overset{\displaystyle O}{\|}}{C}-OH$$

10.112 Ester hydrolysis is the reversible reaction in which water reacts with an ester in the presence of an acid catalyst to produce a carboxylic acid and an alcohol.

Saponification is the basic cleavage of an ester linkage that produces the salt of a carboxylic acid and an alcohol.

Chapter 11 Carbohydrate Structure and Function (Section 11.1)

11.2 **a.** cellulose — structural material in plants

b. sucrose, table sugar — an energy source in our diet

c. glycogen — form of stored energy in animals

d. starch — form of stored energy in plants

11.4 **a.** table sugar–carbohydrate, disaccharide

b. 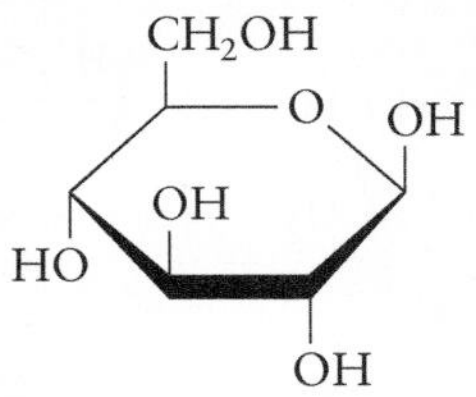 (ring structure: CH_2OH, O, OH, OH, HO, OH) — carbohydrate, monosaccharide

c. starch — carbohydrate, polysaccharide

d. glycogen — carbohydrate, polysaccharide

11.6 Carbohydrates are polyhydroxy aldehydes and ketones, or substances that yield such compounds on hydrolysis. In other words, carbohydrates contain alcohol, aldehyde, and ketone functional groups or will contain those functional groups upon hydrolysis.

Carbohydrate Stereochemistry (Section 11.2)

11.8

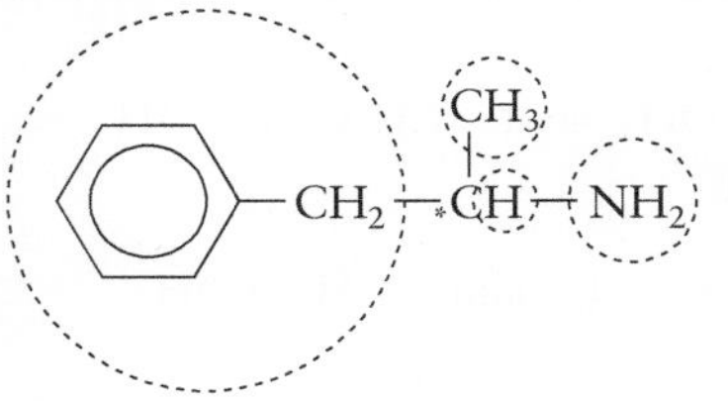

$$C_6H_5-CH_2-\overset{\overset{\displaystyle CH_3}{|}}{\overset{*}{C}H}-NH_2$$

The chiral carbon atom is marked with an asterisk. The four attached groups are:

1. a hydrogen
2. an amino group
3. a methyl group
4. a $C_6H_5-CH_2-$ group

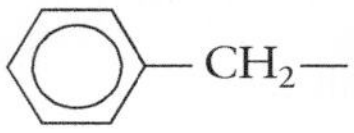

11.10 **a.** yes **b.** yes

c. no **d.** no

11.12 2 chiral carbons; 4 stereoisomers

11.14 **a.**

$$\overset{\overset{\displaystyle OH}{|}}{CH_2}-\overset{\overset{\displaystyle OH}{|}}{CH}-\overset{\overset{\displaystyle OH}{|}}{CH}-\overset{\overset{\displaystyle O}{\|}}{C}-CH_2-OH$$

2 chiral carbon atoms; 4 stereoisomers

b.

$$\overset{\overset{\displaystyle OH}{|}}{CH_2}-\overset{\overset{\displaystyle OH}{|}}{CH}-\overset{\overset{\displaystyle OH}{|}}{CH}-\overset{\overset{\displaystyle OH}{|}}{CH}-CHO$$

3 chiral carbon atoms; 8 stereoisomers

11.16 **a.**

$$CH_3-\underset{\underset{\displaystyle Br}{|}}{\overset{\overset{\displaystyle Br}{|}}{C}}-CH_2-\overset{\overset{\displaystyle OH}{|}}{CH}-COOH$$

(chiral: the CH carbon) 1 chiral carbon atom; a pair of enantiomers

b.

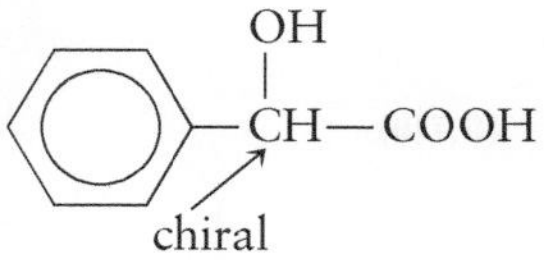

$$C_6H_5-\overset{\overset{\displaystyle OH}{|}}{CH}-COOH$$

(chiral: the CH carbon) 1 chiral carbon atom; a pair of enantiomers

c. 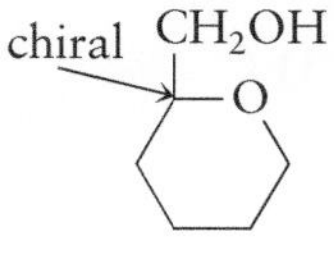 chiral CH_2OH on ring carbon adjacent to O (six-membered ring containing O) 1 chiral carbon atom; a pair of enantiomers

11.18 First, identify the chiral carbon, indicated by the asterisk (*). The stereochemistry at the chiral carbon can be represented by drawing the substituent to the main carbon chain either on a dashed or a wedged bond.

a. 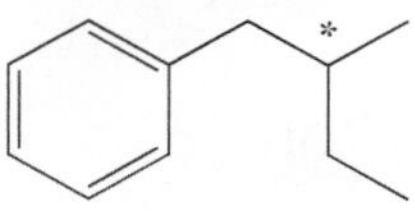 yields 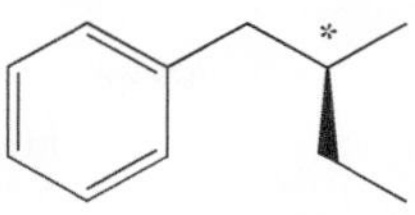 and 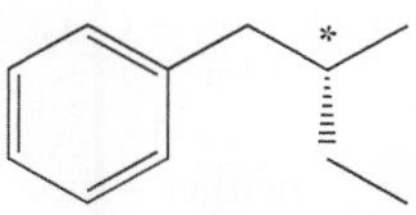 as its two enantiomers

b. 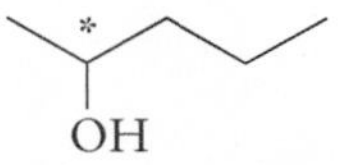 OH yields 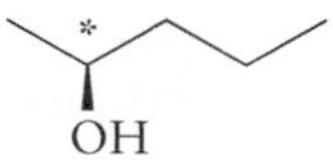 OH and 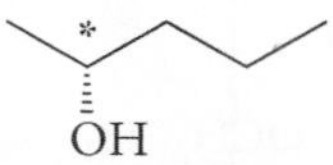 OH as its two enantiomers

11.20 With multiple chiral centers, the enantiomer switches the stereochemistry at *each* chiral center from wedge to dash and vice versa.

a. The enantiomer of (OH, O, H, HS) would be (OH, O, H, HS)

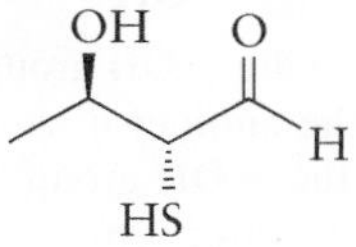

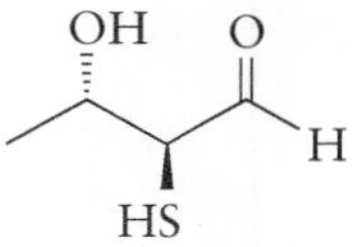

b. The enantiomer of (HO, OH, O, OH, H_3C) would be (HO, OH, O, OH, H_3C)

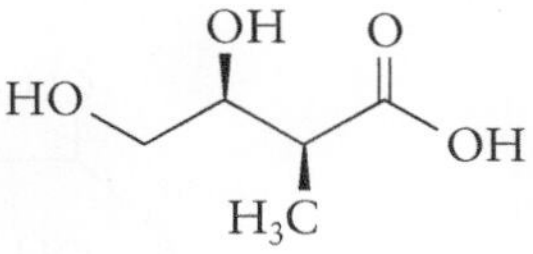

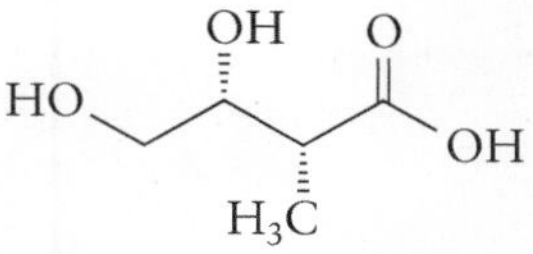

Fischer Projections, Cyclization, and Haworth Projections (Section 11.3)

11.22 a. 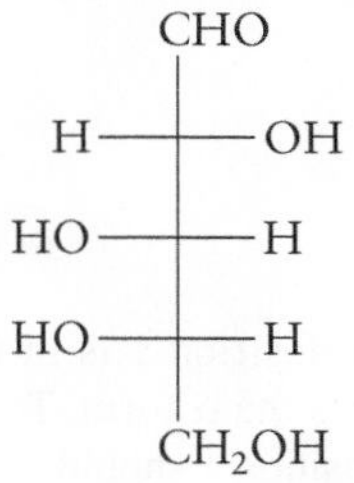 CHO; H—OH; HO—H; HO—H; CH_2OH

b. 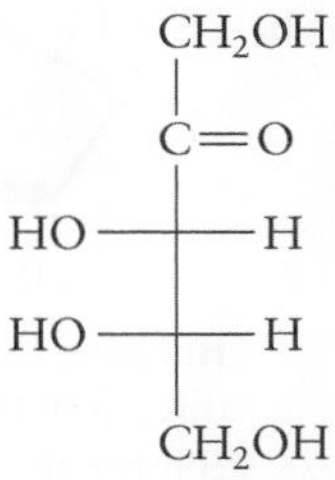 CH_2OH; C=O; HO—H; HO—H; CH_2OH

11.24 a. 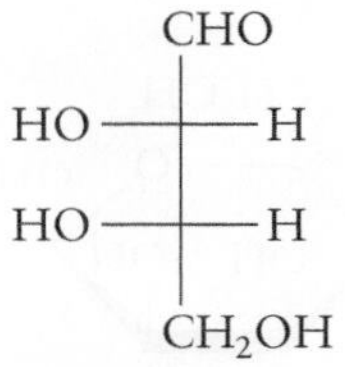 CHO; HO—H; HO—H; CH_2OH — L-form

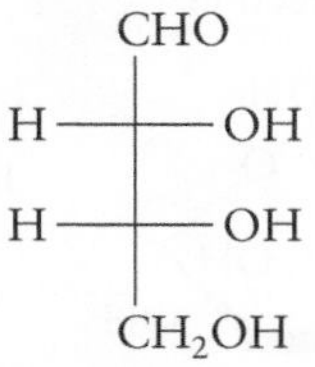

CHO; H—OH; H—OH; CH_2OH

b. 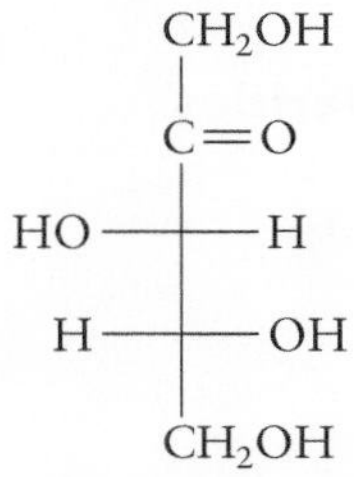 CH_2OH; C=O; HO—H; H—OH; CH_2OH — D-form

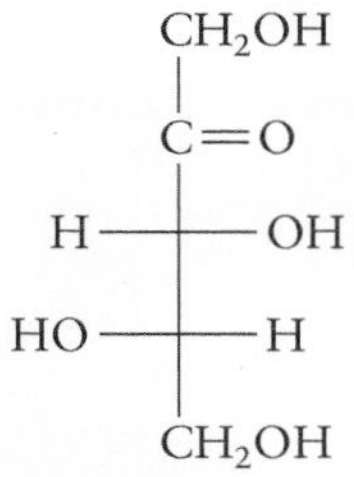

CH_2OH; C=O; H—OH; HO—H; CH_2OH

11.26 a. 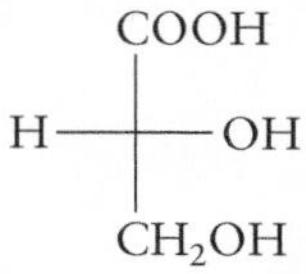 COOH; H—OH; CH_2OH — D-form

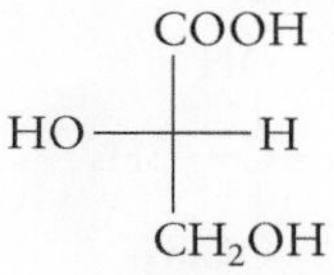

COOH; HO—H; CH_2OH — L-form

b. 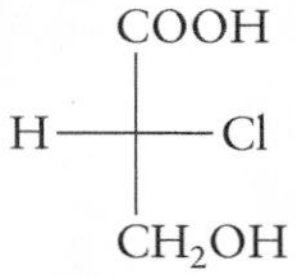 COOH; H—Cl; CH_2OH — D-form

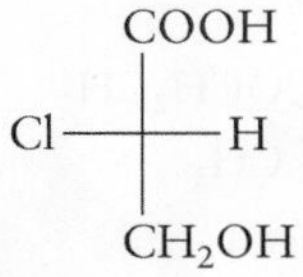

COOH; Cl—H; CH_2OH — L-form

11.28 The study of chiral molecules is important in biochemistry because living organisms consist largely of chiral substances. The stereochemistry of a biological compound plays a key role in determining the receptors with which it will interact, and enzymes will be able to catalyze its reactions, as well as what its physiological effect on an organism will be.

11.30 a. same

b. same

c. rotate plane-polarized light in different directions

d. same

11.32 *False*. The D and L signify the spatial relationships at chiral carbons relative to the reference compound, glyceraldehyde. Both D and L compounds can rotate either dextrorotatory or levorotatory, so they do not signify the direction of how plane-polarized light is rotated by the molecule.

11.34 d

11.36 b, c

11.38 a. $CH_3-\underset{H}{\overset{CH_3}{C}}-OCH_3$

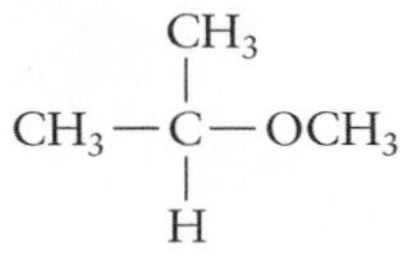

neither – ether

b. $CH_3CH_2\overset{OCH_3}{CH}-OH$

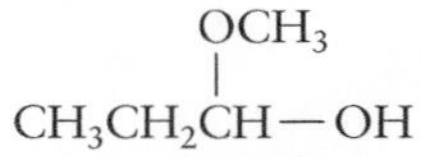

hemiacetal

c. $CH_3-\underset{OH}{\overset{OH}{C}}-CH_2CH_3$

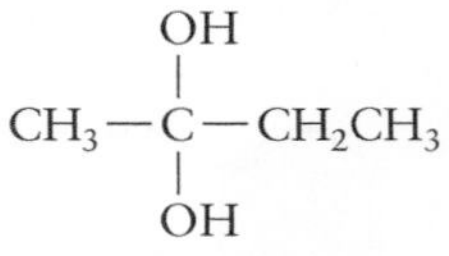

neither – diol

d. $CH_3\underset{CH_3}{\overset{OH}{C}}-OCH_2CH_3$

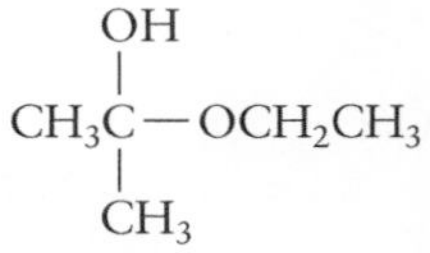

hemiketal

11.40 a. $CH_3-\underset{OCH_2CH_3}{\overset{OCH_2CH_3}{C}}-CH_3$

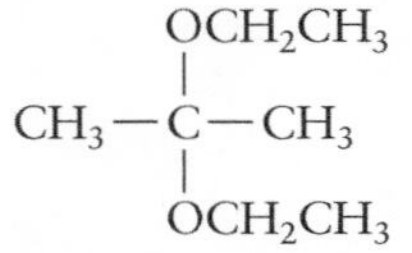

ketal

b. O—CHCH$_3$, O

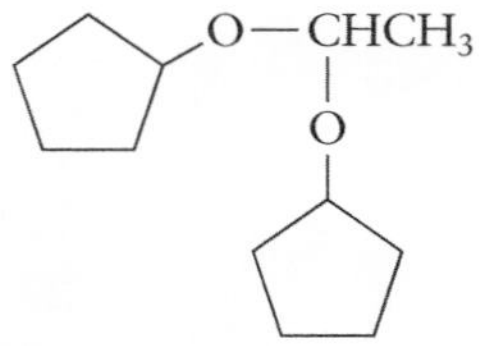

acetal

c. $CH_3CH_2-\underset{OCH_3}{\overset{OH}{C}}-CH_3$

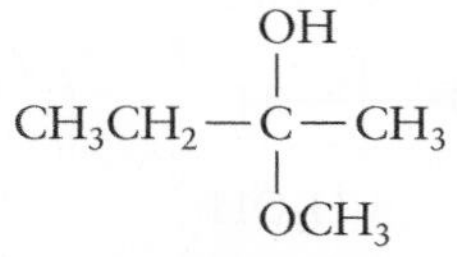

neither – hemiketal

11.42 a. O, OCH_2CH_3, CH_3

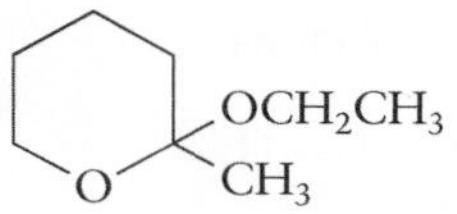

ketal

b. CH_2CH_3, OH

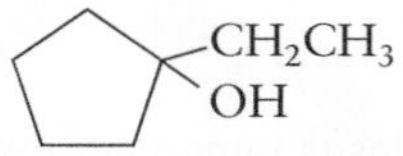

none of these (alcohol)

c. $\underset{OCH_3}{CH}-OH$

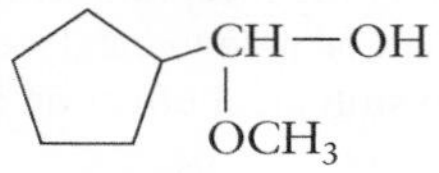

hemiacetal

11.44 a. The hemiacetal within the following molecule is highlighted in blue.

HO, O, OH, HO, OH, OH

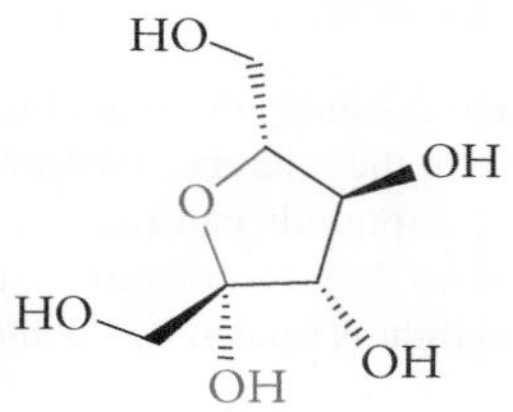

b. The hemiacetal within the following molecule is highlighted in blue.

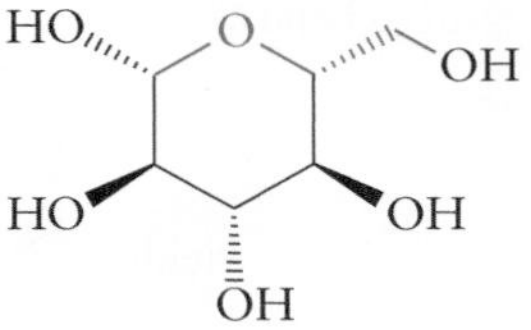

HO, O, OH, HO, OH, OH

11.46 a. CH_2OH, O, OH, OH, HO, OH

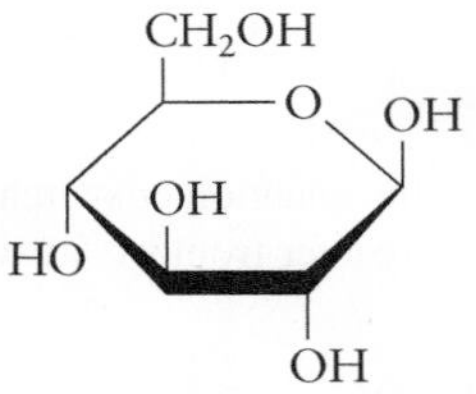

Because the —OH group at position 1 is in the up direction, the anomer drawn is the β form. To draw the α form, the —OH group at position 1 should be drawn in the down direction:

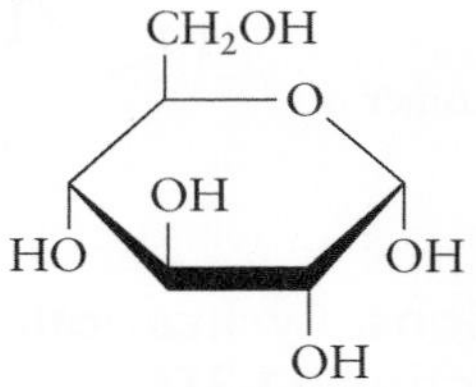

CH_2OH, O, OH, HO, OH, OH

b. $HOCH_2$, O, CH_2OH, OH, OH, OH

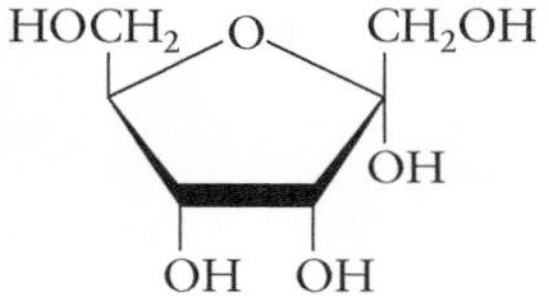

Because the —OH group at position 1 is in the down direction, the anomer drawn is the α form. To draw the β form, the —OH group at position 1 should be drawn in the up direction.

11.48 CH_2OH, O, HO, OH, OH, OH — α-talose

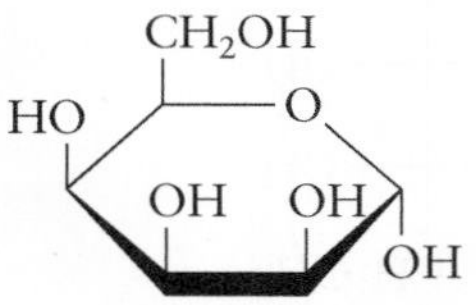

CH_2OH, O, HO, OH, OH, OH — β-talose

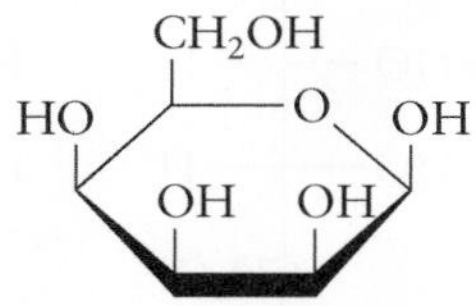

Classification of Monosaccharides (Section 11.4)

11.50 a. ketose

b. aldose

c. aldose

11.52 a. pentose

b. hexose

c. tetrose

11.54 a.

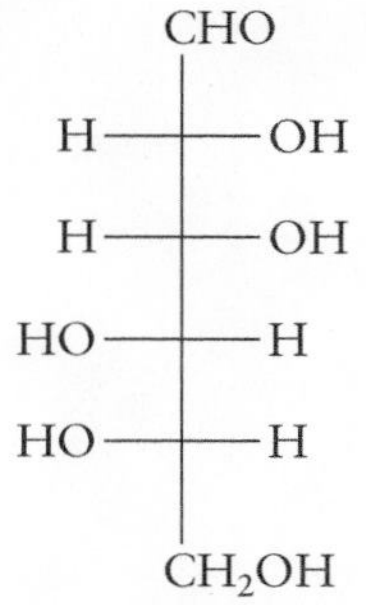

CHO
H—OH
H—OH
HO—H
HO—H
CH_2OH
aldohexose

b.

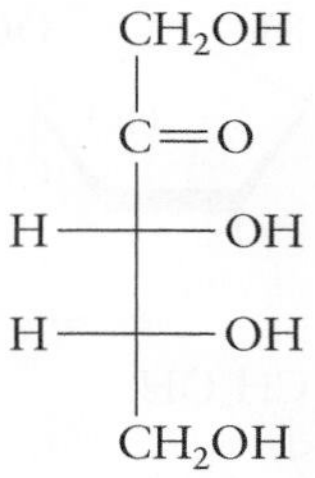

CH_2OH
C=O
H—OH
H—OH
CH_2OH
ketopentose

11.56 a.

CHO
HO—H
HO—H
CH_2OH

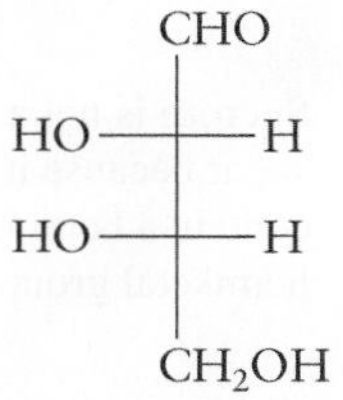

An aldotetrose must have: (1) —CHO at the top, (2) 4 total carbon atoms, and (3) —CH_2OH at the bottom of the structure. This aldotetrose has the L configuration.

b.

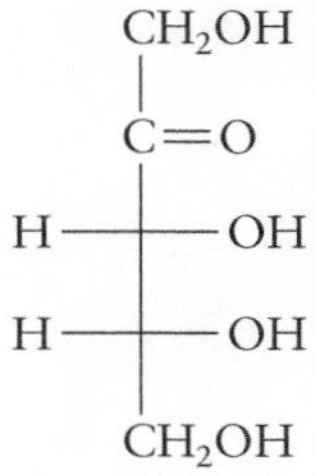

CH_2OH
C=O
H—OH
H—OH
CH_2OH

A ketopentose must have: (1) a —CH_2OH bonded to a carbonyl carbon atom at the top, (2) 5 total carbon atoms, and (3) a —CH_2OH at the bottom of the structure. This ketopentose has the D configuration.

Important Monosaccharides (Section 11.5)

11.58 D-Glucose can be injected directly into the bloodstream to serve as an energy source because glucose is the sugar transported by the blood to body tissues to satisfy energy

(Solution for 11.58 continued)

requirements. Glucose has a high solubility in aqueous solutions, and it dissolves readily in water and in blood; therefore, it poses no risks of toxicity itself, and will not form dangerous solid occlusions in the blood vessels.

11.60 Fructose is a ketohexose.

11.62 a. The natural sources of glucose are honey and fruits, as well as sucrose, which is a disaccharide containing glucose.

b. The natural source of fructose is fruit.

c. The natural source of galactose is milk, which contains lactose, a disaccharide of galactose and glucose.

Properties of Monosaccharides (Section 11.6)

11.64

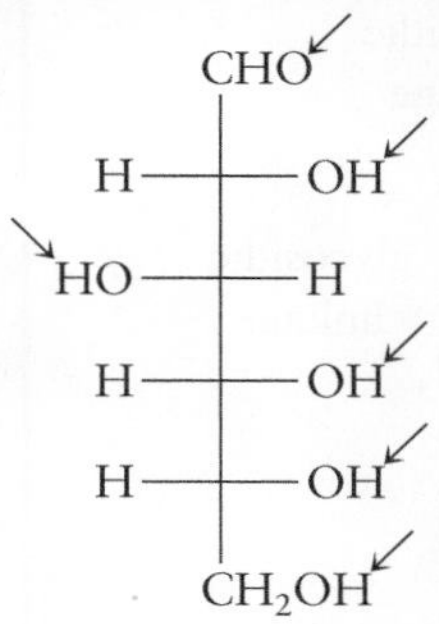

CHO ↙
H—OH ↙
↘ HO—H
H—OH ↙
H—OH ↙
CH_2OH ↙

The arrows indicate the sites on the molecule where water can hydrogen-bond to glucose.

11.66 The cyclic compound β-D-galactose can react with Cu^{2+} and be classified as a reducing sugar because as it undergoes mutarotation in a solution, the ring opens, an aldehyde is formed, and the compound can react with Benedict's reagent, which is an oxidizing agent.

11.68 All will give a positive Benedict's test.

11.70 a.

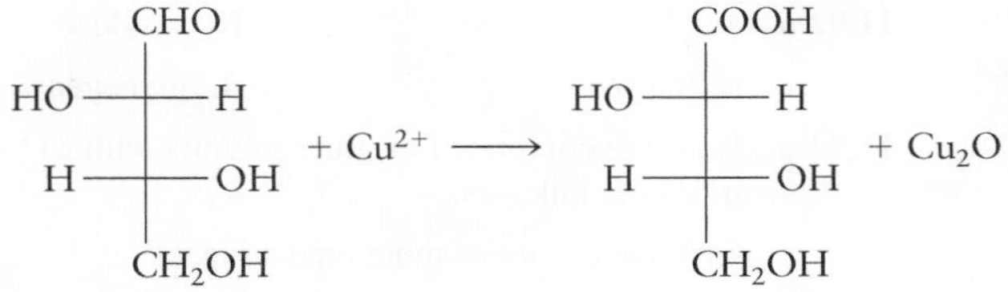

CHO / HO—H / H—OH / CH_2OH + Cu^{2+} ⟶ COOH / HO—H / H—OH / CH_2OH + Cu_2O

b.

CH_2OH, O, OH, HO, OH, OH (ring) + CH_3—OH $\xrightarrow{H^+}$ CH_2OH, O, OH, HO, OCH_3, OH (ring)

+ CH_2OH, O, OCH_3, OH, HO, OH (ring) + $2H_2O$

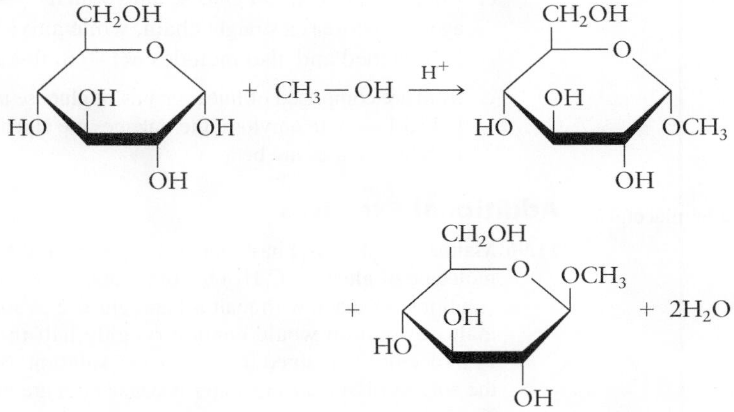

c. 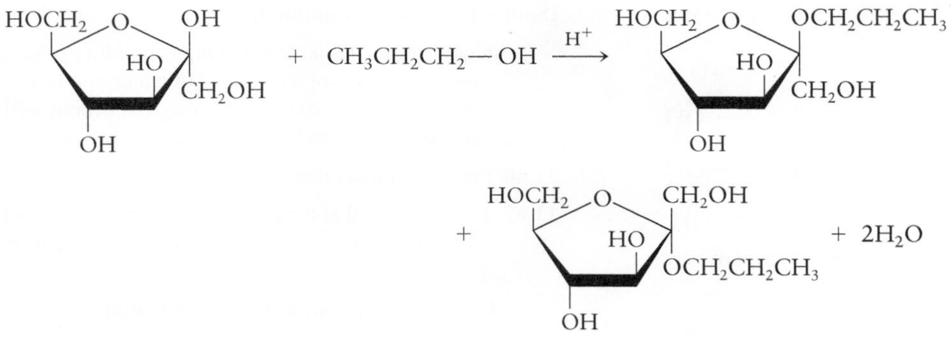 $HOCH_2$ O OH HO CH_2OH OH + $CH_3CH_2CH_2—OH$ $\xrightarrow{H^+}$ $HOCH_2$ O $OCH_2CH_2CH_3$ HO CH_2OH OH + $HOCH_2$ O CH_2OH HO $OCH_2CH_2CH_3$ OH + $2H_2O$

11.72 a. 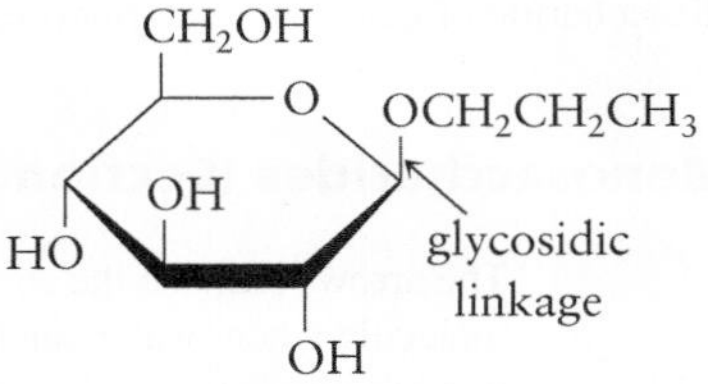 CH_2OH O $OCH_2CH_2CH_3$ OH HO OH glycosidic linkage

b. 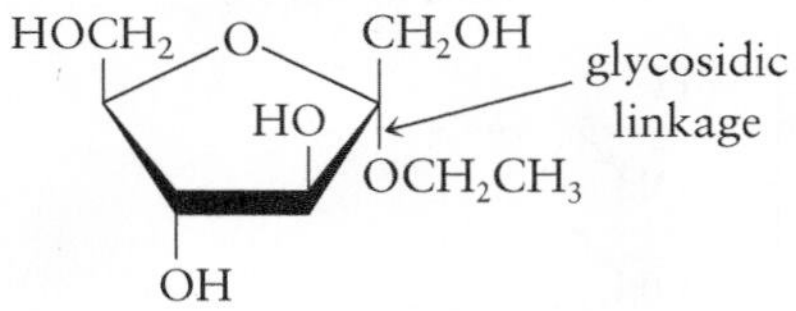 $HOCH_2$ O CH_2OH HO OCH_2CH_3 OH glycosidic linkage

11.74 a. $CH_3CH_2CH_2OH$

b. CH_3CH_2OH

Disaccharides (Section 11.7)

11.76 a. sucrose b. maltose

c. lactose d. maltose

e. sucrose f. sucrose

11.78 The hemiacetal group of a lactose molecule is able to react with Benedict's reagent because the ring containing the hemiacetal group is not "locked." In solution the hemiacetal group can undergo mutarotation, opening the ring into an open-chain aldose that can react with Benedict's reagent.

11.80 a. glucose, fructose b. glucose, glucose

c. galactose, glucose

11.82 a. $1 \rightarrow 4$ b. $1 \rightarrow 6$

c. $1 \rightarrow 4$

11.84 a. β b. α

c. β

11.86 a. yes b. yes

c. yes

11.88 Lactose is a reducing sugar because it contains a hemiacetal group.

a. 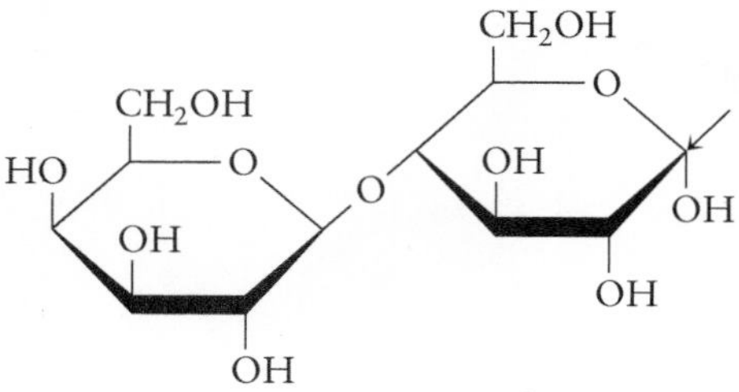 HO CH_2OH O OH OH O CH_2OH O OH OH OH

(Solution 11.88 continued)

b. 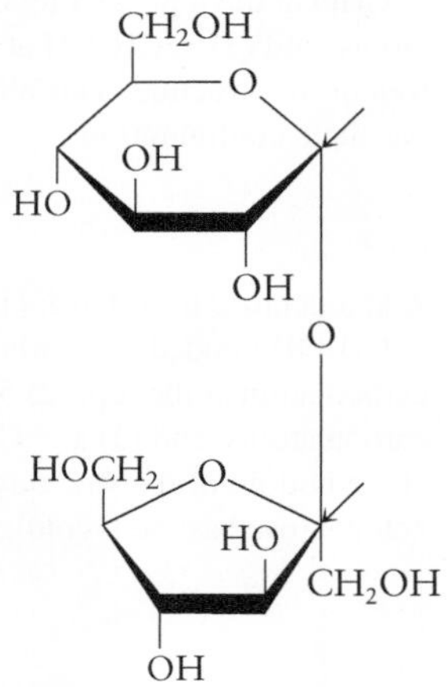 CH_2OH O OH HO OH O $HOCH_2$ O HO CH_2OH OH

Sucrose is not a reducing sugar because it does not contain a hemiacetal or a hemiketal group.

11.90 In the case of melibiose, the anomeric carbon on glucose contains an intact —OH group that can be oxidized by an oxidizing agent. In other words, melibiose can reduce an oxidizing agent, and is thus a reducing sugar.

Polysaccharides (Section 11.8)

11.92 a. amylose, cellulose b. amylose

c. cellulose d. glycogen

11.94 a. Both are comprised of glucose units with $\alpha(1 \rightarrow 4)$ and $\alpha(1 \rightarrow 6)$ linkages.

Glycogen contains more branching.

b. Both are comprised of glucose units with $\alpha(1 \rightarrow 4)$ linkages. Amylose is a straight chain, while amylopectin is branched and also includes $\alpha(1 \rightarrow 6)$ linkages.

c. Both are comprised of linear chains of glucose units, linked $1 \rightarrow 4$. In amylose, the linkages are alpha, while in cellulose they are beta.

Additional Exercises

11.96 Maltose ($C_{12}H_{22}O_{11}$) has approximately twice the mass per molecule of glucose ($C_6H_{12}O_6$); therefore, if two 10% (w/v) solutions are made with maltose and glucose as solutes, the maltose solution would contain roughly half the number of molecules contained in the glucose solution. Neither of the solutes dissociates in water because both are molecular. The boiling point of a solution increases with the number of solute particles in solution; therefore, the glucose solution would contain more molecules of solute and would have the higher boiling point.

11.98 $100 \text{ g sucrose} \left(\frac{1 \text{ mole sucrose}}{342 \text{ g sucrose}}\right)\left(\frac{1 \text{ mole glucose}}{1 \text{ mole sucrose}}\right)\left(\frac{180 \text{ g glucose}}{1 \text{ mole glucose}}\right) = 52.6 \text{ g glucose}$

Chemistry for Thought

11.100 $\frac{2.80 \times 10^5 \text{ amu}}{1 \text{ molecule starch}}\left(\frac{1 \text{ molecule linked glucose}}{171 \text{ amu linked glucose}}\right) = 1.64 \times 10^3 \text{ glucose molecules}$

Note: The molecular mass of glucose is 180 amu, but when glucose links to form a molecule of starch, 1 molecule of water (18.0 amu) is lost for every 2 glucose units or 0.5 molecules of water per glucose unit; therefore, the molecular mass of a linked glucose unit is 180 = 9.0 = 171 amu.

11.102 Aspartame (Nutrasweet) contains calories and yet is used in diet drinks. A drink can contain aspartame as a sweetener and yet be low in calories because the amount needed to sweeten a diet drink is so small that very few calories are added to the drink.

11.104 Foods that would be expected to give a positive starch test include bread, crackers, pasta, and rice.

11.106 Amylose is a straight-chain glucose polymer similar to cellulose. Paper manufactured with amylose instead of cellulose would have less longevity because amylose forms loose spiral structures, whereas cellulose forms extended straight chains that can be aligned side by side to form well-organized, water-insoluble fibers in which the hydroxy groups form numerous hydrogen bonds with the neighboring chains. These hydrogen bonds confer rigidity and strength to the overall structure of the paper.

Chapter 12
Amino Acid Structure and Classification (Section 12.1)

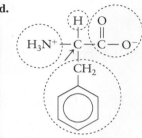

12.2 a. $H_3N^+-CH(CH(OH)-CH_3)-COO^-$ (circled groups: H, H_3N^+, $C(=O)-O^-$, $CH-CH_3$ / OH)

threonine

b. $H_3N^+-CH(CH_2-COO^-)-COO^-$ (circled groups: H, H_3N^+, $C(=O)-O^-$, CH_2 / COO^-)

aspartate

c. $H_3N^+-CH(CH_2-OH)-COO^-$ (circled groups: H, H_3N^+, $C(=O)-O^-$, CH_2 / OH)

serine

d. $H_3N^+-CH(CH_2-C_6H_5)-COO^-$ (circled groups: H, H_3N^+, $C(=O)-O^-$, CH_2 / benzene ring)

phenylalanine

12.4 a. threonine — neutral and polar

b. aspartate — acidic and polar

c. serine — neutral and polar

d. phenylalanine — neutral and nonpolar

Reactions of Amino Acids (Section 12.2)

12.6 A *zwitterion* is a dipolar ion that carries both a positive and a negative charge as a result of an internal acid–base reaction in an amino acid molecule.

12.8 a. $H_3N^+-CH(CH_2-CH_2-S-CH_3)-COO^-$

b. $H_3N^+-CH(CH(OH)-CH_3)-COO^-$

c. $H_3N^+-CH(CH_2-SH)-COO^-$

12.10 a. $H_2N-CH(CH_2-SH)-COO^-$

cysteine

b. $H_2N-CH(CH_3)-COO^-$

alanine

12.12 a. $H_3\overset{+}{N}-CH(CH(CH_3)-CH_3)-COOH$

valine

b. $H_3\overset{+}{N}-CH(CH(OH)-CH_3)-COOH$

threonine

12.14

$$H_3N^+—CH(CH_2CHCH_3CH_3)—C(=O)—O^- + H_3N^+—CH(CH(OH)—CH_3)—C(=O)—O^- \rightarrow H_3N^+—CH(CH_2CHCH_3CH_3)—C(=O)—NH—CH(CHCH_3OH)—C(=O)—O^- + H_2O$$

leucine; threonine; Leu-Thr

$$H_3N^+—CH(CH(OH)—CH_3)—C(=O)—O^- + H_3N^+—CH(CH_2CHCH_3CH_3)—C(=O)—O^- \rightarrow H_3N^+—CH(CH(OH)—CH_3)—C(=O)—O^-—NH—CH(CH_2CHCH_3CH_3)—C(=O)—O^- + H_2O$$

threonine; leucine; Thr-Leu

12.16

$$\overset{+}{H_3N}—CH(CH(CH_3)_2)—C(=O)—NH—CH(CH_2COO^-)—C(=O)—NH—CH(CH_2SH)—C(=O)—O^-$$

Val-Asp-Cys

12.18 alanine–cysteine–valine

12.20

peptide bonds

$$\overset{+}{H_3N}—CH(CH_3)—C(=O)—NH—CH(CH_2SH)—C(=O)—NH—CH(CH—CH_3(CH_3))—COO^-$$

N-terminal amino acid; C-terminal amino acid

Protein Structure (Section 12.3)

12.22 Even though two polypeptides have the same amino acid composition, the sequence of amino acids may be different. A different order of amino acids would give rise to a different three-dimensional structure and thus different functions.

12.24

$$—NH—CH(R)—C(=O)—NH—CH(R')—C(=O)—NH—CH(R'')—C(=O)—NH—CH(R''')—C(=O)—$$

12.26 The formation of secondary structures (α-helix and the β-pleated sheet) arises from hydrogen-bonding between amide groups and is independent from the primary structure.

12.28 The formation of salt bridges can occur only if the side chain is charged. The amino acids with side-chain groups that can form salt bridges are histidine, lysine, arginine, aspartate, and glutamate.

12.30 c, d

12.32 Quaternary protein structure is the arrangement of subunits that form a larger protein.

12.34 The quaternary structure of hemoglobin is four subunits (two alpha chains containing 141 amino acid residues each and two beta chains containing 146 residues each) that each contain a heme group (a planar ring structure centered around an iron atom) located in crevices near the exterior of the molecule. The molecule is nearly spherical, with the four subunits held together rather tightly by hydrophobic forces.

Protein Functions (Section 12.4)

12.36 The role of enzymes in the body is to catalyze chemical reactions.

12.38 **a.** structure
b. transport
c. catalysis

Classes of Enzymes (Section 12.5)

12.40 **a.** hydrolase **b.** transferase
c. lyase **d.** ligase

12.42

	Enzyme	Reaction Catalyzed
a.	decarboxylase	removal of carboxyl group from compounds
b.	phosphatase	hydrolysis of phosphate ester linkages
c.	peptidase	hydrolysis of peptide linkages
d.	esterase	formation of ester linkages

Enzyme Activity (Section 12.6)

12.44 Enzyme catalysis of a reaction is superior to normal laboratory conditions because enzymes are specific in the type of reaction they catalyze and enzyme activity can be regulated by the cell.

12.46 Many types of enzymes are needed because enzymes can only catalyze specific reactions and a body requires many reactions to occur in order to maintain life.

12.48 Enzyme activity in an experiment is observed by any method that allows the rate of product formation or reactant usage to be determined. The disappearance or appearance of a characteristic color is an example.

12.50 The equation shows that the enzyme (E) and substrate (S) establish an equilibrium with the enzyme–substrate complex (ES). This is a reversible reaction. The enzyme–substrate complex can break apart into the enzyme and substrate or react to produce the enzyme and the product (P). The reaction to produce the product is not a reversible reaction.

12.52 a. $^{-}OOC-CH_2-OH$ and $HO-$

b. $CH_3CH_2-COO^-$ and Fe^{2+}

c. $CH_3CH_2CH_2CH_2CH_3$ and (phenyl group)

12.54 Covalent bonds are strong bonds — too strong to permit the eventual release of the substrate.

Factors Affecting Enzyme Activity (Section 12.7)

12.56 The rate of an enzyme-catalyzed reaction reaches a maximum when the substrate concentration is raised beyond the saturation point. At this point, the enzyme becomes the limiting reagent and the substrate is in excess.

12.58 With more enzyme present, the V_{max} will be higher than it should be.

12.60 Enzyme structures are unique, so enzyme properties are also unique.

Enzyme Inhibition and Regulation (Section 12.8)

12.62 Irreversible enzyme inhibition involves the formation of a covalent bond between an irreversible inhibitor and the enzyme, which renders the enzyme inactive. This is not reversible because the bond between enzyme and inhibitor is extremely strong. Reversible enzyme inhibition involves the reversible binding, which may or may not be covalent, of a reversible inhibitor to an enzyme. An equilibrium is established; therefore, the inhibitor can be removed from the enzyme by shifting the equilibrium.

12.64

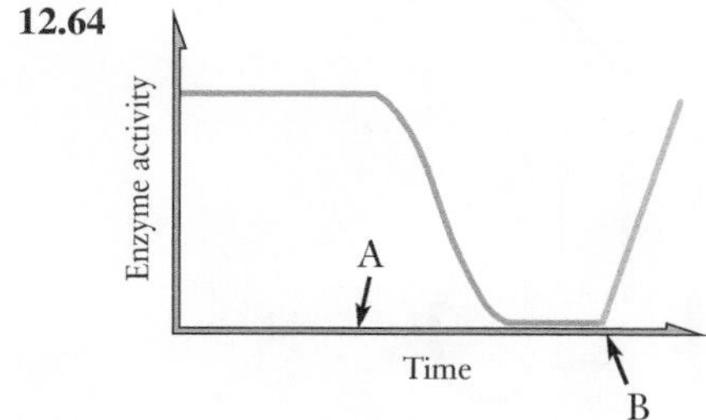

12.66 a. reversible — competitive

b. reversible — noncompetitive

c. irreversible — reversible noncompetitive

d. irreversible — reversible noncompetitive

Enzymes in Medical Applications and Disease (Section 12.9)

12.68 A medical professional might run a diagnostic test detecting specific enzymes in the blood serum in the following scenarios:

1. A patient has experienced an injury causing major tissue/cellular damage.
2. The medical professional is testing a patient for the presence of cancer.
3. The medical professional is diagnosing diseases of the heart, liver, pancreas, prostate, or bones.

Additional Exercises

12.70

Conjugate acid: $H_3\overset{+}{N}-\underset{R}{\overset{H}{C}}-COOH$

Conjugate base: $H_2N-\underset{R}{\overset{H}{C}}-COO^-$

Chemistry for Thought

12.72 The ethyl ester of alanine melts 200 degrees below the melting point of alanine because the ethyl ester of alanine is unable to form a zwitterion, and consequently, it is unable to make ionic bonds to other molecules of the same type. The weak interparticle forces between the molecules of the ethyl ester of alanine correspond to a low melting point, while the strong interparticle forces between the alanine zwitterions correspond to a melting point more like an ionic compound than an organic compound.

Chapter 13
The Flow of Genetic Information (Section 13.1)

13.2 RNA is found primarily in the cytoplasm of eukaryotic cells, though it may be found in other locations, too.

13.4 *Transcription* is the transfer of necessary information from a DNA molecule onto a molecule of messenger RNA. *Translation* is the decoding of the messenger RNA into an amino acid sequence, resulting in the synthesis of a specific protein.

The Components of Nucleic Acids (Section 13.2)

13.6 In DNA, the sugar is 2-deoxyribose. In RNA, the sugar is ribose.

13.8 a. guanine purine

b. thymine pyrimidine

c. uracil pyrimidine

d. cytosine pyrimidine

e. adenine purine

13.10 a. deoxyribose

b. adenine, guanine, cytosine, uracil

c. phosphate

13.12

thymidine 5′-monophosphate

13.14 a. false **b.** false

c. true **d.** false

The Structure of DNA (Section 13.3)

13.16 Two DNA molecules that contain the same number of nucleotides may differ in the order as well as relative numbers of the different types of bases.

13.18 The data obtained from the chemical analysis of DNA that supported the idea of complementary base pairing in DNA proposed by Watson and Crick was that in all DNA the percentages of adenine and thymine were always equal to each other, as were the percentages of guanine and cytosine.

13.20 The double helix is established because the nucleotide base pairs hydrogen-bond with one another, making up the "rungs" of the double helix "ladder." Due to their complementary structures, adenine hydrogen-bonds with thymine and cytosine hydrogen bonds with guanine. See Figure 13.10.

13.22 a. 16 hydrogen bonds

b. 15 hydrogen bonds

13.24 Strand A (CAGTAG) because it has more hydrogen bonds

13.26 a. 14%

b. 36%

c. 14%

13.28 3′ CTAAGT 5′ complementary strand

13.30 For the two strands, the amount of A = T and the amount of G = C. Thus, the complementary strand has 21% T, 29% G, 36% C, and 14% A.

DNA Replication (Section 13.4)

13.32 Semiconservative replication is a replication process in which daughter DNA molecules contain one strand from the parent and a new strand that is complementary to the strand from the parent.

13.34 The function of the enzyme helicase in the replication of DNA is to catalyze the separation and unwinding of the nucleic acid strands at a specific point along the DNA helix.

13.36 The enzymes involved in DNA replication are helicase, DNA polymerase, and DNA ligase.

13.38 The synthesis of a DNA daughter strand growing toward a replication fork grows continuously. A daughter strand growing away from a replication fork has Okazaki fragments separated by nicks.

13.40 a. GTACGT
CATGCA

b. Old strand 5′ GTACGT 3′
New strand 3′ CATGCA 5′ } daughter DNA molecule

New strand 5′ GTACGT 3′
Old strand 3′ CATGCA 5′ } daughter DNA molecule

13.42 Using this diagram will help. Remember to write the answers in the 5′ to 3′ direction.

Daughter Q		Daughter R	
5′	3′	5′	3′
A	T	A	T
A	T	A	T
C	G	C	G
T	A	T	A
G	C	G	C
G	C	G	C
3′	5′	3′	5′
New strand	Parent strand	New strand	Parent strand

a. CCAGTT

b. CCAGTT

c. AACTGG

13.44 Response c is correct. The repeating nucleotide units of a DNA strand are linked together by 3′ to 5′ phosphodiester linkages.

Ribonucleic Acid (RNA) (Section 13.5)

13.46 While the DNA secondary structure is a double helix with consistent hydrogen-bonding across the complimentary base pairs, RNA does not have a set secondary structure. However, there are portions of RNA (stem-loop regions) where the single strand folds back on itself to initiate hydrogen-bonding interactions between complementary base pairs.

13.48 a. rRNA

b. mRNA

c. tRNA

13.50

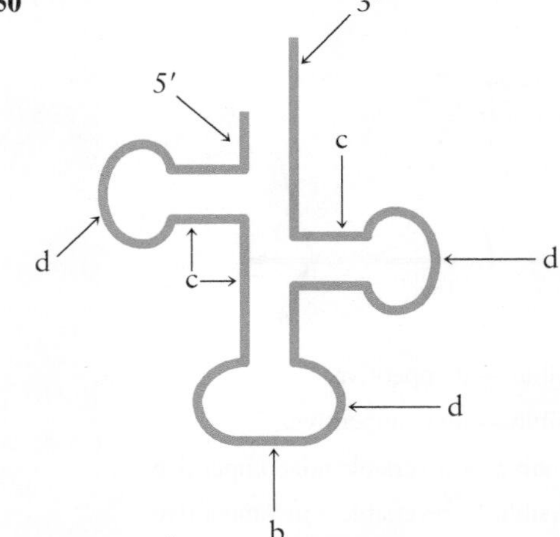

Transcription: RNA Synthesis from DNA (Section 13.6)

13.52 3′ CGGUAUAAC 5′ mRNA

13.54 Written 5′ → 3′ DNA strand is GATTCGCAA

13.56 mRNA written 5′ → 3′ is CGUA

Once the introns are removed, the exon TACG remains.

The Genetic Code and Protein Synthesis (Section 13.7)

13.58 A codon is a sequence of three nucleotide bases that represents a "code word" on the mRNA molecules. Each codon provides genetic code information for a specific amino acid.

13.60 **a.** 5′ UAC 3′ valine

b. 5′ GGG 3′ proline

c. 5′ GUG 3′ histidine

d. 5′ UGG 3′ proline

13.62 glutamic acid–aspartic acid–leucine–glycine–glycine

13.64 All codons are three letters.

13.66 **a.** False. Each codon is composed of three bases, not four.

b. True

c. False Most, but not all, codons represent an amino acid. Some codons represent stop signals.

d. False Each living species is thought to share the same genetic code, not have its own unique genetic code.

e. True

f. False On the contrary, UAG, UAA, and UGA are the known stop signals for protein synthesis.

13.68 A polysome is the configuration of several ribosomes simultaneously translating the same mRNA.

13.70 The stages of protein synthesis are: (1) initiation of the polypeptide chain, (2) elongation of the chain, and (3) termination of polypeptide synthesis.

13.72 The A site is the aminoacyl site located on the mRNA–ribosome complex next to the P site, where an incoming tRNA carrying the next amino acid will bond. The P site is the peptidyl site and is the location on the ribosome that the initiating codon must occupy for the initiation process to begin.

13.74 A genetic mutation is a change in the genetic code that results from an incorrect sequence of bases on a DNA molecule.

13.76 A genetic mutation that causes the mRNA sequence GCC to be replaced by CCC means that the resulting polypeptide will contain the amino acid proline instead of the amino acid alanine.

13.78 **a.** serine–glycine–isoleucine–tryptophan

b. threonine–glycine–isoleucine–tryptophan

c. serine–glycine–isoleucine–tryptophan

Viruses (Section 13.8)

13.80 Viruses "hijack" the cell machinery by forcing the host organism to replicate the virus's genetic material, thereby translating viral proteins in order to produce new virions. This allows the virus to propagate and multiply.

13.82 Prominent differences between an RNA virus, a DNA virus, and a retrovirus are the ways their genetic information is stored and expressed. An RNA virus contains 1 or 2 strands of RNA genome that serves as the direct template for viral protein translation. DNA viruses contain single- or double-stranded DNA genomes that are first transcribed into RNA in the nucleus, and then translated into viral proteins in the cytoplasm of the host cell. Retroviruses also contain RNA genomes, but the genetic information is first reverse transcribed into DNA, then integrated into the host cell genome, before undergoing transcription and translation into viral proteins.

13.84 Retroviruses rapidly evolve because the reverse transcriptase enzyme is highly error prone and introduces many DNA mutations when the genomic RNA is converted into DNA.

13.86 Most commonly employed vaccines use inactivated (attenuated) forms of the pathogen to inoculate patients. Newer mRNA vaccines inoculate patients with the genetic material of a part of the target virus, instead of the fully inactivated form.

Section 13.9 Recombinant DNA (Section 13.9)

13.88 Recombinant DNA contains segments of DNA from two different organisms.

13.90 Substances likely to be produced on a large scale by genetic engineering are human insulin (needed by people with diabetes who are allergic to insulin from pigs or cattle), human growth hormone (for the treatment of dwarfism), interferon (needed to fight viral infections), hepatitis B vaccine (for protection against hepatitis), malaria vaccine (for protection against malaria), and tissue plasminogen activator (needed to promote the dissolving of blood clots).

13.92 CRISPR refers to a series of repeating patterns of DNA nucleotides discovered in bacteria. These repeating patterns in bacterial DNA originated as a primitive immune system to recognize invading viruses.

13.94 Because the CRISPR-Cas 9 system can be used to target and edit very specific regions of human DNA, it could be used as potential cures for cystic fibrosis and sickle-cell disease.

Additional Exercises

13.96 In general, hydrolysis reactions are used to break nucleotides polymers apart, as is the case when introns are excised from pre-mRNA, and condensation reactions are used to join nucleotides, as is the case when exons are spliced together to form mature mRNA.

Chemistry for Thought

13.98 Some benefits of genetic engineering include improved availability of vaccines, enzymes, and hormones for humans and improved production of agricultural food products. Some concerns about genetic engineering include the long-term health effects of overusing these enzymes and the effects of the agricultural hormone residues on humans.

13.100 AZT, a drug used to fight HIV, is believed to act as an enzyme inhibitor. The type of enzyme inhibition most likely caused by this drug is competitive inhibition. AZT is similar to the substrate of the virus enzyme and thus most likely would be a competitor for the enzyme.

13.102 The three-dimensional protein structure develops as a result of backbone and side chain interactions, as well as interactions between the protein and its environment. These interactions occur as a natural consequence of the primary structure of the protein.

Chapter 14
Section 14.1 The Classification of Lipids

14.2 Lipids are a form of stored energy and a structural component of the human body. Some hormones in the human body are lipids.

14.4 **a.** nonsaponifiable

b. saponifiable

c. saponifiable

d. saponifiable

14.6 a and c

Section 14.2 Fatty Acids

14.8

polar, hydrophilic → $HO-\overset{\overset{\large O}{\|}}{C}$; nonpolar, hydrophobic → $(CH_2)_{12}CH_3$

$$HO-\overset{O}{\overset{\|}{C}}-(CH_2)_{12}CH_3$$

14.10 A micelle is a spherical cluster of fatty acid ions in which the nonpolar chains extend toward the interior of the structure away from water and the polar carboxylate groups face outward in contact with the water. The nonpolar chains in a micelle are held together by weak dispersion forces.

14.12 Insoluble lipids form complexes with proteins called lipoprotein aggregates (chylomicrons). These complexes pass into the lymph system and then into the bloodstream. They are modified by the liver into smaller lipoprotein particles, which allows most lipids to be transported to various parts of the body by the bloodstream.

14.14 **a.** saturated solid

b. unsaturated liquid

c. unsaturated liquid

d. saturated solid

14.16 Unsaturated fatty acids have lower melting points than saturated fatty acids because they contain double bonds that form kinks and prevent the unsaturated fatty acids from packing together as effectively as the saturated fatty acids. Weaker intermolecular forces result in lower melting points.

14.18 **a.** saturated

b. polyunsaturated, 4 double bonds, omega-6 fatty acid

c. monounsaturated

d. saturated

Section 14.3 The Structures of Fats and Oils

14.20

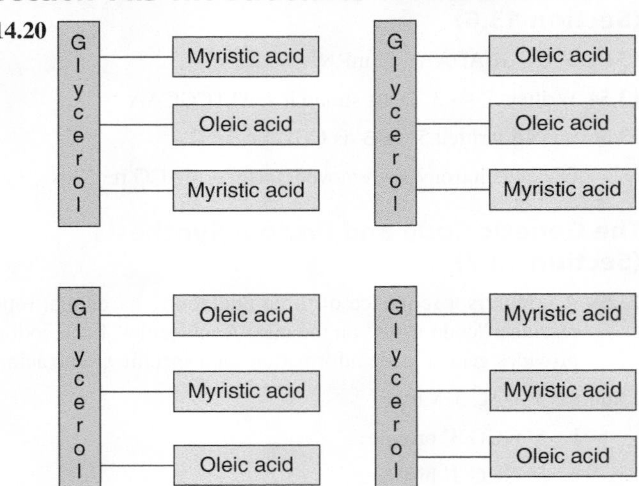

14.22

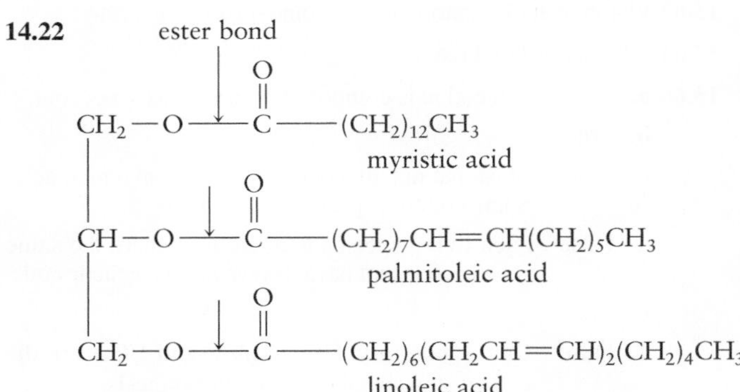

14.24 Fats and oils are both triglycerides that contain a glycerol backbone and three fatty acid chains. Fats contain more saturated than unsaturated fatty acids and oils contain more unsaturated than saturated fatty acids.

14.26 Triglycerides from animal sources (other than fish) tend to have more saturated fatty acids. Triglycerides from plants or fish tend to have more unsaturated fatty acids.

14.28 Triglyceride B has a greater percentage of unsaturated fatty acids (arachidonic + palmitoleic), so it has the lowest melting point.

Section 14.4 Chemical Properties and Reactions of Fats and Oils

14.30 The process used to prepare a number of useful products such as margarines and cooking shortenings from vegetable oils is hydrogenation. In this process, some of the double bonds in the unsaturated fatty acids are hydrogenated by reaction with hydrogen gas in the presence of a catalyst. Because fewer double bonds are present, the melting point of the mixture increases and the vegetable oils become more solid in consistency.

14.32

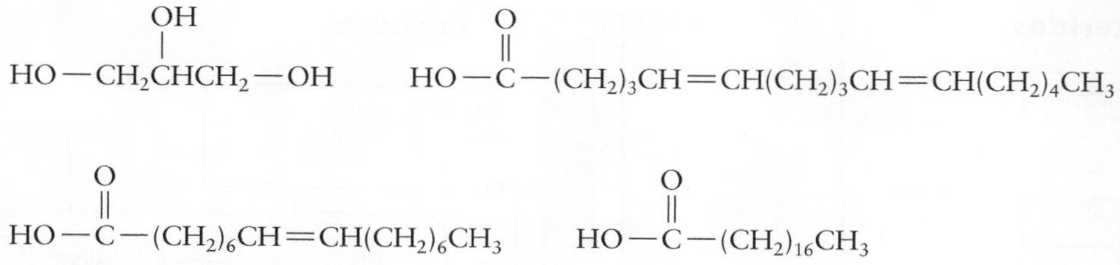

$$HO-CH_2\overset{\displaystyle OH}{\overset{|}{C}}HCH_2-OH \qquad HO-\overset{\displaystyle O}{\overset{\|}{C}}-(CH_2)_3CH{=}CH(CH_2)_3CH{=}CH(CH_2)_4CH_3$$

$$HO-\overset{\displaystyle O}{\overset{\|}{C}}-(CH_2)_6CH{=}CH(CH_2)_6CH_3 \qquad HO-\overset{\displaystyle O}{\overset{\|}{C}}-(CH_2)_{16}CH_3$$

14.34

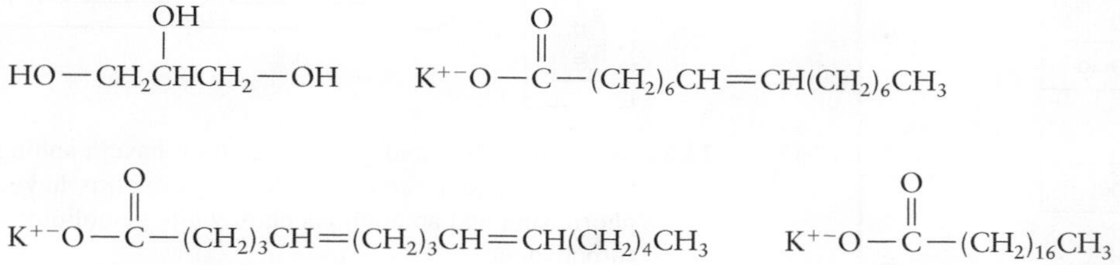

$$HO-CH_2\overset{\displaystyle OH}{\overset{|}{C}}HCH_2-OH \qquad K^{+-}O-\overset{\displaystyle O}{\overset{\|}{C}}-(CH_2)_6CH{=}CH(CH_2)_6CH_3$$

$$K^{+-}O-\overset{\displaystyle O}{\overset{\|}{C}}-(CH_2)_3CH{=}(CH_2)_3CH{=}CH(CH_2)_4CH_3 \qquad K^{+-}O-\overset{\displaystyle O}{\overset{\|}{C}}-(CH_2)_{16}CH_3$$

14.36

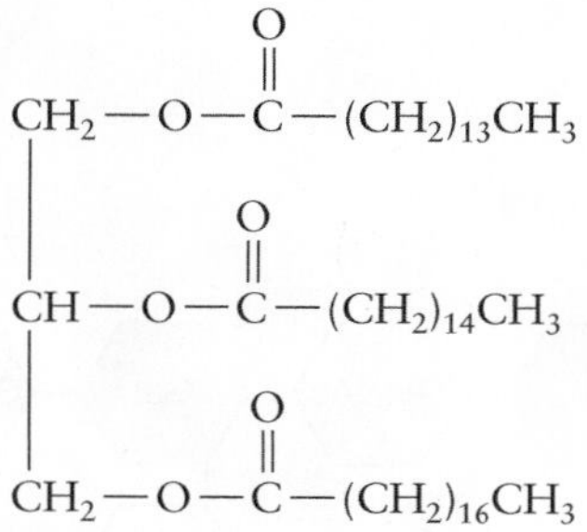

$$\begin{array}{l} CH_2-O-\overset{\displaystyle O}{\overset{\|}{C}}-(CH_2)_{13}CH_3 \\ | \\ CH-O-\overset{\displaystyle O}{\overset{\|}{C}}-(CH_2)_{14}CH_3 \\ | \\ CH_2-O-\overset{\displaystyle O}{\overset{\|}{C}}-(CH_2)_{16}CH_3 \end{array}$$

c.

$$\begin{array}{l} CH_2-O-\overset{\displaystyle O}{\overset{\|}{C}}-(CH_2)_{12}CH_3 \\ | \\ CH-O-\overset{\displaystyle O}{\overset{\|}{C}}-(CH_2)_{14}CH_3 \\ | \\ CH_2-O-\overset{\displaystyle O}{\overset{\|}{C}}-(CH_2)_{20}CH_3 \end{array}$$

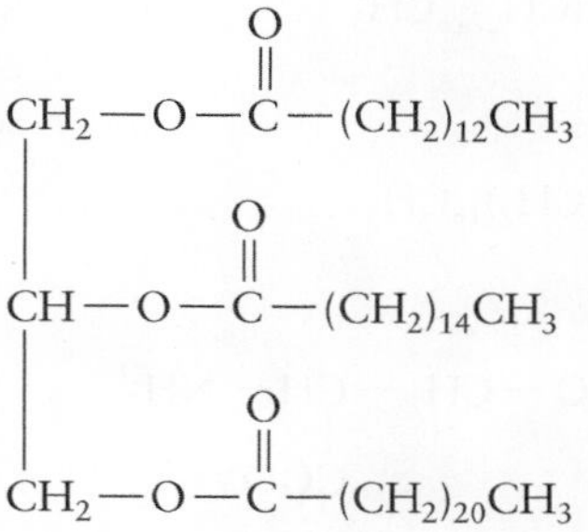

14.38 three

14.40 a.

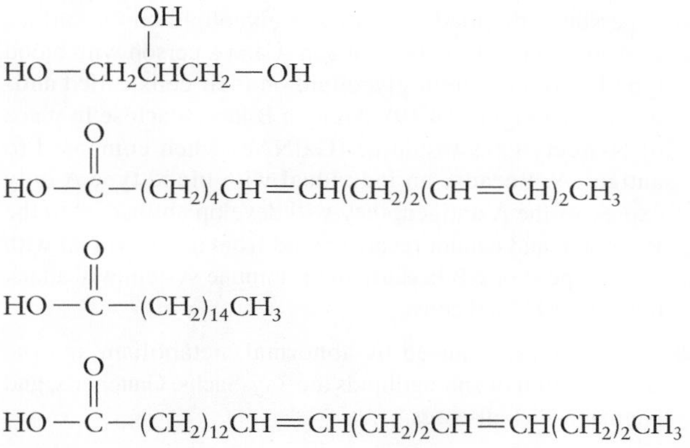

$$HO-CH_2\overset{\displaystyle OH}{\overset{|}{C}}HCH_2-OH$$

$$HO-\overset{\displaystyle O}{\overset{\|}{C}}-(CH_2)_4CH{=}CH(CH_2)_2(CH{=}CH)_2CH_3$$

$$HO-\overset{\displaystyle O}{\overset{\|}{C}}-(CH_2)_{14}CH_3$$

$$HO-\overset{\displaystyle O}{\overset{\|}{C}}-(CH_2)_{12}CH{=}CH(CH_2)_2CH{=}CH(CH_2)_2CH_3$$

b.

$$HO-CH_2\overset{\displaystyle OH}{\overset{|}{C}}HCH_2-OH$$

$$Na^{+-}O-\overset{\displaystyle O}{\overset{\|}{C}}-(CH_2)_4CH{=}CH(CH_2)_2(CH{=}CH)_2CH_3$$

$$Na^{+-}O-\overset{\displaystyle O}{\overset{\|}{C}}-(CH_2)_{14}CH_3$$

$$Na^{+-}O-\overset{\displaystyle O}{\overset{\|}{C}}-(CH_2)_{12}CH{=}CH(CH_2)_2CH{=}CH(CH_2)_2CH_3$$

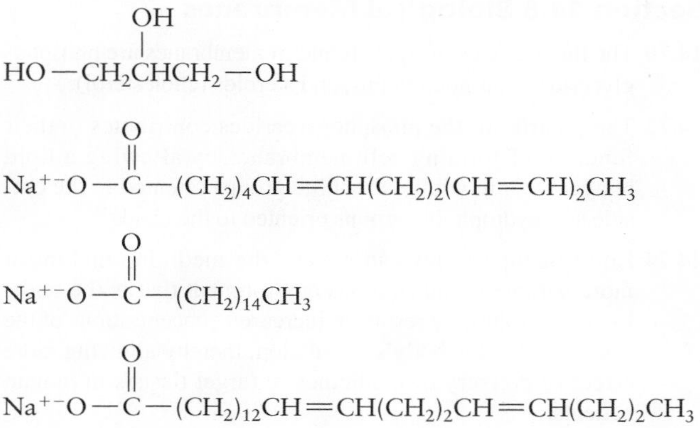

14.42 a. glycerol + 3 fatty acids (more saturated than unsaturated)

b. glycerol + 3 fatty acids (more unsaturated than saturated)

c. glycerol + 3 salts of fatty acids (soaps)

d. glycerol + 3 salts of fatty acids (soaps)

Section 14.5 Waxes

14.44 Waxes and fats are esters of long-chain fatty acids; however, waxes do not contain the glycerol backbones that fats do. Waxes contain one fatty acid chain, while fats contain three fatty acid chains.

14.46

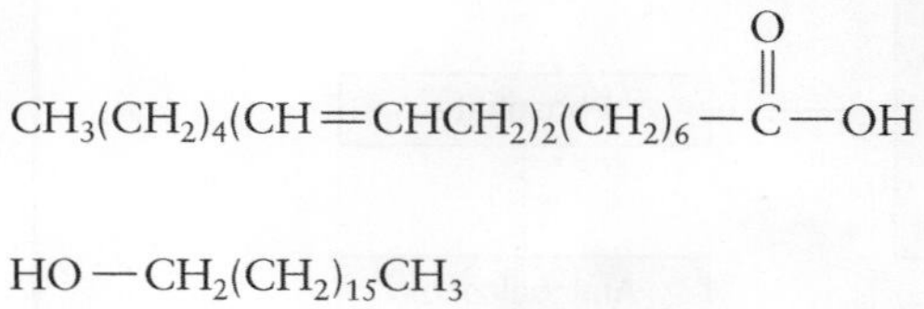

$$CH_3(CH_2)_4(CH{=}CHCH_2)_2(CH_2)_6-\overset{\displaystyle O}{\overset{\|}{C}}-OH$$

$$HO-CH_2(CH_2)_{15}CH_3$$

14.48 Waxes are protective coatings on feathers, fur, skin, leaves, and fruits. They also occur in secretions of the sebaceous glands to keep skin soft and prevent dehydration.

Section 14.6 Phosphoglycerides

14.50

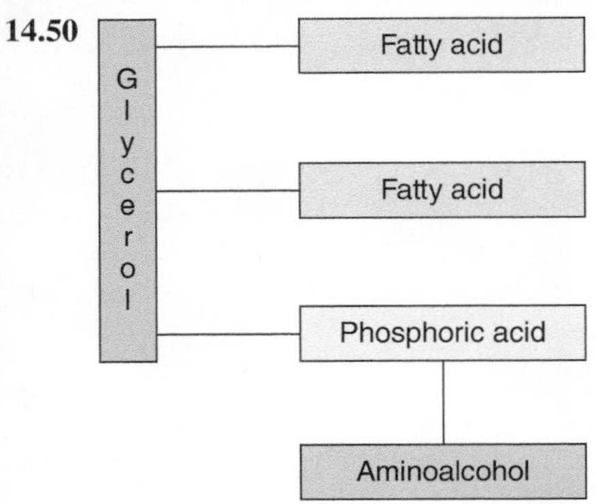

14.52

$$
\begin{array}{l}
CH_2-O-\overset{\displaystyle O}{\overset{\|}{C}}-(CH_2)_{16}CH_3 \\
| \\
CH-O-\overset{\displaystyle O}{\overset{\|}{C}}-(CH_2)_{16}CH_3 \\
| \\
CH_2-O-\overset{\displaystyle O}{\overset{\|}{P}}-O-CH_2-CH-NH_3^+ \\
\qquad\qquad\quad | \qquad\qquad\quad\;\; | \\
\qquad\qquad\;\; O^- \qquad\qquad COO^-
\end{array}
$$

This is a cephalin because it contains serine.

14.54 Lecithins are phosphoglycerides that contain the aminoalcohol choline. Cephalins are phosphoglycerides that contain ethanolamine or serine.

14.56 Lecithins are important structural components in cell membranes as well as emulsifying micelle-forming agents.

Section 14.7 Spingolipids

14.58 Sphingolipids include sphingomyelins and glycolipids. Both of these subclasses contain the sphingosine backbone and a fatty acid.

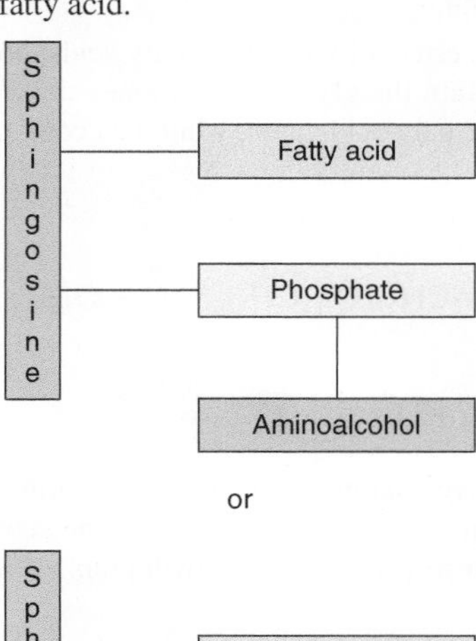

or

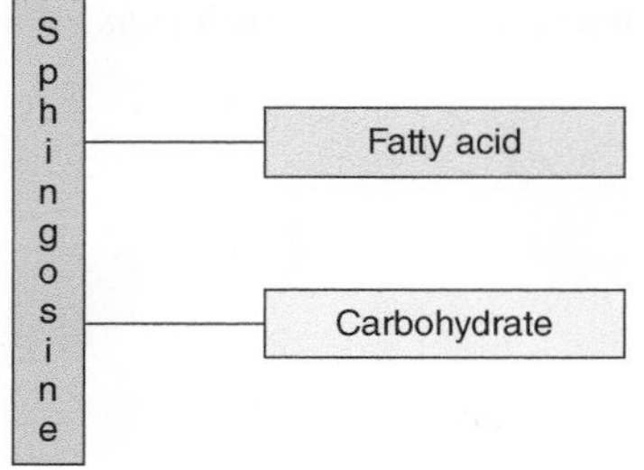

14.60

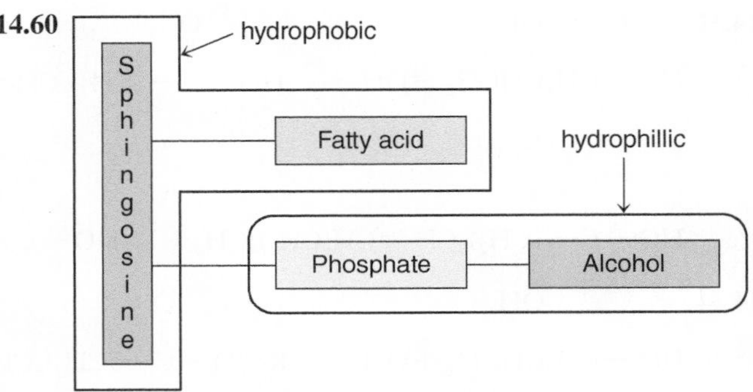

14.62 Sphingomyelins and glycolipids both have a sphingosine backbone and 1 fatty acid. Sphingomyelins have phosphoric acid and an aminoalcohol, while glycolipids have a carbohydrate.

14.64

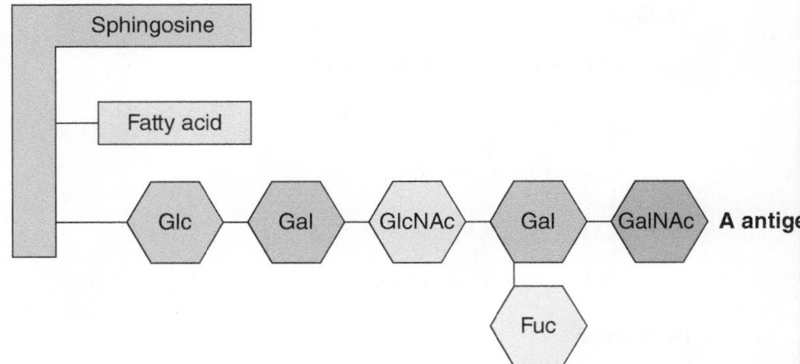

14.66 A person with blood type A has a glycolipid on the surface of their blood cells called antigen A and a person with blood type B has a different glycolipid on their cells called antigen B (see Figure 14.19). Antigen B has galactose in place of N-Acetylgalactosamine (GalNAc) when compared to antigen A. Because an individual with blood type A only expresses the A antigen, they will develop antibodies to the B antigen and cannot receive blood from an individual with blood type B or AB because their immune system will attack the donated blood cells.

14.68 Three diseases caused by abnormal metabolism and the accumulation of sphingolipids are Tay-Sachs, Gaucher's, and Niemann-Pick diseases.

Section 14.8 Biological Membranes

14.70 The three classes of lipids found in membranes are phosphoglycerides, sphingomyelins, and steroids (cholesterol).

14.72 The polarity of the phosphoglycerides contributes to their function of forming cell membranes by allowing a lipid bilayer to form with hydrophilic groups oriented to the outside and hydrophobic groups oriented to the inside.

14.74 Lipid nanoparticles can encase the medicine making it more soluble in the aqueous environment that is the body. Increased solubility results in increased concentrations of the drug entering the body's circulation, thereby allowing more effective delivery of medicines to target tissues in human patients.

Section 14.9 Steroids and Hormones

14.76 All steroids contain a 4 fused ring system composed of 3 six membered rings and 1 five membered ring arranged as shown in the following structure.

14.78 Prostaglandins are characterized as cyclic compounds synthesized from the 20-carbon unsaturated fatty acid arachidonic acid. They contain a carboxylic acid group, various ring substituents, and side-chain double bonds.

14.80 Bile salts aid in the digestion of lipids by emulsifying lipids and breaking the large globules into many smaller droplets in order to increase the surface area available for hydrolysis reactions.

14.82 With a large hydrophobic region and an ionic region, sodium glycocholate functions like a soap, emulsifying cholesterol.

14.84 The major component in most gallstones is cholesterol.

14.86 The two groups of adrenocorticoid hormones are mineralocorticoids and glucocorticoids. Mineralocorticoids regulate the concentration of ions in body fluids. An example of a mineralocorticoid is aldosterone, which regulates the retention of sodium and chloride ions in urine formation. Glucocorticoids regulate carbohydrate metabolism. An example of a glucocorticoid is cortisol, which helps to increase glucose and glycogen concentrations in the body.

14.88 Athletes use anabolic steroids to promote muscular development without excessive masculinization. The side effects of anabolic steroid use range from acne to deadly liver tumors. In males, anabolic steroid use can cause testicular atrophy, a decrease in sperm count, and temporary infertility.

14.90 It is suggested that some people restrict cholesterol intake in their diet because high levels of cholesterol in the blood are linked to atherosclerosis (hardening of the arteries).

14.92 Body processes regulated in part by prostaglandins are menstruation, conception, uterine contractions during childbirth, blood clotting, inflammation, and fever, to name only a few.

14.94 Therapeutic uses of prostaglandins include inducing labor, therapeutic abortion, treating asthma, inhibiting gastric secretions, and treating peptic ulcers.

Additional exercises

14.96 Complex lipids are more predominant in cell membranes than simple lipids because the cell membranes have a lipid bilayer that is composed of lipids with polar head groups and nonpolar tails. Phospholipid molecules have a structure that is more conducive to forming a lipid bilayer than a fat, oil, or wax. Simple lipids do not form as effective a barrier as the complex lipids.

Chemistry for Thought

14.98 Gasoline is soluble in nonpolar solvents, but it is not classified as a lipid because it does not come from a living organism.

14.100 As a new plant begins to grow from its seed, it derives energy from the oils (a low density, but calorie-rich source) because the plant cannot photosynthesize until it has grown leaves.

Chapter 15
The Basics of Nutrition (Section 15.1)

15.2 Macronutrients are nutrients needed by the body in gram quantities every day. Micronutrients are nutrients needed by the body in milligram or microgram quantities every day.

15.4 $2500 \cancel{\text{mg}} \times \dfrac{1 \text{ tsp}}{4250 \cancel{\text{mg}}} = 0.59 \text{ tsp}$

15.6 $300 \cancel{\text{g carbohydrate}} \times \dfrac{1 \text{ potato}}{26 \cancel{\text{g carbohydrate}}} = 11.54 \text{ potatoes}$

or 10 potatoes if 300 is considered to have just one significant figure.

15.8 $30 \text{ g fiber} \times \dfrac{1 \text{ potato}}{2 \text{ g fiber}} = 15 \text{ potatoes}$

or 20 potatoes to one significant figure.

15.10 Carbohydrates are used for energy and materials for the synthesis of cell and tissue components. Lipids are a concentrated source of energy; they store fat-soluble vitamins and help carry them through the body, and they provide essential fatty acids. Lipids are also an essential component of cell membranes. Proteins are used to make new tissues; maintain and repair cells; synthesize enzymes, hormones, and other nitrogen-containing compounds in the body; and provide energy.

15.12 **a.** carbohydrates **b.** proteins
c. lipids **d.** lipids

15.14 **a.** protein **b.** carbohydrate
c. lipid

15.16 **a.** carbohydrates, lipids, proteins
b. carbohydrates, lipids, proteins
c. carbohydrates, proteins
d. carbohydrates, lipids, proteins

15.18 Linoleic acid is called an essential fatty acid for humans because humans cannot synthesize this compound. It must be obtained from the diet.

15.20 $65 \text{ g fat}\left(\dfrac{9 \text{ cal}}{1 \text{ g}}\right) = 585 \text{ cal from fat}$

$20 \text{ g fat}\left(\dfrac{9 \text{ cal}}{1 \text{ g}}\right) = 180 \text{ cal from saturated fat}$

$300 \text{ g carbohydrate}\left(\dfrac{4 \text{ cal}}{1 \text{ g}}\right) = 1.20 \times 10^3 \text{ cal from carbohydrate}$

Note: The significant figures from the calorie per gram conversion were ignored in these calculations.

15.22 **a.** fat-soluble **b.** water-soluble
c. water-soluble **d.** water-soluble

15.24 fat-soluble

15.26 Large doses of fat-soluble vitamins are potentially dangerous because excess fat-soluble vitamins will be stored in the body's fat reserves and stay in the body. Fat-soluble vitamins can be harmful in large quantities; however, large doses of water-soluble vitamins will be excreted, and any effect due to a higher concentration will be only temporary.

15.28 a. vitamin C (ascorbic acid)

b. vitamin B_1 (thiam)

c. vitamin B_{12} (cobalamin)

d. niacin

15.30 A major mineral is a mineral found in the body in quantities greater than 5 g. A trace mineral is a mineral found in the body in quantities less than 5 g.

An Overview of Metabolism (Section 15.2)

15.32 A metabolic pathway is a sequence of reactions used to produce one ultimate product or accomplish one overall process. The product(s) of one reaction is/are used up in the next reaction in the sequence so that all the reactions are essential to the overall process.

15.34 Catabolism is an energy-producing process; anabolism is an energy-consuming process.

15.36 a. formation of ATP — Stage 3

b. digestion of fuel molecules — Stage 1

c. consumption of O_2 — Stage 3

d. generation of acetyl CoA — Stage 2

15.38 The main purpose of the catabolic pathway is to produce ATP.

ATP: The Energy Currency of Cells (Section 15.3)

15.40 ATP represents adenosine triphosphate.

ADP represents adenosine diphosphate.

AMP represents adenosine monophosphate.

15.42 $\Delta G^{\circ\prime}$ represents the free energy change for a reaction under normal physiological conditions (normal body temperature and the appropriate pH and concentration for the tissue involved) in kcal/mol. A negative $\Delta G^{\circ\prime}$ value indicates an exergonic reaction. A positive $\Delta G^{\circ\prime}$ value indicates an endergonic reaction.

15.44 Anabolic processes require energy, so they should have a $+\Delta G^{\circ\prime}$.

Catabolic processes liberate energy, so they should have a $\Delta G^{\circ\prime}$.

Carbohydrate Metabolism and Blood Glucose (Section 15.4)

15.46 a. glucose **b.** glucose, galactose

c. fructose, glucose **d.** glucose

15.48 Blood sugar level is the concentration of glucose in the blood. Normal fasting level is the concentration of glucose in the blood after an 8–12 hour fast.

15.50 a. Blood sugar level is below normal fasting level.

b. Blood sugar level is above normal fasting level.

c. Blood sugar level is above 108 mg/100 mL, and sugar is not completely reabsorbed from the urine by the kidneys.

d. Blood sugar level is above the renal threshold, and glucose appears in the urine.

15.52 Insulin could be used to lower higher levels of blood glucose (hyperglycemia).

An Overview of Cellular Respiration (Section 15.5)

15.54 The main purpose of glycolysis is to provide energy for the body from carbohydrates.

15.56 Glycolysis occurs in the cellular cytoplasm.

15.58 The oxidizing agent in glycolysis is NAD^+.

15.60 A high concentration of glucose-6-phosphate inhibits glycolysis because hexokinase, the enzyme that catalyzes the conversion of glucose to glucose 6-phosphate, is inhibited by glucose 6-phosphate. This feedback inhibition decreases the rate of glycolysis when a high concentration of glucose 6-phosphate is present.

15.62 Pyruvate reacts with coenzyme A and NAD^+ inside the mitochondria to produce acetyl CoA, NADH, and CO_2 during Stage 2 of cellular respiration. Equation 15.5 shows that the pyruvate dehydrogenase complex is the enzyme responsible for catalyzing this reaction.

15.64 a. acetyl CoA **b.** CO_2

c. NADH, $FADH_2$

15.66 a. The enzymes at the three control points of the citric acid cycle are citrate synthetase (Step 1), isocitrate dehydrogenase (Step 3), and the α-ketoglutarate dehydrogenase complex (Step 4).

b. The citric acid cycle produces both NADH and ATP. If the cell has high levels of NADH and ATP, then in all likelihood, the energy needs of the cell are or will be adequately met. Therefore, when the NADH levels are high, NADH serves as an inhibitor to slow the production of NADH and $FADH_2$. When ATP levels are high, the entry of acetyl CoA into the citric acid cycle is reduced. Energy production is lowered because the requirements of the cell are already being met. If the cell has low levels of NADH and ATP, the cell's energy requirements are probably not being met. If levels of ATP are low, levels of ADP are usually high. High ADP levels activate the cycle and stimulate the production of more ATP. Similarly, if the levels of NADH are low, this compound is not available to inhibit the control point enzymes and the cycle is stimulated to produce more NADH and ATP.

c. Two of the control-point enzymes are activated by ADP. This benefits the cell because a high level of ADP indicates a low level of ATP, so the cell can respond immediately to low ATP levels. If the cycle were simply stimulated by a feedback mechanism involving only high or low levels of ATP, the cellular response to low levels of ATP (and low energy) would be slower. Because ADP stimulates the cycle, the cell can respond immediately to its presence rather than waiting for the level of ATP to fall and its activity as an inhibitor to be reversed. A second benefit occurs because ADP is converted directly into ATP. Thus, the substance which stimulates the pathway is the direct substrate of that pathway, and conversion between the two compounds is, therefore, quick.

15.68 **a.** within the cytoplasm of the cell

b. within mitochondria

c. within mitochondria

d. within mitochondria

15.70 The cytochromes in the electron transport chain pass electrons along the chain to oxygen atoms and provide H^+ ions to the oxygen, resulting in the formation of water.

15.72 Iron deficiency might impact the electron transport chain, thus reducing the utilization of oxygen and the production of energy.

15.74 32 ATP equivalents

Glycogen Metabolism and Gluconeogenesis (Section 15.6)

15.76 The entry point for fructose is dihydroxyacetone phosphate or glyceraldehyde-3-phosphate. The entry point for galactose is glucose, which enters in the form of glucose-6-phosphate.

15.78 It is important that the liver store glycogen because one of the functions of the liver is to maintain a relatively constant level of blood glucose.

15.80 The principal site for gluconeogenesis is the liver.

15.82 Lactate, glycerol, and certain amino acids undergo gluconeogenesis. Amino acids are derived from proteins; glycerol is derived from lipids; and lactate has a variety of sources, including pyruvate, which is produced from carbohydrates.

15.84 The Cori cycle is the metabolic pathway in which lactate produced in the muscles is ultimately converted to glucose in the liver. Muscle activity quickly uses up stored ATP, and more energy must be produced by the breakdown of glycogen. The Cori cycle provides a mechanism for energy to be generated under anaerobic conditions and returns an energy source (glucose) to the muscles.

15.86 **a.** Glucagon increases the blood sugar levels by enhancing the breakdown of glycogen in the liver, while insulin reduces the blood sugar level by enhancing the absorption of glucose from the blood into cells of active tissue.

b. Glucagon inhibits glycogen formation, while insulin increases the rate of glycogen formation.

Lipid Metabolism and Biosynthesis (Section 15.7)

15.88 Fat mobilization is the process whereby epinephrine stimulates the hydrolysis of triglycerides (in adipose tissue) into fatty acids and glycerol, which enter the bloodstream.

15.90 Glycerol enters the glycolysis pathway as dihydroxyacetone phosphate.

15.92 The two fates of glycerol after it has been converted to an intermediate of glycolysis are to provide energy to cells or be converted to glucose.

15.94 A fatty acid is prepared for catabolism by a conversion into fatty acyl CoA. This activation occurs outside the mitochondria in the cytoplasm of the cell.

15.96 Ketone bodies are the compounds synthesized within the liver from excess acetyl CoA. They are acetoacetate, β-hydroxybutyrate, and acetone.

15.98 Ketone bodies are formed in the liver. The brain, heart, and skeletal muscles can use these compounds to meet energy needs.

15.100 In ketoacidosis, the kidneys excrete excessive amounts of water in response to the low blood pH.

15.102 Acid–base balance can be restored by using bicarbonate to neutralize the acids.

15.104 Acetyl CoA supplies the carbon atoms for fatty acid synthesis.

15.106 There is a transport system that helps to move citrate from the mitochondria to the cytoplasm. In other words, there are molecules that assist the passage of citrate through the membrane.

15.108 Two polyunsaturated fatty acids, linoleic and linolenic, cannot be synthesized by the body and yet are required for good health.

Amino Acid Metabolism and Biosynthesis (Section 15.8)

15.110 The sources of amino acids in the pool are digested dietary proteins, turnover of body proteins, and synthesis in the liver.

15.112 Other than protein synthesis, amino acids are used for the production of purine and pyrimidine bases of nucleic acids, heme structures for hemoglobin and myoglobin, choline and ethanolamine for phospholipids, and neurotransmitters such as acetylcholine and dopamine.

15.114 Urea is synthesized in the liver.

15.116

$$NH_4^+ + HCO_3^- + 3ATP + 2H_2O + \begin{array}{c} COO^- \\ | \\ CH_2 \\ | \\ CH-NH_3^+ \\ | \\ COO^- \end{array} \longrightarrow H_2N-\overset{\overset{\displaystyle O}{\|}}{C}-NH_2 + \begin{array}{c} H \quad\quad COO^- \\ \diagdown \quad \diagup \\ C=C \\ \diagup \quad \diagdown \\ {}^-OOC \quad\quad H \end{array} + 2ADP + 2P_i + AMP + PP_i$$

15.118 The amino acids isoleucine, phenylalanine, tryptophan, and tyrosine can be both glucogenic and ketogenic.

15.120 The lack of phenylalanine hydroxylase causes PKU.

15.122 During fasting or starvation, glucogenic amino acids are used to synthesize glucose through the process of gluconeogenesis after 12–18 hours without food.

15.124 The essential amino acid that makes tyrosine nonessential is phenylalanine.

15.126 The amino acids synthesized from glutamate are proline, glutamine, and arginine.

Appendix C

Solutions to Learning Checks

Chapter 1

1.1 A change is chemical if new substances are formed and physical if no new substances are formed.

a. Chemical: The changes in taste and odor indicate that new substances form.

b. Physical: The handkerchief doesn't change, and the evaporated water is still water.

c. Chemical: The changes in color and taste indicate that new substances form.

d. Chemical: The gases and smoke released indicate that new substances form.

e. Physical: The air is still air, as indicated by appearance, odor, and so on.

f. Physical: On being boiled, the water forms steam, which is the same substance present before the change.

1.2 **a.** Carbon tetrachloride contains two different elements covalently bonded together in a compound.

b. Molecules containing only sulfur atoms are elements.

c. Glucose contains three different elements covalently bonded together in a larger molecule, so it is a compound.

1.3 **a.** Salads are mixtures that not are not uniform throughout, so they are heterogeneous mixtures.

b. Carbon monoxide consists of two elements covalently bonded together in a pure substance called a compound.

1.4 Substitute the given radius and π values into the equation:

$$A = \pi r^2 = (3.14)(3.5\text{ cm})^2 = (3.14)(12.25\text{ cm}^2) = 38\text{ cm}^2$$

1.5 Because 1 mL = 1000 μL,

$$5.0\ \cancel{\text{mL}} \times \frac{1000\ \mu\text{L}}{1\ \cancel{\text{mL}}} = 5000\ \mu\text{L}$$

1.6 Because 1 kg = 1000 g,

$$0.819\ \cancel{\text{kg}} \times \frac{1000\text{ g}}{1\ \cancel{\text{kg}}} = 819\text{ g}$$

1.7 Use Equation 1.4 to convert 37.0 °C to °F.

$$°\text{F} = \left(\frac{9}{5}\right)(°\text{C}) + 32 = \left(\frac{9}{5}\right)(37.0\ °\text{C}) + 32 = 98.6\ °\text{F}$$

1.8 **a.** Scientific notation is used. In nonscientific notation, the number is written 588.

b. Scientific notation is not used. In scientific notation, the number is written 4.39×10^{-4}.

c. Scientific notation is used. In nonscientific notation, the number is written 0.0003915.

d. Scientific notation is not used. In scientific notation, the number is written 9.870×10^3.

e. Scientific notation is not used. In scientific notation, the number is written 3.677×10^1.

f. Scientific notation is not used. In scientific notation, the number is written 1.02×10^{-1}.

1.9 Perform multiplications and divisions separately for the nonexponential and exponential terms, then combine the results into the final answer.

a. $(2.4 \times 10^3)(1.5 \times 10^4) = (2.4 \times 1.5)(10^3 \times 10^4) = (3.6)(10^{3+4}) = 3.6 \times 10^7$

b. $$\frac{6.3 \times 10^5}{2.1 \times 10^3} = \left(\frac{6.3}{2.1}\right)\left(\frac{10^5}{10^3}\right) = (3.0)(10^{5-3}) = 3.0 \times 10^2$$

c. $$\frac{(1.7 \times 10^4)(2.9 \times 10^3)}{3.4 \times 10^2} = \left(\frac{1.7 \times 2.9}{3.4}\right)\left(\frac{10^4 \times 10^3}{10^2}\right) = (1.5)\left(\frac{10^{4+3}}{10^2}\right) = (1.5)\left(\frac{10^7}{10^2}\right) = (1.5)(10^{7-2}) = 1.5 \times 10^5$$

1.10 The primary challenge is to interpret correctly the significance of zeros. Leading zeros to the left of nonzero numbers are not significant. Trailing zeros are significant only when a decimal is present; all zeros between nonzero digits are significant.

a. 2, trailing zero is not significant, 2.5×10^2

b. 4, 1.805×10^1

c. 3, leading zeros are not significant, 1.08×10^{-2}

d. 2, 3.7×10^1

e. 1, leading zeros are not significant, 1×10^{-3}

f. 4, trailing zero is significant, 1.010×10^2

1.11 **a.** Answer will be rounded to 2 significant figures to match 0.0019.

$$(0.0019)(21.39) = 0.04064 = 4.1 \times 10^{-2}$$

b. Answer will be rounded to 2 significant figures to match 4.1.

$$\frac{8.321}{4.1} = 2.0295 = 2.0$$

c. Answer will be rounded to 3 significant figures to match 0.0911 and 3.22.

$$\frac{(0.0911)(3.22)}{(1.379)} = 0.21272 = 0.213, \text{ or } 2.13 \times 10^{-1}$$

1.12 In each case, the sum or difference is rounded to have the same number of places to the right of the decimal as the least number of places in the terms added or subtracted.

a. Answer will be rounded to have one place to the right of the decimal to match 3.2.

$$8.01 + 3.2 = 11.21 = 11.2$$

b. Answer will be rounded to have no places to the right of the decimal to match 3000.

$$3000. + 20.3 + 0.009 = 3020.309 = 3020$$

c. Answer will be rounded to have two places to the right of the decimal to match both 4.33 and 3.12.

$$4.33 - 3.12 = 1.21$$

d. Answer will be rounded to have two places to the right of the decimal to match 2.42.

$$6.023 - 2.42 = 3.603 = 3.60$$

1.13 Two factors result from the relationship 1 g = 1000 mg. The factors are:

$$\frac{1 \text{ g}}{1000 \text{ mg}} \quad \text{and} \quad \frac{1000 \text{ mg}}{1 \text{ g}}$$

The first factor is used because it cancels the mg unit and generates the g unit:

Step 1: 1.1 mg

Step 2: 1.1 mg $= ? \text{ g}$

Step 3: $1.1 \cancel{\text{mg}} \times \frac{1 \text{ g}}{1000 \cancel{\text{mg}}} = ? \text{ g}$

Step 4: $\frac{(1.1)(1 \text{ g})}{(1000)} = 0.0011 \text{ g} = 1.1 \times 10^{-3} \text{ g}$

The number of significant figures matches the number in 1.1 because the 1 and 1000 are exact numbers.

1.14 Two factors, 1 km/1000 m and 0.621 mi/1 km, are used to convert meters to miles, and two factors, 60 s/1 min and 60 min/1 h, are used to convert seconds in the denominator to hours.

$$10.0 \frac{\cancel{\text{m}}}{\cancel{\text{s}}} \times \frac{1 \cancel{\text{km}}}{1000 \cancel{\text{m}}} \times \frac{0.621 \text{ mi}}{1 \cancel{\text{km}}} \times \frac{60 \cancel{\text{s}}}{1 \cancel{\text{min}}} \times \frac{60 \cancel{\text{min}}}{1 \text{ h}} = 22.356 \text{ mi/h} = 22.4 \text{ mi/h}$$

1.15 Debra lost 184 lb − 167 lb = 17 lb.

$$\text{Percent} = \frac{\text{part}}{\text{total}} \times 100\% = \frac{17 \text{ lb}}{184 \text{ lb}} \times 100\% = 9.2\%$$

1.16 **a.** $\% = \frac{\text{part}}{\text{total}} \times 100\%; = \frac{\$988}{\$1200} \times 100\% = 82.3\%$

b. $\% = \frac{\text{part}}{\text{total}} \times 100\%$

$$90.4\% = \frac{\text{part}}{83} \times 100\%$$

$$\frac{(90.4\%)(83)}{100\%} = \text{part}$$

$= \text{Number voting to not take final}$

$= 75.0$

Number voting to take exam = 83 − 75 = 8.

1.17 **a.** Density $(d) = \frac{m}{V} = \frac{1.04 \text{ g}}{1 \text{ mL}}$

The volume is 1.20 L or 1200 mL.

$$d = \frac{m}{V} \text{ or } m = (d)(V)$$

$m = \frac{1.04 \text{ g}}{1 \cancel{\text{mL}}} \times 1200 \cancel{\text{mL}} = 1248 \text{ g}$ or 1250 g (three significant figures)

b. The sample mass is 98.5 g, and we wish to convert this to cm^3. Rearrange the equation for density, $d = m/V$, to isolate V: $V = m/d$.

$$98.5 \cancel{\text{g}} \times \frac{1.0 \text{ cm}^3}{2.7 \cancel{\text{g}}} = 36.48 \text{ cm}^3 \text{ (calculator answer)}$$

$$= 3.6 \times 10^1 \text{ cm}^3 \text{ (properly rounded answer)}$$

1.18 **a.** The sample mass is equal to the difference between the mass of the container with the sample inside and the mass of the empty container:

$$m = 64.93 \text{ g} - 51.22 \text{ g} = 13.71 \text{ g}$$

The density of the sample is equal to the sample mass divided by the sample volume:

$$d = \frac{m}{V} = \frac{13.71 \text{ g}}{10.00 \text{ mL}}$$

$$= 1.371 \text{ g/mL (properly rounded answer)}$$

b. The identity can be determined by calculating the density of the sample from the data and comparing the density with the known densities of the two possible metals. The volume of the sample is equal to the difference between the cylinder readings with the sample present and with the sample absent:

$$V = 25.2 \text{ mL} - 21.2 \text{ mL} = 4.0 \text{ mL}$$

According to Table 1.3, 1 mL = 1 cm^3, so the sample volume is equal to 4.0 cm^3. The density of the sample is equal to the sample mass divided by the sample volume:

$$d = \frac{m}{V} = \frac{35.66 \text{ g}}{4.0 \text{ cm}^3}$$

$$= 8.9 \text{ g/cm}^3 \text{ (properly rounded answer)}$$

A comparison with the known densities of nickel and chromium allows the metal to be identified as nickel.

Chapter 2

2.1 The periodic table and Figure 2.4 are used.

a. Xe: noble gas

b. As: metalloid

c. Hg: transition metal

d. Ba: alkaline earth metal

e. Br: halogen

2.2 The atomic number, Z, equals the number of protons: $Z = 4$. The mass number, A, equals the sum of the number of protons and the number of neutrons: $A = 4 + 5 = 9$. According to the periodic table, the element with an atomic number of 4 is beryllium, with the symbol Be. The isotope symbol is $^{9}_{4}\text{Be}$.

2.3 **a.** $^{234}_{90}\text{Th}$

b. $^{234}_{90}\text{Th} \rightarrow {}^{0}_{-1}\beta + {}^{234}_{91}\text{Pa}$

c. In general: $^{A}_{Z}X \rightarrow {}^{0}_{-1}\beta + {}^{A}_{Z+1}Y$

When a β particle is emitted, the daughter has the same mass number and an atomic number higher by 1 than that of the decaying nucleus.

2.4 The elapsed time of 79.8 hours is 6 half-lives $\left(\frac{79.8}{13.3} = 6\right)$. The fraction remaining is $\frac{1}{2} \times \frac{1}{2} \times \frac{1}{2} \times \frac{1}{2} \times \frac{1}{2} \times \frac{1}{2} = \frac{1}{64}$ of the original diagnostic dose.

2.5 The subshell filling order is obtained from **Figure 2.32**. All subshells except the last one are filled to their capacities as follows: *s* subshells—filled with 2, *p* subshells—filled with 6, and *d* subshells—filled with 10.

a. Element 9 requires 9 electrons. Full electron configuration: $1s^2 2s^2 2p^5$. Abbreviated electron configuration: [He] $2s^2 2p^5$.

b. Mg is element 12 and so needs 12 electrons.

Full electron configuration: $1s^2 2s^2 2p^6 3s^2$. Abbreviated electron configuration: [Ne] $3s^2$.

c. The element in Group VIA(16) and Period 3 is S, which is element 16. Therefore, it needs 16 electrons. Full electron configuration: $1s^2 2s^2 2p^6 3s^2 3p^4$. Abbreviated electron configuration: [Ne] $3s^2 3p^4$.

2.6 **a.** Atomic radii increase (right to left across a period and top to bottom down a group, **Figure 2.37**). Cs is in Period 6 and Group IA(1) so it is the largest, followed by Fe in Period 4 and Group VIIIB(8), then Al in Period 3 and Group IIIA(13), and the smallest is F in Period 2 and Group VIIA(17).

b. According to **Figure 2.37**, the first ionization energies decrease from the top to the bottom in a group, and decrease from right to let across a period. The trend is the exact opposite of the atomic radii; thus, F in Period 2 and Group VIIA(17) has the highest first ionization energy, followed by Al in Period 3 and Group IIIA(13), then Fe in Period 4 and Group VIIIB(8), and Cs in Period 6 and Group IA(1).

c. Electronegativity decreases right to left across a period and decreases going down a group (**Figure 2.37**). The trend is the exact opposite of the atomic radii, but the same as first ionization energy. So, F in Period 2 and Group VIIA(17) has the highest first ionization energy, followed by Al in Period 3 and Group IIIA(13), then Fe in Period 4 and Group VIIIB(8), and Cs in Period 6 and Group IA(1).

Chapter 3

3.1 **a.** Lithium contains 3 electrons, with one of them classified as a valence-shell electron: $[\text{He}]2s^1$ and Li·.

b. Bromine contains 35 electrons, with 7 of them classified as valence-shell electrons: $[\text{Ar}]4s^2 3d^{10} 4p^5$ and :B̤̈r·.

c. Strontium contains 38 electrons, with 2 classified as valence-shell electrons: $[\text{Kr}]5s^2$ and Ṡr·.

d. Sulfur contains 16 electrons, with 6 classified as valence-shell electrons: $[\text{Ne}]3s^2 3p^4$ and :S̤̈·.

3.2 **a.** The change that will actually take place is the one that involves the fewest electrons. Thus, Li will lose 1 electron and become a cation rather than gaining 7 electrons:

$$\text{Li} \rightarrow \text{Li}^+ + e^-$$

b. F will gain 1 electron to form an anion rather than losing 7 electrons:

$$\text{F} + e^- \rightarrow \text{F}^-$$

c. Fe will lose 2 electrons to form a cation rather than gaining 6 electrons:

$$\text{Fe} \rightarrow \text{Fe}^{2+} + 2e^-$$

3.3 The complete series of ions isoelectronic with Kr is as follows:

$$\text{As}^{3-}, \text{Se}^{2-}, \text{Br}^-, \text{Rb}^+, \text{Sr}^{2+}, \text{In}^{3+}$$

3.4 **a.** Mg, a metal in Group IIA(2), will lose 2 electrons:

$$\text{Mg} \rightarrow \text{Mg}^{2+} + 2e^-$$

O, a nonmetal in Group VIA(16), will gain 2 electrons:

$$\text{O} + 2e^- \rightarrow \text{O}^{2-}$$

The positive and negative ions are combined in the lowest numbers possible to give the compound formula. The combining requirement is that the total number of positive charges from the positive ion must equal the total number of negative charges from the negative ion. In the case of Mg^{2+} and O^{2-}, one of each ion satisfies the requirement, and the formula is MgO.

b. K, a Group IA(1) metal, will lose 1 electron:

$$\text{K} \rightarrow \text{K}^+ + 1e^-$$

S, a Group VIA(16) nonmetal, will gain 2 electrons:

$$\text{S} + 2e^- \rightarrow \text{S}^{2-}$$

In the formula, $2K^+$ will be needed for each S^{2-}. Thus, the compound formula is K_2S.

c. Ca, a Group IIA(2) metal, will lose 2 electrons:

$$\text{Ca} \rightarrow \text{Ca}^{2+} + 2e^-$$

Br, a Group VIIA(17) nonmetal, will gain 1 electron:

$$\text{Br} + 1e^- \rightarrow \text{Br}^-$$

In the formula, $2Br^-$ will combine with each Ca^{2+}: $CaBr_2$.

3.5 For CaO, the formula weight is also equal to the sum of the atomic weights of the atoms in the formula:

$$\begin{aligned} \text{FW} &= (1)(\text{at. wt. Ca}) + (1)(\text{at. wt. O}) \\ &= (1)(40.1 \text{ amu}) + (1)(16.0 \text{ amu}) \\ \text{FW} &= 56.1 \text{ amu} \end{aligned}$$

3.6 **a.** sodium iodide

b. magnesium fluoride

c. Charge balance requires that $2Br^-$ combine with each Co^{2+}: $CoBr_2$. The name is cobalt(II) bromide.

d. Charge balance requires that each S^{2-} combine with each Ca^{2+}: CaS. The name is calcium sulfide.

3.7 **a.** Each H contributes 1 electron, each O contributes 6, and the N contributes 5. The 24 total electrons can satisfy all octets only if two pair are shared between the N and one of the O atoms:

H:Ö:N::Ö: or H—O—N=O (with :Ö: / O bonded below N)

b. Each H contributes 1 electron, the C contributes 4, and the O contributes 6. The total of 12 electrons cannot satisfy all octets unless two pair are shared between the C and O atoms:

```
H:C::Ö:        or        H—C=O
  ..                        |
  H                         H
```

3.8 **a.** Each P contributes 5 valence-shell electrons, and each F contributes 7. There are a total of 40 electrons available:

```
        :F̤:
    ..   |   ..
   :F—P—F:
    ..  / \  ..
     :F:  :F:
      ..   ..
```

Phosphorus must expand its octet to accommodate binding to five fluorine atoms.

b. Each S contributes 6 valence-shell electrons, each O contributes 6, and each H contributes 1 valence electron. There are a total of 32 electrons available:

```
        Ö:
        ||
   HÖ—S—ÖH
    ..  ||  ..
        O:
        ..
```

Sulfur must expand its octet to rid sulfur of its unfavorable formal charge.

3.9 For H_2S, the molecular weight is the sum of the atomic weights of the atoms in the formula:

$$\begin{aligned} MW &= (2)(\text{at. wt. H}) + (1)(\text{at. wt. S}) \\ &= (2)(1.01\ \text{amu}) + (1)(32.1\ \text{amu}) \\ MW &= 34.1\ \text{amu} \end{aligned}$$

3.10 **a.** sulfur trioxide

b. boron trifluoride

c. disulfur heptoxide

d. carbon tetrachloride

3.11 **a.** The P provides 5 electrons, each O provides 6, and the 3^- charge indicates an additional 3 electrons. The total number of electrons is therefore 32, which is enough to satisfy all octets without any multiple bonds:

```
[    :Ö:    ]3−           [      Ö:      ]
[ :Ö:P̤:Ö: ]      or       [      ||      ]
[    :Ö:    ]             [ :Ö⁻—P—Ö⁻: ]
[    ..     ]             [      |       ]
                          [     :Ö:⁻     ]
```

b. The N provides 5 electrons, each H provides 1, and the + charge indicates that 1 must be subtracted. The total number of electrons is therefore 8, which is enough to satisfy the octet of N, and each H requirement of 2 electrons without any multiple bonds:

```
[    H    ]+            [  H   H   H  ]
[ H :N̤: H ]     or      [   \  |  /   ]
[    H    ]             [      N+     ]
                        [      |      ]
                        [      H      ]
```

3.12 **a.** Ca is a Group IIA(2) metal and so forms Ca^{2+} ions. Formula: $CaHPO_4$; name: calcium hydrogen phosphate.

b. Mg is a Group IIA(2) metal and so forms Mg^{2+} ions. Formula: $Mg_3(PO_4)_2$; name: magnesium phosphate.

c. K is a Group IA(1) metal and so forms K^+ ions. Formula: $KMnO_4$; name: potassium permanganate.

d. The ammonium ion, NH_4^+, acts as the positive ion in this compound. Formula: $(NH_4)_2Cr_2O_7$; name: ammonium dichromate.

Chapter 4

4.1 **a.** Step 1: 98.6 g O

Step 2: 98.6 g O $= ?$ mol O

Step 3: $98.6\ \text{g O} \times \dfrac{1\ \text{mol O}}{16.00\ \text{g O}} = ?\ \text{mol O}$

Step 4: $98.6\ \cancel{\text{g O}} \times \dfrac{1\ \text{mol O}}{16.00\ \cancel{\text{g O}}} = 6.16\ \text{mol O}$

b. Step 1: 1 atom O

Step 2: 1 atom O $= ?$ g O

Step 3: $1\ \text{atom O} \times \dfrac{16.00\ \text{g O}}{6.02 \times 10^{23}\ \text{atoms O}} = ?\ \text{g O}$

Step 4:

$1\ \cancel{\text{atom O}} \times \dfrac{16.00\ \text{g O}}{6.02 \times 10^{23}\ \cancel{\text{atoms O}}} = 2.66 \times 10^{-23}\ \text{g O}$

4.2 **a.** Step 1: 63.9 g $C_5H_{10}O_4$

Step 2: 63.9 g $C_5H_{10}O_4$ $= ?$ mol $C_5H_{10}O_4$

Step 3:

$63.9\ \text{g}\ C_5H_{10}O_4 \times \dfrac{1\ \text{mol}\ C_5H_{10}O_4}{134.13\ \text{g}\ C_5H_{10}O_4} = ?\ \text{mol}\ C_5H_{10}O_4$

Step 4:

$63.9\ \cancel{\text{g}\ C_5H}_{10}O_4 \times \dfrac{1\ \text{mol}\ CO_2}{134.13\ \text{g}\ \cancel{C_5H}_{10}O_4} = 0.476\ \text{mol}\ C_5H_{10}O_4$

b. Step 1: 1 molecule $C_5H_{10}O_4$

Step 2: 1 molecule $C_5H_{10}O_4$ $= ?$ g $C_5H_{10}O_4$

Step 3:

$1\ \text{molecule}\ C_5H_{10}O_4 \times \dfrac{134.13\ \text{g}\ C_5H_{10}O_4}{6.02 \times 10^{23}\ \text{molecules}\ C_5H_{10}O_4} = ?\ \text{g}\ C_5H_{10}O_4$

Step 4:

$1\ \cancel{\text{molecule}\ C_5H_{10}}O_4 \times \dfrac{134.13\ \text{g}\ C_5H_{10}O_4}{6.02 \times 10^{23}\ \cancel{\text{molecules}\ C_5H_{10}}O_4} = ?\ \text{g}\ C_5H_{10}O_4$

$= 2.23 \times 10^{-22}\ \text{g}\ C_5H_{10}O_4$

4.3 The relationships between moles of atoms in 1 mol of molecules is given by the subscripts of the atoms. Thus, 1 mol of glucose molecules would contain 6 mol of C atoms, 12 mol of H atoms, and 6 mol of O atoms. One-half mol of glucose would contain half as many moles of each atom, or 3 mol of C atoms, 6 mol of H atoms, and 3 mol of O atoms.

4.4 $\% = \frac{\text{part}}{\text{total}} \times 100\%$

In each compound, the part will be the mass of carbon associated with some mass of compound (the total). The mass relationships are obtained by assuming a sample size equal to 1.00 mol of each compound.

CO_2: 1.00 mol CO_2 molecules = 1.00 mol C atoms + 2.00 mol O atoms or, using atomic weights,

$$44.01 \text{ g } CO_2 = 12.01 \text{ g C} + 32.00 \text{ g O}$$

$$\% \text{ C} = \frac{(12.01 \text{ g C}) \times 100\%}{(44.01 \text{ g } CO_2)} = 27.29\% \text{ C}$$

CO: 1.00 mol CO molecules = 1.00 mol C atoms + 1.00 mol O atoms

$$28.01 \text{ g CO} = 12.01 \text{ g C} + 16.00 \text{ g O}$$

$$\% \text{ C} = \frac{(12.01 \text{ g C})}{(28.01 \text{ g CO})} \times 100\% = 42.88\% \text{ C}$$

4.5 **a.** $2SO_2 + O_2 \rightarrow 2SO_3$

b. $N_2 + 3Mg \rightarrow Mg_3N_2$

4.6 **a.** The mass of NH_3 possible from H_2 is calculated because N_2 is provided in excess and therefore will not limit the amount of NH_3 produced.

$$2.00 \cancel{\text{mol } H_2} \times \frac{34.0 \text{ g } NH_3}{3 \cancel{\text{mol } H_2}} = 22.7 \text{ g } NH_3$$

b. $$10.0 \cancel{\text{g } H_2} \times \frac{1 \cancel{\text{mol } H_2}}{2.02 \cancel{\text{g } H_2}} \times \frac{1 \cancel{\text{mol } N_2}}{3 \cancel{\text{mol } H_2}} \times \frac{28.0 \text{ } N_2}{1 \cancel{\text{mol } N_2}} = 46.2 \text{ g } N_2$$

4.7 **a.** $$\% \text{ yield} = \frac{\text{actual yield}}{\text{theoretical yield}} \times 100\% = \frac{17.5 \text{ g}}{21.0 \text{ g}} \times 100\% = 83.3\%$$

b. The theoretical yield must be calculated.

$$510. \cancel{\text{g } CaCO_3} \times \frac{56.08 \text{ g CaO}}{100.1 \cancel{\text{g } CaCO_3}} = 285.722 \text{ g CaO} = 286 \text{ g CaO}$$

The % yield is then calculated using this theoretical yield:

$$\% \text{ yield} = \frac{\text{actual yield}}{\text{theoretical yield}} \times 100\% = \frac{235 \text{ g}}{286 \text{ g}} \times 100\% = 82.2\%$$

4.8 **a.** Reactants: Zn = 0, $H^+ = +1$; products: $Zn^{2+} = +2$, $H_2 = 0$

Zn is oxidized, so it is the reducing agent.

H^+ is reduced, so it is the oxidizing agent.

b. In KI: K = +1, I = −1, and $Cl_2 = 0$

In KCl: K = +1, Cl = −1, and $I_2 = 0$

I in KI is oxidized, so KI is the reducing agent.

Cl_2 is reduced, so Cl_2 is the oxidizing agent.

c. In IO_3^-: O = −2 and I = +5

In HSO_3^-: H = +1, O = −2, S = +4, and $I^- = -1$

In HSO_4^-: H = +1, O = −2, and S = +6

S in HSO_3^- is oxidized, so HSO_3^- is the reducing agent.

I in IO_3^- is reduced, so IO_3^- is the oxidizing agent.

4.9 **a.** The O.N. of O changes from 0 to −2, and the O.N. of C changes from −2.7 to +4. The reaction is redox. Because a hydrocarbon reacts with oxygen gas to produce carbon dioxide and water vapor, the reaction is combustion.

b. The O.N. of H does not change. The O.N. of O changes from −1 (in a peroxide) to −2 (in water) and 0 (in O_2). This is an example in which the same element is both oxidized and reduced. The reaction is redox. Because one substance changes into two substances, the reaction is a decomposition.

c. No oxidation numbers change. The reaction is nonredox. This is a double-replacement (metathesis) reaction.

d. The O.N. of P changes from 0 to +5, and the O.N. of O changes from 0 to −2. The reaction is redox. Because two substances combine to form a single substance, the reaction is a combination.

e. The O.N. of Na does not change. The O.N. of I changes from −1 to 0, and the O.N. of Cl changes from 0 to −1. The reaction is redox. Because the Cl simply replaces the I in the compound, this is a single-replacement reaction.

Chapter 5

5.1 **a.** **Draw the Lewis structure:**
For BF_3, the Lewis structure is

:F̤—B(—F̤:)—F̤:

Count the electron domains:
The central boron atom has three electron domains—namely, the three single bonds to fluorine.

Determine the electron-domain geometry:
When there are three electron domains around the central atom, the electron-domain geometry is trigonal planar.

Determine the molecular geometry:
Because all three electron domains are bonding domains, the bond angle is 120°, and the molecular geometry is also trigonal planar.

b. **Draw the Lewis structure:**
For SO_4^{2-}, the Lewis structure is

:Ö:⊖—S(=Ö:)(=Ö:)—Ö:⊖

Count the electron domains:
The central sulfur atom has four electron domains—namely, the two single bonds to oxygen and the two double bonds to oxygen.

Determine the electron-domain geometry:
When there are four electron domains around the central atom, the electron-domain geometry is tetrahedral.

Determine the molecular geometry:
Because all four electron domains are bonding domains, the bond angle is 109.5°, and the molecular geometry is also tetrahedral.

c. Draw the Lewis structure:

For CS_2, the Lewis structure is

$$:\ddot{S}=C=\ddot{S}:$$

Count the electron domains:
The central carbon atom has two electron domains—namely, the two double bonds to sulfur.

Determine the electron-domain geometry:
When there are two electron domains around the central atom, the electron-domain geometry is linear.

Determine the molecular geometry:
Because both electron domains are bonding domains, the bond angle is 180°, and the molecular geometry is also linear.

d. Draw the Lewis structure:

For NO_3^-, the Lewis structure is

$$\ddot{O}:\ \Vert\ N^+ \text{ bonded to } {}^{\ominus}:\ddot{O}: \text{ and } :\ddot{O}:^{\ominus}$$

Count the electron domains:
The central nitrogen atom has three electron domains—namely, the two single bonds to oxygen and the one double bond to oxygen.

Determine the electron-domain geometry:
When there are three electron-domains around the central atom, the electron-domain geometry is trigonal planar.

Determine the molecular geometry:
Because all three electron domains are bonding domains, the bond angle is 120°, and the molecular geometry is also trigonal planar.

5.2 **a.** No polarization because both atoms are identical

b. Br is located higher in the group and so is more electronegative than I:

$$\overset{\delta^+}{I}-\overset{\delta^-}{Br}$$
$$+\!\!\longrightarrow$$

c. Br is farther toward the right in the periodic table and so is more electronegative than H:

$$\overset{\delta^+}{H}-\overset{\delta^-}{Br}$$
$$+\!\!\longrightarrow$$

5.3 The values of ΔEN are used.

a. $\Delta EN = 4.0 - 0.8 = 3.2$. Bond is classified as ionic.

b. $\Delta EN = 3.5 - 3.0 = 0.5$. Bond is classified as polar covalent.

c. $\Delta EN = 3.0 - 1.5 = 1.5$. Bond is classified as polar covalent.

d. $\Delta EN = 3.5 - 0.8 = 2.7$. Bond is classified as ionic.

5.4 **a. Draw the Lewis structure:**
For NH_3. the Lewis structure is

$$H-\ddot{N}-H \text{ with } H \text{ below } N$$

Determine the polarity:
The electron-domain geometry is tetrahedral, but the molecular geometry of NH_3 is trigonal pyramidal. The electronegativity values for H and N are 2.1 and 3.0, respectively. Each bond, then, is polarized toward nitrogen, resulting in a partial negative charge on the nitrogen and partial positive charges on the hydrogen atoms. NH_3 has a permanent dipole and is therefore polar.

Determine the major intermolecular force:
NH_3 will exhibit London dispersion forces (as all molecules do), and it will exhibit dipole–dipole interactions because it is polar. However, neither of these are the strongest intermolecular force present because NH_3 molecules can form hydrogen bonds when the hydrogen of one molecule interacts with the lone pair on the nitrogen of another molecule. As a result, the strongest intermolecular force is hydrogen bonding.

b. Draw the Lewis structure:
The Lewis structure of O_2 consists of two oxygens, each with two lone pairs, connected via a double covalent bond.

$$:\ddot{O}=\ddot{O}:$$

Determine the polarity:
The electronegativity of each oxygen atom is 3.5, so the O—O bond is nonpolar.

Determine the major intermolecular force:
Because O_2 is nonpolar, the only intermolecular forces exhibited by the molecule are London dispersion forces.

c. Draw the Lewis structure:
The Lewis structure of H_2S consists of one sulfur with two lone pairs, connected via a single covalent bonds to two hydrogen atoms.

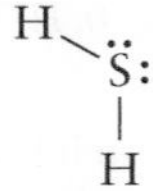

Determine the polarity:
The electron-domain geometry is tetrahedral, but the molecular geometry of H_2S is bent. The electronegativity values for H and S are 2.1 and 2.5, respectively. Each bond, then, is polarized toward sulfur, resulting in a partial negative charge on the sulfur and partial positive charges on the hydrogen atoms. H_2S has a permanent dipole and is therefore polar.

Determine the major intermolecular force:
All molecules exhibit London dispersion forces, but a polar molecule like H_2S also exhibits dipole–dipole interactions, which are stronger, making them the major force. Unlike water, the hydrogens bonded to sulfur cannot form hydrogen bonds.

d. **Draw the Lewis structure:**
The Lewis structure of H_2O consists of one oxygen with two lone pairs, connected via single covalent bonds to two hydrogen atoms. KCl dissociates into ions in this aqueous environment.

K^+ H–Ö:–H Cl^-

Determine the polarity:
The electron-domain geometry for water is tetrahedral, but the molecular geometry of H_2O is bent. The electronegativity values for H and O are 2.1 and 3.5, respectively. Each bond, then, is polarized toward oxygen, resulting in a partial negative charge on the oxygen and partial positive charges on the hydrogen atoms. H_2O has a permanent dipole and is therefore polar.

Determine the major intermolecular force:
All molecules exhibit London dispersion forces, but a polar molecule like H_2O also exhibits dipole–dipole interactions. However, neither of these are the strongest intermolecular force present. Instead, the strongest force can be demonstrated when the K^+ ion interacts with the partial negative on the oxygen atom and the Cl^- interacts with the partial positive on the hydrogens atoms, resulting in ion–dipole forces.

5.5 a. The energy stored in the chemical bonds of ATP is an example of potential energy. When ATP is hydrolyzed to ADP, the energy released is converted to kinetic energy to do work.

b. The energy stored in the covalent bonds of CCl_4 is potential energy.

c. When chemicals are broken down, the potential energy in their bonds is converted to kinetic energy to do work.

5.6 a. DMSO has indefinite shape and intermediate attractive forces, so it must be a liquid.

b. Carbon monoxide has a very low density, no definite shape, and weak attractive forces, so it must be a gas.

5.7 a. Evaporation is endothermic.

b. Freezing is exothermic.

c. Sublimation is endothermic.

5.8 The lower the forces between particles, the higher the vapor pressure will be.

a. The only forces acting between the molecules in each case are dispersion forces. These are stronger between the heavier nitrogen molecules, so helium has the higher vapor pressure.

b. Molecules of HF form strong hydrogen bonds with one another, whereas the only forces between neon molecules are weak dispersion forces. Thus, neon has the higher vapor pressure.

5.9 **Determine the attractive forces in diethyl ether and F_2:**
An assessment of diethyl ether shows it is polar and has an overall dipole moment that points in the direction of oxygen. The strongest intermolecular forces are dipole–dipole interactions in diethyl ether. F_2, on the other hand, is nonpolar. The strongest intermolecular forces are London dispersion forces.

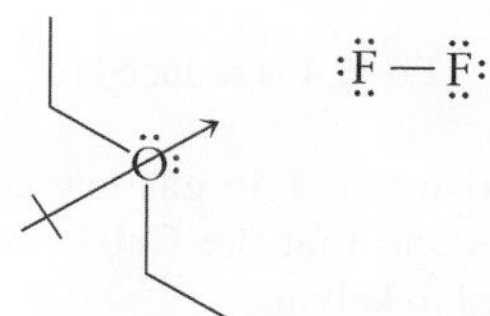

:F̤–F̤:

Determine which compound has the higher boiling point:
Dipole–dipole interactions are stronger intermolecular forces than London dispersion forces, so it takes more energy to change diethyl ether from a liquid to a gas and it has the higher boiling point.

5.10 Use Equation 5.1 to calculate the amount of heat absorbed when the ice melts to liquid water at 0.0 °C:

$$\text{Heat released} = (\text{mass})(\text{heat of fusion})$$

$$q = m \cdot \Delta H_{\text{fus}} = (25.0\ \cancel{g})\left(\frac{80\ \cancel{cal}}{1\ \cancel{g}}\right)\left(\frac{4.184\ J}{1\ \cancel{cal}}\right)$$

$$= 8368\ J = 8400\ J$$

5.11 Heat absorbed = (mass)(specific heat)(temp. change). The specific heat of helium is obtained from Table 5.8, and 1 kg = 1000 g.

$$\text{Heat absorbed} = (1000\ g)(1.25\ cal/g \cdot {}^\circ C)(700\ {}^\circ C - 25\ {}^\circ C)$$
$$= 8.44 \times 10^5\ cal$$

Chapter 6

6.1 $KE = \frac{1}{2}mv^2 = \frac{1}{2}(3.00\ g)(10.0\ cm/s)^2 = 1.50 \times 10^2\ g\ cm^2/s^2$

$KE = \frac{1}{2}mv^2 = \frac{1}{2}(3.00\ g)(20.0\ cm/s)^2 = 6.00 \times 10^2\ g\ cm^2/s^2$

6.2 The factors used come from the information in Table 6.2.

a. $670\ \cancel{torr} \times \dfrac{1\ atm}{760\ \cancel{torr}} = 0.882\ atm$

b. $670\ \cancel{torr} \times \dfrac{14.7\ psi}{760\ \cancel{torr}} = 13.0\ psi$

c. $670.\ \cancel{torr} \times \dfrac{760\ mmHg}{760\ \cancel{torr}} = 670.\ mm\ Hg$

6.3 a. $K = {}^\circ C + 273 = 27 + 273 = 300\ K$

b. $K = {}^\circ C + 273 = 0. + 273 = 273\ K$

c. ${}^\circ C = K - 273 = 0.00 - 273 = -273\ {}^\circ C$

d. ${}^\circ C = K - 273 = 100. - 273 = -173\ {}^\circ C$

6.4 **a. Step 1:** Examine the question for what is given and what is unknown.

Given: initial $V = 650$ L — initial $T = 28$ °C

Unknown: final P = ? — final $V = 650$ L

initial $P = 3.00$ atm — final $T = 20$ °C

Step 2: Identify a formula that links the given and unknown terms.

$$\text{Eq. 6.4: } = \frac{P_iV_i}{T_i} = \frac{P_fV_f}{T_f}$$

Because the chamber is rigid, note that the volume does not change, so $V_i = V_f$

$$\text{Eq. 6.4 is reduced to } \frac{P_i}{T_i} = \frac{P_f}{T_f}$$

Note also that T in gas law equations is expressed in kelvins and that the Celsius temperatures need to be changed to kelvins.

$$28\text{ °C} + 273 = 301\text{ K}$$
$$20\text{ °C} + 273 = 293\text{ K}$$

Step 3: Enter the data into the formula.

$$\frac{3.00\text{ atm}}{301\text{ K}} = \frac{P_f}{293\text{ K}}$$

The desired quantity, P_f, can be isolated by multiplying both sides of the equation by 293 K.

$$\frac{(3.00\text{ atm})(293\text{ K})}{301\text{ K}} = \frac{(P_f)(293\text{ K})}{293\text{ K}}$$

or

$$P_f = \frac{(3.00\text{ atm})(293)}{301} = 2.92\text{ atm}$$

b. Equation 6.4 is used. Temperatures must be converted to kelvins:

$P_i = 1.90$ atm — $P_f = 1.00$ atm

$V_i = 10.0$ L — V_f = ?

$T_i = 30.$°C $= 303$ K — $T_f = 210.2$ °C $= 483.2$ K

$$\frac{P_iV_i}{T_i} = \frac{P_fV_f}{T_f}$$

$$\frac{(1.90\text{ atm})(10.0\text{ L})}{303\text{ K}} = \frac{(1.00\text{ atm})(V_f)}{483.2\text{ K}}$$

Solve for V_f:

$$V_f = \frac{(1.90\text{ atm})(10.0\text{ L})(483.2\text{ K})}{(1.00\text{ atm})(303\text{ K})} = 30.3\text{ L}$$

6.5 To use the ideal gas law, all quantities must have units to match those of R. The only unit that needs to be changed is the temperature:

$$\text{K} = \text{°C} + 273 = 30.\text{°C} + 273 = 303\text{ K}$$
$$PV = nRT$$

Solve for P:

$$P = \frac{nRT}{V}$$
$$= \frac{(2.15\text{ mol})\left(0.0821\frac{\text{L atm}}{\text{mol K}}\right)(303\text{ K})}{(12.6\text{ L})}$$
$$= 4.24\text{ atm}$$

6.6 Dalton's law says $P_t = P_{O_2} + P_{N_2} + P_{H_2O}$. The total pressure of the sample is the atmospheric pressure of 742 torr. Therefore,

$$742\text{ torr} = 141\text{ torr} + 581\text{ torr} + P_{H_2O}$$

and

$$P_{H_2O} = 742 - 141 - 581 = 20.$$

6.7 Like dissolves like, so oil will dissolve best in oily solvents such as light mineral oil or dish soap.

6.8 **a.** Total ionic: $2Na^+(aq) + 2I^-(aq) + Cl_2(aq) \rightarrow 2Na^+(aq) + 2Cl^-(aq) + I_2(aq)$

Na^+ is a spectator ion.

Net ionic: $2I^-(aq) + Cl_2(aq) \rightarrow 2Cl^-(aq) + I_2(aq)$

b. Total ionic: $Ca^{2+}(aq) + 2Cl^-(aq) + 2Na^+(aq) + CO_3^{2-}(aq) \rightarrow 2Na^+(aq) + 2Cl^-(aq) + CaCO_3(s)$

Na^+ and Cl^- are spectator ions.

Net ionic: $Ca^{2+}(aq) + CO_3^{2-}(aq) \rightarrow CaCO_3(s)$

c. Total ionic: $Ba^{2+}(aq) + 2OH^-(aq) + 2H^+(aq) + SO_4^{2-}(aq) \rightarrow 2H_2O(\ell) + BaSO_4(s)$

There are no spectator ions, so the net ionic equation is the same as the total ionic equation.

6.9 **a.** The data may be substituted directly into Equation 6.8:

$$\text{M} = \frac{1.25\text{ mol solute}}{2.50\text{ L solution}} = 0.500\frac{\text{mol solute}}{\text{L solution}}$$

The solution is 0.500 molar, or 0.500 M.

b. In this problem, the volume of solution must be converted to liters before the data are substituted into Equation 6.8:

$$(225\text{ mL solution})\left(\frac{1\text{ L}}{1000\text{ mL}}\right) = 0.225\text{ L solution}$$

$$\text{M} = \frac{0.486\text{ mol solute}}{0.225\text{ L solution}} = 2.16\frac{\text{mol solute}}{\text{L solution}}$$

The solution is 2.16 molar, or 2.16 M.

c. In this problem, the volume of solution must be converted to liters and the mass of solute must be converted to moles before the data can be substituted into Equation 6.8:

$$(100\text{ mL solution})\left(\frac{1\text{ L}}{1000\text{ mL}}\right) = 0.100\text{ L solution}$$

$$(2.60\text{ g NaCl})\left(\frac{1\text{ mol NaCl}}{58.44\text{ g NaCl}}\right) = 0.0445\text{ mol NaCl}$$

In the last calculation, the factor $\frac{1\text{ mol NaCl}}{58.44\text{ g NaCl}}$ comes from the calculated formula weight of 58.44 amu for NaCl. The data are now substituted into Equation 6.8

$$\text{M} = \frac{0.0445\text{ mol NaCl}}{0.100\text{ L solution}} = 0.445\frac{\text{mol NaCl}}{\text{L solution}}$$

The solution is 0.445 molar, or 0.445 M.

6.10 **a.** To calculate % (w/w), the mass of solute contained in a specific mass of solution is needed. The mass of solute is 0.900 g, and the mass of solution is equal to the solvent mass (100.0 g) plus the solute mass (0.900 g):

$$\% \text{ (w/w)} = \frac{\text{solute mass}}{\text{solution mass}} \times 100\% = \frac{0.900 \cancel{g}}{100.9 \cancel{g}} \times 100\%$$
$$= 0.892\% \text{ (w/w)}$$

To calculate % (w/v), the number of grams of solute must be known along with the number of milliliters of solution. The mass of solute is 0.900 g, and the solution volume is 100 mL:

$$\% \text{ (w/v)} = \frac{\text{g of solute}}{\text{mL of solution}} \times 100\% = \frac{0.900 \text{ g}}{100 \text{ mL}} \times 100\%$$
$$= 0.900\% \text{ (w/v)}$$

b. The given quantity is 30.0 mL of beverage, and the desired quantity is milliliters of pure alcohol. The % (v/v) provides the necessary factor:

$$30.0 \cancel{\text{mL beverage}} \times \frac{45 \text{ mL alcohol}}{100 \cancel{\text{mL beverage}}} = 14 \text{ mL alcohol}$$

6.11 Equation 6.12 is used to calculate the volume of 6.00 M NaOH solution needed.

$$(C_c)(V_c) = (C_d)(V_d)$$
$$(6.00 \text{ M})(V_c) = (0.250 \text{ M})(500. \text{ mL})$$
$$V_c = \frac{(0.250 \cancel{M})(500. \text{ mL})}{(6.00 \cancel{M})} = 20.8 \text{ mL}$$

The solution is prepared by putting 20.8 mL of 6.00 M NaOH solution into a 500-mL flask, adding pure water to the mark, and making certain the resulting solution is well mixed.

6.12 **Step 1:** Compare the osmotic pressures of the solution and the cell.
The concentration of glucose is lower outside the cell, so the solution is hypotonic. Thus, the osmotic pressure inside the cell is less than the osmotic pressure outside the cell.

Step 2: Determine the direction of net flow.
The osmotic pressure inside the cell is less, so there is a net flow of water via osmosis into the cell.

Step 3: Describe the change in cell volume.
The net flow of water into the cell will cause the cell to swell. In extreme cases, it will burst.

Chapter 7

7.1 **a.** When the concentration of Fe^{2+} is increased, more effective collisions occur, and more products are obtained over the same time period. This means the rate of the reaction will increase.

b. An decrease in temperature decreases the kinetic energy of the reactants and causes fewer effective collisions. As a result, the rate of the reaction will decrease.

c. The addition of an inhibitor to a reaction reduces the amount of product produced in the same period of time, thus decreasing the reaction rate.

7.2 **a.** $K_{eq} = [Ba^{2+}][SO_4^{2-}]$

b. $K_{eq} = 1.5 \times 10^{-9}$, so you can think of the equilibrium constant as the ratio of 0.0000000015 to 1. This means that there are 0.0000000015 set of ions (Ba^{2+} and SO_4^{2-}) for every 1 $BaSO_4$. As a result, solid $BaSO_4$ predominates at equilibrium.

7.3 **a.** Heat is added to an endothermic system. To restore equilibrium, heat must be used up, so the equilibrium shifts to the right, and more ions will be present at equilibrium in the warmer mixture.

b. The equilibrium will shift to the left in order to use up the added NH_4^+. At the new equilibrium position, less NO_3^- will be present.

7.4 **a.** $\underset{\text{acid}}{HC_2H_3O_2} + \underset{\text{base}}{H_2O} \rightleftarrows \underset{\text{acid}}{H_3O^+} + \underset{\text{base}}{C_2H_3O_2^-}$

conjugate pair (H_2O / H_3O^+)
conjugate pair ($HC_2H_3O_2$ / $C_2H_3O_2^-$)

b. $\underset{\text{base}}{NO_2^-} + \underset{\text{acid}}{H_2O} \rightleftarrows \underset{\text{base}}{OH^-} + \underset{\text{acid}}{HNO_2}$

conjugate pair (H_2O / OH^-)
conjugate pair (NO_2^- / HNO_2)

7.5 **a.** The anhydrous compound is called hydrogen iodide. The name of the water solution is obtained by dropping *hydrogen* from the anhydrous compound name and adding the prefix *hydro-* to the stem *iod*. The *-ide* suffix on the *iod* stem is replaced by the suffix *-ic* to give the name *hydroiodic*. The word *acid* is added, giving the final name *hydroiodic acid* for the water solution.

b. The anhydrous compound is called hydrogen bromide. The name of the water solution is obtained the same way it was for HI(aq) in part a, using the stem *brom* in place of *iod*. The resulting name for the water solution is *hydrobromic acid*.

c. The anhydrous compound is called hydrogen fluoride. The name of the water solution is obtained the same way it was for HI(aq) in part a, using the stem *fluor* in place of *iod*. The resulting name for the water solution is *hydrofluoric acid*.

7.6 The removal of the two H^+ ions leaves behind the CO_3^{2-} polyatomic ion. From Table 3.5, this ion is the carbonate ion. According to Rules 1–4, the *-ate* suffix is replaced by the *-ic* suffix to give the name *carbonic acid*.

7.7 **a.** If $[OH^-] = 1 \times 10^{-5}$ M, then $[H_3O^+] = 1 \times 10^{-9}$ M; 1×10^{-5} is a larger number than 1×10^{-9}, so the solution is basic.

b. $[H_3O^+] = 1 \times 10^{-5}$ M, then $[OH^-] = 1 \times 10^{-9}$ M; 1×10^{-5} is the larger number, so the solution is acidic.

7.8 **a.** pH = 6

b. If $[OH^-] = 1 \times 10^{-3}$ M, then $[H_3O^+] = 1 \times 10^{-11}$ M and pH = 11.

7.9 **a.** If pH = 1.0, then $[H_3O^+] = 1 \times 10^{-1}$ M and $[OH^-] = 1 \times 10^{-13}$ M.

b. If pH = 10.0, then $[H_3O^+] = 1 \times 10^{-10}$ M and $[OH^-] = 1 \times 10^{-4}$ M.

7.10 **a.** H_2SO_4 is the stronger acid. In diprotic and triprotic acids, the second and third protons are sequentially harder to remove, so the corresponding acid is weaker.

b. The stronger base is hydride as indicated in Table 7.5.

c. The conjugate base of a strong acid is weak, and the conjugate base of a weak acid is strong. HNO_3 is a strong acid, so NO_3^- is weak base, and CH_3COOH is a weak acid, so CH_3COO^- is a strong base.

7.11 The pattern is liters solution $A \rightarrow$ mol B, and the pathway is liters NaOH solution $\rightarrow$ mol NaOH $\rightarrow$ mol H_3PO_4. In combined form, the steps in the factor-unit method are:

Step 1: 0.0141 L NaOH solution

Step 2: 0.0141 L NaOH solution = mol H_3PO_4

Step 3: $0.0141\ \cancel{L}\ \cancel{\text{NaOH solution}} \times \dfrac{0.250\ \cancel{\text{mol NaOH}}}{1\ \cancel{L}\ \cancel{\text{NaOH solution}}} \times \dfrac{1\ \text{mol } H_3PO_4}{3\ \cancel{\text{mol NaOH}}} = \text{mol } H_3PO_4$

Step 4: $(0.0141)\left(\dfrac{0.250}{1}\right)\left(\dfrac{1\ \text{mol } H_3PO_4}{3}\right) = 0.00118\ \text{mol } H_3PO_4$

The molarity of the H_3PO_4 solution is calculated:

$$M = \frac{\text{mol } H_3PO_4}{\text{L } H_3PO_4 \text{ solution}} = \frac{0.00118\ \text{mol } H_3PO_4}{0.0250\ \text{L } H_3PO_4 \text{ solution}} = \frac{0.0472\ \text{mol } H_3PO_4}{\text{L } H_3PO_4 \text{ solution}} = 0.0472\ \text{M}$$

7.12 **a.** Because the acid (HCOOH) and conjugate base ($HCOO^-$) concentrations are the same, the pH will equal the pK_a for the acid. This is shown by the following calculation:

$$\begin{aligned} pH &= pK_a + \log\frac{[B^-]}{[HB]} \\ &= 3.74 + \log\frac{[HCOO^-]}{[HCOOH]} \\ &= 3.74 + \log\frac{0.22\ \text{mol/L}}{0.22\ \text{mol/L}} \\ &= 3.74 + \log 1 \\ &= 3.74 + 0 = 3.74 \end{aligned}$$

b. The acid is H_2SO_3, and the conjugate base is HSO_3^- (produced when the $NaHSO_3$ salt dissociates).

$$\begin{aligned} pH &= pK_a + \log\frac{[B^-]}{[HB]} \\ &= 1.82 + \log\frac{(0.10\ \text{mol/L})}{(0.25\ \text{mol/L})} \\ &= 1.82 + \log 0.40 \\ &= 1.82 - 0.40 \\ &= 1.42 \end{aligned}$$

c. The acid is H_2CO_3 and the conjugate base is HCO_3^-. The desired pH is 7.40. Substitution into Equation 7.6 gives:

$$\begin{aligned} pH &= pK_a + \log\frac{[HCO_3^-]}{[H_2CO_3]} \\ 7.40 &= 6.37 + \log\frac{[HCO_3^-]}{[H_2CO_3]} \\ 7.40 - 6.37 &= \log\frac{[HCO_3^-]}{[H_2CO_3]} \\ 1.03 &= \log\frac{[HCO_3^-]}{[H_2CO_3]} \end{aligned}$$

The log of the desired ratio is 1.03. The antilog of 1.03 correctly rounded is 11. So the concentration of HCO_3^- must be 11 times the concentration of H_2CO_3 in the blood to maintain the desired pH.

Chapter 8

8.1 Look for the presence of carbon atoms. Organic molecules contain carbon.

a. Organic

b. Inorganic

8.2 **a.**

$CH_3-CH(OH)-CH_2-CH_2-CH_3$

OH

b.

$CH_3-CH(CH_3)-CH_2-C(=O)-OH$

O

OH

8.3

alcohol

aromatic

amide

alkene

H

N

O

O

ether

capsaicin

8.4 **a.**

H H O—H alcohol

H—C=C—C—H

alkene H

structural isomer

b.

H H O

H—C=C—C—H

alkene aldehyde

no relationship

c. 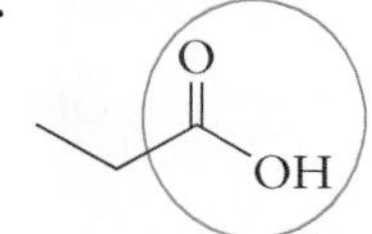

no relationship

d. 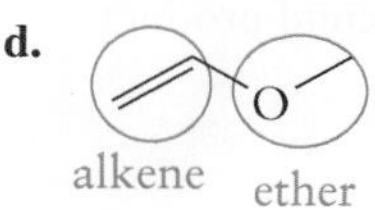

structural isomer

8.5 **a.** Same molecule: In both molecules, there is a continuous chain of five carbons with a branch at position 2.

b. Same molecule: In both molecules, there is a continuous chain of four carbons with a branch at position 2

$$\underset{1}{CH_3}-\underset{2}{CH}(CH_3)-\underset{3}{CH_2}-\underset{4}{CH_3} \qquad \underset{1}{CH_3}-\underset{2}{CH}(\overset{3}{CH_2}-\overset{4}{CH_3})-CH_3$$

8.6 **a.** The chain is numbered from the right to give 2-methylpentane.

b. The chain is numbered from the right to give 4-isopropyl-2-methylheptane.

8.7 **a.**

$$CH_3-C(CH_3)_2-CH_2-CH(CH_3)-CH_3$$

b.

$$CH_3-CH(CH_3)-CH(CH_2-CH_3)-CH(CH_3)-CH_2-CH_2-CH_3$$

8.8 The name 1-ethyl-2-methylcyclopentane is correct, whereas 2-ethyl-1-methylcyclopentane is incorrect because the ring numbering begins with the carbon attached to the first group alphabetically.

8.9 **a.** (1) Trans because the two Br's are on opposite sides of the ring.

(2) Cis; both Cl's are on the same side.

(3) Cis; the two groups are on the same side.

b. In showing geometric isomers of ring compounds, it helps to draw the ring in perspective. Cis should have both substituents on the same side of the ring:

Cl Cl (ring drawn in perspective, both Cl on the same side)

8.10 In each of these alkenes, the double-bonded carbons occur at positions 1 and 2.

a. 3-bromo-1-propene

b. 2-ethyl-1-pentene

c. 3,4-dimethylcyclohexene

8.11 **a.**

$$\underset{cis}{(CH_3)_2CH(H)C=C(H)CH_3} \qquad \underset{trans}{(CH_3)_2CH(H)C=C(CH_3)H}$$

b. This structure does not exhibit *cis–trans* isomerism because there are two methyl groups attached to the left double-bonded carbon.

8.12 Each chain is correctly numbered from the right.

a. 2-pentyne **b.** 5-methyl-2-hexyne

8.13 **a.** $CH_3-CH(CH_3)-CH_2-CH_3$

b. (cyclopentane with CH_3 and H) = (methylcyclopentane, CH_3)

8.14 **a.** Markovnikov's rule predicts that H will attach at position 1 to give:

$$CH_3CH_2CH_2\overset{OH}{CH}-\overset{H}{CH_2} = CH_3CH_2CH_2\overset{OH}{CH}CH_3$$

b. (cyclopentane with H and OH) = (cyclopentanol, OH)

8.15

$$-CH_2\underset{CN}{CH}-CH_2\underset{CN}{CH}-CH_2\underset{CN}{CH}-CH_2\underset{CN}{CH}-$$

8.16 **a.** Numbers or the term *meta* may be used:

1,3-diethylbenzene or *m*-diethylbenzene

b. Numbers must be used when there are three groups: 1,2,3-tribromobenzene.

Chapter 9

9.1 **a.** (benzene ring: OH at 1, CH_3CHCH_3 at 2 and at 6; positions 3, 4, 5 unsubstituted) 2,6-diisopropylphenol

b.

$$\underset{\underset{6\ \ 7}{CH_2CH_3}}{\overset{5}{CH_3}\overset{OH}{\underset{4}{CH}}}\overset{3}{CH}\overset{2}{CH_2}\overset{1}{CH}(CH_3)CH_3$$

As printed: $CH_3CHCHCH_2CHCH_3$ with OH on the carbon numbered 3, CH_2CH_3 (6, 7) on carbon 4, and CH_3 on carbon 1 — 2,5-dimethyl-4-heptanol

c. CH_3 (on C-2 of cyclopentane ring, OH on C-1) 2-methylcyclopentanol

d.
$$\begin{array}{c} CH_3 \\ | \\ CH_2CHCH_2CH_2 \\ |\qquad\qquad | \\ OH \qquad\quad OH \end{array}$$
2-methyl-1,4-butanediol

9.2 In each case, count the number of carbon atoms attached to the hydroxy-bearing carbon.

a. Primary **b.** Secondary **c.** Tertiary

9.3 Methanol molecules form hydrogen bonds, whereas molecules of butane do not.

9.4 **a.** Because hydrobromic acid is a strong acid, it will protonate the oxygen of isopropyl alcohol.

$$(CH_3)_2CH{-}O{-}H + H{-}Br \rightleftharpoons (CH_3)_2CH{-}\overset{+}{O}H_2 + Br^-$$

b. The ethyne anion is a strong base, so it will deprotonate the oxygen on isopropyl alcohol.

$$(CH_3)_2CH{-}O{-}H + Na^+ \,^-C{\equiv}CH \rightleftharpoons HC{\equiv}CH + (CH_3)_2CH{-}O^- + Na^+$$

9.5 **a.** There are two ways of removing water from this structure, but they both produce the same product:

$$CH_3CH{=}CHCH_2CH_3$$

b. Dehydration will occur in the direction that gives the highest number of carbon groups attached to the double-bonded carbons:

1-methylcyclohexene (CH_3 on the double-bonded carbon of a cyclohexene ring)

9.6 This reaction is similar to Equation 9.4, too, so two molecules of alcohol will come together to form an ether.

$$C_5H_9{-}O{-}H \;\; H{-}O{-}C_5H_9 \longrightarrow C_5H_9{-}O{-}C_5H_9 + H_2O$$

9.7 **a.**

$$C_6H_5{-}\overset{O}{\overset{\|}{C}}{-}H \qquad C_6H_5{-}\overset{O}{\overset{\|}{C}}{-}OH$$

first product second product

b. cyclohexanone ($C_6H_{10}{=}O$) $+ H_2O$

c. The alcohol is tertiary, so it does not react.

9.8 **a.** ethyl phenyl ether

b. diisopropyl ether

9.9 **a.** propylamine

b. ethylpropylamine

9.10 In each case, count the number of carbon atoms directly attached to the nitrogen.

a. Secondary **b.** Secondary **c.** Tertiary

9.11 **a.**

$$CH_3{-}\overset{\displaystyle H}{\underset{\displaystyle CH_3}{\overset{|}{\underset{|}{N^+}}}}{-}CH_2CH_3 + OH^-$$

b. $C_6H_5{-}NH_3^+ + OH^-$

c. $C_5H_9{-}NH_3^+\,CH_3CH_2COO^-$

d. $C_6H_5{-}\underset{\displaystyle CH_3}{\underset{|}{NH_2^+}}\,Cl^-$

Chapter 10

10.1 **a.** The aldehyde contains five carbon atoms, so the name is pentanal.

b. The carbonyl carbon atom is position 1. The name is 2,5-dimethylcyclopentanone.

c. The carbonyl carbon atom is position 3. The name is 4-methyl-3-hexanone

10.2 **a.**

$$CH_3CH_2\overset{\displaystyle OH}{\overset{|}{C}}HCH_2CH_3$$

b.

$$C_6H_5{-}CH_2\overset{\displaystyle OH}{\overset{|}{C}}H_2$$

10.3 **a.**

$$CH_3\overset{\displaystyle CH_3}{\overset{|}{C}}HCH_2{-}\overset{O}{\overset{\|}{C}}{-}OH$$

b. This is a ketone, so it does not undergo oxidation.

10.4 **a.**

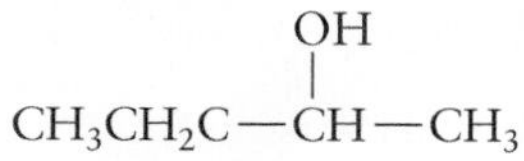

$$CH_3CH_2C(OH)H{-}CH{-}CH_3$$

(structure: $CH_3CH_2C{-}CH{-}CH_3$ with OH on the CH carbon)

b. cyclopentane ring–CH_2OH

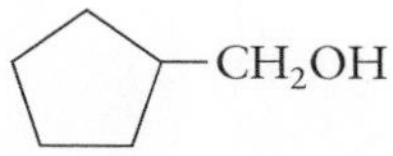

10.5 **a.** The longest chain beginning with the carboxyl group has four carbon atoms. The name is 2-methylbutanoic acid.

b. The carboxyl group is at position 1. The name is 3,4-dibromobenzoic acid.

c. This is the potassium salt of butanoic acid. The name is potassium butanoate.

10.6

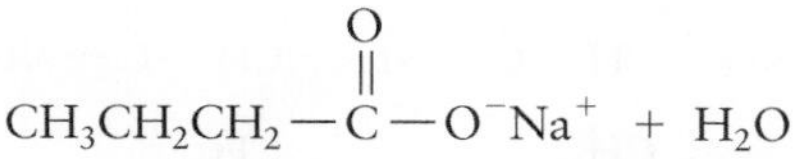

$$CH_3CH_2CH_2{-}\overset{\overset{\displaystyle O}{\|}}{C}{-}O^-Na^+ \ + \ H_2O$$

10.7 **a.** The common name for the acid used to prepare this ester is formic acid. The alkyl group attached to oxygen is isopropyl. Thus, the common name is isopropyl formate. The IUPAC name for formic acid is methanoic acid. Thus, the IUPAC name for the ester is isopropyl methanoate.

b. The common and IUPAC names are identical when the carboxylic acid is benzoic acid. The name of the ester is methyl benzoate.

c. This amide is derived from butanoic acid. The name is butanamide.

d. The common name for the acid, butyric acid, is found in Table 10.4. The common name for the amide is *N*-methylbutyramide. The IUPAC name is *N*-methylbutanamide.

10.8 **a.**

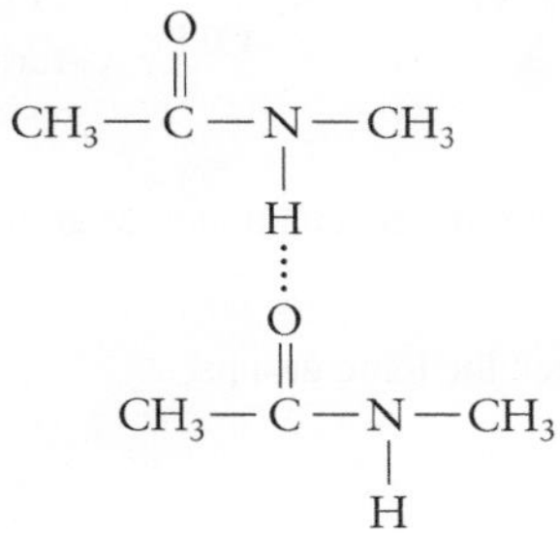

$$CH_3{-}\overset{\overset{\displaystyle O}{\|}}{C}{-}\underset{\underset{\displaystyle H}{|}}{N}{-}CH_3$$

(H ⋯ O hydrogen bond to)

$$CH_3{-}\overset{\overset{\displaystyle O}{\|}}{C}{-}\underset{\underset{\displaystyle H}{|}}{N}{-}CH_3$$

b. $H{-}O{-}H$ ⋯ $O{=}C(CH_3){-}\underset{\underset{\displaystyle H}{|}}{N}{-}CH_3$ (water H bonded to carbonyl O)

or

$$CH_3{-}\overset{\overset{\displaystyle O}{\|}}{C}{-}\underset{\underset{\displaystyle H}{|}}{N}{-}CH_3$$ (N–H ⋯ O of $H{-}O{-}H$)

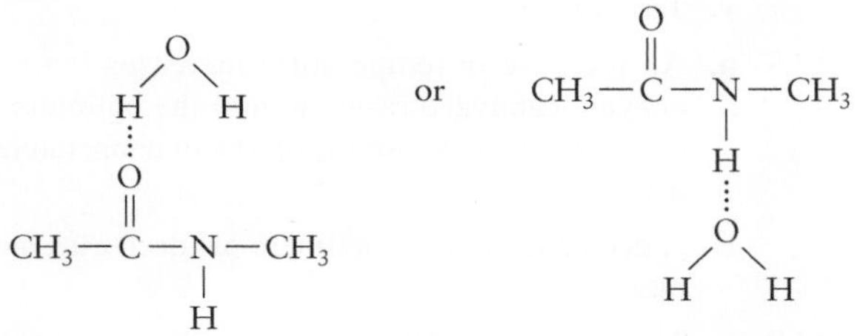

10.9 **a.**

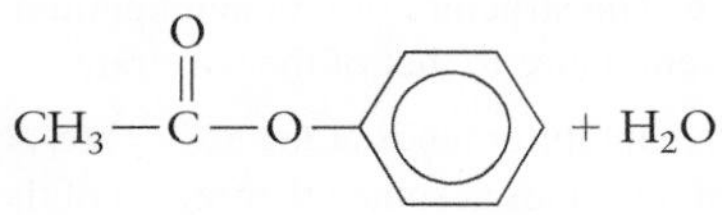

$$CH_3{-}\overset{\overset{\displaystyle O}{\|}}{C}{-}O{-}C_6H_5 \ + \ H_2O$$

b.

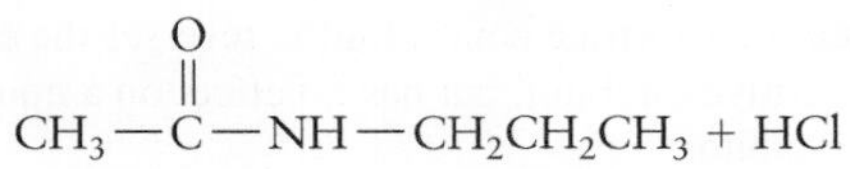

$$CH_3{-}\overset{\overset{\displaystyle O}{\|}}{C}{-}NH{-}CH_2CH_2CH_3 + HCl$$

10.10 **a.**

$$C_6H_5{-}\overset{\overset{\displaystyle O}{\|}}{C}{-}OH \ + \ HO{-}CH_2CH_2CH_3$$

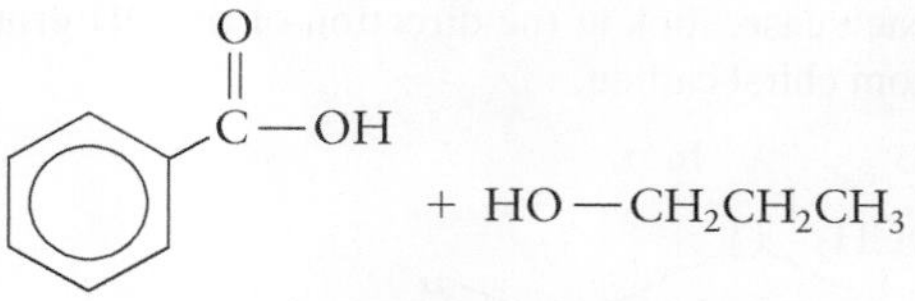

b. Under basic conditions, amide hydrolysis produces an amine and the salt of a carboxylic acid.

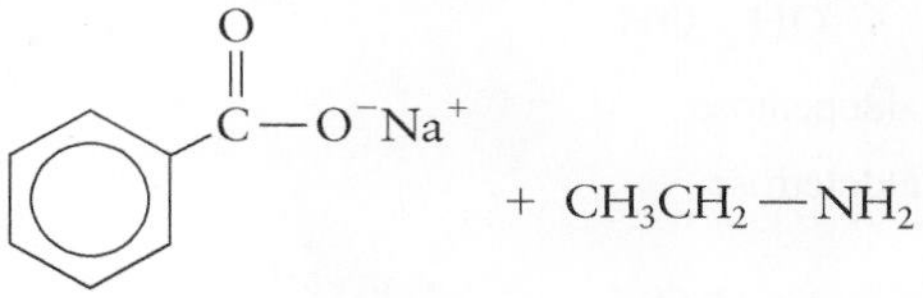

$$C_6H_5{-}\overset{\overset{\displaystyle O}{\|}}{C}{-}O^-Na^+ \ + \ CH_3CH_2{-}NH_2$$

10.11 **a.**

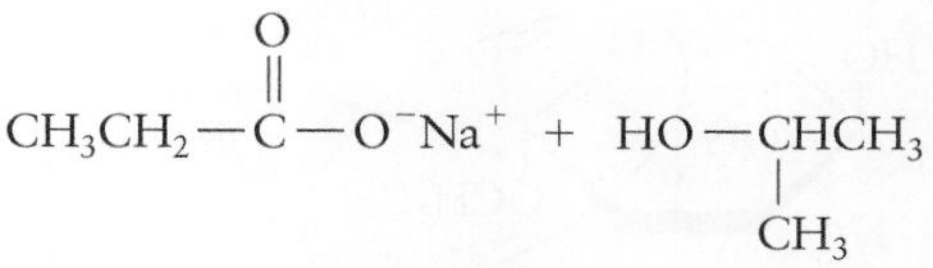

$$CH_3CH_2{-}\overset{\overset{\displaystyle O}{\|}}{C}{-}O^-Na^+ \ + \ HO{-}\underset{\underset{\displaystyle CH_3}{|}}{CH}CH_3$$

b.

$$C_6H_5{-}\overset{\overset{\displaystyle O}{\|}}{C}{-}O^-K^+ \ + \ HO{-}CH_2CH_3$$

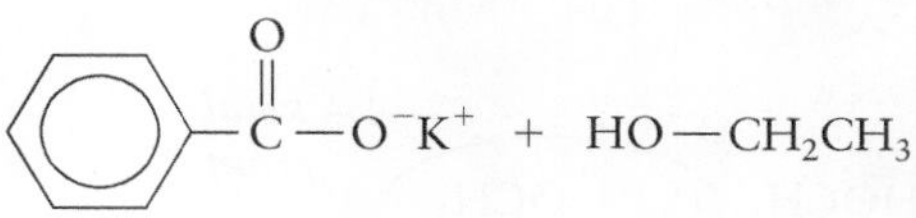

Chapter 11

11.1 In each case, look for the presence of four different groups attached to the colored atom.

a. The carbon is not chiral because of the two attached CH_2OH groups.

b. Yes

c. The carbon is not chiral because it is attached to two H's.

d. Yes

11.2

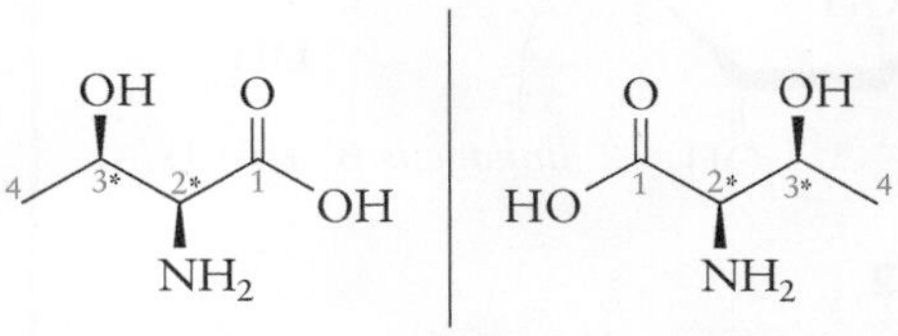

(Skeletal structures of threonine and its mirror image: left, OH on carbon 3* and NH_2 on carbon 2*, with C1 as COOH (O, OH); right, HO–C1(=O), 2*, 3*, 4, with OH and NH_2 on wedges.)

Mirror

The second and third carbons of threonine (denoted with *) are chiral, because they are each bonded to four different groups. In the left representation, the hydrocarbon chain is drawn in the plane of the page, and we arbitrarily chose to represent both the —OH and —NH_2 with a wedge. Thus, the enantiomer (right) must be drawn with the hydrocarbon in the plane of the page, but rotated 180 degrees with the —OH and —NH_2 on a wedge in order to make a mirror image.

11.3 The chiral carbon atom is the one with the attached nitrogen. The chiral carbon atom is placed at the intersection of the two lines, and the COOH is placed at the top:

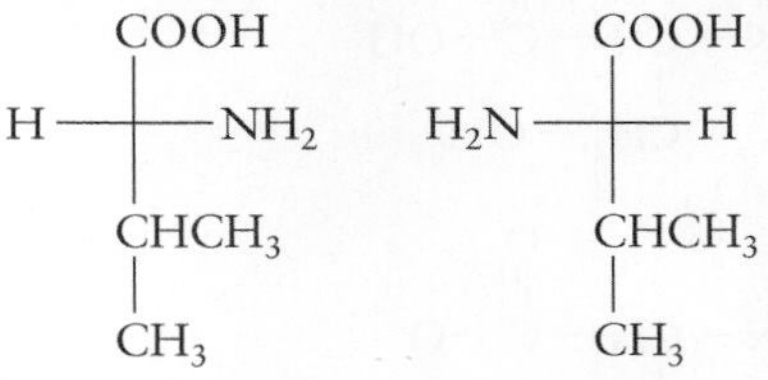

D form	L form
COOH (top); H (left), NH_2 (right); $CHCH_3$ below; CH_3	COOH (top); H_2N (left), H (right); $CHCH_3$ below; CH_3

D form　　L form

11.4 In each case, look at the direction of the OH group on the bottom chiral carbon.

a. D b. L

11.5 $HOCH_2$ O α OH OH OH

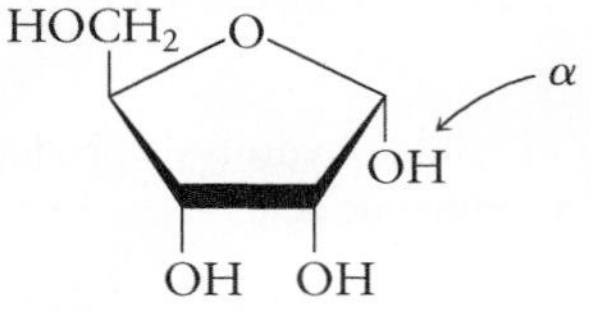

11.6 a. aldopentose

b. aldotetrose

11.7 a.

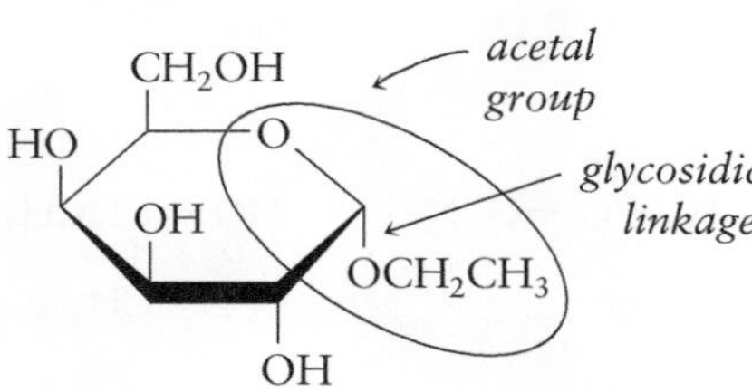

b.

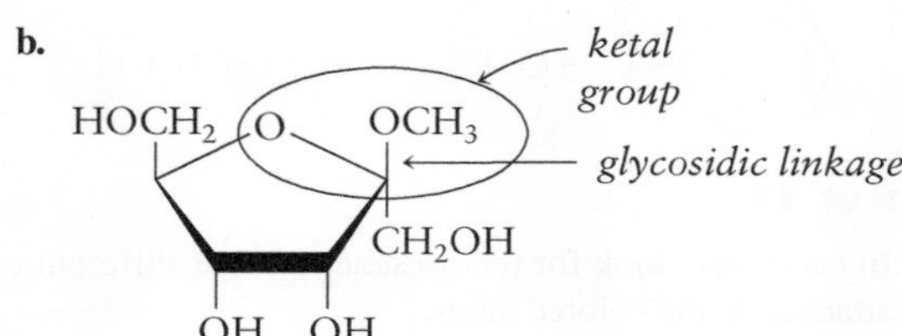

11.8 CH_2OH O OH CH_2OH 4 OH O O OH 1 HO OH OH

linkage is β (1⟶4)

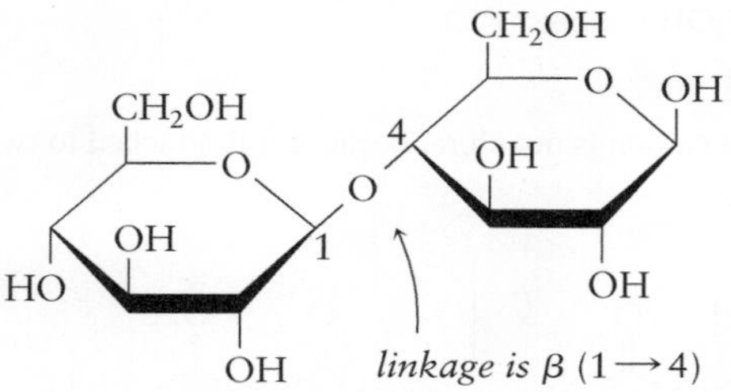

Chapter 12

12.1 $H_3\overset{+}{N}$ O O^- $\overset{+}{NH_3}$

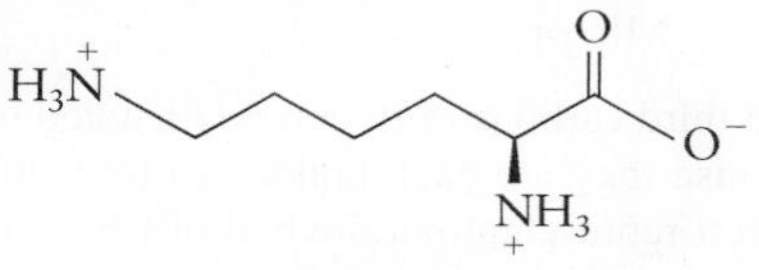

lysine

12.2 a. Acidic polar

b. Neutral nonpolar

c. Basic polar

d. Neutral polar

12.3 a.

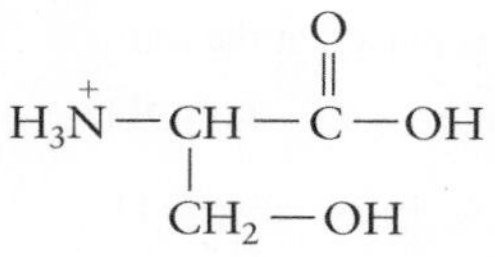

$H_3\overset{+}{N}-CH(CH_2-OH)-C(=O)-OH$

b.

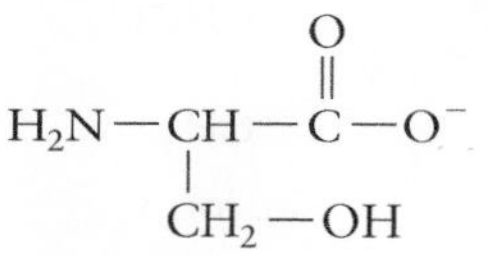

$H_2N-CH(CH_2-OH)-C(=O)-O^-$

12.4

$H_3\overset{+}{N}-CH(CH_2-COO^-)-C(=O)-NH-CH(CH_2C_6H_5)-C(=O)-O^-$

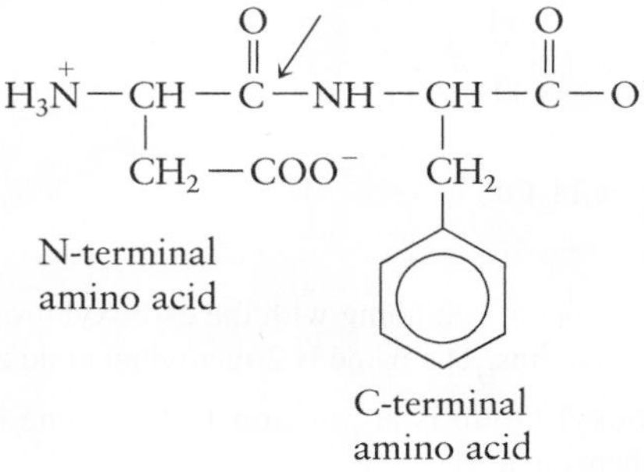

N-terminal amino acid

C-terminal amino acid

12.5 a.

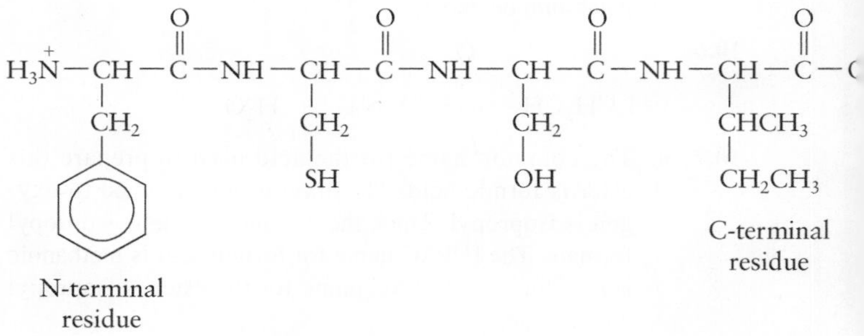

$H_3\overset{+}{N}-CH(CH_2C_6H_5)-C(=O)-NH-CH(CH_2SH)-C(=O)-NH-CH(CH_2OH)-C(=O)-NH-CH(CHCH_3CH_2CH_3)-C(=O)-O^-$

N-terminal residue

C-terminal residue

b. By convention the peptide backbone is shown in the plane of the page with the side chains alternating as wedge and dash.

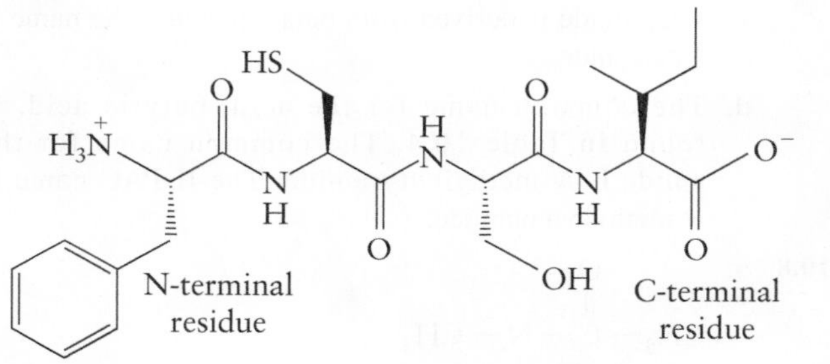

HS $H_3\overset{+}{N}$ O H N O N H O N H O O^- OH O

N-terminal residue

C-terminal residue

12.6 a. Hydrophobic attractions between nonpolar groups

b. Hydrogen bonding

c. Salt bridges between the ionic groups

12.7 a. Quaternary

b. Secondary

c. Tertiary

12.8 a. An increase in temperature increases the rate of the enzyme-catalyzed reaction until the optimum temperature is reached. Above the optimum temperature, the rate decreases.

b. A decrease from the optimum pH decreases the reaction rate.

12.9 a. The structure of a competitive inhibitor resembles that of the substrate. The structure of a noncompetitive inhibitor bears no resemblance to that of the substrate.

b. A competitive inhibitor binds at the active site. A noncompetitive inhibitor binds at some other region of the enzyme.

c. Increasing substrate concentration reverses the effect of a competitive inhibitor, but has no effect on a noncompetitive inhibitor.

Chapter 13

13.1 **a.** purine, both DNA and RNA

b. pyrimidine, DNA

13.2

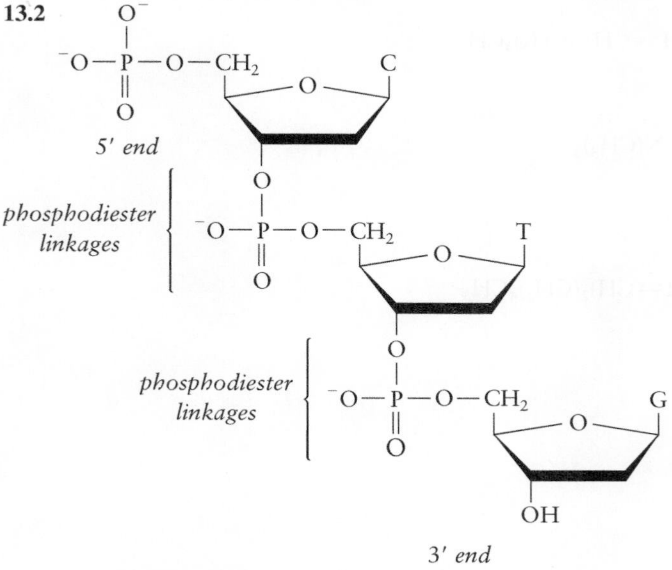

[Structure: trinucleotide with 5′ end phosphate $^-O-P(=O)(O^-)-O-CH_2$ on sugar bearing C; phosphodiester linkages $^-O-P(=O)-O-CH_2$ to sugar bearing T; phosphodiester linkages $^-O-P(=O)-O-CH_2$ to sugar bearing G with OH at 3′ end]

5′ end

phosphodiester linkages

phosphodiester linkages

3′ end

13.3 5′ T-T-A-C-G 3′ original strand

3′ A-A-T-G-C 5′ complementary strand

The complementary strand written in the 5′ to 3′ direction is C-G-T-A-A.

13.4 5′ A-T-T-A-G-C-C-G 3′ DNA template

3′ U-A-A-U-C-G-G-C 5′ mRNA

The mRNA written in the 5′ to 3′ direction is C-G-G-C-U-A-A-U.

13.5 fMet-His-His-Val-Leu-Cys

UAG is the STOP codon that signals for chain termination.

Chapter 14

14.1 With one carbon–carbon double bond, vernolic acid is monounsaturated. Monounsaturated fatty acids, according to Table 14.1, are liquids at room temperature.

14.2 One of several possibilities is the following structure:

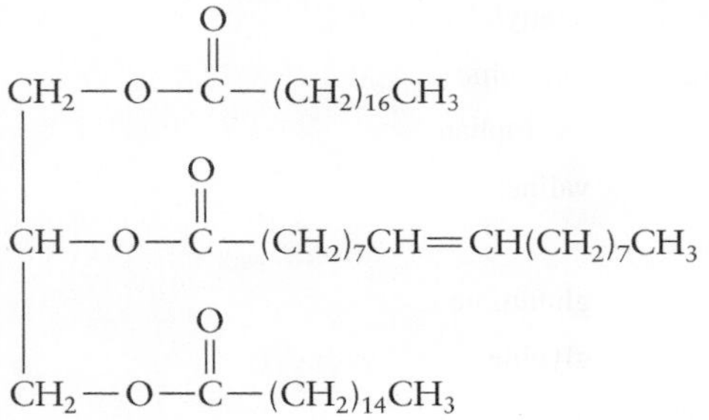

$CH_2-O-C(=O)-(CH_2)_{16}CH_3$ (stearic acid component)

$CH-O-C(=O)-(CH_2)_7CH=CH(CH_2)_7CH_3$ (oleic acid component)

$CH_2-O-C(=O)-(CH_2)_{14}CH_3$ (palmitic acid component)

14.3 CH_2-OH $HOOC-(CH_2)_7CH=CH(CH_2)_7CH_3$

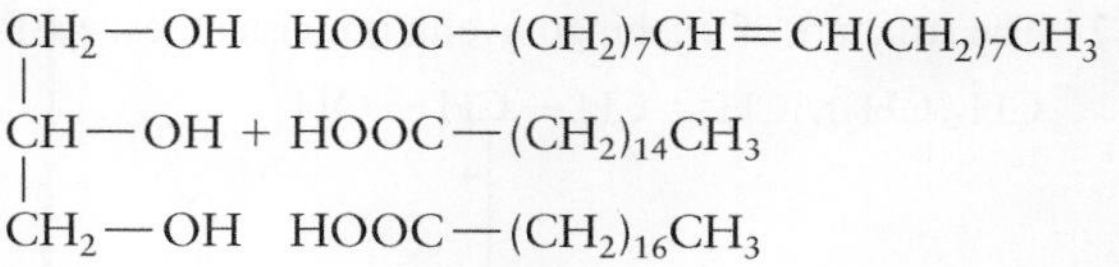

$CH-OH$ + $HOOC-(CH_2)_{14}CH_3$

CH_2-OH $HOOC-(CH_2)_{16}CH_3$

14.4 CH_2-OH $K^{+-}OOC-(CH_2)_{16}CH_3$

$CH-OH$ + $2K^{+-}OOC-(CH_2)_7CH=CH(CH_2)_7CH_3$

CH_2-OH

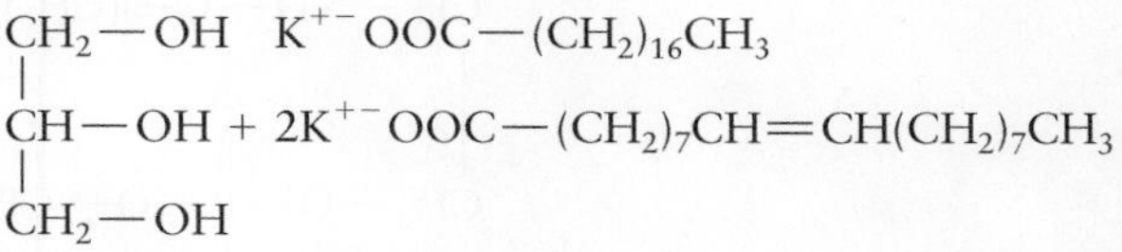

14.5 During complete hydrogenation, all CH=CH groups become CH_2-CH_2 groups:

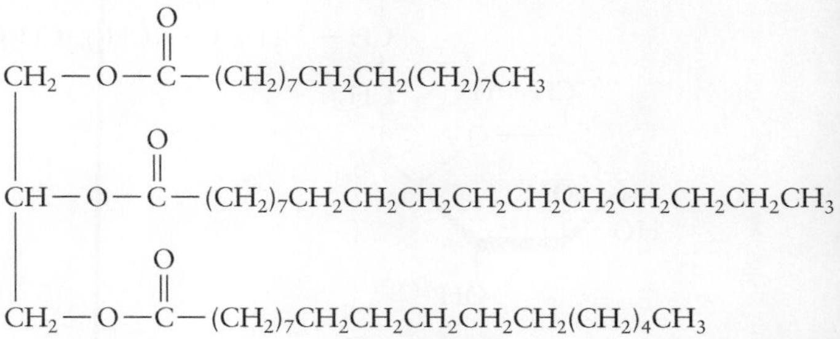

$CH_2-O-C(=O)-(CH_2)_7CH_2CH_2(CH_2)_7CH_3$

$CH-O-C(=O)-(CH_2)_7CH_2CH_2CH_2CH_2CH_2CH_2CH_2CH_2CH_2CH_3$

$CH_2-O-C(=O)-(CH_2)_7CH_2CH_2CH_2CH_2CH_2(CH_2)_4CH_3$

Combining all the CH_2 groups gives:

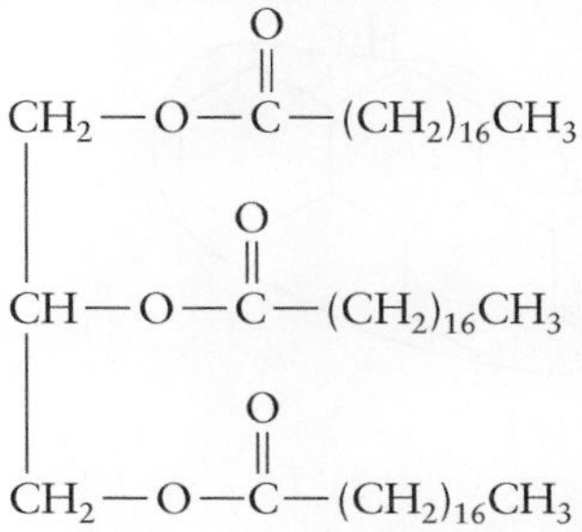

$CH_2-O-C(=O)-(CH_2)_{16}CH_3$

$CH-O-C(=O)-(CH_2)_{16}CH_3$

$CH_2-O-C(=O)-(CH_2)_{16}CH_3$

14.6

$CH_2-O-C(=O)-(CH_2)_{14}CH_3$

$CH-O-C(=O)-(CH_2)_7CH=CH(CH_2)_7CH_3$

$CH_2-O-P(=O)(O^-)-OCH_2CH_2-NH_3^+$

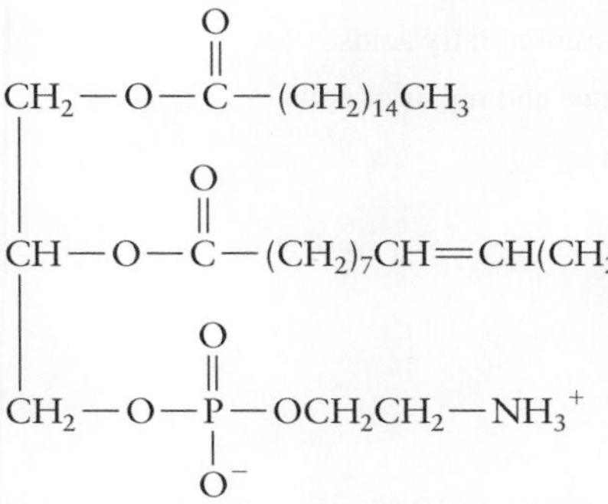

14.7 The structure of a sphingomyelin is constant, except for the fatty acid component.

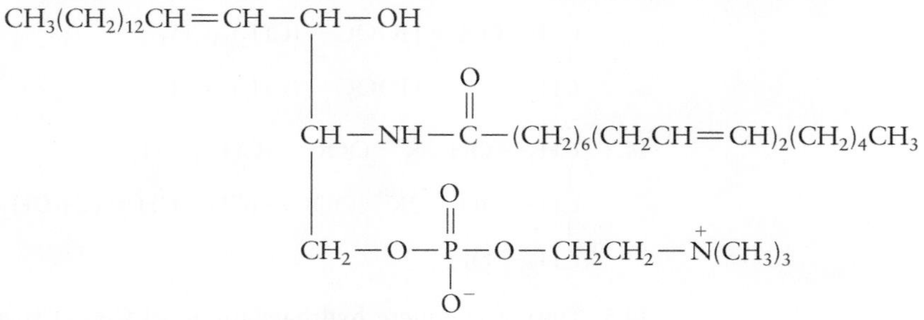

$CH_3(CH_2)_{12}CH{=}CH{-}CH{-}OH$

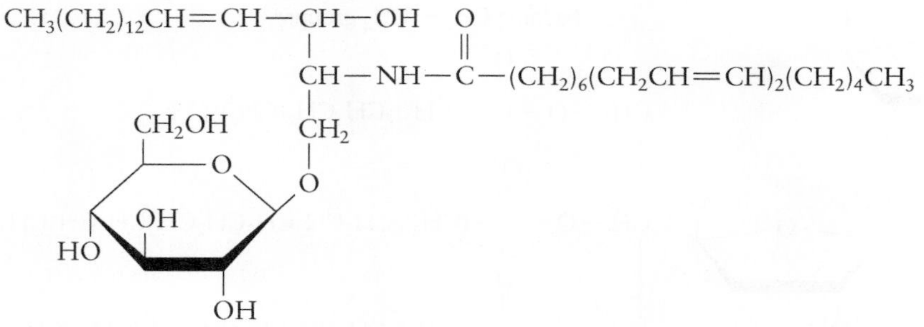

$CH{-}NH{-}C({=}O){-}(CH_2)_6(CH_2CH{=}CH)_2(CH_2)_4CH_3$

$CH_2{-}O{-}P({=}O)(O^-){-}O{-}CH_2CH_2{-}\overset{+}{N}(CH_3)_3$

14.8 $CH_3(CH_2)_{12}CH{=}CH{-}CH{-}OH$

$CH{-}NH{-}C({=}O){-}(CH_2)_6(CH_2CH{=}CH)_2(CH_2)_4CH_3$

CH_2

O

CH_2OH, O, OH, HO, OH

14.9 CH_3

$C{=}O$

H_3C

H_3C

O

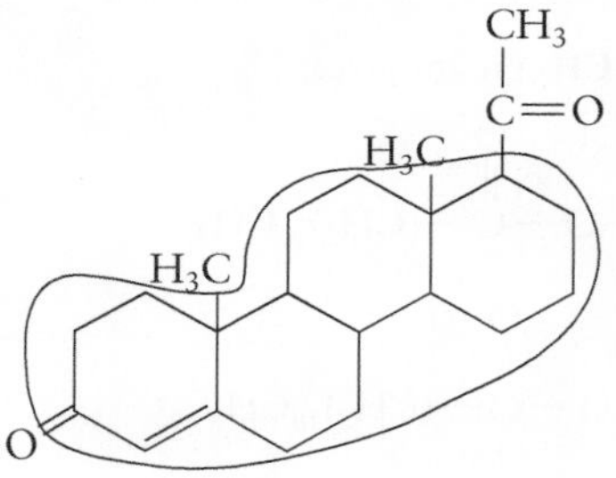

Chapter 15

15.1 **a.** Provide elements for body synthesis

b. Provide essential fatty acids

c. Maintenance and repair of cells

15.2 **a.** potassium

b. calcium

c. iodine

15.3 **a.** Stage 3

b. Stage 1

c. Stage 2

15.4 Glycolysis consists of 10 reactions that begin cellular respiration. The goal of cellular respiration is to produce large amounts of ATP to provide the body with a source of chemical energy. When the body has produced ATP in excess of what it needs, ATP itself will inhibit the pathway that ultimately produces more ATP to reduce the cellular concentration.

15.5 The citric acid cycle consists of 8 reactions that produce electron carriers NADH and $FADH_2$ for the electron transport chain. The electron transport chain is part of oxidative phosphorylation, a process by which our bodies produce large amounts of ATP. When ATP is hydrolyzed, and thus depleted, it produces ADP. Increased ADP is a signal to the body that it is time to make more ATP. Thus, ADP is able to activate the citric acid cycle and trigger the cascade of events that leads to the production of ATP.

15.6 acetoacetate and β-hydroxybutyrate

15.7 **a.** histidine, phenylalanine, isoleucine, threonine, leucine, tryptophan, lysine, valine, methionine

b. alanine, glutamine, arginine, glycine, asparagine, proline, aspartate, serine, cysteine, tyrosine, glutamate

Glossary

absolute specificity The characteristic of an enzyme that it acts on one and only one substance.

absolute zero The temperature at which all motion stops; a value of 0 on the Kelvin scale.

acetal A compound that contains the functional group

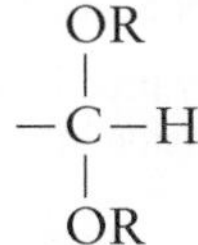

acetone breath A condition in which acetone can be detected in the breath.

acid chloride An organic compound that contains the —C(=O)—Cl functional group.

acid dissociation constant (K_a) The equilibrium constant for the dissociation of an acid.

acidic solution A solution in which the concentration of H_3O^+ is greater than the concentration of OH^-. Also, a solution in which pH is less than 7.

acidosis A low blood pH.

acidosis An abnormally low blood pH.

activation energy Energy needed to start some spontaneous processes. Once started, the processes continue without further stimulus or energy from an outside source.

active site The location on an enzyme where a substrate is bound and catalysis occurs.

acute radiation syndrome The condition that can result from short-term exposure to intense radiation.

addition polymer A polymer formed by the linking together of many alkene molecules through addition reactions.

addition reaction A reaction in which a compound adds to a multiple bond.

adipose tissue A kind of connective tissue where triglycerides are stored.

aerobic In the presence of oxygen.

α-helix The helical structure in proteins that is maintained by hydrogen bonds.

alcohol A compound in which an –OH group is connected to a hydrocarbon.

aldehyde A compound that contains the —C(=O)—H group; the general formula is R—C(=O)—H, which can be written more concisely as RCHO.

alkaloids A class of nitrogen-containing organic compounds obtained from plants.

alkalosis An abnormally high blood pH.

alkane A hydrocarbon that contains only carbon–hydrogen bonds and carbon–carbon single bonds.

alkene A hydrocarbon containing one or more carbon–carbon double bonds.

alkyl group A group differing by one hydrogen from an alkane.

alkyne A hydrocarbon containing one or more carbon–carbon triple bonds.

allosteric regulation When a substance binds to an enzyme at a location other than the active site and alters the catalytic activity.

alpha (α) amino acid An organic compound containing both an amino group and a carboxylate group, with the amino group attached to the carbon next to the carboxylate group.

alpha particle The particle that makes up alpha rays. It is identical to the helium nucleus and is composed of two protons and two neutrons.

amide An organic compound having the functional group —C(=O)—N(—)—

amine An organic compound derived by replacing one or more of the hydrogen atoms of ammonia with alkyl or aromatic groups, as in RNH_2, R_2NH, and R_3N.

amino acid pool The total supply of amino acids in the body.

amino acid residue The parts of amino acids involved in peptide, polypeptide, or protein chain formation by the removal of water.

amphetamines A class of drugs structurally similar to epinephrine, used to stimulate the central nervous system.

anabolism All reactions involved in the synthesis of biomolecules.

anion An atom that has acquired a net negative charge by gaining one or more electrons.

anomeric carbon An acetal, ketal, hemiacetal, or hemiketal carbon atom that gives rise to two stereoisomers.

anomers Stereoisomers that differ in the three-dimensional arrangement of groups at the carbon of an acetal, ketal, hemiacetal, or hemiketal group.

antibiotic A substance produced by one microorganism that kills or inhibits the growth of other microorganisms.

antibody A substance that helps protect the body from invasion by viruses, bacteria, and other foreign substances.

anticodon A three-base sequence in tRNA that is complementary to one of the codons in mRNA.

antioxidant A substance that prevents another substance from being oxidized.

apoenzyme A catalytically inactive protein formed by removal of the cofactor from an active enzyme.

aqueous solution A solution in which the solvent is water. The state of an aqueous solution is indicated by (aq) in chemical reactions.

aromatic hydrocarbon Any organic compound that contains a benzene ring.

Arrhenius acid Any substance that provides H^+ ions when dissolved in water.

Arrhenius base Any substance that provides OH^- ions when dissolved in water.

atomic mass unit (amu) A unit used to express the relative masses of atoms. One amu is equal to 1.661×10^{-24} g, which is 1/12 of the mass of a single carbon-12 atom.

atomic number The number of protons in the nucleus of an atom. Symbolically represented by Z.

atomic orbital A volume of space around atomic nuclei in which electrons of the same energy may reside at any given time. Groups of orbitals with the same *n* value form subshells. Each orbital can hold two electrons.

atomic radius The distance extending from the center of the nucleus of the atom to the location of the valence shell electrons.

atomic symbol The full representation of an element, which communicates both the name (abbreviation) of the element, and also information about the number of subatomic particles present.

atomic weight The mass of the weighted average of the atoms (naturally occurring isotopes) of an element.

atom The limit of chemical subdivision for matter.

Avogadro's law Equal volumes of gases measured at the same temperature and pressure contain equal numbers of molecules.

Avogadro's number The number of atoms or molecules in one mole of a substance, equal to 6.022×10^{23}.

balanced equation An equation in which the number of atoms of each element in the reactants is the same as the number of atoms of each element in the products.

base unit of measurement A specific unit from which other units for the same quantity are obtained by multiplication or division.

basic or **alkaline solution** A solution in which the concentration of OH^- is greater than the concentration of H_3O^+. Also, a solution in which pH is greater than 7.

Benedict's reagent A mild oxidizing solution containing Cu^{2+} ions used to test for the presence of aldehydes.

beta particle The particle that makes up beta rays. It is identical to an electron but is produced in the nucleus when a neutron is changed into a proton and an electron.

binary compound A compound made up of two different elements.

biochemistry A study of the compounds and processes associated with living organisms.

biological unit of radiation A radiation measurement unit indicating the damage caused by radiation in living tissue.

biomolecule A general term referring to organic compounds essential to life.

blood sugar level The amount of glucose present in blood, normally expressed as milligrams per 100 mL of blood.

boiling point The temperature at which the vapor pressure of a liquid is equal to the prevailing atmospheric pressure.

bond polarization A result of shared electrons being attracted to the more electronegative atom of a bonded pair of atoms.

β-oxidation process A pathway in which fatty acids are broken down into molecules of acetyl CoA.

Boyle's law A gas law describing the inverse relationship between the pressure and volume of a gas at constant temperature.

β-pleated sheet A secondary protein structure in which protein chains are aligned side by side in a sheetlike array held together by hydrogen bonds.

branched alkane An alkane in which at least one carbon atom is not part of a continuous chain.

Brønsted-Lowry acid Any hydrogen-containing substance that is capable of donating a proton (H^+) to another substance.

Brønsted-Lowry base Any substance that is capable of accepting a proton (H^+) from another substance.

buffer A solution with the ability to resist changing pH when acids (H^+) or bases (OH^-) are added.

buffer capacity The amount of acid (H^+) or base (OH^-) that can be absorbed by a buffer without causing a significant change in pH.

carbohydrate A polyhydroxy aldehyde or ketone, or substance that yields such compounds on hydrolysis.

carbonyl group The $-\overset{\overset{\displaystyle O}{\|}}{C}-$ group.

carboxylate ion The $R-\overset{\overset{\displaystyle O}{\|}}{C}-O^-$ ion that results from the dissociation of a carboxylic acid.

carboxyl group The $-\overset{\overset{\displaystyle O}{\|}}{C}-OH$ group.

carboxylic acid An organic compound that contains the $-\overset{\overset{\displaystyle O}{\|}}{C}-OH$ functional group.

carboxylic ester or **ester** An organic compound having the functional group

$$-\overset{\overset{\displaystyle O}{\|}}{C}-OR$$

catabolism All reactions involved in the breakdown of biomolecules.

catalyst A substance that changes (usually increases) reaction rates without being used up in the reaction.

cation An atom that has acquired a net positive charge by losing one or more electrons.

cellular respiration The entire process involved in the use of oxygen by cells.

central dogma of molecular biology The well-established process by which genetic information stored in DNA molecules is expressed in the structure of synthesized proteins.

cephalin A phosphoglyceride containing ethanolamine or serine.

change in Gibbs free energy (ΔG) The difference in the product and reactant energies in a chemical reaction.

Charles's law A gas law describing the direct relationship between the volume and temperature of a gas at constant pressure.

chemical changes Changes matter undergoes that involve changes in composition.

chemical kinetics The study of the rates of chemical reactions.

chemical properties Properties that matter demonstrates when attempts are made to change it into new substances.

chemical thermodynamics The study of the changes in energy that take place when processes such as chemical reactions occur.

chemiosmosis The mechanism wherein protons flow across the inner mitochondrial membrane during operation of the electron transport chain, thus providing the energy for ATP synthesis.

chiral A descriptive term for compounds or objects that cannot be superimposed on their mirror image.

chiral carbon A carbon atom with four different groups attached.

chromosome A tightly packed bundle of DNA and protein that is involved in cell division.

chylomicron A lipoprotein aggregate found in the lymph and the bloodstream.

cis- Two groups attached to the *same* side of an organic molecule whose rotation is restricted (as applied to cis–trans isomers).

citric acid cycle A series of reactions in which acetyl CoA is oxidized to carbon dioxide, and reduced forms of coenzymes FAD and NAD^+ are produced.

codon A sequence of three nucleotide bases that represents a code word on mRNA molecules.

coenzyme An organic molecule required by an enzyme for catalytic activity.

cofactor A nonprotein molecule or ion required by an enzyme for catalytic activity.

cohesive forces The attractive forces between particles; they are associated with potential energy.

cold spot Tissue from which a radioactive tracer is excluded or rejected.

collision model of chemical kinetics The idea that a chemical reaction can occur only when the reactant molecules, atoms, or ions collide in the proper orientation and exceed the activation energy.

combination reaction A chemical reaction in which two or more substances react to form a single substance.

combined gas law A gas law that describes the pressure, volume, and temperature behavior of a gas sample when the amount of the gas is held constant.

combustion A chemical reaction in which hydrocarbons are oxidized in the presence of oxygen.

common catabolic pathway The reactions of the citric acid cycle plus those of oxidative phosphorylation.

competitive inhibitor An inhibitor that competes with substrate for binding at the active site of the enzyme.

complementary DNA strands The two strands in double-helical DNA align such that adenine in one strand hydrogen-bonds to thymine in the other, and guanine in one strand hydrogen-bonds to cytosine in the other.

complete protein A protein that contains all the essential amino acids in a sufficient quantity needed for the body.

complex carbohydrates The polysaccharides amylose and amylopectin; collectively called starch.

complex lipid An ester-containing lipid with more than two types of components: an alcohol, fatty acids, and other components.

compound A pure substance consisting of two or more elements, with a definite composition, that can be broken down into simpler substances only by chemical methods.

compressibility The change in volume of a sample resulting from a pressure change acting on the sample.

concentration The relationship between the amount of solute and the specific amount of solution in which it is contained.

condensation An exothermic process in which a gas or vapor is changed to a liquid.

condensed structural representation A structural molecular formula showing the general arrangement of atoms but without all the covalent bonds.

conformations The different arrangements of atoms in three-dimensional space achieved by rotation about single bonds.

conjugate acid–base pair A Brønsted-Lowry acid and its conjugate base.

conjugate base The species remaining when a Brønsted-Lowry acid donates a proton.

conversion factors Fractions obtained from numerical relationships between quantities.

Cori cycle The process in which glucose is converted to lactate in muscle tissue, lactate is reconverted to glucose in the liver, and glucose is returned to the muscle.

covalent bond The attractive force that results between two atoms that are both attracted by a shared pair of electrons.

covalent compound A compound formed when two nonmetals share electrons to form a chemical bond.

crenation (plasmolysis) The shrinking of a cell after exposure to a hypertonic solution due to a net movement of water out of the cell via osmosis.

crystal lattice A rigid three-dimensional arrangement of particles.

C-terminal residue The amino acid on the end of a chain that has an unreacted or free carboxylate group.

cycloalkane An alkane in which carbon atoms form a ring.

cytochrome An iron-containing enzyme located in the electron transport chain.

Daily Reference Values (DRVs) A set of standards for nutrients and food components (such as fat and fiber) that have important relationships with health; used on food labels as part of the Daily Values.

Daily Values (DVs) Reference values developed by the FDA specifically for use on food labels. The Daily Values represent two sets of standards: Reference Daily Intakes (RDIs) and Daily Reference Values (DRVs).

Dalton's law of partial pressures The total pressure exerted by a mixture of gases is equal to the sum of the partial pressures of the gases in the mixture.

daughter nuclei The new nuclei produced when unstable nuclei undergo radioactive decay.

decomposition reaction A chemical reaction in which a single substance reacts to form two or more simpler substances.

degenerate A code in which different codons specify the same amino acid.

dehydration reaction A reaction in which water is chemically removed from a compound.

density The number obtained when the mass of a sample of a substance is divided by its volume.

deoxyribonucleic acid (DNA) A nucleic acid found primarily in the nuclei of cells.

deposition The exothermic process in which a gas is changed directly to a solid without first becoming a liquid.

derived unit of measurement A unit obtained by multiplication or division of one or more basic units.

dextrorotatory Rotates plane-polarized light to the right.

dialysis A process in which solvent molecules, other small molecules, and hydrated ions pass from a solution through a membrane.

dialyzing membrane A semipermeable membrane with pores large enough to allow solvent molecules, other small molecules, and hydrated ions to pass through.

diatomic molecules Molecules that contain two atoms.

diffusion The movement of individual molecules of a substance from an area of higher concentration to an area of lower concentration.

dimensional analysis The systematic, mathematical method of converting a quantity from one unit to another.

dimer Two identical molecules bonded together.

dipeptide A compound formed when two amino acids are bonded by an amide linkage.

dipole–dipole interactions The attractive force that exists between the positive end of one polar molecule and the negative end of another.

diprotic acid An acid that gives up two protons (H^+) per molecule when dissolved.

disaccharide A carbohydrate formed by the combination of two monosaccharide units.

disruptive forces The forces resulting from particle motion; they are associated with kinetic energy.

dissolving The process of solution formation when one or more solutes are dispersed in a solvent to form a homogeneous mixture.

DNA helicase An enzyme responsible for unwinding the DNA double strand to form the replication fork.

DNA ligase An enzyme that joins together Okazaki fragments on a daughter DNA strand.

DNA polymerase A family of enzymes responsible for building daughter DNA strands complementary to parent template strands during replication.

double bonds The covalent bonds resulting from the sharing of two pairs of electrons.

double-replacement reaction A chemical reaction in which two compounds react and exchange partners to form two new compounds.

effective collision A collision that causes a reaction to occur between the colliding molecules.

electrolyte A solute that when dissolved in water forms a solution that conducts electricity.

electron A stable subatomic particle with an electric charge of -1 and a mass of 9.07×10^{-28} g that orbits the center of an atom.

electron capture A mode of decay for some unstable nuclei in which an electron from outside the nucleus is drawn into the nucleus, where it combines with a proton to form a neutron.

electron configurations The detailed arrangement of electrons indicated by a specific notation, such as $1s^2 2s^2 2p^4$.

electron domain Any lone pair of electrons or shared electrons resulting in a single, double, or triple bond.

electron-domain geometry The general shape that a molecule assumes when repulsions are minimized between electron-domains.

electronegativity A measure of the tendency of an atom to attract electrons.

electron transport chain A series of reactions in which protons and electrons from the oxidation of foods are used to reduce molecular oxygen to water.

elemental symbol A one- or two-letter designation assigned to an element based on the name of the element, consisting of one capital letter or a capital letter followed by a lowercase letter.

element A pure substance consisting of only one kind of atom in the form of homoatomic molecules or individual atoms.

elimination reaction A reaction in which two or more covalent bonds are broken and a new multiple bond is formed.

enantiomers Stereoisomers that are nonsuperimposable mirror images.

endothermic A process that absorbs heat.

end point of a titration The point at which the titration is stopped on the basis of an indicator color change or pH meter reading.

enthalpy The sum of internal energies plus the product of pressure and volume.

enzyme A biomolecule that catalyzes chemical reactions.

enzyme activity The rate at which an enzyme catalyzes a reaction.

enzyme inhibitor A substance that decreases the activity of an enzyme.

enzyme international unit (IU) A quantity of enzyme that catalyzes the conversion of 1 μmol of substrate per minute under specified conditions.

equilibrium concentrations The unchanging concentrations of reactants and products in a reaction system that is in a state of equilibrium.

equilibrium constant A numerical relationship between reactant and product concentrations in a reaction at equilibrium.

equilibrium expression An equation relating the equilibrium constant and reactant and product equilibrium concentrations.

equivalence point of a titration The point at which the unknown solution has exactly reacted with the known solution. Neither is in excess.

essential amino acid An amino acid that cannot be synthesized within the body at a rate adequate to meet metabolic needs.

essential fatty acid A fatty acid needed by the body but not synthesized within the body.

esterification The process of forming an ester.

ether A compound that contains a $-\overset{|}{\underset{|}{C}}-O-\overset{|}{\underset{|}{C}}-$ functional group.

eukaryotic cell A cell containing membrane-enclosed organelles, particularly a nucleus.

evaporation or **vaporization** An endothermic process in which a liquid is changed to a gas.

exact numbers Numbers that have no uncertainty; numbers from defined relationships, counting numbers, and numbers that are part of simple fractions.

exon A segment of a eukaryotic DNA molecule that is coded for amino acids.

exothermic A process that liberates heat.

expanded structural representation A structural molecular formula showing all the covalent bonds.

fat A triglyceride that is solid at room temperature.

fat mobilization The hydrolysis of stored triglycerides, followed by the entry of fatty acids and glycerol into the bloodstream.

fatty acid A long-chain carboxylic acid found in fats.

feedback inhibition A process in which the end product of a sequence of enzyme-catalyzed reactions inhibits an earlier step in the process.

fermentation A reaction of sugars, starch, or cellulose to produce ethanol and carbon dioxide.

fibrous proteins Proteins whose secondary structure is made up of cable-like coiled coils.

first ionization energy The energy required to remove the first outermost electron from a neutral atom.

Fischer projection A method of depicting three-dimensional shapes for chiral molecules.

fluid-mosaic model A model of membrane structure in which proteins are embedded in a flexible lipid bilayer.

formula unit The simplest whole number ratio of ions that make up an ionic compound.

formula weight The sum of the atomic weights of the ions present in the formula unit of an ionic compound.

frequency The number of waves that pass a fixed point per unit of time.

functional group A unique reactive combination of atoms that differentiates molecules of organic compounds of one class from those of another.

furanose ring A five-membered sugar ring system containing an oxygen atom.

gamma ray A high-energy ray that is like an X ray, but with a higher energy.

gas laws Mathematical relationships that describe the behavior of gases as they are mixed, subjected to pressure or temperature changes, or allowed to diffuse.

Geiger-Müller tube A radiation-detection device operating on the principle that ions form when radiation passes through a tube filled with low-pressure gas.

gene An individual section of a DNA molecule that is the fundamental unit of heredity.

globular protein A spherical protein that usually forms stable suspensions in water or dissolves in water.

glucogenic amino acid An amino acid whose carbon skeleton can be converted metabolically to an intermediate used in the synthesis of glucose.

gluconeogenesis The synthesis of glucose from noncarbohydrate molecules.

glucosuria A condition in which elevated blood sugar levels result in the excretion of glucose in the urine.

glycogenesis The synthesis of glycogen from glucose.

glycogenolysis The breakdown of glycogen to glucose.

glycolipid A complex lipid containing a sphingosine, a fatty acid, and a carbohydrate.

glycolysis A series of reactions by which glucose is oxidized to pyruvate.

glycoside Another name for a carbohydrate containing an acetal or ketal group.

glycosidic linkage or **glycosidic bond** A covalent bond between carbon and oxygen that joins the –OR group to the carbohydrate ring.

group of the periodic table A vertical column of elements that have similar chemical properties.

half-life The time required for one- half the unstable nuclei in a sample to undergo radioactive decay.

Haworth structure A method of depicting three-dimensional carbohydrate structures.

heat of fusion The amount of heat energy required to melt exactly 1 g of a solid substance at constant temperature.

heat of vaporization The amount of heat energy required to vaporize exactly 1 g of a liquid substance at constant temperature.

hemiacetal A compound that contains the functional group

$$\begin{array}{c} \text{OH} \\ | \\ -\text{C}-\text{H} \\ | \\ \text{OR} \end{array}$$

hemolysis The swelling of a cell after exposure to a hypotonic solution due to a net movement of water into the cell via osmosis.

Henderson-Hasselbalch equation A relationship between the pH of a buffer, pK_a, and the concentrations of acid and salt in the buffer.

heteroatomic molecules Molecules that contain two or more kinds of atoms.

heterocyclic ring A ring containing at least one atom of an element other than carbon.

heterogeneous matter Matter with properties that are not the same throughout the sample.

heterogeneous nuclear RNA (hnRNA) RNA produced when both introns and exons of eukaryotic cellular DNA are transcribed.

homoatomic molecules Molecules that contain only one kind of atom.

homogeneous matter Matter that has the same properties throughout the sample.

homologous series Compounds of the same functional class that differ by a $-CH_2-$ group.

hormone A chemical messenger secreted by specific glands and carried by the blood to a target tissue, where it triggers a particular response.

hot spot Tissue in which a radioactive tracer concentrates.

hydrated ions Ions in solution that are surrounded by water molecules.

hydration The addition of water to a multiple bond.

hydrocarbon An organic compound that contains only carbon and hydrogen.

hydrogenation A reaction in which the addition of hydrogen takes place. A precious metal catalyst (e.g., Pt, Ni, or Pd) is typically needed to make the reaction work.

hydrogen bonding The attraction between a hydrogen atom that is bonded to a F, N, or O atom and any F, N, or O atom on another molecule or a distant region of the same molecule.

hydronium ion The ion resulting from the addition of a proton to water.

hydrophobic Molecules or parts of molecules that repel (are insoluble in) water.

hydroxy group The –OH functional group.

hyperglycemia A higher-than-normal blood sugar level.

hypertonic solution A solution that contains a higher concentration of solutes compared to another solution.

hyperventilation Rapid, deep breathing.

hypoglycemia A lower-than-normal blood sugar level.

hypotonic solution A solution that contains a lower concentration of solutes compared to another solution.

hypoventilation Slow, shallow breathing.

ideal gas law A gas law that relates the pressure, volume, temperature, and number of moles in a gas sample. Mathematically, it is $PV = nRT$.

immiscible A term used to describe liquids that are insoluble in each other.

immunization A medical treatment to prevent an individual from developing a disease upon exposure to its source.

induced-fit theory A theory of enzyme action proposing that the conformation of an enzyme changes to accommodate an incoming substrate.

inhibitor A substance that decreases reaction rates.

inner-transition metals The lanthanides (elements 57–71) and actinides (elements 89–103).

insoluble substance A substance that does not dissolve to a significant extent in a solvent.

integrase The viral enzyme responsible for splicing retroviral DNA into the host cell DNA.

intermolecular forces The attractive forces that exist *between* molecules.

intramolecular forces The attractive forces that exist *within* molecules.

intron A segment of a eukaryotic DNA molecule that carries no codes for amino acids.

ion–dipole interaction The attraction between a charged ion (cation or anion) and a polar molecule.

ionic bond The attractive force that holds together ions of opposite charge.

ionic compound A substance consisting of ions of opposite charge held together by an electrostatic attraction.

ion product of water The equilibrium constant for the dissociation of pure water into H_3O^+ and OH^-.

isoelectric point The characteristic solution pH at which an amino acid has a net charge of 0.

isoelectronic Literally "same electronic." The term is used to describe atoms or ions that have identical electron configurations.

isomerism A property in which two or more compounds have the same molecular formula but different arrangements of atoms.

isotonic solution A solution that contains the same concentration of solutes as another solution.

isotopes Atoms that have the same atomic number but different mass numbers. That is, they are atoms of the same element that contain different numbers of neutrons in their nuclei.

ketoacidosis A low blood pH due to elevated levels of ketone bodies.

ketogenic amino acid An amino acid whose carbon skeleton can be converted metabolically to acetyl CoA or acetoacetyl CoA.

ketone A compound that contains the $-\overset{|}{\underset{|}{C}}-\overset{\overset{O}{\|}}{C}-\overset{|}{\underset{|}{C}}-$ group; the general formula is $R-\overset{\overset{O}{\|}}{C}-R'$, or RC(O)R′.

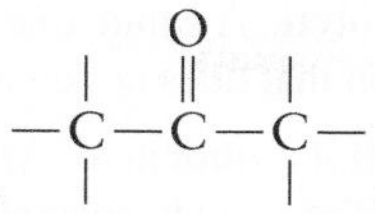

ketone bodies Three compounds—acetoacetate, β-hydroxybutyrate, and acetone—formed from acetyl CoA.

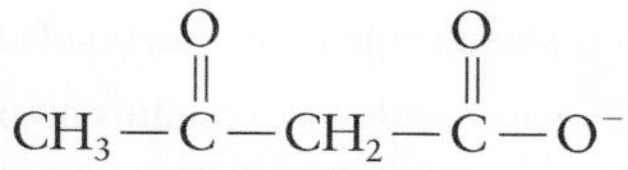

$$CH_3-\overset{\overset{O}{\|}}{C}-CH_2-\overset{\overset{O}{\|}}{C}-O^-$$

acetoacetate

$$CH_3-\overset{\overset{OH}{|}}{CH}-CH_2-\overset{\overset{O}{\|}}{C}-O^-$$

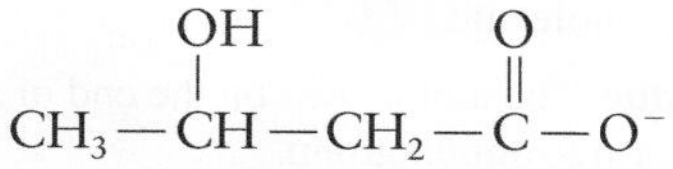

β-hydroxybutyrate
(not a ketone)

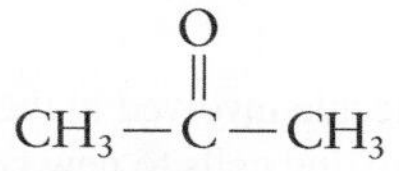

$$CH_3-\overset{\overset{O}{\|}}{C}-CH_3$$

acetone

ketonemia An elevated level of ketone bodies in the blood.

ketonuria The presence of ketone bodies in the urine.

ketosis A condition in which ketonemia, ketonuria, and acetone breath exist together.

kinetic energy The energy a particle has as a result of its motion.

lattice site The individual location occupied by a particle in a crystal lattice.

law of conservation of matter Atoms are neither created nor destroyed in chemical reactions.

Le Châtelier's principle The position of an equilibrium shifts in response to changes made in factors of the equilibrium.

lecithin A phosphoglyceride containing choline.

levorotatory Rotates plane-polarized light to the left.

Lewis dot structure A representation of an atom or ion in which the elemental symbol represents the atomic nucleus and all but the valence-shell electrons. The valence-shell electrons are represented by dots arranged around the elemental symbol.

Lewis structure A diagram that shows the bonding between atoms of a molecule, as well as the nonbonding pairs of electrons that may exist in the molecule.

lipid A biological compound that is soluble only in nonpolar solvents.

lipid bilayer A structure found in membranes, consisting of two sheets of lipid molecules arranged so that the hydrophobic portions are facing each other, and the hydrophilic portions are positioned to interact with water.

lock-and-key theory A theory of enzyme specificity proposing that a substrate has a shape fitting that of the enzyme's active site, as a key fits a lock.

London dispersion forces Very weak attractive forces acting between the particles of all matter. They result from momentary nonsymmetric electron distributions in molecules or atoms.

lone pairs Nonbonding pairs of electrons that serve to depict the completion of an atom's octet.

macronutrient A substance needed by the body in relatively large amounts.

major mineral A mineral found in the body in quantities greater than 5 g.

Markovnikov's rule In the addition of H—X to an alkene, the hydrogen becomes attached to the carbon atom that is already bonded to more hydrogens.

mass A measurement of the amount of matter in an object.

mass number The total number of protons and neutrons in a nucleus.

matter Anything that has mass and occupies space.

melting point The temperature at which a solid changes to a liquid; the solid and liquid have the same vapor pressure.

messenger RNA (mRNA) RNA that carries genetic information from the DNA in the cell nucleus to the site of protein synthesis in the cytoplasm.

metabolic pathway A sequence of reactions used to produce one product or accomplish one process.

metabolism The sum of all reactions occurring in an organism.

metalloids Elements that form a narrow diagonal band in the periodic table between metals and nonmetals. They have properties somewhat between those of metals and nonmetals.

metals Elements found in the left two-thirds of the periodic table. Common properties of metals are: high thermal and electrical conductivities, high malleability and ductility, and a luster.

micelle A spherical cluster of molecules in which the polar portions of the molecules are on the surface and the nonpolar portions are located in the interior.

micronutrient A substance needed by the body only in small amounts.

mineral A metal or nonmetal used in the body in the form of ions or compounds.

mitochondrion A cellular organelle in which reactions of the common catabolic pathway occur.

mixture A blend of matter that can theoretically be physically separated into its separate pure components.

molarity (M) A solution concentration expressed as the number of moles of solute contained in one liter of solution.

molar mass The mass of 1 mol of a substance in grams.

molecular equation An equation written with each compound represented by its formula.

molecular formula A representation of the molecule of a covalent compound, consisting of the symbols of the elements found in the molecule. Elements present in numbers greater than one have the number indicated by a subscript that follows the elemental symbol.

molecular geometry The three-dimensional shape of bonded atoms in a molecule.

molecular weight The sum of the atomic weights of the elements in the molecular formula of a covalent compound.

molecule The smallest particle of a pure substance that has the properties of that substance and is capable of a stable independent existence. Alternatively, a molecule is the limit of physical subdivision for a pure substance.

mole The number of particles (atoms or molecules) contained in a sample of an element or a compound with a mass in grams equal to the atomic or molecular weight, respectively. Numerically, 1 mol is equal to 6.022×10^{23} particles.

monomer The starting material that becomes the repeating units of polymers.

monoprotic acid An acid that gives up only one proton (H^+) per molecule when dissolved.

monosaccharide A simple carbohydrate most commonly consisting of three to six carbon atoms.

mutagen A chemical that induces mutations by reacting with DNA.

mutation Any change resulting in an incorrect base sequence on DNA.

net ionic equation An equation that contains only unionized or insoluble materials and ions that undergo changes as the reaction proceeds. All spectator ions are eliminated.

neurotransmitter A substance that acts as a chemical bridge in nerve impulse transmission between nerve cells.

neutral A term used to describe any water solution in which the concentrations of H_3O^+ and OH^- are equal. Also, an aqueous solution with pH = 7.

neutralization reaction A reaction in which an acid and base react completely, leaving a solution that contains only a salt and water.

neutron A stable subatomic particle with an electric charge of 0 and a mass of 1.67×10^{-24} g that resides in the center of an atom.

noble gas configurations or **abbreviated electron configurations** The arrangement of electrons that starts with the noble gas prior to the valence shell and ends with specific notation representing the electrons in the valence shell, for example, $[He]2s^22p^4$ instead of $1s^22s^22p^4$.

noncompetitive inhibitor An inhibitor that binds to the enzyme at a location other than the active site.

nonelectrolyte A solute that when dissolved in water forms a solution that does not conduct electricity.

nonessential amino acid An amino acid that can be synthesized within the body in adequate amounts.

nonmetals Elements found in the right one-third of the periodic table. They often occur as brittle, powdery solids or gases and have properties generally opposite to those of metals.

nonpolar covalent bond A covalent bond in which the bonding pair of electrons is shared equally by the bonded atoms.

nonpolar molecule A molecule that contains no polarized bonds, or a molecule containing polarized bonds in which the resulting charges are distributed symmetrically throughout the molecule.

N-terminal residue The amino acid on the end of a chain that has an unreacted or free amino group.

nuclear reaction A process where two atomic nuclei or an atomic nucleus and another subatomic particle collide, resulting in one or more new nuclei.

nucleic acid A biomolecule involved in the transfer of genetic information from existing cells to new cells.

nucleic acid backbone The sugar–phosphate chain that is common to all nucleic acids.

nucleotide The repeating structural unit (the monomer) of polymeric nucleic acids DNA and RNA.

nucleus The central core of atoms that contains protons, neutrons, and most of the mass of atoms.

nutrition An applied science that studies food, water, and other nutrients and the ways living organisms use them.

octet rule A rule for predicting electron behavior in reacting atoms—namely, atoms will lose, gain, or share sufficient electrons to achieve an outer electron arrangement identical to that of a noble gas. This arrangement usually consists of eight electrons in the valence shell.

oil A triglyceride that is liquid at room temperature.

Okazaki fragment A DNA fragment produced during replication as a result of strand growth in a direction away from the replication fork.

oligosaccharide A carbohydrate formed by the combination of three to ten monosaccharide units.

optically active molecule A molecule that rotates the plane of polarized light.

optimum pH The pH at which enzyme activity is highest.

optimum temperature The temperature at which enzyme activity is highest.

organelle A specialized structure within a cell that performs a specific function.

organic chemistry The study of carbon-containing compounds.

organic compound A compound that contains the element carbon.

osmosis The process in which solvent molecules flow through a semipermeable membrane into a solution.

osmotic pressure The hydrostatic pressure required to prevent the net flow of solvent through a semipermeable membrane into a solution.

oxidation numbers or **oxidation states** The number assigned to an element within a compound after its atoms have lost or gained electrons.

oxidation Originally, the term referred to a process involving a reaction with oxygen. Today, it means a number of things, including a process in which electrons are given up, hydrogen is lost, or an oxidation number increases.

oxidative deamination An oxidation process resulting in the removal of an amino group.

oxidative phosphorylation A process wherein the electron transport chain is coupled with chemiosmosis to form ATP from ADP and P_i.

oxidizing agent The substance in a redox reaction that oxidizes another substance. The oxidizing agent itself is reduced.

palindrome A section of DNA in which the two strands have the same sequence but run in opposite directions.

partial pressure The pressure an individual gas of a mixture would exert if it were in the container alone at the same temperature as the mixture.

peptide An amino acid polymer of short chain length.

peptide bond The amide linkage between amino acids that results when the amino group of one acid reacts with the carboxylate group of another.

peptide linkage or **peptide bond** The amide linkage between amino acids that results when the amino group of one amino acid reacts with the carboxylate group of another.

peptidyl transferase A protein enzyme located in the large ribosomal subunit that catalyzes amide bond formation between amino acids in a nascent polypeptide.

percent yield The percentage of the theoretical amount of a product that is actually produced by a reaction.

period of the periodic table A horizontal row of elements.

phenol A compound in which an –OH group is connected to a benzene ring. The parent compound is also called phenol.

phenyl group A benzene ring with one hydrogen removed, C_6H_5–.

phosphate ester An organic compound having the following general form:

$$-O-\overset{\overset{\displaystyle O}{\|}}{\underset{\underset{\displaystyle OH}{|}}{P}}-OH$$

phosphoglyceride A complex lipid containing glycerol, fatty acids, phosphoric acid, and an aminoalcohol component.

phospholipid A phosphorus-containing lipid.

pH The negative logarithm of the molar concentration of H_3O^+ in a solution.

physical changes Changes matter undergoes without changing composition.

physical properties Properties of matter that can be observed or measured without trying to change the composition of the matter being studied.

physical unit of radiation A radiation measurement unit indicating the activity of the source of the radiation; for example, the number of nuclear decays per minute.

pK_a The negative logarithm of K_a.

plasmid Circular, double-stranded DNA found in the cytoplasm of bacterial cells.

polar covalent bond A covalent bond that shows bond polarization; that is, the bonding electrons are shared unequally.

polarity The distribution of electrical charge between atoms bonded together.

polar molecule A molecule that contains polarized bonds and in which the resulting charges are distributed nonsymmetrically throughout the molecule.

polyatomic ions Covalently bonded groups of atoms that carry a net electrical charge.

polyatomic molecules Molecules that contain more than three atoms.

polycyclic aromatic compound A derivative of benzene in which carbon atoms are shared between two or more benzene rings.

polymerase chain reaction (PCR) A laboratory technique used to amplify small amounts of DNA.

polymer A very large molecule made up of repeating units.

polymerization A reaction that produces a polymer.

polypeptide An amino acid polymer of intermediate chain length containing up to 50 amino acid residues.

polyribosome or **polysome** A complex of mRNA and several ribosomes.

polysaccharide A carbohydrate formed by the combination of many monosaccharide units.

polyunsaturated A term usually applied to molecules with several double bonds.

position of equilibrium An indication of the relative amounts of reactants and products present at equilibrium.

positron A positively charged electron.

potential energy The energy that is stored in particles due to their position, composition, or arrangement.

pressure A force per unit area of surface on which the force acts. In measurements and calculations involving gases, it is often expressed in units related to measurements of atmospheric pressure.

primary alcohol An alcohol in which the –OH group is attached to CH_3 or to a carbon attached to one other carbon atom.

primary amine An amine having one alkyl or aromatic group bonded to nitrogen, as in R—NH_2.

primary protein structure The linear sequence of amino acid residues in a protein chain.

products of a reaction The substances produced as a result of the reaction taking place. They are written on the right side of the equation, representing the reaction.

prokaryotic cell A simple unicellular organism that contains no nucleus and no membrane-enclosed organelles.

prostaglandin A substance derived from unsaturated fatty acids with hormone-like effects on a number of body tissues.

protein An amino acid polymer made up of more than 50 amino acids.

protein turnover The continuing process in which body proteins are hydrolyzed and resynthesized.

proton A stable subatomic particle with an electric charge of $+1$ and a mass of 1.67×10^{-24} g that resides in the center of an atom.

pure substance Matter that has a constant composition and fixed properties.

pyranose ring A six-membered sugar ring system containing an oxygen atom.

quaternary ammonium salt An ionic compound containing a positively charged ion in which four alkyl or aromatic groups are bonded to nitrogen, as in R_4N^+.

quaternary protein structure The arrangement of subunits that form a larger protein.

radical or **free radical** An electron-deficient particle that is very reactive.

radioactive decay A process in which an unstable nucleus changes energy states and in the process emits radiation.

radioactive nuclei Nuclei that undergo spontaneous changes and emit energy in the form of radiation.

radioisotope An isotope of an element that emits nuclear radiation.

reactants of a reaction The substances that undergo chemical change during the reaction. They are written on the left side of the equation, representing the reaction.

reaction enthalpy The difference in enthalpy between the products and the reactants in a chemical reaction.

reaction rate The speed of a reaction.

recombinant DNA DNA of an organism that contains genetic material from another organism.

reducing agent The substance in a redox reaction that reduces another substance. The reducing agent itself is oxidized.

reducing sugar A sugar that can be oxidized by Cu^{2+} solutions.

reduction Originally, the term referred to a process in which oxygen was lost. Today, it means a number of things, including a process in which electrons are gained, hydrogen is accepted, or an oxidation number decreases.

Reference Daily Intakes (RDIs) A set of standards for protein, vitamins, and minerals; used on food labels as part of the Daily Values.

relative specificity The characteristic of an enzyme that it acts on several structurally related substances.

renal threshold The blood glucose level at which glucose begins to be excreted in the urine.

replication fork A point where the double helix of a DNA molecule unwinds during replication.

replication The process by which an exact copy of a DNA molecule is produced.

representative elements Elements found within Groups 1–2 and 13–18 in the periodic table.

restriction enzyme A protective enzyme found in some bacteria that catalyzes the cleaving of all but a few specific types of DNA.

reverse transcriptase The viral enzyme responsible for converting a viral genomic RNA molecule into a molecule of DNA.

reverse transcription The transfer of genetic information from a viral genomic RNA molecule to a molecule of DNA.

ribonucleic acid (RNA) A nucleic acid found mainly in the cytoplasm of cells.

ribosomal RNA (rRNA) RNA that constitutes about 65% of the material in ribosomes, the sites of protein synthesis.

ribosome A subcellular particle that serves as the site of protein synthesis in all organisms.

RNA polymerase A family of enzymes responsible for building nascent RNA strands from complementary DNA templates during transcription.

saponification The basic cleavage of an ester linkage.

saturated hydrocarbon Another name for an alkane.

saturated solution A solution that contains the maximum amount possible of dissolved solute in a stable situation under the prevailing conditions of temperature and pressure.

scientific notation A way of representing numbers consisting of a product between a nonexponential number and 10 raised to a whole-number exponent that may be positive or negative.

scintillation counter A radiation-detection device operating on the principle that certain substances give off light when struck by radiation.

secondary alcohol An alcohol in which the carbon bearing the –OH group is attached to two other carbon atoms.

secondary amine An amine having two alkyl or aromatic groups bonded to nitrogen, as in R_2NH.

secondary protein structure The arrangement of protein chains into patterns as a result of hydrogen bonds between amide groups of amino acid residues in the chain. The common secondary structures are the α-helix and the β-pleated sheet.

semiconservative replication A replication process that produces DNA molecules containing one strand from the parent and a new strand that is complementary to the strand from the parent.

semipermeable membrane A biological or synthetic membrane that will allow only certain molecules to pass through via diffusion in the case of solutes and osmosis in the case of solvent particles.

shell The location and energy of electrons around a nucleus are designated by a value for n, where $n = 1, 2, 3$ ....

side reactions Reactions that do not give the desired product of a reaction.

significant figures The numbers in a measurement that represent the certainty of the measurement, plus one number representing an estimate.

simple carbohydrates Monosaccharides and disaccharides; commonly called sugars.

simple ion An atom that has acquired a net positive or negative charge by losing or gaining electrons, respectively.

simple lipid An ester-containing lipid with just two types of components: an alcohol and one or more fatty acids.

single-replacement reaction A redox chemical reaction in which an element reacts with a compound and displaces another element from the compound.

skeletal structural representation A structural molecular formula showing carbon atoms as points, with hydrogens bonded to carbon assumed. Other atoms (e.g., nitrogen, oxygen, and chlorine) are shown explicitly, as well as hydrogens bonded to these noncarbon atoms.

soap A salt of a fatty acid often used as a cleaning agent.

solubility The maximum amount of solute that can be dissolved in a specific amount of solvent under specific conditions of temperature and pressure.

soluble substance A substance that dissolves to a significant extent in a solvent.

solute One or more substances present in a solution in amounts less than that of the solvent.

solutions Homogeneous mixtures of two or more pure substances.

solvent The substance present in a solution in the largest amount.

specific heat The amount of heat energy required to raise the temperature of exactly 1 g of a substance by exactly 1°C.

spectator ions The ions in a total ionic reaction that are not changed as the reaction proceeds. They appear in identical forms on the left and right sides of the equation.

sphingolipid A complex lipid containing the aminoalcohol sphingosine.

standard atmosphere The pressure needed to support a 760-mm column of mercury in a barometer tube.

standard conditions (STP) A set of specific temperature and pressure values used for gas measurements.

standard position for a decimal In scientific notation, the position to the right of the first nonzero digit in the nonexponential number.

state of equilibrium A condition in a reaction system when the rates of the forward and reverse reactions are equal.

stereoisomers Compounds with the same structural formula but different spatial arrangements of atoms.

steroid A compound containing four rings fused in a particular pattern.

stoichiometry The study of mass relationships in chemical reactions.

straight-chain alkane Any alkane in which all the carbon atoms are aligned in a continuous chain.

strong acids and **strong bases** Acids and bases that dissociate (ionize) completely when dissolved to form a solution.

structural isomers Compounds that have the same molecular formula but in which the atoms bond in different patterns.

sublimation The endothermic process in which a solid is changed directly to a gas without first becoming a liquid.

subshell A subdivision of a shell that is designated by *s*, *p*, or *d*.

substrate The substance that undergoes a chemical change catalyzed by an enzyme.

subunit A polypeptide chain having primary, secondary, and tertiary structural features that is a part of a larger protein.

supersaturated solution An unstable solution that contains an amount of solute greater than the solute solubility under the prevailing conditions of temperature and pressure.

tertiary alcohol An alcohol in which the carbon bearing the –OH group is attached to three other carbon atoms.

tertiary amine An amine having three alkyl or aromatic groups bonded to nitrogen, as in R_3N.

tertiary protein structure A specific three-dimensional shape of a protein resulting from interactions between R groups of the amino acid residues in the protein.

titration An analytical procedure in which one solution (often a base) of known concentration is slowly added to a measured volume of an unknown solution (often an acid). The volume of the added solution is measured with a buret.

torr The pressure needed to support a 1-mm column of mercury in a barometer tube.

total ionic equation An equation written with all soluble ionic substances represented by the ions they form in solution.

trace mineral A mineral found in the body in quantities less than 5 g.

tracer A radioisotope used medically because its progress through the body or localization in specific organs can be followed.

transaminase An enzyme that catalyzes the transfer of an amino group.

transamination The enzyme-catalyzed transfer of an amino group to a keto acid.

transcription The transfer of genetic information from a DNA molecule to a molecule of messenger RNA.

transfer RNA (tRNA) RNA that delivers individual amino acid molecules to the site of protein synthesis.

transition metals All metallic elements in the central block (Groups 3–12).

translation The conversion of the code carried by messenger RNA into an amino acid sequence of a protein.

translocation The movement of the ribosome from one codon to the next along the mRNA in the 5′ to 3′ direction during the elongation phase of protein translation.

trans- Two groups attached to the *opposite* side of an organic molecule whose rotation is restricted (as applied to cis–trans isomers).

triatomic molecules Molecules that contain three atoms.

triglyceride or **triacylglycerol** A triester of glycerol in which all three alcohol groups are esterified.

triple bonds The covalent bonds resulting from the sharing of three pairs of electrons.

triprotic acid An acid that gives up three protons (H^+) per molecule when dissolved.

turnover number The number of molecules of substrate acted on by one molecule of enzyme per minute.

universal gas constant The constant that relates pressure, volume, temperature, and number of moles of gas in the ideal gas law.

unsaturated hydrocarbon A hydrocarbon that contains one or more carbon–carbon multiple bonds.

urea cycle A metabolic pathway in which ammonium ions are converted to urea.

vaccine A medical treatment utilizing parts of a virus or bacteria in order to instigate an immune response (antibody production) against a disease to prevent future illness during secondary exposure to the virus or bacteria.

valence electrons The electrons found in the outermost (highest-energy) shell of an atom

vapor pressure The pressure exerted by vapor that is in equilibrium with its liquid.

vector A carrier of foreign DNA into a cell.

virus An infectious agent made of nucleic acids (RNA or DNA) that can affect all forms of life and force the host organism to replicate its genetic material.

vitamin An organic nutrient that the body cannot produce in the small amounts needed for good health.

volume The amount of three-dimensional space an object or a substance occupies.

volume/volume percent A concentration that expresses the volume of solute contained in 100 volume units of solution.

VSEPR theory A theory based on the mutual repulsion of electron pairs. It is used to predict molecular shapes.

wavelength The distance between any given point and the same point in the next wave.

wax An ester of a long-chain fatty acid and a long-chain alcohol.

weak (or **moderately weak**) **acids** and **bases** Acids and bases that dissociate (ionize) less than completely when dissolved to form a solution.

weight A measurement of the gravitational force acting on an object.

weight/volume percent A concentration that expresses the grams of solute contained in 100 mL of solution.

weight/weight percent A concentration that expresses the mass of solute contained in 100 mass units of solution. The mass units of the solute and solution must be identical.

zwitterion A dipolar ion that carries both a positive and a negative charge as a result of an internal acid–base reaction in an amino acid molecule. The overall net charge on the zwitterion is neutral.

Index

B

C

D

E

G

H

I

J

K

L

M

N

O

P

Q

R

S

T

U

V

Table of Atomic Weights and Numbers

Name	Symbol	Atomic Number	Atomic Weight	Name	Symbol	Atomic Number	Atomic Weight
Actinium	Ac	89	(227)	Mendelevium	Md	101	(260)
Aluminum	Al	13	26.98	Mercury	Hg	80	200.6
Americium	Am	95	(243)	Molybdenum	Mo	42	95.96
Antimony	Sb	51	121.8	Moscovium	Mc	115	(288)
Argon	Ar	18	39.95	Neodymium	Nd	60	144.2
Arsenic	As	33	74.92	Neon	Ne	10	20.18
Astatine	At	85	(210)	Neptunium	Np	93	(237)
Barium	Ba	56	137.3	Nickel	Ni	28	58.69
Berkelium	Bk	97	(247)	Nihonium	Nh	113	(284)
Beryllium	Be	4	9.012	Niobium	Nb	41	92.91
Bismuth	Bi	83	209.0	Nitrogen	N	7	14.01
Bohrium	Bh	107	(270)	Nobelium	No	102	(259)
Boron	B	5	10.81	Oganesson	Og	118	(294)
Bromine	Br	35	79.90	Osmium	Os	76	190.2
Cadmium	Cd	48	112.4	Oxygen	O	8	16.00
Calcium	Ca	20	40.08	Palladium	Pd	46	106.4
Californium	Cf	98	(251)	Phosphorus	P	15	30.97
Carbon	C	6	12.01	Platinum	Pt	78	195.1
Cerium	Ce	58	140.1	Plutonium	Pu	94	(244)
Cesium	Cs	55	132.9	Polonium	Po	84	(209)
Chlorine	Cl	17	35.45	Potassium	K	19	39.10
Chromium	Cr	24	52.00	Praseodymium	Pr	59	140.9
Cobalt	Co	27	58.93	Promethium	Pm	61	(145)
Copernicium	Cn	112	(285)	Protactinium	Pa	91	231.0
Copper	Cu	29	63.55	Radium	Ra	88	(226)
Curium	Cm	96	(247)	Radon	Rn	86	(222)
Darmstadtium	Ds	110	(281)	Rhenium	Re	75	186.2
Dubnium	Db	105	(268)	Rhodium	Rh	45	102.9
Dysprosium	Dy	66	162.5	Roentgenium	Rg	111	(280)
Einsteinium	Es	99	(252)	Rubidium	Rb	37	85.47
Erbium	Er	68	167.3	Ruthenium	Ru	44	101.1
Europium	Eu	63	152.0	Rutherfordium	Rf	104	(265)
Fermium	Fm	100	(257)	Samarium	Sm	62	150.4
Flerovium	Fl	114	(289)	Scandium	Sc	21	44.96
Fluorine	F	9	19.00	Seaborgium	Sg	106	(271)
Francium	Fr	87	(223)	Selenium	Se	34	78.96
Gadolinium	Gd	64	157.3	Silicon	Si	14	28.09
Gallium	Ga	31	69.72	Silver	Ag	47	107.9
Germanium	Ge	32	72.63	Sodium	Na	11	22.99
Gold	Au	79	197.0	Strontium	Sr	38	87.62
Hafnium	Hf	72	178.5	Sulfur	S	16	32.07
Hassium	Hs	108	(277)	Tantalum	Ta	73	180.9
Helium	He	2	4.003	Technetium	Tc	43	(98)
Holmium	Ho	67	164.9	Tellurium	Te	52	127.6
Hydrogen	H	1	1.008	Tennessine	Ts	117	(294)
Indium	In	49	114.8	Terbium	Tb	65	158.9
Iodine	I	53	126.9	Thallium	Tl	81	204.4
Iridium	Ir	77	192.2	Thorium	Th	90	232.0
Iron	Fe	26	55.85	Thulium	Tm	69	168.9
Krypton	Kr	36	83.80	Tin	Sn	50	118.7
Lanthanum	La	57	138.9	Titanium	Ti	22	47.87
Lawrencium	Lr	103	(262)	Tungsten	W	74	183.8
Lead	Pb	82	207.2	Uranium	U	92	238.0
Lithium	Li	3	6.94	Vanadium	V	23	50.94
Livermorium	Lv	116	(293)	Xenon	Xe	54	131.3
Lutetium	Lu	71	175.0	Ytterbium	Yb	70	173.0
Magnesium	Mg	12	24.31	Yttrium	Y	39	88.91
Manganese	Mn	25	54.94	Zinc	Zn	30	65.38
Meitnerium	Mt	109	(276)	Zirconium	Zr	40	91.22

A value in parentheses is the mass number of the isotope of longest half-life for radioactive elements.